IFMBE Proceedings

Volume 84

The IFMBE Proceedings Book Series is an official publication of *the International Federation for Medical and Biological Engineering* (IFMBE). The series gathers the proceedings of various international conferences, which are either organized or endorsed by the Federation. Books published in this series report on cutting-edge findings and provide an informative survey on the most challenging topics and advances in the fields of medicine, biology, clinical engineering, and biophysics.

The series aims at disseminating high quality scientific information, encouraging both basic and applied research, and promoting world-wide collaboration between researchers and practitioners in the field of Medical and Biological Engineering.

Topics include, but are not limited to:

- Diagnostic Imaging, Image Processing, Biomedical Signal Processing
- Modeling and Simulation, Biomechanics
- Biomaterials, Cellular and Tissue Engineering
- Information and Communication in Medicine, Telemedicine and e-Health
- Instrumentation and Clinical Engineering
- Surgery, Minimal Invasive Interventions, Endoscopy and Image Guided Therapy
- Audiology, Ophthalmology, Emergency and Dental Medicine Applications
- Radiology, Radiation Oncology and Biological Effects of Radiation

IFMBE proceedings are indexed by by SCOPUS, EI Compendex, Japanese Science and Technology Agency (JST), SCImago.

Proposals can be submitted by contacting the Springer responsible editor shown on the series webpage (see "Contacts"), or by getting in touch with the series editor Ratko Magjarevic.

More information about this series at http://www.springer.com/series/7403

Almir Badnjevic · Lejla Gurbeta Pokvić
Editors

CMBEBIH 2021

Proceedings of the International Conference on Medical and Biological Engineering, CMBEBIH 2021, April 21–24, 2021, Mostar, Bosnia and Herzegovina

Springer

Editors
Almir Badnjevic
Verlab Ltd Sarajevo and Faculty of Pharmacy
University of Sarajevo
Sarajevo, Bosnia and Herzegovina

Lejla Gurbeta Pokvić
Verlab Ltd Sarajevo and International Burch University Sarajevo
Sarajevo, Bosnia and Herzegovina

ISSN 1680-0737 ISSN 1433-9277 (electronic)
IFMBE Proceedings
ISBN 978-3-030-73908-9 ISBN 978-3-030-73909-6 (eBook)
https://doi.org/10.1007/978-3-030-73909-6

This Springer imprint is published by the registered company Springer Nature Switzerland AG
The registered company address is: Gewerbestrasse 11, 6330 Cham, Switzerland

Preface

The International Conference on Medical and Biological Engineering in Bosnia and Herzegovina (CMBEBIH) is an exciting, informative and inspiring conference with the aim of sharing ideas, experiences, best practices and the latest advancements in biomedical engineering and all related fields, while staying true to the conference motto: “Stay focused.”

Our mission is to promote biomedical engineering worldwide, especially in Southeast Europe, as well as to establish and strengthen scientific collaborations.

We bring together leading researchers, academic scientists and engineers together with outstanding speakers in the field of biomedical engineering. We offer different forum for discussion on the latest findings, innovative solutions and emerging challenges this field is facing with, in order to improve the quality of healthcare and life in general.

The 2021 edition of CMBEBIH is a continuation of the extensive work conducted by the Bosnia and Herzegovina Medical and Biological Engineering Society and the company Verlab, both in Sarajevo, to foster a fast development of Medical and Biological Engineering in Bosnia and Herzegovina and Southeast European countries. The previous CMBEBIH gave great results with 8 eminent keynote speakers from all around the world, 81 oral presentations, 41 poster presentations, international workshops, EAMBES General Assembly and Council Meeting.

Research and development activities in biomedical engineering are impacting science, technology and health care. They are helping us better understand human physiology and function at multiple levels. They are providing scientists and doctors with improved tools and techniques for disease detection, prevention and treatment and for better managing healthcare processes and services.

This book gathers the proceedings of CMBEBIH 2021, including 105 papers accepted after a rigorous peer review process. The book has been organized in different thematic sections to reflect the scientific sessions held at the conference. Topics include biomedical signal processing, biomechanics, biosensors and bioinstrumentation; artificial intelligence and machine learning in health care; health informatics, clinical engineering and health technology assessment; medical physics, biomedical imaging and radiation protection. Further topics include biomaterials, molecular, cellular and tissue engineering, bio-micro/nano-technologies, as well as cardiovascular, respiratory and endocrine systems engineering, pharmaceutical engineering and COVID-19-related research.

Some papers were presented during international workshops on clinical engineering and health technology assessment, showing important developments achieved by Institutes in Bosnia and Herzegovina and in Southeastern Europe.

Besides the proceedings papers, seven keynote lectures were organized and moderated by prominent experts and professionals in the area. A brief summary is shown below:

The first keynote was held by Prof. Dr. Thomas Penzel (Germany), Scientific Chair of the Interdisciplinary Sleep Medicine Center at Charité-Universitätsmedizin Berlin, Germany. The title of his lecture was "*The future of sleep medicine technologies.*"

Another interesting lecture was about the latest research on COVID-19, held by Prof. dr. Ivan Đikić, Director of the Institute of Biochemistry II at Goethe University Frankfurt.

Prof. Dr. Mladen Poluta, Health Technology Director, from the Western Cape Department of Health (South Africa) talked about "*Global Clinical Engineering – The New Normal.*"

A keynote lecture entitled "*COVID-19 Pandemic Driving Biomedical Innovations: Standards, Regulations, and Patient Safety*" was held by Prof. Dr. Carole C. Carey, the Chair of the IEEE EMBS Standards Committee (USA).

Prof. Dr. Adnan Mehonić from University College London (UK) spoke about "*The future of energy-efficient AI and neuromorphic engineering.*"

Last but not least, Prof. Dr. Anne Humeau-Heurtier from the University of Angers, France, held a very interesting lecture about the use of artificial neural networks in medicine, with a special focus on the analysis respiratory system. The title of the lecture was *Entropy-based measures for the biomedical field.*

The latest advancements in the field of Clinical Engineering were presented by Prof. Dr. Tom Judd, the President of Clinical Engineering Division in International Federation of Medical and Biological Engineering. The title of his keynote speech was "*Clinical Engineering Making a Difference Globally in Partnership with WHO 2020-2021: A Review of Best Practices from Digital Health, COVID19 Equipment & Vaccine Support, and More.s*"

We express our sincere gratitude to the kind support and effort of international organizations such as IFMBE, EAMBES, IEEE and ESEM for endorsing the conference, as well as our publisher Springer Nature. Furthermore, we would like to acknowledge various industry sponsors, which offered their support to the conference.

Heartfelt thanks to the conference participants and authors of the chapters of this book, for their outstanding contributions. We wish them all the best with their ongoing and future research in advancing medical and biological engineering. We wish to come again together with in the next CMBEBIH and create new opportunities for professional growth and networking between fellow and young scientists, engineers and other enthusiastic people. We strongly believe that this conference has created long-lasting memories, fostered networking and supported both new and ongoing collaborations among its participants.

Sincerely,

April 2021

Almir Badnjevic
CMBEBIH 2021 Conference Chair

Organization

Conference Organization Board

Conference Chair

Almir Badnjevic	Bosnia and Herzegovina Medical and Biological Engineering Society

Conference Co-chairs

Lejla Gurbeta Pokvić	Verlab Ltd Sarajevo, Bosnia and Herzegovina
Tamer Bego	Faculty of Pharmacy Sarajevo, University of Sarajevo, Bosnia and Herzegovina
Dubravka Šimić	Faculty of Pharmacy Mostar, University of Mostar, Bosnia and Herzegovina
Ranko Skrbic	Medical Faculty Banja Luka, University of Banja Luka, Bosnia and Herzegovina

Honorary Chairs

Ervin Sejdic	University of Pittsburgh, USA
Ratko Magjarevic	University of Zagreb, Croatia

Program Committee

Abdulhamit Subasi	Effat University, Jeddah, Saudi Arabia
Adnan Beganović	University Clinical Center Sarajevo, Bosnia and Herzegovina
Aida Šapčanin	University of Sarajevo, Bosnia and Herzegovina
Aleksandar Karać	University of Zenica, Bosnia and Herzegovina
Aljo Mujčić	University of Tuzla, Bosnia and Herzegovina

Amina Kozarić	University Sarajevo, Bosnia and Herzegovina
Anne Humeau-Heurtier	University of Angers, France
Antonio Pedotti	Politecnico di Milano, Italy
Baki Karaböce	Ulusal Metroloji Enstitüsü (UME) TÜBITAK, Turkey
Božidar Ferek	Medtronic, Croatia
Christopher James	University of Warwick, UK
Damijan Miklavčić	University of Ljubljana, Slovenia
Damir Marjanović	International Burch University Sarajevo, Bosnia and Herzegovina
Dejan Jokić	International Burch University Sarajevo, Bosnia and Herzegovina
Dragan Primorac	University of Zagreb, Croatia
Dušanka Bošković	University of Sarajevo
Emir Žunić	University of Sarajevo, Bosnia and Herzegovina
Edhem Hasković	University of Sarajevo, Bosnia and Herzegovina
Enisa Omanović Mikličanin	University of Sarajevo, Bosnia and Herzegovina
Ernesto Iadanza	Florence University, Italy
Ervin Sejdić	University of Pittsburg, USA
Fahir Bečić	University of Sarajevo, Bosnia and Herzegovina
Goran Ristić	University of Niš, Serbia
Igor Lačković	University of Zagreb, Croatia
Isak Karabegović	Academy of Sciences and Arts of Bosnia and Herzegovina
Jasmin Kevrić	International Burch University, Bosnia and Herzegovina
Lana Nezić	University of Banja Luka, Bosnia and Herzegovina
Leandro Pecchia	University of Warwick, UK
Lejla Mehmedović	University of Tuzla, Bosnia and Herzegovina
Leonardo Bocchi	University of Florence, Italy
Mario Medvedec	University Hospital Center Zagreb, Croatia
Miroslav Končar	Oracle Healthcare Zagreb, Croatia
Mirsada Hukić	Academy of sciences and Arts of Bosnia and Herzegovina
Mirza Dedić	University of Sarajevo, Bosnia and Herzegovina
Monia Avdić	International Burch University, Bosnia and Herzegovina
Nenad Filipović	University of Kragujevac, Serbia
Paulo de Carvalho	University of Coimbra, Portugal
Radovan Stojanović	University of Montenegro, Montenegro
Safija Herenda	University of Sarajevo, Bosnia and Herzegovina
Siniša Car	University of Zagreb, Croatia
Sven Lončarić	University of Zagreb, Croatia
Tarik Uzunović	University of Sarajevo, Bosnia and Herzegovina

Thomas Penzel	Charite University Hospital Berlin, Germany
Tomislav Pribanić	University of Zagreb, Croatia
Vedran Bilas	University of Zagreb, Croatia
Vukoman Jokanović	Vinča Institute, University of Belgrade, Serbia
Yves Lemoigne	IFMP Ambilly France & CERN, Switzerland
Zdenka Babić	University of Banja Luka, Bosnia and Herzegovina

Local Organizing Committee

*Members of Bosnia and Herzegovina Medical and Biological Engineering Society

Members

Lemana Spahić
Nejra Gurbeta
Elma Hasković
Sumeja Hadžalić
Emina Pita

Conference Webpage

http://www.cmbebih.com/

Bosnia and Herzegovina Medical and Biological Engineering Society (DMBIUBIH)

The Bosnia and Herzegovina Medical and Biological Engineering Society (DMBIUBIH) is a non-profit organization which was established in 2014 to encourage and promote medical and biological engineering in Bosnia and Herzegovina. The members of the society are university professors, medical doctors, engineers and technicians in the disciplines of medicine and engineering, as well as students, youth with interest in this field. It serves as the lead society and professional home for medical and biological engineering in Bosnia and Herzegovina. DMBIUBIH activities include participation in the formulation of public policy and the dissemination of information through publications and forums. The organization promotes and enhances knowledge and education in medical and biological engineering in Bosnia and Herzegovina Society through organized scientific meetings and diversity initiatives. DMBIUBIH Society is a national representative in IFMBE and EAMBES.
http://www.dmbiubih.org/

Organized by

Bosnia and Herzegovina Medical and Biological Engineering Society, Bosnia and Herzegovina

Co–organized by

Faculty of Pharmacy Sarajevo, University of Sarajevo, Bosnia and Herzegovina
Faculty of Pharmacy Mostar, University of Mostar, Bosnia and Herzegovina
Medical Faculty Banja Luka, University of Banja Luka, Bosnia and Herzegovina

Endorsed by

International Federation for Medical and Biological Engineering (IFMBE)
European Alliance for Medical and Biological Engineering & Science (EAMBES)
European Foundation for Education in Biomedical Engineering (ESEM—Educating Students in Engineering and Medicine)

Technical Co-sponsors

IEEE Bosnia and Herzegovina Section

Sponsored by

Verlab doo Sarajevo, Bosnia and Herzegovina
Teh doo Sarajevo, Bosnia and Herzegovina
ZOONO, South East Europe
Privredna banka Sarajevo dd Sarajevo, Bosnia and Herzegovina
Minerva Medica Sarajevo, Bosnia and Herzegovina
Mehmedbašić – Zavod za ginekologiju, perinatologiju i neplodnost Sarajevo, Bosnia and Herzegovina
Inel doo Mostar, Bosnia and Herzegovina
Hrvatske telekomunikacije d.d. Mostar, Bosnia and Herzegovina
Ljekarne Lupriv Mostar, Bosnia and Herzegovina

Contents

Biomedical Signal Processing

Effect of Seasonal Changes on Serum Vitamin D Concentration

Lulzana Shabani[1](✉), Teuta Shabani-Leka[2], Mimoza Bafqari-Bakiji[3], Ibadete Denjalli[1], and Sanije Berisha[3]

[1] Faculty of Mathematics and Natyral Science, Department of Biology – Biochemistry, University of Tetovo, Tetovo, Macedonia
lulzana.shabani@unite.edu.mk

[2] Faculty of Medical Sciences, University of Tetovo, Tetovo, Macedonia

[3] Clinical Hospital of Tetovo, Tetovo, Macedonia

Abstract. Vitamin D is essential for the formation and maintenance of strong and healthy bones. Vitamin D deficiency can occur from inadequate sun exposure, insufficient food intake, decreased absorption, abnormal metabolism or vitamin D resistance. This study aims to assess the status of 25 (OH) D in serum in different seasons in the population of the Polog region during the period of the year 2019, January-December. Serum 25 (OH) D concentration was measured by the Chemiluminescent method. The study included 997 subjects of different ages and genders who routinely determined the serum 25 (OH) D concentration. The largest number of subjects were tested in the period October-December, based on the obtained results the minimum values of 25 (OH) D was 4.87 ng/ml, where as at the age group >60 years with an average of 15.686 ± 7.6 ng/ml. During the fall and winter months, both sexes are at high risk of vitamin D deficiency corresponding to low serum 25 (OH) D levels.

Keywords: Vitamin D · 25 (OH) D · Chemiluminescent · Seasonal changes · Different age groups

1 Introduction

Vitamin D (cholecalciferol) is synthesized in the skin under the influence of ultraviolet B (UVB) rays from sunlight or can be obtained while consuming fish or plant sources (ergocalciferol). Vitamin D is a precursor to the synthesis of steroid hormones that undergoes chemical conversion in the liver and kidneys: during the first reaction 25 (OH) D is created, while in the second reaction the main bioactive form is synthesized, 1,25-dihydroxyvitamin D (1,25 (OH) 2 D) (1). 1,25 (OH) 2 D, acts through vitamin D receptors (VDR), which regulate proliferation in the basal layer of the epidermis and help sequentially differentiate keratinocytes so that they form the upper layers of the epidermis [2]. Serum 25 (OH) D concentration is a good clinical indicator of vitamin D level because it indicates the total amount of synthesis in the skin and dietary intake of vitamin D [3]. People need vitamin D to maintain the body healthy and fight infections.

A. Badnjevic and L. Gurbeta Pokvić (Eds.): CMBEBIH 2021, IFMBE Proceedings 84, pp. 3–9, 2021.
https://doi.org/10.1007/978-3-030-73909-6_1

Vitamin D is an important factor in regulating bone metabolism. But its actions are not limited to the skeletal system, but extend to several skeletal organs such as the brain, heart, prostate, intestines and immune cells [4, 5].

Some key factors such as season, strength of UV rays, duration of exposure to UV rays, age, place of residence, vitamin intake in the form of supplements, physical activity and the amount of pigment in the skin affect the production of vitamin D3. [6–8] Seasonal changes result in changes in the serum 25- (OH) D level. Serum 25- (OH) D level in winter months particulary in young women is lower compared to summer and spring. Vitamin D deficiency is common in the elderly, such as in places with changing temperatures during the seasons, especially in early spring due to reduced synthesis in the skin during the winter months [9, 10], in people whom stay throughout time inside the home (due to age or inability to move) [11] as well as in patients with various medical conditions.[12, 13] The prevalence of vitamin D deficiency is even higher in elderly patients with fragility fractures, ranging from 55%–91.6% [14].

2 The Aim of This Survey

This survey includes patients who routinely determined 25 (OH) D concentrations during various periods of the year. The aim was to analyze the changes of 25 (OH) D concentration between the time periods January-March, April-June, July-September and October-December.

3 Material and Methods

A total of 997 people from the Polog region were included in this retrospective study. Data collection was performed in the period January December 2019, in the laboratory of the Clinical Hospital of Tetovo. The age of the patients included in the sturvey was from 0 years to 82 years old, with an average age of 50.29 ± 18.453.

Serum 25 (OH) D concentration was measured by the Chemiluminiscent method with the ADVIA Centaur apparatus by the DCLIA (Direct Chemiluminescence immunoassay) method. Data collected from the subjects included in the study were divided by gender, age group and by the time period when the measurement of serum 25 (OH) D concentration was performed.

Reference values of serum vitamin D concentration are different in pediatric and adult ages (for pediatric ages the deficit is taken as <15 ng/ml, while in adults < 20 ng/ml).

Microsoft Office Excel (2007) and SPSS version 26 were used to process and present the results of this survey. The data are presented as mean standard deviation (SD) and relative frequencies (%). The value of $p < 0.05$ was taken as statistically reliable.

4 Results

From the results obtained during the survey, resulted that a total of 997 patients participated, of which 226 (22.64%) patients were male and 771 (77.31%) patients were female (Fig. 1).

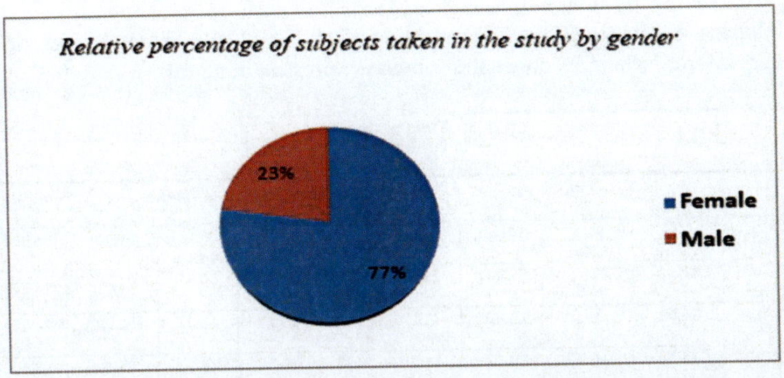

Fig. 1. Relative percentage of subjects taken in the study by gender

The number of patients who have determined the concentration of 25 (OH) D for the period January-December 2019 is high with a relatively high gender difference, where in females it is believed that from the slightest exposure to the sun have appeared concerns and with this is required and the analysis of this parameter (Tables 1 and 2).

Table 1. Distribution of subjects by gender, age group and time of determination of vitamin D level by number

Time period / Age	January-March		April-June		July-September		October-December	
	Male	Female	Male	Female	Male	Female	Male	Female
1-15 years old	6	3	9	4	2	1	15	12
16-30 years old	2	13	9	21	2	8	7	30
31-45 years old	14	33	7	39	4	20	16	84
45-60 years old	7	41	17	74	4	27	21	115
>65 years old	11	38	24	87	13	25	36	96

In the period January-March, 16.8% (n = 168) of the subjects of this study of both sexes were analyzed. The lowest values of 25 OHD were found in females in the age group 46–60 years of 6.85 ng/ml. During this period no statistically significant difference was found between males and females, except for the age group 46–60 years, where $p = 0.0473$. Daily intake of vitamin D is an important factor in maintaining vitamin D in

Table 2. Tabular presentation of the concentration of 25 (OH) D distributed by age group for the period January-March, where its concentration was expressed in ng/ml

Time period JANUARY-MARCH											
	Male					**Female**					
Age groups	**n**	**Min value**	**Max value**	**Mean**	**SD**	**N**	**Min value**	**Max value**	**Mean**	**SD**	**p**
0-15 years old	6	18.58	72.72	35.118	19.667	3	21.32	60.51	43.203	19.991	0.581
16-30 years old	2	16.61	30.0	24.805	7.346	13	15.11	40.87	21.587	7.602	0.585
31-45 years old	14	11.58	46.77	28.622	11.204	33	10.10	78.71	23.30	14.418	0.225
46-60 years old	7	15.45	39.32	25.43	8.166	41	6.85	44.90	18.852	7.848	**0.0473**
>60 years old	11	9.28	37.89	20.751	8.11	38	7.94	87.62	26.521	16.728	0.276

the serum at sufficient levels and protection against vitamin D deficiency, especially on days with less sunlight (Table 3).

Table 3. Tabular presentation of the concentration of 25 (OH) D distributed by age group for the period April-June, where its concentration was expressed in ng/ml

Time period APRIL-JUNE											
	Male					**Female**					
Age groups	**n**	**Min value**	**Max value**	**Mean**	**SD**	**n**	**Min value**	**Max value**	**Mean**	**SD**	**p**
0-15 years old	9	21.98	69.46	39.76	13.94	4	17.42	43.54	27.192	12.073	0.148
16-30 years old	9	16.81	51.90	30.068	12.673	21	14.25	91.68	24.644	16.010	0.376
31-45 years old	7	13.37	67.70	29.530	18.558	39	10.48	126.02	27.402	25.066	0.831
46-60 years old	17	10.10	36.36	23.184	7.411	74	7.72	101.49	22.61	13.13	0.862
>60 years old	24	6.38	46.77	24.417	10.383	87	7.23	66.94	21.509	10.457	0.229

The lowest mean value of 25 OHD 21.509 ± 10.457 ng/ml during this period was encountered in the female gender of the older age group (over 60 years), where the number of examined patients was relatively high. Compared to the previous period, there is a slight increase in the averages in more age groups in both sexes. From the table above it is observed that there is no significant difference between males and females in any age group (Table 4).

This is the most radiant period of the year and the number of subjects analyzed in this period is significantly smaller compared to other time periods 10.6% (n = 106). The highest mean value was found in the second female age group 33.9. 11.47 ng/ml, followed by the fourth male group 31.2 ± 13.5 ng/ml. In the period July-September no

Table 4. Tabular presentation of the concentration of 25 (OH) D distributed by age group for the period July-September, where its concentration was expressed in ng/ml

Time period JULY-SEPTEMBER											
	Male					Female					
Age groups	n	Min value	Max value	Mean	SD	n	Min value	Max-value	Mean	SD	p
0-15 years old	2	13.39	46.50	29.945	23.412	2	21.90	39.58	30.74	12.501	0.970
16-30 years old	2	20.62	39.18	29.900	13.123	7	14.51	47.14	33.977	11.474	0.677
31-45 years old	4	18.41	39.76	28.605	9.105	20	11.52	56.11	29.252	9.815	0.904
46-60 years old	4	17.11	49.80	31.272	13.585	27	10.23	59.85	25.785	12.549	0.425
>60 years old	13	12.81	33.90	24.083	6.868	25	11.08	38.02	19.536	6.591	0.054

statistically significant difference was observed between males and females in any age group (Table 5).

Table 5. Tabular presentation of the concentration of 25 (OH) D distributed by age group for the period October-December, where its concentration was expressed in ng/ml

Time period OCTOBER-DECEMBER											
	Male					Female					
Age groups	n	Min value	Max value	Mean	SD	n	Min value	Max value	Mean	SD	p
0-15 years old	15	12.47	50.76	25.589	9.535	12	8.39	76.28	27.108	17.609	0.776
16-30 years old	7	8.56	28.0	19.198	5.908	30	8.90	34.66	20.072	6.942	0.760
31-45 years old	16	8.89	37.36	19.086	7.658	84	5.15	71.34	18.516	11.317	0.847
46-60 years old	21	10.32	60.13	21.760	10.615	115	6.62	45.39	16.500	7.871	**0.0088**
>60 years old	36	4.87	38.86	15.686	7.685	96	5.71	45.15	16.397	8.016	0.647

The highest percentage of subjects were analyzed in the period October-December, with a percentage of 43.32%. Serum values of 25 OHD in this time period are significantly lower compared to spring and summer period. For each age group the average is significantly lower compared to other time periods. The most endangered age group is the oldest male where the lowest serum value of 25 OHD (4.87 ng/ml) with an average of 15.8 ± 7.6 was encountered, followed by the female of the same age group with an average of 16.3 ± 8.0.

From the statistical processing of data in table no. 5 there is a statistically significant difference between men and women in the age group 46–60 years, where $p = 0.0088$.

5 Conclusion

This was a 12-month survey to determine the seasonal changes of vitamin D concentration. Our results show a large prevalence of vitamin D deficiency and insufficiency in all seasons, with a higher prevalence of deficiency in winter, where the highest values are encountered during the winter period with the lowest of 25 (OH) D concentration, compared to summer and autumn in the population of Polog region. Several studies have found that vitamin D levels increase in summer and decrease in winter [26–28] due to vitamin D dependence on sunlight. However, vitamin D deficiency has also been reported in sunny seasons. The results of our study showed that the mean values of 25 (OH) D were ≤39 ng/ml in males during the four seasons, where as ≤43 ng/ml in females also during the four seasons. In Europe, vitamin D deficiency is estimated to be 40.4%, where ethnic groups with greater skin pigmentation were found to have a higher prevalence of vitamin D deficiency [29]. At high northern latitudes (above 40° N) even with adequate sun exposure, dermal generation of vitamin D is low or absent in winter and thus increases the demand for dietary consumption [30]. The age group at which the lowest values of vitamin D were encountered was the age group > 60 years in the period October-December, when solar radiation is absent. There is a statistically significant difference between males and females in the time periods January-March and October. December, at the age group 46–60 years old.

References

1. Atherton, K., Berry, D.J., Parsons, T., Macfarlane, G.J., Power, C., Hypponen, E.: Vitamin D and chronic widespread pain in a white middle-aged British population: evidence from a cross-sectional population survey. Ann. Rheum. Dis. **68**(6), 817–822 (2009). PMID: 18697776
2. Holick, M.F.: Vitamin D, and bone health. J. Nutr. **126**, 1159S–1164S (1996)
3. El Baba, K., Zantout, M.S., Akel, R., Azar, S.T.: Seasonal variation of vitamin D and HbA 1c levels in patients with type 1 diabetes mellitus in the Middle East. Int. J. Gen. Med. **4**, 635–638 (2011)
4. Holick, M.F.: Vitamin D, deficiency. N. Engl. J. Med. **357**, 266–281 (2007)
5. Bikle, D.D.: Vitamin D and the skin: physiology and pathophysiology. Rev. Endocr. Metab. Disord. **13**, 3–19 (2012)
6. Costanzo, P.R., Elías, N.O., Kleiman Rubinsztein, J., et al.: Ultravio-let radiation impact on seasonal variations of serum 25hydroxy-vitamin D in healthy young adults in Buenos Aires. Medicina (B Aires) **71**, 336–342 (2011)
7. Pilz, S., Kienreich, K., Stückler, D., Meinitzer, A., Tomaschitz, A.: Associations of sun exposure with 25-hydroxyvitamin D and parathyroid hormone levels in a cohort of hypertensive patients: the Graz endocrine causes of hypertension (GECOH) study. Int. J. Endocrinol. **2012**, 732636 (2012)
8. Looker, A.C., Dawson-Hughes, B., Calvo, M.S., Gunter, E.W., Sahyoun, N.R.: Serum 25-hydroxyvitamin D status of adolescents and adults in two seasonal subpopulations from NHANES III. Bone **30**, 771–777 (2002)
9. Nurmi, I., Kaukonen, J.P., Lüthje, P., et al.: Half of the patients with an acute hip fracture suffer from hypovitaminosis D: a prospective study in southeastern Finland. Osteoporos Int. **16**, 2018–2024 (2005)

10. Lips, P.: Vitamin D deficiency and secondary hyperparathyroid-ism in the elderly: consequences for bone loss and fractures and therapeutic implications. Endocr. Rev. **22**, 477–501 (2001)
11. Thomas, M.K., Lloyd-Jones, D.M., Thadhani, R.I., et al.: Hypovita-minosis D in medical inpatients. N. Engl. J. Med. **338**, 777–783 (1998)
12. Moniz, C., Dew, T., Dixon, T.: Prevalence of vitamin D inadequacy in osteoporotic hip fracture patients in London. Cur.r Med. Res. Opin. **21**, 1891–1894 (2005)
13. Beringer, T., Heyburn, G., Finch, M., et al.: Prevalence of vitamin D inadequacy in Belfast following fragility fracture. Curr. Med. Res. Opin. **22**, 101–105 (2006)
14. LeBoff, M.S., Kohlmeier, L., Hurwitz, S., Franklin, J., Wright, J., Glowacki, J.: Occult vitamin D deficiency in postmenopausal US women with acute hip fracture. JAMA **281**, 1505–1511 (1999)
15. Springbett, P., Buglass, S., Young, A.R.: Photoprotection and vita-min D status. J. Photochem. Photobiol. B **101**, 160–168 (2010)
16. Linos, E., Keiser, E., Kanzler, M., et al.: Sun protective behaviors and vitamin D levels in the US population: NHANES 2003–2006. Cancer Causes Control **23**, 133–140 (2012)
17. Thandrayen, K., Pettifor, J.M.: Maternal vitamin D status: implications for the development of infantile nutritional rickets. Endocrinol. Metab. Clin. North Am. **39**, 303–320 (2010)
18. Wortsman, J., Matsuoka, L.Y., Chen, T.C., Lu, Z., Holick, M.F.: Decreased bioavailability of vitamin D in obesity. Am. J. Clin. Nutr. **72**, 690–693 (2000)
19. Oudshoorn, C., van der Cammen, T.J., McMurdo, M.E., van Leeu-wen, J.P., Colin, E.M.: Ageing and vitamin D deficiency: effects on calcium homeostasis and considerations for vitamin D supplementation. Br. J. Nutr. **101**, 1597–1606 (2009)
20. Patel, S., Barron, J.L., Mirzazedeh, M., et al.: Changes in bone mineral parameters, vitamin D metabolites, and PTH measurements with varying chronic kidney disease stages. J. Bone Miner. Metab. **29**, 71–79 (2011)
21. Williams, S., Malatesta, K., Norris, K.: Vitamin D and chronic kidney disease. Ethn. Dis. **19**(4 Suppl. 5) S5-8-11 (2009)
22. Lo, C.W., Paris, P.W., Clemens, T.L., Nolan, J., Holick, M.F.: Vitamin D absorption in healthy subjects and in patients with intestinal malabsorption syndromes. Am. J. Clin. Nutr. **42**, 644–649 (1985)
23. Kennel, K.A., Drake, M.T., Hurley, D.L.: Vitamin D deficiency in adults: when to test and how to treat. Mayo Clin. Proc. **85**, 752–757 (2010)
24. Wang, S.: Epidemiology of vitamin D in health and disease. Nutr. Res. Rev. **22**, 188–203 (2009)
25. Greene-Finestone, L.S., Berger, C., de Groh, M., et al.: 25-Hydroxyvitamin D in Canadian adults: biological, environmental, and behavioral correlates. Osteoporosis international: a journal established as result of cooperation between the European Foundation for Osteoporosis and the National Osteoporosis Foundation of the USA, vol. 22, pp. 1389–1399 (2011)
26. Smith, M.: Seasonal, ethnic and gender variations in serum vitamin D3 levels in the local population of Peterborough. Biosci. Horiz. Int. J. Student Res. **3**, 124–131 (2010)
27. Klingberg, E., Oleröd, G., Konar, J., Petzold, M., Hammarsten, O.: Seasonal variations in serum 25-hydroxy vitamin D levels in a Swedish cohort. Endocrine **49**, 800–808 (2015)
28. Cashman, K.D., Dowling, K.G., Škrabáková, Z., et al.: Vitamin D deficiency in Europe: pandemic? Am. J. Clin. Nutr. **103**, 1033–1044 (2016)
29. Holick, M.F.: Vitamin D: a millenium perspective. J. Cell Biochem. **88**, 296–307 (2003). https://doi.org/10.1002/jcb.10338

Diagnosis of Neuromuscular Disorders Using TQWT and Random Subspace Ensemble Classifier

Abdulhamit Subasi[1(✉)] and Emine Yaman[2]

[1] Effat University, College of Engineering, Jeddah 21478, Saudi Arabia
absubasi@effatuniversity.edu.sa
[2] International University of Sarajevo, Sarajevo, Bosnia and Herzegovina
eyaman@ius.edu.ba

Abstract. Electromyographic (EMG) signals are the elements of the neuromuscular disorders diagnosis, whereas machine learning techniques are utilized as a computer aided decision support system in the diagnosis of neuromuscular disorders. On the other hand, ensemble learners improve the accuracy of the weak learners through weighted combination of multiple classifier models. Therefore, this study uses tunable Q wavelet transform (TQWT) to extract features from the raw EMG, while the Random Subspace ensemble classifier is employed to classify the EMG signals. Hence, the proposed Random Subspace ensemble classifier model with TQWT feature extraction achieved better performance with k-fold cross validation. Experimental results show the feasibility of Random Subspace ensemble classifier model for diagnosis of neuromuscular disorders. Results are promising and showed that the SVM and ANN with Random Subspace ensemble method archived an accuracy of 99%.

Keywords: Electromyography (EMG) · Tunable-Q Wavelet Transform (TQWT) · Random Subspace ensemble classifier

1 Introduction

It is imperative to bear in mind that the human muscular system consists of two primary parts: nervous system and muscular system. Hence, various muscle fibers or nerves can be the cause of the neuromuscular disorders and it is crucial to know the exact cause of the disorder. Electromyography (EMG) is a method introduced for diagnosing neuromuscular disorders on the basis of cell action in the course of muscle activity [1–4]. Performance of feature extraction algorithms and classifiers can significantly affect the accuracy of EMG signal classification. Therefore, through resolution of problems encountered, the performance of the EMG signal classification system can be enhanced. Recently, numerous approaches are presented for extracting appropriate features to classify EMG data [4–7]. Nevertheless, it is still quite challenging to design an accurate diagnostic system. Hence, it is necessary to carry out EMG signals for systematic treatment, to have efficient

A. Badnjevic and L. Gurbeta Pokvić (Eds.): CMBEBIH 2021, IFMBE Proceedings 84, pp. 10–19, 2021.
https://doi.org/10.1007/978-3-030-73909-6_2

EMG signal classification. As a result, scientists have developed numerous quantitative EMG analysis algorithms which are computer-based [1–4, 8–13].

There are important stages in biomedical signal analysis. The most critical of these is the feature extraction stage. This study uses tunable Q wavelet transform (TQWT) for feature extraction from the EMG data, and then statistical values of every sub-band of TQWT is calculated to reduce dimension. Finally, those statistical values of TQWT coefficients serve as inputs to classifier. Since, in many cases single classifiers are not effective, ensemble classifiers can be employed for EMG signals classification in detecting neuromuscular disorders. Ensemble learners use multiple learners to achieve more satisfactory results. Recent researches have shown that ensemble learning algorithms work more effectively than individual algorithms [14–16]. Precisely, ensemble classifiers take care majority of learners, based on various features subset extracted from the original dataset randomly. Moreover, these learning methods tolerate utilization of common classifiers which are applied for diagnosing several disorders. Hence the contribution of this research is to offer the Random Subspace classifier for diagnosis of the neuromuscular disorder with TQWT feature extraction. In this respect, various individual classifiers are employed with Random Subspace ensemble classifier for purpose of improving the classification performance for diagnosing the neuromuscular disorders.

Subasi [1] employed neuro-fuzzy computing methods with three characteristic extraction techniques namely AR, DWT and WPE. The ANFIS classification method yield a classification accuracy of 95% with AR + DWT features. Moreover, Subasi [19] employed SVM method for EMG signals classification utilizing normal, myopathic and neurogenic EMG signals with an evolutionary approach. The proposed evolutionary SVM has accomplished a classification accuracy of 97% through the use of 10-fold cross-validation. Other study [19] suggested the fuzzy support vector machines (FSVM) modelling combined with discrete wavelet transform (DWT) for obtaining better performance. By using cross-validation, 97.67% accuracy was achieved. Another study [2] has utilized of PSO-SVM for EMG signals classification. This PSO-SVM classification system achieved notable improvements regarding the classification accuracy which was 97.41%.

Gokgoz and Subasi [7] studied Multiscale Principal Component Analysis (MSPCA) denoising method impact on EMG signal classification. Multiple Single Classification (MUSIC) feature extraction method was used on EMG signals for classification into normal, ALS or myopathic. They proved that denoising EMG signals with MSPCA has efficiently enhanced classification accuracy. After denoising with MSPCA precision is 92.55% for SVM, 90.02% for ANN, 82.11% for k-NN. Parsaei and Stashuk [21] employed K-means clustering and supervised classification employing certainty-based algorithm. This model has 86.4% accuracy for simulated and 96.4% for real data. Bozkurt et al. [5] employed autoregressive (AR) parametric techniques and subspace-based methods for EMG recordings composed of normal, myopathic, and neurogenic disorders. A feedforward error backpropagation artificial neural network (FEBANN) and combined neural network (CNN) have been utilized and highest performance has been obtained with the eigenvector method. The average accuracy was roughly 94% for CNN and 93.3% for FEBANN.

Sengur et al. [23] proposed a deep learning-based method for effectively classifying ALS and normal EMG signals. They used different time frequency methods combined with convolutional neural network to classify EMG signals. They used ALS and normal EMG signals and achieved 96.80% accuracy with CWT and CNN. Hazarika et al. [24] presented real-time feature extraction and fusion method for automatic classification of electromyogram signals with amyotrophic lateral sclerosis (ALS), myopathy (MYO) and normal (NOR) using Discrete wavelet transform and canonical correlation analysis (CCA) methods. Collected discriminant characteristics are given to k-NN classifier and 98.80% accuracy is achieved with two-fold crossvalidation.

The remaining part of this study contains the following chapters explains materials and methods, and introduces techniques employed in each stage of EMG signal classification procedure and subjects. Section 3 provides a broad experimental study on Random Subspace ensemble classifier employed for EMG signal classification. This section also provides for a performance comparison with the state-of-the-art techniques. The final chapter consists of the conclusion of the study.

2 Materials and Methods

2.1 Subjects and Data Acquisition

Concentric needle-electrode has been utilized to record EMG signals from biceps brachii muscles. Signals have samples at 20 kHz for 5s with 12-bit resolution and band-pass which is filtered at 5 Hz to 10 kHz. EMG data was recorded from 7 control subjects, 3 of which were male and 4 female, their age in the range from ten to fourty-three (mean age ± standard departure: 30.2 ± 10.8 years), as well as from seven myopathic subjects four of which were males and three females, their age in between from seven to fourty-six (mean age ± standard departure: 21.5 ± 13.3 years) and from 13 neuropathic subjects, 8 of which were male and 5 female, their age in the range from seven to fifty-five (mean age ± standard departure (S.D.): 25.1 ± 17.2 years) [8].

2.2 Feature Extraction Using TQWT and Dimension Reduction

It is important to reduce the parameters on the data by using certain methods for better results. We call feature extraction to reduce the parameters with certain methods. Studies have shown that better accuracy is obtained on feature extracted data. This also applies to EMG data classification. Wavelet transform that decompose signal into fundamental functions, referred to as wavelets can be used to decompose a signal which has higher time resolution [33]. The Tunable-Q Wavelet Transform (TQWT) is one of imperative tool for oscillatory signals analysis [26]. It can be simply adjusted as a function of the targeted application. It has some tunable parameters which are Q, r and j. Here, Q represents the Q-factor, r represents the oversampling rate and levels of decomposition are represented by j. The amount of oscillations of the wavelet are adjusted by Q. r controls the unnecessary changes to define the wavelet temporal localization while conserving its form [26, 27].

Due to the fact that the feature vector which are produced from the waveletbased feature extraction algorithms cannot be used as classifier inputs because they are too big, certain dimension reduction techniques must be employed to derive less numbers out of characteristics of wavelet coefficients. Therefore, there are six statistical features implemented for the purposes of dimension reduction which are Mean absolute values of the coefficients in each sub-band, average power of the coefficients in each sub-band, standard deviation of the coefficients in each sub-band, ratio of the absolute mean values of coefficients of adjacent sub-bands, skewness of the coefficients in each sub-band, kurtosis of the coefficients in each sub-band.

2.3 The Random Subspace Method

Random Subspace ensemble learning introduced by Ho [31] modifies practice data in the feature space. In this way, r-dimensional random subspace of the original p-dimensional characteristic space is obtained. Altered practice set composes of rdimensional practice objects to build classifiers in random subspaces and to employ their votes in the final prediction. The final decision has been brought based majority voting. Random Subspace ensemble algorithm employs random subspaces for aggregating and building classifiers. This approach is especially appealing when number of practice objects is smaller than data dimensionality. Hence this allows construction of classifiers in random subspaces and solves problem of small sample size. Dimension of subspace is less than the original characteristics space. However, number of practice objects stays the same. It increases relative size of relative training. Therefore, in the case of a data with adequate redundancy, better classifiers could be constructed in subspaces which are produced randomly as compared to the original feature data. Collective resolution from these classifiers is as precise as compared to an individual classifier build by employing original practice set in total characteristic space [29].

3 Results and Discussion

In order to classify EMG signals, this study utilizes statistical values which are derived for every sub-band of TQWT of EMG signal pattern. Since appropriate classification algorithm must tolerate anticipated differences, some patterns of features must be obtained from EMG data. Furthermore, Random Subspace ensemble classification algorithms which have different configuration are used for obtaining more accurate EMG signal recognition.

3.1 Performance Evaluation Metrics

It is not a good idea to take performance in the training set as a performance indicator in the independent test set. Estimating performance based on limited data is not very reliable. The most commonly used method to estimate the error rate is crossvalidation with 10 folds, if the data set is small [28].

Some statistics are used to evaluate the classification results. They are accuracy, F-measure, area under the curve (AUC) and kappa statistics. Accuracy demonstrates the classification performance of the offered model in percentage. Generally, it means the effectiveness of the applied methods. The relation between sensitivity and specificity of algorithm was examined by AUC. F-measure is a complex measure that is good for more sensitive algorithms [30]. The Kappa statistic measure considers the anticipated form through taking it off from predictor's successes and showing outcome as ratio of the whole for perfect predictor [28].

3.2 Experimental Results

In this research, statistical values derived for every sub-band of TQWT are utilized for diagnosing neuromuscular disorders. Features are obtained from every signal frame in order to generate a complete feature pattern which is employed for EMG signal pattern characterization. Then, ensemble machine learning method is applied to achieve precise results for EMG classification. Typical pattern classification problem occurs when EMG signals are classified into groups with similar shape. In order to develop classifiers, this study extracts feature vectors by using TQWT for every EMG signal frame. Statistical properties from TQWT sub-bands are also employed to calculate feature vectors. Finally, every feature vector is used for producing EMG signal patterns.

During experiments, improved classification accuracy through the use of statistical values are achieved utilizing TQWT and Random Subspace ensemble classifier. 10-fold cross validation is applied because it prevents bias that could be potentially present when choosing certain training and test sets. Tables 1 and 2 contain summary of classifiers' performance for the EMG data. It can be seen that all methods achieved quite good performance accordance to total classification accuracy, AUC, F-measure, Kappa statistics. LAD Tree has lowest output, its result being 89.92% from an individual classifiers' accuracy (see Table 1). However, LAD Tree had accuracy of 90.21% with Random Subspace method (see Table 2). Nonetheless, SVM had the best performance of 98.79% classification accuracy from individual classifiers. Besides, highest output is reached by ANN and SVM – namely, 99%, when Random Subspace ensemble classifier was introduced. Accuracy of weak classifiers such as those of C4.5, Random Tree, REP Tree, LAD Tree have significantly improved.

Table 1. Performance of single classifiers for EMG signal classification

	Accuracy (%)	F – measure	ROC Area	Kappa
ANN	98.25	0.982	0.997	0.974
k-NN	93.79	0.938	0.98	0.907
SVM	98.79	0.988	0.993	0.982
Random forest	98.38	0.984	0.999	0.976
C4.5	96.71	0.967	0.981	0.951
Random tree	93.33	0.933	0.95	0.900
REP tree	93.33	0.933	0.95	0.943
LAD tree	89.92	0.899	0.907	0.849

Table 2. EMG signal classification results for random subspace ensemble classifier

	Accuracy (%)	F – measure	ROC Area	Kappa
ANN	99.00	0.99	0.998	0.985
k-NN	94.58	0.946	0.992	0.919
SVM	99.00	0.99	0.998	0.985
Random forest	98.38	0.984	0.999	0.976
C4.5	98.38	0.984	0.999	0.976
Random tree	97.67	0.977	0.997	0.965
REP tree	97.33	0.973	0.998	0.960
LAD tree	90.21	0.902	0.967	0.853

The F-measure of the methods presented in the Table 1 for single classifiers. Following the Random Subspace ensemble learning methods, up to 0.99 is achieved.

AUC of the methods presented in the Table 1. The best AUC (0.999) was reached by Random Forest and C4.5 using Random Subspace ensemble.

Kappa results of the methods shown in the Table 1. Following the use of Random Subspace ensemble learning technique, Kappa value is increased up to 0.985.

The most important point of EMG Signal Classification efficiency is which of the input variables and methods chosen for the classification. Another important norm for deriving the most crucial features from the EMG is the way of selecting signal processing methods [33]. For EMG signal classification, the most suitable parameters should be used as input to the model. Therefore, the statistical properties inferred for each subband of TQWT are chosen and each subband must be relevant to classify the nonlinear dynamics underlying muscle movements and to allow prediction of the growth and regularity of the complication of EMG [34].

3.3 Discussion

The research objective is to develop a framework to assist in diagnosis and treatment of neuromuscular diseases through use of EMG signals. It must be noted that the best result is obtained through SVM, in cases when individual classifiers are used to classify EMG data. In addition, ANN and SVM were shown to be algorithms with most success when using the Random Subspace ensemble method for success rate enhancement. For the purpose of creating a classifier, the input variable selection plays a very crucial role. For this reason, statistical characteristics for every sub-band of TQWT have been employed being the input of the classifier. This is why clinicians need to comprehend all the requirements prior to actual use. Ensemble learning methods have more successful performance when detecting neuromuscular disorders, judging by final outcome of the study at hand.

The adjustment of parameters is performed according to different logic for each classification method. Namely, in random forest method, the only key parameter which is adjusted is the number of trees. In addition, Researchers still do not have enough information about some parameters. Different machine learning approaches are designed since development of appropriate algorithm is hard to achieve. In this respect, ANN has been proven to have the most success and are considered proper from the time the beginning of computer supported analysis. Since there were almost no studies that cover diagnosis of neuromuscular disorders through TQWT feature extraction method and Random Subspace learning method, this study shows the potential that the combination of these two methods have in the field of neuromuscular disorders diagnosing through the processing of EMG data. Moreover, by applying Random Subspace ensemble method, the classification performance of weak algorithms like LAD tree, REP tree, Random Tree, C4.5 are greater than that of the rest of classifiers like ANN, k-NN and SVM. Consequently, the outcomes of this study are having a satisfactory rate of success of 99% in comparison to other literature examples.

Previous studies on the classification of EMG data using different feature extraction methods are listed in Table 3. Subasi et al. in a study developed a method that use AR method to extract important features and WNN as classifier and achieved 90.7% accuracy. Katsis et al. [9] tried to classify EMG signals using Fuzzy k-means algorithm for feature extraction and SVM for classification and have reached 86.14% success rate. SVM was used by Kaur et al. [1] and achieved 95.90% accuracy. Subasi employed different feature extraction methods like AR and DWT and various types of classifiers and reached maximum 97.67% success rate. Kamali et al. [8, 32] used Time and time-frequency domain features for the first and Time-Frequency features for the second model and SVM classifier. They reported 91% success rate from first study and 97% success rate from second study. Gokgoz and Subasi [6] used Random Forest tree approach with the DWT feature extraction. There are different studies in this area and the comparison of these studies are given in Table 3. When you examine the results, you will see that our method works better than other studies.

Table 3. Classification accuracy accomplished by previous studies.

The study reference	Feature extraction method	Classifier	Classification accuracy
[4]	AR	WNN	90.7%
[17]	Fuzzy k-means	SVM	86.14%
[18]	Peaks of MUAPs	SVM	95.90%
[10]	Time domain features	Adaptive fuzzy k-NN	93.5%
[1]	AR + DWT	ANFIS	95%
[11]	DWT	evolutionary SVM	97%
[19]	DWT	FSVM	97.67%
[2]	DWT	PSO-SVM	97.41%
[32]	Time and time-frequency domain features	SVM	91%
[8]	Time-Frequency	SVM	97%
[7]	MUSIC	SVM	92,55
[6]	DWT	Random forest	96.67%
[5]	MUSIC	Combined neural network (CNN)	94%
[20]	Normalized weight vertical visibility algorithm	SVM	98.36%
[22]	EEMD	LDA	98%
[25]	EMD	LS-SVM	96.33%
[23]	CWT	CNN	96.80%
Proposed method	TQWT	Random subspace ensemble with SVM	99%

4 Conclusion

In biomedical research, neuromuscular disorders diagnosis through electromyography (EMG) signals is a very important topic because the several MUAPs' classifications and their proper identification are the requirement to reach the successful treatment of patients. The paper offers a novel framework for recognition of MUAP by applying the application of statistical values of sub-band of tunable Q wavelet transform (TQWT). Different pathological variations in EMG signals are represented through statistical properties which are extracted from every sub-band TQWT. This algorithm is more successful for identification different types of MUAP than are other literature examples. Namely, output indices of MUAP detection are satisfactory because of the new combination of tunable Q wavelet transform and Random Subspace ensemble method together. Existing literature covers many studies on ensemble learning methods classification problems none of which have unfortunately been used for diagnosing neuromuscular disorders. Based on the obtained results, method proposed in this study is more successful than individual classifiers. Moreover, this framework can be easily integrated into any type of computer-based diagnosis framework.

Acknowledgement. The authors use this opportunity to thank Dr. Mustafa Yilmaz, University of Gaziantep, Department of Neurology, for EMG data used here.

Funding. The study is funded by Effat University, Decision Number UC#7/28 Feb. 2018/10.244h, Jeddah, Saudi Arabia.

References

1. Subasi, A.: Classification of EMG signals using combined features and soft computing techniques. Appl. Soft Comput. **12**(8), 2188–2198 (2012)
2. Subasi, A.: Classification of EMG signals using PSO optimized SVM for diagnosis of neuromuscular disorders. Comput. Biol. Med. **43**(5), 576–586 (2013)
3. Begg, R., Lai, D.T., Palaniswami, M.: Computational Intelligence in Biomedical Engineering. CRC Press , Boca Raton (2007)
4. Subasi, A., Yilmaz, M., Ozcalik, H.R.: Classification of EMG signals using wavelet neural network. J. Neurosci. Methods **156**(1), 360–367 (2006)
5. Bozkurt, M.R., Subasi, A., Koklukaya, E., Yilmaz, M.: Comparison of AR parametric methods with subspace-based methods for EMG signal classification using standalone and merged neural network models. Turkish J. Electr. Eng. Comput. Sci. **24**(3), 1547–1559 (2016)
6. Gokgoz, E., Subasi, A.: Comparison of decision tree algorithms for EMG signal classification using DWT. Biomed. Signal Process. Control **18**, 138–144 (2015)
7. Gokgoz, E., Subasi, A.: Effect of multiscale PCA de-noising on EMG signal classification for diagnosis of neuromuscular disorders. J. Med. Syst. **38**(4), 31 (2014)
8. Kamali, T., Boostani, R., Parsaei, H.: A multi-classifier approach to MUAP classification for diagnosis of neuromuscular disorders. IEEE Trans. Neural Syst. Rehabil. Eng. **22**(1), 191–200 (2014)
9. Katsis, C.D., Exarchos, T.P., Papaloukas, C., Goletsis, Y., Fotiadis, D.I., Sarmas, I.: A two-stage method for MUAP classification based on EMG decomposition. Comput. Biol. Med. **37**(9), 1232–1240 (2007)
10. Rasheed, S., Stashuk, D., Kamel, M.: A software package for interactive motor unit potential classification using fuzzy k-NN classifier. Comput. Methods Programs Biomed. **89**(1), 56–71 (2008)
11. Subasi, A.: Medical decision support system for diagnosis of neuromuscular disorders using DWT and fuzzy support vector machines. Comput. Biol. Med. **42**(8), 806–815 (2012)
12. Dobrowolski, A.P., Wierzbowski, M., Tomczykiewicz, K.: Multiresolution MUAPs decomposition and SVM-based analysis in the classification of neuromuscular disorders. Comput. Methods Programs Biomed. **107**(3), 393–403 (2012)
13. Baraka, A., Shaban, H., Abou El-Nasrand, M., Attallah, O.: Wearable accelerometer and sEMG-based upper limb BSN for Tele-rehabilitation. Appl. Sci. **9**(14), 2795 (2019). https://doi.org/10.3390/app9142795
14. Brown, G.: Ensemble learning. In: Encyclopedia of Machine Learning, pp. 312–320. Springer, Heidelberg (2011)
15. Kuncheva, L.I.: Combining Pattern Classifiers: Methods and Algorithms. Wiley, Hoboken (2004)
16. Valentini, G., Dietterich, T.G.: Bias-variance analysis of support vector machines for the development of SVM-based ensemble methods. J. Mach. Learn. Res. **5**(Jul), 725–775 (2004)
17. Das, R., Sengur, A.: Evaluation of ensemble methods for diagnosing of valvular heart disease. Expert Syst. Appl. **37**(7), 5110–5115 (2010)

18. Kaur, G., Arora, A.S., Jain, V.: Multi-class support vector machine classifier in EMG diagnosis. WSEAS Trans. Signal Process. **5**(12), 379–389 (2009)
19. Kaur, G., Arora, A., Jain, V.: EMG diagnosis via AR modeling and binary support vector machine classification. Int. J. Eng. Sci. Technol. **2**(6), 1767–1772 (2010)
20. Artameeyanant, P., Sultornsanee, S., Chamnongthai, K.: An EMG-based feature extraction method using a normalized weight vertical visibility algorithm for myopathy and neuropathy detection. SpringerPlus **5**(1), 2101 (2016)
21. Parsaei, H., Stashuk, D.W.: EMG signal decomposition using motor unit potential train validity. IEEE Trans. Neural Syst. Rehabil. Eng. **21**(2), 265–274 (2013)
22. Naik, G.R., Selvan, S.E., Nguyen, H.T.: Single-channel EMG classification with ensemble-empirical-mode-decomposition-based ICA for diagnosing neuromuscular disorders. IEEE Trans. Neural Syst. Rehabil. Eng. **24**(7), 734–743 (2016)
23. Khan, M., Singh, J., Tiwari, M.: A multi-classifier approach of EMG signal classification for diagnosis of neuromuscular disorders. Int. J. Comput. Appl. **133**(4) (2016)
24. Hazarika, A., Dutta, L., Boro, M., Barthakur, M., Bhuyan, M.: An automatic feature extraction and fusion model: application to electromyogram (EMG) signal classification. Int. J. Multimedia Inf. Retrieval 1–14 (2018)
25. Mishra, V.K., Bajaj, V., Kumar, A., Sharma, D., Singh, G.: An efficient method for analysis of EMG signals using improved empirical mode decomposition. AEU Int. J. Electron. Commun. **72**, 200–209 (2017)
26. Vetterli, M., Herley, C.: Wavelets and filter banks: theory and design. IEEE Trans. Signal Process. **40**(9), 2207–2232 (1992)
27. Patidar, S., Pachori, R.B.: Classification of cardiac sound signals using constrained tunable-Q wavelet transform. Expert Syst. Appl. **41**(16), 7161–7170 (2014)
28. Bayram, I., Selesnick, I.W.: Frequency-domain design of overcomplete rationaldilation wavelet transforms. IEEE Trans. Signal Process. **57**(8), 2957–2972 (2009)
29. Kalmegh, S.: Analysis of WEKA data mining algorithm REPTree, Simple CART and RandomTree for classification of Indian news. Int. J. Innov. Sci. Eng. Technol. **2**(2), 438–446 (2015)
30. Sokolova, M., Japkowicz, N., Szpakowicz, S.: Beyond accuracy, F-score and ROC: a family of discriminant measures for performance evaluation. Presented at the Australasian Joint Conference on Artificial Intelligence, pp. 1015–1021. Springer, Heidelberg (2006)
31. Bradley, A.P.: The use of the area under the ROC curve in the evaluation of machine learning algorithms. Pattern Recogn. **30**(7), 1145–1159 (1997)
32. Kamali, T., Boostani, R., Parsaei, H.: A hybrid classifier for characterizing motor unit action potentials in diagnosing neuromuscular disorders. J. Biomed. Phys. Eng. **3**(4), 145 (2013)
33. Subasi, A., Yaman, E.: EMG signal classification using discrete wavelet transform and rotation forest. Presented at International Conference on Medical and Biological Engineering (CMBEBIH), vol. 73, pp. 29–35. Springer, Cham (2019). https://doi.org/10.1007/978-3-030-17971-7_5
34. Yaman, E., Subasi, A.: Comparison of bagging and boosting ensemble machine learning methods for automated EMG signal classification. BioMed Res. Int. 2019, Article ID 9152506, 13 p. (2019). https://doi.org/10.1155/2019/9152506

Analysis of Autonomic Nervous System Biosignals

Magdalena Krbot Skorić[1,2](✉), Ivan Adamec[1], Mario Cifrek[2], and Mario Habek[1,3]

[1] University Hospital Center Zagreb, Department of Neurology, Referral Center for Autonomic Nervous System Disorders, Zagreb, Croatia
[2] Faculty of Electrical Engineering and Computing, University of Zagreb, Zagreb, Croatia
[3] School of Medicine, University of Zagreb, Zagreb, Croatia

Abstract. Autonomic nervous system (ANS) regulates many important systems in the human body, such as blood pressure, heart rate, digestion, respiratory system and many others. Dysfunction of the ANS is related to disorders in different systems in the human body, and it can be recognizable in the ANS biosignals. Digital signal processing (DSP) of the ANS biosignals provides information about the functioning of the autonomic nervous system, which is very useful in everyday clinical practice. There is a large variety of different methods for ANS signals analysis, with different features extracted. Due to the increase in the collectible amount of data, there is a growing demand for the automatization of the signal analysis process, which could reduce human error, save time and improve the quality and the accuracy of the calculated results.

Keywords: Autonomic nervous system · Signal processing · Heart rate variability (HRV) · Baroreflex sensitivity index (BRS)

1 Introduction

The autonomic nervous system (ANS) regulates many important systems in the human body, such as blood pressure, heart rate, digestion, respiratory system and many others. These systems have in common that they are mostly independent of human will. According to its function, the ANS could be divided into the sympathetic nervous system (SNS) and the parasympathetic nervous system (PNS). SNS is activated in situations where the body needs to prepare for the possible threat, "fight or flight" situations.

In these situations, the main purpose is to optimize body functions for possible danger, such as pupil dilatation and more light entering the eyes and an increase in blood pressure and heart rate with bronchial dilatation leading to greater blood oxygenation. As opposite, PNS is responsible for "rest and digest" situations, where the body needs to recover and the only purpose is to optimize body functions for basic maintenance, such as a decrease in blood pressure and heart rate which enables conservation of energy [1].

Dysfunction of the ANS is related to disorders in different systems in the human body, and it can be recognizable in the ANS biosignals. Digital signal processing (DSP) of the ANS biosignals provides information about the functioning of the ANS, which

A. Badnjevic and L. Gurbeta Pokvić (Eds.): CMBEBIH 2021, IFMBE Proceedings 84, pp. 20–27, 2021.
https://doi.org/10.1007/978-3-030-73909-6_3

is very useful in everyday clinical practice. The purpose of this manuscript is to present possible methods for the analysis of the ANS biosignals. In Sect. 2 ANS testing and data acquisiton will be briefly described. Section 3 presents time and frequency domain characteristics of the ANS biosignals and Sect. 4 presents additional methods of the ANS biosignals analysis.

2 ANS Testing and Data Acquisition

The basis of the testing of the ANS is the response of different systems that are under its influence (like heart rate and blood pressure) to different well-defined challenges. Because of this, there is no unique test, and usually, a battery of different autonomic tests, that combines different approaches, is used in clinical practice [2].

One of the most frequently examined parts of the ANS is the cardiovascular autonomic nervous system, because it has easily recordable biosignals – heart rate (HR) and blood pressure (BP). Basic cardiovascular biosignals are presented in Fig. 1.

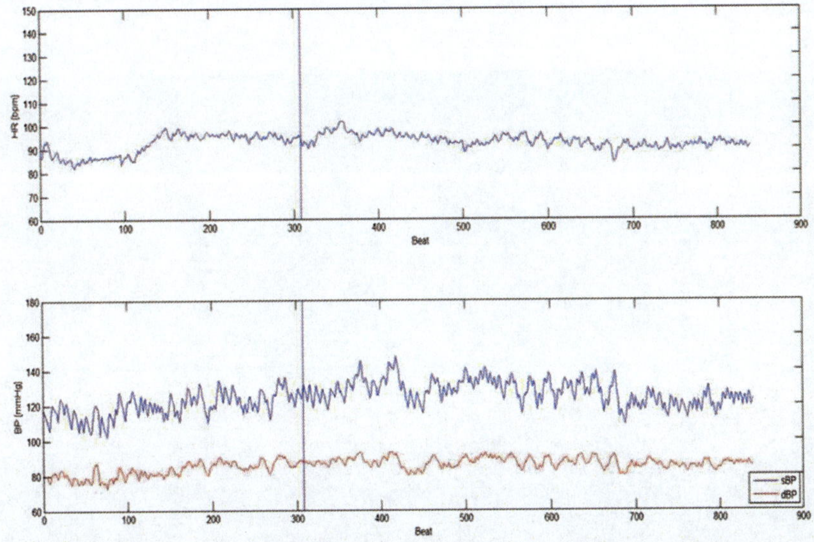

Fig. 1. Presentation of the basic ANS biosignals, in the form of heart rate (HR) and blood pressure (BP), a property of the University Hospital Center Zagreb

2.1 Types of ANS Testing

ANS recordings are usually performed in laboratory conditions, however, they can also be performed bedside. A typical laboratory for ANS recordings is presented in Fig. 2. Continuously monitoring of blood pressure and heart rate is performed with pressure cuff and ECG electrodes, with a patient in the supine position. Different measurement procedures are usually performed: heart rate and blood pressure response to Valsalva maneuver

– a measure of parasympathetic and sympathetic function; heart rate response to deep breathing test – measure of parasympathetic function; and blood pressure response to passive tilt or active standing – a measure of sympathetic function [3, 4]. Additionally, a sudomotor function can be tested with the quantitative sudomotor axon reflex sweat test (QSART) or with the sympathetic skin response (SSR), which presents the momentary change of the electrical potential of the skin.

2.2 Duration of Recordings

Different clinical procedures require a different amount of information, and because of this, recordings of different duration can be performed. Usually, recordings are divided into short-term and long-term recordings. Short-term recordings usually have a duration from 0.5 to 5 min and are related to specific conditions [5]. Long-term recordings are usually defined as 24-h Holter recordings, performed with wearable sensors in subjects freely moving under daily conditions. Different durations of recordings acquire different interpretations, and additional information is necessary for detailed analysis. Some of the variables important for long-term recordings interpretation are physical activity and posture [6].

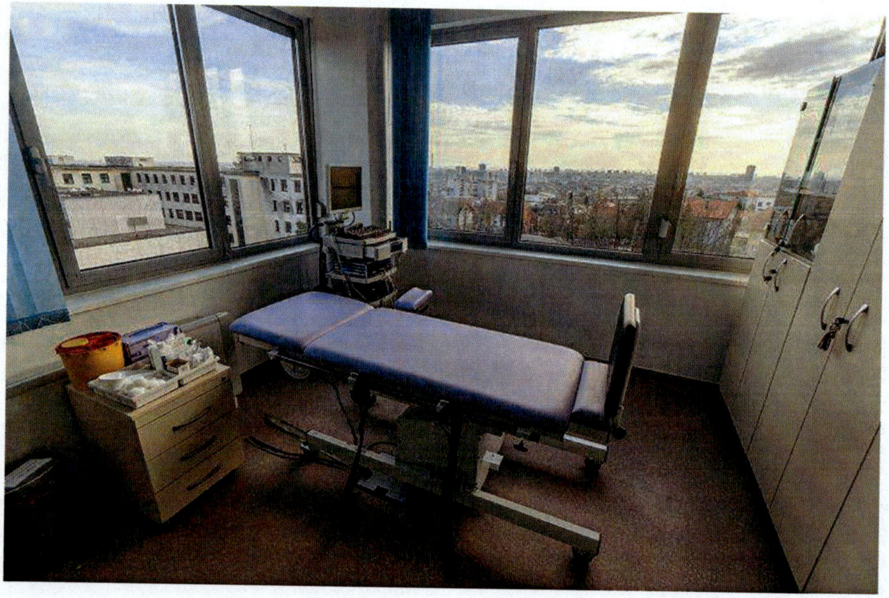

Fig. 2. Laboratory for autonomic nervous system testing, Department of Neurology, University Hospital Center Zagreb

2.3 Characteristics of the Specific Types of Recordings

When recordings are performed in laboratory conditions, all variables can be well controlled, such as patient's activity and environmental conditions (temperature…).

Analysis of the data can be performed in real time and possible difficulties, that could influence the quality of recorded material, can be corrected in real time through active cooperation with the patients. But, this type of examination is time consuming and it enables recording only a limited number of patients with a limited amount of data.

Because of that, various possibilities are examined, to see how a bigger amount of the ANS data could be collected. One of the examples is presented in the work of Komazawa et al., where they collected information from more than 27 000 subjects [7]. With the smartphone camera, that detects the pulse waveform from luminance changes of blood flow on the tip of the finger, the information for HRV analysis were collected. The application enables the collection of a large amount of data at any time, it acquires data about the brightness of the skin and derives a pulse wave based on luminance change. Acquired data is analyzed and the information about the sympathetic and parasympathetic activity is presented on the user's mobile phone.

When performing recordings in non-laboratory conditions (such as a mobile application), it is more convenient to record a bigger number of patients and collect a large amount of data. Also, recording in non-laboratory conditions could provide information about the functioning of the ANS through daily activities. But there are also some disadvantages to this type of recordings. There is no active cooperation with the patients and no control over recording environmental conditions. The quality of data depends on the patient's cooperability and responsibility, and only off-line analysis is available, with no insight into the data quality during the recordings. Data acquisition is dependent on the relationship between the sensor and the skin, and this relationship can vary through daily activities. The smartphone and similar applications have a much lower sampling rate than the laboratory devices (30 Hz vs 1000 Hz), and this also influences the quality of the recorded signal [8].

The quality of the recorded signal can be improved with the measurement of the quality information feedback for the participants, user can discard corrupted data and repeat the measurement, but in most cases, this part of the application is not included [9].

Differences in the collection of the data can provide different insights into the ANS activity, and regarding the type of study design, different characteristics of the signal are appropriate.

3 Time and Frequency Domain Characteristics

Recorded signals can be analyzed through different procedures. Basic cardiovascular biosignals, heart rate and blood pressure, have different time and frequency domain characteristics.

3.1 Heart Rate

The highest emphasis in everyday clinical practice is the analysis of the heart rate (HR). The analysis is usually performed in the time domain in the form of heart rate variability (HRV): standard deviation of RR interval (SDRR), a standard deviation of NN intervals (SDNN), root mean square for standard deviation (RMSSD); and in the frequency domain in the form of spectral analysis: Power Spectral Density (PSD), low frequency (LF, 0.04–0.15 Hz), high frequency (HF, 0.16–0.4 Hz) and LF/HF ratio [10]. There is a large number of time-domain and frequency-domain measures, but the most important question is which of these variables have clinical meaning, and according to that the variables for analysis should be selected [11].

HR signals by their nature can be non-stationary and non-linear, and because of that, the non-linear analysis is also useful in quantification of the structure and complexity of the HR signals [12]. Different metrics, such as detrended fluctuation analysis (DFA), the approximate entropy and correlation dimension, are variables of interest for this type of analysis.

3.2 Blood Pressure

The other important ANS variable is blood pressure (BP). It can be analyzed in the time domain in the form of systolic blood pressure (sBP), diastolic blood pressure (dBP), pulse pressure (PP) or mean arterial pressure [13]. Also, analysis can be performed in the frequency domain in the form of power spectral density (Fast Fourier Transform – FFT), mean and total power, LF band (0.04 to 0.15 Hz), HF band (0.4 to 1.5 Hz) [14].

4 Additional Methods of ANS Signal Analysis

With the growth of the amount of data and different recording procedures, there is a need for additional methods for DSP of ANS biosignals.

4.1 Heart Rate Variability (HRV)

One of the most commonly used methods is heart rate variability (HRV) analysis. The method has well defined protocol and clear mathematical definitions [15]. The most frequently used variable is SDNN (standard deviation of the normal-to-normal interval), and it presents a marker of overall HRV with the contribution of both, sympathetic and parasympathetic nervous system activity. The type of HRV analysis is dependent on the duration of the recordings, which can be divided into short-term (5 min) and long term (24 h) recordings [11]. Data for analysis can be collected with different recording devices (hardwired/wireless photoplethysmography; ring, hand, or chest-based sensors) and analysis can be performed on-line or on a server [16]. Longitudinal HRV monitoring is suitable for "big data" analysis and there are machine learning models for risk prediction based on HRV [17]. The disadvantage of the method is that there are different types of software for its analysis and there can be disagreement between different methods for estimation of HRV [18].

4.2 Baroreflex Sensitivity Indices (BRS)

The baroreflex sensitivity (BRS) index is an important cardiovascular indicator that allows the quantification and analysis of the body's adaptability to hemodynamic changes [19]. Newly introduced BRS indices are BRSa1, α-BRSa, β-BRSa and vagal index

BRSv. BRS indices are calculated from the systolic blood pressure (sBP) curve during the Valsalva maneuver [20, 21]. Indices are usually calculated manually, which is a time-consuming process and dependent on the subjective assessment, which improved the risk of human error. To reduce the subjective human error, reduce the necessary time and improve the precision of the calculated indices, the semi-automated application for calculation was introduced [22]. Semi-automated results showed a statistically significant high association with manually calculated indices, and the results of the application are presented in Fig. 3. Subjective assessment and the likelihood of human error are considerably reduced by using semi-automated calculation, and the application is time saving so it enables analysis of a large amount of data.

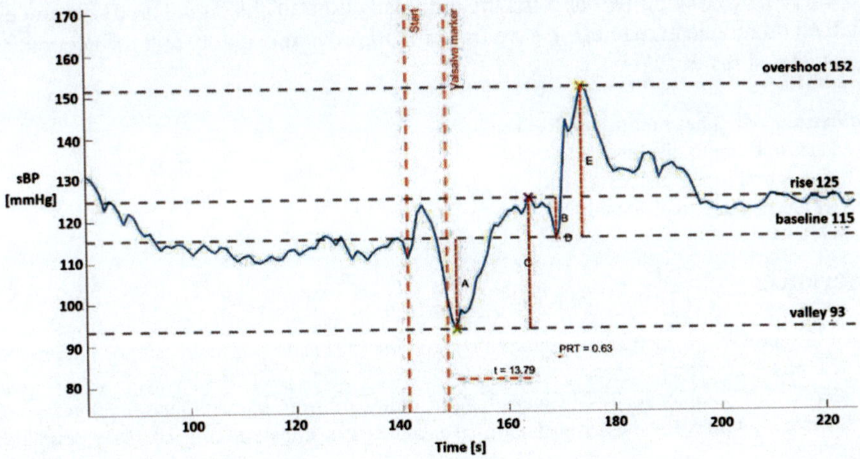

Fig. 3. Example of BRS indices calculation, sBP – systolic blood pressure, the property of University Hospital Center Zagreb

4.3 Blood Pressure and Pulse Rate Variability

Similar to HRV, blood pressure (BP) variability can also be calculated and it varies depending on the duration of BP recordings [23]. Very short-term BP variability presents standard deviation or fluctuation of BP in various frequency bands, mostly recorded in laboratory conditions. Short-term BPV is expressed in the form of standard deviation, coefficient of variation (CoV), 24-h weighted SD, and average real variability (ARV), mostly recorded through the 24 h ambulatory BP monitoring (ABPM). Long-term BPV is related to ABPM over 48 h or home BP monitoring.

(HBPM) over several days or weeks, but it has questionable reliability due to the long duration of the recordings.

Pulse rate variability (PRV) is calculated by pulse oximeter photoplethysmography (PPG), a method that detects blood volume variations in the body tissue, and it provides an estimation of the heart rate. PRV is calculated from the peak to peak time intervals of the PPG signal [24]. PRV could substitute HRV in some non-stationary situations with the usage of wearable and portable medical devices, and because of its robust signal in some situations, PRV has more practical value than HRV signal [25]. A disadvantage of the method is that PRV shows considerable variations regarding the measurement sites [26].

5 Conclusion

Autonomic nervous system testing can be performed in different conditions and there is a large variety of different types of signal analysis, regarding the parameters of the signals and type of signal acquisition. Due to the increase in the collectible amount of data, there is a growing demand for the automatization of the signal analysis process, which could reduce human error, save time and improve the quality and the accuracy of the calculated results.

Disclosure. MKS has nothing to disclose.
IA has nothing to disclose.
MC has nothing to disclose.
MH has nothing to disclose.

References

1. Goldstein, D.S.: Principles of autonomic medicine (2017). https://neuroscience.nih.gov/publications
2. Ziemssen, T., Siepmann, T.: The investigation of the cardiovascular and sudomotor autonomic nervous system-a review. Front Neurol. **10**, 53 (2019). https://doi.org/10.3389/fneur.2019.00053
3. Freeman, R.: Assessment of cardiovascular autonomic function. Clin. Neurophysiol. **117**(4), 716–730 (2006). ISSN 1388-2457
4. Novak, P.: Quantitative autonomic testing. J. Vis. Exp. (53), 2502 (2011). https://doi.org/10.3791/2502. PMID: 21788940; PMCID: PMC3196175
5. Sztajzel, J.: Heart rate variability: a noninvasive electrocardiographic method to measure the autonomic nervous system. Swiss Med. Wkly. **134**(35–36), 514–522 (2004)
6. Hayano, J., Yuda, E.: Pitfalls of assessment of autonomic function by heart rate variability. J. Physiol. Anthropol. **38**, 3 (2019). https://doi.org/10.1186/s40101-019-0193-2
7. Komazawa, M., Itao, K., Kobayashi, H., Luo, Z.: On human autonomic nervous activity related to behavior, daily and regional changes based on big data measurement via smartphone. Health **8**, 827–845 (2016). https://doi.org/10.4236/health.2016.89088
8. Peng, R.C., Zhou, X.L., Lin, W.H., Zhang, Y.T.: Extraction of heart rate variability from smartphone photoplethysmograms. Comput. Math. Methods Med. **2015**, 516826 (2015). https://doi.org/10.1155/2015/516826. Epub 2015 Jan 12
9. Liu, I., Ni, S., Peng, K.: Enhancing the robustness of smartphone photoplethysmography: a signal quality index approach. Sensors **20**(7), E1923 (2020). https://doi.org/10.3390/s20071923
10. Bong, S.Z., Murugappan, M., Yaacob, S.: Analysis of electrocardiogram (ECG) signals for human emotional stress classification. In: Ponnambalam, S.G., Parkkinen, J., Ramanathan, K.C. (eds.) Trends in Intelligent Robotics, Automation, and Manufacturing. IRAM 2012. Communications in Computer and Information Science, vol. 330. Springer, Heidelberg (2012)

11. Shaffer, F., Ginsberg, J.P.: An overview of heart rate variability metrics and norms. Front. Public Health **5**, 258 (2017). https://doi.org/10.3389/fpubh.2017.00258
12. Germán-Sallóa, Z., Germán-Sallób, M.: Non-linear methods in HRV analysis. Procedia Technol. **22**, 645–651 (2016). https://doi.org/10.1016/j.protcy.2016.01.134
13. Kengne, A.P., Czernichow, S., Huxley, R., Grobbee, D., Woodward, M., Neal, B., Zoungas, S., Cooper, M., Glasziou, P., Hamet, P., Harrap, S.B., Mancia, G., Poulter, N., Williams, B., Chalmers, J.: ADVANCE Collaborative Group. Blood pressure variables and cardiovascular risk: new findings from ADVANCE. Hypertension. **54**(2), 399–404 (2009). https://doi.org/10.1161/HYPERTENSIONAHA.109.133041. EPUB 2009 May 26
14. Andriessen, P., Schoffelen, R.L., Berendsen, R.C., de Beer, N.A., Oei, S.G., Wijn, P.F., Blanco, C.E.: Noninvasive assessment of blood pressure variability in preterm infants. Pediatr. Res. **55**(2), 220–3 (2004). Epub 2003 Nov 19
15. Heart rate variability. Standards of measurement, physiological interpretation, and clinical use. Task Force of the European Society of Cardiology and the North American Society of Pacing and Electrophysiology. Eur. Heart J. **17**(3), 354–381 (1996)
16. Drury, R.L.: Heart rate variability in an integrated hardware/software system using artificial intelligence to provide assessment, intervention and performance optimization support. In: Theodoros, A. (ed.) Autonomic Nervous System Monitoring. IntechOpen Limited, London (2019)
17. Chiew, C.J., Liu, N., Tagami, T., Wong, T.H., Koh, Z.X., Ong, M.E.H.: Heart rate variability based machine learning models for risk prediction of suspected sepsis patients in the emergency department. Medicine **98**(6), e14197 (2019). https://doi.org/10.1097/MD.0000000000014197
18. Vest, A.N., Da Poian, G., Li, Q., Liu, C., Nemati, S., Shah, A.J., Clifford, G.D.: An open source benchmarked toolbox for cardiovascular waveform and interval analysis. Physiol. Meas. **39**(10), 105004 (2018). https://doi.org/10.1088/1361-6579/aae021
19. Schrezenmaier, C., Singer, W., Swift, N.M., Sletten, D., Tanabe, J., Low, P.A.: Adrenergic and vagal baroreflex sensitivity in autonomic failure. Arch. Neurol. **64**, 381–6 (2007). https://doi.org/10.1001/archneur.64.3.381
20. Palamarchuk, I.S., Baker, J., Kimpinski, K.: Non-invasive measurement of baroreflex during Valsalva maneuver: evaluation of alpha and beta-adrenergic components. Clin. Neurophysiol. **127**, 1645–51 (2016). https://doi.org/10.1016/j.clinph.2015.10.047
21. Palamarchuk, I.S., Baker, J., Kimpinski, K.: Non-invasive measurement of adrenergic baroreflex during Valsalva maneuver reveals three distinct patterns in healthy subjects. Clin. Neurophysiol. **127**, 858–63 (2016). https://doi.org/10.1016/j.clinph.2015.04.057
22. Višnjić, N., Moštak, I., Mutak, T., Pucić, D., Junaković, A., Adamec, I., Krbot Skorić, M., Habek, M.: Automated calculation of baroreflex sensitivity (BRS) indices. Eur. J. Neurol. **27**, 266 (2020). https://doi.org/10.1111/ene.14307
23. Chadachan, V.M., Ye, M.T., Tay, J.C., Subramaniam, K., Setia, S.: Understanding short-term bloodpressure-variability phenotypes: from concept to clinical practice. Int. J. Gen. Med. **11**, 241–254 (2018). https://doi.org/10.2147/IJGM.S164903. eCollection 2018.
24. Dehkordi, P., Garde, A., Karlen, W., Wensley, D., Ansermino, J.M., Dumont, G.A.: Pulse rate variability compared with heart rate variability in children with and without sleep disordered breathing. In: Conference Proceedings IEEE Engineering in Medicine and Biology Society, pp. 6563–6566 (2013). https://doi.org/10.1109/EMBC.2013.6611059
25. Chou, Y., Zhang, R., Feng, Y., Lu, M., Lu, Z., Xu, B.: A real-time analysis method for pulse rate variability based on improved basic scale entropy. J. Healthc. Eng. (2017). https://doi.org/10.1155/2017/7406896
26. Yuda, E., Yamamoto, K., Yoshida, Y., Hayano, J.: Differences in pulse rate variability with measurement site. J. Physiol. Anthropol. **39**(1), 4 (2020). https://doi.org/10.1186/s40101-020-0214-1

Evaluation of Vectorcardiogram Perspectives in Education and Clinical Practice

Ajdin Fejzic[1], Amina Tihak[1], Dusanka Boskovic[1](✉), and Orhan Lepara[2]

[1] Faculty of Electrical Engineering, University of Sarajevo, Sarajevo, Bosnia and Herzegovina
{afejzic2,atihak1,dboskovic}@etf.unsa.ba
[2] Faculty of Medicine, University of Sarajevo, Sarajevo, Bosnia and Herzegovina
orhan.lepara@mf.unsa.ba

Abstract. The paper presents evaluation of benefits of using a vectorcardiogram (VCG) application in education and clinical practice. In contrast to commonly used visualizations of the ECG signal, spatial presentation of vectorcardiogram is not frequently used, although this approach offers more sensitive and precise insights in cardiac electrical dynamics. In education spatial analysis of cardiac electrical activity is important for understanding the relation between the cardiac abnormalities and the characteristic shapes of the ECG. Technology advancements facilitated VCG related research, including development of novel VCG descriptors, but lack of standardization and established methodology prevented the analyses of VCG to become more popular. The educational version of the VCG application is demonstrated to potential users: medical and engineering students, and their feedback regarding the application benefits for education and clinical practice is collected. The evaluation results are analyzed and activities to raise interest and promote VCG usage are discussed.

Keywords: Vectorcardiography (VCG) · Electrocardiography (ECG)

1 Introduction

In this paper the vectorcardiogram (VCG) educational application is presented, the current and improved version of the eLearning application with integrated visualizations of both ECG and VCG presented in [1]. The objective of the paper is to evaluate: (1) benefits of introducing the VCG application as a teaching tool for both engineering and medical students, and (2) possibilities for more common usage of the VCG tools in clinical practice. This work is a part of broader research of multidisciplinary team aimed to raise interest in the VCG usage.

The structure of the paper is as follows: the next Section provides background information on benefits of using the VCG. Section 3 describes the VCG application and provides sample screens illustrating the VCG for the Normal Sinus Rhythm (NSR), the Myocardial Infarction (MI) and stable angina recording. Section 4 presents the attitude of the prospective users towards the usage of the VCG tool in both academic and clinical environment, collected by user experience survey. The evaluation results are analyzed and discussed. The final Section presents conclusions and recommendations for the future work.

A. Badnjevic and L. Gurbeta Pokvić (Eds.): CMBEBIH 2021, IFMBE Proceedings 84, pp. 28–36, 2021.
https://doi.org/10.1007/978-3-030-73909-6_4

2 Background Information

Understanding the connection between the cardiac cycle, cardiac electrical dynamics, and corresponding ECG tracing presents a fundamental competence both for medical and bioengineering professionals. The measurement and analysis of the ECG presents a good example of the biggest challenge for biomedical engineering (BME) education: the necessity to bring together two different and demanding fields: (1) medicine with fundaments in biology and chemistry; and (2) engineering with a basis in math and physics [2].

The vectorial analysis of heart potentials is based on the convention to represent the generated potential of a heart at a particular instant as a summated vector called instantaneous mean vector [3]. Visualizing the electrical forces generated by heart by means of a continuous movement of cardiac vector is named vectorcardiogram [4].

The orthogonal 3-lead VCG signals records the cardiac electrical activity along axes of the coordinate system defined by three orthogonal planes of the body: frontal, transverse, and sagittal. The VCG signals are accordingly projected onto different planes or visualized in a 3D space. Several lead systems for measuring the VCG were developed, and the orthogonal Frank lead system is recognized as the most common way to measure VCG. A heart cycle is represented by three loops corresponding to P, QRS, and T wave activities [5]. Decrease in clinical interest for VCG motivated researchers to transform 3-lead VCG into the 12-lead ECG, such as Dower transform [6], enables 12-lead ECG diagnosis to those who possessed VCG equipment [7]. Revival of interest in VCG analyses motivated researchers in developing ECG to VCG transformations, and thorough review of transformations in both directions is presented in [8].

Next is the brief review of clinical advantages of VCG usage followed by VCG in education.

2.1 VCG Clinical Perspective

Comparing vector loops with 12-lead ECG traces demonstrated that the largest amplitudes in the leads are not reached at the same time and that the largest amplitude in the heart vector is missed whenever the direction of the heart vector is not parallel to one of the lead vectors at that very moment [7]. The same authors identified several new descriptors of cardiac abnormality that could be further explored by enhancing the usage of the standard 12-lead electrocardiogram with the usage of the VCG.

The 12-lead ECG is more widely used than the 3-lead VCG because physicians are accustomed to using it in clinical applications. It has thus proved its worth, and it is considered to be the Gold Standard. However, all of this knowledge is outdated, and even in that, only a small fraction of the data is actually used by physicians depending on their experience, expertise, and sometimes memorization of ECG signals for various cardiac disorders. The vectorcardiogram overcomes not only the information loss of one or two ECG signals, but also the dimensional problems caused by the 12-lead ECG signals [5].

Recently, the VCG has seen some revival. A number of recent studies have shown that the beat-to-beat variability of the average VCG dipole during repolarization is closely associated with the risk of ventricular arrhythmias in a variety of different cardiomyopathies. VCG related metrics have also been used to detect the existence of a MI scar [9].

The VCG as a monitoring tool for electrical cardiac operation, unlike the ECG, indicates the course of ventricular depolarization and repolarization in the 3D environment. It provides supplementary details for routine 12-lead ECG assessment. VCG provides a higher diagnostic sensitivity compared to ECG in acute MI when combined with left anterior fascicular block, and has a stronger association with the echocardiogram, compared to ECG, in the determination of left ventricular mass. The VCG presents a higher sensitivity to assess the severity of congenital aortic valve stenosis, while in patients with congenital pulmonary valve VCG features has a strong correlation with stenosis. The meaning of the right systolic pressure ventricle and maximum spatial presence the vector on the right side of the horizontal plane [10].

The degree of planarity of the VCG loop can differentiate healthy individuals from patients with ST elevation MI (STEMI) that is compatible with basic understanding of the electrophysiology of the human heart [11].

A recent study showed that VCG adds complexities to the ECG phenomenon in athletes [12].

2.2 VCG in Education

It is acknowledged that VCG can assist in the understanding the physiologic and physiopathologic mechanisms in conduction disease, when training for ECG interpretation competencies. However, to make full use of VCG, additional expertise and special tools are required [13]. Detailed review of VCG advantages for the cardiac diagnosis is presented in [10], and the authors recognize while VCG is still evolving it will always have didactic significance. During the research, it was noticed that there was a large number of applications that could be successfully used in teaching, but there were no indications that this has been done. A VCG application with colorized P, QRS and T wave activities [5] incorporates additional dynamic attributes of spatiotemporal VCG signal such as speed, phase angles, octant numbers, and curvature which improves the automatic assessment of cardiovascular diseases with the use of VCG signals, but it is mainly used for clinical investigations of specific cardiovascular conditions. Authors of the 3D Heart, an application for visual training for ECG analysis described in [14] employ 3D visualizations with a variety of different activation simulations for better understanding of cardiac electrical activities, but follow up for application usage as an educational tool were not documented.

3 Visualizing Vectorcardiogram

The visualization of the VCG is integrated with simultaneous display of the respective ECG traces for three separate leads. The spatial position of the heart vector within the 3D coordinate system follows the ECG charting with heart vector vertex drawing a vectorcardiogram of interest. Visualization tool is implemented in MATLAB, and provides users with sophisticated signal processing capabilities and focusing on the specific parts of interest.

The application is able to process ECG signals from the PhysioNet [15] databases, and in this paper we focus on signals from the PTB Diagnostic ECG Database [16], provided by the National Metrology Institute of Germany, as a compilation of digitized ECG recordings to be used for research, algorithmic benchmarking and teaching. The database is composed of 549 records from 290 subjects, of which 81 women and 209 men, aged 17 to 89 years. Each record includes 15 simultaneously measured signals: the conventional 12 ECG leads and the 3 Frank ECG leads (V_x, V_y, V_z). Each signal is digitized with 1000Hz sampling rate and 16 bit resolution over a range of ± 16.384 mV. The visualizations of the integrated VCG and ECG display enables evaluation of the VCG usage in diagnosing more than 250 pathological conditions including: myocardial infraction, cardiomyopathy, bundle branch block, dysrhythmia, myocardial hypertrophy, valvular heart disease, myocarditis, and comparing them against the healthy controls.

Sample screens of the animated VCG are presented in Fig. 1 where colored vector can be seen as it draws a path with the tip.

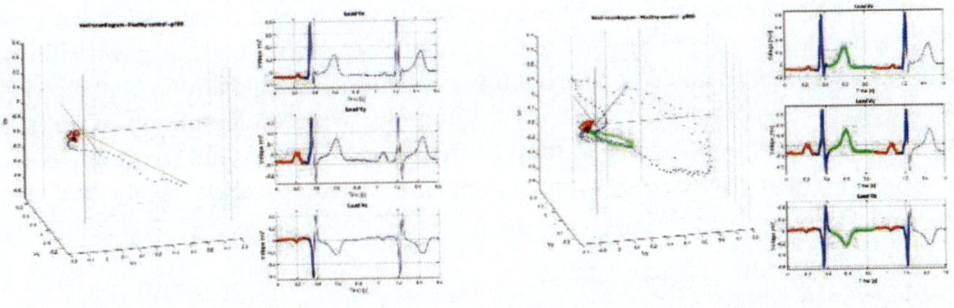

(a) (b)

Fig. 1. Vectorcardiogram and Frank leads XYZ visualization in MATLAB

The tool allows animation to be paused and possibility to observe the formation of VCG. User can review the signals with different health conditions. Different 3D patterns allows the classification of heart health problems. Illustration in Fig. 2 shows an ECG signal lasting two cardiac cycles and a corresponding vectorcardiogram of a healthy control.

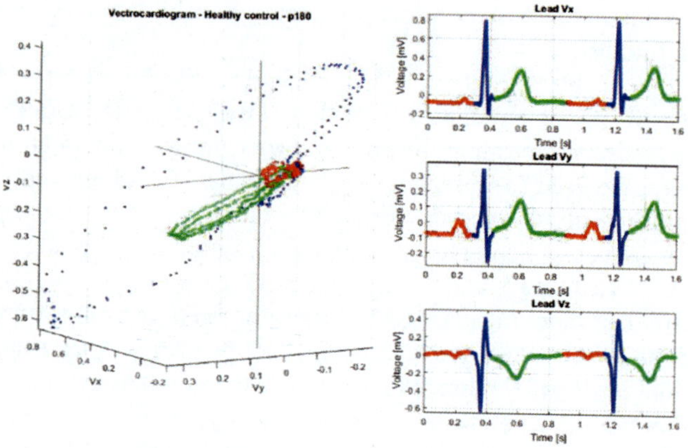

Fig. 2. VCG and Frank leads signals: healthy control; patient 180; signal "s0476_rem"

The next illustrations show VCG lasting two cardiac cycles with diagnosed MI in Fig. 3 and stable angina in Fig. 4. Observing the two graphs, a clear distinction can be made considering the irregularity of the shape of the VCG when compared with the NSR shown in Fig. 2. The problem of MI contains a clear spatial disposition of values visible between the two loops of the cardiac cycle (Fig. 3), while the condition of the stable angina can be clearly distinguished by the absence of "P loop" and irregular shape of the "T loop" named after the eponymous parts of the cardiac cycle (Fig. 4).

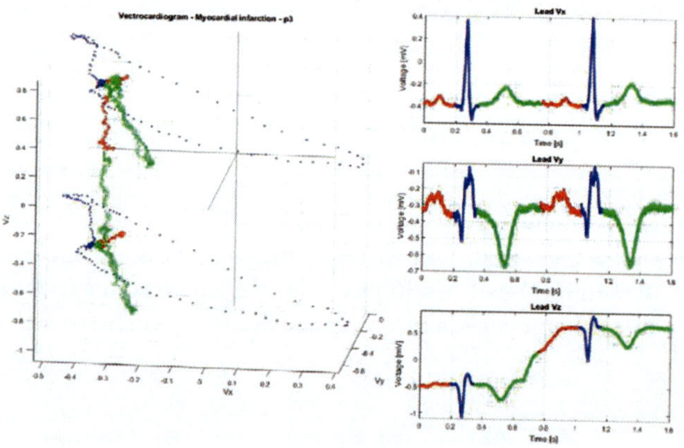

Fig. 3. VCG and Frank leads Vy and Vx animation: MI; patient 3; signal "s0017lrem"

Another important distinction that can be confirmed by observing the previous graphical representations is the "QRS loop" of the VCG (blue color), which in NSR is in an almost planar arrangement, while in MI a slight curvature can be noticed. For a condition of stable angina, the absence of mentioned planarity is clearly visible.

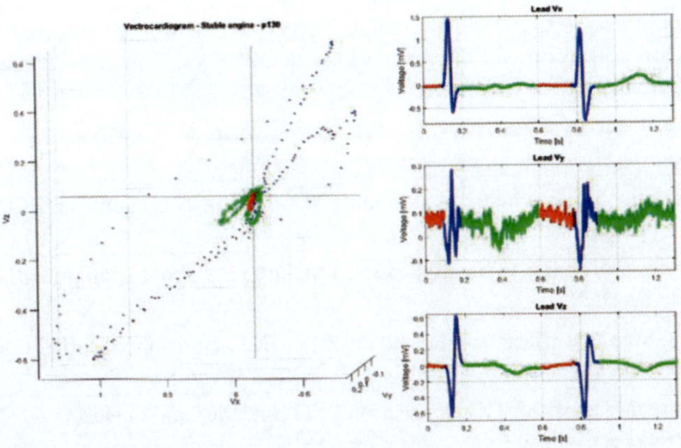

Fig. 4. VCG and Frank leads Vy and Vx animation: Stable angina; patient 130; signal "s0166_rem"

4 User Experience Evaluation

Teaching the BME topics can significantly benefit from using visualizations and animations of underlying phenomena, thus helping understand spatial and temporal links between anatomic parts, physiological processes and physical quantities [1].

With objective to evaluate prospective users' attitude towards significance of using.

VCG tool both in education and clinical practice a user experience survey was designed to collect feedback after reviewing VCG tool. The study was roughly based on the Unified Theory of Acceptance and Use of Technology – UTAUT model [17], with limited focus on Performance Expectancy and Facilitating Conditions, leaving Effort Expectancy and Social Influence for future evaluations. Performance Expectancy has been linked to VCG benefits for education and clinical practice. Survey included 5-point Likert scale questions exploring the following: (a) perceived educational outcome - EO, (b) perceived benefits of VCG in education - EDU, (c) perceived benefits of VCG in clinical practice - CP, and (d) Facilitating Conditions - FC.

We have invited medical (28) and electrical engineering students (15), all of them with previous knowledge of cardiac cycle and ECG measurement and analyses. The results are presented in Table 1.

Questions related to perceived benefits of VCG in education – EDU and in clinical practice – CP, are indicating (above 4) the users trust in the potential of the VCG application. The lower grades for questions related to achieved educational insights are in compliance with users' previous knowledge, but we would like to emphasize high rate for the Question 1: application facilitated better understanding the relation between ECG signal and cardiac electrical activity. Survey results showed that students are interested in VCG usage and recognized the VCG importance. The results for separate student groups did not show differences, except for the last question linked to Facilitating Conditions where medical students expressed higher expectations of large investments and additional training ($\mu_{MED} = 3.9$ vs. $\mu_{EE} = 3.2$).

Table 1. Results of User Experience evaluation: perceived educational outcome - EO, perceived benefits of VCG in education - EDU, perceived benefits of VCG in clinical practice - CP, and facilitating conditions - FC

Question	μ	σ
1. (EO) VCG application has helped me to understand the relation between ECG signal and cardiac electrical activity	4.18	0.63
2. (EO) Before reviewing the VCG application I had no idea about spatial cardiac electrical activities	3.18	1.29
3. (EO) I had no idea that characteristic waves of the ECG signal correspond to the spatial loops of the VCG	3.35	1.25
4. (EO) Before reviewing the VCG application I did not fully understand the cardiac electrical activities	3.20	1.27
5. (EDU) A three-dimensional VCG contains more information than a standard ECG	4.02	0.95
6. (EDU/CP) VCG is superior as a tool for presentation of the physiological heart dynamics	4.12	0.70
7. (EDU) The VCG application has a great potential in teaching	4.31	0.77
8. (CP) VCG can be useful in a diagnostics because it allows more accurate interval duration measurement	4.29	0.74
9. (CP) The combined use of VCG and ECG can facilitate the training of using VCG in diagnostics	4.49	0.65
10. (CP) VCG may indicate pathologies that are not noticeable/visible in the ECG signal	4.08	0.79
11. (FC) Usage of the VCG application in a clinical practice would require large investments and extensive additional training	3.63	1.07

In addition to raising interest and promoting VCG usage, the application initiated interesting and fruitful discussion between multidisciplinary student cohorts.

5 Conclusion

The proposed Vectorcardiogram application exploits real physiological signals data, while providing fully functional VCG visualization tool.

This application can be used for research purposes of both, electrical engineers and medical students, to provide full demonstration of dynamic 3D animations of vectorcardiogram and to learn new ways of classifying the condition of the heart muscle. Platform is great for students of electrical engineering to improve and better the data processing including the ability to extract valuable data in different formats. Additionally, graphical interpretation enables implementation of different kinds of classifications based solely on VCG shape, or combining it with observation of geometrical shapes connected to the parts of ECG signals. Another positive feature of the application is reflected in the ability to choose relevant parts and to display VCG of different duration of the ECG signals. The users evaluated with the highest grade potential of combined use of VCG and ECG, and enabling the training of VCG usage in diagnostics. This is the important approach in ensuring that the wealth of knowledge already available in ECG analyses is transferred and extended with the ECG analyses.

The future work will integrate the interface with displaying the standard ECG descriptors, such as duration of characteristic intervals. Simultaneous visualization in both VCG and ECG will facilitate the knowledge transfer between the domains. After the improvements, the advanced user evaluation study including knowledge assessment will be conducted, for evaluation of the effectiveness of the VCG application as an educational tool.

References

1. Sljivo, S., Boskovic, D., Lepara O.: Vectorcardiogram eLearning application. In: CMBEBIH 2019: Proceedings of the International Conference on Medical and Biological Engineering, Banja Luka, Bosnia and Herzegovina, 16–18 May 2019, vol. 73, p. 81. Springer, Cham (2019)
2. Bošković, D., Badnjević, A.: Opportunities and challenges in biomedical engineering education: focus on Bosnia and Herzegovina. In: 2015 4th Mediterranean Conference on Embedded Computing (MECO), pp. 407–410. IEEE (2015)
3. Hall, J., Guyton, A.: Textbook of Medical Physiology, 11th edn. Saunders Elsevier, Philadelphia (2005)
4. Wilson, F.N., Johnston, F.D.: The vectorcardiogram. Am. Heart J. **16**, 14–28 (1938)
5. Yang, H., Bukkapatnam, S.T.S., Komanduri, R.: Spatiotemporal representation of cardiac vectorcardiogram (VCG) signals. Biomed. Eng. Online (2012)
6. Dower, G.E.: The ECGD: a derivation of the ECG from VCG leads. J. Electrocardiol. **17**, 189 (1984)
7. Man, S., Maan, A.C., Schalij, M.J., Swenne, C.A.: Vectorcardiographic diagnostic & prognostic information derived from the 12-lead electrocardiogram: historical review and clinical perspective. J. Electrocardiol. **48**(4), 463–475 (2015)
8. Jaros, R., Martinek, R., Danys, L.: Comparison of different electrocardiography with vectorcardiography transformations. Sensors **19**, 3027 (2019)
9. Gemmell, P.M., Gillette, K., Balaban, G., et al.: A computational investigation into rate-dependant vectorcardiogram changes due to specific fibrosis patterns in non-ischæmic dilated cardiomyopathy. Comput. Biol. Med. **123**, 103895 (2020)
10. Pérez Riera, A.R., Uchida, A.H., Filho, C.F., Meneghini, A., Ferreira, C., Schapacknik, E., Dubner, S., Moffa, P.: Significance of vectorcardiogram in the cardiological diagnosis of the 21st century. Clin. Cardiol. **30**(7), 319–23 (2007)
11. Ray, D., Hazra, S., Goswami, D.P., Macfarlane, P.W., Sengupta, A.: An evaluation of planarity of the spatial QRS loop by three dimensional vectorcardiography: Its emergence and loss. J. Electrocardiol. **50**(5), 652–660 (2017)
12. Thomas, J.A., Perez-Alday, E.A., Junell, A., Newton, K., Hamilton, C., Li-Pershing, Y., German, D., Bender, A., Tereshchenko, L.G.: Vectorcardiogram in athletes: the sun valley ski study. Ann. Noninvasive Electrocardiol. **24**(3), e12614 (2019)
13. Antiperovitch, P., Zareba, W., Steinberg, J.S., Bacharova, L., Tereshchenko, L.G., Farre, J., Nikus, K., Ikeda, T., Baranchuk, A.: Proposed in-training electrocardiogram interpretation competencies for undergraduate and postgraduate trainees. J. Hosp. Med. **13**(3), 185–193 (2017). https://doi.org/10.12788/jhm.2876. Epub 2017 Nov 8. PMID: 29154379

14. Olson, C.W., Lange, D., Chan, J.K., Olson, K.E., Albano, A., Wagner, G.S., Selvester, R.H.: 3D heart: a new visual training method for electrocardiographic analysis. J. Electrocardiol. **40**(5), 457.e1–7 (2007)
15. Goldberger, A.L., Amaral, L.A.N., Glass, L., Hausdorff, J.M., Ivanov, P., Mark, R.G., Mietus, J.E., Moody, G.B., Peng, C.-K., Stanley, H.E.: PhysioBank, PhysioToolkit, and PhysioNet: components of a new research resource for complex physiologic signals. Circulation **101**(23), e215–e220 (2000)
16. Bousseljot, R., Kreiseler, D., Schnabel, A.: Nutzung der EKG-Signaldatenbank CARDIODAT der PTB über das Internet. Biomedizinische Technik, Band 40, Ergänzungsband 1, S 317 (1995)
17. Venkatesh, V., Morris, M.G., Davis, G.B., Davis, F.D.: User acceptance of information technology: toward a unified view. MIS Q. **27**, 425–478 (2003)

An Improved Model for the Assessment of Cutaneous Microcirculation in Type 1 Diabetes

Eva Rossi[1], Cosimo Aliani[1], Piergiorgio Francia[1], Roberto Anichini[2], and Leonardo Bocchi[1](✉)

[1] Department of Information Engineering, University of Florence, Florence, Italy
leonardo.bocchi@unifi.it
[2] UOs Diabetologia and Diabetic Foot Unit USL Toscana Centro, Pistoia, Italy

Abstract. Patients with Type I Diabetes may develop some complications during their lifetime as a result of impaired metabolic control; among others, some of these common complications concerns microvascular function. The possibility of evaluating the cutaneous microcirculation is considered an important translational model because it could offer a important opportunity to non-invasive evaluation of diabetes-related vascular disorders.

In our previous works we demonstrated the alteration occurring in the vascular system reflect themselves in the shape of the peripheral waveform. In particular, we demonstrated the decomposition of the waveform as a sum of Gaussian curves provides a set of parameters that can be fed into a classifier that successfully discriminate between younger and older subjects.

In the present work, we investigated the application of a different decomposition model, based on a sum of exponential pulses, for the analysis of microvascular alterations correlated with the presence of Type I Diabetes.

The processing pipeline follow the basic architecture of our previous work: a pre-processing step segments each pulsation of the input signal and the removes the long-term trend. Afterward, a Least Squares fitting algorithm determines the set of optimal model parameters that best approximate each single cardiac cycle.

Each vector of model parameters constitutes a compact representation of the pulse waveform; thus, any structural difference in the pulse waveform between controls and patient produces a corresponding alteration of model parameters.

Finally, an example classifier is used to explore the feasibility of discriminating pathological and normal subjects by looking only at the peripheral waveform.

Keywords: Microcirculation · Diabetes · Pulse analysis

1 Introduction

Type 1 diabetes (T1D) is a dreaded autoimmune disease that usually leads to an absolute insulin deficiency and consequent hyperglycemia [2]. Epidemiological studies showed that the incidence and prevalence of T1D are steadily rising worldwide even if they may change from region to region of the world and within the same country [10, 15].

A. Badnjevic and L. Gurbeta Pokvić (Eds.): CMBEBIH 2021, IFMBE Proceedings 84, pp. 37–46, 2021.
https://doi.org/10.1007/978-3-030-73909-6_5

While on the one hand, the incidence of T1D is particularly high in patients younger than 15 years and in particular in children under the age of 4–6, on the other hand, the life expectancy of these patients increases [3, 5, 11].

As consequence, despite continuous therapeutic advances, patients with T1D may develop some feared complications during their lifetime as a result of impaired metabolic control [3, 5].

One of these common complications concerns microvascular function and on clinical perspective it occurs mainly as retinopathy, neuropathy, and nephropathy in addition to skin microcirculation [5, 11, 16].

A combination of factors is considered responsible of skin microvascular dysfunction that, in turn, showed some important relationship with others microvascular complications (i.e. retinopathy and neuropathy). Among these, the excess of non-enzymatic glycosylation of collagen that produces advanced glycation end products (AGEs) has an important role. AGEs increase the collagen cross-links altering the mechanical properties of fibers and so increased vascular stiffness and reduced vessel wall distensibility [16, 19].

In this regard, the possibility of evaluating the cutaneous microcirculation is considered an important translational model because it could offer an important opportunity to non-invasive evaluation of diabetes-related vascular disorders [8, 9].

In our previous work [18], we adopted a model based on a sum of Gaussian for describing the pulse waveform. This model is statistically robust and widely used in literature [4], although it has been recently demonstrated sub-optimal, at least for photopletismographc signals [6].

Moreover, Gaussian curve presents a non-causality behavior that, although practically negligible, could produce unrealistic results. Based on this observation, we analyzed models that are suitable to represent the output of a physiological system. If we approximate the physiological model with a linear system, the expected output consists of a sum of weighted exponential curves. Thus, the most natural choice involves the formulation of the model based on this family of functions. Model parameters are then used to develop a Bayesian classifier for discriminating subject age.

2 Methods

Methods are divided in three sections: *data acquisition and preprocessing* describes how signals were acquired, which instrument was used and the preprocessing steps that were executed before data analysis, *Pulse wave modelling* describes how the model was created and fitted to plethysmographic signals while *classification problem* describes how the created signals in the previews sections were used to train and test a supervised classifier based on Naive Bayes.

2.1 Data Acquisition and Preprocessing

The experiments were performed on the same data set used in our previous work [17], that is briefly summarized here. The data set is composed of 47 T1D patients, attending the Diabetes Unit of the San Jacopo General Hospital (Pistoia, Italy), that were evaluated between December 2016 and March 2018. The control group consists of 63 healthy individuals without any diagnosed pathology; all participants were non-smokers, and control subjects have no known history of cardiovascular disease.

All data collection was performed in accordance with the guidelines of the Declaration of Helsinki: detailed information on the involved procedures and their purpose was provided to the subjects, which gave their written informed consent prior to the measurement sessions.

The collected data was anonymized, and each record is associated with subject age, gender, and BMI; in case of diabetic subjects, T1D duration, T1Drelated complications, and longitudinal records of glycosylated haemoglobin (HbA1c) levels were retrieved from each patient's health record.

Perfusion was recorded on the plantar side of right hallux, with a Periflux 5000 Laser Doppler Flowmetry (LDF) system (Perimed, Sweden). In order to reduce the waveform distortion, the smoothing constant of the instrument was set at its lowest possible value, 0.03s. All acquisition were performed in the same room, at a controlled temperature of 22 °C, after an acclimatization period of 10 min.

The perfusion waveform is pre-processed in accordance with out previous work [18]. In brief, the waveform is upsampled at 100 Hz, before the application of a low pass filter, for reducing noise present in the input signal, using a cutoff frequency of 30 Hz. Afterward, a local maximum detector identifies each local maximum as a candidate systolic peak, and defines the diastolic valley as absolute minimum between each pair of maxima. A refinement step eliminates spurious detection using a set of heuristic rules that concern the relative amplitude and distance of the peaks, and the variability of neighboring cycles. The resulting set of diastolic valleys allows to segment the waveform in single cardiac cycles.

2.2 Pulse Wave Modelling

Several mathematical description of the pulse waveform have been proposed, and a recent work compares the performances of the most frequently used descriptors [6] in describing photoplethysmography imaging.

However, LDF signal has a larger amplitude than photoplethysmography imaging, thus allowing to develop more detailed models. In accordance with this observation, we recently proposed [1] a new model that represents the waveform as a sum of exponential pulses.

This type of representation has the theoretical advantage of representing the response of a linear system to a sequence of ideal pulses. The single exponential response, corresponding to an input pulse at time t_0, is defined as:

$$d(t) = \begin{cases} 0 & \text{for } t < 0 \\ D_{0e^{(t-t_0/\tau)}} & \text{for } t \geq t_0 \end{cases} \tag{1}$$

where D_0 represent the maximum output, at time $t = t_0$, and τ is the time constant of the system. However, this representation presents a discontinuity for $t = t_0$, that is physiologically implausible. The definition of a model composed of a sum of exponential curves thus requires the introduction of mathematical constraints in order to produce a continuous curve.

In order to solve this issue, avoiding any discontinuity in the resulting function, we define an "exponential pulse" as:

$$P_{A,k_1,k_2}(t) = \begin{cases} 0 & \text{for } t < 0 \\ A(e^{-k_1 t} - e^{-k_2 t}) & \text{for } t \geq 0 \end{cases} \tag{2}$$

that is defined by its amplitude A, the rising time constant $t_r = 1/k_1$ and its falling time constant $t_f = 1/k_2$. The definition of the curve ensures its continuity in the origin, and the curve presents a "pulse like" shape, i.e. the curve tends to zero with large values of t.

The simplest model that is able to provide a good approximation of the experimental pulse waveform is composed of three exponential pulses:

$$m(t) = \sum_{i=1}^{3} P_{A_i,k_{1,i},k_{2,i}}(t - t_i) \tag{3}$$

where: A_i, $k_{1,i}$ and $k_{2,i}$ are the parameters of the i-th pulse, and t_i is the beginning time of the relative wave. Each cycle is thus associated with a twelve components parameter vector $p_n = \{A_i, k_{1,i}, k_{2,i}, t_i\}_{i=1,2,3}$, where n is the cycle index.

In this model, the first pulse yields an accurate fit of the systolic pulse, the second one, that is a negative pulse, adapts to the "falling" part of the systolic pulse, while the third pulse models the reflected wave, after the dicrotic notch. The model has the advantage of producing a signal with (practically) limited support, thus being suitable to describe a single cardiac cycle.

It is worth noting that the support of the resulting model, actually, extends after the current cardiac cycle, with the typical values of the estimated time constants. In our fitting procedure we disregard this effect, as the fitting estimation is carried out independently for each cycle.

The estimation of the optimal model parameters is based on the classical Levenberg-Marquardt optimization algorithm for non-linear least squares curve fitting problems [13]. The variability ranges and the initial seed value for each of component of the parameter vector are essential for the stability of the fitting procedure. Both the variability ranges and the seed values were estimated based on a set of heuristic rules from the principal characteristics of the signal, namely the position of the dicrotic notch, the maximum signal amplitude, and the duration of the cardiac cycle.

The objective function of the fitting procedure is the squared difference between the model and the pulse waveform over the cardiac cycle. Therefore, the fitting procedure does not take into account if the support of the model response extends before or after the cardiac cycle. Given the shape of the model response, we did not observe any cycle where the fitting procedures produces a response starting before the beginning of the cycle; however, in almost any case, the support of the model extends beyond the end of the corresponding cycle.

From a physiological perspective, this effect is perfectly logic: in case of an extra-systole, i.e. an increased length between two consecutive beats, the peripheral flow does not fall to zero, but continues its exponential decay until the next heart cycle starts.

2.3 Discrimination Between Diabetic Patients and Controls

While the analysis of the differences in each model parameter between patients and controls gives a first insight on the effects of diabetes in microvascular circulation, we developed a simple classifier to validate the feasibility of using the whole parameter vector for discriminate between physiological and pathological conditions.

The proposed model yields a parameter vector for each cardiac cycle of the subject. On the other hand, the presence of long term oscillations in the vascular flow makes difficult to consider the shape of a single pulse for performing this classification. Indeed, in our previous work, we already verified that the sequence of pulse shapes is correlated with ageing [14].

Thus, two strategies are available for identify the presence of alterations correlated with the pathology: the first one is based on the classification of each cardiac cycle as *normal* or *diabetic*, and later we estimate the subject classification according to the presence of *diabetic* pulses. On the contrary, the second strategy involves the estimation of an "average" cardiac cycle, obtained by simple averaging of all parameters vectors, that can be used to perform the subject classification.

Cycle Based Classification. As stated above, in this approach we build a classifier aiming at discriminating each cycle as normal or possible diabetic. Thus, we created a dataset containing about 28000 cardiac cycles, individually labelled according to the subject class, that were used to train a Bayesian classifier. Although the number of samples is quite large, a random split of this dataset in a training and a test set is expected to introduce a classification bias because it is very probable that pulses of each patient are similar each other.

Thus, we developed the classifier using a "leave-one-subject-out" approach: namely, we removed from the data set all the cycles corresponding to the same subject, train the classifier, and use it to predict the class of the cycles of the subject not included in the data set. As usual, we obtain the overall performance of the classifier by averaging the results over all subjects.

Mean Pulse Classification. This approach analyzes the statistical properties of the subject's cardiac cycles. Each subject s is summarized by a 24-component descriptor $v_s = \{\text{mean}(p_n), \text{std}(p_n)\}$; the descriptor thus includes information about the average values and the standard deviations of the components of p_n, with n ranging over all subject's heart beats.

In this case, there are no correlation between different samples in the data set, but the number of components of the feature vector is comparable with the number of subjects; therefore, it is advisable to use again the leave-one-out method for testing the performance of the classifier.

Thus, we removed one of the subjects, in turn, from the dataset, and we trained a Naive Bayes [7, 12] classifier over the remaining subjects. Finally, the classifier was tested on the removed subject. The average performance over all iterations defines the overall performances of the method.

3 Results

A preliminary validation of the applicability of the proposed method to diabetic subjects is based on the qualitative and quantitative analysis of the results of the fitting procedure. As it can be seen in Fig. 1, the model is able to closely match the experimental data, yielding a comparable quality of the fit for controls and for diabetic patients. It is worth noting that the inclusion of diabetic patients in the experiment did not require tuning of the fitting procedure with respect to our previous work; this includes also the variability ranges and the seed point of the model parameters.

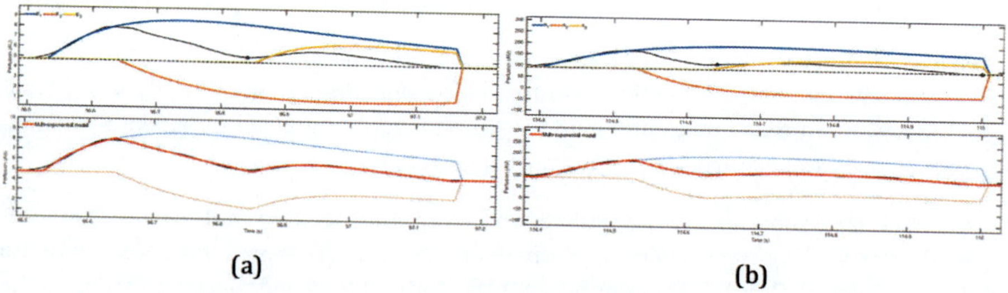

Fig. 1. Sample fitting results: (a) control subject, (b) diabetic subject.

The good performance of the fitting procedure is confirmed by the good values of quality of fit, defined as:

$$R = \frac{1}{S}\sum_{s=1}^{S}\frac{1}{n_i}\sum_{j=1}^{n_i}(1 - r_j) \tag{4}$$

where S is the total number of subjects, n_i is the number of samples of the i-th signal and r_j is the residue of the j-th sample; we measured values of $R = 96\%$ in control subjects and $R = 95\%$ in diabetic subjects.

The statistical distribution of model parameters indicates a significant difference exists in the shape of the pulse waveform between normal and diabetic subjects, as shown in Fig. 2. Numerical values of the mean and standard deviation of each model parameter, for each class, are shown in Table 1.

Table 1. Mean and standard deviation of model parameters of all cardiac cycles in the control and pathological group, respectively. The * indicates statistically significant differences ($p < 0.05$)

Parameter	Control	Pathological	p-value
$A1$	214 ± 214	255 ± 241	<0.001*
$A2$	195 ± 224	221 ± 213	<0.001*
$A3$	231 ± 259	262 ± 297	<0.001*
$k1,1$	3.5 ± 2.7	3.1 ± 2.4	<0.001*
$k1,2$	7.7 ± 5.1	7.0 ± 4.8	<0.001*
$k1,3$	9.4 ± 9.3	8.4 ± 8.1	<0.001*
$k2,1$	9.1 ± 5.4	8.4 ± 5.1	<0.001*
$k2,2$	4.0 ± 5.5	3.8 ± 5.2	0.002*
$k2,3$	9.7 ± 9.5	9.6 ± 9.3	0.39
$dt1$	0.05 ± 0.04	0.05 ± 0.05	1
$dt2$	0.20 ± 0.07	0.21 ± 0.08	<0.001*
$dt3$	0.32 ± 0.1	0.33 ± 0.1	<0.001*
$dt2 - dt1$	0.15 ± 0.07	0.16 ± 0.08	<0.001*
$dt3 - dt1$	0.27 ± 0.1	0.28 ± 0.1	<0.001*

The single pulse classification produces unsatisfactory results, with and accuracy of about 58% of correctly classified pulses. However, the integration of the classification results at patient level improves the outcome; the evaluation, for each subject, of the percentage of pulses classified as diabetic is the most straightforward method to perform such integration. This procedure allows to create a ROC curve of the classifier performance at patient level, as shown in Fig. 3. The optimal performance corresponds to an accuracy of 71%, with sensitivity equal to 57% and specificity of 78%, as shown in the confusion matrix reported in Table 2.

Table 2. Confusion matrix of the integration, at subject level, of the single pulse classifier

	Classified as		
	Control	Pathological	Total
Control	43 (78%)	12 (22%)	55
Pathological	12 (43%)	16 (57%)	28

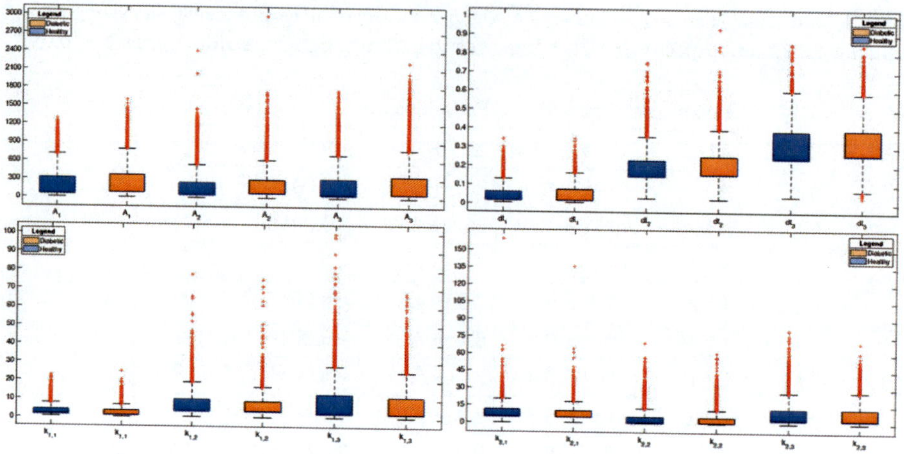

Fig. 2. Boxplot of model's parameters of *healthy* and *diabetic* subjects

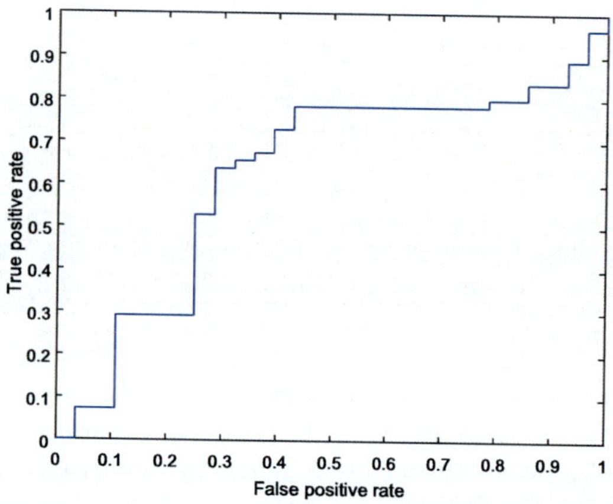

Fig. 3. ROC curve of the classification results at patient level

The mean pulse classification yields similar results; the leave-one-out method obtains a classification accuracy equal to about 64%, with sensibility and specificity of 60% and 65%, respectively (Table 3).

Table 3. Confusion matrix of the mean pulse classification

	Classified as		
	Control	Pathological	Total
Control	36 (65%)	19 (35%)	55
Pathological	11 (40%)	17 (60%)	28

4 Conclusion

The evaluation and monitoring of skin microvascular alterations are becoming increasingly important in the management of the T1D patients.

It is worthy of attention how the proposed model is able to closely match the experimental data, yielding a comparable quality of the fit for both controls and diabetic patients. Moreover, the statistical distribution of model parameters indicates a significant difference in the shape of the pulse wave form between subjects with diabetes and controls. The classification of both single pulse and mean pulse produces satisfactory results showing a performance with a good accuracy, sensitivity and specificity.

The procedure used in this study provides a preliminary indication of the suitability of this methodological framework as a low-cost, non-invasive evaluation for the early detection and monitoring of microvascular disorders related to T1D.

Acknowledgment. This work was supported by Fondazione Cassa di Risparmio di Firenze, Florence, Italy (grants number 2015.0914 and 2017.0800).

References

1. Aliani, C., Rossi, E., Francia, P., Bocchi, L.: An improved model for detectingvascular age from peripheral pulse (2021, under review)
2. American Diabetes Association: 2. classification and diagnosis of diabetes: Standards of medical care in diabetes—2020. Diab. Care **43**(Suppl. 1), S14–S31 (2020). https://doi.org/10.2337/dc20-S002
3. American Diabetes Association: Standards of medical care in diabetes—2020 abridged for primary care providers. Clin. Diab. **38**(1), 10–38 (2020). https://doi.org/10.2337/cd20-as01
4. Couceiro, R., Carvalho, P., Paiva, R.P., Henriques, J., Quintal, I., Antunes, M., Muehlsteff, J., Eickholt, C., Brinkmeyer, C., Kelm, M., Meyer, C.: Assessment of cardiovascular function from multi-Gaussian fitting of a finger photoplethysmogram. Physiol. Meas. **36**, 1801–1825 (2015). https://doi.org/10.1088/0967-3334/36/9/1801
5. DiMeglio, L.A., Evans-Molina, C., Oram, R.A.: Type 1 diabetes. Lancet **391**(10138), 2449–2462 (2018). https://doi.org/10.1016/s0140-6736(18)31320-5
6. Fleischhauer, V., Ruprecht, N., Sorelli, M., Bocchi, L., Zaunseder, S.: Pulse decomposition analysis in photoplethysmography imaging. Physiol. Meas. **41**(9), 095009 (2020). https://doi.org/10.1088/1361-6579/abb005
7. Hastie, T., Tibshirani, R., Friedman, J.: The Elements of Statistical Learning: Datamining, Inference and Prediction, 2nd edn. Springer, New York (2009). https://doi.org/10.1007/978-0-387-84858-7
8. Holowatz, L.A., Thompson-Torgerson, C.S., Kenney, W.L.: The human cutaneouscirculation as a model of generalized microvascular function. J. Appl. Physiol. **105**(1), 370–372 (2008). PMID: 17932300. https://doi.org/10.1152/japplphysiol.00858.2007
9. Ince, C.: The rationale for microcirculatory guided fluid therapy. Curr. Opin. Crit. Care **20**, 301–308 (2014). https://doi.org/10.1097/mcc.0000000000000091
10. Maahs, D.M., West, N.A., Lawrence, J.M., Mayer-Davis, E.J.: Epidemiology of type 1 diabetes. Endocrinol. Metab. Clin. N. Am. **39**(3), 481–497 (2010). Type I Diabetes. https://doi.org/10.1016/j.ecl.2010.05.011

11. Madonna, R., Balistreri, C.R., Geng, Y.J., De Caterina, R.: Diabetic microangiopathy: pathogenetic insights and novel therapeutic approaches. Vasc. Pharmacol. **90**, 1–7 (2017). https://doi.org/10.1016/j.vph.2017.01.004
12. Manning, C.D., Raghavan, P., Schütze, H.: Introduction to Information Retrieval. Cambridge University Press, Cambridge (2008). https://doi.org/10.1017/S1351324909005129
13. Moré, J.J.: The Levenberg-Marquardt algorithm: implementation and theory. In: Numerical Analysis, pp. 105–116. Springer, Heidelberg (1978). https://doi.org/10.1007/BFb0067700
14. Morelli, A., Sorelli, M., Francia, P., Bocchi, L.: Modelling of microcirculatory dynamics with auto-regressive models. In: Jarm, T., Cvetkoska, A., Mahnič-Kalamiza, S., Miklavcic, D. (eds.) 8th European Medical and Biological Engineering Conference, pp. 824–832. Springer, Cham (2021). https://doi.org/10.1007/978-3-030-64610-3_92
15. Ogurtsova, K., da Rocha Fernandes, J., Huang, Y., Linnenkamp, U., Guariguata, L., Cho, N., Cavan, D., Shaw, J., Makaroff, L.: Idf diabetes atlas: global estimates for the prevalence of diabetes for 2015 and 2040. Diabetes Res. Clin. Pract. **128**, 40–50 (2017). https://doi.org/10.1016/j.diabres.2017.03.024
16. Santesson, P., Lins, P.E., Kalani, M., Adamson, U., Lelic, I., von Wendt, G., Fagrell, B., Jörneskog, G.: Skin microvascular function in patients with type 1 diabetes: an observational study from the onset of diabetes. Diab. Vasc. Dis. Res. **14**(3), 191–199 (2017). PMID: 28467200. https://doi.org/10.1177/1479164117694463
17. Sorelli, M., Francia, P., Bocchi, L., De Bellis, A., Anichini, R.: Assessment of cutaneous microcirculation by laser doppler flowmetry in type 1 diabetes. Microvasc. Res. **124**, 91–96 (2019). https://doi.org/10.1016/j.mvr.2019.04.002
18. Sorelli, M., Perrella, A., Bocchi, L.: Detecting vascular age using the analysis ofperipheral pulse. IEEE Trans. Biomed. Eng. **65**(12), 2742–2750 (2018). https://doi.org/10.1109/tbme.2018.2814630
19. Vithian, K., Hurel, S.: Microvascular complications: pathophysiology and management. Clin. Med. **10**(5), 505–509 (2010). https://doi.org/10.7861/clinmedicine.10-5-505

Respiratory System Dynamical Mechanical Properties: Modeling in Time and Frequency Domain

Sara Deumić(✉), Neira Crnčević, and Ivana Zolota

International Burch University, Sarajevo, Bosnia and Herzegovina
{sara.deumic,neira.crncevic,ivana.zolota}@stu.ibu.edu.ba

Abstract. In order to satisfy the energy requirements of the body, the respiratory system takes up oxygen from the environment and discharges carbon dioxide. The mechanical function of the lungs can be examined by analyzing the lung parameters and its overall function with the aim of recognizing early signs of conditions and tracking disease development. Because of the complexity of the respiratory system, models which mimic the mechanical behaviour of the respiratory system were developed. These models contain parameters which can be assessed and changed to match the behaviour of the model of interest. One of these models is the parallel model which features two mechanically different regions called linear compartments. In this paper, we developed two models representing the parallel bicompartment model, which were made in both the time and frequency domain. After making these models, we were able to obtain plots which show the relationship between pressure, flow and volume represented in time and frequency domain.

Keywords: Lungs · Respiratory system · Parallel lung model · Respiratory mechanics · Lung dynamics · Matlab

1 Introduction

The main function of the respiratory system is to take up oxygen from the environment and discharge carbon dioxide with the aim of satisfying the energy requirements of the body [1]. The pulmonary gas exchange represents a continuous process which involves ventilation, diffusion and perfusion. The three physical processes are connected together through the conservation of mass. According to this, within the lungs, each oxygen molecule which is inhaled but not exhaled will diffuse from the alveolar gas to the blood [2]. In order to support oxygen absorption and the release of carbon dioxide, 5–8 L of air per minute are moved in and out of the lungs and approximately three tenths of a litre of oxygen are transferred to the blood from alveoli each minute. Approximately the same volume of carbon dioxide is transferred from the blood to the alveoli and is then exhaled. The function of the respiratory muscles is to support breathing in and out [3].

The main goal of examining the mechanical function of the respiratory system is to assess the lung parameters and its overall function as well as to recognize early signs of

A. Badnjevic and L. Gurbeta Pokvić (Eds.): CMBEBIH 2021, IFMBE Proceedings 84, pp. 47–57, 2021.
https://doi.org/10.1007/978-3-030-73909-6_6

conditions and abnormalities which could negatively affect the patient's health. This is very important in ventilated patients in cases where we need to evaluate if the respiratory function has improved or deteriorated and if the setting for the ventilator match the patient's needs [4]. Because of the complexity of the respiratory system and its many elements, clinicians wanted to learn more about the mechanics in physiology. Because of this, there was a need to create simple models that mimic the mechanical behaviour of the respiratory system to some extent. The models created contained parameters which could be evaluated in order to match the behaviour of the model to the experimental data [5].

The parallel gas redistribution model is composed of an alveolar compartments arranged in a parallel way which are connected by separate airways to the trachea [5]. Each of these compartments represent a mechanically different region of the respiratory system. They have its corresponding pressure, volume, elastance and resistance. The two resistances of the two compartments connected in parallel are combined into the airway resistance [6]. In order to understand this model, the parallel model can be represented as two balloons so that each balloon has its own local airway. Both of them are connected to a common airway [7]. The parameters used in this system include R and E. R is the resistance and E is the elastance of the compartments. One of the drawbacks of the parallel model is its difficult implementation in practice. This is due to the fact that it is not easy to deal with a large number of parameters when it comes to physiological interpretations [8].

This paper aims to get a better insight into the chosen model of the respiratory system dynamical mechanical properties and to evaluate it by doing mathematical procedures in both time and frequency domain and representing results through graphs.

2 Methods and Materials

Eight original, full versioned articles and two books in English language related to the respiratory system dynamical mechanical properties (modeling in time and frequency domain) were used. The papers as well as the books were obtained using the following scientific databases: Google Scholar, PubMed, and Research Gate.

2.1 Respiratory System Physiology

The complexity of the respiratory system can be seen in the large number of its components that need to work in a synchronized way to achieve the maximum performance. Each component has a specific role and a small change in any of these components could potentially disrupt the homeostasis of the entire respiratory system. In Fig. 1, the diagram of the simplified version of the respiratory system is represented.

The respiratory system consists of the nose and the nasal cavity, mouth, sinuses, larynx (voice box), pharynx (throat), diaphragm, trachea, lungs, bronchi, bronchioles, alveoli (air sacs) and the capillaries. The breathing processes begins when air is inhaled into the mouth or the nose. It will travel through the back of the throat to the trachea. This is the windpipe which separates into air bronchial tubes. The function of the epiglottis is to prevent food entrance to the trachea as well as the lungs. During breathing, it is

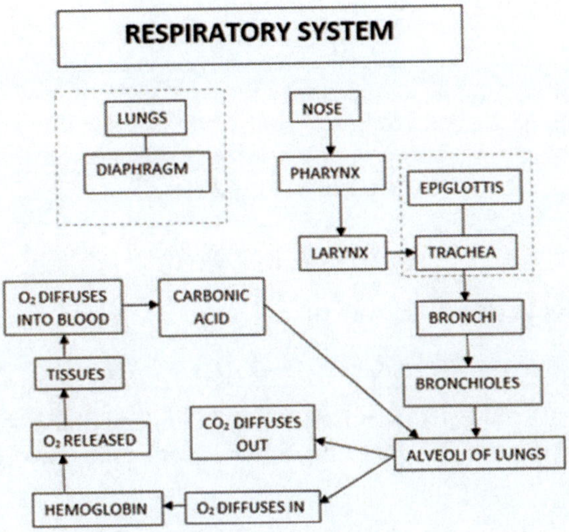

Fig. 1. Diagram of the respiratory system

open allowing air to enter the larynx. The bronchial tubes represent air passages. In order to achieve maximum lung performance, the airways need to be open without any inflammation or extra mucus. The bronchial tubes pass through the lungs and divide into bronchioles, the smaller air passages, that end in air sacs called the alveoli. A mesh of blood vessels (capillaries) surrounds the alveoli. At this point, the oxygen obtained from the inhaled air will pass into the bloodstream. The blood containing the oxygen will then go to the heart and from that point, it will be pumped to the cells across the body. After using the oxygen, the cells generate carbon dioxide which will be transferred to the lungs by the blood. During the process of exhalation, the carbon dioxide will be removed from the body [9].

Table 1. Parameters used for the simulation

Parameters	R_1	R_2	R_c	E_1	E_2
Values	2	4	1	10	5

Table 1 gives the parameters values used for the parallel two compartment model. These parameters will be used in the following equations for the respiratory system dynamical mechanical properties. R_1, R_2, and R_C are measured in $cmH_2O \cdot s \cdot 1^{-1}$, E_1 and E_2 are measured in cmH_2O/l [10].

2.2 Equations for the Time Domain

Elastic pressure is described by:

$$P_1(t) = E_1 V_1(t) \tag{1}$$

$$P_2(t) = E_2 V_2(t) \tag{2}$$

for the first and second compartment.
The resistive pressure is given by:

$$P_{j1}(t) = R_1 \dot{V}_1 + P_1(t) \tag{3}$$

$$P_{j2}(t) = R_2 \dot{V}_2 + P_1(t) \tag{4}$$

The equation for the central airway is:

$$P(t) = R_c\big(\dot{V}_1(t) + \dot{V}_2(t)\big) + P_j(t) \tag{5}$$

Substituting P_1(t) and P_2(t) from Eqs. 1 and 2 into Eqs. 3 and 4, respectively and then substituting P_{j1}(t) and P_{j2}(t) into Eq. 5, the equations for two compartments are obtained:

$$P(t) = (R_c + R_1)\dot{V}_1(t) + R_c\dot{V}_2(t) + E_1 V_1(t) \tag{6}$$

$$P(t) = (R_c + R_2)\dot{V}_2(t) + R_c\dot{V}_2(t) + E_2 V_2(t) \tag{7}$$

From Eqs. 6 and 7, and V can be expressed as:

$$\dot{V}_1(t) = \frac{1}{R_c + R_1} \cdot P(t) - \frac{R_c}{R_c + R_1} \cdot \dot{V}_2(t) - \frac{E_1}{R_c + R_1} \cdot V_1(t) \tag{8}$$

$$\dot{V}_2(t) = \frac{1}{R_c + R_2} \cdot P(t) - \frac{R_c}{R_c + R_2} \cdot \dot{V}_1(t) - \frac{E_2}{R_c + R_2} \cdot V_2(t) \tag{9}$$

2.3 Equations for the Frequency Domain

The first compartment can be described by:

$$\begin{aligned} R_2\dot{P}(t) + E_2 P(t) = {} & [R_1 R_2 + R_c(R_1 + R_2)]\ddot{V}_1 \\ & + [(R_2 + R_c)E_1 + (R_1 + R_c)E_2]\dot{V}_1(t) + E_1 E_2 V_1(t) \end{aligned} \tag{10}$$

and the second compartment by:

$$\begin{aligned} R_1\dot{P}(t) + E_1 P(t) = {} & [R_1 R_2 + R_c(R_1 + R_2)]\ddot{V}_2 \\ & + [(R_2 + R_c)E_1 + (R_1 + R_c)E_2]\dot{V}_2(t) + E_1 E_2 V_2(t) \end{aligned} \tag{11}$$

By applying Laplace method, the transfer functions of Eqs. 10 and 11 were obtained:

$$\frac{V_1(s)}{P(s)} = \frac{R_2 s + E_2}{[R_1 R_2 + R_c(R_1 + R_2)]s^2 + [(R_2 + R_c)E_1 + (R_1 + R_c)E_2]s + E_1 E_2} \tag{12}$$

$$\frac{V_2(s)}{P(s)} = \frac{R_1 s + E_1}{[R_1 R_2 + R_c(R_1 + R_2)]s^2 + [(R_2 + R_c)E_1 + (R_1 + R_c)E_2]s + E_1 E_2} \tag{13}$$

The variables from the equations represent [10]:

- $P_{1,2}$ – the elastic pressure of the first and second compartment
- P_j – the resistive pressure, the pressure at the bifurcation between central and peripheral airways
- P – pressure of the central airway
- $\dot{V}_{1,2}$ – flow of the first and second compartment
- $V_{1,2}$ – volume of the first and second compartment

2.4 Developed Models

In order to simulate the behaviour of the respiratory system, two models were developed. Models were made in the time domain (Fig. 2) and the frequency domain (Fig. 5). Both models contain 2 subsystems named Compartment 1 and Compartment 2 representing two compartments of the parallel model.

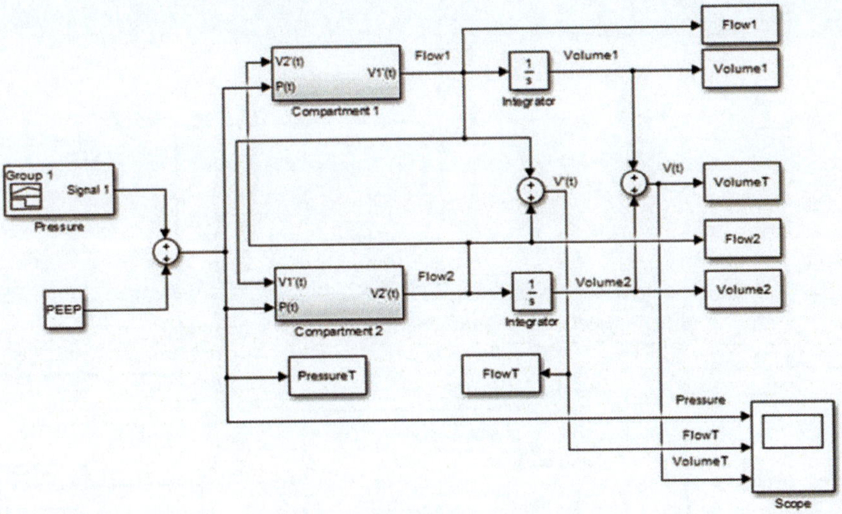

Fig. 2. Model for the time domain

The transfer function for the two compartments in Figs. 6 and 7 are obtained by applying the values from Table 1 into Eqs. 12 and 13, respectively. Compartments 1 and 2 in time (Figs. 3 and 4) and frequency (Figs. 6 and 7) domain models are combined into the entire model (Figs. 2 and 5) of the system by adding the flow and volume of each compartment. The flow and volume of both the compartments and the entire models are obtained and given in the Results section.

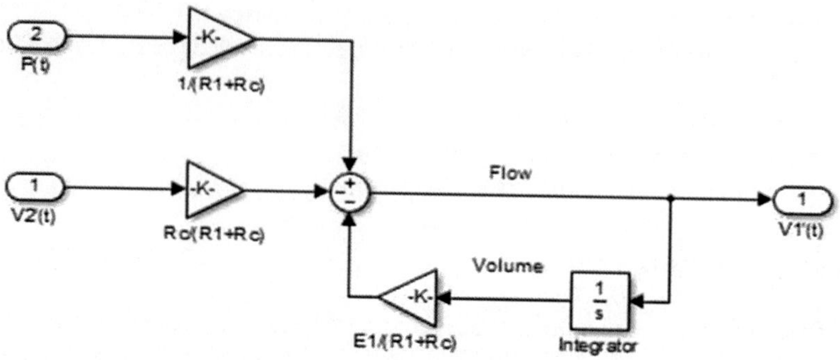

Fig. 3. First compartment of the time domain model

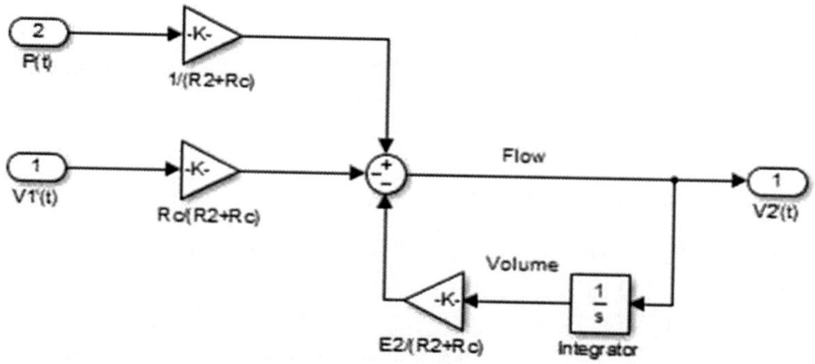

Fig. 4. Second compartment of the time domain model

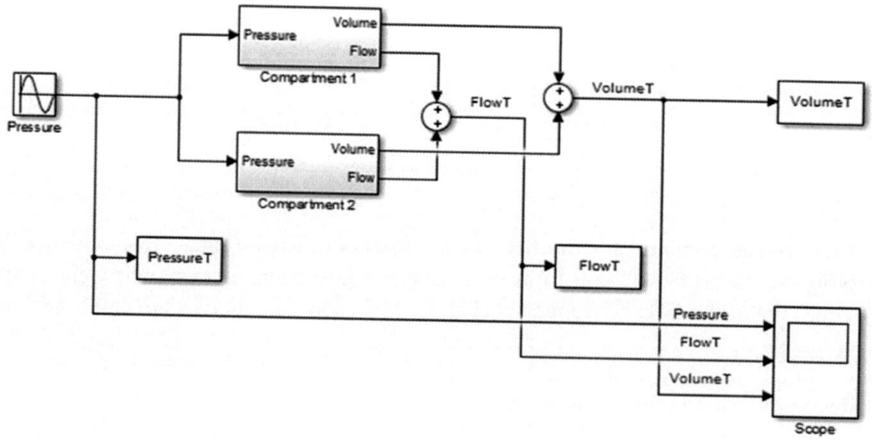

Fig. 5. Model for the frequency domain

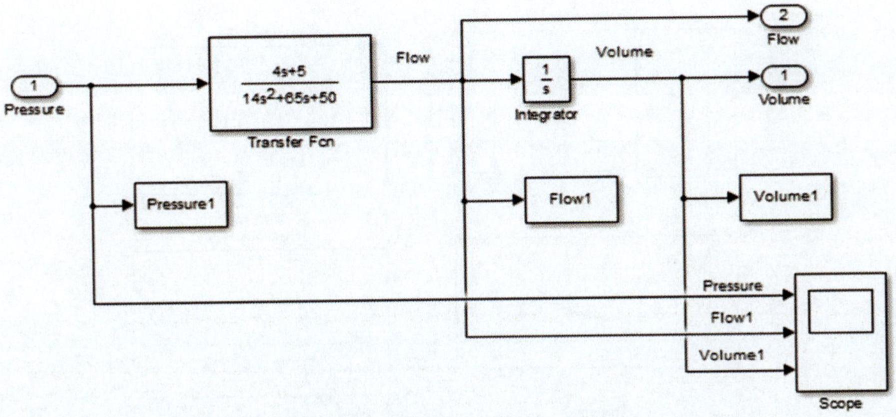

Fig. 6. First compartment of the frequency domain model

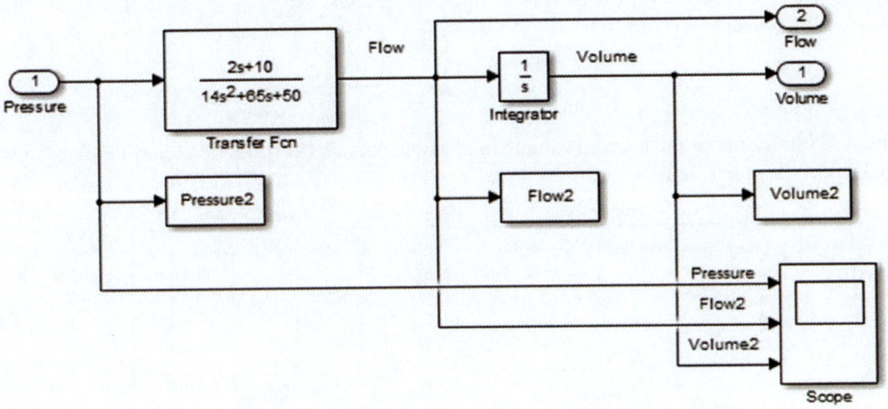

Fig. 7. Second compartment of the frequency domain model

3 Results

Figure 8 represent the relationship between the pressure, flow, and volume in regards to time and frequency domain analysis obtained from the simulation of the two models (Figs. 9 and 10).

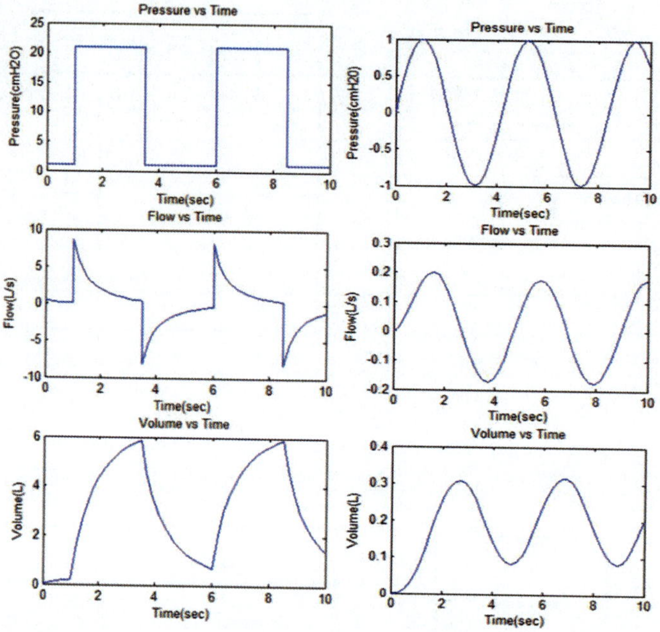

Fig. 8. Plots for pressure, flow and volume for the time domain (left) and frequency domain (right) analysis of the entire models

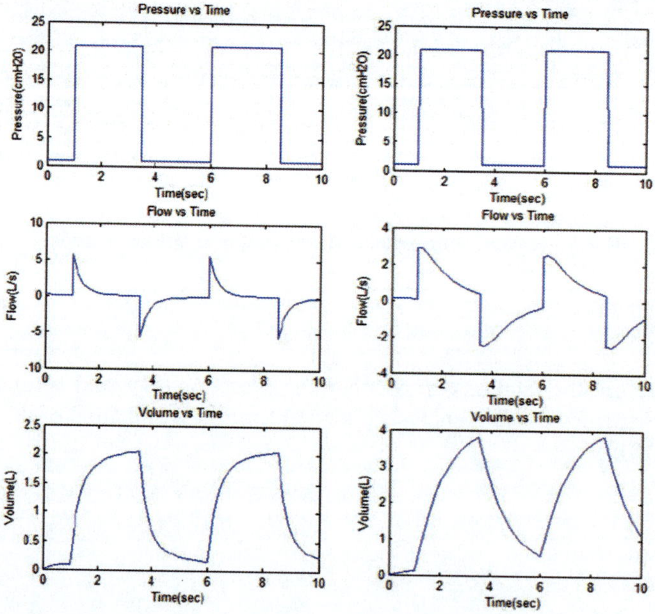

Fig. 9. Plots for pressure, flow and volume for the first compartment (left) and the second compartment (right) of the time domain model

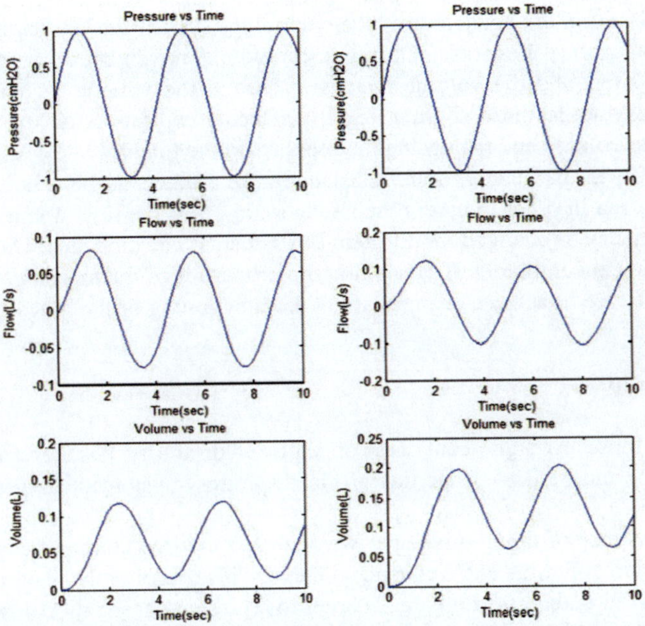

Fig. 10. Plots for pressure, flow and volume for the first compartment (left) and the second compartment (right) of the frequency domain model

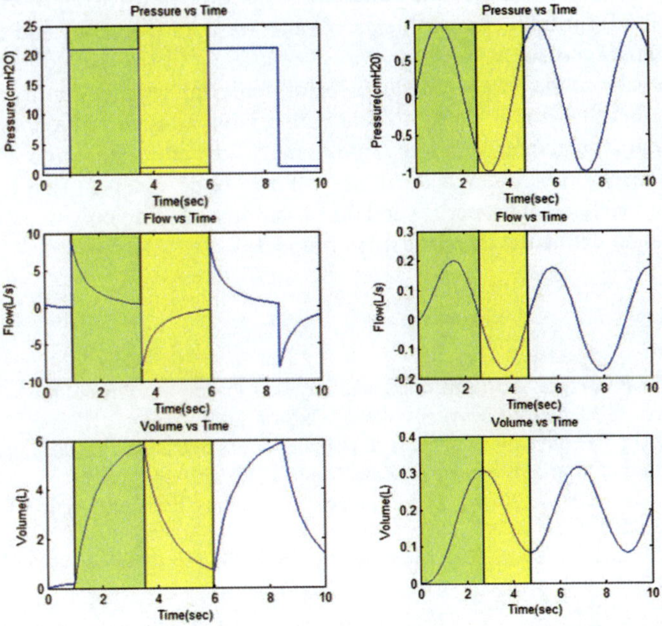

Fig. 11. Plots for pressure, flow and volume representing inspiration (green box) and expiration (yellow box)

Figure 11 shows the behaviour of the respiratory system during inspiration (green box) and expiration (yellow box). During inspiration, a movement of air into the lungs, expands the lungs and the volume increases. Due to the volume increase, the pressure decreases and becomes slightly negative. During expiration, air exists the lungs, decreasing the volume and increasing the pressure in the lungs.

By changing the parameters of the resistance and elastance, the behaviour of the lungs can be monitored in order to assess their functioning. For the time domain model, the pressure signal can be changed for different assessments. The models can be used to represent mechanical ventilators. By changing the properties of the mechanical ventilator, it is possible to see how these changes affect the functioning of the lungs.

4 Conclusion

Lung disease remains a significant cause of death and disability. It is therefore important to estimate the mechanical properties of the respiratory system of lung diseases and disorders.

For the purpose of doing this paper we analysed the two-compartment model, the parallel model in both time and frequency domain. This model of the respiratory system is firstly analysed in the time domain. In order to represent chosen model, several equations have been used for time and frequency domain. To analyse the behaviour of the respiratory system in the frequency domain the equations in time domain were needed to be transferred into the frequency domain. This was done by using transfer function (Laplace method) for both compartments. Results for both frequency and time domain were represented by models.

The drawback of the parallel model is the difficult implementation in practice. This is due to the fact that it is not easy to deal with a large number of parameters when it comes to physiological interpretations. Additionally, for better estimation of mechanical properties of respiratory system in the future it should be considered to explore the respiratory system mechanical properties and the possibility of using only one measurement collection to also define the three-parameter model.

References

1. Hlastala, M.P., Berger, A.J.: Overview and Physical Principles. Physiology of Respiration, 2nd edn, pp. 1–20. Oxford University Press, Oxford (2001)
2. Wagner, P.D.: The physiological basis of pulmonary gas exchange: implications for clinical interpretation of arterial blood gases. Eur. Respir. J. **45**, 227–243 (2015)
3. Otis, A.B., Fenn, W.O., Rahn, H.: Mechanics of breathing in man. J. Appl. Physiol. **2**(11), 592–607 (1950)
4. Restrepo, R.D., Serrato, D.M., Adasme, R.: Assessing respiratory system mechanical function. Clin. Chest Med. **37**(4), 615–632 (2016)
5. Similowski, T., Bates, J.H.: Two-compartment modelling of respiratory system mechanics at low frequencies: gas redistribution or tissue rheology? Eur. Respir. J. **4**(3), 353–8 (1991)
6. Bates, J.H., Rossi, A., Milic-Emili, J.: Analysis of the behavior of the respiratory system with constant inspiratory flow. J. Appl. Physiol. **58**(6), 1840–1848 (1985)

7. Silva, C.A., Ventura, C.E.: A simple model illustrating the respiratory system's time constant concept. Adv. Physiol. Educ. **30**, 129–130 (2006)
8. Carvalho, A.R., Zin, W.A.: Respiratory system dynamical mechanical properties: Modeling in time and frequency domain. Biophys. Rev. **3**(2), 71 (2011)
9. Patwa, A., Shah, A.: Anatomy and physiology of respiratory system relevant to anaesthesia. Indian J. Anaesth. **59**(9), 533–541 (2015)
10. Bates, J.H., Baconnier, P., Milic, E.: A theoretical analysis of interrupter technique for measuring respiratory mechanics. J. Appl. Physiol. **64**(5), 2204–2214 (1988)

Simulating Passage and Absorption of Bolus with an Emphasis on the Small Intestine Using a MultiComponent Model

Dženan Kovačić[1,2](✉), Dado Latinović[1], and Hannah Abigail Boone[1]

[1] Faculty of Engineering and Natural Sciences, Department of Genetics and Bioengineering, International Burch University, Francuske Revolucije bb, 71000 Sarajevo, Bosnia and Herzegovina
dzenan.kovacic@stu.ibu.edu.ba

[2] Sharklab ADRIA: Center for Marine and Freshwater Biology, 27 Chris, Triq Il-Kappella Tal-Lunzjata, In-Naxxar NXR 6031, Malta

Abstract. Modeling of food transport and absorption inside the small intestine is essential to better understand the digestive process as a whole. The models described by our paper present a method to visualize transport of a mass of food representing a bolus from the pylorus to the edge of the ileum. We have developed an integrated model to demonstrate passage of the bolus through the small intestine as well as the absorption of nutrients from it. We also aim to briefly discuss the importance of intestinal modeling, the models that have been developed including those we have based our own models on, and limitations of intestinal modeling as opposed to in vivo studies. Finally, we discuss future directions that intestinal modeling researchers can explore, such as inclusion of other organs within the digestive system and more specific modeling applying to different types of foods.

Keywords: Intestinal modeling · Food passage · Food absorption

1 Introduction

Worldwide, digestive health or lack thereof is becoming an increasingly concerning issue [1]. Obesity and its associated health issues are on the rise, with the World Health Organization reporting as of 2016 that 13% of the world's adult population fell into the obese category, out of 39% of the global population which were classified as overweight [2]. Additionally, autoimmune diseases, which affect over 50 million people in the United States alone, all seem to have an underlying intestinal dysbiosis, inflammation, or permeability component that is a defining aspect of autoimmunity pathogenesis [3]. Due to the global concerns of declining digestive health globally, it is crucial to fundamentally understand the digestive process.

1.1 Anatomy and Histology of the Digestive System

The digestive system is composed of several organs with a wide variety of functions, but the portion where the majority of nutrient absorption occurs is the small intestine

A. Badnjevic and L. Gurbeta Pokvić (Eds.): CMBEBIH 2021, IFMBE Proceedings 84, pp. 58–68, 2021.
https://doi.org/10.1007/978-3-030-73909-6_7

[4]. The small intestine has an average length of 3–5 m and is composed of three major parts: the duodenum, the jejunum, and the ileum [5] (Figs. 1 and 2).

Major anatomy of the digestive system.

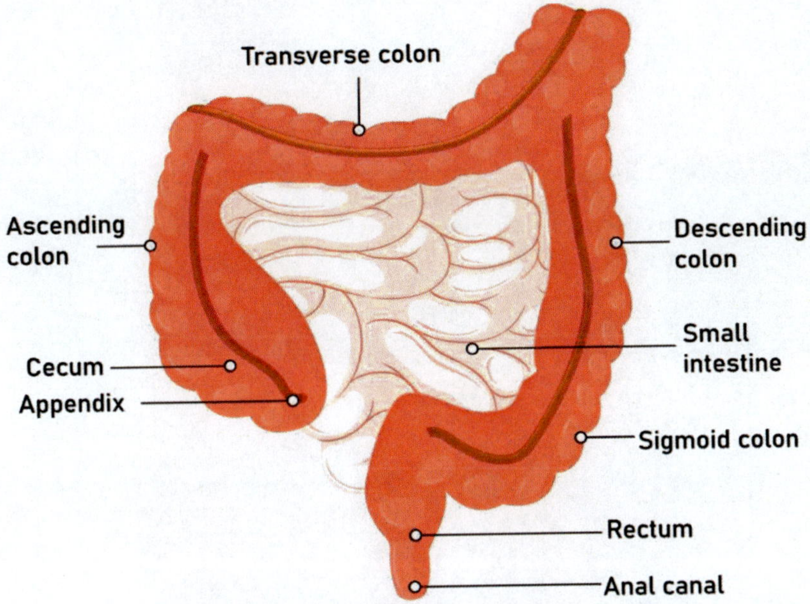

Fig. 1. Major structures within the digestive system.

On a histological level, the small intestine has four tissue layers. The most external is the serosa, followed by the muscularis externa, and then the submucosa [6]. The innermost layer is the mucosa, which contains small hair-like projections known as villi which are the main structure for absorption within the small intestine. Pathologies that damage the small intestine generally affect some component within the mucosa of the small intestine, whether in the form of lining perforations (as in the case of intestinal permeability), ulcers, or clumping of villi [3, 4, 6]. Food passage through the small intestine can be visualized in the form of a bolus, which is first formed in the mouth and then first passes through the esophagus and stomach, and has a standard size corresponding to the relative bite size of an individual [7, 8]. Correct visualization of bolus passage through the small intestine is important for understanding drug, as well as food molecule, absorption inside the intestines [5] (Fig. 3).

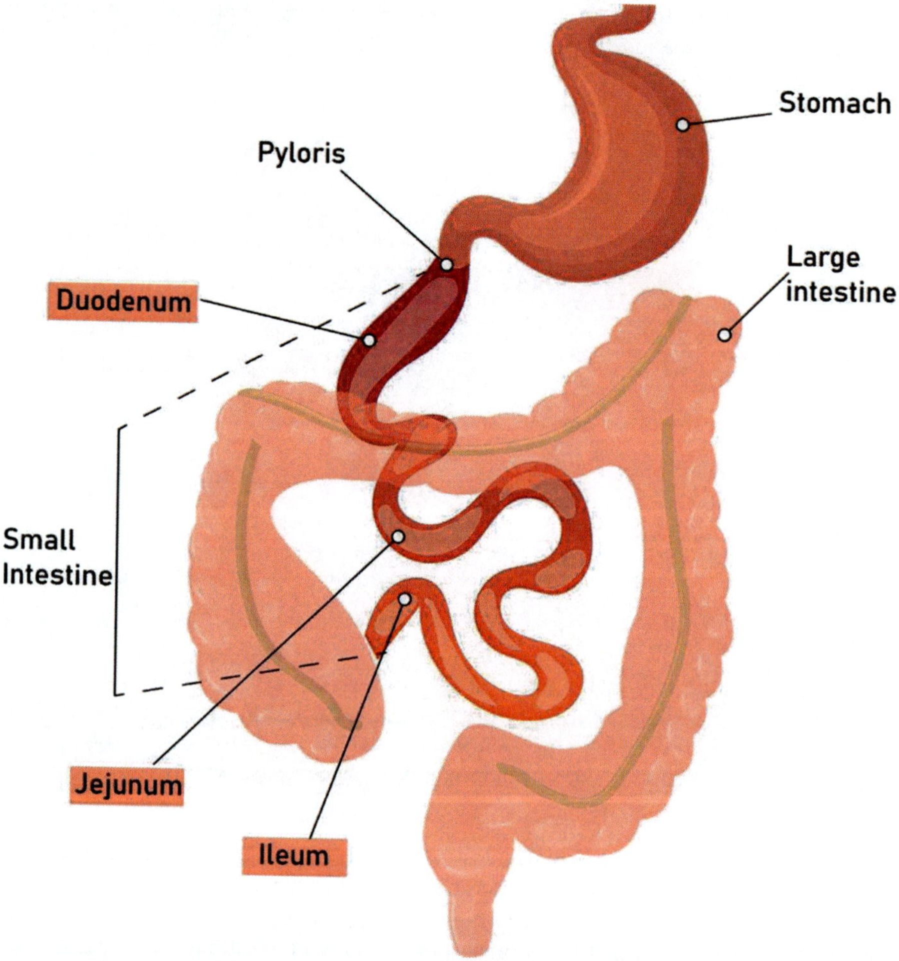

Fig. 2. The three compartments of the small intestine – duodenum, jejunum, and ileum.

1.2 Modeling of Transport and Absorption of Food Through the Small Intestine

Previous modeling attempts have explored various methods for visualizing the digestive process through the small intestine [1]. The model developed by Taghipoor *et al.* relied on a framework of ordinary differential equations in a set of four different models to model different aspects of the digestive process [7]. An updated version of that model factored in the role of dietary fiber and its associated ordinary differential equations [9]. Other models have explored the friction and adhesivity that the mucosal lining has on a small endoscopic robot [10–12]. Studies conducted by Lawrence *et al.* demonstrate that in *in silico* intestinal models, a multi-compartment approach is necessary, as a single-compartment approach does not yield a complete picture of bolus movement through the small intestine [1, 13]. Lastly, a more recent model created by Moxon, Gousetti, and

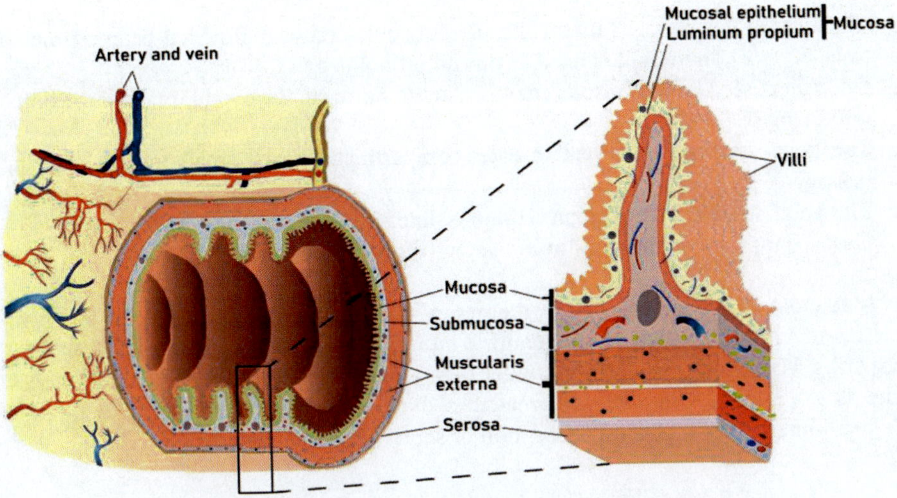

Fig. 3. Histological organization of the small intestine.

Bakalis allowed for simulation of nutrient mass transfer, gastric emptying, and starch hydrolysis using enzyme kinetics described by the Michaelis-Menten equation, in three different models [1].

In this paper, we present an integrated model that shows the passage and absorption of a soft, starch-rich bolus through the small intestine. This model builds upon the work of researchers Taghipoor *et al.* as well as Moxon, Gousetti, and Bakalis. Similar to previous models, our model depicts the small intestine as a one-dimensional cylindrical system with flow in one direction. The model has seven compartments to account for the 3–5 m average size of the small intestine [5, 6].

It should be stated that digestion is an incredibly complex process that cannot, at this time, be fully depicted with models, and the small intestine, although only a single organ within the massive digestion system, is histologically complex and is the site of countless chemical reactions that take place every second [1, 6]. Therefore, the aim of continual exploration of intestinal modeling is to move forward in understanding the mechanics of digestion *in silico*, in order to improve the modelling-based *in vitro* and *in vivo* approaches that advance the field of clinical gastroenterology.

2 Materials and Methods

In order to build this model, several basic assumptions had to be made, and the following conditions satisfied:

1. For simplification purposes, the small intestine will be modelled as a cylindrical structure divided into compartments where absorption of the bolus will take place irrespective of the sum surface of the intestinal villi.

2. The mass of the average human bite size for soft starch-rich food has been estimated to be 12 g [8], thereby it is used as our initial bolus mass.
3. Once processed by the various enzymes in the stomach, the overall mass of the bolus will increase.
4. The bolus shall be absorbed in each compartment of the small intestine during passage.
5. The bolus will be a Newtonian (compressible) fluid.
6. Gastric emptying is the rate-limiting step (Fig. 6) [7].

Mathematical modelling of our theoretical framework was achieved by inputting a series of ordinary differential equations into custom and pre-configured MATLAB scripts by MathWorks™, and the model diagrams were generated using SIMULINK (Fig. 5). An integrated model was created to show both transport of the bolus through the small intestine and general absorption of sequestered metabolites (Fig. 4).

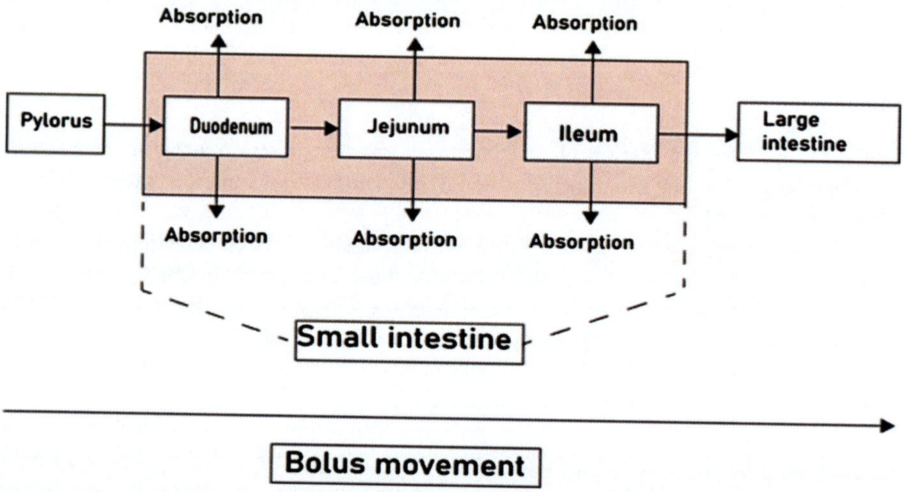

Fig. 4. Block Diagram of Bolus Movement and Absorption in the Small Intestine. The figure demonstrates the movement of the bolus through the small intestine. The bolus moves from the pylorus, which is designated as the beginning of the duodenum, and through the jejunum into the ileum and then out to the large intestine. Along the way, glucose and other metabolites are absorbed by the villi of the small intestine. Our model neglects to include absorption and movement in the stomach and large intestine in order to simplify modeling [1, 5, 7].

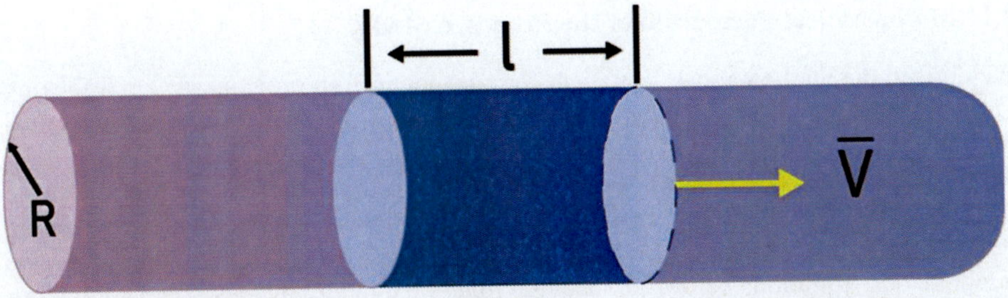

Fig. 5. Cylindrical representation of the small intestine. Surface area of the villi have been excluded for simplification purposes. R – radius; v – velocity, l – length of cylinder.

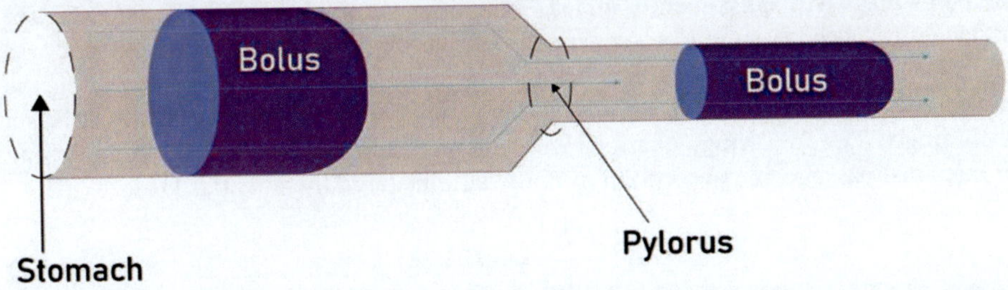

Fig. 6. Schematic representation of the stomach-small intestine transition during gastric emptying. The bolus is compressed once it enters the pylorus, and the overall volume is increased due to enzymatic digestion.

Parameters for transport and food absorption (e.g., glucose) were defined based on previous studies and models. The parameters we selected for our model are summarized in Table 1:

Table 1. Parameters for transport of bolus and absorption of nutrients in the small intestine [1, 5, 7, 8, 14, 15].

Parameter	Meaning	Value	Unit
Kt	Gastric emptying	2	h
Tsi	Mean small intestinal transit time	2	h
t	Total time	4	h
Peff	Human effective permeability	1.5E-05	m/s
R	Radius of small intestine	0.18	m
Mn	Amount of bolus in each compartment	13	g
Ms	Initial bolus size entering pylorus	12	g
a	Absorption segments	7	-
ρ	Density	1	kg/L
v(t)	Average velocity (bolus)	1.7E-04	m/s
l	Length of small intestine	3	m

2.1 Mathematical Formalism of the Proposed Model

We derived the mathematical formalism for our model from previous research seeking to model intestinal passage and absorption of metabolites [1, 7, 16]. The mathematical formalism behind our theoretical model may be applied both integrally and singularly, depending on whether passage of the bolus is being observed in the context of the entire digestive system, or at the resolution of a single compartment (e.g.: small intestine, colon, passage from the stomach to the small intestine). In our SIMULINK model, we have integrated the equations below into a single functional model that accurately computes absorbance of the bolus through the small intestine as a function of time. Of course, the absorption rate will depend on the initial bolus mass, due to the fact that increasing the bolus mass will subsequently increase the gastric emptying time. Human effective permeability (*Peff*) was used as a constant while building the MATLAB model, and was set in accordance to previous studies. Though the initial bolus mass was taken as 12 g, the same logic may be applied for other food, the parameters of which would have to be set accordingly. The following equations were adopted from previous work on modelling of intestinal passage and absorption of orally administered therapeutics [16].

$$\frac{dM_n}{dt} = -K_t M_n \tag{1}$$

Equation (1) is a simple differential equation that describes the process of gastric emptying, which is the rate-limiting step in the passage of food through the gastrointestinal system, with K_t being the gastric emptying rate and M_n representing the initial bolus mass [16].

$$\frac{dMn}{dt} = K_t M_{n-1} - K_t M_n \tag{2}$$

Equation (2) is the differential equation that shows the relationship between gastric emptying time with the amount of bolus present in each compartment [16].

$$\frac{dMn}{dt} = K_t \sum_{n=1}^{a} M_n \tag{3}$$

Equation (3) summarizes the inputs with respect to the absorption that occurs in the small intestine, where a denotes the number of small-intestine compartments that are being integrated, however in our case a was set to seven. Again, the rate-limiting step is the gastric emptying rate [7, 16].

2.2 SIMULINK Model

Figure 7 shows the SIMULINK model for modeling the small intestine containing three compartments which represent the three segments of the small intestine. SIMULINK coding blocks were added to expand upon work done in previous intestinal models. As was seen in Table 1, initial bolus size as it exits the stomach was set to 12 g, which is consistent with models built describing oral digestion [8]. The bolus size inside the small intestine was estimated to be 13 g, with the extra gram accounting for the presence of enzymes and gastric secretions. The density of the bolus would of course vary with the type of food consumed, however; based on previous research we chose to set density at a standard 1 kg/L [14].

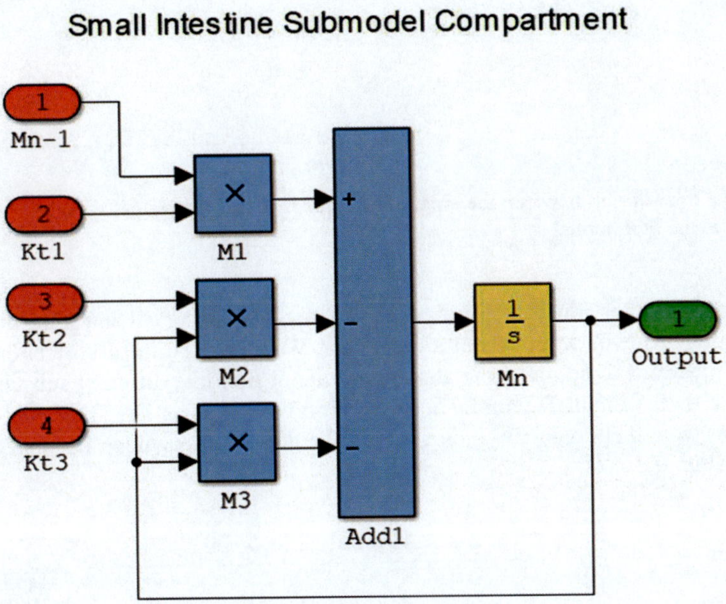

Fig. 7. SIMULINK model of the small intestine for calculating passage of bolus and its absorption.

3 Results

Our MATLAB model calculated the relationship between the amount of food ingested and the time required for the food to pass the entirety of the digestive tract, starting from the pylorus, and ending with the ileum (Fig. 8).

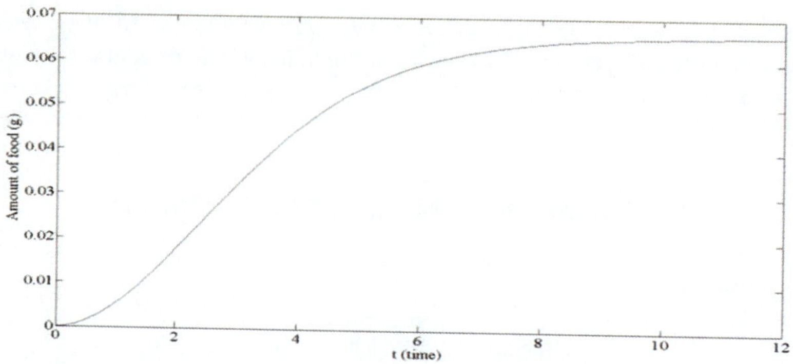

Fig. 8. The relationship between the amount of bolus ingested (in grams) versus time in hours, according to the built model.

Furthermore, for illustrative purposes, we divided the small intestine into seven distinct compartments where absorption occurs. We then used the aforementioned differential equations to compute the absorption rate of multiple different substances as a function of their weight. The initial concentration was taken as 12 g; however, the concentrations were then manually increased to 20 g due to mass gained from solubilizing enzymes (Fig. 9).

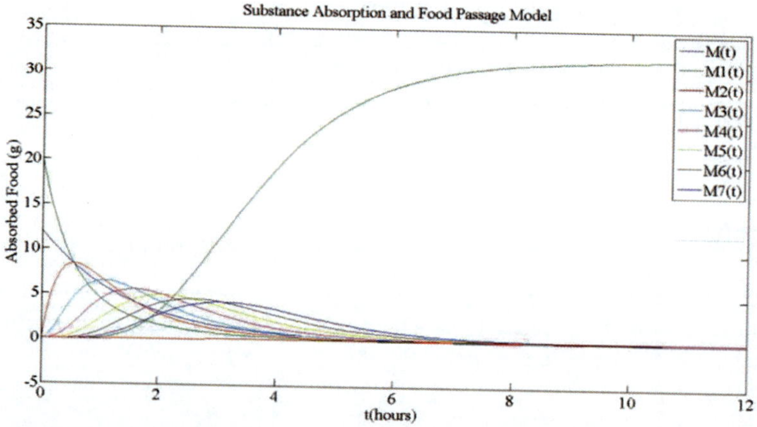

Fig. 9. Results obtained from simulating the absorption in the context of seven absorption compartments of the small intestine.

The model we created uses a combination of ordinary differential equations to describe velocity of the bolus through the small intestine and absorption of glucose, based on previous models developed by researchers [1, 7, 16]. Our seven-compartment model is consistent with research from previous studies, which showed that multicompartment models show a more complete picture of the small intestine as opposed to single-compartment models [13]. The bolus in our model is assumed to be completely solubilized, however; more complex previous models have accounted for non-soluble material in the bolus such as fiber [9]. Our model also falls short in that it does not account for the effect that digestive enzymes will have on the chemical composition of the bolus, thereby neglecting to account for bolus degradation as in previous models [1, 7], only on the addition of mass to the bolus as it passes through the small intestine. Nevertheless, our model is distinct from previous models because it attempts to show both transport and absorption simultaneously. Previous mathematical models show them as separate systems of equations [7].

4 Conclusion

We have created an integrated model to show both transport of food and nutrient absorption through the small intestine, the components of which may individually be used in accordance with pre-determined values of the parameters summarized in Table 1. It should be noted that the bolus (digestate) was taken as a Newtonian compressible fluid, and was modelled in accordance with the passage of Newtonian fluids through a perfect cylinder, using a system of ordinary differential equations. Our model is simplified to include just the three regions of the small intestine, however including seven absorption regions within that specific gastrointestinal segment. This model could be expanded in order to be used for modeling the absorption and passage of digestate through the colon.

Future research should aim to create an integrative *in silico* model that includes all parts of the digestive system and accounts for the introduction of different types of foods into the system. Another avenue that is crucial to be explored in future models is the concept of intestinal permeability. Our model accounts for human effective permeability to a degree, but with illnesses like autoimmune disease on the rise, it will be invaluable to have systems for modeling how foods interact with the lining of the small intestine, as well as the inflammatory responses that can be generated when food molecules escape through damaged tight junctions out of the small intestine into the surrounding environment [3]. As researchers with a desire to enhance understanding of biological processes for better clinical outcomes, we will continue to develop and explore different approaches to our model, and we encourage fellow researchers to do the same.

Conflicts of Interest. The authors declared no conflicts of interest for this paper, including but not limited to research grants, employment to a certain affiliation, or any other sort of funding.

References

1. Moxon, T.E., Gouseti, O., Bakalis, S.: In silico modelling of mass transfer & absorption in the human gut. J. Food Eng. **1**(176), 11020 (2016)
2. WHO. Obesity and overweight [Internet]. World Health Organization (2020). https://Obesity and overweight. Accessed 29 Dec 2020

3. Leech, B., McEwen, B., Sekyere, E.O.: Diet, digestive health, and autoimmunity: the foundations to an autoimmune disease food pyramid—part 2. Altern. Complement. Ther. **26**(4), 158–167 (2020)
4. Collins, J.T., Nguyen, A., Badireddy, M.: Anatomy, abdomen and pelvis, small intestine. StatPearls (2020)
5. Stoll, B.R., Batycky, R.P., Leipold, H.R., Milstein, S., Edwards, D.A.: A theory of molecular absorption from the small intestine. Chem. Eng. Sci. **55**(3), 473–489 (2000)
6. Bass, L.M., Wershil, B.K.: Anatomy, histology, embryology and developmental anomalies of the small and large intestine. In: Sleisenger and Fordtran's Gastrointestinal and Liver Disease, 10th edn, p. 1649. Saunders, Elsevier Inc., Philadelphia (2016)
7. Taghipoor, M., Lescoat, P., Licois, J.R., Georgelin, C., Barles, G.: Mathematical modeling of transport and degradation of feedstuffs in the small intestine. J. Theor. Biol. **7**(294), 114–121 (2012)
8. Bornhorst, G.M., Singh, R.P.: Bolus formation and disintegration during digestion of food carbohydrates. Compr. Rev. Food Sci. Food Saf. **11**(2), 101–118 (2012)
9. Taghipoor, M., Barles, G., Georgelin, C., Licois, J.R., Lescoat, P.: Digestion modeling in the small intestine: Impact of dietary fiber. Math. Biosci. **1**(258), 101–112 (2014)
10. Terry, B.S., Passernig, A.C., Hill, M.L., Schoen, J.A., Rentschler, M.E.: Small intestine mucosal adhesivity to in vivo capsule robot materials. J. Mech. Behav. Biomed. Mater. **1**(15), 24–32 (2012)
11. Zhang, C., Liu, H., Tan, R., Li, H.: Modeling of velocity-dependent frictional resistance of a capsule robot inside an intestine. Tribol. Lett. **47**(2), 295–301 (2012)
12. Zhang, C., Liu, H., Tan, R., Li, H.: Interaction model between capsule robot and intestine based on nonlinear viscoelasticity. Proc. Inst. Mech. Eng. Part H J. Eng. Med. **228**(3), 287–296 (2014)
13. Lawrence, X.Y., Crison, J.R., Amidon, G.L.: Compartmental transit and dispersion model analysis of small intestinal transit flow in humans. Int. J. Pharm. **140**(1), 111–118 (1996)
14. Ferrua, M.J., Singh, R.P.: Modeling the fluid dynamics in a human stomach to gain insight of food digestion. J. Food Sci. **75**(7), R15162 (2010)
15. Bourreau, J., Hernot, D., Bailhache, E., Weber, M., Ferchaud, V., Biourge, V., Martin, L., Dumon, H., Nguyen, P.: Gastric emptying rate is inversely related to body weight in dog breeds of different sizes. J. Nutr. **134**(8), 2039S-2041S (2004)
16. Lawrence, X.Y., Amidon, G.L.: A compartmental absorption and transit model for estimating oral drug absorption. Int. J. Pharm. **186**(2), 119–12 (1999)

Detection of Asthma Inflammatory Phenotypes Using Artificial Neural Network

Selma Delić(✉), Tijana Cvjetković, Medina Čajo, Ismet Fatih Čančar, Adna Čolak, Nejra Ćenanović, and Emina Direk

Faculty of Pharmacy, University of Sarajevo, Sarajevo, Bosnia and Herzegovina
selmadelic@ffsa.unsa.ba

Abstract. Asthma is a common disease that usually starts in childhood. There are more than 300 million people with asthma. In 2020, patients with asthma were at higher risk of developing COVID-19 infection. Previous studies showed benefits of using artificial neural networks when it comes to detection of asthma. This paper included 300 persons out of which 200 of them had disease and 100 which were healthy. Forced Vital Capacity (FVC), Forced Expiratory Volume (FEV1), eosinophiles, FeNO and IgE were measured. The neural network was made and it detected asthma by analyzing these parameters. It had the accuracy of 92.2%. This shows a great potential of neural network use when it comes to the detection of asthma and this could be an instrument used in clinical practice.

Keywords: Asthma · Artificial Neural Network · Artificial intelligence

1 Introduction

Asthma is common disease that starts usually in childhood. It is ranked as 16th among the leading causes of years lived with disability [1]. Also, it is 28th among the leading causes of burden disease, measured by disability-adjusted life years. It is estimated that around 300 million people have asthma worldwide. By 2025, probably around 100 million people more will be affected [2]. It is characterized by a various range of severity, from a very mild, occasional wheeze to acute, life-threatening airway closure. It can be associated with other features of atrophy, such as eczema and hay fever. It leads to multiple hospital admissions and it increases healthcare costs. Hyper-responsiveness, the key feature, can be triggered by many factors. The main problem when asthma is not treated promptly, is that it can have high mortality. Asthma is influenced by genetics and environmental exposure. Triggers for asthma can be viral respiratory tract infections, exercise, gastroesophageal reflux disease, chronic sinusitis, environmental allergens, tobacco smoke, obesity, emotional factors, use of aspirin etc. [1]. In 2020, patients with asthma are at higher risk for developing severe COVID-19 infection. COVID-19 infection is characterized by abnormal, overactive innate immunity and cytokine storm as possible pathological mechanism that can contribute to progression that can cause lethal outcome. Asthmatic patients could be at higher risk due to different antiviral immune responses and exacerbation that can be elicited by common respiratory viruses [3].

A. Badnjevic and L. Gurbeta Pokvić (Eds.): CMBEBIH 2021, IFMBE Proceedings 84, pp. 69–75, 2021.
https://doi.org/10.1007/978-3-030-73909-6_8

There is an increased and chronic airway inflammation when it comes to asthma. Airway sampling done by bronchoscopy shows that sputum has abundant eosinophils and activated T cells. Lung histology demonstrates disordered musocasal epithelium. It is well known that noxious stimuli, including potential allergens, can initiate an acute inflammatory response. That acute inflammatory response is self-limited and it represents non-resolving inflammation. It is linked to asthma and other chronic inflammatory diseases [4].

Asthma is suspected when a patient has a positive medical history of recurrent dry coughing which happens especially at night, wheezing, shortness of breath and tightness. Asthma therapy has an aim to achieve maximum freedom from symptoms and maximum quality of life [5].

Zhang et al. (2014) states that eosinophilic blood counts can accurately predict eosinophilic type of asthma when a patient has persistent uncontrolled asthma despite treatment. Blood counts can be a useful marker for monitoring uncontrolled asthma [6]. Nadif et al. (2016) reports that blood neutrophil counts can be used to classify adults with asthma. Results suggest that blood eosinophilia and neutrophilia may be associated with two particular and specific genotypes of asthma in adults. This can also be related to different characteristics of asthma such as symptoms, lung function and activity [7]. Patients with atopic asthma, have elevated IgE levels and they have good response to anti-IgE therapy [8].

It has been shown that artificial intelligence (AI) can be used for asthma screening and diagnosis, patient classification, asthma management and monitoring and for asthma treatment, as it is used in other fields of medicine [9–15]. For example, a wearable sensor was used to record signals of respiratory sounds. These sounds were subsequently transferred to the smartphone. After signal manipulations, Hidden Markov Model (HMM) was used for respiratory sound classification. It aimed to detect wheezing. This model had 94,91% of accuracy, 89,34% of sensitivity and 96,28% specificity [16]. This shows that there is a great potential of AI in chronic diseases such as asthma. Gurbeta et al. (2018) presented an automated expert diagnostic telehealth system that can diagnose asthma and some other respiratory diseases [17]. Some other diagnostic systems with really high accuracy were developed. An example is expert diagnostic system developed by Badnjevic et al. (2018) that had sensitivity of 96.45% and specificity of 98.71% [18]. Asthma can be also classified based on fuzzy rules that are determined with a combination of spirometry (SPIR) and Impulse Oscillometry System (IOS) test result [19].

The aim of this study is to detect asthma inflammatory phenotypes using Artificial Neural Network with full blood count parameters as a dataset.

2 Materials and Methods

This research was done on a dataset made out of 300 samples. Out of these 300 samples, 200 were for patients with asthma and 100 samples were for healthy patients, i.e. control group. The database consisted of 7 input parameters that are: Forced Vital Capacity (FVC), Forced Expiratory Volume (FEV1), eosinophiles, FeNO, IgE, age, and sex were measured. Based on 8 parameters ANN in this paper predicts whether the patient suffers from asthma or not.

The normal range for FVC starts from >80%. For FEV1 normal range is from 7080% [20]. Usually, FEV1/FVC ratio is measured and in asthma patients, and it is below 70% [21]. When it comes to eosinophiles, the normal range is 0,22–0,32 x^9/L and it is important to measure them when it comes to eosinophilic asthma. Eosinophilic count can be used as a predictor for treatment [22]. IgE has the normal range between 72–900 IU/L and people with asthma can have elevated IgE levels [23] (Table 1).

Table 1. Database for development of neural network.

	Total number of sampels	Number of women	Number of men	Average women age	Average man age
Healthy subjects	100	50	50	38.5 ± 13. 6	36.6 ± 12. 8
Patients with viral infection	200	102	98	41.4 ± 17. 2	44.0 ± 15. 6

The distribution of the dataset for development of the Artificial Neural Network (ANN) is presented in Table 3. Since this is relatively small dataset, distribution of 70:30% for training and validation was used [24].

Artificial Neural Network (ANN) was developed to perform prediction of renal diseases based on diagnostic parameters. Developed ANN has 7 input. The ANN performs classification of samples into two classes: healthy and patients with disease (Table 2).

Table 2. Dataset distribution.

Total number of samples 300	Healthy 100	Disease 200
Training 210	70	140
Validation 90	30	60

For the development of the ANN in this research, standard architecture for pattern recognition was used. That is feedforward neural network which consists of twolayers. This architecture is designed since according to the application experts, that kind of ANN is sufficient to properly perform the classification of input data [9, 18].

The input layer is made of network inputs. Input layer is followed by a hidden layer which consists of parallelly connected neurons. The network was tested on training performance with different number of neurons. Since, this classification is basically highly non-linear data classification, sigmoid function was used for hidden layer [19].

During the design of the network, different number of neurons was used to measure the performance. According to the Mean Square Error (MSE), appropriate number of neurons was chosen. After training, validation of the network performance was done using confusion matrix.

According to these results, specificity, sensitivity and accuracy of the network are calculated according to Eqs. (1), (2) (3):

$$sensitivity = \frac{true\ positive}{true\ positive + false\ neagative} \tag{1}$$

$$specificity = \frac{true\ negative}{true\ negative + false\ positive} \tag{2}$$

$$accuracy = \frac{true\ negative + true\ positive}{true\ negative + true\ positive + false\ negative + false\ positive} \tag{3}$$

3 Results

According to previous research and expert recommendation [9, 18], of classification of prediabetes and diabetes type 2 in this paper, two-layer feedforward artificial neural network, with sigmoid transfer function in hidden layer was developed.

Training performance with the different numbers of neurons is shown in Figs. 1, 2 and 3. As it can be seen, the best performance was achieved with 4 neurons in hidden layer. Accuracy of training with 4 neurons was 92.3%, while accuracy of training performance with 18 and 24 neurons in hidden layer was 89.9% and 76.4% respectively.

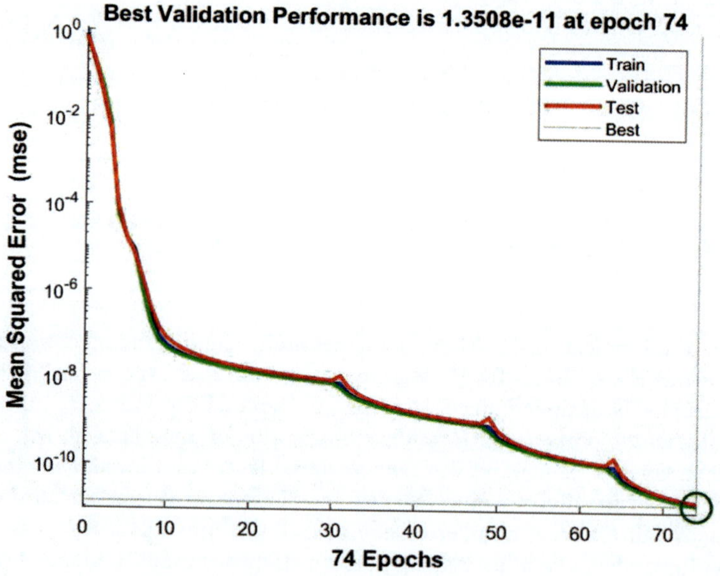

Fig. 1. Performance of training with 4 neurons

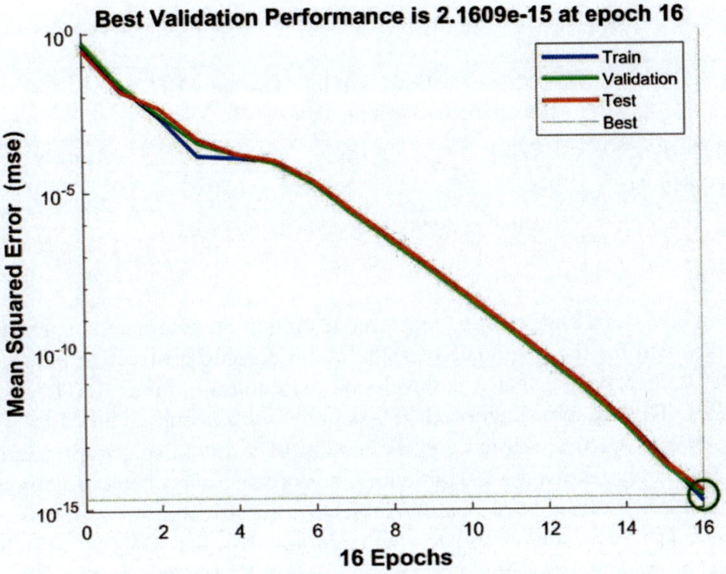

Fig. 2. Performance of training with 18 neurons

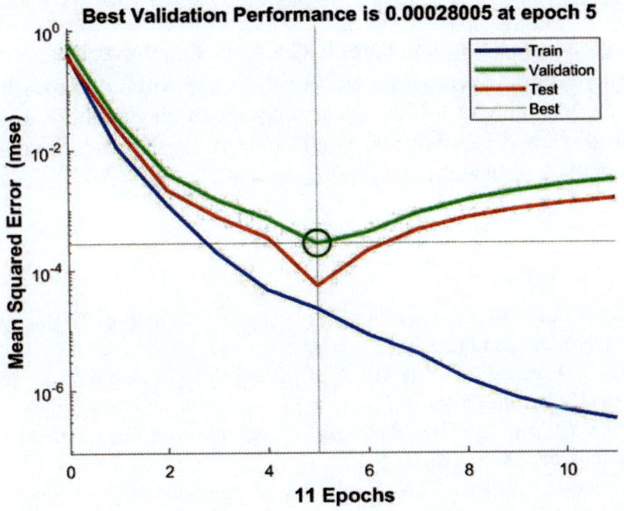

Fig. 3. Performance of training with 24 neurons

Validation performance for neural network with 4 neurons is shown in Table 3. As it can be seen, neural network was trained in the classification of patients with asthma with specificity of 93.3%. Overall accuracy of architecture is 92.2%.

Table 3. Artificial Neural Network performance in classification to class "Asthma"

	GROUP 1 Healthy ANN prediction	Group 2 Disease ANN prediction	
GROUP 1 Healthy 30	TP 27	FN 3	Accuracy [%] 92.2%
Group 2 Disease 60	FP 4	TN 56	
	Sensitivity [%] 90.0	Specificity [%] 93.3	

Gurbeta et al. (2018) developed real-time testing of an automated expert diagnostic telehealth system for the diagnosis of asthma and chronic obstructive pulmonary disease (COPD). The system that was developed uses android, java, MATLAB and PHP technologies. This system diagnosed 780 patients with accuracy an of 97.73% [17]. Badnjević et al. (2016) presented a classification of asthma using implemented fuzzy systems. This study involved 1250 patients. The software classified correctly asthma in 91.89% of cases. In the case of healthy subjects correctly, the software was correct in 95% of cases [19]. The studies showed better results than our ANN but still this shows that there is great potential when it comes to the use of ANN in asthma patients.

4 Conclusion

Asthma is one of the most common chronic diseases. Sometimes it cannot be controlled with medications. There is a great potential for AI use when it comes to asthma. Our ANN showed the accuracy of 92.2% when it comes to diagnosing asthma. This could be a great tool in clinical practice that would help the right time diagnosis of asthma. This would contribute to health care systems as well.

References

1. Hashmi, M.F., Tariq, M., Cataletto, M.E.: Asthma. In: StatPearls [Internet]. StatPearls Publishing, Treasure Island (2020)
2. Dharmage, S.C., Perret, J.L., Custovic, A.: Epidemiology of asthma in children and adults. Front. Pediatr. **18**(7), 246 (2019)
3. Liu, S., Zhi, Y., Ying, S.: COVID-19 and asthma: reflection during the pandemic. Clinic. Rev. Allerg. Immunol. **59**, 78–88 (2020)
4. Levy, B.D., Vachier, I., Serhan, C.N.: Resolution of inflammation in asthma. Clin. Chest Med. **33**(3), 559–570 (2012)
5. Horak, F., Doberer, D., Eber, E., et al.: Diagnosis and management of asthma - statement on the 2015 GINA guidelines. Wien Klin Wochenschr. **128**(1516), 541–554 (2016)
6. Zhang, X.Y., Simpson, J.L., Powell, H., et al.: Full blood count parameters for the detection of asthma inflammatory phenotypes. Clin. Exp. Allergy **44**(9), 1137–1145 (2014)
7. Nadif, R., Siroux, V., Boudier, A., et al.: Bloof granulocyte patterns as predictors of asthma phenotypes in adults from the EGEA study. Eur. Respir. J. **48**(4), 1040–1051 (2016)
8. Patel, T.R., Sur, S.: IgE and eosinophils as therapeutic targets in asthma. Curr. Opin. Allergy Clin. Immunol. **17**(1), 42–49 (2017)

9. Catic, A., et al.: Application of neural networks for classification of patau, edwards, down, turner and klinefelter syndrome based on first trimester maternal serum screening data, ultrasonographic findings and patient demographics. BMC Med. Genomics **11**, 19 (2018). https://doi.org/10.1186/s12920-018-0333-2
10. Badnjevic, A., Cifrek, M., Koruga, D., Osmankovic, D.: Neuro-fuzzy classification of asthma and chronic obstructive pulmonary disease. BMC Med. Inf. Decis. Making J. **15**(Suppl. 3), S1 (2015). https://doi.org/10.1186/1472-694715-S3-S1
11. Alic, B., et al.: Classification of stress recognition using artificial neural network. In: IEEE 5th Mediterranean Conference on Embedded Computing (MECO), 12–16 June 2016, Bar, Montenegro
12. Sejdinovic, D., et al.: Classification of prediabetes and type 2 diabetes using artificial neural network. In: CMBEBIH 2017. IFMBE Proceedings, vol. 62, pp. 685–689. Springer, Singapore (2016). https://doi.org/10.1007/978-981-10-4166-2_103
13. Badnjević, A., Pokvić, L.G., Hasičić, M., Bandić, L., Mašetić, Z., Kovačević, Ž., Kevrić, J., Pecchia, L.: Evidence-based clinical engineering: machine learning algorithms for prediction of defibrillator performance. Biomed. Signal Process. Control **54**, 101629 (2019). https://doi.org/10.1016/j.bspc.2019.101629. ISSN 1746-8094
14. Divovic Mustafic, L., et al.: Diagnosis of severe aortic stenosis using implemented expert system. In: Badnjevic, A., Škrbić, R., Gurbeta Pokvić, L. (eds.) CMBEBIH 2019. IFMBE Proceedings, vol. 73. Springer, Cham (2020)
15. Sarić, R., et al.: FPGA-based real-time epileptic seizure classification using artificial neural network. Biomed. Signal Process. Control **62**, 102106 (2020). https://doi.org/10.1016/j.bspc.2020.102106. ISSN 1746-8094
16. Exarchos, K.P., Beltsiou, M., Votti, C.A.: Artificial intelligence techniques in asthma: a systematic review and critical appraisal of the existing literature. Eur. Respir. J. **56**(3), 2000521 (2020)
17. Gurbeta, L., Badnjevic, A., Maksimovic, M., Omanovic-Miklicanin, E., Sejdic, E.: A telehealth system for automated diagnosis of asthma and chronical obstructive pulmonary disease. J. Am. Med. Inform. Assoc. **25**(9), 1213–1217 (2018)
18. Badnjevic, A., Gurbeta, L., Custovic, E.: An expert diagnostic system to automatically identify asthma and chronic obstructive pulmonary disease in clinical settings. Sci. Rep. **8**(1), 11645 (2018)
19. Badnjevic, A., Gurbeta, L., Cifrek, M., Marjanovic, D.: Diagnostic of asthma using fuzzy rules implemented in accordance with international guidelines and physicians experience. In: IEEE 39th International Convention on Information and Communication Technology, Electronics and Microelectronics (MIPRO), Opatija, Croatia (2016)
20. Barreiro, T.J., Perillo, I.: An approach to interpreting spirometry. Am. Fam. Physician. **69**(5), 1107–1114 (2004)
21. Gallucci, M., Carbonara, P., Pacilli, A.M.G., et al.: Use of symptoms scores, spirometry, and other pulmonary function testing for asthma monitoring. Front. Pediatr. **5**(7), 54 (2019)
22. Kostikas, K., Brindicci, C., Patalano, F.: Blood eosinophils as biomarkers to drive treatment choices in asthma and COPD. Curr. Drug. Targets **19**(16), 18821896 (2018)
23. Mahesh, P.A.: Evaluation of asthma severity: relevance of total serum IgE, sputum and peripheral eosinophilia. Lung India **34**(3), 290–291 (2017)
24. Rau, S.H., Hsu, C.Y., Lin, Y.A., et al.: Development of a web-based liver cancer prediction model for type II diabetes patients by using an artificial neural network. Comput. Methods Programs Biomed. **125**, 58–65 (2016)

Biomechanics, Biosensors and Bioinstrumentation

Development of a Reliable Spiroximeter for Covid-19 Patients' Telemonitoring

Emil Valchinov[1](✉), Konstantinos Rotas[1], Athanasios Antoniou[1], Aris Dermitzakis[2], and Nicolas Pallikarakis[1]

[1] Biomedical Technology Unit, University of Patras, Patras, Greece
emil@upatras.gr
[2] Aeon PC, Patras, Greece

Abstract. This paper presents a non-invasive device, for assessing two vital COVID-19 related parameters: respiratory function and patients' blood oxygenation in a simple manner, capable of detecting early symptoms, thus enabling proactive treatment rather than responsive. Its operation is supported by a smartphone application, via easy wireless Bluetooth connectivity and a platform for remote monitoring. The proposed device computes the mass flow by measuring the differential pressure in a venturi tube using a factory calibrated and temperature compensated differential pressure sensor. The latter is directly connected via a two-wire digital interface to the main control electronic board, and the obtained data are linearized. The venturi tube was preferred over a laminar flow restrictor, due to its simplicity and smooth easy to swipe surfaces despite its higher non-linear response characteristics. It is detachable from the whole device allowing easy disinfection or replacement, thus providing enhanced infection protection. The blood oxygenation and pulse rate are measured by an extremely sensitive and stable factory calibrated module, which includes internal LEDs, photodetectors, optical elements, and low-noise electronics with ambient light rejection. The device is entirely controlled by the smartphone and has no moving mechanical parts, displays, or buttons, thus offering high reliability and low production cost. The preliminary results showed that the proposed design solution is robust, user friendly and can deliver accurate and consistent respiratory and oximetry test results at the point of care, without the need for calibration.

Keywords: Spirometer · Oximeter · IMU · Venture tube · Mass flow · Differential pressure · Bluetooth · Smartphone

1 Introduction

The unfolding COVID-19 pandemic has led to a global crisis that threatens to become a health, economic, and humanitarian disaster. It has presented considerable challenges to global health services and influences almost every aspect of medical practice and policy. Home monitoring for patients without serious symptoms is the cornerstone of the strategy for dealing with the current pandemic. This approach helps to prevent the spread of the disease and reduces the workload of the health system. However, it requires daily

A. Badnjevic and L. Gurbeta Pokvić (Eds.): CMBEBIH 2021, IFMBE Proceedings 84, pp. 79–87, 2021.
https://doi.org/10.1007/978-3-030-73909-6_9

monitoring of patients who remain at home through frequent telephone communication and evaluation of symptoms, and/or visits by specialized physicians and nurses [1–3].

Therefore, a non-invasive device, for assessing the two most vital COVID-19 related parameters: respiratory function and patients' blood oxygenation in a simple manner and at home, will provide early diagnosis, enabling proactive treatment rather than responsive especially in cases of the vulnerable population suffering from respiratory diseases. Moreover, it will promote shifting the focus from hospitals and clinics to patients' homes thus reducing unneeded hospital visits and associated risks of contamination and limiting disease outspread.

The device proposed by this work concerns an easy to use, low-cost device for home patient monitoring, assessing respiratory function and blood oxygenation non-invasively, providing remote data access to the treating physician. It provides exhaled air volume and respiration pattern, heart rate, and blood oxygen saturation in less than 30 s. Its operation is supported by a smartphone application, via easy wireless Bluetooth connectivity and a web platform for remote monitoring, which allows the measurement, analysis, and transmission in real-time patients' key vital data.

2 Materials and Methods

2.1 Hardware Architecture

The proposed handheld device combines two main measuring parts: a spirometer and an oximeter, thus allowing to measure, collect, and analyze vital COVID-19 related parameters as shown in the main block diagram in Fig. 1. At present, spirometers employ the following main flow sensor technologies [4–9].

- Turbine and Pitot Tube (rotating vane) flowmeters: low cost, applying simple technology but accuracy is affected by bearing degradation and requires frequent calibration. Suffers from the so-called "lag-before-start" and "spin-after-stop" effects. Readings are reduced at low temperatures but not affected by barometric pressure.
- Differential pressure flowmeters (pneumotachometers): two types - Lilly (mesh screen) & Fleisch (array of capillaries). Feature no mechanical moving parts with extreme accuracy, good dynamic response, and range but easily get dirty. They are very sensitive to temperature, humidity, and atmospheric pressure and need proper compensation and calibration. Fleisch type is currently the most reliable with the longest useful life and is used as a reference system.
- Ultrasonic flowmeters: feature no mechanical moving parts with extreme accuracy but are sensitive to temperature, tube resistance, turbulence, shear effect and the signal at zero flow (baseline) tends to drift over time. However, lately, they are gaining popularity due to technology advancements that allow overcoming some of the drawbacks.

The proposed device computes the mass flow by measuring the differential pressure due to flow across a flow restriction as shown in Fig. 2. The flow restriction is implemented by a modified venturi tube, where the differential pressure is measured employing a small flow through a state-of-the-art differential pressure sensor, thus creating the so-called bypass configuration, allowing measuring the flow in the tube with high precision.

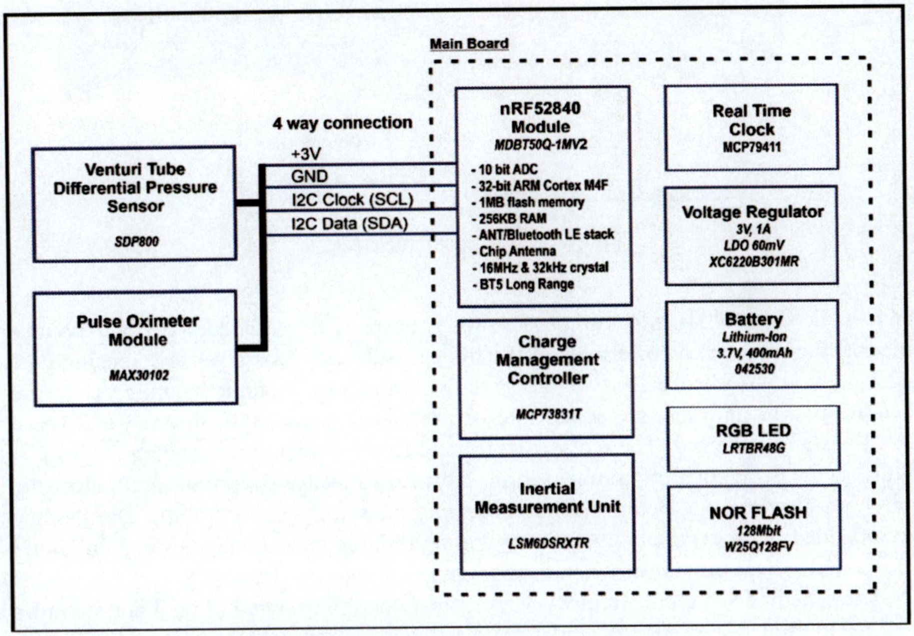

Fig. 1. Block diagram of the proposed device.

The selected sensor [10] provides outstanding accuracy, long-term stability, and a wide range of measurement speeds. It is factory calibrated, temperature compensated and features fast response time, no offset, no drift, and low sensor-to-sensor variation which is especially suited for mass production. The venturi tube was preferred over a laminar flow restrictor (honeycomb structure) type of orifice due to its simplicity and smooth easy to swipe surfaces despite its higher non-linear response characteristics and the need for nonlinearity error compensation [11]. It is designed to be detachable from the whole device allowing easy disinfection or replacement with a new one, thus providing enhanced infection protection. The device is designed to be used with a single-use inline antibacterial and antiviral filter [12], which delivers extra safety. The overall implemented design solution retains low production cost, making it accessible for personal use while increasing the reliability of the device by avoiding mechanical parts that are prone to failures and require a dedicated enclosure.

The embedded pulse oximeter module MAX30102 [13] provides clear SpO2 and heart-rate measurements in LED reflective solution. The module is factory calibrated, extremely sensitive, stable with fast response time, and includes internal LEDs, photodetectors, optical elements, and low-noise electronics with ambient light rejection. It provides a complete system solution to ease the design process for mobile and wearable

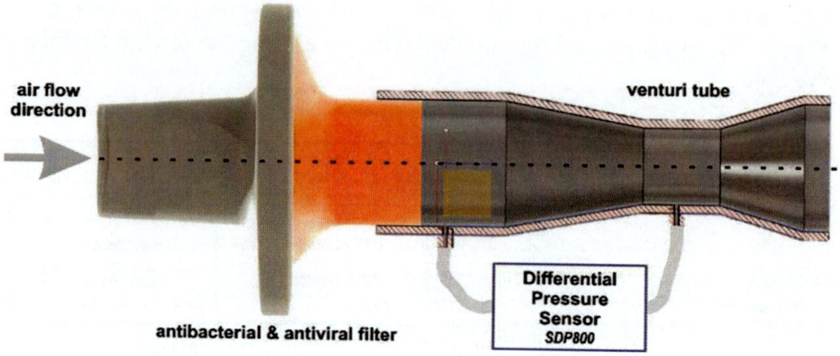

Fig. 2. Flow measurement setup.

devices. The MAX30102 does all the sensing by utilizing the embedded LEDs to bounce light off the arteries and arterioles in the finger's subcutaneous layer and sensing how much light is absorbed with its photodetectors, known as photoplethysmography. The data are passed onto the embedded microcontroller (Cortex M4F) through I2C serial data lines and analyzed. The Cortex M4F handles all algorithmic calculations, digital filtering, pressure/position compensation, advanced R-wave detection, and automatic gain control in order to determine heart rate and blood oxygen saturation. The module is embedded in the ergonomic handle providing reliable measurements since the finger lies naturally over the sensor when the patient is holding the device.

The schematic of the main electronic control board is shown in Fig. 3 and includes also an Inertial Measurement Unit (IMU) used for monitoring of the device's position/inclination during expiration and smart power management. The latter is supported by dedicated algorithms of the embedded firmware. This architecture provides robust detection and compensation of motion artifacts in captured samples and allows to correlate the physical activity with the saturation level (%SpO2) which enables detailed motion analysis during the tests.

The device is entirely controlled by the smartphone and doesn't have any buttons or displays, except for a tiny RGB LED which indicates the current mode of operation. It wakes up when touched/moved and goes to sleep after a specified period of motion inactivity or battery voltage below the minimum threshold. The embedded rechargeable Lithium-Ion battery is charged through a standard micro USB 2.0, AB-type receptacle.

Its main submodules are IMU, NOR-type Flash, RTC, voltage regulator, microcontroller, and 2.4 GHz radio. The last two are implemented with an MDBT50Q module [14], based on the ultra-low-power system-on-chip (SoC) BLE nRF52840 built around a 32-bit ARM Cortex M4F CPU, thus avoiding the use of an extra microcontroller, reducing the part count and keeping the system dimensions small.

2.2 Firmware and Application Program Design

A custom application firmware was developed for the device board, supporting ondemand initiating and terminating sessions of sampling the differential pressure sensor,

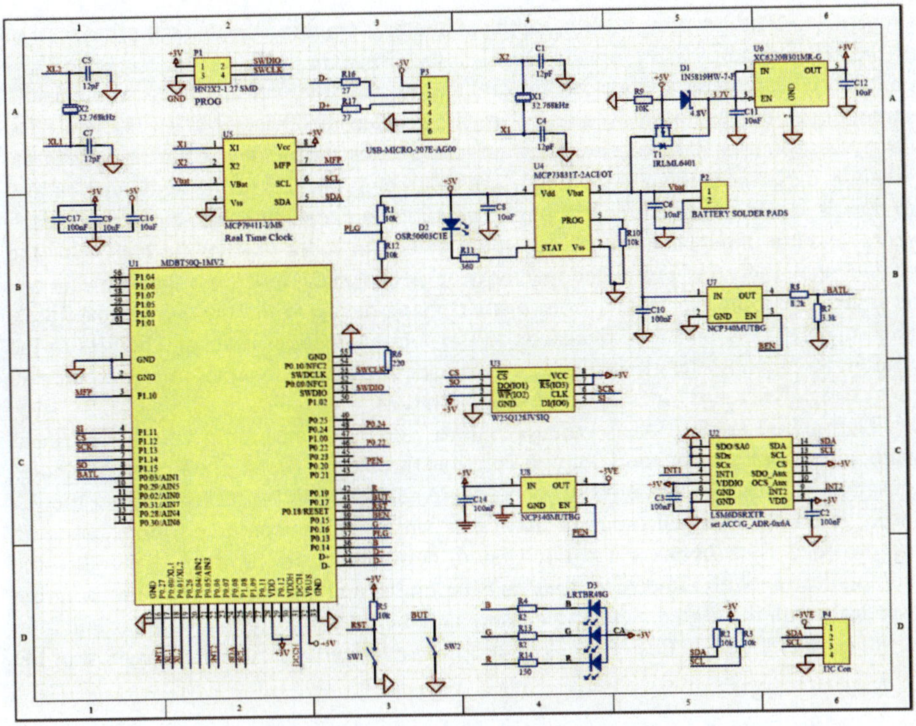

Fig. 3. Schematic of the main control board.

the IMU, and the pulse oximeter. The sampled data are streamed in real-time to a connected smartphone device for further processing. For sampling the differential pressure, the sensor is set in continuous sampling mode, with mass flow compensated differential pressure and no averaging of the sampled values. In this mode, the differential pressure module updates its measurements, which are also temperature compensated, with a rate of 2 kHz. After a short delay, required for the module to produce valid values in a sampling session, the firmware schedules reading the sampled values from the sensor with the highest frequency permitted for a normal operation, which is 1 kHz. Using a higher frequency (e.g. 2 kHz) could interfere with other important and periodic operations of the device such as Bluetooth connectivity related functions, sampling of other sensors, or memory I/O operations. Then the average value over the received samples in an interval of 5 ms are computed, for an effective sampling frequency of 200 Hz. The averaged values are stored in packets scheduled to be sent via Bluetooth to the paired smartphone device, following a custom-designed communication protocol that ensures full use of the available Bluetooth bandwidth.

The oximeter sensor is also sampled on-demand at a predefined interval. SpO2 results are reported as the percentage of hemoglobin that is saturated with oxygen. These data are processed, filtered, and compensated as required on the Cortex M4F microcontroller and then streamed in packets separate from the pressure sensor samples, providing updates to a dedicated channel (Bluetooth "characteristic") which the smartphone can subscribe to. The sampling session for the oximeter is combined with a sampling session of the differential pressure sensor. Besides, the firmware supports sampling the IMU unit for accelerometer and gyroscope data and sending the measurements to the user's smartphone. The sampling frequency for the accelerometer is set at 40 Hz while for the gyroscope it is 104 Hz. The firmware also checks for the device's charging status, monitors the embedded battery voltage level, and sends updates on the remaining battery charge. It is also designed for low power operation and can switch to sleep mode each sensor module independently or the whole device in order to save the battery of the device, upon detecting prolonged inactivity or on-demand. Updates to the firmware can be issued Over the Air (OTA) from a smartphone device with Bluetooth capabilities. Such updates may contain expansions to the communication protocol, changes to the configuration parameters to test and tweak the device's functionality, improvements on processing algorithms, or fixes to operational issues.

The mobile application is written in Java with Android Studio v4.4.1 for Linux, using object-oriented methodology. It contains three main parts. The first part handles the graphical user interface (GUI) and uses Android SDK functions to draw the 2D result graph. The second program part deals with wireless communication, based on the provided BLE libraries and our custom communication protocol. It handles the communication with the device, transmission faults, and interrupts. The third program part deals with the digital signal processing and includes algorithms for the calculation of values FVC, FEV1, PEF, FEV2575 and the nonlinearity error compensation of the flow.

3 Results and Discussion

The first working prototype of the proposed device was developed according to the design described above where the casing was 3D printed as shown in Figs. 4 and 6. It underwent preliminary laboratory testing in order to be validated and calibrated. All the measurements obtained from the prototype were compared to values read from precision flow and pulse oximeter analyzers [15, 16], which are very accurate, and hence a good benchmark.

The prototype was calibrated using a three-liter calibration syringe and Flow Analyzer PF-300, where the venturi tube and the analyzer's transducers were connected in series. The sampling frequency of the prototype was set to 1 kHz for the purpose. Based on the acquired raw data from both prototype and analyzer, the best linearization method was derived using curve fitting with MATLAB simulations. Optimum results were achieved applying rational polynomial curve fitting with N4/D4 terms and averaging filter with 3 sample depth for 200 Hz sampling frequency as shown in Fig. 5.

The preliminary analysis of the linearized data showed that values are almost identical to the ones measured by the flow analyzer, stable in different environmental conditions, and covering the full range of spirometry flow rates. However, further testing is needed to define the exact flow/volume accuracy and dynamic flow resistance of the developed prototype.

After each measurement, the android app calculates and displays the Flow-Volume and Flow-Time graph, and calculates Forced Vital Capacity (FVC), Forced Expiratory

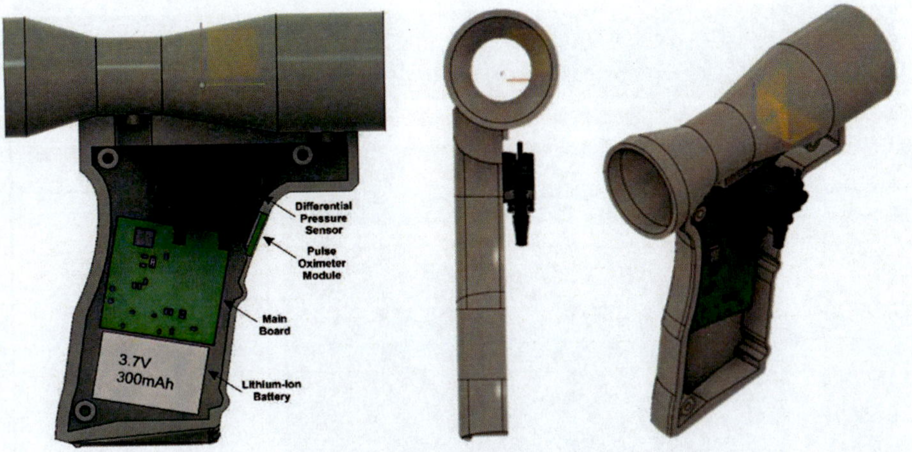

Fig. 4. Design of the prototype casing.

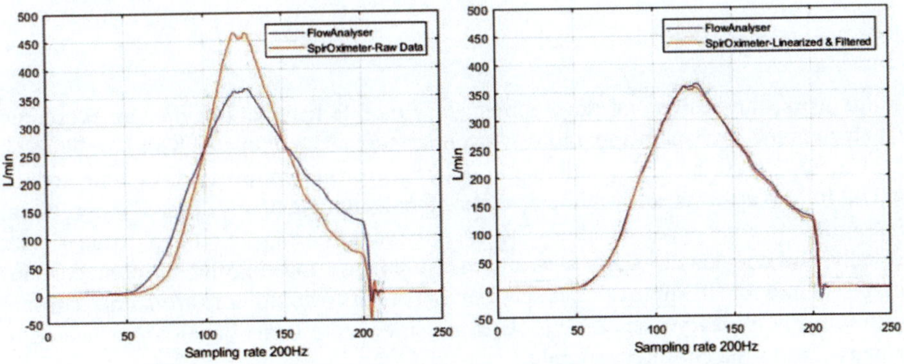

Fig. 5. Screenshot comparing analyzer's and prototype's flow.

Volume in 1 s (FEV1), Peak Expiratory Flow (PEF), and Forced Expiratory Flow 25–75% (FEF 25/75) based on the obtained expiratory data. A screenshot of the Android App together with a photo of the prototype is shown in Fig. 6. Future app development envisages interactive software displays, animated incentive screens, and real-time graphs aiming to enhance user's experience, encouraging child and adult patients to achieve optimal test results. Furthermore, customized solutions including EMR/EHR integration might be included to meet user individual needs.

The use of the Venturi tube proved very robust and fully adequate to perform the spirometry tests using linearization with rational polynomial curve fitting instead of lookup tables. The proposed design solution has no moving parts which make the instrument particularly reliable and highly economic, with low dead space and air resistance. It features a good dynamic response without any latency time, does not get clogged, and is detachable from the handle for fast exchange with a sterile or new one.

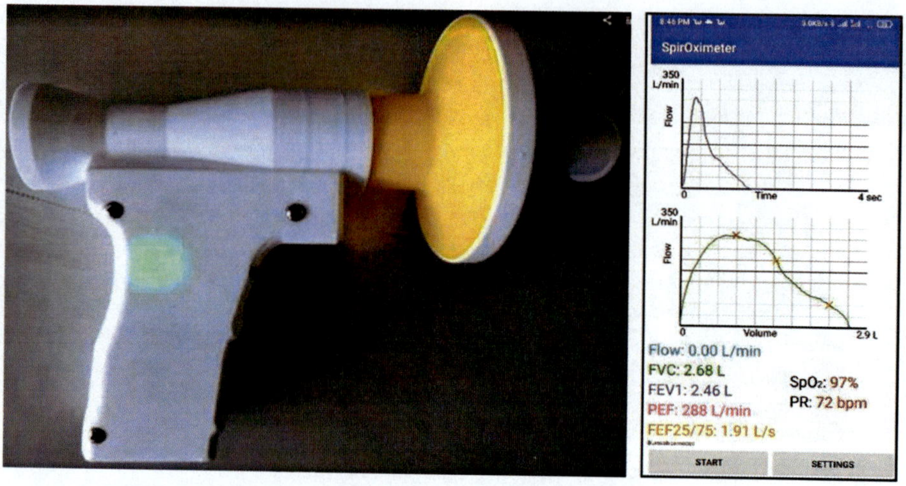

Fig. 6. Photo of the prototype and a screenshot of the Android App.

4 Conclusion

In this work, the design and development of a mobile handheld device are presented which combines two main apparatuses: a spirometer and an oximeter. This combination has been proved to be highly suitable to detect early symptoms of COVID-19, reduce patient follow-up costs, and anticipate the eventual outcome of any clinical progression, thus promoting personalized care. The overall system design is reliable, robust, user friendly, and delivers clinically accurate measurements, meeting the harmonized standards required for CE approval. It is suitable for patients who suffer from asthma, chronic obstructive pulmonary disease, and other respiratory problems which are expected to increase due to the current pandemic.

The preliminary results showed that the proposed design solution can deliver accurate and consistent respiratory and oximetry test results at the point of care without the need for calibration. The wide use of this fully integrated smartphone-driven device can facilitate evidence-based medical decisions for the patient, expedite the diagnosis process, improve the quality of care, and significantly reduce healthcare costs.

References

1. British Thoracic Society: British Thoracic Society Advice for Managing Interstitial Lung Disease Patients during COVID-19 pandemic, March 2020. https://www.brit-thoracic.org.uk/media/455101/bts-management-advice-for-ild-patients-v10-23-march-2020.pdf
2. Hull, J.H., Lloyd, J.K., Cooper, B.G.: Lung function testing in the COVID-19 endemic. Lancet Respir. Med. **8**(7), 666–667 (2020). https://doi.org/10.1016/S2213-2600(20)30246-0
3. Mo, X., et al.: Abnormal pulmonary function in COVID-19 patients at time of hospital discharge. Eur. Respir. J. **55**(6), 2001217 (2020). https://doi.org/10.1183/13993003.01217-2020

4. Yeh, M.P., Adams, T.D., Gardner, R.M., Yanowitz, F.G.: Turbine flowmeter vs. Fleisch pneumotachometer: a comparative study for exercise testing. J. Appl. Physiol. **63**(3), 1289–1295 (1987). https://doi.org/10.1152/jappl.1987.63.3.1289
5. Jones, D.P.: Biomedical Sensors. Momentum Press, New York (2010). Print ISBN: 978-160650-056-9
6. Buess, C., Pietsch, P., Guggenbühl, W., Koller, E.A.: A comprehensive review on accuracy in ultrasonic flow measurement using reconfigurable systems and deep learning approaches. AIP Adv. **10**, 105221 (2020). https://doi.org/10.1063/5.0022154
7. Mortimer, K.M., Fallot, A., Balmes, J.R., Tager, I.B.: Evaluating the use of a portable spirometer in a study of pediatric asthma. Chest **123**(6), 1899–1907 (2003). https://doi.org/10.1378/chest.123.6.1899
8. Skloot, G.S., Edwards, N.T., Enright, P.L.: Four-year calibration stability of the EasyOne portable spirometer. Respir. Care **55**(7), 873–877 (2010)
9. Walters, J., Wood-Baker, R., Walls, J., Johns, D.: Stability of the EasyOne ultrasonic spirometer for use in general practice. Respirology **11**(3), 306–310 (2006). https://doi.org/10.1111/j.1440-1843.2006.00842.x
10. Differential Pressure Sensors SDP800-500. https://www.sensirion.com/en/flow-sensors/differential-pressure-sensors/sdp800-proven-and-improved/. Accessed 20 Jan 2021
11. Murugan, S., Umayal, S.P., Srinivasan, K., Aruna, M.: Nonlinearity error compensation of Venturi flow meter using evolutionary optimization algorithms. IJIRST **3**(7) (2016)
12. MADA Spirometry Filters - 83 Series. https://www-nz.getzhealthcare.com/products/productspeciality/diagnostic-and-treatment-equipment/spirometry/mada-spirometry-filters-83-series#next. Accessed 20 Jan 2021
13. High-Sensitivity Pulse Oximeter and Heart-Rate Sensor for Wearable Health MAX30102. https://www.maximintegrated.com/en/products/interface/sensor-interface/MAX30102.html
14. Bluetooth low energy module MDBT50Q-1MV2. https://www.raytac.com/product/ins.php?index_id=24. Accessed 20 Jan 2021
15. Flow Analyser PF-300. https://www.imtanalytics.com/en/Analyser/FlowAnalyser. Accessed 20 Jan 2021
16. Fluke ProSim SPOT Light SpO2 Pulse Oximeter Analyzer. https://www.flukebiomedical.com/products/biomedical-test-equipment/patient-monitor-simulators/prosim-spot-lightspo2-pulse-oximeter-analyzer. Accessed 20 Jan 2021

Development of a Portable Device for Urodynamic Data Acquisition Suitable for Home Use

Gregor Nikolić[1](✉), Andraž Stožer[2], and Iztok Kramberger[1]

[1] Faculty of Electrical Engineering and Computer Science, Institute for Electronics and Telecommunications, Laboratory for Electronic and Information Systems, University of Maribor, Koroška cesta 46, 2000 Maribor, Slovenia
gregor.nikolic@um.si

[2] Faculty of Medicine, Institute of Physiology, University of Maribor, Taborska ulica 8, 2000 Maribor, Slovenia

Abstract. We use well-known urodynamic examination procedures and measurement devices to identify Lower urinary tract symptoms (LUTS), usually in the clinical environment. Some home-use devices do exist but have their deficits in ease of use and improve the quality of life (QoL). This article presents a newly developed urodynamic data acquisition device named UroDy suitable for home use. The UroDy device records the voided volume and the urine flow rate, elapsed time to the maximum flow, and the total amount of voiding time. The new proposed battery-powered device architecture and its usage concept ensure an easy-to-use approach for elderly users. It is easily washable, with sufficient battery capacity to perform 91 measurements in an estimated time from 3 to 7 days without additional charging. We performed preliminary measurements of voided volume, flow rate, and urine temperature. Further clinical studies are now essential to validate and evaluate all the proposed system architecture and usage concepts.

Keywords: Smart ageing · Healthcare · Smart healthcare · Non-invasive uroflowmetry · Urodynamics · Low Urinary Tract Symptoms (LUTS) · Biometric data acquisition

1 Introduction

The world population is aging, and projections suggest that this trend will continue until 2050, with a slowdown afterwards. Different medical services and increased waiting times are directly related to this [1, 2]. Such challenges are also evident in the Republic of Slovenia, where the population is aging faster than in most European countries [3]. For these reasons, we must strive to develop devices that address longterm challenges, such as population aging, access to health care, shortening waiting times, and, ultimately, improving human QoL. Research shows that such solutions still do not achieve the goals of improving the QoL of the elderly directly [4]. This article focuses on developing a mobile system for urodynamic data acquisition suitable for home use, enabling measurements and preventive monitoring of LUTS.

A. Badnjevic and L. Gurbeta Pokvić (Eds.): CMBEBIH 2021, IFMBE Proceedings 84, pp. 88–96, 2021.
https://doi.org/10.1007/978-3-030-73909-6_10

There are many different solutions and services designed to improve QoL. Research shows that these do not significantly improve our QoL directly as users, but mostly indirectly through others. Such devices' development does not always pay attention to whether the solutions improve the user's life directly, if only in ease of using such devices. They are mostly solutions and services aimed at the people around the user, such as caregivers, doctors, and others, and the user is typically indirectly involved [4]. We must design and develop solutions for the elderly with the thought of whether they will actually help the user and whether they will want to use it in the first place. Research shows that the elderly do not always accept technology, especially if it monitors them or reports directly to others. But they favor such technology insofar as the data are transferred directly to those who care for them or their health, such as health care providers [5]. With this in mind, we designed the UroDy device presented in this article, which would allow older women and men to monitor their urine voiding data and would be easy enough to use.

The most common form of LUTS in women is urinary incontinence (UI) and overactive bladder (OAB), in men most often in the form of benign prostatic hyperplasia (BPH) [4, 6]. Patients receive the instruction to come with a full bladder and a completed voiding diary to perform urodynamic examinations in a hospital environment. Ideally, each patient would be in line as soon as they arrived at the facility with a full bladder for urodynamic examination. Such a patient had retained urine during the time before arrival to the facility, during admission, and then in the waiting room. The patient must then typically sit on a toilet bowl that mimics a real toilet bowl. Such examination and recordings of their urine flow can be inconvenient and contrary to the International Continence Society (ICS) guidelines and study results [8]. ICS also highlights the problem of measurement delay between the actual voiding of urine and the start of measurements. The urine leaving the body first passes to the interface, usually the funnel, and only afterwards to the measuring container [9]. In this case, the funnel represents a delay and, at the same time, a transformation of the waveform of urine excretion; it acts as a kind of filter behind which quick artifacts can hide. The critical aspect is the usability of such devices to be simple enough for immediate use for the elderly. Discomfort and stress directly affect the performed measurements. Further, in urinary incontinence there is an unmet need to capture events when a person suddenly has an urge to void in the home environment, without discomfort and stress [10]. We have considered the above facts when designing the UroDy device.

The goal of the designed UroDy device is to capture standard urodynamic parameters, use the device as a diagnostic tool and monitoring tool during possible pharmacological treatments, with the desire that the user could handle the device as easily as possible without the need for special prior knowledge. According to the above, the proposed system consists of only one part; the specimen container that users position into the toilet bowl, which is independently electrically powered by a built-in battery and is waterproof for easy rinsing under tap water. Thus, the user does not need to assemble anything before taking measurements. The user would use the designed device within the time period prescribed by the doctor. Figure 1 presents the concept of using the device. When we perform diagnostics, the frequency voiding diary duration varies from 3 to 7 days [11]. With widespread devices for home urodynamic measurements, we will be able to determine the amount of time needed for patient improvement during treatment. Through further clinical studies performed using this device, we will be able to even better estimate the time required and the amount of the data necessary to assess treatment progress more reliably.

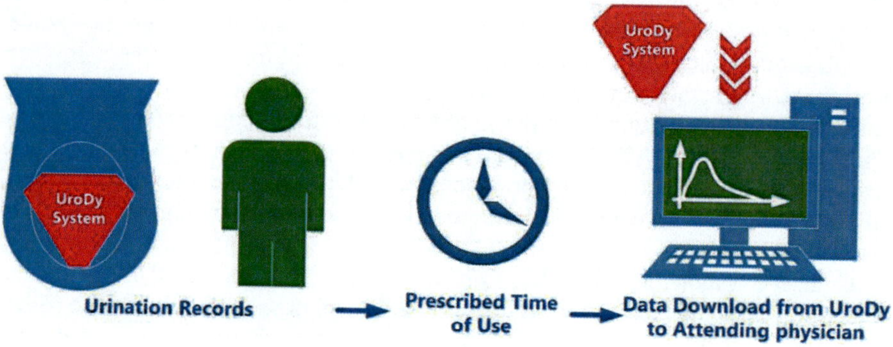

Fig. 1. After the patient's prescribed time of use of the UroDy device, we download data to the patient-doctor for later analysis and examination of recorded data.

The extreme case of use of such a device manifests itself in a patient with symptoms of OAB, where a person suffering from OAB on average performs 13 urinations in a time frame of 24 h [12]. The UroDy device design considers this extreme of use.

2 Methods and Materials

The current state of the art in urodynamic measurements is mainly represented by methods suitable for tests and research in the clinical environment. There are only a few methods or devices suitable for use in the home environment [10]. Research results point out the need for devices as stand-alone units or supplements to other pre-existing devices used for diagnosing LUTS [13]. Currently available devices for measuring urodynamics at home are mainly specimen containers and stopwatches, through which patients measure the time and amount of voided urine. In this way, the doctor obtains at least some quantitative urodynamic data, such as the amount of voided urine and the average flow, given that the patient is highly compliant and has performed the measurements correctly, and also filled in the voiding diary. Inherently, accuracy is always questionable in such cases, as research shows large deviations of the data recorded by patients in the voiding diary and the actual data [14]. We found many smartphone apps on the market that can replace the paper version of the voiding diary, and one electronic device called iU-Flow for home use to record the urine volume. The device is for toilet bowl use that connects wirelessly to a smartphone and contains a digital voiding diary [15]. The device consists of several parts, and one has to use the device with the smartphone. Moreover, a device called Minze records the voided urine volume and flow waveform. It connects wirelessly to a smartphone, and it works by the gravimetric method. Its parts must be assembled before use and this can be impractical for the elderly [13]. Another device called FloRite is a disposable device, intended for use within a 24-h, and allows to measure the volume of voided urine as well as the waveform of urination, and also includes a voiding diary [16]. The device is in the form of a cup, and there are no data available if it is suitable for women and how to use it comfortably.

2.1 Measuring Principle

The method typically used to acquire urodynamic data is the gravimetric method, i.e., weighing urine. For this method, the device must consist of more than one part if one wants to satisfy specific parameters. Therefore, we designed and evaluated a novel capacitive approach presented in Figs. 2 and 3 that meets all the requested parameters, while simultaneously allowing for washability.

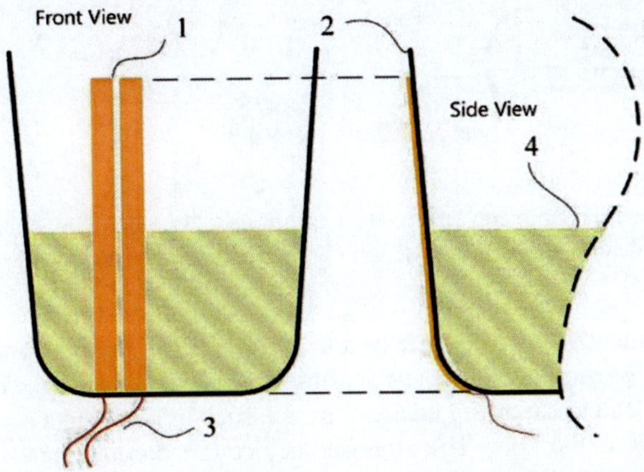

Fig. 2. Copper tape electrodes (1) are placed on the outside of the inner specimen container (2) connected to the UroDy device (3) and are not in direct contact with the collected fluid (4).

The use of the capacitive method for urodynamic data acquisition compared to the existing established non-invasive techniques used in medical institutions was confirmed recently [6]. The study pointed out that additional improvements are needed to achieve the desired functionality; therefore, in this article, we propose a new system architecture. We found that the urine temperature impacts the measured total volume during the evaluation of the proposed measurement principle. Therefore, the urine temperature needs to be measured and taken into account to assure accurate volume measurements. To avoid measurement errors and to provide the capability to measure small volumes, the shape of the specimen container and the electrode positions are crucial. We designed the container's body in a way to reduce splashing of the urine and avoid urination onto the measuring electrodes.

2.2 New Proposed Architecture

We designed the presented UroDy device to be as easy as possible to use so that an older, technology-unfamiliar person can use the device without any assembly or hard-

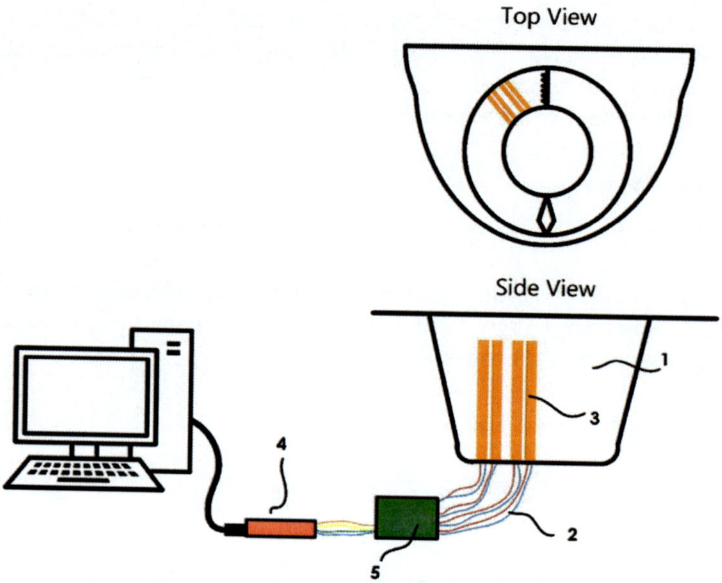

Fig. 3. The specimen container (1), attached electrodes (3), and wire connection (2) to a capacitive-to-digital converter, processing unit (5) with data transfer to a computer (4).

to-understand manuals. The proposed system is suitable for cleaning under running tap water and considers as many factors as possible that may occur in case of incorrect use and which could lead to capturing inaccurate data, such as cleaning a container, moving it around, or a fall on the floor. The proposed device is recording standard urodynamic parameters [9], such as the volume of the voided urine; indirectly from the volume, we calculate the maximum flow value. We calculate the flow waveform from the recorded waveform of the voided volume. We also calculate the voiding time, flow time, maximum and average flow rate, and the elapsed time from beginning to the maximum flow rate from the obtained data. We record time from the urge to start voiding urine; this is when the person sits on the toilet bowl to the moment that urine comes from the person's body; i.e., when urine is detected in the device. This recorded time is a piece of new information for further studies of the correlation of this elapsed time with LUTS.

The proposed system has a battery capacity that allows at least 13 measurements per day in the extreme period for performing measurements for seven days [11]. That means that the user can take at least 91 measurements without recharging the battery, with each measurement lasting 120 s, according to the recommendations [17]. With this approach, we want to achieve no need to recharge the device by the elderly user. Figure 4 presents the hardware concept of the proposed system. To achieve a power consumption as low as possible, we use a high-performance ultra-low-power microcontroller. The device has an integrated memory with a capacity of 128 Mbit to store the planned measurements. Built-in Li-Ion battery is charged and the data transferred from UroDy device to a computer using a micro USB port. Urination itself is detected by the three units of capacitive sensors that are connected to the microcontroller. Each set of electrodes consist of a reference electrode and a measuring electrode made on a polyamide flexible printed circuit board. Measuring electrodes mounted on the outer walls of the inner specimen container of the measuring system offset from each other by 120° are assuring correct measurements due to possibility of tilted device position. Figure 5 shows the physical representation of the partly composed inner specimen container of the whole system for better understanding. For easier handling by the elderly, the system does not have buttons or screens, but instead an RGB light indicator for user feedback. With the help of different colours, we can inform the user about the state of the UroDy system, the beginning of the measurement, the end of the measurement, an empty battery, etc. To detect various states or movements of the measuring system we use a 3-axis accelerometer.

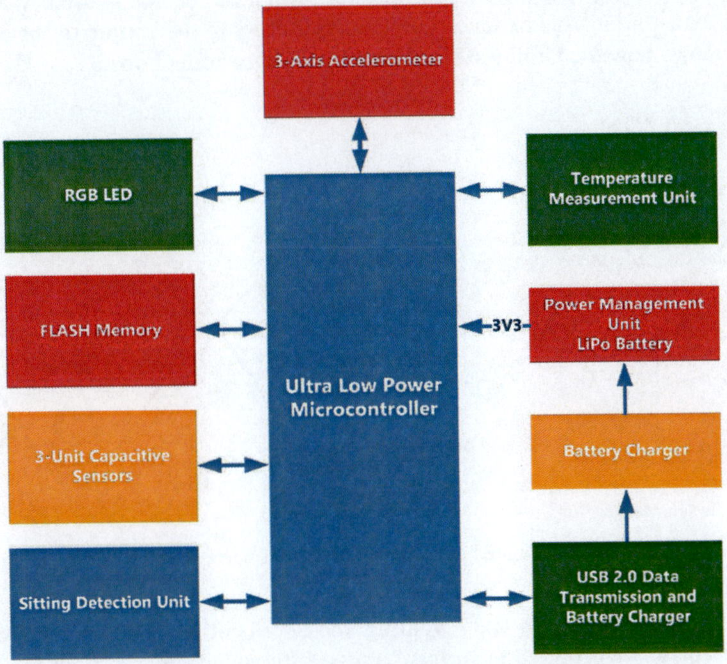

Fig. 4. Block representation of the designed embedded hardware

With an accelerometer, we measure acceleration in all three axes with a three-axis 12bit digital accelerometer. The accelerometer measures accelerations in the measuring range ±2 G. The accelerometer contains an advanced position detection algorithm that can detect all six specimen containers' orientation. With the help of these data, we can distinguish between different states and determine the beginning of measurement and the transition to the state of energy-saving or the shutdown. With these data, we want to achieve that the user only has to grab the UroDy device and position it into the toilet bowl before urinating to minimize the influence on the natural course of events during urination and to maximize ease of use. In the center of the bottom of the UroDy device is positioned a Platinum Thin Film (PTF) temperature sensor for temperature calibration and urine temperature recordings. Its size is only 2.3 mm × 1.1 mm, and it has a fast time response and low thermal inertia, allowing fast temperature measurements. Depending on the sensor class selection, we can achieve a high accuracy of the measured temperature, with an error of ± 0.16 °C to ± 0.17 °C over a measuring range from 35 °C to 41 °C.

3 Results

Based on the new proposed architecture, UroDy device was realized within the hardware and specimen container design. The designed inner specimen container is in the shape of a triangle with flattened corners, as presented in Fig. 5. One of three corners positioned in front of the sitting user minimizes splashing when the device is used. When urine leaves the body, it hits this flattened corner, and it flows to the bottom of the container. Figure 1 shows how the UroDy system is oriented in the toilet bowl.

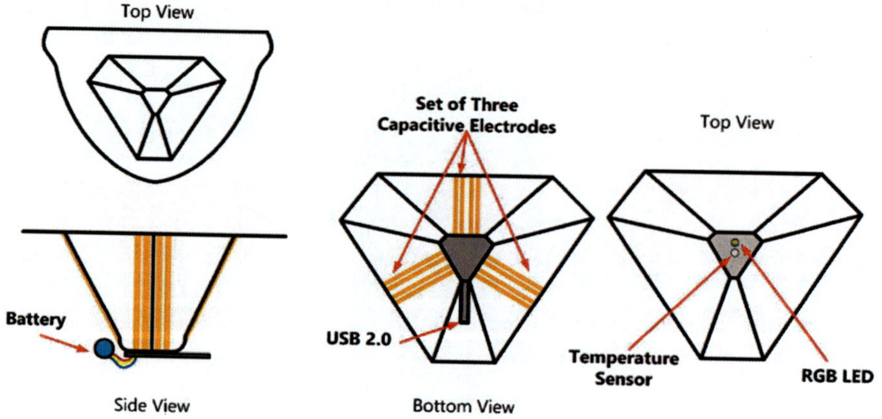

Fig. 5. Partially assembled designed inner part of the specimen container of the UroDy device.

Figure 6 presents preliminary voiding measurement results obtained with the designed UroDy. We acquired the urination waveform and calculated flow, total voided volume, maximum flow rate, time to maximum flow, voiding and flow time, and urine temperature.

In the beginning, in Fig. 6, we can see the small undershoot signal in volume and flow that results when only one electrode is in contact with the urine, and the other two are still intact. This volume is the so-called dead volume needed for correct measurements, and we minimized it with the proposed design of the specimen container. We can see the changes in the flow rate when a person urinates into the UroDy device. The different flow rates and changes in the waveform are the results of a person's urination.

4 Conclusion

This article presents the newly proposed UroDy device architecture and specimen container design, including preliminary measurements that show great potential in the final

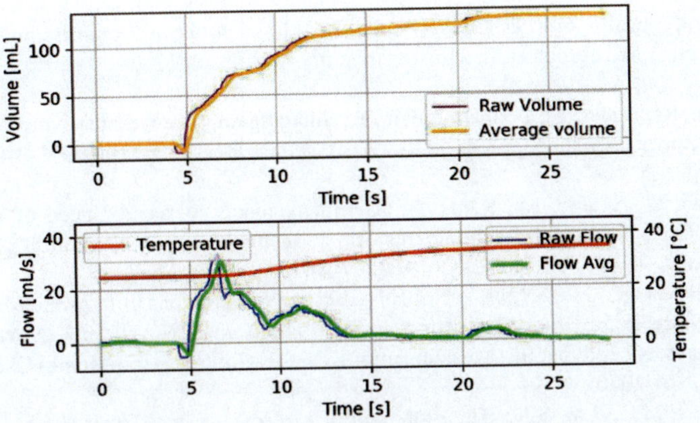

Vol:	128	mL
VT:	20,7	s
FT:	18,2	s
Qmax:	33	mL/s
TQmax:	1,89	s
Qavg:	7,033	mL/s
TrAvg:	10	samp

Fig. 6. Volume and flow voiding measurement with temperature measurement

product. The designed system directly and immediately captures the voiding urine without intermediate containers or funnels that would affect the voided urine waveform. The presented preliminary measurements with the proposed device can be performed in a relaxed home environment. We still have to determine the individual states, such as the device moving or falling on a floor, positioning it into the toilet bowl, etc. To determine all the individual states of the UroDy device and to perform measurements only when required, we need to analyze the limit values of accelerations for each case scenario when the elderly user uses the device. We will carry out a clinical usability and suitability study to determine if it is simple enough for elderly users. We already identified some potential improvements in the design, such as wireless charging and wireless data transfer capability. The wireless communication capability using a smartphone and the voiding diary application can be used to send data directly to the cloud and provide immediate availability for further analysis. Still, it is essential to properly balance power consumption, usability, and ease of use for elderly users. We have to keep in mind that including modern technologies into the device can lead to resistance in the elderly population.

We are continually developing various support devices and algorithms that help speed up and accurately diagnose different patient conditions. A useful support device or algorithm must have many pre-measured data. From this perspective and in the case of massive usage of such devices, the designed system shows great potential to create urodynamic databases, including users with LUTS and healthy users.

Conflicts of Interest. The authors declare no conflict of interest.

References

1. Michel, J.-P., Beattie, B.L., Martin, F.C., Walston, J.D.: Oxford Textbook of Geriatric Medicine. Oxford University Press, Oxford (2018)
2. ESPN Thematic Report on Inequalities in access to healthcare, p. 20.

3. Draguet, V.: Country health profiles 2017. Public Health - European Commission, 20 November 2019. https://ec.europa.eu/health/state/country_profiles/2017_en. Accessed 28 Nov 2020
4. Baraković Husić, J., Baraković, S., Cero Dinarević, E.: Smart ageing: are we succeeding? In: CMBEBIH 2019. Springer, Cham, pp. 387–393 (2020). https://doi.org/10.1007/978-3-030 17971-7_58
5. Demiris, G., Hensel, B., Skubic, M., Rantz, M.: Senior residents' perceived need of and preferences for 'smart home' sensor technologies. Int. J. Technol. Assess. Health Care **24**, 120–124 (2008). https://doi.org/10.1017/S0266462307080154
6. Acta Medico-Biotechnica - 13 2020/01 - A Novel Capacitive Method for Urodynamic Measurements That is Suitable for Home Use. https://actamedbio.mf.um.si/en/article/A+Novel+Capacitive+Method+for+Urodynamic+Measurements+That+is+Suitable+for+Home+Use_1277. Accessed 28 Nov 2020
7. Lee, S.W.H., Chan, E.M.C., Lai, Y.K.: The global burden of lower urinary tract symptoms suggestive of benign prostatic hyperplasia: a systematic review and metaanalysis. Sci. Rep. **7**(1), 7984 (2017). https://doi.org/10.1038/s41598-017-06628-8
8. Gammie, A., Drake, M.J.: The fundamentals of uroflowmetry practice, based on international continence society good urodynamic practices recommendations. Neurourol. Urodyn. **37**(S6), S44–S49 (2018). https://doi.org/10.1002/nau.23777
9. Rosier, P.F.W.M., et al.: International continence society good urodynamic practices and terms 2016: urodynamics, uroflowmetry, cystometry, and pressure-flow study. Neurourol. Urodyn. **36**(5), 1243–1260 (2017). https://doi.org/10.1002/nau.23124
10. Alothmany, N., Mosli, H., Shokoueinejad, M., Alkashgari, R., Chiang, M., Webster, J.G.: Critical review of uroflowmetry methods. J. Med. Biol. Eng. **38**(5), 685–696 (2018). https://doi.org/10.1007/s40846-018-0375-0
11. Yap, T.L., Cromwell, D.C., Emberton, M.: A systematic review of the reliability of frequency-volume charts in urological research and its implications for the optimum chart duration. BJU Int. **99**(1), 9–16 (2007). https://doi.org/10.1111/j.1464410X.2006.06499.x
12. Nixon, A., et al.: A validated patient reported measure of urinary urgency severity in overactive bladder for use in clinical trials. J. Urol. (2005). https://www.auajournals.org/doi/abs/10.1097/01.ju.0000165461.38088.7b. Accessed 29 Nov 2020
13. Abelson, B., Majerus, S., Sun, D., Gill, B.C., Versi, E., Damaser, M.S.: Ambulatory urodynamic monitoring: state of the art and future directions. Nat. Rev. Urol. **16**(5), 291–301 (2019). https://doi.org/10.1038/s41585-019-0175-5
14. Stone, A.A., Shiffman, S., Schwartz, J.E., Broderick, J.E., Hufford, M.R.: Patient noncompliance with paper diaries. BMJ **324**(7347), 1193–1194 (2002). https://doi.org/10.1136/bmj.324.7347.1193
15. Assist: "iUFlow" iUFlow. https://iuflow.com/shop/order-iuflow/. Accessed 29 Nov 2020
16. FloRite^{TM}. https://www.flometrica.co.il/products/florite. Accessed 29 Nov 2020
17. Gammie, A., et al.: International continence society guidelines on urodynamic equipment performance. Neurourol. Urodyn. **33**(4), 370–379 (2014). https://doi.org/10.1002/nau.22546

Comparison Between the Kinematic Behaviour of Different Prototypes of Prosthetic Leg with Actuated Knee and Ankle Joints

Zlata Jelačić(✉)

Faculty of Mechanical Engineering, University of Sarajevo, Sarajevo, Bosnia and Herzegovina
jelacic@mef.unsa.ba

Abstract. Background and Aim: The biggest challenge with today's lower leg prosthetic devices is to climb the stairs in a biologically inspired way. When climbing the stairs, due to the large moments in the knee joint, great forces in the knee occur. Because of the magnitude of these forces, which can be up to six times the weight of a person, it is necessary to use a hydraulic drive with an external power source in the knee and ankle joint. However, most commercially available above-knee prosthetic legs only have actuation in the form of damping in the knee joint. Hence, an amputee is forced to use a healthy leg to advance to each step first while amputated leg follows.

Technique: In this paper the kinematic behaviour of three different prototypes is analysed and compared with the movement of a healthy leg. The analysed prototypes each have linear hydraulic actuators in both the knee and the ankle joint. The kinematic analysis focuses on the first step of the stair ascent while progressing with the prosthetic leg first. Prototype P1 has one-way aggregate for combined actuation of both cylinders. Prototype P2 has two-way hydraulic aggregate for combined actuation of the knee and the ankle joint. And finally, the prototype P3 has separate actuation of the knee and ankle joint with two-way hydraulic aggregate.

Discussion: Obtained diagrams of kinematics motion show improvement for prototype P3 in comparison to older versions of prosthesis. It can be seen that the prosthetic leg with separately actuated knee and ankle joints shows better results in comparison to the movement of subjects without amputation while climbing the stairs.

Clinical relevance: Accurate investigation into the prosthetic biomechanics is crucial in order to develop a device that will make life easier for the final user, the amputees. The overview presented in this paper will guide future developers into designing an even better prosthetic leg.

Keywords: Active prosthetics · Stair climbing · Prototype development · Hydraulic actuators

1 Introduction

The aim of the hydraulic above-knee prosthetic leg project is to develop an above-knee prosthesis that would enable transfemoral above-knee amputees to perform various

A. Badnjevic and L. Gurbeta Pokvić (Eds.): CMBEBIH 2021, IFMBE Proceedings 84, pp. 97–106, 2021.
https://doi.org/10.1007/978-3-030-73909-6_11

activities that are integral part of their every day's life. The main goal, and the hardest one to achieve, is to enable amputees to climb the stairs using the hydraulic above-knee prosthesis and perform it with ease and as naturally as nonamputees.

Advanced microprocessor-controlled prostheses that are available on the market enable amputees to walk on levelled and inclined ground, to descend the stairs and even to ride a bicycle and do various other activities. In order for amputees to climb the stairs in a way that would be, by its kinematics and dynamics, approximate to natural climbing of non-amputees, external power source is needed to substitute large number of muscles needed to perform such an activity [1]. However, climbing the stairs can now be achieved, but it is usually performed in an unnatural "step by step", where (unilateral) amputees progress to the next step with their sound leg, followed by pulling their prosthesis to the same step. This means that progressing to a next stair is always made with same leg.

In order to develop an actively driven prosthesis, the standard Endolite SFEUK above-knee prosthesis was modified in the first phase of the study. This prosthesis was chosen because of the possibility of modification and relatively low purchase price. The SFEUK Endolite, in its original version, has damping in the knee joint with a rigidly tied foot to the prosthesis body.

During the first experimental tests, it was observed that the difficulty of climbing the stairs was due to the great forces and moments to which the joints of the leg, primarily the knee joint, were exposed, as they progressed to the next stair. This realization was the starting point for research on the above-knee prosthesis, which would make it possible for amputee to climb the stairs in a biologically inspired way.

2 Development of Prosthetic Prototype with Active Knee Joint

The original task was to develop a proper knee joint and to simulate the work of muscles that allow these forces and moments to be mastered. To achieve this, it was necessary to select an adequate source of power. At the time when the study on the development of the ankle prosthesis was initiated (ca. 2000), it was not possible to find an electrical actuator of the proper power-to-size ratio. On the other hand, the pneumatics could not provide a sufficient driving force, so the available hydraulic actuators were chosen to drive the joints.

In the first study [2], the calculation of the forces and moments that occur in the knee when climbing stairs was made. The results were obtained using a simplified human wire model for five standard groups of male population, classified by their height and weight, and for six different standard slope steps (Table 1). According to these data, the driving force of a linear hydraulic actuator for knee movement was calculated, which was designed and incorporated into the prosthesis. Subsequently, the prosthesis was experimentally tested, first under laboratory conditions, where the functionality and safety of the actuator were tested, and then with the amputee, where climbing the stairs was simulated by lifting a subject onto the platform. Testing with the amputees showed that the hydraulic actuator was appropriately selected and could produce enough force to lift a person up the stairs.

Without going into the constructive elements and details of the staircase construction as well as the materials from which they are made, the calculation of the knee and ankle joint coordinates will be made based on the ratio of the height of the staircase and the length of the tread (h/b) which define certain patterns determined by the analysis of the human step and other tests.

The ratio h/b is the tangent of the angle of the stair inclination, which normally has a slope of 30°–36°. Stairs with a slope of arms below 30° are mild, and stairs with the inclination above 36° are steep. The exact division of stairs with relation to the slope:

- Stairs of mild inclination with the ratio h/b = 15/32 (cm), for $\alpha = 25°$
- Stairs of a mild inclination with the ratio h/b = 16/30 (cm), for $\alpha = 28°$
- Steps of normal inclination with the ratio h/b = 17/29 (cm), for $\alpha = 30°$
- Stairs of normal inclination where the ratio h/b = 18/27 (cm), for $\alpha = 33°40'$
- Stairs of steep inclination with the ratio h/b = 19/26 (cm), for $\alpha = 36°$ - Stairs of steep inclination with the ratio h/b = 20/25 (cm), for $\alpha = 36°40'$

It was used to calculate the coordinates of the joints of the knee and the feet.

Table 1. Forces and moments in the knee joint for different groups of male population and different slopes.

Stairs	*GROUPS OF MALE POPULATION*				
	I	II	III	IV	V
	58,4 kg	*68,4 kg*	*75,7 kg*	*81,0 kg*	*92,6 kg*
	162,7 cm	*169,1 cm*	*173,5 cm*	*176,9 cm*	*184,2 cm*
h/b	*TOTAL TORQUES (MA)*				
A 15/32	183,569	233,254	267,915	288,655	329,980
B 16/30	188,696	239,278	271,406	296,411	335,144
C 17/29	191,169	244,998	278,128	300,148	340,201
D 18/27	193,581	247,741	281,358	307,334	345,151
E 19/26	198,221	250,407	284,498	310,781	349,991
F 20/25	200,446	252,993	287,549	314,129	354,721

In the following study [3], instead of using a platform simulating a staircase, real stairs were used, and an experiment was performed on stair climbing of a subject with above-knee amputation using a prosthesis with an integrated hydraulic actuator in the knee joint (Fig. 1).

It was decided to concentrate on the fourth group of male population with an average mass of 81 kg and a height of 176.9 cm, as it is the most common group in the population of people with transfemoral amputation (Table 1).

Stair climbing was investigated and biomechanical analysis of stair climbing of subjects with and without amputation was made. The test results of subject without amputation were compared with climbing data of a subject with amputation using a prototype of the above-knee prosthesis (Fig. 1). This research provided a significant base of kinematic features necessary for analysing stair climbing.

An experiment with a subject with amputation using the above-knee prosthetic prototype showed that climbing using a prosthesis with a hydraulic knee actuator could be accomplished, however, some irregularities in the climbing process were observed (Fig. 1).

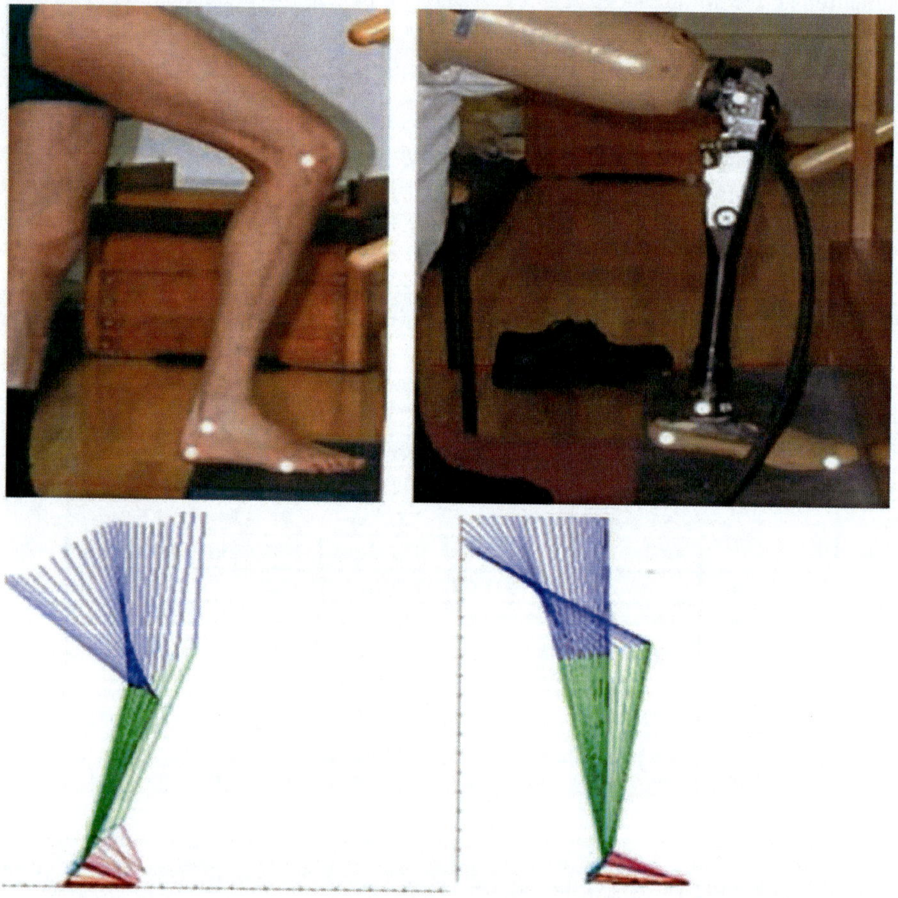

Fig. 1. Examination of stair climbing of a subject without amputation (left) and a subject with amputation who wore above-knee prosthesis with the actuated knee joint (right) [3]

3 Development of Prosthetic Prototype with Active Knee and Ankle Joint

From the results obtained from these experiments, it was concluded that in addition to the knee joint, which is a key factor in coping with the forces and moments that occur when climbing the stairs, another factor significantly affects climbing, namely the ankle joint. The ankle joint is a key factor in the damping of the ground reaction force, as well as the accumulating and releasing the energy at certain points of walking, which is very important for walking, but crucial for achieving natural movement while climbing the stairs. In one of the experiments, it was discovered that the lack of actuated ankle joint on the prosthesis can be potentially dangerous, as a component of the ground reaction force appears, which tends to push back the prosthetic user, which may end up in falling down the stairs.

It was concluded that in order to achieve a natural biological gait when climbing stairs, the knee and ankle must be integrated into the prosthesis and be actively driven by an external power source, which would replace muscle strength.

Subsequent research focused on the integration of a hydraulic ankle actuator in the prototype of the above-knee prosthesis. The position of the ankle joint was determined according to the biological ankle, and a location that was feasible was chosen for the placement of the hydraulic actuator. Using a human wire model, the forces and moments in the ankle were calculated and the appropriate ankle actuator selected and integrated into the prosthesis design [4] (Fig. 2).

The modified Endolite prosthesis, with linear actuators in the knee and ankle of the leg, was experimentally tested in laboratory conditions, and diagrams were obtained, which showed that the prosthesis performed much better with an additional hydraulic linear actuator in the ankle (Fig. 2).

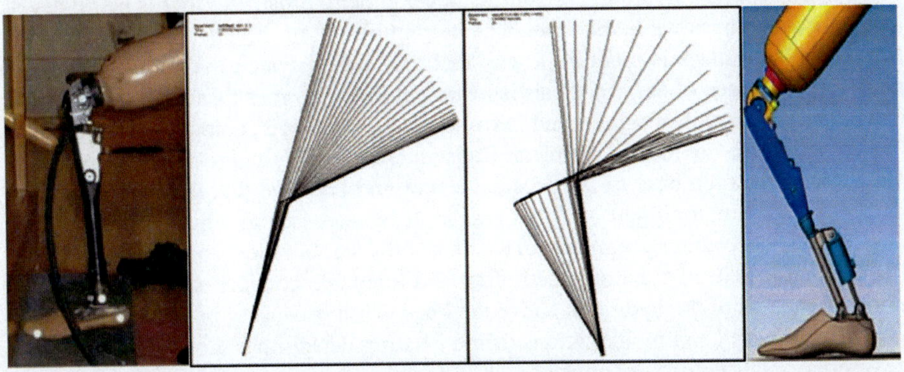

Fig. 2. Comparison of the operation of the prosthesis with the hydraulic actuators only in the knee (left) and with the actuators fitted both in the knee and in the ankle (right) [4]

The simplified wire model of the human body is only an approximation of the human body, and any calculation based on such a model can be significantly different from the calculation with the true human body model. Therefore, a subsequent study was conducted in such a way that the kinematic data of a previous study on subjects without amputation were transferred to a CAD model of a prosthesis with two hydraulically driven joints (knee and ankle), to mimic human walking when climbing stairs.

Using this information, a detailed biomechanical analysis of the prosthetic knee and ankle joints was performed, and the position of the hydraulic ankle actuator optimized.

The time period during which the foot of a healthy person is in contact with the substrate (platform) was also analysed, the dependence of the vertical force of the substrate reaction on the suspension period was analysed and the trajectory of individual characteristic parts of the leg was analysed.

The structural characteristics and values of the required stroke of the linear ankle actuator were obtained as well as the installation dimensions, which can serve as a guide for the prototype of a prosthesis with two linear actuators.

In study [5], a transfemoral prosthesis was designed and tested in which the knee and ankle joints were driven by dual stroke linear actuators that received power from a two-way, battery-powered hydraulic unit. To allow more natural movement of the prosthetic foot and prosthesis when climbing, a prosthetic foot was developed with the possibility of passive bending in the metatarsophalangeal joint. The designed prototype of the hydraulically driven ankle prosthesis and prosthetic foot was tested in working conditions on a subject with a hip amputation while climbing stairs, and the test results were compared with the results of analogue climbing of healthy subjects. As part of the research, a sensor system for the recognition of leg movements was developed, which was adapted and used as measuring equipment in the experimental part of the research.

4 Movement Analysis of the Ankle and Artificial Foot

In the process of new hydraulic prosthesis development, prosthetic foot is being developed in order to improve human gait. The current design of the prosthetic foot and the above-knee prosthesis is made in a way that axis line between the base of the prosthetic foot and knee of the prosthesis is a straight line that goes through central part of above-knee prosthesis, when viewed in frontal plane. However, components of biological human leg are not ideally collinear. Consequently, using the prosthesis with ideally straight axis between base of the prosthetic foot and knee would result in certain discomfort and less natural gait. For that reason, it is necessary to adjust components of prosthetic leg according to segments of user's healthy leg. In order to achieve this feature, it is decided to redesign the prosthetic foot and to enable connection of the prosthetic foot with the rest of the hydraulic above the knee prosthesis using prosthetic adapter.

The goal is to keep modifications of the existing design to a minimum and to find the optimal solution for this connection, keeping already given and defined measures of the foot and prosthesis design. Also, tendency is to avoid using custom made adapter and use standard adapter available on the market.

An important part of the foot construction is the part through which the foot connects to the rest of the above-knee prosthesis and forms the ankle joint. Since the biological foot is not a symmetrical body, the main problem with the structural realisation of this part is its position in the frontal plane.

The research within this section showed that the part that connects the foot to the rest of the prosthesis is located at approximately 2/7 to 2/6 (between 30% and 35%) of the width of the foot, closer to the inside of the foot.

The previous research was based on the prototype P1 of the prosthetic leg which has two one-piece linear hydraulic actuators in the knee and ankle joints, with a single one-way aggregate to power both joints. In that research, a prosthetic foot with the ability to bend in the metatarsophalangeal joint was developed. That prototype represented the starting point for this research (Fig. 3).

As a part of the research which includes this paper, two new physical prototypes of the above-knee prosthesis were developed with active knee and ankle joints. In order to develop a prosthetic prototype which is able to fine-tune the movement, we used the hydraulic system with one two-way hydraulic engine and two linear hydraulic actuators with dual movement in the knee and ankle. That was the first prototype P2 in this study (Fig. 3). The experimental data and the comparison with the biomechanics of the lower extremity of healthy subjects (Fig. 4), showed that the same drive, that is, the combined external power system for both joints, did not meet the required performance and new prototype was developed.

The P3 prototype has separate drive and control for the knee and ankle joints and consists of two twoway hydraulic engines and two dual-way linear hydraulic actuators, one for the knee joint and the ankle joint (Fig. 3). The following figure shows the results in the case of the flexible foot design with combined and separate external power source for the knee and the ankle joint.

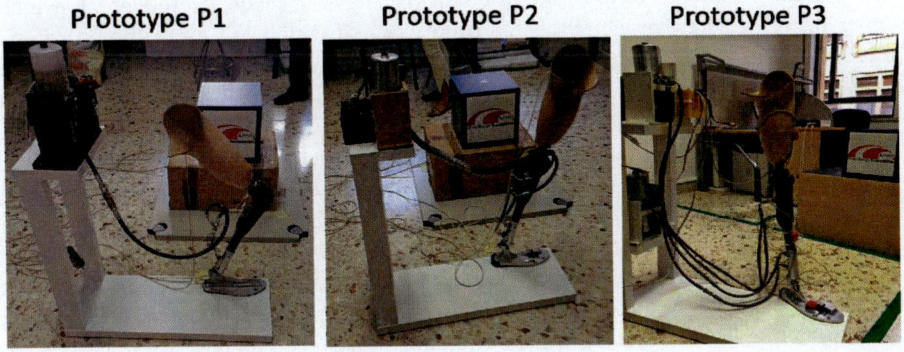

Fig. 3. Different prototypes of the prosthetic leg: P1, P2 and P3.

5 Kinematic Analysis of the Position and Trajectory of Different Prototypes

The amputee's climbing period can be divided into four stages. The first phase is establishing the first contact of the prosthetic foot with the step in which the subject touches the stairs with the toes. The phase which will be singled out in this research is the progression to the next stair in which the analysed leg pushes the subject up. This is denoted as the E3 movement.

Climbing positions and trajectories along the steps of healthy subjects and different prototypes of the above-knee prosthesis are the result of the processing of measurement videos. The recording of the trakSTAR Ascension Technology measurement equipment was used and the prosthesis positions in different simulations of stair climbing were derived using conventional drawing and design toolset. The paths are made by measuring the coordinates of characteristic leg points and importing them in the analysis software. This method made it possible to describe the kinematics of the movements of different prototypes and compare them to the movement characteristics of the subjects without amputation.

For the ease of analysis and comparison, the movement of the healthy leg and the prosthetic prototype has been analysed for the E3 movement. Five points were followed: hip, knee, ankle, heel and metatarsophalangeal joint.

In the results of previous research, the paths were analysed while an amputee was using the prosthetic device. In that case, the subject could correct the prosthetic leg position by himself and hence influence the measurement results. This was specifically the case for prototype P1, which was analysed here as well, with the combined one-way hydraulic power system for both joints.

Therefore, it was decided that, for the sake of objective comparison between different prototypes, we would analyse the autonomous movement of the prosthetic leg without any assistance and intervention of the user. By doing this, we not only analyse the kinematics but also the control system of the prosthetic device.

Different prototypes have different control systems in place. Prototype P1, which was result of previous research, has an on-off switch which can move the prosthesis between the fully stretched and bent leg, respectively. The combined one-way hydraulic installation either fills the cylinders with oil or empties them. Prototype P2 has a control system with sensors that use the reference angle values of the healthy leg in real time, in combination with combined bidirectional hydraulic system. Prototype P3, which was also result of this research as P2, uses the same control system but with separate hydraulic system for the knee and ankle joint. Prototypes P2 and P3 follow in the experiments the prescribed movements in real time [5].

The results showed that prototype P1 was unable to bend in the ankle joint even though it is externally powered. The one directional hydraulic system working on the on-off principle doesn't provide the required movement. The biggest discrepancies are registered in the ankle joint and the prosthetic foot, where no bending in the metatarsophalangeal joint was registered (Fig. 4). As the prototype P1 was unable to bend the prosthesis on its own and provide the necessary swing phase, the prosthetic user has to compensate for this by tossing the hip joint backwards.

To avoid this, it is necessary for the knee and the prosthesis to be hydraulically operated bidirectionally, which is applied to P2 and P3 prototypes. The combined hydraulic system for both the knee and the ankle joint in the P2 prototype causes the prosthetic leg to bend in the ankle and the toes. However, the prosthetic leg doesn't swing and could cause he user to trip.

This problem is solved with prototype P3 where the separate power system exists for both joints. This prototype shows the movement which is closest to the natural one [6].

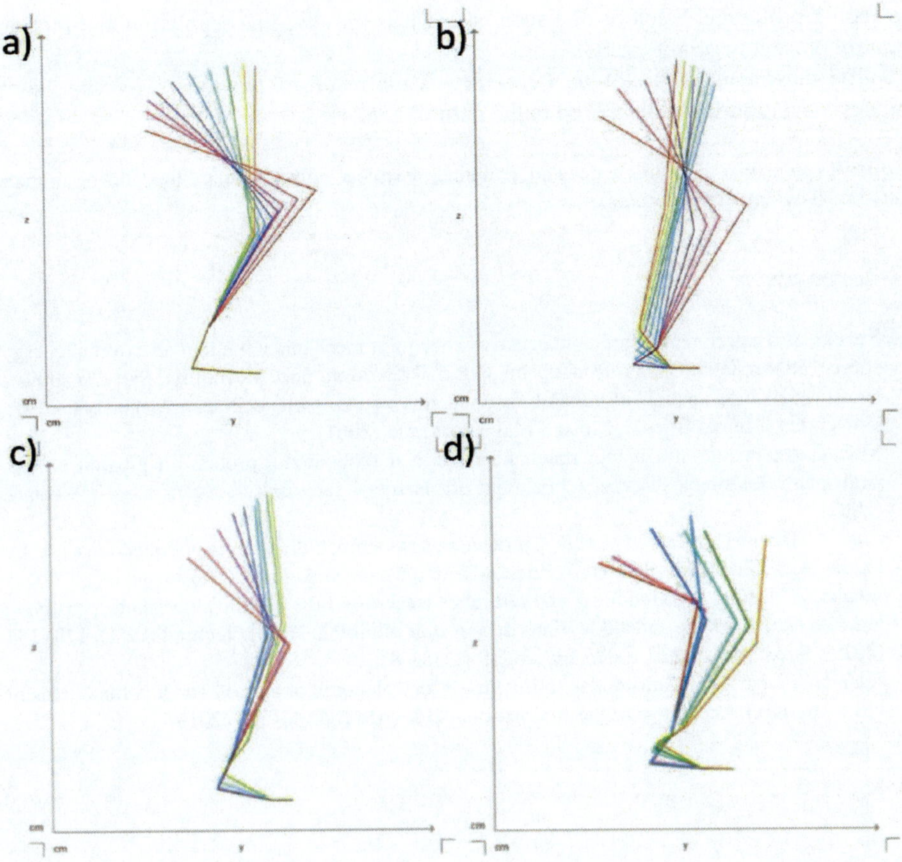

Fig. 4. Autonomous pre-programmed movement of prototype P1 (a) and P2 (b); (c) Movement of non-amputee subject; (d) Real-time movement of prototype P3.

6 Conclusion

Based on the above, it can be seen that the design of the prosthesis P1, which has the same drive for unidirectional linear actuators in the knee and ankle, doesn't provide fine regulation of the ankle joint, present in the case of the newly developed prototypes P2 and P3.

Based on the above, it can be concluded that the most natural separation of the feet from the base, and the trajectory of the angle change in the foot and the ankle joint at the time of separation from the substrate, as compared to healthy subjects, is recorded for prosthesis P3, which has separate drive for knee and ankle joints with two-way aggregates and bidirectional linear actuators.

Movement of the prototype P2, having a single two-way hydraulic drive with bidirectional linear actuators in the knee and ankle, is less natural but quite good compared to the P1 prototype. What can be seen is that it is missing fine regulation and motion control present in prototype P3.

The movement of prototype P1, which is the result of previous research, shows biggest deviation when compared to the natural one.

Ethical Approval. This article does not contain any studies with human participants or animals performed by any of the authors.

References

1. Suzuki, S.: Experimental study on an active knee joint mechanism without an external energy source. Nihon Kikai Gakkai Ronbunshu, C Hen/Trans. Jpn. Soc. Mech. Eng. Part C. (2009)
2. Huđec, M.: Prilog analizi hidrauličkih komponenti za natkoljenične proteze. Mašinski fakultet, Univerzitet Džemal Bijedić, Mostar, Magistarski rad (2001)
3. Vučina, A.: Prilog istraživanju kinematike sila u natkoljeničnoj protezi pri gibanju na stepenicama. Doktorska disertacija. Fakultet strojarstva i računarstva, Sveučilište u Mostaru (2002)
4. Žiga, A.: Dizajn stopala natkoljenične proteze sa ugrađenim hidrauličkim pokretačima u članku i koljenu. Mašinski Fakultet, Univerzitet u Zenici, Magistarski rad (2006)
5. Jelačić, Z., Dedić, R.: Modelling and reference tracking of the robotic above-knee prosthetic leg with actuated knee and ankle joints during stair climbing. Health Technol. **10**(1), 119–134 (2019). https://doi.org/10.1007/s12553-019-00383-8
6. Jelačić, Z, Dedić, R.: Multimodal optimal trajectory planning of a prosthetic leg. In: Annals of DAAAM and Proceedings of the International DAAAM Symposium (2019)

Integrated Intrabody Communication Node Based on OOK Modulation

Antonio Stanešić[1](✉), Željka Lučev Vasić[1], Yueming Gao[2], Min Du[2], and Mario Cifrek[1]

[1] Faculty of Electrical Engineering and Computing, University of Zagreb, Zagreb, Croatia
[2] Faculty of Physics and Information Engineering, Fuzhou University, Fujian, China

Abstract. The goal of this paper is to present the development of an integrated implementation of the transmitter and receiver pair which are designed with intrabody communication (IBC) purposes in mind and uses on-off-keying (OOK) modulation. The system is implemented using Programmable System-on-Chip (PSoC) microcontrollers produced by Cypress, using a minimal number of external components.

Keywords: Intrabody communication · On-off keying · Programmable systemon-chip

1 Introduction and Background

Rapid miniaturization of electronic systems (especially embedded systems) naturally leads to greater inclusion of electronics into various biomedical fields. One of the most prominent places for electronics in the biomedical field is diagnostics, with biomonitoring systems as perfect examples of such systems. Biomonitoring systems are used to monitor various vital and physiological parameters and features present in humans and animals alike.

A typical wireless biomonitoring system communicates via Bluetooth (either its Classic variant or its Low-Energy subvariants), ZigBee, ANT, WiFi, or similar protocols. Such sensor networks are usually made of one receiver (hub) which receives all of the data packages transmitted from various transmitters (sensor nodes) from their place-of-interest [1]. Such protocols and sensor networks can cause a significant amount of electromagnetic radiation, which leads to heating and subsequently irritation of the patient's tissue,) [2]. Other problems of utilizing conventional sensor networks that employ common wireless protocols are security problems and problems of frequency-band congestion. Since such communication nodes rely on emitting waves into space rather than using conductive materials for transmission purposes, eavesdropping is a common problem which has to be mitigated by complex higher-level algorithms. On the other hand, problems of frequency-band congestion are becoming more-and-more serious due to a larger number of devices operating in the same frequency spectrum, leading to failed transmissions and, thus, longer transmission time (impacting battery life and aforementioned radiation as well).

A. Badnjevic and L. Gurbeta Pokvić (Eds.): CMBEBIH 2021, IFMBE Proceedings 84, pp. 107–115, 2021.
https://doi.org/10.1007/978-3-030-73909-6_12

Enumerated problems can be solved by employing a non-standard communication method called intrabody communication. The core idea of such a new paradigm is to use the human body as a signal propagation medium. Since the human body is conductive, lower carrier frequencies can be used, which inherently reduces both energy consumption (prolonged battery life) and electromagnetic radiation (less heating and irritation to the end user) [3, 4]. Using the human body for signal transmission also provides an inherent security mechanism, since it disables non-contact eavesdropping [5]. It occupies a different spectrum (and simultaneously does not radiate the signal significantly into space) than common protocols (mainly 2.4 GHz based technologies), thus it has no coexistence problems, unlike Bluetooth, WiFi, etc.

The communication is typically established between a transmitter which is used to generate, encode and modulate the sampled sensor data, and a receiver which serves as a hub which demodulates, decodes and forwards received signals from multiple sensor nodes [4]. Data transfer characteristics of an IBC system depend on the type of tissue used as a communication channel, the signal path, type of modulation and signal frequency [1]. Since tissue conductivity increases with the carrier frequency, higher data rates can be obtained, thus reducing the required time for transmissions and retransmissions; while on the other hand it inherently increases the consumption. Thus, optimization and re-evaluation are often necessary to find a “sweet spot” of optimal frequency for the desired data rate.

Programmable System-on-Chip (PSoC) is a family of microcontrollers developed and produced by Cypress Semiconductor, which include a CPU core and mixed signal arrays of configurable integrated analog and digital peripherals [7]. The SoC is built around an ARM core and includes analog peripherals such as various amplifiers, comparators, digital-to-analog and analog-to-digital converters, etc.; digital peripherals which are highly configurable and present an opportunity to create digital logic devices of medium size (due to included interconnected datapath and complex programmable logic devices (CPLDs). Digital peripherals can be created from scratch by writing proprietary Verilog-like code. All of the PSoC components are arbitrarily interconnectable and routable to any available GPIO. Such a high level of programmability offers a rapid prototyping environment not limited only to solving digital, but also complex analog problems.

The goal of this paper is to present the development of an integrated implementation of both the transmitter and receiver pair designed for intrabody communication (IBC) purposes. The transmitter and receiver use on-off-keying (OOK) modulation to achieve communication. The system is implemented using Programmable System-onChip (PSoC) microcontrollers produced by Cypress Semiconductor.

2 PSoC-Based IBC System

One of the key decisions to make when evaluating possible approaches for an IBC system is to choose an appropriate type of modulation. The chosen type of modulation also draws implications for the demodulation process – e.g. phase-shift key modulations require coherent demodulation which is much more complex and thus expensive. One of the simplest modulation variants for implementation is amplitude shift keying (ASK).

In ASK, digital data is modulated using discrete amplitude levels for each symbol – for just two symbols (0 and 1) there is one discrete amplitude level assigned to each. The most common form of ASK is on-off keying (OOK), where one symbol is assigned to the signal of zero amplitude (thus, no signal is present, this is the "off" state), while the other symbol is assigned to the signal of predefined amplitude (thus, the signal is present, this is the "on" state).

The selection of the modulation method is, in part, dictated by the selection of the system platform. Coherent demodulation would require significant additions to the basic PSoC system. Thus, OOK is a good candidate for implementation.

The final IBC system implemented in this work functions primarily as COM-2COM communication system (which uses the human body as a communication canal). The system (Fig. 1.) consists of two main parts: the transmitter and the receiver. The transmitter receives the data sent via terminal (via USB), while the receiver sends all demodulated data to the terminal of the other PC (also via USB). To avoid significant capacitive coupling, small-footprint laptops were used as the first and the second PC, while unplugged.

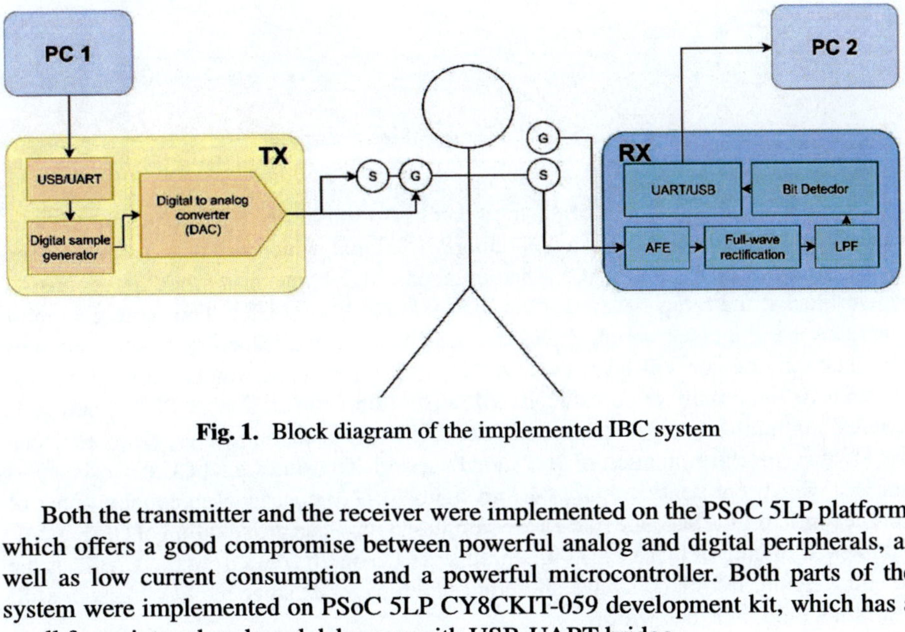

Fig. 1. Block diagram of the implemented IBC system

Both the transmitter and the receiver were implemented on the PSoC 5LP platform, which offers a good compromise between powerful analog and digital peripherals, as well as low current consumption and a powerful microcontroller. Both parts of the system were implemented on PSoC 5LP CY8CKIT-059 development kit, which has a small footprint and on-board debugger with USB-UART bridge.

2.1 Transmitter

Transmitter receives the digital data from the terminal of the PC1 via the onboard USB-to-UART converter and, subsequently, UART component inside the SoC. This data is used to modulate the generated carrier. Placing an inverter on this data line will switch the assignment of "on" and "off" periods: while normally "on" state is assigned to the logical

1, if an inverter is used "on" state will be assigned to the logical 0. The implementation of the transmitter is shown in Fig. 2.

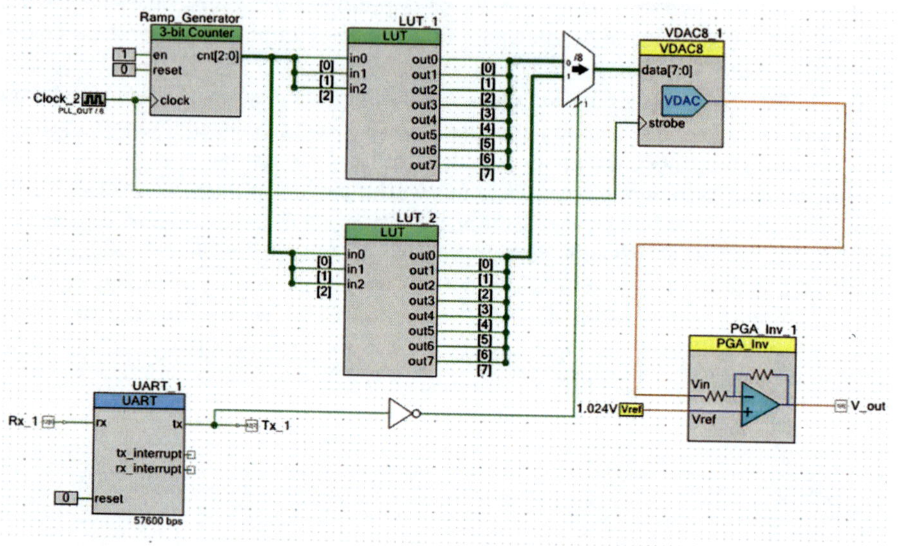

Fig. 2. Implementation of the transmitter

The goal of the digital part of the modulator is to synthesize the samples for the sine wave carrier, as well as for the corresponding "off" signal, which will be transferred to the digital-to-analog converter (DAC). The digital part of the transmitter modulator consists of two blocks: the ramp generator and the lookup tables (LUT). The ramp generator is created using a clock which drives the basic up-counter. On every positive edge of the clock the counter will increment by 1. The counter will automatically reset upon reaching its maximum value (after overflow). Its implementation in PLD macrocells ensures minimum resource occupation. Two LUTs are used, one for the generated sine signal, other for the emulation of "no signal" period. The sine signal LUT consists of 8 samples which represent the sine; the "no signal" LUT consists of an equal number of samples with just the baseline mid-range samples to replicate the situation of "no signal". Samples from both of these LUTs are multiplexed using the data from the UART as the deciding signal. Both LUTs have an output width of 8 bits since the DAC has an 8-bit resolution (and thus, bus width).

The analog output stage consists of a digital-to-analog converter and an output amplifier. The DAC is configured to work in a high-speed mode; DAC bus is used instead of the data bus (register transfer mode) for sample transfers since this provides optimal speed. Range of 1 V is used due to constraints of the subsequent analog amplifier.

The DAC component does not include any kind of an output buffer, thus an external buffer is required to ensure stability of the output voltage range. An inverting amplifier with unity gain is used for this purpose. PSoC amplifiers have a "power elevel" parameter which determines the gain-bandwidth product of the amplifier and effects the power consumption. Here, the "power level" parameter was set to "highpower", in order to prevent attenuation of the high-frequency signals generated by the transmitter. The inverting amplifier has been used to save resources – the voltage reference (here 1.024 V has been used) should be buffered if it is not connected to a high-impedance input of the amplifier.

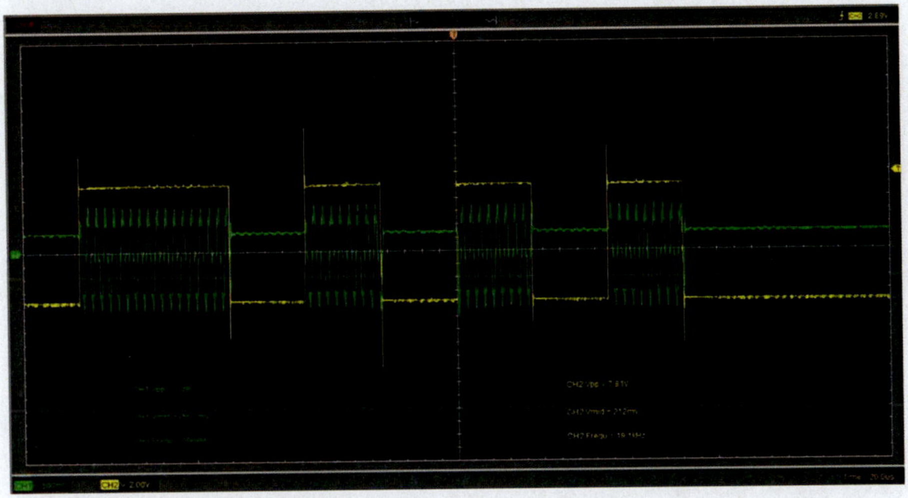

Fig. 3. Generated OOK signal (green) and digital data used for modulation (yellow)

An example of generated OOK signal is presented in Fig. 3.

The generated carrier frequency is determined by two factors:

- The set clock frequency at the clock generator
- The set number of samples for carrier generation

Since samples are changing at the rate of f_C, the generated carrier frequency is:

$$f_{ca}\ \ ie = \frac{f_c}{N_{ape}}$$

The number of samples and the clock frequency can be easily adjusted. In this particular implementation, 8 samples were used, as well as the clock frequency of 4 MHz, in order to achieve a carrier frequency of 500 kHz. Hardware was tested for carrier frequencies up to 3 MHz and it performed as expected.

Figure 4 shows a generated sine carrier of 1.5 MHz and 1 V amplitude.

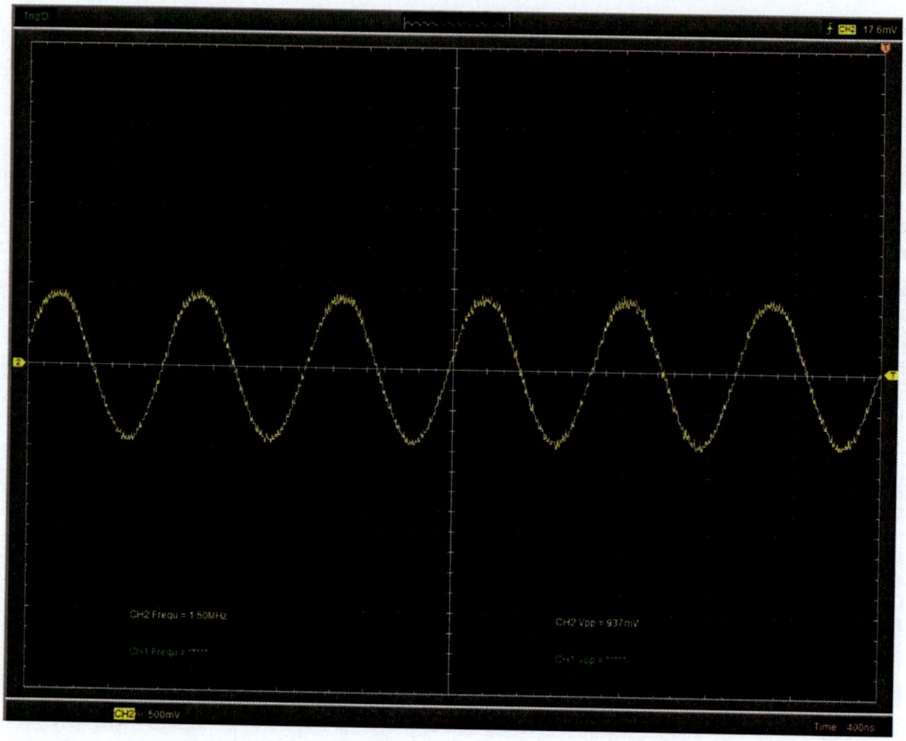

Fig. 4. Generated sine carrier of 1.5 MHz, 1 V amplitude

2.2 Receiver

The receiver electrodes detect a signal sent by the transmitter through the human body. The receiver consists of the analog conditioning part, digital conditioning part and the communication block (Fig. 5). The first, analog stage, starts with a high-pass filter. These are the only external components to the system, and they are necessary primarily to filter out the 50 Hz network mains interference. A high pass filter is created using a CR-network, where the used resistance is 1 kΩ while the used capacitance is 47 nF, resulting in a cutoff frequency of 3.4 kHz. To separate the high-pass filter from the rest of the integrated circuitry, as well as to amplify the received signal, a non-inverting programmable gain amplifier is used.

The next step is the full-wave rectification of the signal. For this purpose, an upmixer is used where the local oscillator is generated by a comparison of the input signal with the predefined threshold or reference. The signal is low-pass filtered after this step (cutoff-frequency is predefined, 76 kHz – higher than the data rate, lower than the carrier frequency; the low-pass filter is formed using an RC network inside of the PSoC), which effectively returns the signal to the baseband (from the frequency domain point of view). The signal can be amplified to simplify later steps, but the signal conditioning in the analog domain is completely done at this point.

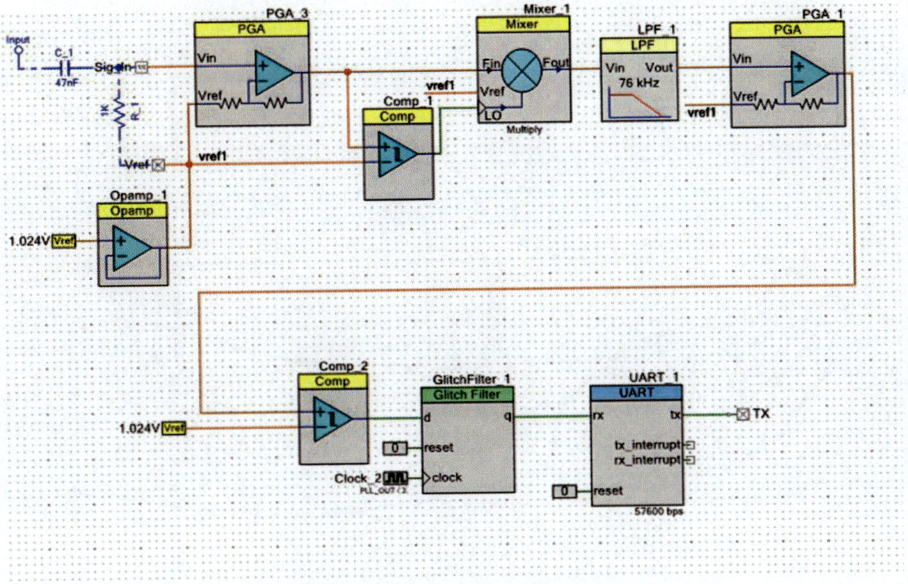

Fig. 5. Implementation of the receiver

In order to preserve resources, a comparator with a fixed threshold is used as an emulation of a 1-bit ADC. Analog signal resembling the digital signal is, thus, completely digitized and susceptible for final digital-domain conditioning. This is performed using a glitch filter, which is used to filter out short glitches (in cases where analog signal spiked above/below the threshold during otherwise stable voltage level. The digital signal is passed to the UART component and the UART component sends it to the terminal of the PC2 via an onboard UART-USB bridge.

3 Measurement Setup and Results

Each stage of both the transmitter and the receiver were tested using a batterypowered oscilloscope in order to confirm the correct waveforms of the respective modulator/demodulator stages. The chosen carrier frequency was 500 kHz [9], while the transfer was tested at rates of 57600 bps, 38400 bps and 19200 bps.

Afterward, both the transmitter and the receiver were connected to the human body using adhesive Ag/AgCl Ambu® WhiteSensor 0215M electrodes. Used electrode configuration was AB [9, 10]: both electrodes (S - signal, G - ground) of the transmitter (TX) were connected to the patient as well as the receiver (RX) signal (S) electrode, while the receiver ground (G) electrode was oriented towards the environment (not connected to the human body, but floating 2 cm above it), as shown in Fig. 6. Care was taken that the human body was not in a direct galvanic contact neither with the transmitter nor with the receiver boards, so that the electrodes were the only point of galvanical contact between the boards and the human body. The transmitter and the receiver were connected to their respective battery-powered laptops, which were a few meters apart.

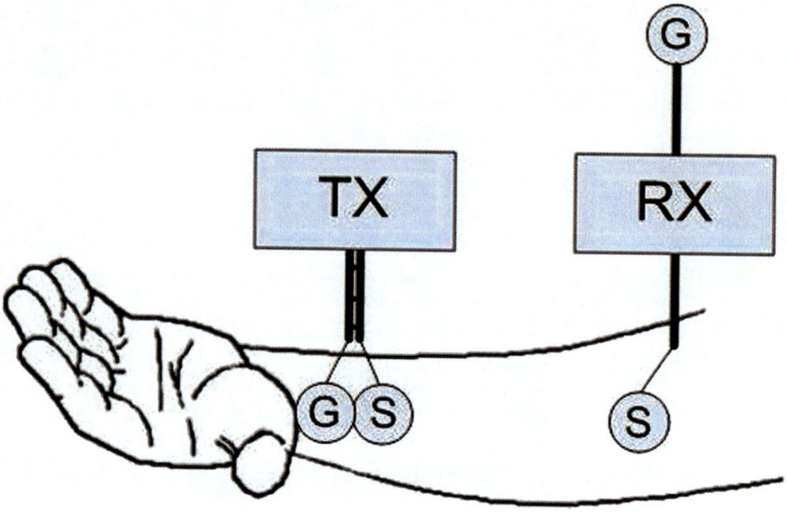

Fig. 6. AB electrode configuration [11]

USB extension cables were used to extend the distance between laptops. Both laptops and the implemented system were placed far from any conducting objects, to minimize the influence of the environment on the measurement results.

The measurement test was successful, and valid data was sent between the two laptops – from the transmitter to the receiver. Error rates were a bit higher for the highest data rate, mostly due to the fact that the data rate and the carrier frequency became comparable.

4 Conclusion

The goal of this paper was the development of an integrated intrabody communication system based on the PSoC microcontroller platform. Aside from the development kit, only two passive external components were used, which would normally be included on the PCB. Much of the resources which the SoC possesses were left unused, especially the valuable analog-to-digital converters (both precision-oriented delta-sigma and speed oriented successive approximation register converters). This demonstrates the viability of a completely encapsulated sensor node inside the PSoC system, which joins the sensor interface and the IBC communication interface. It is also possible to realize the presented implementation using discrete components, thus enabling another choice of SoC. The developed system achieved the desired functionality successfully, with occasional transfer errors which can be corrected by a higher-level protocol.

Conflicts of Interest. The authors declare no conflict of interest.

References

1. Lučev, Ž., Krois, I., Cifrek, M.: Intrabody communication in biotelemetry. In: Lay-Ekuakille, A., Mukhopadhyay, S.C. (eds.) Wearable and Autonomous Biomedical Devices and Systems for Smart Environment – Issues and Characterization, vol. 75 of Lecture Notes in Electrical Engineering, pp. 351–368. Springer, Heidelberg (2010)
2. Grilec, F., Stanešić, A., Vasić, Ž.L., Gao, Y., Du, M., Cifrek, M.: Single-chip intrabody communication node. In: Badnjevic A. (eds) CMBEBIH 2017. IFMBE Proceedings, vol. 62. Springer, Singapore (2017). https://doi.org/10.1007/978-981-10-4166-2_46
3. Zimmerman, T.G.: Personal area networks: near-field intrabody communication. IBM Syst. J. **35**(3–4), 609–617 (1996)
4. Sen, S., Maity, S., Das, D.: To safeguard sensitive data, turn flesh and tissue into a secure wireless channel. In: IEEE Spectrum December 2020, pp. 44–49 (2020)
5. Das. D, Maity. S., Sen. S., Chatterjee, B.: Enabling covert body area network using electro-quasistatic human body communication. In: Nature, Scientific Reports (2019)
6. Grilec, F., Lučev Vasić, Ž., Ni, W., Gao, Y., Du, M., Cifrek, M.: Wireless intrabody communicating sensor node realized using PSoC microcontroller. In: 39th International Convention on Information and Communication Technology, Electronics and Microelectronics (MIPRO), pp. 426–429. IEEE Conference Publications (2016)
7. Cypress Programmable System-on-Chip (PSoC). https://www.cypress.com/products/32-bit arm-cortex-m-psoc
8. CY8CKIT-059 PSoC® 5LP Prototyping Kit Guide, Doc. #: 001–96498 Rev. *G. https://www.cypress.com/file/157971/download
9. Lučev, Ž, Krois, I., Cifrek, M.: A capacitive intrabody communication channel from 100 kHz to 100 MHz. IEEE Trans. Instrum. Meas. **61**(12), 3280–3289 (2012)
10. Naranjo-Hernández, D., Callejón-Leblic, A., Lučev Vasić, Ž., Seyedi, M., Gao, Y.: Past results, present trends, and future challenges in intrabody communication. In: Wireless Communications and Mobile Computing, pp. 1–39. Hindawi Limited (2018). https://doi.org/10.1155/2018/9026847
11. Lučev Vasić, Ž.: Intrabody communication based on capacitive method. Doctoral thesis, University of Zagreb, Faculty of Electrical Engineering and Computing (2014)

Wearable Sensor for Home-Based Biofeedback Therapy for Migraine

Alma Secerbegovic(✉), Mustafa Spahic, Amir Hasanbasic, Vedad Mesic, Haris Hadzic, and Aljo Mujcic

Faculty of Electrical Engineering, University of Tuzla, Tuzla, Bosnia and Herzegovina
alma.secerbegovic@untz.ba

Abstract. The rapid technological development of wearable devices is aimed at enhancing the existing healthcare services and helping chronic patients at home with diseases like diabetes, arrhythmias, and mental and neurological conditions such as migraine. Consumers are now being able to manage their own health with wearable devices, smartwatches, and wearable monitors. As one of the potential applications that can be implemented with wearable devices and smartphones is biofeedback therapy for migraine. Since biofeedback is considered as one of the established alternative therapies for chronic migraine conditions, home-based biofeedback is becoming a necessity especially during challenging periods. In this paper, we have developed and designed a prototype for a finger-based wearable device for the measurement of skin temperature, skin conductance, heart rate, and heart rate variability. Preliminary testing results are reported, with a clear indication of changes in skin temperature and skin conductance are visible during biofeedback treatment. Evaluation of performance in skin conductance measurement is performed when compared to the commercially available device.

Keywords: Wearable device · Biofeedback · Migraine · Skin conductance

1 Introduction

Wearable devices have prospered significantly in recent years, driving the market for wellness, fitness, and health monitoring applications [1, 2]. Increasing demand for connected wearable devices exists in various industries, especially during the global pandemic and crisis recently witnessed. Real-time monitoring with additional personalized treatment or therapy has become the desirable healthcare service that hospitals and insurance companies will offer to patients in the near future. Wearable devices allow for real-time monitoring of the patients while minimizing the number of hospital visits and readmissions. However, wearable devices when used by adults in hospitals are still in validation and testing phases, with unclear improvement outcomes for different studies [3]. On the other hand, wearable devices for personal-based health monitoring give insight to the patient of its current health parameters at home, which is satisfactory for monitoring of healthy individuals.

A. Badnjevic and L. Gurbeta Pokvić (Eds.): CMBEBIH 2021, IFMBE Proceedings 84, pp. 116–123, 2021.
https://doi.org/10.1007/978-3-030-73909-6_13

Research studies have been performed for non-invasive real-time monitoring applications including a wide range of sensors for the recording of different physiological signals such as electrocardiogram, activity monitoring, heart rate, and heart rate variability, respiration rate, electroencephalogram, skin conductance, or electrodermal activity (EDA), oxygen saturation, etc. The main goal of the wearable device is to provide patients with real-time monitoring by collecting and processing physiological signals. With the availability of smartwatches and small-sized wearable sensors, the demand for meaningful usage is on the rise, from preventative medicine to personalized health monitoring.

Migraine is one of the common illnesses that is closely related to mental health, and it is considered one of the underrated and undertreated diseases. One of the effective treatments for migraine, besides oral medications is biofeedback [4]. Biofeedback is a non-invasive therapy that monitors the body's responses and allows the patient to learn and control their physiology with mindfulness-based techniques and relaxation exercises. However, biofeedback sessions are typically conducted in hospital surroundings.

With the rapid development of wearable sensors and smartphones, homebased biofeedback sessions can be put into action. In this paper, we have created and designed a prototype for the wearable sensor that would be used in home-based biofeedback therapy along with a smartphone device. This work focuses on presenting preliminary testing results of the prototype and the evaluation of performance when compared to the commercially available system.

2 Biofeedback Therapy for Migraine

Migraine is a neurological disease characterized by severe throbbing, usually reoccurring pain, disrupting an individual's ability to lead a normal life. Chronic migraine is a condition that affects all age groups - from children to adults, causing significant social and economic burden. Pain-relieving and preventing medications are typically used as oral medications for the treatment of migraine. Alternative therapies, besides traditional oral medications, have shown reductions in the occurrence of migraines when used for an extended period. This type of treatment such as relaxation training and biofeedback, along with pharmacological therapies have proven to be more successful in dealing with migraine compared to either of the interventions separately [5, 6]. Research studies show that biofeedback can reduce the frequency, intensity, and duration of migraine and tension-type headaches [4, 7]. Biofeedback therapy is usually conducted in a hospital setting, where medical-grade devices are used for the measurement of common physiological parameters. Surface electrodes are placed on the specific parts of the body to provide information to the user of its current health state. The biofeedback practitioner follows the physiological responses of the body on a computer screen and guides the patient to a relaxed state. Heart rate, breathing rate, skin temperature, and skin conductance are the most commonly assessed biofeedback parameters. Within real-time monitoring, patients can learn to control their thoughts, sensations, and behaviors with biofeedback. The main goal of biofeedback therapy is to change the activity of the sympathetic nervous system by monitoring and subtly controlling body characteristics. For example, relaxation, or lowering the activity of the sympathetic nervous system, can be induced by decreasing muscle tension, slowing respiration, slowing heart rate, or increasing body temperature. Relaxation training is mostly used in combination with biofeedback. Practices such as slow diaphragmatic breathing, visualization of relaxing and positive images as well as the focus on feelings of heaviness and warmth in different parts of the body are often applied in biofeedback therapy.

The main disadvantage of biofeedback therapy and relaxation training is that it requires time commitment over an extended period and a certain level of effort from the patient. These sessions usually last around an hour and training sessions may include several weekly visits to the biofeedback clinic. Although significant evidence exists to support the use of biofeedback, a large number of patients do not receive the treatment. Certain obstacles exist due to the implementation, cost, and availability of the treatment [8]. Therefore it is necessary to investigate alternative methods for delivering biofeedback treatment for patients at home.

Recently, digital technology has been incorporated into medical delivery practice with mobile health (m-Health) applications for smartphones. A large number of applications are focused on creating mobile headache diaries for tracking symptoms, frequency, duration, pain intensity, etc. [9]. Recently biofeedback smartphone applications have emerged showing that gamified biofeedback apps may lead to better health [10, 11]. With the rapid development of miniaturized sensors and the ubiquitous presence of smartphones, it is possible to provide biofeedback training and therapy to the patient based on mobile phone technology. However, smartphones have no sensors for measuring skin conductance and body temperature, which are sensitive to the physiological responses during the biofeedback treatment [4].

Previous research studies have proposed wearable devices for the measurement of physiological parameters for biofeedback purposes. E.g. Empatica is a smartwatch-type of the sensor with the ability to collect signals such as photoplethysmograph, electrodermal activity, 3-axis accelerometer, and temperature [12]. SensoRing is a ring sensor for the measurement of skin conductance, temperature, heart rate, and locomotion [13], while Breeze presents a wearable pendant placed around the neck that measures breathing and sends biofeedback in realtime [14]. Smartwatch application that maps heart rate and heart rate variability measurement to the emotional state of the patient is presented in [15].

In this paper, we have proposed a prototype design for a wireless wearable device for biofeedback treatment for migraine patients. The proposed device measures skin temperature, oxygen saturation, and heart rate as well as skin conductance. The body's response to high levels of anxiety and stress can be manifested as cold hands. This can be easily explained that as the stress level increases, the body's natural response is to reduce blood flow in the peripheral extremities such as hands. However, when an individual is in a relaxed state, the consequence is returned warming of the fingers as the blood flow is established. Emotional arousal or stress can affect the pores on the surface of the skin, which leads to the increased secretion of the sweat glands. Measuring the skin conductance across the areas with the highest density of the sweat glands can help with the detection of changes in emotional states. Highly emotional states such as anger or stress are represented as increases in skin conductance, while satisfaction and relaxation show the opposite response.

3 Sensor Prototype Design

In this work, we have created a prototype device with the main goal to measure the physiological parameters of the patient while providing feedback from the smartphone. The wearable device is designed to be the minimal size and weight in order to be convenient for usage during extended periods of time.

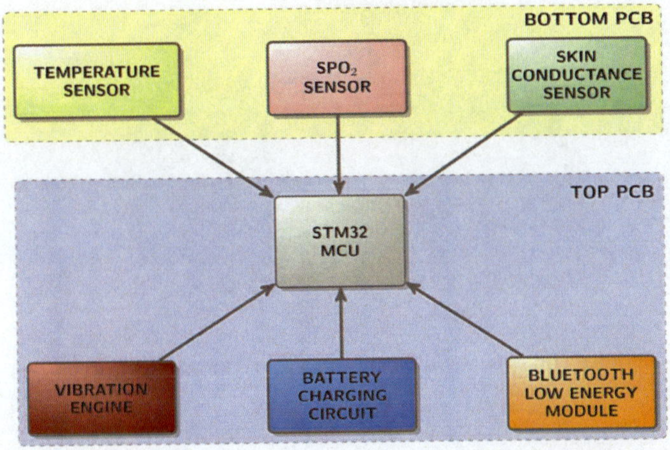

Fig. 1. Block diagram of the proposed hardware architecture

The hardware architecture of the developed prototype biofeedback sensor for multiple physiological parameters measurement is discussed in this section. The main requirements of the proposed platform are to measure finger skin temperature (ST), skin conductance, or electrodermal activity (EDA) along with heart rate, heart rate variability, and oxygen saturation. A wearable device also contains a battery charging circuit along with a rechargeable Lithium-ion battery. The vibration engine component is additionally provided in order to be used as a feedback indicator from the smartphone application. Whenever the user is required to breathe for a certain number of breaths per minute, vibrations from the vibration engine will signalize the periods for the inhales and exhales. Block diagram of the prototype device is shown in Fig. 1.

The proposed sensor prototype consists of two separate printed circuit boards (PCB), bottom and top PCB. The bottom board is a sensor board, where all the sensors from acquisition of physiological parameters are placed. Incorporated sensor components are skin temperature sensor, oxygen saturation sensor and skin conductance sensor. Skin conductance is acquired with two pads, which measure resistance of the skin and the obtained signal is filtered and conditioned by analog circuit on top PCB board. Skin temperature is measured with sensitive TMP116N sensor. It provides a 16-bit temperature result with a resolution of 0.0078°C and an accuracy of up to ±0.2°C with no calibration. It also has an alert feature which was used to notify and wake the main microcontroller unit from the "deep sleep" mode when the device is worn. The MAX30102 is an integrated pulse oximetry and heart-rate monitor module. It includes internal LEDs, photodetectors, optical elements, and low-noise electronics with ambient light rejection. Communication is established through a standard I2C-compatible interface. The module can be shut down through software with zero standby current, allowing the power rails to remain powered at all times. The bottom printed circuit board is shown in Fig. 2a, where the sensor placements are annotated.

Fig. 2. Bottom printed circuit board with temperature, oxygen saturation and skin conductance sensors

Top PCB consists of a powerfull microcontroller unit (STM32F030F4Px), wireless Bluetooth Low Energy (BLE) module that enables communication with smartphone or PC and battery charging circuit (Fig. 3). Since one of the main requirements is power optimization, the device is placed in "deep sleep" mode whenever the temperature sensor is outside of range of normal human body temperature. The preliminary version of the device sends all the raw data collected from the sensors to smartphone device via Bluetooth.

Fig. 3. 3D model of the top printed circuit board

The processing of the signals will be conducted on the smartphone device, where the user can see the physiological data in real-time.

4 Preliminary Testing Results

The prototype device was tested for one individual during relaxation and activation session, as experienced in biofeedback treatment. The created scenario starts with two minutes of normal breathing relaxation session. The rise in skin temperature of the finger can be visually identified in Fig. 4 at the beginning of the recording. The activation session involved the mental workload operation, where the patient was required to subtract an odd number from a very large number for one minute. Afterwards, the patient was instructed to think about stressful event for another minute. This change in emotional state is visible on skin temperature graph, as a drop of finger temperature in the middle of the session. The final session included paced breathing and relaxation, which has lead to the relaxation and the rise of the finger temperature.

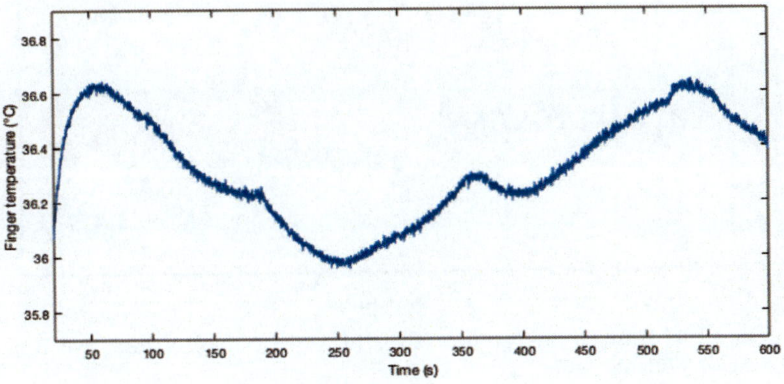

Fig. 4. Skin temperature collected with prototype device during relax and stress scenario

Figure 5 shows skin conductance during relaxation and activation scenarios. Sudden rise and drop in skin conductance level is characterized with higher stimulation of the nervous system, as can be seen in the middle interval, when activation session is conducted.

In order to validate the skin conductance measurement where the electrodes are placed on the forefinger, we have synchronously measured skin conductance with commercially available device Bitalino [16]. Bitalino captures skin conductance with two electrodes placed on the palm, where the sweat glands are highly placed. Figure 6 shows the obtained measures for skin conductance during stressful scenario. Measured values had relatively high correlation coefficient during the experiment, ranging from 0.71 to 0.82.

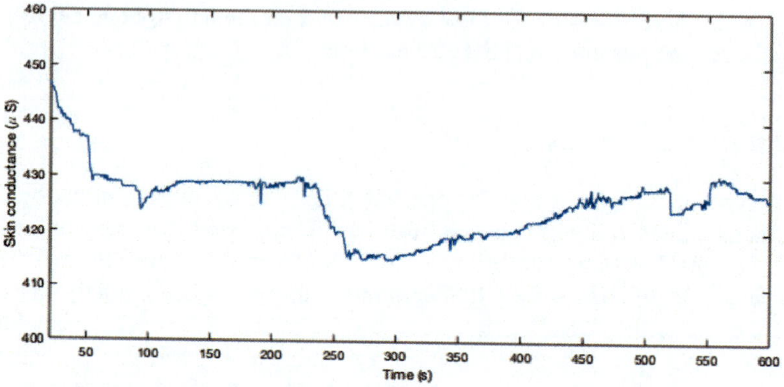

Fig. 5. Skin conductance collected with prototype device during relax and stress scenario

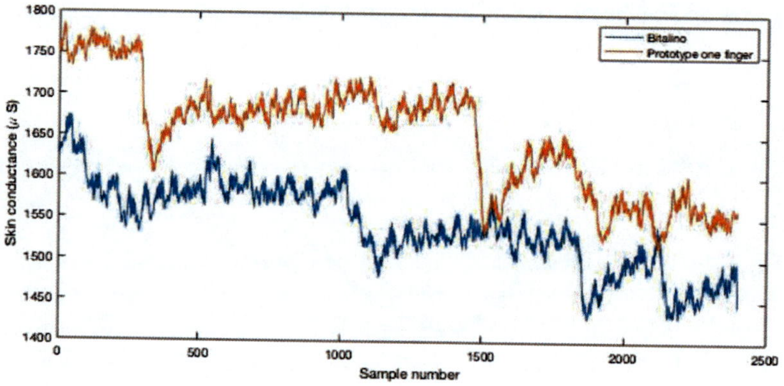

Fig. 6. Comparison between protype one finger EDA sensor measurement and palm-based measurement with Bitalino [16]

5 Conclusion

In this work, we have presented a preliminary prototype-version of the wearable sensor, intended for home-based biofeedback treatment of the migraine. Biofeedback presents an efficient method in reducing pain intensity and numerous occurrences of migraine attacks. However, it is mostly performed in specialized biofeedback clinics. The proposed sensor is aimed at measuring typical physiological parameters in biofeedback treatment such as skin temperature, skin conductance, heart rate, heart rate variability, and oxygen saturation. In this preliminary study, we have tested the prototype for two parameters, skin conductance, and temperature. Experimental results are promising and show that the skin conductance measurements are highly correlated with the reference module. In the next phase of our work, we will be incorporating oxygen saturation measurements and iOS application development for testing in real biofeedback scenarios.

References

1. Metcalf, D., Milliard, S.T.J., Gomez, M., Schwartz, M.: Wearables and the internet of things for health: Wearable, interconnected devices promise more efficient and comprehensive health care. IEEE Pulse **7**(5), 35–39 (2016)
2. Market Analysis Report, Wearable Technology Market Size, Share & Trends Analysis Report By Product (Wrist-wear, Eye-wear & Head-wear, Foot-wear, Neck-wear, Body-wear), By Application, By Region, and Segment Forecasts, 2020–2027 (2020). https://www.grandviewresearch.com/industry-analysis/wearable-technology-market
3. Zheng, Y.L., et al.: Unobtrusive sensing and wearable devices for health informatics. IEEE Trans. Biomed. Eng. **61**(5), 1538–1554 (2014)
4. Schwartz, M.S., Andrasik, F. (eds.): Biofeedback: A Practitioner's Guide. Guilford Publications (2017)
5. Holroyd, K.A., Penzien, D.B.: Pharmacological versus non-pharmacological prophylaxis of recurrent migraine headache: a meta-analytic review of clinical trials. Pain **42**, 1–13 (1990)
6. Powers, S.W., Kashikar-Zuck, S.M., Allen, J.R., et al.: Cognitive behavioral therapy plus amitriptyline for chronic migraine in children and adolescents. JAMA **310**(24), 2622–2630 (2013)
7. Raggi, A., Grignani, E., Leonardi, M., et al.: Behavioral approaches for primary headaches: recent advances. Headache **56**(6), 913–925 (2018)
8. Minen, M.T., et al.: Factors related to migraine patients' decisions to initiate behavioral migraine treatment following a headache specialist's recommendation: a prospective observational study. Pain Med. **19**(11), 2274–2282 (2018)
9. Hundert, A.S., Huguet, A., McGrath, P.J., Stinson, J.N., Wheaton, M.: Commercially available mobile phone headache diary apps: a systematic review. JMIR mHealth uHealth **2**(3), e36 (2014)
10. Hunter, J.F., Olah, M.S., Williams, A.L., Parks, A.C., Pressman, S.D.: Effect of brief biofeedback via a smartphone app on stress recovery: randomized experimental study. JMIR Serious Games **7**(4), e15974 (2019)
11. Dillon, A., Kelly, M., Robertson, I.H., Robertson, D.A.: Smartphone applications utilizing biofeedback can aid stress reduction. Front. Psychol. **7**, 832 (2016)
12. Garbarino, M., Lai, M., Bender, D., Picard, R.W., Tognetti, S.: Empatica E3—a wearable wireless multi-sensor device for real-time computerized biofeedback and data acquisition. In: 2014 4th International Conference on Wireless Mobile Communication and HealthcareTransforming Healthcare Through Innovations in Mobile and Wireless Technologies (MOBIHEALTH), pp. 39–42. IEEE (2014)
13. Mahmud, M.S., Wang, H., Fang, H.: SensoRing: an integrated wearable system for continuous measurement of physiological biomarkers. In: 2018 IEEE International Conference on Communications (ICC), pp. 1–7. IEEE (2018)
14. Frey, J., Grabli, M., Slyper, R., Cauchard, J.R.: Breeze: sharing biofeedback through wearable technologies. In: Proceedings of the 2018 CHI Conference on Human Factors in Computing Systems, pp. 1–12 (2018)
15. Qin, C.Y., Choi, J.H., Constantinides, M., Aiello, L.M., Quercia, D.: Having a heart time? a wearable-based biofeedback system. In: Proceedings of the 22nd International Conference on Human-Computer Interaction with Mobile Devices and Services Adjunct, 2020 (2020)
16. Da Silva, H.P., Guerreiro, J., Lourenço, A., Fred, A.L., Martins, R.: BITalino: a novel hardware framework for physiological computing. In: PhyCS, pp. 246–253 (2014)

The Analysis of Biochemical Markers for the Diagnosis of an Acute Myocardial Infarction Using Artificial Neural Network

Jahić Muamera, Jelačić Neira, Jovičić Tanja, Jusufović Selma, Kajmaković Amir(✉), and Kapić Amna

Farmaceutski fakultet Univerziteta u Sarajevu, Sarajevo, Bosnia and Herzegovina
amirkajmakovic@ffsa.unsa.ba

Abstract. Cardiovascular diseases are the most frequent conditions worldwide. When the cardiac muscle is not getting enough oxygen from the bloodstream, it leads to acute myocardial infarction (AMI). The aim of this paper is to analyze specific biomarkers that can indicate previous AMI using artificial neural network. Specific biomarkers that we used are CK-MG, cTnT, cTnI, myoglobin, homocysteine and CRP. In this study participated 300 examinees. The accuracy of this analysis was not at the satisfactory level. Since the specificity rate was low, identified true negative cases may not be really negative. Even though the selected biomarkers are proven to be accurate in most of the studies similar to this one, limitation of this study is low accuracy.

1 Introduction

Cardiovascular disease (CVD) is a group of serious health conditions and it is classified as the leading cause of death in the world. The annual mortality of this group of diseases is expected to reach 23.6 million by 2030. CVD is a group of diseases that includes coronary heart disease (CHD), cerebrovascular disease, peripheral arterial disease, rheumatic heart disease, deep vein thrombosis, and pulmonary embolism, according to the World Health Organisation. CHD is defined as a disease of the blood vessels supplying the heart muscle. This condition is also referred to as ischemic heart disease (IHD) and acute myocardial infarction (AMI) [1].

Myocardial infarction is a syndrome where myocardial cells die, due to imbalances between myocardial oxygen supply and demand and it can be secondary to acute atherosclerotic plaque disruption, or alterations in myocardial oxygen supply and/or demand in the absence of acute atherothrombosis. The clinical definition of AMI denotes the presence of acute myocardial injury. This injury is easily detected by abnormal cardiac biomarkers in the setting of evidence of acute myocardial ischemia [2]. Ischemia due to decreased coronary artery flow causes serious damage like deterioration of ventricular function and myocardial necrosis. Such enzymes as ALT, AST, LDH, CK, and troponin have been indicators or biomarkers for detecting AMI [3]. A significant role in the regulation of muscle contraction plays cardiac troponin complex. This complex is composed of three isoforms, namely, troponin C, troponin I and troponin T. Troponin I

A. Badnjevic and L. Gurbeta Pokvić (Eds.): CMBEBIH 2021, IFMBE Proceedings 84, pp. 124–129, 2021.
https://doi.org/10.1007/978-3-030-73909-6_14

and troponin T have been verified to be specific and sensitive biomarkers of myocardial damage because they are expressed in the cardiac muscle. This is particularly important in asymptomatic patients, when combined with other biomarkers and examinations [4]. CK is an enzyme that is found primarily in the both cardiac and skeletal muscle. This enzyme has 3 isoenzymes: MM, MB and BB. The CK-MB rises in the serum at 4-9 h after the onset of chest pain. Its peak is approximately 24 h and returns to baseline values at 48-72 h. CK-MB has one advantage over the troponins and that is the early clearance that helps in the detection of reinfarction. Thus, the serum level of troponin along with the level of the CK-MB fraction is assessed for the diagnosis of myocardial infarction [5].

Creatine kinase (CK) is an enzyme that is located inside of cells and it presents itself in greatest amounts in skeletal muscle, myocardium, and brain; smaller amounts occur in other visceral tissues. Disruption of cell membranes due to hypoxia or other injury releases CK from the cellular cytosol into the systemic circulation. Elevated serum levels of CK have been used as a sensitive but non- specific test for myocardial infarction, that means that it can also be used as biomarker [6].

Myoglobin is a cytoplasmic protein present in the heart and skeletal muscles. Myoglobin is not found in any tissue other than the muscle, but it can be present in the bloodstream as a result of muscle damage. It is a sensitive marker for AMI, but has limited specificity. It is rapidly released from the myocardium during the injury and is rapidly excreted from the kidneys within 24 h. Myoglobin rises in the first 30 min in the early period after the onset of an acute event and it is an important biomarker for early detection and/or exclusion of cardiac damage. It is elevated in all AMI patients within 6–10 h and peaks at the 12th hour. Although a relatively sensitive parameter, its diagnostic specificity is limited. Its values grow due to various skeletal muscle injuries, which is the main reason why its use is effective mainly in the exclusion of AIM, but unfortunately not in its confirmation or differentiation from unstable angina pectoris [3].

Currently, creatine kinase–MB isoenzyme (CK-MB), myoglobin and cardiac troponins (cTnI & cTnT) are used in the diagnosis of AMI. However, these cardiac marker proteins are also not satisfactory for the diagnosis of AMI within first 6 h of the onset of chest pain. Myoglobin compared to CK-MB (isoenzyme), cardiac troponin I, or cardiac troponin may provide the best diagnostic efficacy in assessing the risk for AMI. Myoglobin is most useful as a marker when used in combination with other cardiac markers in the rapid diagnosis of AMI, especially in patients with atypical chest pain or nonspecific ECG changes [7]. Patients can be asymptomatic, that is why these parameters are very significant.

Artificial neural network (ANN) is a model that simulates biological neural network using a reduced set of concepts from biological neural systems. It seems that ANNs were not used before in analysis of biomarkers that are used to determine previous AMI [8]. However, although this type of data analysis has only recently been used in practice, many scientists have already used it in their analysis [9–17].

The aim of this study is to analyze specific biomarkers that can indicate previous AMI using artificial neural network.

2 Materials and Methods

The study, reported in this paper, was conducted in November 2020. A total of 300 respondents participated in this study. Of these, 200 were patients who had acute myocardial infarction, while 100 subjects were healthy controls. For all patients, between the event of an acute myocardial infarction and the taking of a blood sample, necessary information for this study, passed between 12 h, up to a maximum of 10 days from the event. This information was used from the history chart of the patient since it is known that after 10 days, markers of acute myocardial infarction return to their initial state.

According to the research topic, the evaluation criteria included: cardiac creatinine kinase, cardiac troponins (cTnT, cTnI), myoglobin, homocysteine, and C-reactive protein. The selected parameters directly and unambiguously indicate the occurrence of acute myocardial infarction, which is why they were chosen for this study. Table 1 lists the reference values for the selected markers.

Table 1. Reference values of selected parameters

Parameter	CKMB (IU/L)	cTnT (ng/L)	cTnI (ng/mL)	Myoglobin (ng/mL)	Homocysteine (μmol/L)	CRP (mg/L)
Reference range	5–25	<14	0.04	25–72	<15	<10

The dataset is presented in Table 2. As already mentioned, in this study participated 300 subjects. 52% of them were female, and 48% male. Most of the examinees were from the age range 56 and older, 54,66%.

Table 2. The dataset

	Female	Male	
Gender	52%	48%	
	Health	Disease	
Health status	100	200	
	18–35	36–55	56+
Age	11,66%	33,66%	54,66%

The distribution of the dataset for development of the Artificial Neural Network (ANN) is 70:30% for training and validation was used [19]. Validation was performed on 120 samples.

Accuracy, sensitivity and specificity of the network are calculated according to Eqs. (1), (2) (3):

$$sensitivity = \frac{true\ positive}{(true\ positive\ +\ false\ negative)} \tag{1}$$

$$specificity = \frac{true\ negative}{(true\ negative\ +\ false\ positive)} \tag{2}$$

$$accuracy = \frac{(true\ negative\ +\ true\ positive)}{(true\ negative + true\ positive + false\ negative + false\ positive)} \tag{3}$$

3 Results

For the purpose of the study, for diagnosis of an acute myocardial infarction based on biochemical markers, two-layer feedforward artificial neural network, with sigmoid transfer function in hidden layer was developed [19, 20].

During neural network development different number of neurons was tested. Results of training with different number of neurons is presented in Table 3. Training was performed using Levenberg – Marquadt training algorithm.

Table 3. Training performance with different number of neurons in hidden layer

Number of neurons	Performance
2	2.5514e–08
5	2.8069e–12
10	1.2931e–11
15	8.4121e–10
20	**1.6902e–12**

Accuracy of training for feedforward neural network with 20 neurons and sigmoid transfer function in hidden layer was 91.8%. The validation of performance of feedforward neural network with 20 neurons and sigmoid transfer function in hidden layer is presented in Table 4.

Table 4. Distribution of dataset

n = 120	GROUP 1 Healthy ANN prediction	Group 2 Disease ANN prediction	
GROUP 1 Healthy 84	65	19	77.5%
Group 2 Disease 36	8	28	
	77.3%	77.8%	

As shown in Table 4 ANN prediction has 70.23% sensitivity, and 30.55% specificity. The accuracy of this method in our study is 58.33%.

Since only the sensitivity rate is reasonably high, that means that cases identified as positive, are most likely really positive. The specificity rate being low (30,55%) means that not all negative cases are really negative.

Review article, written by Mythili et al. discusses relevant inflammation biomarkers indicators of AMI. They suggest that these biomarkers have become the frontline diagnostic tool. They can be very useful in rapid diagnosis and thereby saving lives [5].

4 Conclusion

AMI is cardiac event that doesn't last long, so if patient is asymptomatic, it can go undetected. Biomarkers, such as CK-MG, cTnT, cTnI, myoglobin, homocysteine and CRP, can be very much useful in determining this event even a few hours or days after it happened. Results have shown that accuracy of this ANN analysis is not satisfactory. Accuracy rate is quite low so it leaves a room for mistake. Sensitivity rate is only validation parameter that is reasonably high, which means that more than 70% positive cases are really positive.

References

1. Sjogren, B., Bigert, C., Gustavsson, P.: Chapter 16- Cardivascular Disease. Handbook on the Toxicology of Metales (Fourth Edition). vol. I, pp. 313–331 (2015)
2. Sandoval, J., Jafee, A.S.: Type 2 myocardial infraction: JACC review topic of the week. J. Am. Coll. Cardiol. **73**(15), 1846–1860 (2019)
3. Aydin, S., et al.: Biomarker in acute myiocardialinfration:current perspectives. Vasc. Health Risk Mang. **15**, 1–10 (2019)
4. Wang, X.Y., et al.: The biomarkers for acute myocardial infraction and heart failure. BioMed Res. Int. **14** (2020)
5. Mythili, S., Malathi, N.: Diagnostic markers of acute myocardial infraction. Biomed. Rep. **3**, 743–748 (2015)
6. Walker, H.K., et al.: Clinical Methods: The History, Physical, and Laboratory Examinations. 3rd edition. Boston:Butterworths (1990)
7. Pyati, A.K., et al.: Heart- type fatty acid binding protein. a better cardiac biomarker than CK-MB and myoglobin in the early diagnosis of acute myocardial infraction. J Clin. Diagn. Res. **9**(10), BC08-BC11 (2015)
8. Walczak, S., Cerpa, N.: Artificial neural networks. encyclopedia of physical science and technology, 631–645 (2003)
9. Catic, A., et al.: Application of Neural networks for classification of Patau, Edwards, Down, turner and Klinefelter Syndrome based on first trimester maternal serum screening data, ultrasonographic findings and patient demographics. BMC Med. Genomics **11**, 19 (2018)
10. Sarić, R., et al.: FPGA-based real-time epileptic seizure classification using artificial neural network. Biomed. Sig. Process. Control, **62**, 102106 (2020). ISSN 1746–8094
11. Badnjevic, A., Gurbeta, L., Cifrek, M., Marjanovic, D.: Classification of asthma using artificial neural network. In: IEEE 39th International convention on information and communication technology, electronics and microelectronics (MIPRO), 30 May to 03. June 2016 Opatija, Croatia (2016)
12. Aljovic, A., et al.: Artificial neural networks in the discrimination of alzheimer's disease using biomarkers data. In: IEEE 5th Mediterranean Conference on Embedded Computing (MECO), 12–16 June 2016, Bar, Montenegro (2016)

13. Alic, B., et al.: Classification of stress recognition using artificial neural network. In: IEEE 5th Mediterranean Conference on Embedded Computing (MECO), 12–16 June 2016, Bar, Montenegro (2016)
14. Veljovic, E., et al.: Artificial neural network and docking study in design and synthesis of xanthenes as antimicrobial agents. In: CMBEBIH 2017. IFMBE Proceedings, vol. 62, pp 617–626, Springer, Singapore (2017). https://doi.org/10.1007/978-981-10-4166-2_93
15. Sejdinovic, D., et al.: Classification of prediabetes and type 2 diabetes using artificial neural network. In: CMBEBIH 2017. IFMBE Proceedings, vol. 62, pp 685–689, Springer, Singapore (2017). https://doi.org/10.1007/978-981-10-4166-2_103
16. Hrvat, F., et al.: Artificial Intelligence for prediction of medical device performance: infusion and perfusor pumps case study. In: 9th Mediterranean Conference on Embedded Computing MECO 2020 (2020)
17. Šećkanović, A., et al.: Review of artificial intelligence application in cardiology. In: 9th Mediterranean Conference on Embedded Computing MECO 2020
18. Rau, H.-H., Hsu, C.-Y., Lin, Y.-A., Atique, S., Fuad, A., Wei, L.-M., Hsu, M.-H.: Development of a web-based liver cancer prediction model for type II diabetes patients by using an artificial neural network. Comput. Meth. Programs Biomed. **125**, 58–65 (2016)
19. Anifowose, F., Khoukhi, A., Abdulraheem, A.: Investigating the effect of training–testing data stratification on the performance of soft computing techniques: an experimental study. J. Exper. Theor. Artif. Intell. **29**(3), 517–535 (2017). https://doi.org/10.1080/0952813X.2016.1198936
20. Aggarwal, C.: Book: Neural Networks and Deep Learning. Springer. https://link.springer.com/book/10.1007/978-3-319-94463-0

Influence of Probiotic Bacteria on Mechanical Properties of Nickel-Titanium Alloys Used in Orthodontics

Ines Musa Trolic[3](✉), Stjepan Spalj[1], Sven Karlovic[2], and Goran Bosanac[2]

[1] University of Rijeka, Rijeka, Croatia
[2] University of Zagreb, Zagreb, Croatia
[3] University of Mostar, Mostar, Bosnia and Herzegovina

Fixed orthodontic appliances are made from different metal alloys. The main properties of materials used in oral cavity are biocompatibility and their resistance to mechanical stress. Orthodontic appliances are surrounded by complex and very aggressive environment of oral cavity. Human saliva is composed of inorganic salts, organic acids, enzymes, different bacteria and gastric secretions [1].

Bacteria in the oral cavity produce biofilm and it was found that pH in a 24 h old biofilm can be 4.8 [2]. This conditions can cause corrosion of alloys used in orthodontic treatment what can cause hypersensitivity because of realeasing metal ions in the organisam and also can affect a mechanical properties of the appliance [3–5].

Because of the design of orthodontic appliance maintance of oral hygien in patients with fixed orthodontic appliances is aggravated. The most often consequences of the poor oral hygiene in the orthodontic patients are carries, white spot lesions, gingivitis. In addition to the usual methods of maintaining oral hygiene, it has been recommended use of probiotic supplements to the orthodontic patients with poor oral hygiene because this supplements showed good results in treatment of pathological conditions such a candidiasis, periodontitis, gingivitis [6–10].

Live microorganisms that have health benefits when consumed are called probiotics. For improving health in the oral cavity the most used bacteria is Lactobacillus because they have ability to produce reutericyclin, reutericin and reuterin [11–15]. Lactobacillus reuteri is the one of the probiotic bacteria which is used in the commercially available oral supplement BioGaia Prodentis (BioGaia AB, Stockholm, Sweden). This supplement contains two strains of probiotic bacteria *L. reuteri*, DSM 17938 and ATCC PTA 5289. *L.reuteri*.

Microbiologically induced corrosion is the type of corrosion induced by the activity of the different types of the microorganisms [16–19]. It is known that the *Bacteroides corrodens* and *Streptococcus mutans* can cause microbiological corrosion of dental materials in the oral cavity [18–23]. The ability of bacteria to adhere at metallic surfaces and create a thick biofilm conditions the occurence of corrosion. Bacteria can be directly involved in corrosion process or they can changed surroundings around them and cause corrosion of the affected materials [15]. Corrosion can affect mechanical properties of the orthodontic archwires. Probiotic bacteria show certain corrosion potential [24]. But there is no relevant research about corrosion of orthodontic materials caused by the probiotic.

A. Badnjevic and L. Gurbeta Pokvić (Eds.): CMBEBIH 2021, IFMBE Proceedings 84, pp. 130–137, 2021.
https://doi.org/10.1007/978-3-030-73909-6_15

The aim of the research was to explore influence of probiotic bacteria Lactobacillus reuteri on the mechanical properties of tested orthodontic archwires such a modulus of flexibility and resilience.

Our hypotesis was that corrosion will be increased at archwires exposed to the saliva with the addition of L. reuteri then in archwires exposed just to the artificial saliva. We expected that corrosion will be seen as deterioration of flexibility and resilience, more in unloading than loading.

1 Materials and Methods

NiTi archwires were exposed to artificial saliva and artificial saliva with Lactobacillus reuteri in vitro.

Three types of commercially available orthodontic archwires with dimensions 0.508 × 0.508 mm (0.020 × 0.020 in.) were tested – three Nickel-Titanium alloys (composition Ni = 50,4%; Ti = 49,6%) – uncoated surface (BioForece Sentallo®), nitrided surface (IonGuard®) and rhodium coated surface (High Aestheti ® (Dentsply GAC, Bohemia, USA).

Tani-Zucchi formula of artificial saliva were used: 1.5 g/l KCl, 1.5 g/l NaHCO3, 0.5 g/l NaH2PO4xH2O, 0.5 g/l KSCN, 0.9 g/l of lactic acid, pH 4.8 [20]). Influence of the probiotic bacteria were tested using artificial salive mixed with a culture of L. reuteri. Strains DSM 17938 and ATCC PTA 5289 were used in this research.

Corrosion by probiotic bacteria strains was conducted in the following manner: samples of each archwire, 2.5 cm in length were cut from the straight ends of archwires. Thin plastic tubes with stoppers were used for testing. In every tube were 6 ml of experimental solution, artificial saliva or artificial salive with probiotic bacteria and sample of tested archwire. For bacterial nutrition, broth media was aded through experiment and it was mainly composed of D (+) glucose (20 g/l). To ensure viability of bacteria every 48 h new tubes with new saliva and bacteria were used and tested wires were transfered into them. The tubes wich contains samples with pure saliva were not changed during the experiment. The wires were immersed 28 days. To the end of the experiment test tubes were placed in the incubator. The temperature in the incubator was 37 °C. A Gaspac system CO2 generator (BD BBL, Becton, Dickinson and Company, New Jersey, USA) was used for achieving anaerobic atmosphere. Unexposed archwires were used as control. On the tested archwires was created visible biofilm so that was the proof of bacteria viability. At the same time viability assay was performed in parallel experiment. After a week of incubation tested archwires, each 1,5 cm long, were removed from experimental media with L. reuteri. Archwires were washed with phosphate-buffered saline, PBS (Sigma-Aldrich-Chemie, Steinheim, Germany) three times to eliminate unbound bacteria. After washing they were transferred to plastic tubes with 5 ml PBS and sonificated in a water bath (Bactosonic, Bandelin, Berlin, Germany) at 40 kHz for 1 min. L. reuteri were counted. Lactobacilli were placed on MRS agar for 3 days at 37 °C. An average bacterial count of 3000 on NNiTi (range 1000–5000), 1667 on NiTi (range 1000–3000) and <100 on RhNiTi (limit of detection was 100). To ensure L. reuteri viability the culture was constantly kept in Petri dishes on De Man, Rogosa and Sharpe (MRS) agar in Petri dishes. A GENbox anaer (BioMérieux, Marcy-l'Étoile, France) was used to achieve anaerobic atmosphere necessery for survival of anaerobic L. reuteri.

Three-point bending test was used for testing mechanical properties of the archwires. The support span of Texture Analyser TA.HD.plus (Stable Micro Systems, Godalming, UK) was set to 12 mm and loaded with a low-force cell (5 kg, factory calibrated), according to current standards. During the measurements the temperature of 37 °C was set in a thermal chamber. Temperature as set with the help of a thermal coil and regulated automatically with thermal sensor. Archwires were loaded to a deflection of 3.1 mm and then unloaded to zero deflection at a cross-head speed of 0.0167 mm/s. The Texture Exponent software program (Stable Micro Systems) was used for collecting load in Newtons (N) and deflection in millimeters (mm) for each archwire, every 5 ms. Figure 1. shows representative curve. From the hysteresis loop loading and unloading elastic modulus (*E*) and yield strength (YS) were read off. Springback ratio as a measure of flexibility (YS/E) and modulus of resilience as a measure of resilience ($YS^2/2E$) was calculated. Decrease of both values indicates deterioration of elastic properties.

Statistical analysis was done using commercial software IBM SPSS 22 (IBM Corp, Armonk, USA). Oneway ANOVA with Student–Newman–Keuls post hoc test was used

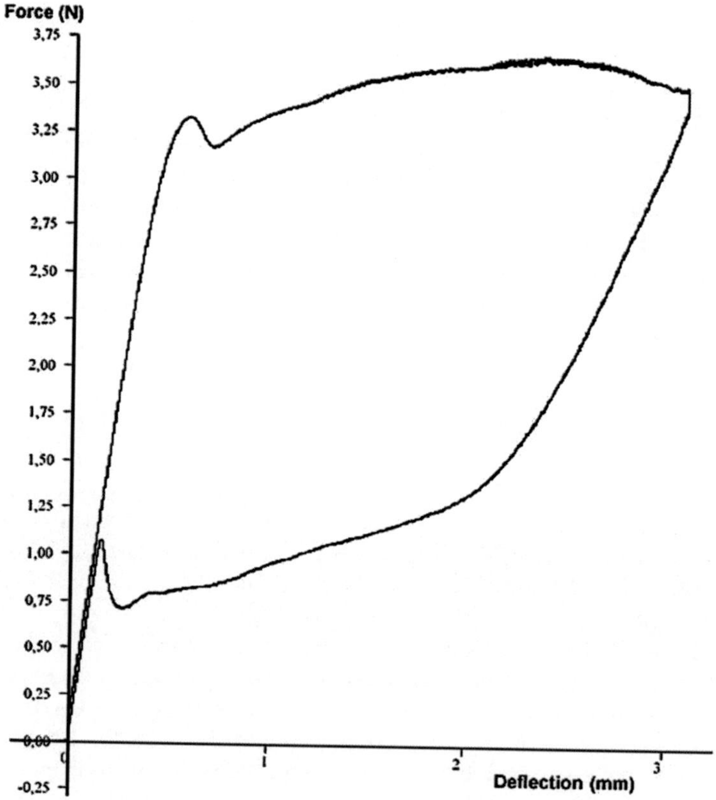

Fig. 1. Representative force-deflection curve

for each type of archwire to explore does probiotic bacteria influence mechanical caracteristics in comparison to as-received condition and negative control (artificial saliva). Effect size, i.e., the magnitude of the difference between groups, was quantified η2. The following criteria were used for interpretation: η2 = 0.02–0.13 = small effect size, 0.13–0.26 = medium and >0.26 large effect size.

2 Results

Exposure to the artificial saliva and probiotic bacteria decreased elastic modulus (E) and increased yield strength (YS) of the NiTi alloys independentlly of the coating of the archwire with large effect size ($p \leq 0.044$; $\eta2 = 0.406–0.980$; Fig. 2). Also flexibility and resilience are decreased with large effect size ($p \leq 0.001$; $\eta2 = 0.809–0.998$; Table 1). Bacteria don't change examined properties of the NiTi alloys above the influence of the saliva on the NiTi archwires.

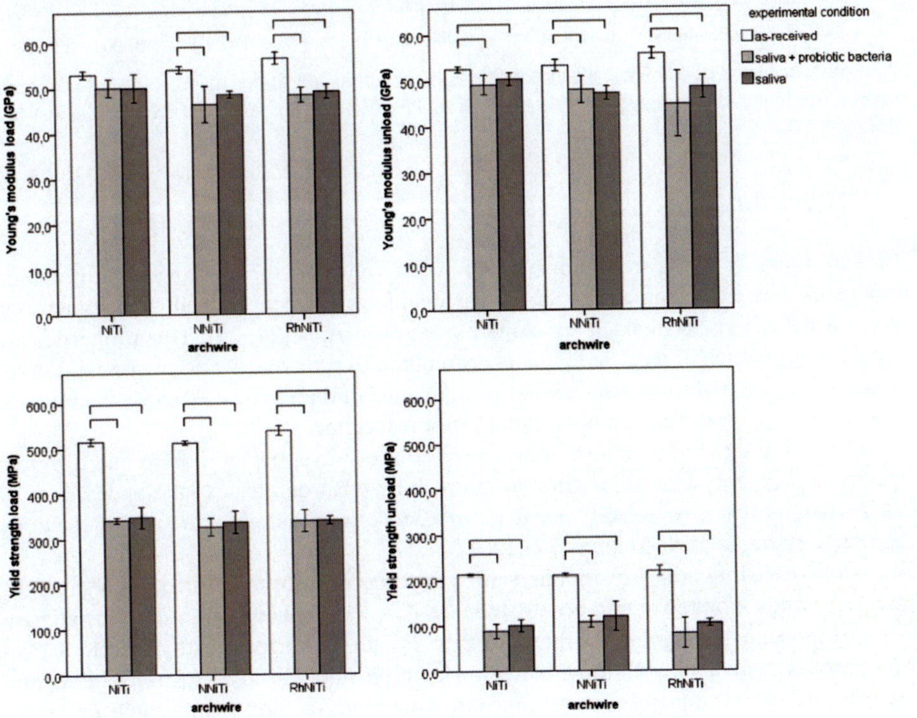

Fig. 2. Young's modulus of elasticity and yieald strength in relation to archiwire type and exposure

Table 1. Flexibility and resilience in relation to archiwire type and exposure

Type	Experimental condition					
	Parameter	As-received	Bacteria + saliva	Saliva	p	η2
NiTi	Flexibility load	9.72 ± 0.21[a]	6.83 ± 0.26[b]	6.97 ± 0.37[b]	<0.001	0.970
	Flexibiltiy unload	4.04 ±0.06[a]	1.78 ± 0.26[b]b	1.99 ± 0.28[b]	<0.001	0.995
	Resilience load	2.50 ± 0.08[a]	1.17 ± 0.05[b]	1.21 ± 0.13[b]	<0.001	0.998
	Resilience unload	0.43 ± 0.01[a]	0.08 ± 0.03[b]	0.10 ± 0.03[b]	<0.001	0.988
NNiTi	Flexibility load	9.50 ± 0.11[a]	7.09 ± 0.90[b]	6.95 ± 0.42[b]	<0.001	0.852
	Flexibiltiy unload	3.95 ± 0.08[a]	2.37 ± 0.19[b]	2.55 ± 0.70[b]	0.001	0.809
	Resilience load	2.43 ± 0.03[a]	1.16 ± 0.20[b]	1.17 ± 0.15[b]	<0.001	0.961
	Resilience unload	0.42 ± 0.01[a]	0.14 ± 0.01[b]	0.16 ± 0.08[b]	<0.001	0.910
RhNiTi	Flexibility load	9.55 ± 0.10[a]	7.01 ± 0.32[b]	6.94 ± 0.18[b]	<0.001	0.981
	Flexibiltiy unload	3.92 ± 0.16[a]	1.83 ± 0.62[b]	2.16 ± 0.11[b]	<0.001	0.908
	Resilience load	2.57 ± 0.05[a]	1.19 ± 0.13[b]	1.18 ± 0.05[b]	<0.001	0.989
	Resilience unload	0.43 ± 0.04[a]	0.08 ± 0.05[b]	0.11 ± 0.01[b]	<0.001	0.959

p-significance level of ANOVA, η2-effect size. Experimental conditions that do not share the same superscript letters differ significantly according to ANOVA and Student-Newman-Keuls post-hoc test.

3 Discussion

Nickel – titanium alloys are the most common used alloys in the first stage of orthodontic treatment. Their mechanical properties are ideal for alignment of teeth: shape memory, low stiffness, high springback and superelastic properties [24, 25]. This properties are ideal for alignamen of teeth in the nivelation phase of the orthodontic treatment. Forces used in this phase of treatment have to be light and continous to avoid side effects of the orthodontic treatment such a bone loss or root resorption.

In the oral cavity this alloys are exposed to the saliva, bacteria, food and different mechanical activity like a chewing, toothbrushing what can lead to the appearance of the corrosion. It was reported that saliva can affect ultimate stress, modulus of elasticity and yield strength of Ni-Ti archwires [26].

Alloys used in dentistry can be changed by different types of corrosion such as galvanic, general, pitting, pointed or microbially induced [27]. General corrosion affecting all metals, and it occurs throughout the metal surface. Localized corrosion affects certain areas more intensely than others. Pitting corrosion is a type of localized corrosion. It penetrates deep into the metal, and only one pit can lead to material fracture. Pitting corrosion on the NiTi surface is mostly manifested in the weak spots of the surface [28, 29].

Our previous research indicates that L. reuteri could induce the occurrence of localized corrosion but only to a small degree [30].

Orthodontic tooh movement depends of the applied force. Nickel Titanium archwires have the ability to return in their original shape after aplication of the force and this propertie is used in orthodontic to produce ideal force for tooth movement. After loading of the NiTi archwires they return in to initial shape but the unloading force is always lesser than loading force because of energy loss. This unloading phase is the most important propertie of the archwire because it is the working phase when the teeth are moved into ideal position.

Any type of corrosion can affect mechanical properties of the materials. Changes in mechanical properties affects the unloading force and thus the outcome of the treatment. It is known that saliva solely can change th properties of the archwires in loadin and unloading [31].

Our hypotesis was that corrosion will be greater in the archwires exposed to the artificial salive with adittion of L. reuteri then in archwires exposed to the saliva solely and that this microbiologically induced corrosion can cause changes in mechanical properties oof tested archwires. Our hypothesis has not been confirmed.

According to our research bacteria L. Reuteri doesn't influence significant on mechanical properties like a flexibility and resilience of tested archwires made of NiTi. Changes which are caused by artificial saliva in regards to unexposed tested archwires were not modiefied by adding probiotc bacteria *L. reuteri.*

4 Conclusion

By the changes caused by saliva, probiotic bacteria do not induce significant additional corrosion of NiTi alloys that would be visible as a change in mechanical properties.

References

1. Humphrey, S.P., Williamson, R.T.: A review of saliva: normal composition, flow, and function. J. Prosthet. Dent. **85**, 162–169 (2001)
2. Arneberg, P., Giertsen, E., Emberland, H., Ogaard, B.: Intra-oral variations in total plaque fluoride related to plaque pH A study in orthodontic patients. Caries Res. **31**, 451–456 (1997)
3. Bass, J.K., Fine, H.F., Cisneros, G.J.: Nickel hypersensitivity in the orthodontic patient. Am. J. Orthod. Dentofacial Orthop. **103**, 280–285 (1993)
4. Kerosuo, H., Kullaa, A., Kerosuo, E., Kanerva, L., Hensten-Pettersen, A.: Nickel allergy in adolescents in relation to orthodontic treatment and piercing of ears. Am. J. Orthod. Dentofacial Orthop. **109**, 148–154 (1996)
5. Setcos, J.C., Babaei-Mahani, A., Silvio, L.D., Mjor, I.A., Wilson, N.H.: The safety of nickel containing dental alloys. Dent. Mater. **22**, 1163–1168 (2006)
6. Stamatova, I., Meurman, J.H.: Probiotics: health benefits in the mouth. Am. J. Dent. **22**, 329–338 (2009)
7. Kazor, C.E., Mitchell, P.M., Lee, A.M., Stokes, L.N., Loesche, W.J., Dewhirst, F.E., et al.: Diversity of bacterial populations on the tongue dorsa of patients with halitosis and healthy patients. J. Clin. Microbiol. **4**, 558–563 (2003)
8. Burton, J.P., Drummond, B.K., Chilcott, C.N., Tagg, J.R., Thomson, W.M., Hale, J.D., et al.: Influence of the probiotic Streptococcus salivarius strain M18 on indices of dental health in children: a randomized double-blind, placebo-controlled trial. J. Med. Microbiol. **62**, 875–884 (2013)
9. Kõll-Klais, P., Mändar, R., Leibur, E., Marcotte, H., Hammarström, L., Mikelsaar, M.: Oral lactobacilli in chronic periodontitis and periodontal health: species composition and antimicrobial activity. Oral. Microbiol. Immunol. **20**, 354–361 (2005)

10. Schlagenhauf, U., Jakob, L., Eigenthaler, M., Segerer, S., Jockel-Schneider, Y., Rehn, M.: Regular consumption of Lactobacillus reuteri-containing lozenges reduces pregnancy gingivitis: an RCT. J. Clin. Periodontol. **43**, 948–954 (2016)
11. Vishnu, H.P.: Probiotics and oral health. In: Virdi, M.S. (ed.) Oral health care – pediatric, research, epidemiology and clinical practices. InTech, pp. 195–204 (2012)
12. Deepak, T., Manjunath, M., Pewa, S.: Antibiotics are passe: take a look at probiotics. World J. Dent. **1**, 109–111 (2010)
13. Kabuki, T., Saito, T., Kawai, Y., Uemura, J., Itoh, T.: Production, purification and characterization of reutericin 6, a bacteriocin with lytic activity produced by Lactobacillus reuteri LA6. Int. J. Food. Microbiol. **34**, 145–156 (1997)
14. Gänzle, M.G., Höltzel, A., Walter, J., Jung, G., Hammes, W.P.: Characterization of reutericyclin produced by Lactobacillus reuteri LTH2584. Appl. Environ. Microbiol. **66**, 4325–4333 (2000)
15. Talarico, T.L., Casas, I.A., Chung, T.C., Dobrogosz, W.J.: Production and isolation of reuterin, a growth inhibitor produced by Lactobacillus reuteri. Antimicrob Agents Chemother. **32**, 1854–1858 (1988)
16. Beech, I.B., Sunner, J.: Biocorrosion: towards understanding interactions between biofilms and metals. Curr. Opin. Biotech. **15**, 181–186 (2004)
17. Geiser, M., Avci, R., Lewandowski, Z.: Microbially initiated pitting on 316L stainless steel. Int. Biodeterior. Biodegrad. **49**, 235–243 (2002)
18. Papadopoulou, K., Eliades, T.: Microbiologically-influenced corrosion of orthodontic alloys: a review of proposed mechanisms and effects. Aust. Orthod. J. **25**, 63–75 (2009)
19. Chang, J.C., Oshida, Y., Gregory, R.L., Andres, C.J., Barco, T.M., Brown, D.T.: Electrochemical study on microbiology-related corrosion of metallic dental materials. Biomed. Mater. Eng. **13**, 281–295 (2003)
20. Graber, T.M., Vanarsdall, R.L., Vig, K.W.L.: Orthodontics: Current Principles and Techniques. Elsevier, New York (2005). str. 361–69
21. Souza, J.C.M., et al.: Corrosion behaviour of titanium in the presence of Streptococcus mutans. J. Dent. **41**, 528–534 (2013)
22. Bahije, L., et al.: Behavior of NiTi in the presence of oral bacteria. Corrosion Streptococcus Mutans Int. Orthod. **9**, 110–119 (2011)
23. Zhang, S.M., Qiu, J., Tian, F., Guo, X.K., Zhang, F.Q., Huang, Q.F.: Corrosion behavior of pure titanium in the presence of Actinomyces naeslundii. J. Mat. Sci. Mater. Med. **24**, 1229–1237 (2013)
24. Miura, F., Mogi, M., Ohura, Y., Hamanaka, H.: The super-elasitc property of the Japanese NiTi alloy wire for use in orthodontics. Am. J. Orthod. Dentofac Orthop. **90**(1), 1–10 (1986)
25. Burstone, C.J., Qin, B., Morton, J.Y.: Chinese NiTi wire – a new orthodontic alloy. Am. J. Or-thod. **87**(6), 445–452 (1985)
26. Harris, E.F., Newman, S.M., Nicholson, J.A.: Ni-tinol arch wire in a simulated oral environment: changes in mechanical properties. Am. J. Orthod. Dentofacial Orthop. **93**(6), 508–513 (1988)
27. Ramazanzadeh, B.A., Ahrari, F., Sabz Eliades, T., Athanasiou, A.E.: In vivo aging of orthodontic alloys: implications for corrosion potential, nickel release, and biocompability. Angle Orthod. **72**, 222–237 (2002)
28. Hu, T., Xin, X.C., Wu, S.L., Chu, C.I., Lu, J., Cuan, I., et al.: Corrosion behavior on orthopedic NiTi alloy with nanocrystalline/amorphous surface. Mater. Chem. Phys. **126**, 102–107 (2011)
29. Mirjalili, M., Momeni, M., Ebrahimi, N., Moayed, M.H.: Comparative study on corrosion behavior of Nitinol and stainless steel orthodontic wires in simulated saliva solution in presence of fluoride ions. Mater. Sci. Eng. C Mater. Biol. Appl. **33**, 2084–2093 (2013)

30. Musa Trolic, I., Serdarevic, N., Todoric, Z., Budimir, A., Spalj, S., Otmacic, C.H.: Corrosion of orthodontic archwires in artificial saliva in the presence of Lactobacillus reuteri. Surf. Coat. Technol. **370**, 44–52 (2019)
31. Hosseinzadeh Nik, T., Ghadirian, H., Ahmadabadi, M.N., Shahhoseini, T., Haj-Fathalian, M.: Effect of saliva on load-deflection characteristics of superelastic nickel-titanium orthodontic wires. J. Dent. (Tehran). **9**, 171–179 (2012)

Immunosensors: Recent Advances and Applications

Selena Hadžić, Anja Trkulja, and Iman Alihodžić(✉)

International Burch University, Francuske Revolucije bb, 71000 Sarajevo, Bosnia and Herzegovina
iman.alihodzic@stu.ibu.edu.ba

Abstract. Immunosensors are solid-state devices that involve the coupling of immunochemical reactions to appropriate transducers. In this review we took a systematic approach to the recent developments in the immunosensor field and summarized all of the data available. The aim was to provide detailed introduction to the updated methods in the field and create a comparison among them. This systematic review was written according to the guidelines proposed by Preferred Reporting Items for Systematic Reviews and Meta-Analyses (PRISMA). In order to collect data, multiple databases were searched and different keywords were used. We also involved several inclusion and exclusion criteria to be sure that our materials fit our article. Our results were presented in the table which shows a comparison between different immunosensor types, their characteristics and applications. Below are results explained in words. Today, immunosensor research is highly focused on the development of PoC devices, since they represent the future and a better alternative to current methods. Recent advances in immunosensors allow for better usage in clinical environments that subsequently result in better and faster analysis and diagnosis. Such advances combined with the low production cost are very promising and will firmly assure the presence of immunosensors in the future.

Keywords: Immunosensor · Antibody · Antigen

1 Introduction

Immunosensors are solid-state devices that involve the coupling of immunochemical reactions to appropriate transducers [29]. Their rapid development in recent decades occurred due to the demand for fast and simple analytical methods that have a wide spectrum of applications. Immunosensors rely on the interaction of specific immunochemical recognition of antibodies immobilized on a transducer to antigens [3].

The history of all biosensors dates back to 1956 when Leland C. Clark, acknowledged as one of the founding fathers of the field of biosensors, described the first biosensor, developed to detect glucose levels in serum samples using a membrane-bound biologically sensitive element. This work evolved from his work on the oxygen electrode, widely known as the Clark electrode [27]. Immunosensors are a special type of biosensor, developed to detect specific antigens or antibodies. They convert the signal generated

A. Badnjevic and L. Gurbeta Pokvić (Eds.): CMBEBIH 2021, IFMBE Proceedings 84, pp. 138–151, 2021.
https://doi.org/10.1007/978-3-030-73909-6_16

by antigen-antibody binding events into a measurable signal, which can be further analyzed [28].The design of immunosensors mainly consists of three parts: a biological recognition element, a physicochemical transducer, and an electronic part. The biological recognition element are usually antibodies or antibody derivatives, which are either integrated or associated with the transducer. The electronic part is used to digitalize the physicochemical output signal from the transducer devices. These can be electrochemical, optical, or microgravimetric devices [3]. Gizeli and Lowe [1] state that the design of an immunosensor should possess the following elements: the ability to detect and quantify the antigens (antibodies), the capacity to transform the binding event without externally added reagents, the ability to repeat the measurement on the same device, and the capacity to detect the specific binding of the antigens (antibodies) in real samples [1]. Recent advances in the development of these sensors using nanoparticles and polymeric nanomaterials have enabled a wider range of functions and approaches in diagnostic and prognostic research. There are several challenges that engineers face while designing an immunosensor. The primary challenge that all immunosensors must overcome is to enable specific recognition of target molecule by immobilizing antibodies and antibody fragments, as well as proteins or oligonucleotides. The second challenge is their size. Efforts have been made to shrink antibody-based tools through the dissociation of antibodies into smaller antigen-binding fragments, making them easier to work with [28]. In this review, we took a systematic approach to the recent developments in the immunosensor field. Motivated by the lack of information currently available for this topic, we summarized all of the data available to us via several search engines and in several journals. The goal was to provide an introduction to the updated methods in the field and create a comparison between them, which can be used as a starting point for future research.

2 Methods and Materials

The systematic review was written adhering to the guidelines proposed by Preferred Reporting Items for Systematic Reviews and Meta-Analyses (PRISMA) [30]. Appropriate studies published between January 1st, 2000 and December 31st, 2019 were searched on the following databases; PubMed, ScienceDirect, Scopus, SpringerLink, ResearchGate and ACS Publications by combining keywords such as "immunosensors" with more specific keywords such as "label-free","fluorescence", "antibody", "antigen", "immobilization", "silver nanoparticles" and "applications". Data were collected based on the following inclusion criteria: 1. the study is an original research article, 2. the study is published in a peer-reviewed journal, 3. the study is published in English and 4. The study is not published earlier than Jan 1st, 2000, nor later than Dec 31st, 2019. The 2 exclusion criteria were defined as follows: 1. The study is a non-original article (review articles, conference proceedings, letters to the editor, commentaries, technical notes, etc.), 2. Articles with no abstract, 3. Incomplete study (no results) and 4. Duplicate publications of the same study (type I and type II). Data collection was done in duplicate to prevent selection bias. After removing 5545 duplicates from 10122 identified articles using Mendeley reference manager software, 4577 titles and abstracts were reviewed by three independent researchers to determine if they fit the exclusion criteria. Following the title and abstract review, 4467 articles were excluded for fitting the exclusion criteria. The full-article review was performed on the remaining 110 articles, and an additional 78 articles were excluded from the study for not fitting all of the inclusion criteria. Finally, the remaining 32 articles were included in the study and underwent systematic review. The study flow chart, presenting the identification of eligible studies with the inclusion and exclusion criteria, is shown in the figure below (Fig. 1 and Table 1).

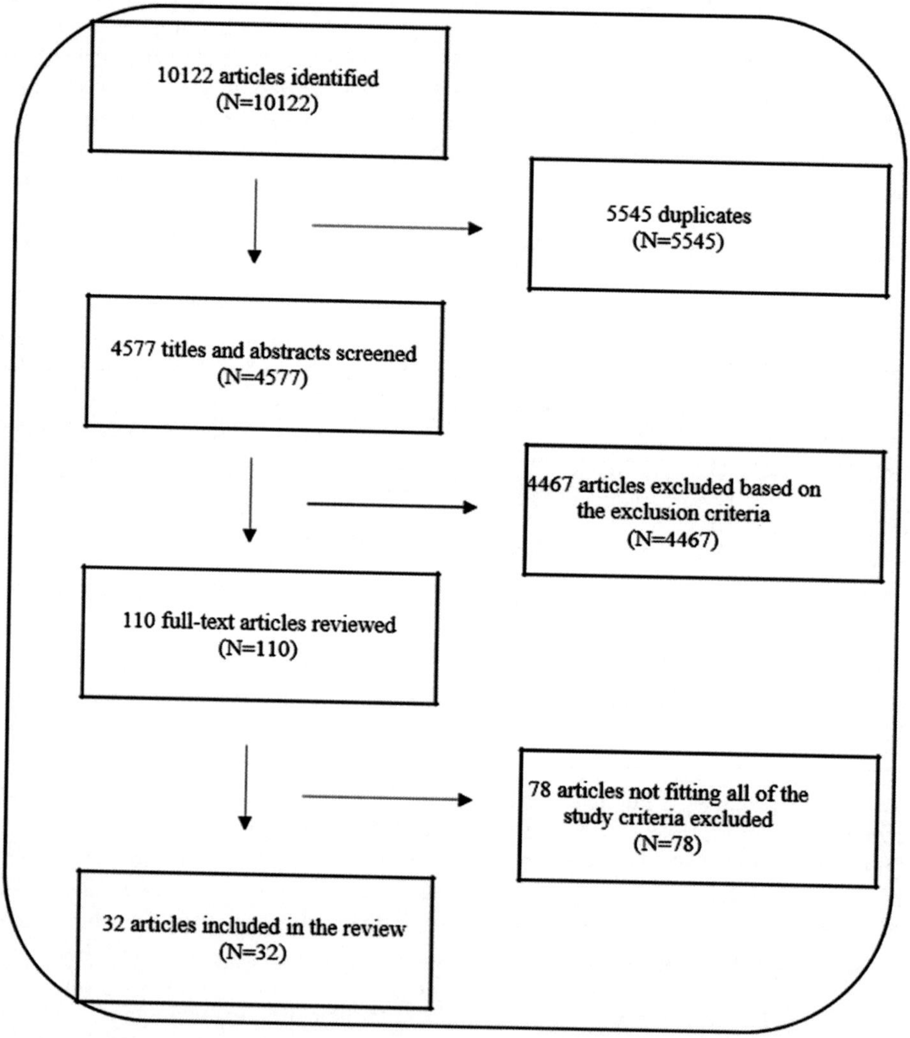

Fig. 1. Study flow chart: identification of eligible studies with inclusion and exclusion criteria

Table 1 Summary of articles in systematic literature review

Target	Sensor	Substrate	Label	Sensitivity	Application	References
Label-free strategies						
Au substrate						
Anti-capsid protein VP1A	Impedance spectroscopy	Au	Label-free	-	Detection of anti-capsid protein VP 1 Ab for enetrovirus detection	Mejri et al. (2009)
Bovine insulin	SPR	Au	Label-free	100 ng mL-1	Detection of bovine insulin	Lee et al. (2005)
Cancer antigen125 (CA 125)	SPR	Au	Label-free	-	Detection of CA 125	Dutra et al. (2007)
Influenza virus	BioDVD	Au	Label-free	10 nM	Typing and sub-typing	Gopinath et al. (2009)
Fibrinogen	Nanostructure	Au	Label-free	0.1 ng mL-1	Ag–Ag interactions	Hiep et al. (2010)
Enzyme alanine aminotransferase (ALT)	Quartz crystal microbalance (QCM) and electrochemical impedance spectroscopy (EIS)	Au	Label-free	QCM-60 ng/ mL, EIS-92 ng/mL	Detection of ALT	Wu et al. (2011)
Cardiac troponin	SPR	AuNP	Label-free	0.1 ng mL-1	Detection of cardiac troponin	Li et al. (2013)
BSA	Piezoelectric-cantilever	Borosilicate	Label-free	1 pg mL-1	Sensing Ab–Ag interaction	Maraldo and Mutharasan (2007)
IgG	Quartz crystal microbalance	Magnetic nanocompostie	Label-free	15 ng mL-1	Ag–Ag interactions	Zhang et al. (2008)
PSMA	BioCD	Silica	Label-free	4 ng mL-1	Detection of PSMA	Wang et al. (2011
Hemagluttinin (HA)	SPR	Dextran	Label-free	-	Detection of HA	Leonard et al. (2011
Palytoxin	SPR	Dextran	Label-free	0.52 ng mL-1	Detection of Palytoxin	Yakes et al. (2011
Glycosylated haemoglobin (HbA1c)	Electrochemical	Glassy carbon	Label-free	4.5%	Detection of HbA1c in Serum	Liu et al. (2012)
Influenza-H1/p53 protein	SERS	SAM of 4-aminothiophenol	Label-free	-	Multiplexdetectionof influenza-H1/p53 protein	Kho et al. (2012)

67

(*continued*)

Table 1 (*continued*)

Anti-hepatitis C virus (HCV) NS5A protein	Amperometric measurement	Silk fibrion	Label-free	28 µA μg^{-1} mL	Detection of Anti-hepatitis C virus (HCV) NS5A protein	Moraes et al. (2013)
Strategies with – substrate	**labeling or tags Au**					
IgG	Impedance spectroscopy	Au	Biotin	2 mg mL^{-1}	Detection of IgG	Pei et al. (2001)
IgG	Sandwich SPR	Au	Au	1 ng mL^{-1}	Detection of IgG	Wang et al. (2009)
IgE	SPR	Au	Au	1 fM	Sensitive detection of IgE	Kim et al. (2010)
IgE	electrochemical	Au	Au and methylene blue	0.52 ng mL^{-1}	Detection of IgE	Wang et al. (2010)
Benzo[a]pyrene (BaP)	Amperometric immunosensor	Au	Enzymatic	5.6 ng mL^{-1}	Detection of BaP in H_2O	Ahmad and Moore (2012)
Bovine leukemia virus antigen gp51	Surface enhanced Raman scattering	Magnetic AuNP	Au	0.95 mg mL^{-1}	Detection of gp51 in milk	Baniukevic et al. (2013)
Influenza virus	Prism-free SPR	Au	Fluorescent	0.2 HAU/ml	Detectionofinfluenza	Nomura et al. (2013
Anti-Brucella sp. antibodies	Fluorescence	Glass	Cyanine-5	0.005 mg mL^{-1}	Detection ofanti-Brucella sp. antibodies	Silva et al. (2004)
IgG	BioCD	Silicon	Fluorescent	30 pg mL^{-1}	Ag–Ag interactions	Wang et al. (2008)
C reactiveprotein (CRP)	Sandwich	AR chips Epoxy	Dye	10 µg mL^{-1}	CRP detection in serum	Pultar et al. (2009)
HA	Waveguide	SiO_2	Dye	1 nM	Detection of influenza	Gopinath et al. (2010)
Carcinoembryon-ic antigen	Chemiluminescence	Mesoporous silica nanoparticles	HRP	0.1 ng/mL	Detection of Carcinoembryonic antigen	Chen et al. (2011
Iduronate-2-sulfate sulfatase	ELISA	Polystyrene	Enzymatic	8 ng mL^{-1}	Immuno-quantification	Sosa et al. (2011)

(*continued*)

Table 1 (*continued*)

IgG against Helicobacter pylori	Chemiluminescence	3-aminopropyl-modified controlled poreglass(AP-CPG)	Alkaline phosphatase (AP)	0.17 U mL^{-1}	Detection of IgG against *Helicobacter pylori* in serum sample	Seia et al. (2012)
Human chorionic gonadotrophic antigen (HCG-Ag)	Fluorescence	Nanoporous PtRu	Quantum dot (QD) CdTe	-	Detection of HCG-Ag	Zhang et al. (2012)
HA	Waveguide	SiO_2	Au	8×10^3 pfu/ml	Detection of influenza	Gopinath et al. (2013b)
Chloramphenicol	Lateral flow immunoassay	Nitrocellulose	Europium	0.25 ng mL^{-1}	Detection of chloraphenicol	Xia et al.(2013)

3 Results

3.1 Major Types of Immunosensors

Major division of immunosensors is determined according to the type of its transducer. Consequently, three main types of immunosensors are used today: electrochemical, optical, and microgravimetric [3].

3.1.1 Electrochemical Immunosensors

As the major group of immunosensors, electrochemical immunosensors possess several advantages: high sensitivity, low cost, and portable design. They work on a principle of electrochemical detection of the labeled immunoagents (tags) such as enzymes, metal ions, or other electroactive compounds [3]. Electrochemical immunosensors work on a principle of immobilization of a recognition element (i.e. antigen or antibody) on the electrode surface. In an electrochemical format, the enzymatic reaction that occurs leads to the production of a molecule which is electroactive, meaning that it gives an electrochemical signal in presence of certain inputs [4]. There are five kinds of transducers used in electrochemical immunosensors: potentiometric, conductometric, capacitive, impeditive, and amperometric. Potentiometric immunosensors depend on a change in potential that occurs between an indicator and a reference electrode as a result of a particular interaction between an antibody and its antigen [3]. Conductometric transducers are the oldest type of electrochemical transducers and are not widely used. Their functioning principle is based on the fact that almost all enzymatic reactions involve either consumption or production of charged species and, thus, lead to a global change in the ionic composition of the tested sample [5]. Capacitance and impedance immunosensors are highly sensitive and widely used. They are based on the principle that the electrolyte capacitance of an electrode depends on the thickness and dielectric behavior of the dielectric layer on the electrode surface and the solid/solution interface. Amperometric immunosensors are the most popular immunosensing devices. They are based on the measurement of the currents resulting from the electrochemical oxidation or reduction of electroactive species at a certain constant voltage [3].

3.1.2 Optical Immunosensors

Optical immunosensors take advantage of the optical phenomena (e.g. adsorption, fluorescence,luminescence) at the sensing surface for biochemical sensing design. The optical immunosensors are divided into two major types: direct optical immunosensors and indirect immunosensors depending upon the use of labeled signaling molecules. They can also be divided into surface plasmon resonance sensors, fluorescence immunosensors,and chemiluminescence sensors [3].

3.1.3 Microgravimetric Immunosensors

The detection principle of these immunosensors is based on adsorbate recognition where the selective binding causes changes in mass loading and interfacial properties. Those are recognized by a corresponding shift in the oscillation frequency. These immunosensors have two types of sensing formats: gas phase and solution phase sensing. Apart from this division, immunosensors can also operate either as direct or indirect ones. Direct immunosensors detect the analytes in real-time. They achieve this by using transducers that directly detect the physical or chemical effects, forming an immunocomplex at the interfaces, using no additional labels (label-free). On the other hand, indirect immunosensors use one or multiple labeled bio-reagents during the detection processes. The transducers then detect the signals from the labels (tags). It was shown that indirect immunosensors could have higher sensitivity [3].

3.2 Functionalization of the Sensor Surface

Direct immobilization of molecules of interest onto the sensing surfaces can be done either by chemical means or physical adsorption. The latter is achieved through complex formation with other molecules or by direct molecule attachment to the surface. The chemical reactions allowing for adsorption of Ab on the surface include hydrophobic interactions, Van der Waals forces, hydrogen bonding, and electrostatic interactions [6]. The crucial criterion of a faithful outcome development on a given sensing plate is surface functionalization, albeit physical adsorption is regarded as non-specific since there is no reaction between the residues, the majority of the times. Furthermore, the reactions resulting from the interaction of different residues in Chemical immobilization offer higher specificity. The general method of chemical immobilization is based on the reactions involving amino acid residues of previously immobilized proteins or Abs and the ability of the groups on the sensing surface to react with amines (such as periodate, glutaraldehyde, isothiocyanate, N-hydroxysuccinimide) and active ester [6]. Given the ability of Au to form semi-covalent bonds with thiol-modified Ab, the immobilization of Abs on Au surface is attainable through the attachment of thiol groups to the Abs, thus earning Au the status of the common platform of sensor development [7]. Alkyl silanes possess reactive groups including aldehydes, thiols, amines, and carboxyl acids which are used in various sensor plates [8]. The fact that silicon materials are used frequently in sensor production allows for amine coupling to persist as one of the most favored functionalization techniques [9].

Namely, surfaces with amine functionalization are able to link with aldehyde groups, e.g. glutaraldehyde (Glu), which has two aldehyde groups forming a peptide bond with the amines, resulting in protein immobilization on the said amine-containing surface [10]. Atomic force microscopy images indicate that the surface chemistry of Glu-modified surfaces is optimal for smaller sized molecules [11]. Recently, with the aims of attaining a highly sensitive and specific detection technique, polyethylene glycol (PEG) polymers were used on a Glu-modified surface [12]. Moreover, Self-assembled monolayers (SAMs) proved to be significant in both surface modification and specific reactions on the sensing surface [13]. They are assembled upon spontaneous absorption of molecules containing the functional groups on the surface, chiefly on the surface of Au. The immobilization of Ab onto the surface is ensured through the interactions of the functional groups e.g. COOH. Thus, the surface properties (e.g. charge of the surface) can be preserved through the utilization of a SAM, in order not to alter the immobilized Abs orientation [14]. COOH-SAMs were employed in Surface Plasmon Resonance (SPR) prism-free sensor surface, for immobilization of anti-influenza Abs on an Au surface. Various sizes of gold nanoparticles (AuNP) conjugated with streptavidin were utilized to evaluate the aptness of surface modifications for different molecules. Out of the surface chemistries tested, N, N0-carbonyldiimidazole (CDI) was found to be the most apt for molecules with higher molecular weight (<20 nm).Therefore, influenza viral (100 nm in size) detections havebeen performed using Ab [12]. In spite of various surface chemistries that have been proposed, the Ab-based sensor sensitivity remains largely dependent on the orientation of the immobilized Abs. Unsolved issues preventing Abs from establishing a stable interaction with Ag have resulted in very few commercial implementations of immunosensors [15]. These issues encompass immobilization, conformation, quantity, density, and orientation, all of them factors dependent on the immobilization technique, sensing surface, and surface characteristics [6]. The performance of any given sensor is expected to increase with the proper orientation of Abs immobilized on the surface [14]. Regarding orientation on the sensing surface, there are three possible ways for the immobilization of Abs. (1) exposure of the fragment antigen-binding region (Fab); (2) non-exposure of the Fab; and (3) partial exposure of the fab; (Fig). Abs with proper orientation are usually immobilized through linkers, e.g. N-hydroxysuccinimide (NHS) and N-ethyl-N0-(3-dimethylaminopropyl) carbodiimide hydrochloride (EDC), via linkage to the carboxyl-terminus of the Abs, on an amine-containing surface. The method, based on EDC-NHS coupling, converts Ab into a highly reactive NHS ester intermediate, which forms a peptide bond with the amine-functionalized surface, facilitating the immobilization of Ab. In this manner, highly specific and sensitive binding is achieved due to the Ag-binding region of the Ab being exposed. It can be further evaluated on Glu or Ab-immobilized surfaces interacting with the Ag.

Capturing the fragment crystallizable (Fc) region of a given Ab is another way to achieve the proper orientation, by utilizing another probe or an aptamer on the sensing surface [16]. In a similar fashion, the proteins A and G can serve as the capturing agent for properly orienting the Abs, since they bind the Fc region of Abs. Analysis of the immobilized Abs on amine- and carboxyl- (positively and negatively charged, respectively) terminated SAMs, that IgG1 had better orientation on amine-terminated SAMs [14]. However, it was found that this was due to the surface charges of Ab acting as the controlling agent of the Ab orientation since IgG2 was found to have equal adsorption on both surfaces. Light assisted antibody immobilization utilizing UV irradiation, was implemented to demonstrate the absorption of irradiation by tryptophan residues and its disruptive effect on the disulfide bridges of the nearby thiol groups, leading to the proper orientation of Abs on the surface due to the interaction of thiol groups with the Au surface [17]. Furthermore, polymers such as dextran and poly (ethylene glycol)-b-poly (acrylic acid) (PEG-bPAAc) and pentaethylenehexamine-terminated PEG (N6-PEG)] (PEG) have a significant role in proper molecule orientation, so by blocking the polymers on the SiO2 surface, proper orientation was actualized and the sensitivity was increased up to thousand-fold [12]. In order to enhance its signal-to-noise ratio and maximize its performance, a quality immunosensor should minimize biofouling. For this purpose, multiple blocking agents enhance the sensor's specificity through proper ligand alignment to the sensing surface. Still, the usual blocking agents (e.g. casein, BSA, ethanolamine) are not apt for every sensing surface. Alternatively, the alleviation of non-specificity can be carried out by several polymers that can reduce non-specific binding. Such polymers include synthetic polymers, and densely packed PEG tethered chain surfaces [6, 18]. Moreover, PEG dual chains, a combination of a long and a short PEG, significantly reduce non-specific adsorption of proteins, evident in the upsurge insensitivity [19]. Different PEG block copolymers have been designed including PEG-b-polycations, PEG-b-polyanions, and PEG-b-hydrophobic polymers, all of which are versatile materials for immunosensing applications.

3.3 Signal Enhancement

3.3.1 Emission Assisted Signaling

Fluorescent and chemiluminescent labeling represent the two most common methods of augmenting sensitivity, the former being the most common one due to its non-destructive and sensitive properties. Its sources are fluorophores (e.g. fluorescein isothiocyanate, rhodamine, coumarin, and cyanine) which allow monitoring of Ag-Ab interactions in real-time. Using monoclonal mouse anti-goat IgG, conjugated with cyanine-5 in an immune-homogeneous assay with fluorescent Abs, IgG was detected by optical fibers [20]. Moreover, the authors demonstrated that anti-Brucella sp. Abs are detectable in 0.005–0.11 mgml_1 range as the proof of concept. Quantification of IgG Abs against Helicobacter pylori was done by utilizing secondary Abs conjugated with the alkaline-phosphatase enzyme, coupled with laser-induced fluorescence. The detection limit of the portable system is 0.17 UmL_1 [21]. Furthermore, another chemiluminescence-based portable system utilizes horseradish peroxidase (HRP) to catalyze the substrate luminol, forming 3aminophthalate. Alternatively, Quantum Dot (QD) is more suitable when a system with a longer fluorescence lifetime, less susceptibility to environmental quenching, and higher quantum yield is needed, than fluorophores-based systems. QD's narrow emission spectra and broad excitation range with high and near fluorescence quantum yield, all stem from its long fluorescent lifetime.

3.3.2 Nanoparticle Assisted Signal Amplifications

Out of the various metal nanoparticles (NPs) in the development of immunosensors, AuNP and AgNP (silver nanoparticles) are the most commonly used ones. The NP-conjugated immunosensing encompasses Au, platinum NPs, platinum NPs, europium NPs, and AgNPs, and inorganic NPs [12]. The sensitivity is improved by increasing the particle size, which in turn increases the surface area, providing accommodation for more Abs [14]. An analysis of Ab attachment to AuNPs of the following sizes: 14, 16, 35, and 38 nm, found that to coat the smaller NPs (16 nm), a lower amount of Ab is required. Moreover, the AuNP-conjugated Ab typically lower migration rate, and higher molecular size than that of a bare AuNP. A 106 – fold increase in sensitivity was recorded for the aptamer-IgE-Abs reactions when using AuNP conjugates [22]. 10nm AuNP-conjugated Abs of influenza virus were used for the analysis of the influenza subtypes, with the subtypes detected at a concentration of 8 x 103 pfu/ml [12]. Another investigation in which HRP and Ab against carcinoembryonic Ag were immobilized on the mesoporous silica nanoparticles surface, detection of Ag up to 0.1–40 ng/mL was enabled [14]. Biotinylated Ab against 6 troponin and immobilized on the surface of luminolfunctionalized AuNP, is used for detection of that very protein (Li et al. 2013). A linkage was formed between the with the biotinylated Ab and the streptavidin-coated AuNPs on the modified Au surface electrodes, resulting in an interaction with the cardiac troponin. The detection limit reached 0.1 ng/mL-1 [23]. Despite the ability of larger NPs to improve the system's sensitivity, it has been demonstrated that they are more prone to non-specificity than their smaller counterparts [14]. Thus the report of non-specificity associated with AuNPs of a larger diameter. Immunochromatography test (ICT) is a rapid diagnostic system for direct visualization, at the basis of which are metal NPs. In such a system, the binding of Ag to Ab-conjugated metal NPs results in amplification of signal accompanied by colored bands production. Home diagnostic tests are commonly ICT-based. The first ICT-based home pregnancy test was designed to detect human chorionic gonadotrophin (HCG)). A europium NP-based lateral flow immune assay was recently proposed for the detection of chloramphenicol. QD CdTe, along with Ab against HCG-Ag was immobilized on the surface of nanoporous PtRu alloy, for enhanced signal amplification. Glu linked with Ab was used for the immobilization on the chitosan-coated Fe3O4 magnetic NPs' surface. The NPs' magnetic properties allowed for the assembly and separation of Ab on the indium thin oxide glass surface. The detected signal amplification was approx. 4.67-fold higher than in the unamplified QD [24].

4 Conclusion

Immunosensor research is focused on the development of PoC devices to allow prompt management of a particular issue whether commercial, medical, environmental, or security by non-trained personnel [25]. They represent the future and a better alternative to current methods. Recent advances allow for better usage in clinical environments that subsequently result in better and faster analysis and diagnosis. The main improvements focus on biocompatibility and sensitivity of transducers [26]. The significantly reduced production cost and portability allow them to be used in rural areas with scarce resources [14]. As stated by Holford et al. (2012), for any immunosensor to play a significant role in monitoring analytes, it must be of low unit cost, subjectable to automation, reusable, and easy to use. The before mentioned advances combined with the cheap production cost allow for a future proof method of analysis with a large room for improvement that has substantial space for growth which will surely assure the presence of immunosensors in the future.

Declaration of Conflict of Interest. The authors declare no conflict of interest in preparing this article.

References

1. Gizeli, E., Lowe, C.R.: Immunosensors. Curr. Opin. Biotechnol. **7**(1), 66–71 (1996, in press). https://doi.org/10.1016/s0958-1669(96)80097-8
2. Davey, J., Lord, M.: Essential Cell Biology. Oxford University Press, Oxford (2003)
3. Wang, H., Shen, G., Yu, R.: Aspects of recent development of immunosensors. Electrochem. Sens. Biosens. Their Biomed. Appl. 237–260 (2008, in press). https://doi.org/10.1016/b978-012373738-0.50011-8
4. Ricci, F., Adornetto, G., Palleschi, G.: A review of experimental aspects of electrochemical immunosensors. Electrochim. Acta **84**, 74–83 (2012). https://doi.org/10.1016/j.electacta.2012.06.033.inpress
5. Jaffrezic-Renault, N., Dzyadevych, S.: Conductometric microbiosensors for environmental monitoring. Sensors, **8**(4), 2569–2588 (2008, in press). https://doi.org/10.3390/s8042569
6. Yoshimoto, K., Nishio, M., Sugasawa, H., Nagasaki, Y.: Direct observation of adsorption-induced inactivation of antibody fragments surrounded by mixed-PEG layer on a gold surface. J. Am. Chem. Soc. **132**(23), 7982–7989 (2010, in press). https://doi.org/10.1021/ja910372e
7. Gopinath, S.C.B., Awazu, K.K.R., Fons, P., Tominaga, J., Kumar, P.: A sensitive multilayered structure suitable for biosensing on the BioDVD platform. Analyt. Chem. **81**(12), 4963–4970 (2009, in press). https://doi.org/10.1021/ac802757z
8. Mukundan, H., et al.: Waveguide-based biosensors for pathogen detection. Sensors **9**(7), 5783–5809 (2009, in press). https://doi.org/10.3390/s90705783
9. Awazu, K., Rockstuhl, C., Fujimaki, M., Fukuda, N., Tominaga, J., Komatsubara, T., Ikeda, T., Ohki, Y.: High sensitivity sensors made of perforated waveguides. Opt. Express **15**(5), 2592 (2007, in press). https://doi.org/10.1364/oe.15.002592
10. Gopinath, S.C.B., Awazu, K., Fujimaki, M.: Detection of influenza viruses by a waveguide-mode sensor. Anal. Methods **2**(12), 1880 (2010, in press). https://doi.org/10.1039/c0ay00491j
11. Gopinath, S.C.B., Awazu, K., Fujimaki, M., Shimizu, K., Mizutani, W., Tsukagoshi, K.: Surface functionalization chemistries on highly sensitive silica-based sensor chips. The Analyst **137**(15), 3520 (2012, in press). https://doi.org/10.1039/c2an35159e
12. Lakshmipriya, T., Fujimaki, M., Gopinath, S.C.B., Awazu, K., Horiguchi, Y., Nagasaki, Y.: A high-performance waveguide-mode biosensor for detection of factor IX using PEG-based blocking agents to suppress non-specific binding and improve sensitivity. Analyst **138**(10), 2863 (2013, in press). https://doi.org/10.1039/c3an00298e
13. Iyer, S.S., Anderson, A.S., Reed, S., Swanson, B., Schmidt, J.G.: Synthesis of orthogonal end functionalized oligoethylene glycols of defined lengths. Tetrahedron Lett. **45**(22), 4285–4288 (2004, in press). https://doi.org/10.1016/j.tetlet.2004.04.004
14. Chen, S., Liu, L., Zhou, J., Jiang, S.: Controlling antibody orientation on charged self-assembled monolayers. Langmuir **19**(7), 2859–2864 (2003, in press). https://doi.org/10.1021/la026498v

15. Luppa, P.B., Sokoll, L.J., Chan, D.W.: Immunosensors—principles and applications to clinical chemistry. Clin. Chim. Acta **314**(1–2), 1–26 (2001, in press). https://doi.org/10.1016/s00098981(01)00629-5
16. Hiep, H.M., Saito, M., Nakamura, Y., Tamiya, E.: RNA aptamer-based optical nanostructured sensor for highly sensitive and label-free detection of antigen–antibody reactions. Anal. Bioanal. Chem. **396**(7), 2575–2581 (2010, in press). https://doi.org/10.1007/s00216010-3488-z
17. Ventura, B.D., Schiavo, L., Altucci, C., Esposito, R., Velotta, R.: Light assisted antibody immobilization for bio-sensing. Biomed. Opt. Express **2**(11), 3223 (2011, in press). https://doi.org/10.1364/boe.2.003223
18. Kamimura, M., Miyamoto, D., Saito, Y., Soga, K., Nagasaki, Y.: Design of poly(ethylene glycol)/streptavidin coimmobilized upconversion nanophosphors and their application to fluorescence biolabeling. Langmuir **24**(16), 8864–8870 (2008, in press). https://doi.org/10.1021/la801056c
19. Uchida, K., Otsuka, H., Kaneko, M., Kataoka, K., Nagasaki, Y.: A reactive poly(ethylene glycol) layer to achieve specific surface plasmon resonance sensing with a high S/N ratio: the substantial role of a short underbrushed PEG layer in minimizing nonspecific adsorption. Anal. Chem. **77**(4), 1075–1080 (2005, in press). https://doi.org/10.1021/ac0486140
20. Zeng, X., Shen, Z., Mernaugh, R.: Recombinant antibodies and their use in biosensors. Anal. Bioanal. Chem. **402**(10), 3027–3038 (2011, in press). https://doi.org/10.1007/s00216-011-5569-z
21. Seia, M.A., Pereira, S.V., Fontán, C.A., Vito, I.E.D., Messina, G.A., Raba, J.: Laser-induced fluorescence integrated in a microfluidic immunosensor for quantification of human serum IgG antibodies to Helicobacter pylori. Sens. Actuat. B: Chem. **168**, 297–302 (2012, in press)
22. Kim, S., Lee, J., Lee, S.J., Lee, H.J.: Ultra-sensitive detection of IgE using biofunctionalized nanoparticle-enhanced SPR. Talanta **81**(4–5), 1755–1759 (2010, in press). https://doi.org/10.1016/j.talanta.2010.03.036
23. Li, F., Yu, Y., Cui, H., Yang, D., Bian, Z.: Label-free electrochemiluminescence immunosensor for cardiac troponin I using luminol functionalized gold nanoparticles as a sensing platform. Analyst **138**(6), 1844 (2013, in press). https://doi.org/10.1039/c3an36805j
24. Zhang, Y., et al.: Magnetic beads-based electrochemiluminescence immunosensor for determination of cancer markers using quantum dot functionalized PtRu alloys as labels. Analyst **137**(9), 2176 (2012, in press). https://doi.org/10.1039/c2an16170b
25. Holford, T.R., Davis, F., Higson, S.P.: Recent trends in antibody based sensors. Biosens. Bioelectron. **34**(1), 12–24 (2012, in press). https://doi.org/10.1016/j.bios.2011.10.023
26. Balahura, L.-R., Van Staden, R.-I., Van Staden, J., Aboul-Enein, H.Y.: Advances in immunosensors for clinical for clinical applications. J. Immunoassay Immunochem. **40**(1), 40–51 (2018, in press). https://doi.org/10.1080/15321819.2018.1543704
27. Heineman, W.R., Jensen, W.B. Leland, C. Clark, Jr. (1918–2005) Biosens. Bioelectron **21**, 1403–1404 (2006). https://doi.org/10.1016/j.bios.2005.12.005.
28. Li, Z., Chen, G.: Current conjugation methods for immunosensors. Nanomaterials **8**(5), 278 (2018). https://doi.org/10.3390/nano8050278
29. Cristea, C., Florea, A., Tertiș, M., & Săndulescu, R.: Immunosensors. In: Rinken, T. (ed.), Biosensors - Micro and Nanoscale Applications, pp. 166–202 (2015). https://doi.org/10.5772/60524
30. Moher, D., Liberati, A., Tetzlaff, J., Altman, D.G.: Preferred reporting items for systematic reviews and meta-analyses: the PRISMA statement. J. Clin. Epidemiol. **62**(10), 1006–1012 (2009). https://doi.org/10.1016/j.jclinepi.2009.06.005
31. Mejri, M., et al.: Electrochemical impedance spectroscopy for specific detection of enterovirus. Sens. Lett. **7**(5), 896–899 (2009). https://doi.org/10.1166/sl.2009.1168

32. Lee, W., Oh, B.K., Lee, W.H., Choi, J.W.: Immobilization fo antibody fragment for immunosensor application based on surface plasmon resonance. Colloids Surfaces B: Interfaces **40**(3–4), 143–148 (2005). https://doi.org/10.1016/j.colsurfb.2004.10.021
33. Dutra, R.F., Mendes, R.K., Lins da Silva, V., Kubota, L.T.: Surface plasmon resonance immunosensor for human cardiac troponin T based on self-assembled monolayer. J. Pharm. Biomed. Anal. **43**(5), 1744–1750 (2007). https://doi.org/10.1016/j.jpba.2006.12.013
34. Wu, J., Park, J.P., Dooley, K., Cropek, D.M., West, A.C., Banta, S.: Rapid development of new protein biosensors utilizing peptides obtained via phage display. PLoS ONE **6**(10), e24948 (2011). https://doi.org/10.1371/journal.pone.0024948
35. Maraldo, D., Mutharasan, R.: Optimization of antibody immobilization for sensing using piezoelectrically excited-millimeter-sized cantilever (PEMC) sensors. Sens. Actuat. B: Chem. **123**(1), 474–479 (2007). https://doi.org/10.1016/j.snb.2006.09.034
36. Zhang, Y., et al.: A reusable piezoelectric immunosensor using antibody-adsorbed magnetic nanocomposite. J. Immunol. Methods **332**(1–2), 103–111 (2008). https://doi.org/10.1016/j.jim.2007.12.019
37. Wang, X., Zhao, M., Nolte, D.D., Ratliff, T.L.: Prostate specific antigen detection in patient sera by fluorescence-free BioCD protein array. Biosens. Bioelectron. **26**(5), 1871–1875 (2011). https://doi.org/10.1016/j.bios.2010.02.009
38. Leonard, P., Hearty, S., O'Kennedy, R.: Measuring protein-protein interactions using biacore. In: Walls D., Loughran S. (eds.), Protein chromatography. Methods in Molecular Biology (Methods and Protocols), vol. 681, pp. 403–418 (2011). https://doi.org/10.1007/978-1-60761913-0_22
39. Yakes, B.J., DeGrasse, S.L., Poli, M., Deeds, J.R.: Antibody characterization and immunoassays for palytoxin using an SPR biosensor. Anal. Bioanal. Chem. **400**, 2865–2869 (2011). https://doi.org/10.1007/s00216-011-5019-y
40. Liu, G., Khor, S.M., Iyengar, S.G., Gooding, J.J.: Development of an electrochemical immunosensor for the detection of HbA1c in serum. Analyst **137**(4), 829–832 (2012). https://doi.org/10.1039/C2AN16034J
41. Kho, K.W., Dinish, U.S., Kumar, A., Olivo, M.: Frequency shifts in SERS for biosensing. ACS Nano **6**(6), 4892–4902 (2012). https://doi.org/10.1021/nn300352b
42. Moraes, M.L., Lima, L.R., Silva, R.R., Cavicchioli, M., Ribeiro, S.J.: Immunosensor based on immobilization of antigenic peptide NS5A-1 from HCV and silk fibroin in nanostructured films. Langmuir **29**(11), 3829–3834 (2013). https://doi.org/10.1021/la304404v
43. Pei, R., Cheng, Z., Wang, E., Yang, X.: Amplification of antigen-antibody interactions based on biotin labeled protein-streptavidin network complex using impedance spectroscopy. Biosens. Bioelectron. **16**(6), 355–361 (2001). https://doi.org/10.1016/S09565663(01)00150-6
44. Wang, R., et al.: Interdigitated array microelectrode based impedance immunosensor for detection of avian influenza virus H5N1. Talanta **79**(2), 159–164 (2009). https://doi.org/10.1016/j.talanta.2009.03.017
45. Wang, J., Munir, A., Li, Z., Zhou, H.S.: Aptamer-Au NPs conjugatesaccumulated methylene blue for the sensitive electrochemical immunoassay of protein. Talanta **81**(1–2), 63–67 (2010). https://doi.org/10.1016/j.talanta.2009.11.035
46. Ahmad, A., Moore, E.: Electrochemical immunosensor modified with selfassembled monolayer of 11-mercaptoundecanoic acid on gold electrodes for detection of benzo[*a*]pyrene in water. Analyst **137**(24), 5839–5844 (2012). https://doi.org/10.1039/C2AN35236B
47. Baniukevic, J., Boyaci, I.H., Bozkurt, A.G., Tamer, U., Ramanavicius, A., Ramanaviciene, A.: Magnetic gold nanoparticles in SERS-based sandwich immunoassay for antigen detection by well oriented antibodies. Biosens. Bioelectron. **43**(15), 281–288 (2013). https://doi.org/10.1016/j.bios.2012.12.014

48. Nomura, K., Gopinath, S.C.B., Lakshmipriya, T., Fukuda, N., Wang, X., Fujimaki, M.: An angular fluidic channel for prism-free surface-plasmon-assisted fluorescence capturing. Nature Commun. **4**, 2855 (2013). https://doi.org/10.1038/ncomms3855
49. Silva, M., Cruz, H., Rossetti, O., Arese, A., Oliva, A.: Development of an optical immunosensor based on the fluorescence of Cyanine-5 for veterinarian diagnostics. Biotech. Lett. **26**, 993–997 (2004). https://doi.org/10.1023/B:BILE.0000030046.96039.e8
50. Pultar, J., Sauer, U., Domnanich, P., Preininger, C.: Aptamer-antibody on-chip sandwich immunoassay for detection of CRP in spiked serum. Biosens. Bioelectron. **24**(5), 1456–1461 (2009). https://doi.org/10.1016/j.bios.2008.08.052
51. Chen, L., Zhang, Z., Zhang, P., Zhang, X., Fu, A.: An ultra-sensitive chemiluminescence immunosensor of carcinoembryonic antigen using HRP-functionalized mesoporus silica nanoparticles as labels. Sens. Actuat. B: Chem. **155**(2), 557–561 (2011). https://doi.org/10.1016/j.snb.2011.01.007
52. Sosa, A.C., et al.: Development of a sandwich enzyme linked immunosorbent assay (ELISA) for the quantification of iduronate-2-sulfate sulfatase. J. Immunol. Methods **368**(1–2), 64–70 (2011). https://doi.org/10.1016/j.jim.2011.03.004
53. Gopinath, S.C., Awazu, K., Fujimaki, M., Shimizu, K., Shima, T.: Observation of immunogold conjugates on influenza viruses using waveguide-mode sensors. PLoS ONE **8**(7), e69121 (2013). https://doi.org/10.1371/journal.pone.0069121
54. Xia, X., et al.: A highly sensitive europium nanoparticle-based lateral flow immunoassay for detection of chloramphenicol residue. Anal. Bioanal. Chem. **405**, 7541–7544 (2013). https://doi.org/10.1007/s00216013-7210-9

Artificial Intelligence and Machine Learning in Healthcare

Prediction of Multi-drug Resistance in Escherichia Coli Using Machine Learning Algorithms

Emina Imamović[1](✉), Amar Deumić[1,7], Abdulrahman Khouly[1], Kursat Talha Pisil[1], Elida Avdić[2], Mirsada Hukić[3,4], Sanja Jakovac[5,6], and Monia Avdić[1]

[1] International Burch University, Sarajevo, Bosnia and Herzegovina
emina.imamovic@stu.ibu.edu.ba
[2] Cantonal Hospital Mostar, Mostar, Bosnia and Herzegovina
[3] Academy of Sciences and Arts of Bosnia and Herzegovina, Sarajevo, Bosnia and Herzegovina
[4] Institute for Biomedical Diagnostics and Research Nalaz, Sarajevo, Bosnia and Herzegovina
[5] Department of Microbiology and Molecular Diagnostics, University Clinical Hospital Mostar, Mostar, Bosnia and Herzegovina
[6] Faculty of Medicine, University of Mostar, Mostar, Bosnia and Herzegovina
[7] Verlab - Medical Devices Verification Laboratory, Sarajevo, Bosnia and Herzegovina

Abstract. Extensive usage of antibiotics has led to the development of antibiotic resistance in bacteria. Multidrug resistant bacteria (MDR), or bacteria that are resistant to more than three classes of antibiotics, are becoming a major issue today. Standardized antibiotics for infection treatment are prescribed, as advised by European Committee for Antimicrobial Susceptibility Testing (EUCAST), however they may differ slightly from region to region depending on the resistance profiles of the bacteria there. Hence, the use of Machine Learning (ML) algorithms in prediction of MDR in bacteria would aid physicians to interpret the results of antibiograms. This paper presents the results of comparison of four different ML algorithms for prediction of antibiotic resistance of *Escherichia coli* (*E. coli*). For the analysis, a database of 94 samples was used. All samples were collected in three different medical centers in Bosnia and Herzegovina (Mostar and Sarajevo). The Naïve Bayes based classifier shows the best potential to be used for prediction of MDR of *E. coli* since it resulted in 86.20% accuracy of prediction, 86.20% precision and 81.81% sensitivity. The results of this study are promising since they can be used in healthcare institutions for creating MDR databases as well as contribute to application of novel strategies in cases where MDR is predicted.

Keywords: Machine learning · Prediction · Antimicrobial resistance · Bacterial infection · *Escherichia coli*

1 Introduction

Extensive usage and misusage of antibiotics has led to the development of antibiotic resistance in bacteria. Multi-drug resistance (MDR) is a term that defines bacteria resistant to multiple classes of antibiotics [1]. The emergence of MDR bacteria has led to

A. Badnjevic and L. Gurbeta Pokvić (Eds.): CMBEBIH 2021, IFMBE Proceedings 84, pp. 155–163, 2021.
https://doi.org/10.1007/978-3-030-73909-6_17

antibiotic-treatment failure and is the cause of over 700,000 deaths annually [2, 3]. The lack of alternative solutions to treat multi-drug resistant bacteria leads to an increase of untreatable infections [4]. It is predicted that, if the usage of antibiotics and treatment of diseases are not changed, the number of deaths caused by antibiotic resistant bacteria will go up to 10 million per year by 2050 [5].

Traditionally, an essential step before recommending antibiotic therapy is performing antimicrobial susceptibility testing using Kirby-Bauer disk diffusion method. In this test, zones of inhibition, after overnight incubation, are compared to standards recommended by European Committee for Antimicrobial Susceptibility Testing (EUCAST) and Clinical and Laboratory Standards Institute (CLSI) [6–8].

On the other hand, the use of artificial intelligence (AI) in microbiology has a potential to revolutionize the field and shorten the time needed to draw conclusions from standardized tests by processing vast amount of data in a short period of time. The quality and efficiency of patient care has significantly improved with introduction of AI tools in medical diagnosis [9–13] and treatment [14–18]. Machine learning (ML) is a field of AI used for dataset analysis, with a general idea of training an algorithm to predict an output [11].

The application of ML algorithms for solving the problem of antibiotic resistance has been investigated earlier. Genomes of antibiotic-resistant *Pseudomonas aeruginosa* isolates were used by Khaledi et al. [18] to train machine learning classifiers and produce models that identify resistance biomarkers to four commonly prescribed antibiotics. Obtained results showed high sensitivity and predictive values that could be used to improve diagnostic performance. Whole genome sequencing was also used in the study performed by Kim et al. [19] that developed a new bioinformatics tool used to build models for prediction of antibiotic resistance. Prediction models achieved high accuracies through nested cross validation and showed potential for use in healthcare systems. Also, Hicks et al. [20] performed meta-analysis of *Neisseria gonorrhoeae, Klebsiella pneumoniae* and *Acinetobacter baumannii* datasets by employing three different machine learning algorithms in order to predict resistance of the aforementioned bacteria. Their findings emphasize the importance of incorporating biological and epidemiological knowledge when generating a machine learning model, since that allows modeling for individual drugs, pathogens and clinical populations. *E. coli* was also the subject of research in the study of Sharaha et al. [21], which combined infrared microscopy together with machine learning to examine *E. coli* susceptibility to antibiotics. The results of their study showed high accuracy, which could be used to reduce the time needed to identify antibiotic resistance of *E. coli* and help determine optimum choice of antibiotic to be applied.

Problem of antibiotic resistance is developing rapidly, therefore new methods for testing antibiotic sensitivity are needed to improve the quality of treatment in healthcare institutions. The aim of our study is to predict multi-drug resistance of *E. coli* strains based on the results of their antibiogram using multiple machine learning algorithms: Naïve Bayes, Logistic Regression and J48, and to see which yields the highest accuracy, precision and sensitivity.

2 Materials and Methods

2.1 Dataset

The dataset used in the study was obtained from three different medical centers in Bosnia and Herzegovina: University Clinical Hospital of Mostar, Cantonal Hospital of Mostar and Center for diagnostics and research Nalaz Sarajevo. These three medical centers are located into two geographical regions of Bosnia and Herzegovina. All samples in the dataset were acquired in the period of 2016–2017 by medical specialists. A total of 3140 of *Escherichia coli* isolates from urine specimens were analyzed in order to generate the database that is analyzed in this research.

Each sample in the database represents bacteria antibiogram. The identification and characterization of bacteria was done by performing biochemical tests established as a routine for this purpose (oxidase, indole, lactose and beta-glucuronidase tests) [22]. Antibiotic susceptibility testing was done using Kirby-Bauer disk diffusion test [6] and was later compared to EUCAST standards [8].

In the study, antimicrobial resistance of Escherichia coli isolates was tested for 18 antibiotics as follows: *(1) Amoxicillin-clavulanic acid, (2) Amoxicillin, (3) Ampicillin, (4) Cefuroxime, (5) Ceftazidime, (6) Cefixime, (7) Ceftriaxone, (8) Cefotaxime, (9) Cefepime, (10) Imipenem, (11) Meropenem, (12) Colistin, (13) Amikacin, (14) Gentamicin, (15) Ciprofloxacin, (16) Levofloxacin, (17) Nitrofurantoin and (18) Trimethoprim / Sulfamethoxsazole.*

Collected database was pre-processed in terms that samples from the same patients in reoccurring infections were not taken into consideration if they had the same antibiogram, in order to prevent overlapping and overfitting. Also, filtering was applied to remove the duplicated samples. At the end, the total number of samples used for building predictors based on machine learning algorithms was 94 samples. Table 1. shows the division of the used dataset according to resistance and sensitivity of samples to different antibiotic classes.

Table 1. Multi-drug resistance/sensitivity to different antibiotic classes

Escherichia coli	MDR	Not MDR
Sample (Total number = 94)	36 38.29 (%)	58 61.71 (%)

In this database, the classification is performed to two classes, as it can be seen from Table 1: MDR or Not MDR. A sample was classified as MDR if it was resistant to three or more antibiotic classes and it was classified as Not MDR if it was not resistant to three or more antibiotic classes.

According to standard principles applied in similar studies [18, 24], overall dataset was divided into training and validation datasets. Since this is relatively small dataset, the distribution of 70/30 was used, according to the practice [18, 24]. The training dataset

Table 2. Data distribution for system development

Escherichia coli	MDR	Not MDR
Total no. of samples 94 (100%)	36 (38.29%)	58 (61.71%)
Training dataset 65 (69.14%)	25 (38.46%)	40 (61.54%)
Training 45 (69.69%)	31 (69.56%)	14 (30.44%)
Testing 20 (30.31%)	14 (70.00%)	6 (30.00%)
Validation 29 (30.86%)	11 (38.29%)	18 (61.71%)

was further divided into training and testing subsets. Distribution of the dataset per these categories is shown in Table 2.

2.2 Classifier for Determining Multi-drug Resistance Based on Machine Learning Algorithms

In this research machine learning algorithms are used to predict multi-drug resistance of *E. coli*. The prediction is based on the results of antibiogram. For this purpose, three different machine learning algorithms were tested: Naïve Bayes, Logistic Regression and J48. The aim of this research was to evaluate the performance of 3 different classifiers and to identify the one with the highest accuracy, precision and sensitivity. Due to their wide usage in prediction/classification problems, these machine learning algorithms were selected to be used in this study as well.

For assessing the performance of developed classifiers, performance (class probability), accuracy, precision and sensitivity were calculated follows:

$$accuracy = \frac{number\ of\ correct\ predctions}{total\ number\ of\ sample} * 100\% \tag{2.1}$$

$$precision = \frac{true\ positive\ samples}{(true\ positive + false\ positive)sample} * 100\% \tag{2.2}$$

$$sensitivity = \frac{true\ positive\ samples}{(true\ positive + false\ positive)sample} * 100\% \tag{2.3}$$

3 Results

The classifiers were developed based on the database containing 65 samples, as shown in Table 2. The results of performance of different ML based classifiers for training and testing subsets are shown in Table 3.

Table 3. Performance of training and testing of classifiers (training dataset)

Classifier	Training	Testing
Naïve Bayes	87.0%	86.2%
Logistic Regression	93.8%	79.3%
J48	89.2%	75.9%

Table 4. Performance evaluation for J48 classifier

<table>
<tr><th>Total no. of samples = 29</th><th colspan="3">J48 Classifier</th></tr>
<tr><td rowspan="3">Database samples</td><td>Classes</td><td>Not MDR</td><td>MDR</td></tr>
<tr><td>Not MDR</td><td>17</td><td>1</td></tr>
<tr><td>MDR</td><td>6</td><td>5</td></tr>
<tr><td>Performance measure</td><td>Accuracy = 75.86%</td><td>Precision = 75.90%</td><td>Sensitivity = 45.45%</td></tr>
</table>

As it can be observed in the table above, the training of the classifiers yielded higher accuracy than the testing of the classifiers. This was expected outcome due to the restrictions of the used dataset, mainly because only 65 samples were used. The maximum variation in classifier performance during training was 6.8% meaning that the highest accuracy was achieved by Logistic Regression during training process, and the minimum accuracy was obtained by Naïve Bayes based classifier. As for testing part of the process, Naïve Bayes classifier has the highest accuracy of 86.2% so it was expected for this classifier to have the best performance during performance validation process. Classifier performance evaluation using validation dataset is presented in Tables 4, 5 and 6.

Table 5. Performance evaluation for Logistic Regression classifier

Total no. of samples = 29	**Logistic Regression classifier**		
Database samples	**Classes**	Not MDR	MDR
	Not MDR	16	2
	MDR	4	7
Performance measure	Accuracy = 79.31%	Precision = 79.30%	Sensitivity = 63.63%

Table 6. Performance evaluation for Naïve Bayes classifier

Total no. of samples = 29	**Naïve Bayes classifier**		
Database samples	**Classes**	Not MDR	MDR
	Not MDR	16	2
	MDR	2	9
Performance measure	Accuracy = 86.20%	Precision = 86.20%	Sensitivity = 81.81%

As it can be observed in Tables 4, 5 and 6, when compared by accuracy, the best performing classifier was one based on Naïve Bayes algorithm which was expected during development phase. The overall accuracy of this classifier was 86.20%, precision 86.20% and sensitivity 81.81%. Our results are a bit lower in accuracy when compared to results from previously mentioned studies. Lower accuracy could be due to the restriction of the database and lower number of samples used in the study.

4 Discussion and Conclusion

Multidrug resistant bacteria (MDR) pose a rapidly developing problem in the world today. Their emergence resulted in antibiotic-treatment failure and this is the cause of many deaths annually. The interpretation of results of antibiograms and additional information about resistance rates, including presence of MDR strains, in certain regions would aid in raising awareness of the rational use of antibiotics. This is especially significant in treatment of urinary tract infections, were recurrent infections are very common and hence resistance rates towards antibiotics are very high. Consequently, development of Machine Learning (ML) algorithms for prediction of MDR in bacteria would aid physicians in raising awareness of presence of MDR in certain regions and aid in rational prescription of antibiotics.

In this study, when the accuracy, precision and sensitivity were compared between different ML algorithms used to perform prediction of multi-drug resistance in *E. coli*, the best results were obtained with Naïve Bayes algorithm. This expert system showed 86.20% overall accuracy, 86.20% precision and 81.81% sensitivity.

The results obtained in this study are comparable to results from previously mentioned studies. Khaledi et al. [18] reported 0.87 overall F1-score for predicting antimicrobial resistance of *Pseudomonas aeruginosa*. Kim et al. [19] reported 91.1% mean accuracy for all antibiotic-pathogen combinations they examined to predict antimicrobial resistance using whole genome sequencing data. Hicks et al. [20] reported 92.5% overall accuracy for prediction of *K. pneumoniae* resistance and 97.5% overall accuracy for prediction of *A. baumanii* resistance to antibiotics. Sharaha et al. [21], who also investigated prediction of *E. coli* resistance, reported 89% sensitivity for determination of the optimum choice of antibiotic for treating *E. coli*.

The lower accuracy rate in our study, compared to the aforementioned studies, can be attributed to a lower number of samples since all the repeating samples were excluded from the study and use of different laboratory methodologies for creation of the database.

This study analyzes prediction of multi-drug resistance of *E. coli* by utilizing three different ML algorithms, which is an improvement when compared with previous papers that investigated only one algorithm. For future aspects, this study could be improved by collecting a larger number of samples, since that would improve overall accuracy of the algorithm being used. Furthermore, ML algorithms could be used for prediction of other aspects of antibiotic resistance that are emerging in the world today.

Conflict of Interest Declaration. Authors have nothing to declare.

References

1. Zaman, S.B., Hussain, M.A., Nye, R., Mehta, V., Mamun, K.T., Hossain, N.: A review on antibiotic resistance: alarm bells are ringing. Cureus **9**(6), e1403 (2017). https://doi.org/10.7759/cureus.1403
2. Ragheb, M.N., Thomason, M.K., Hsu, C., Nugent, P., Gage, J., Samadpour, A.N., Kariisa, A., Merrikh, C.N., Miller, S.I., Sherman, D.R., Merrikh, H.: Inhibiting the evolution of antibiotic resistance. Mol. Cell **73**(1), 157–165 (2019)

3. Alekshun, M.N., Levy, S.B.: Molecular mechanisms of antibacterial multidrug resistance. Cell **128**(6), 1037–1050 (2007)
4. Worthington, R.J., Melander, C.: Combination approaches to combat multidrugresistant bacteria. Trends Biotechnol. **31**(3), 177–184 (2013)
5. World Health Organization: Antimicrobial resistance: global report on surveillance. World Health Organization (2014)
6. Hudzicki, J.: Kirby-Bauer disk diffusion susceptibility test protocol (2009)
7. Jenkins, S.G., Schuetz, A.N.: Current concepts in laboratory testing to guide antimicrobial therapy. Mayo Clin. Proc. **87**(3), 290–308 (2012). https://doi.org/10.1016/j.mayocp.2012.01.007
8. Cusack, T.P., Ashley, E.A., Ling, C.L., Rattanavong, S., Roberts, T., Turner, P., Wangrangsimakul, T.D.A.B., Dance, D.A.: Impact of CLSI and EUCAST breakpoint discrepancies on reporting of antimicrobial susceptibility and AMR surveillance. Clinical Microbiology and Infection (2019)
9. Pokvic, L.G., Spahic, L., Badnjevic, A.: Chapter 23: Implementation of industry 4.0 in transformation of medical device maintenance systems. In: Handbook of Research on Integrating Industry 4.0 in Business and Manufacturing, pp. 512–532. IGI Global (2020)
10. Mustafić, L.D., et al.: Diagnosis of severe aortic stenosis using implemented expert system. In: Badnjevic, A., Škrbić, R., Gurbeta Pokvić, L. (eds,) CMBEBIH 2019. CMBEBIH 2019. IFMBE Proceedings, vol 73. Springer, Cham (2020)
11. Deo, R.C.: Machine learning in medicine. Circulation **132**(20), 1920–1930 (2015)
12. Alić, B., Gurbeta, L., Osmanovic, A., Badnjević, A.: Machine learning techniques for classification of diabetes and cardiovascular diseases. In: 2017 6th Mediterranean Conference on Embedded Compu-ting (MECO), Bar, Montenegro, 2017, pp. 1–4 (2017). https://doi.org/10.1109/MECO.2017.7977152
13. Catic, A., Gurbeta, L., Kurtovic-Kozaric, A., Mehmedbasic, S., Badnjevic, A.: Application of Neural Networks for classification of Patau, Edwards, Down, Turner and Klinefelter Syndrome based on first trimester maternal serum screening data, ultrasonographic findings and patient demographics. BMC Med. Genom. **11**, 19 (2018). https://doi.org/10.1186/s12920-018-0333-2
14. Badnjevic, A., Gurbeta, L., Custovic, E: An expert diagnostic system to automatically identify asthma and chronic obstructive pulmonary disease in clinical settings. Sci. Rep. (2018)
15. Aljovic, A, Badnjevic, A, Gurbeta, L.: Artificial neural networks in the discrimination of Alzheimer's disease using biomarkers data. In: IEEE 5th Mediterranean Conference on Embedded Computing (MECO), 12–16 June 2016, Bar, Montenegro 10
16. Gurbeta, L., Izetbegović, S., Badnjević-Čengić, A.: Inspection and testing of infant incubators. In: Badnjević A., Cifrek M., Magjarević R., Džemić Z. (eds.) Inspection of Medical Devices. Series in Biomedical Engineering. Springer, Singapore (2018)
17. Badnjevic, A., Cifrek, M., Gurbeta, L., Feric, E.: Classification of chronic obstructive pulmonary diseases based on neuro-fuzzy software. In: The Book: "Chronic Obstructive Pulmonary Disease (COPD): Clinical Symptoms, Emerging Treatment Strategies and Impact on Quality of Life". Nova Science Publisher (2016). ISBN 978–1–63484–500–7
18. Granulo, E., Becar, L., Gurbeta, L., Badnjevic, A.: Telemetry system for diagnosis of asthma and Chronical Obstructive Pulmonary Disease (COPD. In: Ahmed, M.U., Begum, S., Raad, W. (eds.). Internet of Things Technologies for HealthCare, pp. 113–118. Springer (2016). https://doi.org/10.1007/978-3-319-51234-1_18
19. Khaledi, A., Weimann, A., Schniederjans, M., Asgari, E., Kuo, T.H., Oliver, A., Cabot, G., Kola, A., Gastmeier, P., Hogardt, M., Jonas, D.: Predicting antimicrobial resistance in Pseudomonas aeruginosa with machine learning-enabled molecular diagnostics. EMBO Molecul. Med. **12**(3), e10264 (2020)

20. Kim, J., Greenberg, D.E., Pifer, R., Jiang, S., Xiao, G., Shelburne, S.A., Koh, A., Xie, Y., Zhan, X.: VAMPr: VAriant Mapping and Prediction of antibiotic resistance via explainable features and machine learning. PLoS Comput. Biol. **16**(1), e1007511 (2020)
21. Hicks, A.L., Wheeler, N., Sánchez-Busó, L., Rakeman, J.L., Harris, S.R., Grad, Y.H.: Evaluation of parameters affecting performance and reliability of machine learningbased antibiotic susceptibility testing from whole genome sequencing data. PLoS Comput. Biol. **15**(9), e1007349 (2019)
22. Sharaha, U., Rodriguez-Diaz, E., Sagi, O., Riesenberg, K., Salman, A., Bigio, I.J., Huleihel, M.: Fast and reliable determination of Escherichia coli susceptibility to antibiotics: Infrared microscopy in tandem with machine learning algorithms. J. Biophoton. **12**(7), e201800478 (2019)
23. Holding, A.J., Collee, J.G.: Chapter I Routine biochemical tests. In: Methods in Microbiology, vol. 6, pp. 1–32. Academic Press (1971)
24. Alic, B., Sejdinovic, D., Gurbeta, L., Badnjevic, A.: Classification of stress recognition using artificial neural network. In: IEEE 5th Mediterranean Conference on Embedded Computing (MECO), 12–16 June 2016, Bar, Montenegro

Intrusion Detection in Smart Healthcare Using Bagging Ensemble Classifier

Abdulhamit Subasi[1], Shahad Algebsani[1], Wafa Alghamdi[1], Emir Kremic[2](✉), Jawaher Almaasrani[1], and Najwan Abdulaziz[1]

[1] College of Engineering, Effat University, Jeddah 21478, Saudi Arabia
absubasi@effatuniversity.edu.sa
[2] Federal Institute of Statistics,
Bosnia and Herzegovina, 71000 Sarajevo, Bosnia and Herzegovina

Abstract. Critical infrastructure systems such as ehealthcare, which are crucial to government, society, and commerce, are directly related to people's lives. In order to offer extremely secured critical infrastructure systems, a reliable, scalable, and robust threat detection system must be established to effectively alleviate cyber threats. Moreover, big data from threat monitoring systems cause severe trials for cyber-attack since increasing number of devices in the structure and the quantity of monitoring data accumulated from IoT based healthcare devices need to prevent cyberattacks. Intrusion detection systems, or IDSs, have become an important component in maintaining the security on a corporate network. The popularity of Internet usage comprises many risks of network attacks. Intrusion detection aims to discover unusual access or attacks to secure internal networks. To build our proposed intrusion detection system, Machine learning techniques are utilized with benchmark attack data. In this paper, Bagging ensemble classifier for intrusion detection in a smart healthcare environment is proposed. Bagging ensemble classifier with Random Forest achieved better performance (97.67% accuracy) as compared to its counterpart classifier models for intrusion detection in a smart healthcare with a benchmark data. Experimental results have revealed the practicality of Bagging ensemble classifier by accomplishing a better performance for the cyber security threats in a smart healthcare environment.

Keywords: Bagging ensemble classifier · Smart healthcare · Cybersecurity · Anomaly detection · Internet of Things (IoT)

1 Introduction

Internet of Things (IoT) brings considerable advantages for Smart Healthcare (SH) applications. Trends bring a smart healthcare concept with a crucial challenge of cyber security threats in the smart healthcare. Actually, as a critical infrastructure, the smart healthcare is more sensitive to cyber-attacks, since the control and monitoring of smart healthcare can be carried out with standard internet-based protocols. The most ubiquitous risk to the end user is obstacle of privacy of the data and malicious control of personal equipment and devices. Hence, security is one of the major concerns hindering the rapid

A. Badnjevic and L. Gurbeta Pokvić (Eds.): CMBEBIH 2021, IFMBE Proceedings 84, pp. 164–171, 2021.
https://doi.org/10.1007/978-3-030-73909-6_18

and large scale application and establishment of the IoT [1]. Smart IoT devices for healthcare and domestic life provide extraordinary services and enhanced quality of life [2].

The increasing cost of healthcare and the occurrence of chronic diseases around the world instantly require the revolution of healthcare from a hospital-centered system to a patient-centered environment, with a focus on disease management as well as their wellbeing. Security is a main problem in which networks are arranged at large scales. IoTbased smart healthcare structures related to human data. Data collected from wearable sensors is vulnerable to highest privacy concerns. In IoT-based smart healthcare structures, since most of the equipment and their infrastructures are wireless in nature, security and privacy are the main concern. An IP-enabled sensor in a Medical Sensor Network (MSN) can send medical data to a remote healthcare service and routed through an untrusted network infrastructure, such as Internet. Furthermore, since security and privacy of patients are key issue in smart healthcare, many efforts achieved in designing Smart e-Health Gateways. Hence, the required security incident to the medical sensors, smart gateways eliminate the need for authentication and authorization of remote healthcare centres from the sensors and any malicious action can be detected before entering to a medical constrained domain [3].

Computer system security is described as the way toward ensuring three variables: confidentiality, integrity, and availability (CIA). Network security is ending up noticeably dynamically more vital because of the gigantic development of computer systems use and system applications [4]. In this way, an Intrusion detection system (IDS) is a security system that screens computer systems and system activity and examines the movement for conceivable hostile attacks starting from outer clients to the interior attackers. The objective of intrusion detection systems (IDS) is to identify assaults from intruders with high precision to secure inner systems [5]. An effective intrusion detection system needs to diminish false positives/cautions rate. In the meantime, it should be effective at getting assaults, keeping up high detection rate. One of the principle issues with IDS is the overhead, which can turn out to be restrictively high. The recognition reaction time is another significant issue in the IDS. Computer systems have a dynamic nature it might be said that data and information inside them are ceaselessly evolving [4].

Smart cities use IT solutions to maintain the services such as transportation, health care and more. Smart healthcare faces unique and escalating cyber threats. The smart grid technology offers a two-way dialogue where electricity and information can be exchanged between the utility and its customers [6]. In addition, smart healthcare demands the transformation of healthcare from a hospitalcantered system to a patient-cantered environment, with a focus on disease management as well as their wellbeing.

That might require the usage of wireless sensor networks [7]. For that reason, smart city cyber security in the context of smart healthcare became a hot topic. It may therefore be advantageous to further investigate the security issues and challenges of the IoT-based smart healthcare. However, only few studies have examined this topic. Therefore, further research on this topic in needed. Smart city security concerns need to tackle new challenges for a safe, reliable, effective and secure operation of the smart healthcare. It should be highlighted that current security techniques are inappropriate, impractical, incompatible, or basically insufficient that should be substituted by innovative techniques to guarantee the security of the extremely massive and complicated dynamic smart healthcare environment. A smart healthcare can be considered as the combination of physical system elements and cyber system infrastructure comprising hardware, software, and communication constraints. Furthermore, in a smart healthcare, both cyber and physical system securities are essential and concern of security issues in cyber domain cannot capture the whole picture [8].

In this study, the security issues and challenges of the IoT-based Smart Healthcare in a smart city concept is examined. In Sect. 2 we briefly describe Security issues and challenges in Smart Healthcare. In Sect. 3, Bagging ensemble classifier is described. Experimental results are given in Sect. 4. Section 5 is conclusion.

2 Security Issues and Challenges in Smart Healthcare

Because of direct involvement of people in IoT-based social services applications, giving vigorous and secure information correspondence among sensors, actuators and patients, are critical. Abuse or protection concerns may confine individuals to use IoT-based secure applications. Routine security mechanisms including existing cryptographic arrangements, secure rules, and protection verification cannot be utilized because of asset requires security level prerequisites, and IoT-based medical services frameworks. To alleviate the previously mentioned threats, robust system security foundations for a short and long-term correspondence are required. There are noteworthy security answers for current remote systems which are not specifically appropriate to IoT based applications. Security convention offers validation, key trade, and assurance of utilization information. A protected and proficient validation and approval engineering for IoT-based frameworks utilizing smart e-healthcare passages as a part of a conveyed design is introduced [3].

In the models of medical services IoT, not just information can be gathered by keen gadgets (sensors) and transmitted to end-clients, however end-clients can control, and manage medical sensors through the Web. Since patients' wellbeing information is utilized for empowering applications and administrations in public services, IoT must present secure end-to-end communication between endclients and medical sensors to ensure the exchange of health information. Healthcare security seeks to achieve privacy and security of patient information to increase confidence with the hospital. The problem in hospital is that some people hide or change their personal data because of the lack of confidence in the security and protection of patient information at the hospital. The patient should have their basic rights to know if their information secure or not.

3 Bagging Ensemble Classifier

Machine learning is an optimized programming model to address a performance criterion built on historical experiences or training data. The optimization model can be utilized to predict the future or to get knowledge from the data. In order to solve the optimization problem with the massive amount of data, an effective training algorithm is needed. Once the model is trained and learned effectively , the representation and solution algorithm for the inference should be as effective as possible. The effectiveness of the learning algorithm is depends on its predictive accuracy, despite its space and time complexity [9].

Bagging (Bootstrap aggregating) is the simple ensemble modeling which uses very basic methods for the creation of the base model and the aggregation. Bootstrap samples are utilized in the training sets to create base models and then they are merged by unweighted voting. It gives an enhancement compared to single models created by using the same algorithm if the algorithm is unstable. There is no enhancement or even minor reduction of quality of prediction for stable algorithms. Modeling algorithm has no specific requirements, except the instability. If it makes it more stable, it can be simplified compared to what would be usually employed in the case of single model creation. The over fitting of the ensemble is not affected by the over fitting of the base model because their aggregation will cancel it out. Bagging algorithm stabilizes the unstable algorithms. Single unstable models maybe sensitive to changes depending on a particular training set. For the models, which are only different in their training samples, it is impossible to have repeating training and assessment cycles several times in order to produce low-variance performance evaluations that can be used for model selection. The bagging ensemble performance can be improved by increasing the number of base models up to a certain level after it stabilizes [10].

4 Experimental Results

It is essential to compare the performance of different machine learning schemes in terms of the error rate. Taking a larger sample for the training purpose increases the classifier performance, but the returns might begin to lessen once reached a certain level. Also, taking a larger sample for the testing purpose enhances the accuracy of the error estimate. Since the real problem happens when there is not a massive amount of data available, the holdout procedure is carried to hold over a certain amount of data for testing, and reminder data used for the training. A more general procedure called the repeated holdout procedure is carried out by repeating the training and testing stages several times with multiple random samples. Another simple statistical method called cross-validation by deciding a fixed number of folds of the data. Each fold employed for testing and the reminder for training, and this procedure repeated in a way that every fold has been employed exactly once for testing [11]. In this study 10-fold cross validation is employed during the experiments.

There are several expressions that must be taken into consideration when computing the classifier performance. The true positives (TP) and true negatives (TN) denote to the examples, which are accurately classified and rejected respectively. A false positive (FP) is once the example is wrongly classified as positive when it is negative. A false negative (FN) is once the example is incorrectly categorized as negative when it is positive [11]. The most common measure, accuracy, differentiate between the number of correct labels of different classes [12].

$$\text{Accuracy} = \frac{\text{TP} + \text{TN}}{\text{TP} + \text{FN} + \text{TN} + \text{FP}} \times 100\% \tag{1}$$

Table 1. Performance of classifiers for smart healthcare intrusion detection.

	Single				Bagging			
	Accuracy (%)	F – measure	AUC	Kappa	Accuracy (%)	F – measure	AUC	Kappa
ANN	92.70	0.925	0.943	0.744	94.06	0.938	0.974	0.784
k-NN	95.76	0.957	0.957	0.855	95.86	0.958	0.966	0.859
SVM	91.81	0.912	0.802	0.687	91.90	0.913	0.837	0.691
Random Forest	97.00	0.970	0.996	0.896	97.67	0.977	0.997	0.921
C4.5	97.12	0.971	0.971	0.902	97.65	0.976	0.996	0.902
Random Tree	96.80	0.968	0.950	0.893	97.49	0.975	0.993	0.916
REP Tree	97.24	0.972	0.984	0.908	97.47	0.975	0.994	0.914

F-measure is another performance measure and defined by:

$$F - measure = 2 \times \frac{\Pr ecision \times \text{Recall}}{\Pr ecision + \text{Recall}} \tag{2}$$

$$\Pr ecision \frac{TP}{TP + FP}, \ \text{Recall} \frac{TP}{TP + FN} \tag{3}$$

where

ROC analysis, known as receiver operating characteristic, is graphical representation for evaluating the performance of a classifier regardless of class distribution or error costs. It is defined as a plot of the test sensitivity on the vertical coordinate axis and the 1-specificity on the horizontal coordinate at various threshold settings. The diagonal line for a model predicts a chance but cannot be considered as a discriminating model. Ideally, the curve will climb up towards the top-left corner that means the classifier is correctly predicting the instances, but unfortunately this is not the case in real practices. The model would be able to disseminate between positives and negatives is when the curve is further from the diagonal line. There is a useful statistic function that can be calculated from this curve, like the Area Under the Curve (AUC). It computes the entire two-dimensional area underneath the ROC curve as a whole and measures the performance across all possible classification thresholds [11].

Kappa Statistic is an essential statistical measure of the inter-rater agreement between two raters. The two raters evaluate the dataset independently to check the performance while correcting for an agreement that occurs by chance. It computes the agreement between predicted and observed categorizations of a dataset. The Kappa statistic varies from 0 to 1, where 0 means the agreement equivalent to a random chance and 1 means the perfect agreement. The calculation formula for two raters shown below where P is the relative observed agreement among raters and P_e is the hypothetical probability of chance agreement [13].

$$K = \frac{P_0 = P_e}{1 - P_e} = 1 - \frac{1 - P_0}{1 - P_e} \tag{4}$$

In this paper, we employed a benchmark data in a smart grid environment to evaluate the performances of the models to test the accuracies, F-measure, area under the ROC curve and kappa statistics of each algorithm. Various machine learning algorithms are employed for classification with a 10-fold cross-validation.

Several machine learning methods such as Artificial Neural Networks (ANN), k-Nearest Neighbor (k-NN), Support Vector Machine (SVM), Random Forest (RF), C4.5 Decision Tree, Random Tree, REP Tree, and Bagging ensemble methods are employed for intrusion detection in the smart healthcare and the results are presented in Table 1. REP Tree achieved the maximum classification accuracy with 97.24%, C4.5 is the second one with 97.12%. Random Forest and Random Tree achieve 97%, and 96.68% respectively. SVM is the last one with 91.81% accuracy among the single classifiers.

F-measure is another performance measure and achieved similar results as the classification accuracy. It can be seen from Table 1 that, F-measure and the total classification accuracy are close to each other that represents the reliability of results. Hence, the F-measure results are above 0.912 for the classifiers which is coincident with total accuracies of the classifiers. Moreover, the performance of the machine learning algorithms is assessed by the area under the ROC curve (AUC). AUC of Random Forest is the best (0.996) and followed by REP Tree (0.84) then C4.5 (0.971), and SVM is the last (0.802). Kappa statistic accomplished the results similar to the total classification accuracy. Thus, REP Tree algorithm achieved the best result with a kappa value of 0.908; then C4.5 achieved the second-best result with 0.902 kappa value. Random Forest is in the third position. Again, SVM is the least effective classifier with a kappa value of 0.802.

With Bagging, Random Forest achieved the highest classification accuracy with 97.67%, C4.5 is the second one with 97.65%. Random Tree achieved 97.49%, REP Tree achieved 97.47% and SVM was the last one with 91.90% accuracy. F-measure which is another performance measure achieved results similar to the classification accuracy. It is revealed from Table 1 that all the decision tree-based classifiers accomplished higher F-measure value than other classifiers with Bagging. Furthermore, the performance of the Bagging ensemble classifier is evaluated by the area under the ROC curve (AUC). All classifiers achieved an AUC value of above 0.966 except SVM which is 0.837. Bagging achieved the kappa statistic similar to the AUC. Therefore, Random Forest classifier achieved the best kappa value which is 0.921; then the second-best result is accomplished by Random Tree with 0.916 kappa statistics. Again, the least effective classifier is SVM as with a kappa value of 0.691.

The performance of classifiers in terms of accuracy, Fmeasure, area under the ROC curve and kappa statistic values revealed that the Bagging ensemble classifier is effective for intrusion detection in the smart healthcare. Actually, the improvements for the proposed model is very small, it is crucial when detecting intrusion in a very sensitive environments such as smart healthcare. The results are consistent for the all machine learning techniques are revealed that the REP Tree model can yield reliable results with high accuracy as a single classifier. On the other hand, Random Forest achieved best classification accuracy with Bagging ensemble classifier.

5 Conclusions

Intrusion detection in the Smart Healthcare is a new research area which has been concerned quickly by the industry, academia and government. In this study, Intrusion detection in the Smart Healthcare is presented. The novel intrusion detection architecture based on Bagging ensemble classifier proposed for attack prevention in the Smart Healthcare. Intrusion detection is still under progress in the Smart Healthcare, since the information security should be taken into consideration. As the Smart Healthcare communication network infrastructure includes heterogeneous network architectures and devices, scalability, delay limitations on different time scales, and diversified capabilities of embedded devices, it is unrealistic to consistently achieve robust security over the Smart Healthcare. Therefore, the Smart Healthcare necessitates fine-grained security solutions dedicated mainly for diverse network requests, producing an intrusion detection framework for the Smart Healthcare in the future [7]. Moreover, it is crucial to accomplish an intrusion detection system for the Smart Healthcare to stop denial-of-service attacks and false-data injection. In this paper, an intrusion detection framework for the Smart Healthcare is proposed to stop intrusions by utilizing Bagging ensemble classifier. It is realized that the proposed Bagging ensemble classifier increased the accuracy of other classifiers. In this study, an effective anomaly-based intrusion detection system (IDS) is suggested by using Bagging ensemble classifier. The analysis is done utilizing 20% of NSL-KDD dataset without attribute selection. The proposed approach altogether outperforms, over state-of-the-art techniques as far as accuracy, F-measure, AUC and Kappa statistics. In future study, we will study on-depth the security of a keyelement of the smart cities where we focus on how we can securely integrate smart healthcare, equipped with smart sensors and smart body sensors by collecting real sensor data so that end-user could securely participate in the smart healthcare environment.

References

1. Bekara, C.: Security issues and challenges for the iotbased smart grid. Procedia Comput. Sci. **34**, 532–537 (2014)
2. Elmaghraby, A.S., Losavio, M.M.: Cyber security challenges in Smart Cities: safety, security and privacy. J. Adv. Res. **5**(4), 491–497 (2014)
3. Moosavi, S.R., et al.: End-to-end security scheme for mobility enabled healthcare Internet of Things. Future Gener. Comput. Syst. (2016)
4. Kevric, J., Jukic, S., Subasi, A.: An effective combining classifier approach using tree algorithms for network intrusion detection. Neural Comput. Appl. **28**(1), 1051–1058 (2016). https://doi.org/10.1007/s00521-016-2418-1
5. Rights, R.F.: SANS institute InfoSec reading room. Risk **1**, 27 (2001)
6. Kumar, P., Lee, H.-J.: Security issues in healthcare applications using wireless medical sensor networks: a survey. Sensors **12**(1), 55–91 (2012)
7. Wang, W., Lu, Z.: Cyber security in the Smart Grid: survey and challenges. Comput. Netw. **57**(5), 1344–1371 (2013)
8. Anwar, A., Mahmood, A.N.: Cyber security of smart grid infrastructure. ArXiv Prepr. ArXiv14013936 (2014)
9. Alpaydin, E.: Introduction to Machine Learning. MIT Press, Cambridge (2009)
10. Cichosz, P.: Data Mining Algorithms: Explained Using R. Wiley, Chichester (2014)

11. Witten, I.H., Frank, E., Hall, M.A.: Data mining: practical machine learning tools and techniques. Elsevier, Burlington (2011)
12. Sokolova, M., Japkowicz, N., Szpakowicz, S.: Beyond accuracy, F-score and ROC: a family of discriminant measures for performance evaluation. In: Presented at the Australasian Joint Conference on Artificial Intelligence, pp. 1015–1021 (2006)
13. Cohen, J.: A coefficient of agreement for nominal scales. Educ. Psychol. Meas. **20**, 37–46 (1960)

A Machine Learning Approach to Predict the Sepsis Status: Analyzing the Connection Between Relevant Laboratory Values and Other Physiological Measurements Obtained in Intensive Care Unit

Sebahattin Babur[1(✉)], Sanam Moghaddamnia[2], and Mehmet Recep Bozkurt[1]

[1] Faculty of Engineering, Electrical-Electronics Engineering, Sakarya University, Esentepe Kampüsü, Üniversite Cd., 54050 Serdivan, Sakarya, Turkey
sebahattin.babur@tau.edu.tr

[2] Faculty of Engineering, Electrical-Electronics Engineering Şahinkaya Cad., Turkish-German University, 106 - 34820 Beykoz, Istanbul, Turkey

Abstract. Sepsis is a life-threatening disease caused by the body's response to an infection, which leads to tissue damage, organ failure and death. According to literature, about 1.7 million Americans each year suffer from sepsis with approximately 270,000 fatal cases, and more than one third of mortalities in hospitals are associated with sepsis. It is estimated that worldwide about 30 million people suffer from sepsis each year, and 6 million patients with sepsis die. Above all, the cost of sepsis care is extremely high and constitutes a health risk mostly in developing countries. Having regard to the above, early diagnosis and therapy of sepsis play a key role in increasing the chance of cure for many patients. In this regard, many clinicians and researchers have proposed new definitions for sepsis (Singer et al. 2016), however the subject of early diagnosis methods and their practical usability and limits for sepsis remains to a large extent unexplored. Therefore, a cost-effective self-diagnostic technic has been proposed in this work, which is based on the XGBoost algorithm and deep convolution neural network model for sepsis prediction. Septic data were used for the model training, and the highest accuracy and the area under the ROC curve (AUC) obtained to evaluate the performance of the model were 96.49% and 70.78%, respectively. The novelty of this study lies in the investigations that unlike other studies are performed using vital signs determined independently from laboratory measurements. It has shown that the proposed model is faster and can provide more accurate results, and thus has great potential to be used as an efficient technique in predicting sepsis.

Keywords: Sepsis · Convolutional neural network (CNN) · Clinical data · Regression · Machine learning · Xgboost

1 Introduction

The immune system has a well-developed protection mechanism against infections that can be caused by bacteria, viruses or fungi. When the immune system cannot provide

A. Badnjevic and L. Gurbeta Pokvić (Eds.): CMBEBIH 2021, IFMBE Proceedings 84, pp. 172–183, 2021.
https://doi.org/10.1007/978-3-030-73909-6_19

a special defense against infection, it causes a sepsis in the body, which leads to a damage of inflammatory chemicals to create a mass effect [1]. Sepsis is defined as a highly complex and fatal syndrome with different clinical indications and creates a challenging environment for detection and treatment [2]. According to international knowledge, sepsis is defined as fatal organ damage resulting from the body's response to an infection [2]. The cornerstone of organ damage is the subsequent hypoperfusion caused by the metabolic needs of the tissue and the inflammatory state of the body. It is caused by insufficient blood supply to an organ or extremity, such as the brain or hand in general.

Cardiac dysfunction due to inflammation and redistribution of systemic blood volume play a key role, exacerbated by the use of oxygen from damaged tissue. [3]. Organs frequently damaged by sepsis are kidneys, lungs, liver, heart, central nervous and hematological systems [2]. Infections affecting sepsis are resistant to antibiotics and lead to rapid deterioration of health conditions [4]. Sepsis symptoms include fever, irregular heart rate, blood vessel leakage, inflammation, and coagulation difficulties [5]. Sepsis can be classified as sepsis 1, 2 or 3. Sepsis 1 is known to occur as a result of a systemic inflammatory response to known or suspected infections. Sepsis 2 diagnosis is based on inflammatory, haemodynamic, organ diseases and tissue perfusion parameters, whereas sepsis 3 is diagnosed according to organ lifethreatening due to its unstable response to infection [6].

Early diagnosis and timely management of sepsis play an important role in decreasing mortality rates. Blood cultures are examined and biomarkers such as procalcitonin (PCT), C-reactive protein (CRP), cell-free DNA (cfDNA) are used as standards for early diagnosis of sepsis [7, 8]. However, using blood cultures as a diagnostic tool exhibits some drawbacks [9]. Their systems developed to address them are costly and their integration into clinical algorithms is challenging.

Therefore, it is important to develop a cost-effective, automated diagnostic tool for early diagnosis of sepsis. In this way, it reduces the time it takes to diagnose sepsis in environments with high patient density and opens the way for treatment in a short time [10].

Indicator abstraction and selection processes are naturally automated in deep learning techniques, facilitating classifications, so deep learning models are increasingly used in the detection of various diseases [11–16]. In this research, temporal convolutional network is used for prediction of sepsis. The proposed method aims to predict sepsis quickly with high accuracy. Further information is provided in the next sections. Section 2, methodology, presents vector features in a table. The correlation matrix with the relationship of all the features in the data set is shown in the appendix. We conducted statistical analysis between laboratory values to be used in sepsis and invasive measurements using correlation matrices and created the regression model. Sections 3 and 4 discuss results and offers a comparison with other studies, while Sect. 5 concludes the article with recommendations for future research.

2 Methodology

2.1 Data Acquired and Pre-processing Method

We draw on the data from the open source dataset published for the PhysioNet Computing in Cardiology 2019 Challenge [17]. These include data from 2 hospitals: Beth Israel Deaconess Medical Center (hospital system A) and Emory University Hospital (hospital system B), which contains 1790 and 1142 septic records, respectively. The data are based on sepsis-3 criteria and each patient's record contains 40 features. It consists of 8 vital signs features recorded hourly, 26 laboratory measurements and 6 demographic variables. The attributes are presented in Table 1 along with their explanations.

Table 1. Clinical time series data provided for the Challenge: vital signs (rows 1–8), laboratory values (rows 9–34), demographics (rows 35–40), and outcome (row 41)

Row	Measurement	Description
1	**HR**	**Heart rate (beats/min)**
2	O2Sat	Pulse oximetry (%)
3	**Temp**	**Temperature (°C)**
4	SBP	Systolic BP (mm Hg)
5	MAP	Mean arterial pressure (mm Hg)
6	DBP	Diastolic BP (mm Hg)
7	**Resp**	**Respiration rate (breaths/min)**
8	Etco2	End tidal carbon dioxide (mm Hg)
9	BaseExcess	Excess bicarbonate (mmol/L)
10	Hco3	Bicarbonate (mmol/L)
11	Fio2	Fraction of inspired oxygen (%)
12	pH	pH
13	Paco2	Partial pressure of carbon dioxide from arterial blood (mm Hg)
14	Sao2	Oxygen saturation from arterial blood (%)
15	AST	Aspartate transaminase (IU/L)
16	**BUN**	**Blood urea nitrogen (mg/dL)**
17	Alkalinephos	Alkaline phosphatase (IU/L)
18	**Calcium**	**Calcium (mg/dL)**
19	Chloride	Chloride (mmol/L)
20	Creatinine	Creatinine (mg/dL)
21	**Bilirubin direct**	**Direct bilirubin (mg/dL)**
22	Glucose	Serum glucose (mg/dL)
23	Lactate	Lactic acid (mg/dL)
24	Magnesium	Magnesium (mmol/dL)

(*continued*)

Table 1. (*continued*)

Row	Measurement	Description
25	Phosphate	Phosphate (mg/dL)
26	Potassium	Potassium (mmol/L)
27	**Bilirubin total**	**Total bilirubin (mg/dL)**
28	TroponinI	Troponin I (ng/mL)
29	Hct	Hematocrit (%)
30	Hgb	Hemoglobin (g/dL)
31	PTT	Partial thromboplastin time (s)
32	WBC	Leukocyte count (count/L)
33	**Fibrinogen**	**Fibrinogen concentration (mg/dL)**
34	Platelets	Platelet count (count/mL)
35	Age	Age (yr)
36	Gender	Female (0) or male (1)
37	Unit 1	Administrative identifier for ICU unit (medical ICU); false (0) or true (1)
38	Unit 2	Administrative identifier for ICU unit (surgical ICU); false (0) or true (1)
39	HospAdmTime	Time between hospital and ICU admission (hours since ICU admission)
40	**ICULOS**	**ICU length of stay (hours since ICU admission)**
41	SepsisLabel	For septic patients, SepsisLabel is 1 if $t \geq$ tsepsis—6 and 0 if $t <$ tsepsis—6. For nonseptic patients, SepsisLabel is 0

Since the missing data in the data set are too many, it is preferred not to calculate values such as median and mean of all samples and use them for missing data. First, the correlation matrix was created by cleaning out the missing data. Parametric statistical analysis was used due to the assumption (central limit theorem) that the sampling distribution will be distributed normally in case of a sample size greater than 30 in each group [18]. Therefore, the Pearson Correlation Coefficient matrix was created. The results show that 8 vital symptom features, 26 laboratory measurements and 6 demographic features were related to one another. These correlation coefficients were useful in identifying the most important vital sign feature for sepsis, laboratory measurement and demographic characteristics. These features are HR, Temp, Resp, Blood urea nitrogen (BUN), Calcium, Biluribin_direct, Bilurubin_total, Fibrinogen and ICU length-of-stay (ICULOS). This research aimed to predict the features obtained from blood culture with other invasively measurable features to estimate the distribution of laboratory measurement values and to alleviate the missing values problem.

It is important to examine the correlation between laboratory measurements and vital signs features. The dual distribution graphs between BUN and other vital features, which have the highest correlation with these measurements, are presented in Fig. 1. It showed that there is a positive and negative distribution between laboratory measurements and vital signs features.

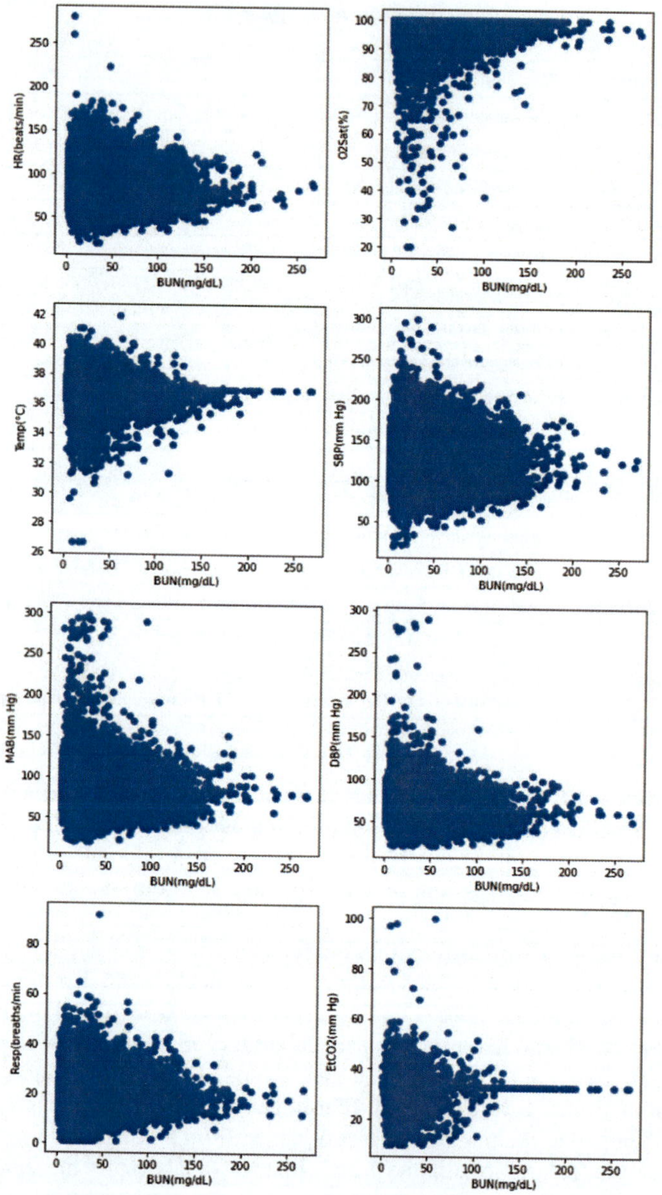

Fig. 1. Scatter plot between 8 recorded vital signs and BUN laboratory measurement results.

We have used different regression methods for comparison purposes in order to determine the missing ones of the determined laboratory measurements by using vital sign features. The results obtained from the Linear Regression, Lasso, MLP Regressor,

KNeighborsRegressor, DecisionTreeRegressor and Support Vector Regression methods are presented in Table 2.

As can be seen here, the method with the lowest mean square error value for the BUN value was MLP, MLP for Calcium, SVR for Bilirubin, for direct it was SVR, Bilurubin, for total it was KNeighborsRegressor, for WBC, it was SVR, for Fibrinogen it was Lasso.

Table 2. Linear Regression, Lasso, MLP Regressor, KNeighbors Regressor, DecisionTree Regressor and Support Vector Regression methods

Method	BUN Mean Squared Error	Calcium Mean Squared Error	Bilirubin direct Mean Squared Error	Bilirubin total Mean Squared	WBC Mean Squared Error	Fibrinogen Mean Squared Error
Linear Regression (LR)	13.121	3.405	3.218	2.986	5.588	106.011
LASSO	13.267	3.410	0.958	3.160	5.555	**105.569**
Multi-Layer Perceptron Regressor	**12.981**	**3.393**	2.371	2.988	5.751	121.047
KNeighbors Regressor	14.056	3.665	1.100	**2.485**	5.947	114.684
DecisionTree Regressor	19.806	4.765	1.153	3.094	7.569	168.877
Support Vector Regression	13.358	3.867	**0.380**	3.237	**5.526**	116.958

2.2 Convolutional Neural Network

Deep learning models are neural networks consisting of multiple layers and parameters that assist classification tasks [19]. Deep models ranging from convolutional neural networks (CNN) [20] to long short-term memory (LSTM) [21] and automatic encoders [22] are frequently used to detect arrhythmia [11, 12, 23] and schizophrenia [13]. In this study, a 6-layer CNN structure was created with 4 intermediate layers and input/output layers. A total of 6,049 parameters were calculated (Fig. 2).

```
Model: "sequential"
_________________________________________________________________
Layer (type)                 Output Shape              Param #
=================================================================
dense (Dense)                (None, 36)                684
_________________________________________________________________
dense_1 (Dense)              (None, 36)                1332
_________________________________________________________________
dense_2 (Dense)              (None, 36)                1332
_________________________________________________________________
dense_3 (Dense)              (None, 36)                1332
_________________________________________________________________
dense_4 (Dense)              (None, 36)                1332
_________________________________________________________________
dense_5 (Dense)              (None, 1)                 37
=================================================================
Total params: 6,049
Trainable params: 6,049
Non-trainable params: 0
_________________________________________________________________
```

Fig. 2. Convolutional Neural Network model summary.

2.3 Extreme Gradient Boosting (XGBoost)

Chen and Guestrin's Extreme Gradient Boosting (XGBoost) is a machine learning technique for classification and regression problems [24]. In this study, max depth, n_estimators and learning rate parameters were determined as 21, 300, and 0.05, respectively.

2.4 Training, Testing and Validation of Model

For each patient, there are selected features and labels that indicate the onset of sepsis. These labels are 1, starting from the beginning of sepsis and throughout the section with sepsis, and 0 in other cases. 80% of the data were used for training and 20% for testing [25]. The area under the ROC curve (AUROC) was calculated to validate the model. The calculated metrics are divided into 4 categories: Accuracy, Sensivity, Specificity, and AUC metrics are calculated once for the test set [17] (Tables 3 and 4).

Table 3. Selected features

Feature number	Data columns	Non-null count	Dtype
1	HR	22806	float64
2	GO2Sat	22806	float64
3	Temp	22806	float64

(continued)

Table 3. (*continued*)

Feature number	Data columns	Non-null count	Dtype
4	SBP	22806	float64
5	MAB	22806	float64
6	DBP	22806	float64
7	Resp	22806	float64
8	EtCO2	22806	float64
9	BUN	1739	float64
10	Calcium	3767	float64
11	Bilirubin_direct	80	float64
12	Bilirubin_total	549	float64
13	WBC	1743	float64
14	Fibrinogen	515	float64
15	Age	22806	int64
16	Gender	22806	float64
17	HospAdmTime	22806	float64
18	ICULOS	22806	int64
19	Sepsis Label	22806	int64

Table 4. Train/Test validation dataset.

Train/Test validation	Samples	Features
Sepsis Train Set	18244	18
Sepsis Test Set	4562	18
Toplam	22806	18

3 Results

Table 5 shows the results of the classification based on performance metrics. Data from septic patients are used for training in the model. To evaluate the performance of the model, the highest accuracy and area under the ROC curve (AUC) are 96.49% and 70.78%, respectively. The precision and sensitivity of the proposed model are 98.85% and 42.70%, respectively and The model has reached 70.78% AUC values.

Table 5. Results for classification based methods.

Classification methods	Accuracy(%)	Sensivity(%)	Specificity(%)	AUC(%)
Convolutional Neural Network (CNN)	**96.49**	**98.85**	**42.70**	**70.78**
Extreme Gradient Boosting (XGBoost)	96.64	99.79	25	62.39

4 Discussion

The proposed model in Fig. 3, achieved the highest classification accuracy and AUC value of 96.49%. Since the study was in the initial phase, the estimation of the determined laboratory measurements obtained from the clinical data was determined by estimating the NaN values. The results of 6 measurements were reported, while other studies' reporting ignored the given NaN values. Kamaleswaran et al. [26] study on comparing logistic regression and random forest classifiers shows that our proposed method is successful with 70% AUC using only clinical data.

In order to train the model, a data set that contains 22806 samples was used in this study. Large data sets from both classes are needed to train deep learning models. Since the difference between Sepsis and non-Sepsis people in the data set is high, it does not give stable results when training is done. Therefore, accuracy charts and a classifier using unbalanced data sets do not reflect the actual classification performance, because this leads to a misleading interpretation of the sensitivity of the model [27]. ROC charts, in contrast, measure the classification performance of the model more accurately because some of the true positives among the positive predictions are calculated in these charts [27].

The process of estimating lab results from clinical data by regression takes time, so the pre-treatment and training phases may take longer than traditional pre-processing. When the sample number of this process is low, the Mean Squared Error and standard deviation values are high. However, since obtaining laboratory results is more costly and time consuming, the used method shows that it is advantageous both during the process and in terms of calculation.

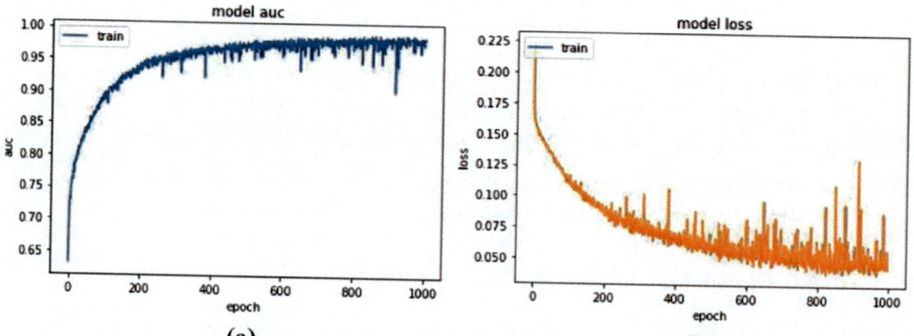

(a) (b)

Fig. 3. (a) Training auc vs epoch graph and (b) loss vs epoch graph.

5 Conclusion and Future Work

Sepsis is a condition that occurs when a person's immune system releases large amounts of inflammatory chemicals to fight infection caused by pathogens. The release of large amounts of chemicals causes fluid to accumulate in a person's tissues, leading to organ dysfunction and eventually septic shock. Prediction of sepsis status is imperative to completely avoid sepsis and its permanent effects on the body. Deep learning models are increasingly using traditional machine learning techniques. In this study, deep convolution network is used to predict sepsis. Some of the septic data were used for training in the model. The highest accuracy and area under the ROC curve (AUC) obtained to assess the performance of the model were 96.49% and 70.78%, respectively. Our experiments have shown that the proposed model is an effective self-diagnostic tool with high accuracy and AUC can be used to predict sepsis. For future research, we intend to improve the performance of the model by training it with more diverse data including sepsis 1, 2 and 3 criteria, using a large data set. With more balanced data, the model can learn better and hence achieve a higher accuracy and AUC value. In datasets which have many features, choosing the most appropriate features may be the solution to increase performance.

Appendix

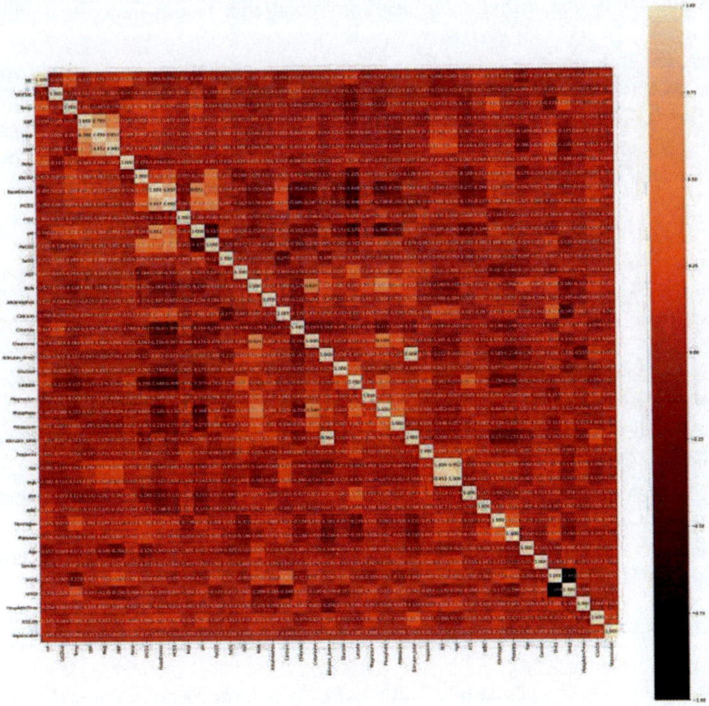

References

1. Berg, D., Gerlach, H.: Recent advances in understanding and managing sepsis. F1000 Res. **7**(1570), 1–8 (2018)
2. Caraballo, C., Jaimes, F.: Organ dysfunction in sepsis: an ominous trajectory from infection to death. Yale J. Biol. Med. **92**(4), 629–640 (2019)
3. Howell, M.D., Donnino, M., Clardy, P., Talmor, D., Shapiro, N.I.: Occult hypoperfusion and mortality in patients with suspected infection. Intensive Care Med. **33**(11), 1892–1899 (2007)
4. Vincent, J.L.: The clinical Challenge of sepsis identification and monitoring. PLoS Med. **13**(5), 1–10 (2016)
5. Lynn, B., Rn, M., Cncc, B., Morgan, B.L.: Critical Care Concepts Sepsis Overview October 22, 2013 1, 2017, pp. 1–20, no. C (2017)
6. Gül, F., Arslantaş, M.K., Cinel, I., Kumar, A.: Changing definitions of sepsis, Turk Anesteziyoloji ve Reanimasyon Dern. Derg. **45**(3), 129–138 (2017)
7. Marik, P.E.: Don't miss the diagnosis of sepsis! Crit. Care **18**(5), 12–14 (2014)
8. Sinha, M., Jupe, J., Mack, H., Coleman, T.P., Lawrence, S.M., Fraley, I.: Towards detection directly from whole Blood: current and emerging technologies for rapid diagnosis of microbial infections without. Clin. Microbiol. Rev. **31**(2), 1–26 (2018)
9. Pierrakos, C., Vincent, J.L.: Sepsis biomarkers: a review. Crit. Care **14**(1), 1–18 (2010)
10. Brown, S.M., Jones, J., Kuttler, K.G., Keddington, R.K., Allen, T.L., Haug, P.: Prospective evaluation of an automated method to identify patients with severe sepsis or septic shock in the emergency department. BMC Emerg. Med. **16**(1), 1–7 (2016)
11. Acharya, U.R., et al.: Automated identification of shockable and non-shockable life-threatening ventricular arrhythmias using convolutional neural network. Future Generat. Comput. Syst. **79**, 952–959 (2018)
12. Yildirim, O., Baloglu, U.B., Tan, R.S., Ciaccio, E.J., Acharya, U.R.: A new approach for arrhythmia classification using deep coded features and LSTM networks. Comput. Methods Progr. Biomed. **176**, 121–133 (2019)
13. Oh, S.L., Vicnesh, J., Ciaccio, E.J., Yuvaraj, R., Acharya, U.R.: Deep convolutional neural network model for automated diagnosis of schizophrenia using EEG signals. Appl. Sci. **9**(14), 2870 (2019)
14. Jahmunah, V., et al.: Computer-aided diagnosis of congestive heart failure using ECG signals – a review. Phys. Med. **62**(March), 95–104 (2019)
15. Shu Lih Oh, M.K., Jahmunah, V., Ooi, C.P., Tan, R.-S., Ciaccio, E.J., Yamakawa, T., Tanabe, M.: Classification of Heart Sound Signals Using a Novel Deep WaveNet Model 2020 no. June (2020)
16. Lih, O.S., et al.: Comprehensive electrocardiographic diagnosis based on deep learning, Artif. Intell. Med. **103** (2020)
17. Reyna, M.A., et al.: Early prediction of sepsis from clinical data: the PhysioNet/computing in Cardiology Challenge 2019. Crit. Care Med. **48**(2) (2020)
18. Kul, S.: Uygun istatistiksel test seçim kilavuzu/guideline for suitable statistical test selection. Plevra Bülteni **8**(2), 26 (2014)
19. Lih, O.S., et al.: Comprehensive electrocardiographic diagnosis based on deep learning. Artif. Intell. Med. **103**, 101789 (2020)
20. Indolia, S., Goswami, A.K., Mishra, S.P., Asopa, P.: Conceptual understanding of convolutional neural network - a deep learning approach. Procedia Comput. Sci. **132**, 679–688 (2018)
21. Cascade-correlation, R., Chunking, N.S.: Long short-term memory. Neural Comput. **9**(8), 1–32 (1997)

22. Bengio, Y.: Autoencoders, unsupervised learning, and deep architectures Pierre. Found. Trends® Mach. Learn. **2**(1), 1–127 (2009)
23. Murat, F., Yildirim, O., Talo, M., Baloglu, U.B., Demir, Y., Acharya, U.R.: Application of deep learning techniques for heartbeats detection using ECG signals-analysis and review. Comput. Biol. Med. **120**, 103726 (2020)
24. Chen, T., Guestrin, C.: Xgboost: a scalable tree boosting system. In: Proceedings of the 22nd ACM SIGKDD International Conference on Knowledge Discovery and Data Mining (2016)
25. Duda, R.O., Hart, P.E., Stork, D.G.: Pattern Classification. Wiley, New York (2001)
26. Kamaleswaran, R., Akbilgic, O., Hallman, M.A., West, A.N., Davis, R.L., Shah, S.H.: Applying artificial intelligence to identify physiomarkers predicting severe sepsis in the PICU, Pediatr. Crit. Care Med. **19**(10), (2018). https://doi.org/10.1097/PCC.0000000000001666
27. Saito, T., Rehmsmeier, M.: The precision-recall plot is more informative than the ROC plot when evaluating binary classifiers on imbalanced datasets. PLoS ONE **10**(3), 1–21 (2015)

Analysis of Predictive Parameters in Prediction of the Occurrence of Myocardial Infarction Using Artificial Neural Networks

Merima Bukva(✉), Ajla Bešlija, Lejla Bihorac, Melika Brčkalija, Semira Budimović, and Nejra Buljubašić

Faculty of Pharmacy, University of Sarajevo, Zmaja od Bosne 8, 71000 Sarajevo, Bosnia and Herzegovina
merimabukva@ffsa.unsa.ba

Abstract. One of the most common conditions in population is myocardial infarction, which refers to the tissue death of the heart muscle caused by ischemia - lack of oxygen delivery to the myocardial tissue. This type of acute coronary syndrome can be described with a sudden or short-term change in symptoms related to blood flow to the heart. This paper presents the development of a neural network for prediction of the risk for occurrence of myocardial infarction based on diagnostic parameters. Neural network was developed based on 300 samples, 100 were healthy individuals, and 200 individuals with already detected myocardial infarction. Developed neural network has 8 input parameters: age, sex, HDL-C level, total cholesterol level, systolic blood pressure (treated/untreated), diabetes status and smoking habits. Neural network, based on these 8 input parameters, provides prediction of the risk score for occurrence of myocardial infarction. Based on the predicted risk score the patient can be classified in one of three categories: high, medium and low. Architecture with 4 neurons in hidden layer and sigmoid transfer function showed the best accuracy – 84.4%.

Keywords: Myocardial infarction · Cardiovascular disease risk · Artificial neural network

1 Introduction

Myocardial infarction is one of the leading causes of death in the world [1]. It is a disease caused by a reduction or cessation of blood flow in a part of the heart, leading to necrosis of the heart muscle. It most often occurs as a result of a blood clot in the epicardial artery. In the heart muscle, the blood supply must be equal to the muscle's need for oxygen. The disease can also be defined as the cell death of cardiac myocytes, caused by ischemia [2]. We find this topic particularly interesting and valuable, since the cardiovascular diseases are the leading cause of death worldwide, taking 17.9 million people every year, so there is absolutely a great need for improvement in assessing the patient in a timely and more accurate manner [3].

A. Badnjevic and L. Gurbeta Pokvić (Eds.): CMBEBIH 2021, IFMBE Proceedings 84, pp. 184–190, 2021.
https://doi.org/10.1007/978-3-030-73909-6_20

Due to the modern lifestyle and unhealthy habits, risk factors such as smoking, psychological stress, obesity and low physical activity come forward as the main elements connected to the myocardial infarction [4]. In all probability, risk factors such as hypertension and abnormal lipid profile are the most relevant factors.

Low-density lipoprotein (LDL) cholesterol acts as a precursor for oxidized form (OxLDL) which contributes to the forming of atherosclerosis and its progression which, most often, leads to the heart failure [5]. Contrary to that, high-density lipoprotein (HDL) cholesterol is inversely associated with cardiovascular diseases, as it catches the "bad" cholesterol and carries it to liver [6]. Analogous, body weight, more precisely obesity, can be directly connected to the higher risk of heart-associated morbidity and mortality [7]. Similarly, the association between the systolic pressure values higher than the recommended ones, and the likelihood of fatal outcome in cardiac arrest, has been proven in multiple studies already, since the chronic hypertension causes cardiac hypertrophy, and therefore higher oxygen demand [8, 9]. As regards to the age factor, in developed countries, myocardial infarction tends to affect people between the sixth and seventh decade of life, whereas in less developed places this scale lowers to a younger age, approximately one decade earlier [10]. In addition, there has been solid evidence that the gender plays an important role when it comes to the susceptibility to coronary diseases. Concerning the women, multiple studies have shown that heart attack occurs 6–10 years later than in men due to the protective role of estrogen, which plays a central factor in sustaining healthy lipid profile throughout the life. Still, the risk is equalized after the menopause [11]. Unhealthy habits, such as smoking, increase the risk of coronary disorder, especially in men. The greatest danger regarding the tobacco smoking lies in the accelerated development of atherosclerosis by causing damage to the vascular endothelium and promoting the formation of blood clots [11, 12]. It is also important to mention the effect that chronic conditions, such as diabetes have to the overall condition of the organism. Specifically, untreated diabetes leads to high blood glucose levels, which causes an increase in reactive oxygen species, atherosclerotic changes and macrovascular complications [13].

In this study, our goal is to make a different approach to the prediction of the cardiovascular diseases based on the usage of the neural networks as novel strategies in this department. These artificial intelligence based models show great potential for the utilization in medicine sphere [14]. Application of neural networks for prediction of the appearance and outcome of different diseases has been used in healthcare already. For example, in a study by Ćatić et al. [15], neural networks are used for the prenatal diagnosis and early detection of Patau, Edwards, Down, Turner and Klinefelter syndrome. There are several other works connected to the respiratory system, asthma, chronical obstructive pulmonary disease and pulmonary function prediction [16–21], cardiovascular system disease prediction [22–24] and metabolic disorders like prediabetes, diabetes and metabolic syndrome prognosis [25, 26], all of which have artificial networks as their foundation. For instance, Badnjević et al. [18, 20] explored the usage of ANNs in manipulating the process of diagnosing and classification for asthmatic patients. Neural networks are also actively used for the prediction of machines' performances such as defibrillators [27] and other medical devices. In a research by Kovačević et al., ANNs are used for the prognosis of the infant incubator performances [28], and in a study by

Hrvat et al., ANNs are used for the predicament of the infusion and perfusor pumps machinery [29]. Artificial intelligence is used on a genetic level as well, with studies such as the one by Mrkulić et al. where scientists made a correlation of the likelihood of leukemia gene expression with occurring mutations [30].

2 Materials and Methods

Detailed information about the dataset used in this study is presented in the Table 1.

Table 1. Properties of the dataset used for development of artificial neural network

Class	Myocardial infarction	Healthy	TOTAL
Male	95	46	141
Average age (years)	54,5	46,15	50,32
Female	105	54	159
Average age (years)	51	51,57	51,28
Total number of patients per class	200	100	300

As it can be seen from Table 1, this research was done using the database which consisted of 300 samples. Out of those 300 samples, 100 were healthy individuals with no health conditions that could be connected to the development of the cardiovascular diseases, and 200 individuals with disturbed levels of parameters of interest, such as HDL-cholesterol, total cholesterol and systolic blood pressure. Besides that, information about age, gender, smoking habits and diabetes was collected. Based on these parameters, the risk of myocardial infarction occurrence was predicted by using neural networks.

The distribution of the dataset for development of the Artificial Neural Network (ANN) is presented in Table 2. Since this is relatively small dataset, distribution of 70:30% for training and validation was used [31].

Table 2. Training and validation dataset distribution

Total number of samples 300300	Healthy	Myocardial infarction
No. of samples	100	200
Training 210	70	140
Validation 90	30	30

Artificial Neural Network (ANN) was developed in order to perform prediction of risk for the occurrence of myocardial infarction based on eight input parameters as mentioned before. The output of the network is calculation of the risk score for occurrence of myocardial infarction. Based on the risk score there are three risk categories to which patient can be classified: high, medium and low. These three categories are formed based on the Framingham Risk Score system that is traditionally used [32].

Framingham Risk Score system is a tool which, after dividing individuals by gender and age, evaluates parameters linked to the development of cardiovascular disease and its outcome: systolic blood pressure, HDL cholesterol, total cholesterol, as well as diabetes and smoking which are well-known risk factors. It works on the principle of adding points to estimate the risk of the cardiovascular event. In general, absolute risk percentage is classified as low if it is $<10\%$, intermediate if it is 10–20% and high risk if it is $>20\%$.

The ANN used in this research is constructed of two layers and based on the standard architecture for pattern recognition. This particular construction has been used because it shows the best relation to our research demands, considering that our input data includes multiple parameters, as previously mentioned [15, 17]. The input layer is made of network inputs. Input layer is followed by a hidden layer which consists of parallelly connected neurons. The network was tested on training performance with different number of neurons. Since, this classification is basically highly non-linear data classification, sigmoid function was used for hidden layer [20]. During the design of the network, different number of neurons was used in order to measure the performance. According to the Mean Square Error (MSE), appropriate number of neurons was chosen. After training, validation of the network performance was done using confusion matrix. Training was performed using Levenberg–Marquadt training algorithm.

3 Results

According to the previous research and expert recommendation [15, 17], for prediction of risk for myocardial infarction in this paper, two-layer feedforward artificial neural network, with sigmoid transfer function in hidden layer was developed. For this research, ANN with 8 inputs and 1 output was developed. Input factors are systolic blood pressure, systolic blood pressure treatment, HDL cholesterol, total cholesterol, diabetes, smoking, age and sex.

Training was performed using Levenberg–Marquadt training algorithm. During training of the neural network, different number of neurons was tested. System was tested for number of neurons from 2 to 30 with increment of 1 neuron in each step. Training performance was varying between 75.3% for 25 neurons to 86.9% achieved with 4 neurons. Finally, architecture with 4 neurons in hidden layer and sigmoid transfer function was chosen for the validation step. Mean error for prediction of risk for myocardial infarction during training process was 1.67 absolute value.

Performance of the neural network during training with 4 neurons is shown in the Fig. 1.

The results of the validation of neural network with 4 neurons is presented in the Table 3.

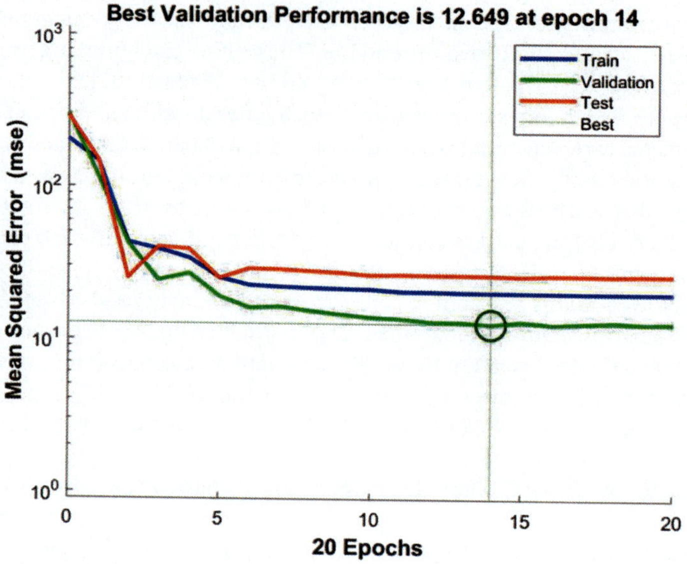

Fig. 1. Performance with 4 neurons

Table 3. Artificial Neural Network performance during validation

n = 90	Group 1 Low risk	Group 2 Middle risk	Group 3 High risk	
GROUP 1 LOW risk 38	30	6	2	**Accuracy** [%] 84.4
Group 2 Middle risk 9	0	7	2	
GROUP 3 High risk 43	1	3	39	
	Accuracy Group 1 [%]	Accuracy Group 2 [%]	Accuracy Group 3 [%]	
	78.9%	77.8%	90.7%	

4 Conclusion

Cardiovascular diseases, especially myocardial infarction, are one of the leading causes of death. In this paper, we present an assessment system in predicting the outcome of a heart attack based on the parameters of interest that can have impact in development of cardiovascular diseases. By creating a database and using ANN we can predict potential risk of myocardial infarction due to decrease prevalence of mortality. We develop two-layer feedforward artificial neural network, with sigmoid transfer function in hidden layer for our research. Our ANN is based on the eight input parameters and it calculates risk score and sorts it in three categories: high, medium and low according to the Framingham risk score. In the end of the training final accuracy for all groups was 84.4%. In this research we worked on a quite small sample, but we are considering developing more sophisticated and multifunctional ANN in the future. That ANN would be based on this ANN and new data that we found out in this research. Our goal in the future is to raise public awareness of the importance of monitoring and later successfully treating variable parameters, primarily by changing lifestyles and then using appropriate medications to reduce or avoid the incidence of a heart attack.

Conflict of Interest. The authors have no conflicts of interest to disclose.

References

1. Thygesen, K., Alpert, J.S., White, H.D.: Universal Definition of Mycardial Infarction. J. Am. Coll. Cardiol. **50**(22), 2173–2195 (2007)
2. Saleh, M., Ambrose, J.A.: Understanding myocardial infarction. F1000Research **7**, 1378 (2018)
3. GBD 2015 Mortality and Causes of Death Collaborators. Global, regional, and national life expectancy, all-cause mortality, and cause-specific mortality for 249 causes of death, 1980–2015: a systematic analysis for the Global Burden of Disease Study 2015. Lancet. **388**(10053):1459–1544 (2016)
4. Jayaraj, J.C., Davatyan, K., Subramanian, D.D., Priya, J.: Epidemiology of Myocardial Infarction. IntechOpen (2018)
5. Pirillo, A., Norata, G.D., Catapano, A.L.: LOX-1, OxLDL, and atherosclerosis. Mediators Inflamm. **2013**, 152786 (2013)
6. Voight, B.F., Peloso, G.M., Orho-Melander, M., et al.: Plasma HDL cholesterol and risk of myocardial infarction: a mendelian randomisation study. Lancet **380**(9841), 572–580 (2012)
7. Yusuf, S., Hawken, S., Ounpuu, S., et al.: Obesity and the risk of myocardial infarction in 27,000 participants from 52 countries: a case-control study. Lancet **366**(9497), 1640–1649 (2005)
8. Pedrinelli, R., Ballo, P., Fiorentini, C., et al.: Hypertension and acute myocardial infarction. J. Cardiovascular Med. **13**(3), 194–202 (2012)
9. Picariello, C., Lazzeri, C., Attanà, P., et al.: The impact of hypertension on patients with acute coronary syndromes. Int. J. Hyperten. **2011**, 563657 (2011)
10. Abed, M.A., Eshah, N.F., Moser, D.K.: Risk profile of myocardial infarction in young versus older adults. Heart Lung. **47**(3), 226–230 (2018)
11. Pedersen, L.R., et al.: Risk factors for myocardial infarction in women and men: a review of the current literature. Curr Pharm Des. **22**(25), 3835–3852 (2016)
12. Lisowska, A., Makarewicz-Wujec, M., Filipiak, K.J.: Risk factors, prognosis, and secondary prevention of myocardial infarction in young adults in Poland. Kardiol Pol. **74**(10), 1148–1153 (2016)
13. Huang, D., Refaat, M., Mohammedi, K., Jayyousi, A., Al Suwaidi, J., Abi, K.C.: Macrovascular complications in patients with diabetes and prediabetes. Biomed Res Int. **2017**, 7839101 (2017)
14. Weng, S.F., Reps, J., Kai, J., Garibaldi, J.M., Qureshi, N.: Can machine-learning improve cardiovascular risk prediction using routine clinical data? PLoS ONE **12**(4), e0174944 (2017)
15. Catic, A., et al.: Application of Neural Networks for classification of Patau, Edwards, Down, Turner and Klinefelter Syndrome based on first trimester maternal serum screening data, ultrasonographic findings and patient demographics. BMC Med. Genomics **11**, 19 (2018)
16. Gurbeta, L., et al.: A telehealth system for automated diagnosis of asthma and chronical obstructive pulmonary disease. J Am Med Inform Assoc. **25**(9), 1213–1217 (2018)

17. Badnjevic, A., Gurbeta, L., Custovic, E.: An expert diagnostic system to automatically identify asthma and chronic obstructive pulmonary disease in clinical settings. Sci Rep. **8**(1), 11645 (2018)
18. Badnjevic, A., Gurbeta, L., Cifrek, M., Marjanovic, D.: Diagnostic of asthma using fuzzy rules implemented in accordance with international guidelines and physicians experience. In: IEEE 39th International convention on information and communication technology, electronics and microelectronics (MIPRO), 30. May to 03. June 2016. Opatija, Croatia
19. Granulo, E., et al.: Telemetry system for diagnosis of Asthma and Chronical Obstructive Pulmonary Disease (COPD). In: The 3rd EAI International Conference on IoT Technologies for HealthCare, 18–19 October 2016, Vasteras, Sweden
20. Badnjevic, A., Gurbeta, L., Cifrek, M., Marjanovic, D.: Classification of asthma using artificial neural network. In: IEEE 39th International Convention on Information and Communication Technology, Electronics and Microelectronics (MIPRO), 30 May to 03 June 2016. Opatija, Croatia (2016)
21. Badnjevic, A., Gurbeta, L., Cifrek, M., Pecchia, L.: Pre-classification process symptom questionnaire based on fuzzy logic for pulmonary function test cost reduction. In: CMBEBIH 2017. IFMBE Proceedings, vol. 62. Pp. 608–616, Springer, Singapore (2017)
22. Šećkanović, A., et al.: Review of artificial intelligence application in cardiology. In: 9th Mediterranean Conference on Embedded Computing MECO 2020 (2020)
23. Mustafić, L.D., et al.: Diagnosis of severe aortic stenosis using implemented expert system. In: Badnjevic, A., Škrbić, R., Gurbeta Pokvić, L. (eds.) CMBEBIH 2019. IFMBE Proceedings, vol. 73. Springer, Cham (2019)
24. Alić, B., et al.: Machine learning techniques for classification of diabetes and cardiovascular diseases. In: 6th Mediterranean Conference on Embedded Computing (MECO), Bar, Montenegro, pp. 1–4 (2017)
25. Sejdinovic, D., et al.: Classification of prediabetes and Type 2 Diabetes using Artificial Neural Network. In: CMBEBIH 2017 (2017)
26. Alic, B., et al.: Classification of metabolic syndrome patients using implemented expert system. In: CMBEBIH 2017. IFMBE Proceedings, vol. 62, pp 601–607, Springer, Singapore (2017)
27. Badnjević, A., Gurbeta Pokvić, L., Hasičić, M., Bandić, L., Mašetić, Z., Kovačević, Ž., Kevrić, J., Hifziefendić, J., Pecchia, L.: Evidence-based clinical engineering: machine learning algorithms for prediction of defibrillator performance. In: Biomedical Signal Processing and Control, vol. 54, p. 101629, September 2019
28. Kovačević, Ž., et al.: Prediction of Medical Device Performance Using Machine Learning Techniques: Infant incubator case study. Health and Technology (2020)
29. Hrvat, F., et al.: Artificial Intelligence for prediction of medical device performance: Infusion and perfusor pumps case study. In: 9th Mediterranean Conference on Embedded Computing MECO 2020 (2020)
30. Mrkulić, F., et al.: A correlation of leukemia genes overexpression and point mutations in different tissues. In: Badnjevic, A., Škrbić, R., Gurbeta Pokvić, L. (eds.) CMBEBIH 2019. CMBEBIH 2019. IFMBE Proceedings, vol 73. Springer, Cham
31. Rau, H.H., et al.: Development of a web-based liver cancer prediction model for type II diabetes patients by using an artificial neural network. Comput. Methods Programs Biomed. **125**, 58–65 (2016)
32. Continuing Medical Implementation. Risk Calculators. FRAMINGHAM RISK SCORE. Dostupno na: https://www.cvtoolbox.com/cvtoolbox1/risk_calc/rc.html?fbclid=IwAR2vGV7_izlZUILSEmFjXyPQb5dGf-V8NxH6ZIfTqvCqKA1-iiYfuUb0cvc

Application of Neural Network in the Kidney Living Donor Selection Criteria Using Biomarkers Data

Memnuna Hasanović, Ena Hasković(✉), Alisa Hebibović, Azra Herić, Amila Hodžić, and Nura Hodžić

Faculty of Pharmacy, University of Sarajevo, Zmaja od Bosne 8, 71 000 Sarajevo, Bosnia and Herzegovina
enahaskovic@ffsa.unsa.ba

Abstract. Selecting a suitable kidney donor is demanding and significant for both the recipient and the living donor. We chose to apply Artificial Neural Network (ANN) to the living kidney donor selection criteria using biomarkers' data, and their application in patients' classification when it comes to managing the waiting list. For the development of the ANN used for classification a database containing following biomarkers was used: HIV (human immunodeficiency virus), HCV (hepatitis C virus), HBV (hepatitis B virus), active ischemic heart disease, recent history of cancer, glomerular filtration rate, body mass index, impaired glucose tolerance, dyslipidemia, diabetes mellitus, hematuria. Based on these biomarkers, an ANN was developed in order to classify possible living kidney donors into two categories: eligible or ineligible. The database consisted of 300 patients: 100 healthy and 200 non-healthy patients. Training and validation performances were done on a neural network with 30, 12, and 2 neurons. As the training and validation performances of the neural network with 12 neurons in the hidden layer were the highest, it is suggested that this neural network be used for predicting whether a person is eligible for a kidney donor.

Keywords: Living kidney donor · Solid organ transplantation · ANN · KDPI

1 Introduction

Solid organ transplantation (SOT) offers life-saving treatment for diseases considered terminal of those associated with a significant impairment in a patients' quality of life [1]. Worldwide need for solid organs for saving lives is growing at a rapid pace. According to the Global Observatory on Donation & Transplantation (GODT), 11469 solid organs were transplanted in 2012, which only met approximately 10% of the global needs.

Two options for treating kidney failure are dialysis and kidney transplantation [2]. Living donor (LD) kidneys may be obtained through a directed donation, a non-directed donation, or an exchange program in which one donor kidney is swapped with another, more compatible kidney. Deceased donor (DD) kidneys are designated by the Kidney

A. Badnjevic and L. Gurbeta Pokvić (Eds.): CMBEBIH 2021, IFMBE Proceedings 84, pp. 191–198, 2021.
https://doi.org/10.1007/978-3-030-73909-6_21

Donor Profile Index (KDPI), ranging from 0% to 100%. The KDPI scoring system integrates factors including hypertension, diabetes, ethnicity, and donor age. Most kidneys are from individuals declared brain dead, designated as donation after brain death. In recent guidelines from Public Health Service, increased-risk donor kidneys include kidneys from donors who have tested negative for human immunodeficiency virus (HIV), hepatitis C virus (HCV), and hepatitis B virus (HBV), but have an increased risk for exposure due to behavioral risk factors [3]. Guideline recommendations for the medical assessment of living kidney donor candidates are: measured glomerular filtration rate (mGFR), cystoscopy, kidney biopsy, and additional cardiac investigations. Identifying cardiac risk factors in the donor evaluation process includes assessment for hypertension, obesity, diabetes, dyslipidemia and smoking history. Chest X-rays and electrocardiograms are also standard [4]. Absolute contraindications to kidney transplantation are: active ischemic heart disease or severe cardiomyopathy, active infection, recent history of cancer other than nonmelanoma skin cancer, cirrhosis or advanced liver fibrosis, active substance abuse or dependance, active psychosis and incorrigible noncompliance [5].

Artificial Neural Networks (ANN) find significant applications in medicine because they can provide decision support that effectively reduces the workload for clinicians [6]. ANNs are systems for information processing. They represent a series of algorithms, leading to conclusion, that work in a similar way like human neuron cells [7]. Aware of the importance of their usage in pharmacy and medicine, and inspired by many publications [8–15] we developed ANN for living kidney donor selection using biomarkers data. The main usage of these systems is in patients' classification when it comes to managing the waiting list, considering that as one of the greatest problems transplant centers are faced with.

2 Material and Methods

For the development of the ANN a database containing following biomarkers was used: (1) HIV (human immunodeficiency virus), (2) HCV (hepatitis C virus), (3) HBV (hepatitis B virus), (4) active ischemic heart disease, (5) recent history of cancer, (6) glomerular filtration rate, (7) body mass index, (8) impaired glucose tolerance, (9) dyslipidemia, (10) diabetes mellitus, and (11) hematuria.

A database consisted of 300 samples, out of which 100 corresponded to healthy patients and 200 to non-healthy patients. Each sample was represented with these 11 biomarkers, so an ANN was developed in order to classify possible living kidney donors into two categories: eligible or ineligible. The dataset distribution for the ANN development is presented in Table 1. As this is a relatively small dataset (Table 1.), a distribution of 70:30% for training and validation was used [16].

For the development of the ANN in this research, the standard architecture for pattern recognition was used. That is a feedforward neural network which consists of two-layers. This architecture is designed because, according to the application experts, that kind of an ANN is sufficient to properly perform the classification of input data [17, 18].

The input layer is made of network inputs. Input layer is followed by a hidden layer which consists of parallelly connected neurons. The network was tested on training performance with different numbers of neurons. Since this classification is basically a

Table 1. Dataset distribution

Total number of samples	Eligible	Ineligible
349	148	201
Training 143	103	140
Validation 106	45	61

highly non-linear data classification, transig function was used for the hidden layer [19]. During the design of the network, a different number of neurons was used in order to measure the performance. According to the Mean Square Error (MSE), an appropriate number of neurons was chosen. After training, the validation of the network performance was done using a confusion matrix.

According to these results, specificity, sensitivity and accuracy of the network are calculated according to Eqs. (1), (2) (3):

$$sensitivity = \frac{true\ positive}{(true\ positive + false\ negative)} \tag{1}$$

$$specificity = \frac{true\ negative}{(true\ negative + false\ positive)} \tag{2}$$

$$accuracy = \frac{(true\ negative + true\ positive)}{(true\ negative + true\ positive + false\ negative + false\ positive)} \tag{3}$$

3 Results

Based on previous research and expert recommendation [17], for the classification of eligible kidney donor prediabetes and diabetes type 2 in this paper, we have developed a two-layer feedforward artificial neural network, with transig transfer function in the hidden layer.

Table 2. Artificial Neural Network performance in classification

n = 106	**Eligible**	**Ineligible**	
Eligible 45	16	29	**Accuracy [%]** 78.3
Ineligible 61	54	7	
	Sensitivity [%] 88.5	**Specificity [%]** 64.4	

During development architecture with 30, 12 and 2 neurons in hidden layer were tested. Training accuracy for the feedforward neural network with 12 neurons and transig transfer function in the hidden layer was 89.3%. Training was performed using Levenberg – Marquadt training algorithm. Training performance of the developed neural network is presented in Fig. 1.

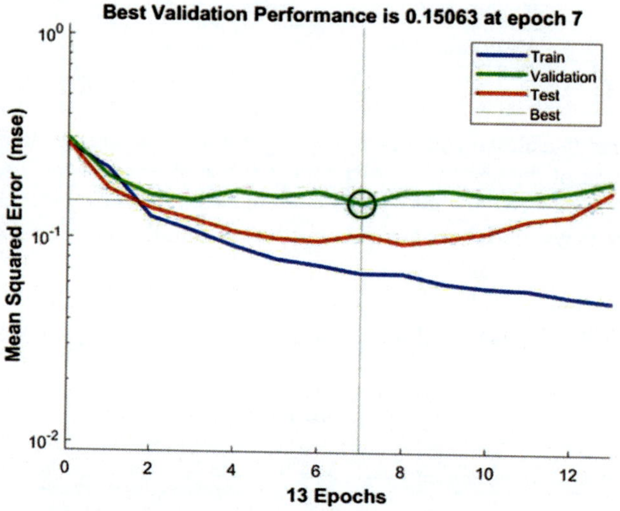

Fig. 1. Training performance of the network with 12 neurons in hidden layer

The performance validation of the feedforward neural network with 12 neurons and transig transfer function in hidden layer is presented in Table 2.

Table 3. Artificial Neural Network performance in classification

n = 106	**Eligible**	**Ineligible**	
Eligible 45	16	29	**Accuracy** [%] 80.2
Ineligible 61	56	5	
	Sensitivity [%] 91.8	**Specificity** [%] 64.4	

Training accuracy of the feedforward neural network with 2 neurons and transig transfer function in the hidden layer was 87.2%. Training was performed using Levenberg – Marquadt training algorithm. Training performance of the developed neural network is presented in Fig. 2.

Performance validation of the feedforward neural network with 2 neurons and transig transfer function in the hidden layer is presented in Table 3.

Training accuracy of the feedforward neural network with 30 neurons and transig transfer function in the hidden layer was 88.8%. Training was performed using Levenberg

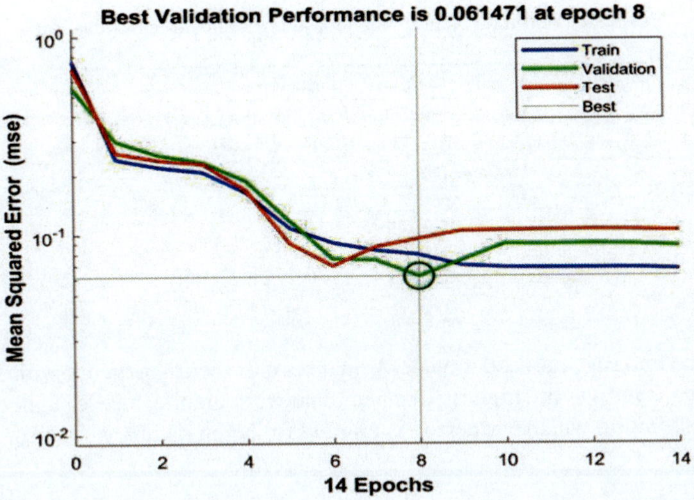

Fig. 2. Training performance of the network with 2 neurons in hidden layer

– Marquadt training algorithm. Training performance of the developed neural network is presented in Fig. 3.

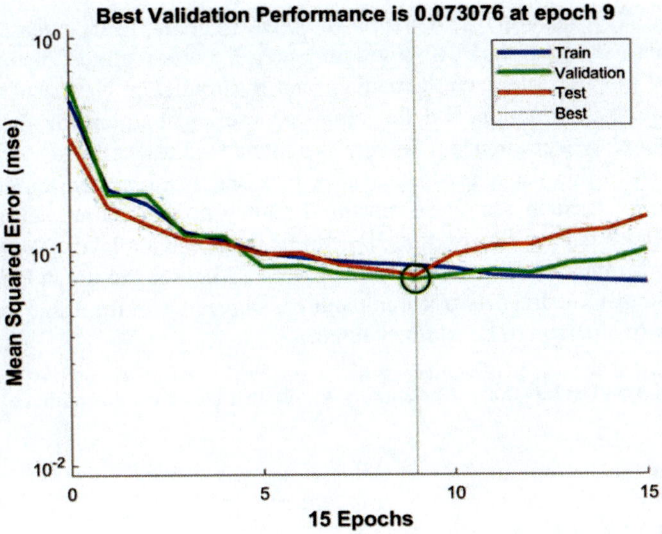

Fig. 3. Training performance of the network with 30 neurons in hidden layer

Performance validation of the feedforward neural network with 30 neurons and transig transfer function in hidden layer is presented in Table 4.

Table 4. Artificial Neural Network performance in classification

n = 106	Eligible	Ineligible	
Eligible 45	24	21	**Accuracy** [%] 74.5
Ineligible 61	58	3	
	Sensitivity [%] 95.1	**Specificity** [%] 46.6	

Since the training and validation performance of the neural network with 12 neurons in the hidden layer was the highest, this neural network architecture is suggested for the usage for predicting whether a person is eligible for being a kidney donor or not.

4 Conclusion

This paper presents the application of neural networks in the criteria for selecting kidney donors using biomarker data. We are witnessing that today, transplants of all solid organs are very common, all because they are a life-saving treatment. The causes of kidney failure are different. This paper describes some of the biomarkers that determine whether someone is a suitable donor or not. Some of these biomarkers that have been processed are HIV, hepatitis C, hepatitis B, active ischemic heart disease, history of cancer, glomerular filtration, BMI, blood pressure, diabetes mellitus, history of smoking, impaired glucose tolerance, hematuria, and dyslipidemia. Neural networks have helped us process information, and they represent a series of algorithms that lead to the conclusion that they work in a similar way as a human neuron cell.

The results in this paper prove that an expert diagnostic system based on neural networks can be used in search for potential kidney donors. Three neural networks were used – those having 12, two and 30 neurons in the hidden layer. The training and validation performance of the neural network with 12 neurons in the hidden layer was the highest, thus this neural network architecture is suggested for the usage for predicting the eligibility of a person to be a kidney donor.

Conflict of Interest Declaration. The authors declare that they have no conflict of interest.

References

1. Black, C.K., Termanini, K.M., Aguirre, O., Hawksworth, J.S., Sosin, M.: Solid organ transplantation in the 21st century. Ann. Transl. Med. **6**(20), 409 (2018). https://doi.org/10.21037/atm.2018.09.68
2. Scheuher, C.: A review of organ transplantation: heart, lung, kidney, liver, and simultaneous liver-kidney. Crit. Care Nurs. Q. **39**(3), 199–206 (2016). https://doi.org/10.1097/CNQ.0000000000000115

3. Voora, S., Adey, D.B.: Management of kidney transplant recipients by general nephrologists: core curriculum 2019. Am. J. Kidney Dis. **73**(6), 866–879 (2019). https://doi.org/10.1053/j.ajkd.2019.01.031
4. Lam, N.N., Lentine, K.L., Garg, A.X.: Renal and cardiac assessment of living kidney donor candidates. Nat. Rev. Nephrol. **13**(7), 420–428 (2017). https://doi.org/10.1038/nrneph.2017.43
5. Scandling, J.D.: Kidney transplant candidate evaluation. Semin. Dial. **18**(6), 487–484 (2005). https://doi.org/10.1111/j.1525-139X.2005.00094.x
6. Šarić, R., et.al.: FPGA-based real-time epileptic seizure classification using Artificial Neural Network. Biomed. Signal Process. Control **62**, 102106 (2020). https://doi.org/10.1016/j.bspc.2020.102106. ISSN 1746-8094
7. Penny, W., Frost, D.: Neural networks in clinical medicine. Med. Decis. Making **16**(4), 386–398 (1996). https://doi.org/10.1177/0272989X9601600409
8. Badnjević, A., Ćifrek, M., Koruga, D., Osmanković, D.: Neuro-fuzzy classification of asthma and chronic obstructive pulmonary disease. BMC Med. Inform. Decis. Making J. **15**(Suppl 3), S1 (2015). https://doi.org/10.1186/1472-6947-15-S3-S1
9. Gurbeta, L., et al.: A telehealth system for automated diagnosis of asthma and chronical obstructive pulmonary disease. J. Am. Med. Inform. Assoc. **25**(9) (2018). https://doi.org/10.1093/jamia/ocy055
10. Badnjević, A., et al.: Evidence-based clinical engineering: machine learning algorithms for prediction of defibrillator performance. Biomed. Signal Process. Control **54**, 101629 (2019). https://doi.org/10.1016/j.bspc.2019.101629. ISSN 1746-8094
11. Kovačević, Ž., Gurbeta, P.L., Spahić, L., Badnjević, A.: Prediction of medical device performance using machine learning techniques: infant incubator case study. Health Technol., 151–155(2019). https://doi.org/10.1007/s12553-019-00386-5
12. Badnjević, A., Ćifrek, M., Koruga, D.: Classification of Chronic Obstructive Pulmonary Disease (COPD) using integrated software suite. In: IFMBE XIII Mediterranean Conference on Medical and Biological Engineering and Computing (MEDICON), 25–28. Chronic Obstructive Pulmonary Disease (COPD): Clinical Symptoms, Emerging Treatment Strategies and Impact on Quality of Life, pp. 1–26 (2013)
13. Avdakovic, S., et al.: Diagnosis of epilepsy from EEG signals using global wavelet power spectrum. In: 6th European Conference of the International Federation for Medical and Biological Engineering (MBEC), IFMBE Proceedings, vol. 45. Springer, Cham (2014). https://doi.org/10.1007/978-3-319-11128-5_120
14. Divović Mustafić L., et al.: Diagnosis of severe aortic stenosis using implemented expert system. In: CMBEBIH 2019. IFMBE Proceedings, vol. 73. Springer, Cham (2020). https://doi.org/10.1007/978-3-030-17971-7_23
15. Dícz-Sanmartín, C., Sarasa Cabezuelo, A.: Application of artificial intelligence techniques to predict survival in kidney transplantation: a review **9**(2), 572(2020). https://doi.org/10.3390/jcm9020572
16. Siao-Hsien, R., Chien-Yeh, H., Lin, Y.-A., Suleman, A., Anis, F., Wei, L.-M., Hsu, M.-H.: Development of a web-based liver cancer prediction model for type II diabetes patients by using an artificial neural network. Comput. Methods Programs Biomed. **125**, 58–65 (2016). https://doi.org/10.1016/j.cmpb.2015.11.009, ISSN 0169-2607
17. Ćatic, A., et al.: Application of Neural Networks for classification of Patau, Edwards, Down, Turner and Klinefelter Syndrome based on first trimester maternal serum screening data, ultrasonographic findings and patient demographic. BMC Med. Genomics **11**, 19 (2018). https://doi.org/10.1186/s12920-018-0333-2

18. Badnjević, A., Gurbeta, L., Čustović, E.: An expert diagnostic system to automatically identify asthma and chronic obstructive pulmonary disease in clinical settings. Sci. Rep. **8**, 1–9 (2018)
19. Badnjević, A., Gurbeta, L., Ćifrek, M., Marjanović, D.: Classification of asthma using artificial neural network. In: IEEE 39th International Convention on Information and Communication Technology, Electronics and Microelectronics (MIPRO), Opatija, Croatia (2016)

Diagnosis of Different Types of Hyperbilirubinemia Using Artificial Neural Network

Drače Amina(✉), Duraković Murveta, Džafić Amel, Džananović Elmedina, Džanko Meliha, and Džubur Alma

Faculty of Pharmacy, University of Sarajevo, Sarajevo, Bosnia and Herzegovina
aminadrace@ffsa.unsa.ba

Abstract. Hyperbilirubinemia is a common condition in newborn children, as they are inapt to easily discharge bilirubin which induces its concentration in the blood and within other fluids in the body. This condition can be a serious signal for a more severe underlying disease. Its frequency is justification enough for the further development of early diagnostic techniques. The modernization of clinical diagnosis methods has paved the way for the application of artificial intelligence. This is often an enhancement in accuracy and affordability in comparison to traditional diagnosis methods and tools such as bilirubin meters. The main purpose of this paper is to show the possibility of using artificial neural networks (ANNs) in the classification of different types of hyperbilirubinemia. The neural network which was developed in this research is two-layer feedforward artificial neural network with 13 input parameters: *gender, age, skin color, mucosa color, sclera color, urine color, stool color, bilirubin in urine, urobilinogen in urine, urobilinogen in a stool, total bilirubin level, conjugated bilirubin level, unconjugated bilirubin level, and liver enzyme (ALT and ALP) levels.* This work outlines the testing of samples from 360 patients who are either healthy or have a syndrome associated with hyperbilirubinemia. The study successfully assesses the sensitivity and accuracy of this diagnostic approach and evaluates it against the data set from the afore mentioned sample.

Keywords: Hyperbilirubinemia · Artificial neural network · Neonatal hyperbilirubinemia

1 Introduction

Bilirubin is the ultimate breakdown product of hemoglobin formed from heme by the opening of the heme ring at the α-carbon bridge. Serum total bilirubin categorized into unconjugated (UBIL) and conjugated bilirubin (CBIL) [1, 2]. When glucuronides are not excreted, for example, due to obstruction of the bile ducts, the concentration of CBIL in the serum increases. On the other hand, if we have increased bilirubin synthesis, the overloaded liver will not be able to metabolize (conjugate) the entire amount of bilirubin and the concentration of UBIL in the serum will increase. This is the point of

A. Badnjevic and L. Gurbeta Pokvić (Eds.): CMBEBIH 2021, IFMBE Proceedings 84, pp. 199–207, 2021.
https://doi.org/10.1007/978-3-030-73909-6_22

the conjugation process. Based on this, hyperbilirubinemia can be classified as *unconjugated hyperbilirubinemia* (hemolytic jaundice, Gilbert's syndrome, and CrigelNajjar syndrome) and as *conjugated hyperbilirubinemia* (cholestatic jaundice, hepatocellular jaundice, Dubin-Johnson syndrome, and Rotor syndrome) [3].

Hyperbilirubinemia is one of the most prevalent clinical diseases. This condition is characterized by elevated levels of bilirubin (greater than 20 μmol/L) [4]. These elevated levels are usually sign of an underlying disease that affects the liver and biliary tract. Also, it can be a result of a variety of inherited causes or can result from hematologic diseases [5]. Research shows that around 6% of term newborns develop icterus, and 2% of them exceed a total serum bilirubin (BIT) concentration that is above 340 μmol/L. Additionally, recent data suggest that encephalopathy happens to one baby per 50,000100,000 live births [6]. Grohman et al. [7] state that the first routine method that is done includes the skin test for newborns. A skin test is done with a nonchemical photometric device. Results are trusted up to 250 μmol/L bilirubin values. If the skin test result exceeds 200 μmol/L then standard laboratory methods should be used [7]. It is important to measure BIT, and it can be used as a significant predictor that newborns will develop significant hyperbilirubinemia. It will also determine newborns who need phototherapy [8]. There are different syndromes related to hyperbilirubinemia such as: Gilbert syndrome, Crigler-Najjar syndrome, Dubin Johnson syndrome, and Rotor syndrome.

Certain problems can occur during the usual laboratory methods for diagnosis of different types of hyperbilirubinemia, for example: using algorithms that are different, different reference intervals for parameters which can cause false positive or negative results. Beside this common diagnostic procedure, other diagnostic tools based on artificial intelligence can be investigated. Application of artificial intelligence is not uncommon in healthcare nowadays. Successful diagnostic procedures based on artificial neural networks have already been developed in the field of genetics [9], respiratory diseases such as asthma and chronical obstructive pulmonary disease [10, 11], Alzheimer disease [12] and other cardiovascular diseases and conditions [13–18]. This approach has been implemented in the field of hyperbilirubinemia detection. For instance, Falk [19] used machine learning approach for jaundice detection using color corrected smartphone images for the determination of hyperbilirubinemia. Smartphone cameras are used to capture skin induces color values and it is a low-cost alternative to expensive bilirubin meters. Falk reported in his master thesis that his neural network had 75% fewer false negative predictions [20]. Danuhawer et al. [21] investigated enhanced early prediction of clinically relevant neonatal hyperbilirubinemia with machine learning and concluded that earlier detection can be performed when it comes to clinically relevant hyperbilirubinemia when machine learning is used [21].

This paper presents the use of the artificial neural network in the diagnosis of different types of hyperbilirubinemia following the parameters listed in Materials and Methods, with the main goal to facilitate the diagnosis and minimize previously mentioned problems during diagnosis.

2 Materials and Methods

During this study, a database which consisted of 360 samples was used.

Table 1. Dataset overview

Total number of samples	360
Healthy	150
Hyperbilirubinemia syndrome 1. *Gilbert syndrome* 2. *Crigler-Najjar syndrome* 3. *Dubin Johnson syndrome,* 4. *Rotor syndrome* 5. *Cholestatic jaundice* 6. *Hepatocellular jaundice*	210 Each symptom has 35 samples in the database

Detailed information in the database is presented in the Table 1.

For each sample in the database, following parameters were recorded: (1) gender, (2) age, (3) skin color, (4) mucosa color, (5) sclera color, (6) urine color, (7) stool color, (8) bilirubin in urine, (9) urobilinogen in urine, (10) urobilinogen in stool, (11) total bilirubin, (12) conjugated bilirubin, (13) unconjugated bilirubin (UBIL), (14) alanine aminotransferase and (15) alkaline phosphatase. Based on these 15 parameters, each patient is classified into two distinctive categories, as presented in Table 1. The color of the skin, mucosa color, and sclera color are characteristic of each disease: in healthy subjects the skin color is normal, and in patients suffering from any syndrome of hyperbilirubinemia there is yellowish skin color.

The distribution of the dataset for the development of the Artificial Neural Network (ANN) is presented in Table 2. Since this is a relatively small dataset, a distribution of 70:30% for training and validation was used [22].

Table 2. Dataset distribution

Total number of samples 360 (100%)	Healthy 150	Disease 210
Training 252 (70%)	105 (70%)	147 (70%)
Validation 108 (30%)	45 (30%)	63 (30%)

The ANN which is developed performs the classification of patients who are suffering from hyperbilirubinemia and to perform detection of the type of hyperbilirubinemia.

According to these results, specificity, sensitivity, and accuracy of the network are calculated according to Eqs. (1), (2) (3):

$$sensitivity = \frac{true\,positive}{(true\,positive + false\,negative)} \tag{1}$$

$$specificity = \frac{true\,negative}{(true\,negative + false\,positive)} \tag{2}$$

$$accuracy = \frac{(true\,negative + true\,positive)}{(true\,negative + true\,positive + false\,negative + false\,positive)} \tag{3}$$

3 Results

The block diagram of the system which was developed in this study is presented in the Fig. 1.

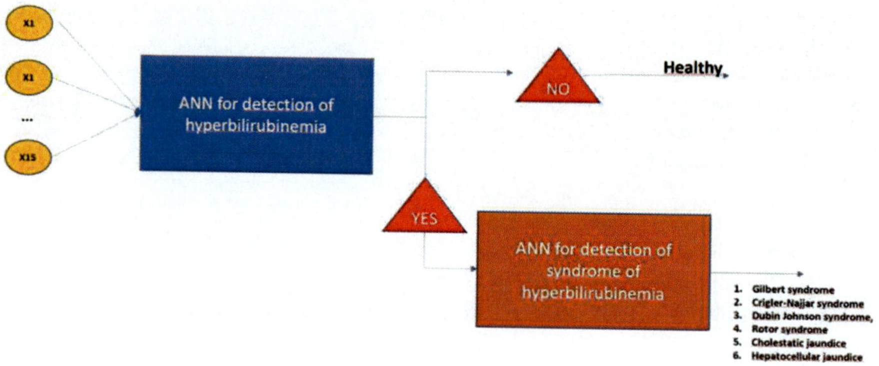

Fig. 1. Block diagram of the developed system

Both developed neural networks are two-layer feedforward architecture with logsig transfer function in the hidden layer. First ANN was developed for classification of patients between healthy and diseased and the second one was developed to classify the exact syndrome that patient suffers from.

For the development of the first ANN number of neurons in hidden layer was changed. Table 3 presents the results of the training process for the first neural network. As it can be seen from the table, the best performance was achieved with the neural network with 15 neurons in the hidden layer. During the training Levenberg – Marquart algorithm was used and for the transfer function, logsig function was used.

Table 3. Training performance

Number of neurons	Training mean square error
2	6.0614e−14
5	5.2300e−15
10	4.4887e−05
15	**1.3823e−08**
30	7.2888e−04
50	3.8747e−08

Training performance for a neural network with 15 neurons in the hidden layer is presented in Fig. 2.

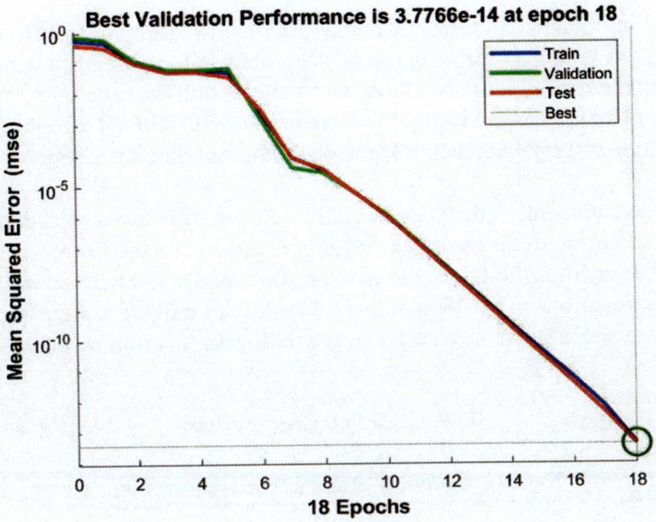

Fig. 2. Training performance of the neural network with 15 neurons in the hidden layer

The validation of the performance of a feedforward neural network with 15 neurons and logsig transfer function in the hidden layer is presented in Table 4. As it can be seen from the Table 3, the overall accuracy of the developed neural network is 93.5% while specificity for classification to the group disease is 93.75.

Table 4. Artificial Neural Network performance during validation

n = 120	**Group 1 Healthy** ANN prediction	**Group 2 Disease** ANN prediction	Accuracy
Group 1 Healthy 45	TP 42	FN 3	**93.5%**
Group 2 Disease 63	FP 4	TN 59	
Sensitivity	93.3%		
Specificity	93.7%		

During development of the second neural network 147 samples from training dataset were used. These samples were subsequently distributed into training, testing and validation dataset in the ratio 70:15:5%. Out of 147 samples Gilbert syndrome, Crigler-Najjar syndrome and Dubin Johnson syndrome were represented with 24 samples, while Rotor syndrome, Cholestatic jaundice and Hepatocellular jaundice were represented with 25 samples.

For the development of the second ANN number of neurons in hidden layer was changed. Table 5 presents the results of the training process for the second neural network. As it can be seen from the table, the best performance was achieved with the neural network with 5 neurons in the hidden layer. During the training Levenberg – Marquart algorithm was used and for the transfer function, logsig function was used.

Table 5. Training performance

Number of neurons	Training mean square error
2	6.8414e−11
5	**4.3300e−15**
10	8.4887e−05
15	1.7852e−08
30	2.1228e−04
50	3.8747e−07

The validation of the performance of a feedforward neural network with 5 neurons and logsig transfer function in the hidden layer is presented in Table 6.

As it can be seen from Table 6, developed ANN shows great sensitivity in terms of detection of the specific syndrome. The lowest sensitivity for classification was for Dubin Johnson syndrome while for Crigler-Najjar syndrome it was the highest. The accuracy of developed architecture can be improved by training with a larger dataset.

Table 6. Artificial Neural Network performance during validation

n = 63	**GROUP 1 Gilbert syndrome** ANN prediction	**GROUP 2 Crigler-Najjar syndrome** ANN prediction	**GROUP 3 Dubin Johnson syndrome** ANN prediction	**GROUP 4 Rotor syndrome** ANN prediction	**GROUP 5 Cholestatic jaundice** ANN prediction	**GROUP 6 Hepatocellular jaundice** ANN prediction
GROUP 1 Gilbert syndrome 10	7	–	1	-	1	1
GROUP 2 Crigler-Najjar syndrome 10	1	8			1	
GROUP 3 Dubin Johnson syndrome 10	2		6		1	1
GROUP 4 Rotor syndrome 11	1	1		8	1	
GROUP 5 Cholestatic jaundice 11		1			9	
GROUP 6 Hepatocellular jaundice 11		2				8
	Accuracy for class [%] 70%	Accuracy for class [%] 80%	Accuracy for class [%] 60%	Accuracy for class [%] 72.7%	Accuracy for class [%] 81.8%	Accuracy for class [%] 72.7

4 Conclusion

This research paper presents the ability of neural networks to select and classify patients with suspected hyperbilirubinemia. Results have enabled us to diagnose and classify disease, chose proper treatment, and skip out on excessive testing which can often consume a lot of resources. The best performance was achieved with the neural network with 15 neurons in the hidden layer, with accuracy of 93.5%. This paper has proven that diagnostic approaches that are based on neural networks can be effective for the classification of syndromes that are closely related to hyperbilirubinemia.

Conflict of Interest. The authors have no conflicts of interest to declare.

References

1. Fevery, J.: Bilirubin in clinical practice: a review. Liver Int. **28**(5), 592–605 (2008)
2. Memon, N., Weinberger, B., Hegyi, T., et al.: Inherited disorders of bilirubin clearance. Pediatr. Res. **79**, 378–386 (2016)
3. Čepelak, I., Čvorišćes D.: Štrausova medicinska biokemija. Medicinska naklada, Zagreb (2009)
4. Ullah, S., Rahman, K., Hedayati, M.: Hyperbilirubinemia in neonates: types, causes, clinical examinations, preventive measures and treatments: a narrative review article. Iran J. Public Health **45**(5), 558–568 (2016)
5. Sullivan, J.I., Rockey, D.C.: Diagnosis and evaluation of hyperbilirubinemia. Curr. Opin. Gastroenterol. **33**(3), 164–170 (2017)
6. Guidelines for detection, management and prevention of hyperbilirubinemia in term and late preterm newborn infants (35 or more weeks' gestation) - Summary. Paediatr. Child Health **12**(5), 401–418 (2007)
7. Grohmann, K., Roser, M., Rolinski, B., et al.: Bilirubin measurement for neonates: comparison of 9 frequently used methods. Pediatrics **117**(4), 1174–1183 (2006)
8. Alpay, F., Sarici, S.U., Tosuncuk, H.D., et al.: The value of first-day bilirubin measurement in predicting the development of significant hyperbilirubinemia in healthy term newborns. Pediatrics **106**(2), E16 (2000)
9. Catic, A., et al.: Application of Neural Networks for classification of Patau, Edwards, Down, Turner and Klinefelter Syndrome based on first trimester maternal serum screening data, ultrasonographic findings and patient demographics. BMC Med. Genomics **11**, 19 (2018). https://doi.org/10.1186/s12920-018-0333-2
10. Gurbeta, L., et al.: A telehealth system for automated diagnosis of asthma and chronical obstructive pulmonary disease. J. Am. Med. Inform. Assoc. **25**, 1213–1217 (2018)
11. Badnjevic, A., Gurbeta, L., Custovic, E.: An expert diagnostic system to automatically identify asthma and chronic obstructive pulmonary disease in clinical settings. Sci. Rep. **8**, 1–9 (2018)
12. Aljovic, A., Badnjevic, A., Gurbeta, L.: Artificial neural networks in the discrimination of alzheimers disease using biomarkers data. In: IEEE 5th Mediterranean Conference on Embedded Computing (MECO), 12–16 June 2016, Bar, Montenegro (2016)
13. Šećkanović, A., et al.: Review of artificial intelligence application in cardiology. In: 9th Mediterranean Conference on Embedded Computing. MECO 2020 (2020)
14. Mustafić, L.D., et al.: Diagnosis of severe aortic stenosis using implemented expert system. In: Badnjevic, A., Škrbić, R., Gurbeta Pokvić, L. (eds.) CMBEBIH 2019. IFMBE Proceedings, vol. 73. Springer, Cham (2019)
15. Alić, B., et al.: Machine learning techniques for classification of diabetes and cardiovascular diseases. In: 6th Mediterranean Conference on Embedded Computing (MECO), Bar, Montenegro, pp. 1–4 (2017)
16. Sejdinovic, D., et al.: Classification of prediabetes and Type 2 Diabetes using Artificial Neural Network. In: CMBEBIH 2017 (2017)
17. Alic, B., et al.: Classification of metabolic syndrome patients using implemented expert system. In: CMBEBIH 2017. IFMBE Proceedings, vol 62, pp. 601–607. Springer, Singapore (2017)
18. Rijad Sarić, et al.: FPGA-based real-time epileptic seizure classification using Artificial Neural Network. Biomed. Signal Process. Control **62**, 102106 (2020). https://doi.org/10.1016/j.bspc.2020.102106, ISSN 1746-8094
19. Falk, H.H.: A machine learning approach for jaundice detection using color corrected smartphone images. Norwegian University of Science and Technology (2018)

20. Daunhawer, I., Kasser, S., Koch, G., et al.: Enhanced early prediction of clinically relevant neonatal hyperbilirubinemia with machine learning. Pediatr. Res. **86**, 122–127 (2019)
21. Anifowose, F., Khoukhi, A., Abdulraheem, A.: Investigating the effect of training–testing data stratification on the performance of soft computing techniques: an experimental study. J. Exp. Theoret. Artif. Intell. **29**(3), 517–535 (2017). https://doi.org/10.1080/0952813X.2016.1198936

Check for updates

Machine Learning Techniques for Prediction of Psoriatic Arthritis Development in Patients with Psoriasis

Habibović Lejla(✉), Hamidović Azra, Habibović Nihada, Hadžić Dženana, Halilović Neira, and Halilović Samila

Faculty of Pharmacy, University of Sarajevo, Sarajevo, Bosnia and Herzegovina
lejlahabibovic@ffsa.unsa.ba

Abstract. Psoriatic arthritis (PsA) is a chronic inflammatory disease which 30% of psoriasis patients develop, but due to its symptoms and a high prevalence of undiagnosed patients, it is important to diagnose it early, or even better predict it. In this study, using a database of 450 patients, classified in low, medium and high risk patients, and machine learning (ML) algorithms such as Naïve Bayes (NB) and Random Forest (RF), we built classifiers for classification of risk for developing PsA in patients with psoriasis. Input parameters, based on which the risk of PsA development is predicted were obesity, smoking, alcohol consumption, location and the surface of psoriatic lesions, family history of PsA, nail lesions, and since it was a relatively small dataset the distribution of the samples for training and validation was 70:30% . During the training process of the classifiers, the one based on the Random Forest ML algorithm had the highest accuracy (99.82%), while the k-NN algorithm had the lowest accuracy (89.6%). Based on these results, we can see that we have a great framework and basis for future development, because even though the results are satisfying, by using a much bigger and varied dataset, we can achieve a better accuracy and develop a ML algorithm for predicting the risk of PsA development which physicians would be able to rely on while diagnosing a patient with psoriasis.

Keywords: Psoriatic arthritis · Psoriasis · PASI score · Machine learning · Prediction

1 Introduction

Psoriasis is an immune-mediated inflammatory skin disorder that affects 1–3% of the population [1]. Psoriasis varies greatly in severity, from an intermittent condition with a few localized patches of skin involvement, to a widespread serious skin disease with severe pruritus, extensive skin involvement, complications and disability. It is considered to be an incurable, chronic condition that has a variable course, periodically improving and worsening [2]. Approximately one third of patients with psoriasis will develop psoriatic arthritis (PsA) which is a chronic inflammatory musculoskeletal disease [3]. The prevalence of PsA is rare in the general population (<0.5%), but although it occurs in 30%

A. Badnjevic and L. Gurbeta Pokvić (Eds.): CMBEBIH 2021, IFMBE Proceedings 84, pp. 208–216, 2021.
https://doi.org/10.1007/978-3-030-73909-6_23

of psoriasis patients, there is a high prevalence of undiagnosed PsA in those patients [4, 5]. Joint pain, swelling, stiffness and movement limitations are the main signs and symptoms [4]. PsA is also associated with comorbidities such as osteoporosis and cardiovascular disease, representing a relevant health issue that has a great impact on the patients quality of life [6]. Early diagnosis is important for effective management of the disease, because if it's left untreated or if the treatment is delayed due to difficulties in diagnosis PsA can cause irreparable joint damage [4]. The results of a study conducted in 2015 were that even a 6-month delay from symptom onset to the first visit with a rheumatologist contributes to the development of more severe joint erosion and inflammation, and worse long-term physical function [7].

Although it is not clear why psoriatic arthritis develops in patients with psoriasis, there are risk factors that indicate the possible development of psoriatic arthritis in patients with psoriasis. Studies have shown that a single variable doesn't predict transition to PsA but complex interaction of multiple features underlines this progression [8]. Available literature suggests that there are many risk factors that can lead to PsA in psoriatic patients and some of them are: obesity, smoking, alcohol consumption, location and the surface of psoriatic lesions, family history of PsA, nail lesions and other [9, 10].

In this paper the accent will be on the already mentioned risk factors. Obesity is the most common health problem in the world and it has been a risk factor for PsA in patients with psoriasis. The pathophysiological link is not clear, but one of the explanations is based on systematic inflammation induced by adipose tissue [11]. Smoking and alcohol consumption increased the risk of developing PsA in psoriatic patients. The association between those two factors and PsA is not researched enough, but the oxidative stress caused by smoking and liver and heart diseases related to the alcohol consumption can be accepted as a reasonable explanation for the relationship between PsA and those risk factors [9, 11]. The heritability of PsA is estimated to be around 80%, higher than that reported for psoriasis in general, so genetic factors, nail lesions and severe psoriasis are, by far, the highest risk factors for the transition of psoriasis to PsA [5, 8, 10].

Machine Learning (ML) provides methods, techniques, and tools that can help solving diagnostic and prognostic problems in a variety of medical domains [12]. ML is being used for the analysis of the importance of clinical parameters and their combinations for prognosis, prediction of disease progression, therapy planning and support, for the overall patient management and it's exposing novel ideas to practitioners, and specialists [12–20]. These are also capable of autonomous diagnosis of different diseases under clinical regulation [13]. With the ML models, it can also be possible to improve the quality of medical data, reduce fluctuations in patient rates, and save when it comes to medical costs [13].

To reduce the joint damage caused by psoriatic arthritis and managing it effectively requires early detection, as we already mentioned, which is why the ML models could help in the prediction of the disease and prevention of health deterioration, which are the main purposes of this study.

2 Material and Methods

The parameters for the dataset for machine learning training are based on the numerous studies we found on PubMed, Google Scholar, ResearchGate. Multiple studies have identified risk factors for PsA development in the general population and patients with psoriasis and some of these factors are presented in Table 1. One of the most important parameters we used was the PASI score. The Psoriasis Area and Severity Index (PASI) is the most frequently and widely used psoriasis assessment tool for measuring the severity of psoriatic lesions based on area coverage and plaque appearance in clinical practice and therapeutic decision making [21–23]. According to the strength of the association of parameters from Table 1 for PsA development, we classified risk factors for PsA development as low, medium and high.

Table 1. Parameters for the classification of risk for development of psoriatic arthritis in patients with psoriasis

Parameter	Parameter value
Gender	Male Female
Age	18–70
Smoking	Current smoker Past smoker Never
Alcohol consumption	Daily Social None
BMI	Normal 18,5–24,9 Overweight 25–29,9 Obese > 30
Family history of PsA	Yes No No data
Nail changes	None Nail pitting Nail onycholysis Nail hyperkeratosis
PASI score	<10 mild 10–20 moderate >20 severe

Low risk parameters are patients with normal BMI [24, 25], none smokers [25, 26], no alcohol consumption [27, 28], PASI score less than 10 [27, 29], and these parameters together represent datasets for low risk patients. Medium risk parameters are overweight patients [24, 25], smokers [25, 26], alcohol consumers [27, 28], PASI score

10–20 [27, 29], and for medium risk patient datasets we combined medium and low risk parameters. High risk parameters are obese patients [24, 30], family history of PsA [27, 29, 31], nail changes [27, 29, 32], and a PASI score over 20 [27, 29]. For high-risk patient datasets we used a combination of high-risk parameters with medium and low risk parameters.

The database consisted of 450 patients in total of which 150 were low risk patients, 150 medium risk patients and 150 high risk patients as it can be seen in Table 2.

Table 2. Database used for prediction of psoriatic arthritis development in patients with psoriasis

Class	Low risk patients	Medium risk patients	High risk patients	Total
Male	75	75	75	225
Average age (years)	44,53	33,57	48,24	42,11
Female	75	75	75	225
Average age (years)	45,09	49,33	53,30	49,24
Total number of patients per class	150	150	150	450
Average age (years)	44,81	41,45	50,80	45,69

The distribution of the dataset for development of the machine learning algorithms is presented in Table 3. Since this is relatively small dataset, distribution of 70:30% for training and validation was used [33].

Table 3. Dataset distribution

Total number of samples	Low risk patients	Medium risk patients	High risk patients
Training 315	105	105	105
Validation 135	45	45	45

For building classifier for classification of PsA, following machine learning algorithms were used: Naïve Bayes (NB), Decision Tree (DT), Random Forest (RF), and kNearest Neighbor (k-NN) algorithm.

For evaluating performance of developed classifier following was used:
Accuracy and precision can be calculated according to the formula [33]:

$$Accuracy = \frac{TP + TN}{TP + FP + TN + FN} \tag{1}$$

$$Precision = \frac{TP}{TP + FP} \tag{2}$$

3 Results

Figure 1 presents the results of training of classifiers based on 4 different machine learning algorithms. As it can be seen from the figure, during training process, classifier based on Random Forest algorithms achieved the highest accuracy of 99.82%. The lowest accuracy was achieved by k-NN algorithm and it was 89.6%. Average training time was around 4.76 s.

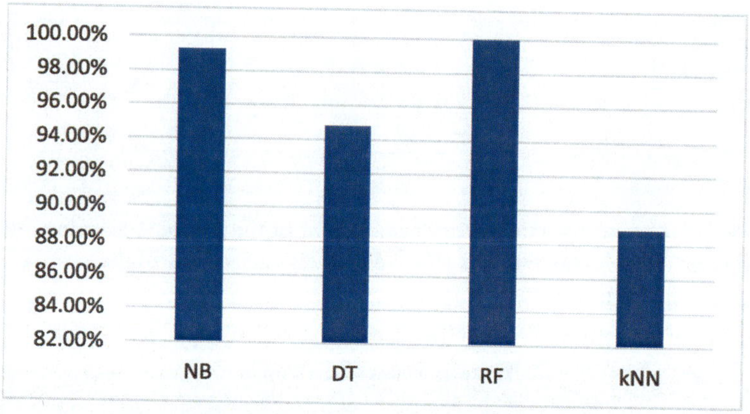

Fig. 1. Training performance of classifiers

Validation was performed on 135 samples. Each classifier was validated and the results of validation are presented in the Tables 4, 5, 6 and 7 below.

As it can be seen from Fig. 2 in comparison to the other ML classifiers, the Random Forest based one has the best performance.

Table 4. Validation performance of NB based classifier

	Low risk patients	Medium risk patients	High risk patients	Accuracy
Low risk patients 45	42	2	1	92.6%
Medium risk patients 45	3	40	2	
High risk patients 45	0	2	43	
Total samples 135	93.3	88.9	95.6	

Table 5. Validation performance of DT based classifier

	Low risk patients	Medium risk patients	High risk patients	Accuracy
Low risk patients 45	40	3	2	93.3
Medium risk patients 45	1	42	2	
High risk patients 45	1	0	44	
Total samples	88.9%	93.3	97.8	

Table 6. Validation performance of RF based classifier

	Low risk patients	Medium risk patients	High risk patients	Accuracy
Low risk patients 45	44	0	1	97.0%
Medium risk patients 45	1	43	1	
High risk patients 45	1	0	44	
Total samples	97.8	95.6	97.8	

Table 7. Validation performance of k-NN based classifier

	Low risk patients	Medium risk patients	High risk patients	Accuracy
Low risk patients 45	43	1	1	94.8%
Medium risk patients 45	0	44	1	
High risk patients 45	4	0	41	
Total samples	95.6	97.8	91.1	

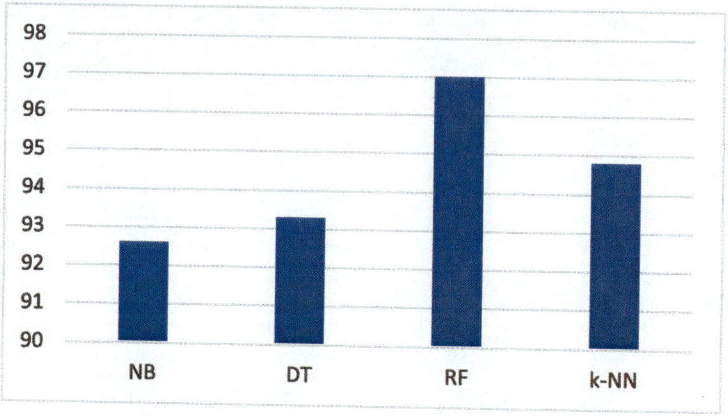

Fig. 2. Comparison of the ML classifier performance

4 Conclusion

Early diagnosis of PsA in patients with psoriasis is crucial in treating and preventing the manifestation of more severe PsA symptoms, which may ultimately lead to serious and irreversible joint damage. In this paper, we used 4 different machine learning algorithms: Naïve Bayes (NB), Decision Tree (DT), Random Forest (RF), and k-Nearest Neighbor (k-NN) algorithm. A database was created with 450 patients with psoriasis, which were classified into low, medium and high risk for the development of PsA based on the strength of the risk parameters associated with the development of PsA.

During the training process, the classifier based on Random Forest (RF) algorithm achieved the highest accuracy of 99.82%, while the lowest accuracy was achieved by k-NN algorithm and it was 89.6%. Random forest (RF) had the best validation performance with an accuracy of 97.0% while Naïve Bayes (NB) had an accuracy of 92.6%.

Even though these results are satisfying, they can be improved by training a network with a larger data set.

ML can significantly help in predicting the development of PsA in patients with psoriasis. Although the algorithms were trained on a small sample, it could be a good basis for developing more sophisticated ML algorithms, which could be used in every-day practice for diagnosis, making a decision on the treatment, monitoring of patients condition, providing adequate therapy for patients and, in some cases, prevention of PsA in patients with psoriasis.

Conflict of Interest Declaration. The authors declare that they have no conflict of interest.

References

1. Napolitano, M., Caso, F., Scarpa, R., Megna, M., Patrì, A., Balato, N., Costa, L.: Psoriatic arthritis and psoriasis: differential diagnosis. Clin. Rheumatol. **35**(8), 1893–1901 (2016)
2. LiverTox: Clinical and Research Information on Drug-Induced Liver Injury [Internet]. National Institute of Diabetes and Digestive and Kidney Diseases, Bethesda, MD (2012)

3. Jalali-najafabadi, F., Dand, N., Ho, P., Smith, C.H., Barker, J.N., McHugh, N., Warren, R.B., Barton, A., Bowes, J.: Application of machine learning methods for prediction modelling of psoriatic arthritis in patients with psoriasis. Ann. Rheum. Dis. **77**(Suppl 2) (2018)
4. Patrick, M.T., Stuart, P.E., Raja, K., et al.: Genetic signature to provide robust risk assessment of psoriatic arthritis development in psoriasis patients. Nat. Commun. **9**, 4178 (2018)
5. Cretu, D., Lisa, G., Liang, K., Soosaipillai, A., Diamandis, E.P., Chandran, V.: Differentiating psoriatic arthritis from psoriasis without psoriatic arthritis using novel serum biomarkers. Arthritis Care Res. (Hoboken) **70**(3), 454–461 (2018)
6. Ocampo, D.V., Gladman, D.: Psoriatic arthritis. F1000Res. 8: F1000 Faculty Rev-1665 (2019)
7. Haroon, M., Gallagher, P., FitzGerald, O.: Diagnostic delay of more than 6 months contributes to poor radiographic and functional outcome in psoriatic arthritis. Ann. Rheum. Dis. **74**, 1045–1050 (2015)
8. Scher, J.U., Ogdie, A., Merola, J.F., et al.: Preventing psoriatic arthritis: focusing on patients with psoriasis at increased risk of transition. Nat. Rev. Rheumatol. **15**, 153–166 (2019)
9. Busse, K., Liao, W.: Which psoriasis patients develop psoriatic arthritis? Psoriasis Forum **16**(4), 17–25 (2010)
10. Eder, L., Haddad, A., Rosen, C.F., Lee, K.A., Chandran, V., Cook, R., Gladman, D.D.: The incidence and risk factors for psoriatic arthritis in patients with psoriasis: a prospective cohort study. Arthritis Rheumatol. (Hoboken, N.J.) **68**(4), 915–923 (2016)
11. Ogdie, A., Gelfand, J.M.: Clinical risk factors for the development of psoriatic arthritis among patients with psoriasis: a review of available evidence. Curr. Rheumatol. Rep. **17**(10), 64 (2015)
12. Magoulas, G.D., Prentza, A.: Machine learning in medical applications. In: Machine Learning and Its Applications, pp. 300–307. Springer, Heidelberg (2001)
13. Battineni, G., Sagaro, G.G., Chinatalapudi, N., Amenta, F.: Applications of machine learning predictive models in the chronic disease diagnosis. J. Pers. Med. **10**(2), 21 (2020)
14. Hodzic, J., et al.: Overview of next-generation sequencing platforms used in published draft plant genomes in light of genotypization of immortelle plant (helichrysium arenarium). Med. Arch. **71**(4), 288–292 (2017)
15. Catic, A., et al.: Application of Neural Networks for classification of Patau, Edwards, Down, Turner and Klinefelter Syndrome based on first trimester maternal serum screening data, ultrasonographic findings and patient demographics. BMC Med. Genomics **11**(1), 19 (2018)
16. Gurbeta, L., Badnjevic, A., Maksimovic, M., Omanovic-Miklicanin, E., Sejdic, E.: A telehealth system for automated diagnosis of asthma and chronical obstructive pulmonary disease. J. Am. Med. Inform. Assoc. **25**(9), 1213–1217 (2018)
17. Badnjevic, A., Gurbeta, L., Custovic, E.: An expert diagnostic system to automatically identify asthma and chronic obstructive pulmonary disease in clinical settings. Sci. Rep. **8**(1), 11645 (2018)
18. Badnjevic, A., Gurbeta, L., Hasicic, M., Bandic, L., Masetic, Z., Kovacevic, Z., Kevric, J., Pecchia, L.: Evidence-based clinical engineering: machine learning algorithms for prediction of defibrillator performance. Biomed. Signal Process. Control **54**, 101629 (2019)
19. Kovačević, Ž., et al.: Prediction of medical device performance using machine learning techniques: infant incubator case study. Health Technol. **10**(2), 151–155 (2020)
20. Sarić, R., et al.: FPGA-based real-time epileptic seizure classification using Artificial Neural Network. Biomed. Signal Process. Control **62**, 102106 (2020)
21. British Association of Dermatologists 2020. Psoriasis Area and Severity Index (PASI) worksheet. https://www.bad.org.uk/page.aspx?sitesectionid=1353. Accessed Dec 2020
22. Bożek, A., Reich, A.: The reliability of three psoriasis assessment tools: psoriasis area and severity index, body surface area and physician global assessment. Adv. Clin. Exp. Med. **5**, 851–856 (2017)

23. Llamas-Velasco, M., de la Cueva, P., Notario, J., Martínez-Pilar, L., Martorell, A., Moreno-Ramírez, D.: Moderate psoriasis: a proposed definition. Actas Dermosifiliogr. **108**(10), 911–917 (2017)
24. Li, W., Han, J., Qureshi, A.A.: Obesity and risk of incident psoriatic arthritis in US women. Ann. Rheum. Dis. **71**(8), 1267–1272 (2012)
25. Love, T.J., Zhu, Y., Zhang, Y., Wall-Burns, L., Ogdie, A., Gelfand, J.M., Choi, H.K.: Obesity and the risk of psoriatic arthritis: a population-based study. Ann. Rheum. Dis. **71**(8), 1273–1277 (2012)
26. Nguyen, U.D.T., Zhang, Y., Lu, N., Louie-Gao, Q., Niu, J., Ogdie, A., Gelfand, J.M., LaValley, M.P., Dubreuil, M., Sparks, J.A., Karlson, E.W., Choi, H.K.: Smoking paradox in the development of psoriatic arthritis among patients with psoriasis: a population-based study. Ann. Rheum. Dis. **77**(1), 119–123 (2018)
27. Eder, L., Haddad, A., Rosen, C.F., Lee, K.A., Chandran, V., Cook, R., Gladman, D.D.: The incidence and risk factors for psoriatic arthritis in patients with psoriasis: a prospective cohort study. Arthritis Rheumatol. **68**(4), 915–923 (2016)
28. Wu, S., Cho, E., Li, W.Q., Han, J., Qureshi, A.A.: Alcohol intake and risk of incident psoriatic arthritis in women. J. Rheumatol. **42**(5), 835–840 (2015)
29. Taylor, W., Gladman, D., Helliwell, P., Marchesoni, A., Mease, P., Mielants, H., CASPAR Study Group: Classification criteria for psoriatic arthritis: development of new criteria from a large international study. Arthritis Rheum. **54**(8), 2665–2673 (2016)
30. Eder, L., Polachek, A., Rosen, C.F., Chandran, V., Cook, R., Gladman, D.D.: The development of psoriatic arthritis in patients with psoriasis is preceded by a period of nonspecific musculoskeletal symptoms: a prospective cohort study. Arthritis Rheumatol. **69**(3), 622–629 (2017)
31. Tey, H.L., Ee, H.L., Tan, A.S., Theng, T.S., Wong, S.N., Khoo, S.W.: Risk factors associated with having psoriatic arthritis in patients with cutaneous psoriasis. J. Dermatol. **37**(5), 426–430 (2010)
32. Wilson, F.C., Icen, M., Crowson, C.S., McEvoy, M.T., Gabriel, S.E., Kremers, H.M.: Incidence and clinical predictors of psoriatic arthritis in patients with psoriasis: a population-based study. Arthritis Rheum. **61**(2), 233–239 (2009)
33. Rau, H.H., Hsu, C.Y., Lin, Y.A., Atique, S., Fuad, A., Wei, L.M., Hsu, M.H.: Development of a web-based liver cancer prediction model for type II diabetes patients by using an artificial neural network. Comput. Methods Programs Biomed. **125**, 58–65 (2016)

Laboratory Diagnosis of Viral Infection Using Artificial Network

Djoja Mirna, Foco Amna, Glamoc Medina, Gljiva Amina, Gudic Lamija, Gutosic Emina(✉), and Dzudzevic Rudaba

Facutly of Pharmacy, University of Sarajevo, Sarajevo, Bosnia and Herzegovina
eminagutosic@ffsa.unsa.ba

Abstract. The viral infection is not synonymous with a disease but it can result in disease of varying severity. Viruses are a cause of viral infection and they can be classified as RNA or DNA viruses. Symptoms of virus infection include fatigue, cough, fever, myalgia, rhinorrhea, and diarrhea. Laboratory diagnosis of viral infection is based on the value of parameters such as white blood cells count (WBC), neutrophils (N), leukocytes (L), red blood cells (RBC), hemoglobin (HB), red cells distribution width (RDW), platelets (PLT) and C-reactive protein (CRP). In order to be more accurate in the diagnosis of viral infection, artificial neural network (ANN) can be used. For ANN development 400 participants were included, 150 healthy subjects and 250 with the viral infection. The standard architecture of ANN for pattern recognition was used and it had two-layers. The results showed 95% accuracy, 97.8% sensitivity, and 93.3% specificity, so it can be concluded that this ANN can be used for the determination of viral infection. It would be a good tool in clinical practice when there is a need to determine whether the patient has a viral or bacterial infection.

Keywords: Viral infection · Virus · Artificial neural network · Laboratory diagnosis

1 Introduction

The viral infection is not synonymous with disease. Many viral infections are subclinical, but some result in disease of varying severity that is typically accompanied by characteristic clinical signs in the affected host. Lots of factors can contribute to disease development. The outcome of the virus–host encounter is essentially the product of the virulence of the infecting virus and the susceptibility of the host. [1].

Viruses can be classified due to differences in their genetic material so there are RNA or DNA viruses. Also, they can be classified because of the differences in their configuration and as well as the orientation of their encoded genes. They can be distributed across several segments. Sometimes they are packaged together in a virion or often in separate virus particles [2]. The most common symptoms of virus infection are fatigue, cough, fever, myalgia, rhinorrhea, and diarrhea [3].

Lots of hematological markers can be used in the diagnosis of viral infection. In the last decade, the importance of platelets (PLT) in virus infection has been shown.

A. Badnjevic and L. Gurbeta Pokvić (Eds.): CMBEBIH 2021, IFMBE Proceedings 84, pp. 217–223, 2021.
https://doi.org/10.1007/978-3-030-73909-6_24

Platelets are anucleate blood cells and they are playing a crucial role in the maintenance of homeostasis. Viruses are modulating their activity. Elevated platelet counts (thrombocytosis) are associated with an increased risk of thrombotic complications and low platelet counts (thrombocytopenia) are associated with an increased risk of bleeding [4]. Han et al. (2020) state that PLT levels were decreased in suspected influenza patients. Decreased PLT levels could differentiate influenza virus infections from others [5]. For example, COVID-19 patients often have mild thrombocytopenia and at the same time they have increased PLT consumption [6]. Red blood cell levels are higher in infection caused by influenza virus. Also, levels of hemoglobin are higher in this type of infection. They can be used as differential diagnostic indicators to compare infection caused by the influenza virus versus bacteria. It is explained by the fact that influenza hemagglutinin readily binds to sialic acid receptors on the membranes of red blood cells. That causes red blood cells to clump together or agglutinate. Bacterial infections on the other side cause intravascular hemolysis leading to sepsis which reduces the red blood cell count. Reduction in red blood cell count can be a sign of bacterial infection [5]. Lymphocytes represent around 20 to 40% of white blood cells (WBC). Lymphocytosis is state where there is an increase in WBC of more than 40% in the presence of a normal absolute white cell count. That state can be caused by several viral infections such as Epstein-Barr Virus (EBV), Cytomegalovirus, and Influenza virus [7]. However low white blood cell count called lymphopenia is associated with severity of COVID-19 infection [8]. Poly-morphonuclear leukocytes called neutrophils represent the first immune cells to the site of injury and microbial infection. They have a crucial role in controlling bacterial and fungal infections. They are also shown to play important roles in viral pathogenesis. Depletion of neutrophils is characteristic state of Influenza A infection. These decreased levels can be associated with viral infection [9]. Neutrophil/lymphocyte ratio is used as a marker for COVID-19 and it is suspected that it can predict severity of pneumonia [10]. One of the signs of the inflammation process, C-reactive protein (CRP) is an acute inflammatory protein that can increase up to 1,000-old at sites of infection or inflammation. It is synthesizing primarily in liver hepatocytes but also by smooth cells muscles, macrophages, endothelial cells, lymphocytes, and adipocytes [11]. Sasaki et al. (2002) in their study have concluded that levels of CRP are slightly high or even normal in viral infection [12]. Elevated CRP indicates bacterial infection but it was reported that virus infections also can cause elevated CRP [13]. Studies have shown that higher levels are associated with mortality due to COVID-19 infection [14].

Artificial intelligence tools such as neural networks and machine learning have already been used in healthcare, from management of medical devices [15–18] to classification and prediction of several diseases [19–27]. When it comes to epidemiology, it can be used for variety of epidemiological assessments, such as Ebola or even COVID-19 infections [28].

The aim of this study is development of artificial neural network for the classification of patients with viral disease. This system is envisioned as diagnostic aid tool for primary healthcare practice when often medical specialist is not available. Such tool increases chance for early diagnosis and effective treatment.

2 Material and Methods

The neural network developed in this study is based on the following input parameters: sex, age, white blood cells count (WBC), neutrophils (N), leukocytes (L), red blood cells (RBC), hemoglobin (HB), red cells distribution width (RDW), platelets (PLT) and C-reactive protein (CRP). It was developed based on the dataset which consisted of information about 400 participants. Out of the total number of participants included in the study, 250 were participants with viral disease and 150 were healthy subjects. Detailed information about the database is presented in Table 1.

Table 1. Database used for development of neural network

	Total number of samples	Number of women	Number of men	Average women age	Average man age
Healthy subjects	150	69	81	39.8 ± 14.9	41.1 ± 14.5
Patients with viral infection	250	135	115	40.8 ± 15.59	39.3 ± 14.9

The network does the diagnosis of the viral infection for the patient based on these set of parameters that are obtained by Standard Operating Procedures for the hematology parameters. Parameters are determined in blood after venipuncture in whole blood, serum or plasma.

The distribution of the dataset for development of the Artificial Neural Network (ANN) is presented in Table 2. Since this is a relatively small dataset, a distribution of 70:30% for training and validation was used [29].

Table 2. Dataset distribution

Total number of samples 400	Healthy 150	Disease 250
Training 280	105	175
Validation 120	45	75

For the development of the ANN in this research, standard architecture for pattern recognition was used. That is a feedforward neural network which consists of two-layers. This architecture is designed since according to the application experts, that kind of ANN is sufficient to properly perform the classification of input data.

The input layer is made of network inputs. Input layer is followed by a hidden layer which consists of parallel connected neurons. The network was tested on training performance with different number of neurons. Since, this classification is highly non-linear data classification, sigmoid function was used for the hidden layer.

During the design of the network, a different number of neurons was used to measure the performance. According to the Mean Square Error (MSE), appropriate number of neurons was chosen. After training, validation of the network performance was done using the confusion matrix.

According to these results, specificity, sensitivity, and accuracy of the network are calculated according to Eqs. (1), (2) (3):

$$sensitivity = \frac{true\ positive}{(true\ positive + false\ negative)} \tag{1}$$

$$specificity = \frac{true\ negative}{(true\ negative + false\ positive)} \tag{2}$$

$$accuracy = \frac{(true\ negative + true\ positive)}{(true\ negative + true\ positive + false\ negative + false\ positive)} \tag{3}$$

3 Results

According to previous research and expert recommendation, prediction of viral infection in this paper, two-layer feedforward artificial neural network, with logsig transfer function in the hidden layer was developed.

Table 3 presents the results of training process for the neural network when a number of neurons in hidden layer was changed. As can be seen from the table the best performance was achieved with the neural network with 12 neurons in hidden layer. During the training Levenberg – Marquart algorithm was used and for the transfer function, logsig function was used.

Table 3. Training performance

Number of neurons	Training mean square error
2	0.0037
5	0.0040
10	0.0053
12	**8.2156e–08**
15	3.5385e–04
30	0.0016
50	0.0057

Training performance for neural network with 12 neurons in hidden layer is presented in Figure 1.

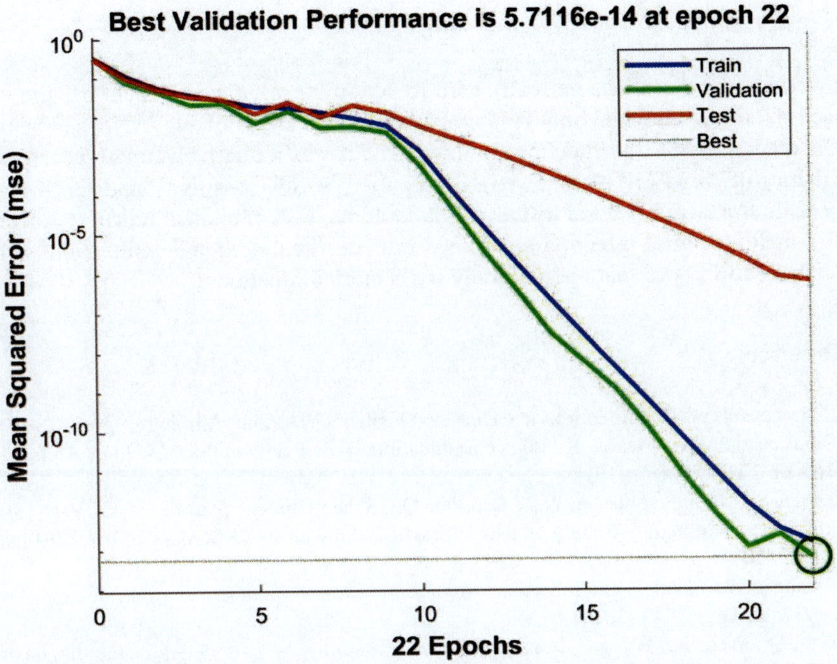

Fig. 1. Training performance of neural network with 12 neurons in hidden layer

The validation of the performance of the feedforward neural network with 12 neurons and logsig transfer function in hidden layer is presented in Table 4. As it can be seen from the Table 4, overall accuracy of the developed neural network is 95 % while specificity for classification to the group disease is 93.3 %.

Table 4. Artificial Neural Network performance during validation

n = 120	**Group 1 Healthy** ANN prediction	**Group 2 Disease** ANN prediction	
GROUP 1 Healthy 45	TP 44	FN 1	**Accuracy** [%] 95%
Group 2 Disease 75	FP 5	TN 70	
	Sensitivity [%] 97.8	**Specificity** [%] 93.3	

4 Conclusion

It is well known that it can be really hard to determine whether a patient has the viral or bacterial infection when only laboratory parameters are available. ANN can be used for the detection of viral infection. In this study, it was a relatively small dataset with distribution of 70:30%. ANN showed 95% accuracy, 97.8% sensitivity, and 93.3% specificity. This shows really good results and it could be used in clinical practice. Accurate determination of viral infection could lead to better treatment and could avoid wrong prescribing antibiotics that are used only for bacterial infections.

References

1. Pathogenesis of Viral Infections and Diseases. Fenner's Veterinary Virology, pp. 47–78 (2017)
2. Simmonds, P., Aiewsakun, P.: Virus classification - where do you draw the line? Arch. Virol. **163**(8), 2037–2046 (2018)
3. Zayet, S., Kadiane-Oussou, N.J., Lepiller, Q., et al.: Clinical features of COVID-19 and influenza: a comparative study on Nord Franche-Comte cluster. Microbes Infect. **22**(9), 481–488 (2020)
4. Assinger, A.: Platelets and infection - an emerging role of platelets in viral infection. Front. Immunol. **18**(5), 649 (2014)
5. Han, Q., Wen, X., Wang, L., et al.: Role of hematological parameters in the diagnosis of influenza virus infection in patients with respiratory tract infection symptoms. J. Clin. Lab. Anal. **34**(5), e2319 (2020)
6. Wool, G.D., Miller, J.L.: The impact of COVID-19 disease on platelets and coagulation. Pathobiology **13**, 1–13 (2020). https://doi.org/10.1159/000512007
7. Hamad, H., Mangla, A.: Lymphocytosis, StatPearls [Internet]. Treasure Island (FL): StatPearls Publishing, 7 October 2020 (2020). https://www.ncbi.nlm.nih.gov/books/NBK549819/. Accessed 19 Nov 2020
8. Zhao, Q., Meng, M., Kumar, R., et al.: Lymphopenia is associated with severe coronavirus disease 2019 (COVID-19) infections: a systemic review and meta-analysis. Int. J. Infect. Dis. **96**, 131–135 (2020)
9. Drescher, B., Bai, F.: Neutrophil in viral infections, friend or foe? Virus Res. **171**(1), 1–7 (2013)
10. Feng, Z., Yu, Q., Yao, S., et al.: Early prediction of disease pro- gression in COVID-19 pneumonia patients with chest CT and clinical characteristics. Nat. Commun. **11**, 4968 (2020)
11. Sproston, N.R., Ashworth, J.J.: Role of C-reactive protein at sites of inflammation and infection. Front. Immunol. **13**(9), 754 (2018)
12. Sasaki, K., Fujita, I., Hamasaki, Y., et al.: Differentiating between bacterial and viral infection by measuring both C-reactive protein and 2'-5'-oligoadenylate synthetase as inflammatory markers. J. Infect. Chemother. **8**(1), 76–80 (2002)
13. Jeon, J.S., Rheem, I., Kim, J.K.: C-Reactive protein and respiratory viral infection. Korean J. Clin. Lab. Sci. **49**, 15–21 (2017)
14. Sahu, B.R., Kampa, R.K., Padhi, A., et al.: C-reactive protein: a promising biomarker for poor prognosis in COVID-19 infection. Clin. Chim. Acta **509**, 91–94 (2020)
15. Badnjević, A., Gurbeta Pokvić, L., Hasičić, M., Bandić, L., Mašetić, Z., Kovačević, Ž., Kevrić, J., Hifziefendić, J., Pecchia, L.: Evidence-based clinical engineering: machine learning algorithms for prediction of defibrillator performance. Biomed. Signal Process. Control **54**, 101629 (2019)

16. Kovačević, Ž., et al.: Prediction of medical device performance using machine learning techniques: infant incubator case study. Health Technol. **10**, 151–155 (2020)
17. Spahić, L., Kurta, E., Čordić, S., et al.: Machine learning techniques for performance prediction of medical devices: infant incubators. In: Badnjevic, A., Škrbić, R., Gurbeta Pokvić, L. (eds.) CMBEBIH 2019. CMBEBIH 2019. IFMBE Proceedings, vol. 73, pp. 483–490. Springer (2020)
18. Hrvat, F., et al.: Artificial intelligence for prediction of medical device performance: infusion and perfusor pumps case study. In: 9th Mediterranean Conference on Embedded Computing MECO 2020 (2020)
19. Sarić, R., Jokić, D., Beganović, N., et al.: FPGA-based real-time epileptic seizure classification using Artificial Neural Network. Biomed. Signal Process. Control **62**, 102106 (2020)
20. Badnjevic, A., Gurbeta, L., Cifrek, M., Marjanovic, D.: Classification of asthma using artificial neural network. In: IEEE 39th International Convention on Information and Communication Technology, Electronics and Microelectronics (MIPRO), Opatija, Croatia, 30 May–03 June 2016
21. Badnjevic, A., Gurbeta, L., Cifrek, M., Pecchia, L.: Pre-classification process symptom questionnaire based on fuzzy logic for pulmonary function test cost reduction. In: CMBEBIH 2017. IFMBE Proceedings, vol. 62, pp. 608–616. Springer, Singapore (2017)
22. Šećkanović, A., et al.: Review of artificial intelligence application in cardiology. In: 9th Mediterranean Conference on Embedded Computing MECO 2020 (2020)
23. Mustafić, L.D., et al.: Diagnosis of severe aortic stenosis using implemented expert system. In: Badnjevic, A., Škrbić, R., Gurbeta Pokvić, L. (eds.). CMBEBIH 2019. IFMBE Proceedings, vol. 73. Springer, Cham (2019)
24. Alić, B., et al.: Machine learning techniques for classification of diabetes and cardiovascular diseases. In: 6th Mediterranean Conference on Embedded Computing (MECO), Bar, Montenegro, pp. 1–4 (2017)
25. Sejdinovic, D., et al.: Classification of prediabetes and Type 2 diabetes using artificial neural network. In: CMBEBIH 2017 (2017)
26. Alic, B., et al.: Classification of metabolic syndrome patients using implemented expert system. In: CMBEBIH 2017. IFMBE Proceedings, vol. 62, pp 601–607. Springer, Singapore (2017)
27. Catic, A., Gurbeta, L., Kurtovic-Kozaric, A., et al: Application of Neural Networks for classification of Patau, Edwards, Down, Turner and Klinefelter Syndrome based on first trimester maternal serum screening data, ultrasonographic findings and patient demographics. BMC Med. Genomics **11**, 19 (2018)
28. Vaishya, R., Javaid, M., Khan, I.H., Haleem, A.: Artificial Intelligence (AI) applications for COVID-19 pandemic. Diabetes Metab. Syndr. **14**(4), 337–339 (2020)
29. Anifowose, F., Khoukhi, A., Abdulraheem, A.: Investigating the effect of training–testing data stratification on the performance of soft computing techniques: an experimental study. J. Exp. Theor. Artif. Intell. **29**(3), 517–535 (2017). https://doi.org/10.1080/0952813X.2016.1198936

Using Artificial Neural Networks in Diagnostics of Familial Combined Hyperlipidaemia Based on Levels of Certain Blood Parameters and Risk Assessment of Developing Cardiovascular Disease

Šahinović Berina, Šehić Faruk, Šerak Rijad(✉), Šero Aiša, Škrijelj Melisa, Špago Ajla, Špago Merima, and Almir Badnjević

Faculty of Pharmacy, University of Sarajevo, Sarajevo, Bosnia and Herzegovina
rijad.serak@ffsa.unsa.ba

Abstract. Hyperlipidaemia is a condition that refers to an extremely high level of lipids in the blood. Cholesterol and triglyceride levels are highest in people with hereditary hyperlipidaemias, which are divided into five types. In this article we will discuss type IIb hyperlipidaemia. It is an inherited disorder that results in accelerated atherosclerotic early death, most commonly due to myocardial infarction. The aim of this study was to predict on the basis of blood lipid parameters, the risk of occurrence and development of cardiovascular disease, and to assist doctors in the proper diagnosis of this disease. For the purposes of this paper, artificial neural networks were used. Parameters used in the diagnosis of Familial hyperlipidaemia were: cholesterol (mmol/l), triglycerides (mmol/l), LDL (mmol/l), HDL (mmol/l), Apo B100 (mg/dl), systolic blood pressure and smoking status. The artificial neural network has an accuracy of 97.2%, reliably predicts the disease which means it can be used to help diagnose Familial hyperlipidaemia.

Keywords: Hyperlipidaemia · Artificial neural network · Cardiovascular disease

1 Introduction

Familial combined hyperlipidaemia (FCHL) also known as hyperlipidaemia type IIb is the most common type of primary dyslipidaemia. Prevalence of FCHL in the general population is 1-3% and ranges from 20–38% in patients who have previously suffered from myocardial infarction (MI) [1].

FCHL is defined by fluctuations in serum lipid profile and represents a heterogeneous condition which is associated with certain metabolic disorders that correlate with FCHL such as obesity, insulin resistance and type 2 diabetes mellitus (T2D) [1]. Patients diagnosed with FCHL are at higher risk of developing cardiovascular disease. Certain studies showed that in patients diagnosed with FCHL, the profile of lipoprotein fractions was modified, notably the properties of very low-density lipoprotein (VLDL) as well as low-density lipoprotein (LDL), and high-density lipoprotein (HDL) [2].

A. Badnjevic and L. Gurbeta Pokvić (Eds.): CMBEBIH 2021, IFMBE Proceedings 84, pp. 224–229, 2021.
https://doi.org/10.1007/978-3-030-73909-6_25

Using artificial neural networks (ANNs) for disease diagnosis represents a relatively novel and increasingly interesting area of research thanks to easily accessible large quantities of medical data. Deep neural networks try to create model abstractions from large-scale data to conceptualize various types of data (e.g. text, sound, images) [3]. There have been many studies on hyperlipidaemia diagnosis based on blood and urine parameters, which provide a good basis that can be implemented in deep learning techniques in order to create a faster and robust diagnostic tool compared to existing traditional methods [4].

Previous studies combining deep learning techniques with diagnosis of hyperlipidaemia are scarce [4, 5] and use blood and urine parameters as well as electronic health records (EHRs) of patients. However, they are aimed at diagnosing all types of hyperlipidaemia and do not assess the risk of developing potential cardiovascular disease in patients with FCHL.

In this article we will be focusing on the diagnosis of FCHL using artificial neural networks which is based on measuring the levels of cholesterol, tri-glycerides, different lipoprotein fractions (LDL, HDL, apoB) in the blood and comparing them to pre-established concentration levels for healthy individuals and diagnosed patients. In addition, we will be assessing the potential risk factor of developing cardiovascular disease depending on the severity of FCHL and combination of factors such as age and cigarette smoking, among others [6].

2 Materials and Methods

Parameters used in diagnosis of FCHL were: cholesterol (mmol/l), triglycerides (mmol/l), LDL (mmol/l), HDL (mmol/l), Apo B100 (mg/dl), systolic blood pressure and smoking status.

Although there are many more parameters such as the interpretation of electropherogram results, visual examination of blood plasma and electrophoretic mobility, we decided not to present those variables, because we concluded the aforementioned seven parameters are enough in our case.

The dataset for this study consisted of 480 samples. Out of the overall number, 240 samples were from patients diagnosed with FCHL and 240 were from healthy individuals. This dataset is from November 2020 [7, 8].

As indicated in Table 1, the dataset is represented by data from patients with both diagnosed FCHL and of healthy individuals that presented no symptoms. [9]. Average age of males was 58, 76 years while the average age of female patients was 60,56.

This dataset was used for creating an automated model for diagnosing patients with FCHL and assessing their risk of developing cardiovascular disease. Although the dataset is relatively small, appropriate neural network architectures can be used for this purpose [10, 11].

Based on SCORE – European High Risk Chart [13] we estimated the risk of cardiovascular disease using parameters such as: age, gender, total cholesterol (mmol/l), systolic blood pressure and smoking status [12].

The dataset for this study consisted of 480 samples. Out of the 480 samples, 240 are presented with the risk of cardiovascular disease while the other 240 are healthy individuals.

Table 1. Properties of the dataset used for development of artificial neural network

Class	Patients with FCHL	Healthy	Total
Male	113	131	244
Average age (years)	57,54	59,98	58,76
Female	127	109	236
Average age (years)	61,85	59,28	60,56
Total	240	240	480

Table 2. Distribution of patients by classes for risk in developing cardiovascular disease.

Risk of cardiovas cular disease (%)	< 1%	1%	2%	3% –4%	5%–9%	10%–14%	15% and over	Total
Male	2	22	8	13	32	19	22	118
Female	22	15	13	20	40	6	6	122
Patients per class	24	37	21	33	72	25	28	240

As indicated in Table 2, patients diagnosed with FCHL are distributed across different classes ranked for their risk in developing cardiovascular disease. Each class has been ranked according to the SCORE Chart of the European Guidelines on CVD prevention in Clinical Practice.

3 Results

The distribution of the dataset for development of the Artificial Neural Network (ANN) is presented in Table 3. Since this is a relatively small dataset, distribution of 70:30% for training and validation was used [12].

Table 3. Dataset distribution

Total number of samples 480	Healthy 240	Disease 240
Training 336	168	168
Validation 144	72	72

According to previous research and expert recommendation [13], twolayer feed-forward artificial neural network, with sigmoid transfer function in hidden layer was developed. Neural network architecture consists of inputs: sex, age, cholesterol (mmol/l), triglycerides (mmol/l), LDL (mmol/l), HDL (mmol/l), Apo B100 (mg/dl), systolic blood pressure and smoking status. During development, different number of neurons was tested. The results of training for different number of neurons in hidden layer are presented in the Table 4 below (Fig. 1).

Table 4. Training and validation of neural network

Number of neurons	MSE
2	1.4402e−11
5	**3.9074e−13**
10	1.3832e−06
11	3.2916e−05
15	5.8533e−09
20	2.1347e−10

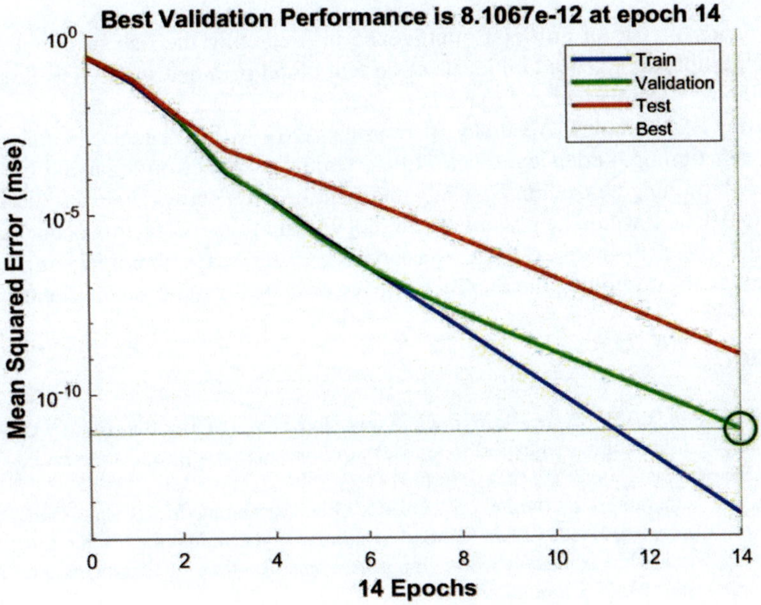

Fig. 1. Training performance with 5 neurons

Based on the training results, ANN with 5 neurons in hidden layer was chosen for diagnostics of Familial combined hyperlipidaemia based on levels of certain blood parameters and risk assessment of developing cardiovascular disease. Results of validation are presented in the Table 5 below.

Table 5. Artificial Neural Network performance during validation

n = 144	**Group 1 Healthy** ANN prediction	**Group 2 Disease** ANN prediction	
GROUP 1 Healthy 72	TP 71	FN 1	**Accuracy** [%] 97.9
Group 2 Disease 72	FP 2	TN 70	
	Sensitivity [%] 98.6	**Specificity** [%]	

4 Conclusion

In this paper the use of artificial intelligence in predicting the risk of Familial combined hyperlipidaemia based on levels of certain blood parameters and risk assessment of developing cardiovascular disease is presented. The developed neural network was made using 480 samples. Accuracy of training ANN with five neurons and sigmoid transfer functioning hidden layer was 97.9%. Training was performed using Levenberg – Marquadt training algorithm. By using and training such networks, in addition to collecting real data, it would be possible to predict which of the risk factors occur the most. This could have a preventive effect in the early stages of disease, thus achieving the most important goal - improving the quality of life of both healthy and sick patients.

References

1. Vargas-Vázquez, J.E., Cortés-Arroyo, G., Tapia-González, I., Cruz-Bautista, C.A.: Aguilar-Salinas, Familial combined hyperlipidemia: current knowledge, perspectives and controversies. Rev. Invest. Clin. **70**(5), 224–236 (2018)
2. Puig, N., Miñambres, I., Benítez, S., Gil, P., Grau-Agramunt, M., Rivas-Urbina, A., Pérez, A., Sánchez-Quesada, J.L.: Familial combined hyperlipidemia (FCH) patients with high triglyceride levels present worse lipoprotein function than FCH patients with isolated hypercholesterolemia. Biomedicines **8**(1), 6 (2020)
3. Cao, C., et al.: Deep learn- ing and its applications in biomedicine. Genomics Proteomics Bioinform. **16**(1), 17–32 (2018)
4. Zhang, Q., Liu, Y., Liu, G., Zhao, G., Qu, Z., Yang, W.: An automatic diagnostic system based on deep learning, to diagnose hyperlipidemia. Diabetes Metab. Syndr. Obes. **12**, 637–645 (2019)
5. Liu, Y., Zhang, Q., Zhao, G., Liu, G., Liu, Z.: Deep learning-based method of diagnosing hyperlipidemia and providing diagnostic markers automatically. Diabetes Metab. Syndr. Obes. **13**, 679–691 (2020)

6. Hodzic, J., et al.: Overview of next-generation sequencing platforms used in published draft plant genomes in light of genotypization of immortelle plant (helichrysium are- narium). Med Arch. **71**(4), 288–292 (2017). https://doi.org/10.5455/medarh.2017.71.288-292
7. Catic, A., et al.: Application of Neural Networks for classification of Patau, Edwards, Down, Turner and Klinefelter Syndrome based on first trimester maternal serum screening data, ultrasonographic findings and patient demographics. BMC Med. Genomics **11**, 19 (2018). https://doi.org/10.1186/s12920-018-0333-2
8. Gurbeta, L., et al.: A telehealth system for automated diagnosis of asthma and chronical obstructive pulmonary disease. J. Am. Med. Inform. Assoc. **25**, 1213–1217 (2018)
9. Badnjevic, A., et al.: An expert diagnostic system to automatically identify asthma and chronic obstructive pulmonary disease in clinical settings. Sci. Rep. **8**, 1–9 (2018)
10. Badnjević, A., Pokvić, L.G., Hasičić, M., Bandić, L., Mašetić, Z., Kovačević, Ž, Kevrić, J., Hifziefendić, J., Pecchia, L.: Evidence-based clinical engineering: machine learning algorithms for prediction of defibrillator performance. Biomed. Signal Process. Control **54**, 101629 (2019)
11. Kovačević, Ž., et al.: Prediction of medical device performance using machine learning techniques: infant incubator case study health and technology. Health Technol **10**(1), 151–155 (2019)
12. Rau, H.-H., Hsu, C.-Y., Lin, Y.-A., Atique, S., Fuad, A., Wei, L.-M., Hsu, M.-H.: Development of a web-based liver cancer prediction model for type II diabetes patients by using an artificial neural network. Comput. Methods Programs Biomed. **125**, 58–65 (2016). https://doi.org/10.1016/j.cmpb.2015.11.009, ISSN 0169-2607
13. Piepoli, M.F., et al.: 2016 European Guidelines on cardiovascular disease prevention in clinical practice: The Sixth Joint Task Force of the European Society of Cardiology and Other Societies on Cardiovascular Disease Prevention in Clinical Practice (constituted by representatives of 10 societies and by invited experts) Developed with the special contribution of the European Association for Cardiovascular Prevention & Rehabilitation (EACPR). Eur. Heart J. **37**(19), 2315–2381 (2016)

Machine Learning Techniques for Assessment of Stress and Burnout Syndrome in Pharmacists

Aida Begić(✉), Almedina Alibegović, Nejira Aličajić, Amina Alihodžić, Aida Aljović, Naida Bašić, and Lejla Bureić

Facutly of Pharmacy, University of Sarajevo, Sarajevo, Bosnia and Herzegovina
aidabegic@ffsa.unsa.ba

Abstract. Stress can be defined as a physical, mental or emotional factor that causes bodily or mental tension. Burnout syndrome is increasingly recognized among healthcare professionals, including pharmacists. The operational definition of burnout has three dimensions: emotional exhaustion (EE), which is commonly considered the core symptom of the burnout syndrome; depersonalization (DP) is the development of negative, cynical attitudes towards patients; and, finally, lack of personal accomplishment (PA), defined as the tendency to believe that one is no longer effective in working with patients. The goal of this paper was to develop Machine Learning Techniques for stress assessment and burnout syndrome in pharmacists. For developing classifier for classification of burnout syndrome, following machine learning algorithms were used: (1) Naïve Bayes (NB), (2) Decision Tree (DT), (3) Random Forest (RF), and (4) k-Nearest Neighbor (k-NN) algorithm. During training process, classifier based on k-NN algorithms achieved the highest accuracy of 99.64%. The lowest accuracy was achieved by Naïve Bayes algorithm and it was 97.20%.

Keywords: Health-system workers · Machine learning techniques · Stress · Burnout syndrome

1 Introduction

The term stress refers to the human body's adaptive response to a stressors such as a challenge, disturbance, or environmental demand. Stress can be defined as a physical, mental or emotional factor that causes bodily or mental tension [1]. Adaptation to stress is supervised by a complex neuroendocrine, cellular and molecular infrastructure called the stress system that is located in the central nervous system and on the periphery of the organism.

Symptoms of stress and sometimes chronic changes, occur due to bodily functions that change in the stress states. The way how an individual will deal with stressors is influenced by genetics, experience and behavior. Although several epidemiological studies were performed in order to evaluate this phenomenon, there is a lack of data on the prevalence of burnout, due to different definitions of the syndrome and heterogeneity of assessment methods.

A. Badnjevic and L. Gurbeta Pokvić (Eds.): CMBEBIH 2021, IFMBE Proceedings 84, pp. 230–238, 2021.
https://doi.org/10.1007/978-3-030-73909-6_26

This lack of evidence was also pointed out by a recent review, demonstrating a remarkable variability in published prevalence of burnout, ranging from 0% to 80.5%, and highlighting the necessity to develop a common definition for burnout and standardized measurement tools [2].

Health-care, as an industry, places numerous pressures on healthcare providers, including the challenges of clinical work, time constraints, competing demands, lack of control over work processes and scheduling, and conflicting relationships with leadership [3]. Pharmacists' job satisfaction may vary according to practice setting, with community pharmacists likely experiencing lower job satisfaction and, consequently, higher rates of burnout symptoms, compared with pharmacists practicing in other settings [4]. Frontline healthcare professionals, who prioritize the care of patients each day, may not realize the effect and burden that working in these professions places on one's own health and cognition. For example, poor provider well-being and moderate-to-high levels of burnout are associated with poorer patient safety outcomes and increased medical errors [5].

During the COVID-19 pandemic, healthcare workers and population in general experience higher levels of stress than usually. Not being able to cope with it in the right way, stress can cause condition called "Burnout syndrome". Burnout is generally considered a response by a subject to chronic work-related stress in an attempt to adapt or protect oneself from it [6]. Lack of control, social support and work-life balance, dysfunctional workplace dynamics are just one of few causes of job burnout. Consequences that may follow burnout syndrome are usually fatigue, insomnia, and even heart diseases, high blood pressure, diabetes type 2 in perspective. The operational definition of burnout has three dimensions: emotional exhaustion (EE), which is commonly considered the core symptom of the burnout syndrome; depersonalization (DP) is the development of negative, cynical attitudes towards patients; and, finally, lack of personal accomplishment (PA), defined as the tendency to believe that one is no longer effective in working with patients. The most common tool to calculate burnout level is the Maslach Burnout Inventory Human Services Survey (MBI-HSS) [6, 7]. That is a psychological assessment that is the leading measure of burnout for professionals.

With the rise of application of artificial intelligence in healthcare [8–15] it is interesting to investigate how these tools can be used for assessment of stress and burnout syndrome. The prediction and assesment of stress level using artificial neural networks (ANNs) have been investigated previously. Ladstätter et al. [16] have investigated the Application of artificial neural networks to a study of nursing burnout and concluded that the usage of ANNs in the field of burnout has been justified due to their superior ability to capture non-linear relationships, which is relevant for theory development. Also, Sejdinović et al. [17] successfully implemented Artificial Neural Network for Stress Recognition for drivers using based on measurement of physiological parameters.

Following this research, the aim of this paper is to present the development of machine learning classifier for assessment of stress and burnout syndrome in pharmacist. Machine learning (ML) [18] is the scientific discipline that focuses on how computers learn from data. It arises at the intersection of statistics, which seeks to learn relationships from data, and computer science, with its emphasis on efficient computing algorithms. This technology seems to be very promising and further development opens a new and exciting area in various fields of biomedical sciences.

2 Methods and Materials

An electronic questionnaire based on the Maslach Burnout Inventory Human Services Survey (MBI-HSS) [6, 7] was distributed to a target population of health-system workers (pharmacists) via e-mail. This questionnaire was sent in order to assess study participants for burnout. Final classification was done by psychologist.

Study participant was classified with burnout syndrome based on MBI-HSS if feelings of emotional exhaustion, depersonalization, and reduced personal accomplishment were detected. Table 1. shows how participant of this study were classified into classes which were further used for development of classifier based on ML algorithms. The age distribution of the participants is shown in Fig. 1.

Table 1. Study participant classification

Total number of samples	400
Burnout syndrome	300
Healthy	100
No. of male participants	150
No. of female participants	150

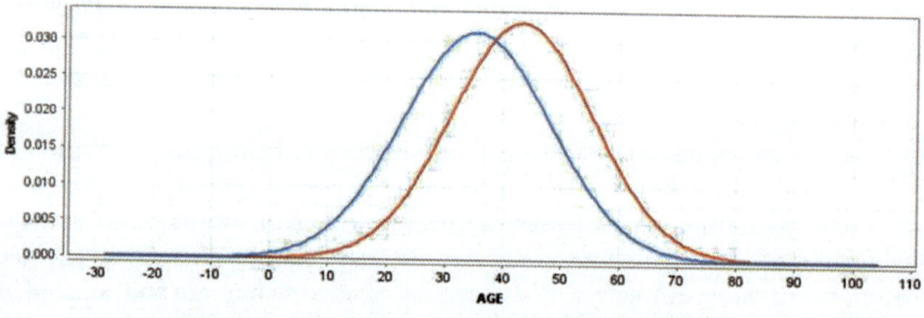

Fig. 1. Age distribution of the participants in the study

During the survey, for each participant following information were collected: sex, age, smoking status, alcohol consumption, while following parameters were assessed: EE (emotional exhaustion), D (depersonalization), PA (lack of personal accomplishment) according to the MBI-HSS.

Abovementioned database was further used for development of classifier based on machine learning algorithm which can be used for assessment of burnout syndrome based on those input parameters. The distribution of the dataset for development of the classifier based on machine learning algorithms is presented in Table 2. Since this is relatively small dataset, distribution of 70:30% for training and validation was used [19].

Table 2. Dataset distribution per training and validation subset

	Group 1 Burnout syndrome	Group 2 Healthy	Total number of samples
Training	212	68	280
Validation	88	32	120
Total	**300**	**100**	**400**

For developing classifier for classification of burnout syndrome, following machine learning algorithms were used: (1) Naïve Bayes (NB), (2) Decision Tree (DT), (3) Random Forest (RF), and (4) k-Nearest Neighbor (k-NN) algorithm. For evaluating performance of developed classifier accuracy and precision were used. They are calculated according to the formula:

$$Accuracy = \frac{TP + TN}{TP + FP + TN + FN} \quad (1)$$

$$Precision = \frac{TP}{TP + FP} \quad (2)$$

Where:

- TP – true positive
- TN – true negative
- FP – false positive
- FN – false negative

3 Results

The development of classifiers based on ML algorithms was done in RapidMiner studio [20]. The results of training performance for classifiers based on 4 different machine learning algorithms are presented in Fig. 2. Training was performed on 280 samples who were subsequently distributed according to the 70:30 ration using random classification. As it can be seen from the figure, during training process, classifier based on k-NN algorithms achieved the highest accuracy of 99.64%. The lowest accuracy was achieved by NB algorithm and it was 97.20%. Average training time was around 5 s.

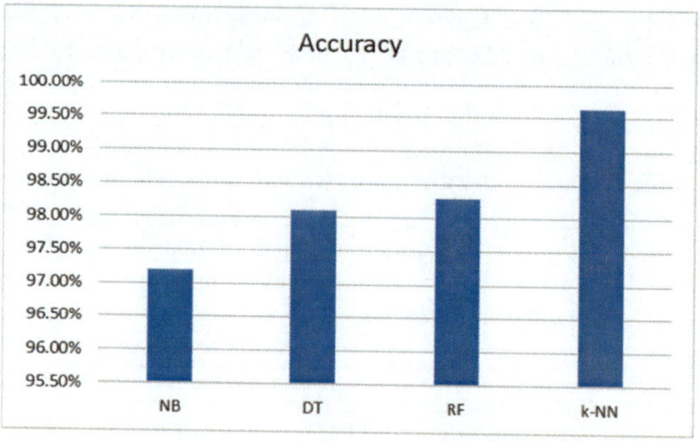

Fig. 2. Training performance of classifiers

After training, for all developed classifiers validation was done. It was performed on 120 samples as shown in Table 2. Each classifier was validated and the results of validation are presented in the Table 3, 4, 5 and 6 presented below.

Table 3. Validation performance of NB based classifier

Dataset/Naïve Bayes	Group 1 Burnout syndrome Classification	Group 2 Healthy Classification	Accuracy 91.6%
Group 1 Burnout syndrome 88	82	6	
Group 2 Healthy 32	4	28	
Total 120 samples	Sensitivity 93.2%	Specificity 87.5%	

Figure 3 shows sensitivity and accuracy of classifiers during validation process. Based on these results it can be seen that k-NN based classifier achieved the highest accuracy during validation. It was 97.5%. the lowest accuracy was achieved by classifier based on Random Forest and it was 88.3%.

The analysis of the database showed that examined parameters age and alcohol consumption do not support the classification and prediction of burnout syndrome. It was shown that D, PA, EE have more significant impact to the classification of the burnout syndrome. Table 7 presents the results of validation of k-NN based classifier when only abovementioned three inputs are used. The improvement in accuracy for 0.8% is achieved.

Table 4. Validation performance of DT based classifier

Dataset/Decision Tree	Group 1 Burnout syndrome Classification	Group 2 Healthy Classification	Accuracy 95.0%
Group 1 Burnout syndrome 88	84	4	
Group 2 Healthy 32	2	30	
Total 120 samples	Sensitivity 94.5%	Specificity 93.8%	

Table 5. Validation performance of RF based classifier

Dataset/Random Forest	Group 1 Burnout syndrome Classification	Group 2 Healthy Classification	Accuracy 88.3%
Group 1 Burnout syndrome 88	82	6	
Group 2 Healthy 32	8	24	
Total 120 samples	Sensitivity 93.2%	Specificity 75.0%	

Table 6. Validation performance of k-NN based classifier

Dataset/k-Nearest Neighbour	Group 1 Burnout syndrome Classification	Group 2 Healthy Classification	Accuracy 97.5%
Group 1 Burnout syndrome 88	85	3	
Group 2 Healthy 32	2	30	
Total 120 samples	Sensitivity 96.6%	Specificity 93.8%	

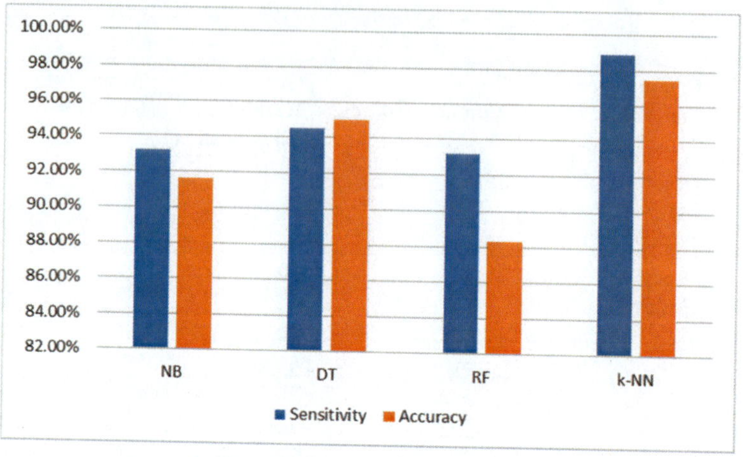

Fig. 3. Classifier validation performance and accuracy

Table 7. Validation performance of k-NN based classifier when 3 inputs are used

	Group 1 Burnout syndrome	Group 2 Healthy	Accuracy 98.3%
Group 1 Burnout syndrome 88	86	2	
Group 2 Healthy 32	0	32	
Total 120 samples	Sensitivity 97.7%	Specificity 100.0%	

4 Conclusion

Burnout syndrome has an increasing prevalence among healthcare workers. Early recognition is useful in many ways, and it can be achieved by using machine learning algorithms.

In our research it was shown that the classifier based on k-NN algorithms achieved the highest accuracy of 99.64%. Artificial intelligence has a broad spectrum of usage in medicine.

Machine learning, deep learning, and cognitive computing are promising and can potentially change the way in which medicine is practiced and the way how some of the diseases and syndromes are detected and diagnosed.

References

1. Russell, E., Koren, G., Rieder, M., Uum, S.V: Hair cortisol as a biological marker of chronic stress: current status, future directions, and unanswered questions. Psychoneuroendocrinology **37**(5), 589–601 (2012)
2. Vinnikov, D., Tulekov, Z., Akylzhanov, A., Romanova, Z., Dushpanova, A., Kalmatayeva, Z.: Age and work duration do not predict burnout in firefighters. BMC Public Health **19**(1), 308 (2019)
3. Lyndon, A.: Burnout among health professionals and its effect on patient safety. https://psnet.ahrq.gov/per-spectives/perspective/190/burnout-among-health-professionals-and-its-effect-on-patient-safety. Accessed 25 June 2017
4. Maio, V., Goldfarb, N.I., Hartmann, C.W.: Pharmacists' job satisfaction: variation by practice setting. P T., vol. 29, pp. 184–190 (2004)
5. Shanafelt, T.D., Balch, C.M., Bechamps, G., et al.: Burnout and medical errors among American surgeons. Ann. Surg. **251**, 995–1000 (2010)
6. Gil-Monte PR. El síndrome de quemarse por el trabajo (burnout). Una enfermedad laboral en la sociedad del bienestar. Madrid: Pirámide; 2005.
7. Balyen, L., Peto, T.: Promising artificial intelligence-machine learning-deep learning algorithms in ophthalmology. Asia Pac. J. Ophthalmol. (Phila) **8**(3), 264–272 (2019)
8. Spahić, L., et al.: Machine learning techniques for performance prediction of medical devices: infant incubators. In: Badnjevic, A., Škrbić, R., Gurbeta Pokvić, L. (eds.) CMBEBIH 2019. CMBEBIH 2019. IFMBE Proceedings, vol. 73. Springer, Cham (2020)
9. Alić, B., Gurbeta, L., et al.: Machine learning techniques for classification of diabetes and cardiovascular diseases. In: 2017 6th Mediterranean Conference on Embedded Computing (MECO), Bar, Montenegro (2017)
10. Rijad Sarić, et al.: FPGA-based real-time epileptic seizure classification using Artificial Neural Network. Biomed. Signal Process. Control 62, 102106 (2020).https://doi.org/10.1016/j.bspc.2020.102106, ISSN 1746-8094
11. Badnjevic, A., Cifrek, M., Koruga, D., Osmankovic, D.: Neuro-fuzzy classification of asthma and chronic obstructive pulmonary disease. BMC Med. Inform. Decis. Making J. **15**(Suppl 3), S1 (2015). https://doi.org/10.1186/1472-6947-15-S3-S1
12. Gurbeta, L., et al.: A telehealth system for automated diagnosis of asthma and chronical obstructive pulmonary disease. J. Am. Med. Inform. Assoc. **25**(9), 1213–1217 (2018). https://doi.org/10.1093/jamia/ocy055
13. Badnjević, A., et al.: Evidence-based clinical engineering: machine learning algorithms for prediction of defibrillator performance. Biomed. Signal Process. Control **54**, 101629 (2019).https://doi.org/10.1016/j.bspc.2019.101629, ISSN 1746-8094
14. Kovačević, Ž., et al.: Prediction of medical device performance using machine learning techniques: infant incubator case study. Health Technol. (2019). https://doi.org/10.1007/s12553-019-00386-5
15. Rau, H.-H., Fuad, A., Hsu, C.-Y., Wei, L.-M., Lin, Y.-A., Hsu, M.-H., Atique, S.: Development of a web-based liver cancer prediction model for type II diabetes patients by using an artificial neural network. Comput. Methods Programs Biomed. (2015)
16. Ladstätter, F., Garrosa, E., Badea, C., Moreno, B.: Application of artificial neural networks to a study of nursing burnout. Ergonomics **53**(9), 1085–1096 (2010)
17. Alic, B., Sejdinovic, D., Gurbeta, L., Badnjevic, A.: Classification of stress recognition using artificial neural network. In: IEEE 5th Mediterranean Conference on Embedded Computing (MECO), Bar, Montenegro, 12–16 June 2016
18. Mello, R.F., Ponti, M.A.: Machine Learning: A Practical Approach on the Statistical Learning Theory. Spinger

19. Anifowose, F., Khoukhi, A., Abdulraheem, A.: Investigating the effect of training–testing data stratification on the performance of soft computing techniques: an experimental study. J. Exp. Theor. Artif. Intell. **29**(3), 517–535 (2017). https://doi.org/10.1080/0952813X.2016.1198936
20. Rapid Miner Studio. https://rapidminer.com/

Artificial Intelligence in Type 2 Diabetes

Musić Enisa, Nefić Belma, Neslanović Adin, Njemčević Enisa, Nikočević Ines, and Odobašić Nejla(✉)

Faculty of Pharmacy, University of Sarajevo, Sarajevo, Bosnia and Herzegovina
{belmanefic,nejlaodobasic}@ffsa.unsa.ba

Abstract. The incidence of type 2 diabetes mellitus has increased significantly in recent years. This study aimed to identify the artificial intelligence applications for type 2 diabetes mellitus care. For the development of this system 600 samples consisting of information about Fasting Plasma Glucose [FPG], Glucose Tolerance Test [OGTT] and blood test called HbA1c were used. Based on the training results, Artificial Neural Network [ANN] with 10 neurons in hidden layer was chosen for diagnostics of Type 2 diabetes [T2D] and was trained with 420 samples. Testing of developed neural network was performed with 180 samples for validation. Out of 420 samples, developed network was accurate in 97.2% cases for the classification of T2D.

Keywords: Diabetes type 2 · Fasting plasma glucose · HbA1c · Artificial neural network

1 Introduction

Type 2 diabetes mellitus [T2D] is a serious, chronic and demanding disease which affects life in many ways.Today is one of the most common diseases. [1] Mostly known as adult-onset, diabetes is characterized by high blood sugar, peripheral insulin resistance, impaired regulation of hepatic glucose production and declining β-cell function, eventually leading to β-cell failure [2].

Common symptoms include increased thirst, frequent urination and weight loss. The gland affected by type 2 diabetes is the pancreas which produces insulin-which allows glucose from the bloodstream to enter the body's cells where it's used for energy. In type 2 diabetes - not enough insulin is produced or the body cannot use it properly or both. That results is "too much" glucose in the bloodstream [3]. The symptoms of T2D can develop more slowly than the ones of T1D, making the condition harder to spot. That's why DT2 can progress asymptomatically or patients with diabetes don't notice symptoms [1]. The difference between type 1 and type 2 diabetes mellitus is that type 1 diabetes doesn't produce insulin and type 2 diabetes doesn't respond to insulin as well as it should - and later while the disease is progressing not enough insulin is made. [4] The basic factors for the rapidly rising incidence of diabetes mellitus are sedentary lifestyle and dietary habits.

Increased risk for diabetes is primarily associated with age, ethnicity, family history of diabetes, smoking, obesity and physical inactivity [1]. Extremely high blood sugar can

A. Badnjevic and L. Gurbeta Pokvić (Eds.): CMBEBIH 2021, IFMBE Proceedings 84, pp. 239–245, 2021.
https://doi.org/10.1007/978-3-030-73909-6_27

lead to a potentially deadly condition where body can't process sugar - hyperglycemic hyperosmolar nonketotic syndrome [HHNS] [5].

T2D is a prevalent illness that causes major vascular, renal and neurologic complications. Possible complications include kidney failure, leg amputation, vision loss and nerve damage. Adults with diabetes also have two to threefold increased risk of heart attacks and strokes. In pregnancy, poorly controlled diabetes increases the risk of fetal death and other complications [6]. Recently, elevated HbA1c level has been considered as one of the leading risk factors for developing micro-vascular and macro-vascular complications. Prevention and treatment of diabetes and its complications are of paramount importance [1].

WHO estimates that diabetes was the seventh leading cause of death in 2016. In high-income countries, the premature mortality rate due to this disease from 2000 to 2010 stopped but then increased in 2010–2016. By contrast, the probability of dying from any one of the four main noncommunicable diseases [cardiovascular diseases, cancer, chronic respiratory diseases, or diabetes] between the ages of 30 and 70 decreased by 18% globally between 2000 and 2016 [7]. The National Diabetes Statistics Report 2020 [USA] analyzed health data through 2018; cases of type 1 and type 2 diabetes have significantly increased among US youth.

For adults diagnosed with diabetes from 2008 through 2018, 15% were smokers, 89% were overweight, 38% were physically inactive, 37% had chronic kidney disease [stages 1 through 4] and fewer than 25% with moderate to severe chronic kidney disease [stage 3 or 4] were aware of their condition [8]. Diagnosis of diabetes is done by blood tests such as fasting plasma glucose, oral glucose tolerance test, or HbA1C [2]. Fasting blood glucose is a test which determine how much glucose is in a blood sample after overnight fast. The normal range for blood glucose is 70 to 100 mg/dL. Levels between 100 and 126 mg/dL are referred to as impaired fasting glucose or pre-diabetes. Diabetes is typically diagnosed when fasting blood glucose levels are 126 mg/dL or higher [9].

In people without diabetes, fasting blood glucose concentration is modestly and non-linearly associated with risk of vascular disease [10]. A hemoglobin A1c [HbA1c] test measures the amount of blood sugar [glucose] attached to hemoglobin. HbA1c results are given in percentages: normal [HbA1c below 5.7%], prediabetes [HbA1c between 5.7% and 6.4%], diabetes [HbA1c of 6.5% or higher] [11]. A glucose tolerance test may be used to diagnose type 1 diabetes mellitus, type 2 diabetes mellitus, and gestational diabetes mellitus. It is a blood test that involves taking multiple blood samples over time, usually 2 h. The test can show whether a person can metabolize a standardized measured amount of glucose [12].

Nowadays, many multidisciplinary studies are being done to support the prevention and treatment of T2D, and it has become a worldwide research priority [13]. Thus, the surveillance, prevention and control of diabetes and its complications using two or more disciplines are becoming increasingly important. As each discipline has its own potential and added value, the combination of them might be a solution for T2D [14, 15].

Artificial Neural Networks [ANN] are currently the next promising area of interest. Already it's successfully applied to various areas of medicine such as diagnostic systems, biochemical analysis, image analysis and drug development [16, 17]. Artificial neural networks are an advanced method for estimating outcomes and prioritizing risk factors

[18]. The ANN technique is an advanced modeling technique based on brain neurons that has been widely used in recent years and can be helpful for diagnosing, estimating, and predicting various diseases [17, 18]. Our aim is to present a diagnostic model that can predict and determine T2D using an ANN model.

2 Materials and Methods

This research was done on the data-set made out of 600 samples. Out of these 600 samples, 300 were for patients with T2D and 300 samples were for healthy patients. The database is consisted of 3 input parameters: fasting plasma glucose [FPG], oral glucose tolerance test, and HbA1c.

The fasting blood glucose test is commonly used to detect diabetes mellitus. As it can be seen from Table 1, according to the results of FPG test, patient is classified as healthy if the FPG is below 5.6 mmol/L.

Table 1. Parameters for the classification

Criteria	Normal	T2D
FPG	<5.6 mmol/L	≥7 mmol/L
HbA1c	<6.0%	>6.5%
OGTT	<7.8 mmol/L	≥11.1 mmol/L

A fasting blood sugar level from 100 to 125 mg/dL [5.6 to 6.9 mmol/L] is considered prediabetes. Diabetes is typically diagnosed when fasting blood glucose levels are 126 mg/dL or higher [19]. Fasting blood sugar [FBS] test measures blood sugar levels or a glycosylated hemoglobin, also called a hemoglobin A1C test. According to the National Institute for health and care excellence, all patients who have HbA1c above 6.5% are classified as diabetic. The criteria for prediabetes is between 6,0% and 6.4%, while the ones classified as healthy have HbA1c below 6,0% [19].

The glucose tolerance test identifies abnormalities in the way your body handles glucose after a meal - often before your fasting blood glucose level becomes abnormal. A normal blood glucose level is lower than 140 mg/dL [7.8 mmol/L]. A blood glucose level between 140 and 199 mg/dL [7.8 and 11 mmol/L] is considered impaired glucose tolerance, or prediabetes. A blood glucose level of 200 mg/dL [11.1 mmol/L] or higher may indicate diabetes [20].

Artificial Neutral Network [ANN] was development in order to perform prediction of diabetes mellitus type 2 based on diagnostic parameters (Table 2).

The distribution of the data-set for development of the Artificial Neural Network [ANN] is presented in Table 3. Since this is relatively small data-set, distribution of 70:30% for training and validation was used [21].

According to previous research and expert recommendation [22, 23] two-layer feed forward artificial neural network, with sigmoid transfer function in hidden layer was

Table 2. Database for development of neural network

Class	T2D	Healthy	Total
Male	152	151	303
Average age [years]	46,36	44,42	45,39
Female	148	149	297
Average age [years]	48,43	44,92	46,68
Total number of patients per class	300	300	600

Table 3. Data-set distribution

Total number of samples 600	Healthy 300	Disease 300
Training 420	210	210
Validation 180	90	90

developed. Neural network architecture is consisted of 5 inputs: Sex, age, OGTT, FGP, HbA1c.

According to these results, specificity, sensitivity and accuracy of the network are calculated according to Eqs. (1), (2) (3):

$$sensitivity = \frac{true\ positive}{(true\ positive + false\ negative)} \tag{1}$$

$$specificity = \frac{true\ negative}{(true\ negative + false\ positive)} \tag{2}$$

$$accuracy = \frac{(true\ negative + true\ positive)}{(true\ negative + true\ positive + false\ negative + false\ positive)} \tag{3}$$

3 Results

During development, different number of neurons were tested. The results of training for different number of neurons in hidden layer are presented in the Table 4 below.

Table 4. Training and validation of neural network

Number of neurons	MSE	No of iterations
1	3.6784e−10	16
5	7.9635e−11	18
10	1.6071e−10	19
20 neurons	1.9441e−10	19

Training was performed using Levenberg – Marquadt training algorithm. Training performance of the developed neural network is presented in the Fig 1.

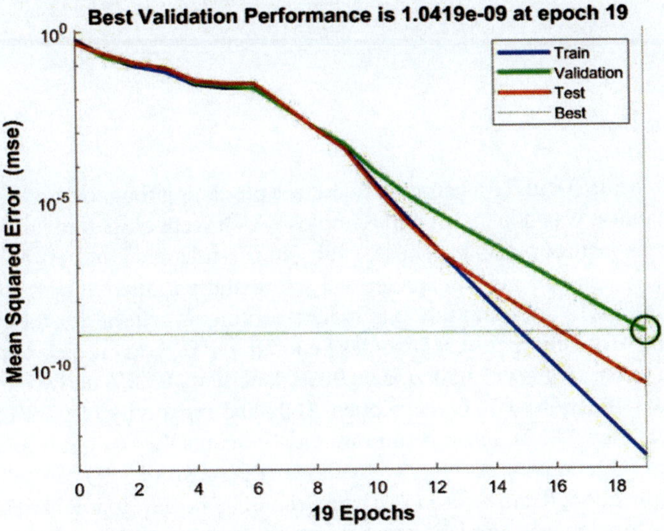

Fig. 1. Training performance with 10 neurons

Based on the training results, ANN with 10 neurons in hidden layer was chosen for diagnostics of Type 2 diabetes [T2D]. Results of validation are presented in Table 5 below.

These results indicate that in 97.8% of patients, DM2 will be diagnosed using an artificial neural network. The percentage of specificity shows that 96.6% of healthy patients will be negative on DM2. The total accuracy of this method, artificial neural network, for diagnosing DM2 is 97.2%. This is promising results, since such expert system can be used as an aid to healthcare professional at primary healthcare level in situations when medical specialist is not available.

Table 5. Artificial Neural Network performance in classification to class "Diabetes type 2"

n = 180	**Group 1 Healthy** ANN prediction	**Group 2 Disease** ANN prediction	
GROUP 1 Healthy 90	TP 88	FN 2	**Accuracy** [%] 97.2
Group 2 Disease 90	FP 3	TN 87	
	Sensitivity [%] 97.8	**Specificity** [%] 96.6	

4 Conclusion

This paper presents Artificial Neural Network for predicting diagnosis of diabetes mellitus. ANN method was done with 420 Samples which were classified based on results of tests such as fasting plasma glucose, oral glucose tolerance and HbA1C, into two groups - healthy patients and T2D patients. Fasting plasma glucose, oral glucose tolerance test, and HbA1c are important parameters serving as criteria for the diagnosis of T2D according to the International Diabetes Federation [IDF]. In the case of prediabetes. classification sensitivity is calculated to be 90%, specificity 95,5% and accuracy 94,1%.

For the classification of T2D, developed ANN had sensitivity of 97,8%, specificity 96,6% and accuracy 97,2%. These results are satisfying, but they can be improved trough network with a larger data set. ANN as a method in diagnosis of T2D can be improved to enlarge its application. It can be used as a background in the development of a device that could help doctors in more rapid diagnostics, monitoring and possible adequate therapy.

References

1. Sami, W., Ansari, T., Butt, N.S., Ab Hamid, M.R.: Effect of diet on type 2 diabetes mellitus: a review (2017)
2. Mahler, R.J., Adler, M.L.: Type 2 diabetes mellitus: update on diagnosis, pathophysiology, and treatment. J. Clin. Endocrinol. Metab. **84**, 1165–1171 (1999)
3. Hormone Health Network: Type 2 diabetes. https://www.hormone.org/diseases-and-conditions/diabetes/type-2diabetes
4. Healthline: Type 1 and Type 2 Diabetes. https://www.healthline.com/health/difference-between-type-1-andtype-2-diabetes
5. WebMD: High Blood Sugar and Diabetes. https://www.webmd.com/diabetes/guide/diabetes-hyperglycemia
6. World Health Organization: Diabetes. https://www.who.int/health-topics/diabetes#tab=tab_1

7. World Health Organization: Global report on diabetes. https://apps.who.int/iris/bitstream/handle/10665/204871/9789241565257_eng.pdf;jsessionid=222040F48428DA1F9B18D960B275D994?sequence=1
8. National Diabetes Statistics Report 2020: Estimates of Diabetes and Its Burden in the United States. https://www.cdc.gov/diabetes/pdfs/data/statistics/national-diabetesstatistics-report.pdf
9. MedicineNet: Medical definition of fasting blood glucose. https://www.medicinenet.com/fasting_blood_glucose/definition.htm
10. Sarwar, N., Gao, P., Kondapally Seshasai, S.R., Gobin, R., Kaptoge, S., Di Angelantonio, E., Ingelsson, E., Lawlor, D.A., Selvin, E., Stampfer, M., Stehouwer, C.D.A., Lewington, S., Pennells, L., Thompson, A., Sattar, N., White, I.R., Ray, K.K., Danesh, J.: Diabetes mellitus, fasting blood glucose concentration, and risk of vascular disease: a collaborative meta-analysis of 102 prospective studies (2010)
11. MedlinePlus: Hemoglobin A1C [HbA1c] Test. https://medlineplus.gov/lab-tests/hemoglobin-a1c-hba1c-test/
12. Eyth, E., Basit, H., Smith, C.J.: Glucose Tolerance Test (2020)
13. Badnjević, A., Gurbeta Pokvić, L., Hasičić, M., Bandić, L., Mašetić, Z., Kovačević, Ž., Kevrić, J., Hifziefendić, J., Pecchia, L.: Evidence-based clinical engineering: machine learning algorithms for pre- diction of defibrillator performance. Biomed. Signal Process. Control **54**, 101629 (2019)
14. Abhari, S., Niakan Kalhori, S.R., Ebrahimi, M., Hasannejadasl, H., Garavand, A.: Artificial intelligence applications in type 2 diabetes mellitus care. Focus on machine learning methods (2019)
15. Kovačević, Ž., et al.: Prediction of medical device performance using machine learning techniques: infant incubator case study health and technology. Health and Technology
16. Sejdinovic, D., et al.: Classification of prediabetes and Type 2 Diabetes using Artificial Neural Network. In: CMBEBIH 2017 (2017)
17. Sarić, R., et al.: FPGA-based real-time epileptic seizure classification using Artificial Neural Network. Biomed. Signal Process. Control **62**, 102106 (2020)
18. Borzouei, S., Soltanian, A.R.: Application of an artificial neural network model for diagnosing type 2 diabetes mellitus and determining the relative importance of risk factors (2018)
19. Healthline:Blood Glucose Test. https://www.healthline.com/health/glucosetestblood?fbclid=IwAR29NqdrZoChm3hszz6yjecfbaYOFVH3vsBYKksY7Nbd6shwpuWfpdldqg
20. Mayo clinic: Glucose tolerance test. https://www.mayoclinic.org/tests-procedures/glucose-tolerancetest/about/pac-20394296
21. Rau, H.-H., Hsu, C.-Y., Lin, Y.-A., Atique, S., Fuad, A., Wei, L.-M., Hsu, M.-H.: Development of a web-based liver cancer prediction model for type II diabetes patients by using an artificial neural network. Comput. Methods Programs Biomed **125**, 58–65 (2016). https://doi.org/10.1016/j.cmpb.2015.11.009, ISSN 0169-2607
22. Catic, A., et al.: Application of Neural Networks for classification of Patau, Edwards, Down, Turner and Klinefelter Syndrome based on first trimester maternal serum screening data, ultrasonographic findings and patient demographics. BMC Med. Genomics **11**, 19 (2018). https://doi.org/10.1186/s12920-018-0333-2
23. Badnjevic, A., et al.: An expert diagnostic system to automatically identify asthma and chronic obstructive pulmonary disease in clinical settings. Sci. Rep. **8**, 1–9 (2018)

Check for updates

A Fuzzy Model for the Risk of Urinary Tract Infections Prediction Using Microscopic and Chemical Urine Analysis

Ismar Šahović[1], Monia Avdić[1], Mirsada Hukić[2,3], and Zerina Mašetić[1](✉)

[1] International Burch University, Francuske Revolucije bb, 71000 Sarajevo, Bosnia and Herzegovina
zerina.masetic@ibu.edu.ba
[2] Institute for Biomedical Research and Diagnostics NALAZ, Sarajevo, Bosnia and Herzegovina
[3] ANUBIH, Academy of Science and Arts Bosnia and Herzegovina, Sarajevo, Bosnia and Herzegovina

Abstract. Urinary Tract Infections (UTIs) are the most common type of bacterial infection and the physical, chemical and microscopic analysis of urine, followed by culture is a commonly conducted procedure in microbiological laboratories. Even though physical, chemical and microscopic analysis are more time efficient, culture is still considered the gold standard in diagnosing UTI, although its results and the results of antibiogram are obtained within a period of days. The aim of this research is to create a fuzzy model to predict the risk of urinary tract infections using microscopic and chemical urine analysis. A dataset containing 595 samples was used to create a fuzzy model using a Mamdani-type inference system with 5 inputs (nitrite, leukocyte esterase, bacteria, white blood cells and red blood cells) and 1 output (prediction of infection). The accuracy of the model achieved was 86%, after the contaminated samples were excluded. The recall value for predicting the absence of UTI samples was the highest, 89%, while the recall value for predicting the presence of UTI was 71%. The lowest recall achieved was for samples that could present both, the absence and the presence of UTI, with a value of 29.6%. The model was successful and revealed adequate accuracy, however future experiments should use more parameters.

Keywords: Urinary tract infections (UTIs) · Fuzzy model · Urine analysis · Nitrite · Leukocyte esterase · Bacteria · White blood cells (WBC) · Red blood cells (RBC)

1 Introduction

Urinary tract infections and recurrent urinary tract infections, caused by the reinfection with the original bacterial isolate, are amongst the most common outpatient infections amongst women and consequently they represent a major burden for the healthcare system as well as a personal burden that has a negative impact on the quality of life (QoL) [1, 2]. Urine, as the most submitted specimen in microbiology laboratories, is subjected to

A. Badnjevic and L. Gurbeta Pokvić (Eds.): CMBEBIH 2021, IFMBE Proceedings 84, pp. 246–255, 2021.
https://doi.org/10.1007/978-3-030-73909-6_28

microscopic, physical and chemical analysis and subsequently the specimen is cultured [3]. Results of the microscopic and chemical analysis of urine specimens are obtained in a short period of time, however culture requires a 24 h period for bacterial growth and additional 24 h–48 h for species identification and ordination of antibiotic therapy according to the obtained antibiogram results [4, 5]. Results of culture are reported as potencies of 10 Colony Forming Units (CFU), where 10^4 is considered probable absence of infection, while 10^4–10^5 in asymptomatic patients requires taking another specimen and repeating the procedure, whereas a CFU above 10^5 is considered infection [6].

Amongst the physical parameters, urine transparency is amongst the most significant in prediction of UTI, while from the chemical parameters' presence of nitrites and leukocyte esterase are, amongst other parameters, the most indicative of infection [4, 5].

Regarding the microscopic examination of the urine specimen, absence of white blood cells (WBC) and bacteria in a clean catch uncentrifuged urine specimen is indicative of absence of infection. Contrary one or more bacterial cells per oil-immersion field is indicative of 10^5 or more bacteria per milliliter in the specimen and additionally presence of WBC further implies presence of infection [6]. Variations within the number of bacterial cells and WBC per oil-immersion field are apparent in different CFU obtained in the results of culture.

Consequently, in our previous work we have designed a fuzzy system that aims to predict the risk of UTI based on input parameters that include transparency and microscopic examination of urine specimens (presence of bacteria, WBC and Red Blood Cells (RBC)) [7]. Considering the significance of presence of nitrites and leukocyte esterase in the prediction of UTI, conducted within chemical urinalysis, in this study we aimed to design a fuzzy model with an increase in input parameters to yield higher sensitivity in the prediction of UTI.

The fuzzy model, which is a qualitative tool used to explain system behavior through a natural language [8], will aid physicians in the prediction of UTIs. Fuzzy models operate on information known as fuzzy sets and fuzzy logic and can place multiple values through the same variable.

The automated process of UTI predictions have been researched in various previous studies. However, fuzzy logic as the methodology for the UTIs prediction hasn't been applied much.

Ramgopal et al. [9] derived and validated a model for the febrile infants at low risk for SBIs identification, using supervised machine learning approaches. They used 4 features (urinalysis, white blood cell count, absolute neutrophil count, and procalcitonin), and showed that the random forest model achieved the highest specificity (74.9, 95%) and sensitivity of 98.6% in the validation group. Ozkan, Koklu and Sert [10] aimed to develop the automated model for UTIs diagnosis using a decision tree (DT), support vector machine (SVM), random forest (RF) and artificial neural network (ANN). The highest classification accuracy was achieved with ANN with the result of 98.3%, using pollacuria, suprapubic pain symptoms and erythrocyte features. Papageorgiou [11] worked on the development of the model for making the decisions for UTI treatment, called fuzzy cognitive maps (FCMs) to handle the uncertainty and missing information. The developed model gave a front-end decision on antibiotics' suggestion for the treatment of UTI which could be helpful references for physicians and patients.

The rest of the work is organized as follows: Sect. 2 will explain the data used in the study and methods applied for creating the fuzzy model. Section 3 will present and discuss the obtained results, while the conclusions will be given in Sect. 4.

2 Material and Methods

2.1 Data Collection

The study included a total of 595 specimens that were collected from a medical center in Bosnia and Herzegovina Institute for Biomedical Diagnostics and Research NALAZ in the period from January 2018 - December 2018. Criteria for inclusion into the study was that more than one of three successive urine samples was not taken into consideration.

The obtained urine specimens were subjected to physical, chemical and microscopic examination followed by culture. Physical examination was conducted by visual inspection of the specimen. Chemical examination was conducted using combi screen 11 SYS, where results for nitrite and leukocyte esterase were considered significant. Microscopic examination was carried out by microscopic inspection of a drop of wellmixed, uncentrifuged urine under the oil immersion objective.

The features obtained for the further study are nitrite, leukocyte esterase, bacteria, white blood cells (WBC) and red blood cells (RBC). Table 1 explains the features.

Table 1. Input parameters of the fuzzy model

Parameter	Explanation
Nitrite	Bacteria that enter the urinary tract turn nitrate in the urine into nitrite and this is a sign of UTI
Leukocyte Esterase	Esterase is a type of enzyme that is produced by leukocytes. Urinalysis tests use this parameter to test for infection as white blood cells are a strong indicator of UTI. This parameter also allows for detecting lysed leukocytes
Bacteria	Bacteria in urine is responsible for UTIs. If this parameter is found during urinalysis, this is a strong indicator that infection is present
WBC	Presence of WBC indicates urine infection
RBC	RBC indicates an infection, viral or bacterial, inflammation of the kidney or bladder, blood disorder or bladder or kidney cancer

There is only one output variable, and this expresses the model's prediction for the infection, representing the presence of UTI, absence of UTI, or output that could either present presence or absence of UTI.

The distribution of the data, based on the output is shown in Table 2.

Table 2. The distribution of input samples per output class

Output class	Input samples
Absence of UTI	300
"Maybe" UTI	27
Presence of UTI	268

2.2 Fuzzy Model

The urine sample features nitrite, leukocyte esterase, bacteria, white blood cells (WBC) and red blood cells (RBC) are used as the input to the fuzzy model for automated UTIs prediction, as shown in Fig. 1. The model is created using MATLAB, which enables the usage of fuzzy logic toolbox to create if-then rules and to explain how a system behaves. To create the model, we used Mamdani's fuzzy inference system (FIS) [12]. Mamdani's fuzzy inference is the most common method and the output membership functions are fuzzy sets. After aggregation, each of the fuzzy sets require defuzzification. Moreover, it uses membership functions, if-then rules and fuzzy logic operators.

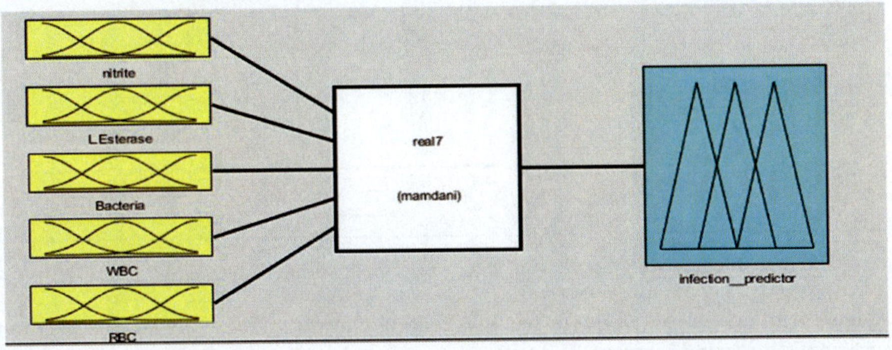

Fig. 1. Fuzzy model inference system with nitrite, leukocyte esterase, bacteria, WBC and RBC features

Input values must undergo fuzzification using membership functions to define how well they fit into relevant fuzzy sets.

Figure 2 represents the Membership function of the input value nitrite, which describes the amount of nitrite present in the urine sample. The membership function represents 3 categories, and these are presented in the dataset as negative, trace and positive. Value 1 represents negative, value 2 represents trace and value 3 represents positive.

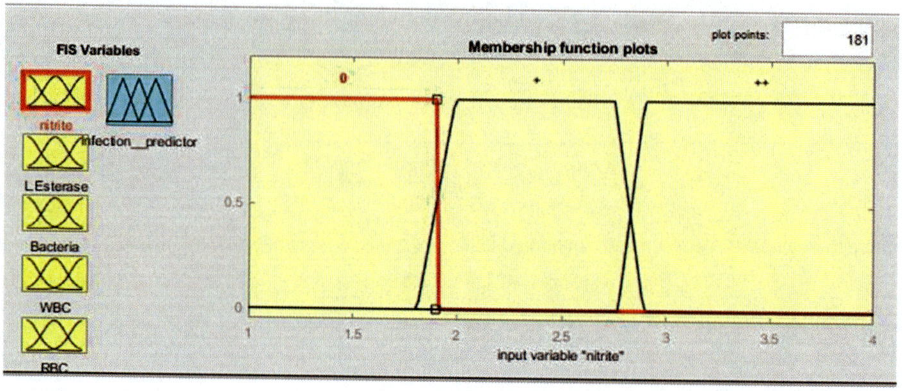

Fig. 2. Membership functions plot for fuzzy input variable nitrite

For the feature leukocyte esterase, which represents the amount of leukocyte esterase present in the urine sample, the membership function represents 5 categories, and these are presented in the dataset as negative, trace, positive, very positive and extremely positive. These are numerically expressed from 1 to 5 respectively, as shown in Fig. 3.

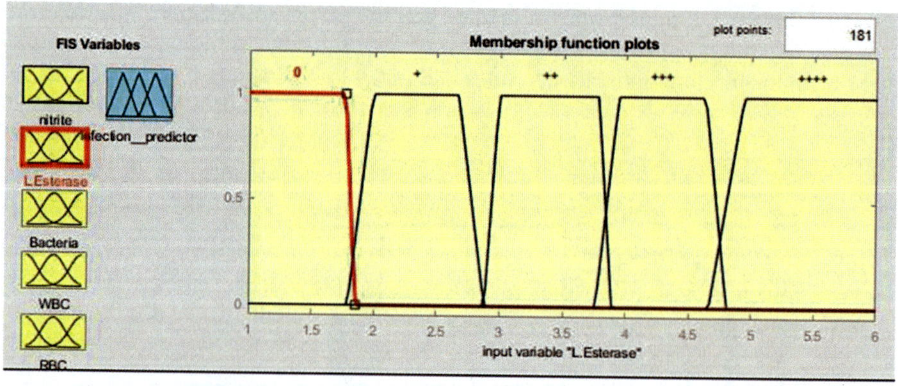

Fig. 3. Membership functions plot for fuzzy input variable leukocyte esterase

The membership functions for bacteria feature shows three categories, represented by negative, trace and positive. The following categories are represented by the number 1 to 3. For WBC features, membership functions represent values of white blood cells from 1 to 25. The first category includes the values from 1 to 5, the second category includes 2 to 3, the third includes 4 to 10.5, the fourth includes 7.5 to 15.5, the fifth includes 13 to 20.5 and the sixth category includes 18.5 to 25. The following categories are represented by the number 1 to 6.

Last membership functions represent the number of RBC in the urine sample. ranged from 0 to 20. The first category includes the values from 0 to 1.9, the second category includes 1.6 to 2.5, the third includes 2.2 to 4, the fourth includes 3.4 to 10.4, the fifth includes 7.5 to 17.2 and the sixth category includes 13.2 to 20. The following categories are represented by the number 1 to 6.

Next, the fuzzy rule set is created. Membership functions expressed as words combined describe the fuzzy system [12]. Our fuzzy system is made of a block consisting of the linguistic values for 5 inputs, such as: negative, trace and positive, and block representing the linguistic term for output variables: negative, maybe and positive. Following fuzzification, 5 inputs are taken and passed through the fuzzy inference system mapping out an output. This fuzzy model uses the Mamdani-type fuzzy inference system with a default AND operator [12].

The fuzzy model has one output which describes whether the patient has an infection by observing the 5 input variables. The defuzzification stage transforms the fuzzy output from the fuzzy inference system into a crisp output. To get this crisp output, a representative value needs to be chosen as the final output. This model uses the centroid calculation to return the center of area under the curve. Defuzzification process is shown in Fig. 4.

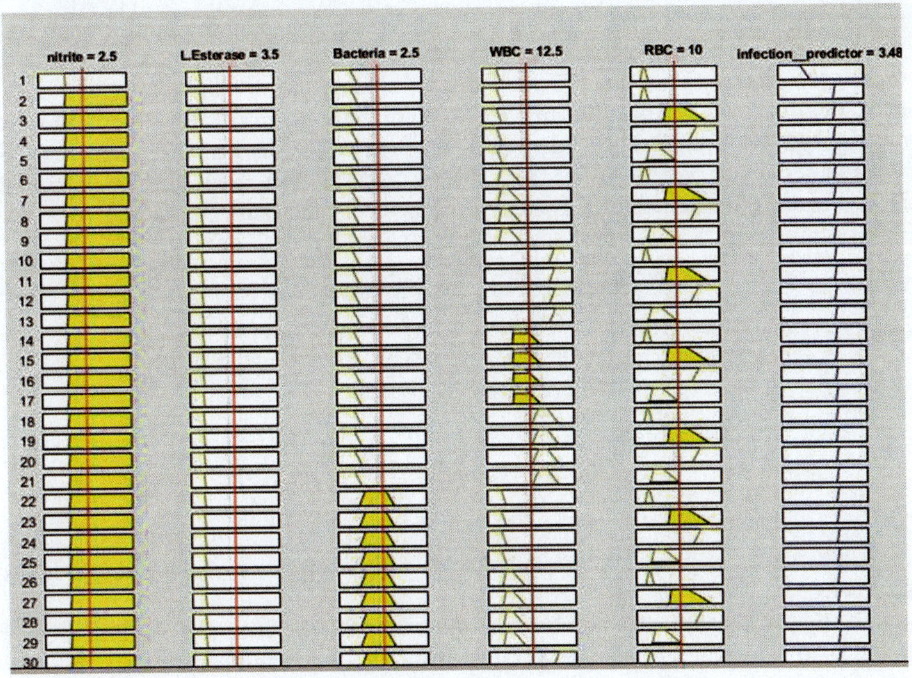

Fig. 4. Defuzzification process to transform the fuzzy output to a crisp value

3 Results and Discussion

The results of the study will be evaluated using the accuracy and recall performance measures. The accuracy is the performance metric that shows the number or percentage of correctly classified input instances within the whole dataset. It can be represented in terms of true positive (TP), true negative (TN), false positive (FP) and false negative (FN) outcomes, following the formula (1) [13]:

$$Accuracy = \frac{TP + TN}{TP + TN + FP + FN} \tag{1}$$

The sensitivity or recall represents the number or percentage of actual positive input instances classified correctly, represented as (2) [13]:

$$Recall = \frac{TP}{TP + FN} \tag{2}$$

The proposed fuzzy model achieved the prediction accuracy of 78%, when the whole dataset was considered. However, we excluded samples that contained all input factors, which indicate the contamination, from the original dataset, which led to the accuracy of the model to 86%.

The recall value for the prediction of UTI presence in the input samples was 71%, indicating that the model successfully predicted 189 out of 268 samples. The highest recall value was achieved for the prediction of UTI absence, with the result of 89%, indicating that 266 out of 300 samples were correctly classified. Finally, the lowest recall value was achieved in prediction of the "maybe" cases, or those that could be classified either as presence of UTI or absence of UTI. The recall value was around 30%, where 8 out of 27 instances were correctly predicted (Table 3).

Table 3. Recall values per output class

Classes	All instances	Correctly classified	Recall (%)
No UTI	300	266	89%
"Maybe" UTI	27	8	30%
UTI	268	189	71%

Additionally, we measured the accuracy with the presence of each single input parameter.

The first parameter is the number of nitrites found in the urine sample. Presence of nitrite in urine indicates bacteria as nitrate, a naturally occurring substance, is converted into nitrite by bacteria [14]. Our fuzzy model produces a high accuracy value for the samples in which the nitrite parameter is positive, approximately 99.6%. Other studies, such as dos Santos, Weber and Perez [15] revealed the high accuracy with the nitrite features, concluding that if the patient has a high nitrite number, this leads to the presence of UTI more than 99%.

Leukocyte esterase as the second parameter in our study, is a specific protein of white blood cells (leukocytes), which reacts with agents on the dipstick to produce a blue color. The main advantage of this test is its ability to detect esterase and confirm presence of leukocytes even when they are lysed. Our fuzzy model showed the accuracy of 92% in detection of leukocyte esterase samples in UTI. In dos Santos, Weber and

Perez [15], the accuracy with the leukocyte esterase feature, the accuracy achieved was 89.5%, slightly lower compared to the accuracy achieved in this study, still indicating the leukocyte esterase as the important in UTI prediction.

Bacteria are organisms that are normally not present in the urine, however a small number of bacteria can be found in the urine of people that do not have an UTI. An increased level of bacteria in urine also increases the chances of the patient having an UTI. The main cause of bacteria in urine would be all factors that are affecting the flow of urine. In other words, bacteria number is increased when there is retention of urine. Our model showed the accuracy of 85% when bacteria number was classified as many. In previous studies [15], authors achieved better classification accuracy, 92.9%, compared to the accuracy achieved here, when bacteria feature is considered, concluding that microscopic-urinary-sediment analysis for bacteria in the urine is found to be very important in the UTI prediction.

Leukocytes or WBC are cells that are produced as a response to an infection. Most of the time leukocytes in urine indicate UTI. Leukocytes can also be present if a patient has kidney stones, kidney infection, urinary system block age and holding urine. The model showed the accuracy of 75% in detection of UTI using leukocytes above 20 as the only parameter.

Finally, Erythrocytes or RBC are usually not present in urine. Presence of these substances in urine can indicate an abnormality. Red blood cells in urine can be caused by infections such as: urinary tract infection, bladder, kidney or prostate infection. The proposed model achieved an accuracy of 70% in predicting UTI when RBCs were above 5. Usually this is not a strong parameter unless combined with other parameters because of the aforementioned reasons.

These results are compared in Fig. 5.

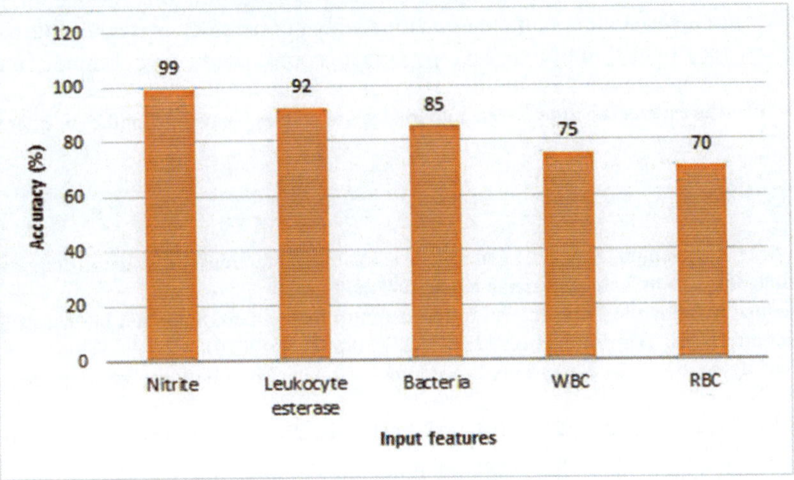

Fig. 5. Accuracy results per input parameters

The Fig. 5 compares the accuracies with the presence of each single input parameter, increased at the input, in the conjunction with other parameters. Based on the achieved accuracies, it is noticeable that the highest impact on the UTI prediction has the presence of the nitrite in the urine. Additionally, the presence of leukocyte esterase ensures the presence of UTI with 92% accuracy, which leads to the conclusion of the importance of this feature for the UTI detection and further prediction. The lowest impact on the prediction of UTI from the person's urine has the presence of RBC among given parameters. However, it could also indicate the presence of UTI, when blood is noticed in the urine sample.

In review research papers regarding the analysis of urine and the prediction of the UTIs, the five mentioned features are with the highest correlation with UTIs [4, 16]. This means that if the presence of these features is noticed in the person's urine, it is most likely that they could be suffering from UTIs.

4 Conclusions

As UTIs are amongst the most common infections, due to the severity of symptoms that may develop if treatment is not applied timely, routine testing, that takes 2 to 3 days, represents a major issue.

Within the routine testing results of microscopic, chemical and physical analysis are indicative of infection, however each of these parameters separately may be caused by other diseases of the urinary tract and pathophysiological conditions.

Therefore, the aim of this study was to develop a system that would take into consideration the presence of the parameters that are most indicative of UTI and all the possible combinations that arise in such infections to predict CFU as the final output. The accuracy of the model of 86% can be considered as successful, however, due to the rise of antibiotic resistance, an increase in the number of parameters is advisable to reach higher accuracy rates if antibiotic therapy is to be coordinated using computer models.

Conflict of Interest Declaration. The authors declare that they have no conflict of interest.

References

1. Kodner, C., Gupton, E.K.T.: Recurrent urinary tract infections in women: diagnosis and management. Am. Fam. Physician **82**(6), 638–643 (2010)
2. Medina, M., Castillo-Pino, E.: An introduction to the epidemiology and burden of urinary tract infections. Ther. Adv. Urol. **11**, 1756287219832172 (2019)
3. Washington, J.A. (ed.): Laboratory Procedures in Clinical Microbiology. Springer, Heidelberg (2012)
4. Simerville, J.A., Maxted, W.C., Pahira, J.J.: Urinalysis: a comprehensive review. Am. Fam. Physician **71**(6), 1153–1162 (2005)
5. Strasinger, S.K., Di Lorenzo, M.S.: Urinalysis and body fluids. FA Davis Company (2014). ISBN 9780803639201
6. Vandepitte, J., Verhaegen, J., Engbaek, K., Rohner, P., Piot, P., Heuck, C.C., Heuck, C.C.: Basic laboratory procedures in clinical bacteriology. World Health Organization (2003)

7. Ibrisimovic, M.A., Karlı, G., Balkaya, H.E., Ibrisimovic, M., Hukic, M.: A fuzzy model to predict risk of urinary tract infection. In: CMBEBIH 2017, p. 289293. Springer, Singapore (2017)
8. Sugeno, M., Yasukawa, T.: A fuzzy-logic-based approach to qualitative modeling. IEEE Trans. Fuzzy Syst. **1**(1), 7–31 (1993). https://doi.org/10.1109/TFUZZ.1993.390281
9. Ramgopal, S., Horvat, C.M., Yanamala, N., Alpern, E.R.: Machine learning to predict serious bacterial infections in young febrile infants. Pediatrics **146**(3), e20194096 (2020). https://doi.org/10.1542/peds.2019-4096
10. Ozkan, I.A., Koklu, M., Sert, I.U.: Diagnosis of urinary tract infection based on artificial intelligence methods. Comput. Methods Programs Biomed. (2018). https://doi.org/10.1016/j.cmpb.2018.10.007
11. Papageorgiou, E.I.: Fuzzy cognitive map software tool for treatment management of uncomplicated urinary tract infection. Comput. Methods Programs Biomed. **105**(3), 233–245 (2012)
12. Foroutan, E., Maleki, J.: Risk Assessment for Urban Area Using Fuzzy Logic. International Society for Photogrammetry and Remote Sensing, (Freksa 1994) (2001). www.isprs.org/proceedings/2011/Gi4DM/PDF/OP31.pdf
13. Avdić, M., Mašetić, Z., Sayed, A.E., Odobašić, L., Hukić, M.: A novel approach in determination of biofilm forming capacity of bacteria using random forest classifier. In: Badnjevic, A., Škrbić, R., Gurbeta Pokvić, L. (eds.) CMBEBIH 2019. IFMBE Proceedings, vol. 73. Springer, Cham (2020). https://doi.org/10.1007/978-3-030-17971-7_42
14. Penders, J., Fiers, T., Delanghe, J.R.: Quantitative Evaluation of Urinalysis Test Strips (2002)
15. dos Santos, J.C., Weber, L.P., Perez, L.R.R.: Evaluation of urinalysis parameters to predict urinary-tract infection. Braz. J. Infect. Dis. **11**, 479–481 (2007). www.bjid.com.br
16. Wilson, M.L., Gaido, L.: Laboratory diagnosis of urinary tract infections in adult patients. Med. Microbiol. 9 (2004). https://academic.oup.com/cid/article-abstract/38/8/1150/441696

Machine Learning Techniques for Predicting Breast Cancer Based on Biomarkers

Mediha Salić, Nejra Samardžić(✉), Nejla Selmanović, Irma Sinanović, Muhamed Sirćo, and Belma Suljević

Univerzitet u Sarajevu, Farmaceutski fakultet, Sarajevo, Bosnia and Herzegovina
nejrasamardzic@ffsa.unsa.ba

Abstract. Worldwide, one of the most common malignancies is breast cancer, which makes about 10% of all malignancy cancers. This paper presents how machine learning techniques can be used in early diagnosis of breast cancer (BC). A database for prediction of possibility of BC appearance was formed by using eight parameters that include: age, body mass index, glucose, leukocytes count, volatile organic compounds, adiponectin, leptin and neutrophils. Obesity has shown a high level of breast cancer risk increase so the Body mass index (BMI) was incorporated as one of the predictors. The paper included 100 healthy volunteers and 250 breast cancer patients. Different classifiers were tested and among them are Naive Bayes based (NB), Decision Tree (DT), Random Forest (RF) and k-nearest neighbors algorithm (kNN). All methods had accuracy over 90% and the highest was Decision Tree with accuracy rate of 99.05%. Results suggest that parameters used in this paper can be used as a cheap and effective breast cancer predictor by Machine Learning Techniques which can increase number of early discovered patients, the ease of diagnosis and decrease the number of cases with metastasis and mortality.

Keywords: Machine learning · Prediction · Breast cancer · Biomarkers

1 Introduction

Breast cancer (BC) is still the most commonly diagnosed cancer in women worldwide. According to the World Health Organization (WHO) there were approximately 2 500 000 cases of breast cancer in 2020. However, due to improvement in medical science and technology, mortality is decreasing. On the other hand, costs per year are in a constant increase [1].

According to The Merck Manual of Diagnosis and Therapy (MSD manuals) breast cancer occurs when cells in the breast become abnormal and divide uncontrollably. It starts in the lobules or the ducts and can invade locally or spread through the lymph nodes or bloodstream. Metastatic BC may affect other organs such as lungs, liver, brain or even bone [2].

Typically there are no symptoms when the tumor is small, which is why screening is an important strategy for early detection so it can provide greater opportunities for a

A. Badnjevic and L. Gurbeta Pokvić (Eds.): CMBEBIH 2021, IFMBE Proceedings 84, pp. 256–263, 2021.
https://doi.org/10.1007/978-3-030-73909-6_29

positive outcome in the future. Even though painless lump is the most common sign, sometimes cancer can spread to underarm lymph nodes and be present without knowing it [3, 4].

Previously clinicians have used basic software programs, such as Microsoft Excel, SPSS and STATA in analyzing factors influencing breast cancer. These conventional statistical methods are not quite adaptable in identifying new variables as well as generating creative and integrative visualizations so their drawback has led to the wide usage of various machine learning (ML) approaches [5].

Machine learning still draws heavily from statistics and probability, but it is fundamentally more powerful because it allows inferences or decisions to be made that could not otherwise be made using conventional statistical methodologies. For instance, many statistical methods are based on multivariate regression or correlation analysis. While generally very powerful, these approaches assume that the variables are independent and that data can be modeled using linear combinations of these variables. When the relationships are nonlinear and the variables are interdependent (or conditionally dependent) conventional statistics usually flounders. It's in these situations where machine learning tends to shine. Among the more commonly noted problems was an imbalance of predictive events with parameters (too few events, too many parameters), overtraining, and a lack of external validation or testing. Nevertheless, it was clear that machine learning methods, relative to simple statistical methods, could substantially (15–25%) improve the accuracy of cancer susceptibility and cancer outcome prediction [6].

Machine learning is a branch of Artificial intelligence that incorporate a variety of statistical and optimization techniques that allow computers to "learn" from past examples. This process is similar to traditional models: there is an outcome, covariates and a statistical function linking the two. Advantage of using machine learning is in handling enormous numbers of predictors and combining them in nonlinear and interactive ways [6, 7]. These features are well-suited to medical applications and are frequently used in healthcare [8–14].

Previous studies have focused on: differentiation of benign and malignat tumors based on genetic programming and machine learning algorithms, prediction of breast cancer based on data which can be collected in routine blood analyses and using machine learning for personalized breast cancer risk prediction comparison with the BCRAT and BOADICEA models [3, 15, 16].

Using Machine Learning Techniques in medicine, can help detect cancer in early stages. This can prevent further development of the disease, reduce mortality and more severe forms of the disease like metastases. Also, the costs of treatment are significantly lower in terms of therapy and hospital treatment. This study was designed to show all benefits and how Machine Learning Techniques can be used for predicting breast cancer. Using Machine Learning Techniques we aim to predict possibility of breast cancer appearance based on higher or lower value of eight biomarker predictors: age, body mass index, glucose, adiponectin, leptin, volatile organic compounds, leukocyte count and neutrophils [17–23].

The purpose of this study is to review Artificial Intelligence application in predicting possibility of breast cancer based on several biomarkers.

2 Materials and Methods

A database for developing Artificial Neural Network for prediction of possibility of BC appearance was formed by using 8 parameters that include: age, body mass index, glucose, leukocytes count, volatile organic compounds, adiponectin, leptin and neutrophils. The paper included 100 healthy volunteers and 250 breast cancer patients.

Table 1. Properties of the dataset used for development of artificial neural network

Class	Breast cancer	Healthy	Total
Male	/	/	/
Average age (years)	/	/	/
Female	250	100	350
Average age (years)	56,06	51,05	53,55
Total number of patients in class	250	100	350

Obesity (BMI > 30.0) has been shown to increase a risk of breast cancer so we incorporate Body mass index (BMI) as one of the predictors. Results of studies had shown that women had elevated breast cancer risk with increasing weight and BMI [6]. In the study "Molecular Links between Central Obesity and Breast Cancer." by Zimta, Alina-Andreea et al. it was shown that adipose microenvironment in obese people has similarities with tumor microenvironment in respect to associated cellular composition, and that the adipose tissue that functions as an endocrine organ, in obese people, produces a high level of tumor-promoting hormones, such as leptin or estrogen and low level of tumor-suppressor hormone adiponectin [18].

Leukocytes have a major role in immune system response against cancer as they recognize and infiltrate tumors. Expansion of leukocytes in the blood is defined as leukocytosis. Leukocytosis is common among patients suffering from cancer and those who are at risk of developing it, when leukocytes count exceeds 11,000 cells per mm^3 (normal level in adults is 4500 to 11000 cells per mm^3). Studies have found leukocytes count increase in diabetes mellitus patients as high glucose levels (normal level is 3.8–6.1 mmol/L) increase tumor cell proliferation and high leukocytes count can induce inflammation leading to the progression of cancer and increase mortality rate [19].

Also, there was a study that found association between leukocytes count and random glucose level (RGL). In female cancer patients, leukocytes count was higher in those with RGL > 150 mg/dL, and the lowest count was detected in those with RGL < 70 mg/mL. For RGL test, a result of 200 mg/dL or above indicates that a person may have diabetes [20].

Human breath contains volatile organic compounds (VOCs) that are biomarkers of breast cancer, because the disease is accompanied by increased oxidative stress which generates volatile organic compounds (VOCs) that are excreted in breath. Reference value for 2-propanol is 60–600 ppg. [21].

Leptin is a hormone that is secreted more intensely in obese people who are associated with the appearance of breast cancer [22]. Normal level of leptin in adults is 14, 1 to 37 ng/ml.

Adiponectin is measured in human serum in the range of 2 to 20 μg/ml and the data suggests that plasma adiponectin does not change with age. However, in contrast to leptin, adiponectin decreases with increasing body mass index, and circulating levels have been found to decrease in obese individuals as compared with nonobese individuals. In stark contrast with adiponectin, plasma leptin concentration increases with body mass index [24–26].

Based on the study of the inflammatory parameters of complete blood count in breast cancer cases, parameter of neutrophil counts over 2,56 $\times$ 104/mm^3, proved to be an independent parameter in predicting breast cancer [27].

The distribution of the dataset for development of the classifier based on machine learning algorithms (ML) is presented in Table 2. Since this is a relatively small dataset, distribution of 70:30% for training and validation was used [9].

Table 2. Dataset distribution per training and validation subset

Total number of samples 350	Healthy 100	Disease 250
Training 245	70	175
Validation 105	30	75

Total study had 350 samples, and 70% of those were used for training of algorithms, while 30% are used for validation. Training and validation pool was equally distributed among healthy and disease groups in validation phase, among 105 participants, 75 were in group disease, while 30 were in healthy group.

For building classifier for classification of burnout syndrome, following machine learning algorithms were used: Naïve Bayes (NB), Decision Tree (DT), Random Forest (RF) and k-Nearest Neighbor (k-NN) algorithm.

In the Tables 1, 2, 3 and 4, data about accuracy, specificity and sensitivity is given for each algorithm. According to these results, those values are calculated in Eqs for sensitivity, specificity and accuracy.

It has to be clarified that True Positive (TP in tables) is assigned to the person who is sick and which AI recognizes as sick. On the other hand, True negative is healthy person recognized by AI as healthy. False positive is a person which is healthy, but classified by AI as sick, while false negative is the person which is sick but by algorithm is classified as healthy.

3 Results

In the fourth section, results for different classifiers are given. We have tested for different classifier and among them are Naive Bayes based (NB), Decision Tree (DT), Random Forest (RF) and k-nearest neighbors algorithm (k-NN). All four classifiers gave promising results with more than 90% of accuracy recorded in each of them. The tables include data about total samples, which were 105 in our study for the validation part (Tables 5 and 6).

Table 3. Validation performance of NB based classifier

	Group 1 healthy	Group 2 disease	Accuracy
Group 1 Healthy *prediciton*	TN 24	FN 2	98.09%
Group 2 Disease *prediciton*	FP 0	TP 79	
	Specificity 100%	Sensitivity 97.5%	

Table 4. Validation performance of DT based classifier

	Group 1 healthy	Group 2 disease	
Group 1 Healthy *prediciton*	TN 29	FN 0	Accuracy 99.05%
Group 2 Disease *prediciton*	FP 1	TP 75	
	Specificity 96.67%	Sensitivity 100%	

It can be noticed that Decision Tree (DT) algorithm has highest accuracy rate of 99.05%. This means that in the case of DT only one case was wrongfully assumed by AI and that one was predicted to be sick, while real patient was healthy. That brings us that sensitivity in this case was 100% (no patients were considered healthy while they were sick) and specificity of 96.67%. Worst results were obtained by RF based classifier, with still quite high accuracy of 91.42%. In this case, specificity was significantly lower then sensibility (81.4% compared to 94.9%). resulted in highly big number (5) of those predicted by AI to be sick, while they were healthy. However, as already mentioned all models had very high accuracy which was higher than 90%. In the Fig. 1, we can see accuracy, sensitivity and specificity for all four models given.

Table 5. Validation performance of RF based classifier

	Group 1 healthy	Group 2 disease	Accuracy
Group 1 Healthy *prediciton*	TN 22	FN 4	91.42%
Group 2 Disease *prediciton*	FP 5	TP 74	
	Specificity 81.4%	Sensitivity 94.9%	

Table 6. Validation performance of k-NN based classifier

	Group 1 healthy	Group 2 disease	Accuracy
Group 1 Healthy *prediciton*	TN 29	FN 2	96.67%
Group 2 Disease *prediciton*	FP 1	TP 73	
	Specificity 96.67%	Sensitivity 96.67%	

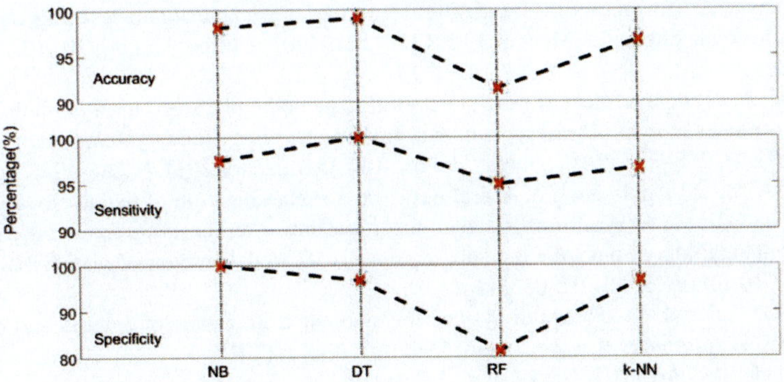

Fig. 1. Training performance of classifiers in terms of accuracy, sensitivity and specificity

4 Conclusion

In this study we have attempted to demonstrate usage of machine learning technique that are being applied to cancer prediction and prognosis based on eight biomarkers. Even though machine learning has been used primarily as an aid to cancer diagnosis and detection, It has only been relatively recently that cancer researchers have attempted to apply machine learning towards cancer prediction and prognosis. The best result showed Decision Tree (DT) algorithm with highest accuracy rate of 99.05%. The use of computers (and machine learning) in disease prediction and prognosis is a part of a growing trend towards personalized, predictive medicine. This movement towards predictive medicine is important, not only for patients (in terms of lifestyle and qualityof-life decisions) but also for physicians (in making treatment decisions) as well as health economists and policy planners (in implementing large scale cancer prevention or cancer treatment policies).

References

1. Paho.org. (2020). https://www.paho.org/hq/index.php?option=com_docman&view=download&category_slug=4-cancer-country-profiles-2020&alias=51561-global-cancer-profile-2020&Itemid=270&lang=fr. Accessed 3 Dec 2020
2. Breast Cancer - Gynecology and Obstetrics - MSD Manual Professional Edition. MSD Manual Professional Edition (2020). https://www.msdmanuals.com/professional/gynecology-and-obstetrics/breast-disorders/breast-cancer. Accessed 3 Dec 2020
3. Patrício, M., Pereira, J., Crisóstomo, J., et al.: Using Resistin, glucose, age and BMI to predict the presence of breast cancer. BMC Cancer **18**, 29 (2018). https://doi.org/10.1186/s12885017-3877-1
4. IARC Working Group on the Evaluation of Cancer-Preventive Interventions. Breast cancer screening. Lyon (FR): International Agency for Research on Cancer (2016). PMID: 31553546
5. Ganggayah, M.D., Taib, N.A., Har, Y.C., Lio, P., Dhillon, S.K.: Predicting factors for survival of breast cancer patiens using machine larning techniques (2019)
6. Cruz, J.A., Wishart, D.S.: Applications of machine learning in cancer prediction and prognosis. Cancer Inform. (2006). https://doi.org/10.1177/117693510600200030
7. Obermeyer, Z., Emanuel, E.J.: Predicting the future - big data, machine learning, and clinical medicine. New England J. Med. **375**,13, 1216–1219 (2016). https://doi.org/10.1056/NEJMp1606181
8. Hodzic, J., et al.: Overview of next-generation sequencing platforms used in published draft plant genomes in light of genotypization of immortelle plant (helichrysiumarenarium). Med. Arch. **71**(4), 288–292 (2017). https://doi.org/10.5455/medarh.2017.71.288-292
9. Catic, A., et al.: Application of neural networks for classification of patau, edwards, down, turner and klinefelter syndrome based on first trimester maternal serum screening data, ultrasonographic findings and patient demographics. BMC Med. Genomics **11**, 19 (2018). https://doi.org/10.1186/s12920-018-0333-2
10. Gurbeta, L., et al.: A telehealth system for automated diagnosis of asthma and chronical obstructive pulmonary disease. J. Am. Med. Inf. Assoc. (2018)
11. Badnjević, A., Pokvić, L.G., Hasičić, M., Bandić, L., Mašetić, Z., Kovačević, Z., Kevrić, J., Hifziefendić, J., Pecchia, L.: Evidence-based clinical engineering: Machine learning algorithms for prediction of defibrillator performance. Biomed. Signal Process. Control **54**, 101629 (2019)
12. Kovačević, Z., et al.: Prediction of medical device performance using machine learning techniques: Infant incubator case study Health and Technology. Health and Technology
13. Sarić, R., Jokić, D., Beganović, N., et al.: FPGA-based real-time epileptic seizure classification using Artificial Neural Network. Biomed. Signal Process. Control **62**, 102106 (2020). ISSN 1746-8094

14. Ming, C., Viassolo, V., Probst-Hensch, N., et al.: Machine learning techniques for personalized breast cancer risk prediction: comparison with the BCRAT and BOADICEA models. Breast Cancer Res **21**, 75 (2019). https://doi.org/10.1186/s13058-019-1158-4
15. Dhahri, H., Al Maghayreh, E., Mahmood, A., Elkilani, W., Nagi, M.F.: Automated Breast Cancer Diagnosis Based on Machine Learning Algorithms. Saudi Arabia, Tunisia, Jordan (2019)
16. Shah, R., Rosso, K., Nathanson, S.D.: Pathogenesis, prevention, diagnosis and treatment of breast cancer. World J. Clin. Oncol. **5**(3), 283–298 (2014). https://doi.org/10.5306/wjco.v5.i3.283. PMID: 25114845; PMCID: PMC4127601
17. Zimta, A.-A., et al.: Molecular links between central obesity and breast cancer. Int. J. Mol. Sci. **20**, 21 5364 (2019). https://doi.org/10.3390/ijms20215364
18. Margolis, K.L., Rodabough, R.J., Thomson, C.A., Lopez, A.M., McTiernan, A.: Prospective Study of Leukocyte Count as a Predictor of Incident Breast, Colorectal, Endometrial, and Lung Cancer and Mortality in Postmenopausal Women (2007)
19. Almehmadi, M.M.: Association Between Random Glucose Level and Leukocytes Count in Female Cancer Patients. UAE (2020)
20. Phillips, M., Cataneo, RN., Saunders, C., Hope, P., Schmitt, P., Wai, J.: Volatile biomarkers in the breath of women with breast cancer (2010)
21. Jardé, T., Perrier, S., Vasson, M.P., Caldefie-Chézet, F.: Molecular mechanisms of leptin and adiponectin in breast cancer. Eur J Cancer. **47**(1), 33–43 (2011). https://doi.org/10.1016/j.ejca.2010.09.005. PMID: 20889333
22. Utility of Peripheral Blood Parameters in Predicting Breast Cancer Risk. Asian Pacific J. Cancer Prevent. **16**(6), 2409–2412 (2015)
23. Cleary, M.P., Grossmann, M.E., Ray, A.: Effect of obesity on breast cancer development. Vet Pathol. **47**(2), 202–213 (2010). https://doi.org/10.1177/0300985809357753. Epub 2010 Feb 1 PMID: 20124008
24. Ryan, A.S., et al.: Plasma adiponectin and leptin levels, body composition, and glucose utilization in adult women with wide ranges of age and obesity. Diabetes Care **26**(8), 2383–2388 (2003)
25. Jardé, T., et al.: Involvement of adiponectin and leptin in breast cancer: clinical and in vitro studies. EndcorRelat Cancer. **16**(4), 1197–1210 (2009). https://doi.org/10.1677/ERC-090043. Epub 2009 Aug 6 PMID:19661131
26. Ercan, T., et al.: Utility of peripheral blood parameters in predicting breast cancer risk. Asian Pac. J. Cancer Prev. **16**(6), 2409–2412 (2015)
27. Rau, H.H., et al.: Development of a web-based liver cancer prediction model for type II diabetes patients by using an artificial neural network. Comput. Methods Programs Biomed. **125**, 58–65 (2016)

Machine Learning Techniques for Prediction of Liver Fibrosis Based on Biomarkers

Amila Suljić, Ines Konjević, Lamija Smaka(✉), Saadet Leyla Suleymanoglu, Muamera Subašić, and Nermina Sofić

Univerzitet u Sarajevu, Farmaceutski fakultet, Sarajevo, Bosnia and Herzegovina
lamijasmaka@ffsa.unsa.ba

Abstract. Liver fibrosis is a condition caused by severe chronic liver tissue damage. The aim of this study is to confirm whether tested parameters are good predictors of liver fibrosis in respondents using machine learning techniques. We used alanine aminotransferase (ALT), aspartat aminotransferase (AST), bilirubin, albumin, gama- glutamil transpeptidase (GGT), protormbine time (PT), tumor necrosis factors- TNF-α, TGF-β, type IV collagen and laminin as biomarkers. The total number of patients were 602, both sexes within the range of 18–75 years. The data was analised using Machine Learning classifiers as folows. Naive Bayes (NB), Decision Tree (DT), Random Forest (RF) and KNearest Neighbors (k-NN). During training process, classifier based on random forest (RF) algorithms achieved the highest accuracy of 100%. Based on dataset outcome, above mentioned parameters are very good predictors of liver fibrosis when machine learning techniques are used to predict it.

Keywords: Liver fibrosis · Machine learning techniques · Prediction

1 Introduction

Liver is the organism's main laboratory that plays a crucial role in several processess in the body, such as metabolism, detoxification, protein synthesis, digestive enzymes production, as well as regulation of red blood cells (RBC) and glucose synthesis and storage [1].

Liver fibrosis is a condition characterized by excessive accumulation of extracellular matrix proteins (ECM), such as collagen. It results from chronic damage to the liver tissue, in conjunction with the accumulation of ECM proteins. Such accumulation distorts the hepatic architecture by forming a fibrous scar. Moreover, the subsequent development of nodules of regenerating hepatocytes leads to cirrhosis, resulting in hepatocellular dysfunction and increased intrahepatic resistance to blood flow. Consequently, hepatic insufficiency and portal hypertension occure.The main causes of liver fibrosis in industrialized countries include chronic hepatitis C (HCV) infection, alcohol abuse, and nonalcoholic steatohepatitis (NASH) [2].

Determining the degree of hepatic fibrosis is paramount in the assessment of patients with chronic liver disease, as it is the key determinant of prognosis. The severity of

A. Badnjevic and L. Gurbeta Pokvić (Eds.): CMBEBIH 2021, IFMBE Proceedings 84, pp. 264–269, 2021.
https://doi.org/10.1007/978-3-030-73909-6_30

fibrosis dictates the need for therapy, it's duration, screening and surveillance strategies, in addition to defining the treatment response [3]. Liver fibrosis is a reversible process, unless it is progressive and leads to cirrhosis [4]. For a long time, liver biopsy was the only method of determining hepatic fibrosis. However its well-acknowledged limitations have led to the search for alternative, non-invasive methods for fibrosis assessment that include clinical and serum biomarkers. Over the past decade, the number of serum biomarkers for liver fibrosis has increased significantly and they are slowly becoming a part of routine clinical practice [3].

Biomarkers of fibrosis are typically divided into indirect and direct markers of fibrogenesis and fibrinolysis. Indirect markers, followed in this article, include simple liver function tests, such as alanine and aspartate aminotransferases (ALT, AST), bilirubin, albumin, gammaglutamyl transpeptidase (GGT) and pro-thrombin time.Direct markers are more directly involved in the molecular pathogenesis of fibrogenesis and fibrinolysis and they include pro-inflammatory and pro-fibrotic cytokines, such as tumour necrosis factor-α (TNF-α) and transforming growth factor-β (TGF-β), type IV collagen and laminin.

These biomarkers are commonly combined with clinical risk factors for hepatic fibrosis, such as age, gender or diabetes [3].

Machine learning (ML) is a form of artificial intelligence. ML techniques represent a broad range of algorithms that perform intelligent predictions based on a data set. In twenty-first century, the growth of dataset demands increase in computer competence. Nowdays, the artificial intelligence techniques have attained what appears to be a human level of semantic understanding and information extraction, and sometimes the ability to detect abstract patterns with greater accuracy than human experts. There are multiple variety in ML techniques. Which technique one should use is determined by a particular problem, the characteristics of the data and the type of desired outcome [5]. Today, many scientists in different types of research use ML techniques [6–12].

The aim of this paper is to confirm whether above mentioned parameters are good predictors of liver fibrosis in respondents using machine learning techniques. Above mentioned biomarkers are the most efficient, and the least invasive way of determining liver fibrosis.

2 Materials and Methods

The parameters that were used for making of our database are: (1) Alanine Transaminase-ALT, (2) Aspartate Transaminase - AST, (3) Bilirubin, (4) Albumin, (5) Gamma-Glutamyl Transferase - GGT, (6) Prothrombin time- PT, (7) Tumor Necrosis Factor Alpha - TNF-α, (8) Transforming Growth Factor Beta - TGF-β, (9) Type IV Collagen, (10) Laminin, (11) Sex, and (12) Age. Reference values are shown in Table 1.

The database consists of 602 examinees, both sexes within the range of 18–75 years (Table 2). 301 of the examinees are among the healty group, while the other 301 examinees are among the ailed group. There are 148 healhty males and 153 healthy females, the average age for the healthy group being 46.9. There are 148 men with liver fibrosis, as well as 153 women with the ailment, the average age for the diseased group being 46.1, as seen in the Table 2.

Table 1. Reference values

ALT	AST	Bilirubin	Albumin	GGT
0-45 IU/L[1]	0-35 IU/L[1]	2-17 μmol/L[1]	40-60 g/L[1]	0-30 IU/L[1]
PT	**TNF- α**	**TGF-β**	**Type IV collagen**	**Laminin**
10,9- 12,5s[1]	≤8,1 pg/ml[2]	1 - 27 ng/ml[3]	≤6 ng/ml[4]	50-180 ng/ml[5]

Table 2. Dataset

Class	Liver fibrosis	Healthy	Total
Male	148	148	296
Female	153	153	306
Total number of patients per class	301	301	602
Average age	46.1	46.9	46.5

3 Results

Values above normal for ALT and AST values indicate the existence of a malady. Bilirbuin values, in liver fibrosis tend to stay within the normal range, while the values of albumin are slightly reduced. For GGT, values above 30 IU/L are also indicative of a malady, specifically affecting the liver. Elevated values of PT indicate severe liver damage. Higher values, above the normal, for TNF-α and TGF-β indicate an impaired liver. Type

IV Collagen values above the normal, though they are usually only slightly elevated, point to an existing disorder. Values above 180 ng/ml for Laminin also point to the existence of a pathological process.

As shown in Fig. 1 in this study following machine learning algorithms (classifiers) are used: Naive Bayes (NB), Decision Tree (DT), Random Forest (RF) and K-Nearest Neighbors (k-NN). The validation performance of these algorithms was calculated in Tables 3, 4, 5 and 6.

During training process, classifiers based od RF had the highest accuracy of 100%. Second highest accuracy had DT, 97,23%, and the k-NN had the accuracy of 96,70%. The lowest accuracy had NB classifier, 95,58%.

Table 5 shows that only RF based classifier had 100% sensitivity and specificity, as well as accuracy. Since sensitivity rate is high in every classifier, that proves that the value of true positive examinees is very high, while the value of false negative is very low. That means that parameters that we have chosen are an excellent indicator of liver firbosis.

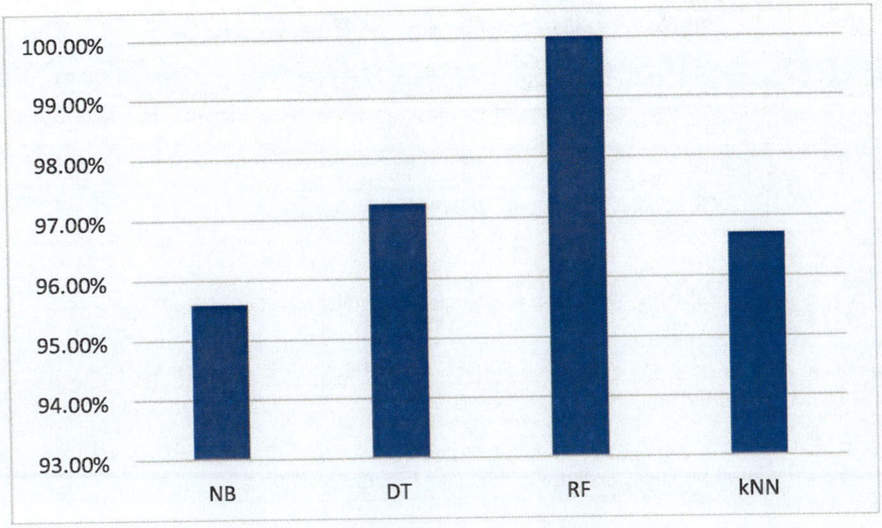

Fig. 1. Training performance of classifiers

Table 3. Validation performance of NB based classifier

	Healthy	Disease	Accuracy
Healthy	88	3	95.58%
Healthy	5	85	
Total samples	Sensitivity 96,7%	Specificity 94,44%	

Table 4. Validation performance of DT based classifier

	Healthy	Disease	Accuracy
Healthy	79	1	97.23%
Disease	4	97	
Total samples	Sensitivity 98,75%	Specificity 96,03%	

Table 5. Validation performance of RF based classifier

	Healthy	Disease	Accuracy
Healthy	94	0	100.00%
Disease	0	87	
Total samples	Sensitivity 100%	Specificity 100%	

Table 6. Validation performance of k-NN based classifier

	Healthy	Disease	Accuracy
Healthy	90	3	96.70%
Disease	2	86	
Total samples	Sensitivity 96,77%	Specificity 97,72%	

4 Conclusion

Liver, as organism's main laboratory, is a very sensitive organ. It easily gets damaged. Liver fibrosis is a result of chronically damaged liver tissue. It is very important to spot this damage in time, so it can be treated. The most simple way to detect liver firbosis is to track her biomarkers. The results of validation performance of RF based classifier showed 100% accurancy, so it can be concluded that this kind of prediction of liver fibrosis based on biomarkers should be use in medical practice.

References

1. Lala, V., Goyal, A., Bansal, P., Minter, D.A.: Liver Function Tests. Updated 2020 Jul 4]. In: StatPearls [Internet]. Treasure Island (FL): StatPearls Publishing, January 2020. https://www.ncbi.nlm.nih.gov/books/NBK482489/?fbclid=IwAR00m1GLpaZ6dQB_gcOXxXV_2DKDrN3mLN6D54JSddPmGVDbw8fq3JxEpAo%20REF%201
2. Li, G., Wu, W., Zhang, X., Huang, Y., Wen, Y., Li, X., Gao, R.: Serum levels of tumor necrosis factor alpha in patients with IgA nephropathy are closely associated with disease severity. BMC Nephrol. (2018). https://www.ncbi.nlm.nih.gov/pmc/articles/PMC6236996/
3. Kyrtsonis, M.C., Repa, C., Dedoussis, G.V., Mouzaki, A., Simeonidis, A., Stamatelou, M., Maniatis, A.: Serum transforming growth factor-beta 1 is related to the degree of immunoparesis in patients with multiple myeloma. Med Oncol. (1998). https://pubmed.ncbi.nlm.nih.gov/9789221/
4. Sakugawa, H., Nakayoshi, T., Kobashigawa, K., et al.: Clinical usefulness of biochemicalmarkers of liver fibrosis in patients with nonalcoholic fatty liver disease. World J. Gastroenterol. (2005). https://www.ncbi.nlm.nih.gov/pmc/articles/PMC4205412/
5. Wang, L., Zhang, Y., Deng, Y., Li, X., Zheng, X., Wang, F., Zou, Y.: Clinical and laboratory characterizations of hepatic capillariasis. Acta Trop. (2019). https://pubmed.ncbi.nlm.nih.gov/30710532/
6. Lala, V., Goyal, A., Bansal, P., et al.: Liver Function Tests. StatPearls [Internet]. Treasure Island (FL): StatPearls Publishing (2020). https://www.ncbi.nlm.nih.gov/books/NBK482489/

7. Bataller, R., Brenner, D.A.: Liver fibrosis. J. Clin. Invest. **115**(2), 209–218 (2005). https://www.ncbi.nlm.nih.gov/pmc/articles/PMC546435/
8. Adams, L.A.: J. Gastroenterol. Hepatol. **26**(5), 802–809 (2001)
9. https://www.medscape.com/viewarticle/741737_1
10. Aydın, M.M., Akçalı, K.C.: Liver fibrosis. Turk J Gastroenterol. **29**(1), 14–21 (2018)
11. https://www.ncbi.nlm.nih.gov/pmc/articles/PMC6322608/
12. Nichols, J.A., Herbert Chan, H.W., Baker, M.A.B.: Machine learning: applications of artificial intelligence to imaging and diagnosis. Biophys. Rev. **11**(1), 111–118 (2018). https://doi.org/10.1007/s12551-018-0449-9
13. Badnjevic, A., Gurbeta, L., Custovic, E.: An expert diagnostic system to automatically identify asthma and chronic obstructive pulmonary disease in clinical settings. Sci. Rep. (2018)
14. Badnjević, A., Pokvić, L.G., Hasičić, M., Bandić, L., Mašetić, Z., Kovačević, Z., Kevrić, J., Hifziefendić, J., Pecchia, L.: Evidence-based clinical engineering: machine learning algorithms for prediction of defibrillator performance. Biomed. Signal Process. Control **54**, 101629 (2019)
15. Kovačević, Z., Pokvić, L.G., Spahić, L., Badnjević, A.: Prediction of medical device performance using machine learning techniques: Infant incubator case study Health and Technology. Health and Technology
16. Alić, B., Gurbeta, L., Osmanovic, A., Badnjević, A.: Machine learning techniques for classification of diabetes and cardiovascular diseases. In: 2017 6th Mediterranean Conference on Embedded Computing (MECO), Bar, Montenegro, pp. 1–4 (2017)
17. Spahić, L., Kurta, E., Ćordić, S., Bećirović, M., Pokvic, L.G., Kovacevic, Z., Izetbegovic, S., Badnjevic, A.: Machine learning techniques for performance prediction of medical devices: infant incubators. In: Badnjevic, A., Škrbić, R., Gurbeta Pokvić, L. (eds.) CMBEBIH 2019. CMBEBIH 2019. IFMBE Proceedings, vol. 73. Springer, Cham (2019)
18. Hrvat, F., Spahić, L., Pokvić, L.G., Badnjević, A.: Artificial Intelligence for prediction of medical device performance: Infusion and perfusor pumps case study. In: 9th Mediterranean Conference on Embedded Computing MECO 2020 (2020)
19. Šećkanović, A., Šehovac, M., Spahić, L., Ramić, I., Mamatnazarova, N., Kacila, M., Badnjević, A., Gurbeta, L.: Review of artificial intelligence application in cardiology. In: 9th Mediterranean Conference on Embedded Computing MECO 2020 (2020)

Application of Artificial Intelligence Tools in Classification and Diagnosis of Heart Disease: General Review

Muhamed Karajić[1](✉), Edin Begić[2,3], Emina Hrvat[2], and Lejla Gurbeta Pokvić[4,5]

[1] Faculty of Information Technology, University Džemal Bijedić of Mostar, Mostar, Bosnia and Herzegovina
muhamed.karajic@edu.fit.ba
[2] General Hospital, Sarajevo, Bosnia and Herzegovina
[3] School of Medicine, Sarajevo School of Science and Technology, Sarajevo, Bosnia and Herzegovina
[4] International Burch University, Sarajevo, Bosnia and Herzegovina
[5] Medical Device Inspection Laboratory Verlab, Sarajevo, Bosnia and Herzegovina

Abstract. To identify heart diseases several contributory risk factors and information about patient must be taken into account. Diagnosis of heart disease requires medical specialist that are often timely unavailable or costly for some patients in remote and rural areas, but also in overly populated urban areas. For this reason, over the years, new approaches based on Artificial Intelligence (AI) tools have been investigated for the purpose of classification and diagnosis of heart diseases. Such expert systems enable diagnosis of heart disease based on multiple input parameters and with fewer resources avoiding human limitations in parallel analysis of large quantity of data. The growing computing power of everyday phones, tablets, and computers brings a new approach in using AI for diagnostic purposes of this kind. In respect to this, this paper gives a review of application of AI tools in classification and diagnosis of heart disease. In this paper, papers published in the period 2010–2020 dealing with usage of AI in diagnosis of various types of heart diseases have been reviewed.

Keywords: Artificial intelligence · Artificial neural networks · Machine learning · Heart disease · Heart failure · Review

1 Introduction

Cardiovascular diseases are the leading cause of death in modern medicine. In highly developed countries, the prevalence of cardiovascular diseases is also increasing with increasing age of the population [1]. During the 1950s, deaths were recorded in one-third of patients with acute myocardial infarction (AMI), and in the 21st century, in-hospital mortality was reduced to half that number [1]. AMI is the second leading cause of heart failure (HF), which is a one of the greatest risk factor leading to mortality [2]. Complications that occur as a result of acute myocardial infarction reduce the quality of life and affect work ability [2].

A. Badnjevic and L. Gurbeta Pokvić (Eds.): CMBEBIH 2021, IFMBE Proceedings 84, pp. 270–278, 2021.
https://doi.org/10.1007/978-3-030-73909-6_31

In everyday clinical practice, cardiologists encounter coronary artery disease, valvular malformations, heart failure and arrhythmias, and the algorithm of diagnosis and therapy is the basis of their work [3, 4].

Diagnostic and therapeutic algorithm of heart diseases including acute myocardial infarction is performed through taking anamnestic data, physical examination (inspection, palpation, auscultation), electrocardiogram (ECG) recording, laboratory processing, chest X-ray, echocardiography and etc. [3]. The use of magnetic resonance imaging (MRI) and computed tomography (CT) is not part of the routine treatment of patients with acute coronary syndrome (ACS), although they have their place in the assessment of ischemic heart disease. In the diagnosis of acute myocardial infarction, in addition to clinical symptoms (pain with propagation in the neck, axilla, back or abdomen, discomfort, suffocation, shortness of breath, malaise, hiccups), electrocardiographic changes are essential in the evaluation and differential diagnosis in patients with ACS. ECG changes go in three directions: ST segment elevation or newly formed left bundle branch block (LBBB), without ST segment elevation (NSTEMI) or UA (ST segment depression, inversion or increase in T wave amplitude, transient ST segment elevation). [5–8] Heart Failure (HF) is a clinical condition that results from any structural or functional damage to the ventricles with an inadequate blood supply to the organs (reduced stroke volume of the heart muscle and/ or increased intracardiac pressure at rest or during exercise), according to their needs. [4, 5] HF is associated with increased morbidity and mortality and carries a significant burden on the health system. [4, 5] Incidence of ACS, large number of patients with HF (cardiomyopathies are also included), large number of patients with arrhythmia (idiopathic or acquired), patients with valve defects, and patients after revascularization or other surgical treatment, as well as a large number of consultative examinations by cardiologists, in daily clinical work, burdened daily work of cardiologist.

The main clinical symptoms of heart disease are dyspnea, chest pain, symptoms related to heart rhythm disorders, palpitations, cyanosis and syncope. The three main symptoms of heart failure in children are dyspnea and edema, which, like the others, are not strictly specific only for cardiovascular diseases. Nowadays, the situation in healthcare sector is that patients demand more from the healthcare providers, and during the provision of health services, the volume of generated data increase. This is why, currently, healthcare sector is at another major shift. This shift is referred to the application on the artificial intelligence (AI) for providing diagnosis to the patients. Diagnosis tools based on AI offer a number of advantages in respect to the traditional approaches using analytics and clinical decision-making techniques because it enables continuous improvement. AI based tools have the ability to constantly learn and improve performance as they are fed with more real-time based data. This property of AI tools in healthcare enables unprecedented insights into diagnostics, care processes, treatment variability, and patient outcomes.

Application of AI in healthcare can vary from diagnostic applications, to patient treatment and support. So, nowadays, new links between genetic codes can be found by AI tools, also, robots can perform surgeries, diseases can be recognized based on AI powered image analysis, and etc. When speaking of AI tools, mostly researchers are using Artificial Neural Networks (ANNs) and Machine Learning Algorithms (MLs). The application of ANNs and MLs evolved in healthcare, but their full real-time application in healthcare environment is still limited due to the ethical considerations which haven't been yet resolved. Up to now, beside diagnosis of heart diseases, researchers have been using these tools in classification and diagnosis of chronic diseases such as asthma [9, 10] or diabetes [11, 12]. Also, along with diagnostic approach to disease, researchers have been using these tools [13, 14].

Based on this, the aim of this study is to provide review of application of AI tools in diagnosis and classification of heart diseases. Mainly, in this paper applications of ANNs and five ML algorithms have been investigated. The purpose of this review is to give comprehensive conclusion on which AI tools show the most promising results in the classification and diagnosis of heart diseases.

2 Materials and Methods

This review study investigates application of artificial neural networks and machine learning algorithms for classification and diagnosis of heart disease. Following AI tools were investigated: (1) feedforward neural networks, (2) feedback neural networks, machine learning algorithms: (3) Support Vector Machine (SVM), (4) Decision Tree (DT), (5) Logistic Regression (LR), (6) Bagging Algorithm (BA), (7) Gradient Boosted Tree (GBT), (8) Naïve Bayes (NB), (9) Deep Learning (DL), (10) Generalized Linear Model (GLM), (11) Random Forest (RF) and (12) K-Nearest Neighbors (KNN).

Papers were selected based on selection criteria which was following:

- Paper is published in PubMed, Scopus and Web of Science database,
- Paper is published in English language,
- Keywords: *heart disease, classification, diagnosis, machine learning, neural network, artificial intelligence, Support Vector Machine, Decision Tree, Logistic Regression, Bagging Algorithm, Gradient Boosted Tree, Naïve Bayes, Deep Learning, Generalized Linear Model, Random Forest, K-Nearest Neighbors.*
- Paper reports following performance indicators: accuracy, sensitivity, specificity.

First author provided paper selection, while other authors reviewed selected paper and contributed in the information analysis. Following findings were discarded: not a scientific article (review, dissertation/thesis, book chapter, comparative study, etc.).

Following aspects were addressed in our review: the accuracy of expert system, the sensitivity of expert system, the specificity of expert system, the AI method, dataset, and number of input features.

3 Results

In 2020, Google released revealed that AI significantly dominates among highly cited publications. [15] Also, the PubMed database shows that publications dealing with heart disease and artificial intelligence are of interest for research community [16].

During this review study, total number of papers taken into consideration was 61, but only 23 papers had all necessary information presented [17–40].

The results show, that an average accuracy of expert systems based on AI tools, both ANN and ML, for classification and diagnosis of heart disease, was 88%. When

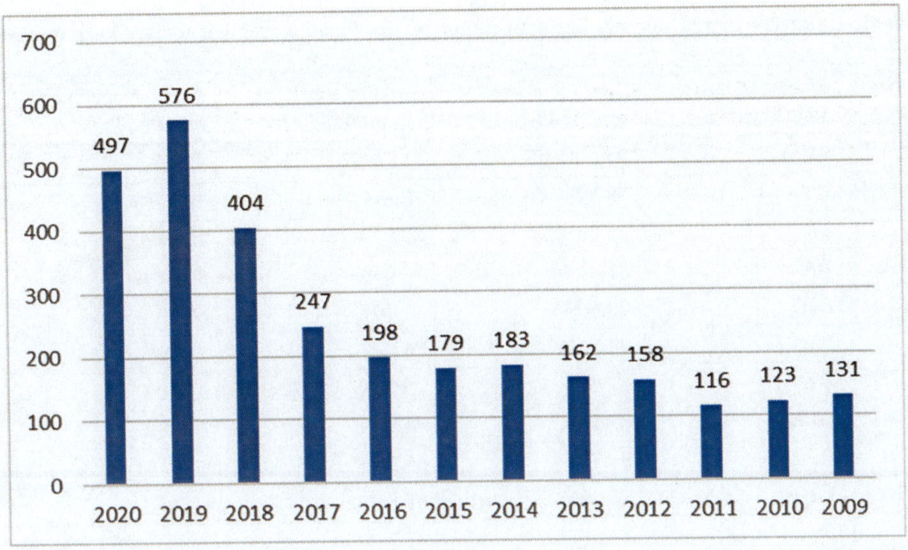

Fig. 1. Number of publications in PubMed database containing keywords artificial intelligence and heart disease

reporting the ability of the system to correctly recognize patients suffering from this disease, it can be seen that an average accuracy of 88.94% was achieved. Developed systems can better differentiate healthy patients, since it was calculated that an average specificity of observed systems is 90.07%.

Average accuracy for expert systems based on artificial neural networks, both feed-forward and feedback architecture were 84.35% accuracy with an average of 80.13% specificity and 85.93% sensitivity. It was seen that feedforward architectures are mostly used for this application. It was noted that reinforced ANN gives better result with proportionately higher accuracy, specificity and sensitivity (97.78%, 95.24% and 100% respectively).

Table 1. provides review of 10 machine learning algorithms that are commonly used for these purposes.

As it can be seen from Table 1, the highest average achieved accuracy is for DL algorithm. The highest average sensitivity of expert system was achieved also when DL algorithm was used. Table 1 shows that the highest average specificity for expert system is based on RF algorithm.

Observed papers report expert systems based on AI tools with different number of features. Mainly, number of features vary between 11 and 13 input features based on which diagnosis and classification is provided. Commonly used features are: age, gender (sex), constrictive pericarditis, resting blood pressure, serum cholesterol, fast blood sugar, resting electrocardiographic, maximum heart rate achieved, exercise induced angina, old peak, the slope of the peak exercise ST segment, number of major vessels and the heart status.

The first paper published in PubMed database containing keywords artificial intelligence and congestive heart failure was published in 1989. Up to 2005 this topic was not

Table 1. Review of machine learning application for classification and diagnosis of heart disease

No.	Machine learning algorithm	Average accuracy reported [%]	Average sensitivity reported [%]	Average specificity reported [%]
1.	SVM	84.00%	75.00%	85.00%
2.	DT	78.87%	70.31%	76.51%
3.	LR	76.08%	60.93%	78.68%
4.	BA	81.41%	74.93%	86.64%
5.	GBT	78.30%	80.70%	60.00%
6.	NB	78.71%	76.62%	73.89%
7.	DL	87.40%	95.00%	33.30%
8.	GLM	85.10%	94.90%	20.00%
9.	RF	85.76%	65.36%	90.04%
10.	k-NN	86.78%	85.54%	80.45%

really investigated since in average 4 papers per year were published. In the past three years, this topic is gaining interest therefore this review study can be very informative for determining next steps in developing expert systems for diagnosis of congestive heart failure based on static and dynamic patient assessment.

4 Discussion

Information technology is a part of every segment of cardiology, from emergency cardiology to day-to-day clinical work, and has shown its role in times of epidemiological crises. [41–45] In our study, the highest achieved accuracy was recorded for the DL algorithm. This algorithm will be used more in the future due to the trend in cardiology which is going towards an increasing number of collaborations, multicentre studies, on which conclusions are drawn, and important information for clinical practice are obtained.

Atherosclerosis forms the basis of about 80% of cardiovascular diseases (2). Given the high percentage of potentially fatal complications of arterial thrombosis (the process progresses through life, before finally manifesting as an acute ischemic event), a correct understanding of the pathogenesis of this disease is of great importance for determining the optimal modality of prophylaxis and therapy. Taking into account all possible risk factors that lead to the occurrence of a cardiovascular incident will lead to a broad and comprehensive analysis of the patient, and the collection of a large number of data. All treatment strategies are part of the programs of regional and national societies, and the guidelines for the strategy themselves are obtained from registers that collect actual unselected data. Pharmacological optimization of therapy in relation to the patient's profile, dosing in relation to the latest knowledge from the most important studies, continuous education of medical staff, and selection of a certain pharmacological substance in relation to the patient's profile, taking into account all its characteristics (even genetic polymorphisms that cause an enhanced effect or drug resistance) should be an imperative in clinical practice. [46] The stated path in the treatment of the patient will have to be based on the conclusions from the meta-analysis, which will enable further use of DL algorithms. Although the DT algorithms have a satisfactory rate of accuracy, their use is still not accepted in cardiac practice, although it can be used in differential diagnosis in the very beginning of symptoms occurrence. The use of AI represents modern clinical practice, it seems that it is still in the field of cardiology at the very beginning, with great potential, and should become a part of clinical work, which will contribute to better treatment of the patient.

5 Conclusion

This general review summarizes the current state of the application of artificial intelligence tools in diagnosis of heart diseases.

During the analysis of the papers, it was concluded that only few of them have been actually practically implemented in healthcare settings. This review shows that research in this field is at satisfying level, and that at this point, shift from research into real-time application needs to happen with support of qualified technical and medical staff in healthcare institutions.

Conflict of interest. Authors declare no conflict of interests.

References

1. Kumar, A., Cannon, C.P.: Acute coronary syndromes: diagnosis and management, part I. Mayo Clin. Proc. **84**(10), 917–938 (2009)
2. Thygesen, K., Alpert, J.S., Jaffe, A.S., Chaitman, B.R., Bax, J.J., Morrow, D.A.: Executive group on behalf of the joint european society of cardiology (ESC)/American college of cardiology (ACC)/American heart association (AHA)/World heart federation (WHF) Task force for the universal definition of myocardial infarction. fourth universal definition of myocardial infarction. Circulation, **138**(20), e618–e651 (2018)
3. McMurray, J.J., Adamopoulos, S., Anker, S.D., Auricchio, A., Böhm, M.: ESC Committee for practice guidelines. ESC guidelines for the diagnosis and treatment of acute and chronic heart failure 2012: the task force for the diagnosis and treatment of acute and chronic heart failure 2012 of the european society of cardiology. developed in collaboration with the heart failure association (HFA) of the ESC. Eur. Heart J. **33**(14), 1787–847 (2012)
4. Malik, A., Brito, D., Chhabra, L.: Congestive heart failure 10 Aug 2020. StatPearls [Internet]. Treasure Island (FL): StatPearls Publishing (2020)
5. Pieske, B., Tschöpe, C., de Boer, R.A., Fraser, A.G., Anker, S.D., Donal, E., et al.: How to diagnose heart failure with preserved ejection fraction: the HFAPEFF diagnostic algorithm: a consensus recommendation from the heart failure association (HFA) of the European society of cardiology (ESC). Eur. J. Heart Fail. **22**(3), 391–412 (2020)
6. Elliott, P., Andersson, B., Arbustini, E., Bilinska, Z., Cecchi, F., Charron, P., et al.: Classification of the cardiomyopathies: a position statement from the European society of cardiology working group on myocardial and pericardial diseases. Eur. Heart J. **29**(2), 270–276 (2008)
7. Elliott, P.M.: Classification of cardiomyopathies: evolution or revolution? J. Am. Coll. Cardiol. **62**, 2073 (2013)

8. Jain, P., Patel, P.A., Fabbro, M.: Hypertrophic cardiomyopathy and left ventricular outflow tract obstruction: expecting the unexpected. J. Cardiothorac. Vasc. Anesth. **32**(1), 467–477 (2018)
9. Gurbeta, L., Badnjevic, A., Maksimovic, M., Omanovic-Miklicanin, E., Sejdic, E.: A telehealth system for automated diagnosis of asthma and chronical obstructive pulmonary disease. J. Am. Med. Inform. Assoc. **25**(9), 12131217 (2018)
10. Badnjevic, A., Gurbeta, L., Custovic, E.: An expert diagnostic system to automatically identify asthma and chronic obstructive pulmonary disease in clinical settings. Sci. Rep. **8**, 11645 (2018)
11. Sejdinovic, D., Gurbeta, L., Badnjevic, A., Malenica, M., Dujic, T., Causevic, A., et al.: Classification of prediabetes and type 2 diabetes using artificial neural network, CMBEBIH. IFMBE Proc. **62**, 685–689 (2017)
12. Brunetti, N.D., Scalvini, S., Acquistapace, F., Parati, G., Volterrani, M., Fedele, F., et al.: Telemedicine for cardiovascular disease continuum: a position paper from the Italian society of cardiology working group on telecardiology and informatics. Int. J. Cardiol. **184**, 452–458 (2015)
13. Badnjevic, A., Gurbeta Pokvic, L., Hasicic, M., Bandic, L., Masetic, Z., Kovacevic, Z., et al.: Evidence-based clinical engineering: machine learning algorithms for prediction of defibrillator performance. Biomed. Sig. Process. Control **54**, (2019)
14. Kovacevic, Z., Gurbeta Pokvic, L., Spahic, L., Badnjevic, A.: Prediction of medical device performance using machine learning techniques: infant incubator case study. Health Technol. **10**, 151–155 (2020)
15. Google Scholar reveals its most influential papers for (2020). https://www.natureindex.com/news-blog/googlescholar-reveals-most-influential-papers-research-citations-twenty-twenty
16. PubMed database. https://pubmed.ncbi.nlm.nih.gov/
17. Abushariah, M., Alqudah, A., Adwan, O., Yousef, R.: Automatic heart disease diagnosis system based on artificial neural network (ANN) and adaptive neuro-fuzzy inference systems (ANFIS) approaches. J. Softw. Eng. Appl. **7**, 1055–1064 (2014)
18. Alic, B., Gurbeta, L., Osmanovic, A., Badnjevic, A.: Machine learning techniques for classification of diabetes and cardiovascular diseases. In: 6th Mediterranean Conference on Embedded Computing (MECO), Bar, Montenegro, pp. 1–4 (2017)
19. Spahic, L., Kurta, E., Cordic, S., Becirovic, M., Gurbeta, L., Kovacevic, Z., et al.: Machine learning techniques for performance prediction of medical devices: infant incubators. In: Badnjevic, A., Skrbic, R., Gurbeta Pokvic, L. (eds) CMBEBIH 2019. CMBEBIH 2019. IFMBE Proceedings, vol. 73, pp. 483–490 (2019)
20. Mohan, S., Thirumalai, C., Srivastava, G.: Effective heart disease prediction using hybrid machine learning techniques. IEEE Access **7**, 8154281554 (2019)
21. Das, R., Turkoglu, I., Sengur, A.: Diagnosis of valvular heart disease through neural networks ensembles. Comput. Methods. Programs Biomed. **93**(2), 185–91 (2009)
22. Olaniyi, E.O., Oyedotun, O.K., Khashman, A.: Heart diseases diagnosis using neural networks arbitration. Int. J. Intell. Syst. Appl. (IJISA), **7**(12), 75–82 (2015)
23. Zebardast, B., Ghaffari, A., Masdari, M.A.: New generalized regression artificial neural networks approach for diagnosing heart disease. Int. J. Innov. Appl. Stud. **4**(4), 679–689 (2013)
24. Xing, Y., Wangg, J., Zhao, Z., Gao, Y.: Combination data mining methods with new medical data to predicting outcome of coronary heart disease. In: 2007 International Conference on Convergence Information Technology - ICCIT 2007, pp. 868–872 (2007)
25. Uyar, K., Ilhan, A.: Diagnosis of heart disease using genetic algorithm based trained recurrent fuzzy neural networks. Precedia Comput. Sci. **120**, 588–593 (2017)

26. Cako, S., Njegus, A., Matic, V.: Effective diagnosis of heart disease presence using artificial neural networks. In: Sinteza 2017 - International Scientific Conference on Information Technology and Data Related Research, pp. 3–8 (2017)
27. Shouman, M., Turner, T., Stocker, R.: Using decision tree for diagnosing heart disease patients. In: Proceedings of the 9-th Australasian Data Mining Conference (AusDM 2011), Ballarat, Australia, pp. 23–29 (2011)
28. Cui, S., Li, K., Ang, L., Liu, J., Cui, L., Song, X., et al.: Plasma phospholipids and sphingolipids identify stent restenosis after percutaneous coronary intervention. JACC Cardiovasc. Interv. **10**(13), 1307–1316 (2017)
29. Haq, A.U., Li, J.P., Memon, M.H., Nazir, S., Sun, R.A.: Hybrid intelligent system framework for the prediction of heart disease using machine learning algorithms. Mob. Inf. Syst. (2018)
30. Obasi, T., Shafiq, M.O.: Towards comparing and using machine learning techniques for detecting and predicting heart attack and diseases. In: 2019 IEEE International Conference on Big Data (Big Data), Los Angeles, CA, USA, pp. 2393–2402 (2019)
31. Terrada, O., Cherradi, B., Raihani, A., Bouattane, O.: Classification and prediction of atherosclerosis diseases using machine learning algorithms. In: 2019 5th International Conference on Optimization and Applications (ICOA), Kenitra, Morocco, pp. 1–5 (2019)
32. Gupta, N., Ahuja, N., Malhotra, S., Bala, A., Kaur, G.: Intelligent heart disease prediction in cloud environment through ensembling. Expert Syst. **34**, (2017)
33. Pouriyeh, S., Vahid, S., Sannino, G., De Pietro, G., Arabnia, H., Gutierrez, J.: A comprehensive investigation and comparison of machine learning techniques in the domain of heart disease. In: 2017 IEEE Symposium on Computers and Communications (ISCC), Heraklion, pp. 204–207 (2017)
34. Nilashi, M., Ahmadi, H., Manaf, A.A., Rashid, T.A., Samad, S., Shahmoradi, L., et al.: Coronary heart disease diagnosis through self-organizing map and fuzzy support vector machine with incremental updates. Int J Fuzzy Syst. **22**, 1376–1388 (2020)
35. Hongzong, S., Tao, W., Xiaojun, Y., Huanxiang, L., Zhide, H., Mancang, L., et al.: Support vector machines classification for discriminating coronary heart disease patients from non-coronary heart disease. West Indian Med. J. **56**(5), 451–457 (2007)
36. Jabbar, M.A.: Prediction of heart disease using k-nearest neighbor and particle swarm optimization. Biomed. Res. **28**(9), 4154–4158 (2017)
37. Jabbar, M.A., Samreen, S.: Heart disease prediction system based on hidden Naïve bayes classifier. In: 2016 International Conference on Circuits, Controls, Communications and Computing (I4C), Bangalore, pp. 1–5 (2016)
38. Bashir, S., Qamar, U., Javed, Y.M.: An ensemble based decision support framework for intelligent heart disease diagnosis. In: International Conference on Information Society (i-Society 2014), London, pp. 259–264 (2014)
39. Shylaja, S., Muralidharan, R.: Comparative analysis of various classification and clustering algorithm. CiiT Int. J. Biometrics Bioinform. **10**(4), 74–77 (2018)
40. Kolukisa, B., Hacilar, H., Goy, G., Kus, M., Bakir-Gungur, B., Aral, A.: Evaluation of classification algorithms, linear discriminant analysis and a new hybrid feature selection methodology for the diagnosis of coronary artery disease. In: 2018 IEEE International Conference on Big Data (Big Data), Seattle, WA, USA, pp. 2232–2238 (2018)
41. Naser, N., Tandir, S., Begic, E.: Telemedicine in cardiology - perspectives in Bosnia and Herzegovina. Acta Inform. Med. **25**(4), 263–266 (2017)
42. Masic, I., Begic, E.: Mobile clinical decision support systems in our hands - great potential but also a concern. Stud. Health Technol. Inform. **226**, 63–6 (2016)
43. Mandzuka, M., Begic, E., Boskovic, D., Begic, Z., Masic, I.: Mobile clinical decision support system for acid-base balance diagnosis and treatment recommendation. Acta Inform. Med. **25**(2), 121–125 (2017)

44. Escobar-Curbelo, L., Franco-Moreno, A.I.: Application of telemedicine for the control of patients with acute and chronic heart diseases. Telemed. J. E. Health **25**(11), 1033–1039 (2019)
45. Eberly, L.A., Khatana, S.A.M., Nathan, A.S., Snider, C., Julien, H.M., Deleener, M.E., Adusumalli, S.: Telemedicine outpatient cardiovascular care during the COVID-19 pandemic: bridging or opening the digital divide? Circulation **142**(5), 510–512 (2020)
46. Begic, E., Obradovic, S., Jankovic, S., Romanovic, R., Djenic, N., Dzudovic, B., et al.: Increased C-reactive protein is associated with major adverse cardiovascular events after STEMI. Erciyes Med J **42**(3), 276–280 (2020)

Diagnosis of Hyperthyroidisim Using Artificial Neural Networks

Hodžić Mubina(✉), Huseinspahić Lamija, Husović Lejla, Ikanović Emina, Islamović Minela, and Isović Amina

Faculty of Pharmacy, University of Sarajevo, Sarajevo, Bosnia and Herzegovina
mubinahodzic@ffsa.unsa.ba

Abstract. Hyperthyroidism is a pathological disease in which there is an increase in the synthesis and secretion of hormones through the thyroid gland. Hyperthyroidism can be overt or sub-clinical. Overt hyperthyroidism is characterized by low serum concentrations of thyroid-stimulating hormone (TSH) and elevated concentrations of thyroid hormones: thyroxine (T4), triiodothyronine (T3), or both. Subclinical hyperthyroidism is characterized by low serum TSH levels, but normal T4 and T3 concentrations occur in the blood serum. Possible treatments for hyperthyroidism include radioactive iodine, anti-thyroid medications, beta-blockers, surgery (thyroidectomy). In this study, we developed a model of an artificial neural network for the diagnosis of hyperthyroidism. Parameters of 330 patients were entered into the database, of which 130 were healthy and 200 sick subjects, taking into account the sex and age of the subjects. The neural network was developed based on the free fractions of hormones T3 and T4 (f-T3 and f-T4 T3), T4, TSH, as well as anti-TPO and antiTG antibodies. During the training of the neural network, different numbers of neurons were tested. The system was tested for a number of neurons from 2 to 20 with different increments of neurons in each step. Accuracy of training for feedforward neural network with 18 neurons and sigmoid transfer function in the hidden layer was 94.9%. The development of accurate Artificial Neural Network models can and will progressively be able to change not only the diagnostics of hyperthyroidism, but all aspects of diagnostics in the future.

Keywords: Artificial neural network · Model · Hyperthyroidism · Diagnostics · Parameters

1 Introduction

Hyperthyroidism is a pathological disease in which there is an increase in the synthesis and secretion of hormones through the thyroid gland. Hyperthyroidism can be overt or subclinical. Overt hyperthyroidism is characterized by low serum concentrations of thyroid-stimulating hormone (TSH) and elevated concentrations of thyroid hormones: thyroxine (T4), triiodothyronine (T3), or both. Subclinical hyperthyroidism is characterized by low serum TSH levels, but normal T4 and T3 concentrations occur in the blood serum [1].

A. Badnjevic and L. Gurbeta Pokvić (Eds.): CMBEBIH 2021, IFMBE Proceedings 84, pp. 279–287, 2021.
https://doi.org/10.1007/978-3-030-73909-6_32

Clinical manifestations depend on several factors of the concomitant disease, the age and gender of the patient, the duration of the disease, and the cause [1]. Women are about 20 times more likely to suffer from hyperthyroidism than men [2]. The disease has a significant genetic component, which includes welldocumented clinically documented familial forms of graves' disease (BG). Studies on twins have shown that genetic factors contribute between 70 and 80%. When people with a sensitive genetic background are exposed to certain environmental factors, such as smoking, excessive iodine intake, stress, the mother's postpartum period, immunological changes caused by the microbiome, and other unknown conditions, the production of autoantigens against TSH receptors begins [3].

Symptoms of overt hyperthyroidism include rapid heartbeat, anxiety, heat intolerance, weight loss, fatigue, muscle weakness, and, in women, irregular menstruation. Clinical data may include tachycardia, eyelid congestion and heat, tremor, and wet skin. Symptoms and signs of subclinical hyperthyroidism are very vague and non-specific if present [4]. Treatment for hyperthyroidism has not changed much over the past few decades [1]. Hyperthyroidism caused by the overproduction of thyroid hormones can be treated in several ways: with thyroid medications (methimazole and propylthiouracil), removal of radioactive thyroid iodine, or surgical thyroidectomy. The method of treatment depends on the main diagnosis, the presence of contraindications to a particular method of treatment, the severity of hyperthyroidism, and the patient's propensity [5].

As the 4th industrial revolution is already underway, and tools of artificial intelligence are nowadays part of everyday activities, it is interesting to see how these tools are affecting the healthcare sector as well. Tools such as machine learning algorithms including artificial neural networks have been used for healthcare practices for some time now, for diagnosis and disease classification, for data management, management of medical devices, drug development, ultrasound diagnostics, X-rays, biochemical analysis, and etc. [6–19]

In this paper, the development of an artificial neural network model in the diagnosis of hyperthyroidism will be discussed. With the aim of developing the best model of ANN for the diagnosis of hyperthyroidism, relevant works in the field of diagnosis of thyroid disease with ANN have been studied. [21–25]

2 Materials and Methods

2.1 Data Collection

The neural network model developed in this paper is based on the database which consisted of information taken from 330 patients of which 130 were healthy and 200 sick subjects, taking into account the sex and age of the subjects. Besides that, the following information was recorded in the database: the free fractions of hormones T3 and T4 (f-T3 and f-T4), T3, T4, TSH, as well as anti-TPO and anti-TG antibodies as presented in Table 1.

Table 1. Normal values for selected parameters

Parameter	Normal (reference value)
f-T3	2.8–6.5 pmol/L
f-T4	11.5–22.7 pmol/L
T3	1.3–2.6 nmol/L
T4	58–161 nmol/L
TPO	<5.61 IU/ml
TG	4.11 IU/ml

2.2 Data Classification

The diagnosis of hyperthyroidism is made on the basis of higher levels of T4, T3, and suppressed TSH. In the early stage, TSH is not lowered because it takes some time, which is also the case in a few subjects. The earliest sign of hyperthyroidism is high T3 or its free fraction. Detailed information about the database is presented in Table 2.

Table 2. Properties of the dataset used for development of artificial

	Disease	Healthy	Total
Male	100	54	154
Average age	43,5	42,1	42,8
Female	100	76	176
Average age	39,7	41,9	40,08

The distribution of the dataset for the development of the Artificial Neural Network model (ANN) is presented in Table 3. Since this is a relatively small dataset, a distribution of 70:30% for training and validation was used [18].

Table 3. Dataset distribution

Total number of samples 330	Healthy	Disease
Training 231	91	140
Validation 99	39	60

Code and cross-reference validation operations to verify that data were included and it was taken care of the distribution of instances.

The Artificial Neural Network (ANN) model was developed in order to perform the prediction of renal diseases based on diagnostic parameters. Developed ANN model has 7 input parameters: gender, age, creatinine clearance (ml/min), urea (g/h), albumin (mg/h), glucose (mmol/L) and cystatin C (mg/l). The ANN model performs the classification of samples into two classes: healthy and patients with renal disease.

For the development of the ANN model in this research, standard architecture for pattern recognition was used. That is feedforward neural network which consists of two-layers. This architecture is designed since according to the application experts, that kind of ANN is sufficient to properly perform the classification of input data [19].

The input layer is made of network inputs. The input layer is followed by a hidden layer which consists of parallel-connected neurons. The network was tested on training performance with a different number of neurons. Since this classification is basically highly non-linear data classification, the sigmoid function was used for the hidden layer.

The appropriate number of neurons were selected by trying out different numbers, using MSE (Mean Square Error) as a fitting function. During the design of the network, different number of neurons was used in order to measure the performance. After training, validation of the network performance was done using a confusion matrix.

According to these results, specificity, sensitivity, and accuracy of the network are calculated according to Eqs. (1), (2) (3) [20]:

$$sensitivity = \frac{true\ positive}{(true\ positive\ +\ false\ negative)} \tag{1}$$

$$specificity = \frac{true\ negative}{(true\ negative\ +\ true\ positive)} \tag{2}$$

$$accuracy = \frac{(true\ negative\ +\ true\ positive)}{(true\ negative\ +\ true\ positive\ +\ false\ negative\ +\ false\ positive)} \tag{3}$$

The developed ANN model showed satisfactory results of specificity and sensitivity.

3 Results and Discussion

According to previous research and expert recommendation [6–19], for the classification of hyperthyroidism in this paper, two-layer feedforward artificial neural network model, with sigmoid transfer function in the hidden layer was developed.

During neural network development, different number of neurons was tested. Results of training with a different number of neurons is presented in Table 4. The training was performed using Levenberg – Marquadt training algorithm.

Table 4. Training performance with different number of neurons in hidden layer

Number of neurons	Performance
2	6.6012e−14
5	2.1917e−13
10	5.8837e−13
13	2.8454e−13
15	3.2779e−14
18	3.1614e−14
20	6.2114e−14

Training performance neural network with varying number of neurons in the hidden layer is presented in Fig. 1, 2, 3, 4, 5, 6 and 7.

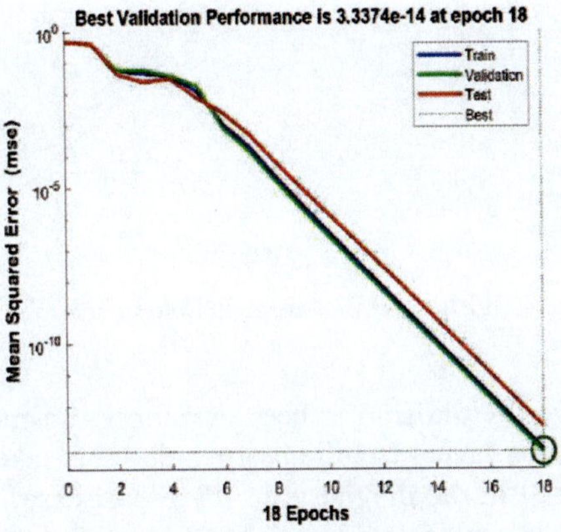

Fig. 1. 2 neurons in hidden layer

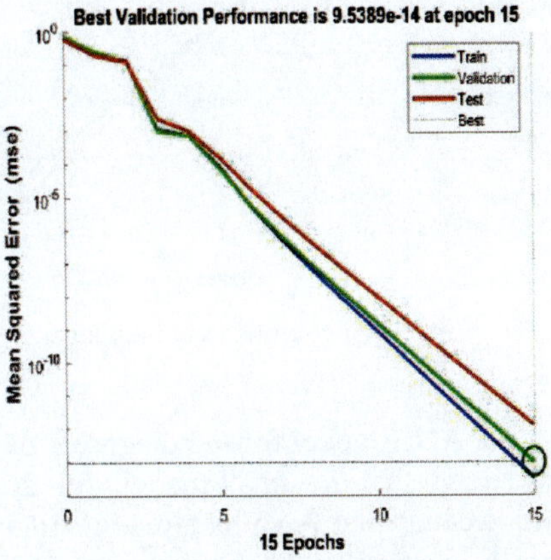

Fig. 2. 5 neurons in hidden layer

Accuracy of training for feedforward neural network with 18 neurons and sigmoid transfer function in the hidden layer was 94.9%. The validation of the performance of a feedforward neural network with 18 neurons and sigmoid transfer function in the hidden layer is presented in Table 5. Code and crossreference validation operations to verify that data were included. The selection of ANN was based on the training results. ANN with 20 neurons in the hidden layer showed the most satisfactory results.

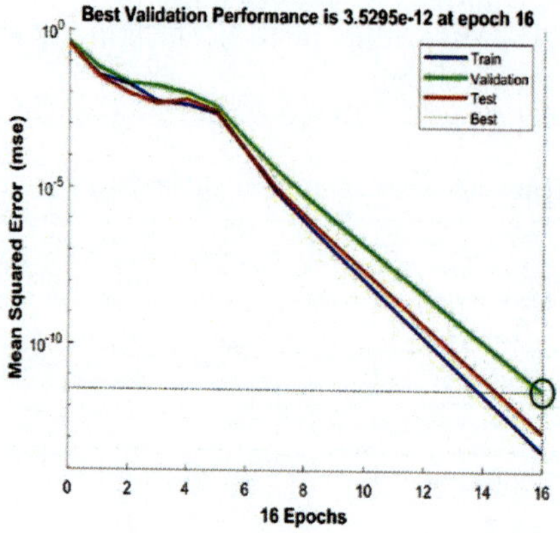

Fig. 3. 10 neurons in hidden layer

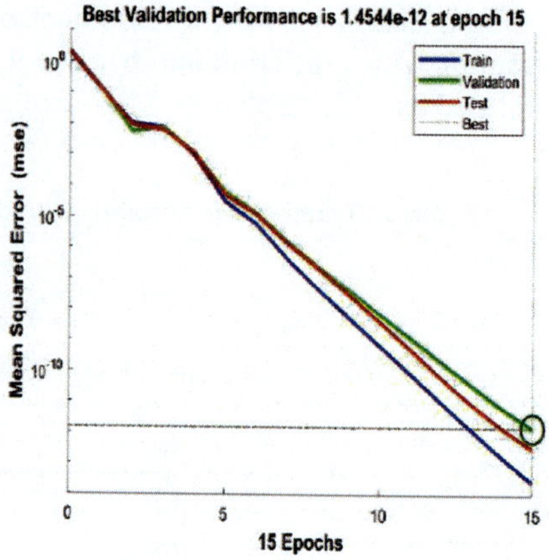

Fig. 4. 13 neurons in hidden layer

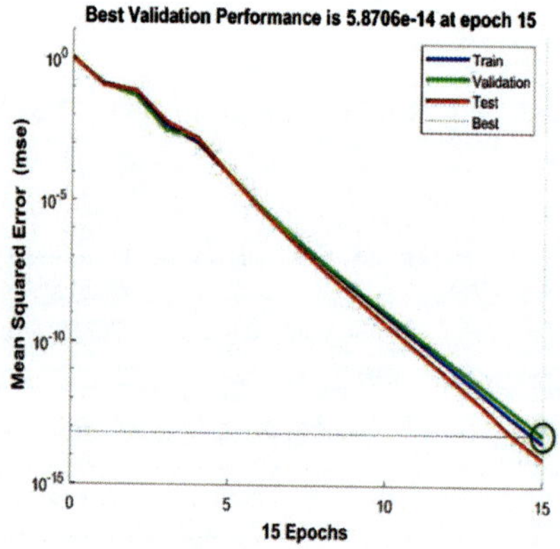

Fig. 5. 15 neurons in hidden layer

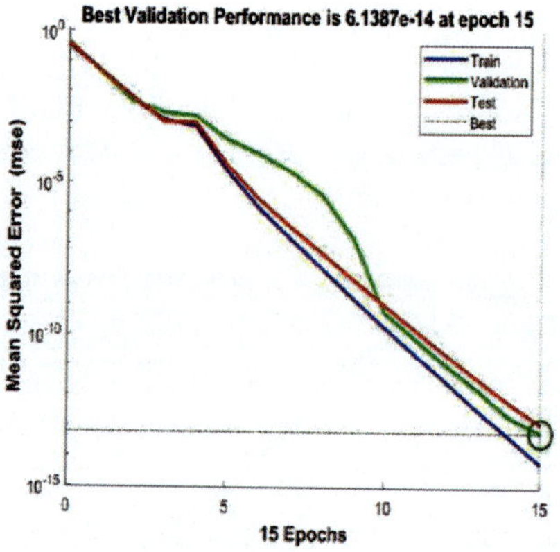

Fig. 6. 18 neurons in hidden layer

An ANN model for the diagnosis of hyperthyroidism has been developed. Testing has shown that the ANN model with 20 hidden layers has the best results. It is taken into account that ANN is already widely used in the diagnosis of thyroid disease. A search of relevant databases found 56 clinical studies in which ANN is used in the diagnosis of thyroid disease. However, only one paper was related to their application in hyperthyroidism outcome, and none in its diagnosis [21]. Therefore, this proposed model of ANN for the diagnosis of hyperthyroidism represents a significant contribution to the development of adequate ANN for the diagnosis of this very common disease. Further research is needed.

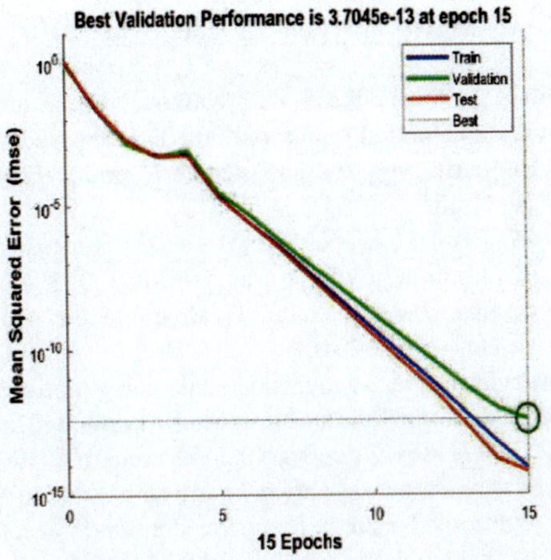

Fig. 7. 20 neurons in hidden layer

Table 5. Artificial Neural Network performance in classification

n = 99	Healhty	Disease	
Healthy 39	38	1	**Accuracy** [%] 94.9%
Disease 60	4	56	
	Sensitivity [%] 97.4	**Specificity** [%] 93.3	

4 Conclusion

In this study, we monitored hyperthyroidism using neural networks. We examined 330 patients, of which 130 patients were healthy and 200 patients were ill.

We monitored the performance of the model, changing the number of neurons in the hidden layer, and found that the smallest error was when we had 18 neurons in the hidden layer. Sensitivity in our case is 97.4%, specificity 93.3%, and accuracy 94.9%.

The significance of this work is reflected in the possibility of diagnosing hyperthyroidism using artificial neural networks, which are widely used in all branches of medicine. In this way, it is easier for doctors and enables much faster diagnosis of various diseases, including thyroid diseases.

References

1. De Leo, S., Lee, Y.S., Baverman, E.L.: Hyperthyroidism, Lancet (2016). https://doi.org/10.1016/S0140-6736(16)00278-6

2. Nygaard, B.: Hyperthyroidism (primary). BMJ Clin. Evid. (2010). https://pubmed.ncbi.nlm.nih.gov/21418670/
3. Vejrazkova, D., Vcelak, J., Vaclavikova, E., Vankova, M., Zajickova, K., Duskova, M., Vrbikova, J., Bendlova, B.: Genetic Predictors of the Development and Recurrence of Graves Disease, Institute of Endocrinology, Prague, Czech Republic (2018). www.biomed.cas.cz/physiolres/pdf/67/67_S431.pdf
4. Sharma, M., Aronow, W., Patel, L., Gandhi, K., Desai, H.: Hyperthyroidism. Med. Sci. Monit. (2011). https://www.ncbi.nlm.nih.gov/pmc/articles/PMC3539526/
5. Kravets, I.: Hyperthyroidism: Diagnosis and Treatment. Stony Brook University School of Medicine, Stony Brook, New York (2016)
6. Hodzic, J., et al.: A overview of next-generation sequencing platforms used in published draft plant genomes in light of genotypization of immortelle plant (helichrysium arenarium). Med. Arch. **71**(4), 288–292 (2017) https://doi.org/10.5455/medarh.2017.71.288-292
7. Catic, A., et al.: Application of neural networks for classification of Patau, edwards, down, turner and klinefelter syndrome based on first trimester maternal serum screening data, ultrasonographic findings and patient demographics. BMC Med. Genomics, **11**, 19 (2018). https://doi.org/10.1186/s12920-018-0333-2
8. Gurbeta, L., et al.: A telehealth system for automated diagnosis of asthma and chronical obstructive pulmonary disease. J. Am. Med. Inform. Assoc. **29**(9), 1213–1217 (2018)
9. Badnjevic, A., Gurbeta, L., Cifrek, M., Marjanovic, D.: Diagnostic of asthma using fuzzy rules implemented in accordance with international guidelines and physicians experience. In: IEEE 39th International convention on information and communication technology, electronics and microelectronics (MIPRO), 30 May to 03 June. Opatija, Croatia (2016)
10. Badnjevic, A., Gurbeta, L., Custovic, E.: An expert diagnostic system to automatically identify asthma and chronic obstructive pulmonary disease in clinical settings. Sci. Rep. **8**(1), 1–9 (2018)
11. Badnjevića, A., Pokvić, L.G., Hasičić, M., Bandić, L., Mašetić, Z., Kovačević, Z., Kevrić, J., Hifziefendić, J., Pecchia, L.: Evidence-based clinical engineering: machine learning algorithms for prediction of defibrillator performance. Biomed. Sign. Process. Control **54**, 101629 (2019)
12. Kovačević, Z., et al.: Prediction of medical device performance using machine learning techniques: Infant incubator case study Health Technol. **10**(1), 151–155 (2020)
13. Sarić, R., et al.: FPGA-based real-time epileptic seizure classification using artificial neural network. Biomed. Sign. Process. Control, **62**, 102106 (2020). ISSN 1746–8094. https://doi.org/10.1016/j.bspc.2020.102106
14. Mustafić, L.D., et al.: Diagnosis of severe aortic stenosis using implemented expert system. In: Badnjevic, A., Škrbić, R., Gurbeta Pokvić, L., (eds) CMBEBIH 2019. CMBEBIH 2019. IFMBE Proceedings, vol 73. Springer, Cham (2020)
15. Mrkulić, F., et al.: Correlation of leukemia genes overexpression and point mutations in different tissues. In: Badnjevic, A., Škrbić, R., Gurbeta, Pokvić, L. (eds) CMBEBIH 2019. CMBEBIH 2019. IFMBE Proceedings, vol 73. Springer, Cham (2020)
16. Hrvat, F., et al.: Artificial Intelligence for prediction of medical device performance: infusion and perfusor pumps case study. In: 9th Mediterranean Conference on Embedded Computing MECO 2020 (2020)
17. Šećkanović, A., et al.: Review of artificial intelligence application in cardiology. In: 9th Mediterranean Conference on Embedded Computing MECO 2020 (2020)
18. Anifowose, Fatai., Khoukhi, Amar, Abdulraheem, Abdulazeez: Investigating the effect of training–testing data stratification on the performance of soft computing techniques: an experimental study. J. Exper. Theor. Artif. Intell. **29**(3), 517–535 (2017). https://doi.org/10.1080/0952813x.2016.1198936

19. Aggarwal, C.: Book: Neural Networks and Deep Learning. Springer. https://link.springer.com/book/10.1007/978-3-319-94463-0
20. Yerushalmy, J.: Statistical problems in assessing methods of medical diagnosis with special reference to x-ray techniques. Public Health Rep. **62**(2), 1432–1439 (1947). https://doi.org/10.2307/4586294.JSTOR4586294.PMID20340527
21. Orunesu, E., Bagnasco, M., Salmaso, C., Altrinetti, V., Bernasconi, D., Del Monte, P., Pesce, G., Marugo, M.: Mela GS. use of an artificial neural network to predict graves' disease outcome within 2 years of drug withdrawal. Eur. J. Clin. Invest. **34**(3), 210–217 (2004). https://doi.org/10.1111/j.1365-2362.2004.01318.x.PMID:15025680
22. Liang X, Yu J, Liao J, Chen Z. Convolutional Neural Network for Breast and Thyroid Nodules Diagnosis in Ultrasound Imaging. Biomed Res Int. 2020 Jan 10;2020:1763803. doi:
23. Liu, C., Xie, L., Kong, W., Lu, X., Zhang, D., Wu, M., Zhang, L., Yang, B.: Prediction of suspicious thyroid nodule using artificial neural network based on radiofrequency ultrasound and conventional ultrasound: a preliminary study. Ultrasonics, **99,** 105951 (2019). https://doi.org/10.1016/j.ultras.2019.105951Epub2019Jun24.PMID:31323562
24. Sanyal, P., Mukherjee, T., Barui, S., Das, A., Gangopadhyay, P.: Artificial intelligence in cytopathology: a neural network to identify papillary carcinoma on thyroid fine-needle aspiration cytology smears. J. Pathol. Inform. **3,** (9), 43 (2018). https://doi.org/10.4103/jpi.jpi_43_18.PMID:30607310;PMCID:PMC6289006
25. Fragopoulos, C., Pouliakis, A., Meristoudis, C., Mastorakis, E., Margari, N., Chroniaris, N., Koufopoulos, N., Delides, A.G., Machairas, N., Ntomi, V., Nastos, K., Panayiotides, I.G., Pikoulis, E., Misiakos, E.P.: Radial basis function artificial neural network for the investigation of thyroid cytological lesions. J. Thyroid Res. **24** (2020), 5464787. https://doi.org/10.1155/2020/5464787PMID:33299540;PMCID:PMC7707952

Using Artificial Intelligence in Prediction of Osteoporosis

Pajević Amila, Pašalić Nejra, Piljug Nejra, Pinjić Adis, Planinić Matej(✉), and Pojata Amina

Faculty of Pharmacy, University of Sarajevo, Sarajevo, Bosnia and Herzegovina
matejplaninic@ffsa.unsa.ba

Abstract. Osteoporosis is a bone disease that makes the bones weaker, making them fragile and more likely to break. The artificial neural network [ANN] represents a system which uses risk factors, biochemical tests, and x-ray scans for development of a model, that is used for diagnosing and prediction of osteoporosis. The parameters used in this study, with the aim of predicting the risk of osteoporosis in patients, are: age, sex, hip fracture, spinal fracture, joint fracture, T-score, glucocorticoids use, smoking, calcium, and vitamin D. Reference values for serum calcium are 1,9−2,6 mmol/L, and for vitamin D 2580 ng/mL. T-score result less than −2.5 indicates osteoporosis. During development, different number of neurons was tested. During training, mean square error [MSE] for neural network with 2 neurons was 1.1075e−12, with 5 neurons it was 1.1075e−12, with 10 neurons it was 1.5541e−06, with 15 neurons it was 2.1113e−12. The optimum training performance was achieved by a neural network with 2 neurons in hidden layer. Based on the training results, ANN with 2 neurons in hidden layer was chosen for prediction of osteoporosis risk in patients.

Keywords: Osteoporosis · Artificial neural networks · Risk of osteoporosis

1 Introduction

Osteoporosis is described as a bone disease that develops when bone mineral density [BMD] and bone mass [BM] decreases, or when the microarchitectural deterioration of bone tissue commences. That being said, the changes in the bone can lead to a decrease in bone strength, which increases bone fragility and the susceptibility to fracture, primarily the hip, spine and other skeletal sites [1, 2]. This can result in devastating physical, psychosocial and economic consequences, which are often overlooked, because osteoporosis is clinically "silent", meaning there are typically no symptoms, before manifesting as fracture [3].

The disease can affect people of all ethnic backgrounds, and with the aging of the world population, osteoporosis is becoming an increasingly prevalent public health problem. It can occur at any age, but the risk for developing the disease increases as the individual, male or female, gets older [4]. More than 200 million people are currently suffering from osteoporosis. In their lifetime 1 in 5 men and 1 in 3 women over the age

A. Badnjevic and L. Gurbeta Pokvić (Eds.): CMBEBIH 2021, IFMBE Proceedings 84, pp. 288–295, 2021.
https://doi.org/10.1007/978-3-030-73909-6_33

of 50 years will experience osteoporosis linked fractures according to recent statistics from the International Osteoporosis Foundation [5].

The essential element in preventing osteoporosis is the achievement of normal peak bone mass, which can be managed by appropriate calcium and vitamin D intake, and a well balanced exercise program.

Since the fracture risk is higher in older women than in older men, postmenopausal women should be checked for signs of osteoporosis during their routine physical examinations [2, 6].

Many risk factors are associated with this disease, including low peak bone mass, the use of certain drugs [e.g. glucocorticoids], hormonal factors, low intake of calcium and vitamin D, low physical activity, smaller body size, and a personal or a family history of fracture. It is of great worth to take these factors into account, when assessing the risk of fracture and further treatment [1, 2, 6].

Osteoporosis is defined by a threshold value for BMD that is 2.5 standard deviations or more below the sex-specific young adult mean, whilst "severe" or "established" osteoporosis is defined by the presence of one or more documented fragility fractures. [2]

The World Health Organization and the International Osteoporosis Foundation recommend that the reference technology for the diagnosis of osteoporosis is dual-energy X-ray absorptiometry [DXA] applied to the femoral neck, which is the preferred site because of its higher predictive value for fracture risk. Combining the clinical assessment of osteoporotic risk factors and objective measures of BMD, can lead to improved identification of patients who will benefit from intervention and can potentially reduce mortality and morbidity associated with fractures. [2, 6]

According to recent studies, machine learning [ML] is becoming more and more reliable in solving healthcare tasks, even being better than humans in specific tasks, such as diagnosing disease. In healthcare, it is mainly used in diagnosis and treatment recommendations, patient engagement and adherence, and administrative activities [7].

ML is a statistical technique, with many versions to it, which trains a system with an adequate model, to use and "learn" data. The system can then predict the estimated output for each new input using the model [7, 8].

Since there are multiple physical, psychosocial and economic consequences of osteoporosis, building a model using artificial intelligence that could predict risk groups, could ease the burden of said consequences on society.

There are already multiple models, one of which is developed by artificial neural networks [ANN], and acts as a tool for prediction and diagnosis of osteoporosis, using extracted features from X-ray images, clinical blood biochemical examination, and risks related to the disease. The results of the study have concluded that the ANN is an effective diagnostic system of osteoporosis [9].

In this paper we use artificial intelligence to predict risk of osteoporosis in patients, using risk factors related to the disease, information extracted from X- ray images and results gathered from bone mineral density testing [10–31].

2 Materials and Methods

This paper presents the use of a neural network, with the aim of predicting the risk of osteoporosis in patiens. Dataset contains of 100 healthy and 150 unhealthy patients. The parameters we used are: age, sex, hip fracture, spinal fracture, joint fracture, T-score, glucocorticoids use, smoking, calcium and vitamin D. Reference values for serum calcium are 1,9–2,6 mmol/L, and for vitamin D 25–80 ng/mL. T-score represents a way of presenting bone mineral density results, and a T- score less than −2.5 indicates osteoporosis.

The distribution of the dataset for development of the Artificial Neural Network [ANN] is presented in Table 1. Since this is relatively small dataset, distribution of 70:30% for training and validation was used. [32].

Table 1. Dataset distribution

Total number of samples 250	Healthy 100	Unhealthy 150
Training 175	70	105
Validation 75	30	45

3 Results

According to previous research and expert recommendation [11], two-layer feedforward artificial neural network, with sigmoid transfer function in hidden layer was developed. Neural network architecture consist of inputs: Sex, Hip fracture, spinal fracture, joint fracture T-score, glucocorticoids, smoking, calcium [serum] [mmoL/L], vitamin D [ng/mL]. ANN has one output parameter and that is Prediction of Osteoporosis Risk in Patients. It can have two values: Yes – there is a certain risk for development of osteoporosis and No – there is no risk for development of ostheoporosis [13].

During development, different number of neurons was tested. The results of training for different number of neurons are presented in Figures below. During training mean square error [MSE] for neural network with 2 neurons was 1.1075e−12, with 5 neurons it was 1.1075e−12, with 10 neurons it was 1.5541e−06, with 15 neurons it was 2.1113e−12 (Figs. 1, 2 and 3)

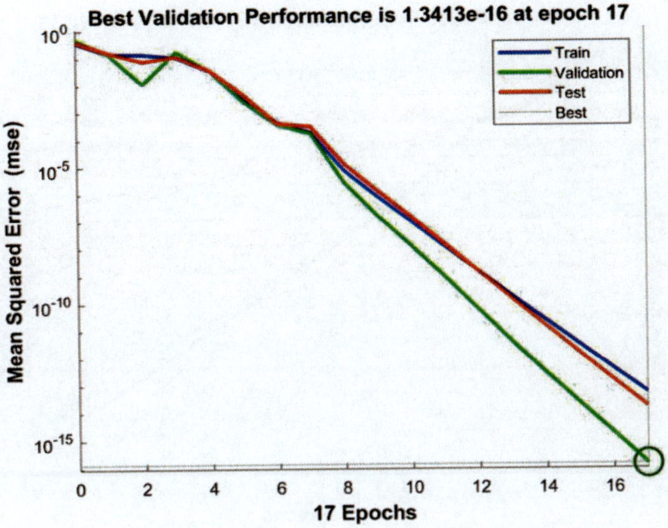

Fig. 1. Training performance with 2 neurons

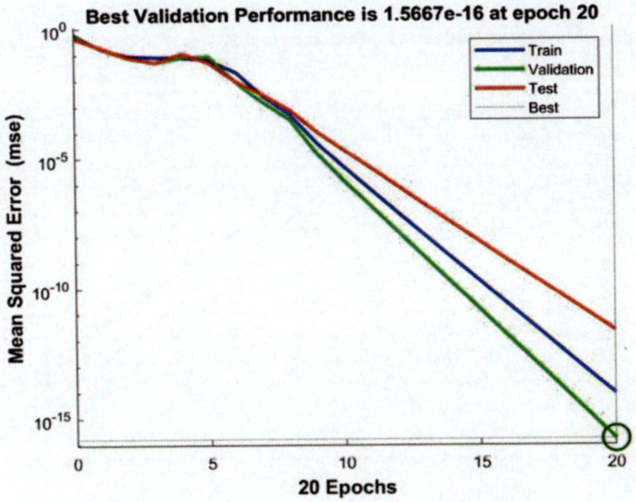

Fig. 2. Training performance with 5 neurons

Based on the figures shown above, it can be concluded that the optimum training performance was achieved by neural network with 2 neurons in hidden layer.

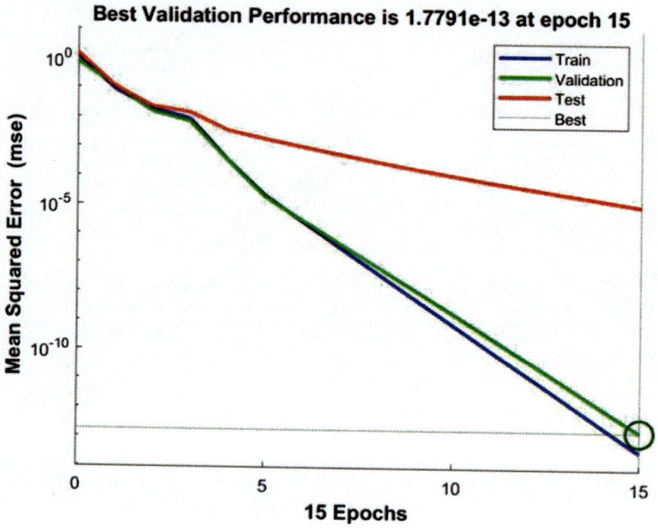

Fig. 3. Training performance with 10 neurons

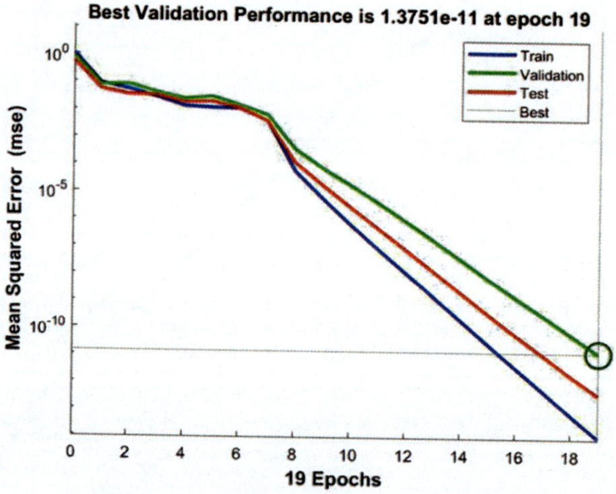

Fig. 4. Training performance with 15 neurons

Based on the training results, ANN with 2 neurons in hidden layer was chosen for prediction of Osteoporosis Risk in Patients. Validation of the developed neural network is given in the table below (Table 2).

Table 2. Artificial neural network performance during validation

n = 75	Group 1 healthy ANN prediction	Group 2 disease ANN prediction	
Group 1 Healthy = 30	29	1	**Accuracy** [%] 98.7
Group 2 Disease = 45	0	45	
	Sensitivity [%] 96.7	**Specificity** [%] 100%	

4 Conclusion

In this paper, the use of artificial intelligence in predicting the risk of osteoporosis is presented. Osteoporosis is a chronic progressive metabolic disease of the skeletal system characterized by a decrease in bone mineral density as well as by the damage to its microarchitecture. The developed neural network was made using 250 samples, of which it was trained with 175 samples and validated with 75. Based on the training results, ANN with 2 neurons in hidden layer was chosen for prediction of osteoporosis risk in patients with accuracy of 98.7%. By using and training such networks, in addition to collecting real data, it would be possible to predict which of the risk factors occur the most. This could have a preventive effect in the early stages of osteoporosis, thus achieving the most important goal - improving the quality of life of both healthy and sick individuals.

References

1. National Institute of Arthritis and Musculoskeletal and Skin Diseases. Osteoporosis Overview. 2019. https://www.bones.nih.gov/health-info/bone/osteoporosis/overview. Accessed 19 Nov 2020
2. Compston, J., Cooper, A., Cooper, C., et al.: UK clinical guideline for the prevention and treatment of osteoporosis. Arch. Osteoporos. **12**, 43 (2017)
3. Bethel, M.: MD Resident Physician, Department of Internal Medicine. Georgia Regents University. Osteoporosis, Medscape (2020)
4. Nancy, E.: Lane. Epidemiology, etiology, and diagnosis of osteoporosis. Am. J. Obstet. Gynecol. **194**[2 Suppl], S3–11 (2006)
5. Sözen, T., Özışık, L., Başaran, N.C.: An overview and management of osteoporosis. Euro. J. Rheumatol. **4**[1], 46–56 (2017)
6. Nuti, R., Brandi, M.L., Checchia, G., Di Munno, O., Dominguez, L., Falaschi, P., Fiore, C.E., Iolascon, G., Maggi, S., Michieli, R., Migliaccio, S., Minisola, S., Rossini, M., Sessa, G., Tarantino, U., Toselli, A., Isaia, G.C.: Guidelines for the management of osteoporosis and fragility fractures. Intern. Emerg. Med. **14**[1], 85–102 (2019)
7. Davenport, T., Kalakota, R.: The potential for artificial intelligence in healthcare. Future Healthcare J. **6**(2), 94–98 (2019)

8. Zhao, Y., Li, Y., Zhang, X., Geng, G., Zhang, W., Sun, Y.: A survey of networking applications applying the software defined networking concept based on machine learning. IEEE Access (2019). https://doi.org/10.1109/access.2019.2928564
9. Xinghu, Y., Ye, C., Xiang, L.: Application of artificial neural network in the diagnostic system of osteoporosis. Neurocomputing **214**, 376–381 (2016)
10. Hodzic, J., Gurbeta, L., Omanovic-Miklicanin, E., Badnjevic, A., et al.: Overview of nextgeneration sequencing platforms used in published draft plant genomes in light of genotypization of immortelle plant [helichrysium arenarium]. Med Arch. **71**(4), 288–292 (2017). https://doi.org/10.5455/medarh.2017.71.288-292
11. Catic, A., et al.: Application of neural networks for classification of patau, edwards, down, turner and klinefelter syndrome based on first trimester maternal serum screening data, ultrasonographic findings and patient demographics. BMC Med. Genomics, **11**(1), 19 (2018). https://doi.org/10.1186/s12920-018-0333-2
12. Gurbeta, L., et al.: A telehealth system for automated diagnosis of asthma and chronical obstructive pulmonary disease. J. Am. Med. Inform. Assoc. **25**(9), 1213–1217 (2018)
13. Badnjevic, A., et al.: An expert diagnostic system to automatically identify asthma and chronic obstructive pulmonary disease in clinical settings. Sci. Rep. **8**(1), 1–9 (2018)
14. Badnjevića, Almir., et al.: Evidence-based clinical engineering: machine learning algorithms for prediction of defibrillator performance. Biomed. Sign. Process. Control **54**, (2019)
15. Kovačević, Z., et al.: Prediction of medical device performance using machine learning techniques: infant incubator case study Health and Technol. **10**(1), 151–155 (2020)
16. Sarić, R., et al.: FPGA-based real-time epileptic seizure classification using artificial neural network. Biomed. Sign. Process. Control, **62**, 102106 (2020). ISSN 1746-8094. https://doi.org/10.1016/j.bspc.2020.102106
17. Badnjevic A, et al.: Diagnostic of asthma using fuzzy rules implemented in accordance with international guidelines and physicians experience. In: IEEE 39th International Convention on Information And Communication Technology, Electronics and Microelectronics [MIPRO], 30 May to 03 June 2016. Opatija, Croatia
18. Badnjevic, A., Gurbeta, L, Cifrek M, Marjanovic, D.: Classification of asthma using artificial neural network. In: IEEE 39th International convention on information and communication technology, electronics and microelectronics [MIPRO], 30. May to 03. June 2016. Opatija, Croatia
19. Aljovic A, et al.: Artificial neural networks in the discrimination of alzheimer's disease using biomarkers data. In: IEEE 5th Mediterranean Conference on Embedded Computing [MECO], 12–16 June 2016, Bar, Montenegro
20. Alic, B., et al.: Classification of stress recognition using artificial neural network. In: 5th Mediterranean Conference on Embedded Computing [MECO], 12–16 June 2016, Bar, Montenegro IEEE (2016)
21. Granulo, E., et al.: Telemetry system for diagnosis of asthma and chronical obstructive pulmonary disease [COPD]. In: The 3rd EAI International Conference on IoT Technologies for HealthCare, 18–19 October 2016, Vasteras, Sweden (2016)
22. Veljovic, E., et.al.: Artificial neural network and docking study in design and synthesis of xanthenes as antimicrobial agents. In: CMBEBIH 2017. IFMBE Proceedings, vol 62, pp 617–626, Springer, Singapore (2017). https://doi.org/10.1007/978-981-10-4166-2_93
23. Alic, B., et.al.: Classification of metabolic syndrome patients using implemented expert system. In: CMBEBIH 2017. IFMBE Proceedings, vol 62, pp 601–607, Springer, Singapore (2017). https://doi.org/10.1007/978-981-10-4166-2_91
24. Badnjevic, A.: Pre-classification process symptom questionnaire based on fuzzy logic for pulmonary function test cost reduction. In: CMBEBIH 2017. IFMBE Proceedings, vol 62, pp 608–616, Springer, Singapore (2017). https://doi.org/10.1007/978-981-10-4166-2_92

25. Sejdinovic, D., et al.: Classification of prediabetes and type 2 diabetes using artificial neural network. In: CMBEBIH 2017 IFMBE Proceedings, vol 62. pp 685–689, Springer, Singapore (2017). https://doi.org/10.1007/978-981-10-4166-2_103
26. Alić, B., Gurbeta, L., et al.: Machine learning techniques for classification of diabetes and cardiovascular diseases. In: 2017 6th Mediterranean Conference on Embedded Computing [MECO], Bar, Montenegro, pp. 1–4 (2017). https://doi.org/10.1109/meco.2017.7977152
27. Mustafić, L.D., et al.: Diagnosis of severe aortic stenosis using implemented expert system. In: Badnjevic, A., Škrbić, R., Gurbeta Pokvić, L. [eds]. In: CMBEBIH 2019. IFMBE Proceedings, vol 73. Springer, Cham (2019)
28. Mrkulić, F., Gurbeta, L., Omanović-Mikličanin, E., Bego, T., Hasanefendić, B., Badnjević, A.: Correlation of leukemia genes overexpression and point mutations in different tissues. In: Badnjevic, A., Škrbić, R., Gurbeta Pokvić, L., [eds] CMBEBIH 2019 IFMBE Proceedings, vol. 73. Springer, Cham (2019)
29. Spahić, L., et al.: Machine learning techniques for performance prediction of medical devices: infant incubators. In: Badnjevic, A., Škrbić, R., Gurbeta Pokvić, L. [eds] CMBEBIH 2019 IFMBE Proceedings, vol 73. Springer, Cham (2020)
30. Hrvat, F., Spahić, L., Pokvić, L.G., Badnjević, A.: Artificial intelligence for prediction of medical device performance: infusion and perfusor pumps case study. In: 9th Mediterranean Conference on Embedded Computing MECO 2020 (2020)
31. Šećkanović, A., Šehovac, M., Spahić, L., Ramić, I., et al.: Review of artificial intelligence application in cardiology. In: 9th Mediterranean Conference on Embedded Computing MECO 2020 (2020)

Check for updates

Machine Learning Techniques for Risk Assessment and Diagnosis of Diabetes Mellitus

Lejla Mehić, Sumeja Muhić, Amina Mujagić, Almedina Mujčinović(✉), Amra Mujić, and Sabina Murto

Faculty of Pharmacy, University of Sarajevo, Sarajevo, Bosnia and Herzegovina
almedinamujcinovic@ffsa.unsa.ba

Abstract. Diabetes represents a group of diseases that are characterized by hyperglycemia resulting from defects in insulin secretion, insulin action, or both. Machine Learning algorithms are increasingly used for risk assessment and diagnosis of various disease, so this paper represents development of machine learning-based model used for risk assessment and diagnosis of diabetes mellitus. For development of this system 300 samples consisting of information about Fasting Plasma Glucose (FPG), Oral Glucose Tolerance Test (OGTT) and blood test called HgbA1c were used. All samples were obtained from several healthcare institutions in Bosnia and Herzegovina, and diagnosis of diabetes mellitus and healthy patients in this dataset were established by medical professionals. Machine learning techniques for risk assessment and diagnosis of diabetes mellitus, were trained with 210 samples. Testing of developed system was performed with 90 samples for validation. The developed system shows 98,89% accuracy while classifying diabetes.

Keywords: Machine learning · Diabetes mellitus · Risk assessment

1 Introduction

Diabetes represents a group of diseases that are characterized by hyperglycemia resulting from defects in insulin secretion, insulin action, or both [1]. Diabetes shows up as an outcome of several pathogenic processes. These processes include autoimmune destruction of the β-cells of the pancreas with consequent insulin deficiency and abnormalities that results with resistance to insulin action. Deficient insulin action results from inadequate insulin secretion and/or diminished tissue responses to insulin at one or more points in the complex pathways of hormone action [1]. Sometimes it is not very clear which abnormality is the primary cause of the hyperglycemia, is it impairment of insulin secretion or defect in insulin action. Symptoms of marked hyperglycemia include polyuria, polydipsia, weight loss, sometimes polyphagia, and blurred vision. Acute, life-threatening consequences of uncontrolled diabetes are hyperglycemia with ketoacidosis or nonketotic hyperosmolare syndrome. Long-term complications of diabetes include retinopathy with possible loss of vision, nephropathy leading to renal failure, peripheral neuropathy with risk of foot ulcers, amputations etc. Hypertension and lipoprotein metabolism abnormalities are often found in people with diabetes [1].

A. Badnjevic and L. Gurbeta Pokvić (Eds.): CMBEBIH 2021, IFMBE Proceedings 84, pp. 296–302, 2021.
https://doi.org/10.1007/978-3-030-73909-6_34

The majority of cases of diabetes fall into two board etiopathogenetic categories: Type 1 (T1DM) and Type 2 diabetes (T2DM). In T1DM an absolute deficiency causes a disease. In the other, much more prevalent category, T2DM this state represents a combination of resistance to insulin reaction and an inadequate compensatory insulin secretory response. In this type of diabetes an asymptomatic period can be shown that carbohydrate metabolism is not normal by measurement of plasma glucose in the fasting state or after a challenge with an oral glucose load [1].

The current gold standard for diagnosis diabetes is measurement of glucose in venous plasma. The oral glucose tolerance test (OGTT) should be used in individuals with fasting plasma glucose 6.1−6.9 mmol/l to determine glucose tolerance status. HbA1c is suitable as a primary diagnostic tool for excluding diabetes in some cases.

Digital technologies and electronic healthcare, along with Big Data that they generate and the artificial intelligence methods which analyze them, represent a major opportunity to reevaluate diabetes, as they are expected to have a major impact on all aspects of diabetes, including its prevention, diabetes care and management. However, these are still early days in the delivery of health services and information that using the internet and related technologies, just as smartphone apps and connected devices (named mHealth) and also for social media, telemedicine and electronic health records that can change the diabetes landscape all the way. Furthermore, data generated online and by digital technologies, that authors suggest to be grouped under the term "digitosome", constitute a powerful potential for identifying new digital markers and patterns of risk. When this is combined with clinical data, it can improve diabetes management and quality of life, and prevent diabetes-related complications. For example, it has been suggested that Twitter might be used for analyzing disease related symptoms and medication use. Therefore, combing unstructured data like Tweets with clinical data would be a new approach against what is now available in clinical settings, making this new shape of research a promising area, accommodated between computer science, epidemiology and medical research [9].

The aim of our paper is to present use Machine Learning Techniques for risk assessment and diagnosis of prediabetes and diabetes by measuring glucose by classical methods and tests.

2 Materials and Methods

Total number of participants were 300, of which there were 200 with diagnoses and 100 healthy subjects. Different parameters will be observed.

When performing diagnosis of diabetes mellitus there are used several blood tests to determine the amount of glucose in the blood. In this article, we used informations from several tests such as fasting plasma glucose (FPG), oral glucose tolerance test (OGTT) and HgbA1c [1]. Fasting plasma glucose test checks fasting blood sugar levels. This test is usually done before breakfast, after at least 8 h with no food or drinks (except water) [1]. Interpretation of OGT test is based on plasma glucose results before and 2 h after a 75 g oral glucose load [10]. Parameter HgbA1C gives an integrated index of glycaemia ovet the entire 120-day lifespan of hte red blood cell [1].

Parameters such as polyuria, polydipsia, BMI (body mass index) and fatigue were also considered.

Table 1. Data used for development of machine- learning-techniques

	Total number of samples	Number of women	Number of men	Average woman age	Average men age
Healthy subjects	**100**	**41**	**59**	**31,07**	**28,02**
Subjects whit diagnoses	**200**	**99**	**101**	**32,38**	**34,01**

From Table 1, it can be seen if patient is classified as healthy, with prediabetes or with diabetes. According to the results, patients are classified as diabetic if they have HgbA1c above 6.5% [1] or FPG above 7.0 mmol/L [9] or OGTT above 11.0 mmol/ [10]. Parameters used for the classification of diabetes are shown in the Table 2.

Table 2. Parameters for the classification of diabetes

	HgbA1c	Fasting plasma Glucose (FPG)	OGTT
Normal/Healthy	<5.7%	<6,1 mmol/L	<7.8 mmol/L
Prediabetes	5.7% – 6.4%	6.1–7.0 mmol/L	7.8–11 mmol/L
Diabetes	≥6.5%	≥7.0 mmol/L	≥11.0 mmol/L

The dataset for this article consisted of 300 samples. The distribution of the dataset for development of the classifier based on machine learning algorithms (ML) is presented in Table 3. Since this is relatively small dataset, distribution of 70:30% for training and validation was used [11].

Table 3. Dataset distribution per training and validation subset

Total number of samples 300	Group 1 disease 200	Group 2 healthy 100
Training 210	140	70
Validation 90	60	30

For building classifier for classification of burnout syndrome, following machine learning algorithms were used: Naïve Bayes (NB), Decision Tree (DT), Random Forest (RF), and k-Nearest Neighbor (k-NN) algorithm.

For evaluating performance of developed classifier following was used:

Accuracy and precision can be calculated according to the formulas:

$$Accuary = \frac{TP + TN}{TP + FP + TN + FN} \tag{1}$$

$$P\,eci\,i = \frac{TP}{TP + FP} \tag{2}$$

Where:

- TP – true positive
- TN – true negative
- FP – false positive
- FN – false negative

3 Results

The development of classifiers based on ML aglorithms was done in RapidMiner studio. The results of training performance for classifiers based on 4 different machine learning algorithms are presented in Fig. 2. Training was performed on 210 samples who were subsequently distributed according to the 70:30 ration using random classification [11, 12].

Figure 1. presents the results of training of classifiers based on 4 different machine learning algorithms. As it can be seen from the figure, during training process, classifier based on Naïve Bayes algorithms achieved the highest accuracy of 98.89%. The lowest accuracy was achieved by RF and it was 95.56%.

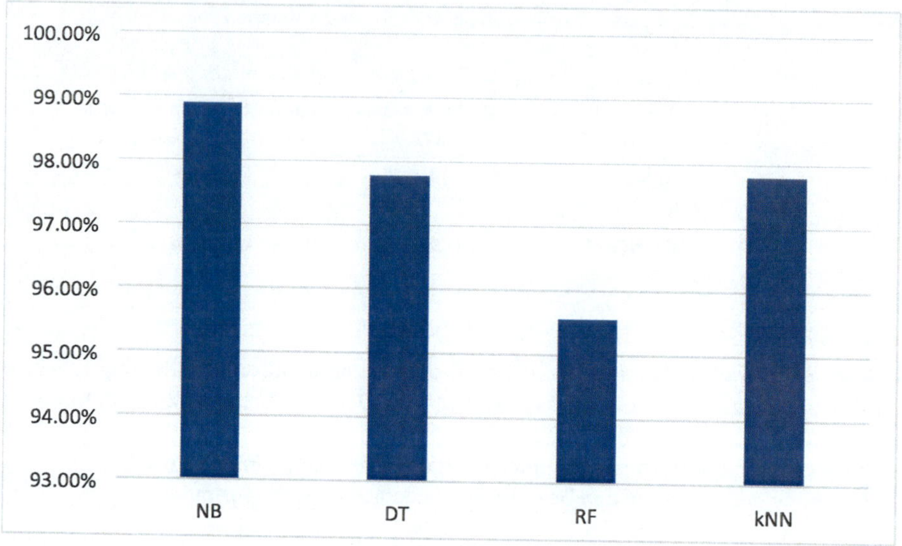

Fig. 1. Training performance of classifiers

Validation was performed on 90 samples. Each classifier was validated and the results of validation are presented in the Tables 4, 5, 6 and 7.

Table 4. Validation performance of NB based classifier

Class	Healthy	Disease	Accuracy
Healthy = 30	29	1	97.7%
Disease = 60	1	59	
Total samples	96.67	98.3	

Table 5. Validation performance of DT based classifier

Class	Healthy	Disease	Accuracy
Healthy = 30	30	0	97.78%
Disease = 60	2	58	
Total samples	100%	96.67	

Table 6. Validation performance of RF based classifier

Class	Healthy	Disease	Accuracy
Healthy = 30	24	6	92.22%
Disease = 60	1	59	
Total samples	80	98.3	

Table 7. Validation performance of k-NN based classifier

Class	Healthy	Disease	Accuracy
Healthy = 30	28	2	96.67%
Disease = 60	1	59	
Total samples	93.3	98.3	

4 Conclusion

This paper presents a classification of patients with prediabetes and diabetes using machine learning techniques. There are two types of diabetes, T1DM and T2DM and it is important to have a system which will enable classification of these conditions. Machine learning techniques for risk assessment and diagnosis of diabetes mellitus, were trained with 210 samples. Testing of developed system was performed with 90 samples for validation. During training process, classifier based on Naïve Bayes algorithms achieved the highest accuracy of 98,89%. The lowest accuracy was achieved by RF and it was 95,56%. Definitely this method of diagnosis is of great importance for both medical staff and the patient, therefore it should be developed

References

1. American Diabetes Association. Diagnosis and Classification of Diabetes Mellitus (2013)
2. Spahić, L., et al.: Machine learning techniques for performance prediction of medical devices: infant incubators. In: Badnjevic, A., Škrbić, R., Gurbeta Pokvić, L. (eds) CMBEBIH 2019. CMBEBIH 2019. IFMBE Proceedings, vol 73. Springer, Cham (2020)
3. Alić, B.L., et al.: Machine learning techniques for classification of diabetes and cardiovascular diseases. In: 2017 6th Mediterranean Conference on Embedded Computing (MECO), Bar, Montenegro (2017)
4. Sarić, R., et.al.: FPGA-based real-time epileptic seizure classification using artificial neural network. Biomed. Sig. Process. Control, **62**, 102106 (2020). ISSN 1746–8094. https://doi.org/10.1016/j.bspc.2020.102106
5. Badnjevic, A., Cifrek, M., Koruga, D., Osmankovic, D.: Neuro-fuzzy classification of asthma and chronic obstructive pulmonary disease. BMC Med. Inform. Dec. Making J. **15**(Suppl 3), 1–9 (2015). https://doi.org/10.1186/1472-6947-15-s3-s1
6. Gurbeta, Lejla., et al.: A telehealth system for automated diagnosis of asthma and chronical obstructive pulmonary disease. J. Am. Med. Inform. Assoc. **25**(9), 1213–1217 (2018). https://doi.org/10.1093/jamia/ocy055

7. Badnjević, A., et.al.: Evidence-based clinical engineering: machine learning algorithms for prediction of defibrillator performance. Biomed. Sig. Process. Control, **54**, 101629, (2019). ISSN 1746–8094. https://doi.org/10.1016/j.bspc.2019.101629
8. Kovačević, Ž., et al.: Prediction of medical device performance using machine learning techniques: infant incubator case study. Health Technol. (2019). https://doi.org/10.1007/s12553-019-00386-5
9. Fagherazzi, G., Ravaud, P.: Digital diabetes: perspectives for diabetes prevention, management and research. Diab. Metab. (2018)
10. Cloete, L., Mitchell, B.G., Morton, D.: Protocol: investigating the effectiveness and cost benefit of a lifestyle intervention targeting type 2 diabetes in Australia. BMC Endocr. Disord. **19**(1). (2019)
11. Anifowose, F., Khoukhi, A., Abdulraheem, A.: Investigating the effect of training–testing data stratification on the performance of soft computing techniques: an experimental study. J. Exper. Theor. Artif. Intell. **29**(3), 517–535 (2017). https://doi.org/10.1080/0952813X.2016.1198936
12. Aggarwal, C.: Book: Neural Networks and Deep Learning. Springer. https://link.springer.com/book/10.1007/978-3-319-94463-0

Utilization of Machine Learning Techniques for Identification of *Escherichia Coli* Based on Results of Bauer Kirby Antibiotic Susceptibility Testing

Amel Spahić[1](✉), Zerina Mašetić[1], Irma Mahmutović-Dizdarević[2], and Monia Avdić[3]

[1] Department of Information Technologies, International Burch University, Sarajevo, Bosnia and Herzegovina
amel.spahic@stu.ibu.edu.ba

[2] Department of Biology, Faculty of Science, University of Sarajevo, Sarajevo, Bosnia and Herzegovina

[3] Department of Genetics and Bioengineering, International Burch University, Sarajevo, Bosnia and Herzegovina

Abstract. Rise of antibiotic resistance has become a major issue worldwide. The most common procedure for the evaluation of antibiotic susceptibility, in use today, is the Bauer Kirby disk diffusion testing, however the procedure is time consuming and therapy is ordinated 48–72 h upon obtaining the patient's sample. This is problematic especially in treatment of urinary tract infections (UTI) where the symptoms are severe, and therapy should be ordained as soon as possible. The aim of our work was to utilize machine learning techniques to identify *Escherichia coli* as the most frequent causative agent of urinary tract infection based on the results of the antibiotic susceptibility testing. The study included a total of 162 antibiograms obtained from the Institute for Biomedical Research and Diagnostics NALAZ collected in the period of 2018–2019. Using the Multi-layer Perceptron classifier, the precision, sensitivity, and accuracy for identification of *E. coli* and its distinction from other causative agents of urinary tract infection was over 93%. These results indicate that machine learning techniques can be of great value to aid in diagnostic procedures and microbiology is a field where the use of AI can be exploited to a much greater extent.

Keywords: Bacteria identification · Escherichia coli · Antibiotic susceptibility · Machine learning · Random forest · J48 · Multi-layer perceptron · Logistic regression

1 Introduction

Antibiotic resistance has become a major challenge with the potential to rise to the level of emerging as a global crisis [1]. Antibiotics as small molecules, whose origin initially most likely was as signaling molecules in bacterial communication, still must be utilized in adequate inhibitory and microbicidal doses as the only effective mode of

A. Badnjevic and L. Gurbeta Pokvić (Eds.): CMBEBIH 2021, IFMBE Proceedings 84, pp. 303–312, 2021.
https://doi.org/10.1007/978-3-030-73909-6_35

treatment of most infections [2]. Antibiotics can be defined as "a substance produced by a microorganism (as a bacterium or a fungus) and in dilute solution having the capacity to inhibit the growth of or kill another microorganism (such as a disease or germ)" [3]. Antibiotic resistance of bacteria can be a natural trait or may be acquired through rendering the bacteria resistant to antibiotics through mutation or acquiring the resistance gene from another source [4]. Inherited resistance may be due to traits such as the structure of the cell wall or lack of a transport system for the antibiotic and many others [4].

Prediction of antibiotic resistance patterns has been employed in many studies with the aim to contribute to solving this issue.

Up to now genome-based prediction of antibiotic resistance has been employed by determining the presence of antibiotic resistance genes and single nucleotide polymorphisms (SNP) against approaches that deploy machine learning and statistical models [5]. An in silico antibiogram for eight drugs has been established in the work of Van Camp et al. 2020, where the NCBI BioSample Database was utilized to train and cross-validate eight XGBoost-based machine learning models [6].

In another work from Kim et al., 2020, VAMPr: VAriant Mapping and Prediction of antibiotic resistance via explainable features and machine learning exhibited promising results and yielding an accuracy of 91% for all antibiotic – pathogen combinations [7].

Bradly et al. 2015 utilized the Mykrobe predictor for prediction of resistance in Staphylococcus aureus and Mycobacterium tuberculosis yielding an accuracy rate comparable to the gold standard for 12 antibiotics, with sensitivity/specificity of 99.1%/99.6%, respectively [8].

In the work of Chowdhury et al., 2020 the accuracy of predicting antibiotic resistance in gram negative bacteria was 93% to 99% by employing game theory to choose protein characteristics that should be utilized in their machine learning model [9].

All the mentioned antibiotic resistance predictions were based upon data of genome sequences which requires the use of sophisticated methodologies and hence is costly, whereas the antibiogram data using Bauer Kirby Disk diffusion testing is vast and affordable for most labs.

Employment of antibiotic susceptibility testing and avoidance of overuse of antibiotics may be only part of the solution and it holds obtaining an antibiogram in the clinical microbiology lab, preceded by cultivation of the microorganism and subsequent biochemical identification [10].

Each of these processes requires an incubation period of 18–24 h, adding up to a total of 48–72 h before administering antibiotic therapy - which is very problematic in cases where the symptoms of infection are severe.

Employment of methodologies to shorten this time frame and at the same time does not contribute to enhancing the problem of antibiotic resistance would be of great benefit to the field and to patients.

Considering the empiric use of therapy in different regions as well as medical guidelines exposure to antibiotics in certain hospitals is uniform, hence the resistance patterns in certain microbial species may be similar due to being part of the same ecological niche. Deploying this fact and feeding the plethora of data about antibiotic susceptibility testing to machine learning algorithms the aim of our paper is to predict the bacterial species based on their antibiotic resistance pattern.

2 Materials and Methods

The study included a total of 162 antibiograms obtained from the Institute for Biomedical Research and Diagnostics NALAZ collected in the period of 2018–2019. Species identification was carried out using standardized biochemical testing, whereas the choice of antibiotics was in accordance with the European Committee on Antimicrobial Susceptibility Testing (EUCAST) standard from that period [11]. Criteria for inclusion into the study was that only one of three successive urine samples from the same patient was taken into consideration.

Kirby-Bauer disk diffusion test is used as the method for antibiotic susceptibility testing for a total of 16 different antibiotics: (1) *Trimethoprim/ Sulfamethoxazole*, (2) *Ciprofloxacin*, (3) *Ceftriaxone*, (4) *Ceftazidime*, (5) *Ampicillin*, (6) *Amoxicillin*, (7) *Amikacin*, (8) *Cefuroxime*, (9) *Ertapenem*, (10) *Nitrofurantoin*, (11) *Cefixime*, (12) *Gentamicin*, (13) *Fosfomycin + glucose-6-phosphate*, (14) *Vancomycin*, (15) *Teicoplanin*, (16) *Linezolid*.

2.1 Data Pre-processing

For the diversity of the dataset and getting more accurate model, several antibiograms of different bacterial species are used, as follows: *Escherichia coli*, *Klebsiella pneumoniae*, *Proteus mirabilis*, *Pseudomonas aeruginosa*, *Streptococcus agalactiae*, *Enterococcus* spp (Fig 1).

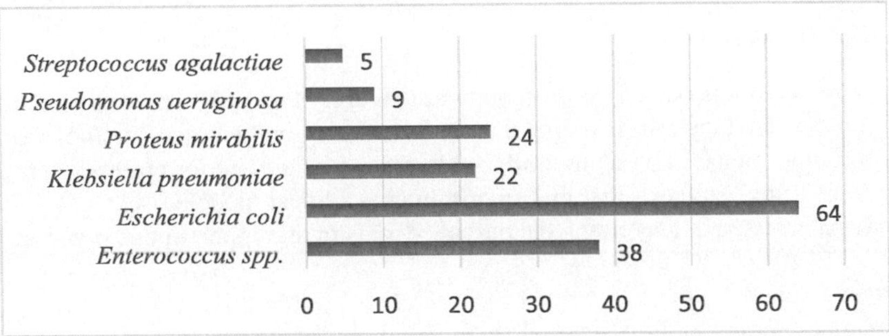

Fig. 1. Dataset composition

Machine learning model should identify *Escherichia coli* based on antibiograms from different bacteria, so the pre-processing was done by adding a new column representing two classes for the output:

- *Y* – indicates *E. coli*.
- N – indicates other bacteria.

Table 1. Pre-processed dataset

	Escherichia coli	
	Y	N
Samples	64	98
(Total: 162)	39.506%	60.494%

The data were reprocessed to best represent the real-life experience, so the dataset contains antibiograms whose results have a frequency greater than 1. Total number of samples used on the machine learning algorithm is 162.

Having a relatively small dataset, k-fold cross validation method as a test option is used. The dataset is divided into k folds, where k-1 folds are used as a training set, while 1-fold is used as a testing set. The process is repeated k times, each time having a different fold for the testing. The error of these k-models is averaged. This is done for each of the model parameters to be tested, and the model with the lowest error is selected [12].

2.2 Machine Learning Techniques

There are several ML approaches to classify a bacteria's susceptibility to certain groups of bacteria. It is sometimes necessary to pre-process data before implementing ML techniques, so that these techniques can take advantage of the data information. In specific, four classification methods will be examined in this paper: logistic regression, decision tree, random forest, and artificial neural network [13].

2.3 Logistic Regression

Logistic regression is one of the most popular ML algorithms which makes predictions using a nonlinear function. If output data is expected to be binomial (yes/no) this is a good algorithm to use. These are used for classification and not for regression, but the name "regression" suggests that this algorithm uses a linear model [14].

Parameters used in generating the output of logistic regression in this paper are:

- Ridge: 1.0E−8
- Maximum number of iterations: −1 (default −1, until convergence)

2.4 Decision Tree

Decision trees are machine learning algorithms that are used for predictive modeling as well as classification. In the terms of the volume of input data, they do not require too much information and they are a good fit for this paper. Representation of these types of algorithms is a binary tree.

This paper uses an optimized version of C4.5 algorithm –J48. This algorithm passes through the decision tree, visits each node, and selects the optimal split using GainRatio as the splitting criterion [15].

Parameters used in generating the output:

- J48 pruned tree as a classifier model
- Number of leaves: 4
- Size of the tree: 7
- Confidence factor: 0.25

The optimal accuracy is reached when the size of tree is 7 and number of leaves 4.

2.5 Artificial Neural Network

Neural networks consist of a huge number of processing units called neurons, nodes, processing units or cells. Each is connected to another by related weight. Neural network consists of input nodes which receive data from outside and form input layer, output nodes which send data out of the network and form output layer and hidden nodes whose signals remain in the network and form hidden layer. Here, the network is created with one input layer, consisting of 20 inputs and 1 hidden layer, each consisting of 11 nodes, as these parameters give the optimal classification accuracy [16].

2.6 Random Forest

Random Forest is a classification method which combines multiple tree predictors such that each tree depends on a value of randomly chosen vector, distributed to all trees in the forest. Two measures of the accuracy of individual classifiers are strength and correlation, respectively. Random forest with random features is formed by choosing a small group of input variables on each node, randomly. Here, a random forest consisted of 100 trees, with 0 random features. Different values of trees and random features were used, using the classification accuracy as fitting function [17].

3 Results and Discussion

The aim of this research is to use antibiograms to create a machine learning model that will predict whether the results of the antibiogram indicates bacterium *E. coli* or not.

For the performance of these classifiers, several indicators are used: accuracy, precision, and sensitivity, which can be obtained using the following formulas [18]:

Equation 1: $Accuracy = Correct\ Predictions/Number\ of\ Samples$

Equation 2: $Precision = TP/(TP + FP)$

Equation 3: $Recall = TP/(TP + FN)$

Model accuracy is the output metric that shows the number of instances correctly classified within the input range, or the percentage of instances correctly classified.

Beside the *accuracy* result, there are additional benchmarks like *true positive (TP), false positive (FP), true negative (TN), false negative (FN).*

The above-mentioned classifiers are used to predict the presence of *Escherichia coli* from the given antibiogram: *Decision tree* (DT), *Random Forest* (RF), *Logistic Regression* (LR), and *Artificial Neural Networks* (ANN).

Each classifier uses the same set of data and cross-validation sampling procedure with 10 folds.

The performance of the classifiers measured through accuracy is shown in Table 2.

Table 2. Machine learning classifier performance.

Classifier	Correctly classified instances	Incorrectly classified instances
DT	87.037%	12.963%
LR	83.333%	16.667%
RF	87.654%	12.346%
ANN	93.210%	6.790%

All selected classifiers showed good performance when it comes to the detection of *Escherichia coli* bacteria among five different types of bacteria and based on the antibiograms. However, ANN outstood, achieving the highest accuracy during the classification process (93.2%), while LR has the lowest result (83.3%), among four classifiers. DT and RF classifiers achieved almost the same results, with DT misclassifying 1 instance more, compared to RF.

ANNs are very capable in detecting all possible interactions between predictor variables, and therefore are of high importance in such problems.

Additionally, the confusion matrices, with accuracy, precision and recall values of all predictors are shown in Tables 3, 4, 5 and 6 and evaluated.

Table 3. Decision tree classifier performance

Samples: 162	Decision tree		
Confusion matrix	**Classes**	**Y**	**N**
	Y	58	6
	N	15	83
Performance	**Accuracy**	**Precision**	**Recall**
	87.04%	87.80%	87.00%

Table 4. Random forest classifier performance

Samples: 162	Random forest		
Confusion matrix	**Classes**	**Y**	**N**
	Y	59	5
	N	15	83
Performance	**Accuracy**	**Precision**	**Recall**
	87.65%	88.60%	87.70%

Table 5. Logistic regression classifier performance

Samples: 162	Logistic regression		
Confusion matrix	**Classes**	**Y**	**N**
	Y	59	5
	N	22	76
Performance	**Accuracy**	**Precision**	**Recall**
	83.33%	88.50%	83.30%

Table 6. Artificial neural network classifier performance

Samples: 162	Artificial neural network		
Confusion matrix	**Classes**	**Y**	**N**
	Y	60	4
	N	7	91
Performance	**Accuracy**	**Precision**	**Recall**
	93.21%	93.30%	93.20%

On the complete dataset of 162 samples, DT achieved 87.04% classification accuracy. Considering the confusion matrix in the Table 3, it misclassified 6 cases of *Escherichia coli* bacteria. Moreover, it falsely classified 15 cases of other bacteria as *Escherichia coli.* DT achieved 87% of the recall value which states the percentage of correctly classified *Escherichia coli* cases among all cases classified as *Escherichia coli* bacteria. Precision, with the value of 87.8% indicates the percentage of correctly classified *Escherichia coli* cases among all actual cases of *Escherichia coli.*

Similarly, RF classifier was 87.65% successful in differentiating among presence and absence of *Escherichia coli* bacteria. Additionally, it achieved slightly better precision and recall rate, of 88.06% and 87.7%, respectively, as shown in Table 4. RF classified one more instance of *Escherichia coli* bacteria correctly compared to DT, but the rate of misclassification of other bacteria is the same as for DT.

Table 5 shows the confusion matrix, accuracy, precision, and recall results, obtained by Logistic Regression. Compared to the other three classifiers, it achieved the lowest classification accuracy. However, it achieved a higher precision rate than DT classifier, which is of higher importance in detecting the *Escherichia coli* bacteria.

Finally, the best performance, in terms of all performance evaluators was achieved by Artificial Neural Networks. In total, it misclassified 11 instances from 162 instances in total. It achieved high precision and recall values, which means that it is capable of recognizing *Escherichia coli* bacteria among all other bacteria with a very high rate.

As it can be seen from Tables 2, 3, 4, 5 and 6, ANN method gives the best results when considering all three statistical indices. These results show that it can be applicable in the detection of *Escherichia coli* bacteria among the mentioned bacteria with the 93.21%, 93.30%, and 93.20% for accuracy, precision, and recall measures, respectively. The performance of the ANN depends much on the values of weights, with the aim to minimize the prediction error. Therefore, the weight values were selected, considering the accuracy, recall and prediction as the fitting functions. The characteristics of the classifiers and the proper selection of the parameters determine the classifier's performance.

In our work up to now, we have used fuzzy logic to detect UTI [19] and we have used machine learning algorithms to predict the biofilm category of causative agents of UTI [20]. On the other hand, studies conducted on antibiotic resistance up now were directed towards prediction of the resistance patterns where the input data was mainly based upon resistance genes as in the papers of [2–9]. The aim of our work was to exploit phenotypic data on susceptibility testing using Bauer Kirby disk diffusion method, that is available in vast amounts, to predict the bacterial species that are the causative agents of UTI, as urine is the most submitted specimen in microbiology laboratories. The contribution to the field is that we utilized the reverse logic compared to studies done up to now, where the output data was the bacterial resistance/susceptibility pattern to antibiotics [21], whereas in our work the susceptibility profile was only the input data to predict the species as the final output. Further on in this study we employed logic utilized by microbiologists for species identification in cases where species identification is not in accordance with the obtained antibiogram and broadened it with the aim to shorten the time needed to carry out diagnostic procedures. The drawback of our study was a low number of samples, since only samples with an antibiogram where 12 antibiotics were tested were taken into consideration. Hence in our future work we aim to increase the number of samples and include also samples where less than 12 antibiotics were routinely tested, as well as separate data into categories depending on the fact if the patient is hospitalized or if it is an out-patient since the resistance patterns differ to a great extent in these ecological niches.

Overall, this study only opens a plethora of possibilities of the application of ML techniques on microbiological databases to upgrade the field of diagnostics in medicine in general.

4 Conclusion

Phenotypic testing of bacterial susceptibility to antibiotics yields vast data where the information is further utilized by physicians to ordinate therapy. The obtained data can be further exploited using machine learning techniques to shorten the time needed to identify bacteria species and in doing so contribute to the diagnostics in the field. This is of great value especially in infections where the symptoms are severe and timely application of antibiotics is of great value for the patient but also to prevent the infection from ascending into the higher portions of the urinary system.

The results of this study indicate that using Artificial Neural Network, it is possible to achieve an accuracy, precision, and sensitivity above 93% for the identification of *E. coli,* amongst other causative agents of UTI, which is very promising considering that the results of antibiotic susceptibility testing were only used as input parameters. From this it is evident that microbiology is a field where machine learning techniques can be applied to different databases and obtain results which will aid the diagnostics in the field of medicine.

References

1. Bush, K., Courvalin, P., Dantas, G., Davies, J., Eisenstein, B., Huovinen, P., Lerner, S.A.: Tackling antibiotic resistance. Nat. Rev. Microbiol. **9**(12), 894–896 (2011)
2. Bulgakova, V.G., Vinogradova, K.A., Orlova, T.I., Kozhevin, P.A., Polin, A.N.: Action of antibiotics as signalling molecules. Antibiotiki i khimioterapiia = Antibiot. chemother. [sic], **59**(1–2), 36 (2014)
3. Davies, J.: Are antibiotics naturally antibiotics? J. Ind. Microbiol. Biotechnol. **33**(7), 496–499 (2006)
4. Todar, K.: Bacterial resistance to antibiotics (page 3). Todar's Online Textbook Bacteriol. 4 (2011)
5. Su, M., Satola, S.W., Read, T.D.: Genome-based prediction of bacterial antibiotic resistance. J. Clin. Microbiol. **57**(3 (2019)
6. Van Camp, P-J., Haslam, D.B., Porollo, A.: Prediction of antimicrobial resistance in gram-negative bacteria from whole-genome sequencing data. Front. Microbiol. **11**, 1013 (2020)
7. Kim, J., Greenberg, D.E., Pifer, R., Jiang, S., Xiao, G., Shelburne, S.A., Zhan, X.: VAMPr: variant mapping and prediction of antibiotic resistance via explainable features and machine learning. PLoS Comput. Biol. **16**(1), (2020)
8. Bradley, P., Gordon, N.C., Walker, T.M., Dunn, L., Heys, S., Huang, B., Piazza, P.: Rapid antibiotic-resistance predictions from genome sequence data for Staphylococcus aureus and Mycobacterium tuberculosis. Nat. Commun. **6**(1), 1–15 (2015)
9. Chowdhury, A.S., Call, D.R., Broschat, S.L.: Antimicrobial resistance prediction for gram-negative bacteria via game theory-based feature evaluation. Sci. Rep. **9**(1), 1–9 (2019)
10. Vandepitte, J., Engbaek, K., Rohner, P., Piot, P., Heuck, C.C., et al.: Basic laboratory procedures in clinical bacteriology/ J. Vandepitte … [et al, 2nd ed. World Health Organization. (2003). https://apps.who.int/iris/handle/10665/42696
11. The European Committee on Antimicrobial Susceptibility Testing. Breakpoint tables for interpretation of MICs and zone diameters, version 9.0 (2019)
12. Berrar, D., Cross-validation. Encycl. Bioinform. Computat. Biol. **1**, 542–545 (2019)
13. Top 10 Machine Learning Algorithms - Is AI Scary in 2020?. https://techjury.net/blog/machine-learning-algorithms/. Accessed 30 Dec 2020

14. Maimon, O., Rokach, L.: Data Mining and Knowledge Discovery Handbook. Springer Science + Business Media, New York (2005)
15. Quinlan, J.R.: C4.5: Programs for Machine Learning, Morgan Kaufmann Publishers, Inc., Burlington (1993)
16. Krose, B., van der Smagt, P.: An Introduction to Neural Networks, 8th ed., University of Amsterdam (1996)
17. Breiman, L.: Random forests. Mach. Learn. **45**(1), 5–32 (2001)
18. Sulistyaningrum, D.R., Rasyida, A., Setiyono, B.: Rice disease classification based on leaf image using multilevel support vector machine (SVM). J. Phys.: Conf. Ser. **1490** 012053 (2020). https://doi.org/10.1088/1742-6596/1490/1/012053
19. Ibrisimovic, M.A., Karlı, G., Balkaya, H.E., Ibrisimovic, M., Hukic, M.: A fuzzy model to predict risk of urinary tract infection. In: CMBEBIH 2017, pp. 289–293. Springer, Singapore (2017)
20. Avdić, M., Mašetić, Z., El Sayed, A., Odobašić, L., Hukić, M.: A novel approach in determination of biofilm forming capacity of bacteria using random forest classifier. In International Conference on Medical and Biological Engineering, pp. 273–279. Springer, Cham (2019)
21. Deumić, A., Imamović, E., Ramić, I., Pokvić, L.G., Avdić, M., Jakovac, S., Hukić, M.: Decision support system based on artificial neural network for prediction of antibiotic sensitivity of causative agents of urinary tract infection in certain geographical regions. In: European Medical and Biological Engineering Conference, pp. 314–323. Springer, Cham, November 2020

Application of Artificial Intelligence Techniques to Predict Effects of Cigarette Smoking on Hematological Parameters and Attributable Diseases

Krekić Nejra, Kruško Lejla(✉), Kubat Nermana, Korić Amina, Lauš Božana, Lelo Amina, and Badnjević Almir

Faculty of Pharmacy, University of Sarajevo, Sarajevo, Bosnia and Herzegovina
lejlakrusko@ffsa.unsa.ba

Abstract. Consequences of cigarette smoking, on human health are significant and, in many cases, fatal. Smoking affects all hematological parameters acutely and chronically. The objective of this study was to determine the magnitude of the adverse effects of smoking cigarettes on the biochemical properties of blood in the male population from age 18 to 55 by using artificial intelligence techniques such as artificial neural network. A total of 300 healthy adult male subjects were enrolled in this study including smokers ($n = 200$) and non-smokers as control ($n = 100$) in the age range from 18 to 55 years. Database for artificial neural network was created by using six parameters: age, WBC-White blood cells, RBC-Red blood cells, HB-Hemoglobin, HTC-Hematocrit and MCVMean Corpuscular Volume. The smokers had higher levels of WBC (white blood cells), RBC (red blood cells), HB (hemoglobin), HCT (hematocrit) and MCV (mean corpuscular volume). Our results have shown that continuous cigarette smoking has significant effects on hematological parameters (e.g. hemoglobin, hematocrit, WBC count) and these alterations could be associated with a higher risk for developing atherosclerosis, polycythemia vera, chronic obstructive pulmonary disease and/or cardiovascular diseases.

Keywords: Hematological parameters · Blood cells count · Artificial neural network · Cigarette smoking

1 Introduction

Cigarette smoking is one of the greatest avoidable causes of premature death and disability in the world. It is the most common form of tobacco use worldwide, but all forms of tobacco are harmful, and there is no safe level of exposure to tobacco [1]. Smoking is also a major cause of atherosclerotic disease and is considered one of the three major risk factors for coronary heart disease along with high blood pressure and cholesterol disorders. A cigarette smoker is exposed to a number of harmful substances including nicotine, free radicals, carbon monoxide and other gaseous products. All these substances potentially affect atherogenesis and thrombosis. Moreover, cigarette smoking

A. Badnjevic and L. Gurbeta Pokvić (Eds.): CMBEBIH 2021, IFMBE Proceedings 84, pp. 313–318, 2021.
https://doi.org/10.1007/978-3-030-73909-6_36

accelerates pathogenesis in different type of cancers such as lung, pancreas, breast, liver and kidney [2]. Similarly, it also enhances pH in stomach that resulted in peptic ulcers and gastric diseases and also plays an inhibitory role on ulcer healing [3]. The nicotine induces formation of a clot in the coronary arteries, it weakens the vascular activity, and increases endothelium dysfunction. Increase in the level of carboxy-hemoglobin may cause hypoxia, and it is also responsible for subendothelial edema considering that it alters the vascular permeability and accumulation of lipids. Smoking has both acute and chronic effect on hematological parameters [4]. A correlation was established between smoking, and white blood cell (WBC) counts. Relatively higher WBC counts were detected in smokers and also smoking has been suggested to increase the levels of hematological parameters as hemoglobin (Hb) concentration and red blood cell (RBC) [5]. Some scientists suggested that increase in hemoglobin level in blood of smokers could be a compensatory mechanism. However, some were of view that smoking does not increase in hemoglobin level in all smokers and this relates to tolerance potential of individual to different kind of diseases. Moreover, episodic duration of smoking and age of individual might have changed the adverse effects of smoking on blood characteristics of human being [6]. However, common methods that are used to collect and compare data on changes in hematological parameters might not be trustworthy enough. That being said, we tried to collect and analyze data by using artificial intelligence technique such asartificial neural network(ANN). By definition an ANN is a mathematical representationof the human neural architecture, reflecting its "learning" and "generalization" abilities. ANNs are widely applied in research because they canmodel highly nonlinear systems in which the relationship among the variables is unknown or very complex [7].

Due to the substantial plasticity of input data, ANNs have proven useful in the analysis of blood and urine samples of diabetic patients, diagnosis of tuberculosis, classification of epileptic seizures or even classification of classification of Patau, Edwards, Down, Turner and Klinefelter Syndrome [8, 9].

The application of artificial intelligence (AI) in hematology it not new at all. However, it increasingly becomes part of the measurement of hematological parameters and subsequently also influences decisionmaking [10]. In the field of hematology the first devices based on AI have been applied to the routine laboratory data management. New tools concern the differential diagnosis in specific diseases, based on neural networks trained with data from peripheral blood analysis [11].

The aim of the present study was to assess the extent of adverse effects of cigarette smoking on biochemical characteristics of blood in male population and attributable diseases using artificial neural network.

2 Materials and Methods

Neural network presented in this study is developed based on the dataset which consisted of of 300 healthy adult male subjects, including smokers (n = 200) and non-smokers as control (n = 100) in the age range from 18 to 55 years.

The smokers were regularly consuming minimum of 10 cigarettes per day for at least 3 years. The clinical data, medical history and other relevant information were collected from subjects by personal interview. Information on smoking habits and the amount of tobacco consumed were collected by a selfcontrolled survey to be filled in by the participants. Informed consent was acquired from all the participants, and the study was approved by ethical committee of the university, prior to the start of study. The enrolled subjects did not have any serious health problems, nor history of drug use. None of the subjects have donated or received blood in last 6 months.

The blood samples were collected in the morning time from 8.00 to 9.00 am after an overnight fasting for at least 10 h. Venous blood samples were drawn in K3 EDTA (15%) Becton Dickinson Vacuum tubes and mixed gently. Complete blood cell count (CBC) was analyzed by CELL-DYN 3700 fully automatic hematological analyzer. The Hematological analyzer was calibrated by standardized commercially available calibrators' kit.

The distribution of the dataset for development of the Artificial Neural Network (ANN) is presented in Table 1. Since this is relatively small dataset, distribution of 70:30% for training and validation was used [12, 13]

Table 1. Dataset distribution

Total number of samples 300	Healthy 100	200
Training	70	140
210		
Validation 90	30	60

According to these results, specificity, sensitivity and accuracy of the network are calculated according to Eqs. (1), (2) (3):

$$sensitivity = \frac{true\ positive}{(true\ positive + false\ negative)} \tag{1}$$

$$specificity = \frac{true\ negative}{(true\ negative + false\ positive)} \tag{2}$$

$$accuracy = \frac{(true\ negative + true\ positive)}{(true\ negative + true\ positive + false\ negative + false\ positive)} \tag{3}$$

Artificial Neural Network (ANN) was developed to detect diseases based on hematological parameters. Developed ANN has 6 parameters: age, WBCwhite blood cells, RBC- red blood cells, HB- hemoglobin), HCT- hematocrit and MCV-mean corpuscular volume.

3 Results

According to previous research and expert recommendation [14], two-layer feed forward artificial neural network, with sigmoid transfer function in hidden layer was developed. During development, different number of neurons was tested. The results of training for different number of neurons in hidden layer are presented in the Table 2 (Fig. 1).

Table 2. Training and validation of neural network

Number of neurons	MSE
5 neurons	0.0420
15 neurons	**0.0295**
25 neurons	0.0230

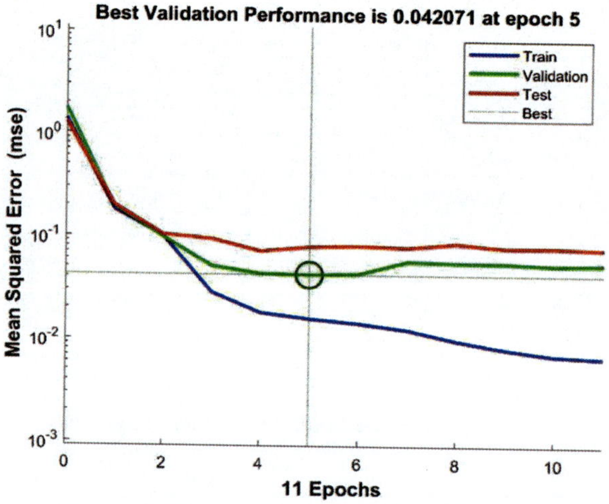

Fig. 1. Training performance with 15 neurons

Based on the training results, ANN with 15 neurons in hidden layer was chosen for diagnostics of Effects of cigarette smoking on hematological parameters and attributable diseases. Results of validation are presented in Table 3.

These results indicate that in 90% of patients (smokers) the diseases caused by effects of smoking on hematological parameters will be predicted by artificial neural network (ANN). Specificity values show that 93.3% of non-smokers will not have these diseases caused by smoking. The total accuracy of this artificial neural network for diagnosing diseases caused by smoking is 92.2%. These results show that this expert system can be used as an aid to healthcare professionals for maximum benefit to patients and the health-care system. The results obtained in this research follow the good practices established in previous similar studies [15–21].

Table 3. Artificial Neural Network with 15 neurons performance

n = 90	**Group 1 Non smokers** ANN prediction	**Group 2 Smokers** ANN prediction	
Group 1 Non smokers 30	27	3	**Accuracy** [%] 92.2
Group 2 Smokers 60	4	57	
	Sensitivity [%] 90	**Sensitivity** [%] 90	

4 Conclusion

Our finding show that cigarette smoking has effects on hematological parameters. WBC, RBC, HB, HCT and MCV counts are significantly higher in smokers. This study suggests that these biomarkers might be associated with a greater risk for developing atherosclerosis, polycythemia vera or cardiovascular diseases in smokers as compared to non-smokers.

The implications of these relationship range from helping to guide the medical work-up in a smoker with leukocytosis or erythrocytosis to broadening our understanding of how smoking leads to atherosclerosis, polycythemia vera, chronic obstructive pulmonary disease and cardiovascular disease. Reduction in smoking does improve measures of some of these biomarkers, demonstrating that these biomarkers are sensitive to change in smoking intake.

References

1. Kume, A., Kume, T., Masuda, K., Shibuya, F., Yamzaki, H.: Dose-dependent effect of cigarette smoke on blood biomarkers in healthy volunteers: observations from smoking and nonsmoking. J. Health Sci. **55**(2), 259–264 (2009)
2. Gitte, R.N.: Effect of cigarette smoking on plasma fibrinogen and platelet count. Asian J. Med. Sci. **2**, 181–184 (2011)
3. Li, L.F., Chan, R.L., Lu, L., Shen, J., Zhang, L., Wu, W.K., Wang, L., Hu, T., Li, M.X., Cho, C.H.: Cigarette smoking and gastrointestinal diseases: the causal relationship and underlying molecular mechanisms (review). Int. J. Mol. Med. **34**(2), 372–380 (2014). https://doi.org/10.3892/ijmm.2014.1786. Epub 2014 May 22 PMID: 24859303
4. Malenica, M., Prnjavorac, B., Bego, T., et al.: Effect of Cigarette smoking on haematological parameters in healthy population. Med. Arch. **71**(2), 132–136 (2017). https://doi.org/10.5455/medarh.2017.71.132-136
5. Inal, B., Hacıbekiroglu, T., Cavus, B., Musaoglu, Z., Demir, H., Karadag, B.: Effects of smoking on healthy young men's hematologic parameters. North ClinIstanb. **31**(1), 19–25 (2014). https://doi.org/10.14744/nci.2014.39974.PMID:28058297;PMCID:PMC5175019

6. Asif, M., Karim, S., Umar, Z., Malik, A., Ismail, T., Chaudhary, A., Al-Qahtani, Mohammed, R.M.: Effect of cigarette smoking based on hematological parameters: comparison between male smokers and non-smokers. Turkish J. Biochemistry/Turk BiyokimyaDergisi. **38** (2013). https://doi.org/10.5505/tjb.2013.68077
7. Amato, F., López-Rodríguez, A., Peña-Méndez, E.M., Vaňhara, D., Hampl, P., Josef, A.H.: Artificial neural networks in medical diagnosis. J. Appl. Biomed. 11. 47–58 (2013). https://doi.org/10.2478/v10136-012-0031-x
8. RijadSarić, D., Beganović, N., Pokvić, L.G., Badnjević, A.: FPGAbased real-time epileptic seizure classification using artificial neural network, Biomed. Sign. Process. Control, **62** (2020). 102106, ISSN 1746-8094
9. Catic, A., Gurbeta, L., Kurtovic-Kozaric, A., Mehmedbasic. S., Badnjevic, A.: Application of neural networks for classification of Patau, Edwards, down, turner and klinefelter syndrome based on first trimester maternal serum screening data, ultrasonographic findings and patient demographics. BMC Med. Genomics **11**(19) (2018). https://doi.org/10.1186/s12920018-0333-2
10. Zini, G.: Artificial intelligence in hematology. Hematology. **10**(5), 393–400 (2005). https://doi.org/10.1080/10245330410001727055. PMID:16203606
11. Kaestner, L.: Artificial intelligence meets hematology. TransfusApher Sci. **2**, 102986 (2020). https://doi.org/10.1016/j.transci.2020.102986 Epub ahead of print. PMID:33153901
12. Rau, S-H., Hsu, C-Y., Lin, Y-A., Suleman, A., Fuad, A., Wei, L-M., Hsu, M-H.: Development of a web-based liver cancer prediction model for type II diabetes patients by using an artificial neural network. Comput. Meth. Programs Biomed. **125**, 58–65 (2016). ISSN 0169-2607
13. Anifowose, F., Khoukhi, A., Abdulraheem, A.: Investigating the effect of training–testing data stratification on the performance of soft computing techniques: an experimental study. J. Exper. Theor. Artif. Intell. **29**(3), 517–535 (2017). https://doi.org/10.1080/0952813X.2016.1198936
14. Aggarwal, C.: Book: Neural Networks and Deep Learning. Springer https://link.springer.com/book/. https://doi.org/10.1007/978-3-319-94463-0
15. Almina, S., et al.: Review of Artificial Intelligence application in cardiology. In: 9th Mediterranean Conference on Embedded Computing MECO 2020
16. Badjei., A., Gureta, L., Cifrek, M., Marjaoi, D.: Diagnostic of Astha using fuzzy rules implemented in accordance with international guidelines and physicians experience. In: IEEE 39th International convention on information and communication technology, electronics and microelectronics (MIPRO), 30 May to 03. June Opatija, Croatia (2016)
17. Badnjevic, A., Gurbeta, L., Custovic, E.: An expert diagnostic system to automatically identify asthma and chronic obstructive pulmonary disease in clinical settings. Sci. Rep. **8**(1), 1–9 (2018)
18. Badnjević, A., Pokvić, L., Gurbeta, H.: Mehrija, B., Lejla, M., Zerina, K., Živorad, K., Jasmin, P.L.: Evidence-based clinical engineering: Machine learning algorithms for prediction of defibrillator performance, Biomed. Sign. Process. Control, **54**, 101629 (2019)
19. Živorad, K., et al.: Prediction of medical device performance using machine learning techniques: Infant incubator case study Health Technol
20. Rijad, S., et al.: FPGA-based real-time epileptic seizure classification using artificial neural network, Biomed. Sign. Process. Control. **62**, 102106 (2020). ISSN 1746-8094. https://doi.org/10.1016/j.bspc.2020.102106
21. Mustafić, L., et al.: Diagnosis of severe aortic stenosis using implemented expert system. Badnjevic, A., Škrbic, R., Gurbeta, PL. (eds.) CMBEBIH 9. CMBEBIH 9. IFMBE Proceedings, vol 73. Springer, Cham (2020)

Health Informatics

Comparison of the Wavelet Denoising Methods for Denoising of Phonocardiogram Signal

Dželila Mehanović(✉), Zerina Mašetić, Dino Kečo, and Jasmin Kevrić

International Burch University, Sarajevo, Bosnia and Herzegovina
dzelila.mehanovic@ibu.edu.ba

Abstract. Heart diseases are the number one cause of death all over the world. Many deaths are caused due to late detection of heart diseases. During the process of heart sound recording, beside heart sound, environment noise is being recorded too. In this work signal denoising was performed using wavelet denoising method. Furthermore, parameter values are compared to find the best combination. Signal to noise ratio (SNR), mean square error (MSE), root mean square error (RMSE) and peak signal to noise ratio (PSNR) were calculated to evaluate results. The highest mean value of the signal-to-noise ratio was 85.38 and it is obtained by Daubechies wavelet method with order 16, 'sqwtolog' threshold selection rule and threshold rescaling 'one'. Wavelet denoising is appropriate method for phonocardiogram analysis and could be useful in heart disease detection.

Keywords: Heart disease · Phonocardiogram (PCG) · Signal denoising · Wavelet denoising · SNR

1 Introduction

We live in the era of big data, where huge amounts are generated on daily basis. On the other side, there are many problems related to health that could be solved with a system that is going to interpret data correctly in an efficient manner. One of these problems is heart disease detection.

Term heart disease refers to the number of disorders of the heart. Heart diseases are the number one cause of deaths all over the world and thus the serious problem in healthcare. Around 610 000 people in the United States die from heart disease every year, which is one death in every four deaths. Most of these deaths are caused by a heart attack, while around 735 000 people experience heart attack yearly [1].

In their work [2], authors mentioned the problem of the remote monitoring of the heart rate. Emergencies of the telemedicine brought the need for remote monitoring of the heart rate. Among several solutions, phonocardiogram (PCG) signals seem to be the simplest and cheapest to introduce. Phonocardiogram (PCG) is a mechanical recording of the heart sounds and murmurs. These sounds are results of vibrations caused by the closure of heart valves [3].

Based on collected information about numbers of people with heart disease and deaths caused by it, this is noticeably the area of human life which requires to be improved when it comes to the faster and more reliable heart disease detection.

A. Badnjevic and L. Gurbeta Pokvić (Eds.): CMBEBIH 2021, IFMBE Proceedings 84, pp. 321–329, 2021.
https://doi.org/10.1007/978-3-030-73909-6_37

Because of the fact PCG is among the simplest and cheapest solutions, we decided to see whether it is possible to denoise PCG to apply heart disease detection. Furthermore, wavelet denoising methods are used by many authors for signal analysis. In [4] wavelet denoising is applied on electroencephalogram (EEG) signals, while authors in [5] applied wavelet methods with heart signals to detect disease. In this work we used wavelet denoising methods, that is *wden* [13] function from Matlab to see if it is applicable with PCG signals. After that, we compared results obtained by different parameter values to find the best option.

The rest of the work is organized as follows: Sect. 2 gives an overview of related works. Section 3 provides details about the data used and describes methods applied for signal denoising and noise removal. Results are described in Sect. 4. Finally, we conclude our work in Sect. 5.

2 Literature Review

The problem of phonocardiogram analysis for reliable heart disease detection and noise reduction is addressed in various research studies. Some of those are mentioned here.

Salman et al. [6] compared several signal denoising methods: wavelet transform (WT), total variance (TV) and empirical mode decomposition (EMD). Authors applied Majorization-Minimization algorithm with regularization factor equal to 0.3 to solve optimization problems. It is shown that the EMD method performs better than others because its denoised signal is the most similar to original signal. Also, EMD method results in larger SNR and smaller RMSE and PRD values. For S1 and S2 heart sound signals, values of SNR, RMSE and PRD obtained by EMD are 10.0415 dB, 0.037586 and 31.4716% respectively. Other two methods, WT and TV, resulted in the following values: 9.825 and 7.9689 for SNR, 0.038535 and 0.047716 for RMSE, 32.2665% and 39.9537% for PRD.

Sujadevi et al. [7] presented denoising of large-scale phonocardiogram dataset using variational mode decomposition (VMD). Modes are generated using identified spectral components. As it is discovered in PCG signal decomposition, initial modes contain signal characteristics and rest of the modes contain noise. Performance of the algorithm is tested on large-scale dataset which encompasses multiple databases and different noise conditions are applied. As an example of results, SNR values obtained for PCG signals with normal characteristics are approximately in the range of 17–19, 21–24, 2229, 29–34 when applied sound levels of 5 dB, 10 dB, 15 dB and 20 dB.

Jain and Tiwari [8] presented an adaptive method for threshold value estimation which is based on statistical parameters and non-linear mid function. This function keeps large value coefficients unchanged, sets small coefficients to zero and shrinks moderate coefficient values by the threshold. It has been shown that the proposed mid function gives higher SNR and fit values than hard and soft functions. SNR and fit values are 4.03 ± 0.15 and 85.09 ± 0.21 for the proposed method, and 3.07 ± 0.21, 82.59 ± 0.23 and 3.43 ± 0.25, 62.11 ± 0.41 for soft and hard functions.

Table 1 displays a summary of experiments mentioned above.

In the related work, to denoise a signal, authors tried different methods such as wavelet transform, total variance, variational and empirical mode decomposition, and

Table 1. Overview of related works

#	Authors	Preprocessing methods	Results
1	Amy H. Salman, Nur Ahmadit, Richard Mengko, Armein Z. R. Langi, Tati L. R. Mengko	Wavelet transform, total variance and empirical mode decomposition	SNR = 9.825, 7.9689, 10.0415 dB, RMSE = 0.038535, 0.047716, 0.037586, PRD = 32.2665%, 39.9537%, 31.4716%
2	Sujadevi V.G, Soman K.P, Sachin Kumar S, Neethu Mohan, Arunjith A.S.	Variational mode decomposition	SNR = 17–19, 21–24, 22–29, 29–34 when applied sound levels of 5 dB, 10 dB, 15 dB and 20 dB
3	Mohan, Arunjith A.S.	Non-linear mid function	SNR = 4.03 ± 0.15, 85.09 ± 0.21

nonlinear mid function. The highest SNR values are achieved by non-linear mid function. Also, a variational and empirical mode decomposition outperformed wavelet and total variance methods.

While heart rate is recorded, there are different sources of noise that need to be removed to achieve a valid heartbeat. The main objective of this research is to evaluate performance of wavelet denoising methods and to find the best parameter values to denoise PCG signal.

3 Methodology

3.1 Dataset

The dataset provided by PASCAL Classifying Heart Sound Challenge which is available at [9] is used in this study. Data set contains 312 files in wav format with lengths from 1 s to 30 s. These recordings can contain noise in the last second when the stethoscope is removed from the body, or during recording caused by breathing or by friction between clothes and stethoscope. Sounds could be classified as normal, murmur and extrasystole.

Normal sounds are healthy heart sounds and should have a "lub dub lub dub" pattern where lub is S1 and dub is S2. Murmur sounds contain noise between lub and dub periods or dub and lub periods. This noise can mean some type of heart disease. Extrasystole sounds are heart sounds that contain additional heartbeats or skipped heartbeats. These sounds do not mean presence of heart disease, but they could be one of the symptoms to detect disease on time [9].

Table 2 summarizes the distribution of each sound type in the dataset.

3.2 Signal Denoising

Beside useful signal information, sound can contain noise caused by different sources. Therefore, it is necessary to remove or reduce noisy parts from the data and increase

Table 2. Distribution of sound types in dataset

Sound type	# of instances
Normal	200
Murmur	66
Extrasystole	46

signal power. To denoise signals, we concentrated on the wavelet method [10]. Wavelet is defined as a wave oscillator whose amplitude starts and ends at zero, with the increment in between.

The performance is measured and evaluated through several performance metrics:

a. Signal to noise ratio (SNR)

$$SNR = \frac{P_{signal}}{P_{noise}} \tag{1}$$

Signal to noise ratio represents ratio of signal and noise power in provided signal. The higher value of SNR means better results, since it means that there is more signal than noise. The value one means power of the signal and power of the noise are equal. The negative values mean that noise power is higher than signal power [12].

b. Mean squared error (MSE) represents the average of square errors [12].

$$MSE = n^1 \sum_{k=1}^{n} (Yk - Y'k)^2 \tag{2}$$

c. Root mean square error (RMSE) represents the root of mean square error [12].

$$RMSE = \sqrt{MSE} \tag{3}$$

d. Peak signal-to-noise-ratio (PSNR) represents the ratio of maximum power of signal and noise power (dB) [14].

$$PSNR = 20\log_{10}(\text{MAX PIXEL}) * \log_{10}(\text{MSE}) \tag{4}$$

3.3 Wavelet Denoising

A signal denoising process is presented in this section. Firstly, signal-to-noise-ratio is calculated using Matlab function *snr* [11] which takes a vector of sample data as a parameter and returns signal-to-noise ratio measured in decibels (dB).

From Fig. 1, it can be noticed that SNR value is less than zero, which means it is necessary to perform signal denoising to remove noise and improve signal quality. To denoise signal, Matlab function *wden* is used with the following parameters: sample data, threshold selection rule, type of thresholding, threshold rescaling, level of wavelet decomposition and name of the wavelet.

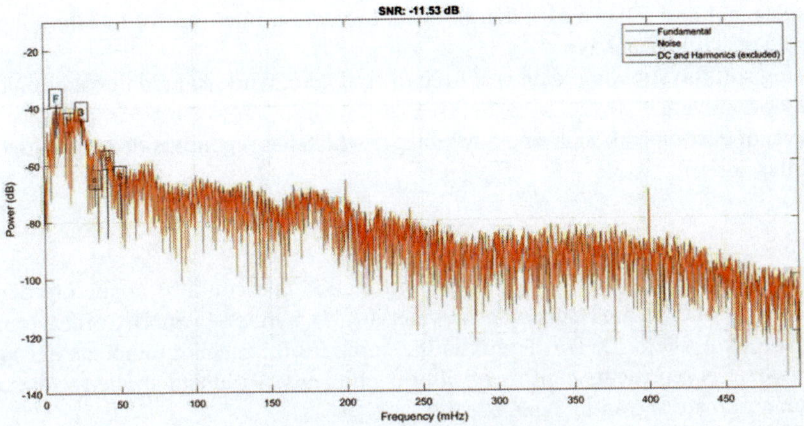

Fig. 1. SNR of the noisy signal

Table 3. Threshold selection rule

Value	Description
rigrsure	Estimates soft threshold using a quadratic loss function
sqtwolog	2ln (length(y))
heursure	Combines rigrsure and sqtwolog
minimaxi	Used to obtain minimum and maximum mean square error

Table 4. Types of thresholding

Value	Description
s	Soft thresholding
h	Hard thresholding

Table 5. Types of threshold rescaling

Value	Description
one	There is no rescaling applied
sln	Coefficients of the first level are used to estimate the level of the noise and perform rescaling
mln	Estimation of noise level is level dependent and rescaling is performed according to that

Tables 3, 4 and 5 list values that threshold selection rule, type of thresholding and threshold rescaling can have.

Using soft thresholding, data is shifted toward zero, while in hard thresholding data is kept or removed.

Level of decomposition is a positive integer which has recommended value given by the expression (5).

$$\text{fix}(\text{log2}(\text{length}(y))) \tag{5}$$

There are multiple variations of wavelet function that could be applied to denoise signals. Some of them are Daubechies wavelets (dbN), Symlets (symN), Coiflets (coifN), Fejér-Korovkin filters (fkN) where N is the number of vanishing moments except for the fk, where N represents number of filter coefficients. Valid and applied values of N for each group are shown in Table 6.

Table 6. Valid and applied values of N for wavelet denoising families

Type	Value of N	Applied Value of N
Daubechies wavelets	1–45	4, 8, 16, 32, 45
Symlets	2–45	4, 8, 16
Coiflets	1–5	2, 3, 5
Fejér-Korovkin filters	4, 6, 8, 14, 18, 22	4, 8, 22

For all other listed parameters, threshold selection rule, type of thresholding and threshold rescaling, all presented variations were applied. For each of these combinations signal-to-noise ratio, mean square error (MSE), root mean square error (RMSE) and peak signal-to-noise ratio (PSNR) were calculated. These values are calculated using Matlab functions *snr*, *psnr*, *rms* and *imms*.

4 Results and Discussion

Table 7 shows best mean values of SNR, MSE, RMSE and PSNR for normal, murmur and extrasystole datasets. Firstly, the highest SNR values were found. Then the values of other parameters were listed, to present the best combination of parameters for each wavelet method. Finally, mean values of MSE, RMSE and PSNR obtained by those combinations of parameters were found.

The highest mean SNR values for all three types of heart sounds are obtained by db16 wavelet transform method, sqwtolog threshold selection rule, type one of threshold rescaling. Threshold type is interchanged between soft and hard, but after comparing their results, we found out that results are very close to each other, so this parameter does not have a high impact on results.

There were combinations of parameters and methods that obtained very low results, such as fk wavelet denoising methods. Therefore, we believe that processing of many

Table 7. Best results for normal, murmur and extrasystole heart sounds

Wavelet method	Threshold selection rule	Thresholding type	Threshold rescaling	Mean value of SNR	Mean Value of MSE	Mean value of RMSE	Mean value of PSNR	Sound type
db16	**sqtwolog**	**h**	**one**	**85.38**	**0.010128**	**0.006948**	**21.66**	**Normal**
sym16	**sqwtolg**	**h**	**one**	**73.35**	**0.024099**	**0.014957**	**19.21**	**Murmur**
db16	**sqwtolog**	**s**	**one**	**84.99**	**0.010753**	**0.005105**	**21.83**	**Extrasystole**

different combinations of parameters, helped us to not miss the best combinations and to achieve higher results than previous works.

The obtained results are presented in the Table 8 and could be compared to the results presented in previous works listed in Table 1. It is noticeable that we achieved higher values of SNR than those in previous works. In the [8], authors achieved the highest SNR 85.09 ± 0.21, while our highest mean SNR value is equal to 85.38. Again, if we look at SNR values of individual sounds, we were able to find SNR values greater than 85.38 such as 86.56 and higher.

Table 8. Comparison of the obtained results with previous studies

SNR	MSE	RMSE	PSNR
85.38	0.010128	0.006948	21.66

According to the results of the present study and understanding of PCG, the following can be emphasized:

- Wavelet denoising is proved to be suitable for PCG signal analysis and processing.
- The optimal parameters for wavelet method, threshold selection rule, thresholding type and threshold rescaling are obtained by testing various combinations of them, which led to obtaining the quality signal without much noise.

5 Conclusion

In this work, the analysis of a dataset which contains 312 instances divided into three categories: normal, extrasystole and murmur heart sounds was performed. Normal heart sounds are those that are not affected by any kind of heart diseases. Murmur heart sounds represent the presence of heart disease. Extrasystole sounds could indicate the presence of heart diseases as they could contain early symptoms of diseases.

Signal-to-noise ratio values calculated on the initial sounds were below zero. To denoise signal, we applied a wavelet method, more precisely we used the Matlab function *wden* which implements wavelet denoising.

Various combinations of parameter values for this function were tested. Since this function takes four parameters, wavelet method, threshold selection rule, thresholding type and threshold rescaling, and there are multiple variations of them, we had a huge dataset of results for each of these heart sounds.

To evaluate performance, signal-to-noise ratio, peak signal-to-noise ratio, mean square error and root mean square error are calculated for each sound type. After we compared values returned by different combinations of these parameters, we found that the highest mean value of signal-to-noise ratio 85.38 is returned by db16 wavelet method, hard thresholding, sqwtolog threshold selection rule and threshold rescaling parameter equal to 'one'.

Additionally, we were able to conclude that this combination of parameters succeeded in eliminating the noise best. Also, we found out that thresholding type is not an important parameter when we analyze results. Finally, we found an answer to the question if it is possible to denoise phonocardiogram, and we proved that phonocardiogram can be successfully denoised and necessary noises could be removed.

Conflict of Interest Declaration. The authors declare that they have no conflict of interest.

References

1. Heron, M.: Deaths: leading causes for 2017. Natl. Vital Stat. Rep. **68**(6), 1–77 (2019)
2. Samieinasab, M., Sameni, R.: Fetal phonocardiogram extraction using single channel blind source separation. In: 2015 23rd Iranian Conference on Electrical Engineering (2015). https://doi.org/10.1109/iraniancee.2015.7146186
3. Abbas, K., Abbas, A.K., Bassam, R.: Phonocardiography signal processing. Synth. Lect. Biomed. Eng. **4**(1), 1–194 (2009). http://dx.doi.org/10.2200/s00187ed1v01y200904bme031
4. Kevric, J., Subasi, A.: The impact of Mspca signal de-noising in real-time wireless brain computer interface system. Southeast Euro. J. Soft Comput. **4**(1) 2015. http://dx.doi.org/10.21533/scjournal.v4i1.90
5. Ramović, A., Bandić, L., Kevrić, J., Germović, E., Subasi, A.: Wavelet and teager energy operator (TEO) for heart sound processing and identification. In: IFMBE Proceedings, pp. 495–502 (2017). http://dx.doi.org/10.1007/978-981-10-4166-2_76
6. Salman, A.H., Ahmadi, N., Mengko, R., Langi, A.Z.R., Mengko, T.L.R.: Performance comparison of denoising methods for heart sound signal. In: 2015 International Symposium on Intelligent Signal Processing and Communication Systems (ISPACS). (2015). http://dx.doi.org/10.1109/ispacs.2015.7432811
7. Sujadevi, V.G., Soman, K.P., Sachin Kumar, S., Mohan, N., Arunjith, A.S.: Denoising of phonocardiogram signals using variational mode decomposition. In: 2017 International Conference on Advances in Computing, Communications and Informatics (ICACCI) (2017). https://doi.org/10.1109/icacci.2017.8126043
8. Jain, P.K., Tiwari, A.K.: An adaptive thresholding method for the wavelet based denoising of phonocardiogram signal. Biomed. Sign. Process. Control **38**, 388–399 (2017)
9. Classifying Heart Sounds Challenge. http://www.peterjbentley.com/heartchallenge/#taskoverview. Accessed 17 Jan 2021
10. Walnut, D.F.: An Introduction to Wavelet Analysis. Springer Science & Business Media (2013)
11. Signal-to-noise ratio - MATLAB snr - MathWorks Benelux. https://nl.mathworks.com/help/signal/ref/snr.html. Accessed 17 Jan 2021

12. Ai, Q., Liu, Q., Meng, W., Xie, S.Q.: Advanced Rehabilitative Technology: Neural Interfaces and Devices. Academic Press (2018)
13. Automatic 1-D denoising - MATLAB wden. https://www.mathworks.com/help/wavelet/ref/wden.html. Accessed 17 Jan 2021
14. Jude, D., Hemanth, D., Gupta, A., Balas, V.E.: Intelligent Data Analysis for Biomedical Applications: Challenges and Solutions. Academic Press (2019)

Electronic Health Records System for Efficient Healthcare Services

Izabela Mitreska(✉), Ninoslav Marina, and Dijana Capeska Bogatinoska

University for Information Science and Technology "St. Paul the Apostle", Ohrid, Republic of North Macedonia
izabela.mitreska@cse.uist.edu.mk, {rector, dijana.c.bogatinoska}@uist.edu.mk

Abstract. There has been a steady and significant growth of necessity for digitalization of healthcare services. That can be done by implementing different Electronic Health Records systems which will increase the interoperability between health systems from different medical institutions. In this paper, we focus on the possibility of using Electronic Health Records system, which is used by healthcare providers to submit a collection of patient data in order to detect chronically ill patients in an effective and timely manner. The proposed Electronic Health Records system manages different medical data such as diagnoses, therapies, measurements and remarks, available in a digital form to authorized health care providers. The main objective of the system is to allow the patients to review their own medical records and to actively engage them in their own healthcare. A survey is created to examine the rate of satisfaction among different users of the proposed system. The obtained results from the survey satisfied our expectations and proved the efficiency of implementing Electronic Health Records System.

Keywords: Electronic Health Records Systems · Physicians · Patients

1 Introduction

The constantly increasing usage of new technology on a global scale introduced the digitalization of the healthcare services by implementing different Electronic Health Records (EHR) systems. EHR systems are electronic versions of medical and treatment history of patients that improve the health surveillance as well as informing clinical decision making [1]. Possession of the complete picture filled with medical treatment information allows physicians to distinguish chronically ill patients and identify the proper diagnose intended to provide medical treatment. Early intervention of health related issues is fundamental for effective treatment of the medical complications. Granting adequate diagnoses and treatments, medical personnel gain the opportunity to safeguard people's lives in an effective and timely manner.

Over the past years, the usage of software for patient and healthcare services management has been drastically increased [2]. As a consequence, the benefits of existing EHR systems became better understood and the performance gains associated with EHR adoption were clarified.

A. Badnjevic and L. Gurbeta Pokvić (Eds.): CMBEBIH 2021, IFMBE Proceedings 84, pp. 330–338, 2021.
https://doi.org/10.1007/978-3-030-73909-6_38

A key characteristic of the adoption of the EHR systems is the 24/7 data access to the medical records by the medical personnel within a given medical organization. Additionally, if the system has the functionality, medical checkup appointments can be scheduled by the patient himself or herself. By providing access to their own medical records, the patients can review and track their own medical history at any given time. With the implementation of such system we reject the traditional paper-based medical records and instead we embrace digital medical records. That allows for data backups and in case of a disaster or system failure, the medical records can be easily restored. Last but not least, the medical history of each patient can be examined by any physician because such system would guarantee a medical records update instantly. This results with more detailed timeline of the patient medical history and allows better healthcare services. The aforementioned characteristics of EHR systems show the performance gains of implementing such system.

The increased rate of adoption of EHR systems at hospitals demonstrate the efficiency with which patients are treated. The advantages are enormous because hospitals want to deliver quality healthcare for their patients without severe cost overruns. Implementing an EHR system is an effective way to increase demand on patient-centered care and to improve the resident-patient [3].

The overriding reason to use EHR system is to have all of the information we need for patient care at one place, reduce medical errors and increase support for reimbursement for healthcare providers. However, EHR systems can also help different branches of medicine, which will result with more efficient national healthcare system.

On the other hand, EHR systems have some disadvantages which are worth considering. Many medical institutions use different EHR systems which result with different data formats and result with lack of interoperability between those medical institutions. Additionally, physicians have different medical and personal history, and depending on that they might prefer the traditional paper-based medical records over the modern digitalized medical records [4, 5].

The EHR system represented in this paper provides different features for efficient healthcare management, i.e. health tracking, diagnoses, applied therapies, physicians reports etc. The main feature of the proposed system is to increase the interoperability among different medical institutions, to increase the efficiency of the physicians and to actively engage the patients in their own healthcare. The proposed system provides a feature for the patients to log in and to review and track their own medical history at any given time. This allows the patients to instantly view their medical checkup results or medical history instead of waiting hours or even days for the same information to be provided by the physician.

A survey is created to examine the rate of satisfaction of the application User Interface and User Experience among the users, both physicians and patients, of the proposed system in this paper. Moreover, the survey would help us to improve the proposed EHR system by adding additional features.

The structure of the paper is as follows. Section 2 surveys the related concepts on EHR systems. Section 3 explicate the quality of the system, followed by Sect. 4 that describes the system architecture of the proposed EHR system. Using some programming skills, we evaluate the proposed EHR system presented in Sect. 5. Lastly, Sect. 6 contains the discussion of the results from the survey and the paper is concluded in Sect. 7.

2 Related Concepts

The concept of EHR systems has been exploited in many different settings, including both low and middle income countries. The benefits of the EHR systems are significant, but the process of addressing potential challenges were critical for success. The implementation of such systems allow for early intervention as well as detections of all risks to vulnerable populations that will facilitate current health needs and mitigate future risks [1].

Although the EHR systems were implemented in 2001, paper-based records remain to be in use. However, the difference in age, gender, job title, previous computer experience and education level show that EHR systems were quicker, more secure and more accurate over the traditional paper-based records [6].

The usage of the EHR systems greatly increase the precision and comprehensiveness of medical data, which will enhance standards and disease prevention capabilities. Databases consisted of medical records make data to be more easily shared between providers and organizations [7]. Patient perspectives are essential to the success of the implementation and development of the EHR systems. Many discussions related to the EHR systems represent the importance of clarity, simplicity and usability of patient's records, as well as the level of transparency and patient's trustworthiness [8].

The invention of such systems improve the worth and effectiveness of the health care. The high satisfaction of medical data collected in this systems results in the functionalities available for prescribing drugs [9].

EHR systems was considered as a key factor for improving healthcare quality and safety. Physicians and nurses expect many benefits from the use of EHR systems that will improve the positive impact on quality care [10].

3 Why EHR Systems

Permitting different data of the patient to be kept independently of the category of cure that need to be obtained explicate the quality of the system. This provides an opportunity medical personnel to gain data on each occasion and hoard the evidence for their patient after their examination.

EHR systems can be owned by a private company or by a government that offer healthcare services. Their stability and consistence is one of the main reasons to provide satisfaction on the workplace for the medical personnel. Stable system, optimized with efficient workflow and pure data representation provides an opportunity to medical coworkers to be pleased and moreover focused in their work. On the contrary, a slow system with a bad user experience might lead in frustration and workplace disappointment.

4 System Architecture of the Proposed EHR System

The system was developed using *Angular* as a framework, *JavaScript* and *TypeScript* as a programming languages. The user interface is designed using *HTML* and *CSS*. *Firebase* provides *API* for data storage, data retrieving and user authentication. The user interface is developed using *Angular* as a fronted framework and *TypeScript* as a programming language fully supported by *Angular*. The user interface communicates with *Firebase* via the provided *API*. The architecture of the system is represented in the diagram below (see Fig. 1):

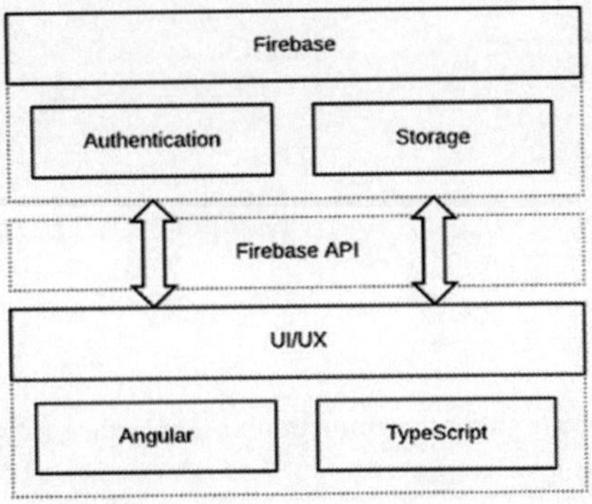

Fig. 1. System architecture of the proposed EHR system

5 Overview of the Proposed EHR System

The EHR system presented in this paper is a digital version of a patient's data.

This system makes it possible for the health data to be available instantly and securely for the registered users. In order to be organized the work of the medical professionals and patients, two types of user and computer interact were classified:

- Physician User Interface
- Patient User Interface

The *Physician User Interface* demonstrates data with the following categorization:

- *Dashboard page* – represents summarized medical information: total number of examinations, total number of patients and total number of diagnoses.
- *New patient page* – popup which comprises a form with: *input elements* – first name, last name, patient unique citizen number (EMBG), date of birth, gender, blood type, address and phone and *two buttons* – submit and cancel.
- *Search Patient page* – single-line box containing dedicated function of accepting the input of the user to be searched for in a database.
- *Patient Profile page* – constitutes the whole examination history for the current patient.

- *Patient examination page* – illustrates a personalized health data of a patient: examination information, remarks, diagnoses, therapies and measurements.
- *New examination page* – popup that holds a form with date that is *disabled* or filled with *default values* – patient, physician and date and *empty input* – department (Fig. 2, 3).

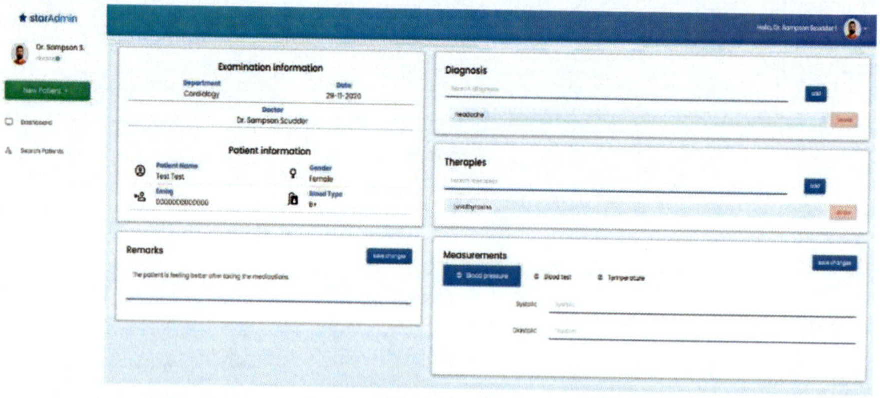

Fig. 2. EHR system – patient examination user interface (physician interface)

- *Patient's Medical History* – were patients could see all information about them.

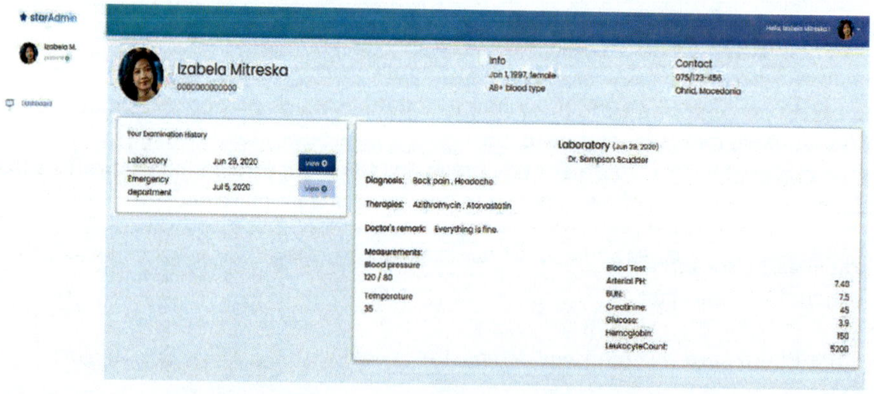

Fig. 3. EHR system – patient profile user interface

The EHR system workflow for patient and physician is shown on the diagram bellow (see Fig. 4):

The *patient user interface* demonstrates data which is classified in one single part:

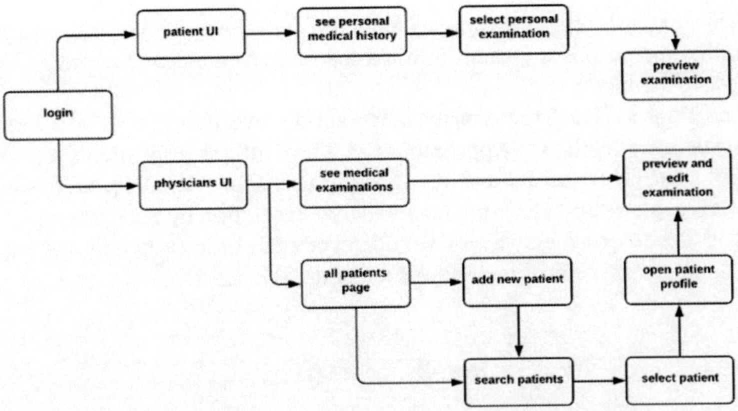

Fig. 4. EHR system workflow for patient and physicians

6 Results

Based on the results from the survey, the proposed EHR system achieved our expectations. The results that we get from the survey prove that the proposed EHR system is perfect medical records system to be implemented in the future.

Fifty-three participants, physicians and patients, took part in the survey. All participants received the survey through their emails and they answered all of the questions which were asked. The survey contains 10 questions related to medical personnel and patients (see Table 1):

Table 1. Questions from the survey

#	Question
1.	Does the system meet the expectations regarding medical practices?
2.	Does the system follow medical standards regarding privacy and efficiency in healthcare?
3.	How easy is the system for daily usage?
4.	How fast do you receive the requested medical records?
5.	How well is the system integrated with the existing work flows?
6.	How likely are you to recommend the system to others?
7.	Do you have any remarks regarding medical practices?
8.	Select the feature that you want to be added in the future
9.	What goals did we help you achieve with the proposed EHR system?
10.	Do you think that the system is useful for small or bigger hospitals?

The first and the second question in the survey focus on how well does the system meet the expectations of the users, physicians and patients, and how well is it following the medical standards regarding privacy and efficiency. Over 75.5% of the respondent's answered that the proposed system fulfilled their expectations and it was following all of the medical standards.

The questions 3, 4 and 5 focus on the meaningful usage, ease of use and the integration with the existing work flows. Approximately, 73.6% of the respondents answered that the system is extremely useful and very easy to use. More than 90% were satisfied by the fast data response and clean medical data representation by the system.

Most of the respondents (87%) would recommend the proposed system to their colleagues and other medical institutions (see Fig. 5):

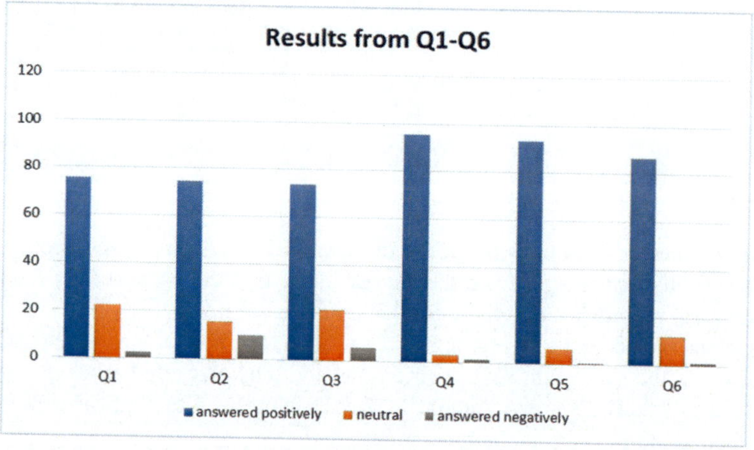

Fig. 5. Results from Q1–Q6

Only 5% of the responses had remarks regarding the responsiveness of the application and usage on smaller devices, such as tablets and smartphones. The most requested feature for implementation in the future is a phone call integration for better communication and implementation of push notifications (see Fig. 6):

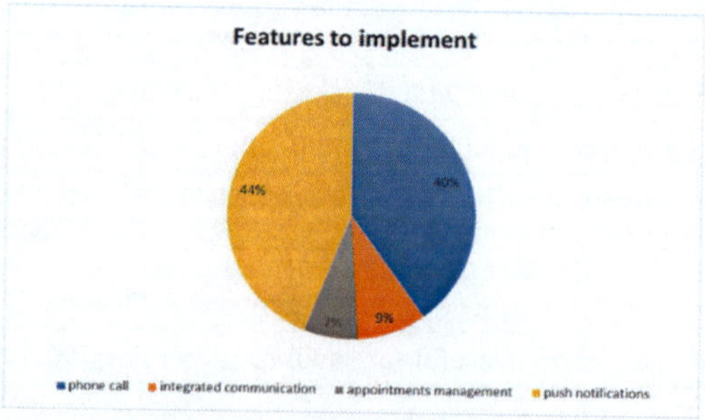

Fig. 6. Features to implement

The last two questions were related to the goals that can be achieved with the proposed system and the general opinion whether the system was useful for small medical institutions, big medical institutions or both. The results showed that the system allowed them to manage the medical records more efficiently. On the other hand 11.3% answered that such system would benefit small medical institutions and 88.7% favored the big medical institutions, such as regional hospitals or clinics (see Fig. 7):

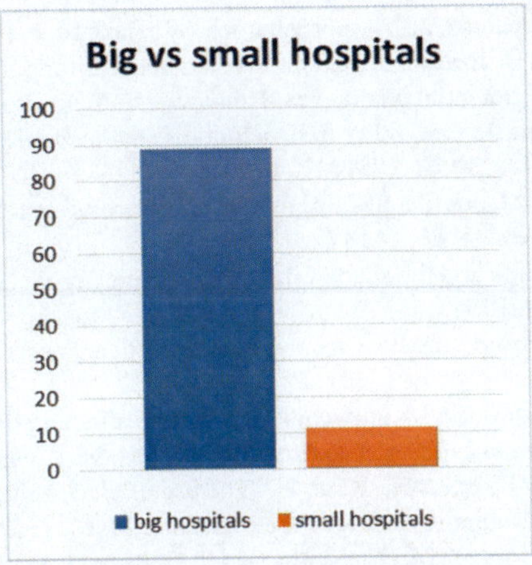

Fig. 7. Big vs small hospitals

Most of the responses were from Macedonia (93.4%). The rest of the responders were from Holland and Germany. All respondents were physicians who worked in small or big hospitals, nurses who took care of the patients in that particular hospital and some of them were patients.

7 Conclusion

The EHR systems made the work of the medical professionals easier and more comfortable. Additionally, the patient records were more accurate and protected due to their sensitive nature. This was the reason why we created this system, which works as intended.

Because of the benefits that the EHR systems provide, many health institutions had adopt some form of such system. However, these systems provide the opportunity for all patients to be part of the process of their healthcare with the help of the medical providers who will help their patients to take the right decision for the benefit of their own health.

The main important step is the creation of the survey which is required to see the rate of satisfaction of the proposed system and also it would help us improve the system by adding different features. Majority of the surveyed medical personnel was satisfied of the proposed system because of the broad specification provided by the system.

In the future, the EHR systems needs to adapt to different changes in the medical industry, i.e. new diseases and new healthcare methods. There are a lot of EHR trends that need to be considered in the upcoming years, especially both, accessibility and standardization.

References

1. Dornan, L., Pinyopornpanish, K., Jiraporncharoen, W., Hashmi, A., Dejkriengkraikul, Angkurawaranon, C.: (2019) Utilization of electronic health records for public health in asia: a review of success factors and potential challenges. Hindawi, BioMed. Res. Int. **9** (2019)
2. Milstein, J.A., Everson, J., Lee, S.D.: EHR adoption and hospital performance: time-related effects. Health Serv. Res. **50**(6), 1751–1771 (2015)
3. Wani, D., Malhota, M.: Does the meaningful use of electronic health records improve patient outcomes? J. Oper. Manage. **60**, 1–18 (2018)
4. Shull, J.G.: Digital health and the state of interoperable electronic health records. JMIR Med. Inform. **7**(4), e12712 (2019)
5. Alpert, J.S.: The electronic medical record in 2016: advantages and disadvantages. Digit. Med. **2**, 48–51 (2016)
6. Msiska, K.E.M., Kumitawa, A., Kumwenda, B.: Factors affecting the utilization of electronic medical records system in Malawian central hospitals. M. M. J. **29**, 247–253 (2017)
7. Kruse, C.S., Stein, A., Thomas, H., Kaur, H.: The use of electronic health records to support population health: a systematic review of the literature. M. M. J. **42**, 214 (2018)
8. Aviran, C.M.H., Brelsford, K.M., McKenna, K.C., Graham, R.D., Lampron, Z.M., Beskow, L.M.: Research use of electronic health records: patients' views on alternative approaches to permission. AJOB Empirical Bioeth. **11**(3), 172–186 (2020)
9. Schopf, T.R., Nedrebq, B., Hufthammer, K.O., Daphu, I.K., Laerum, H.: How well is the electronic health record supporting the clinical tasks of hospital physicians? a survey of physicians at three Norwegian hospitals. BMC Health Serv. Res. **19**, 934 (2019)
10. Benedicits, A.D., Lettieri, E., Gastaldi, L., Masella, C., Urgu, A., Tartaglini, D.: Electronic medical records implementation in hospital: an empirical investigation of individual and organizational determinants. PLOS One, **15**(6), e0234108(2020)

Diagnosis of Iron-Deficiency Anemia Using Artificial Neural Network

Merhunisa Mahir(✉), Lejla Mahmutović, Klara Lovrić, Amna Lenjinac, Benjamin Mahić, and Nerma Mačković

Faculty of Pharmacy, University of Sarajevo, Sarajevo, Bosnia and Herzegovina
merhunisamahir@ffsa.unsa.ba

Abstract. Iron deficiency anemia (IDA) is a type of microcytic anemia, which is presented by the decrease of mean red cell volume (MCV) as a consequence of reduced hemoglobin (Hb) production. When it comes to diagnosing IDA and the decision making process behind it, physicians have to take into account patients disease history, use of supplements, his current state and lab results, which is why we made an artificial neural network (ANN) that can make the diagnosis easier. The dataset for developing the ANN consisted of 400 samples, of which 150 were healthy individuals and 250 had IDA. The results showed that the ANN with 15 neurons in hidden layer was the best for diagnosis of Iron- deficiency Anemia, with high accuracy, specificity and sensitivity. Even though the dataset was relatively small, the results of ANN development were satisfactory, which indicates that with more samples and therefore much bigger datasets, the accuracy of the developed ANN would increase, which would make the diagnosis of IDA faster and easier.

Keywords: Iron-deficiency · Anemia · Artificial Neural Network · Classification

1 Introduction

The fact that over 30% of the world's population is anemic indicates anemia as a growing problem of public health worldwide [1, 2]. The common cause of anemia is iron deficiency (ID) and accounts for 50% of all cases of anemia. Iron-deficiency anemia (IDA) is a type of microcytic anemia because as a consequence of iron deficiency there is a reduced synthesis of heme in the hemoglobin molecule, and thus a decrease in hemoglobin production. Generally, anemias are commonly associated with impaired cognitive function, increased risk of falls in the elderly and increased newborn and maternal mortality. IDA is specially related to chronic kidney disease, chronic heart failure, cancer, and inflammatory bowel disease [3, 4].

The predominant mechanisms for ID are chronic bleedings of various etiologies, most often from the urogenital and digestive tract. The next major cause is iron malabsorption combined with iron-poor diets. Common causes of iron malabsorption include coeliac disease, bypass gastric surgery and *Helicobacter pylori* infection and uncommon causes of malabsorption include gut resection, atrophic gastritis and bacterial overgrowth [4, 5].

Symptoms of IDA are often vague and can be connected with different clinical conditions. Highly suggestive symptoms of IDA include fatigue, lack of endurance and

A. Badnjevic and L. Gurbeta Pokvić (Eds.): CMBEBIH 2021, IFMBE Proceedings 84, pp. 339–346, 2021.
https://doi.org/10.1007/978-3-030-73909-6_39

weakness ascribed to low delivery of oxygen to body tissues and decreased activity of iron-containing enzymes. Important symptoms are also cold intolerance, pagophagia (ice craving), hair loss, and restless legs syndrome. Signs of ID include skin pallor, reduced papillation of the tongue, cheilosis and defects in the nail beds which include Mees lines and spooning of the nails [1, 5, 6].

As symptoms and signs of IDA are often nonspecific, the initial suspicion of IDA usually comes after the results of hematological tests. Early and accurate diagnosis of IDA is important because it can prevent various complications and improve patient's quality of life.

Artificial Neural Networks are defined as computational modeling tools based on the neural structure of the brain [7, 8]. The neural network itself consists of multiple simple processing elements, artificial neurons, arranged in 3 basic types of layers: input, hidden, and output layer. The input layer processes the input data from the system while the output layer is the last layer of the network in which the outputs of neurons are predicting system solutions [7, 8]. The main application of artificial neural networks is searching for dependencies between variables that are not in a linear relationship and predicting a complex relationship between those variables, and the usage of ANN in disease classification is very often nowadays [8–15].

The aim of this research is to develop artificial neural network for diagnosis of anemia caused by lack of iron in blood. The aim of this system is to be used as aid in diagnosis especially in primary healthcare units in rural and remote areas.

2 Materials and Methods

Artificial Neural Network is this study was developed based on database which contains 400 samples. Each sample in this database have following parameters recorded hemoglobin, mean corpuscular hemoglobin (MCH), mean corpuscular volume (MCV), serum ferritin, serum iron, transferrin saturation and serum soluble transferrin receptors.

Table 1. shows the reference values of the abovementioned parameters. If a subject has parameters within these limits then they are considered to be healthy. Any deviation from these parameters is indication for disorder.

Table 1. Parameters used to predict iron deficiency anemia

Parameters	Reference values
Hemoglobin	≥130 g/L (males), ≥120 g/L (females)
Mean corpuscular hemoglobin (MCH)	27–34 pg
Mean corpuscular volume (MCV)	80–95 fl
Serum ferritin	40–300 μg/l (males), 20–200 μg/l (females)
Serum iron	10–30 μg/l
Transferrin saturation	≥16%, 45%≤
Serum soluble transferrin receptors	Low or normal

For a definitive diagnosis of IDA, laboratory evaluation is required. The diagnosis of anemia is made after confirmation of a reduced blood hemoglobin concentration. Anemia is defined as a hemoglobin level <130 g/L in men, but <120 g/L in women, according to the World Health Organization (WHO) [16–18].

A complete blood count (CBC) with red blood cell indices can provide useful and confirmatory data for IDA. Test show reduced mean corpuscular hemoglobin, with values below 27 pg, and reduced mean corpuscular volume (MCV), which corresponds to microcytosis with values below 80 fl [6, 18]. The determination of iron status is somewhat complex and the use of multiple tests that indicate iron status overcomes the limitations of using a single test. In the absence of inflammation, the serum ferritin level is the most specific indicator of IDA and it reflects total body iron stores. The generally accepted cut-off value for serum ferritin level is <30 μg/l [4, 16].

In cases of iron deficiency, serum iron (SI) is reduced below 10 μg/l and total iron-binding capacity (TIBC) is increased, resulting in a considerable reduction in transferrin saturation. Transferrin saturation represents the ratio of SI to TIBC and a transferrin saturation level of less than 16% indicates an iron deficiency.

A recent additional test for diagnosing IDA is the measurement of serum soluble transferrin receptors (sTfR). In the case of IDA, sTfR levels are increased and this test is considered in cases where the diagnosis is unclear. Although thresholds are not clearly defined and values vary according to the method used, the increase of sTfR levels is used to diagnose IDA [3, 17, 18].

Table 2. shows detailed information about the database which was used for development of ANN in this study.

Table 2. Properties of the dataset used for development of artificial neural network

Class	IDA	Healthy	Total
Male	139	84	223
Average age (years)	42,92	40.37	41,65
Female	111	66	177
Average age (years)	42,84	40.46	41,65
Total number of patients per class	250	150	400

The distribution of the dataset for development of the Artificial Neural Network (ANN) is presented in Table 3. Since this is relatively small dataset, distribution of 70:30% for training and validation was used [20].

Table 3. Dataset distribution

Total number of samples	Healthy	Disease
Training = 280	105	175
Validation = 120	45	75
Total = 400	**150**	**250**

Artificial Neural Network classifies input parameters of each patient into two classes: **iron deficiency anemia (disease)** and **no iron deficiency anemia (healthy)**.

3 Results

According to previous research and expert recommendation [21], two-layer feedforward artificial neural network, with sigmoid transfer function in hidden layer was developed. Neural network architecture consists of 9 inputs: gender, age, hemoglobin [g/L], mean corpuscular hemoglobin [pg], mean corpuscular volume [fl], serum ferritin [μg/l], serum iron [μg/l], transferrin saturation [%] and serum soluble transferrin receptors.

During development, different number of neurons were tested. The results of training for different number of neurons in hidden layer are presented in the Table 4.

Table 4. Training and validation of neural network

Number of neurons	MSE	No of iterations for training
5 neurons	1.4855e-10	263
15 neurons	9.1738e-15	14
25 neurons	1.6669e-10	166

Training performance of ANN is presented in Figs. 1–3.

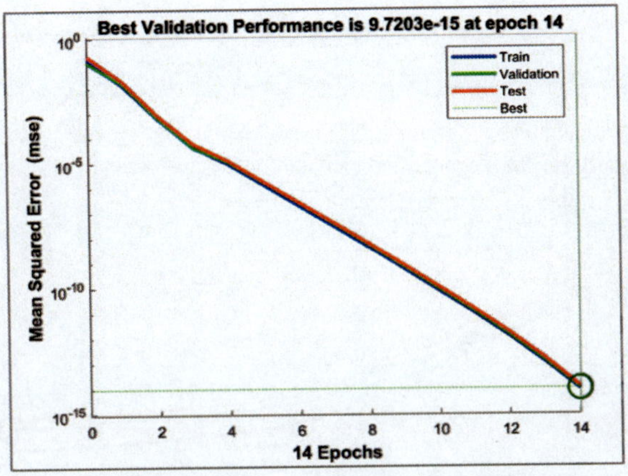

Fig. 1. Training performance with 15 neurons in hidden layer

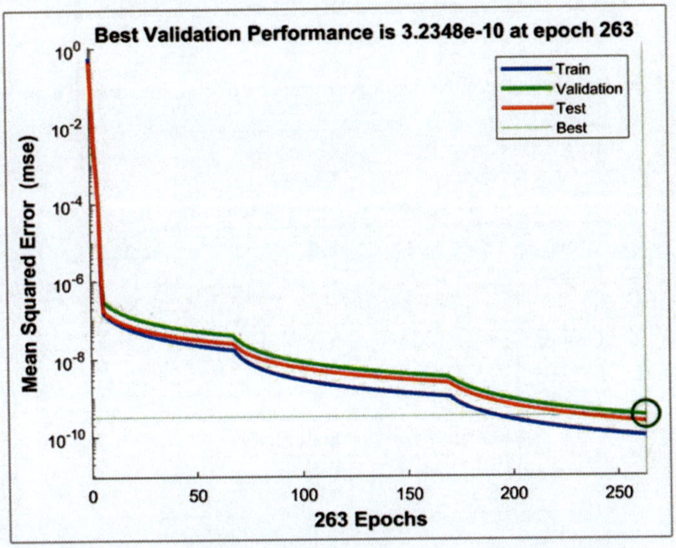

Fig. 2. Training performance with 5 neurons in hidden layer

Based on the training results, ANN with 15 neurons in hidden layer was chosen for Diagnosis of Iron- deficiency Anemia. Results of validation are presented in the Table 5 below.

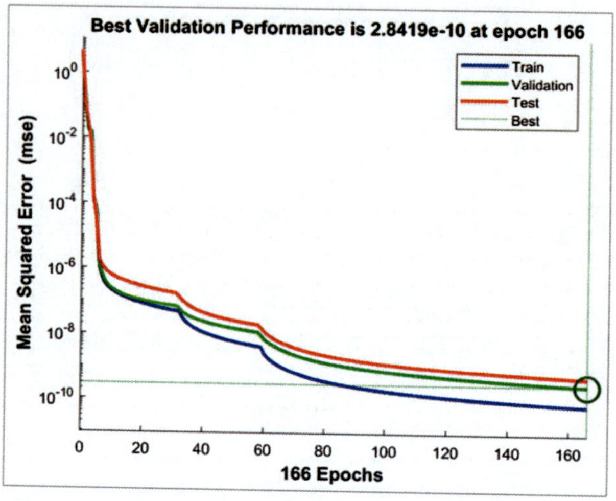

Fig. 3. Training performance with 25 neurons in hidden layer

Table 5. Artificial Neural Network performance in classification to class IDA

n = 120	Group 1 Healthy ANN prediction	Group 2 Disease ANN prediction	
Group 1 Healthy 45	TP 43	FN 2	Accuracy [%] 95.8
Group 2 Disease 75	FP 3	TN 72	
	Sensitivity [%] 95.5	Specificity [%] 96.0	

4 Conclusion

This paper presents the diagnosis of iron deficiency anemia using Artificial Neural Networks. Anemia is a growing problem worldwide, because approximately a third of the world population is anemic. Symptoms of IDA are often vague and nonspecific, so the initial suspicion of IDA usually comes only after the results of hematological tests. Using ANN to develop a device that can be used by physicians for early and accurate prediction of IDA would be a solution to this problem.

The ANN was developed by using 280 samples for training and 120 samples for validation. During the development the ANN with 15 neurons in hidden layer showed the best results, where out of 400 samples, 95,8% were diagnosed correctly, with a specificity of 95.5% and sensitivity of 96%. It is important to mention that this ANN wouldn't be able to substitute a doctor, rather it should be used by one for the best results.

References

1. Auerbach, M., Adamson, J.W.: How we diagnose and treat iron deficiency anemia. Am. J. Hematol. **91**(1), 31–8 (2016)
2. De Franceschi, L., Iolascon, A., Taher, A., Cappellini, M.D.: Clinical management of iron deficiency anemia in adults: systemic review on advances in diagnosis and treatment. Eur. J. Intern. Med. **42**, 16–23 (2017)
3. Hempel, E.V., Bollard, E.R.: The evidence-based evaluation of iron deficiency anemia. Med. Clin. North Am. **100**(5), 1065–75 (2016)
4. Lopez, A., Cacoub, P., Macdougall, I.C., Peyrin-Biroulet, L.: Iron deficiency anaemia. Lancet **387**(10021), 907–16 (2016)
5. DeLoughery, T.G.: Iron deficiency anemia. Med. Clin. North Am. **101**(2), 319–332 (2017)
6. Camaschella, C.: Iron-deficiency anemia. N. Engl. J. Med. **372**(19), 1832–43 (2015)
7. Basheer, I.A., Hajmeer, M.: Artificial neural networks: fundamentals, computing, design, and application. J. Microbiol. Methods **43**(1), 3–31 (2000)
8. Renganathan, V.: Overview of artificial neural network models in the biomedical domain. Bratisl. Lek. Listy **120**(7), 536–540 (2019)
9. Hodzic, J., et al.: Overview of next-generation sequencing platforms used in published draft plant genomes in light of genotypization of immortelle plant (helichrysium arenarium). Med. Arch. **71**(4), 288–292 (2017)
10. Catic, A., et al.: Application of neural networks for classification of patau, edwards, down, turner and klinefelter syndrome based on first trimester maternal serum screening data, ultrasonographic findings and patient demographics. BMC Med. Genomics **11**(1), 19 (2018)
11. Gurbeta, L., et al.: A telehealth system for automated diagnosis of asthma and chronical obstructive pulmonary disease. J. Am. Med. Inform. Assoc. **25**(9), 1213–1217 (2018)
12. Badnjevic, A., et al.: An expert diagnostic system to automatically identify asthma and chronic obstructive pulmonary disease in clinical settings. Sci. Rep. **8**(1), 11645 (2018)
13. Badnjevic, A., et al.: Evidence-based clinical engineering: machine learning algorithms for prediction of defibrillator performance. Biomed. Sign. Process. Control **54**, 101629 (2019)
14. Kovačević, Ž., et al.: Prediction of medical device performance using machine learning techniques: infant incubator case study. Health Technol. **10**(2), 151–155 (2020)
15. Sarić, R., et al.: FPGA-based real-time epileptic seizure classification using artificial neural network. Biomed. Sign. Process. Control **62**, 102106 (2020)
16. Boone, S., Powers, J.M., Goodgame, B., Peacock, W.F.: Identification and management of iron deficiency anemia in the emergency department. J. Emerg. Med. **57**(5), 637–645 (2019)
17. Cappellini, M.D., Musallam, K.M., Taher, A.T.: Iron deficiency anaemia revisited. J. Intern. Med. **287**(2), 153–170 (2020)
18. WHO/UNICEF/UNU. Iron deficiency anaemia assessment, prevention, and control: a guide for programme managers. Geneva, World Health Organization (2001). http://www.who.int/nutrition/publications/en/ida_assessment_prevention_control.pdf. Accessed 25 Nov 2020
19. Rau, H.H., Hsu, C.Y., Lin, Y.A., Atique, S., Fuad, A., Wei, L.M., Hsu, M.H.: Development of a web-based liver cancer prediction model for type II diabetes patients by using an artificial neural network. Comput. Methods Programs Biomed. **125**, 58–65 (2016)

20. Anifowose, F., Khoukhi, A., Abdulraheem, A.: Investigating the effect of training–testing data stratification on the performance of soft computing techniques: an experimental study. J. Exp. Theor. Artif. Intell. **29**(3), 517–535 (2017). https://doi.org/10.1080/0952813x.2016.1198936
21. Aggarwal, C.: Neural Networks and Deep Learning. Springer. https://link.springer.com/book/10.1007/978-3-319-94463-0

Using Artificial Network for Identification of Kidney Cancer

Ajla Turajlić, Hava Turković, Mevlija Tursunović, Amina Vatreš, Nedžma Vehabović, and Minela Viteškić(✉)

Faculty of Pharmacy, University of Sarajevo, Sarajevo, Bosnia and Herzegovina
minelaviteskic@ffsa.unsa.ba

Abstract. Since renal cell carcinoma (RCC) is a very common renal disease and the number of diseases constantly increases, it is very important to prevent this malignity by making an appropriate diagnosis. In this study, artificial neural networks are used to identify kidney cancer based on parameters that have the most detrimental effects on kidney's health: urine volume, urine color, urine sediment, blood in urine, urine pH, creatinine, creatinine clearance, urea, uric acid, LDH, CEA, C3, C4, CH50, haptoglobin, and fibrinogen.

Collected data derives from 300 patients, 100 of them healthy (Group 1) and 200 diseased (Group 2). Group 1 didn't have any abnormality of these parameters. In the case of Group 2, the parameters of every respondent were not normal, which means that the value of parameters is higher or lower than reference values of the same parameters. The result of those abnormalities is kidney cancer.

2, 5, and 15 neurons were tested during the development. An Artificial Neural Network with 15 neurons shows the best results with accuracy training 99,01% and accuracy validation 98,62%. After Artificial Neural Network performance validation accuracy was 93,3%, sensitivity 90%, and specificity 95,0%. Accuracy helps us to separate healthy and diseased and also helps to recognize the stage of the disease. Although the use of ANN has long been present in urology and even diagnostics in kidney diseases, this paper tried to give its contribution in the form of the development of ANN with the best possible parameters of specificity and sensitivity.

Keywords: Renal cell carcinoma · Artificial neural network · Network specificity · Network sensitivity · Network accuracy

1 Introduction

Renal cell carcinoma (RCC) accounts for approximately 90% of all renal malignancies and is the 13th most common malignancy worldwide. In the US, the incidence of kidney cancer increased by an average of 1.1% over 2004–2013, and it was estimated that close to 63,000 new cases were diagnosed in 2016 [1]. The incidence of renal cancer is highest in developed countries with rates generally higher in Europe and North America than in India, Africa, and China [2].

A. Badnjevic and L. Gurbeta Pokvić (Eds.): CMBEBIH 2021, IFMBE Proceedings 84, pp. 347–355, 2021.
https://doi.org/10.1007/978-3-030-73909-6_40

Multiple factors have been implicated in an increased risk of RCC, such as nutrition and diet, but only a few established risk factors have been confirmed. Age is considered to be a risk factor, as the incidences of RCC in Europe and US increase consistently [3]. The estimated risk of RCC for men between the ages of 50 and 59 is around 0.3% compared to 1.3% for older individuals [1]. Age-controlled incidences of RCC also indicate a 2:1 predominance of men compared to women [4].

It is known that the diagnosis of cancer by examination of the urine has potential to improve patient outcomes by means of earlier detection. Due the fact that the urine contains metabolic signatures of many biochemical pathways, this biofluid is ideally suited for metabolomics analysis, especially involving diseases of the kidney and urinary system.

The purpose of this study is to identify kidney cancer using artificial neural network. A similar research method was conducted in February 2018. where the Application of Neural Networks was used to classify Patau, Edwards, Down, Turner and Klinefelter Syndrome based on first trimester maternal serum screening data, ultrasonographic findings and patient demographics [5]. Also, we have another similar research method conduced in August 2018. where an expert diagnostic system was used to automatically identify asthma and chronic obstructive pulmonary disease in clinical settings [6].

Physicochemical parameters that we will observe in our study of healthy and sick subjects are: urine volume, urine color, urine sediment, presence of blood in urine, urine pH, creatinine, creatinine clearance, urea, uric acid, LDH, CEA, C3, C4, CH50, haptoglobin and fibrinogen.

2 Materials and Methods

In a study from November 2020, we collected data from 300 patients. They were grouped into two groups. The first group consisted of 100 healthy individuals, and this group was called healthy control (Group 1). The second group consisted of 200 sick individuals, and they have kidney cancer (Group 2). There are many parameters that can affect the development of kidney cancer, for the purposes of this study we have selected parameters which have the most detrimental effects on kidney's health. Parameters that we used are: urine volume, urine color, urine sediment, blood in urine, urine pH, creatinine, creatinine clearance, urea, uric acid, LDH, CEA, C3, C4, CH50, haptoglobin and fibrinogen.

The parameters were selected in accordance with detailed checks of the latest literature related to the assessment of the clinical condition of the kidney, comparing already used parameters in the design of ANN and other parameters that are important in the clinical diagnosis of kidney disease [7–9].

Group 1 doesn't have any abnormality of these parameters. In case of Group 2, parameters of every respondent was not normal, which means that value of parameters are higher or lower than reference values of the same parameters. Result of those abnormalities is kidney cancer.

In Group 1, 51 respondents were male and 49 were female (100 in total). In Group 2, 102 respondents were male and 98 were female (200 in total). Many studies has shown that risk to get a kidney cancer is higher for males than females.

Table 1. shows reference values for parameters that we used in our study.

Table 1. Reference values

Parameters	Reference value	Min	Average	Max
Urine volume	750–2000 ml	750	1750	2000
Urine color	Colorless-Dark yellow	Colorless	Colorless	Dark yellow
Urine sediment	None	No	No	Yes
Blood in urine	None	No	No	Yes
pH urine	4,5–8,4	4,5	4	8,4
Creatinine	M: 53–97 μmol/l F: 44–80 μmol/l	53 44	101,5 88,5	97 89
Creatinine clearance	M: 100–150 ml/min F: 85–125 ml/min	100	125	150
		85	167,5	125
Urea	1,7–8,3 mmol/l	1,7	5	8,3
Uric acid	M: 214,2–458 μmol/l F: 148,7–404,4 μmol/l	214,1 148,7	443,1 350,9	458 404,4
LDH	< 241 U/l	56	150	270
CEA	< 3,4 μg/l	0,5	3,4	7
C3	0,9–1,8 g/l	0,9	1,05	1,8
C4	0,1–0,4 g/l	0,1	0,25	0,4
CH50	≤ 10 IU/ml	2	7	15
Haptoglobin	0,5–2,2 g/l	0,5	1,35	2,2
Fibrinogen	1,8–3,5 g/l	1,8	3,55	3,5

The distribution of the dataset for development of the Artificial Neural Network (ANN) is presented in Table 3. Since this is relatively small dataset, distribution of 70:30% for training and validation was used [10]. In order to achieve statistically significant results, validation and training were performed on 30% of healthy samples and the remaining 70% on patients with kidney disease. The reason for this distribution is that data for healthy patients is easy to interpret while data for patients with kidney disease shows different variations during training and validation, so more samples were needed Table 2.

During the design of the network, different number of neurons was used in order to measure the performance. According to the Mean Square Error (MSE), appropriate number of neurons was chosen. After training, validation of the network performance was done using confusion matrix.

Table 2. Dataset distribution

Total number of samples 300	Healthy 100	Disease 200
Training	70	140
210		
Validation 90	30	60

According to these results, specificity, sensitivity and accuracy of the network are calculated according to Eqs. (1), (2) (3):

$$sensitivity = \frac{true\ positive}{(true\ positive + false\ negative)} \tag{1}$$

$$specificity = \frac{true\ negative}{(true\ negative + false\ positive)} \tag{2}$$

$$accuracy = \frac{(true\ negative + true\ positive)}{(true\ negative + false\ negative + false\ positive)} \tag{3}$$

3 Literature Review

The use of ANN is wide in the diagnosis of kidney disease, among other things in the diagnosis of kidney tumors. When reviewing the literature data, significant databases such as Google Scholar, EBSCO, Pubmed and Medline were searched. Testing of the proposed algorithms was performed by processing and analyzing scientific studies related to the analysis of significant clinical markers in biological samples [11–13].

4 Results

According to previous research and expert recommendation [14, 15] for classification kidney disease in this paper, two-layer feedforward artificial neural network, with sigmoid transfer function in hidden layer was developed. Neural network architecture consist of 17 inputs:

Sex, Urine volume, Urine color, Urine sediment, Blood in urine, pH urine, Creatinine, Creatinine clearance, Urea, Uric acid, LDH, CEA, C3, C4, CH50, Haptoglobin, Fibrinogen.

It has one output. It classifies patients to two categories: healthy and disease.

During development, different number of neurons was tested. The results of training for different number of neurons in hidden layer are presented in the Table 3 below.

Table 3. Training and validation of neural netowork

Number of neurons	Accuracy training	Accuracy validation
2	97.6190	96.6667
5	98.99	98.76
15	99.01	98.62

Accuracy of training for feedforward neural network with 2 neurons and sigmoid transfer function in hidden layer was 97.6190%. Training was performed using Levenberg – Marquadt training algorithm.

Network performance during training with different number of neurons is shown in the Figures below. The training and validation were done for the second time in this test set Fig. 1.

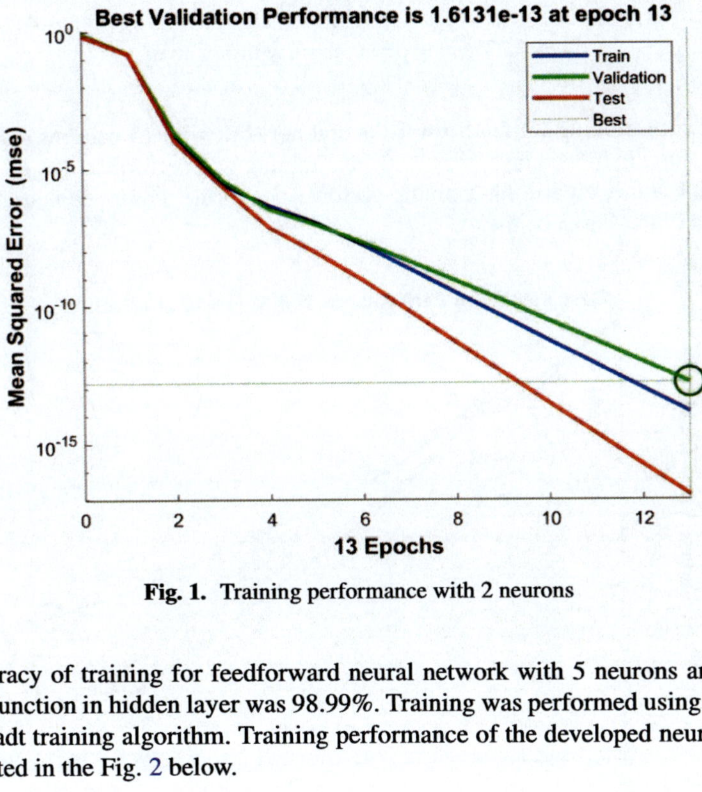

Fig. 1. Training performance with 2 neurons

Accuracy of training for feedforward neural network with 5 neurons and sigmoid transfer function in hidden layer was 98.99%. Training was performed using Levenberg – Marquadt training algorithm. Training performance of the developed neural network is presented in the Fig. 2 below.

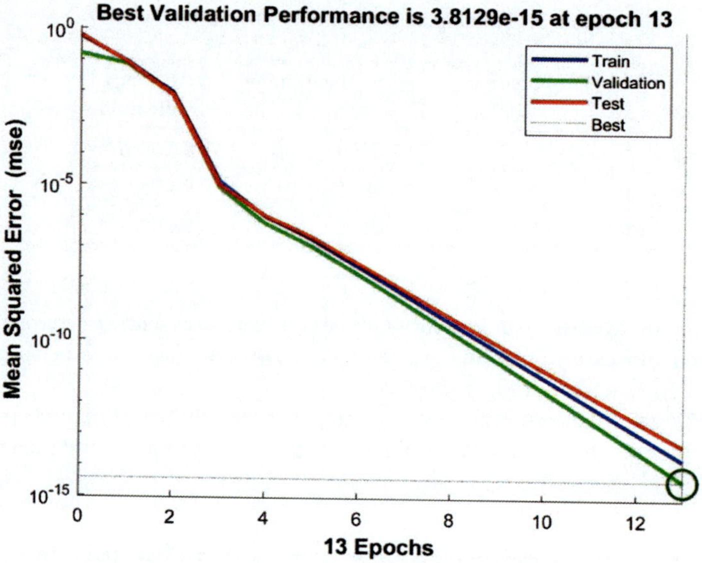

Fig. 2. Training performance with 5 neurons

Accuracy of training for feedforward neural network with 15 neurons and sigmoid transfer function in hidden layer was 99.01%. Training was performed using Levenberg – Marquadt training algorithm. Training performance of the developed neural network is presented in the Fig. 3 below.

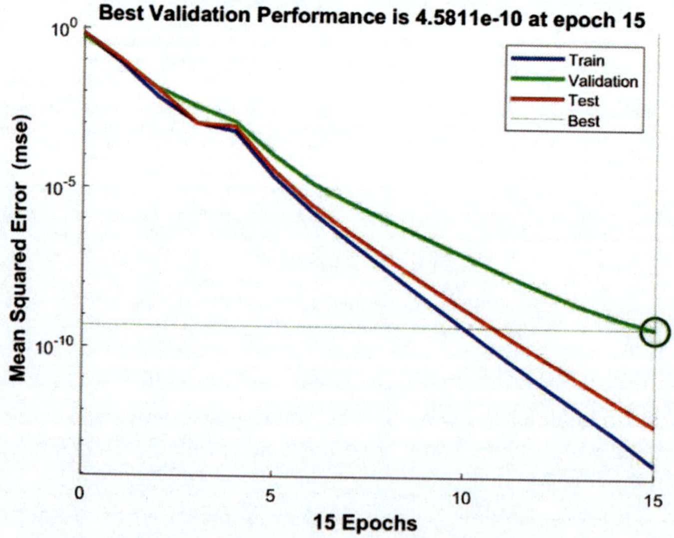

Fig. 3. Training performance with 15 neurons

As it can be seen from the training process, neural network with 15 neurons shows the best results.

Validation of such architecture is shown below (Table 4).

Table 4. Artificial neural network performance validation

n = 90 – number of times validation was performed	Group 1 Healthy ANN predic tion	Group 2 Disease ANN predic-tion	
GROUP 1 Healthy 30	TP 27	FN 3	Accuracy [%] 93.3%
Group 2 Disease 60	FP 3	TN 57	
	Sensitivity [%] 90%	Specificity [%] 95.0	

The results showed better specificity and similar sensitivity compared to the literature data mentioned earlier as shown in Table 5. The information used for this database was collected exclusively for this work from healthy volunteers or with the permission of patients in the period from 2017 to 2019 in Bosnia and Herzegovina, and the authors have an insight into it. Tumor markers CEA, C3, C4, CH50, haptoglobin, and fibrinogen were identified as the most important parameters due to literature data [14, 15].

Table 5. Sensitivity and specificity

	Other ANN	Our ANN
Sensitivity	85%–90%	90%
Speccificity	90%	95%

5 Conclusion

In summary, the diagnosis of cancer by examination of the urine has potential to improve patient outcomes by means of earlier detection. In this study, we used artificial neural network to identify kidney cancer based on parameters, and since the training of neural network with 15 neurons in hidden layer was the highest (accuracy training 99,01% and accuracy validation 98,62%) this neural network architecture is suggested for the usage for prediction of kidney cancer. This is promising results, since such expert system can be used as an aid to healthcare professional at primary healthcare level in situations when medical specialist is not available.

References

1. Siegel, R.L., Miller, K.D., Jemal, A.: Cancer statistics, 2016. CA: Cancer J. Clin. **66**(1),7–30 (2016)
2. Ferlay, J., Shin, H.R., Bray, F., Forman, D., Mathers, C., Parkin, D.M.: Estimates of worldwide burden of cancer in 2008: GLOBOCAN 2008. Int. J. Cancer **127**(12), 2893–2917 (2010)
3. Chow, W.H., Devesa, S.S.: Contemporary epidemiology of renal cell cancer. Cancer J. (Sudbury, Mass) **14**(5), 288–301 (2008)
4. Howlader, N., Noone, A., Krapcho, M., et al.: [Accessed January, 2017]; SEER Cancer Statistics Review, pp. 1975–2013 (2016). http://seer.cancer.gov/csr/1975_2013/
5. Catic, A., et al.: Application of neural networks for classification of patau, edwards, down, turner and klinefelter syndrome based on first trimester maternal serum screening data, ultrasonographic findings and patient demographics. BMC Med. Genomics **11**, 19 (2018). https://doi.org/10.1186/s12920-018-0333-2
6. Gurbeta, L., et al.: A telehealth system for automated diagnosis of asthma and chronical obstructive pulmonary disease. J. Am. Med. Inform. Assoc. **25**(9), 1213–1217 (2018)
7. Niel, O., Bastard, P.: Artificial intelligence in nephrology: core concepts, clinical applications, and perspectives. Am. J. Kidney Dis. **74**(6), 803–810 (2019). https://doi.org/10.1053/j.ajkd.2019.05.020. Epub 2019 Aug 23. PMID: 31451330
8. Biourge, V., Delmotte, S., Feugier, A., Bradley, R., McAllister, M., Elliott, J.: An artificial neural network-based model to predict chronic kidney disease in aged cats. J. Vet. Intern. Med. **34**(5), 1920–1931 (2020). https://doi.org/10.1111/jvim.15892. Epub 2020 Sep 7. PMID: 32893924; PMCID: PMC7517863
9. Rashidi, P., Bihorac, A.: Artificial intelligence approaches to improve kidney care. Nat. Rev. Nephrol. **16**(2), 71–72 (2020). https://doi.org/10.1038/s41581-019-0243-3. PMID: 31873197; PMCID: PMC7591106
10. Rau, H.H., Hsu, C.-Y., Lin, Y.A., Atique, S., Fuad, A., Wei, L.M., Hsu, M.H.: Development of a web-based liver cancer prediction model for type II diabetes patients by using an artificial neural network. Comput. Methods Programs Biomed. **125**, 58–65 (2016). ISSN 01692607, https://doi.org/10.1016/j.cmpb.2015.11.009
11. Checcucci, E., Autorino, R., Cacciamani, G.E., Amparore, D., De Cillis, S., Piana, A., Piazzolla, P., Vezzetti, E., Fiori, C., Veneziano, D., Tewari, A., Dasgupta, P., Hung, A., Gill, I., Porpiglia, F.: Urotechnology and some working group of the young academic urologists working party of the european association of urology. Artificial Intell. Neural Netw. Urol. Curr. Clin. Appl. Minerva Urol. Nefrol. **72**(1), 49–57 (2020). https://doi.org/10.23736/s0393-2249.19.03613-0. Epub 2019 Dec 12. PMID: 31833725
12. Shah, M., Naik, N., Somani, B.K., Hameed, B.M.Z.: Artificial intelligence (AI) in urology-current use and future directions: an iTRUE study. Turk. J. Urol. **46**(Supp. 1), S27–S39 (2020). https://doi.org/10.5152/tud.2020.20117. Epub 2020 May 27. PMID: 32479253; PMCID: PMC7731952
13. Santoni, M., Piva, F., Porta, C., Bracarda, S., Heng, D.Y., Matrana, M.R., Grande, E., Mollica, V., Aurilio, G., Rizzo, M., Giulietti, M., Montironi, R., Massari, F.: Artificial neural networks as a way to predict future kidney cancer incidence in the United States. Clin. Genitourin. Cancer **10**, S1558-7673 (2020). https://doi.org/10.1016/j.clgc.2020.10.008. Epub ahead of print. PMID: 33262083

14. Rani, B.S., Suchitra, M.M., Srinivasa Rao, P.V.L.N., Kumar, V.S.: Serum tumor markers in advanced stages of chronic kidney diseases. Saudi J. Kidney Dis. Transpl. **30**(4), 898–904 (2019). https://doi.org/10.4103/1319-2442.265466. PMID: 31464247
15. Amiri, F.S.: Serum tumor markers in chronic kidney disease: as clinical tool in diagnosis, treatment and prognosis of cancers. Ren. Fail. **38**(4), 530–44 (2016). https://doi.org/10.3109/0886022X.2016.1148523. Epub 2016 Feb 24 PMID: 26907957

Clinical Engineering and Health Technology Assessment

Sentiment Analysis for Performance Evaluation of Maintenance in Healthcare

Lorenzo Mascii[1], Alessio Luschi[2], and Ernesto Iadanza[2](✉)

[1] Biomedical Engineering, University of Florence, Florence, Italy
lorenzo.mascii@stud.unifi.it

[2] Department of Information Engineering, University of Florence, Florence, Italy
{alessio.luschi,ernesto.iadanza}@unifi.it

Abstract. This paper presents a framework which makes use of Sentiment Analysis techniques for retrieving Real World Data (RWD) starting from scheduled and corrective maintenance data. The scope of the analysis is to automatically extract features from maintenance work orders, in order to calculate Key Performance Indicators of maintenance operations on medical devices, for Health Technologies Assessment purposes. Data are extracted from Computerized Maintenance Management System reports of healthcare facilities.

Keywords: Clinical engineering · HTM · KPI · Sentiment analysis · HTA · Medical equipment

1 Introduction

World Health Organization (WHO) defines Health Technology Assessment (HTA) as "the systematic evaluation of properties, effects, and/or impacts of health technology. It is a multidisciplinary process to evaluate the social, economic, organizational and ethical issues of a health intervention or health technology. The main purpose of conducting an assessment is to inform a policy decision making" [1].

In recent years, the analysis of Real-World Data (RWD) has become more and more important in HTA, allowing to generate reliable Real-World Evidence (RWE) which is very useful in terms of HTA analysis and reporting. The main challenge about RWD is their high degree of heterogeneity. They can originate from Randomized Clinical Trials (RCT), epidemiological studies, administrative databases, regulatory or institutional registers and secondary sources. In recent years, social media have proved to be one of the most popular and wide secondary sources of RWD, surely increasing the level of diversity and complexity [2]. Data from social media are public and can be analyzed relatively easily, overcoming many logistical obstacles associated with traditional approaches, while enabling fast and cost-effective data collection, providing a reliable insight into the patient experience. Therefore, the design of effective methods for the automated extraction of useful and aggregated Smart Data (i.e. data enabling accurate and precise decision-making processes) by using data mining processes on various heterogeneous sources is currently needed [3].

A. Badnjevic and L. Gurbeta Pokvić (Eds.): CMBEBIH 2021, IFMBE Proceedings 84, pp. 359–367, 2021.
https://doi.org/10.1007/978-3-030-73909-6_41

Sentiment Analysis is a fairly new field of study, grown in parallel with the spread of social media, for analyzing opinions, evaluations, attitudes and sentiments of people in relation to products, services, organizations, topics and personalities [4].

Maintenance is a crucial subject for medical equipment life cycle management. Evidence-based maintenance consists of the continuous performance monitoring of equipment, starting from the evidence and improvement of its effectiveness by making the required changes. Maintenance service work orders are stored in Computerized Maintenance Management Systems (CMMS) [5–7]. These are software applications which supports technicians and managers to track resources, manage work orders and costs, schedule maintenance plans and check the compliance to quality standards [8–10].

Other studies have developed failure classification models and KPIs for evaluating medical equipment maintenance performance [11]. These methods rely on data extracted and then classified directly by users, with a subsequential high cost in terms of computational evaluation time and resources.

This study aims to apply Sentiment Analysis and Natural Language Processing (NLP) techniques to extract features from maintenance's work orders reports stored in hospital's CMMS (both scheduled and corrective). The extracted features are then used as input to a supervised machine learning (ML) classifier to obtain fault codes and finally to evaluate KPIs, thus improving evidence-based maintenance.

2 Methods

2.1 Sentiment Analysis

According to literature, there are two main approaches to Sentiment Analysis: Lexicon-based approach and Artificial Intelligence (AI) approach.

- The former uses common dictionaries (e.g. SentiWordNet [12]) to compare nouns and adjectives and score them with a positive, neutral or negative weight [13, 14].
- The latter uses AI algorithms to solve NLP problems, and to evaluate the final sentiment score by using classification models [15, 16].

Both techniques require pre-processing normalization and transformation to transform the sentences recovered via text mining [17, 18] into comprehensive and usable input text. Pre-processing task includes:

- Sequence-splitting (extraction of single sequences from analyzed texts)
- Spell-checking
- Case-folding (making each word lowercased or uppercased)
- Word-stopping (automatically excluding given words from further steps basing upon dictionaries)
- Lemming/Stemming (converting all verb forms and tenses to present infinitive and all plurals nouns to singulars)
- Tokenization (converting text into numerical representation)

When using the ML approach, feature extraction and selection are also needed. Typical NLP algorithms for feature extraction tasks are *BoW* (Bag of Words) and *TF-IDF* (Terms Frequency – Inverse Document Frequency) [15]. The former simply counts how many times each word occurs inside the document, while the latter links each word with a statistic measurement estimating the importance of the word itself inside the document. Specifically, the weight w_{ij} of the word t_i inside the document j is evaluated:

$$\frac{N}{w_{ij} = tf_{ij} \times \log df_i}$$

Where:

- tf_{ij} is how many times the word t_i occurs inside the document j
- N is the total number of documents
- df_i is the number of documents which includes the word t_i

Feature selection is the process of selecting the most significant and relevant features from a vast set of features in the given dataset. Feature selection helps in finding the smallest set of features which results in:

- Faster training
- Reducing the complexity of a model and making it easier to interpret
- Building a sensible model with better predicting power
- Reducing over-fitting by selecting the right set of features

This task can be performed with filtered methods, wrappers or embedded methods. With a filtered method the selection is machine-learning-independent and thus can be used with both approaches (supervised and unsupervised). In wrapper methods, the feature selection process is based on a specific ML algorithm that we are trying to fit on a given dataset. With embedded methods, feature selection is done by observing each iteration of model training phase (Fig. 1).

Sentiment Analysis may use ML approaches to obtain a sentiment score for the initial data. In case of supervised ML, a specific algorithm is used to find a statistical model which better represents already classified elements, such as Support Vector Machine (SVM), K-Nearest Neighbors (KNN), Decision-tree, Naïve Bayes (NB), Maximum Entropy or Neural Networks (NN).

In unsupervised ML, users do not need to provide the algorithm with output labels. Instead, the algorithm is able to autonomously discover patterns and information about the selected features.

A Recurrent Neural Network (RNN) is a class of artificial neural networks for unsupervised ML techniques where connections between nodes form a directed graph along a temporal sequence [19]. Convolutional Neural Network (CNN) is another class of deep neural networks, most commonly applied to analyzing visual imagery but which can also be implemented for NLP problems [20].

Figure 2 summarizes the whole described process.

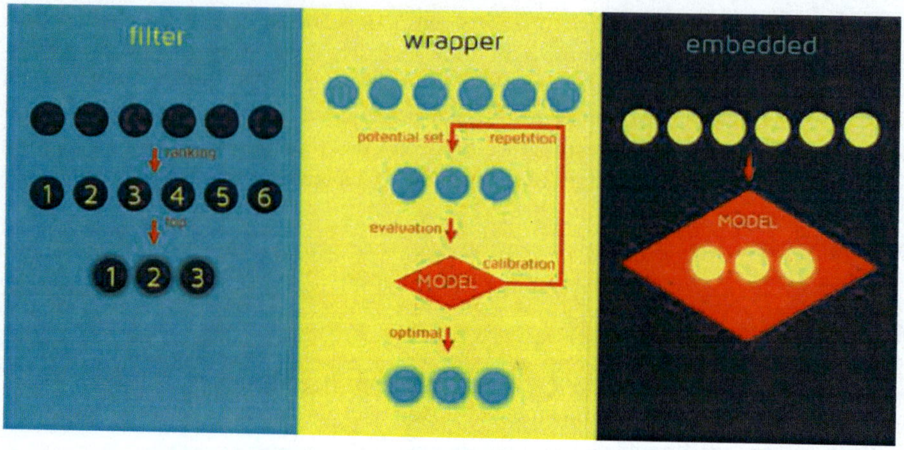

Fig. 1. Schematic representation of feature selection methods.

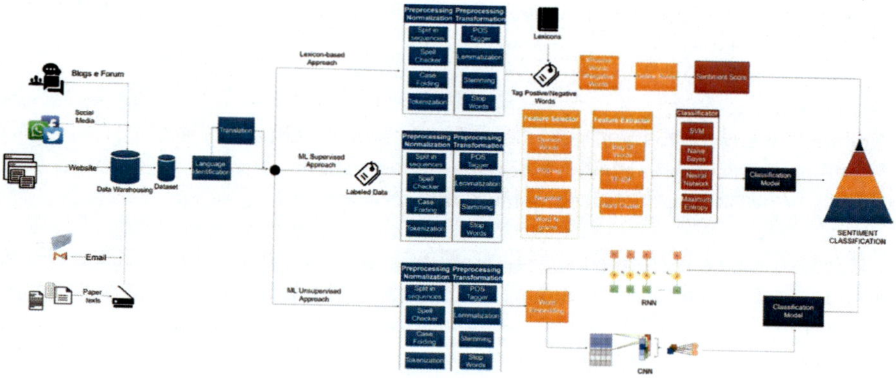

Fig. 2. Different sentiment analysis approaches.

2.2 Failure Classification and KPIs

Tecnicians from the Clinical Engineering service assign a fault code whenever a maintenance activity is performed. The code is chosen among a set of available codes [11]. This method allows evaluating and managing of the efficacy and efficiency of each maintenance operation (Fig. 3).

The UNI EN 15341:2007 standard [21] describes a system for managing KPIs to measure maintenance performance as influenced by key maintenance factors and to assess and improve efficiency and effectiveness. The standard suggests that the KPIs be structured into three groups to measure every aspect of the maintenance process: financial, technological and organizational. Previous studies have taken into account the criteria suggested by the above standard and designed a set of 20 KPIs described in Table 1 [11].

Source	Activity	Code	Code definition	Severity
Equipment	Corrective maintenance (CM)	UPF	Unpreventable failure, evident to user, typically caused by normal wear and tear but is unpredictable	1
		PPF	Preventable and predictable failure, evident to user, typically caused by wear and tear that can be predicted or detected	3
		USE	Failures induced by use, eg, abuse, abnormal wear and tear, accident, or environment issues	1
		SIF	Service-induced failure, ie, failure induced by corrective or scheduled maintenance that was not properly completed or a part that was replaced and had premature failure ("infant mortality")	2
	Scheduled maintenance (SM)	PF	Potential failure, ie, failure is either about to occur or in the process of occurring but has not yet caused equipment to stop working or problems to patients or users	4
		EF	Evident failure, ie, a problem that can be detected but was not reported by the user without running any special tests or using specialized test/measurement equipment	1
		HF	Hidden failure, ie, a problem that could not be detected by the user unless running a special test or using specialized test/measurement equipment	5
	CM and SM	NPF	No problem found, including alleged failures that could not be duplicated ("cannot duplicate" [CND])	0
Accessories	CM or SM	BATT	Battery failure, ie, battery(ies) failed before the scheduled replacement time	1
		ACC	Other accessory failures, excluding batteries, evident to user, typically caused by normal wear and tear	1
Accessory (including rechargeable battery) failures are segregated from equipment failures to help determine the cause of problems and identify solutions.				

Fig. 3. Set of available fault codes for both scheduled and corrective maintenance.

Table 1. List of KPIs.

Index	Name	Typology
KPI 1	Downtime (%) (non-availability time)	Technological
KPI 2	Uptime (%) (availability time)	Technological
KPI 3	MTTR (mean time to restoration)	Technological
KPI 4	MTBF (mean time between failures)	Technological
KPI 5	Class failure ratio (fails per class)	Technological
KPI 6	Global failure rate (defectiveness)	Technological
KPI 7	AFR: age failure rate	Technological
KPI 8	"Negligent" actions (%)	Organizational
KPI 9	"1 day" actions	Organizational
KPI 10	SM with failure (%)	Organizational
KPI 11	SM coverage rate (scheduled maintenance)	Organizational
KPI 12	No problem found (fake faults) (%)	Organizational
KPI 13	No. devices per technician (internal)	Organizational
KPI 14	Time cost of the workforce	Organizational
KPI 15	COSR (cost of service ratio = global maintenance to acquisition cost) (%)	Financial

(continued)

Table 1. (*continued*)

Index	Name	Typology
KPI 16	External maintenance Cost (% with respect to total maintenance cost)	Financial
KPI 17	Internal maintenance cost (% with respect to total maintenance cost)	Financial
KPI 18	Corrective maintenance cost (CM) (% with respect to total maintenance cost)	Financial
KPI 19	Scheduled maintenance cost (SM) (% with respect to total maintenance cost)	Financial
KPI 20	Cost of spare parts (+ consumables) (% with respect to total maintenance cost)	Financial

3 Results

The input of the designed framework is a collection of work orders extracted from the CMMS. The work orders alone cannot be directly used to input the system. In fact, data are unstructured (i.e. they are written in several different ways), resulting in unstructured information. Hence, the first step is to implements a TimeML compliant time-expression tagger [22]. HeidelTime has been chosen to be implemented as temporal tagger to extract time information and normalize them according to TimeML's TimeX3 tag. Once data have been tagged and normalized, common Sentiment Analysis algorithms can be applied.

Preprocessing is initially needed to obtain semantically meaningful words and usable input text to subsequently perform feature extraction and selection on. This is achieved by using Sentiment Analysis preprocessing methods described in Sect. 2.1.

Then TF-IDF algorithm is used for feature extraction. This has been chosen over BoW because more accurate and reliable. However, frequent words may not always have semantic significance and so they might be excluded from the obtained features. Moreover, this technique may easily lead to a huge amount of features which would cause the machine learning process to be spatial and temporal impracticable. *Dropfeature* predictive method is indeed used to give each feature a score (feature importance) to define which features have relevance in computational terms and which not, and so perform a feature selection.

Initial data contained inside the work orders of the CMMS can often be unbalanced, i.e. some classes may occur more frequently than others. This kind of data significantly reduces the performance of machine learning models: classifiers trained with unbalanced data tends to completely ignore the minority classes despite being useful for the final result. Many algorithms actually exist which prevent unbalanced learning by oversampling or undersampling the starting dataset. *Synthetic Minority Oversampling Technique* (SMOTE) has been successfully applied in many cases where unbalanced learning needed to be avoided, and it allows to add new data to the original dataset creating artificial datapoints based upon likeness between existent minorities [23].

Thereafter, machine-learning models are implemented to classify the extracted and filtered features. Support Vector Machine (SVM) supervised classifier has been chosen because it performs better in high-dimensions spaces and has a good cost-to-performance ratio [23, 24].

The desirable output is a classification of the number of maintenance operations for type of fault code, both for scheduled and corrective maintenance. Finally, a subset of the previously described KPIs (especially the technological ones) can be automatically evaluated by using some fault codes or a combination of them (Fig. 4).

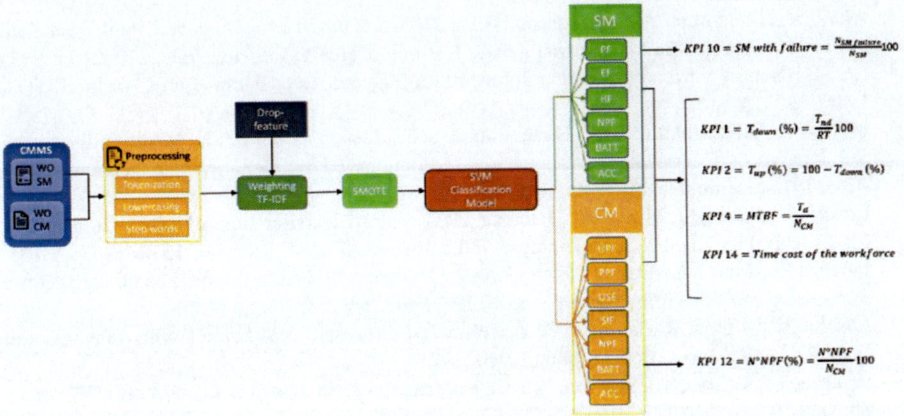

Fig. 4. Designed framework for automatic classification of maintenance's work orders.

4 Conclusion

The proposed HTA framework implements Sentiment Analysis methods to analyze RWD extracted from maintenance's work orders reports manually produced inside CMMS.

The analysis itself is a restricted NLP problem. In fact, the system does not need to comprehend the semantic of each phrase, but just few features correlated to the maintenance classification according to a set of available fault codes. This type of RWD are extremely unstructured and unbalanced, so external algorithms and models (time-expression tagging, SMOTE) are needed to transform them to be suitable for the analysis. Sentiment Analysis' common preprocessing is then performed (lemming, stemming, case-folding, tokenization), followed by feature extraction via TD-IDF algorithm, feature selection via drop-feature algorithm and supervised machine learning classification with SVM.

The output is a multi-class classification of work orders with fault codes, allowing an evidence-based maintenance approach, which monitors the efficacy and efficiency of maintenance operations on medical devices. KPIs are (partly) automatically calculated by the system, allowing top-management to objectively better understand the weak points and improve the process.

Eventually, the designed framework leads to an optimization of costs and resources, and also to improvements in terms of patients' safety and risk management by predicting faults of critical assets.

References

1. WHO – Health technology assessment: https://www.who.int/medical_devices/assessment/en
2. Yingjie, L., Yang, W., Jingfang, L., Li, J., Pengzhu, Z.: Understanding health care social media use from different stakeholder perspectives: a content analysis of an online health community. J. Med. Internet Res. **19**(4), e109 (2017)
3. Akay, A., Dragomir, A., Erlandsson, B.E.: Network-based modeling and intelligent data mining of social media for improving care. J. Biomed. Health Inform. **19**(1), 210–218 (2013)
4. Liu, B.: Sentiment analysis and opinion mining. Synth. Lect. Hum. Lang. Technol. **5**(1), 1–167 (2012)
5. Medenou, D., Fagbemi, L.A., Houessouvo, R.C., Jossou, T.R., et al.: Medical devices in SubSaharan Africa: optimal assistance via a computerized maintenance management system (CMMS) in Benin. Health Technol. **9**, 219–232 (2019)
6. Luschi, A., Marzi, L., Miniati, R., Iadanza, E.: A custom decision-support information system for structural and technological analysis in healthcare. IFMBE Proc. **41**, 1350–1353 (2013)
7. Iadanza, E., Marzi, L., Dori, F., Biffi Gentili, G., Torricelli, M.C.: Hospital healthcare offer a monitoring multidisciplinar approach. IFMBE Proc. **14**, 3685–3688 (2006)
8. Luschi, A., Miniati, R., Iadanza, E.: A web based integrated healthcare facility management system. IFMBE Proc. **45**, 633–636 (2015)
9. Iadanza, E., Luschi, A., Ancora, A.: Bed management in hospital systems. IFMBE Proc. **68**(3), 313–316 (2019)
10. Iadanza, E., Luschi, A., Gusino, R., Terzaghi, F.: Designing a healthcare computer aided facility management system: a new approach. In: International Conference on Medical and Biological Engineering. IFMBE Proceedings, vol. 73, pp. 407–411 (2019)
11. Iadanza, E., Gonnelli, V., Satta, F., et al.: Evidence-based medical equipment management: a convenient implementation. Med. Biol. Eng. Comput. **57**, 2215–2230 (2019)
12. Miller, G., Beckwith, T., Fellbaum, C., et al.: WordNet: an on-line lexical database. Int. J. Lexicogr. **3**(4), 235–244 (1990)
13. Whitelaw, C., Garg, N., Argamon, S.: Using appraisal groups for sentiment analysis. In: Proceedings of the ACM SIGIR Conference on Information and Knowledge Management (CIKM), pp. 625–631 (2005)
14. Vu, L., Le, T.: A lexicon-based method for sentiment analysis using social network data. In: International Conference on Information and Knowledge Engineering (IKE 2017), Las Vegas, USA (2017)
15. Du, J., Xu, J., Song, H., Xiangyu, L., Cui, T.: Optimization on machine learning based approaches for sentiment analysis on HPV vaccines related tweets. J. Biomed. Semant. **8**(9), 1–7 (2017)
16. Duyu, T., Bing, Q., Ting, L.: Deep learning for sentiment analysis: successful approaches and future challenges. WIREs Data Min. Knowl. Discov. **5**(6), 292303 (2015)
17. Mine, G., Elder, J., Fast, A., et al.: Practical Text Mining and Statistical Analysis for Nonstructured Text Data Applications. Academic Press, Cambridge (2012)
18. Heerschop, B., Goossen, F., Hogenboom, A., et al.: Polarity analysis of texts using discourse structure. In: 20th ACM Conference on Information and Knowledge Management (2011)
19. Socher, R., Perelygin, A., Wu, J., Chuang, J., et al.: Recursive deep models for semantic compositionality over a sentiment treebank. In: Proceedings of the 2013 Conference on Empirical Methods in Natural Language Processing, pp. 1631–1642 (2013)

20. Porumb, M., Iadanza, E., Massaro, S., Pecchia, L.: A convolutional neural network approach to detect congestive heart failure. Biomed. Sign. Process. Control **55**, 101597 (2020)
21. UNI EN 15341:2007 Maintenance - Maintenance Key Performance Indicators
22. TimeML – Markup Language for Temporal and Event Expressions. http://www.timeml.org/
23. Zhang, T., Bhatia, A., Pandya, D., et al.: Industrial text analytics for reliability with derivative-free optimization. Comput. Chem. Eng. **135**, 106763 (2020)
24. Hartmann, J., Huppertz, J., Schamp, C., Heitmann, M.: Comparing automated text classification methods. Int. J. Res. Market. **36**(1), 20–38 (2020)

Application of Artificial Intelligence Techniques to Predict Renal Function Based on Diagnostic Parameters

Abdagić Valida, Abdukić Tarik(✉), Adžamija Amra, Alagić Amar, Alajbegović Anisa, and Alešević Lamija

Faculty of Pharmacy, University of Sarajevo, Sarajevo, Bosnia and Herzegovina
tarikabdukic@ffsa.unsa.ba

Abstract. Approximately 10% of the population worldwide is affected by chronic kidney disease (CKD), and millions die each year because they do not have access to affordable treatment. Over 2 million people worldwide currently receive treatment with dialysis or a kidney transplant to stay alive, yet this number may only represent 10% of people who actually need treatment to live. This paper presents the possible modality of diagnosis kidney disease. Testing was done on 400 samples; 200 samples are for chronic kidney disease and 200 for the healthy class of patients. Artificial Neural Network (ANN) was developed in order to perform prediction of renal diseases based on diagnostic parameters. Parameters that we used were gender, age, creatinine clearance (ml/min), urea (g/h), albumin (mg/h), glucose (mmol/L) and cystatin C (mg/l). Although limited number of data were present in this study, such automated systems can be used as assistant tool during diagnosis in real-time clinical settings.

Keywords: Artificial neural network · Renal function · Chronic kidney disease

1 Introduction

Kidneys are organs with multiple functions that include regulation of electrolyte and volume, excretion of nitrogenous waste and elimination of exogenous molecules (drugs), synthesis of a variety of hormones (erythropoietin), metabolism of low molecular weight proteins (insulin), etc. 1. They perform their function through three processes including glomerular filtration (glomerulus), active tubular excretion and active and passive tubular reabsorption (tubules) 2.

Chronic kidney disease (CKD) is a serious health problem in the world with possible unfortunate outcomes such as kidney failure and premature death. In addition to urea and creatinine from the blood, the two most important indicators of kidney disease are eGFR (estimated glomerular filtration rate) and urinary albumin. At the very beginning of kidney disease, there are often no clear signs or symptoms, but it can be detected by simple laboratory tests - determination of protein in the urine and urate, urea and creatinine in the blood. As the disease progresses, other signs appear such as swollen

A. Badnjevic and L. Gurbeta Pokvić (Eds.): CMBEBIH 2021, IFMBE Proceedings 84, pp. 368–376, 2021.
https://doi.org/10.1007/978-3-030-73909-6_42

ankles, fatigue, difficulty concentrating, decreased appetite and an unusual appearance of urine.

Over 850 million people, according to statistics, suffer from chronic kidney disease, acute renal injury (AKI) or are on renal replacement therapy (RTT), which is a worrying figure since it exceeds the number of people with diabetes. The risk of the disease is almost equal in women and men 3. People with diabetes are especially susceptible to developing kidney disease. Therefore, they should be checked regularly.

In the age of 4th industrial revelution, the health sector is one of the areas where artificial intelligence has already begin to play a significant role in improving the existing systems and processes. Already, intelligent systems which a large amount of data such as: diagnostic images, laboratory tests and medical records have been developed and are aiding medical professionals in the everyday activities. Research in this field is one of the most productive ones nowadays. Application of different techniques is being investigated, such as artificial neural networks (ANNs) and machine learning (ML) algorithms.

For instance, Sarić et al. [4] suggest usage of ANNs implemented on FPGA for epileptic seizure classification. Implementation of these systems is not uncommon for genetic testings as well. Catic et al. [5] have developed neural network for classification of prenatal syndroms based on parameters that can be acquired non-invasively from the pregnant woman. Successful implementation of ANNs for diagnosis of other conditions and diseases can be seen in healthcare sector [6–9]. Researchers have investigated application of these tools for detection and classification of chronic kidney disease and other pathological conditions associated with the urinary system [10]. Certain artificial intelligence techniques have made it possible to detect kidney stones, based on the results of laboratory tests such as creatinine, uric acid, glucose, lymphocytes and other blood components. Neural networks and other machine learning techniques have also been applied to identify the stage of chronic kidney disease. Some research suggests a neural network model to detect CKD from patient laboratory data, as well as comparisons with other machine learning models 10.

The aim of this research is to develop artificial neural network for prediction of renal function based on diagnostic parameters that can be easily collected within healthcare environment. The goal for such system is to serve as an aid in the early detection and prevention of chronic kidney disease, especially in rural and remote areas where clinical specialist isn't available.

2 Methods and Materials

For the development of an artificial neural network, database of clinical parameters related to occurence of chronic kidney disease was used. Each sample in the database consisted of information about five input parameters for each patient, and those parameters are: (1) creatinine clearance (ml/min) (2) urea (g/h) (3) albumin (mg/h) (4) glucose (mmol/L) and (5) cystatin C (mg/l). Table 1. shows the values of the mentioned parameters for healthy patients. Increased values of these clinical parameters in blood and urine, indicate acute and/or chronic kidney disease and urinary tract obstruction.

The data set for this study consisted of 400 subjects, out of overall number, 200 samples were for class of patients with chronic kidney disease and 200 for the healthy

Table 1. Parameters for prediction of renal function

Parameters	Reference (Normal) values
(1) Creatinine clearance	125–135 ml/min
(2) Urea	12–20 g/24 h
(3) Albumin	3–30 mg/24 h
(4) Glucose	0,0–0,8 mmol/L
(5) Cystatin C	0,65 mg/L ± 0, 085 =0,56–0,73 (male) 0,83 mg/L ± 0,103 =0,727–0,933 (female)

class of patients. As indicated in Table 2, in first group, 107 were men with an average age of 43, and the remaining 93 were women with an average age of 41. The rest of the patients were from the control group, 98 of them were male with average age of 42, and 102 female with average age of 43 years.

Table 2. Properties of the dataset used for development of artificial neural network

Class	Chronic kidney disease	Healthy	Total
Male	107	98	205
Average age (years)	43,17	42,12	42,64
Female	93	102	195
Average age (years)	41,13	43,07	42,1
Total number of patients per class	200	200	400

The distribution of the dataset for development of the Artificial Neural Network (ANN) is presented in Table 3. Since this is relatively small dataset, distribution of 70:30% for training and validation was used [11].

Table 3. Dataset distribution

Total number of samples 400	Healthy 200	Disease 200
Training 280	140	140
Validation 120	60	60

Artificial Neural Network (ANN) was developed in order to perform prediction of renal diseases based on diagnostic parameters. Developed ANN has 7 input parameters: gender, age, creatinine clearance (ml/min), urea (g/h), albumin (mg/h), glucose (mmol/L) and cystatin C (mg/l). The ANN performs classification of samples into two classes: healthy and patients with renal disease.

For the development of the ANN in this research, standard architecture for pattern recognition was used. That is feedforward neural network which consists of two-layers. This architecture is designed since according to the application experts, that kind of ANN is sufficient to properly perform the classification of input data [12].

The input layer is made of network inputs. Input layer is followed by a hidden layer which consists of parallely connected neurons. The network was tested on training performance with different number of neurons. Since, this classification is basically highly non-linear data classification, sigmoid function was used for hidden layer [13].

During the design of the network, different number of neurons was used in order to measure the performance. According to the Mean Square Error (MSE), appropriate number of neurons was chosen. After training, validation of the network performance was done using confusion matrix.

According to these results, specificity, sensitivity and accuracy of the network are calculated according to Eqs. (1) (2) (3):

$$senstivity = \frac{true\ positive}{(true\ positive + false\ negative)} \tag{1}$$

$$specificity = \frac{true\ negative}{(true\ negative + false\ positive)} \tag{2}$$

$$accuracy = \frac{true\ negative + true\ positive}{(true\ negative + true\ positive + false\ negative + false\ positive)} \tag{3}$$

3 Results

According to previous research and expert recommendation [4–8, 12, 14–16], for classification of renal diseases in this paper, two-layer feedforward artificial neural network, with sigmoid transfer function in hidden layer was developed.

Accuracy of training for feedforward neural network with 2 neurons and sigmoid transfer function in hidden layer was 84.3%. Training was performed using Levenberg – Marquadt training algorithm. Training performance of the developed neural network is presented in the Fig. 1.

The validation of performance of feedforward neural network with 2 neurons and sigmoid transfer function in hidden layer is presented in Table 4.

Accuracy of training for feedforward neural network with 5 neurons and sigmoid transfer function in hidden layer was 88.2%. Training was performed using Levenberg – Marquadt training algorithm. Training performance of the developed neural network is presented in the Fig. 2.

The validation of performance of feedforward neural network with 2 neurons and sigmoid transfer function in hidden layer is presented in Table 5.

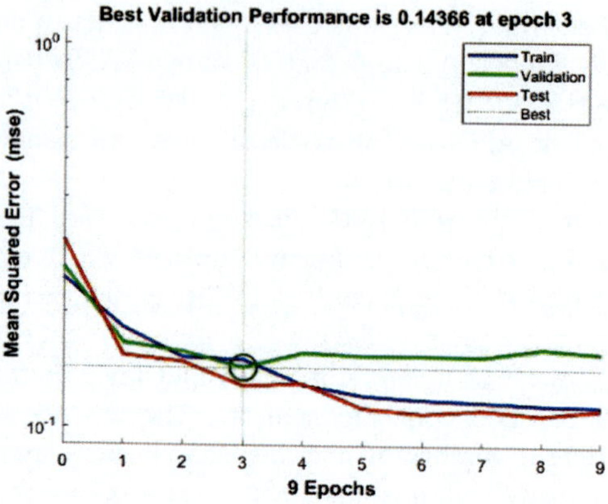

Fig. 1. Training performance of the network with 2 neurons in hidden layer

Table 4. Artificial neural network performance in classification

n = 120	GROUP 1 Healthy ANN prediction	Group 2 Disease ANN prediction	
GROUP 1 Healthy 60	TP 54	FN 6	Accuracy [%] 78.3
Group 2 Disease 60	FP 20	TN 40	
	Sensitivity [%] 90.0	Specificity [%] 66.7	

Accuracy of training for feedforward neural network with 15 neurons and sigmoid transfer function in hidden layer was 92.3%. Training was performed using Levenberg – Marquadt training algorithm. Training performance of the developed neural network is presented in the Fig. 3.

The validation of performance of feedforward neural network with 10 neurons and sigmoid transfer function in hidden layer is presented in Table 6.

Since the training and validation performance of neural network with 15 neurons in hidden layer was the highest this neural network architecture is suggested for the usage for prediction of renal disease.

Using this expert system at primary healthcare level, based on basic diagnostic parameters there is above 80% chance that patient will be diagnosed correctly. This is promising results, since such expert system can be used as an aid to healthcare professional at primary healthcare level in situations when medical specialist is not available.

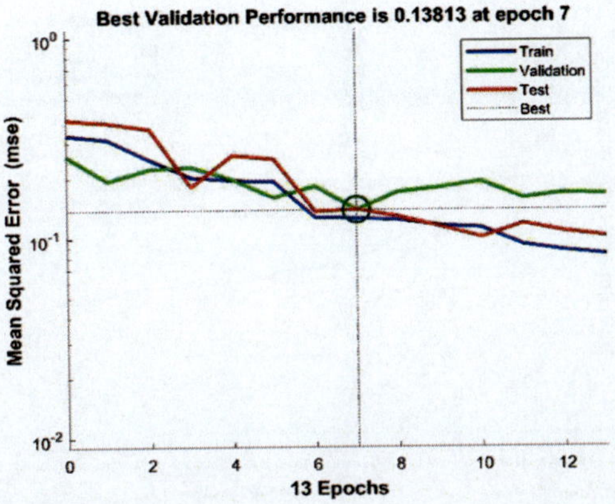

Fig. 2. Training performance of the network with 2 neurons in hidden layer

Table 5. Artificial neural network performance

n = 120	GROUP 1 Healthy ANN prediction	Group 2 Disease ANN prediction	
GROUP 1 Healthy 60	TP 58	FN 2	Accuracy [%] 87.5%
Group 2 Disease 60	FP 13	TN 47	
	Sensitivity [%] 96.6	Specificity [%] 78.3	

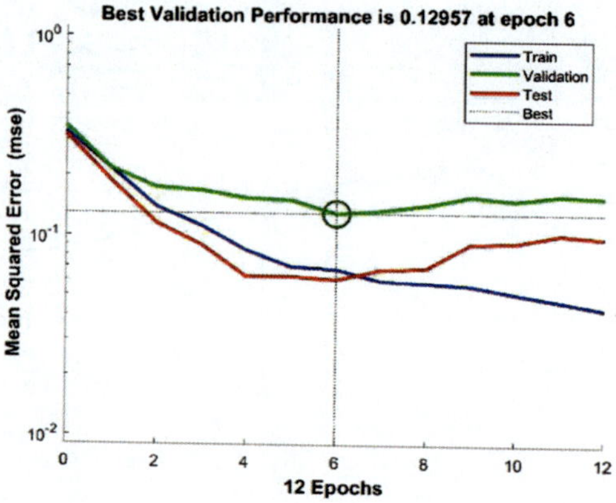

Fig. 3. Training performance of the network with 10 neurons in hidden layer

Table 6. Artificial neural network performance

n = 120	GROUP 1 Healthy ANN prediction	Group 2 Disease ANN prediction	
GROUP 1 Healthy 60	TP 58	FN 2	Accuracy [%] 91.7%
Group 2 Disease 60	FP 8	TN 52	
	Sensitivity [%] 96.6	Specificity [%] 86.6	

4 Conclusion

Chronic kidney disease is a serious health problem that can be predicted by Artificial Intelligence. Artificial Neural Network (ANN) was developed in order to perform prediction of renal disease based on diagnostic parameters: creatinine clearance (ml/min), urea (g/h), albumin (mg/h), glucose (mmol/L) and cystatin C (mg/l).

Training and validation performance of neural network with 15 neurons in hidden layer is suggested for the usage for prediction of renal disease.

Using this expert system at primary healthcare level, based on basic diagnostic parameters, indicate that there is above 80% chance that patient will be diagnosed correctly.

References

1. Bindroo, S., Quintanilla Rodriguez, B.S., Challa, H.J.: Renal Failure. 2020 Aug 13. In: StatPearls [Internet]. Treasure Island (FL): StatPearls Publishing; 2020 Jan – PMID: 30085554. Preuzeto sa. https://pubmed.ncbi.nlm.nih.gov/30085554/. Accessed 28 Nov 2020
2. Murray, I., Paolini, M.A.: Histology, Kidney and Glomerulus. [Updated 2020 May 5]. In: StatPearls [Internet]. Treasure Island (FL): StatPearls Publishing; 2020 Jan - Preuzeo sa. https://www.ncbi.nlm.nih.gov/books/NBK554544/. Accessed 28 Nov 2020
3. Jager, K.J., Kovesdy, C., Langham, R., Rosenberg, M., Jha, V., Zoccali, C.: A single number for advocacy and communication—worldwide more than 850 million individuals have kidney diseases, Nephrology Dialysis Transplantation, vol. 34, no. 11, pp. 1803–1805, November 2019. https://doi.org/10.1093/ndt/gfz174. Accessed 28 Nov 2020
4. Sarić, R., Jokić, D., Beganović, N., Pokvić, L.G., Badnjević, A.: FPGA based real-time epileptic seizure classification using Artificial Neural Network. Biomed. Sign. Process. Control **62**, 102106 (2020). ISSN 1746-8094, https://doi.org/10.1016/j.bspc.2020.102106
5. Catic, A., Gurbeta, L., Kurtovic-Kozaric, A., Mehmedbasic, S., Badnjevic, A.: Application of Neural Networks for classification of Patau, Edwards, Down, Turner and Klinefelter Syndrome based on first trimester maternal serum screening data, ultrasonographic findings and patient demographics. BMC Med. Genomics **11**, 19 (2018). https://doi.org/10.1186/s12920 018-0333-2
6. Badnjevic, A., Gurbeta, L., Cifrek, M., Marjanovic, D.: Classification of asthma using artificial neural network. In: IEEE 39th International Convention on Information and Communication Technology, Electronics and Microelectronics (MIPRO), vol. 30. Opatija, Croatia, May to 03. June 2016
7. Badnjevic, A., Cifrek, M., Koruga, D., Osmankovic, D.: Neuro-fuzzy classification of asthma and chronic obstructive pulmonary disease. BMC Med. Inform. Dec. Making J. **15**(3), 1–9 (2015). https://doi.org/10.1186/1472-6947-15-s3-s1
8. Sejdinovic, D., Gurbeta, L., Badnjevic, A., Malenica, M., Dujic, T., Causevic, A., Bego, T., Divovic, L.: Classification of prediabetes and Type 2 diabetes using artificial neural network. In: CMBEBIH 2017. IFMBE Proceedings, vol. 62, pp. 685–689. Springer, Singapore (2017). https://doi.org/10.1007/978-981-10-4166-2_103
9. Rau, S.-H., Hsu, C.-Y., Lin, Y.-A., Atique, S., Fuad, A., Wei, L.-M., Hsu, M.-H.: Development of a web-based liver cancer prediction model for type II diabetes patients by using an artificial neural network. Comput. Methods Programs Biomed. **125**, 58–65 (2016). ISSN 0169-2607, https://doi.org/10.1016/j.cmpb.2015.11.009
10. Vásquez-Morales, G.R., Martínez-Monterrubio, S.M., Moreno-Ger, P., RecioGarcía, J.A.: Explainable prediction of chronic renal disease in the colombian population using neural networks and case-based reasoning. IEEE Access **7**, 152900152910 (2019). https://doi.org/10.1109/access.2019.2948430
11. Anifowose, F., Khoukhi, A., Abdulraheem, A.: Investigating the effect of training–testing data stratification on the performance of soft computing techniques: an experimental study. J. Exp. Theor. Artif. Intell. **29**(3), 517–535 (2017). https://doi.org/10.1080/0952813X.2016.1198936
12. Badnjevic, A., Gurbeta, L., Cifrek, M., Marjanovic, D.: Classification of asthma using artificial neural network. In: IEEE 39th International Convention on Information and Communication Technology, Electronics and Microelectronics (MIPRO). Opatija, Croatia (2016)
13. Aggarwal, C.: Neural Networks and Deep Learning. Springer. https://link.springer.com/book/10.1007/978-3-319-94463-0
14. Badnjevic, A., Gurbeta, L., Custovic, E.: An expert diagnostic system to automatically identify asthma and chronic obstructive pulmonary disease in clinical settings. Sci. Rep. **8**(1), 1–9 (2018)

15. Gurbeta, L., Badnjevic, A., Maksimovic, M., Omanovic-Miklicanin, E., Sejdic, E.: A telehealth system for automated diagnosis of asthma and chronical obstructive pulmonary disease. J. Am. Med. Inform. Assoc. **25**(9), 1213–1217 (2018)
16. Divovic Mustafic, L., et al.: Diagnosis of severe aortic stenosis using implemented expert system. In: Badnjevic, A., Škrbić, R., Gurbeta Pokvić, L. (eds) CMBEBIH 2019. IFMBE Proceedings, vol. 73. Springer, Cham (2020)

Cardiovascular Disease Risk Assessment in Patients with Metabolic Syndrome Using Artificial Neural Network

Neira Omeragić(✉), Aida Omanović, Minela Omerović, Lejla Osmanagić, Ehlimana Omanović, and Nedžmina Ohran

Faculty of Pharmacy, University of Sarajevo, Sarajevo, Bosnia and Herzegovina
neiraomeragic@ffsa.unsa.ba

Abstract. Coronary heart disease is the most common cause of death in Europe. A risk assessment allows more accurate estimates of risk to guide the prevention of cardiovascular events. The calculation predicts the absolute risk of having a coronary heart disease event (e.g., myocardial infarction, heart failure, death) over a certain period. It is based on an empiric equation and artificial networking that combines major risk factors, such as blood pressure and cholesterol levels. This is highly important because patients who are aware of their risk level are more likely to initiate risk-reducing therapies. Based on the training results, ANN with 15 neurons in the hidden layer was chosen for Cardiovascular Disease Risk Assessment in Patients with Metabolic Syndrome, with an accuracy of 97.7%. The results are showing that this new approach and more sophisticated ANN networks based on an even bigger sample could be a powerful tool in advancing what will medical treatments represent in the future.

Keywords: Cardiovascular disease · Metabolic syndrome · Artificial network

1 Introduction

The process of defining risk factors for any clinical syndrome relies upon the conduction of population-based studies of the condition being examined and combining the results. However, when a cluster of potentially fatal cardiovascular disease risk factors exist in the same patient the situation becomes much more complex (i.e., syndrome X) [1].

Metabolic syndrome (MetS), also known as syndrome X, insulin resistance, etc., is defined by World Health Organization as a pathologic condition characterized by abdominal obesity, insulin resistance, hypertension, and hyperlipidemia [2]. This condition has serious implications on an individual's health and healthcare cost so it is necessary to recognize the rising prevalence of the metabolic syndrome, as through intervention, the progression of the syndrome can be halted and potentially reversed [3].

Early research of this syndrome has focused mainly on the role of insulin resistance but more recently the focus has been shifted on the MetS as an epidemiologic tool related to the risk of developing cardiovascular disease, and therefore traditional cardiovascular disease risk factors have been adopted as the defining features [4]. Given that heart

A. Badnjevic and L. Gurbeta Pokvić (Eds.): CMBEBIH 2021, IFMBE Proceedings 84, pp. 377–385, 2021.
https://doi.org/10.1007/978-3-030-73909-6_43

disease remains a leading cause of death around the world, elucidating mechanisms by which MetS increases cardiovascular risk is essential for developing future treatments and preventing this global epidemic [5].

Although insulin resistance has been excluded from the clinical definition of the MetS, the fact that insulin resistance contributes to many of the features of the MetS like dysglycemia, elevated fatty acid levels and hyperinsulinemia cannot be fully ignored. The epidemiologic literature consistently demonstrates associations of hyperinsulinemia with obesity-related heart disease. Studies of the metabolic physiology of insulin resistance have led to a deeper understanding of the interplay of fuel selection, with excess fatty acids contributing in part to cellular resistance to glucose metabolism but also with impairment in insulin receptor and post-receptor signaling events and abnormalities in mitochondrial number and function all contributing to the overall phenomenon [4].

Artificial neural networking (ANN) is part of computer science that can analyze and identify complex connections between input and output data. The data can be used for diagnosis, treatment, and outcome prediction in various patients [6].

The aim of the study was to assess both risks of developing cardiovascular disease in patients with metabolic syndrome using artificial neural networks and the potential of artificial neural networks in helping to predict and possibly prevent the development of cardiovascular disease [7–11].

2 Materials and Methods

A total of 300 respondents participated in this study, which began in November 2020 and lasted for a month (Table 1). These subjects were divided into two study groups. The first study group consisted of healthy subjects, a total of 100. The second study group consisted of subjects with metabolic syndrome, a total of 200.

Table 1. Total number of respondents and number of respondents by groups

Total number of respondents	Healthy controls	Patients with metabolic syndrome
300	100	200

Criteria for general evaluation of subjects were gender and age, while criteria for the health status of subjects were cigarette consumption (smoker), body mass index (BMI), glucose concentration in the body, blood pressure, cholesterol, and triglyceride values in the body. Reference values for these parameters are listed in Table 2.

Criteria for inclusion of subjects in this study, for a group of healthy subjects were normal reference values of parameters listed in Table 2. Inclusion criteria for patients with metabolic syndrome were one or more of these parameters, which deviated from the reference values listed in Table 2. According to WHO, for a person to be defined with the metabolic syndrome, they must have:

Table 2. Reference (normal) values of parameters for evaluation [12–15]

Parameter	Smoker	BMI (kg/m^2)	Glucose level (mmol/L)	Blood pressure (mmHg)	Cholesterol (mmol/L)	Triglycerides (mmol/L)
Reference (normal) value	Yes/No	18.5–24.90	4–6	120/80	2.16–5.52	0.8–1.92

1. Blood pressure ≥ 140/90 mmHg or taking medication for hypertension.
2. Triglycerides ≥ 150 mg/dL (1,7 mmol/L) or taking medication for triglyceridemia.
3. HDL cholesterol < 35 mg/dL (0,9 mmol/L) for men or < 39 mg/dL (1,0 mmol/L) for women or taking medications for reduced HDL.
4. Waist to hip ratio > 0,9 for men and > 0,85 for women or BMI > 30 kg/m^2 [16].

In the control group, a group of healthy subjects, the exclusion criterion was one or more of the listed parameters, which was outside the reference values.

The risk estimate is based on major risk factors and is calculated using the charts for estimation of cardiovascular disease in men and women (Figs. 1 and 2).

3 Results

The distribution of the dataset for the development of the Artificial Neural Network (ANN) is presented in Table 3. Since this is a relatively small dataset, a distribution of 70:30% for training and validation was used [18].

According to previous research and expert recommendation [19, 20], two-layer feed-forward artificial neural network, with sigmoid transfer function in hidden layer was developed. Neural network architecture consists of 9 inputs: Gender 1–2; Age 18–75; Smoker 1/0; BMI (kg/m2) 18.5–24.9; Glucose 4–6 mmol/L; Blood pressure – Systolic 120 mmHg; Blood pressure – Diastolic 80 mmHg; Cholesterol 2,16–5,52 mmol/L; Triglycerides 0,8–1,92 mmol/L.

According to these results, specificity, sensitivity, and accuracy of the network are calculated according to Eqs. (1), (2) (3):

$$sensitivity = \frac{true\ positive}{(true\ positive + false\ negative)} \tag{1}$$

$$specificity = \frac{true\ negative}{(true\ negative + false\ positive)} \tag{2}$$

$$accuracy = \frac{(true\ negative + true\ positive)}{(true\ negative + true\ positive + false\ negative + false\ positive)} \tag{3}$$

During development, a different number of neurons was tested. The results of training for a different number of neurons in the hidden layer are presented in the table below (Table 4) (Fig. 3).

Step 1: Age

Years	Points	Years	Points
30 to 34	–1	55 to 59	4
35 to 39	0	60 to 64	5
40 to 44	1	65 to 69	6
45 to 49	2	70 to 74	7
50 to 54	3		

Step 2: LDL or TC Level

LDL

mg per dL	mmol per L	Points
< 100	< 2.59	–3
100 to 129	2.59 to 3.35	0
130 to 159	3.36 to 4.13	0
160 to 190	4.14 to 4.92	1
> 190	> 4.92	2

TC

mg per dL	mmol per L	Points
< 160	< 4.14	–3
160 to 199	4.14 to 5.16	0
200 to 239	5.17 to 6.20	1
240 to 279	6.21 to 7.23	2
≥ 280	≥ 7.24	3

Step 3: HDL Level

mg per dL	mmol per L	Points (if LDL used in step 2)	Points (if TC used in step 2)
< 35	< 0.91	2	2
35 to 44	0.91 to 1.15	1	1
45 to 49	1.16 to 1.28	0	0
50 to 59	1.29 to 1.54	0	0
≥ 60	≥ 1.55	–1	–2

Step 4: Blood Pressure

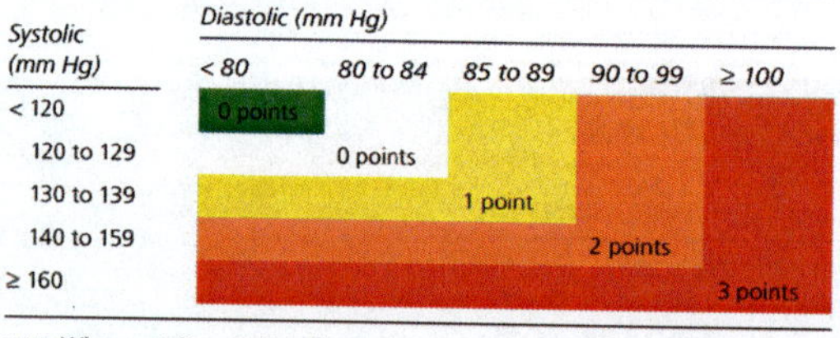

NOTE: *When systolic and diastolic pressures provide different point scores, use the higher score.*

Step 5: Diabetes Mellitus

Present?	Points
No	0
Yes	2

Step 6: Smoking

Smoker?	Points
No	0
Yes	2

Step 7: Total Points

Step 1: Age	____
Step 2: LDL or TC level	____
Step 3: HDL level	____
Step 4: Blood pressure	____
Step 5: Diabetes mellitus	____
Step 6: Smoking	____
Total points	____

Step 8: CHD Risk

Total points	10-year risk if LDL used in step 2 (%)	10-year risk if TC used in step 2 (%)
≤ –3	1	—
–2	2	—
–1	2	2
0	3	3
1	4	3
2	4	4
3	6	5
4	7	7
5	9	8
6	11	10
7	14	13
8	18	16
9	22	20
10	27	25
11	33	31
12	40	37
13	47	45
≥ 14	≥ 56	≥ 53

Step 9: Comparative Risk

Age (years)	Average 10-year CHD risk (%)	Average 10-year risk of hard event* (%)	Low 10-year CHD risk† (%)
30 to 34	3	1	2
35 to 39	5	4	3
40 to 44	7	4	4
45 to 49	11	8	4
50 to 54	14	10	6
55 to 59	16	13	7
60 to 64	21	20	9
65 to 69	25	22	11
70 to 74	30	25	14

**—Hard events exclude angina pectoris.*
†—Low risk as calculated for a man of the same age who does not smoke or have diabetes, and has optimal blood pressure, an LDL level of 100 to 129 mg per dL or TC level of 160 to 199 mg per dL, and an HDL level of 45 mg per dL.

Key

Color	Relative risk	Color	Relative risk
Green	Very low	Orange	High
White	Low	Red	Very high
Yellow	Moderate		

Fig. 1. Chart for estimation of coronary heart disease risk in men [17]

Based on the training results, ANN with 15 neurons in the hidden layer was chosen for Cardiovascular Disease Risk Assessment in Patients with Metabolic Syndrome. Results of validation are presented in the table below (Table 5).

Step 1: Age

Years	Points	Years	Points
30 to 34	−9	55 to 59	7
35 to 39	−4	60 to 64	8
40 to 44	0	65 to 69	8
45 to 49	3	70 to 74	8
50 to 54	6		

Step 2: LDL or TC Level

LDL

mg per dL	*mmol per L*	*Points*
< 100	< 2.59	−2
100 to 129	2.59 to 3.35	0
130 to 159	3.36 to 4.13	0
160 to 190	4.14 to 4.92	2
> 190	> 4.92	2

TC

mg per dL	*mmol per L*	*Points*
< 160	< 4.14	−2
160 to 199	4.14 to 5.16	0
200 to 239	5.17 to 6.20	1
240 to 279	6.21 to 7.23	1
≥ 280	≥ 7.24	3

Step 3: HDL Level

mg per dL	*mmol per L*	*Points (if LDL used in step 2)*	*Points (if TC used in step 2)*
< 35	< 0.91	5	5
35 to 44	0.91 to 1.15	2	2
45 to 49	1.16 to 1.28	1	1
50 to 59	1.29 to 1.54	0	0
≥ 60	≥ 1.55	−2	−3

Step 4: Blood Pressure

Systolic (mm Hg)	*Diastolic (mm Hg)* < 80	80 to 84	85 to 89	90 to 99	≥ 100
< 120	−3 points	0 points	0 points	2 points	3 points
120 to 129	0 points	0 points	0 points	2 points	3 points
130 to 139	0 points	0 points	0 points	2 points	3 points
140 to 159	2 points	2 points	2 points	2 points	3 points
≥ 160	3 points	3 points	3 points	3 points	3 points

NOTE: *When systolic and diastolic pressures provide different point scores, use the higher score.*

Step 5: Diabetes Mellitus

Present?	*Points*
No	0
Yes	4

Step 6: Smoking

Smoker?	*Points*
No	0
Yes	2

Step 7: Total Points

Step 1: Age	____
Step 2: LDL or TC level	____
Step 3: HDL level	____
Step 4: Blood pressure	____
Step 5: Diabetes mellitus	____
Step 6: Smoking	____
Total points	____

Step 8: CHD Risk

Total points	*10-year risk if LDL used in step 2 (%)*	*10-year risk if TC used in step 2 (%)*
≤ −2	1	1
−1	2	2
0	2	2
1	2	2
2	3	3
3	3	3
4	4	4
5	5	4
6	6	5
7	7	6
8	8	7
9	9	8
10	11	10
11	13	11
12	15	13
13	17	15
14	20	18
15	24	20
16	27	24
≥ 17	≥ 32	≥ 27

Step 9: Comparative Risk

Age (years)	*Average 10-year CHD risk (%)*	*Average 10-year risk of hard event* (%)*	*Low 10-year CHD risk† (%)*
30 to 34	<1	<1	<1
35 to 39	<1	<1	1
40 to 44	2	1	2
45 to 49	5	2	3
50 to 54	8	3	5
55 to 59	12	7	7
60 to 64	12	8	8
65 to 69	13	8	8
70 to 74	14	11	8

**—Hard events exclude angina pectoris.*
†—Low risk as calculated for a woman of the same age who does not smoke or have diabetes, and has optimal blood pressure, an LDL level of 100 to 129 mg per dL or TC level of 160 to 199 mg per dL, and an HDL level of 55 mg per dL (1.42 mmol per L).

Key

Color	*Relative risk*	*Color*	*Relative risk*
Green	Very low	Orange	High
White	Low	Red	Very high
Yellow	Moderate		

Fig. 2. Chart for estimation of coronary heart disease risk in women [17]

Table 3. Dataset distribution

Total number of samples 300	Healthy 100	Disease 200
Training 210	70	140
Validation 90	30	60

Table 4. Training and validation of neural network

Number of neurons	MSE
1	1.7886e–12
5	2.2668e–09
10	1.1420e–09
14	1.7138e–10
15	**3.1890e–13**
16	9.0975e–11
20	2.4231e–06

4 Discussion

Developed ANN is trained with 210 samples and validated with 90 samples that were classified based on the presence of the metabolic syndrome in two groups: healthy and disease. In Group 1 which was labeled as "Healthy", ANN prediction showed 28 true positives from a healthy group of 30 samples and 0 false positives from a disease group of 60 samples. In Group 2 which was labeled as "Disease", ANN prediction showed 2 false negatives from a healthy group of 30 samples and 60 true negatives from a disease group of 60 samples. In the case of cardiovascular disease risk assessment in patients with metabolic syndrome, sensitivity is calculated to be 93.3%, specificity 100%, and accuracy 97,7%.

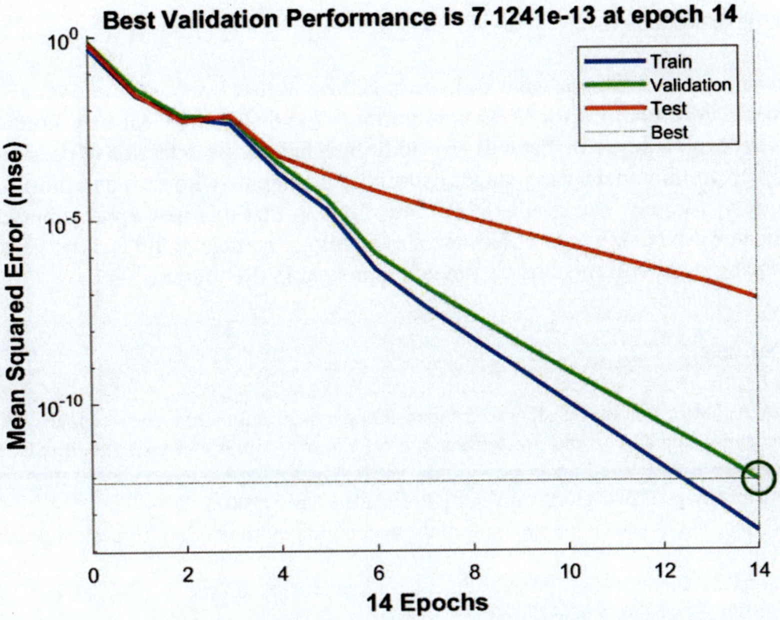

Fig. 3. Training performance with 15 neurons

Table 5. Results of validation

	Group 1 Healthy ANN prediction	*Group 2 Disease* ANN prediction	Accuracy
GROUP 1 Healthy 30	TP 28	FN 2	***97.7%***
Group 2 Disease 60	FP 0	TN 60	
Sensitivity	***93%***		
Specificity	***100%***		

5 Conclusion

The prevalence of cardiovascular disease is getting higher every year and while developing better formulations for CVD is important, it is becoming clear that more crucial thing is to develop a system that will give additional help in the detection of possible risks and help prevention in the early stages especially in patients with serious conditions like metabolic syndrome. The results of the are showing that this new approach and more sophisticated ANN networks based on an even bigger sample could be a powerful tool in advancing what will medical treatments represent in the future.

References

1. Reaven, G.M.: Insulin resistance, compensatory hyperinsulinemia, and coronary heart disease: syndrome X revisited. In: Jefferson, L.S., Cherrington, A.D. (eds.) Handbook of Physiology, Section 7, The Endocrine System, Vol II, The Endocrine Pancreas and Regulation of Metabolism, p. 126. Oxford University Press, New York (2001)
2. Saklayen, M.G.: The global epidemic of the metabolic syndrome. Curr. Hypertens. Rep. **20**(2), 12 (2018)
3. Swarup, S., Goyal, A., Grigorova, Y., et al.: Metabolic syndrome. In: StatPearls. StatPearls Publishing, Treasure Island (2020)
4. Tune, J.D., Goodwill, A.G., Sassoon, D.J., Mather, K.J.: Cardiovascular consequences of metabolic syndrome. Transl. Res. J. Lab. Clin. Med. **183**, 57–70 (2017)
5. Berwick, Z.C., Dick, G.M., Tune, J.D.: Heart of the matter: coronary dysfunction in metabolic syndrome. J. Mol. Cell. Cardiol. **52**(4), 848–856 (2012)
6. Ramesh, A.N., Kambhampati, C., Monson, J.R., Drew, P.J.: Artificial intelligence in medicine. Ann. R. Coll. Surg. Engl. **86**(5), 334–8 (2004)
7. Hodzic, J., Gurbeta, L., Omanovic-Miklicanin, E., Badnjevic, A.: Overview of next-generation sequencing platforms used in published draft plant genomes in light of genotypization of immortelle plant (helichrysium arenarium). Med. Arch. **71**(4), 288–292 (2017)
8. Gurbeta, L., Maksimovic, M., Omanovic-Miklicanin, E., Badnjevic, A., Sejdic, E.: A telehealth system for automated diagnosis of asthma and chronical obstructive pulmonary disease. J. Am. Med. Inform. Assoc. **25**(9), 1213–1217 (2018)
9. Badnjević, A., Pokvić, L.G., Hasičić, M., Bandić, L., Mašetić, Z., Kovačević, Ž., Kevrić, J., Hifziefendić, J., Pecchia, L.: Evidence-based clinical engineering: machine learning algorithms for prediction of defibrillator performance. Biomed. Sign. Process. Control **54**, 101629 (2019)
10. Kovačević, Ž., Pokvić, L.G., Spahić, L., Badnjević, A.: Prediction of medical device performance using machine learning techniques: infant incubator case study health and technology. Health and Technology
11. Sarić, R., Jokić, D., Beganović, N., Pokvić, L.G., Badnjević, A.: FPGA-based real-time epileptic seizure classification using Artificial Neural Network. Biomed. Sign. Process. Control **62**, 102106 (2020). ISSN 1746-8094
12. Purnell, J.Q.: Definitions, classification, and epidemiology of obesity. In: Feingold, K.R., Anawalt, B., Boyce, A., et al. (eds.) South Dartmouth (MA), MDText.com, Inc. (2000)
13. Maphephu, S.M., Kasvosve, I.: Reference intervals for serum total cholesterol, HDL cholesterol and non-HDL cholesterol in Batswana adults. S. Afr. Med. J. **101**(8), 546–7 (2011)

14. Mathew, T.K., Tadi, P.: Blood glucose monitoring. In: StatPearls. StatPearls Publishing, Treasure Island (2020)
15. Brzezinski, W.A.: Blood pressure. In: Walker, H.K., Hall, W.D., Hurst, J.W. (eds.) Clinical Methods: The History, Physical, and Laboratory Examinations. 3rd edn. Butterworths, Boston (1990). Chapter 16
16. World Health Organization. Definition, diagnosis and classification of diabetes mellitus and its complications. Geneva, WHO (1999). https://apps.who.int/iris/handle/10665/66040
17. Viera, A.J., Sheridan, S.L.: Global risk of coronary heart disease: assessment and application. Am. Fam. Phys. **82**(3), 265–74 (2010)
18. Rau, H.H., Hsu, C.Y., Lin, Y.A., Atique, S., Fuad, A., Wei, L.M., Hsu, M.H.: Development of a web-based liver cancer prediction model for type II diabetes patients by using an artificial neural network. Comput. Methods Program. Biomed. **125**, 58–65 (2016). ISSN 0169-2607
19. Catic, A., Gurbeta, L., Kurtovic-Kozaric, A., Mehmedbasic, S., Badnjevic, A.: Application of Neural Networks for classification of Patau, Edwards, Down, Turner and Klinefelter Syndrome based on first trimester maternal serum screening data, ultrasonographic findings and patient demographics. BMC Med. Genomics **11**, 19 (2018)
20. Badnjevic, A., Gurbeta, L., Custovic, E.: An Expert Diagnostic System to Automatically Identify Asthma and Chronic Obstructive Pulmonary Disease in Clinical Settings. Scientific Reports (2018)

Performance Inspection of Patient Monitors According to the Legal Metrology Framework: Bosnia and Herzegovina Case Study

Amar Deumić[1,2](✉), Emina Imamović[1], Lejla Gurbeta Pokvić[1,2], and Almir Badnjević[1,2]

[1] International Burch University, Sarajevo, Bosnia and Herzegovina
amar.deumic@stu.ibu.edu.ba
[2] Verlab – Medical Devices Verification Laboratory, Sarajevo, Bosnia and Herzegovina

Abstract. In Bosnia and Herzegovina, legal metrology framework has been implemented as a tool for ensuring safety and performance accuracy of medical devices. This paper presents the analysis of results for these periodical inspections of patient monitor. The analysis is based on 2118 patient monitor inspection reports from April 2015–December 2020. Inspections were conducted in clinical centers, hospitals and several primary healthcare institutions using etalon with documented traceability to SI units. The results show that in six-year period of implementation of this metrology-based framework for medical device management and assessment at national level the drop of 5.38% of patient monitor fault rate was achieved. Also, all performance measurements are available through information system which provides comprehensive management of these medical devices at institutional and national level.

1 Introduction

Medical professionals are working with sophisticated medical devices but interpretation of obtained values and understanding of their function is very important. Due to the gap in this area patient safety issue is gaining on its importance [1].

Recently, majority of healthcare institutions is undertaking accreditation procedures, depending on national legislation, to ensure quality of treatment provided to their patients. The reason for such activities is increasing competitiveness on national and international market, especially now when major shift from public to private healthcare can be observed. However, even in these situation mechanisms of ensuring patient safety, in terms of medical device reliability achieved through post-market supervision are not well defined. This is proven by numerous incident reports [2, 3] involving variety of medical devices including infant incubators, but importantly emphasized by recent amendments of medical device directives at European level.

One way of overcoming this issue can be by implementation of an interdisciplinary supervision mechanism for medical devices that relies on scientific evidence-based methodology [4, 5]. Such has been done in Bosnia and Herzegovina where Legal Metrology Framework for medical devices with measuring function has been introduced [6].

A. Badnjevic and L. Gurbeta Pokvić (Eds.): CMBEBIH 2021, IFMBE Proceedings 84, pp. 386–392, 2021.
https://doi.org/10.1007/978-3-030-73909-6_44

This framework applies for 10 types of both diagnostic and therapeutic medical devices that are used in healthcare, including patient monitors.

Patient monitor is a medical device which continuously monitors physiological parameters of a patient. It is most commonly used during operational interventions, or for monitoring of patients at intensive care unit (ICU) [7, 8]. Physiological parameters which can be monitored are electrocardiograph (ECG), respiration rate (RR), blood oxygen saturation (SpO2), non-invasive blood pressure (NIBP), invasive blood pressure (IBP), temperature (TEMP) and heart rate (HR) [9]. All these parameters are obtained through biosensors placed on a specific place on a patient's body.

As multi-parameter measuring devices, patient monitors must meet several requirements posed on every phase of device life-cycle, from manufacturing to the usage in real-time setting [10]. During production of patient monitors manufacturing company must comply with ISO 13485 which ensures that manufacturer successfully meets customer's and regulatory requirements. Additionally, all medical devices, including patient monitors, must comply to requirements of the basic safety and essential performance standard issued by International Electrotechnical Commission (IEC) [11]. IEC standard for patient monitors is IEC 80601-2-49:2018 "Medical electrical equipment – Part 2-49: Particular requirements for the basic safety and essential performance of multifunction patient monitoring equipment" has a scope restricted to patient monitors which are intended for connection to a single patient that has two or more physiological monitoring units.

Ensuring safety and performance accuracy during design, approval and manufacturing phase is very well regulated on international and national level, but post-market surveillance and management of medical devices is shown to be comprehensive task both on institutional and national level. Nowadays, post-market surveillance mechanism and device performance evaluation relies on manufacturers reporting incidents and conducting risk minimization strategies [2, 3]. On the other hand, legal metrology framework presents an independent evidence-based methodology for periodical inspection of safety and performance of medical devices. According to legal metrology framework, medical devices such as mechanical ventilators, infant incubators [12], anesthesia machines, defibrillators [13], therapeutic ultrasound [14], dialysis machines [15], infusion and perfusor pumps [16], including patient monitors are periodically inspected by independent inspection body accredited by ISO 17020 international standard [17]. Inspection is done for each physiological parameter that can be measured from patient. Measurements during inspection are taken with etalons which have stated traceability to SI unit. For every inspection a report with all measurements is created and stored in unique electronic database [18]. In this way, an evidence of device performance and safety is created for every single device used.

The objective of this paper is to present results of annual inspection of patient monitors in the period of six years and their impact on patient monitor proper functioning.

2 Materials and Methods

The analysis presented in this paper is based on 2118 patient monitor inspection reports which were acquired during April 2015–December 2020 time frame. Inspection of patient monitors were conducted in clinical centers, hospitals and several primary healthcare institutions by appointed inspection body.

Inspection of patient monitors were conducted by a procedure which is in accordance to ISO 17020 *Conformity assessment – Requirements for the operation of various types of bodies performing inspection*, national regulation (*Rules on metrological and technical requirements for patient monitors, Official Gazzete of B&H, No. 75/14*) and international standards, Table 1.

Table 1. International standards for patient monitors

Parameter	Standard	Parameter to measure	Ref.
Pulse oximetry	ISO 9919:2005	Basic safety and essential performance requirements of multifunction patient monitoring equipment	[17]
ECG and NIBP	IEC 60601-2-27 IEC 60601-2-34 IEC 60601-2-49 ANSI/AAMI SP10:2002/A1:2003	Basic safety and essential performance requirements of multifunction patient monitoring equipment For Patient Monitor, Patient Monitor Simulator is required to check and simulate critical parameters like NIBP, Respiration, ECG, IBP, SpO_2 and also to simulate performance waveforms Standard for electronic or automated sphygmomanometers	[11]
Heart rate (HR)	ANSI/AAMI EC 13: 2002/rev 2007 and IEC 60601-2-27	Standard for cardiac monitors, heart rate meters, and alarms	[11]

Every patient monitor was subject to visual inspection, followed by electrical safety inspection and performance inspection. For electrical safety inspection Fluke Biomedical ESA 620 etalon was used [19] while for performance inspection Fluke Biomedical ProSim 8 etalon was used [20].

During visual inspection, general information about the device is collected (manufacturer, type, serial number), as well as observation of structural properties is done. Device is checked for integrity and functionality according to manufacturer specification.

Electrical safety inspection is conducted according to IEC 60601 – Medical electrical equipment: General requirements for basic safety and essential performance of medical devices. The third step of inspection is performance inspection with the aim of determining measurement error for the device. Table 2 presents the ranges for inspected parameters on patient monitor as well as allowed error values.

Inspection criteria is presented in Table 3.

Table 2. Measuring points for patient monitor parameters

Parameter	HR	RR	SpO2	NIBP
Range	30–300 BPM	20–150 bpm	70–99%	50/80 mmHg 120/80 mmHg 150/100 mmHg 200/150 mmHg
Error	±1% *Relative error*	±2% *Relative error*	±2% *Absolute error*	±5 mmHg *Absolute error*

Table 3. Inspection criteria

Device inspection report result	Visual inspection	Electrical safety inspection	Performance inspection
Pass	✔	✔	✔
Fail	X	Not perfomed	Not perfomed
Fail	✔	X	Not perfomed
Fail	✔	✔	X

3 Results

Evaluating medical device's performance and safety is of great concern and although a lot of research has been done on certain medical devices, evaluation of patient monitors has not been researched in depth. Breteler et al. [21] evaluated HR and RR of four different patient monitor types and used Clarke Error Grid analysis on gathered data to evaluate monitor's performance. They reported that significant variation in measurements of these parameters wasn't detected. For measuring the RR, three of monitor types showed reasonable accuracy while one was outside of acceptable limits. Maglogiannis et al. [22] showed that the highest risk threats come from failure of power supply, failures in software modules and user-based errors.

Our analysis (Table 4) shows that over the course of six years, the overall percentage of accurate devices was 94.85%. This means than an average of 5.15% inspected patient monitors didn't satisfy the inspection criteria and was not acceptable for usage on patients. Analysis of visual inspection reports revealed that majority of failures were due to worn out patient cables, with broken insulations that are used in patient treatment. Such cases were immediately replaced by technical service in healthcare institutions.

Figure 1 shows the comparison of percentage of faulty devices obtained in a six-year period. The highest percentage of faulty devices were detected in 2015 and 2017. The results of 2015 were very alarming since the sample number is significantly smaller than for 2017.

Table 4. Classification of devices in accurate and faulty

Year of inspection	Pass (P)	Fail (F)	Total number of inspected devices
2015	73	7	80
2016	315	22	337
2017	405	44	449
2018	395	12	407
2019	420	10	430
2020	401	14	415
2015–2020	**2009**	**109**	**2118**

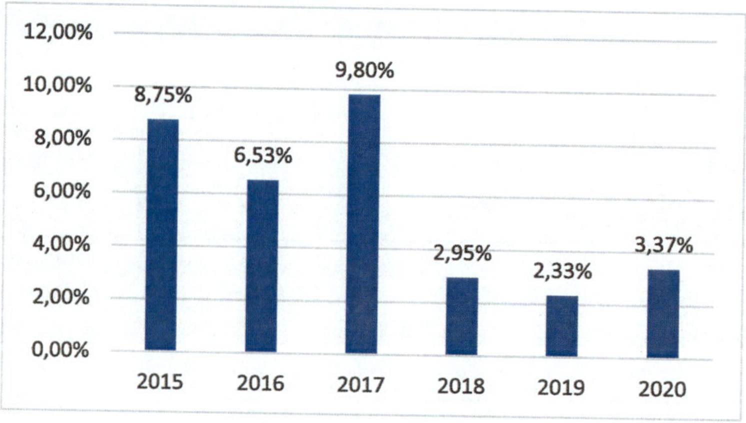

Fig. 1. Percentage of faulty devices

As a result of application of Legal metrology framework for medical devices a drop of 5.38% in faulty rate was detected in six-year period. An average variation of faulty rate of 0.83% can be detected for 2018–2020 period which is smaller in comparison to 2.74% variation in average for 2015–2017 period. This shows that medical device management according to legal metrology framework is beneficial at institutional level, since electrical safety and performance failures are timely detected and taken care of.

All measurement results, for all inspected patient monitors, as mentioned before, are available in digital database and can be managed through information system. In this way, big data concerning patient monitors is created [23]. This opens new opportunities for creating expert systems for prediction of performance of medical devices. It offers new approach in medical device management. For instance, up to now researchers have used different machine learning algorithms to make a prediction whether a specific patient monitor would reach the critical value in the next hour while monitoring patient [24]. Stored data on device electrical safety and performance opens new possibilities for usage of machine learning for prediction of device performance and failures [25, 26].

4 Conclusion

Following the facts presented above, legal metrology framework, consists of three key components: metrology, international standards and medical devices. The overall goal of proposed supervision mechanism is ensuring safety and performance of medical devices. This goal is achieved by implementation of procedures for periodical conformity assessment of medical device safety and performance.

Results presented in this paper show that as a result of application of this framework on example of patient monitors, decrease in device faulty rate is detected. This framework is beneficial not only for institutional level for ensuring safety and accuracy of devices, but also at national level, since real-time data about used devices are available through information system that manages inspections.

Conflict of Interest. The authors hereby declare to have no conflict of interest.

References

1. Wallace, D.R., Kuhn, D.R.: Failure modes in medical device soft-ware: an analysis of 15 years of recall data. Int. J. Reliab. Qual. Saf. Eng. **8**(04), 351–371 (2001)
2. Liebel, T.C., Daugherty, T., Kirsch, A., Omar, S.A., Feuerstein, T.: Analysis: using the FDA MAUDE and medical device recall databases to design better devices. Biomed. Instrum. Technol. **54**(3), 178–188 (2020)
3. Chung, G., Etter, K., Yoo, A.: Medical device active surveillance of spontaneous reports: a literature review of signal detection methods. Pharmacoepidemiol. Drug Saf. **29**(4), 369–379 (2020)
4. Badnjevic, A., Gurbeta, L., Boskovic, D., Dzemic, Z.: Measurement in medicine – past, present, future. Folia Med. Facultatis Med. Univ. Saraeviensis J. **50**(1), 43–46 (2015)
5. Badnjević, A., Gurbeta Pokvić, L., Spahić, L.: Chapter 74 - inspection of medical devices. In: Iadanza, E. (ed.) Clinical Engineering Handbook (Second Edition), pp. 491–497. Academic Press, Elsevier (2020)
6. Badnjević, A., Cifrek, M., Magjarević, R., Džemić, Z. (eds): Inspection of Medical Devices. Series in Biomedical Engineering. Springer, Singapore
7. Badnjević, A., Gurbeta Pokvić, L.: Chapter 76 - intensive care unit. In: Iadanza, E. (ed.) Clinical Engineering Handbook (Second Edition), pp. 503–508, Academic Press, Elsevier (2020)
8. Gilbert, A.: 23 The patient-monitor system in intensive care. Engineering Psychology and Cognitive Ergonomics: Volume 2: Job Design and Product Design, vol. 175 (2017)
9. Ramakrishna, S., Tian, L., Wang, C., Liao, S., Teo, W.E.: Medical Devices: Regulations, Standards and Practices. Woodhead Publishing, Cambridge (2015)
10. Rathi, V.K., Krumholz, H.M., Masoudi, F.A., Ross, J.S.: Characteristics of clinical studies conducted over the total product life cycle of high-risk therapeutic medical devices receiving FDA premarket approval in 2010 and 2011. JAMA **314**(6), 604–612 (2015)
11. International Electrotechnical Commission. Medical Electrical Equipment Part 2–27: Particular Requirements for the Basic Safety and Essential Performance of Electrocardiographic Monitoring Equipment. ICS (2011)
12. Badnjevic, A., Gurbeta, L., Jimenez, E.R., Iadanza, E.: Testing of mechanical ventilators and infant incubators in healthcare institutions. Technol. Health Care **25**(2), 237–250 (2017)

13. Gurbeta, L., Dzemic, Z., Bego, T., Sejdic, E., Badnjevic, A.: Testing of anes-thesia machines and defibrillators in healthcare institutions. J. Med. Syst. **41**, 133 (2017). https://doi.org/10.1007/s10916-017-0783-7
14. Gurbeta, L., Badnjevic, A., Dzemic, Z., Jimenez, E.R., Jakupovic, A.: Testing of therapeutic ultrasound in healthcare institutions in Bosnia and Herzegovina. In: 2nd EAI International Conference on Future Access Enablers of Ubiquitous and Intelligent Infrastructures, pp. 24–25. Belgrade, Serbia (2016)
15. Gurbeta, L., Alic, B., Dzemic, Z., Badnjevic, A.: Testing of dialysis machines in healthcare institutions in Bosnia and Herzegovina. In: Eskola, H., Väisänen, O., Viik, J., Hyttinen, J. (eds.) EMBEC & NBC 2017. IFMBE Proceedings, vol. 65. Springer, Singapore (2017)
16. Gurbeta, L., Alic, B., Dzemic, Z., Badnjevic, A.: Testing of infusion pumps in healthcare institutions in Bosnia and Herzegovina. In: Eskola, H., Väisänen, O., Viik, J., Hyttinen, J. (eds.) EMBEC & NBC 2017. IFMBE Proceedings, vol 65. Springer, Singapore (2017)
17. International Organization for Standardization. Medical electrical equipment — Particular requirements for the basic safety and essential performance of pulse oximeter equipment for medical use (2005)
18. Gurbeta, L., Badnjević, A.: Inspection process of medical devices in healthcare institutions: software solution. Health Technol. **7**(1), 109–117 (2017). https://doi.org/10.1007/s12553-016-0154-2
19. Fluke Biomedical ESA etalon. https://goo.gl/eJuqWE
20. Fluke Biomedical ProSim 8 etalon. shorturl.at/ezCG0
21. Breteler, M.J., KleinJan, E.J., Dohmen, D.A., Leenen, L.P., van Hillegersberg, R., Ruurda, J.P., Kalkman, C.J.: Vital signs monitoring with wearable sensors in high-risk surgical patients: a clinical validation study. Anesthesiology **132**(3), 424–439 (2020)
22. Maglogiannis, I., Zafiropoulos, E., Platis, A., Lambrinoudakis, C.: Risk analysis of a patient monitoring system using Bayesian Network modeling. J. Biomed. Inform. **39**(6), 637–647 (2006)
23. Gurbeta, L., Badnjević, A., Kurta, E.: EVerlab: software tool for medical device safety and performance inspection management. In: Badnjevic, A., Škrbić, R., Gurbeta Pokvić, L. (eds.) CMBEBIH 2019. IFMBE Proceedings, vol. 73. Springer, Cham (2020)
24. Chang, D., Chang, D., Pourhomayoun, M.: Risk prediction of critical vital signs for icu patients using recurrent neural network. In: 2019 International Conference on Computational Science and Computational Intelligence (CSCI), pp. 1003–1006. IEEE (2019)
25. Badnjevića, A., Pokvić, L.G., Hasičić, M., Bandić, L., Mašetić, Z., Kovačević, Z., Kevrić, J., Pecchia, L.: Evidence-based clinical engineering: machine learning algorithms for prediction of defibrillator performance. Biomed. Sign. Process. Control **54**, 101629 (2019)
26. Kovačević, Z., Pokvić, L.G., Spahić, L., Badnjević, A.: Prediction of medical device performance using machine learning techniques: infant incubator case study. Health Technol. **10**(1), 151–155 (2020)

Pharmaceutical Engineering

Transdermal Patches as Noninvasive Drug Delivery Systems

Jasmina Hadžiabdić(✉), Lejla Šejto, Ognjenka Rahić, Amina Tucak, Lamija Hindija, and Merima Sirbubalo

Faculty of Pharmacy, Department of Pharmaceutical Technology, University of Sarajevo, Zmaja od Bosne 8, 71000 Sarajevo, Bosnia and Herzegovina
jasmina.hadziabdic@ffsa.unsa.ba

Abstract. The most common pharmaceutical dosage forms applied to the skin to deliver an exact dose of active substance through the skin into the bloodstream are self-adhesive transdermal drug delivery systems or transdermal patches. There have been many attempts to manufacture transdermal patches that deliver many different active substances, but this process has proved to be quite tedious and difficult, as active substances must meet a long list of strict criteria for their transdermal delivery to be possible. Transdermal patches as physical devices differ in their composition and manufacturing method. They are designed and manufactured as multi-layered, single-dose pharmaceutical preparations. Components of a solid adhesive transdermal patch are: backing layer, liner, reservoir or polymeric matrix with the incorporated active substance, rate controlling membrane (in reservoir-type transdermal patches, mostly), and the pressure-sensitive adhesive. Most of these components are made of different polymers, which allows adjustment of a transdermal patch's mechanical and adhesive properties, among others. Transdermal patches release active substances through various mechanisms. In order for the active substances to be released from a transdermal patch, solubilization of active substances needs to occur, followed by diffusion. This diffusion follows first-order kinetics and is usually represented by the two- or five-compartmental kinetic model of the skin.

Keywords: Transdermal patch · Polymers · Layers

1 Introduction

Transdermal patches (TDPs) are a drug delivery system used and designed to allow drug permeation through the skin layers and into systemic circulation, as an alternative to oral and intravenous drug delivery. These multi-layered noninvasive systems have a number of advantages such as minimizing the first pass metabolism effects by avoiding enterohepatic circulation, allowing for steady state concentration to be maintained, while being applied by the patient, further simplifying dosage and use of these products. However, there is always risk and possibility of skin irritation or hypersensitivity, so caution is necessary [1, 2]. Specially designed TDPs allow complete control of how,

A. Badnjevic and L. Gurbeta Pokvić (Eds.): CMBEBIH 2021, IFMBE Proceedings 84, pp. 395–402, 2021.
https://doi.org/10.1007/978-3-030-73909-6_45

when and what amount of the active substance/drug is released from the patch and delivered into the body. Although these devices vary greatly in appearance, components used in their manufacture all fall into one of two groups: mechanical and chemical [3, 4]. Mechanical components of TDPs are: protective layers (liner and backing layer), the reservoir containing the active substance, adhesive layer and the rate controlling membrane. Chemical components are used to produce the mechanical ones: polymeric matrix, the active substance that is going to be placed inside the reservoir, chemical penetration enhancers, adhesive polymers, and backing layer polymers [4]. Most commonly used are the solid adhesive patches that adhere to the *stratum corneum* tightly [2, 3]. These solid patches have either a reservoir, or a polymeric matrix, into which the active substance is incorporated and from which it is released. Matrix-type adhesive TDP formulations may have one or several "active" layers – in which the active substance located [1, 3]. Based on the time of their development, there are three generations of TDPs. Most of the patches used today are 1st generation, but lately, 3rd generation patches such as those with microneedles, are used to deliver macromolecules and vaccines [2, 3, 5].

2 Components of Solid Adhesive TDPs'

Solid adhesive TDPs are usually manufactured as a reservoir or matrix type of a product. In reservoir systems the active substance may be dissolved or dispersed in a semisolid basis or in a solid polymer matrix separated from the skin by a rate controlling membrane and a pressure-sensitive adhesive. Matrix systems contain the active substance in a solid or semi-solid matrix whose properties control the diffusion pattern to the skin. The matrix system may also be a solution or solid dispersion of the active substance in the pressure-sensitive adhesive [6]. However, there is another type of formulation – the *drug-in-adhesive* (DIA) patch (Fig. 1). Regardless of the production process or product type, all solid adhesive TDPs have similar components and type of skeleton: drug (in matrix, adhesive or reservoir), adhesive, protective layers (backing and liner), while some contain a rate controlling membrane [1, 3, 7].

2.1 Active Substances in TDPs

Over the years, TDPs have been used to deliver many drugs such as fentanyl, pizotifen, meloxicam, pentazocine, propranolol, atenolol, carvedilol, losartan, valsartan, irbesartan, verapamil, diltiazem, buspirone, estrogen, vitamins, tamoxifen and drugs for the treatment attention deficit hyperactivity disorder [8–18]. In order to be a candidate for transdermal delivery via a patch, a drug must meet a large number of very strict criteria, which is why there are so few commercially available TDPs. These criteria are mostly directed at the molecule size, shape and chemical characteristics [1, 19]: solubility in both water and oil phases (>1 mg/mL), melting point < 93 °C, molecular mass < 1000 Da (usually <500 Da), it's saturated aqueous solution should have a pH value of 5–9, doesn't ionize at physiological pH, efficient in dosages < 25 mg/day and has a short half-life, stability and non-irritability in direct contact with skin, doesn't develop tolerance during transdermal use, doesn't cause immunological reactions or bind irreversibly to dermis' structures, doesn't metabolize extensively in the skin [1, 19]. While

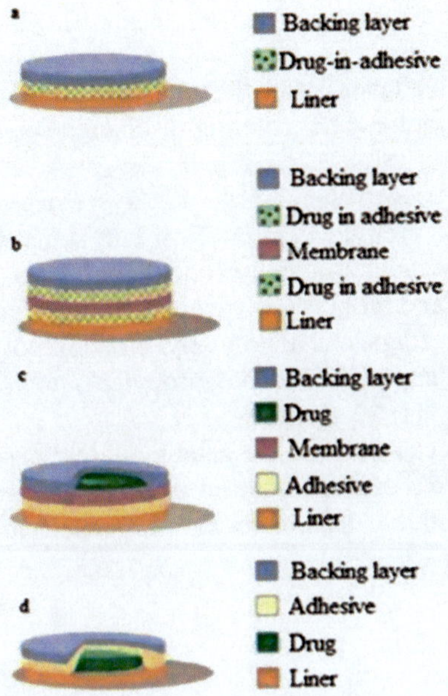

Fig. 1. Most common TDP designs: single layer DIA patch (a), multi-layer DIA patch (b), reservoir system (c), and matrix system patch (d) [4].

all these factors are important, the key limiting factors are molecule size, ionization and partition coefficient [20].

2.2 Polymer Matrix

The polymer matrix is the backbone of a TDP, which, together with the adhesive, controls the release rate of the drug. Polymers used as matrix-formers have effect on cohesive properties of the TDP as well as its mechanical properties and stability [21]. These polymers have to be inert, non-irritating, non-toxic, chemically and thermally stable [1].

Polymers most used as matrix-formers are cross-linked polyethylene glycol networks, acrylic-acid matrices, ethyl cellulose (EC), polyvinyl pyrrolidone (PVP) (EC/PVP mixture with 30% dibutyl phthalate as a plasticizer), hydroxypropyl methylcellulose (HPMC), chitosan (alone or in combination with HPMC), ethyl vinyl acetate copolymer [10, 21, 22].

2.3 Reservoir

The reservoir is placed between the backing layer and rate controlling membrane [22, 23]. Reservoirs in transdermal patches are usually manufactured as hydrogels from polysaccharide materials, alginate hydrogels, ethyl vinyl alcohol hydrogel, polyelectrolyte hydrogels and thermoresponsive hydrogels (poly-N-isopropyl acrylamide) [24]. In a microreservoir system, the chamber is usually made of an elastomeric material such as polydimethylsiloxane [25].

2.4 Adhesives

The purpose of the adhesive layer is to adhere a transdermal patch strongly to the skin under light pressure. It shouldn't be irritating or change surrounding flora of the skin while in contact with it [1].

Chemically, polymers used to manufacture this layer can be divided into three groups: acrylates (methyl-, ethyl-, propyl-, isopropyl-, n-butyl-, hexyl-, heptyl, 2ethylhexyl-, 2-ethylbutyl-, dodecyl-, hexadecyl- and cyclohexyl acrylate), silicones (polydimethylsiloxane with silica resin) and rubbers (poly isobutylene, styrene/isoprene/styrene block copolymer, natural rubber, tragacanth). Polyvinyl alcohol, polyvinyl pyrrolidone, ethylene vinyl acetate copolymer, low density polypropylene, polysulfones, polyamides, and polyesters can also be used [1, 22, 26, 27].

Natural polymers that can be used are gelatin, starch, casein and sodium alginate, and semi-synthetic polymers that can be used as adhesives are sodium carboxymethyl cellulose (Na-CMC) and other cellulose derivatives. These can also be used as backing layers [27].

2.5 Backing Layer

This protective layer of a TDP has to be impermeable to active substances and penetration enhancers, have optimum elasticity and flexibility, and at the same time be affordable and compatible with active substances, penetration enhancers, adhesives and other excipients used [1].

The backing layer is usually made of vinyl, polyethylene and polyester films [1, 29]. Polyvinyl acetate, acrylate esters and EC are commonly used [30]. Natural polymers used for this layer are gelatin, sodium alginate, starch, arabic gum and natural rubber; semi-synthetic polymers used are Na-CMC and other cellulose derivatives [27, 30].

2.6 Liner

During storage, the adhesive layer is covered by a protective layer that is removed before applying the patch to the skin. Although this layer should be considered as part of the packaging, it still comes into direct contact with the drug delivery system, so it must meet very strict criteria regarding inertness and drug, penetration enhancer, and water permeation [22].

Polymers used to manufacture this layer can be divided into two groups: silicone coated liners for use with non-silicone adhesives and fluorocarbon and fluorosilicone coated liners for standard use [3]. It can be made of polyethylene, polyvinyl chloride, polydimethylsiloxane (with or without silicone resin), polyester foil and metallized layers [1, 12].

2.7 Rate Controlling Membrane

The rate controlling membrane is a crucial component of a large number of TDPs. Properties of polymers used to manufacture the rate controlling membrane affect the manner and speed in which an active substance is released, whether it's placed in a reservoir type, or a multi-layer drug-in-adhesive formulation [1, 4, 31].

This layer usually consists of natural polymers, such as chitosan [1, 12]. The membrane can also be a mixture of sodium alginate and tragacanth, and a mixture of PVP and gelatin [33]. Other materials used for this layer are silicone rubber, polyethylene, polyurethane, polysulfones and polyvinylidene fluoride (to make hydrophilic membranes), polytetrafluoroethylene (to make hydrophobic membranes) [12, 23, 28]. Poly2-hydroxyethyl methacrylate membranes can also be used [1].

3 Mechanism and Kinetics of Drug Release

In order to allow their insertion into the TDP in the maximum possible concentration, drugs are usually used in their crystal form. However, since molecules of a crystallized drug can't be transported to the tissues, these molecules must be transformed into their free form. This is done by solubilization of the drug, as water permeates into the polymer matrix over time and dissolves the crystals of drug. Solubilization of drug is followed by the drug flux – the free drug molecules are exposed to bodily fluids. Thus, the polymer absorbs large amounts of liquid and swells in the body fluids. The free drug then diffuses out of the swollen matrix system, travel via systemic circulation and binds to specific cellular receptors, followed by its intake by endocytosis [33].

Skin permeation mechanisms are usually explained by Fick's laws of diffusion. Since these equations are extremely complex (especially with biological variations in systems such as the human body), a simplified approach has been developed. Essentially, TDP, skin, and blood are observed as different compartments. Depending on whether skin and blood are observed as one compartment or as subcompartments – *stratum corneum* and epidermis, and capillaries, a two- to five-compartmental model is created. Most kinetic models for determining drug diffusion and skin permeation use the so-called representative model of the skin (Fig. 2). In this model it is assumed that drug diffusion follows first order kinetics, and that the main barrier for drug permeation through the skin is *stratum corneum*. Recently, another model has been proposed: the multistage modeling of multicomponent transdermal diffusion [34].

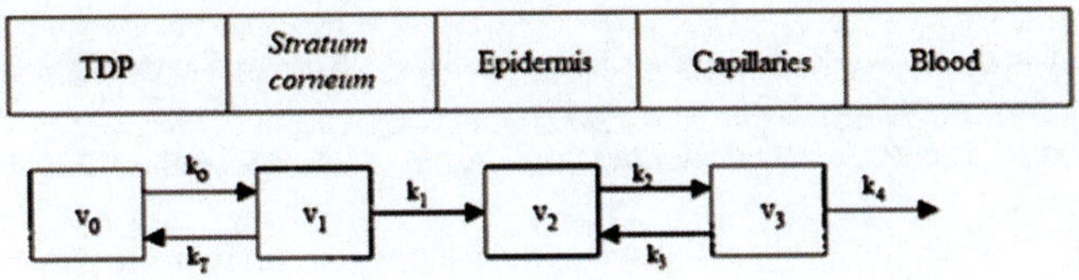

Fig. 2. The "representative" kinetic model of the skin [34].

Equation (1) shows the rate of permeation through the skin C_d and C_r represent permeant concentrations in two different compartments e.g., on the surface of the *stratum corneum* and within the body and P_s is the coefficient of skin permeability for the drug. From this equation it is clear that a constant diffusion rate can only be achieved if C_d is significantly higher than C_r; $C_d >> C_r$ (for example, if the concentration on the surface of the *stratum corneum* would be constantly significantly higher than in the body), and Eq. (1) becomes Eq. (2) [4]:

$$dQ/dt = P_s \times (C_d - C_r) \tag{1}$$

$$dQ/dt = P_s \times C_d \tag{2}$$

Equation (3) allows calculation of the maximum possible speed of drug permeation through the skin [4]:

$$(dQ/dt)max = P_s \times C_s \tag{3}$$

Equation (3) shows that the maximum skin permeation rate depends on the skin permeability coefficient P_s and equilibrium solubility in the *stratum corneum* C_s [4].

4 Conclusion

Transdermal drug delivery via a patch is a good alternative to conventional delivery paths such as oral and parenteral, mainly because it minimizes unwanted or adverse effects, and allows patients to apply the product themselves. TDPs are designed as multi-layered devices that allow controlled and gradual drug release into the skin and subsequently a systemic circulation. *Stratum corneum* acts mostly as a barrier to this delivery, making it difficult to manufacture devices to deliver most of currently known active substances. The main components of a TDP are protective layers (backing layer and liner), reservoir or polymer matrix into which the active substance is incorporated, the rate controlling membrane, and the adhesive. It is possible to adjust all these layers to improve the TDP's mechanical and chemical properties and thus improve drug delivery.

Conflict of Interest. The authors have no conflicts of interest to disclose.

References

1. Saha, A., Debnath, S.: Transdermal Drug Delivery System. LAP Lambert Academic Publishing, Saarbrücken (2012)
2. Prausnitz, M.R., Langer, R.: Transdermal drug delivery. Nat. Biotechnol. **26**(11), 1261–1268 (2008)
3. Miller, K.J.: Transdermal product formulation development. In: Benson, H.A.E., Watkinson, A.C. (eds.) Transdermal and Topical Drug Delivery Principles and Practice, pp. 287–309. John Wiley and Sons Inc, Hoboken (2011)
4. Hafeez, A., Jain, U., Singh, J., Maurya, A., Rana, L.: Recent advances in transdermal drug delivery system (TDDS): an overview. J. Sci. Innov. Res. **2**(3), 695–709 (2013)
5. Choi, S.H., Wang, Y., Conti, D.S., Raney, S.G., Delvadia, R., Leboeuf, A.A., Witzmann, K.: Generic drug device combination products: regulatory and scientific considerations. Int. J. Pharm. **544**(2), 443–454 (2018)

6. Benson, H.A.E.: Skin structure, function, and permeation. In: Benson, H.A.E., Watkinson, A.C. (eds.) Transdermal and Topical Drug Delivery Principles and Practice, pp. 3–14. John Wiley and Sons Inc, Hoboken (2011)
7. Sheth, N.S., Mistry, R.B.: Formulation and evaluation of transdermal patches and to study permeation enhancement effect of eugenol. J. Appl. Pharm. Sci. **1**(3), 96–101 (2011)
8. Sachan, R., Bajpai, M.: Transdermal drug delivery system: a review. Int. J. Res. Dev. Pharm. Life Sci. **3**(1), 2274–2290 (2017)
9. Patel, H.J., Trivedi, D.G., Bhandari, A.K., Shah, D.A.: Penetration enhancers for transdermal drug delivery system: a review. Int. J. Pharm. Innov. **1**(2), 67–80 (2011)
10. Manish, K., Trivedi, V., Shukla, A.K., Dev, S.K.: Effect of polymers on the physicochemical and drug release properties of transdermal patches of atenolol. Int. J. Appl. Pharm. **10**(4), 68–73 (2018)
11. Mounika, K., Reddy, B.V., Reddy, K.N.: Formulation and evaluation of carvedilol transdermal patches with hydrophilic polymers. World J. Pharm. Res. **3**(10), 815–826 (2014)
12. Yewale, C., Tandel, H., Misra, A.: Application of polymers in transdermal drug delivery. In: Misra, A., Shahiwala, A. (eds.) Applications of Polymers in Drug Delivery, pp. 175–211. Smithers Rapra Technology Ltd, Shawbury (2014)
13. Baviskar, D.J., Parik, V.B., Gupta, H.N., Maniyar, A.H., Jain, D.K.: Design and evaluation of patches for transdermal delivery of losartan potassium. PDA J. Pharm. Sci. Technol. **66**(2), 126–135 (2012)
14. Srilakshmi, A., Teja, V.R., Thippesh, M., Lakshmi, K., Salma, U.S., Nagendrababu, B.: Formulation and evaluation of transdermal patches of Irbesartan. Indian J. Res. Pharm. Biotechnol. **5**(3), 212–215 (2017)
15. Serna Jimenez, C.E., del Rio Sancho, S., Calatayud-Pascual, M.A., Balaguer-Fernandez, C., Femenia-Font, A., et al.: Development of antimigraine transdermal delivery systems of pizotifen malate. Int. J. Pharm. **492**(1–2), 223–232 (2015)
16. Ah, Y.C., Choi, J.K., Ki, H.M., Bae, J.H.: A novel transdermal patch incorporating meloxicam: in vitro and in vivo characterization. Int. J. Pharm. **385**(1–2), 12–19 (2010)
17. Verma, P.R.P., Chandak, A.R.: Development of matrix controlled transdermal delivery systems of pentazocine: in vitro/in vivo performance. Acta Pharm. **59**(2), 171–186 (2009)
18. Adhyapak, A., Desai, B.G.: Preparation and in vitro characterization of the transdermal drug delivery system containing tamoxifen citrate for breast cancer. Asian J. Pharm. **5**(1), 41–45 (2011)
19. Peddapalli, H., Ganta, R.P., Boggula, N.: Formulation and evaluation of transdermal patches for antianxiety drug. Asian J. Pharm. **12**(2), 127–136 (2018)
20. Murthy, S.N.: Approaches for delivery of drugs topically. AAPS Pharm. Sci. Tech. **21**(30), 1–2 (2019)
21. Kandavilli, S., Nair, V., Panchagnula, R.: Polymers in transdermal drug delivery systems. Pharm. Technol. **26**(5), 62–81 (2002)
22. Esuendale, D., Gabriel, T.: Cellulosic on transdermal drug delivery system: a review. J. Drug Deliv. Ther. **6**(5), 57–64 (2016)
23. Valenta, C., Auner, B.G.: The use of polymers for dermaln and transdermal delivery. Eur. J. Pharm. Biopharm. **58**(2), 279–289 (2004)
24. Sharma, K., Singh, V., Arora, A.: Natural biodegradable polymers as matrices in transdermal drug delivery. Int. J. Drug Dev. Res. **3**(2), 85–103 (2011)
25. Ochoa, M., Mousoulis, C., Ziaie, B.: Polymeric microdevices for transdermal and subcutaneous drug delivery. Adv. Drug Deliv. Rev. **64**(14), 1603–1616 (2012)
26. Jones, D.: Pharmaceutical applications of polymers for drug delivery. Rapra Rev. Rep. **15**(6), 15–32 (2004)
27. Sugibayashi, K., Morimoto, Y.: Polymers for transdermal drug delivery systems. J. Control. Release **20**(1–2), 177–185 (1994)

28. Patil, S.M., Bagawan, M.G.: A review: transdermal drug delivery system – basic understanding for development. Int. J. Inst. Pharm. Life Sci. **3**(6), 76–92 (2013)
29. Sinko, P.J.: Martin's Physical Pharmacy and Pharmaceutical Sciences, 6th edn. Lipincott Williams and Wilkins, Philadelphia (2010)
30. Arias, M., van Dijk, P.: What is natural rubber and why are we searching for new sources? Front. Young Mind. **7**(100), 1–9 (2019)
31. Pastore, M.N., Kalia, Y.N., Hortsmann, M., Roberts, M.S.: Transdermal patches: history, development and pharmacology. Br. J. Pharmacol. **172**(9), 2179–2209 (2015)
32. Pal, K., Banthia, A., Majumdar, D.K.: Polyvinyl alcohol-gelatin patches of salicylic acid: preparation, characterization and drug release studies. J. Biomater. Appl. **21**(1), 75–91 (2006)
33. Chakravarty, K., Dalal, D.C.: A nonlinear mathematical model of drug delivery from polymeric matrix. Bull. Math. Biol. **81**(1), 105–130 (2018)
34. Guy, R.H., Hadgraft, J.: Feasibility assessment in topical and transdermal delivery: mathematical models and in vitro studies. In: Guy, R.H., Hadgraft, J. (eds.) Transdermal Drug Delivery, 2nd edn., pp. 16–32. CRC Press, New York (2003)

Hydrocephalus: 5-HIAA and HVA in the Cerebrospinal Fluid

Mirsada Salihović[1](✉) and Emin Sofić[2]

[1] Department of Natural Sciences in Pharmacy, Faculty of Pharmacy, University of Sarajevo, Zmaja od Bosne 8, 71 000 Sarajevo, Bosnia and Herzegovina
mirsada.salihovic@ffsa.unsa.ba

[2] Department of Chemistry, Faculty of Science, University of Sarajevo, Zmaja od Bosne 33–35, 71 000 Sarajevo, Bosnia and Herzegovina

Abstract. The choroid plexus form cerebrospinal fluid (CSF) within four ventricles the brain. The composition of CSF is very similar to the of the brain's extracellular, and its composition depends on filtration and diffusion from the blood. Few conditions cause CSF volume and pressure to fall below normal (dehydration may be one), but many conditions can cause it to increase, sometimes under pressure. This may lead to blocked flow CSF from one or more ventricles, which causes an increase in intracranial pressure, and CSF buildup i.e. to hydrocephalus. Hydrocephalus results from one of three possible causes: over secretion of CSF, impaired absorption of CSF or obstruction of CSF pathways. The concentration of metabolites of the monoamines homovanillic acid (HVA) and 5-hydrocephalus acetic acid (5-HIAA) were determined with HPLC-ED in the CSF of 17 patients (children of 1 age) with hydrocephalus. Determined 5-HIAA concentrations are in the range of 0,99–113,89 ng/mL and HVA 0,98–165,92 ng/mL in CSF due to pressure changes occurring in hydrocephalus is increasing the turnover of serotonin and dopamine to its metabolites. The correlation no could be established between the levels of 5-HIAA and HVA and type of hydrocephalus, and the duration of disease.

Keywords: Hydrocephalus · 5-HIAA · HVA · Cerebrospinal fluid

1 Introduction

For more than 30 years, neurotransmitters such as noradrenaline (NA), dopamine (DA), and serotonin (5-HT) have been the main targets of neurochemical research for the development of therapies for neurological and psychiatric diseases Biosynthesis of catecholamines (CAs) DA, NA, and adrenaline (A) is performed in the brain, chromaffin cells of adrenal glands and in synaptic nerve endings. CAs are synthesized through a simple sequence of successive enzymatic reactions. The tyrosine hydroxylase (TH) enzyme catalyzes the synthesis of 3, 4-dihydroxyphenylalanine (DOPA) from tyrosine, which is synthesized from phenylalanine. DA is produced by the conversion of DOPA which is catalyzed by the aromatic amino acid decarboxylase enzyme. CAs break down very

A. Badnjevic and L. Gurbeta Pokvić (Eds.): CMBEBIH 2021, IFMBE Proceedings 84, pp. 403–412, 2021.
https://doi.org/10.1007/978-3-030-73909-6_46

quickly, approximately two minutes in the serum. The major metabolites relevant for laboratory testing are homovanillic acid (HVA), metanephrine (MN), and vanillylmandelic acid (VMA) [1].

5-HT mediates a number of central and peripheral functions, which makes its physiology very complex. Among other things, it is thought to be involved in the modulation of behavior and homeostatic processes in the body: mood regulation, cognitive and motor functions, and circadian and neuroendocrine rhythms such as regulation of breathing, sleep, cardiovascular function, metabolism, temperature, feeding, sexual activity and more [2, 3]. 5-HT is broken down by the action of the MAO enzyme and aldehyde dehydrogenase to 5-hydroxyindoleacetic acid (5-HIAA). Aside from the brain, this acid can be found in the cerebrospinal fluid (CSF), blood and urine [4]. Many literature reports indicate that low concentrations of 5-HIAA in CSF can be the cause of psychomotor agitation and positive schizophrenic symptoms, while high concentrations of 5-HIAA in CSF cause negative symptoms [5, 6].

It's known that CSF has several important roles in the functioning of the CNS. The cerebrospinal fluid protects the brain and spinal cord from serious injury in sudden movements and strokes. In subarachnoid spaces, the cerebrospinal fluid allows the brain to "float" which prevents the weight of the brain from pressing the roots of the cranial nerves and blood vessels against the inner surface of the skull. The cerebrospinal fluid enables the dilution and elimination of products of metabolism of brain cells and other substances coming from the blood into the brain. Furthermore, cerebrospinal fluid enables intracerebral transport [7, 8]. Excess CSF expands the brain chambers and creates pressure in the brain. The pressure created by an increase in the level of CSF associated with hydrocephalus can damage brain tissue and cause greater damage to brain functions.

Hydrocephalus is the enlargement of the cerebral chambers by excess CSF. Manifestations are head enlargement and brain atrophy. Increase in intracranial pressure causes irritability and bulging fontanelle in children. Diagnosis in infants is performed by ultrasound and in older children by CT or MRI [9].

Catecholamine and indolamine metabolites concentrations are usually determined by high-pressure liquid chromatography with electrochemical detection (HPLC-ED), gas chromatography (GC), or gas chromatography with mass spectrometry (GC-MS). For GC or GC-MS, the samples must be derivatized, while the HPLC method can analyze polar compounds without prior derivatization [10]. The advantage of HPLC methods is that they are fast and allow simultaneous determination of various catecholmines and indoleamine and their metabolites in the same sample.

The aim of this study is to determine the concentrations of 5-HIAA and HVA in CSF, by HPLC-ED method, in 17 patients diagnosed with hydrocephalus.

2 Materials and Methods

Biological Material: Liquor.

Reagents and Solutions: Perchloric acid ($HClO_4$) 60%, pro analysis, Merck; 5-HIAA minimum 98%, titration, Sigma; HVA, Sigma; Water Redistilled for Chromatography, Panreac; EDTA titriplex III disodium salt- dihydrate min 99%, pro analysis, Merck;

Sodium acetate anhydrous, for analysis, Merck; Acetic glacial-glacial acid ≥ 99,8%, pro analysis, Panreac; Methanol LiChrosolv, Merck.

Equipment and Accessories: Mixer VWR International, Germany AC 230 V 50 Hz 51 W 100–2400 1/min.; Microcentrifuge, Hettich; Micro 22R; Ultrasonic Water Bath, Sonis, Iskra; Vacuum Filtration Device, Supelco; Pair filter, nylon 66 membranes; 0,45 um × 47 mm, Supelco; HPLC, Shimadzu LCSOL SINGLE-LC EN, Made in Kyoto, Japan; Detector: ED, BAS Liquid Chromatography CC-5E, LC-4C; Pump: LC-20AT; Degasser: DGU-20A3; Oven: CTO-10Asvp; Working Electrode: Glassy carbon; model MF-1000; BASS; Reference electrode: Ag/AgCl; model MW-2030; BASS; Columns: ODS Hypersil 5 μm 200 × 4,6 mm Hewlett Packard.

2.1 HPLC-ED Analysis

Samples were stored at −80 °C. After thawing, the samples were diluted with $HClO_4$ (0,2 mol/L) at a ratio of 1:1. Such dilutions were centrifuged at 15,000 rpm for 30 min at 4 °C. After that, samples were decanted into clean and dry Eppendorf microcentrifuge tubes and re-centrifuged at 15,000 rpm for 10 min at 4 °C and decanted into pure Eppendorf microcentrifuge Tubes. These samples were injected (20 μl) in HPLC for analysis.

Standard Solutions: 5-HIAA and HVA 0,2 mol/L.

Mobile Phase: EDTA 372 mg/L; 0,1 M Sodium acetate (13,61 g/L); 50% Methanol (63,3 ml/L); 7,9 ml Acetate - glacial acid.

Table 1. Data for the calibration curve of standard solutions 5-HIAA and HVA

	5-HIAA	ng/mL		HVA	ng/mL
10	1171498	10,628	10	548965	9,513
	983528	8,959		553693	9,305
75	8550778	76.183	20	1352981	21,090
	8451767	75,304		1337587	20,866
100	11211557	99,820	100	6738233	100,222
	11131136	99,106		6783594	99,561
R	0,9998977		R	0,9998686	
R^2	0,9997954		R^2	0,9997373	
m	24914,1		m	82350	

Conditions for HPLC-ED Analysis: Detector conditions; Range: 50 nA; Potential: + 0,75 V; Filter: 0,02 Hz; Flow rate = 0,9 mL/min; Column: ODS Hypersil 5 μm, 200 × 4,6 mm Hewlett Packard.

Quantification of 5-HIAA and HVA in CSF: standard solutions of 5-HIAA and HVA in 0,2 N $HClO_4$ were analyzed by HPLC-ED method. Table 1 showed concentrations of 5-HIAA and HVA standard solutions and surfaces of peaks. Based on the data presented, a calibration curve for the standard 5-HIAA and HVA solutions was constructed.

Figure 1 showed a chromatogram of a mixture of standards 5-HIAA and HVA in 0,2 N $HClO_4$, whose concentration is 40 ng/mL, which showing retention times for 5-HIAA standards and HVA. By analyzing the linear correlation of 5-HIAA and HVA standards, the concentration of 5-HIAA and HVA in the CSF samples was determined.

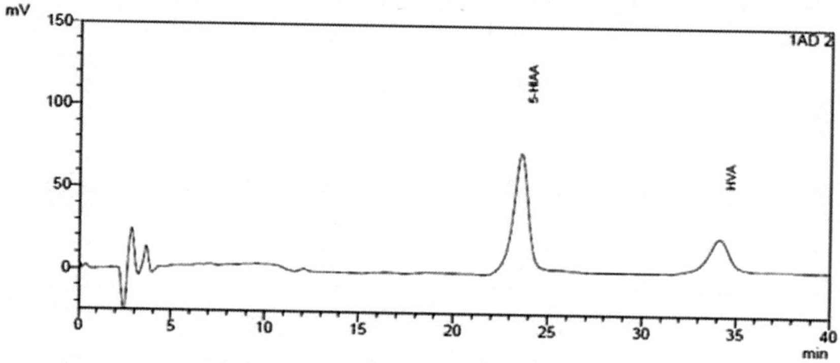

Fig. 1. Chromatogram of 5-HIAA and HVA standard solution (40 ng/mL)

3 Results

The concentration of 5-HIAA and HVA in CSF of children determined with high-pressure liquid chromatography with electrochemical detection using HPLC Shimadzu, Kyoto, Japan with BAS amperometric detector. Studies were carried out on CSF of children belonging to both sexes, with ages from few months to one year old. Patients were with hydrocephalus conditions.

Table 2 are shown the concentrations for 5-HIAA and HVA determined in 17 CSF samples in individuals up to 1 year of age, of different sex, diagnosed with hydrocephalus. The mean concentration of 5-HIAA is 38,13 (SEM $\pm$ 8,31) ng/mL, and the HVA is 47,25 (SEM $\pm$ 9,22) ng/mL. The detection limit for 5-HIAA in the CSF is 4,6 pg and for HVA it is 3,6 pg.

Table 2. The concentration of 5-HIAA and HVA in the CSF of persons of different sex and age of 1, diagnosed with hydrocephalus

Number	Sex	5-HIAA (ng/mL ± SD)	HVA (ng/mL ± SD)
1	Male	19,9 ± 0,90	23,54 ± 0,29
2	Male	7,95 ± 0,68	39,19 ± 0,97
3	Male	23,5 ± 0,75	32,32 ± 0,32
4	Male	9,6 ± 0,15	1,37 ± 0,46
5	Male	77,7 ± 50	59,5 ± 1,12
6	Male	113,1 ± 0,78	38,72 ± 0,38
7	Male	4,36 ± 0,13	44,46 ± 0,57
8	Male	60,88 ± 0,46	115,12 ± 1,13
9	Male	46,00 ± 0,68	165,92 ± 0,98
10	Male	25,68 ± 0,72	36,78 ± 0,86
11	Male	0,52 ± 1,01	15,20 ± 0,63
12	Male	0,38 ± 1,35	12,12 ± 0,88
13	Male	113,86 ± 0,23	29,50 ± 1,16
14	Female	32,73 ± 0,55	51,06 ± 0,77
15	Female	40,13 ± 1,43	48,23 ± 0,94
16	Female	36,42 ± 1,21	46,29 ± 0,32
17	Female	35,43 ± 0,73	43,94 ± 0,91

3.1 Determination of 5-HIAA Concentration in CSF

After HPLC-ED analysis of the CSF samples and the standard solutions of 5-HIAA and HVA, the concentrations of these metabolites were determined. The concentrations of 5-HIAA shown in Fig. 1 were determined based on the calibration curve of the standard solution 5-HIAA (example, Figs. 2 and 3).

On the chromatogram of the CSF sample No. 1, shown in Fig. 3 it can be seen that the retention times for 5-HIAA ($t_R = 23{,}933$) and HVA ($t_R = 34{,}435$) correspond to the retention times of the standard solutions of 5-HIAA and HVA, i.e. for 5-HIAA $t_R = 24{,}268$, and for HVA $t_R = 33{,}556$.

Figure 3 shows that the concentration of 5-HIAA in sample CSF No.1 was determined based on the calibration curve of standard solution 5-HIAA and the peak area of 5-HIAA in sample CSF No.1.

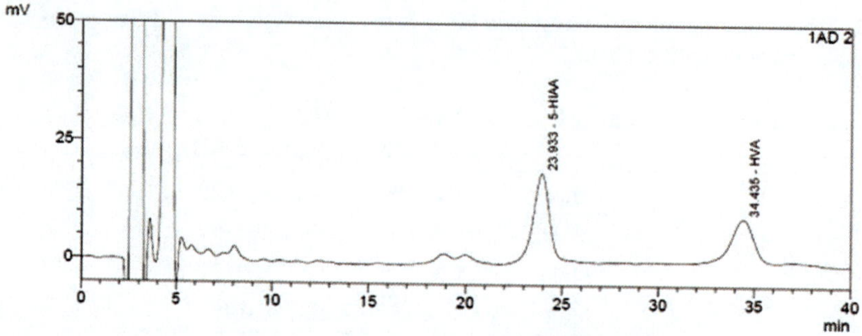

Fig. 2. Chromatogram of liquor sample No.1

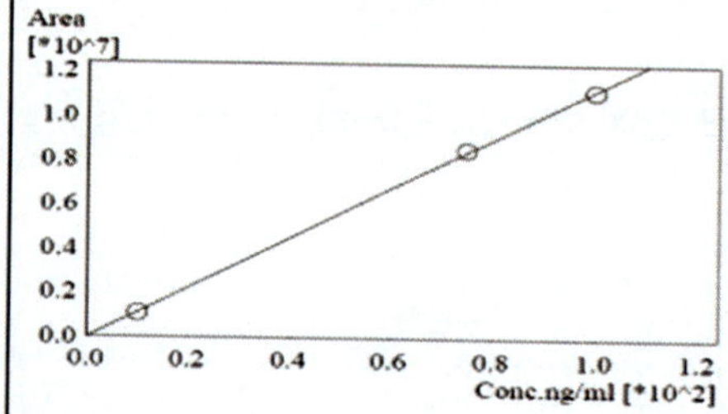

#	Conc.ng/ml	MeanArea	Area
1	10.000	1077513.1	1171498
			983528
2	75.000	8501272.2	8550778
			8451767
3	100.000	11171346.4	11211557
			11131136

Quantitative Results

AD2

ID#	Name	Ret. Time	Area	Height	Conc.ng/ml
1	5-HIAA	23.933	1095012	18789	9.949

Stvarna koncentracija 5-HIAA u uzorku likvora br.20 je 19.898 ng/ml, jer je razrijenjenje R=1:1

Fig. 3. Calibration curve of standard 5-HIAA solution

3.2 Determination of HVA Concentration in CSF

The HVA concentrations shown in Fig. 2 are based on the calibration curve of the standard HVA solution (for example, Figs. 4 and 5).

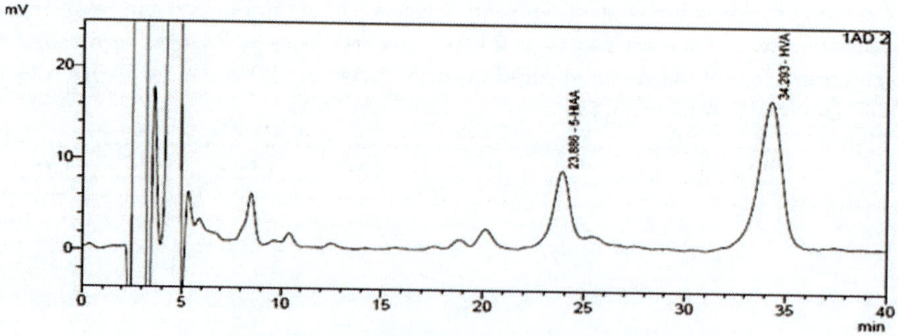

Fig. 4. Chromatogram of liquor sample No.2

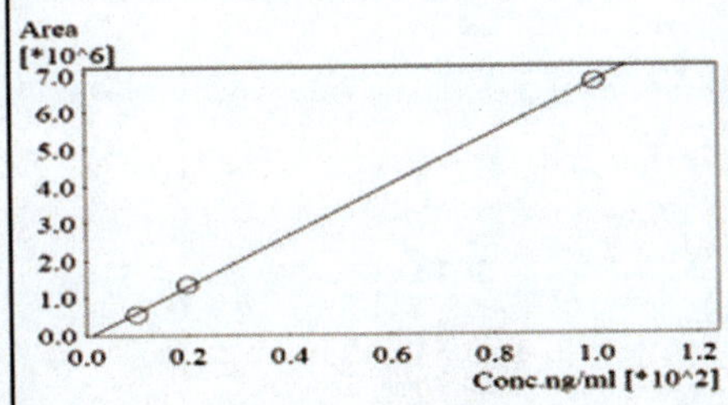

#	Conc.ng/ml	MeanArea	Area
1	10.000	551328.8	548965
			553693
2	20.000	1345284.0	1352981
			1337587
3	100.000	6760913.3	6738233
			6783594

Quantitative Results

AD2

ID#	Name	Ret. Time	Area	Height	Conc.ng/ml
1	HVA	34.293	1259901	15950	19.596

Stvarna koncentracija HVA u uzorku likvora br.16 je 39.192 ng/ml. jer razrijedjenje R= 1:1

Fig. 5. Calibration curve of standard HVA solution

On the chromatogram of sample CSF No.2 showed in Fig. 5 shows that the retention times for 5-HIAA ($t_R = 23{,}886$) and HVA ($t_R = 34{,}293$) correspond to the retention times of standard solutions 5-HIAA and HVA, i.e. 5-HIAA $t_R = 24{,}268$, and for HVA $t_R = 33{,}556$.

Figure 4 shows that the HVA concentration in the CSF sample No.2 was determined based on the calibration curve of the standard HVA solution and the HVA peak area in the CSF sample No.2.

3.3 The Influence of Gender on the Concentration of 5-HIAA and HVA in CSF

The obtained results of the 5-HIAA and HVA concentrations in 17 samples of CSF, subjects different gender were divided into two groups:

- Group 1 - Thirteen male examiners
- Group 2 - Four female respondents

In these two groups, the influence of gender on the concentration of 5-HIAA and HVA in CSF was analyzed. Mean concentration values for 5-HIAA are shown in Fig. 6 and mean concentration values for HVA are shown in Fig. 7.

Figure 6 shows that the mean value of 5-HIAA concentrations in CSF in male subjects is higher than the mean value of 5-HIAA concentrations in CSF in female subjects. Whereas in Fig. 7, the mean value of HVA concentrations in CSF in male subjects is slightly higher than the mean value of HVA concentration in CSF in female subjects.

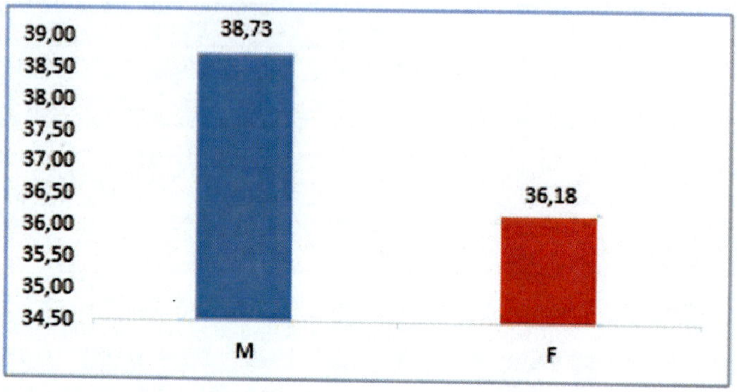

Fig. 6. Mean values concentrations of 5-HIAA female and male subjects

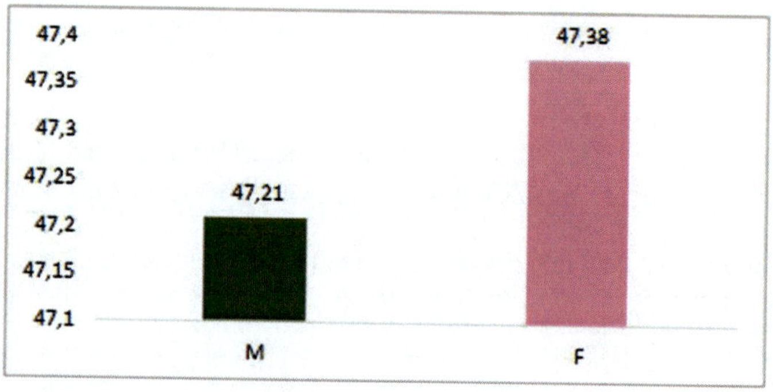

Fig. 7. Mean values concentrations of HVA female and male subjects

4 Discussion

The major metabolites of 5-HT and DA i.e. 5-HIAA and HVA can be found except in the brain and the CSF, plasma, serum, and urine [4]. Since CSF is in close contact with CNS structures, data on biochemical changes in the brain and spinal cord could be collected by CSF analysis, which could be used in the diagnosis and treatment of various diseases [11, 12].

The HVA concentration determined in CSF and plasma is used to study the neurotransmission of DA in neurological and psychiatric diseases [13]. Many literature data indicate that low concentrations of 5-HIAA in CSF may be the cause of psychomotor agitation and positive schizophrenic symptoms, while high concentrations of 5-HIAA in CSF cause negative symptoms [14]. Concentrations of CA and indolamine metabolites are usually determined by HPLC-ED, GC or GC-MS. For GC or GC-MS, samples must be derivatized, while HPLC can analyze compounds without first derivatization [10]. The advantage of HPLC methods is that they are fast and allow simultaneous determination of various CA and indoleamine and their metabolites in the same sample. The method used in this study relies principally on HPLC-ED methods for the determination of biogenic amines and their metabolites in biological samples, as well as in tissue homogenates, cerebrospinal fluid, urine and serum of humans and animals [4, 15]. In our work, concentrations of 5-HIAA and HVA in CSF were determined by HPLC Shimadzu apparatus, Kyoto, Japan, with a BAS amperometric detector and a 20 μl manual volume injector. The obtained results showed that the male subjects have a slightly higher mean concentration of 5-HIAA, and a lower mean value of the HVA concentration in the CSF compared to the female subjects. However, further study of CSF on a larger number of subjects is needed to establish certain irregularities. CA and 5-HT deficiency and excess cause various diseases, such as adrenal carcinoid or other chromaffin tissues [4].

A tumor derived from argentaffin cells intensively produces 5-HT that converts to 5-HIAA, so patients excrete large amounts of 5-HIAA, which in urine can grow more than 100 times normal. Tumors most commonly occur in the intestinal tract, but can also occur in other organs containing argentaffin cells (biliary tract, pancreas, etc.) [16].

5 Conclusions

The concentrations of 5-HIAA and HVA in the CSF of 17 subjects up to one year of age and of different sex were determined. The mean concentration of 5-HIAA is 38,13 ± 8,31 ng/mL and the HVA is 47,25 ± 9,22 ng/ml. The detection limit of 5-HIAA is 4,6 pg and the HVA is 3,6 pg. The provided HPLC-ED method is suitable for the quantification of very low concentrations of 5-HIAA and HVA in one complex biological sample such as CSF. However, the Shimadzu apparatus, made in Kyoto, Japan with an ED detector, BAS has an injector calibrated to 20 μl, which is a major technical drawback of this HPLC-ED system. The mean value concentration of 5-HIAA in CSF in male subjects was 38,73 ± 10,87 ng/mL, and in female subjects was 36,18 ± 1,33 ng/mL. The mean value concentration of HVA in CSF in male subjects was 47,21 ± 12,07 ng/mL, and in female subjects was 47,38 ± 1,31 ng/mL. Male subjects had a slightly higher mean concentration of 5-HIAA and a lower mean value of HVA in the CSF compared to female subjects. Further studies of CSF on a higher number of subjects are needed to establish certain rightness.

References

1. Sasaki, T., Fukushima, T., Ohishi, M., Toyo´oka, T.: Development of 6-hydroxychroman-based derivatization reagent: application to the analysis of 5-hydrxytryptamine and catecholamines by using high-performance liquid chromatography with electrochemical detection. Biomed. Chromatogr. **22**, 888–899 (2008)
2. Meltzer, H.Y., Arora, R.C.: Platelet serotonin studies in affective disorders: evidence for serotogenic abnormality. U: 5-Hydroxytryptamine in Psychiatry, A Spectrum of Ideas, Oxford University Press, pp. 50–89 (1991)
3. Hoyer, D., Clarke, D.E., Fozard, J.R., Hartig, P.R., Martin, G.R., Mylecharane, E.J., Humphrey, P.P.: International Union of Pharmacology classification of receptors for 5-hydroxytryptamine (Serotonin). Pharmacol. Rev. **46**(2), 157–203 (1994)
4. Sofić, E.: Metabolic turnover of biogenic amines in physiological fluids: diagnostic significances. Turk. J. Biochem. **32**(3), 120–129 (2007)
5. Salihović, M., Sofić, E.: 5-HIAA and HVA in the coma cerebri, hydrocephalus and tumor cerebri. In: CMBEBIH 2017, pp. 396–401. Springer, Singapore (2017)

6. Nordström, P., Samuelsson, M., Åsberg, M., Träskman-Bendz, L., Åberg-Wistedt, A., Nordin, C., Bertilsson, L.: CSF 5-HIAA predicts suicide risk after attempted suicide. Suicide Life-Threat. Behav. **24**(1), 1–9 (1994)
7. Klarica, M.: Uloga osmolarnosti likvora u patofiziologiji intrakronijalnog tlaka. Disertacija, Medicinski fakultet Sveučilišta u Zagrebu, Zagreb (1992)
8. Juretic, A., Spagnoli, G.C., Schultz-Thater, E., Sarcevic, B.: Cancer/testis tumour-associated antigens: immunohistochemical detection with monoclonal antibodies. Lancet Oncol. **4**(2), 104–109 (2003)
9. Relkin, N., Marmarou, A., Klinge, P., Bergsneider, M., Black, P.M.: Diagnosing idiopathic normal-pressure hydrocephalus. Neurosurgery **57**(suppl_3), S2–S4 (2005)
10. Lionetto, L., Lostia, A.M., Stigliano, A., Cardelli, P., Simmaco, M.: HPLC-mass spectrometry method for quantitative detection of neuroendocrine tumor markers: Vanillylmandelic acid, homovanillic acid and 5-hydroxyindolacetic acid. Clin. Chim. Acta **398**, 1–4 (2008)
11. Sherman, J.L., Citrin, C.M.: Magnetic resonance demonstration of normal CSF flow. Am. J. Neuroradiol. **7**(1), 3–6 (1986)
12. Saunders, N.R., Habgood, M.D., Dziegielewska, K.M.: Barrier mechanisms in the brain, I. Adult brain. Clin. Exp. Pharmacol. Physiol. **26**(1), 11–29 (1999)
13. Ramirez-Bermudez, J., Ruiz-Chow, A., Perez-Neri, I., Soto-Hernandez, J.L., Flores-Hernandez, R., Nente, F., Montes, S., Rios, C.: Cerebrospinal fluid homovanillic acid is correlated to psychotic features in neurological patients with delirium. General Hosp. Psychiatry **30**, 337–343 (2008)
14. Csernansky, J.G., Wrona, C.T., Bardgett, M.E., Early, T.S., Newcomer, J.W.: Subcortical dopamine and serotonin turnover during acute and subchronic administration of typical and atypical neuroleptics. Psychopharmacol. **110**(1–2), 145–151 (1993)
15. Sperk, G.: Simultaneus determination of serotonin, 5-hidroxyindoleacetic acid, 3, 4-dihydroxyphenylacetic acid and homovanillic acid by hight performance liquid chromatography with electrochemical detection. J. Neurochem. **38**, 840–843 (1982)
16. Csernansky, J.G., King, R.J., Faustman, W.O., Moses, J.A., Poscher, M.E., Faull, K.F.: 5-HIAA in cerebrospinal fluid and deficit schizophrenic characteristics. Br. J. Psychiatry **156**, 501–507 (1990)

Calculation of Hazard Quotient Based on the Content of Heavy Metals in Different Mushrooms

Aida Šapčanin(✉), Mirsada Salihović, Selma Korać, Emina Ramić, Belma Pehlivanović, and Šaćira Mandal

Faculty of Pharmacy, Department of Natural Sciences in Pharmacy, University of Sarajevo, Zmaja od Bosne 8, 71 000 Sarajevo, Bosnia and Herzegovina
aida.sapcanin@ffsa.unsa.ba

Abstract. Since heavy metals may enter the food chain as a result of their uptake by edible mushrooms, it is necessary to assess the levels of heavy metals and to report possible contamination that would represent a health hazard. The aim of this study was to estimate the risk posed to human's health based on the content of heavy metals found in the mushrooms collected in the western parts of Bosnia and Herzegovina. Various wild edible mushroom species (*Macrolepiotaprocera, Agaricuscampester, Amanita caesarea, Boletus edulis, Lycoperdonpyriforme, Agaricusurinascens*) were collected and the content of Zn, Co, Cu, Ni, Fe, Cr, Se, As, Pb and Cd has been measured by flame atomic absorption spectrometry (F-AAS), after lyophilisation and microwave digestion. A hazard quotient (HQ) has been calculated for estimation of health risk. The estimation of hazard quotient of metals in different wild mushrooms from various sites in the western parts of Bosnia and Herzegovina demonstrated alarming results especially when the population of children is concerned. Though the concentration of Co, Cu, Fe and Cr was significantly higher in examined mushrooms, the health risk associated with these metals was also found to be very high.

Keywords: Mushroom · Heavy metals · Hazard quotient · Health risk estimation

1 Introduction

Mushrooms are classified as specific microbiological group of organisms with significant ecological, nutritive and pharmaceutical value [1]. Besides being excellent source of nutrients, mushrooms very often contain various metals whose presence could generate toxic reactions. Presence of heavy metals in edible wild mushrooms can generate various consequences for human health. Such health-related risks are difficult to assess due to the limited and restricted data. Heavy metals are classified according to their nutritive role as essential (Cu, Zn and Fe) and non-essential (Pb, Cd and Cr) metals [1–5]. The presence of non-essential metals is undesirable and the absence of essential metals in human nutrition over long period of time can produce various metabolic processes and trigger different disorders. Low intake of metals can result in physiological removal and

A. Badnjevic and L. Gurbeta Pokvić (Eds.): CMBEBIH 2021, IFMBE Proceedings 84, pp. 413–422, 2021.
https://doi.org/10.1007/978-3-030-73909-6_47

non-significant effects on biochemical processes in organism. Some of the metals, such as Fe, Cu and Zn belong to the group of biogenic elements but their presence in high concentrations is toxic for plants and, if included in the food chain, they also present a great risk to both animal and human health. Some heavy metals in soil, when present in the appropriate concentrations, have stimulating effects and some are present without physiological significance for living organisms. Furthermore, in case of their excessive concentrations, some of heavy metals are toxic and therefore have significant ecological value. Examples of such metals are Pb, Cd and As [4, 5].

As uptake of the edible mushrooms causes presence of heavy metals in the food chain, it is obligatory to determine the concentrations of heavy metals and to estimate possible contamination that could result in the damage and risks on human health [1, 6, 7]. Environmental Protection Agency (EPA) states: „Human health risk assessment is defined as the process to estimate the nature and probability of adverse health effects in humans exposed to chemicals in contaminated environmental media, now or in the future [3, 5, 8, 9]. Therefore, it is necessary to detect and examine possible risk parameters and their dependence on exposure frequency and amount of contaminant. However, it is important to take into consideration the oral reference dose and average body weight when assessing potential risk parameters [9, 10]. Thus, the aim of the present study was to estimate the risks associated with human health based on determining the content of heavy metals present in the mushrooms collected from the western parts of Bosnia and Herzegovina.

2 Material and Methods

Biological Material: Various wild edible mushroom species (*Macrolepiotaprocera, Agaricuscampester, Amanita caesarea, Boletus edulis, Lycoperdonpyriforme, Agaricusurinascens*) were collected from western Bosnia and Herzegovina were examined by flame atomic absorption spectrometry (F-AAS), after lyophilization and microwave digestion. The investigated mushroom species are shown in Fig.

1. The contents of Zn, Co, Cu, Ni, Fe, Cr, Se, As, Pb, and Cd were selected as these are representative trace metals the levels of which in the environment represent a reliable index of environmental pollution, and also as some of them are essential metals and play an important role in biological systems. All elements concentrations were determined on a dry weight basis (Fig. 1).

Calculation of Hazard Quotient: A hazard quotient (HQ) is the ratio of the potential exposure to a substance and the level at which no adverse effects are expected. The screening-level risk associated with consumption of potentially contaminated food can be assessed using the hazard quotient US EPA [8] Hazard quotient for adults (male and female) and children associated with the intake of metals along with mushrooms was assessed by using the following formula:

$$HQ = \frac{(DI) \times (C_{metal})}{(R_f D) \times BW}$$

Daily intake (DI) of mushrooms was taken as 0.100 kg for adults, as this is the minimum mushroom requirement for a balanced diet [9]. Average body weight (BW)

Fig. 1. Investigated mushroom species: (a) *Macrolepiotaprocera*, (b) *Agaricuscampester*, (c) *Amanita caesarea*, (d) *Boletus edulis*, (e) *Lycoperdonpyriforme* and (f) *Agaricusurinascens*

Table 1. Reference oral dose (mg/kg/day) for heavy metals

Heavy metal	Reference oral dose (mg/kg/day)	Reference
Co	0.0003	
Cu	0.0400	
Fe	0.7000	
Cr	0.0030	
Zn	0.3000	[12] and [13]
Ni	0.0200	
Se	0.0050	
Cd	0.0005	
Pb	0.0035	
As	0.0003	

for adult males was taken as 70 kg, for female as 65 kg, and for children as 15 kg [10, 11, 12] Reference oral dose R_fD/mg/kg/day) of selected metals is given in Table 1.

The health risk associated with any pollutant is dependent upon the level of exposure and amount of absorption by the human body. Thus, a hazard quotient is a valid tool to assess the level of risk associated with a particular pollutant. If the level of Hazard quotient is less than 1, the risk associated with exposure to metal is negligible. However, if the level of hazard quotient is higher than 1, the metal may pose serious health hazards.

3 Results and Discussion

Obtained results demonstrated variations of trace metal concentrations found in various wild mushrooms across different areas in the western parts of Bosnia and Herzegovina. It has been noticed that a wide variation of the total content of some metals may be a consequence of the geological evolution of soil from which examined mushrooms were gathered. Based on determining the HQ of metals present in the examined wild mushrooms from the western parts of Bosnia and Herzegovina, the gained results were alarming, especially in respect of the effects on the population of children. Findings indicated both an increase in concentrations of Co, Cu, Fe, and Cr in examined mushrooms as well as an increase of health risks associated with consumption of these metals (Fig. 2, 3, 4, 5 and 6).

Cobalt is an essential biogenic trace element and a component of vitamin B12. Also, it is an essential element of erythropoiesis and it is involved in various physiological mechanisms. The presence of cobalt in organisms is mediated through the food intake only. However, higher concentrations of cobalt indicate toxic reactions [2, 3, 5]. Results from this study showed that the hazard quotient of Co, Cu, Cr and Fe was more than one for all of the examined populations in almost all investigated mushrooms from the western parts of Bosnia and Herzegovina (Fig. 2, 3, 4 and 5).

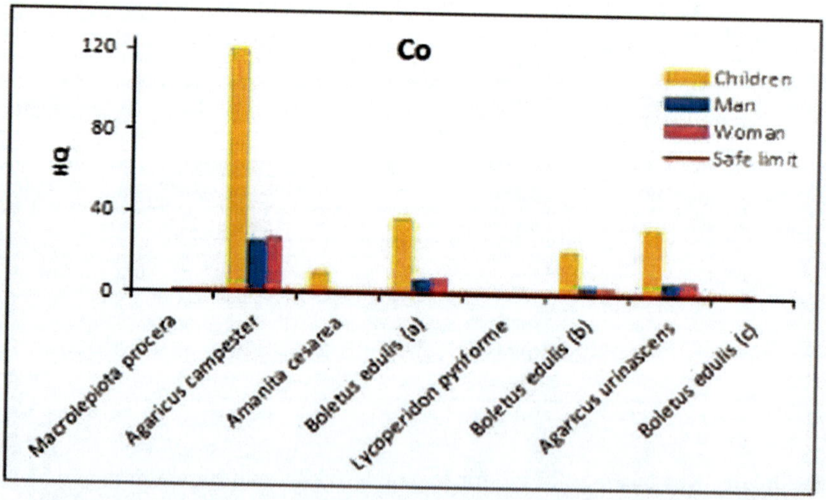

Fig. 2. Hazard quotient for adults and children of Co in investigated mushrooms

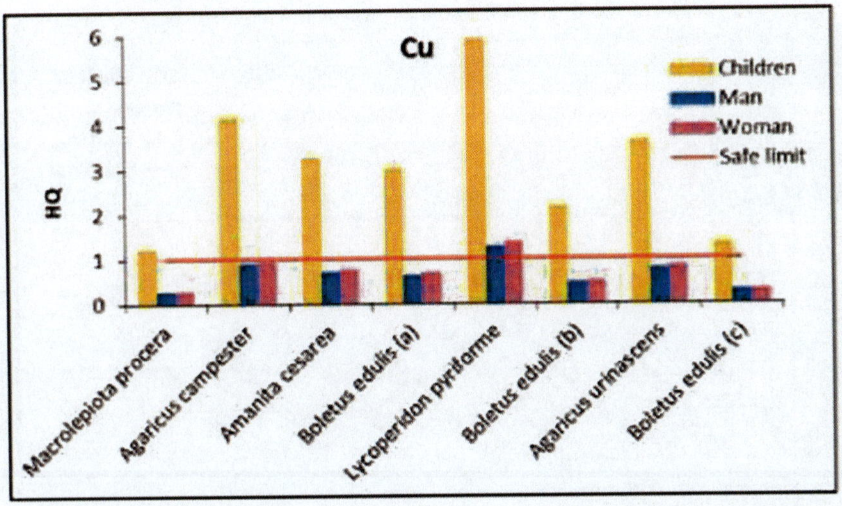

Fig. 3. Hazard quotient for adults and children of Cu in investigated mushrooms

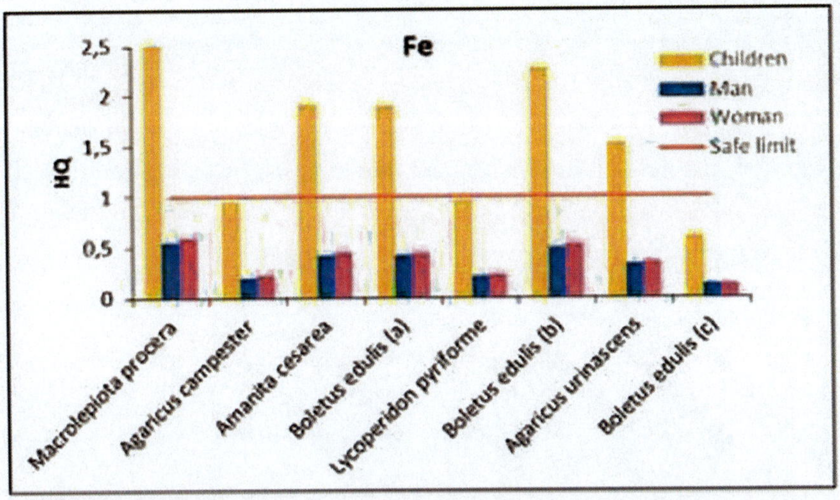

Fig. 4. Hazard quotient for adults and children of Fe in investigated mushrooms

Various studies have demonstrated that high concentrations of Fe are responsible for the deposition of iron oxides [4, 6, 8]. High concentrations of Co in rats induced adverse effects on blood, kidney, and heart [11, 15–17].

According to the International Agency for Research on Cancer (IARC), Cd, Ni, As, and Cr (VI) have been identified as human or animal carcinogens. Results of our study are shown in Fig. 5, 6, 7, 8 and 9.

Low concentrations of Ni are detected in the environment and they might result in lung-associated damages.

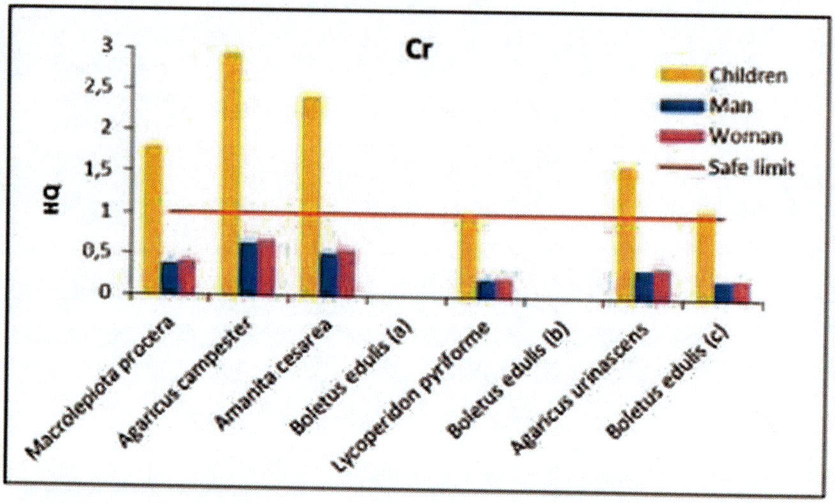

Fig. 5. Hazard quotient for adults and children of Cr in investigated mushrooms

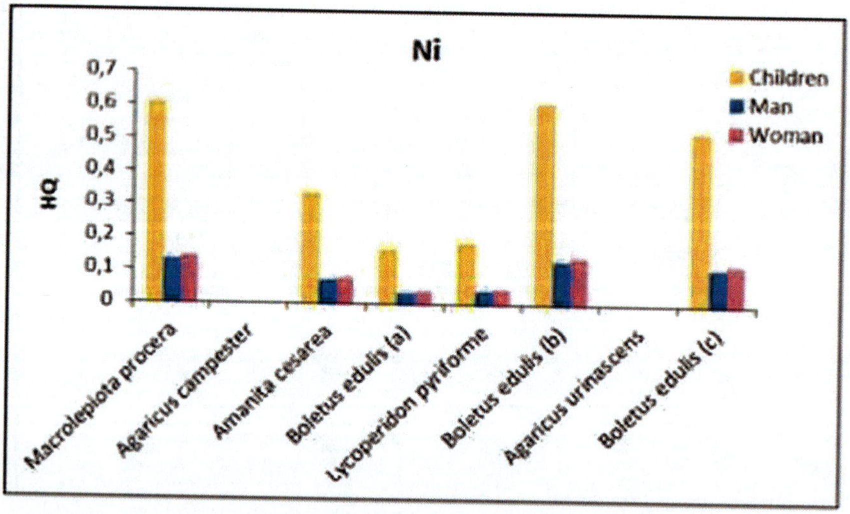

Fig. 6. Hazard quotient for adults and children of Ni in investigated mushrooms

US EPA has declared Cd, Co, Cu, and Fe as potentially harmful metals for their target organs which can lead to serious health disorders. It has been shown that the activity of the endocrine gland is affected by Co and that the activity of the kidney and gastrointestinal tract is affected by Cd, Cu, and Fe [13].

Results from our study indicate significantly higher concentrations of Pb in some examined areas (Fig. 9). The obtained concentrations of Pb are above the approved limit [9, 10]. It is known that higher concentrations of Pb result in sleep disorder and weight loss but they also affect hearing.

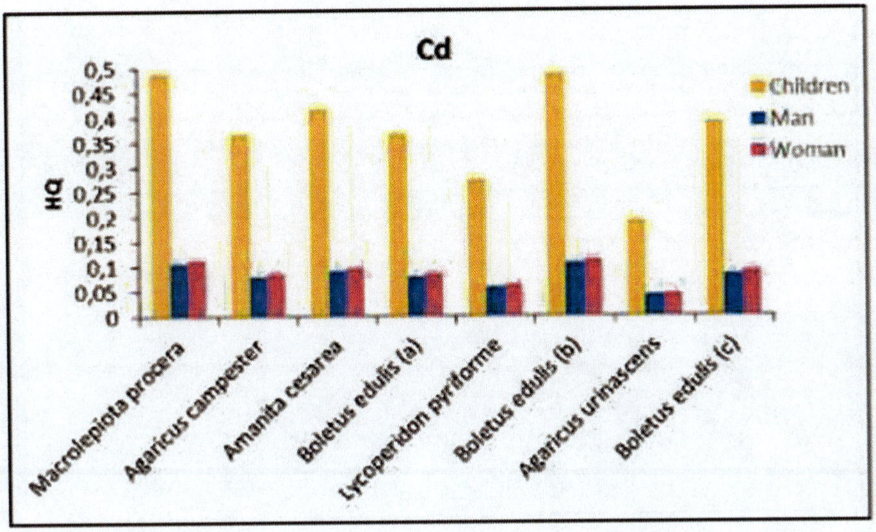

Fig. 7. Hazard quotient for adults and children of Cd in investigated mushrooms

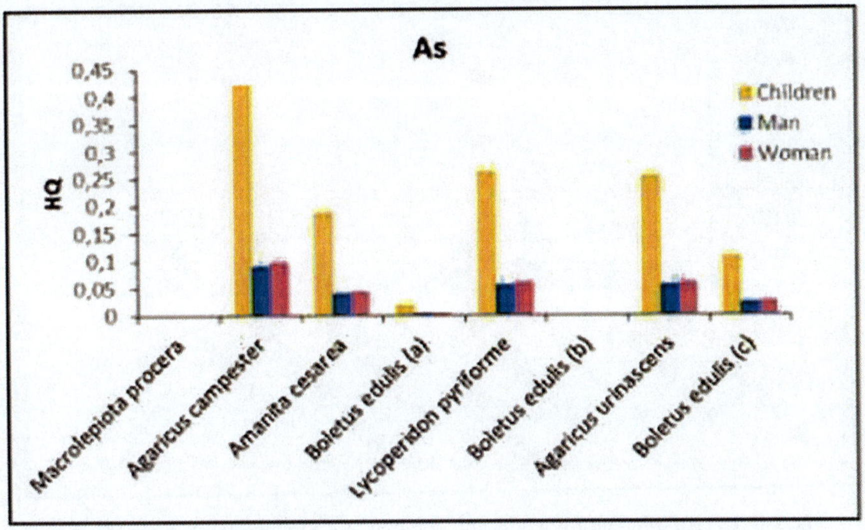

Fig. 8. Hazard quotient for adults and children of As in investigated mushrooms

The results for the essential trace elements Zn and Se are shown in Fig. 10 and 11. Adequate Intake of Se for adults and pregnant women was established by the EFSA Panel for Dietary Products, Nutrition and Allergies [19].

Research has shown that there is a very narrow range of safety of intake of Se, which makes the examination of its content in food even more important [20].

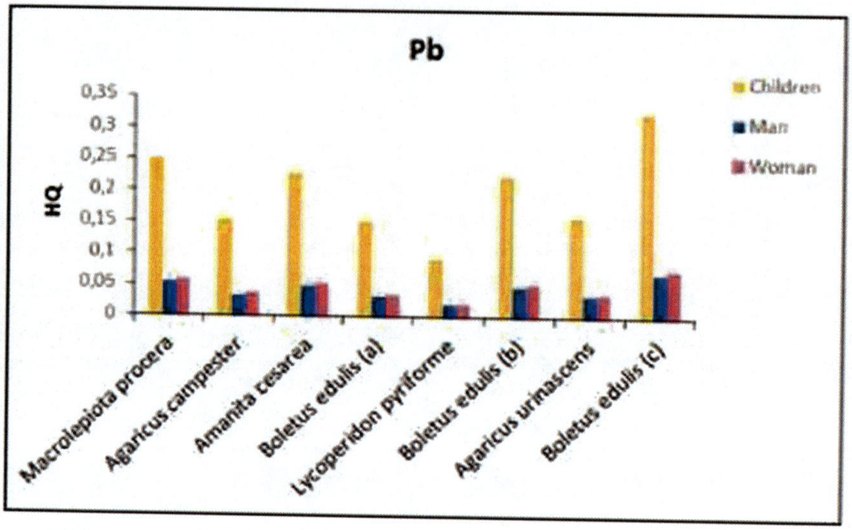

Fig. 9. Hazard quotient for adults and children of Pb in investigated mushrooms

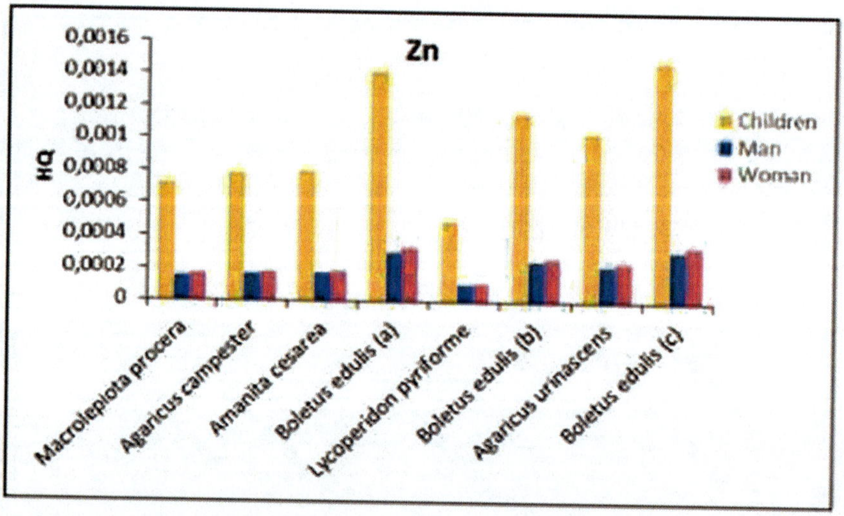

Fig. 10. Hazard quotient for adults and children of Zn in investigated mushrooms

Tolerable Upper Intake Level of Zn for adults, pregnant and lactating women established by the EFSA [21].

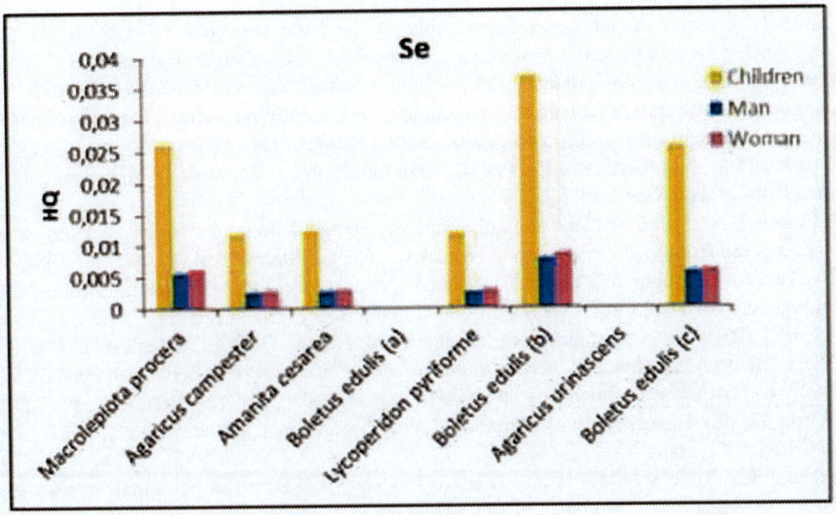

Fig. 11. Hazard quotient for adults and children of Se in investigated mushrooms

4 Conclusions

The present study has generated data associated with risk assessment of consumer's exposure to the heavy metals from wild mushrooms traditionally used especially in rural areas.

The estimation of the hazard quotient of metals in different wild mushrooms from various sites demonstrated alarming results, especially when the population of children is concerned. Though the concentration of Co, Cu, Fe, and Cr was significantly higher in the examined mushrooms, the health risk associated with these metals was also found to be very high.

The average consumption of the studied wild mushrooms, in general, will not expose consumers to the effect of toxic elements, since the studied species do not occur in nature throughout the year and therefore their everyday consumption cannot be expected.

It is suggested that regular monitoring of heavy metals in mushrooms and other food items should be performed for the prevention of high buildup of heavy metals in the human food chain.

Acknowledgement. This work was supported by the Federal Ministry of Education and Science in Bosnia and Herzegovina and was carried out within the project "Chemical compositionand an antioxidative potential of wild mushrooms from Bosnia and Herzegovina" (Grant no. 0101–8183-3/17, dated 01.12.2017). We thank Assist.Prof. Vesna Suljic, PhD for editing the text in English.

References

1. Cheung, P.: The nutritional and health benefits of mushrooms. Nutr. Bull. **35**(4), 292–299 (2010)

2. Järup, L.: Hazards of heavy metal contamination. Br. Med. Bull. **68**, 167–182 (2003)
3. Guerra, F., Trevizam, A.R., Muraoka, T., Marcante, N.C., Caniatti-Brazaca, S.G.: Heavy metals in vegetables and potential risk for human health. Sci. Agric. **69**, 54–60 (2012)
4. Kalač, P.: A review of chemical composition and nutritional value of wild-growing and cultivated mushrooms. J. Sci. Food Agric. **93**(2), 209–218 (2013)
5. Fowler, B.A., Nordberg, M., Friberg, L.: Handbook on the Toxicology of Metals, 3rd edn. Academic Press, New York (2007)
6. Chittaragi, A., Naika, R.: Determination of trace elements on some wild mushroom samples encountered from Western ghats of Karnataka. J. Chem. Pharm. Res. **6**(7), 2124–2135 (2014)
7. Sarikurkcu, C., Copur, M., Yildiz, D., Akata, I.: Metal concentration of wild edible mushrooms in Soguksu National Park in Turkey. Food. Chem. **128**, 731–734 (2011)
8. US EPA. Risk assessment guidance for superfund: Human Health Evaluation (1989)
9. World Health Organization, Environmental Health Criteria Series, EHC 228. Principles and Methods for the Assessment of Risk from Essential Trace Elements. WHO Geneva (2002)
10. World Health Organization, International Programme on Chemical Safety, Publications. IPCS/EHC Series
11. Agency for Toxic Substances and Disease Registry, ATSDR. Toxicological Profiles 12.US EPA. A review of the reference dose and reference concentration processes. EPA/630/P-02 (2002)
12. US EPA. A review of the reference dose and reference concentration processes. EPA/630/P-02 (2002)
13. US EPA. United States Environmental Protection Agency, in Quantitative Risk Assessment Calculations 7–9 (2015)
14. Hadis, K., Nabi, S., Shahrokh, N., Mojtaba, M., Babak, M., Amin, M.K.: The Concentration and probabilistic health risk of potentially toxic elements (PTEs) in edible mushrooms (wild and cultivated) samples collected from different cities of Iran. Biol. Trace. Elem. Res (2020)
15. Svoboda, L., Chrastny, V.: Levels of eight trace elements in edible mushrooms from a rural area. Food. Addit. Contam. **25**, 51–58 (2008)
16. Falandysz, J., et al.: ICP/MS and ICP/AES elemental analysis (38 elements) of edible wild mushrooms growing in Poland. Food. Addit. Contam. **18**, 503–513 (2001)
17. Simonsen, L.O., Harbak, H., Bennekou, P.: Cobalt metabolism and toxicology—a brief update. Sci. Total. Environ. **432**, 210–215 (2012)
18. Nordberg, G., Sandström, B., Becking, G., Goyer, R.A.: Essentiality and toxicity of trace elements: principles and methods for assessment of risk from human exposure to essential trace elements. J. Trace Elem. Exp. Med. **14**, 261–273 (2001)
19. EFSA. Scientific Opinion on dietary reference values for selenium; EFSA Panel on Dietetic Products, Nutrition and Allergies (NDA). EFSA J. **12**(10), 3846 (2014)
20. Falandysz, J.: Review: on published data and methods for selenium in mushrooms. Food Chem. **138**, 242–250 (2013)
21. EFSA. Tolerable upper intake levels for vitamins and minerals; Scientific Committee on Food,Scientific Panel on Dietetic Products, Nutrition and Allergies (2006)

Association Between Serum Concentrations of Free Fatty Acids with Free Iron in Type 2 Diabetes

Sacira Mandal(✉)

Faculty of Pharmacy, Department of Natural Sciences in Pharmacy, University of Sarajevo, Sarajevo, Bosnia and Herzegovina

Abstract. Free fatty acids (FFAs) are represent one of the major risk factor for insulin resistance (IR) and they are involved in pathophysiological processes of Type 2 *Diabetes mellitus* (T2D) but the underlying mechanisms a still unknown. Recent studies have showed important role of elevated iron levels in pathogenesis of T2D and IR. Aim of this study was to find association between serum of FFA concentrations and free iron levels in T2D patients. The study included 51 participants (27 healthy control and 24 newly diagnosed diabetes patients), both gender and the age of 45–70. Results shown that concentrations of serum FFAs and free iron levels were elevated in diabetic patients compare to healthy subjects while statistical significant difference for all tested parameters demonstrated between diabetic and control groups ($p < 0.05$). In addition, there were negative correlations between total iron levels and C18:1 and C20:3 acids concentration ($p < 0.05$) as well as a positive association of C12:0 concentration with iron levels in control group ($p < 0.001$). In diabetic patients, a positive association between C20:5 and iron levels ($p < 0.05$) was observed. These findings suggest that increased serum of FFA and free iron concentrations may have an important role and influence in development of T2D and its incidence, as well as higher risk of further complications of diabetes.

Keywords: Free fatty acid · Free iron · Type 2 diabetes mellitus

1 Introduction

Type 2 *Diabetes mellitus* (T2D) is a common form of diabetes and its prevalence has increased rapidly worldwide. According to International Diabetes Federation (IDF) almost 700 million people would be affected until 2040 [1]. This chronic metabolic disease characterized by defective secretion or action of insulin and accompanied by high glucose and lipid levels in the blood. Because of alterations in lipid and carbohydrates as well protein metabolic pathways, circulating plasma free fatty acids (FFAs) and lipoproteins concentration rises. The long chain fatty acids, polyunsaturated fatty acids (PUFA) represent an important signal molecules, energy source and main constituents of cell membrane [2–6]. They play a key role in the cell and tissue functions by involve in gene expression, cell growth and differentiation, and therefore regulate the β-cells

A. Badnjevic and L. Gurbeta Pokvić (Eds.): CMBEBIH 2021, IFMBE Proceedings 84, pp. 423–432, 2021.
https://doi.org/10.1007/978-3-030-73909-6_48

functionally. Elevated concentration of FFAs (especially saturated fatty acids, SFA) in serum is associated with impaired insulin sensitivity (IS) and action, as well as glucose intolerance. A disturbance in the regulation of free fatty acid metabolism is a main event responsible for insulin resistance (IR), obesity, dyslipidemia and T2D [7–9]. According to the glucose-fatty acid vicious cycle of Randle et al. (1963), preferential oxidation of FFAs over glucose has key roles in IS as well hyperglycemia and leads to metabolic disturbances of diabetes [10].

Recent studies demonstrated that elevated concentrations of either endogenous or exogenous iron were associated with an increased risk of development Type 2 diabetes. In addition, iron metabolism directly or indirectly can affect to the onset and development of T2D [11–13].

Iron (Fe), one of the essential trace elements, is involved in electron and oxygen transfer between cells, participates in genomic synthesis, and involved in regulating the differentiation and growth of living cells. In its free form, released by ferritin by the action of reducers that convert Fe^{3+} to Fe^{2+} ions, iron as a strong oxidant can generate oxidative stress by formation of reactive oxygen species (ROS) (especially detrimental hydroxyl radicals) through the Fenton reaction. These free radicals are strong pro-oxidants, which cause lipid peroxidation of cellular membrane, damage the configuration of proteins and replace nucleic acids in genes [14–17]. In addition, the blood glucose concentrations related are closely to the iron contents in pancreatic β-cells, which are more likely to accumulate iron when compared with other cells. Iron has an important role in pathogenesis of the disease by mediation of β-cell dysfunction and IR, as well in regulation of energy homeostasis by involving in glucose and fatty acids metabolism. Potential role of both endogenous and exogenous iron with PUFAs in the development of T2D and its vascular complications present in Fig. 1 [18–20].

Despite efforts to find underlying mechanisms, few studies have evaluated the association between serum fatty acids and iron in T2D patients. Therefore, the aim of this

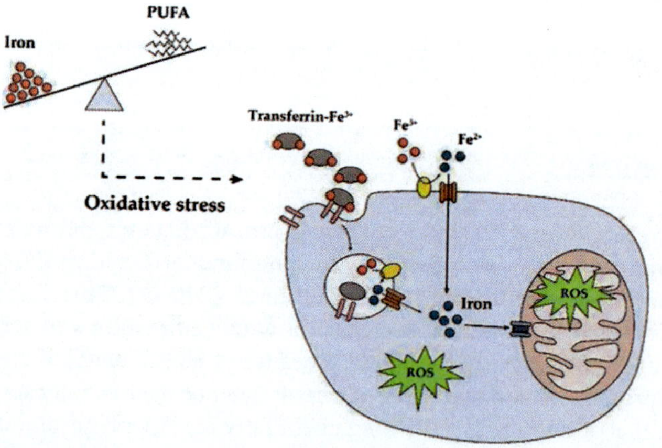

Fig. 1. Potential role of endogenous and exogenous iron with PUFAs in the development of T2D and its related complications

study was examine relationship between free fatty acids and free iron in newly diagnosed Type 2 diabetes in Bosnian population.

2 Materials and Methods

2.1 Subjects

Total of 51 participants of which 24 Type 2 *diabetes mellitus* (DM) patients and 27 healthy controls were included in this study after providing informed consent. Patients had been a newly diagnosed T2DM by endocrinologist and they were recruited from The Endocrine Centre of University Clinical Centre Sarajevo, Sarajevo, BH; under criteria of American Diabetes Association (ADA) (fasting plasma glucose levels were higher than 7.0 mmol/l and postprandial serum glucose was more than 11.1 mmol/. Cut of values for HbA1c values were (<5.7%) for healthy controls, and Type 2 diabetics (>6.5%) [21].

No patient or control had any other known illness and none was taking medications. Controls were healthy subjects who had no family history of DM. The study protocol was approved by the Ethnics Committee of the UKC Sarajevo and complied with ethical principles outlined in World Medical Association Declaration of Helsinki – Ethical Principles for Medical Research Involving Human Subjects (initiated in June 1964, last amendment in October 2000).

2.2 Sample Analysis

Blood samples were drawn after of 12 h fast and centrifuged at 3000 rpm for 10 min at room temperature for serum separation. The fasting serum samples were kept frozen at –70 °C until analysis. Biochemical parameters including fasting glucose, glycated hemoglobin (HbA1c), and lipid profile (total cholesterol, triglycerides, high-density lipoprotein cholesterol, low-density lipoprotein cholesterol, very low-density lipoprotein cholesterol) were measured by standard procedures of International Federation for Clinical Chemistry (IFCC) with an autoanalyzer VITROS 350 Chemistry System (Ortho-Clinical Diagnostic, Rochester, NY, USA). Serum free iron concentration was determined by the Ferrozine method using Autoanalyzator Dimension® RXL clinical-chemical system (Siemens, Erlangen, Germany) [22].

Preparation of sample for fatty acid analysis was done by methylation of serum samples, followed by conversion of FFAs to corresponding fatty acid methyl esters [23]. FFAs with medium and long-chain length, from C12 to C22 carbon atoms with different saturation levels, in serum of newly T2D, and healthy subjects were determined by gas chromatography equipped with mass spectrometer (Agilent 7890A GC + Agilent DB23 capillary column; Agilent 5975C MSD, Agilent Technologies, Inc., Wilmington, DE, USA). The identity of each fatty acid peak was obtained by comparing the retention time of the peak with the retention times of referent standards with known fatty acids composition, and C17:0 as internal standard was used for calculation the concentration of individual FFAs.

2.3 Statistical Analysis

Calculations were performed using SPSS 23.0 for Windows. The differences FFAs, free iron concentrations and biochemical variables between cases and controls were assessed by Mann-Whitney test, while associations FFAs and iron levels were done by the Spearman's coefficient correlation test. Differences were considered statistically significant when the p value was <0.05.

3 Results

The characteristics of newly diagnosed T2D patients and control subjects are presented in Table 1. The significant differences were found between fasting glucose, HbA1c, age, lipid profile and free iron levels. As expected, levels of free iron were higher in T2D patients than controls.

Table 1. Biochemical and clinical characteristics of T2D patients and controls.

Parameters	Cases (N = 24)	Controls (N = 27)
Age, years	64^a (60–68)	48^a (42–54)
Glucose, mmol/L	8.20^a (7.59–8.81)	5.3^a (5.24–5.49)
Total cholesterol, mmol/L	5.34^a (5.02–5.65)	4.53^a (4.12–4.94)
HDL-cholesterol, mmol/L	1.17^a (0.89–1.44)	1.27^a (1.16–1.38)
LDL-cholesterol, mmol/L	3.45^a (3.12–3.78)	2.47^a (2.08–2.86)
VLDLC, mmol/L	1.03^a (0.50–1.56)	0.95^a (0.80–1.10)
Triglycerides, mmol/L	2.08^a (1.77–2.40)	1.68^a (1.28–2.08)
HbA1c, %	7.16^a (6.88–7.44)	5.74^a (5.57–5.90)
Iron, µmol/L	23.38^a (14.44–32.31)	19.74^a (19.52–19.95)

*Data presents as median (lower and upper quartile). HDL - high density lipoprotein cholesterol; LDL - low density lipoprotein choles- terol; VLDLC - very low-density lipoprotein cholesterol; HbA1c - glycated hemoglobin.
[a]Significances were calculated by Mann-Whitney U test ($p < 0.001$).

The composition and profile pattern of the study population for identified and measured individual FFAs showed in Table 2. The significant differences were found between cases and healthy subjects for values of all the determined FFAs ($p < 0.001$). Analyzed fatty acids were decreased in T2D patients, except for lauric, myristoleic, palmitic, stearic, DGLA, DTA, and DPA acids.

The negative association of serum diohomo-γ-linolenic acid (C20:3 n6) concentrations and free iron levels was demonstrated in the study participants (Fig. 2). The Spearman's coefficient correlation was shown a positive association between iron levels and C12:0 ($p < 0.001$) while a negative association iron with C18:1 and C20:3 concentrations ($p < 0.05$) in healthy subjects was observed (Fig. 3). In group of newly diagnosed T2D patients, a positive significant association between EPA (C20:5 n3) with iron concentrations was observed (Fig. 4).

Table 2. Composition and concentration of FFAs.

Free fatty acid (μmol/L)	Cases	Controls
Lauric acid, C12:0	0.02045[c] (8.76E–3–3.2E—2)	0.0362[c] (1.23E–2–6.02E—2)
Myristic acid, C14:0	0.2517[c] (1.75E–1–3.28E–1)	0.2398[c] (1.72E–1–3.08E—1)
Myristoleic acid, C14:1	0.1015[a] (2.02E–2–1.83E–1)	0.00508[a] (2.573E–3–7.61E–3)
Palmitic acid, C16:0	0.1150[c] (9.34E–2–1.37E–1)	0.3625[c] (2.27E–1–4.98E–1)
Palmitoleic acid, C16:1 n7	0.0757[c] (2.71E–2–1.24E–1)	0.0496[a] (–5.158E–3–1.04E—1)
Stearic acid, C18:0	0.0768[c] (5.58E–2–9.79E–2)	0.2428[c] (2.04E–1–2.82E–1)
Oleic acid, C18:1 n9	0.09267[a] (4.39E–2–1.41E–1)	0.1872[a] (7.87E–2–2.30E—1)
Linoleic acid, C18:2 n6	0.1680[c] (1.29E–1–2.07E–1)	0.3634[c] (2.39E–1–4.88E–1)
γ-Linolenic acid, C18:3 n6	0.0298[a] (1.89E–2–4.07E–2)	0.1032[a] (8.62E–2–1.20E–1)
Arachidic acid, C20:0	0.002727[c] (1.96E–3–3.49E–3)	0.007[c] (4.86E–3–1.01E–2)
DGLA, C20:3 n6	0.01351[b] (–4.59E–3–3.16E–2)	0.0119[b] (5.37E–3–1.84E–2)
Arachidonic acid, C20:04 n6	0.041579[c] (3.18E–2–5.14E–2)	0.2012[c] (1.60E–1–2.42E–1)
EPA, C20:5 n3	0.00728[c] (4.09E–3–1.05E–2)	0.0082[c] (7.57E–3–8.84E–3)
Behenic acid, C22:0	0.00119[c] (7.46E–4–1.64E–3)	0.00711[c] (3.98E–3–1.01E–2)
DTA, C22:4 n6	0.00013[c] (2.27E–5–2.42E–4)	0.0001245[c] (6.71E–5–1.82E–4)
DPA, C22:5 n3	.0001426[a] (–3.67E–5–3.22E–4)	.0000258[a] (–3.27E–5–8.43E–5)
DHA, C22:6 n6	0.0000726[b] (–3.17E–5–1.77E–4)	0.0003569[b] (8.16E–5–6.32E–4)

*Data presents as median (lower and upper quartile). EPAeicosapentaenoic acid, DGLA-Dihomo-γ-linolenic acid; DTAdocosatetraenoic acid, DPA-docosapentaenoic acid, DHAdocosahexaenoic acid. Significances were tested by Mann-Whitney U test ([a]$p < 0.05$; [b]$p < 0.01$; [c]$p < 0.001$).

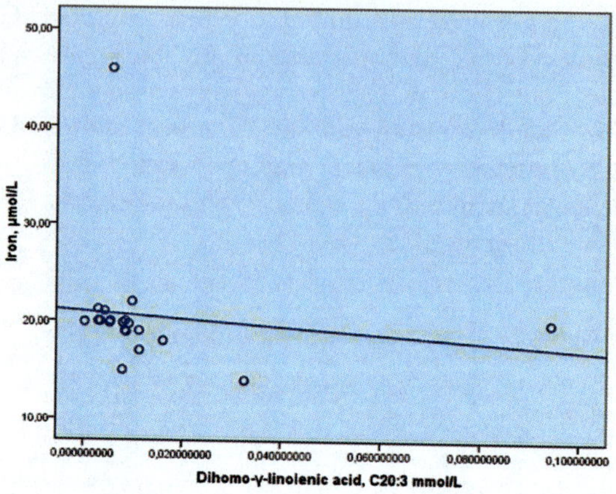

Fig. 2. Correlation between diohomo-γ-linolenic acid concentration and free iron levels in study population ($\rho = -0.524$, $p < 0.05$).

4 Discussion

A number of studies have investigated an association of dietary intake of iron and serum iron with IR and T2D. Most of them demonstrated a significant correlation of serum iron concentrations and its ferritin form with IR and increased risk of development of T2D.

Several other and recent investigations found that elevated concentration of FFAs are as one the major risk factors for developing IR and T2D. The relation of common SFAs and PUFAs (C16:0, C18:0 or C20:4 n6) suggest severe and detrimental effects on signaling gene regulation, growth and differentiation of cells and, therefore, β-cells function. Recent studies demonstrated that individual FFAs association with the pathogenesis of diabetes, specifically with differences in IS in T2D patients [24–27]. Total content of C16:0, C18:1, C18:2, and C22:4 n6 were associated with decrease IS and insulin secretion. Concentration of EPA (C20:5 n3) only was associated with slightly increase IS in T2D patients, which is partially in accordance with results of present study (Fig. 4).

However, a few studies have explored the relationship between FFAs and iron with IR, obesity, and T2D. In this work, the negative associations between serum concentrations of DGLA (C20:3 n6) and free iron were observed (Fig. 2, and Fig. 3C). These results are interesting and might reflect the influence of age and ethnicity on iron and FFAs metabolism in participants. In study of *Tsurutani* et al. they found that a high serum DGLA concentration was positive associated with obesity, body fat accumulation, and IR in T2D patients [28, 29].

Recent research demonstrated that elevated concentrations of both endogenous and exogenous iron were associated directly or indirectly with occurrence and development of T2D. However, the giving results of the association between iron metabolisms with increased risk of T2D were inconsistent because investigation provided in different age, gender, and ethnicity groups. So, data obtained in Asian population, both gender, showed significantly higher serum iron concentrations compare to White population

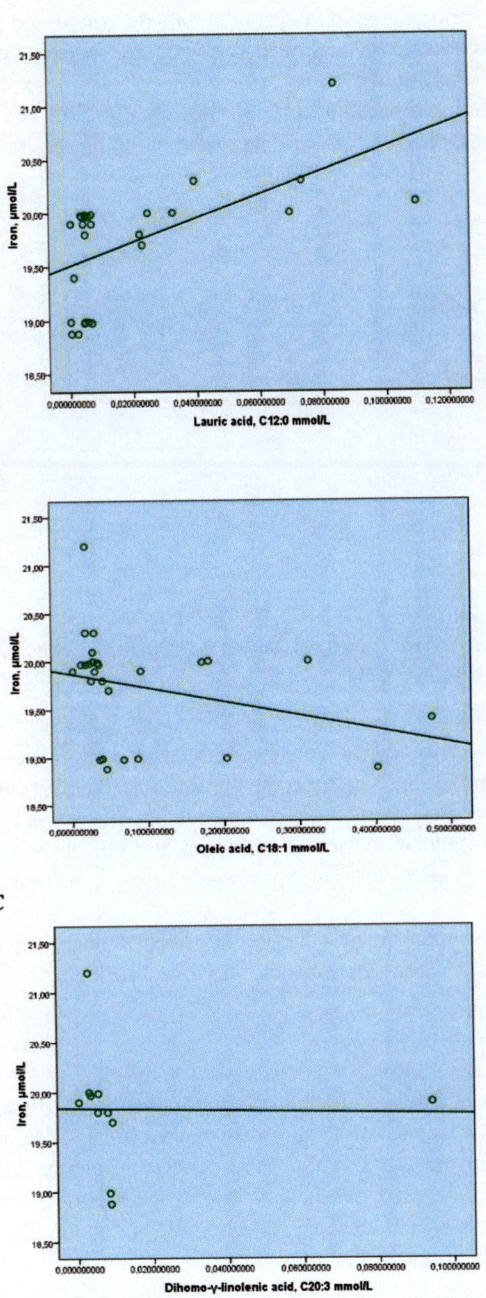

Fig. 3. Correlations between free iron levels and lauric acid (A), oleic acid (B) and diohomo-γ-linolenic acid (C) concentrations in controls. (Spearman's coefficient correlation $\rho = 0.681$, $p < 0.001$; $\rho = -0436$, $p < 0.05$; $\rho = -0.648$, $p < 0.05$, respectively).

[18, 30, 31]. *In vitro* and *in vivo* animal model data demonstrated that increased iron can damage neurons and other cells *via* indirect or direct biochemical pathways. Considering fact that PUFAs have antioxidant effect in an iron-related prooxidant influence of tissue and cells, recent studies found that a higher ratio iron/PUFA ratio was associated with increase risk of developing of T2D and its related vascular complications [32–34].

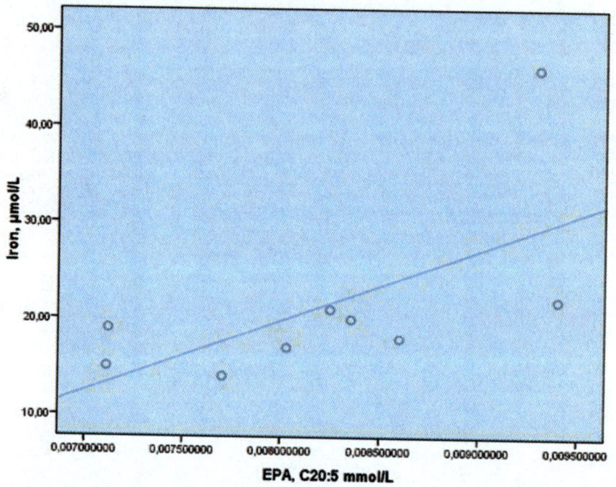

Fig. 4. Correlation between free iron levels and EPA (C20:5n3) concentration in newly diagnosed Type 2 diabetic patients ($\rho = 0.750$ $p < 0.05$).

Previuos studies examined the relation between free iron levels and lipid profile in T2D patients. In some of investigation, the authors found elevation in iron levels of diabetics and significant correlations with lipid profile but others were not demonstrated [35, 36]. In addition, no association between glucose and HbA1c with serum iron concentration were demonstrated in this study, opposite to reported findings of other research [37–39].

However, future examinations including larger cohorts of subjects, as well as studies in other populations are needed to substantiate these findings.

5 Conclusions

The results of this first study performed in population from Bosnia and Herzegovina, in which a significance of association between specific PUFAs with serum free iron concentrations was investigated in the pathogenesis of newly diagnosed T2D. These findings suggest that increased serum of FFA and free iron concentrations may have an important role and influence in development of T2D and its incidence, as well as higher risk of further complications of diabetes.

References

1. International Diabetes Federation (IDF). Definition and diagnosis of diabetes mellitus and intermediate hyperglycemia. IDF Diabetes Atlas — 9th Edition. Diabetes Atlas (2019)

2. Pankow, J.S., Duncan, B.B., Schmidt, M.I., Ballantyne, C.M., Couper, D.J., Hoogeveen, R.C., Golden, S.H.: Fasting plasma free fatty acids and risk of type 2 diabetes. Diabetes Care **27**, 77–82 (2004)
3. Boden, G.: Free fatty acids as target for therapy. Curr. Opin. Endocrinol. Diabetes **11**(5), 258–263 (2004)
4. Cnop, M.: Fatty acids and glucolipotoxicity in the pathogenesis of type 2 diabetes. Biochem. Soc. Trans. **36**(3), 348–352 (2008)
5. Boden, G.: Obesity, insulin resistance and free fatty acids. Curr. Opin. Endocrinol. Diabetes Obes. **18**(2), 139–143 (2011)
6. Ragheb, R., Medhat, A.M.: Mechanisms of fatty acid-induced insulin resistance in muscle and liver. J. Diabetes Metab. **2**, 127–132 (2011)
7. Oh, Y.S., Bae, G.D., Baek, D.J., Park, E.-Y., Jun, H.-S.: Fatty acidinduced lipotoxicity in pancreatic beta-cells during development of type 2 diabetes. Front. Endocrinol. **9**(384), 1–10 (2018)
8. Wolf, G.: Role of fatty acids in the development of insulin resistance and type 2 diabetes mellitus. Nutr. Rev. **66**(10), 597–600 (2008)
9. Brown, A.E., Walker, M.: Genetics of insulin resistance and the metabolic syndrome. Curr. Cardiol. Rep. **18**(75), 1–8 (2016)
10. Randle, P.J., Garland, P.B., Hales, C.N., Newsholme, E.A.: The glucose fatty-acid cycle. Its role in insulin sensitivity and the metabolic disturbances of diabetes mellitus. Lancet **1**(7285), 785–789 (1963)
11. Sanjeevi, N., Freeland-Graves, J., Beretvas, S.N., Sachdev, P.K.: Trace element status in type 2 diabetes: a meta-analysis. J Clin Diagn Res. **12**(5), 1–8 (2018)
12. Arija, V., et al.: Excess body iron and the risk of type 2 diabetes mellitus: a nested case–control in the PREDIMED (PREvention with MEDiterranean Diet) study. Br. J. Nutr. **112**, 1896–1904 (2014)
13. Simcox, J.A., McClain, D.A.: Iron and diabetes risk. Cell Metab. **17**(3), 329–341 (2013)
14. Fernández-Real, J.M., McClain, D., Manco, M.: Mechanisms linking glucose homeostasis and iron metabolism toward the onset and progression of type 2 diabetes. Diabetes Care **38**, 2169–2176 (2015)
15. Orban, E., Schwab, S., Thorand, B., Huth, C.: Association of iron indices and type 2 diabetes: a meta-analysis of observational studies. Diabetes Metab Res Rev. **30**(5), 372–394 (2014)
16. Podmore, C., et al.: Association of multiple biomarkers of iron metabolism and type 2 diabetes: the EPICInterAct Study. Diabetes Care **39**, 572–581 (2016)
17. Rajpathak, S.N., Crandall, J.P., Wylie-Rosett, J., Kabat, G.C., Rohan, T.E., Hu, F.B.: The role of iron in type 2 diabetes in humans. Biochim. Biophys. Acta **1790**(7), 671–681 (2009)
18. Backe, M.B., Moen, I.W., Ellervik, C., Hansen, J.B., MandrupPoulsen, T.: Iron regulation of pancreatic beta-cell functions and oxidative stress. Annu. Rev. Nutr. **36**, 241–273 (2016)
19. Liu, J., Li, Q., Yang, Y., Ma, L.: Iron metabolism and type 2 diabetes mellitus: a meta-analysis and systematic review. J. Diabetes Investig. **11**(4), 946–955 (2020)
20. Shaaban, M.A., Dawod, A.E., Nasr, M.A.: Role of iron in diabetes mellitus and its complications. Menoufia Med. J. **29**, 11–16 (2016)
21. American Diabetes Association 2. Classification and Diagnosis of Diabetes: Standards of Medical Care in Diabetes-2018. Diabetes Care. **41**(Suppl 1): S13–S27 (2018)
22. Smith, F.E., Herbert, J., Gaudin, J., Hennessy, D.J., Reid, G.R.: Serum iron determination using ferene triazine. Clin Biochem. **17**(5), 306–310 (1984)
23. Lepage, G., Roy, C.R.: Specific methylation of plasma nonesterified fatty acids in a one-step reaction. J. Lipid Res. **27**, 114–120 (1986)
24. Clore, J.N., Allred, J., White, D., Li, J., Stillman, J.S.: The role of plasma fatty acid composition in endogenous glucose production in patients with type 2 diabetes mellitus. Metabolism **51**(11), 1471–1477 (2002)

25. Clore, J.N., Stillman, J.S., Li, J., O'Keefe, S.J.D., Levy, J.R.: Differential effect of saturated and polyunsaturated fatty acids on hepatic glucose metabolism in humans. Am. J. Physiol. Endocrinol. Metab. **287**, 358–365 (2004)
26. Mandal, S., Causevic, A., Dzudzevic-Cancar, H., Semiz, S.: Free fat-ty acid profile in type 2 diabetic subjects with different control of glycemia. In: IFMBE Proceedings, vol. 62, pp. 781–786 (2017)
27. Huang, J., Jones, D., Luo, B., Sanderson, M., Soto, J., Abel, E.D.: Iron overload and diabetes risk: a shift from glucose to fatty acid oxidation and increased hepatic glucose production in a mouse model of hereditary hemochromatosis. Diabetes **60**(1), 80–87 (2011)
28. Yamashita, K., et al.: Dihomo-gamma-linolenic acid levels and obesity in patients with type 2 diabetes. Diabetology International **6**(3), 206–211 (2014). https://doi.org/10.1007/s13340-014-0187-6
29. Tsurutani, Y., Inoue, K., Sugisawa, C., Saito, J., Omura, M., Nishikawa, T.: Increased serum dihomo-linolenic acid levels are associated with obesity, body fat accumulation, and insulin resistance in Japanese patients with type 2 diabetes. Intern. Med. **57**, 2929–2935 (2018)
30. Wallace, D.F.: The regulation of iron absorption and homeostasis. Clin. Biochem. Rev. **37**(2), 51–62 (2016)
31. Aregbesola, A., Voutilainen, S., Virtanen, J.K., Mursu, J., Tuomainen, T.-P.: Gender difference in type 2 diabetes and the role of body iron stores. Annals Clin. Biochem. **54**(1), 113–120 (2017)
32. Huth, C., Beuerle, S., Zierer, A., Heier, M., Herder, C., Kaiser, T., et al.: Biomarkers of iron metabolism are independently associated with impaired glucose metabolism and type 2 diabetes: the KORA F4 study. Eur. J. Endocrinol. **173**(5), 643–653 (2017)
33. Kim, K., Song, Y., Oh, T.J., Choi, S.H., Jang, H.C.: Association between iron intake and diabetic peripheral neuropathy in type 2 diabetes: significance of iron intake and the ratio between iron intake and polyunsaturated patty acids intake. Nutrients **12**(11), 1–13 (2020)
34. Swaminathan, S., Fonesca, V.A., Alam, M.G., Shah, S.V.: The role of iron in diabetes and its complications. Diabetes Care **30**(7), 1926–1933 (2007)
35. Mandal, Š: Serum iron concentration and lipid profile in type 2 diabetes patients. Bull. Chemists Technol. Bosnia and Herzegovina **54**, 7–12 (2020)
36. Wolide, A.D., Zawdie, B., Alemayehu, T., Tadesse, S.: Association of trace metal elements with lipid profiles in type 2 diabetes mellitus patients: a cross sectional study. BMC Endocrine Disorders **17**(64), 1–7 (2017)
37. Stechemesser, L., Eder, S.K., Wagner, A., Patsch, W., Feldman, A., Strasser, M., et al.: Metabolomic profiling identifies potential pathways involved in the interaction of iron homeostasis with glucose metabolism. Mol. Metab. **6**, 38–47 (2017)
38. Misra, G., Bhatter, S.K., Kumar, A., Gupta, V., Khan, M.Y.: Iron profile and glycaemic control in patients with type 2 diabetes mellitus. Med. Sci. **4**(22), 1–8 (2016)
39. Gohel, M., Sirajwala, H.B., Chacko, A.: Serum free iron concentration in patients with type 2 diabetes mellitus with good and poor control and its correlation with glycemic control. Int. J. Diabetes Res. **2**(2), 33–38 (2013)

Potentially New Synergistic Combination of Curcumin and Rosuvastatin: An *in Vitro* Study

Belma Pehlivanović[1](✉), Kenan Čaklovica[2], Aida Šapčanin[3], Dina Lagumdžija[1], Naida Omerović[1], Nermina Žiga Smajić[1], Selma Škrbo[1], and Fahir Bečić[1]

[1] Department of Clinical Pharmacy, Faculty of Pharmacy, University of Sarajevo, Zmaja od Bosne 8, 71000 Sarajevo, Bosnia and Herzegovina
belma.pehlivanovic@ffsa.unsa.ba

[2] Department of Food Safety, Veterinary Faculty, University of Sarajevo, Zmaja od Bosne 90, 71000 Sarajevo, Bosnia and Herzegovina

[3] Department of Natural Sciences, Faculty of Pharmacy, University of Sarajevo, Zmaja od Bosne 8, 71000 Sarajevo, Bosnia and Herzegovina

Abstract. Due to the increased rate of infectious diseases, an insufficient number of therapeutics and antibiotic resistance, the detection of potential sources of novel therapeutic agents is of crucial importance. As the current medicines are simply not sufficient, further studies have been based on a combinational approach that evaluates the synergistic pharmacological activity of plantderived compounds and other medicines. Curcumin, a natural polyphenol extracted from plant *Curcuma longa* L., and rosuvastatin, synthetic lipid-lowering medicine, pose a wide range of pharmacological properties. So far, the combined activity of curcumin and rosuvastatin has not been investigated. The aim of this study was to examine the potential synergistic antimicrobial, antioxidant and anti-inflammatory *in vitro* activity of a combination of curcumin and rosuvastatin. *In vitro* agar well diffusion method was used to determine antibacterial activity against Gram-positive and Gram-negative bacteria as well as the antifungal activity. Antioxidant activity was evaluated by a commonly used *in vitro* assay for scavenging free radicals. *In vitro* assay of inhibition of protein denaturation was applied for the determination of anti-inflammatory activity. Results have demonstrated that a combination of curcumin and rosuvastatin suppressed the growth of all tested strains of microorganisms, therefore implying significant antimicrobial activity. Results have also suggested that a combination of curcumin and rosuvastatin shows a dose-dependent activity in scavenging free radicals and inhibiting protein denaturation. Therefore, the present *in vitro* study emphasizes the potential application of a combination of curcumin and rosuvastatin as antimicrobial, antioxidant and anti-inflammatory agents. However, current findings are limited to *in vitro* assays and further research should be conducted involving *in vivo* models.

Keywords: In vitro · Curcumin · Rosuvastatin · Antimicrobial · Anti-inflammatory · Antioxidative activity

A. Badnjevic and L. Gurbeta Pokvić (Eds.): CMBEBIH 2021, IFMBE Proceedings 84, pp. 433–440, 2021.
https://doi.org/10.1007/978-3-030-73909-6_49

1 Introduction

In the last two decades, in vitro studies have become one of the most important aspects of modern drug discovery across different fields of biomedical science [1]. The need of the pharmaceutical industry for efficient and precise determination of pharmacological parameters of existing and novel drugs increased the use and modifications of various in vitro models [2]. Because some in vitro techniques can provide rapid, more relevant and precise data than in vivo studies, their application is of crucial importance in early preclinical trials [3]. Lately, the scientific community is facing an urgent need for novel therapeutic agents due to the increased rate of infection diseases, antibiotic resistance and an insufficient number of medicines [4]. Future efforts on detecting novel therapeutic agents are based on a combinational approach that evaluates the pharmacological activity ofnatural substances and synthetic drugs. The investigation of synergistic potential of combining natural with the synthetic product can resultinan advance of pharmacotherapy, improved pharmacological activity and reduction of side effects [5].

Curcumin [(1E,6E)-1,7-bis(4-hydroxy-3-methoxyphenyl) hepta-1,6-diene-3,5-dione)] is natural, yellow-orangesubstancethat can be extracted from the rhizome of plant CurcumalongaL.,Zingiberaceae. For many years, it has been used as a spice, food preservative, colouring agent and traditional Ayurvedic medicine for different diseases. It has been reported that curcumin possesses significant antitumour, antioxidant, anti-inflammatory, antiviral, antimicrobial, lipid-lowering, hepatoprotective and neuroprotective properties [6, 7]. Rosuvastatin is a synthetic statin, a selective inhibitor of enzyme 3-hydroxy-3-methyl-glutaryl-coenzyme A reductase (HMG-CoA reductase), which is applied for the treatment of hyperlipidemia and reduction of cardiovascular risk [8]. Also, rosuvastatin has been marked as an „ideal statin" due to its unique pharmacological properties [9]. Compared to the other hypolipemic drugs, an additional advantage of rosuvastatin is based on the possession of specific, non-lipid pleiotropic properties [10]. Various studies have shown that curcumin and rosuvastatin possess a wide range of pharmacological properties and that their activity is mediated through similar targets and molecular mechanisms [11–15]. Both curcumin and rosuvastatin are marked as highly pleiotropic substances, which could be considered promising for new therapeutic indications [6, 7, 10, 16, 17]. Possible reasons for this consideration may be attributed to their anti-inflammatory and antioxidant activity [8, 18, 19].

So far, the combined activity of curcumin and rosuvastatin has not been investigated and there are no available data on their potential synergistic activity. Therefore, the aim of the present study was to examine potential synergistic antimicrobial, antioxidant and anti-inflammatory in vitro activity of a combination of curcumin and rosuvastatin.

2 Materials and Methods

Curcumin was obtained from Sigma-Aldrich (CAS No: 458–37-7, St. Louis, MO, the USA), and rosuvastatin calcium standard (99.1% purity; Lot No: BM14000510) was obtained from ILKO ILAC SANAYII VE TICARET A.S. Ethanol (puriss. p.a., $\geq 99.8\%$) and 1,1-diphenyl-2-picrylhydrazyl radical (DPPH) were purchased from Sigma-Aldrich, and nutrient agar and antibiotic discs were purchased from Laboratorios Conda S.A.

Dimethyl sulfoxide (DMSO) was obtained from Semikem, BiH. All other chemicals were of the highest analytical grade available. Spectrometer Lambda 25 UV/VIS, PerkinElmer, was used for measuring the absorbance, and incubator LabLine Imperial III, Barnstead, the USA, for the incubation of Petri plates.

2.1 *In vitro* Antimicrobial Activity

Agar well diffusion assay was applied for the evaluation of antibacterial activity of two Gram-positive bacteria (*Staphylococcus aureus* ATCC6538, *Listeria monocytogenes* ATCC35152) and two Gram-negative bacteria (*Pseudomonas aeruginosa* ATCC9027, *Escherichia coli* ATCC8739). The same assay was used for the assessment of antifungal activity against *Candida albicans* (ATCC10231). Curcumin and rosuvastatin were dissolved in DMSO and prepared as a single test solution at the following concentrations: 0.10, 0.25, 0.35 and 0.50 mg/ml. During the preparation of inoculums, suspensions of microorganisms were adjusted on approximately 1.5 × 108 CFU/ml and cultured on nutrient agar. In every Petri plate, four wells of 6 mm diameter were created with sterile borer and in each well, the total volume of 100 μl of test solutions was applied. Also, the volume of 100 μl of DMSO was applied into well and served as a negative control. Chloramphenicol (30 μg/disc), a standard antibiotic, was applied as a positive control. After application of test substances, plates containing bacteria were incubated at 37 °C for 18–24 h and plates with fungi at 25 °C for 48 h. After the incubation period, the diameter of zone of inhibition (mm) was measured and recorded for each plate. Obtained values of the diameter of zone of inhibition were proportional to the degree of antimicrobial sensitivity of test solutions. The experiment was performed in triplicate and results were presented as Mean ± Standard deviation (SD).

2.2 *In vitro* Antioxidant Activity

The most commonly used DPPH radical scavenging assay was applied for the evaluation of antioxidant activity. Both curcumin and rosuvastatin were dissolved in DMSO and prepared as a single test solution at the following concentrations: 0.10, 0.25, 0.50, 0.75 and 1.00 mg/ml. The solution of ascorbic acid was prepared at the same concentration and used as a positive control. The reaction mixture consisted of 2.5 ml of the test solution and 1.0 ml of ethanolic solution of DPPH at the concentration of 3 × 10(4) mol/L. The test solutions were stored ina dark place and kept at the room temperature for 20 min, after which the absorbance was measured at 517 nm. The absorbance of the ethanolic solution of DPPH was also measured and served as a control. Antioxidant activity was expressed as Radical Scavenging Capacity (RSC) and calculated using the following formula: RSC (%) = 100 − (Asample × 100/Acontrol). The experiment was performed in triplicate and results were presented as Mean ± Standard deviation (SD).

2.3 *In vitro* Anti-inflammatory Activity

To assess anti-inflammatory activity, convenient in vitro assay of heat-induced inhibition of protein denaturation was applied. Both curcumin and rosuvastatin were dissolved in DMSO and prepared as a single test solution at the following concentrations: 0.10, 0.25, 0.50, 0.75 and 1.00 mg/ml.Acetylsalicylic acid was used as a positive control and prepared at the same concentrations. The assay was performed according to Ullah et al. [20] The reaction mixture contained 2.0 ml of the test solution, 2.8 ml of phosphate buffer saline with an adjusted pH value of 6.4 and 0.2 ml of egg albumin. The control mixture contained 2.0 ml of distilled water instead of the test solution. The mixtures were incubated for 15 min at 37 ± 2 °C and then heated for 5 min at 70 °C. After cooling down at the room temperature, the absorbance was measured at 660 nm. Anti-inflammatory activity was expressed as the percentage of inhibition of protein denaturation and calculated using the following formula:

Inhibition (%) = 100 × (1 − Abssample/Abscontrol). The experiment was performed in triplicate and results were presented as Mean ± Standard deviation (SD).

3 Results and Discussion

3.1 *In vitro* Antimicrobial Activity

Results of the in vitro agar well diffusion assay showed that investigated test solutions of a combination of curcumin and rosuvastatin exhibited strong antibacterial activity against both Gram-positive and Gram-negative bacteria as well as the antifungal activity (Table 1). The strongest antibacterial activity was observed at the concentration of 0.50 mg/ml against *E. coli, L. monocytogenes, S. aureus and P. aeruginosa*, resulting in the maximum zone of inhibition of 13.31 mm, 11.96 mm, 11.31 mm and 9.30 mm, respectively. It is interesting to notice that tested combination of curcumin and rosuvastatin demonstrated stronger antibacterial activity against *S. aureus and P. aeruginosa* than a standard antibiotic, chloramphenicol. Application of DMSO, which served as a negative control, demonstrated no inhibition of microbial growth. An increase in the concentration of test solutions resulted in a larger diameter of zones of inhibition. Therefore, results indicated that the antimicrobial activity of a tested combination of curcumin and rosuvastatin is dependent upon the concentration. These findings are consistent with those previously published for curcumin and rosuvastatin separately [16, 21]. However, there are no available data on the antimicrobial activity of a combination of curcumin and rosuvastatin.

Table 1. Zones of inhibition (mm) obtained through the agar well diffusion method

Test solutions	The diameter of zone of inhibition (mm)				
	Escherichia coli (ATCC8739)	Pseudomonas aeruginosa (ATCC9027)	Staphylococcus aureus (ATCC6538)	Listeria monocytogenes (ATCC 35152)	Candida albicans (ATCC10231)
C+R (0.10 mg/ml)	12.66 ± 0.16	8.32 ± 0.33	9.69 ± 0.16	10.07 ± 0.13	9.03 ± 0.09
C+R (0.25 mg/ml)	13.00 ± 0.11	8.73 ± 0.09	9.72 ± 0.33	10.30 ± 0.19	10.00 ± 0.11
C+R (0.35 mg/ml)	13.00 ± 0.11	8.70 ± 0.11	10.30 ± 0.19	11.07 ± 0.11	10.69 ± 0.16
C+R (0.50 mg/ml)	13.31 ± 0.20	9.30 ± 0.19	11.31 ± 0.12	11.96 ± 0.38	11.00 ± 0.11
CHL(30 μg/disc)	14.68 ± 0.13	8.05 ± 0.23	10.03 ± 0.09	18.00 ± 0.11	NA
NYS (100 μg/disc)	NA	NA	NA	NA	15.71 ± 0.20
DMSO	ND	ND	ND	ND	ND

Results are presented as Mean ± Standard deviation (n = 3).
NA = not applied; ND = not detected; C + R = combination of curcumin and rosuvastatin;
CHL = chloramphenicol; NYS = nystatin.

3.2 *In vitro* Antioxidant Activity

Results obtained by the free radical scavenging capacity of test solutions of ascorbic acid and a combination of curcumin and rosuvastatin are presented in Table 2. Test solutions of ascorbic acid, which were applied as a positive control, demonstrated a stronger antioxidant activity than a combination of curcumin and rosuvastatin. This was expected because ascorbic acid is a well-known and documented antioxidant. The percentage of RSC ranged from 50.11% to 68.51% and 48.70% to 59.87% for ascorbic acid and a combination of curcumin and rosuvastatin, respectively. Also, it has been noticed that percentage of RSC increases with an increase in the concentration of test solutions. Therefore, results indicated a dose-dependent antioxidant response for both ascorbic acid and a combination of curcumin and rosuvastatin. Previous in vitro studies have shown that statins exhibit the radical scavenging effects on multiple radicals [22]. Also, it has been demonstrated that statins could reduce the effects of hydroxyl radicals in in vitro conditions [16, 23]. A recent study has revealed that the free radical scavenging activity of curcumin is based on the presence of phenol and CH2 groups of the ß-diketone moiety [24]. However, there are no available data on the free radical scavenging capacity of a combination of curcumin and rosuvastatin.

Table 2. Percentage (%)of the free radical scavenging capacity of test solutions

Concentration of test solutions (mg/ml)	RSC (%)	
	Ascorbic acid	Combination of curcumin and rosuvastatin
1.00	68.51 ± 0.30	59.87 ± 0.13
0.75	65.11 ± 0.04	56.31 ± 0.70
0.50	60.03 ± 1.11	51.39 ± 0.07
0.25	55.99 ± 0.07	50.11 ± 0.15
0.10	50.11 ± 0.03	48.70 ± 0.10

Results are presented as Mean ± Standard deviation (n=3); RSC = radical scavenging capacity

3.3 *In vitro* Anti-inflammatory Activity

Results of the in vitro assay of heat-induced inhibition of protein denaturation are given in Table 3. The highest percentage of protein denaturation inhibition for solutions of acetylsalicylic acid and a combination of curcumin and rosuvastatinwas at theconcentration of 1.00 mg/ml. Test solutions of a combination of curcumin and rosuvastatin showed the mean inhibition of egg albumin denaturation of 90.65%, 90.17%, 89.30%, 89.11% and 88.46%, at the concentrations of 1.00 mg/ml, 0.75 mg/ml, 0.50 mg/ml, 0.25 mg/ml and 0.10 mg/ml, respectively. When compared to the obtained percentage of protein denaturation inhibition for solutions of acetylsalicylic acid, it can be stated that a combination of curcumin and rosuvastatin exhibited similar in vitro anti-inflammatory activity. Furthermore, our results revealed that an increase in the concentrations leads to an increase in protein denaturation inhibition, therefore indicating that anti-inflammatory activity of a combination of curcumin and rosuvastatin is dependent upon the concentration. Findings from this study are in accordance with our results for individually examined substances curcumin and rosuvastatin [16, 21].

Table 3. Percentage (%) of inhibition of protein denaturation of test solutions

Concentration of test solutions (mg/ml)	Inhibition %	
	Acetylsalicylic acid	Combination of curcumin and rosuvastatin
1.00	96.84 ± 0.24	90.65 ± 0.20
0.75	91.27 ± 0.03	90.17 ± 0.10
0.50	87.59 ± 0.21	89.30 ± 0.09
0.25	85.12 ± 0.89	89.11 ± 0.78
0.10	85.03 ± 0.10	88.46 ± 0.19

Results are presented as Mean ± Standard deviation (n = 3).

4 Conclusion

This study has presented the first in vitro investigation of antimicrobial, antioxidant and anti-inflammatory activity of a combination of curcumin and rosuvastatin so far. Our results have demonstrated strong antibacterial and antifungal activity against selected microorganisms as well as the significant antioxidant and anti-inflammatory activity. Furthermore, results of the present study have revealed that an increase in the concentrations leads to an increase in activity and implied that antimicrobial, antioxidant and anti-inflammatory activity of a combination of curcumin and rosuvastatin is dependent upon the concentration. Therefore, our study emphasises a potentially new combination of substances that can be considered for further research involving in vivo models. However, findings of this study are limited to a narrow interval of concentrations of curcumin and rosuvastatin, which were obtained by the most commonly used in vitro assays. Consideration should be given to providing results with a wider interval of concentrations in different types of in vitro assays. Also, further research should be conducted on in vivo models.

References

1. Carvalho, C., Varela, S.A.M., Bastos, L.F., et al.: The relevance of in silico, in vitro and non-human primate based approaches to clinical research on major depressive disorder. Altern Lab Anim. **47**(3–4), 128–139 (2019)
2. Pehlivanović, B.: In vitro models to determine the pharmacokinetic parameters. Folia Med. Fac. Med. Univ. Saraeviensis **50**(1), 40–42 (2015)
3. Haddadi, M., Javad, M.M., Mohseni, S., Golnaz, M.: In vitro ADME screening instead of in vivo studies in preclinical safety. Biomed. J. Sci. Tech. Res. **24**(4), 1837118376 (2020)
4. Freedman, D.H.: Hunting for New Drugs with AI. The pharmaceutical industry is in a drug-discovery slump. How much can AI help? Nature **576**, S49–S53 (2019)
5. Bałan, B.J., et al.: Synergistic Activity for Natural and Synthetic Inhibitors of Angiogenesis Induced by Murine Sarcoma L-1 and Human Kidney Cancer Cells. In: Pokorski, M. (ed.) Clinical Research and Practice. AEMB, vol. 1020, pp. 91–104. Springer, Cham (2017). https://doi.org/10.1007/5584_2017_17
6. Aggarwal, B.B., Sung, B.: Pharmacological basis for the role of curcumin in chronic diseases: an age-old spice with modern targets. Trends Pharmacol. Sci. **30**(2), 85–94 (2009)
7. Pulido-Moran, M., Moreno-Fernandez, J., Ramirez-Tortosa, C., Ramirez-Tortosa, M.: Curcumin and Health. Molecules **21**(3), 264 (2016)
8. Alisi, I.O., Uzairu, A., Abechi, S.E., Idris, S.: Evaluation of the antioxidant properties of curcumin derivatives by genetic function algorithm. J. Adv. Res. **12**, 4754 (2018)
9. McTaggart, F.: Comparative pharmacology of rosuvastatin. Atherosclerosis **4**(1), 9–14 (2003)
10. Fabijanić, D.: Pleiotropic effects of statins. Medicus **19**(2), 163–169 (2010)
11. Beevers, C.S., Huang, S.: Pharmacological and clinical properties of curcumin. Botanics Targets Therapy **1**, 5–18 (2011)
12. Collins, R., Reith, C., Emberson, J.: Interpretation of the evidence for the efficacy and safety of statin therapy. Lancet **388**(10059), 2532–2561 (2016)
13. Panahi, Y., Ahmadi, Y., Teymouri, M., Johnston, T.P., Sahebkar, A.: Curcumin as a potential candidate for treating hyperlipidemia: a review of cellular and metabolic mechanisms. J. Cell Physiol. **233**(1), 141–152 (2018)

14. Pehlivanović, B., Bečić, F.: Curcumin: phytochemical therapy in the treatment of hyperlipidemia. Bull. Chemists Technol. Bosnia and Herzegovina **52**, 1116 (2019)
15. Zhao, J.F., Ching, L.C., Huang, Y.C., Chen, C.Y., Chiang, A.N., Kou, Y.R., et al.: Molecular mechanism of curcumin on the suppression of cholesterol accumulation in macrophage foam cells and atherosclerosis. Mol. Nutr. Food Res. **56**(5), 691–701 (2012)
16. Pehlivanović, B., Čaklovica, K., Lagumdžija, D., Žiga Smajić, N., Bečić, F.: *In vitro* evaluation of pleiotropic properties of rosuvastatin. Int. J. Pharm. Sci. Res. **12**(2), 1000–1006 (2021)
17. Oesterle, A., Laufs, U., Liao, J.K.: Pleiotropic effects of statins on the cardiovascular system. Circ. Res. **120**, 229–243 (2017)
18. Banfi, C., Baetta, R., Gianazza, E., Tremoli, E.: Technological advances and proteomic applications in drug discovery and target deconvolution: identification of the pleiotropic effects of statins. Drug Discov. Today. **22**, 848–869 (2017)
19. Jeenger, M.K., Shrivastava, S., Yerra, V.G., Naidu, V.G., Ramakrishna, S., Kumar, A.: Curcumin: a pleiotropic phytonutrient in diabetic complications. Nutrition. **31**(2), 276–282 (2015)
20. Ullah, A.H.M., et al.: Evaluation of antinociceptive, in-vivo & in-vitro anti-inflammatory activity of ethanolic extract of Curcuma zedoaria rhizome. Complement. Alternat. Med. **14**, 346–354 (2014)
21. Pehlivanović, B., et al.: Curcumin: natural antimicrobial and anti-inflammatory agent. J. Pharm. Res. Int. **32**(43), 1–8 (2020)
22. Umeda, R., Takanari, H., Ogata, K., et al.: Direct free radical scavenging effects of watersoluble HMG-CoA reductase inhibitors. J. Clin. Biochem. Nutr. **64**(1), 20–26 (2019)
23. Rajkiran, M., Deepa, K.: Pleiotropic antioxidant potential of rosuvastatin in preventing cardiovascular disorders. Eur. J. Pharmacol. **711**(1–3), 57–62 (2013)
24. Sökmen, M., Akram Khan, M.: The antioxidant activity of some curcuminoids and chalcones. Inflammopharmacology **24**(2–3), 81–86 (2016). https://doi.org/10.1007/s10787-016-0264-5

Check for updates

Influence of Electrolytes on Acetylsalicylic Acid Concentration

Elma Hasković(✉) and Sumeja Hadžalić

Faculty of Pharmacy, University of Sarajevo, Zmaja od Bosne 8, 71 000 Sarajevo, Bosnia and Herzegovina

Abstract. In this paper is presented the anaylisis of acetylsalicylic acid (aspirin) using the spectrophotometric method. The subject of this paper is the study of the influence of electrolytes on the concentration of aspirin from different manufacturers. Spectrophotometric determination of acetylsalicylic acid was performed, based on the principle that acetylsalicylic acid is converted into Fe (III) - a salicylate complex that absorbs light in the visible region of the spectrum. Based on the obtained results, we conclude that electrolytes have an effect on the concentration of acetylsalicylic acid.

Keywords: Acetylsalicylic acid · Spectrophotometry · Electrolytes

1 Introduction

A drug is chemical substance that affects physiological function in a specific way. A medicinal product containing a substance or combination of substances manufactured and intended for the treatment or prevention of disease in humans or animals, for diagnosis, improvement or alteration of physiological functions, and for the attainment of other medically justified objectives. The basic division of drugs is into natural, semi-synthetic and synthetic, more precisely into drugs (of plant, animal and mineral origin), inorganic and organic chemical preparations, galenic preparations and serums and vaccines [1].

In most cases, the drug works by binding to specific sites in the cell called receptors. As the receptor is distant from the site of administration, it is important that the drug is transported from the site of administration to the site of action. The drug binds to the receptor, and the binding strength may be different, which determines the duration of the effect of the drug. It is desirable that the drug reaches the desired site of action, ie. to have a selective effect, thus reducing the possibility of complications due to additional side effects. One or more drugs can act on the human body, and this effect is called pharmacodynamics [2].

When the body acts on the drug, the processes of absorption, metabolism and excretion of the drug occur.These processes are described by pharmacokinetics. The drug can be ingested in several ways, and which route of administration is most appropriate for a particular patient depends on the pharmacological properties of the drug, as well as the therapy used [2].

A. Badnjevic and L. Gurbeta Pokvić (Eds.): CMBEBIH 2021, IFMBE Proceedings 84, pp. 441–448, 2021.
https://doi.org/10.1007/978-3-030-73909-6_50

Acetylsalicylic acid, better known by its trade name as aspirin, belongs to a small number of therapeutic agents that are well known to both health professionals and patients. Aspirin is one of the most intensively studied drugs, but it is also one of the most commonly used and cheapest drugs in the world [3].

It belongs to a group of non-steroidal anti-inflammatory drugs (NSAID). It is a synthetic derivate of saliciln glycoside found in plants of the famies Salix (willow) and Spiraea (whey).

Salicylic acid (2-hydroxybenzoic acid) was first descovered in the mid19th century, when the glycoside salicin was converted to the acid form. However, this form was difficult to take orally because it caused irritation in the oral cavity and stomach and tasted bad. In 1897, Arthur Eichengun solved this problem by synthesizing acetylsalicylic acid (2acetoxybenzoic acid), a form of aspirin as we use today [4].

It is sometimes used in smaller doses to prevent heart attack in people who are at risk of developing blood clots [5]. Aspirin is one of the most versatile drugs known in medicine. The first synthesis of aspirin was performed by Felix Hoffman in 1897 in the labaratory of the Freidrich Bayer factory in Elberfeld, Germany [3]. Aspirin synthesis is classified as an esterification reaction. In this reaction, acetic anhydrate, an acetic aid derivate, is added to the salicylic acid as the reactant, which, in the presence of an 85% phosphotic acid (H_3PO_4) catalyst (sometimes sulfuric acid, H_2SO_4), gives acetylsalicylic acid and as a by-product, ace (acetylsalcylic acid). During the reaction itself, acetate anhydrate binds to the phenolic group of salycilic acid (OH–) turning it into an acetyl group (ROH → R-OCOCH$_3$) Fig. 1.

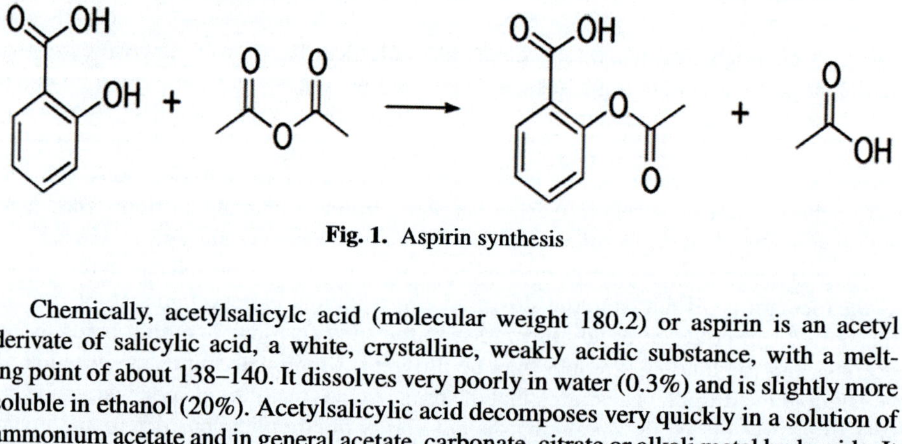

Fig. 1. Aspirin synthesis

Chemically, acetylsalicylc acid (molecular weight 180.2) or aspirin is an acetyl derivate of salicylic acid, a white, crystalline, weakly acidic substance, with a melting point of about 138–140. It dissolves very poorly in water (0.3%) and is slightly more soluble in ethanol (20%). Acetylsalicylic acid decomposes very quickly in a solution of ammonium acetate and in general acetate, carbonate, citrate or alkali metal hydroxide. It is stabile in dry air, hile it is gradually hydrolized by the action of moisture from the air on acetic acid and salicylic acid [6]. An electrolyte is a substance (solution or solid) that is electrically conductive due to the movement of free ions, ie. possesses the proprety of iconic conductivity. An easy way to obtain electrolytes is by dissolving or diluting an acid, base or salt in a suitable solution, usually water. In this case, the molecules of the acid, base or salt dissociate into two ions whose charges are of equal amounts, but opposite in polw or sign. Positively charged ions (cations) are the metal part of a salt, and in the case of an acid it is a hydrogen atom. Negative ions (anions) are an acid residue in acids and salts, and in bases it is the hydroxyl group OH^- [7]. All substances that conduct electricity by ions, when dissolved or dissolved in water, are called electrolytes, and all other substances that do not conduct electricity under the same conditions are called non-electrolytes. Electrolyte solutions show a greater decrease in the freezing point and melting point as well as a higher osmotic pressure than corresponds to their molecular weight [8, 10].

2 Materials I Methods

Material: acetylsalicylic acid (Bayer, Hemofarm, Bosnalijek); NaOH p.a. (Centrohem d.o.o., Stara Pazova); iron (III) – chloride hexahydrate ($FeC_{13} \cdot 6H_2O$), (Lach – Ner,Czech Republic); $CaCl_2$ p.a. (Fisher Chemical, Wien, Austria); $MgCl_2$ p.a. (Fisher Chemical, Wien, Austria); KCl p.a. (Fisher Chemical, Wien, Austria).

Method: The instrument on which the measurments where made is called spectrophotometer (PerkinElmer instruments). It is used during spectrophotometric determining of aspirin a procedure based on the properties of complex being performed from aspirin to absorb light at wavelenght of 545 nm. In solution, aspirin by itself does not absorb light in the visible area, but when translated to iron (III) salicylate complex, absorption characteristics of the colored solution, can be used to determine the quantitative concentration of aspirin.

The general procedure is called spectrophotometric analysis and has many important applications in chemistry. In the base medium, aspirin hydrolyzes and forms dianion salicylate (Fig. 2) [9–12].

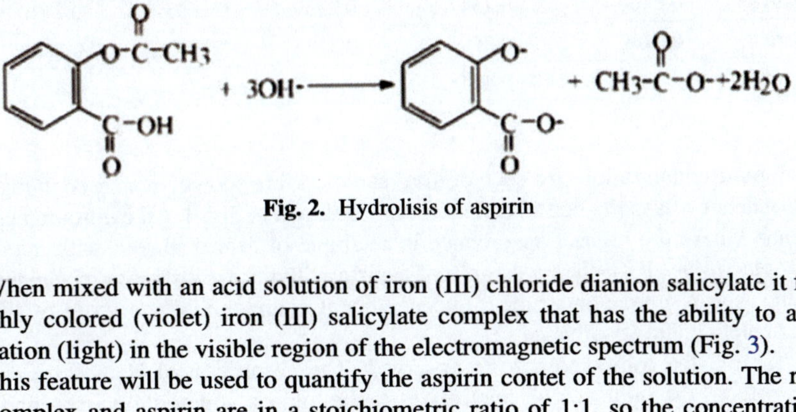

Fig. 2. Hydrolisis of aspirin

When mixed with an acid solution of iron (III) chloride dianion salicylatc it forms a highly colored (violet) iron (III) salicylate complex that has the ability to absorb radioation (light) in the visible region of the electromagnetic spectrum (Fig. 3).

This feature will be used to quantify the aspirin contet of the solution. The resulting complex and aspirin are in a stoichiometric ratio of 1:1, so the concentration of the complex is equal to the conentration of aspirin. In this paper several solutions of known concentration of salicylate complex were prepared (0,01 mg/mL; 0,02 mg/mL; 0,03 mg/mL; 0,04 mg/mL; 0,05 mg/mL) and the absorbance of each of them was determined at the wavelength of maximum absorption. After that a concentration of different electrolytes was added to the solution ($CaCl_2$, $MgCl_2$, KCl) which corresponds to the concentration in the extracellular fluid of 103 mmol/l. Lambert – Beer's law graphically shows absorbances as a function of concentration [9, 13, 14].

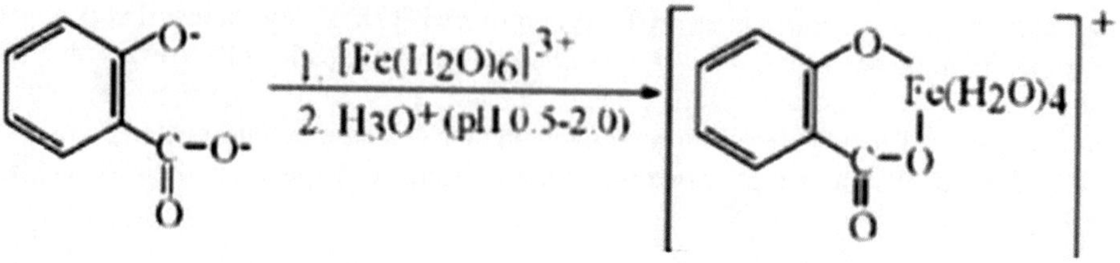

Fig. 3. Reaction of formation of iron (III) salicylate complex

3 Results and Discussion

During the analysis and quantitative determination of the content of acetysalicylic acid by spectrophotometric method, three aspirin tablets from different manufacturers were used, which can be found on our market, and that are: Aspirin, Bayer; Midol protect, Hemofarm and Bospyrin, Bosnalijek. Results are presented in table and graph.

Table 1. Aspirin (Bayer) results without and with electrolyte influence.

	Concentration mg/ml	A without electrolyte	A [$MgCl_2$] 103 mmol/l	A [KCl] 103 mmol/l	A [$CaCl_2$] 103 mmol/l
Blank sample	0,0000	0,0086	– 0,0095	0,1259	0,0061
Standard 1	0,0100	0,0908	0,0659	0,0662	0,0809
Standard 2	0,0200	0,1744	0,1511	0,1456	0,1523
Standard 3	0,0300	0,2547	0,2359	0,2603	0,2420
Standard 4	0,0400	0,3501	0,3225	0,3275	0,3188
Standard 5	0,0500	0,4248	0,4152	0,4261	0,4205

In Table 1 conentrations for each of the 5 standards are shown, expressed in mg/mL, the dependence of absorbance an concentration is shown in Fig. 4. All calibration curves are lienar, where the highest apsorbance is a sample of aspirin Bayer, with the influence of electrolyte KCl, then a sample of aspirin without the influence of electrolyte, while the aspirin sample under the influence of $MgCl_2$ and the influence of $CaCl_2$ have approximately equal absorbance.

In Table 2 the following can be observed: The absorption of Hemofarm aspirin tablets without the influence of electrolytes is the highest and continuously increases with increasing concentration.

The absorbance and the effect of $CaCl_2$ are the higher compared to other electrolytes. This table shows the graph shown in Fig. 5.

In Fig. 5, the calibration curves are linear, where the sample of Hemofarm aspirin without the influence of elecrolytes has the highest absorbance, followed by the sample of aspirin under the influence of $CaCl_2$ and the sample of aspirin under the influence of $MgCl_2$, and KCl solution has the least effect, reulting in the lowest absorbance.

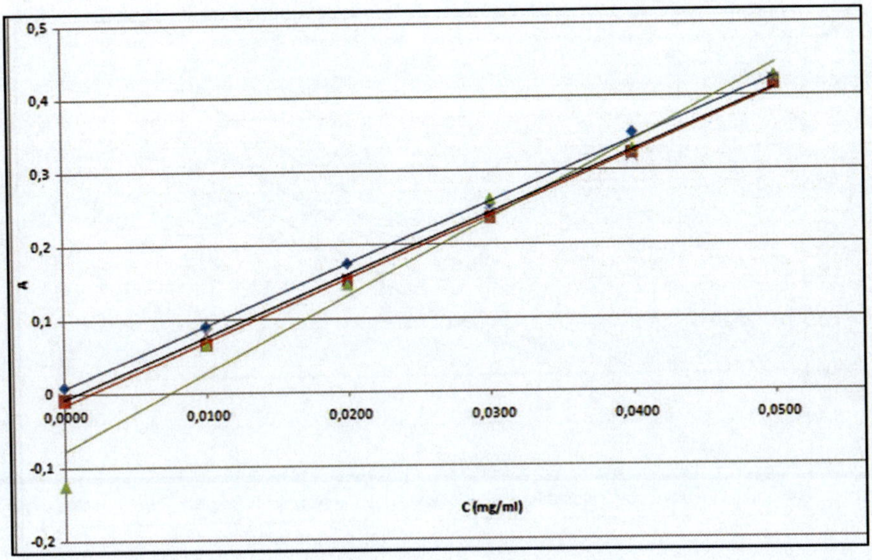

Fig. 4. Graphic representationof the calibration cure of Bayer aspirin samples, with and without electrolyte influence (blue line - sample without electrolyte influence, red line - sample with $MgCl_2$ influence, green line - sample with KCl influence, red line - sample ith $CaCl_2$ influence).

Table 2. Results of aspirin absorption (Hemofarm) without and with the influence of electrolytes.

Solution	Concentration mg/ml	A without electrolytes	A [$MgCl_2$] 103 mmol/l	A [KCl] 103 mmol/l	A [$CaCl_2$] 103 mmol/l
Blank sample	0,0000	0,0056	– 0,0233	– 0,0221	– 0,0211
Standard 1	0,0100	0,0685	0,0342	0,0382	0,0372
Standard 2	0,0200	0,1506	0,0997	0,0997	0,1205
Standard 3	0,0300	0,2293	0,1702	0,1751	0,2047
Standard 4	0,0400	0,3197	0,2474	0,2427	0,2756
Standard 5	0,0500	0,3965	0,3214	0,2924	0,3563

In Table 3, the absorbance of the Bosnalijek aspirin sample without the effect of electrolytes is the highest, ith standard 2 at a concentration of 0.0200 mg/ml indicating that the $CaCl_2$ solution has an effect, resulting in higher absorbance in such a sample. This table shows the graph shown in Fig. 6.

Figure 6 shows the ratio of absorbance and concentration of all 5 standards with and without the influence of electrolytes. All calbration curves are linear, where the sample of Bosnalijek aspirin without the influence of electrolytes has the highest absorbance with increasing concentration. At concentrations of 0.01 and 0.02 mg/ml, it could be said that the aspirin sample under the influence of KCl and the aspirin sample without any electrolyte effect have approximately the same absorbance, however, with increasing

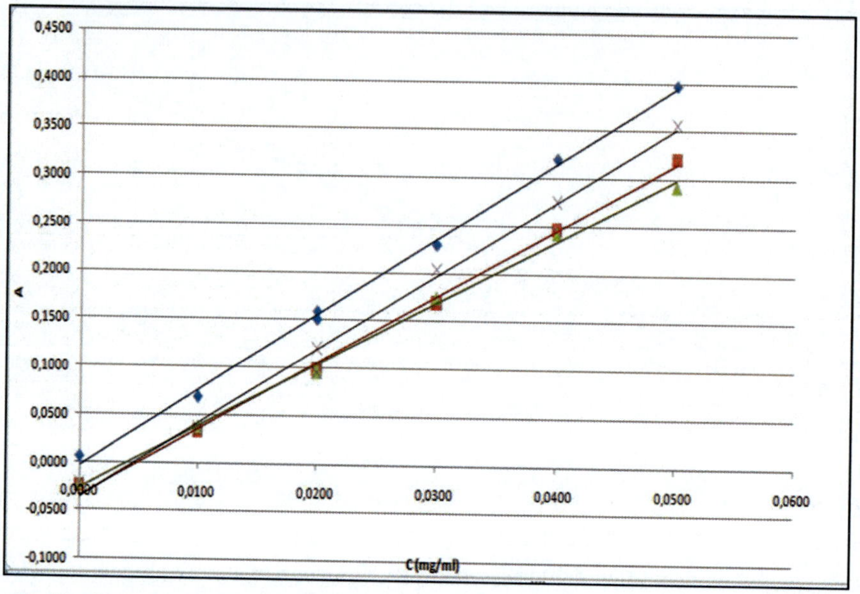

Fig. 5. Graphical display of the calibration curve of Hemofarm aspirin samples, with and without eletrolyte influence (blue line - sample without electrolyte influence, red line - sample ith $MgCl_2$ influence, green line - sample with KCl influence, black line - sample with $CaCl_2$ influence).

Table 3. Results of aspirin absorption (Bosnalijek) without and with the influence of electrolytes

Solution	Concentration mg/ml	A without electrolytes	A [$MgCl_2$] 103 mmol/l	A [KCl] 103 mmol/l	A [$CaCl_2$] 103 mmol/l
Blank sample	0,0000	0,0086	−0,0210	0,0080	−0,0228
Standard 1	0,0100	0,0875	0,0542	0,0838	0,0594
Standard 2	0,0200	0,1705	0,1315	0,1601	0,1939
Standard 3	0,0300	0,2636	0,2291	0,2443	0,2122
Standard 4	0,0400	0,3571	0,3082	0,3396	0,3104
Standard 5	0,0500	0,4378	0,3973	0,4240	0,4029

concentration the absorbance of the aspirin sample under the influence of KCl decreases. The aspirin sample under the influence of $CaCl_2$ has slightly higher absorbance than the aspirin sample under the influence of $MgCl_2$.

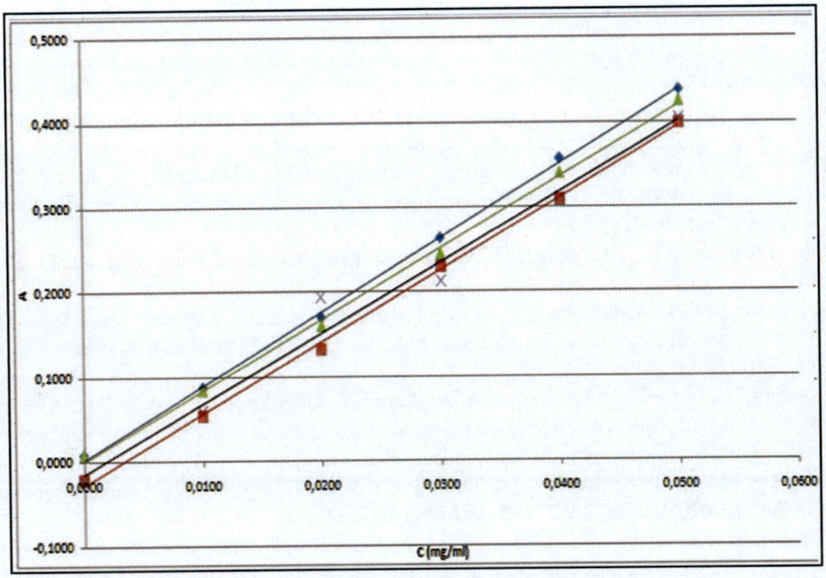

Fig. 6. Graphic display, calibration curves of Bosnalijek aspirin samples, with and without electrolyte influence (blue line - sample without electrolyte influence, red line - sample with $MgCl_2$ influence, green line- sample with KCl influence, black line - sample with $CaCl_2$ influence).

4 Conclusion

The spectrophotometric method of examing the efect of electrolytes on aspirin from different manufacturers led to the conclusion that allthree examined electrolyte absorbances increase with increasing drug concentration. The addition of electrolytes reduces the absorbance, which means that it affects the kinetics of the drug and shows inhibitory properities on the action of aspirin. The effect of potassium ions on aspirin (Bosnalijek) is the smallest, probably due to the valence of potassium and physicochemical properties, compared to divalent ions such as Ca^{2+} and Mg^{2+}, which increase and accelerate the kinetics of aspirin. From all the results we conclude that electrolytes are very important in the body and that they have an impact on the kinetics of the drug.

References

1. Ivaniš, B.: Validacija HPLC metode za odre ivanje sadržaja lidokaina u Lidokain spreju. Diplomski rad. Zagreb: Fakultet kemijskog inženjerstva i tehnologije (2012)
2. Petrović, M.: Kako lijek djeluje i kako se unosi u organizam. Zagreb (2009)
3. Sartori, E.R.: Square-Wave Voltammetric Determination of Acetylsalicylic Acid in Pharmaceutical Formulations using a Boron-Doped Diamond Electrode without the Need of Previous Alkaline Hydrolysis Step, vol. 2, pp. 360–366 (2009)
4. Lovrek, K.: Sinteza i IR spektroskopska analiza aspirina i paracetamola. Diplomski rad. Sveučilište J. J. Strossmayera, Osijek (2012)

5. Anthony, O.: A comparative anaylsis of aspirin from various analgesic formulations using titrimetry Spectroscopic and Hypenated Chromatographic Techniques. J. Mol. Imaging Dynn. **2**, 1–6 (2017)
6. Pavlović, A.N.: Kinetičko odre ivanje nekih analgetika i sedativa. Doktorska disertacija. Univerzitet u Nišu, Prirodno matematički fakultet, Niš (2009)
7. Filipović, I., Lipanović, S.: Opća i anorganska kemija, 9 izd. Školska knjiga, Zagreb (1995)
8. Brčkalić, E.: Elektroliti. Univerzitet u Tuzli, Rudarsko Geološko Gra evinski fakultet, Seminarski rad. Tuzla (2014)
9. Interni praktikum iz fizikalne hemije. Sarajevo: Univerzitet u Sarajevu, Farmaceutski fakultet (2016)
10. Ali Rasool, M.A., Mohammed, A.A., Benas thamer, M.Z.A.J.: Development Method for Determination of Aspirin as Sodium Salicylate by UV-VIS Spectroscopy. Mater. Sci. Eng. **571**, p. 012104 (2019)
11. Ochieng, A., Hemed, S.M., Mataka, A.M., Abdul, J., Ochieng, J.O., Okoli, C.P.: A comparative anaylsis of aspirin from various analgesic formulations using titrimetry spectroscopic and hypenated chromatographic techniques. J. Mol. Imaging Dyn. **7**, 2 (2017)
12. Gabriela, M., Aurel, P.: Studies of different types of aspirin by spectrophotometric methods. ACTA CHEMICA IASI, 22_2, 155–164 (2014)
13. Laghari, M.G.H., Darwis, Y., Memon, A.H.: New spectrophotometric methods for the determination of p-Aminosalicylic acid in tablets. Tropical J. Pharm. Res. **13**(7), 11331139 (2014)
14. Mohauman Mohammad, A.R.: Spectrophotometric determination of sodium salicylate in pharmaceutical preparations by coupling with diazotized para–amino benzoic acid, MOJ Bioequiv Availab. **2**(2), 17–21 (2016)

Effect of Diclofenac and Potassium Ions on Catalase Activity

Safija Herenda[1], Edhem Hasković[2], Denis Hasković[3], Ena Hasković[4], and Svjetlana Dilber[2(✉)]

[1] Department of Chemistry, Faculty of Science, University of Sarajevo, Zmaja od Bosne 33-35, 71 000 Sarajevo, Bosnia and Herzegovina
[2] Department of Biology, Faculty of Science, University of Sarajevo, Zmaja od Bosne 33-35, 71 000 Sarajevo, Bosnia and Herzegovina
[3] Clinical Center, Organizational Unit Clinical Pathology, Cytology and Human Genetics, University of Sarajevo, Bolnička 25, 71 000 Sarajevo, Bosnia and Herzegovina
[4] Clinical Center, Organizational Unit Clinical Chemistry and Biochemistry, University of Sarajevo, Bolnička 25, 71 000 Sarajevo, Bosnia and Herzegovina

Abstract. The aim of this study was to investigate the type of inhibition of the anti - inflammatory drug diclofenac and potassium ions on the activity of the enzyme catalase *in vitro*. All experiments in this work were performed at a temperature of 37 °C, which corresponds to normal human body temperature, by the spectrophotometric method at a wavelength of 405 nm. The blank sample contained the substrate hydrogen peroxide (H_2O_2), catalase, ammonium molybdate and phosphate buffer, and the tested samples in addition to the listed components also contained diclofenac in various concentrations of 0,5065; 1,013; 1,5195; 2,026; 2,5325 and 3,039 mM, as well as a 4,5 mM KCl concentration. Based on the obtained results for the values of kinetic constants V_{max} (maximum rate) and K_M (Michaelis-Menten constant), obtained by the graphical method, we conclude that diclofenac, as well as potassium ions have a partially competitive type of inhibition on enzyme catalase.

Keywords: Inhibition · Enzymes · Catalase · Diclofenac · Potassium ions

1 Introduction

The enzyme is a protein with catalytic properties which, as a catalyst, lowers the activation energy of the reaction (E_a), thus increasing the reaction rate without affecting the equilibrium of the reaction itself. The enzyme lowers the activation energy most often by specific binding to the intermediate, thus stabilizing it [1]. An inhibitor is a substance that reduces the rate of an enzymatically catalyzed reaction, whereby this inhibition can be competitive, non-competitive and competitive. The anti-inflammatory drug (diclofenac) is a benzene acetic acid product, a nonsteroidal anti-inflammatory drug, a cyclooxygenase inhibitor. According to its chemical structure, it is a product of 2-[(2,6 - dichlorophenyl) amino] benzene acetic acid. Sodium and potassium salts of this

A. Badnjevic and L. Gurbeta Pokvić (Eds.): CMBEBIH 2021, IFMBE Proceedings 84, pp. 449–455, 2021.
https://doi.org/10.1007/978-3-030-73909-6_51

drug are soluble in water and the drug is taken orally in the treatment of inflammatory processes and severe pain. It's greatest effect is antiinflammatory activity. In addition, it acts as an analgesic and antipyretic. It is most often used as a treatment for acute and chronic pain, rheumatitis and osteoarthritis. It's action is based on the inhibition of COX activity and the consequent formation of pro-inflammatory mediators such as prostaglandins (PGs) and thromboxanes [2] (Fig. 1).

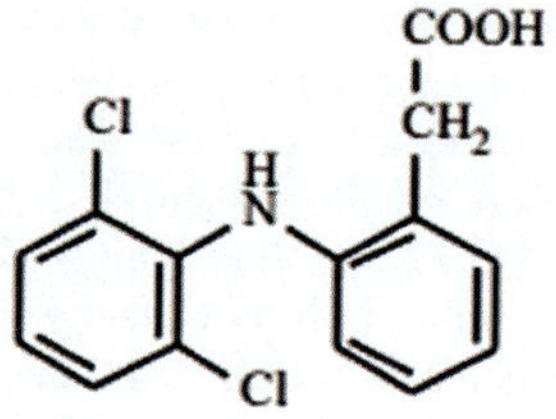

Fig. 1. Structure of diclofenac

Eicosanoids are paracrine hormones, substances that act only on cells that are close to their formation. These fatty acid (MC) products have a number of effects on vertebrate tissues. They are involved in reproductive function, inflammation, fever and pain associated with injuries and disease, in the formation of blood clots and regulation of blood pressure, in the secretion of stomach acid, and in many other processes important for human health or disease. All eicosanoids are derived from arachidonic acid 20:4 ($\Delta^{5,8,11,14}$).

There are three types of eicosanoids: prostaglandins, leukotrienes and thromboxanes. Prostaglandins contain an alkyl chain with a 5-C ring derived from arachidonic acid, named after the gland, the prostate from which they were first isolated. Thromboxanes also have a sixmembered ring containing an ester bond. They are formed by platelets from arachidonic acid, and participate in the formation of blood clots and reduce blood flow to the thrombus. Nonsteroidal anti-inflammatory drugs (NSAIDs) such as aspirin, ibuprofen, meclofenamate as well as diclofenac inhibit the enzyme prostaglandin H_2 synthetase (cyclooxygenase) which catalyzes the early step in the process from arachidonic to prostaglandin and thromboxane [3]. During extensive research aimed at developing anti-inflammatory drugs with improved biological properties, a large number of o-amino-phenyl-acetic acids were synthesized and their pharmacological activity examined.

In the implementation of this broad synthesis program, the goal was to provide a basic principle of a molecule that would meet the following conditions:

1. Acidity constant (pKa) of 4–5,
2. A coefficient of P (n-octanol/aqueous buffer; pH 7.4) with a value of approximately 10,
3. Two aromatic rings assembled together.

The selection of these criteria is based on the structural and physicochemical properties of phenyl-butazone, mefenamic acid and indomethacin, all of which have been shown to be successful in acting against various forms of inflammation and pain. Of the nearly more than 200 analogs, sodium [*o*-[(2,6-dichlorophenyl)-amino] phenyl] acetate (diclofenac sodium) has been shown to be the most interesting in terms of its pharmacological properties. The compound meets the physicochemical target properties, with a pKa of 4,0 and a P coefficient of 13,4. Wherein the X-ray analysis proves a structure in which two aromatic rings are bound together (distortion angle 69°). Two chlorine atoms in the ortho position of one of the phenyl rings cause maximum twisting of this ring [4]. In this work, the pharmacokinetics of diclofenac and potassium ions on catalase activity were monitored by spectrophotometric method in simulated *in vivo* conditions.

2 Materials and Methods

2.1 Materials

Hydrogen peroxide, 30% w/w, Sigma-Aldrich; Bovine liver catalase enzyme, aqueous solution, c100-50 MG, Sigma-Aldrich; Ammonium molybdate, Sigma-Aldrich; Phosphate buffer, KH_2PO_4 and Na_2HPO_4, Fisher Chemical (Wien, Austria); Diclofenac, Bosnalijek, BiH; Potassium chloride (KCl), p. A., Semikem d.o.o, BiH; Spectrophotometer, Multiskan GO with Thermo Scientific SkanIt program, Thermo Fisher Scientific Oy, Finland.

2.2 Method

The spectrophotometric method in the presence of ammonium heptamolybdate, described in the Góth literature [5], was used. The obtained absorbances, using kinetic laws, were translated into appropriate velocities. The Michaelis-Menten equation was used to determine the effect of diclofenac on catalase activity. The kinetic constants maximum velocity (V_{max}) and Michaelis-Menten constant (K_m) were determined graphically, i.e. via the Lineweaver-Burk diagram. The influence of different concentrations of the substance on the activity of the enzyme is observed via the Lineweaver-Burk diagram [6, 7]. Based on the obtained directions, it is possible to determine the appropriate type of inhibition in the observed system. The type of inhibitory activity of diclofenac was determined by determining the kinetic parameters of enzyme inhibition.

3 Results and Discussions

To examine the type of inhibitory effect of different concentrations of diclofenac in the range of hydrogen peroxide substrate concentrations (1,6; 2; 2,4; 4; 5; and 6 mM) on enzyme activity, we used the Michaelis-Menten kinetic model. By applying reciprocal values of substrate concentrations on the abscissa as well as reciprocal values of enzyme reaction rates, linear dependences of these two parameters were obtained.

This further allows the Michaelis-Menten constant and the maximum reaction rate to be calculated based on the Lineweaver – Burk diagram. Figure 2, Table 1.

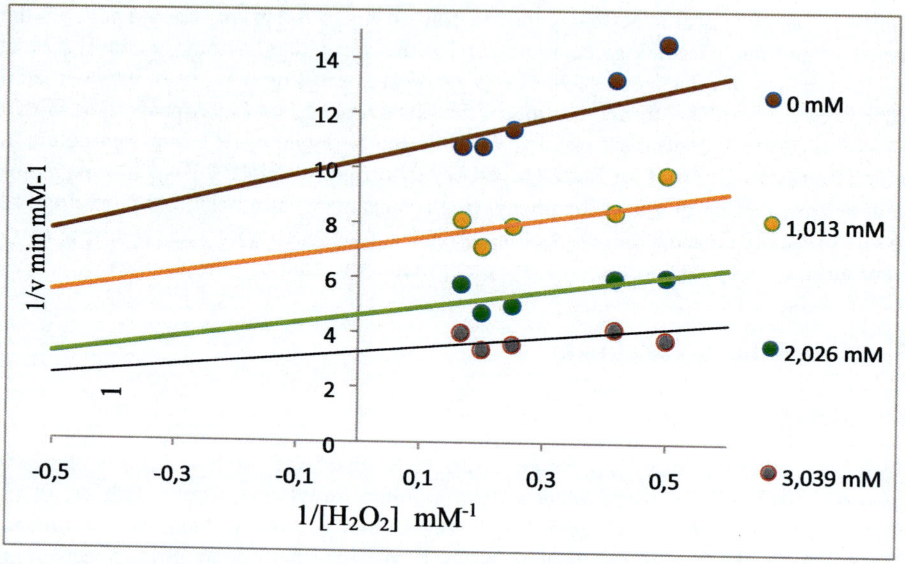

Fig. 2. Lineweaver - Burk diagram for determination of V_{max} and K_m without the presence and with the presence of different concentrations of diclofenac

Using the Lineweaver-Burk equation, we can calculate the kinetic parameters, i.e. the constants K_m and V_{max} for the corresponding concentrations of diclofenac:

$$\frac{1}{\upsilon_0} = \frac{K_m}{V_{max}S}\frac{1}{S} + \frac{1}{V_{max}}\left(1 + \frac{[I]}{K_i}\right)$$

Table 1. K_m and V_{max} values without the presence and with the presence of different concentrations of diclofenac

[diklofenac] (mM)	Vmax (mM min-1)	K_M (mM)
0	0,0978	0,5016
1,013	0,1411	0,47206
2,026	0,2169	0,6253
3,039	0,3061	0,5492

Based on the graphical representation of the influence of diclofenac, it is obvious that it is an acompetitive type of enzyme inhibition, which is indicated by parallel directions for different concentrations of diclofenac. In Fig. 2 we see that at the beginning of the reaction an enzyme-substrate complex (catalase-peroxide) is formed. The hydrogen peroxide decomposition reaction is considered to take place in two steps:

$$H_2O_2 + Fe(III) - E \rightarrow H_2O_2 + O = Fe(IV) - E(.+)$$

$$H_2O_2 + O = Fe(IV){-}E(.+) \rightarrow H_2O_2 + Fe(III) - E + O_2.$$

Fe (III)-E is the center of iron in the heme group. Fe (IV)-E (.+) Is a mesomeric form of Fe (V)-E. Thus, iron is not completely oxidized but receives a "supporting electron" from heme ligand. Heme is designated here as a radical cation (.+). When hydrogen peroxide reaches the active center of catalase, it then interacts with the amino acids asparagine at position 147 (Asn 147) and histidine at position 74 (His 74). On that occasion, the proton, the hydrogen ion H+, is transferred between oxygen atoms. When free oxygen coordinates, the resulting water molecule is released and Fe (IV) = O. After that Fe (IV) = O reacts with the next molecule of hydrogen peroxide and forms Fe (III) -E and water and oxygen are formed. With the presence of phenolate ligand on tyrosine at position 357 (Tyr 357) the reactivity of the iron center can be improved. Here Tyr 357 participates in the oxidation of Fe (III) to Fe (IV) [8]. As the concentrations of diclofenac in the observed system increase (Fig. 2), the values of the constants V_{max} and K_m increase, so we conclude that diclofenac has a partial competitive effect on catalase activity. Partial competitive inhibition takes place according to the following mechanism [9]:

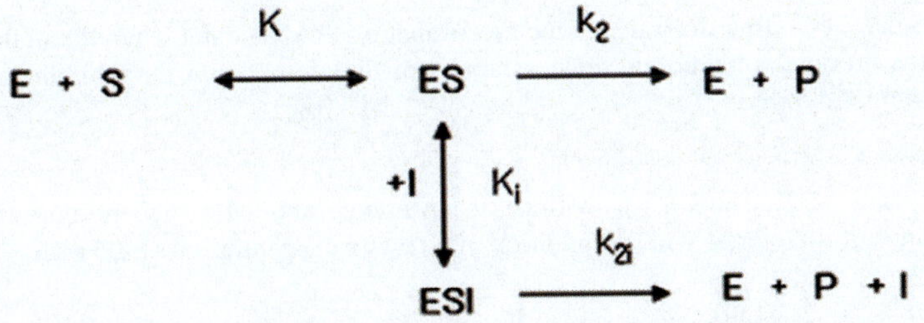

Partial inhibition of catalase suggests that degradation continues in part even after diclofenac has inhibited the enzyme-substrate complex, i.e., the award-winning enzyme-substrate-inhibitor complex, dissociates, in part, into the enzyme, product, and inhibitor [10]. The effect of potassium ion concentration of 4.5 mM on catalase activity is shown in Fig. 3, Table 2.

Using the Lineweaver-Burk equation, we can calculate the kinetic parameters, i.e. the constants K_m and V_{max} for the corresponding concentrations of diclofenac:

$$\frac{1}{\upsilon_0} = \frac{K_m}{V_{max}S}\frac{1}{S} + \frac{1}{V_{max}}\left(1 + \frac{[I]}{K_i}\right)$$

Based on a graphical representation of the effects of diclofenac and potassium ions, it is obvious that it is a competitive type of enzyme inhibition. As the concentrations of diclofenac in the observed system increase, so does the value of the constants V_{max} i K_m. We conclude that diclofenac has a partial competitive effect and with the addition of potassium ions on catalase activity. With the addition of potassium ions, in the concentration present in the human body, the type of inhibition remains the same.

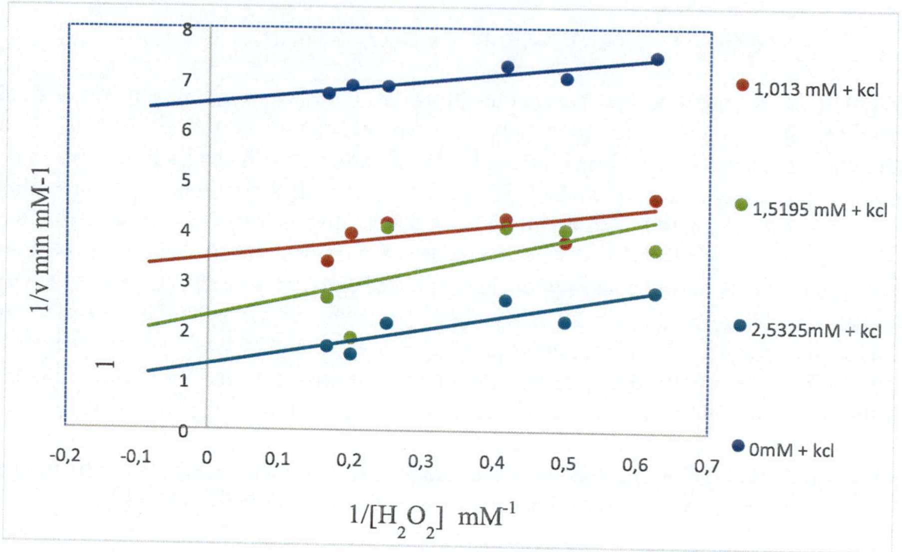

Fig. 3. Lineweaver - Burk diagram for the determination of Vmax and Km without the presence and with the presence of different concentrations of diclofenac with the addition of KCl at a concentra-tion of 4,5 mM

Table 2. Values of K_m and V_{max} without the presence and with the presence of different concentrations of diclofenac with the addition of KCl at a concentration of 4,5 mM

[diklofenac] (mM)	Vmax (mM min-1)	K_M (mM)
0	0,1537	0,2183
1,013	0,2927	0,4729
1,5195	0,4413	1,3381
2,5325	0,7682	1,8048

4 Conclusion

By considering the results obtained by spectrophotometric analysis, we came to the conclusion that the study showed that the enzyme catalase follows the Michaelis-Menten model. Based on the graphical representation of the influence of diclofenac and potassium ions, it is obvious that this is an acompetitive type of enzyme inhibition, which is indicated by parallel directions for different concentrations of diclofenac. Using the Lineweaver-Burk equation, the values of the kinetic parameters V_{max} and K_m were calculated. The values of the constants increase with increasing concentrations of diclofenac, which indicates that it is a partially acompetitive inhibition.

Optimal conditions were achieved for conducting the research at a temperature of 37 °C and pH = 7. Diclofenac inhibits catalase in vitro and affects its activity at different concentrations, which indicates that at the tested concentrations of the substrate it exhibits improved antiinflammatory activity.

References

1. Marangoni, A.G.: Enzyme Kinetics: A Modern Approach. Wiley, Hoboken (2003)
2. Naveed, S., Qamar, F.: UV spectrophotometric assay of Diclofenac sodium available brands. J. Innov. Pharm. Biol. Sci. JIPBS **1**(3), 92–96 (2014). Faculty of Pharmacy Jinnah University for women, Karachi, Safila Naveed
3. Nelson, D., Cox, M.: Principles of Biochemistry. W. H Freeman and Company, New York (2013)
4. Menasse, R., Hedwall, P.R., Kraetz, J., Pericin, C., Riesterer, L., Sallmann, A., Ziel, R., Jaques, R.: Pharmacological propertcies of diclofenac sodium and its metabolites. Scand. J. Rheumatol. **22**, 516 (1978)
5. Góth, L.: A simple method for determination of serum catalase activity and revision of reference range. Clinica Chemica Acta **196**(2–3), 143–151 (1991)
6. Michaelis, L., Menten, M.L.: Die Kinetik der Invertinwirkung. Biochem. Z. **49**, 333–369 (1913)
7. Markus, M., Hess, B., Ottaway, J.H., Cornish- Bowden, A.: The analysis of kinetic data in biochemistry a critical evaluation of methods. FEBS Lett. **63**, 225–230 (1976)
8. Putnam, C.D., Arvai, A.S., Bourne, Y., Tainer, J.A.: Active and inhibited human catalase structures: ligand and NADPH binding and catalytic mechanism. J. Mol. Biol. **296**(1), 295–309 (2000)
9. Whiteley, C.G.: Enzyme kinetics: partial and complete uncompetitive inhibition. Biochem. Educ. **28**(3), 144–147 (2000)
10. Park, J.H., Yoo, I.K., Kwon, O.Y., Ryu, K.: Partial uncompetitive inhibition of horseradish peroxidase by a water-miscible onic liquid [BMIM][MeSO4]. Biotech. Lett. **33**(8), 16571662 (2011)

Current Trends in Cancer Immunotherapy

Amila Hajdarević(✉), Miralem Kmetaš, Faris Begović, Merima Durić, and Ema Vajzović

Students of Faculty of Pharmacy, University of Sarajevo, Sarajevo, Bosnia and Herzegovina
amilahajdarevic@ffsa.unsa.ba

Abstract. Cancer is known as the second cause of death worldwide. Although various methods have been implemented such as chemotherapy, radiotherapy, and surgical resection of tumors, and have shown satisfactory results, we are experiencing an increase in the number of patients who do not benefit from these treatments. Moreover, these treatments may cause severe side effects and greatly impact the patient's quality of life. The insufficient precision and toxicity of accessible therapeutics have encouraged an increased exploration of alternative approaches to treating cancer, namely immunotherapy. Immunotherapy includes monoclonal antibodies, immune checkpoint blockades, cancer vaccines, and dendritic cell vaccines as the latest and most efficient methods in combating cancer. It also provides a place for a more molecularly targeted approach to treating cancers, using biomarkers that will ultimately achieve personalized immunotherapy. Immunotherapy has shown an immense advancement in the treatment of cancers and has the needed potential to become the first therapeutic option in treating most cancers. In this paper, we discuss prominent methods of immunotherapy in cancer treatment along with their prospective furtherance.

Keywords: Immunotherapy · Monoclonal antibodies · Cancer vaccines · Immune checkpoint blockades · Personalized immunotherapy · Dendritic cell vaccines

1 Introduction

Ever since it first became known to doctors and scientists, cancer has been a major challenge to prevail over. Several methods of treating cancer gained popularity and have shown benefits for a large scale of cancer patients. These methods include surgical resection of tumors, chemotherapy, radiotherapy. However, these measures are considered aggressive as they can cause serious side effects that may impact the patient's quality of life.

To reduce the toxic effects of traditional radiotherapy and chemotherapy, scientists started to exploit the body's own immune system to kill cancer cells. This led to the development of a new approach to treating cancer, known as immunotherapy.

Immunotherapy strengthens the body's natural immune system and allows the immune cells to identify cancer cells by recognizing specific antigens found on their

A. Badnjevic and L. Gurbeta Pokvić (Eds.): CMBEBIH 2021, IFMBE Proceedings 84, pp. 456–461, 2021.
https://doi.org/10.1007/978-3-030-73909-6_52

surface. The success of immunotherapy has largely been limited by the cancer cells' capability to avoid the immune system in various ways [1].

Recently developed approaches tend to be involved in stopping the avoidance of cancer cells. Immunotherapy includes several therapies such as monoclonal antibodies, immune checkpoint blockades, cancer vaccines, and dendritic cell vaccines. The advancements in these methods have provided basis for reaching personalized immunotherapy.

Immunotherapy has a great potential of becoming the first therapeutic option when it comes to most cancers. However, similar to any other novelty, some of these treatments have serious side effects such as nausea, fever, and diarrhea [2].

This paper aims to explore the concept of cancer immunotherapy and provide a review of current trends and developments in the area of cancer immunotherapy.

2 Monoclonal Antibodies

The immune system produces antibodies that recognize malignant or pathogenic cells in the body by antigens expressed on their surface. Antibodies attach to the specific antigens and destroy foreign particles [1]. Identification of antigens on cancer cells is the most important step in selecting monoclonal antibodies. This is an extremely challenging task because cancer cells are sensitive to mutations [1].

There are several mechanisms by which monoclonal antibodies destroy cancer cells. One of the main mechanisms is referred to as targeted therapy [1, 3]. Monoclonal antibodies have the ability to recognize antigens on cancer cells and mark them as a target that the immune system will destroy. If some monoclonal antibodies are directly fused to cancer cells, it can lead to cell apoptosis [4].

Monoclonal antibodies can be divided into naked monoclonal antibodies that have a single function and can act directly on cancer cells. The second group is conjugated monoclonal antibodies that deliver radioactive particles or chemotherapy drugs directly to cancer cells. Bispecific monoclonal antibodies are the latest group of antibodies.

They consist of two antibodies that target different specific antigens.

Monoclonal antibodies can also be divided according to genetic development techniques. The first generation of monoclonal antibodies is composed of murine monoclonal antibodies that have not been able to interact with the human immune system. The next generation of chimeric monoclonal antibodies consists of constant human regions and variable mouse regions [5]. Humanized monoclonal antibodies are 95% human, and the remainder is the mouse region that determines complementarity. Finally, we arrive at fully-humanized monoclonal antibodies that are 100%, sequencelike human [6].

Monoclonal antibody-based treatment has proved to be one of the most successful treatments for hematologic malignancies and solid tumors in the last twenty years. This therapy is based on decades of research that has focused on understanding the interaction between cancer cells and immune system cells, and techniques for generating optimal antibodies at tumor target sites [7].

3 Cancer Vaccines

Cancer vaccines are a new generation of vaccines, different from the traditional prophylactic vaccines because they are in charge of attacking cells already present in the body. Two types of cancer vaccines can be specified: vaccines that prevent cancer and vaccines that treat cancer.

Preventive (prophylactic) vaccines prevent viral infections that cause cancer or contribute to cancer development. These vaccines stimulate the immune system to fight viruses before they cause an infection and are administered to healthy people before cancer develops.

Vaccines used to treat cancer work to promote the body's immune system to detect cancer-specific antigens found on the surface of cancer cells, but not healthy cells. This allows the immune system to interfere with cancer cells, stop their growth and proliferation, and ultimately kill these cells. There are several categories of cancer vaccines which include cell vaccines, peptide vaccines and genetic vaccines [1].

Making treatment vaccines that actually benefit a patient is a challenge due to several limitations. The first one is that larger tumors are hard to remove using only vaccines. This is why doctors give a cancer vaccine along with other treatments. The second limitation is the use of vaccines on the older population and people with a weakened immune system (due to other cancer treatments). They will most probably be unable to produce a strong immune response after receiving a vaccine. The third limitation is the possible malfunctioning antigen presentation that could induce tolerance to the antigens contained within the vaccine and jump-start rapid tumor progression [8]. Another thing to mention is the inability to differentiate clones of highly aggressive neoplastically transformed cells because they no longer express cancer cell-specific molecules. In these cases, cancer vaccines are of little use.

For the mentioned reasons, many researchers believe that cancer vaccines may work well for smaller tumors and cancer in its early stages [8].

4 Immune Checkpoint Blockades

Immune checkpoint blockade has been one of the most effective and significant advances in cancer immunotherapy of the past decade [9]. The condition for the formation of an immune response is the presence of molecules on certain immune cells that need to be activated or inactivated, which are called immune checkpoints [10, 11]. They have the role of either promoting or inhibiting T cell activation.

Antibodies that target checkpoint molecules, cytotoxic T lymphocyte antigen 4 (CTLA4), programmed cell death 1 (PD1), and PD1 ligand 1 (PD-L1) got attention because of their ability to use checkpoints to escape from being attacked by the immune system [12, 13]. The goal of immune checkpoint blockade is to remove inhibitory signals of T-cell activation, which enables tumor-reactive T cells to overcome regulatory mechanisms and mount an effective antitumor response.

Despite its success, response rate to immune checkpoint blockades is low in most cases [14]. This has largely to do with a lack of tumor T cell infiltration that distinguishes "cold" tumors from "hot tumors". This is why hot tumors, T-cell inflamed tumors, can benefit from immune checkpoint blockades, but cold tumors may need a different approach [15]. Several methods of turning cold tumors into hot tumors are currently under investigation.

5 Dendritic Cell Vaccines

Dendritic cells have been proven to be the connection between innate and adaptive immune systems. The goal of dendritic cells is to process and present antigens to Tcells. It can also be stated, that dendritic cells are able to play a significant role in cancer immunotherapy. Non-targeted protein and nucleic acid-based vaccines are one of the approaches in using dendritic cells in cancer immunotherapy. Although there are different methods, the most often method is based on dendritic cells, that are ex vivo differentiated from CD14 + monocytes, isolated by leukapheresis [16]. The monocytes are cultured in the presence of macrophage-colony-stimulating factor (GMCSF) and interleukin 4. Despite the fact about gathered data, that shows tolerance and good safety profile of dendritic cell vaccines, clear therapeutic outcomes emerge in less than 15% of patients [17, 18]. The success of this type of immunotherapy depends on the ability of dendritic cells to emigrate from the injection site towards lymph nodes, and the ability to elicit cytotoxic T-cells.

6 Developing Personalized Immunotherapy

Lack of precision tends to be the main cause of why many cancer patients do not obtain an effective treatment from the immunotherapies they are receiving. Thus, with each new novelty in cancer research, precision comes to be the main objective.

Recent studies are focusing on using genomic information for a more molecularly targeted approach when it comes to cancer treatment. This is done using predictive and prognostic biomarkers in cancers that serve as a predecessor to "personalized immunotherapy" [1].

The main premise of these biomarkers is to put a halt to the use of drugs that are not beneficial to the patients and advance their treatment in terms of specificity. They aim to give patients the treatment they will most likely respond to.

One method of doing this is converting "cold" tumors, less immune responsive tumors, to "hot tumors", ones that are more responsive to immune checkpoint blockades [19]. This is done by altering the microenvironment and surrounding tissue of the cold tumor and can advance the use of immune checkpoint blockades on a larger group of patients.

Achieving precision and specificity is still a daunting challenge for scientists and oncologists but might provide a viable solution for numerous cancer patients [20].

7 Conclusion

Immunotherapy has shown great potential in combating cancer. But despite its revolutionary achievements, many cancers still do not respond to immunotherapy. This largely depends on the strength of the body's natural immune system, but also the size and severity of neoplastically transformed cells. In conclusion, to achieve precision in cancer treatment is still a major issue, but the recent use of prognostic and predictive biomarkers in cancer treatment provides significant hope, and therefore there is a need for future research that will further develop and optimize cancer immunotherapy.

References

1. Karachi, A.: Immunotherapy for treatment of cancer. IntechOpen (2018). https://doi.org/10.5772/intechopen.81150
2. Maude, S.L., Barrett, D., Teachey, D.T., Grupp, S.A.: Managing cytokine release syndrome associated with novel T cell-engaging therapies. Cancer J. **20**(2), 119–122 (2014)
3. Vanneman, M., Dranoff, G.: Combining immunotherapy and targeted therapies in cancer treatment. Nat. Rev. Cancer **12**(4), 237–251 (2012)
4. Coulson, A., Levy, A., Gossell-Williams, M.: Monoclonal antibodies in cancer therapy: Mechanisms, successes and limitations. West Indian Med. J. **63**(6), 650–654 (2014)
5. Hansel, T.T., Kropshofer, H., Singer, T., Mitchell, J.A., George, A.J.: The safety and side effects of monoclonal antibodies. Nat. Rev. Drug Discov. **9**(4), 325–338 (2010)
6. Lonberg, N.: Human monoclonal antibodies from transgenic mice. Handb. Exp. Pharmacol. **181**, 69–97 (2008)
7. Andrew, M.S., Allison, J.P., Wolchok, J.D.: Monoclonal antibodies in cancer therapy. Cancer Immun. (12)(1) 14 (2012)
8. Bodey, B., Bodey, B., Jr, Siegel, S.E., Kaiser, H.E.: Failure of cancer vaccines: the significant limitations of this approach to immunotherapy **20**(4), 2665–2676 (2000)
9. Matlung, H.L., et al.: The CD47-SIRPalpha signaling axis as an innate immune checkpoint in cancer. Immunol. Rev. **276**, 145–164 (2017)
10. Bayer, V.R., Davis, M.E., Gordan, R.A., et al.: Immunotherapy. In: Olsen, M.M., LeFebvre, K.B., Brassil, K.J., (eds.) Chemotherapy and Immunotherapy Guidelines and Recommendations for Practice, pp. 149-189. Oncology Nursing Society, Pittsburgh, PA (2019)
11. Kaunitz, G.J., Loss, M., Rizvi, et al.: Cutaneous eruptions in patients receiving immune checkpoint blockade: clinicopathologic analysis of the nonlichenoid histologic pattern. Am J Surg Pathol. **41**(10), 1381–1389 (2017)
12. Chowell, D., et al.: Patient HLA class I genotype influences cancer response to checkpoint blockade immunotherapy. Science **359**, 582–587 (2018)
13. Pitt, J.M., et al.: Resistance mechanisms to immune-checkpoint blockade in cancer: tumorintrinsic and -extrinsic factors. Immunity **44**, 1255–1269 (2016)
14. Bonaventura, P., Shekarian, T., Alcazer, V., et al.: Cold tumors: a therapeutic challenge for immunotherapy. Front Immunol. **10**, 168 (2019). https://doi.org/10.3389/fimmu.2019.00168
15. Park, J.A., Cheung, N.-K.V.: Limitations and opportunities for immune checkpoint inhibitors in pediatric malignancies. Cancer Treat. Rev. **58**, 22–33 (2017). https://doi.org/10.1016/j.ctrv.2017.05.006
16. Calmeiro, J., Carrascal, M.A., Tavares, A.R., Ferreira, D.A., Gomes, C., Falcão, A., Cruz, M.T., Neves, B.M.: Dendritic cell vaccines for cancer immunotherapy: the role of human conventional type 1 dendritic cells. Pharmaceutics **12**, 158 (2020)
17. Anguille, S., Smits, E.L., Lion, E., van Tendeloo, V.F., Berneman, Z.N.: Clinical use of dendritic cells for cancer therapy. Lancet Oncol. **15**, e257–e267 (2014)
18. Constantino, J., Gomes, C., Falcão, A., Cruz, M.T., Neves, B.M.: Antitumor dendritic cell–based vaccines: Lessons from 20 years of clinical trials and future perspectives. Transl. Res. **168**, 74–95 (2016)

19. Karachi, A., Dastmalchi, F., Mitchell, D., Rahman, M.: Temozolomide for immunomodulation in the treatment of glioblastoma. Neuro-Oncology noy07 2 (2018). https://doi.org/10.1093/neuonc/noy07
20. Kakimi, K., Karasaki, T., Matsushita, H., Sugie, T.: Advances in personalized cancer immunotherapy. Breast Cancer **24**(1), 16–24 (2016). https://doi.org/10.1007/s12282-016-0688-1

Smartdrugs: Mechanisms of Action and Ethical Issues

Riana Jaha(✉), Tea Kolak, Hanna Helać, Haris Ćesir, Emina Sarajlić, and Melika Spahić

Students of Faculty of Pharmacy, University of Sarajevo, Sarajevo, Bosnia and Herzegovina
rianajaha@ffsa.unsa.ba

Abstract. Nootropics, also known by the name of "smartdrugs", are a relatively new class of medications that enhance cognitive functions and can be used for treatment of mental illnesses such as narcolepsy, ADHD, dementia and Alzheimer's disease. There are three aspects of bioethics which are of importance: scientific and medical fundaments, long-term effects of pharmacodynamics and pharmacokinetics of mentioned substances as well as philosophy that connects medicine and social science. Few theories describe how they work in particular and this will be described more in detail in further reading. Such substances would be optimal for office workers, students and other people whose jobs rely on their concentration, however, there is a lack of research related to the safety of certain drugs. Moreover, people end up using these substances off-label and for recreational means so this can be a problem. Even though there is evidence of them affecting the nervous system, a lot can be contributed to the placebo effect. We hope for more substantial studies in the future of these molecules and their pharmacology so they could be used for various conditions.

Keywords: Nootropics · Brain performance · Neuroenhancers · ADHD · Cognitive enhacement · Ethical issues · Modafinil · Methylyphenidate

1 Introduction

Humans have always wanted to improve their cognitive functions, from the use of hallucinogens in ancient civilizations who wanted to communicate with their gods, to the common use of caffeine in various forms [1]. Nootropics are a class of substances that have a boosting effect on brain performance. They've been developing for over three decades and are the mainly used method for treating cognitive deficits. These drugs are also called "smart drugs", cognition enhancers or memory enhancing substances. Prescription nootropics have stimulant effects. They can counteract the symptoms of medical conditions such as narcolepsy, dementia, attention deficit hyperactivity disorder (ADHD) or Alzheimer's disease. Prescription nootropics include:

modafinil - a stimulant that addresses sudden drowsiness of narcolepsy, Adderall - contains amphetamines to treat ADHD, methylphenidate - used for symptoms of narcolepsy and ADHD, memantine - treats symptoms of Alzheimer's disease. These

A. Badnjevic and L. Gurbeta Pokvić (Eds.): CMBEBIH 2021, IFMBE Proceedings 84, pp. 462–468, 2021.
https://doi.org/10.1007/978-3-030-73909-6_53

nootropics can have some side effects such as insomnia, high blood pressure, and addiction. Nonprescription nootropics can enhance brain performance or focus but because of the lack of research, we cannot say with certainty that this is true. It is believed that they have impact on thinking, memory, or other mental functions. Nonprescription nootropics do not treat diseases. Presumably, they act by altering the levels of neurotransmitters, enzymes, and hormones that are available to the brain. Some researchers say that people who believe their mental performance has been increased thanks to nootropics are largely being influenced by a placebo effect. The aim of this paper is to present and highlight two aspects of nootropics: how they work and how we should use them.

2 Cognitive Enhancement

Cognitive enhancement is defined as the use of drugs with aim to improve cognitive functions of healthy subjects, particularly memory, intelligence, attention and creativity without any medical indication [1]. These cognitive pharmacological enhancers are licensed for use in three disorders: dementia, ADHD and narcolepsy. Drugs that are used to treat patients with dementia such as memantine hydrochloride or acetylcholinesterase inhibitors are currently not misused. There are still drugs that are misused such as modafinil or the class B drug methylphenidate. Methylyphenidate (commercially known as Ritalin) and modafinil (commercially known as Provigil) are prescribed off-label for refractory depression, Parkinson's disease and narcolepsy.

There is no consistent evidence for neuroenhancement of methylphenidate, even though there is some evidence for memory enhancement. A single dose of modafinil kept up memory, wakefulness and executive functions in individuals moderately sleep-deprived. On the other hand, when repeated, doses couldn't prevent cognitive decrement and only maintained wakefulness and prolonged sleep deprivation. Modafinil is found to have more effect on individuals with lower IQ by causing overconfidence, obstructing the ability to correctly self-assess cognitive performance. Modafinil is proven to have positive effect on mood, pattern recognition, memory for digits and "spatial planning and stop-signal reaction time". However, some studies reported contradictory results such as only raising the anxiety levels of the volunteers given modafinil, with no cognitive enhancement. Both, modafinil and methylyphenidate give different results depending on the studies [2]. It is known that a neuron has structural and functional plasticity termed as synaptic plasticity which is responsible for synaptic remodeling, also known as cellular learning. Some nootropics work mechanisms in cognitive performance are: glutaminergic signaling, cholinergic system, amyloid precursor protein, secondary messenger, miscellaneous [3].

2.1 Glutaminergic Signallng

Glutamate is one of the essential neurotransmitters involved in cognitive processes. Two different types of glutamate receptors are: metabotropic receptors and ionotropics, distributed on the pre- and postsynaptic sites of the neuron. N-methyl-D-aspartate (NMDA) and AMPA receptors are going to be activated by glutamate release. NMDA receptor is in charge of learning and memory, AMPA receptor, on the other hand, is responsible for synaptic transmission. Brain respond will be opening Na^{+}/K^{+} ion channel and depolarizing the cell membrane. The main excitatory neurotransmitter related to cognition function is glutamate which acts by activating NMDA receptor. Its main role is to mediate instant and fast postsynaptic response to glutamate release; this could lead to increased synaptic plasticity. AMPA receptor is located throughout the brain, particularly in hypothalamus, hippocampus, thalamus, basal ganglia, cerebral cortex and cerebellum [3].

2.2 Cholinergic System

Essential neurotransmitter in regulating cognitive functions is the central cholinergic system and it is affiliated with acetylcholine (ACh). The receptor that is in charge of hippocampal synaptic transmission by stimulating the release of glutamate is neuronal nicotinic ACh receptor. Activation of protein kinase C (PKC) causes activation of nicotinic ACh receptor and these reactions sustain the phosphorylation of the receptor and regulate glutamate release. Diminishing of cholinergic function leads to cognitive dysfunction which is treated by stimulation of central cholinergic activity. The main cause of Alzheimer's disease is loss of neuronal cholinergic in the hippocampal area. Use of established nootropics leads to increasing level of ACh and regulation of receptor binding for cholinergic in hippocampus and frontal cortex. The main mechanism of action for nootropics is downregulation of ACh esterase activity and this reduction leads to regulation of ACh expression in the brain. Amyloid protein (amyloid β-protein and neurofibrillary tangles) is involved in dysfunction of the cholinergic system in Alzheimer's disease. APP fragments are involved in synaptic maintenance and formations. This influence leads to memory formation. Upregulation of APP expression is caused by the use of natural nootropics which leads to increasing memory performance and learning. Another fragment of APP is β-amyloid peptide which is accumulated in brain in insoluble or oxidized form. Activated caspase-3 could also be present in the patient's brain. That is cysteine protease that is included in apoptosis induced by the mitochondrion. This apoptosis leads to the neuronal degeneration. Indicator of development of Alzheimer's disease is production of amyloid-β fibril. Stimulation of Ca^{+2} conductance and permeability of lipid membrane is caused by amyloid-β fibril [3].

2.3 Secondary Messenger

Activation of the NMDA receptor is related to LTP and leads to stream of Ca2+. PKC in the dentate gyrus is activated by Ca2+. The administration of PKC activator and nootropic drug improve memory performance, with suggested involvement of similar PKC pathway. When PKC is activated, it goes to specific subcellular sites and gives variety of physiological functions. Besides PKC's role in learning and memory, it improves synaptic plasticity in the brain. PKC activity and diminished calcium/calmodulin-dependent protein kinase II, contribute to downregulation of NMDA receptor, reducing the release of glutamate [3].

2.4 Miscellaneous

Insulin receptor is very important for investigating the effect of nootropic in cognitive purposes. In the hippocampus, high expression of insulin is found in the dentate gyrus and CA1 pyramidal cells. The synaptic activity mediated by NMDA can be modulated by neuron insulin or insulin receptor in the hippocampus. Activity and the synthesis of a number of neurotransmitters is altered as a result. The signaling pathway of cognitive function is going to be stimulated by activation of neuronal insulin, as well as activation of mitogen-activated protein kinase (MAPK), PKC, and phosphatidylinositol-3-kinase (PI3K). Angiotensin receptor also helps the signal transduction in the CNS. Blood pressure is regulated by Ang II by various actions. Some most significant actions are renal actions, vasoconstriction and increased aldosterone biosynthesis. The cognitive and behavioral processes are facilitated by Ang II through its specific receptors and metabolites expressed in animal models. After cognitive deterioration of patient with Alzheimer's disease, Ang II inhibits the release of acetylcholine by inhibiting Ang I receptor. The cognition enhancer influences the cognition function, including the learning, retrieval and consolidation, and memory [3].

3 Addiction Issues

In recent years, smart drugs, also known as "study drugs", have become popular between college students and prescribed medications are now often misused. The term "misuse" is defined as use of prescription drugs not prescribed for the respondent or respondent use only for the experience of the feeling they caused [4]. The main reasons students use study drugs are to enhance alertness or energy, to experiment, to improve academic performance or they use it recreationally. Some studies have shown that misuse of smart drugs have three potential negative consequences: addiction issues, possible medical complications and reactions to high doses. When it comes to addiction problem, over 60 studies have proven that methylphenidate can have the same effect as d-amphetamine or cocaine. It has been noted that methylphenidate can cause depression, vivid dreams, anxiety, exhaustion, etc. Use of smart drugs cause irregular heartbeat and it raises blood pressure. Thus, it is contraindicated if one has cardiovascular-related concerns, depression, anxiety, glaucoma, motor tics, etc. When smart drugs are used in high doses, they can cause emotional lability, aggressiveness, blurred vision, confusion, dizziness and insomnia. Limiting access to prescribed cognitive enhancers and education should be right approach to solving these problems [5].

4 The Ethics Behind Neuroenhancement

In terms of moral guidelines regarding the use of neuroenhancers, due to a widespread (ab)use of such substances, primarily among students and working-class individuals whose jobs require focus and attentiveness beyond normal human capabilities, concerns have been raised. According to a study [6], one in seven healthy, college students in Switzerland uses neuroenhancers to fuel themselves. There are three important aspects of bioethics: scientific and medical fundaments, long-term effects of pharmacodynamics and pharmacokinetics of neuroenhancers, and philosophy that connects medicine and social science. Besides them, we mention the effect it can have on the individual's personality and lowering the disproportionality rates of unequal starting points of different classes [7]. One of the most common drugs used to treat ADHD in children is methylphenidate, better known for its brand name Ritalin. Considering how children are developing and growing, the question is imposed: is it a treatment for an overwhelmed brain or a temporary fix for a permanent issue? How much gray space is in between treating a symptom of psychological distress and an otherwise-manageable problem? [8]. In the context of smart drugs, we cannot make a definite differentiation between a medical enhancer, used preventively and medication used exclusively to treat symptoms, since they can vary, especially at an early age [9]. There are no clear studies that indicate whether the long-term effects of these drugs are detrimental to health. Pharmacovigilance has reported some cases of psychological impact such as addiction and therefore, caution is advised [10]. When discussing whether someone should make their own decision about whether they use the mentioned substances, it is important not to diminish the value of pharmacists. While some things might be a matter of bodily autonomy, we still rely on pharmacists to advise and warn us about precautions with drug intakes. An informed adult is capable of deciding what they wish to take in to improve themselves, but the decision should be in correlation to the benefit/risk ratio. Opponents of this view will argue that explicitly banning neuroenhancers off-the-counter sales directly coincides with human freedom and their right to use them at free will [11].

4.1 Personality and Authenticity

Personality and authenticity are considered vitally important for humans. There is not enough proof that smart drugs change people's personalities, but it is an issue that should be researched. A popular definition of authenticity says that for a person to be authentic, he/she must be "true to himself/herself" and feel that his/her experiences and feelings are really "his/her own." [7] Using smart drugs brings out the question; can someone's success really be considered their own if they achieved it by using enhancers? Critics claim that taking smart drugs might make people feel like their achievements are not theirs and this can lead to the loss of awareness of who a person truly is [12]. It is often considered the way a person would like to be, their selffulfillment and self-creation. Cognitive enhancement can "push" a person towards something they would like to be, by improving their concentration, thinking, focus…

Some critics claim that this would make people more authentic rather than less. Smart drugs could help people become the person they want to be, a better version of themselves [13].

4.2 Issues of justice

We live in a world full of inequalities. Do neuroenhancers work in favor of making unequal societies? It's quite possible. Those who are privileged already have better access to additional classes, training, tutoring… Knowledge is often expensive and it can make one social group more successful than the other. In the academic area, knowledge is the most valuable currency and it is not equally distributed in all social groups [7]. Smart drugs would just make an addition to this problem, because they would cost some money. Easy, no-limited access to these drugs would make the difference gaps between the classes even bigger. Some opinions support the fact that cognitive enhancers can help with reducing the size of the difference gap [7]. Life is often unfair and it doesn't offer the same possibilities to everyone. If used cleverly, smart drugs can act as a chance for those who are disadvantaged or born with less cognitive skills and neutralize the unfairness of life. This could only work if regulated by strict regulations. Smart drugs would only be available to those who are underachieving or experiencing learning difficulties, through a health-care system or in schools.

5 Conclusion

Smart drugs occur naturally in various plant species, but also can be synthesized in laboratories. Suggested mechanisms related to the effect of cognitive enhancement are glutaminergic signaling, cholinergic system, amyloid precursor protein, secondary messenger, miscellaneous. Cognitive enhancement can often be attributed to placebo effect, some researchers claim. The main mechanism of action for nootropics is downregulation of ACh esterase activity and this reduction leads to regulation of ACh expression in the brain. Modern way of living has probably contributed for the search of such substances, since many people's jobs rely on their concentration and wakefulness. Because of this, these drugs end up being misused, even though some haven't been such as memantine hydrochloride. In some cases, nootropic supplements can help boost cognitive function, but anyone who is considering such an option should first consult a healthcare professional. Some studies have shown that nootropic drugs can have an effect on the central nervous system, but there is a lack of larger, controlled, long-term studies, so there is not enough data whether these substances consistently work and are absolutely safe to use. Because of the absence of scientific data, these drugs are used off-label, for treatment of diseases where there is no evidence of their positive effect.

References

1. Frati, P., et al.: Smart drugs and synthetic androgens for cognitive and physical enhancement: revolving doors of cosmetic neurology. Curr. Neuropharmacol. **13**(1), 5–11 (2015). https://doi.org/10.2174/1570159X13666141210221750. PMID: 26074739; PMCID: PMC4462043
2. Jeon, J.P.: Nootropics for Healthy Individuals. The First-Year Papers (2010 - present). Trinity College Digital Repository, Hartford, CT (2015). https://digitalrepository.trincoll.edu/fypapers/61
3. Suliman, N.A., Mat Taib, C.N., Mohd Moklas, M.A., Adenan, M.I., Hidayat Baharuldin, M.T., Basir, R.: Establishing natural nootropics: recent molecular enhancement influenced by natural nootropic. Evid Based Complement Alternat Med. 4391375 (2016). https://doi.org/10.1155/2016/4391375. Epub 2016 Aug 30. PMID: 27656235; PMCID: PMC5021479
4. Sussman, S., Pentz, M.A., Spruijt-Metz, D., et al.: Misuse of "study drugs: " prevalence, consequences, and implications for policy. Subst. Abuse Treat. Prev. Policy **1**, 15 (2006). https://doi.org/10.1186/1747-597X-1-15

5. Bossaer, J.B., Pharm, D., Gray, J.A., Pharm, D., Miller, S.E., Pharm, D., Enck, G., Gaddipati, V.C., Enck, R.E.: The use and misuse of prescription stimulants as "cognitive enhancers" by students at one academic health sciences center, academic medicine, vol. 88, no. 7, pp. 967–971 July 2013. https://doi.org/10.1097/ACM.0b013e318294fc7b
6. Maier, L.J., Liechti, M.E., Herzig, F., Schaub, M.P.: To dope or not to dope: neuroenhancement with prescription drugs and drugs of abuse among Swiss university students. PLoS ONE **8**(11), e77967 (2013)
7. Buyx, A.M.: Smart drugs: Ethical issues. In: Clausen, J., Levy, N. (eds.) Handbook of Neuroethics, pp. 1191–1206. Springer, New York, NY (2015)
8. Singh, I.: Will the "Real Boy" Please behave: Dosing dilemmas for parents of boys with ADHD. Am. J. Bioethics **5**(3), 34–47 (2005)
9. Buyx, A.: Be careful what you wish for? Theoretical and normative aspects of wishfulfilling medicine. Med. Health Care Philos. **11**(2), 133–143 (2008)
10. Heinz, A., Kipke, R., Heimann, H., Wiesing, U.: Cognitive neuroenhancement: false assumptions in the ethical debate. J. Med. Ethics **38**(6), 372–375 (2012)
11. Bostrom, N.: Smart policy: Cognitive enhancement in the public interest. In: Zonneveld, L., Dijstelbloem, H., Ringoir, D. (eds.) Reshaping the Human Condition: Exploring Human Enhancement, pp. 29–36. Rathenau Institute, The Hague (2008)
12. President's Council on Bioethics: Beyond Therapy: Biotechnology and the Pursuit of Happiness. Dana Press, New York (2003)
13. DeGrazia, D.: Enhancement technologies and human identity. J. Med. Philos. **30**, 261–283 (2005)

Pharmacogenomics: Current Trends and Future Perspectives

Azra Ibrahimović(✉), Dalida Adilović, Lamija Brković, Nedžla Bučo, Amra Hadžagić, and Lana Popović

Students of Faculty of Pharmacy, University of Sarajevo, Sarajevo, Bosnia and Herzegovina
azraibrahimovic@ffsa.unsa.ba

Abstract. Pharmacogenetic research provides molecular diagnostic tools that can be used to optimize drug selection and dosage for individual patients. Many clinical trials of oncology drugs are now taken into consideration of pharmacogenomics, germline studies, or acquired genetic factors that determine drug response and toxicity. However, there is a lack of agreement on actual practice on where pharmacogenomic studies should be included in the oncological clinical course of drug development. This paper discusses the recent growth of pharmacogenomics in oncology clinical trials, especially in early-stage studies and several critical issues facing the incorporation of pharmacogenomics into the early development of oncology drugs.

Keywords: Pharmacogenomics · Pgx · Pharmacogenetics · Personalized medicine · Oncology medicine

1 Introduction

The modern way of life, globalization and digitalization have brought with them new improvements in the way of treating diseases. One of them is certainly pharmacogenomics, a branch of pharmacology that involves knowing the genome of each person who requires therapy. The goal of pharmacogenomics is to understand how an individual's genome determines the action of drugs in the body and the occurrence of frequent side effects. In short, pharmacogenomics assists the physician in prescribing the best possible therapy as well as drug dose, while reducing the risk of side effects and overdose as well as in preventing side effects caused by drug interactions.

The aim of this paper is to explain personalized medicine and benefits of this way of treatment, and importance of PGx methods in drug prescription for each individual patient. Finally, this paper focuses on current clinical practice of oncology pharmacogenomics and cancer drug development, as well as future prospectives of this method.

2 Pharmacogenomics in Individualized Cancer Therapy

One of the main components of personalized medicine, personalized care, is focused on the variable response to drugs in the forms of lack of response and adverse reaction,

A. Badnjevic and L. Gurbeta Pokvić (Eds.): CMBEBIH 2021, IFMBE Proceedings 84, pp. 469–473, 2021.
https://doi.org/10.1007/978-3-030-73909-6_54

and the incentive to properly use medications. In view of these findings, the hypothesis proposed that one of the key causes was different genetic backgrounds among individuals, although environmental factors may also lead to various drug reactions. The discipline of pharmacogenetics has arisen as a branch of pharmacology to classify genetic variants from candidate genes linked to drug metabolism, transport or molecular targets/pathways in order to provide a deeper understanding of the relationship between human genetics and drug response. The word pharmacogenetics has progressively developed into pharmacogenomics since entering the postgenomic period in the new century as genome-wide integrative research has been increasingly used. In certain scenarios today, the two words are simply synonymous. Pharmacogenetics and pharmacogenomics are universally referred to as PGx.

So far, PGx consideration or package-insert labeling has been recommended by the US FDA for more than 120 drugs with relationships to more than 50 genes. These medications are widely used to treat cancers and respiratory infections and psychiatric disorders. It is clear that for the following reasons, PGx anti-cancer agents are one of the most actively researched fields. Next, chemotherapeutics are sometimes more severe and often more severe because of the limited therapeutic indices. Second, two related but separate genomic structures (tumor and germline genomes) that are specific to anticancer therapy need to be investigated to enhance drug effectiveness and minimize toxicity. Third, the pipelines for the production of anticancer agents have been very successful, contributing to many more drugs being put on the market or undergoing clinical trial assessments. In addition, if the package-insert label stipulates the use the beneficial effects of PGx may also be applied to Phase IV postmarketing trials.

2.1 Characterization of Heritability for Anticancer Agents

Generally speaking, either natural causes or environmental factors may be linked to the multiple opioid reactions [1]. It will also be important and essential to determine whether and how much genetics plays a role in variable drug responses before performing PGx discovery research. Heritability analyses may be used for this purpose to qualify and measure whether the drug reaction is a heritable trait. Either in twins or in large pedigrees, conventional heritability research is done. Measures of heritability will vary from 0 to 1, with nearly 0 showing that it is less likely to be a heritable trait, while nearly 1 suggests that genetics is responsible for most phenotypic variations.

It is impossible for anticancer agents to conduct heritability tests in humans, because the application of chemotherapy medications to noncancer relatives is illegal [2]. Study approach has also been based on conducting heritability experiments using resources obtained from human beings. Lymphoblastoid cell lines originating from large pedigrees, for example, have been used to study cisplatin and cytotoxicity caused by 5-fluorouracil (5-FU). More recently, Peters et al. performed a study of heritability among 29 Anticancer drugs approved by the FDA that use a total of 125 lymphoblastoid cell lines from 14 broad Caucasian pedigrees. At different therapy doses, cytotoxicity was calculated for each drug. For these 29 drugs, they found heritability ranging from 0.06 to 0.64, indicating the differing functions of genetics in the cellular reaction to anticancer agents. Nineteen out of 29 medications reported a median heritability of more than 0.3, indicating that biology contributed for more than 30% of the total variance in cytotoxicity, which justified more PGx studies to classify genetic variations leading to these heritable characteristics [3].

3 Methods Used in Study PGx

3.1 Genetic Variants

In oncology, genetic modifications, all of which are targets of PGx studies, can be present either in the germline genome as germline alterations or in the tumor genome as somatic mutations. Any observable and heritable variance in the lineage of germ cells, such as genetic polymorphisms in genes that encode drug-metabolizing enzymes, is a germline variation; whereas somatic mutations are unintended modifications that can trigger cancer in a genomic sequence of DNA. Although drug efficacy and toxicity could theoretically be predicted by germline variants, somatic mutations are also used to refine chemotherapy agent selection to improve efficacy. SNPs, nucleotide addition, deletion, tandem repeat, copy number variation and chromosomal translocation are the genetic variations usually tested in PGx. Furthermore, gene expression in PGx is also widely examined for significance in tumorigenesis and reaction to chemotherapy. This term has now been formally adopted by the new FDA description of pharmacogenomics. The SNP is currently the most commonly studied genetic variant, thanks to the rapid technical advancement in the speed, coverage, precision and reduced cost of acquiring SNP genotypes. Their future practical importance also lies in the interest in researching SNPs. These variants may subsequently affect the transcription of genes, translation of genes and splicing of RNA, as well as cell stability of RNA. In a population in which each individual has his or her particular genetic make-up, all of which could contribute to variable drug responses.

3.2 PGx Discovery for Optimizing Anticancer Agents Usage

Advances in PGx have steadily begun to unveil the mystery of interindividual variations in drug responses in the past decade. A substantial number of the advances in the field of anticancer treatment have been made at present. A literature search on PGx in cancer therapy revealed a steady rise in publications in the past 10 years, in line with this pattern. In 2011, the publication figure is almost 2.5 times that of 10 years earlier. Ses PGx studies of anticancer agents have revealed several genetic variants for their possible role in tailoring individualized chemotherapy, or have pointed to certain unique chromosome loci. To date, the drug labels for 30 FDA-approved anticancer agents contain PGx data for 24 biomarkers. Such biomarkers include variations of genes, functional defects, changes in speech, chromosomal anomalies and others. Notice that some of these PGx markers, while others are tumor specific, are germline variants. Ses markers were interpreted differently in multiple parts of the FDA-approved drug labeling on the basis of the extent of support for clinical evidence.

Both PGx markers used in the drug box inserts are thus, of importance. Medical intervention should be taken for such markers suggested as ‘mandatory’ or ‘recommended’. With the accumulation of extra research data, these recommendations can change/evolve.

Improved effectiveness or decreased toxicity have been found when integrating these PGx markers into directing cancer chemotherapy. In addition, numerous credible studies have shown pharmaco-economic benefits.

4 Current Progress and Future Perspectives

The main goal of clinical medicine is to increase the therapeutic effect of drugs and decrease their toxicity level [4]. Pharmacogenomics is included more and more in drug development, because it allows targeting drugs in specific location, in individual patients. Currently, a lot of PGx studies are based on SNP, which, in the future, will provide drug prescription personalised for each patient [1]. Pharmacogenomic treatment approach in clinical practice is still not fully utilized, since there is not enough effective information about the benefits of this way of treatment. It is important to know, that besides the genome profile of individual patient, there are more thing to consider when defining convinient therapy, such as patients' age, gender, weight and also drug interactions.

In past years, a great progress has been made in the field of tumor biology, but also tumor development and response to treatment. Current gene-based PGx approaches face a lot of difficulties, since there is not enough reliable information about tumor drug response, as well as normal tissue toxicity, although there was considerable success where single genes play a big role in overall drug response [5, 6]. Ideally, a genome-wide studies should provide much better results and are the future of pharmacogenomic analysis [7]. Latest genome-wide studies show that new biomarkers are needed for possible future drug targets. Moreover, it is important that all new discoveries are later implemented in clinical practice and drug development.

In the end, pharmacogenomic approach to cancer treatment has full potential for personalising therapy for each patient, regarding optimal drug doses that will fully benefit individual patients as well as minimize the risk of toxicity. There is still a lot of economical, legal, ethical and clinical issues with pharmacogenomic approach to treatment, which need to be solved before PGx is completely incorporated in clinical practice [8]. Also, with technology development in the future, all researches will develope even more and the money needed to invest for those researches will be less with time.

5 Conclusion

Pharmacogenomics determines how your genes affect your body's response to drugs. This way of testing can help your doctor choose the right medicine for you as well as help the amount of medicine to suit your needs. Good communication between doctor and patient is extremely important, and the doctor will be able to determine your potential response to the drug based on gender, weight, age, diet, other medications and environmental factors, such as long-term exposure to cigarette smoke. Remember, your genetic code can also have a big impact on the reaction to a particular drug and its amount. In fact, genetic factors can affect up to 95% of the way you respond to medications.

In the future, large clinical trials will be needed to collect more information about the effectivnes of PGx treatment approach, but to this day a lot of positive effects are shown. One of the priorities for more and more clinicians to prepare and learn about

PGx testing and analyzing, since personalized medicine is indeed the future of health care. Finally, pharmacogenomics show great potential of revolutionzing personalized medicine, including cancer therapy, with the ability of increasing therapy efficiacy, and decreasing toxicity levels.

References

1. Paugh, S.W., Stocco, G., McCorkle, J.R., Diouf, B., Crews, K.R.: Evans Cancer pharmacogenomics. **90**(3), 461–466 (2011)
2. Filipski, K.K., Mechanic, L.E., Long, R.: Freedman Pharmacogenomics in oncology care, April 2014
3. Weng, L., Zhang, L., Peng, Y., Haung, R.S.: Pharmacogenetics and pharmacogenomics: a bridge to individualized cancer trerapy. **14**(3), 315–324 (2013)
4. Xie, H.-G., Frueh, F.W.: Pharmacogenomics steps toward personalized medicine, October 2005
5. Rodríguez-Vicente, A.E., et al.: Pharmacogenetics and pharmacogenomics as tools in cancer therapy. Drug Metabol. Person. Therapy **31**(1), 25–34 (2016)
6. Ruwali, M.: Pharmacogenetics and Cancer Treatment: Progress and Prospects, Molecular Medicine, Sinem Nalbantoglu and Hakima Amri, IntechOpen, January 2019
7. Weng, L., Zhang, L., Peng, Y., Huang, R.S.: Pharmacogenetics and pharmacogenomics: a bridge to individualized cancer therapy. **14**(3), 315–324 (2019)
8. Stadler, Z.K., Gallagher, D.J., Thom, P., Offit, K.: Genome-wide association studies of cancer: principles and potential utility. Oncology (Williston Park). **24**(8), 667 (2010)

Spectrophotometric Determination of Cysteine Based on Complex Reaction Alizarin Red with Cooper

Mirha Pazalja(✉) and Mirsada Salihović

Faculty of Pharmacy, Department of Natural Sciences in Pharmacy, University of Sarajevo, Zmaja od Bosne 8, 71 000 Sarajevo, Bosnia and Herzegovina
mirha.pazalja@ffsa.unsa.ba

Abstract. Cysteine (Cys) is a non-essential amino acid that is important for making protein and other metabolic functions. For determination of Cys have been developed simple, sensitive, easily available, and fast methods. The aim of our study is the determination of Cys in pharmaceutical products by the spectrophotometric method. That method is based on the complexation reaction that occurs between Cu^{2+} and alizarin red (AR). The absorption maximum for AR in Britton-Robinson (BR) buffer solution is obtained at a wavelength of 433 nm. The addition of Cu^{2+} shifts the absorption maximum to 465 nm and changes the color from yellow to red. Determination of Cys in pharmaceutical products (samples 1 and 2) was carried out by spectrophotometric method with recoveries 96,25 and 95%. Comparing our results with the results obtained by the electrochemical method for the same samples can be concluded that is the electrochemical method that is more efficient than the spectrophotometric method. However, the spectrophotometric method in relation to the electrochemical method can be used as a simple, easy workup, low cost, easily applicable, and provides short reaction times.

Keywords: Cysteine · Cooper · Alizarin red · Spectrophotometric · AR – Cu^{2+} complex

1 Introduction

Cysteine (Cys) or 2-amino-3-mercaptopropanoic acid (Fig. 1a) is a sulfur containing amino acid that plays an important role in numerous biological processes. The sulfhydryl group of Cys plays an important role in the biological activity of proteins and enzymes. The benefits of Cys as protection of the brain and liver from alcohol consumption hazards, antioxidant properties due to its reaction with free radicals, and the ability to chelate with heavy metals are well described. There are several applications of Cys: it is a component of some pharmaceutical products, is used in food additives and it is a component in cosmetics [1].

Because of the crucial importance of Cys great efforts have been made to develop simple, sensitive, easily available, and fast methods for its determination. Numerous analytical methods have been developed for the quantitative determination of Cys as

A. Badnjevic and L. Gurbeta Pokvić (Eds.): CMBEBIH 2021, IFMBE Proceedings 84, pp. 474–480, 2021.
https://doi.org/10.1007/978-3-030-73909-6_55

high-performance liquid chromatography (HPLC), capillary electrophoresis, chemiluminescence, spectrofluorimetry, electrochemistry [2–5]. Many of these methods are very popular with equipment available in laboratories, bat several disadvantages can be cited. Those methods use expensive equipment and reagents and usually require extensive sample treatment. In contrast, spectrophotometric methods have advantages such as easy handling, sensitivity, low cost, and high selectivity [6].

Spectrometry is one of the most frequently used analytical methods and finds wide application in chemistry, forensics, agriculture, etc. For analytical purposes, absorption spectroscopy is the most widely used, which measures the dependence of the quantity of absorbed radiation on the wavelength for a particular substance. It can be used in qualitative and quantitative analysis. Spectrophotometric study of various analytes based on complex reaction alizarin red S (ARS) with different metal ions is previously reported [7–10]. According to our knowledge, there are no reports on the spectrophotometric determination of Cys with alizarin red (AR), which differs from ARS in chemical structure. The structural formulas for AR and ARS are shown in Fig. 1b, and c [11].

The aim of our study is the spectrophotometric determination of Cys in pharmaceutical products based on the reaction of AR with Cu^{2+} which gives color compounds as products. The addition of Cys to AR – Cu^{2+} solution produced the AR due to the redox reaction between Cys and Cu^{2+}.

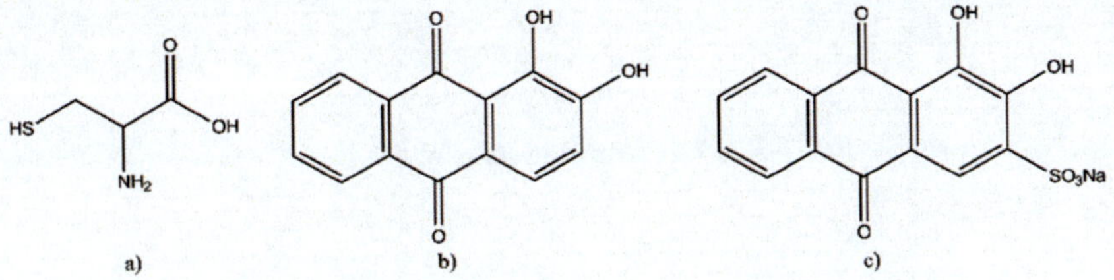

Fig. 1. The structure of (a) Cys, (b) AR and (c) ARS

2 Material and Methods

2.1 Instrumentation

All absorbance measurement was performed on a Shimadzu 1280 spectrophotometer with a 1 cm quartz cuvette. The pH of the buffer solution was measured with a pH meter (Mettler Toledo).

2.2 Reagens and Solutions

All chemicals were of analytical grade and were used without further purification. A solution of AR (Merk) was prepared by dissolving 12 mg of AR in 50 cm^3 of methanol (1 mmol dm^{-3}). A Cu^{2+} solution was prepared by dissolving 25 mg of $CuSO_4 \cdot 5H_2O$ in 100 cm^3 of methanol (1 mmol dm^{-3}). Britton-Robinson (BR) buffer solution was prepared by mixing phosphoric acid (0,04 mol dm^{-3}) acetic acid (0,04 mol dm^{-3}) and boric acid (0,04 mol dm^{-3}) in 1 dm^3 flask and diluting the mixture with water to the mark. Initial pH of 1,90 was adjusted at 7,40 with NaOH (0,20 mol dm^{-3}). Doubly distilled water was used in all experimental work. The concentrations of Cys in pharmaceutical products were determined based on the calibration curve of a standard solution of Cys, shown in Fig. 2. Primarily, the standard stock solution of Cys was freshly prepared by dissolving 12,60 mg Cys in 10 cm^3 methanol, at a concentration of 10 mmol dm^{-3}. From the stock solution, standard solutions were then prepared of concentrations: 2, 3, 4, 5, and 6 mmol dm^{-3}, which were used in this study [12].

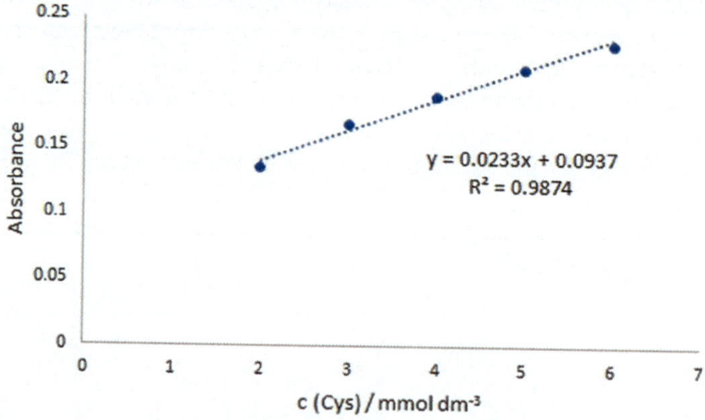

Fig. 2. Calibration curve of Cys

2.3 Procedure

Preparation of an AR – Cu^{2+} Complex Solution

Previous research has shown that it is for the formation of AR – Cu^{2+} complex achieved the best results by mixing Cu^{2+} and AR solutions in a stoichiometric ratio of 1:2 (Cu^{2+}: AR).Therefore a 0,50 cm^3 AR solution was transferred into 10 cm^3 volumetric flask and added 0,25 cm^3 Cu^{2+} solution. After stirring, the resulting mixture stabilized the color for 15 min. Then 1 cm^3 of BR buffer solution (pH = 7,40) was added. Methanol was added to reach the mark on a volumetric flask. The spectra were measured with a UV-Vis spectrophotometer at a wavelength of 300 to 700 nm [8, 13].

Preparation of an AR – Cu^{2+} - Cys Solution

A 0,50 cm^3 AR solution was transferred into 10 cm^3 volumetric flask and added 0,25 cm^3 Cu^{2+} solution, 1,25 cm^3 standard solution of Cys, 1 cm^3 of BR buffer solution of pH 7,40. Methanol was added to reach the mark on a volumetric flask. After stirring, the resulting mixture stabilized the color for 15 min. The spectra were measured with a UV-Vis spectrophotometer at a wavelength of 300 to 700 nm [8].

2.4 Sample Preparation

Two samples (labeled S1 and S2) were capsules from Germany (450 mg cysteine), and Austria (500 mg cysteine). An amount of the capsules equivalent to 4,90 mg of cysteine was weighed into a 10 cm^3 volumetric flask, 3 cm^3 of methanol were added and shaken carefully for about 15 min. The contents were diluted to the mark with methanol. Quantification was performed by the spectrophotometric method described above.

3 Results and Discusion

AR and ARS are organic compounds known as complexing ligands for metal ions such as Ca^{2+}, Mn^{2+}, Al^{3+}, Zn^{2+}, and many others [14–17]. In this study, the spectrophotometric method for the determination of Cys in pharmaceutical products is based on the complexation reaction that occurs between Cu^{2+} and AR. Figure 3 illustrates the absorption spectra of AR, AR in the presence of Cu^{2+} and the spectrum of AR in the presence of Cu^{2+} and Cys. The absorption maximum for AR in BR buffer solution is obtained at a wavelength of 433 nm. The addition of Cu^{2+} shifts the absorption maximum to 465 nm and changes the color from yellow to red. The reaction of forming complex AR – Cu^{2+} occurs at its peri-hydroxy carbonyl group and results in the formation of a $C = O \rightarrow M - O$ coordination bond [18]. While research has shown that, the complexation of aluminum with AR can be occurs at two potential sites 1,2-dihydroxyl (catechol) and the 1-hydroxy-9-keto groups [19]. Wu et al. reported that ARS reagent formed a complex with Cu^{2+} ion in the acetate buffer at pH 5,20 [12]. Some literature indicates that at pH below 7,00 alizarin red occurs in the form of yellow-colored only slightly dissociated molecules and shows little tendency to form complexes [14]. Therefore, BR buffer solution (pH 7,40) was chosen for further work.

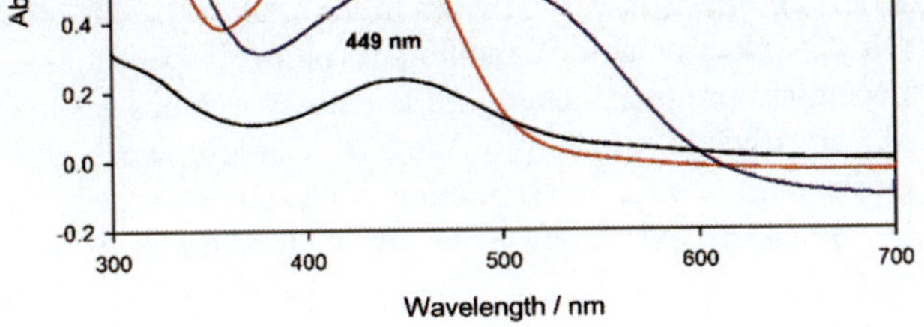

Fig. 3. UV-VIS spectra of AR, AR – Cu^{2+} complex and AR – Cu^{2+} - Cys

The addition of Cys to the AR – Cu^{2+} solution results in a change of the spectrum as shown in Fig. 3. The spectral recovery of AR is detected, and color anew changed from red to yellow. This implies that the addition of Cys decomposes the AR – Cu^{2+} complex and releases AR. As well electrochemical reduction process for alizarin–ion metal complex is used for determination of Pb(II), Cd(II) and Zn(II) by adsorptive cathodic stripping voltammetry [20]. The prediction reaction of metal complex formation AR – Cu^{2+} and decompound are presented in Fig. 4.

Similar complex ARS with Cu^{2+} is reported by several authors [7, 12, 18].

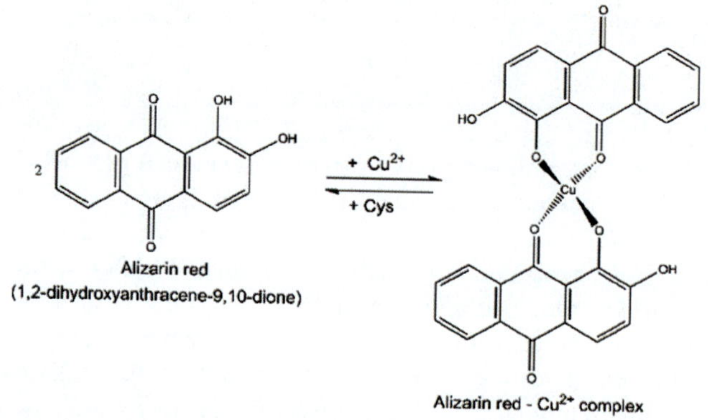

Fig. 4. The prediction reaction of formation AR – Cu^{2+} complex and decompound with the addition of Cys

By adding Cys to the solution of the AR – Cu^{2+} complex, the absorption maximum shifts to 449 nm. The calibration curve and the spectrophotometric quantification of Cys in pharmaceutical products were obtained at 449 nm. As shown in Fig. 4, an increase of Cys concentration causes the increase of absorbance suggesting a good spectral response. Dependence of Cys concentration versus absorbance is given by linear regression $A_{449nm} = 0{,}0233\ c_{Cys}\ (\text{mmol dm}^{-3}) + 0{,}0937$, with correlation factor $R^2 = 0{,}9874$, and the standard calibration curve was linear over the range (2–6 mmol dm^{-3}) (Fig. 5).

This method using AR as complexing ligands for Cu^{2+} was applied to the spectrophotometric determination of Cys in commercially available pharmaceutical products, and results are presented in Table 1. In Table 1 are presented and results of Cys content in the same samples of pharmaceutical products obtained by amperometric Flow-Injection Analysis (FIA) [5]. A comparison of the results of the analysis of the two methods given in Table 1 shows that the electrochemical method has more effective for the determination of Cys in pharmaceutical products. Though, the recovery values for the spectrophotometric method do not diverge much from the FIA method, especially for sample 2. Therefore, the spectrophotometric determination of Cys in pharmaceutical preparations can be used as a simple, fast, routine non-expensive method that does not require special sample preparation.

4 Conclusions

The spectrophotometric method for the determination of cysteine in a commercial sample of pharmaceutical products was proposed. It is based on the complex reaction at room

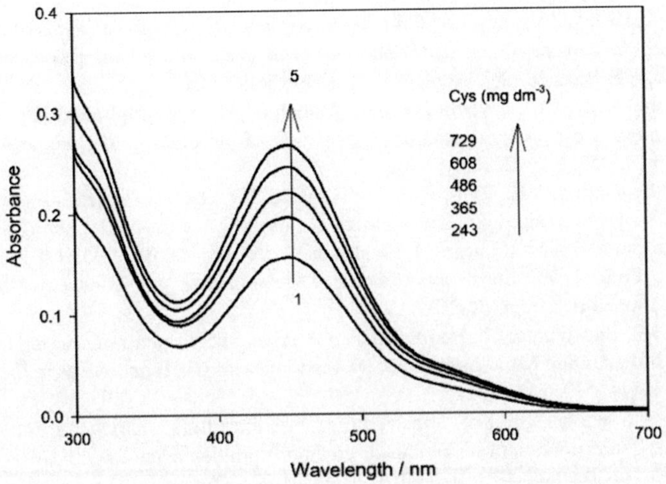

Fig. 5. UV-VIS spectra of AR, AR – Cu^{2+} complex and AR – Cu^{2+} - Cys complex

Table 1. Determination of Cys in a pharmaceutical product using spectrophotometric method and comparation with results FIA amperometric method

Method	Sample	Added Cys (mmol dm^{-3})	Found Cys mmol dm^{-3})	Recovery (%)	Ref
Spectrophotometric with AR-Cu^{2+}	S1	4,00	3,85	96,25	This work
	S2	4,00	3,80	95,00	
FIA amperometric	S1	1,65	1,61	97,60	[4]
	S2	1,65	1,57	95,15	

temperature using Cu^{2+} ions and AS as reagents. Based on the obtained results can be concluded that it is an electrochemical method is more efficient than the spectrophotometric method. However, the spectrophotometric method in relation to the electrochemical method can be used as a simple, easy workup, low cost, easily applicable and provides short reaction times.

References

1. Rehman, T., Shabbir, M.A., Inam-Ur-Raheem, M., Manzoor, M.F., Ahmad, N., Liu, Z.W., Ahmed., M.H., Siddeeg., A., Abid., M., Aadil, R.M.: Cysteine and homocysteine as biomarker of various diseases. Food Sci. Nutr. **8**(9), 4696–4707 (2020)
2. Wei, L., Lu, X., Kang, X., Song, Y.: Determination of glutathione and cysteine in human breast milk by high-performance liquid chromatography with chemiluminescence detection for evaluating the oxidative stress and exposure to heavy metals of lactating women. Anal. Lett. 1–12 (2020)

3. Kang, S.H., Wei, W., Yeung, E.S.: On-column derivatization for the analysis of homocysteine and other thiols by capillary electrophoresis with laser-induced fluorescence detection. J. Chromatogr. B Biomed. Sci. Appl. **744**(1), 149–156 (2000)
4. Hormozi-Nezhad, M.R., Seyedhosseini, E., Robatjazi, H.: Spectrophotometric determination of glutathione and cysteine based on aggregation of colloidal gold nanoparticles. Scientia Iranica **19**(3), 958–963 (2012)
5. Pazalja, M., Kahrović, E., Zahirović, A., Turkušić, E.: Electrochemical sensor for determination of L-cysteine based on carbon electrodes modified with Ru(III) Schiff base complex, carbon nanotubes and Nafion. Int. J. Electrochem. Sci **11**, 10939–10952 (2016)
6. Li, X.D., Zhai, Q.Z.: Spectrophotometric Determination of Calcium with Dibromo*p*-methylsulfonazo. J. Chem. (2020)
7. Jannatin, M., Supriyanto, G., Pudjiastuti, P.: A novel spectrophotometric method for determination of histamine based on its complex reaction with Ni(II) and Alizarin Red S. Indones. J. Chem. **17**(1), 139–143 (2017)
8. Jannatin, M., Ayu Nabila, I.L., Supriyanto, G., Pudjiastuti, P.: A novel spectrophotometric method for determination of histamine based on its complex reaction with Cu(II) and alizarin red S. J. Chem. Technol. Metallur. **52**(6), 1045–1050 (2017)
9. Wu, L., Forsling, W.: Potentiometric and spectrophotometric study of calcium and alizarin red S complexation. Acta Chem. Scand. **46**, 418–422 (1992)
10. Alsamarrai, K.F.: Spectrophotometric assay of lead in human hair samples by using alizarin red (S) in Samarra area. J. Univ. Anbar pure science **5**(3), 5–12 (2011)
11. Thomas, O., Brogat, M.: Organic Constituents. In: UV-Visible Spectrophotometry of Water and Wastewater, pp. 73–138. Elsevier (2017)
12. Wu, F.Y., Liao, W.S., Wu, Y.M., Wan, X.F.: Spectroscopic determination of cysteine with alizarin red S and copper. Spectrosc. Lett. **41**(8), 393–398 (2008)
13. Gahlan, A.A., El-Mottaleb, M.A., Badawy, N.A., Kamale, F.H., Ali, S.H.: Spectrophotometeric studies on binary and ternary complexes of some metal ions with alizarin red S and cysteine. Int. J. Adv. Res **2**(10), 570–584 (2014)
14. Puchtler, H., Meloan, S.N., Terry, M.S.: On the history and mechanism of alizarin and alizarin red S stains for calcium. J. Histochem. Cytochem. **17**(2), 110–124 (1969)
15. Sholehah, H., Susane, H.: Histamine test kit formation from Manganase(II), 5-fluorouracil and Alizarin red S complex compound. Jurnal Pijar Mipa **15**(2), 177–181 (2020)
16. Supian, S.M., Ling, T.L., Heng, L.Y., Chong, K.F.: Quantitative determination of Al(III) ion by using Alizarin Red S including its microspheres optical sensing material. Anal. Methods **5**(10), 2602–2609 (2013)
17. Kaushik, R., et al.: Alizarin red S–zinc(II) fluorescent ensemble for selective detection of hydrogen sulphide and assay with an H_2S donor. RSC Adv. **5**(97), 79309–79316 (2015)
18. Jannatin, M., Ayu Nabila, I.L., Wahyuni, S., Supriyanto, G., Ibrahim, W.A.W.: Rapid spectrophotometric method for histamine determination in fish using alizarin red s and metal. Malaysian J. Anal. Sci. **23**(3), 505–515 (2019)
19. Say-Liang-Fat, S., Cornard, J.P.: Al(III) complexation by alizarin studied by electronic spectroscopy and quantum chemical calculations. Polyhedron **30**(13), 2326–2332 (2011)
20. Pardi, H., Suyani, H., Zein, R., Alif, A., Edelw, T.W.: The development method for a sensitive simultaneous determination of Pb(II), Cd(II) and Zn(II) by adsorptive cathodic stripping voltammetry using Alizarin as a complexing agent. 8(7), 885–898 (2016)

Quantitative Structure-Activity Relationship Study of DPP-4 Enzyme Inhibitors as Drugs in Therapy of Type 2 Diabetes Mellitus

Sanja Rogic, Miljana Nukic, and Zarko Gagic(✉)

Department of Pharmaceutical Chemistry, Faculty of Medicine, University of Banja Luka, Banja Luka, Bosnia and Herzegovina
zarko.gagic@med.unibl.org

Abstract. Dipeptidyl peptidase 4 (DPP-4) is a multi-functional protein that, besides its catalytic activity, also functions as a binding protein and a ligand for a variety of extracellular molecules. One of its most substantial roles is impact on glucose homeostasis by regulating the biological activities of two incretins, glucagon-like peptide 1 (GLP-1) and glucose-dependent insulinotropic polypeptide (GIP). Thus, DPP-4 enzyme inhibitors, collectively called glyptins, have been successfully used in treatment of type 2 diabetes mellitus. 3D-quantitative structure-activity relationship study (3D-QSAR) was conducted on a series of 48 molecules with more than 85% structural similarity to sitagliptin in order to determine the influence of the chemical structure on the biological activity. Structures and data on the DPP-4 inhibitory activity (IC_{50}) of these molecules were collected from the ChemBL database. Molecular structures were previously optimized using the semiempirical PM3 method, and the modeling was performed using PLS regression analysis. Based on the calculated statistical parameters, $R^2 = 0.96$, $Q^2 = 0.71$ and $R^2_{pred} = 0.610$, it can be concluded that the formed 3D-QSAR model has good predictive power and can be used to predict DPP-4 inhibitory activity of structurally related compounds. Variables obtained from QSAR contain information of crucial structural determinants that impact activity and provide guidelines for future design of more potent inhibitors.

Keywords: DPP-4 · Inhibitor · 3D-QSAR · Diabetes mellitus

1 Introduction

Type 2 diabetes mellitus (T2DM, insulin-independent diabetes, senile diabetes) is a chronic metabolic disorder caused by insulin resistance, which may be accompanied by decreased insulin secretion. This type of diabetes is the most common in the population, accounting for around 90% of all diabetes cases. Although it does not belong to the group of infectious diseases, the number of diabetes patients is increasing incessantly. According to the estimates of the IDF (International Diabetes Federation), the total number of patients will have reached around 640 million by the year 2040 [1]. Drugs of different mechanisms of action are used in the therapy of DMT2, including sulfonylurea

A. Badnjevic and L. Gurbeta Pokvić (Eds.): CMBEBIH 2021, IFMBE Proceedings 84, pp. 481–488, 2021.
https://doi.org/10.1007/978-3-030-73909-6_56

derivatives, meglutinids, biguanides, α-glucosidase inhibitors, thiazolidinediones, DPP-4 enzyme inhibitors [2].

Dipeptidyl peptidase-4 inhibitors are relatively new oral antidiabetics that showed good efficacy in treating T2DM [3]. They were originally synthesized as an oral alternative to GLP-1 agonists and analogs (e.g., exenatide), but were found to prolong the action of gastrointestinal incretins, GLP-1, and GIP by reversibly inhibiting their metabolic inactivation by the DPP-4 enzyme [4]. They can be used as a monotherapy, but are often given in combination with metformin to achieve optimal HbA1C levels in these patients [3]. Since 2006, when sitagliptin was discovered, eight more DPP-4 inhibitors have been included in DMT2 therapy: saxagliptin, vildagliptin, alogliptin, teneligliptin, gemigliptin, anagliptin, trelagliptin, and linagliptin. These drugs are available on the market today, but research aimed at obtaining a new inhibitor has not stopped [5].

Dipeptidyl peptidase-4 or T-cell antigen CD 26 catalyzes the cleavage of X-alanine or X-proline dipeptides from the N-terminal end of its substrate [6, 7]. It belongs to the serine protease family, which also includes DPP-6, DPP-8, DPP-9, aminoacyl carboxypeptidase, prolyl carboxypeptidase, fibroblast activating protein α (FAP-α), and cell prolyl peptidase [4, 8]. DPP-4 is found on epithelial and endothelial cells, as well as in many organs including the liver, pancreas, lungs, gastrointestinal tract, kidneys, and placenta [8]. In its free soluble form, it is present in plasma and certain body fluids such as semen and cerebrospinal fluid [9]. DPP-4 is a type 2 transmembrane protein, consisting of a short cytoplasmic and transmembrane domain, and a dominant extracellular domain containing a C-terminal active site, a glycosylated, and cysteinerich region. It exhibits enzyme activity in the form of homodimers, and can also form heterodimeric forms with similar peptidases (such as FAP-α), with each enzyme retaining its activity [8]. It participates in a large number of physiological processes such as cell migration, tissue remodeling, inflammation regulation, glycemic regulation, immune response, cell signaling and others [6]. Due to the current pandemic caused by the Sars-Cov-2 virus, the role of DPP-4 in development of the COVID-19 infection was also examined. In addition to ACE2, DPP-4 has been shown to act on the cell surface as a receptor for the coronavirus, and to participate in the inflammatory response. Therefore, the use of DPP-4 inhibitors could reduce the entry of the virus into the cells, the replication of the virus and a pronounced inflammatory response, ie. cytokine storm [10].

For type 2 diabetes mellitus, the DPP-4 enzyme is important in terms of its indirect effect on glycemic regulation by metabolic inactivation of gastrointestinal incretins [6]. Gastrointestinal incretins, glucagon-like peptide 1 (GLP-1), and glucosedependent insulinotropic peptide (GIP) are secreted in L (GLP-1) and K (GIP) cells of the intestinal mucosa, in response to increased blood glucose levels. They stimulate the secretion of insulin from the β-cells of the pancreas, and reduce the secretion of glucagon from the α-cells, which enables the establishment of normoglycemia. This process is referred to as the incretin effect [11]. In addition to its insulinotropic and glucagonostatic effect, GLP-1 slows down gastric emptying, reduces appetite and enables the regeneration of pancreatic β-cells, which also has a beneficial effect on glycemic regulation [12]. With the action of DPP-4 enzyme, the half-life of GIT incretin is shortened, which is why their effect is not pronounced in patients with DMT2, unless DPP-4 inhibitors are included in the therapy [9, 12, 13].

The aim of this study was to investigate the influence of the chemical structure on the biological activity of DPP-4 inhibitors by developing an appropriate 3D quantitative structure-activity relationship (3D-QSAR) model that could be used to predict activity of structurally similar DPP-4 inhibitors and design new compounds with enhanced antidiabetic activity.

2 Materials and Methods

2.1 Data Set

The data set consists of 48 DPP-4 inhibitor compounds (EL1–EL48), which are structural analogs of sitagliptin, with more than 85% similarity (Fig. 1). The compounds were collected from the ChemBL database, together with their experimentally determined inhibitory activities expressed as IC_{50} [14]. In order for the IC_{50} values to be used for further 3D-QSAR modeling, a negative decimal logarithm of IC_{50} was calculated to obtain pIC_{50}.

Based on the structural characteristics of the compounds, the data set was divided into a training set (35 molecules), from which a 3D-QSAR model was formed, and a test set (13 molecules) was used for external model validation. The range of pIC50 inhibitory activity for the compounds in this set is 5,801–8,699.

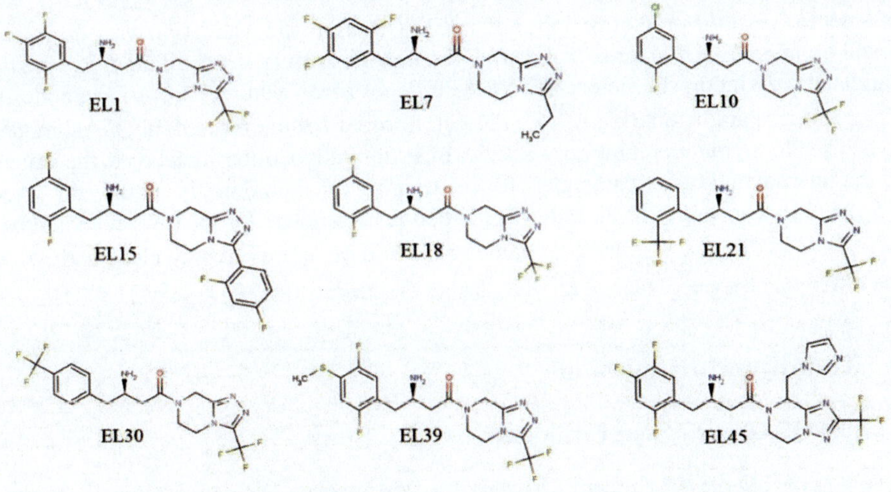

Fig. 1. Chemical structures of selected molecules from the data set.

Prior to model formation, dominant forms of the compounds were determined at pH 7.4 using the MarvinSketch program [15]. Since 3D-QSAR depends on the conformation of the molecules, the obtained dominant shapes were optimized in order to achieve the energy minimum. Optimization was performed using the PM3 semiempirical method (Parameterized Model Revision 3) [16, 17] in Gaussian 98 W software, which was incorporated into the Chem3D Ultra 7.0 program [18, 19].

2.2 Development of the 3D-QSAR Model

The Pentacle program was used to form the 3D-QSAR model [20]. This program uses three-dimensional molecular descriptors independent of alignment (GRid Independent Descriptors (GRIND)), which are calculated based on the Molecular Interactions Field (MIFs). Molecular interaction fields are simplified and presented as variables, and for their calculation the program uses 4 chemical probes that simulate the interaction of the ligand with the target site: DRY-probe represents hydrophobic interaction, O-probe describes hydrogen bond acceptor, N1-probe describes hydrogen bond donor and TIP-probe describes a stern interaction. In order to single out the most important regions, which represent the most favourable interactions of the ligand and the probe, the ALMOND algorithm was used. Two criteria are important for this process: field intensity and distance between selected regions. Selected regions were translated into GRIND variables using the CLACC (Consistently Large Auto and Cross Correlation) algorithm. Using the FFD (Fractional Factorial Design) algorithm, the initial number of variables was reduced until no further improvement of R^2 and Q^2 parameters was possible. The relationship between variables and pIC_{50} was analysed using the PLS (Partial Least Squares) regression method. The number of latent variables (LV) was set to five, and the optimal model containing five latent variables (LV5) was chosen.

2.3 Validation of the 3D-QSAR Model

In order to assess the predictive power of the 3D-QSAR model, validation was performed based on internal and external validation parameters: correlation coefficient R^2, cross-validated correlation coefficient Q^2, RMSEP (Root Mean Square Error of Prediction), R^2_{pred} and r^2_m parameters (r^2_m, r'^2_m, $\bar{r}^2_m$ *i* Δr^2_m). In order for the formed 3DQSAR model to be used to predict the inhibitory activity of structurally similar molecules, the values of the internal validation parameters for the training set should be: $R^2 > 0.6$ and $Q^2 > 0.5$ [21, 22]. The values of the external validation parameters for the test set should be: $R^2_{pred} > 0.5$ [21, 22], r^2_m and r'^2_m parameters should be approximately close and >0.5, the difference between them $\Delta r^2_m < 0.2$, and the mean value ($\bar{r}^2_m$) > 0.5 [23, 24].

3 Results and Discussion

3.1 3D-QSAR Validation Parameters

As a result of 3D-QSAR studies on a data set consisting of 48 DPP-4 inhibitors, an appropriate 3D-QSAR model was formed. The predictive power of the model was checked based on the parameters of internal and external validation. The obtained statistical parameters (Table 1) indicate a good predictive power of the model and ensure that obtained model can be used for prediction of DPP-4 inhibitory activity.

Table 1. Values of internal and external validation parameters of 3D-QSAR model

R^2	Q^2	RMSEP	R^2_{pred}	r^2_m	$\bar{r}^2_m$	r'^2_m	Δr^2_m
0.96	0.71	0.396	0.610	0.5525	0.5274	0.5023	0.0502

3.2 Determination of 3D-Pharmacophore

The correlogram (Fig. 2), which was obtained as a result of QSAR studies, presents the variables in the form of peaks that positively or negatively affect the inhibitory activity of the molecules. The ten most important variables were selected, six of which have a positive impact (var. 281, 457, 347, 210, 168, 446) and the remaining four have a negative effect on the inhibitory activity (var. 671, 569, 289, 624).

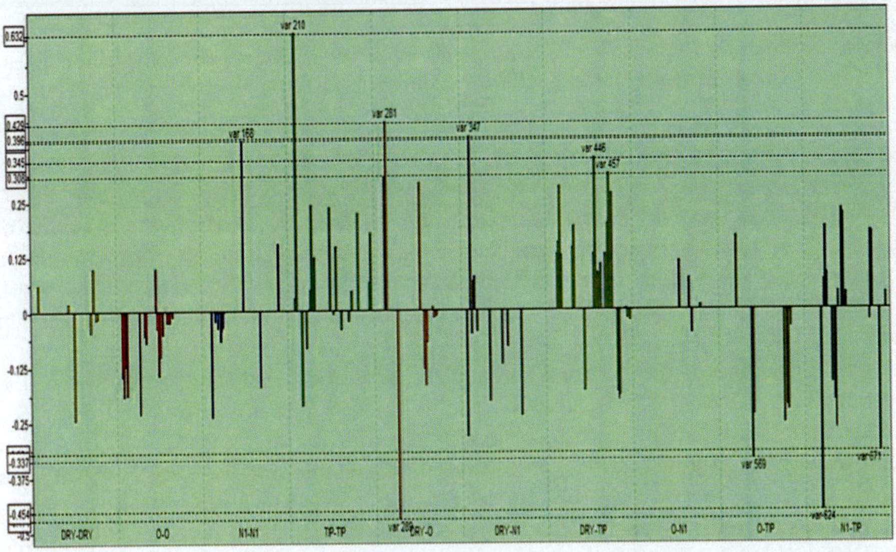

Fig. 2. Correlogram.

Variables That Have a Positive Effect on the Inhibitory Activity. The distance of 2.4A–2.8A between the two stern centers at the C4 position of the benzene ring, marked as v210: TIP-TIP, has the most positive effect on the inhibitory activity of the molecules. In addition to v281, the following variables have a favourable effect on the inhibitory activity of the molecules from the data set: v281: DRY-O which represents the distance of 3.6A-4A between the hydrophobic condensed cycle [6,8-dihydro-5H- [1,2,4] -triazolo [4,3a] pyrazin-7-yl) and a hydrogen bond acceptor i.e. a quaternary ammonium ion at the C3 position of the butane chain; v347: DRY-N1 marks the distance of 2.8 A–3.2 A between the hydrophobic condensed cycle and the nitrogen (N1) from the triazole ring as a hydrogen bond donor; v446: DRY-TIP represents the distance of 15.2 A–15.6 A between the hydrophobic piperazine located within the condensed cycle and the steric center at the C4 benzene position; v457: DRY-TIP marks the distance of 19.6 A–20 A between the triazole as the hydrophobic center in the condensed cycle and the steric center at the C4 position of benzene. V168: N1-N1 is not present in all molecules from the data set, and marks the distance of 12.8 A–13.2 A between two centers that represent hydrogen bond donors. It has a positive effect on inhibitory activity. Selected variables with positive effect on the inhibitory activity are presented in Fig. 3.

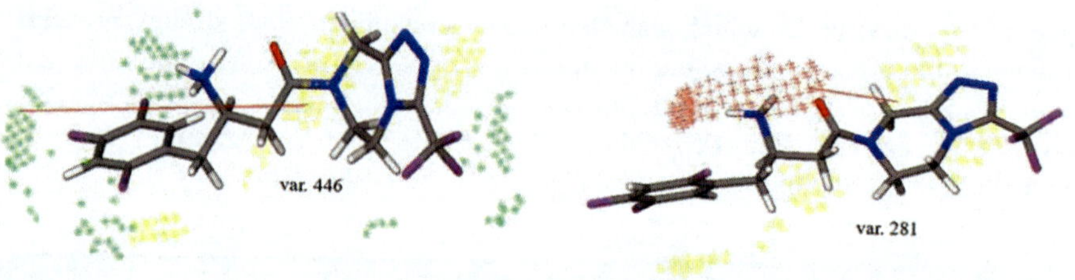

Fig. 3. Variables that have a positive effect on the inhibitory activity of compounds from a data set. The DRY-probe is marked in yellow, the O-probe in red, the N1probe in blue and the TIP-probe in green.

Variables That Have a Negative Effect on the Inhibitory Activity. The distance of 6.8 A–7.2 A between the hydrophobic benzene ring and the quaternary ammonium ion as a hydrogen bond acceptor at the C3 butane chain position, marked as v289: DRYO, has the most negative effect on the inhibitory activity of the molecules in the data set. In addition to this variable, a distance of 10 A–10.4 A between the hydrogen bond acceptor at the C3 position of the butane chain and the steric center at the C4 position of the benzene ring (v569: O-TIP), and a distance of 4.8 A–5.2 A between the nitrogen as a donor hydrogen bonds at the N2 position of the triazole ring and the steric center at the trifluoromethyl group at the C3 position of the triazole ring (v624: N1-TYPE) also have a negative effect on the inhibitory activity. V671: N1-TIP is present only in certain compounds from the data set (EL43, EL18, EL14, EL41, EL30, EL39), and has a negative effect on inhibitory activity. It represents the distance of 23.6 A–24 A between the nitrogen (N2) from the triazole ring as a hydrogen bond donor and the steric center on the benzene ring. In EL41 and EL43, the steric center is located at the C4 position of benzene, unlike EL14, EL18 and EL39 whose steric center is shifted to the C5 position of the benzene ring, due to the absence of electronegative fluorine at the C4 position. In the chemical structure of the EL30 molecule, the steric center is also at the C4 position of benzene, but instead of fluorine there is a trifluoromethyl group, which has an adverse effect on the inhibitory activity. In EL14, pyridine is present at the C3 position of the triazole ring, the nitrogen of which acts as a hydrogen bond donor instead of the nitrogen from the triazole ring. Selected variables with negative effect on the inhibitory activity are presented in Fig. 4.

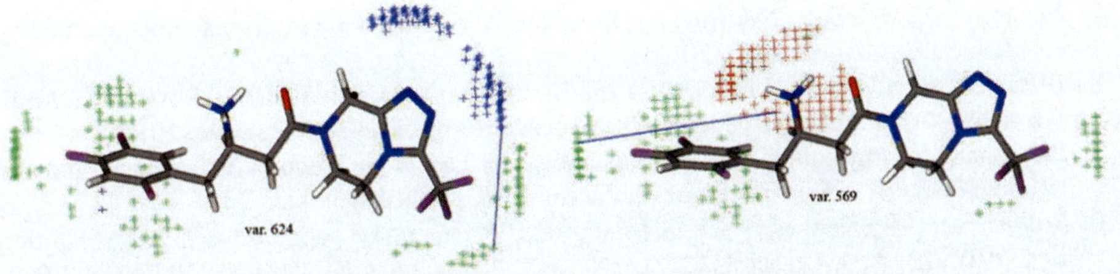

Fig. 4. Variables that have a negative effect on the inhibitory activity of compounds from a data set. The DRY-probe is marked in yellow, the O-probe in red, the N1probe in blue and the TIP-probe in green.

4 Conclusion

A 3D-QSAR study was conducted to evaluate the effect of the chemical structure on the inhibitory activity of DPP-4 inhibitors. Based on the compounds, which were taken from the ChemBL database, together with their IC_{50} inhibitory activities, a 3D-QSAR model was formed using the Pentacle program. The predictive power of the formed model was confirmed on the basis of the values of the internal and external validation parameters, which are within the reference values. Analysis of the variables revealed functional groups that are crucial for inhibitory activity: an electronegative steric group at the C4 position of benzene, a donor of hydrogen bonds in the condensed cycle, and an amine group at the C3 position of the butane chain. In further studies, based on the obtained data, targeted structural modifications can be made in order to obtain new, more potent DPP-4 inhibitors.

Acknowledgement. This work was supported by the Ministry of Scientific and Technological Development, Higher Education and Information Society of Republic of Srpska, Contract # 19.032/961–149/19.

References

1. The International Diabetes Federation Homepage. https://www.idf.org/aboutdiabetes/type2-diabetes.html. Accessed 21 Nov 2020
2. Pathak, R., Bridgeman, M.B.: Dipeptidyl peptidase-4 (DPP-4) inhibitors in the management of diabetes. Pharmacy Ther. **35**(9), 509–513 (2010)
3. Andersen, E.S., Deacon, C.F., Holst, J.J.: Do we know the true mechanism of action of the DPP-4 inhibitors? Diabetes Obes. Metab. **20**(1), 34–41 (2017)
4. Williams, A.D., Lemke, L.T.: Foye's Principles of Medicinal Chemistry, 7 edn. Lippincott Williams & Wilkins, Baltimore (2013)
5. Patel, B.D., Bhadada, S.V., Ghate, M.D.: Design, synthesis and anti-diabetic activity of triazolotriazine derivatives as dipeptidyl peptidase-4 (DPP-4) inhibitors. Bioorg. Chem. **72**, 345–358 (2017)
6. De, S., Banerjee, S., Kumar, S.K.A., Paira, P.: Critical role of dipeptidyl peptidase . Mini Rev. Med. Chem. **19**(2), 88–97 (2019)

7. Mauricio, D.: Molecular Nutrition and Diabetes: A Volume in the Molecular Nutrition Series. Academic Press, Amsterdam (2015)
8. Biftu, T., SinhaRoy, R.: 7.14 - DPP-4 Inhibitors. In: Chackalamannil, S., Rotella D., Ward, S.E. (eds.) Comprehensive Medicinal Chemistry III, pp. 512–555. Elsevier (2017)
9. Deacon, C.F.: Physiology and pharmacology of DPP-4 in glucose homeostasis and the treatment of type 2 diabetes. Front. Endocrinol. **10**, 80 (2019)
10. Solerte, S.B., Di Sabatino, A., Galli, M., Fiorina, P.: Dipeptidyl peptidase-4 (DPP4) inhibition in COVID-19. Acta Diabetol. **57**(7), 779–783 (2020). https://doi.org/10.1007/s00592-020-01539-z
11. Nauck, M.A., Meier, J.J.: Incretin hormones: their role in health and disease. Diabetes Obes. Metab. **1**, 5–21 (2018)
12. Aaboe, K., Krarup, T., Madsbad, S., Holst, J.J.: GLP-1: physiological effects and potential therapeutic applications. Diabetes Obes. Metab. **10**, 994–1003 (2008)
13. Makrilakis, K.: The role of DPP-4 inhibitors in the treatment algorithm of type 2 diabetes mellitus: when to select, what to expect. Int. J. Environ. Res. Public Health **16**, 2720 (2019)
14. ChEMBL Database. https://www.ebi.ac.uk/chembl/
15. Chem Axon Marvin 6.3.0 program (2013). www.chemaxon.com/products.html
16. Stewart, J.J.P.: Optimization of parameters for semiempirical methods I Method. J. Comput. Chem. **10**(2), 209–220 (1989)
17. Stewart, J.J.P.: Optimization of parameters for semiempirical methods II applications. J. Comput. Chem. **10**(2), 221–264 (1989)
18. Gaussian 98 (RevisionA.7) Frisch MJ etal.Gaussian, Inc, PittsburghPA, (1998)
19. Chem3D Ultra7.0, Cambridge Soft Corporation, 100 Cambridge Park Dr, Cambridge, USA (2001). https://www.cambridgesoft.com/
20. Pentacle,Version1.0.6, MolecularDiscoveryLtd., Perugia, Italy (2009)
21. Ojha, P.K., Roy, K.: Comparative QSARs for antimalarial endochins: importance of descriptor-thinning and noise reduction prior to feature selection. Chem. Intell. Lab. Syst. **109**(2), 146–161 (2011)
22. Tropsha, A.: Best practices for QSAR model development, validation, and exploitation. Mol. Inf. **29**, 476–488 (2010)
23. Roy, P., Roy, K.: On some aspects of variable selection for partial least squares regression models. QSAR Comb. Sci. **27**, 302–313 (2008)
24. Ojha, P.K., Mitra, I., Das, R.N., Roy, K.: Further exploring rm2 metrics for validation of QSPR models dataset. Chem. Intell. Lab. Syst. **107**, 194–205 (2011)

Improving the Formulation Aspects of Orodispersible Tablets by Co-processed Excipients: Results of the Latest Studies

Lamija Hindija(✉), Jasmina Hadžiabdić, Amina Tucak, Merima Sirbubalo, and Ognjenka Rahić

Department of Pharmaceutical Technology, Faculty of Pharmacy, University of Sarajevo, Zmaja od Bosne 8, 71 000 Sarajevo, Bosnia and Herzegovina
lamija.hindija@ffsa.unsa.ba

Abstract. Orodispersible tablets (ODTs) improve drug efficiency and patient acceptance and they are suitable for many conditions. Excipients are responsible for providing the best properties of ODTs concerning fast disintegration, acceptable mouthfeel and good mechanical strength. Combining different excipients leads to creating co-processed excipients (CPEs) that have upgraded functionality compared to single excipients, enabling the manufacturing of ODTs with shorter disintegration time and finer dispersions whilst achieving acceptable tablet friability, all together with cost and time savings. This paper aims to describe the wide range of possibilities offered by the use of CPEs according to the results of studies carried out in the past three years. A literature review was done by searching for studies where different CPEs were examined, evaluated and compared, as well as their impact on final formulation properties.

Keywords: Orodispersible tablets · Fast dissolving · Co-processed excipients

1 Introduction

Administration of conventional tablets can be difficult due to swallowing problems in patients with dysphagia or insufficient oral cavity muscle development in children, choking risks, oral infections or other diseases that cause pain during swallowing. These problems can be solved by using orodispersible tablets (ODTs) [1]. ODTs are formulations that disintegrate in the mouth and can be administrated without water. They also provide better bioavailability as they avoid first-pass metabolism [2]. Requirements for fast disintegration and pleasant mouthfeel lead to difficulties in formulation aspects. To achieve fast disintegration and dissolution of the drug in the mouth, tablet matrix needs to be soft and porous so the tableting is done under the lower compression force. That leads to a lack of mechanical strength, final products are brittle and need specific packaging [3].

A. Badnjevic and L. Gurbeta Pokvić (Eds.): CMBEBIH 2021, IFMBE Proceedings 84, pp. 489–498, 2021.
https://doi.org/10.1007/978-3-030-73909-6_57

2 Methods

The type of research was descriptive, comparative and retrospective. Criteria for searching and including studies were pertaining to the publication date that was not older than 2018 and that they specifically investigated the use of CPEs only in ODTs, not conventional tablets or other orodispersible formulations. Original articles published in medical journals and databases such as PubMed, Scopus, Elsevier and Google Scholar were used as sources of information.

3 Excipients for Formulating ODTs

Excipients play a key role in providing optimal properties of the final formulation. Flavorings, sweeteners and dyes improve organoleptic properties [4], lubricants prevent flow problems and hinder tablet mass from adhering to the walls of machines [5]. Surfactants decrease the hydrophobic properties of the drug [6]. Disintegrants have a crucial impact on disintegration effectiveness. They determine disintegration time, hardness and friability of the ODT. Disintegrants have to ensure fast dissolution of the tablet matrix in contact with saliva. Also, they must have good flowability and compressibility to ensure adequate hardness of the final product. Solubility and compressibility of the fillers also affect the disintegration rate [7]. However, using all these excipients may be insufficient to obtain the final product with desired properties. Problems with flowability, compressibility, poor dissolution rate and mutual incompatibility may often occur. The solution to these problems lies in the addition of co-processed excipients (CPEs) in the formulation [8].

3.1 Co-processed Excipients

CPEs are combinations of two or more excipients designed to physically modify their properties in a way not achievable by simple physical mixing and without significant chemical modifications [9]. Combining these excipients includes their physical mixing in form of homogenous dispersion or solution followed by different technologies, such as roller compaction (suitable for hygroscopic and thermolabile substances), wet granulation, hot melt extrusion, spray drying, roller drying, co-transformation, milling, melt granulation and solvent evaporation [10]. Components of CPEs act synergistically and they significantly improve the characteristics of the final product compared to products that contain physical mixtures of single excipients. Optimizing the particle size and size distribution improves flow properties of the powder mass. CPEs also improve compressibility, dilution potential, fill weight variation and lubricant sensitivity [9]. CPEs have better dilution potential than single excipients thus allowing the maintenance of the compressibility properties of the powder blend even when diluted with other excipients. Their brittle character provides low lubricant sensitivity by preventing the formation of a coherent lubricant network. Their high solubility, swelling and wicking capacity provide rapid disintegration. They're physically and chemically stable. Implementation of CPEs in manufacturing processes simplifies the production process, reduces the error rate and enables cost and time savings [10]. High dilution potential of CPEs eliminates the need for many excipients in a formulation [8].

For example, adding lubricant to a mixture of filler and binder can help overcome the problem of long disintegration time by increasing the porosity of tablet matrix and thus reducing the disintegration time. CPEs help to better mask the unpleasant taste of the drug [11, 12].

Commercially Available Co-processed Excipients

Division of the commercially available CPEs can be based on their main components and is shown in Table 1.

Uniform content of the formulation can be achieved if the various, mixed substances have similar shapes, preferably spherical, similar density, size and ratio in the blend. Therefore, some manufacturers offer CPEs with different densities and particle size distribution, like Prosolv® [29].

There are two available grades of SmartEx on market. SmartEx QD-50 has particles with an average diameter of 45–75 μm and provides quicker disintegration, while SmartEx QD-100 particles have a diameter is 85–125 μm and provide harder tablets [28].

Table 1. List of some of the commercially available CPEs and their main components

Trade name	Main components	Group
Ludipress® [13]	– α-lactose monohydrate – povidone – crospovidone	Co-processed lactose
Starlac® [14]	– α-lactose monohydrate - corn starch	
Cellactose® [15]	– α-lactose monohydrate - powdered cellulose	
MicroceLac® 100 [16]	– α-lactose monohydrate – micro crystalline cellulose (MCC)	
Retalac® [17]	– α-lactose monohydrate - hypromellose	
Combilac® [18]	– α-lactose monohydrate – MCC – white maize starch	
Prosolv® ODT G2 [19]	– MCC – colloidal silicum dioxide	Co-processed cellulose
Avicel CE 15®	– MCC - guar gum	
Avicel DG®	– MCC – dibasic calcium phosphate	
Avicel HFE 102® [20]	– MCC – mannitol	

(*continued*)

Table 1. (*continued*)

Trade name	Main components	Group
Compressol S® SM [21]	– sorbitol – mannitol – silicium dioxide	Co-processed sugar alcohols
Pearlitol® [22]	– mannitol – maize starch	
Formaxx® [23]	– sorbitol - calcium carbonate	
Pharmaburst® 500 [24]	– mannitol – silicium dioxide – crospovidone	
F-melt® [25]	– mannitol – MCC – xylitol – crospovidone – dibasic calcium phosphate anhydrate	
Ludiflash® [26]	– mannitol – crospovidone – polyvinyl-acetate	
Disintequik® [27]	– mannitol – lactose monohydrate – crospovidone – dextrose monohydrate	
SmartEx® [28]	– mannitol – low substituted hydroxyl-propyl-cellulose – polyvinyl alcohol	

The more porous the material is, the better wettability and the faster disintegration is. Ludipress® is one of the CPEs that improves dissolution in the oral cavity. On the other hand, Prosolv® and F-Melt® are developed to improve the hardness of tablets [29].

4 Results of the Studies

Drašković et al. [30] used dynamic compaction analysis and percolation theory to evaluate CPE intended for direct compression of ODTs. They evaluated properties of Pharmaburst® 500, Parteck®, Ludiflash® and Disintequik™ ODT. Critical quality attributes (CQAs) were tablet disintegration and mechanical properties. High compactibility and tabletability of excipients were required for sufficient mechanical resistance of formulated ODTs. On the other hand, high porosity with higher extent of elastic deformations, average tabletability and comapctibility were necessary for fast ODT disintegration. Based on the results of the study, porosity, mechanism of consolidation, compactibility and tabletability were the most important factors for CQAs. Pharmaburst® 500 had the highest elastic recovery, the lowest relative density, average compactibility and tabletability, as well as remarkably high dilution capacity and was considered the most convenient CPE for direct compression of ODTs.

Vodáčková et al. [31] investigated the physical properties of Avicel CE 15®, Avicel DG® and Avicel HFE 102®. Avicel CE 15® had mid-range properties, its particles were mid-size and non-spherical, it had high hygroscopicity, growing negative charge, best lubricity, lowest tensile strength, mid-long disintegration time, as well as the best organoleptic characteristics. Avicel DG® had the worst flow properties, smallest and the most asymmetrical particles, lowest hygroscopicity, stable charge, intermediate lubricity and tensile strength and short disintegration time. Avicel HFE 102® had the best flow properties, large symmetrical particles and medium hygroscopy and its charge fluctuated during blending. It had lower lubricity, the highest tensile strength and slow disintegration of tablets. That made Avicel HFE 102® a good candidate for improving the flow properties of poorly flowing substances.

Svačinová et al. [32] conducted a comprehensive study on three CPEs, F-Melt® type C, M and F1, comparing their flow, viscoelastic and compacts properties. After the required testing (AFM imaging, wettability, etc.) they stated that F-Melt F1® and F-Melt M® had similar properties due to their similar composition and were better suited for incorporating hygroscopic compounds than F-Melt C®.

Borkhataria et al. [33] prepared multifunctional, directly compressible CPE for ODT production by spray drying technique. Spray drying was chosen because it provides precise particle size control and forms porous agglomerates suitable for direct compression, which is its advantage over melt granulation. Powder mixtures contained mannitol, MCC, Kyron T-314® (Polacrillin Potassium), maltodextrin, colloidal silicon dioxide and neotame. Results showed that mannitol and MCC mostly affected Carr's index, while Kyron T-314® and MCC reduced disintegration time. Mannitol had the impact on compactibility, flowability and tensile strength of powder. MCC also influenced flowability while maltodextrin had potential effect on tablet tensile strength. This novel developed CPE can successfully improve compressibility, compactibility and flowability and decrease disintegration time as well.

Kothawade et al. [34] investigated the influence of three disintegrants on nimodipine ODTs properties, sodium starch glycolate, crospovidone and co-processed sodium starch glycolate with crospovidone. The formulation containing co-processed sodium starch glycolate with crospovidone, mannitol and MCC showed the shortest disintegration time and 100% drug release within 45 min. Combining the nimodipine complexed with β-CD for solubility enhancement and co-processing of the disintegrants, showed greater disintegrating ability than the simple excipient effect. Formulations were evaluated concerning the type and concentration of superdisintegrant in ODTs with respective to disintegration time, wetting time and *in vitro* drug release by using 2^3 factorial design. Assessed variables were significantly dependent on disintegrants concentration.

Bowles et al. [35] claimed that CombiLac®, F-melt C®, SmartEx QD100® provided acceptable flowability, tabletability and ejection results in addition to low friability, short disintegration time and good dispersibility. That made them a good choice for manufacturing the ODTs by direct compression.

El-Nabarawi et al. [36] compared the ability of nine CPEs to optimize drug delivery and manufacturability of ODTs. They used an equal dose of water-soluble baclofen and poorly water-soluble meloxicam as model drugs. They concluded that F-melt® improved formulation characteristics when ODTs were produced by direct compression. Pharmaburst® and Prosolv® also showed good results in both ODTs evaluation tests containing baclofen or meloxicam.

Lura et al. [37] compared the improvement in properties of orodispersible mini-tablets (ODMTs) loaded with hydrochlorothiazide and enalapril maleate when Ludiflash® and isomalt were added in the formulation. Choosing isomalt allows to select different types and doses of disintegrant and formulations with hydrochlorothiazide showed fast and complete drug release. However, formulation with enalapril maleate showed slower drug release than Ludiflash® when isomalt was added without disintegrant. Attempts were made to simplify the manufacturing process of ODMTs by using a single excipient like isomalt. To improve the flow properties, fumed silica was also added. It increased the content uniformity in some cases but also disturbed required disintegration properties. The key factor for the stability of both formulations, containing Ludiflash® or isomalt, was packaging.

Kaur et al. [38] attempted to enhance bioavailability and improve the dissolution profile of febuxostat ODTs by using a combination of crospovidone and MCC. ODTs prepared by direct compression showed the ability to release more than 80% of the drug within 10 min of the start of dissolution testing and showed better drug release profile compared to the available marketed formulation.

Conceição et al. [39] aimed to develop ODTs of poorly soluble drug carbamazepine by combining hydroxypropyl-β-cyclodextrin (HP-β-CD) and directly compressible CPEs, Prosolv® ODT G2 and F-Melt® type C. HP-β-CD improves wetting and solubility of the poorly soluble drugs by forming inclusion complexes and CPEs promote rapid disintegration and improve flowability of the inclusion complexes. That was confirmed in this study when the flow properties of all materials and main properties of the tablets were investigated. *In vitro* disintegration time of ODTs met the requirements of European Pharmacopeia and drug dissolution was fast and complete, concluding that this combination can improve release properties of ODTs with poorly soluble drugs prepared by direct compression.

Kuldeep et al. [40] formulated ODTs of montelukast by direct compression and evaluated the effect of CPEs over physical mixture of the same excipients. Sodium starch glycolate and mannitol were used to develop CPEs by solvent evaporation and then incorporated in ODTs in different ratios. Hardness, friability, weight variation, drug content, thickness, disintegration and wetting time and *in vitro* drug release were evaluated. Obtained results revealed that formulations with CPE had faster drug release *in vitro* compared to the physical mixture.

The effects of these investigated CPEs on ODTs in studies mentioned above were discussed in Table 2. In studies where more CPEs were compared, the CPE with the most desirable effects was emphasized. In other studies, properties of ODTs containing CPEs were compared with ODTs containing a physical mixture of excipients.

Table 2. Improved properties of ODTs containing CPEs

Co-processed excipients examined in the study	Effects on the final formulation
Pharmaburst® 500	– High elastic recovery – Low relative density – Good compactibility and tabletability – High dilution capacity
Avicel HFE 102®	– Low lubricity – Good flow properties – Large symmetrical particles – Long disintegration time – High tensile strength - Medium hygroscopy
Avicel CE 15®	– Good lubricity – Mid-range flow properties – Mid-size particles – Mid-long disintegration time – Low tensile strength – Good organoleptic properties – High hygroscopy
Avicel DG®	– Intermediate lubricity – Bad flow properties
	– Small particles – Short disintegration time – Intermediate tensile strength - Low hygroscopy
F-Melt ®	– Low hygroscopy – Good tabletability – Good dispersibility – Low friability
Mannitol Microcrystalline cellulose Kyron T-314® Maltodextrin Colloidal silicon dioxide Neotame (spray dried)	– Good compressibility – Good compactibility – Improved flowability – Decreased disintegration time
Co-processed sodium starch glycolate with crospovidone	– Decreased disintegration time – Fast *in vitro* drug release
CombiLac® SmartEx QD100®	– Good flowability – Good tabletability – Low friability – Short disintegration time – Good dispersibility
Ludiflash®	– Fast *in vitro* drug release

(*continued*)

Table 2. (*continued*)

Co-processed excipients examined in the study	Effects on the final formulation
Co-processed crospovidone and microcrystalline cellulose	– Fast *in vitro* drug release – High bioavailability
Prosolv®	– Rapid disintegration – Good flowability – Fast and complete drug dissolution – Fast disintegration time *in vitro*
Co-processed sodium starch glycolate and mannitol (solvent evaporation)	– Fast *in vitro* drug release

Despite all the benefits of using CPEs, there are still some limitations. Ratio of all the individual excipients used is permanent and cannot be adjusted for different purposes. CPEs don't have an official monograph in pharmacopeia. That's the main reason why they are still included in formulations only when significant advantages in tablet compaction are shown over the physical mixture of the excipients [10].

5 Conclusion

ODTs are undoubtedly promising pharmaceutical formulations that offer various advantages over conventional tablets. Some of the drawbacks in their post-manufacturing stability, such as high friability or high hygroscopy can be overcome with adequate excipients. CPEs integrate good properties of a single excipient they're made of and provide improved flowability, compressibility, dilution potential, stability and reduced disintegration time.

However, there's no ideal CPE, their selection is individual for each case and depends on various factors. Desired manufacturing requirements and properties of the final product would determine which one would be the most suitable for the particular purpose. If direct compression is the technology of choice for manufacturing ODTs, convenient CPEs would be Pharmaburst®, Prosolv®, CombiLac® and SmartEx QD100® because they provide good mechanical strength, good flowability and low friability but also short disintegration time. The most suitable CPE for hygroscopic drugs would be F-Melt M or F1 which can also be used for ODTs manufacturing by direct compression. Avicel CE 15® offers the best organoleptic properties and good lubricity, good flowability can be achieved with Avicel HFE 102®, while short disintegration time and low hygroscopy can be accomplished with Avicel DG®.

The increased demand for excipients with improved functionalities without compromising quality, safety and regulatory aspects as well as the desire to simplify the manufacturing processes and reduce costs, will certainly further increase the use of CPEs.

Conflict of Interest. The authors have no conflicts of interest to disclose.

References

1. Ghosh, T., Ghosh, A., Prasad, D.: A review on new generation orodispersible tablets and its future prospective. Int. J. Pharm. Pharm. Sci. **3**, 1–7 (2011)
2. Manivannan, R.: Oral disintegrating tablets: a future compaction. Int. J. Pharm. Res. Dev. **1**, 1–10 (2009)
3. Gittings, S., Turnbull, N., Roberts, C.J., Gershkovich, P.: Dissolution methodology for taste masked oral dosage forms. J. Control Release. **173**, 32–42 (2014)
4. Abed, K.K., Hussein, A.A., Ghareeb, M.M., Abdulrasool, A.A.: Formulation and optimization of orodispersible tablets of diazepam. AAPS PharmSciTech. **11**, 356–361 (2010)
5. Gupta, A., Mishra, A., Gupta, V., Bansal, P.: Recent Trends of Fast Dissolving Tablet - An Overview of Formulation Technology Recent Trends of Fast Dissolving Tablet - An Overview of Formulation Technology (2010)
6. Nayak, A.K., Manna, K.: Current developments in orally disintegrating tablet technology. J. Pharm. Educ. Res. **2**, 21–34 (2011)
7. Panda, S., Latha, V., Pattnaik, S., Maharana, L., Prasad, R.: An update on various excipients employed for orodispersible tablets with a special focus on superdisintegrants **7**, 440–449 (2015)
8. Rojas, J., Buckner, I., Kumar, V.: Co-proccessed excipients with enhanced direct compression functionality for improved tableting performance. Drug Dev. Ind. Pharm. **38**, 1159–1170 (2012)
9. Pawar, S.B., Ahirrao, S.P., Kshirsagar, S.J., City, B.K., Knowledge, B.: Review on novel pharmaceutical coprocessed excipients. Pharm. Reson. **2**, 14–20 (2019)
10. Bin, L.K., Gaurav, A., Mandal, U.: A review on co-processed excipients: current and future trend of excipient technology. Int. J. Pharm. Pharm. Sci. **11**, 1–9 (2019)
11. Soujanya, B., Priya, P.G., Murthy, T.E.G.K.: Coprocessing of excipients: a review on excipient development for improved tabletting performance. Res. J. Pharm. Dos. Forms Technol. **7**, 149–155 (2015)
12. Gohel, M.C., Jogani, P.D.: A review of co-processed directly compressible excipients. J. Pharm. Pharm. Sci. a Publ. Can. Soc. Pharm. Sci. **8**, 76–93 (2005)
13. Ludipress®. https://pharmaceutical.basf.com/global/en/drug-formulation/products/ludipress.html. Accessed 29 Dec 2020
14. StarLac® Product Detail - Meggle Pharma - Excipients & Technology. https://www.meggle-pharma.com/en/lactose/14-starlac.html. Accessed 29 Dec 2020
15. Cellactose® 80 Product Detail - Meggle Pharma - Excipients & Technology. https://www.meggle-pharma.com/en/lactose/12-cellactose-80.html. Accessed 29 Dec 2020
16. MicroceLac® 100 Product Detail - Meggle Pharma - Excipients & Technology. https://www.meggle-pharma.com/en/lactose/13-microcelac-100.html. Accessed 29 Dec 2020
17. RetaLac® Product Detail - Meggle Pharma - Excipients & Technology. https://www.meggle-pharma.com/en/lactose/15-retalac.html. Accessed 29 Dec 2020
18. CombiLac® Product Detail - Meggle Pharma - Excipients & Technology, https://www.meggle-pharma.com/en/lactose/25-combilac.html. Accessed 29 Dec 2020
19. PROSOLV® ODT G2 | Orally Disintegrating Tablet Matrix- JRS Pharma. https://www.jrspharma.com/pharma_en/products-services/excipients/hfe/prosolv-odt-g2.php. Accessed 29 Dec 2020
20. Avicel®. https://www.pharma.dupont.com/pharmaceutical-brands/avicelr-forsolid-dose-forms.html. Accessed 29 Dec 2020
21. Compressol® SM. https://www.spipharma.com/en/products/drug-deliverysystems/compressol-sm/. Accessed 29 Dec 2020

22. PEARLITOL® l Co-Processed Mannitol Starch l. Roquette. https://www.roquette.com/pharma-and-nutraceuticals-coprocessed-mannitolstarch. Accessed 29 Dec 2020
23. ForMaxx Calcium carbonate 70 (calcium carbonate-sorbitol) from MilliporeSigma - Product Description and Details. https://www.americanpharmaceuticalreview.com/25260-Excipients/5821583ForMaxx-Calcium-carbonate-70-calcium-carbonate-sorbitol/. Accessed 29 Dec 2020
24. Pharmaburst® 500 Orally Disintegrating Tablets - SPI Pharma. https://www.spipharma.com/en/products/drug-delivery-systems/pharmaburst-500/. Accessed 29 Dec 2020
25. F-MELT Directly Compressible Excipient System for Fast-Disintegrating Oral Tablets. https://www.fujichemical.co.jp/english/newsletter/newsletter_pharma_0802.html. Accessed 29 Dec 2020
26. Ludiflash® – The taste of success. https://pharmaceutical.basf.com/global/en/drug-formulation/products/ludiflash.html. Accessed 29 Dec 2020
27. Disintequik ODT | Pharma Excipients. https://www.pharmaexcipients.com/product/disintequik-odt/. Accessed 29 Dec 2020
28. SmartEx® - For Pharmaceutical : Shin-Etsu Cellulose. https://www.metolose.jp/en/pharmaceutical/smartexr.html. Accessed 29 Dec 2020
29. Franc, A., Vetchý, D., Vodáčková, P., Kubal̆ák, R., Jendryková, L., Goněc, R.: Co-processed excipients for direct compression of tablets Společně zpracované pomocné látky pro přímé lisování tablet. Čes. slov. Farm. **67**, 175–181 (2018)
30. Drašković, M., Djuriš, J., Ibrić, S., Parojčić, J.: Functionality and performance evaluation of directly compressible co-processed excipients based on dynamic compaction analysis and percolation theory. Powder Technol. **326**, 292–301 (2018)
31. Vodáčková, P., Vraníková, B., Svačinová, P., Franc, A., Elbl, J., Muselík, J., Kubalák, R., Solný, T.: Evaluation and Comparison of Three Types of Spray Dried Coprocessed Excipient Avicel® for Direct Compression. Biomed Res. Int. 2739428 (2018)
32. Svačinová, P., et al.: Comprehensive study of co-processed excipients F- Melts®: Flow, viscoelastic and compacts properties. Powder Technol. **355**, 675–687 (2019)
33. Borkhtaria, C., Gurjar, R., Chavda, J., Manek, R., Patel, K., Patel, D.: Development of multifuncional directly compressible co-processed orodispersible excipient using spray drying technique. Curr. Pharma Res. **9**, 3002–3019 (2019)
34. Kothawade, S., Biyani, S., Gattani, S.: Effect of co-processed superdisintigrant on nimodipine orodispersible tablets using 2 3 factorial design approach effect of co-processed superdisintigrant on nimodipine. Eur. J. Pharm. Med. Res. **6**, 468–478 (2019)
35. Bowles, B.J., et al.: Co-processed excipients for dispersible tablets-part 1: manufacturability. AAPS PharmSciTech. **19**, 2598–2609 (2018)
36. El Nabarawi, M.A., Abd El-Monem, R.A., Al-Samadi, I.E.I.: Effect of coprocess excipients in formulation of ODTs using a model drug. Int. J. Pharm. Sci. Res. **4**, 627 (2019)
37. Lura, A., Luhn, O., Suarez Gonzales, J., Breitkreutz, J.: New orodispersible mini-tablets for paediatric use – a comparison of isomalt with a mannitol based co-processed excipient. Int. J. Pharm. **572**, 118804 (2019)
38. Kaur, M., Mittal, A., Gulati, M., Sharma, D., Kumar, R.: Formulation and in vitro evaluation of fast dissolving tablets of febuxostat using co-processed excipients. Recent Pat. Drug Deliv. Formul. **14**, 48–62 (2020)
39. Conceição, J., Adeoye, O., Cabral-Marques, H., Concheiro, A., Alvarez-Lorenzo, C., Sousa Lobo, J.M.: Orodispersible carbamazepine/hydroxypropyl-β-cyclodextrin tablets obtained by direct compression with five-in-one co-processed excipients. AAPS PharmSciTech **21**(2), 1 (2020). https://doi.org/10.1208/s12249-019-1579-5
40. Kuldeep, K., Himansu, C., Kumar, S.G.: Formulation and evaluation of fast dissolving tablet of montelukast by using co-processed excipients. Res. J. Pharm. Technol. **12**, 5543–5548 (2019)

Application of *In Silico* Methods in Pharmacokinetic Studies During Drug Development

Azra Hamidović, Ena Hasković, Sumeja Muhić, Matej Planinić, Naida Omerović(✉), and Selma Škrbo

Faculty of Pharmacy, University of Sarajevo, Zmaja od Bosne 8, 71 000 Sarajevo, Bosnia and Herzegovina
naida.omerovic@ffsa.unsa.ba

Abstract. Many compounds present potential drug candidates, but only a few of them will be formulated, approved, and registered as a drug for a certain health issue. A search for that compound could be a long, complex, and expensive process, and sometimes it is urgent to develop a drug, for example, during the COVID-19 pandemic. The aim of this study is to provide a theoretical overview of the application of computer-aided methods in pharmacokinetic studies during drug development. Computer-aided *in silico* methods filter a large number of compounds to a few ones with best properties, thus shortening the whole process and making it more efficient. Virtual screening, whether it is ligand-based or structure-based, identifies molecules that are most likely to bind to a receptor or an enzyme. It is becoming a more substantial method during drug development. Quantitative structure-activity relationship (QSAR) and quantitative structure-property relationship (QSPR) methods are based on the mathematical algorithm that predicts chemical features encoded within descriptors (chemical attributes that control the observed response). These studies consist of data preparation, data processing, and data interpretation for a set of substances and are very valuable. *In silico* evaluation of absorption, distribution, metabolism, and excretion (ADME) properties helps predict drug behaviour *in vivo* by combining pharmacokinetic processes into one model. The most important physicochemical properties related to ADME processes are logP and logD, as well as solubility and pKa. The progress of *in silico* methods in drug development is unstoppable, thus their use nowadays is ever-increasing. The future of drug development lies in *in silico* methods, as they are cost-effective and time-saving, with good prediction accuracy.

Keywords: Drug development · *In silico* methods · Pharmacokinetics · Virtual screening · QSAR · ADME properties

1 Introduction

In the process of developing novel drugs, a huge number of potential compounds presents a great challenge, thus it is necessary to single out compounds with best characteristics.

A. Badnjevic and L. Gurbeta Pokvić (Eds.): CMBEBIH 2021, IFMBE Proceedings 84, pp. 499–510, 2021.
https://doi.org/10.1007/978-3-030-73909-6_58

To design a novel drug, scientists need to investigate which part of an organism to target, in what way a novel drug needs to act, and how to apply a drug most effectively. To answer these questions, one can apply pharmaceutical bioinformatics and artificial intelligence. These scientific areas combine biological and chemical processes with informatic technologies, intending to understand how to effectively discover and develop novel drugs, and are important especially during pharmacokinetic studies [1, 2].

Computer-aided drug discovery/design (CADD) methods have made key contributions to drug development. These methods filter a large number of compounds into smaller groups that would be tested experimentally while increasing their activity or optimising their pharmacokinetic properties. They can be classified as structure-based, ligand-based or hybrid methods. Structure-activity relationship (SAR) and structure-property relationship (SPR) models explore the qualitative relationship between molecules' three-dimensional (3D) chemical structure and their activity and properties, respectively. Some models explore the quantitative relationship (QSAR, QSPR), and all of them give valuable information during drug discovery and development [3, 4]. *In silico* evaluation of absorption, distribution, metabolism, and excretion (ADME) properties implies the use of computer modelling to understand and predict pharmacokinetic properties of a drug based on its structure [2].

In this article, we provide a short theoretical overview of the application of computer-aided methods in pharmacokinetic studies during drug development. We will describe virtual screening and its applications, then SAR, SPR, QSAR and QSPR modelling, and *in silico* evaluation of ADME properties.

2 Materials and Methods

Materials used for the writing of this article were obtained from databases such as Medline and PubMed, using scientific literature and websites relevant to our topic. Studies used in this article were published in relevant scientific magazines. Keywords used for research were: drug development, *in silico* methods, pharmacokinetics, virtual screening, QSAR, ADME properties.

3 Results

In this section, we will present notable results of studies conducted regarding virtual screening, QSAR modelling, and *in silico* evaluation of ADME properties.

3.1 Virtual Screening

Virtual screening is a tool that saves thousands of hours and money during drug discovery. It uses algorithms and software packages to distinct and filter a large number of potential drugs in early phases of drug development [5]. Virtual screening methods have rapidly developed because of their proven efficiency. They include docking-based and pharmacophore-based methods [6], as well as similarity searching [7] and machine learning methods [8]. Molecular docking is mostly used when the 3D structure of a target protein is known [9]. However, this method is not always the most successful, as its limitations include target structural flexibility [9] and accurate molecule binding affinity for a target protein, because the solvation and entropic effects are not adequately considered [10]. Solvation and entropic effects are often excluded from virtual screening because of the challenge to determine their referent states and parameterize them because they belong to different energetic categories [10]. Also, most of the virtual screening does not take in consideration other parameters, such as the residence time, but only focuses on binding affinity [11].

A general classification of virtual screening methods is: ligand-based virtual screening (LBVS) and structure-based virtual screening (SBVS). The LBVS methods rely on molecular features, bioactivity parameters, pharmacophore 3D shape matching, and QSAR [5]. For example, support vector machine (SVM) is often used as a machine learning method [8, 12, 13], because of the very low rate in false results [14]. The SVM is a tool that classifies large data into specific groups, based on assigned characteristics [15]. The SBVS methods use the 3D structure of a biological target that already exists in the database (determined either experimentally through X-ray crystallography and nuclear magnetic resonance (NMR) or computationally using homology modelling) to match candidate compounds and select the one with the best complementation to a binding spot [5]. One of the most important steps in SBVS is ranking different parameters to choose the most fitting molecule to a specific target. This step is also known as "scoring" [16, 17]. There are many challenges in this process because not every parameter can be measured and ranked the same way. The most used scoring functions are force field-based, knowledge-based and empirically-based ones [18, 19]. Target structural flexibility can also be challenging because of the complexity of its calculation [20]. However, soft docking (existing in all docking programs) or softening the van der Waals potentials can decrease the number of errors but with a larger risk of false-positive results, because the database of compounds binding to a target can be expanded [21].

When there are many compounds (potential drugs) in the trial, it is very useful to find their common pharmacophore. Pharmacophore is presented as the 3D image of functional groups of a drug, such as hydrogen bond donor (HBD), hydrogen bond acceptor (HBA), cations, anions, aromatics, and hydrophobic area (Hyp) [22, 23]. The 3D structure of a pharmacophore is the base for the prediction of molecule-target interactions. Some of the programs for pharmacophore modelling are Catalyst, DISCO, GASP, Phase, MOE, and LigandScout [6, 8]. Pharmacophore-based virtual screening shortens the drug development process and pharmacokinetic studies [24, 25]. Also, there are many risks in this method, such as insufficient or inaccurate conformations, molecular overlapping, or incorrect binding affinity estimation [26].

Deep learning methods have a large capacity, very efficient extraction ability, and low generalization error during *de novo* drug designing [27, 28]. *De novo* drug designing means designing new compounds with desired properties that fit a target of interest, which used to be a very long and complex process [29, 30]. For this purpose, the Simplified Molecular Input Line Entry Specification (SMILES) and Recurrent Neural Networks (RNN) were created [31]. The SMILES was created to narrow down existing molecular libraries to molecules similar to a molecule in the trial [28, 32]. The RNN are a class of neural networks, which algorithm remembers its input based on the internal memory, creating new molecular libraries that could be examined by machine learning methods and generating molecules of interest fast and easily [33].

New methods include a combination of SBVS and LBVS. This combination can be successful when all structural and chemical properties are known, such as the 3D structure of a target protein and a molecule in the trial [16]. The SBVS is the most used method in virtual screening, but the majority of hits are often not very potent. The LBVS is less used but statistically has more potential hits, thus has shown more successful results. Both LBVS and SBVS are certainly worth developing because they have shown a lot of positive outcomes so far [5].

3.2 QSAR Modelling

Studying the QSAR has prevented complications of drug designing, such as a high cost and a low percentage of clinical approvement of small innovative molecules [34–36]. The idea of QSAR modelling came from a simple concept based on the correlation between chemical nature and response. The QSAR analysis derives information from a small number of substances that are then used to predict responses of a large number of structurally related substances. A QSAR study attempts to find a mathematical correlation between chemical response and its attributes, which represent the intrinsic nature of a molecule that can be quantified. It is then used to predict the chemical behaviour of new or hypothetical molecules. The QSAR method is based on an adequate mathematical algorithm that enables the prediction and exploration of chemical features encoded within quantitative numbers called descriptors. They represent chemical attributes that define and control the observed response and are used to establish a mathematical correlation. There are different types of descriptors: physicochemical, structural, topological, and electronic. Descriptors also allow the modification of existing chemical behaviour by allowing a user to make changes in the chemical structure [34, 37, 38]. The response parameter, such as activity, can be used as a predictor variable for modelling another type of activity endpoint (modelled response). It is important to note that a predictor variable in QSAR modelling should be quantitative for developing a regression model [34, 39].

A QSAR study usually consists of three steps: data preparation, data processing, and data interpretation for a set of substances. The first step involves the data arrangement and conversion to a more suitable form. Quantitative data are derived from predictor or independent variables that define chemical attributes and the endpoint of interest. Endpoints are usually categorized as "activity" (e.g. anti-arrhythmic). Data obtained from endpoints usually exist in two forms, either "a dose-fixed response" or "a response-fixed dose". The first pattern involves measuring the concentration of a substance required to produce a desirable fixed response (e.g. effective concentration in 50% of the population; EC50). The second pattern consists of a chosen response elicited by a fixed dose of a substance. Preferably, response data are used as independent variables. Furthermore, information on the varying concentration of a substance required to produce a fixed response can be used to develop an adequate model. It is important to logarithmically transform response data for two main reasons, the first being the conversion of a large scale into a smaller one, while the second being the ability to construct a sigmoidal curve that represents the dose-response relationship, which renders a linear middle portion that eases modelling. Compounds used for a specific model have to exhibit the same mechanism of action towards the opted response. On the other hand, data from predictor variables are obtained from experiments involving different physicochemical measures and theoretical calculations. Both types of data are then used to form a data matrix, in which rows present various substances, while columns present the response variable. The next step is data processing, which includes several pretreatment operations (data division and removal of intercorrelated features). Data matrix can then be used to develop a model in a linear or nonlinear fashion, combined with a suitable algorithm for selecting features. Various methods can be used for the development of linear correlation models, such as multiple linear regression (MLR) and partial least squares (PLS), while stepwise and genetic algorithms can be used for selecting variables. On the contrary, approaches to nonlinear modelling include artificial neural networks (ANN). This step involves the manipulation of a large amount of data, and such studies must be accompanied by proper statistical tests, which are used to validate and characterize the quality and stability of a developed model. The final step, being the model interpretation, is a crucial one and requires vast knowledge on biochemical aspects of molecules and the observed response [5, 34, 39, 40].

There are several key objectives that QSAR modelling looks to accomplish, such as the prediction of the given response, data categorization, reduction and replacement of animals used in experiments, optimization of lead molecules, and structural refinement of synthetic molecules of interest. Application of this method combined with other *in silico* methods has been of great success during drug development, with some of the examples being approved later on by the US Food and Drug Administration (US-FDA). Notable drugs include zanamivir (1999) used in the treatment of *influenza* virus infection, imatinib (2003) used in the treatment of cancer, captopril (1981) used in the treatment of hypertension, norfloxacin (1986) used in the treatment of bacterial infections, and many more [41].

3.3 *In silico* Evaluation of ADME Properties

For more than a decade, *in silico* prediction of ADME properties of compounds has represented the cornerstone of early phases of drug discovery and development [42–44]. Early ADME studies minimize cost, time, and labour intensiveness of testing and screening by identifying only promising compounds and rejecting those with low success probability [43, 45]. The goal of *in silico* modelling of ADME properties is to ultimately predict the *in vivo* disposition of potential drugs in the human body by merging all pharmacokinetic processes into one inclusive model [42]. *In silico* ADME models have made a lot of progress and we are now in the era of machine learning with many models serving their intended purpose in the industry and complementing experimental methods [46].

In pharmacokinetic studies, physicochemical properties of a drug have a significant effect on its behaviour in the human body (Table 1) [42, 47].

Table 1. Physicochemical properties related to ADME processes or parameters

Physicochemical property	ADME process or parameter	Example
logP and logD	Membrane permeability, absorption, distribution, clearance route	*Ab initio* methods based on the molecular simulation; Property-based methods; Substructure-based methods [48]
Solubility	Dissolution of tablet/capsule, intestinal absorption, oral bioavailability	General solubility equation (GSE) based on the effect of topological polar surface area (TPSA); the solubility of 81% of compounds was predicted accurately [49, 50] Multi-model protocol based on the random forest model, resulting in more accurate results [51]
pK_a	Solubility, lipophilicity, permeability, absorption	*Ab initio* quantum mechanical calculations; Statistical and machine learning approaches; Semi-empirical approaches [48]

Many models have been integrated into different software programs for *in silico* prediction of drug absorption properties from *in vitro* and *in vivo* data. Some of the examples of available software programs are GastroPlus™, SimCYP®, IDEA, and GI-Sim [52–54].

The compartmental absorption transit (CAT) model divides the gastrointestinal tract into series of compartments with different physiological properties and is based on assumptions that instant dissolution, passive permeability (linear absorption kinetics), and minimal absorption occur in the stomach and colon [55]. It was later modified into the Advanced CAT (ACAT) model, which includes nonlinear absorption kinetics, first-pass metabolism effects, and linear transfer kinetics. A commercially available software package, GastroPlus™, was developed based on the ACAT model and has been successfully used to predict and understand the absorption of several compounds, for example, entacapone [56, 57]. The Advanced Dissolution, Absorption and Metabolism (ADAM) model is similar to the ACAT model because it was also developed based on the CAT model. Software that implements this model is SimCYP®. The ADAM model takes into account the heterogeneity of the gastrointestinal tract [58].

The most important pharmacokinetic parameters that describe drug distribution are the volume of distribution (Vd) and the unbound drug fraction (f_u), and are usually predicted in pharmacokinetic studies. When it comes to metabolism-related prediction models, the main focus is on the following studies [48]:

1. the interaction models of enzymes with xenobiotics, which are often used to differentiate whether a xenobiotic is an inhibitor or a substrate of the CYP450 system, and then to evaluate drug-drug interaction possibilities
2. the site of metabolism (SOM) models, which can be used to predict the most probable metabolising sites of xenobiotics (soft spots)
3. the clearance models of the liver, which quantitatively predict the metabolic stability of xenobiotics
4. the metabolite prediction models, which predict all of the potential metabolites of xenobiotics.

Pharmacokinetic parameters that are important in renal drug excretion are the fraction of drug excreted unchanged in urine (f_e) and renal clearance (CL_r) [59].

An overview of recent *in silico* ADME studies is presented in Table 2.

Table 2. Recent *in silico* ADME studies

Pharmacokinetic process	Results	Related source
Absorption	The GastroPlus™ Oral Cavity Compartmental Absorption and Transit (OCCAT™) model was used to simulate the fraction and rate of absorption of zolpidem sublingual tablets. It showed satisfactory performance	[60]
	GI-Sim is suitable for an early risk assessment and guiding the decision-making process during drug development	[61]
	When a high degree of accuracy is needed in predicting the intestinal absorption of incompletely absorbed drugs, GastroPlus and GI-Sim are better than SimCYP	[62]
Distribution	Partial least square (PLS) regression and nonlinear recursive partitioning approaches were used to compare performances of the developed model and those in the commercial software Volsurf. The developed model performed better in predicting the unbound drug fraction (f_u) and the prediction accuracy of the volume of distribution (Vd) was the same	[63]
	Three *in silico* models were developed for predicting the volume of distribution at a steady-state (Vd_{ss}) in a dataset of 1096 compounds. Random forest model was the optimal prediction method	[64]

(continued)

Table 2. (*continued*)

Pharmacokinetic process	Results	Related source
Metabolism	The FlexX/PMF/GOLD approach was able to discriminate CYP450 substrates and nonsubstrates with 94% prediction accuracy	[65]
Excretion	Binary classification model of the fraction of drug excreted unchanged in urine (f_e) based on structure information was calculated in Volsurf and Molconn-Z, as a topological predictor, in a dataset of 130 compounds, which resulted in 65–80% prediction accuracy. Volsurf descriptors had better predictions compared with Molconn-Z ones	[66]
	Binary classification model for the prediction of major clearance pathways of 141 approved drugs was developed, and the final model had 88% prediction accuracy	[67]
	CpathPred was modified and improved by increasing the number of descriptors and including a support vector machine (SVM)	[68, 69]
	A dataset of 754 compounds was used to construct an *in silico* model for the prediction of human clearance, and the model showed better performance compared with the previously published models	[70]
	The prediction system of renal drug excretion focused on f_e and renal clearance (CLr) was based on structure information and developed using freely available software and large datasets (411 and 401 compounds), which can be used as a tool in early phases of drug development	[71]

- absorption – *in silico* programs, such as GI-Sim, proved to be useful in early risk assessment and the decision-making process during drug development
- distribution – adequate accuracy in the prediction of Vd was achieved
- metabolism – highly accurate differentiation of CYP450 substrates and nonsubstrates was shown
- excretion – accurate information on clearance pathways and relevant pharmacokinetic parameters was noted.

4 Conclusion

Chemical compounds represent a valuable part of human necessities considering their applicability in various aspects of human life, from health care and industrial processes to household use. A lot of promising substances fail at late developmental stages and can not reach the clinical stage because of the poor pharmacokinetic properties, which are considered the main reasons for terminating the development process of drug candidates in almost 50% of cases. A lot of attention has been paid to pharmaceutical bioinformatics, artificial intelligence and CADD methods, which could play a crucial role during drug discovery and development. They are cost-effective, time-saving, limit the use of animal models, and help rationalise drug designing. Therefore, virtual screening, QSAR modelling, and *in silico* evaluation of ADME properties are worth investigating and investing in. These methods can speed up the process of drug discovery and development when urgent, for example, in the search of the potential cure for COVID-19 infection.

Conflict of Interest. The authors have no conflicts of interest to disclose.

References

1. Pharmaceutical Bioinformatics [Internet]. Sweden: Uppsala University. Accessed November 2020
2. Mbah, C.J., Okorie, N.H.: Pharmaceutical bioinformatics: its relevance to drug metabolism. Madridge J. Bioinform. Syst. Biol. **1**(1), 19–26 (2018)
3. Sliwoski, G., Kothiwale, S., Meiler, J., Lowe, E.W., Jr.: Computational methods in drug discovery. Pharmacol. Rev. **66**(1), 334–395 (2014)
4. Zhang, S.: Computer-aided drug discovery and development. Methods Mol. Biol. **716**, 23–38 (2011)
5. Lavecchia, A., Di Giovanni, C.: Virtual screening strategies in drug discovery: a critical review. Curr. Med. Chem. **20**(23), 2839–2860 (2013)
6. Kim, K.H., Kim, N.D., Seong, B.L.: Pharmacophore-based virtual screening: a review of recent applications. Expert Opin. Drug Discov. **5**(3), 205–222 (2010)
7. Willett, P.: Similarity-based virtual screening using 2D fingerprints. Drug Discov. Today **11**(23–24), 1046–1053 (2006)
8. Leelananda, S.P., Lindert, S.: Computational methods in drug discovery. Beilstein J. Org. Chem. **12**, 2694–2718 (2016)
9. Chen, Y.C.: Beware of docking! Trends Pharmacol. Sci. **36**(2), 78–95 (2015)
10. Huang, S.Y., Zou, X.: Inclusion of solvation and entropy in the knowledge-based scoring function for protein-ligand interactions. J. Chem. Inf. Model. **50**(2), 262–273 (2010)
11. Copeland, R.A.: The dynamics of drug-target interactions: drug target residence time and its impact on efficacy and safety. Expert Opin. Drug Discov. **5**(4), 305–310 (2010)
12. Liew, C.Y., Ma, X.H., Liu, X., Yap, C.W.: SVM model for virtual screening of LCK inhibitors. J. Chem. Inf. Model. **49**(4), 877–885 (2009)
13. Melville, J., Burke, E., Hirst, J.: Machine learning in virtual screening. Comb. Chem. High Throughput Screen. **12**(4), 332–343 (2009)
14. Ma, X., Jia, J., Zhu, F., Xue, Y., Li, Z., Chen, Y.: Comparative analysis of machine learning methods in ligand-based virtual screening of large compound libraries. Comb. Chem. High Throughput Screen. **12**(4), 344–357 (2009)
15. Ben-Hur, A., Weston, J.: A user's guide to support vector machines. Methods Mol. Biol. **609**, 223–239 (2010)
16. Kitchen, D.B., Decornez, H., Furr, J.R., Bajorath, J.: Docking and scoring in virtual screening for drug discovery: methods and applications. Nat. Rev. Drug Discov. **3**(11), 935–949 (2004)
17. Brooijmans, N., Kuntz, I.D.: Molecular recognition and docking algorithms. Annu. Rev. Biophys. Biomol. Struct. **32**, 335–373 (2003)

18. Krovat, E.M., Steindl, T., Langer, T.: Recent advances in docking and scoring. Curr. Comput. Aided Drug Des. **1**(1), 93–102 (2005)
19. Jain, A.N.: Scoring functions for protein-ligand docking. Curr. Protein Pept. Sci. **7**(5), 407–420 (2006)
20. Evers, A., Hessler, G., Matter, H., Klabunde, T.: Virtual screening of biogenic amine-binding G-protein coupled receptors: comparative evaluation of protein- and ligand-based virtual screening protocols. J. Med. Chem. **48**(17), 5448–5465 (2005)
21. Jiang, F., Kim, S.H.: "Soft docking": matching of molecular surface cubes. J. Mol. Biol. **219**(1), 79–102 (1991)
22. Hou, T., Wang, J., Zhang, W., Wang, W., Xu, X.: Recent advances in computational prediction of drug absorption and permeability in drug discovery. Curr. Med. Chem. **13**(22), 2653–2667 (2006)
23. Dror, O., Shulman-Peleg, A., Nussinov, R., Wolfson, H.J.: Predicting molecular interactions in silico: I. a guide to pharmacophore identification and its applications to drug design. Curr. Med. Chem. **11**(1), 71–90 (2004)
24. Rognan, D.: Chemogenomic approaches to rational drug design. Br. J. Pharmacol. **152**(1), 38–52 (2007)
25. Klabunde, T.: Chemogenomic approaches to drug discovery: similar receptors bind similar ligands. Br. J. Pharmacol. **152**(1), 5–7 (2007)
26. Yang, S.Y.: Pharmacophore modeling and applications in drug discovery: challenges and recent advances. Drug Discov. Today **15**(11–12), 444–450 (2010)
27. LeCun, Y., Bengio, Y., Hinton, G.: Deep learning. Nature **521**(7553), 436–444 (2015)
28. Unterthiner, T., Mayr, A., Klambauer, G., Steijaert, M., Ceulemans, H., Wegner, J.K., et al.: Deep learning as an opportunity in virtual screening. Paper presented at: The Workshop on Deep Learning & Representation Learning (2014)
29. Hartenfeller, M., Schneider, G.: *De novo* drug design. Methods Mol. Biol. **672**, 299–323 (2011)
30. Mullard, A.: The drug-maker's guide to the galaxy. Nature **549**(7673), 445–447 (2017)
31. Olivecrona, M., Blaschke, T., Engkvist, O., Chen, H.: Molecular *de-novo* design through deep reinforcement learning. J. Cheminform. **9**, 48 (2017)
32. Segler, M.H.S., Kogej, T., Tyrchan, C., Waller, M.P.: Generating focussed molecule libraries for drug discovery with recurrent neural networks. ACS Cent. Sci. **4**(1), 120–131 (2018)
33. Zhong, F., Xing, J., Li, X., Liu, X., Fu, Z., Xiong, Z., et al.: Artificial intelligence in drug design. Sci. China Life Sci. **61**(10), 1191–1204 (2018)
34. Todeschini, R., Consonni, V., Gramatica, P.: Chemometrics in QSAR. In: Brown, S., Tauler, R., Walczak, R. (eds.) Comprehensive chemometrics, vol. 4. Oxford: Elsevier, pp. 129–172 (2009)
35. Selassie, C.D.: History of quantitative structure-activity relationships. In: Abraham, D.J. (ed.) Burger's Medicinal Chemistry and Drug Discovery, vol. 1. New York, John Wiley & Sons, pp. 1–48 (2003)
36. Sinko, P.J.: Martin's physical pharmacy and pharmaceutical sciences, 6th ed. Baltimore: Lippincott Williams & Wilkins (2011)
37. Esposito, E.X., Hopfinger, A.J., Madura, J.D.: Methods for applying the quantitative structure-activity relationship paradigm. Methods Mol. Biol. **275**, 131–214 (2004)
38. Pragyan, P., Kesharwani, S.S., Nandekar, P.P., Rathod, V., Sangamwa, A.T.: Predicting drug metabolism by CYP1A1, CYP1A2, and CYP1B1: insights from MetaSite, molecular docking and quantum chemical calculations. Mol. Divers. **18**(4), 865–878 (2014)
39. Tute, M.S.: History and objectives of quantitative drug design. In: Hansch, C., Sammes, P.G., Taylor, J.B., (eds.) Comprehensive medicinal chemistry, vol 4. Oxford: Pergamon Press, pp. 1–31 (1990)

40. Clark, R.D., Daga, P.R.: Building a quantitative structure-property relationship (QSPR) model. Methods Mol. Biol. **1939**, 139–159 (2019)
41. Roy, K., Kar, S., Das, R.N.: QSAR/QSPR Modeling: introduction. In: Roy, K., Kar, S., Das, R.N., (eds.) A primer on QSAR/QSPR modeling. New York: Springer-Verlag Inc, pp. 1–36 (2015)
42. Alqahtani, S.: In silico ADME-Tox modeling: progress and prospects. Expert Opin. Drug Metab. Toxicol. **13**(11), 1147–1158 (2017)
43. Yamashita, F., Hashida, M.: In silico approaches for predicting ADME properties of drugs. Drug Metab. Pharmacokinet. **19**(5), 327–338 (2004)
44. Hall, A.H.: Computer modeling and computational toxicology in new chemical and pharmaceutical product development. Toxicol. Lett. **102–103**, 623–626 (1998)
45. Valerio, L.G., Jr.: Application of advanced in silico methods for predictive modeling and information integration. Expert Opin. Drug Metab. Toxicol. **8**(4), 395–398 (2012)
46. Bhhatarai, B., Walters, W.P., Hop, C.E.C.A., Lanza, G., Ekins, S.: Opportunities and challenges using artificial intelligence in ADME/Tox. Nat. Mater. **18**(5), 418–422 (2019)
47. Honório, K.M., Moda, T.L., Andricopulo, A.D.: Pharmacokinetic properties and in silico ADME modeling in drug discovery. Med. Chem. **9**(2), 163–176 (2013)
48. Wang, Y., Xing, J., Xu, Y., Zhou, N., Peng, J., Xiong, Z., et al.: In silico ADME/T modelling for rational drug design. Q. Rev. Biophys. **48**(4), 488–515 (2015)
49. Jain, N., Yalkowsky, S.H.: Estimation of the aqueous solubility I: application to organic nonelectrolytes. J. Pharm. Sci. **90**(2), 234–252 (2001)
50. Ali, J., Camilleri, P., Brown, M.B., Hutt, A.J., Kirton, S.B.: Revisiting the general solubility equation: in silico prediction of aqueous solubility incorporating the effect of topographical polar surface area. J. Chem. Inf. Model. **52**(2), 420–428 (2012)
51. Chevillard, F., Lagorce, D., Reynès, C., Villoutreix, B.O., Vayer, P., Miteva, M.A.: In silico prediction of aqueous solubility: a multimodel protocol based on chemical similarity. Mol. Pharm. **9**(11), 3127–3135 (2012)
52. Parrott, N., Lave, T.: Prediction of intestinal absorption: comparative assessment of GASTROPLUS and IDEA. Eur. J. Pharm. Sci. **17**(1–2), 51–61 (2002)
53. Sjörgen, E., Westergren, J., Grant, I., Hanisch, G., Lindfors, L., Lennernäs, H., et al.: In silico predictions of gastrointestinal drug absorption in pharmaceutical product development: application of the mechanistic absorption model GI-Sim. Eur. J. Pharm. Sci. **49**(4), 679–698 (2013)
54. Tubic, M., Wagner, D., Spahn-Langguth, H., Bolger, M.B., Langguth, P.: In silico modeling of non-linear drug absorption for the P-gp substrate talinolol and of consequences for the resulting pharmacodynamic effect. Pharm. Res. **23**(8), 1712–1720 (2006)
55. Yu, L.X., Lipka, E., Crison, J.R., Amidon, G.L.: Transport approaches to the biopharmaceutical design of oral drug delivery systems: prediction of intestinal absorption. Adv. Drug Dcliv. Rev. **19**(3), 359–376 (1996)
56. Agoram, B., Woltosz, W.S., Bolger, M.B.: Predicting the impact of physiological and biochemical processes on oral drug bioavailability. Adv. Drug Deliv. Rev. **50**(1), S41-67 (2001)
57. Alqahtani, S., Kaddoumi, A.: Development of a physiologically based pharmacokinetic/pharmacodynamic model to identify mechanisms contributing to entacapone low bioavailability. Biopharm. Drug Dispos. **36**(9), 587–602 (2015)
58. Huang, W., Lee, S.L., Yu, L.X.: Mechanistic approaches to predicting oral drug absorption. AAPS J. **11**(2), 217–224 (2009)
59. Rowland, M., Tozer, T.N., Rowland, M.: Clinical pharmacokinetics and pharmacodynamics: concepts and applications, 4th ed. Baltimore: Lippincott William & Wilkins (2011)

60. Xia, B., Yang, Z., Zhou, H., Lukacova, V., Zhu, W., Milewski, M., et al.: Development of a novel oral cavity compartmental absorption and transit model for sublingual administration: illustration with Zolpidem. AAPS J. **17**(3), 631–642 (2015)
61. Sjögren, E., Thörn, H., Tannergren, C.: Comment on: In silico modeling of gastrointestinal drug absorption: predictive performance of three physiologically based absorption models. Mol. Pharm. **14**(1), 340–343 (2017)
62. Sjögren, E., Thörn, H., Tannergren, C.: In silico modeling of gastrointestinal drug absorption: predictive performance of three physiologically based absorption models. Mol. Pharm. **13**(6), 1763–1778 (2016)
63. Del Amo, E.M., Ghemtio, L., Xhaard, H., Yliperttula, M., Urtti, A., Kidron, H.: Applying linear and non-linear methods for parallel prediction of volume of distribution and fraction of unbound drug. PLoS One **8**(10), e74758 (2013)
64. Lombardo, F., Jing, Y.: In silico prediction of volume of distribution in humans. extensive data set and the exploration of linear and nonlinear methods coupled with molecular interaction fields descriptors. J. Chem. Inf. Model. **56**(10), 2042–2052 (2016)
65. Keseru, G.M.: A virtual high throughput screen for high affinity cytochrome P450cam substrates. implications for in silico prediction of drug metabolism. J. Comput. Aided Mol. Des. **15**(7), 649–657 (2001)
66. Doddareddy, M.R., Cho, Y.S., Koh, H.Y., Kim, D.H., Pae, A.N.: In silico renal clearance model using classical Volsurf approach. J. Chem. Inf. Model. **46**(3), 1312–1320 (2006)
67. Kusama, M., Toshimoto, K., Maeda, K., Hirai, Y., Imai, S., Chiba, K., et al.: In silico classification of major clearance pathways of drugs with their physiochemical parameters. Drug Metab. Dispos. **38**(8), 1362–1370 (2010)
68. Toshimoto, K., Wakayama, N., Kusama, M., Maeda, K., Sugiyama, Y., Akiyama, Y.: In silico prediction of major drug clearance pathways by support vector machines with feature-selected descriptors. Drug Metab. Dispos. **42**(11), 1811–1819 (2014)
69. Wakayama, N., Toshimoto, K., Maeda, K., Hotta, S., Ishida, T., Akiyama, Y., et al.: In silico prediction of major clearance pathways of drugs among 9 routes with two-step support vector machines. Pharm. Res. **35**(10), 197 (2018)
70. Berellini, G., Waters, N.J., Lombardo, F.: In silico prediction of total human plasma clearance. J. Chem. Inf. Model. **52**(8), 2069–2078 (2012)
71. Watanabe, R., Ohashi, R., Esaki, T., Kawashima, H., Natsume-Kitatani, Y., Nagao, C., et al.: Development of an in silico prediction system of human renal excretion and clearance from chemical structure information incorporating fraction unbound in plasma as a descriptor. Sci. Rep. **9**, 18782 (2019)

Evaluation of Tablet Splitting Methods: A Case Study of Propranolol

Irma Mureškić[1](✉), Božana Jevđenić[1], Kanita Muhamedagić[1], Anđelka Račić[2], Biljana Gatarić[2], and Nataša Bubić Pajić[2]

[1] Department of Pharmacy, Faculty of Medicine, University of Banja Luka, Bulevar Vojvode Petra Bojovića 1A, 78000 Banja Luka, Bosnia and Herzegovina

[2] Department of Pharmaceutical Technology and Cosmetology, Faculty of Medicine, University of Banja Luka, Save Mrkalja 14, 78000 Banja Luka, Bosnia and Herzegovina

Abstract. Propranolol is a beta-blocker used for treatment of infantile haemangioma in pediatric population. Due to lack of dosage forms containing adequate doses for children, tablet splitting is sometimes the only option for dose adjustment. The aim of this study was to compare different methods of tablet splitting in regard to mass variation/loss, content uniformity and dissolution properties of propranolol in order to find the optimal approach for dose adjustment for pediatric population. Uncoated tablets of propranolol (40 mg) were split into quarters using different approaches: tablet splitter, knife and by hand. Obtained quarters were subjected to the tests for subdivided scored tablets and content uniformity (Ph. Eur. 9.0), weight loss (FDA) and the dissolution rate (USP42-NF37). Hand scored tablets did not meet the criteria of Ph. Eur. tests for subdivided scored tablets, compared to two other methods. Weight loss was the highest for tablets obtained by the tablet-splitter as well as variations in drug content after splitting, while tablets scored by other methods met the requirements. Regarding dissolution rate after splitting, the acceptance criterion was fulfilled. Similarity with whole tablets' dissolution profiles was found in the case of quarters obtained by using a tablet-splitter. Presented results indicate that different methods for tablet splitting significantly affect the uniformity of doses, directly influencing the drug effect. Considering that the tablets scored by knife meet the requirements of all executed tests, when necessary, this method of tablet splitting is considered to be the most appropriate for dose adjustment in pediatric population.

Keywords: Propranolol · Pediatric population · Tablet splitting · Dissolution test

1 Introduction

Pediatric formulations have recently been an active area of global interest and significant advancements were made in their development [1, 2]. The United States Food and Drug Administration (FDA) and European Medicines Agency (EMA) encouraged the development of pediatric formulations by establishing regulations in order to ensure the availability of high-quality and safe pediatric medicines [3, 4]. However, despite these

A. Badnjevic and L. Gurbeta Pokvić (Eds.): CMBEBIH 2021, IFMBE Proceedings 84, pp. 511–520, 2021.
https://doi.org/10.1007/978-3-030-73909-6_59

efforts, the lack of age-appropriate medicines for children is still present [5]. Furthermore, even available pediatric medicines are not always suitable for children [6] and some of them may cost thousands of dollars per bottle [7]. Thus, the administration of medicines to children still presents unique challenge for both parents and healthcare professionals. In order to provide the adequate dose of medicine to children, it is often required to manipulate/modify the drug formulations designed for adults. The manipulation/modification can be done by pharmacists, who prepare the appropriate medicine from an authorized dosage form and excipients, and by caretaker (doctors, nurses, parents) at the point of administration.

The most common manipulation of dosage form is splitting, crushing or dispersing of tablets [8, 9]. Even though some guidance exists [10, 11], no standard method is available at the moment when it comes to tablet manipulation. Different techniques for tablet splitting may be applied: tablet splitter, hand breaking, kitchen knives or scissors [12]. A several studies were conducted with the aim to compare these techniques and the findings were inconsistent, while some studies claimed that tablet cutter is superior to other techniques [13, 14], others were in favor to hand splitting [12, 15]. Somogyi et al. [16] suggested that in order to choose best splitting techniques, one must take external features of tablet (i.e. shape, score-line, film-coat and size). Commercial tablets are often crushed and used as a source of active ingredient to prepare oral liquids, suspensions or powders, which can be easily administered to children [17].

The healthcare professionals must be aware of the risk that tablet manipulation carries. There is always a possibility of under- or over- optimal dosing due to imprecise tablet splitting, which can affect the total drug exposure. This is especially problematic for drugs with narrow therapeutic windows [18]. In the case of controlled release formulations, drug bioavailability could be significantly modified as a consequence of disturbance of formulation integrity. The changes in the total surface area of the solid dosage form caused by its manipulation, can significantly alter dissolution rate of immediate release tablets [9]. It is also important to keep in mind that the loss of mass during splitting could be potentially very harmful for human health and environment when the tablet contains cytotoxic, mutagenic and reproductiontoxic active substance [19].

Orally administered propranolol is used in pediatric dermatology for the treatment of proliferating infantile haemangioma requiring systemic therapy i.e. life- or function-threatening haemangioma, ulcerated haemangioma with significant pain, haemangiomas at significant risk of ulceration, deformity and/or psychosocial impact [20]. The majority of pediatric dermatologists recommend 2 mg kg^{-1} per day as a therapeutic dose [21]. Additionally, despite the drug is not licensed for all of the following indications, it is prescribed in children in the management of hypertension, arrhythmias, hyperthyroidism, phaeochromocytoma and tetralogy of Fallot, as well as for prophylaxis of migraine [22]. Currently, a commercial preparation of propranolol for infants is authorized in Europe (Hemangiol 3.75 mg/mL oral solution, Pierre Fabre Laboratories, Paris, France) and can be prescribed by the physician. Alternatively, if the commercial preparation is not available or if the cost of authorized formulation is unacceptable for health care system, a prescribed dose can be achieved by manipulating commercially available propranolol tablets [20].

The purpose of this study is to compare different methods of tablet splitting used by parents and caretakers in order to determine whether there is a method that is more likely to deliver the most accurate dose of propranolol for pediatric population.

2 Methods

Uncoated tablets containing 40 mg of propranolol, produced by a single manufacturer, were purchased from the local pharmacy. Tablets were round with flat surfaces, pale pink colored, cross scored on one side. In total, 114 tablets were used for the study.

As described in Sects. 2.1–2.4, in order to estimate effects of tablet splitting on dose uniformity, three different methods of splitting were evaluated: a tablet splitter (Romed, Holland), a kitchen knife, and by hand. After splitting on quarters, obtained tablet portions were then subjected to the test for subdivided scored tablets (thirty samples) and the uniformity of content (ten samples) by following recommendations given by the European Pharmacopoeia (Ph. Eur. 9.0) [23]. Testing for the weight loss after scoring was carried out on fifteen samples according to FDA's guidelines [24].

Additionally, to assess whether tablet splitting affects the drug release, *in vitro* dissolution testing of whole tablets and quarters of tablets obtained by using different methods of splitting, was performed.

2.1 Uniformity of Content Test According to Ph. Eur. 9.0

Content of propranolol in ten randomly selected quarters obtained by different splitting methods was evaluated. The tablets passed the test if not more than one individual content was outside 85–115% of the average content and if all contents were within 75–125% of the average.

To determine the content of propranolol, procedure described in the United States Pharmacopoeia (USP) [25] was used. One tablet was transferred to 100 mL volumetric flask and 5 mL of dilute hydrochloric acid, 1:100 w/w (Lach:Ner, Ltd., The Czech Republic) was added with occasional swirling until the tablets disintegrated. Afterwards, 70 mL of methanol (Lach: Ner, Ltd., The Czech Republic) was added, sonicated for 1 minute and diluted with methanol up to mark. Part of the solution was centrifuged to separate inactive ingredients and diluted with methanol. The absorbance of samples was determined at 290 nm (UV-1800 spectrophotometer, Shimadzu, Japan), using calibration curve method constructed with standard solutions of propranolol hydrochloride, ≥99% (Sigma Aldrich, United States of America). Content of the drug was calculated and compared to requirements specified in the Ph. Eur. 9.0 [23].

2.2 Test for Subdivided Scored Tablets According to Ph. Eur. 9.0

As stated in Ph. Eur. 9.0 [23], thirty tablets were randomly selected and split into four parts, using three mentioned methods. One quarter from each tablet was obtained for the test and weighed individually. The average mass was calculated and individual masses of quarters were compared to it. The tablets have passed the test if not more than one individual mass was outside 85–115% of the average mass, and none of the individual masses were outside the range of 75–125% of the average mass.

2.3 Loss of Mass According to FDA

According to the FDA's guidance for industry for tablet scoring [24], fifteen tablets were used for this test. The tablets were divided into quarters and the masses of whole tablets and quarters obtained after splitting, were measured. The tablets comply with the test if the loss of mass is less than 3% after tablet splitting.

2.4 Dissolution Test According to USP42-NF37

In vitro dissolution test was performed as described in USP [25], in apparatus 1 (Erweka DT 720) at 100 rpm, using 1000 mL of dilute hydrochloric acid, 1:100 w/w (Lach:Ner, Ltd., The Czech Republic) as dissolution medium at constant temperature of 37 ± 0.5 °C. Aliquots of 5 mL were withdrawn after 10, 15, 20 and 30 minutes, with instant substitution of the fresh medium, to maintain sink conditions. Samples were filtered using 0.22 μm membrane filters (Chromafil X-tra PTFE-20/25, Macherey-Nagel, Germany). The filtrates were analyzed spectroscopically ($r^2 = 0.9999$) at 290 nm (UV-1800 spectrophotometer, Shimadzu, Japan). The content of propranolol was calculated and compared to the acceptance criterion stated by USP, which is not less than 75% of label amount in 30 minutes. Similarity of dissolution profiles of propranolol from obtained quarters with whole tablets was evaluated by calculating similarity factors (f_2) according to Eq. (1):

$$f2 = 50 \times \log\left\{\left[1 + \frac{1}{n}\sum_{t=1}^{n}(Rt - Tt)^2\right]^{-0.5} \times 100\right\} \quad (1)$$

where n is the number of time points of sampling, Rt is the amount of drug dissolved from reference product (in this study, whole tablets) at time t, Tt is the amount of drug dissolved from the test product (quarters obtained by different splitting methods) at time t.

3 Results and Discussion

3.1 Uniformity of Content

Average content of propranolol in the split tablets was between 105.26% and 114.52%, depending on the method used for tablet splitting. As shown in Table 1, quarters obtained by using a tablet-splitter did not meet the criteria of the test, given that the content of the drug in three out of ten examined samples was outside the range of 85–115% of the average content. Further, the highest variations in drug content were observed when using a tablet splitter, with content of propranolol ranging from 7.91 mg to 14.73 mg, which is 20.9% less and 47.3% more than the intended dose of 10 mg. In contrast to these results, tablets split by other two methods have met the criteria of the test.

Table 1. Results for uniformity of content

Sample number	Tablet quarters (scored by tablet splitter)		Tablet quarters (split by knife)		Tablet quarters (scored by hand)	
	Drug content (% of label amount)	85–115%of the average content	Drug content (% of label amount)	85–115%of the average content	Drug content (% of label amount)	85–115%of the average content
1	124.75	**No**	113.77	Yes	98.63	Yes
2	114.01	Yes	118.88	Yes	125.13	Yes
3	114.40	Yes	99.55	Yes	90.75	**No**
4	108.02	Yes	103.50	Yes	124.43	Yes
5	113.52	Yes	105.95	Yes	111.43	Yes
6	112.03	Yes	104.63	Yes	124.65	Yes
7	100.57	Yes	104.60	Yes	118.77	Yes
8	86.57	No	95.85	Yes	114.34	Yes
9	102.38	Yes	93.03	Yes	114.27	Yes
10	85.09	**No**	112.86	Yes	122.86	Yes
Average content	106.13		105.26		114.52	
SD	12.65		8.10		11.68	
Pass	**No**		**Yes**		**Yes**	

3.2 Test for Subdivided Scored Tablets

Tablets split by tablet splitter and by knife fulfill the requirements of the test for subdivided tablets. For quarters obtained by using a knife, individual masses varied between 87.45% and 111.96% of the average mass (Table 2). The interval for tablets split by tablet splitter was 86.40–119.40%. Measured individual masses were within the range of 85–115% of the average, except one which was within 75–125% of the average. Hand-split tablets exhibited a large deviation from the average mass, where five samples were outside the range of 85–115% of the average mass, three of which were also outside the range of 75–125% of the average. Therefore, hand split tablets failed this test.

3.3 Loss of Mass

Tablets split by using a knife and by hand met the criteria of the test, with average loss of tablet mass after splitting of 0.95% and 0.61%, respectively. The average loss of mass for tablets split by tablet splitter was 3.55%, with individual losses reaching up to 7.19% (Table 3).

3.4 Dissolution Test

FDA's guidance for industry for tablet scoring [24] states that dissolution properties of scored tablets should meet release requirements of whole tablets. By utilizing different

Table 2. Results of test for subdivided scored tablets

Sample number	Tablet quarters (split by tablet splitter)		Tablet quarters (split by knife)		Tablet quarters (split by hand)	
	Weight of a tablet quarter (g)	85–115% of the average mass	Weight of a tablet quarter (g)	85–115% of the average mass	Weight of a tablet quarter (g)	85–115% of the average mass
1	0.0597	**No**	0.0498	Yes	0.0569	Yes
2	0.0532	Yes	0.0500	Yes	0.0518	Yes
3	0.0542	Yes	0.0552	Yes	0.0647	**No**
4	0.0551	Yes	0.0508	Yes	0.0527	Yes
5	0.0509	Yes	0.0465	Yes	0.0368	**No**
6	0.0507	Yes	0.0550	Yes	0.0564	Yes
7	0.0525	Yes	0.0511	Yes	0.0446	Yes
8	0.0491	Yes	0.0441	Yes	0.0556	Yes
9	0.0495	Yes	0.0570	Yes	0.0559	Yes
10	0.0507	Yes	0.0558	Yes	0.0524	Yes
11	0.0538	Yes	0.0556	Yes	0.0582	Yes
12	0.0566	Yes	0.0520	Yes	0.0599	**No**
13	0.0502	Yes	0.0517	Yes	0.0464	Yes
14	0.0432	Yes	0.0465	Yes	0.0487	Yes
15	0.0472	Yes	0.0571	Yes	0.0457	Yes
16	0.0450	Yes	0.0519	Yes	0.0379	**No**
17	0.0462	Yes	0.0532	Yes	0.0475	Yes
18	0.0500	Yes	0.0480	Yes	0.0535	Yes
19	0.0495	Yes	0.0528	Yes	0.0602	Yes
20	0.0484	Yes	0.0460	Yes	0.0586	Yes
21	0.0503	Yes	0.0490	Yes	0.0534	Yes
22	0.0477	Yes	0.0466	Yes	0.0470	Yes
23	0.0456	Yes	0.0536	Yes	0.0523	Yes
24	0.0486	Yes	0.0526	Yes	0.0410	**No**
25	0.0472	Yes	0.0446	Yes	0.0536	Yes
26	0.0530	Yes	0.0536	Yes	0.0555	Yes
27	0.0470	Yes	0.0531	Yes	0.0540	Yes
28	0.0438	Yes	0.0480	Yes	0.0456	Yes
29	0.0443	Yes	0.0517	Yes	0.0469	Yes
30	0.0558	Yes	0.0479	Yes	0.0493	Yes
Average	0.0500		0.0510		0.0514	
SD	0.0040		0.0037		0.0066	
Pass	**Yes**		**Yes**		**No**	

methods of tablet splitting, the average amount of propranolol dissolved from obtained quarters at the end of *in vitro* dissolution test varied between 102.40% ± 7.26 and 110.02% ± 5.86 of the anticipated drug amount, while 100.05% ± 0.89 of the drug was released from whole tablets (Figure 1). Given that more than 75% of the drug was released in 30 min of the test, the acceptance criterion given by USP is fulfilled.

Table 3. Loss of mass after splitting tablets by using different methods

Sample number	Tablets split by tablet splitter			Tablets split by knife			Tablets split by hand		
	Weight of whole tablet (g)	Weight after splitting (g)	Loss of mass (%)	Weight of whole tablets (g)	Weight after splitting (g)	Loss of mass (%)	Weight of whole tablets (g)	Weight after splitting (g)	Loss of mass (%)
1	0.2023	0.1887	6.72	0.2002	0.1977	1.25	0.1999	0.1977	1.10
2	0.1985	0.1930	2.77	0.1986	0.1946	2.01	0.1910	0.1902	0.42
3	0.2046	0.2009	1.81	0.1959	0.1953	0.31	0.1980	0.1976	0.20
4	0.2014	0.1924	4.47	0.1990	0.1984	0.30	0.2014	0.2002	0.60
5	0.2015	0.1970	2.23	0.1987	0.1978	0.45	0.2009	0.2000	0.45
6	0.2000	0.1960	2.00	0.2047	0.2010	1.81	0.1976	0.1972	0.20
7	0.2027	0.1959	3.35	0.1986	0.1980	0.30	0.1990	0.1983	0.35
8	0.1958	0.1878	4.09	0.1982	0.1970	0.61	0.2015	0.1992	1.14
9	0.2017	0.1955	3.07	0.2004	0.1989	0.75	0.1973	0.1969	0.20
10	0.2030	0.1960	3.45	0.1990	0.1967	1.16	0.1992	0.1989	0.15
11	0.1984	0.1924	3.02	0.2033	0.2010	1.13	0.2010	0.1994	0.80
12	0.2001	0.1918	4.15	0.2014	0.2000	0.70	0.2011	0.1987	1.19
13	0.2013	0.1979	1.69	0.1973	0.1955	0.91	0.1996	0.1974	1.10
14	0.1986	0.1921	3.27	0.1994	0.1951	2.16	0.1986	0.1976	0.50
15	0.2004	0.1860	7.19	0.2028	0.2020	0.39	0.2070	0.2054	0.77
Average loss of mass (%)			3.55			0.95			0.61
SD			1.62			0.63			0.38
Pass			**No**			**Yes**			**Yes**

Based on the calculated similarity factors, only the dissolution profiles of quarters obtained by using a tablet-splitter were shown to be similar with dissolution profiles of propranolol from whole tablets (Table 4).

Dissolution properties of obtained tablet portions meet the requirements given by FDA, confirming that manipulation of tablets has not impacted the drug release. Nevertheless, differences in the amount of drug dissolved can be attributed to variations in the drug content uniformity after splitting tablets by employing different methods.

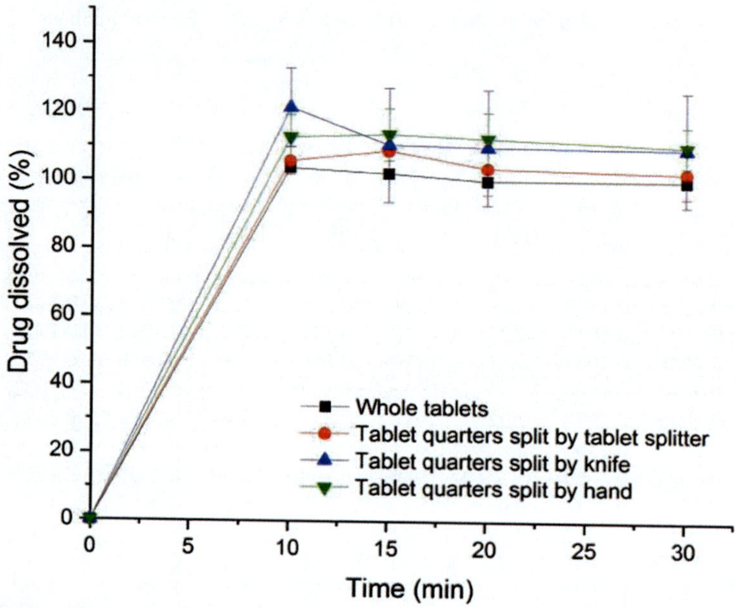

Fig. 1. Dissolution profiles of propranolol from whole tablets and tablet quarters

Table 4. Similarity factors between dissolution profiles of whole tablets and quarters of tablets obtained by different methods of splitting

Sample	Similarity factors
Tablet quarters split by tablet splitter	68.68
Tablet quarters split by knife	46.24
Tablet quarters split by hand	48.34

4 Conclusion

Through a series of test described in official literature, it has been shown that tablet splitting can have a remarkable effect on dose uniformity of propranolol, depending on the splitting method used. Quarter portions of tablets obtained by using a tablet splitter did not meet the requirements for content uniformity and the loss of mass, and hand scored tablets did not comply with the test for subdivided tablets. On the other side, tablets split by using a knife fulfilled the requirements of every conducted test. Dissolution properties of the drug from split portions were not significantly changed, considering that the acceptance criterion stated in USP was fulfilled. Despite the fact that earlier reports indicate that the tablet splitter method was the most convenient, great variations in content uniformity and loss of mass were detected regarding this method in case of propranolol tablets. Although similarity between the dissolution profiles of propranolol from quarters obtained by using a knife with whole tablets was not found, more than 75% of the label amount was released. Also, this method was shown to be the most precise regarding dose uniformity. Therefore, tablet splitting by using a knife could provide optimal doses of propranolol for pediatric patients, with the lowest risks of subdosing or overdosing.

Acknowledgment. The authors would like to acknowledge the financial support from the Ministry for Scientific-Technological Development, Higher Education and Information Society, Republic of Srpska, through the Project 1257026 and Faculty of Medicine, University of Banja Luka.

Conflict of interests declaration. The authors report no conflicts of interests.

References

1. Strickley, R.G.: Pediatric oral formulations: an updated review of commercially available pediatric oral formulations since 2007. J. Pharm. Sci. **108**(4), 1335–1365 (2019)
2. Harris, D., Hermans, E., Klein, S., Wagner-Hattler, L., Walsh, J.: Age-appropriate solid oral formulations for pediatric applications with a focus on multiparticulates and minitablets: Summary of September 2019 EuPFI workshop. Eur. J. Pharm. Biopharm. **153**, 222–225 (2020)
3. European Parliament Union, Regulation (EC) No 1901/2006 of the European Parliament and of the Council of 12 December 2006 on medicinal products for paediatric use and amending Regulation (EEC) No 1768/92, Directive 2001/20/EC, Directive 2001/83/EC and Regulation (EC) No 726/2004. OJEU 378, 1–19 (2006)
4. Food and Drug Administration Safety and Innovation Act (FDASIA), 126 Stat. 993, Pub. L, 112–144 (2012). https://www.gpo.gov/fdsys/pkg/PLAW-112publ144/pdf/PLAW-112publ144.pdf. Accessed 23 Nov 2020
5. del Moral-Sanchez, J.M., Gonzalez-Alvarez, I., Gonzalez-Alvarez, M., NavarroRuiz, A., Bermejo, M.: Availability of authorizations from EMA and FDA for age-appropriate medicines contained in the WHO essential medicines list for children 2019. Pharmaceutics **12**(4), 316 (2020)
6. Schrier, L., et al.: Off-label use of medicines in neonates, infants, children, and adolescents: a joint policy statement by the European Academy of Paediatrics and the European society for Developmental Perinatal and Pediatric Pharmacology. Eur. J. Pediatr. **179**(5), 839–847 (2020). https://doi.org/10.1007/s00431-019-03556-9
7. Meyers, R.: A wish list for drug development in pediatrics. J. Pharm. Sci. **109**(2), 939–943 (2020)
8. Ernest, T.B., et al.: Preparation of medicines for children - a hierarchy of classification. Int. J. Pharm. **435**(2), 124–130 (2012)
9. Watson, C., Webb, E.A., Kerr, S., Davies, J.H., Stirling, H., Batchelor, H.: How close is the dose? Manipulation of 10 mg hydrocortisone tablets to provide appropriate doses to children. Int. J. Pharm. **545**(1–2), 57–63 (2018)
10. Wellchild, Royal College of Paediatrics and Child Health, Neonatal and Paediatric Pharmacists Group. Medicines for Children. https://www.medicinesforchildren.org.uk/part-dose-tablet-or-capsule. Accessed 25 Nov 2020
11. MODRIC: Manipulation of Drugs Required in Children. Guidance for health care professionals. https://alderhey.nhs.uk/application/files/5014/9935/2778/MODRICGUIDELINES.pdf. Accessed 25 Nov 2020

12. van Riet-, D.A., et al.: The accuracy, precision and sustainability of different techniques for tablet subdivision: breaking by hand and the use of tablet splitters or a kitchen knife. Int. J. Pharm. **466**(1–2), 44–51 (2014)
13. Habib, W.A., Alanizi, A.S., Abdelhamid, M.M., Alanizi, F.K.: Accuracy of tablet splitting: comparison study between hand splitting and tablet cutter. Saudi Pharm. J. **22**(5), 454–459 (2014)
14. Verrue, C., Mehuys, E., Boussery, K., Remon, J.P., Petrovic, M.: Tablet-splitting: a common yet not so innocent practice. J. Adv. Nurs. **67**(1), 26–32 (2011)
15. van Reuler, A.V., et al.: Subdivision of aspirin tablets? Use your hands: a study on aspirin tablet subdivision using four different methods. J. Pharm. Pract. Res. **48**(1), 44–48 (2018)
16. Somogyi, O., Meskó, A., Csorba, L., Szabó, P., Zelkó, R.: Pharmaceutical counselling about different types of tablet-splitting methods based on the results of weighing tests and mechanical development of splitting devices. Eur. J. Pharm. Sci. **106**, 262–273 (2017)
17. Helin-Tanninen, M., Naaranlahti, T., Kontra, K., Savolainen, K.: Nifedipine capsules may provide a viable alternative to oral powders for paediatric patients. J. Clin. Pharm. Ther. **32**(1), 49–55 (2007)
18. Zheng, Z., et al.: Preparation and application of subdivided tablets using 3D printing for precise hospital dispensing. Eur. J. Pharm. Sci. **149**, 105293 (2020)
19. Breitkreutz, J., Boos, J.: Paediatric and geriatric drug delivery. Expert Opin. Drug Deliv. **4**(1), 37–45 (2007)
20. Smithson, S.L., Rademaker, M., et al.: Consensus statement for the treatment of infantile haemangiomas with propranolol. Australas J. Dermatol. **58**(2), 155–159 (2017)
21. Wedgeworth, E., Glover, M., et al.: Propranolol in the treatment of infantile haemangiomas: lessons from the European Propranolol in the treatment of complicated haemangiomas (PITCH) taskforce survey. Br. J. Dermatol. **174**(3), 594601 (2016)
22. Martindale: The Complete Drug Reference. 39th edn. Pharmaceutical Press, London (2017)
23. European Pharmacopoeia. 9th edn. Council of Europe, Strasbourg (2016)
24. Guidance for Industry - Tablet Scoring: Nomenclature, labeling, and data for evaluation. Food and Drug Administation, Center for Drug Evaluation and Research (CDER), U.S. Department of Health and Human Services (2013)
25. The United States Pharmacopeia 42, The National Formulary 37. The United States Pharmacopeial Convention, Rockville (2019)

Effects of Microencapsulation on the Release and Permeation of Active Substances from Topical Preparations

Dženita Redžić(✉) and Alisa Elezović

Department of Pharmaceutical Technology, Faculty of Pharmacy, University of Sarajevo, Zmaja od Bosne 8, 71000 Sarajevo, Bosnia and Herzegovina
dzenitaredzic@ffsa.unsa.ba

Abstract. Topical application involves application to a specific place on the skin where the given preparation manifests its effect. The properties of finished topical preparations and the incorporated active substances can often change due to various environmental and intrinsic factors. In order to prevent these changes and at the same time improve the efficacy and stability of the preparation, a large number of microencapsulation methods have been developed for the production of microparticles (microcapsules and microspheres) as a delivery system. The results of the study showed that microencapsulation achieves control of the permeation of substances in such a way that the substances remain on the surface of the skin or within a certain layer of the skin where they exert their action. Controlled release of the substances has also been achieved, thus ensuring better efficacy and reducing toxicity.

Keywords: Topical preparations · Microencapsulation · Controlled release · Permeation

1 Introduction

There are three main functions of the skin, as the largest organ of our body, namely the protective function, the prevention of water loss and the thermoregulatory function. Formulations have been developed that enable topical and transdermal delivery of active substances [1]. Transdermal delivery involves the diffusion of substance, through the layers of the skin, to the systemic circulation to exert its effect. On the other hand, topical delivery of substances is delivery to a specific site on the skin or inside it, where the concentration of the substance that reaches the systemic circulation is minimal [2]. Topical delivery should ensure direct contact of the active substance and the skin while minimizing adverse systemic effects. Topical preparations that can be used are available in the form of ointments, creams, lotions, solutions, suspensions, gels, foams and sprays [3]. The path that a molecule chooses to pass through the stratum corneum and other layers of the skin depends primarily on its physical and chemical properties, but also on the carrier of that molecule. Based on this, efforts are being made to develop systems that ensure adequate delivery of both cosmetic active substances and drugs [4].

A. Badnjevic and L. Gurbeta Pokvić (Eds.): CMBEBIH 2021, IFMBE Proceedings 84, pp. 521–527, 2021.
https://doi.org/10.1007/978-3-030-73909-6_60

The active substances that are most often found in cosmetic preparations are: vitamins (A, E, C, P, F), UV filters, urea, salicylic acid, benzoyl peroxide, D-panthenol, caffeine, quercetin [5]. Examples of active substances within medicinal preparations are: gentamicin sulfate, clotrimazole, nystatin, acyclovir, hydrocortisone, 5-fluorouracil, diclofenac, minoxidil, lidocaine, erythromycin [6]. With the aim of enabling the penetration of substances to a certain layer of the skin, various delivery systems have been developed [4]. The term delivery system means any carrier that carries and transports active substances to a specific location, where they can also control the release of the active substance in order to achieve the desired concentration at a given site of action [5]. Some examples of delivery systems are liposomes, elastic vesicles - transferosomes, micro and nanoparticles, but also formulations such as gels, emulsions or suspensions [4].

2 Microencapsulation

Microencapsulation is the process by which solid, liquid or gaseous particles of micron size are embedded inside an inert shell that insulates them and protects them from external influences. This process produces microparticles (microcapsules and microspheres) that differ in morphology and internal structure [7]. Microcapsules are defined as particles having a size between 1 μm and 1 mm whose core is usually made up of a bioactive substance, and microspheres are actually empty spherical particles, although these two terms are often used interchangeably [4]. Microspheres can still be represented as matrix systems, while microcapsules are reservoir-type systems [8]. There are several reasons why microencapsulation is performed [9]:

- Separation of incompatible components
- Conversion of liquids into solid particles
- Increased stability (protection against oxidation and degradation)
- Masking odors and tastes
- Protection against environmental factors
- Achieving controlled release of substances (continuous or delayed release)
- It allows the release at a specific site.

The interior of the particles is called the nucleus, and the shell that surrounds the nucleus is constructed mainly of polymers [10]. The core may consist of a liquid, solid or gaseous phase depending on the application itself. The shell can be permeable, semipermeable or impermeable [5]. The coating (shell) material can be arranged in one, two or more layers [11]. The characteristics of the microparticles depend on what kind of final product is to be obtained and what are the properties of the substance to be encapsulated. It is possible to encapsulate liquids, solids, proteins, bacterial cells and more, depending on the purpose of the preparation. When producing microparticles, different coating materials are used, with which it is necessary to choose the most suitable excipients, because the stiffness, integrity and degradability of the microparticle are its basic parameters [12].

Examples of the most commonly used coating materials are: vegetable gums (acacia gum, dextran sulfate, carrageenan), cellulose and its derivatives, polymers, proteins (gelatin, collagen) and waxes (paraffin, beeswax) [13]. Polymers can be biodegradable and biopolymers (natural), such as chitosan and copolymers of polylactic and polyglycolic acid (PLGA) for which there is great interest in developing a delivery system for topical preparations. They are natural, non-toxic and nonreactive with human tissue, and are easily degraded unlike other materials that tend to accumulate within the tissue and cause irritation [5].

The methods used to make microparticles can be chemical or physical. A clear division of methods is not easy, as there are some methods that can be classified into both chemical and physical methods [4]. Some of the most important methods are: interfacial polymerization, solvent evaporation/extraction, coacervation, spray drying, and fluid bed coating. Studies conducted to prove the effectiveness of the microencapsulation process have shown that this process has a very high potential to improve the beneficial properties of substances, resulting in new premium products [9]. Accordingly, numerous studies have been conducted to examine the effect of microencapsulation on the controlled release of substances and their permeation within the skin layers.

3 The Effect of Microencapsulation on the Controlled Release of Active Substances

One of the biggest challenges in the pharmaceutical industry is to ensure controlled delivery of substances [14]. If a prolonged effect is achieved, the efficiency of preparation is better and longer, leading to reduction in the application frequency [15].

Retinol as a derivative of vitamin A is often used as the main active ingredient in many skin preparations, primarily for the treatment of acne, psoriasis and the like [16]. Shields et al. [16] examined the effect of microencapsulation of retinol on its release. The capsules are made of silane monomers by a sol-gel polymerization process. The formulations made were compared with formulations containing Microsponge® particles. The release of retinol from silicone particles has been shown to be slower than that of retinol from Microsponge® particles. For formulations in which the retinol concentration is 0.2%, the release rate is slowed by 66.9%, while for a formulation in which the retinol concentration is 0.5%, the release rate is slowed by 14.1%. The faster release from Microsponge® particles is attributed to the larger size and porosity compared to those made from silane monomers whose pores could not be detected by microscopic analysis. These results indicate that it is possible to adjust the release of a substance, slower or faster, depending on the purpose of the formulation. Although Microsponge® technology is one of the market leaders, the good results achieved by encapsulation provide further opportunities for the incorporation of retinol and other molecules within silane microparticles.

Patel et al. [17] examined the properties of oxybenzone, a UV filter, when encapsulated in microspheres made of gelatin by the emulsification method. The results show that different ratios of oxybenzone and gelatin affect its release, so when the ratio of oxybenzone and gelatin is 1:6, the slowest release was achieved, and the fastest was in the ratio 1:2. This means that it is possible to control the release of oxybenzone by encapsulation, and that it primarily depends on the ratio of oxybenzone to the gelatin used. The higher the proportion of gelatin, the higher the density of the capsule matrix, thus prolonging the diffusion path that oxybenzone needs to cross in order to be released. The ability to control the release of oxybenzone is very important for its activity and efficacy as a sunscreen. The formulation with oxybenzone: gelatin in ratio 1:6 also showed the best efficacy. Oxybenzone is separated by microencapsulation, thus there is no direct contact with the skin, and the microspheres create a film that is additional physical barrier for UV rays.

Minoxidil in the form of minoxidil sulfate (MXS) in the form of ethanolic solution and hydrogel is widely used in the treatment of alopecia. However, it is considered that these formulations do not enable adequate delivery of minoxidil to the hair follicles and its release is not maintained. Therefore, efforts have been made to develop a formulation that will improve these properties and reduce the number of applications of these preparations [15]. Gelfuso et al. [15] examined how minoxidil encapsulation affects its release. The microparticles were made of chitosan by spray drying. The results indicate that the release of MXS from the microparticles is maintained for 9 h compared to the solution. 68% of MXS is released from the solution within 2 h, while the encapsulated MXS is released in the same amount within 12 h. Encapsulation has been shown to allow slower and controlled release of MXS relative to solution, due to swelling and gradual diffusion of MXS from the capsules. Since the preparation currently on the market requires application at least twice a day, encapsulation could potentially solve this problem, ensure the continuous release of MXS for a longer period of time and at the same time maintain the therapeutic concentration in the hair follicle. This would reduce the required number of applications, and thus increase patient compliance.

Haddadi et al. [18] examined the effect of microencapsulation on the release of urea from the formulation. The microparticles were made of PLGA by the solvent evaporation method, after which they were incorporated into the O/W and W/O cream and the comparison was made between the creams containing free and encapsulated urea. Encapsulation has been shown to allow controlled release. The maximum level of release of free urea was about 50–60% from O/W cream and 20–25% from W/O cream, while the release of encapsulated form was significantly lower. The release of urea from microparticles was studied during 24 h and during that time the released amount did not reach the plateau. Non-encapsulated urea from the cream is mostly released during the first half hour. After that, very little or no urea is released, so it is considered that such sudden release of free urea is the cause of skin irritation, which could be avoided if urea is encapsulated based on the obtained results.

4 Effect of Microencapsulation on Permeation of Active Substances

Butyl methoxydibenzoylmethane (BMDBM) as the most commonly used UVA filter tends to pass through certain layers of the skin, i.e. to penetrate within the stratum corneum and viable epidermis. UV filters should remain as much as possible on the surface of the skin where they exert their effect, so efforts are being made for the development of formulations that will not penetrate. Scalia et al. [19] examined the permeation of BMDBM in free form, in complex with hydroxypropyl-β-cyclodextrin (HPβ-CD), encapsulated/non-encapsulated in lipid microparticles (LM). The microparticles were prepared by the solution emulsification method. The results show different percentage of BMDBM permeated within the stratum corneum, so the formulation containing free BDMBM shows a permeation of 9.7% ± 2.5, while encapsulated BMDBM shows a significant increase in permeation of 15.1% ± 2.7, as well as BMDBM complexed with cyclodextrin where the percentage of permeation was 17.1% ± 3.2. On the other hand, in the case when the BMDBM/HP-β-CD complex was encapsulated, the permeation was reduced to 6.0% ± 1.5, which was the goal. Under normal conditions, cyclodextrin molecules as well as its complexes are by nature such that they cannot permeate through the skin. However, when a hydrophilic cyclodextrin is in contact with a carrier (in this case a test cream) containing water, it may change percutaneous diffusion due to a change in thermodynamic activity. This further explains the results showing greater permeation of the BMDBM-HP-β-CD complex which is not incorporated within the lipid microparticles and, thus, not separated from the carrier. By encapsulating or incorporating the BMDBM-HP-β-CD complex within lipid microparticles, a reduction in permeation is achieved. This is explained by the fact that the diffusion of complex BMDBM through lipid microparticles relative to the free form is difficult and thus more BMDBM is retained within the particles and prevented from penetrating within the stratum corneum [19]. The UV filter thus stays on the surface where it is its target site of action.

By the same principle, Scalia et al. [20] showed that the incorporation of UV filters, BMDBM and EHMC, within lipid microparticles contributes to reduced permeation.

In this case, the permeation into the stratum corneum was reduced for BMDBM from 30.3% in free form to 22.7% in LMs, and form EHMC from 32.4% to 25.3%, respectively. The release of UV filters from LMs made of different lipids (tristearin, glyceryl behenate and stearic acid) was also examined. The release from LM made of glyceryl behenate and stearic acid has been shown to be the slowest, and the release from LM from tristearin was the fastest. Accordingly, glyceryl behenate and stearic acid were selected as materials for making LM, whereas in the previous study [19] LM was made from tristearin. This could explain why an adequate reduction in permeation was not achieved in the first study, because LMs release tristearins faster than UV filters and thus provide the possibility of greater permeation. It is assumed that different degrees of permeation are caused by different affinity of UV filters for individual lipids. Successful encapsulation of UV filters within LMs achieves their retention on the skin surface, contributes to their protective effect, and reduces the possibility of toxicity [20].

Salicylic acid (SA) is mainly used in the production of cosmetics, but the problem that arises is its insufficient retention and reduced peeling effect, because as a small molecule it tends to easily penetrate deeper layers of the skin. In addition, the irritation it causes, which limits its use, is also a problem [21]. In this *in vitro* transdermal study, the retention of free and encapsulated SA in the skin and on its surface was examined. The microcapsules are made of acacia gum and maltodextrin, by the spray-drying method. In this study, the release of SA was also studied and while 92.1% of pure SA was released within 1 h, this value was less than 65.0% for SA microcapsules. Encapsulation has been shown to also allow a constant permeation rate that was consistently lower than pure SA within 19 h. In this way, it is possible to keep SA longer on the surface of the skin. Retaining microcapsules with SA on the skin surface provides greater efficiency, while reducing the possibility of irritation.

Vitamin C or ascorbic acid is often used to slows down skin aging and as a whitening agent. The positive effects on the skin are reflected in the fact that vitamin C has an antioxidant effect, stimulates collagen synthesis and suppresses melanin. Yet vitamin C is thermolabile substance that is poorly skin permeable which has been attempted to address by encapsulation [22]. The results of Gu et al. [22] show that 26.04 ± 8.38% of vitamin C in solid-lipid microparticles (SLM) was found in the skin compared to the control solution (0.33 ± 0.15%). At the same time, higher percentage of vitamin C incorporated in SLM permeated within the dermis (44.79 ± 5.60%) compared to aqueous solution (8.12 ± 3.27%). This effect is attributed to the macadamia oil used to make SLM because it is rich in unsaturated fatty acids. These acids can interact with intercellular lipids in the *stratum corneum* and thus form special pathways through which vitamin C penetrates more easily. Thus, the topical delivery of vitamin C is facilitated, both in the upper layers of the skin and in the dermis. The amount of vitamin C showed that prolonged permeation was achieved which would mean that SLMs could be used as systems for prolonged delivery. In this case, the effective concentration of vitamin C could be maintained for even longer than 24 h.

5 Conclusion

Attempts have been made to bring release and permeation under control using various encapsulation techniques. Microparticles show good properties in terms of ease of production, relatively high percentage of encapsulation efficiency, but also control of physicochemical parameters of the obtained microcapsules. When manufacturing microparticles, it is necessary to take into account the structural characteristics of the skin, i.e. the lipophilicity and hydrophilicity of its individual layers. Based on the physicochemical characteristics of the active substance, it is necessary to select the most suitable excipients, adjust their share, which will modulate the permeation extent and rate of the active substance depending on its target site of action. In addition, controlled release can be achieved, i.e. delivery of the active substance in sufficient (effective) concentration for prolonged time period. Thus, microencapsulation can control delivery in terms of either retention on the skin surface itself (reduction of skin accumulation, irritation and toxicity) or penetration into specific layers for poorly permeable substances (increased efficiency).

Conflict of Interest Declaration. The authors declare no conflict of interest.

References

1. Lane, M.E.: Skin penetration enhancers. Int. J. Pharm. **447**(1–2), 12–21 (2013)
2. Brown, M.B., Martin, G.P., Jones, S.A., Akomeah, F.K.: Dermal and transdermal drug delivery systems: current and future prospects. Drug Deliv. **13**(3), 175–187 (2006)
3. Gabard, B., Surber, C., Elsner, P., Treffel, P. (eds.): Dermatopharmacology of Topical Preparations. Springer, Heidelberg (2000)
4. Rosen, M.R.: Delivery System Handbook for Personal Care and Cosmetic Products: Technology, Applications, and Formulations. William Andrew Publishing, Norwich (2005)

5. Casanova, F., Santos, L.: Encapsulation of cosmetic active ingredients for topical application – a review. J. Microencapsul. **33**(1), 1–17 (2015)
6. Chang, R.-K., Raw, A., Lionberger, R., Yu, L.: Generic development of topical dermatologic products: formulation development, process development, and testing of topical dermatologic products. AAPS J. **15**(1), 41–52 (2012)
7. Jyothi, N.V.N., Prasanna, P.M., Sakarkar, S.N., Prabha, K.S., Ramaiah, P.S., Srawan, G.Y.: Microencapsulation techniques, factors influencing encapsulation efficiency. J. Microencapsul. **27**(3), 187–197 (2010)
8. Lengyel, M., Kállai-Szabó, N., Antal, V., Laki, A.J., Antal, I.: Microparticles, microspheres, and microcapsules for advanced drug delivery. Sci. Pharm. **87**(3), 20 (2019)
9. Dubey, R., Shami, T.C., Bhasker Rao, K.U.: Microencapsulation technology and applications. Def. Sci. J. **59**(1), 82–95 (2009)
10. Peanparkdee, M., Iwamoto, S., Yamauchi, R.: Microencapsulation: a review of applications in the food and pharmaceutical industries. Rev. Agric. Sci. **4**, 56–65 (2016)
11. Pech-Canul, A., de la, C., Ortega, D., García-Triana, A., González-Silva, N., Solis-Oviedo, R.L.: A brief review of edible coating materials for the microencapsulation of probiotics. Coatings **10**(3), 197 (2020)
12. Tomaro, D.C., Saha, S., Malhotra, M., Kahouli I., Prakash, S.: Microencapsulation for the therapeutic delivery of drugs, live mammalian and bacterial cells, and other biopharmaceutics: current status and future directions. J. Pharm. 1–19 (2013)
13. Singh, M., Dua, J.S., Menra, M., Soni, M., Prasad, D.N.: Microencapsulation and its various aspects: a review. Int. J. Adv. Res. **4**(6), 2094–2108 (2016)
14. Jyoti, J., Kumar, S.: Inovative and novel strategy: microsponges for topical drug delivery. J. Drug Deliv. Ther. **8**(5), 28–34 (2018)
15. Gelfuso, G.M., Gratieri, T., Simão, P.S., de Freitas, L.A.P., Lopez, R.F.V.: Chitosan microparticles for sustaining the topical delivery of minoxidil sulphate. J. Microencapsul. **2**(7), 650–658 (2011)
16. Shields, C.W., White, J.P., Osta, E.G., Patel, J., Rajkumar, S., Kirby, N., et al.: Encapsulation and controlled release of retinol from silicone particles for topical delivery. J. Control. Release **278**, 37–48 (2018)
17. Patel, M., Jain, S.K., Yadav, A.K., Godna, D., Agrawal, G.P.: Preparation and characterization of oxybenzone-loaded gelatin microspheres for enhancement of sunscreening efficacy. Drug Deliv. **13**(5), 323–330 (2006)
18. Haddadi, A., Aboofazeli, R., Erfan, M., Farboud, E.S.: Topical delivery of urea encapsulated in biodegradable PLGA microparticles: O/W and W/O creams. J. Microencapsul. **25**(6), 379–386 (2008)
19. Scalia, S., Coppi, G., Iannuccelli, V.: Microencapsulation of a cyclodextrin complex of the UV filter, butyl methoxydibenzoylmethane: in vivo skin penetration studies. J. Pharm. Biomed. Anal. **54**, 345–350 (2011)
20. Scalia, S., Mezzena, M., Ramaccini, D.: Encapsulation of the UV filters ethylhexyl methoxycinnamate and butyl methoxydibenzoylmethane in lipid microparticles: effect on in vivo human skin permeation. Skin Pharmacol. Physiol. **24**, 182–189 (2011)
21. Huang, Y., Li, Y., Fan, H., Xia, Q.: Preparation and characterization of salicylic acid-loaded microcapsules as delivery systems for cosmetics. Integr. Ferroelectr. **152**(1), 22–28 (2014)
22. Gu, C., Hu, C., Ma, C., Fang, Q., Xing, T., Xia, Q.: Development and characterization of solid lipid microparticles containing vitamin C for topical and cosmetic use. Eur. J. Lipid Sci. Technol. **118**(7), 1093–1103 (2016)

Assessment of Parameters for the Diagnosis of Insulin Resistance Using Artificial Neural Network

Indira Zuko(✉), Hana Turković, Andrea Šumic, Izudin Ugljanin, Anisa Tandir, and Denisa Tahirovic

Faculty of Pharmacy Sarajevo, University of Sarajevo, Sarajevo, Bosnia and Herzegovina
indirazuko@ffsa.unsa.ba

Abstract. Today insulin resistance has high prevalence, not only in elderly, but also in young population. The aim of this paper is to assess parameters in diagnosing insulin resistance using Artificial Neural Network (ANN). We used BMI, FPG, OGTT and A1C test as relevant parameters in diagnosing this disease. In this research we used standard architecture for pattern recognition in ANN method. We proved that this method has high accuracy in assessing above mentioned parameters and it is sufficient to properly perform the classification of input data and to diagnose insulin resistance.

1 Introduction

By its structure, insulin is a peptide hormone. It is secreted by the β cells of the pancreatic islets of Langerhans directly in the bloodflow. Its main role in organism is to maintain normal glucose level by facilitating cellular glucose uptake, regulating carbohydrate, lipid and protein metabolism and promoting cell division and growth through its mitogenic effects [1]. When the biologic response to insulin stimulation of target tissues is impaired states such as hyperglycemia, hypertension, dyslipidemia, visceral adiposity, hyperuricemia, elevated inflammatory markers, etc. occured. That is called insulin resistance. This health condition impairs glucose disposal, resulting in a compensatory increase in β-cell insulin production and hyperinsulinemia and can lead to metabolic syndrome, nonalcoholic fatty liver disease, and type 2 diabetes mellitus [2].

Insulin resistance can be hereditary, acquired and mixed. The majority of the patients with insulin resistance are in acquired category. The causes of acquired insulin resistance can be: aging, excess dysfunctional adipose tissue, physical inactivity, nutritional imbalance, medications like glucocorticoids, anti-adrenergic, protease inhibitors, atypical antipsychotics, and some exogenous insulin, increased sodium diets, glucose toxicity and lipotoxicity from excess circulating free fatty acids [2]. In skeletal muscles, insulin resistance is manifested as reduction in insulin-stimulated glycogen synthesis due to reduced glucose transport. An important role in inducing insulin resistance plays ectopic liquid accumulation. There are multiple defects in insulin signalling that are responsible for impaired glucose metabolism such as inflamatory cytokines and lipid metabolites [3].

A. Badnjevic and L. Gurbeta Pokvić (Eds.): CMBEBIH 2021, IFMBE Proceedings 84, pp. 528–534, 2021.
https://doi.org/10.1007/978-3-030-73909-6_61

For treating insulin resistance, first of all, patient must change his lifestyle. He must eat healthy, nutritional and low-calorie diet. It is known that carbohydrates stimulate excessive insulin demand, and they must be avoided as much as possible. Besides healthy diet, physical activity in recomended. It helps in increasing energy levels and improving muscle insulin sensitivity. Insulin response can be improved by certain medicines, such as: metformin, glucagon-like peptide one inhibitors, sodium-glucose cotransporter two inhibitors, thiazolidinediones and dipeptidyl peptidase-4 inhibitors [2].

One of the parameters used to assess insulin resistance is body mass index (BMI). It is the metric currently in use for defining anthropometric height/weight characteristics and then classifying them into the groups [4]. Another useful parameter in assessing insulin resistance is fasting plasma glucose (FPG). FPG is a parameter that is widely used in diagnostic and screening tests for diabetes. This parameter has more discriminatory power to distinguish between individuals at diabetes risk and those not at risk than post-load glucose values during oral glucose tolerance test (OGTT) and HbA1c test. OGTT is also widely used in clinical practice when determined weather someone suffers from diabetes or not [5]. HbA1c test is an important indicator of long- term glycemic control. It has the ability to reflect the cumulative glycemic history of the preceding two to three months. That is why this parameter is also used in this study for assessing insulin resistance [6].

Artificial neural network is a machine learning technique. It is powerful and flexible. ANN is an analogue to a human brain. In this technique input and output signals are being used. Signals are transmitter from input to output nodes. Before reaching output nodes, input signals are being weighted according to their respective importance. After that the combined signal is processed by activation function [7].

Machine learning techniques are being used more end more every day in biomedical sciences. They can be used in classification of diseases [8, 11–14, 16, 17], in prediction of machine performance [9, 10], in drug synthesis [15], etc.

The aim of this study was to evaluate the parameters: BMI, FPG, OGTT and A1C test, that are used to diagnose insulin resistance, in a selected group of subjects using artificial neural network (ANN). This is a method that is used in medical sciences more and more every day. It can detect relationships between variables in data that human brain may fail to detect.

2 Materials and Methods

This study started in September 2020 and lasted until November 2020. For the mentioned period, data from a total of 300 respondents were collected. Respondents, ie their data were grouped into two groups. The first group consisted of healthy individuals and this group was called healthy control, while the second group consisted of 200 sick individuals, ie subjects with insulin resistance. Given that there are many parameters that can affect the development of insulin resistance, for the purposes of this study we have selected six parameters, which have the most detrimental effects on health.

Besides gender and age, those parameters are:

- body mass index (BMI) - measured according to the formula,
- fasting plasma glucose (FPG) - glucose levels are measured on an empty stomach, at least 8 h after the last meal,
- oral glucose tolerans test (OGTT) - glucose levels are measured 2 h after taking glucose orally,
- A1C test - represents the average blood sugar level measured in the last 3 months.

Body mass index (BMI) was calculated according to a well-known formula:

$$\text{BMI} = \frac{\text{Weight (kg)}}{\text{Height (m}^2\text{)}} \quad (1)$$

The reference range for BMI is: underweight $< 18{,}5$; normal 18,5 - 24,9; overweight 25.0–29.9; obese $> 30{,}0$. The reference range for other three parameters are listed in Table 1.

Table 1. Reference value of FPG, OGTT, A1C

Parameter	FPG (mg/dL)	OGTT (mg/dL)	A1C (%)
Normal	<100	<140	<5,7
Prediabetes	100–125	140–199	5,7–6,4
Diabetes	>126	>200	>6,5

In Table 2 the dataset of participants in this study is shown.

Table 2. Dataset of participants

Gender		
Male	Female	Total
(127) 42,33%	(127) 57,66%	(300) 100%
Health status		
Healthy	Disease	
100	200	300
Age range		
18–29	30–49	50+
71,66%	28,33%	0%

3 Results

Neural network developed in this research consists of 6 inputs that are: gender, age, body mass index, fasting plasma glucose, oral glucose tolerans test and HbA1c test. As mentioned in Introduction part, these inputs (or biomarkers) are widely used to diagnose or predict development of insulin resistance. The network has one output and that is diagnosis of insulin resistance. Output class can have one of two possible values, healthy or disease.

The distribution of the dataset for development of the Artificial Neural Network (ANN) is presented in Table 3. Since this is relatively small dataset, distribution of 70:30% for training and validation was used [18].

Table 3. Dataset distribution

Total number of samples	Healthy	Disease
300	100	200
Training		
210	70	140
Validation		
90	30	60

For the development of the ANN in this research, standard architecture for pattern recognition was used. That is feedforward neural network which consists of two-layers. This architecture is designed since according to the application experts, that kind of ANN is sufficient to properly perform the classification of input data [8].

The input layer is made of network inputs. Input layer is followed by a hidden layer which consists of parallely connected neurons. The network was tested on training performance with different number of neurons. Since, this classification is basically highly non-linear data classification, sigmoid function was used for hidden layer [12].

During the design of the network, different number of neurons was used in order to measure the performance. According to the Mean Square Error (MSE), appropriate number of neurons was chosen. After training, validation of the network performance was done using confusion matrix.

According to these results, specificity, sensitivity and accuracy of the network are calculated according to Eqs. (1), (2) (3):

$$sensitivity = \frac{true\ positive}{(true\ positive + false\ negative)} \tag{2}$$

$$specificity = \frac{true\ negative}{(true\ negative + false\ positive)} \tag{3}$$

$$accuracy = \frac{(true\ negative + true\ positive)}{(true\ negative + true\ positive + false\ negative + false\ positive)} \tag{4}$$

During development, different number of neurons was tested. The results of training for different number of neurons in hidden layer are presented in the Table 4 below.

Table 4. Training and validation of neural network

Number of neurons	Accuracy training	Accuracy validation
2 neurons	96.62	96.8
5 neurons	99.0	99.8
15 neurons	99.05	99.3

As it can be seen from the training process, neural network with 5 neurons shows the best results. Validation of such architecture is shown in Table 5.

Table 5. Artificial Neural Network performance validation

n=120	**GROUP 1 Healthy** ANN prediction	**Group 2 Disease** ANN prediction	
GROUP 1 Healthy 30	TP 28	FN 2	**Accuracy** [%] 96.7%
Group 2 Disease 60	FP 1	TN 59	
	Sensitivity [%] 93.3%	**Specificity** [%] 98.3	

As seen in Table 5 the accuracy of this method is satisfactory, 96,7%. Sensitivity is 93,3% and specificity is 98,9%. That means that every participant identified as true positive (or has an insulin resistance) is really positive, and other way around, every true negative participant is really healthy (or negative to insulin resistance).

4 Conclusion

There are, besides genetics, many causes or insulin resistance. Due to modern lifestyle, prevalence of insulin resistance is getting higher every day. Lifestyle changes are first recommended to treat insulin resistance. This study is using artificial intelligence method-artificial neural network in order to evaluate parameters in diagnosing insulin resistance: BMI, FPG, OGTT and A1C test, in a selected group of subject using Artificial Neural Network. Since the accuracy of this method is high, this method is sufficient to properly perform the classification of input data and to diagnose insulin resistance using above mentioned parameters. This paper provides yet another evidence that ANN method is very useful in assesing diagnostic parameters.

References

1. Wilcox, G.: Insulin and insulin resistance. Clin. Biochemist. Rev. **26**(2), 19–39 (2005)
2. Freeman, A.M.: Insulin resistance, In: StatPearls [Internet]. StatPearls Publishing, Treasure Island (FL) (2020)
3. Sesti, G.: Pathophysiology of insulin resistance. Best Pract. Res. Clin. Endocrinol. Metab. **20**(4), 665–679 (2006)
4. Nuttall, F.Q.: Body mass index: obesity, BMI, and health: a critical review. Nutr. Today **50**(3), 117–128 (2015)
5. Janghorbani, M., Amini, M.: Normal fasting plasma glucose and risk of prediabetes and type 2 diabetes: the Isfahan diabetes prevention study. Rev. Diabetic Stud. RDS **8**(4), 490–498 (2011)
6. Sherwani, S.I., Khan, H.A., Ekhzaimy, A., Masood, A., Sakharkar, M.K.: Significance of HbA1c test in diagnosis and prognosis of diabetic patients. Biomarker Insights **11**, 95–104 (2016)
7. Zhang, Z.: A gentle introduction to artificial neural networks. Ann. Transl. Med. **4**(19), 370 (2016)
8. Catic, A., Gurbeta, L., Kurtovic-Kozaric, A., Mehmedbasic, S., Badnjevic, A.: Application of neural networks for classification of patau, edwards, down, turner and klinefelter syndrome based on first trimester maternal serum screening data, ultrasonographic findings and patient demographics. BMC Med. Genom. **11**, 19 (2018). https://doi.org/10.1186/s12920-018-0333-2
9. Badnjevića, A., Pokvić, L.G., Hasičić, M., Bandić, L., Mašetić, Z., Kovačević, Ž., Kevrić, J., Hifziefendić, J., Pecchia, L.: Evidence-based clinical engineering: Machine learning algorithms for prediction of defibrillator performance. Biomed. Signal Process. Control **54**, 101629 (2019)
10. Kovačević, Z., Pokvić, L.G., Spahić, L., Badnjević, A.: Prediction of medical device performance using machine learning techniques: Infant incubator case study Health and Technology. Health and Technology
11. Sarić, R., Jokić, D., Beganović, N., Pokvić, L.G., Badnjević, A.: FPGA-based real-time epileptic seizure classification using Artificial Neural Network. Biomed. Signal Process. Control **62**, 102106 (2020). ISSN 1746-8094
12. Badnjevic, A., Gurbeta, L., Cifrek, M., Marjanovic, D.: Classification of asthma using artificial neural network. In: IEEE 39th International Convention on Information and Communication Technology, Electronics and Microelectronics (MIPRO), 30 May–03June 2016. Opatija, Croatia (2016)
13. Aljovic, A., Badnjevic, A., Gurbeta, L.: Artificial neural networks in the discrimination of Alzheimer's disease using biomarkers data. In: IEEE 5th Mediterranean Conference on Embedded Computing (MECO), 12–16 June 2016, Bar, Montenegro (2016)
14. Alic, B., Sejdinovic, D., Gurbeta, L., Badnjevic, A.: Classification of stress recognition using artificial neural network. In: IEEE 5th Mediterranean Conference on Embedded Computing (MECO), 12–16 June 2016, Bar, Montenegro (2016)

15. Veljovic, E., Spirtovic – Halilovic, S., Muratovic, S., Osmanovic, A., Badnjevic, A., et.al.: Artificial neural network and docking study in design and synthesis of xanthenes as antimicrobial agents. In: CMBEBIH 2017. IFMBE Proceedings, vol. 62, pp. 617–626. Springer, Singapore (2017). https://doi.org/10.1007/978-981-10-4166-2_93
16. Sejdinovic, D., Gurbeta, L., Badnjevic, A., Malenica, M., Dujic, T., Causevic, A., Bego, T., Divovic, L.: Classification of prediabetes and Type 2 diabetes using artificial neural network. In: CMBEBIH 2017. IFMBE Proceedings, vol. 62, pp. 685–689. Springer, Singapore (2017). https://doi.org/10.1007/978-981-10-4166-2_103
17. Alić, B., Gurbeta, L., Osmanovic, A., Badnjević, A.: Machine learning techniques for classification of diabetes and cardiovascular diseases. In: 2017 6th Mediterranean Conference on Embedded Computing (MECO), Bar, Montenegro, pp. 1–4 (2017)
18. Rau, H.-H., et al.: Development of a web-based liver cancer prediction model for type II diabetes patients by using an artificial neural network. Comput. Methods Programs Biomed. **125**, 58–65 (2016)

RP-HPLC Determination of Lipophilicity in Series of Corticosteroids

Minela Dacić[1](✉), Alija Uzunović[2], Saša Pilipović[2], Larisa Alagić-Džambić[3], and Kemal Durić[4]

[1] Faculty of Pharm and Health, Slavka Gavrančića 17c, Travnik, Bosnia and Herzegovina
[2] Agency for Medicinal Products and Medical Devices, Titova 9, Sarajevo, Bosnia and Herzegovina
[3] Quality Assurance and Quality Control Department, Bosnalijek, Jukićeva 53, 71000 Sarajevo, Bosnia and Herzegovina
[4] Faculty of Pharmacy, University of Sarajevo, Zmaja od Bosne 8, Sarajevo, Bosnia and Herzegovina

Abstract. The adrenal cortex secretes hormones called corticosteroids, which are divided into: mineralocorticoids and glucocorticoids. They are used for the controlled treatment of various dermatological diseases, since they have shown good results. Lipophilicity is a physicochemical feature of a compound that describes its ability to dissolves in fats, oils and non-polar solvents such as hexane or toluene. For drugs it is very important property because it affects their pharmacokinetics, pharmacodynamics and toxicology profile, ie solubility, absorption, distribution, metabolism, elimination, binding to plasma proteins and interaction with the receptor. The main method for determination of lipophilicity is extraction in two different solvents (n-octanol and water), but in this study, we used two different HPLC-UV methods for determination the correlation of this parameter vs retention time. The HPLC-UV manual injection apparatus was used during the study, using a ZORBAX Eclipse XDB-C18, 150 × 4.6 mm, 3.5 µm and Zorbax Phenyl, 250 × 4.6 mm, 5.0 µm columns, and corticosteroids: alclomethasone dipropionate, betamethasone dipropionate, dexamethasone, mometasone, mometasone furoate, hydrocortisone and clobetasole propionate. The results showed that both methods have excellent results in correlation of retention time vs ACD/LogP and retention time vs ClogP but method witch use ZORBAX Eclipse XDB-C18 column are better because of shorter analysis run time.

Keywords: Corticosteroids · HPLC-RP · Lipophilicity

1 Introduction

The biological activity of a drug substance is conditioned by its physicochemical characteristics. The processes of pharmacokinetics and pharmacodynamics depend of the ability of molecules to interact with two different media: non-aqueous (cell membrane) and aqueous (cell interior). Interaction property with aqueous and nonaqueous media

A. Badnjevic and L. Gurbeta Pokvić (Eds.): CMBEBIH 2021, IFMBE Proceedings 84, pp. 535–540, 2021.
https://doi.org/10.1007/978-3-030-73909-6_62

characterized is the notion of lipophilicity [1]. Lipophilicity is usually expressed by logarithm partition coefficient (log P) which describes the equilibrium distribution of the dissolved compound between water and a water-immiscible organic solvent [2]. Due to the need for fast methods that could, in a short time, provide insight into the pharmacokinetic parameters of a large number of molecules, for the purpose of testing the lipophilicity are introduced chromatographic techniques. The retention of the analyte in the reverse-phase liquid and thin layer chromatography was shown to be associated with lipophilicity and that the results correlate well with by the n-octanol/water partition coefficient [3]. This is the reason why we developed and compared two different HPLC-UV methods for determination how the corticosteroid retention time depends of lipophilicity.

For this purpose, we analyzed seven different corticosteroids which results will be showed in next passuses.

2 Materials and Methods

Standards and chemicals:

- Alclometasone dipropionate RM, Batch No: RS-26080717
- Betamethasone dipropionate RM, Batch No: RS-10510217
- Dexamethasone CRM, Sigma-Aldrich, Lot: LRAA9235
- Hydrocortisone CRM, Sigma-Aldrich, Lot: LRAB3713
- Mometasone furoate RM, Merck, Lot: L-000386149-000Y040
- Clobetasole propionate, ointment

Solvents:

- Methanol RS, LabExpert, Lot: 27.2211106
- Acetonitrile HPLC grade, Termo Fischer Scientific, Lot: 1736152
- Formic acid HPLC grade, Kemika Zagreb, Lot: 920244
- Ultrapure water, Genom

First Method

Standard Preparation

Stock standard solutions were prepared by weighing 10 mg of standard in a 10-mL volumetric flask and dissolved in methanol; dilutions were prepared with 50% methanol. After identification of retention times for every single corticosteroid, there was made mix standard which contained all corticosteroids.

Operating Condition

The method is based at the procedure described in work published in JCS [4]. The first method used reversed-phase Zorbax Phenyl, 250 × 4.6 mm, 5.0 μm column at a wavelength of 240 nm. Injecting volume was 20 μL and the oven temperature was 45 °C. The mobile phase was water, HCOOH 0.1% and acetonitrile, HCOOH 0.1% with a flow rate fixed at 1.0 mL/min.

Elution:

Time (min)	Water, HCOOH 0.1% (% v/v)	Acetonitrile, HCOOH 0.1% (%v/v)
0	95	5
2	95	5
7	60	40
25	60	40
45	30	70
46	95	50
50	95	5

Second Method

Standard Preparation

Stock standard solutions were prepared by weighing 10 mg of standard in a 10-mL volumetric flask and dissolved in methanol; dilutions for peak identification and determination were prepared with a mixture of acetonitrile:water:acetic acid = 60%:40%:0.1%. After identification of retention times for every single corticosteroid, there was made mix standard which contained all corticosteroids.

Operating Conditions

The isocratic method used in this investigation was combination of three methods in works: Rahmayuni at all (2018), Matičević (2017) and Uzunović (2006) [5–7]. For this method we used a ZORBAX Eclipse XDB-C18, 150 × 4.6 mm, 3.5 μm column at 254 nm. Injecting volume was 20 μL and the oven temperature was 25 °C. The mobile phase was acetonitrile:water = 55:45 with a flow rate fixed at 1.0 mL/min.

3 Results and Discussion

3.1 First Method

Under the developed HPLC conditions, retention times ranged from 12.16 to 40.10 min and the peaks symmetry factor were in range 0.77 to 1.46. Calculated ACD/LogP were in range from 1.43 to 4.42, while CLogP were in range 1.79 to 4.12.

Levels of correlation between retention times and log P values; and between retention times and ACD/LogP values calculated by chemical softwares (ChemDraw Professional 16.0, ClogP and ACD/logP) are provided in Fig. 1 and 2. [8].

The correlation coefficient for this investigation were 0.9804 and 0.9560 which shows that this method has very good linearity which means: if the drug has bigger lipophilicity, the retention time will be higher. Also, if we know the lipophilicity of some drug, we can predict the retention time for it which is very useful when we don't know which retention time could we expect.

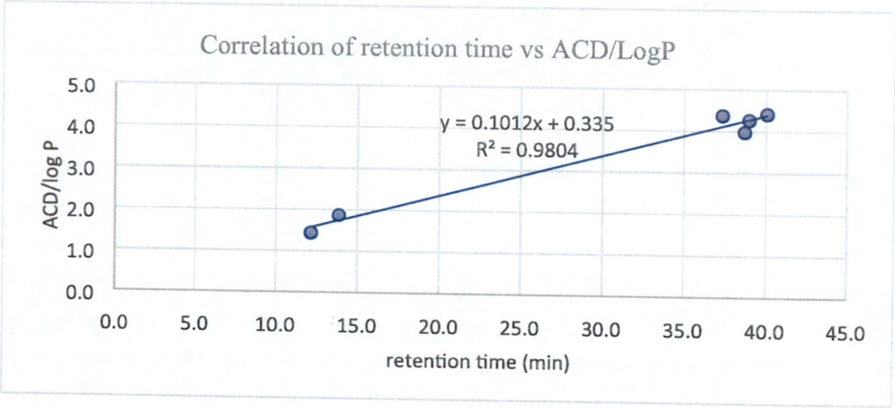

Fig. 1. Correlation of retention time vs ACD/LogP

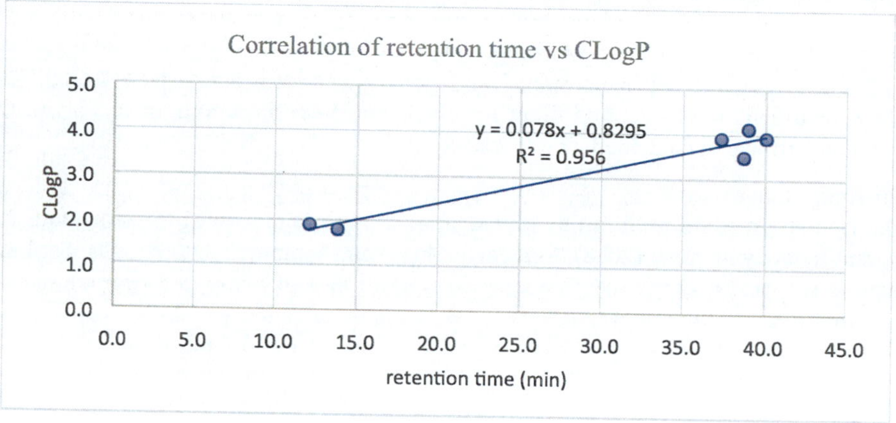

Fig. 2. Correlation of retention time vs CLogP

3.2 Second Method

Under the developed HPLC conditions, retention times ranged from 1.76 to 9.75 min and the peaks symmetry factor were in range 0.93 to 0.98. Calculated ACD/LogP were in range from 1.43 to 4.42, while CLogP were in range 1.79 to 4.12.

Levels of correlation between retention times and log P values; and between retention times and ACD/LogP values calculated by chemical softwares (ChemDraw Professional 16.0, ClogP and ACD/logP) are provided in Fig. 3 and 4.

The correlation coefficient for second investigation were 0.9680 and 0.9584 which shows that this method has very good linearity too. That means: if the drug has bigger lipophilicity, the retention time will be higher. Like in first investigation, if we know the lipophilicity of some drug, we can predict the retention time for it. This method has shorter run time and is easier for performing so it has to be first choice if we have to choose.

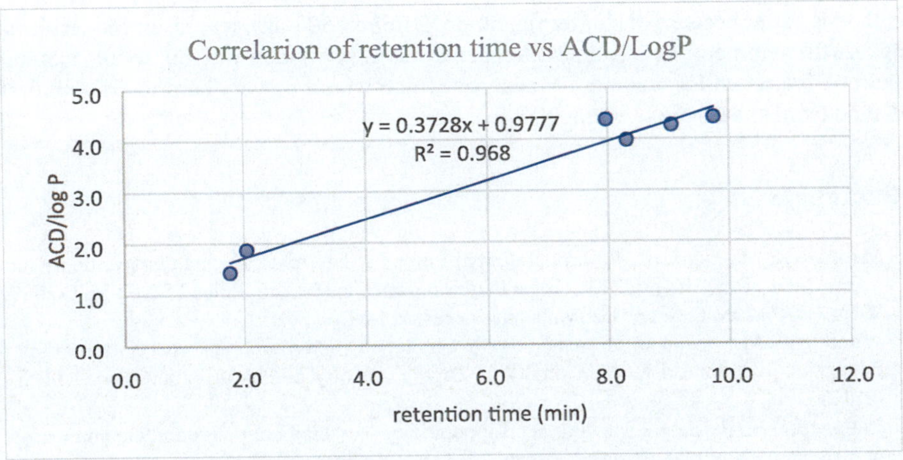

Fig. 3. Correlation of retention time vs ACD/LogP

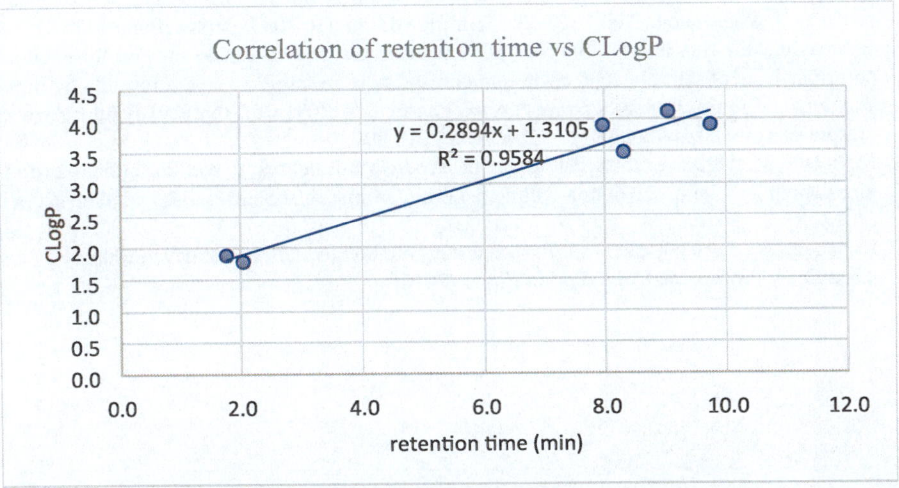

Fig. 4. Correlation of retention time vs CLogP

4 Conclusion

There was analyzed seven different corticosteroids by two different HPLC-UV methods and results of retention times were used for determination of correlation between retention time vs ClogP and retention time vs ACD/LogP. Each method gives excellent results where the correlation coefficient was between 0.9560 and 0.9804. From this correlation coefficients, there is obvious that methods have very good linearity for these parameters and they could be used for predicting a retention times for other corticosteroids using their lipophilicity value. Comparing these methods, there should be told that the second method is better because of shorter run time (12 min vs 50 min), it is easier for performing and the symmetry factors are better; peaks has very good resolution, so this method could be improved for quantitative analysis which would be very hard to do with first method (some peaks are overlapping).

References

1. Marinković, M., Đorđević, S.: Lipophilicity of drugs – determination and therapeutic importance. Med. Rev. **5**(4) (2013). https://www.md-medicaldata.com/files/15-MD-Vol%205%20No%204%20Marinkovic%20Maja.pdf. Accessed 20 Dec 2020
2. Antolković, M.: Primjena kromatografskih tehnika za predvi anje apsorpcije azatioprina i tiogvanina – diplomski rad. University of Zagreb, Faculty of Pharmacy and Biochemistry (2017)
3. Rutkowska, E., Pajak, K., Jóźwiak, K.: Lipophilicity – methods of determination and its role in medicinal chemistry. Acta Pol. Pharm. **70**, 3–18 (2013)
4. Gimeno, P., Françoise Maggio, A., Bancilhon, M., Lassu, N., Gornes, H., Brenier, C., Lempereur, L.: HPLC–UV method for the identification and screening of hydroquinone, ethers of hydroquinone and corticosteroids possibly used as SkinWhitening agents in illicit cosmetic products. J. Chromatogr. Sci. 1–10 (2015). https://doi.org/10.1093/chromsci/bmv147
5. Rahmayuni, E., Harmita, H., Suryadi, H.: Development and validation method for simultaneous analysis of retinoic acid, hydroquinone and corticosteroid in cream formula by high-performance liquid chromatography. J. Appl. Pharm. Sci. **8**(09), 087–092 (2018). https://www.japsonline.com. https://doi.org/10.7324/JAPS.2018.8913
6. Matičević, M.: Utjecaj ekstrakcijskog otapala na odre ivanje odabranih kortikosteroida u prirodnim dermatološkim proizvodima – diplomski rad, Farmaceutskobiokemijski fakultet, Zagreb (2017)
7. Uzunović, A.: Specijalistički rad, Zavod za kontrolu lijekova FBiH, Sarajevo (2006)
8. ChemDraw Professional 16.0, Perkin Elmer (2016)

In Vitro Evaluation of Three Brands of Salicylic Acid Plasters from Bosnian Markets

Alija Uzunović[1], Minela Dacić[2](✉), Saša Pilipović[1], Larisa Alagić-Džambić[3], Kemal Durić[4], and Hurija Džudžević-Čančar[4]

[1] Agency for Medicinal Products and Medical Devices, Titova 9, Sarajevo, Bosnia and Herzegovina
[2] Faculty of Pharm and Health, Slavka Gavrančića 17c, Travnik, Bosnia and Herzegovina
[3] Quality Assurance and Quality Control Department, Bosnalijek, Jukićeva 53, 71000 Sarajevo, Bosnia and Herzegovina
[4] Faculty of Pharmacy, University of Sarajevo, Zmaja od Bosne 8, Sarajevo, Bosnia and Herzegovina

Abstract. Medicated plasters are solid pharmaceutical dosage form consisting of a flexible preparation with an adhesive layer containing one or more active substances. Salicylic acid plasters are a medicated plaster for removal of corns. The plaster contains salicylic acid localized and targeting the corn area, without affecting the surrounding tissue. The aim of this study was to evaluate differences in the dissolution characteristics of salicylic acid plasters commercially available on the local market (formulation I, II and III).

USP Apparatus 5 (Paddle Over Disc) is used for release testing at fixed rotation speed (100 rpm). The plaster was exposed to 300 mL of phosphate buffer solution pH 5.0. All dissolution studies were carried out at 32 ± 0.5 °C and samples were collected at various time intervals (30, 120, 180, 1560 and 2880 min) and analyzed for salicylic acid content using optimized and validated high-performance liquid chromatography (HPLC). The mobile phase consisted of a mixture of methanol: KH_2PO_4 buffer pH 2.0 60: 40 v/v.

Although several apparatus and procedures have been used to study in vitro release characteristics of medical plasters, USP Apparatus 5 could be considered as a discriminatory test that would be able to detect the differences in the dissolution rate of all tested salicylic acid plasters over a wide range sampling intervals (0.5–48 h).

Keywords: Salicylic acid · Plasters · Dissolution · USP Apparatus 5

1 Introduction

Salicylic acid is widely used as the active ingredient in nonprescription liquids, gels and plasters for the home treatment of warts, corns and calluses. It softens the epidermis, allowing the corn to be removed easily, is inexpensive but must be used with caution. Plasters are intended to afford protection and support an occlusion and macerating action and to bring medication into close contact with the skin. There is evidence to suggest

A. Badnjevic and L. Gurbeta Pokvić (Eds.): CMBEBIH 2021, IFMBE Proceedings 84, pp. 541–546, 2021.
https://doi.org/10.1007/978-3-030-73909-6_63

that plasters containing salicylic acid for the treatment of corns have been available for nearly 200 years and continue to be used within the general population as a self-treatment. Corns are a common foot problem, however there is little evidence to indicate which current treatments provide long term resolution [1, 2].

In vitro drug release test from medicated plasters is a widespread method for measuring and determining the release variability between products and these tests are proposed by regulatory bodies. For the dissolution rate test, a paddle over a disc, a paddle over the cell and a rotating cylinder method may be used, which depends on the composition, size and shape of the plaster. The paddle over disk method is considered the USP standard method for testing the release rate of the active drug substance from the medicated plasters. This method is simple, does not require special equipment and is most similar to the methods used to determine the rate of dissolution of the active substance from other dosage forms such as tablets or capsules.

The aim of this study was to compare the dissolution rate from salicylic acid plasters available on the local market (formulation I, II and III). Drug dissolution testing is routinely used to provide critical *in vitro* drug release information for quality control purposes, to assess batch-to-batch consistency and drug development. There are situations where dissolution testing plays a main role: formulation and optimization decisions, during product development, for products where dissolution performance is a critical quality attribute.

2 Materials and Methods

2.1 Chemicals

Salicylic acid (USP reference standard) was purchased from Sigma-Aldrich, Germany, Potassium phosphate dibasic (KH_2PO_4) from Fisher Scientific, USA, H3PO_4 and methanol HPLC grade ($\geq$99,9%) from Honeywell Riedel-de Haën, Germany.

2.2 Salicylic Acid Plasters

The extraction procedure was carried out in an ultrasonic bath with methanol at 25 °C. Mean values for each formulation (n = 10) was obtained after 240 min of sonication. Formulation I contains 4.34 ± 1,04 mg of salicylic acid/plaster, formulation II contains 5.60 ± 0,04 mg of salicylic acid/plaster and formulation III contains 18.77 ± 0,11 mg of salicylic acid/plaster.

2.3 Dissolution Apparatus

USP Apparatus 5 (Paddle Over Disc) is used to measure the salicylic acid dissolution rate from a plaster. The rate of dissolution of salicylic acid from the plasters is measured using a 125 μm stainless steel mesh disc that holds the plaster straight at the bottom of the dispenser vessel. The area from which salicylic acid is released is turned upward. During the test, a distance of 25 ± 2 mm is maintained between the disk surface and the paddle. Each of the six dissolution vessels was filled with 300 mL of 300 mL of KH_2PO_4 buffer pH 5.0 at 32 °C ± 0.5 °C. Rotation of paddle was fixed at 100 rpm. Samples (5 mL of dissolution media) were collected at various time intervals (0.5, 2, 6, 26 and 48 h).

2.4 High Performance Liquid Chromatography (HPLC)

Salicylic acid was determined by High-Performance Liquid Chromatography (HPLC, Shimadzu, Japan) system consisting of solvent delivery system, an autosampler, pump and UV-VIS detector. The performed separation method for analysis is previously described [3, 4]. Salicylic acid was detected at 238 nm. The separation was carried out with a LiChrospher 100, RP-18 endcapped column (125 × 4 mm, 5 μm) at room temperature using a mobile phase consisting of a mixture of methanol: KH_2PO_4 buffer pH 2.0 (60:40 v/v). The injection volume was 10 μl and flow rate 1.8 mL/min. Samples for HPLC analysis were filtered through membrane filter prior to injection.

3 Results and Discussion

In the first step, the HPLC method for the identification and determination of salicylic acid in medicated plasters was validated (specificity, linearity and repeatability of the method). We made two calibration curves because of different concentrations of salicylic acid in three samples. Concentration ranges were 0.002–0.32 mg/ml and 0.00010.008 mg/ml. Sampling was done after different times of dissolution when was continued with HPLC analysis. Results show that sample 1 was released 11.78% of extracted salicylic acid after 48 h, 98.45% and 87.85% after 48 h from sample 2 and 3 respectively.

Based on the retention time for salicylic acid (3.43 min) in the standard solution and samples confirmed the specificity of the analytical procedure using the test conditions. Linearity was confirmed using different concentration (0.0001 to 0.32 mg/mL) of standard solutions of salicylic acid. Using the correlation coefficient (r), linearity ($r \geq 0.995$) was confirmed. The repeatability of the method is confirmed based on the values for a relative standard deviation (RSD) ≤ 2.0% obtained by measuring the intensity of absorbance [mAU] of 0.05 mg/mL salicylic acid solution in 10 parallel measurements. For formulation I, II and III, salicylic acid plasters ($n = 10$) were taken at random and weight individually. Each of the ten samples were weight individually and the average content of salicylic acid calculated (Table 1). The results of content uniformity are summarized in Table 1.

Figure 1 shows the *in vitro* dissolved vs extracted amount of salicylic acid from three medicated plasters tested under the following conditions: volume of dissolution media

Table 1. Content uniformity of salicylic acid plasters (formulation I, II and III)

Sample 1 #	RT	Area	Content SA (mg)	Sample 3 #	RT	Area	Content SA (mg)
1	3,417	82,2	**4,62**	1	3,423	321,7	**18,348**
2	3,42	75,7	**4,25**	2	3,423	358,7	**20,457**
3	3,419	76,7	**4,31**	3	3,425	318,4	**18,160**
4	3,417	108,8	**6,10**	4	3,425	353,5	**20,161**
5	3,422	80,8	**4,54**	5	3,429	319,2	**18,205**
6	3,426	49,3	**2,77**	6	3,43	335,2	**19,117**
7	3,424	73,8	**4,14**	7	3,432	322,2	**18,376**
8	3,424	79,1	**4,44**	8	3,433	317,5	**18,109**
9	3,426	67,3	**3,78**	9	3,436	322,7	**18,405**
10	3,426	79,2	**4,45**	10	3,438	322,4	**18,388**
AVG			**4,34**	**AVG**			**18,77**
STDEV			**8,21E-01**	**STDEV**			**8,59E-01**
RSD %			**18,92**	**RSD %**			**4,58**

Sample 2 #	RT	Area	Content SA (mg)
1	3,422	99,8	**5,700**
2	3,424	101,5	**5,797**
3	3,424	102,1	**5,831**
4	3,426	94	**5,369**
5	3,429	93	**5,312**
6	3,431	93	**5,312**
7	3,43	101,6	**5,802**
8	3,433	93,2	**5,323**
9	3,433	100,7	**5,751**
10	3,436	101,9	**5,819**
AVG			**5,60**
STDEV			**2,38E-01**
RSD %			**4,25**

300 mL, pH of dissolution medium 5.0, speed of paddle rotation 100 rpm and sampling time at the end of 48 h.

For formulation II and III, the amount of active ingredient released from medicated plasters were significantly higher (87.85–98.45%) compared to formulation I (11.78%). Also, for formulation II and III, relative standard deviation (RSD) of content uniformity ($n = 10$) were significantly lower (4.25–4.58%) compared to formulation I (18.92%).

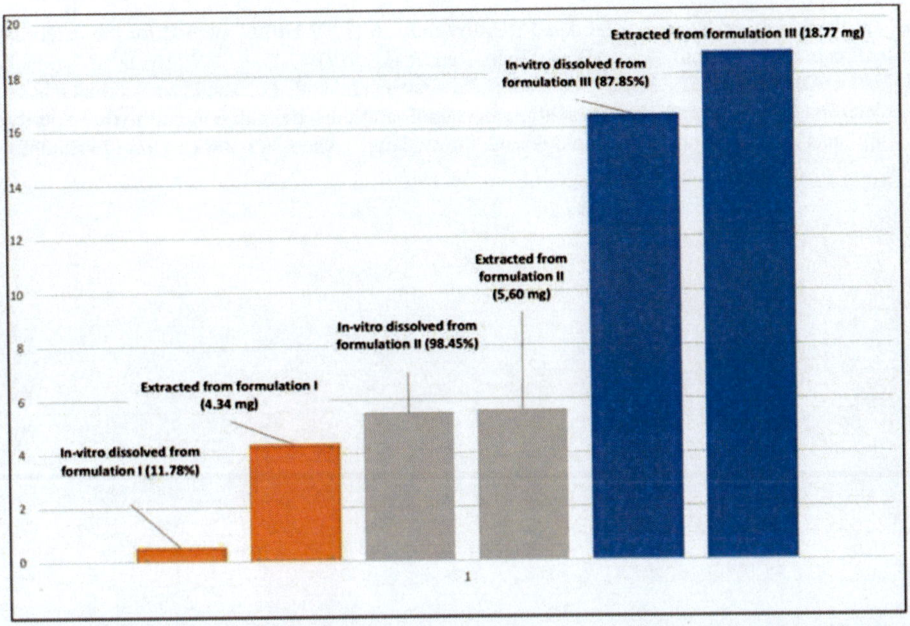

Fig. 1. Comparison of in *vitro* dissolved vs extracted amount of salicylic acid from three medicated plasters (formulation I, II and III)

4 Conclusions

USP Apparatus 5 (Paddle Over Disc method) is considered the USP standard method for testing the release of active substance from salicylic acid plasters. The paddle rotation speed was chosen in accordance with the pharmacopoeia requirements (100 rpm). At pH 5.0 the salicylic acid release rate from three medicated plasters was measured after 0.5–48 h. Formulation I has the lowest concentration of salicylic acid and the worse quality of plasters (the content and content uniformity of salicylic acid plasters). The results showed that the proposed *in vitro* method allows determination of the rate of salicylic acid release from plasters, thereby contributing to the quality control of commercially available salicylic acid plasters. This study highlights the need for constant market monitoring of new products, as well as existing products on market. These tests have shown that, different formulations from the market can be successfully compared and evaluated for quality.

References

1. Farndon, L.J., Vernon, W., Walters, S.J., Dixon, S., Bradburn, M., Concannon, M., Potter, J.: The effectiveness of salicylic acid plasters compared with "usual" scalpel debridement of corns: a randomised controlled trial. J. Foot Ankle Res. **6**, 40 (2013)
2. Stephenson, J., Farndon, L., Concannon, M.: Analysis of a trial assessing the long-term effectiveness of salicylic acid plasters compared with scalpel debridement in facilitating corn resolution in patients with multiple corns. J. Dermatol. **43**(6), 662–669 (2016)

3. Talath, S., Dhaneshwar, S.: Validated stability-indicating RP-HPLC method for the determination of salicylic acid. Am. J. PharmTech. Res. **7**(4) (2017)
4. Akay, C., Degim, I.T., Sayal, A., Aydin, A., Özkan, Y., Gul, H.: Rapid and simultaneous determination of acetylsalicylic acid, paracetamol, and their degradation and toxic impurity products by hplc in pharmaceutical dosage forms. Turk. J. Med. Sci. **38**(2), 167–173 (2008)

Health Sector Pricing and Financing Models: Challenges and Pharmacoeconomic Trends in Bosnia and Herzegovina

Srđan Lučić[1,2](✉), Lejla Gurbeta Pokvić[3,4], Ervina Bečić[1], and Almir Badnjević[1,3]

[1] Faculty of Pharmacy, University of Sarajevo, Sarajevo, Bosnia and Herzegovina
srdjan.lucic@roche.com
[2] Roche, Sarajevo, Bosnia and Herzegovina
[3] Verlab – Medical Devices Verification Laboratory, Sarajevo, Bosnia and Herzegovina
[4] International Burch University, Sarajevo, Bosnia and Herzegovina

Abstract. Scientific progress continues to deliver new treatments that prolong survival and significantly improve the quality of life of patients. Novel treatment options introduce new challenges for healthcare systems, often requiring novel pricing and payments models. It has been common practice that healthcare problems and challenges, both here and in the countries of the region, are addressed mainly by healthcare experts in broad terms, while interaction with experts in other fields has been rare; even when there is interaction, there are problems in achieving a clear common understanding. Indeed, this is not unique for just the health sector. In developed countries, including some countries of this region, important processes in the country and society are considered meticulously from different perspectives, including the economic perspective in particular, since the biggest challenge of the consistent implementation of defined policies is their stable financing. Needless to say, the health policy is among top priorities in every country by reason of its important share in GDP, overall budgets of national governments and the number of beneficiaries and its vital importance for the whole population.

Keywords: Health sector · Pricing models · Financing models · Pharmacoeconomic trends

1 Introduction

Novel treatments as a result of significant scientific advances offer benefits to patients but also pose new challenges that current health sector pricing and financing models struggle to address. Payers sometimes face uncertainty regarding the real-world value and extent of the benefits of treatments at the time of assessment [1]. Medicines may deliver significantly different value across a range of contexts, whereas current pricing and financing processes are often rigid in assigning a single composite value to a medicine. Not all healthcare systems are able to measure the degree to which a product is used

A. Badnjevic and L. Gurbeta Pokvić (Eds.): CMBEBIH 2021, IFMBE Proceedings 84, pp. 547–553, 2021.
https://doi.org/10.1007/978-3-030-73909-6_64

in one indication versus another, or in combination with other products, to provide a payment reflective of the number of patients treated by indication [2].

Thus, novel models of pricing and reimbursement may be required for better patient access that will overcome the hurdles and challenges of the existing models. Indication-based pricing [3] is a mechanism where the price of the product reflects its observed value across indications. Combination-based financing [4] seeks to resolve the additional complexities in assigning value and negotiating prices when different marketing authorization holders are responsible for individual medicines used in combinations. In outcomes-based financing [5], payment for a medicine is conditioned on its real-world performance through a financial structure and is related to observable outcomes. There are a range of outcomes-based payments tailored to particular concerns.

Coverage with evidence development allows access to the product on the condition that additional evidence is provided at a defined point in time, designed to address uncertainty at the time of evaluation. Also referred to as staggered payment [1], these allow for payers to make payments to manufacturers over fixed periods for each patient that receives therapy. Structuring payments this way helps mitigate the high upfront cost that would otherwise be associated with one-off therapies. When the overtime payment is linked to a particular outcome being observed, necessitating the collection of real-world outcomes data, these can also be a way for payers to address uncertainty around the clinical benefit due to the nature of the evidence available at launch. In some circumstances, these may be used to decouple payment for a treatment from the number of patients that receive the medicine. Because the payment is not in relation to the number of patients that underwent treatment, the predicted budget impact associated with treating patients for a area of disease could be more easily anticipated and avail the payers to ensure longterm sustainability. There are existing lump-sum payments, where a given volume is paid with a fixed amount.

In reality, these novel pricing and payment approaches are often combined into a novel pricing and payment model, depending on the challenges related to reimbursing the treatment. For example, anticipating multiple current and potential future indications of a given product, payers and companies have in the past reached multi-indication multi-year agreements [6]. Based on the experience of using these models today, there is already evidence to show how they can accelerate and broaden patient access, and contribute to improving the sustainability of healthcare systems. Nevertheless, countries in Europe have varying levels of experience [7]. To address the challenge of paying for the value of therapies with curative potential, pricing and reimbursement approaches have incorporated over-time payment elements. For example, over-time payments have been implemented in Italy and at the level of individual sick funds in Germany [8, 9]. Models focusing on subscription payments are rarely used in Europe today but there is an example in Denmark [10]. Beyond Europe, their application has been considered and some elements incorporated into agreements for Hepatitis C in some United States of America [11].

2 Challenges and Pharmacoeconomic Trends in Bosnia and Herzegovina

It must be noted that the health sector in Bosnia and Herzegovina (BH) has long been in the special focus of politics, the narrow professional community and the general public with an emphasis on the problems of its funding. It seems that stakeholders' public statements show that the focus is far sharper on revenue than it is on expenditure. Looking back (on the past ten years, for example), there has been a significant growth in health expenditure in BH.

Healthcare expenditures are growing fast and is greater in accordance to the receipts to the health Funds. The debt of the healthcare sector in BH is the reason for poor cash flow coming into the Health Insurance Funds from compulsory health insurance contributions by reason of a small number of employed insured persons and huge debts incurred with respect to unpaid employer contributions, largely by public companies and institutions [12]. This paper provides analysis of challenges and pharmacokinetic trends in healthcare sector in Bosnia and Herzegovina. Five key challenges have been analyzed.

2.1 Elder Population and Depopulation

An indicator of a certain increase in healthcare spending in the future is the current trend in ageing of the population, through the increase in healthcare costs and pharmaceutical consumption by the elderly population. The aging population could be not only one of the key causes of the increase in health care expenditures in BH so far, but also a clear indicator and alarm that warns that health care expenditures in BH will inevitably increase in the future. Healthcare expenses among the elderly, as a rule, are much higher than those among young people, while the distribution of healthcare spending by age is largely asymmetric. Over the last 4 decades, the average age of population in Bosnia and Herzegovina has increased significantly. According to the data of the BH Agency for Statistics [13], an average age of population in 1981 was 29.6 years, which placed BH among the countries with young population. Only 10 years later, according to the results of the 1991 census, an average age was 33 years. According to the latest, 2013 census, an average age is 39.5 years, while the people aged above 65 account for 14% of the total population [14].

Regarding the growing migration trend – people are leaving the country in search for jobs – needless to say, the young, working-age people are leaving the country, which, coupled with a negative natural population growth, has certainly changed the age structure of the population for the worse since the 2013 census [15]. If we use some developed countries in Europe for comparison with Bosnia and Herzegovina in regards to average age, BH has a better average age in comparison to Germany (46). However, Germany has an extremely high immigration rate (apart from the U.S. Germany has the highest number of immigrants on an annual basis, over 1.3 million a year), while BH has a galloping trend of brain drain which is doubled from 2010 to 2014 [14].

A worrying fact as well is a negative natural growth of population. The last positive natural growth of population, though insignificant (150), was recorded in 2008, when the number of deaths was 34,026 and the number of live births 34,176. According to the latest available data for 2018, there is a significant increase in deaths: the number of deaths in 2018 increased to 37,744, and the number of live births drastically declined to 29,467, which is a negative difference of 8,277 persons [15].

2.2 Technological Development and Investements in Healthcare

Although we are no longer among younger nations, there is no overuse of healthcare services. For example, an average number of primary health care visits per person in the BH Federation is 5.3 (in 2015) [15], including visits to primary care physicians and other medical staff, with differences among the cantons. Compared with the countries of the European region, inpatient healthcare indicators are also at lower levels. It is difficult to find precise statistics (summarized) on an increase in the number of healthcare services in BH over the past period due to insufficient transparency and public availability of data, but some indicators show that there has been no major increase in the number of services.

There is no significant impact on healthcare speding in BH by the technological development and investements in healthcare. On the contrary, investments in healthcare have seen a mild decrease. An increase in the number of healthcare services could not have a significant impact on an increase in healthcare spending since BH ranks far below the EU average volume of service provision nor is there a significant increase in the number of health services. Even though the development of technology in healthcare was occuring in our country, we can not conclude that it was a valuable contributor to the spending since there was no great leap forward in the development of technology in healthcare in BH. Also, investments in healthcare have declined noticeably.

2.3 Number of People Employed in the Healthcare Sector in BH

An increase in the number of people employed in the healthcare sector in BH could have a stronger impact on an increase in healthcare spending, and a particular problem in this regard is a high percentage of non-medical staff, medical administration and finances [16]. In view of a small number of physicians per capita in BH, the number of other medical staff, non-medical staff in particular, could be an issue. The share of non-medical staff in the healthcare system of the Federation of Bosnia and Herzegovina (FBH) is around 30%. According to the data of the Health Insurance and Reinsurance Fund of the BH Federation, in 2016, nonmedical staff accounted for 26% of the total health service staff. One year earlier, in 2015, medical staff accounted for 71.4% of the total health service staff and the rest were administrative employees, allied healthcare professionals and staff with lower levels of qualification. The European Union recommends that the share of non-medical staff should not exceed 10% [17]. The Ministry of the Health Federation BH says that over the last several years they have worked towards reducing the size of non-medical staff and that between 2010 and 2015 the number of health service workers went up by nearly 6%, while the number of allied healthcare professionals increased nearly 17% while the number of administrative and technical workers decreased by 3% [18].

2.4 Sustainability of the Health System

Along with depopulation caused by an extremely unfavorable migration trend in BH [19], a negative natural population growth, the strong need for an adequate employment policy and a solution to the financing of the pension system in BH (there is a causal link between all these issues), sustainability of the health system is probably the biggest challenge facing this country. Polemic, discussions, objections, accusations, dissatisfaction and other similar forms of public debate and advocacy of numerous health sector issues in BH emerge continuously, at frequent intervals.

The sector of pharmacy accounts for 1.28% of the total revenue of the whole economy of BH and for 0.85% of the total number of employees (without self-employed individuals) [20]. Over the past three years, 100 leading pharmaceutical industry companies in BH created 534 new jobs. Spending on medicinal products is continuously increasing both in BH and the world.

2.5 Financing of Healthcare System

There are a number of well-known models of financing health insurance, although, roughly, the health financing models are divided into two basic systems: the Bismarck system [21], which is based on compulsory health insurance, based on the principles of solidarity and mutuality, and the Beveridge system [21], in which the health sector is financed directly from the state budget, from the government taxes paid by all citizens. The Bismarck model which used in Bosnia and Herzegovina's healthcare financing system is based on compulsory social and health insurance (most of the healthcare expenses are financed in this manner, however, BH relies on a number of sources of healthcare financing, with a high level out-of-pocket payments).

Bosnia and Herzegovina is a kind of phenomenon in terms of health spending, health spending as a share of GDP and its growth trend. Despite the fact that the level of its development is incomparable to the levels of development of the EU Member States, and although the country is not among highly developed countries, its health spending growth in absolute values is higher than theirs, both percentage wise and as a share of GDP, and its growth - which is much larger than an average growth in Central and Eastern Europe - is the best ranked. Even the share of private health expenditure places the country on top of the list. According to the latest available national accounts' statistics of the BH Agency for Statistics [22], health expenditure in BH in 2013 was BAM 2,525 billion, while GDP was BAM 26,743 billion (9.84%), or BAM 675 per capita. Public spending was BAM 1,784 billion, and private spending was BAM 741,194 billion. In 2014, healthcare costs grew to minimum BAM 2,698 billion, or by BAM 165 million and in 2018 healthcare costs grew to BAM 2,972 billion (public spending 70%, private spending 30%) [22, 23]. The last official WHO demographic data shows following in BH: population 3 323 930; Health expenditure (as a proportion of GDP in 2016) = 9,23% [24].

3 Conclusions

Could be possible to slow down an increase in health spending in Bosnia and Herzegovina by cutting staff in the health sector, notably by reducing the number of nonmedical employees and medical administration and financing on the basis of a detailed analysis of the structure of employees in the health sector in BH, and through long-term policies for the suppression of risk factors to health (poverty, inadequate nutrition, pollution).

It is necessary to slow down the current population ageing trend through long-term strategies for retaining young population (like adequate employment strategies, etc.) and an increased birth rate.

The addresing of healthcare challenges associated with novel treatment and accelerated patient access can be supported through novel payment and pricing models. There is a need for additional enabling factors that would support the use of novel payment and pricing models beyond what the industry can commit to alone. Regulatory and procedural frameworks need to be in place to allow for the use of different novel pricing and payment approaches (such as combination-based pricing and over-time payments). This includes an openness to engage in a dialogue earlier on to ensure predictability for both parties and a willingness to adapt traditional pricing and reimbursement to facilitate the use of novel payment and pricing models when this would broaden and accelerate patient access.

The health sector as a whole and the availability of medicines in particular should be strongly improved in the european path of Bosnia and Herzegovina – in the contex of the fulfilment of the country's obligations on that path.

Conflict of Interest. The authors hereby declare to have no conflict of interest.

References

1. European Federation of Pharmaceutical Industries and Associations (EFPIA). Novel pricing and Payment Models (2020). https://efpia.eu/media/554543/novel-pricing-and-payment-models-new-solutions-to-improve-patient-access300630.pdf
2. Espin, J., et al.: Projecting pharmaceutical expenditure in EU5 to 2021: adjusting for the impact of discounts and debates. Appl. Health Econ Health Policy **16**(6), 803–817 (2018)
3. Yeung, K., Li, M., Carlson, J.J.: Using performance-based risk-sharing arrangements to address uncertainty in indication-based pricing. J. Managed Care Specialty Pharm. **23**(10), 1010–1015 (2017)
4. Cole, A.: What are the Economic Implications of Moving Away from Paying a Single Price for a Single Drug. OHE (2018)
5. Cole, A., Towse, A., Lorgelly, P., Sullivan, R.: Economics of innovative payment models compared with single pricing of pharmaceuticals. OHE Res. Pap. **18**, 4 (2018)
6. Danko, D.: Health technology assessment in middle-income countries: recommendations for a balanced assessment system, vol. 2. Taylor & Francis (2014)
7. European Federation of Pharmaceutical Industries and Associations (EFPIA). Healthier Future. The case for outcomes-based sustainable healthcare (2019)
8. Bruce, F.: Novartis strikes Kymriah outcomes-based deal with German insurers (2019). https://pink.pharmaintelligence.informa.com/PS124927/Novartis-Strikes-Kymriah-OutcomeBased-Deal-With-German-Insurers
9. Schofield, I.: Italy unveils new payment model for Novartis's CAR-T Kymriah (2019). https://pink.pharmaintelligence.informa.com/PS140701/Italy-Unveils-New-Payment-Model-For-Novartiss-CART-Kymriah

10. Bruce, F.: Vertex extends portfolio pricing and reimbursement approach to Denmark (2018). https://pink.pharmaintelligence.informa.com/PS124039/Vertex-Extends-Portfolio-Pricing-and-Reimbursement-ApproachTo-Denmark
11. Ferrario, A., et al.: The implementation of managed entry agreements in central and eastern Europe: findings and implications. Pharmacoeconomics **35**(12), 1271–1285 (2017). https://doi.org/10.1007/s40273-017-0559-4. PMID: 28836222; PMCID: PMC5684278
12. Bajić, B.: Thematic conference "Analysis of the state of integrity in the processes of drug availability in BiH", Healthcare Integrity initiative, Sarajevo, 14 June 2017
13. OECD, International Migration Outlook (2016). https://www.oecd.org/els/mig/international-migration-outlook-2016-statistical-annex.pdf
14. Union for Sustainable Return: Emigration of young people from BiH: Belly for bread, more and more mass. https://www.uzopibih.com.ba/2016-07-07.html
15. Agency for Statistics of BiH, Bosnia and Herzegovina in numbers (2019). https://bhas.gov.ba/data/Publikacije/Bilteni/2020/NUM_00_2019_TB_0_BS.pdf
16. Pobrić, R., Marilović, V.: Analysis of the state of integrity in the processes of drug availability in BiH (2017). https://netconsulting.ba/objavljujemodokument-analiza-stanja-integriteta-u-procesima-dostupnosti-lijekova-u-bih/
17. European Agency for Safety and Health at Work. https://osha.europa.eu/en/safety-and-health-legislation/european-guidelines
18. Journalist's Resource "Medical spending among the U.S. elderly". https://journalistsresource.org/studies/government/health-care/elderly-medicalspending-medicare/
19. Ministry of Security of BiH, Migration Profile of Bosnia and Herzegovina (2019). https://dijaspora.mhrr.gov.ba/wp-content/uploads/2020/07/Migracijski-profil-Bosne-i-Hercegovine-za-2019.-godinu.pdf
20. Valetudo Magazine, Review of the most successful companies in the pharmaceutical industry in Bosnia and Herzegovina (2016)
21. Health Care Systems – Four Basic Models | Physicians for a National Health Program. www.pnhp.org
22. Agency for Statistics of BiH, Statistics of National Health Accounts, Sarajevo, July 2020. https://bhas.gov.ba/data/Pub-likacije/Saopstenja/2020/NHA_01_2018_Y1_0_BS.pdf
23. The sum of calculations of health assets in the Federation for 2014 and the growth of spending of the RS Health Insurance Fund (2013–2014, excluding private out-of-pocket health spending in the RS) according to official data of the Federation and RS Funds for 2014
24. World Health Organization (WHO), Current status of health intervention and technology assessment in the Balkan region (2020). https://apps.who.int/iris/bitstream/handle/10665/336228/WHO-EURO-2020-1303-4105355733-eng.pdf

Biomaterials, Molecular, Cellular and Tissue Engineering

A Study of the $CaCl_2$ Induced *E. Coli* DH5-Alpha Transformation by Heat Shock Accompanied by Vibration

Lejla-Nur Smajović[1], Furkan Enes Oflaz[2], and Ayla Arslan[1,3](✉)

[1] Genetics and Bioengineering Program, Faculty of Engineering and Natural Sciences, International University of Sarajevo, Hrasnička cesta br. 15, 71210 Sarajevo, Bosnia and Herzegovina

[2] Gottfried Schatz Research Center for Cell Signaling, Metabolism and Aging, Medical University of Graz, Neue Stiftingtalstraße 6/6 8010, Graz, Austria

[3] School of Advanced Studies, University of Tyumen, Ulitsa 8 Marta, 2/1, Tyumen 625000, Russia

a.arslan@utmn.ru

Abstract. Chemical methods are commonly used for the induction of Escherichia coli (*E. coli*) competency to allow transformation mediated by heat shock, an essential procedure for gene cloning. Being the most labor and resource efficient chemical method, the 'calcium chloride' ($CaCl_2$) method involves a brief heat shock of bacteria in the presence of Ca^{2+} ions. Several optimizations were made on this original protocol but the effect of vibration or rotational shaking (revolution per minute) accompanied by the heat shock has not been studied. This paper reports the effect of shaking during or after the heat shock of competent DH5-Alpha strain of *E. coli,* induced by $CaCl_2$. Our results show that shaking during heat shock does not have any positive impact on the transformation efficiency. On the other hand, after the heat shock, 180 rpm shaking at 37 °C for 1 h has increased transformation efficiency significantly ($P < 0.05$).

Thus, a post-heat shock incubation with a rotational shaking attains higher yield of transformed cells affecting overall efficiency of transformation of *E. coli* DH5-Alpha.

Keywords: Escherichia coli · DH5-alpha · Chemical transformation · Heat shock · $CaCl_2$ method · Cloning

1 Introduction

Since its discovery by Griffith in 1928, the transforming principle, the process by which the competent bacteria are able to take up foreign DNA, has been adopted to laboratory applications increasingly. This became especially fruitful when bacteria, such as *E. coli*, lacking natural competence, were induced their competence artificially. Artificial induction of competence and transformation is performed by several methods such as

A. Badnjevic and L. Gurbeta Pokvić (Eds.): CMBEBIH 2021, IFMBE Proceedings 84, pp. 557–564, 2021.
https://doi.org/10.1007/978-3-030-73909-6_65

chemical methods and electroporation. While electroporation gives the higher transformation efficiency, it requires special apparatus. Chemical methods, on the other hand, are more practical for general laboratory applications.

The demonstration of transformation in the artificially induced bacteria has led to the development of the first chemical method by Mandel and Higa in 1970 [1]. The so called 'calcium chloride' ($CaCl_2$) method involves a brief heat shock of bacteria in the presence of Ca^{2+} ions. Several optimizations were made on this original protocol which involves the use of variant chemicals, multiple heat shock cycles, use of SOC **(Super Optimal broth with Catabolites repression)** medium instead of LB **(Luria–Bertani)** as growth medium, use of microwave instead of water bath for the heat shock, use of glycerol in the growth medium and the shaking instead of the heat shock [2–6]. In the present study we aim to explore the effect of the shaking (or vibration as measured revolution per minute, or 1/60 Hertz) at specific points during or after the heat shock transformation of competent DH5-Alpha strain of *E. coli*, induced by $CaCl_2$. Shaking has been generated either by thermal shaker and/or rotational incubator. Based on shaking type and stage of induction, each condition has been described and denoted with a number (1-5) and compared to standard heat shock transformation protocol modified from Mandel and Higa [1]. Our results show that shaking during heat shock does not have any positive impact on transformation efficiency. On the other hand, 180 rpm shaking (after heat shock) during incubation at 37 °C for 1 h (Condition 1 and 5) has increased transformation efficiency significantly ($P < 0.05$). Thus, a post heat shock incubation with shaking attains higher yield of transformed cells affecting overall efficiency of transformation procedure. To our knowledge, our study provides the first evidence for the effect of shaking during different stages of heat shock transformation mediated by $CaCl_2$.

2 Methods

2.1 Preparation of Competent *E. Coli* by $CaCl_2$ Method

This protocol is adapted from Sambrook and Russel [7], with some modifications. Under aseptic conditions, an inoculate of *E. coli* DH5-Alpha culture in 5 mL LB medium was prepared by picking one colony and incubating at 37 °C for 24 h with vigorous shaking at 180 rpm. Aliquots of 0.5 mL of the grown culture were added into two sterile conical flasks containing warm 50 mL LB followed by incubation at 37 °C with shaking at 180 rpm. The growth was monitored till the OD_{600} reaches to 0.35–0.4.

When suitable growth has been reached, the cultures were chilled on ice for 10 min and then the cultures were transferred to an ice cold sterile 2 centrifuge tubes (50 mL) before the collection of the cell pellets by centrifugation at 6000 rpm for 5 min at 4 °C.

The cell pellets were resuspended in 12.5 ml of an ice-cold 50 mM $CaCl_2$ solution and incubated on ice for 20 min. Following this, the cell pellets were collected by centrifugation at 6000 rpm for 5 min at 4 °C and resuspend with 1.25 ml ice-cold 50-mM $CaCl_2$ containing 10% glycerol. The cells were then aliquoted as 100 μl and stored at −80 °C for further use. Figure 1 is the demonstration of the steps of the procedure.

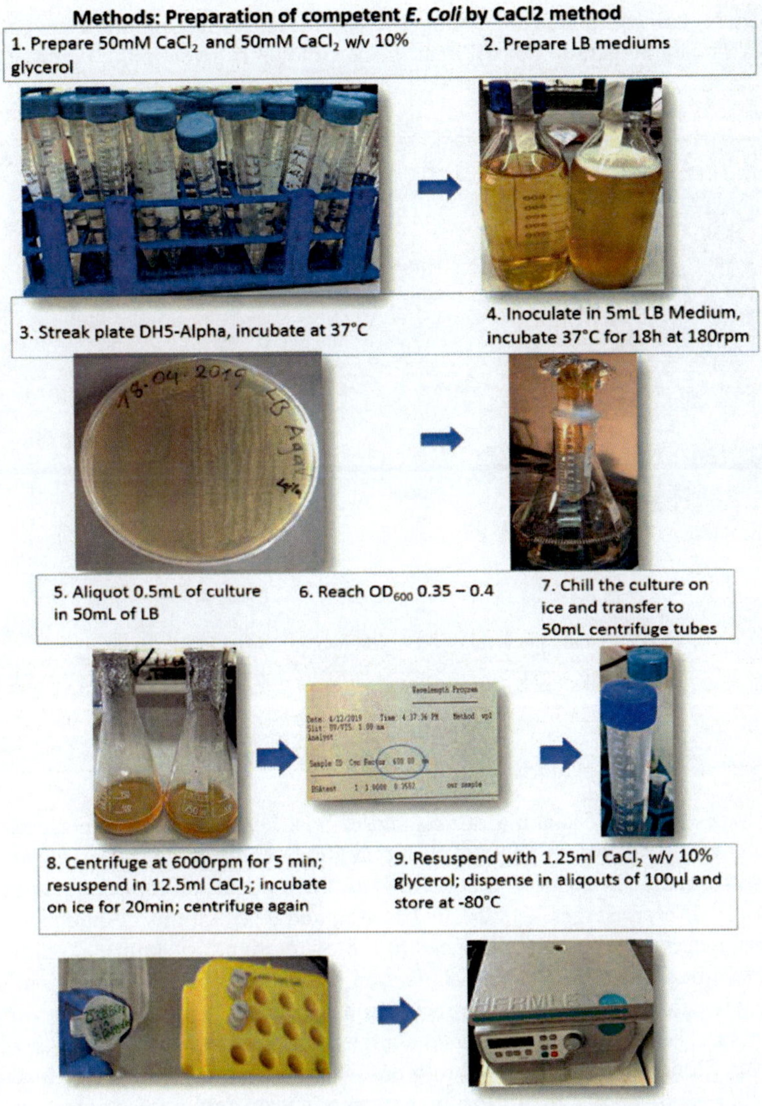

Fig. 1. Preparation of competent bacteria by $CaCl_2$ Method

2.2 Heat Shock Transformation Protocol

Competent bacteria are mixed with plasmid DNA (pcDNA 3.1 (-) plasmid) in ice and incubated in ice for 30 min then transferred to 42 °C water bath for 45 s and placed immediately to ice for 2 min. This is followed by an addition of 100 µl of 37 °C LB media, containing appropriate antibiotic (1 µl of 100 µg/ml ampicillin per 1 mL LB) and subsequent incubation at 37 °C for 1 h. After incubation, the bacteria were plated on 1.5% (w/v) LB agar plates containing 100 μg/mL ampicillin and incubated at 37 °C overnight. The demonstration of the heat shock protocol is given in the Figure 2.

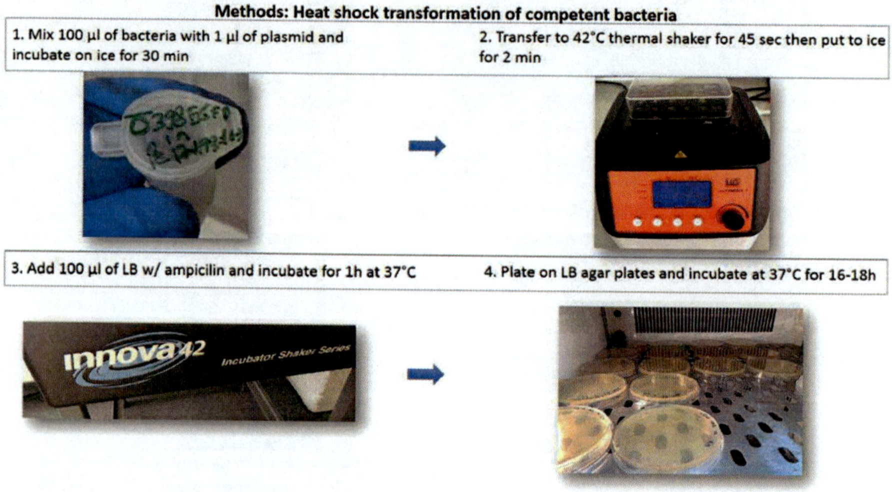

Fig. 2. Heat shock transformation protocol used in the study

2.3 Experimental Test Conditions

In order to test effect of shaking during and/or post heat shock, different conditions (1–5) were set up. These correspond to 5 conditions and the control condition (heat shock transformation without shaking) (Fig. 3). Each condition includes induction of vibrations at different stages. Condition 1 is assigned to vibrations at incubation stage of standard heat shock protocol (induced at 1 h of rotational shaking). Condition 2 is assigned to vibrations at the stage of induction of heat for 45 s at 42 °C. Condition 3 is assigned to vibrations at 2-min ice incubation. Condition 4 is assigned to vibrations induced both at heat and ice stages of standard heat-shock protocol, for 45 s and 2 min, respectively. Condition 5 is assigned to vibrations induced in all stages of heat-shock protocol. For comparisons, the following parameters were kept fixed:

1. Growth phase early log phase (0.35–0.4 OD_{600})
2. Use of 100 μl bacterial aliquots for transformation and transformation efficiency
3. (TrE) calculation;
4. Use of pcDNA 3.1 (-) plasmid only in fixed amounts (about 40 ng)
5. Use of LB media for growth and post heat-shock growth;
6. Use of 45s heat shock for transformation; and
7. Final resuspension volume of 1.25 ml from initial 50 ml of bacteria cultures.

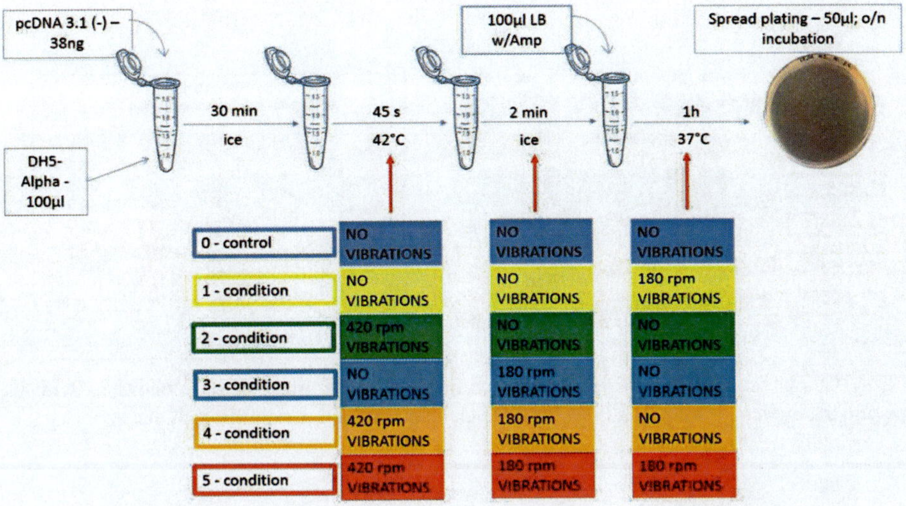

Fig. 3. Schematic representation of the experimental conditions. Five different experimental conditions were tested with respect to control condition, i.e., conventional heat shock transformation

2.4 Statistical Analysis

Experiments are conducted as two sets. Each condition is compared to control condition according to student's t-test. $p < 0.05$ is considered as statistically significant.

3 Results

In this study two different modes of shaking were used: 420 rpm shaking by thermal shaker and 180 rpm shaking by shaker incubator (Fig. 3). The former was used during 42 °C heat shocks while the latter was used during 4 °C (ice) and/or 37 °C incubation (Fig. 4a). Results show that (Fig. 4b), 180 rpm shaking during 37 °C incubation has increased the transformation efficiency significantly ($p < 0.05$). This was indeed expected as it is generally advised to incubate the cells in a shaker incubator after transformation. An example of colonies for each condition is shown in the Fig. 5.

We did not find any effect of shaking during heat shock. As shown in the Fig. 4, none of the conditions which include the shaking during the heat shock only (Conditions 2 and 4) showed a better transformation efficiency compared to control condition (Fig. 4b).

Since the shaking during any stage of the heat shock did not increase the transformation efficiency (as described in the above section), and even produced less transformants than the control condition (Fig. 4b), this then raises the idea that shaking during heat shock would have even negative effect on the transformation efficiency. However a careful analysis of the data show that shaking during heat shock does not seem to have a direct positive or negative effect since the 1st condition (no any shaking except for the incubation stage) and 5th condition (shaking at all stages) have almost the same transformation efficiency (Fig. 4b).

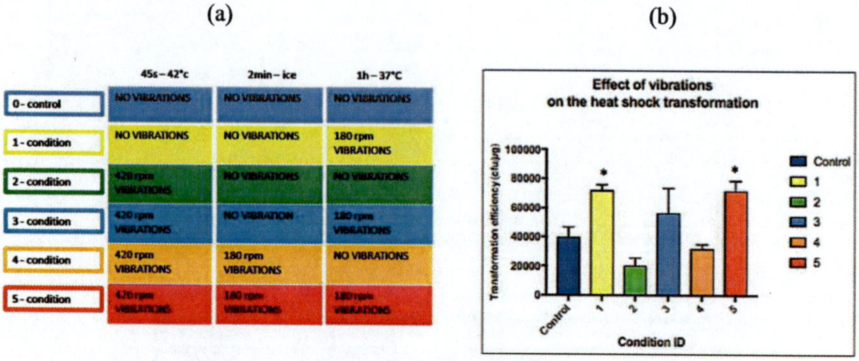

Fig. 4. Comparison of the TrE under test conditions. (a) Summary of test conditions with its specific parameters of vibration (shaking). (b) Comparison of five conditions in heat shock

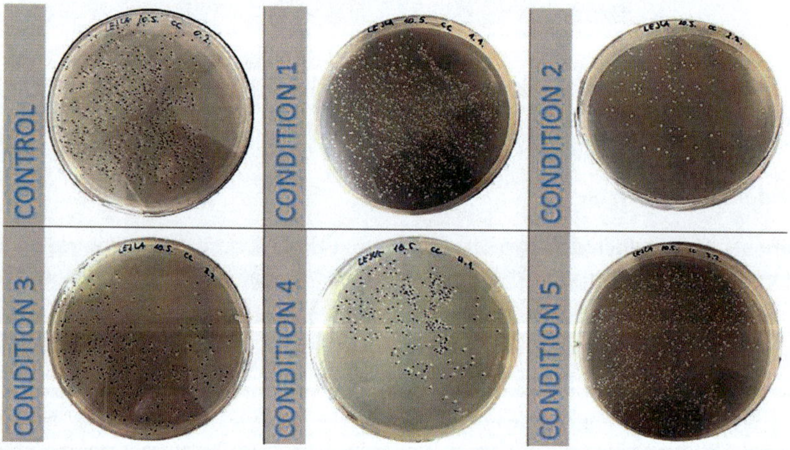

Fig. 5. The colonies grown from the one set of transformation experiment including control and tested conditions described in the Fig. 4. Black dots in some plates represent the marker which is used to count colonies.

4 Discussion and Conclusion

$CaCl_2$ method is the most resource and labor efficient method among the chemical transformation methodologies, thus making it an ideal candidate for further investigations for optimal conditions. In the present study we analyzed the effect of shaking during heat shock transformation of $CaCl_2$ induced DH5-Alpha strain of *E. coli*. Our results show that shaking during heat shock did not improve or worsen the transformation efficiency, however a gentle shaking (180 rpm) during the incubation after the heat shock has improved transformation efficiency significantly.

Accumulating literature suggests various strategies to optimize the transformation efficiency of the conventional heat shock method. These include -but are not limited to- the use of microwave instead of water bath during the heat shock [2], for example.

Moreover factors such as the type of bacterial strain [6, 8], storage period [9], the concentration of calcium chloride and competent bacteria solution [4], rotation speed and duration in centrifugation [4], the type of antibiotic used in the growth media [10] seem to play a role in the transformation efficiency. Although the latter has been found critical during electroporation heat shock, one study showed that the effect of antibiotic exists in the heat shock mediated transformation of $CaCl_2$ induced bacteria (DH1 *E. coli*) as well [9]. Thus, it is not possible to formulate one optimized protocol that fits all bacteria at once. Nevertheless, use of log phase bacteria, a heat shock and an incubation step before plating were found to be beneficial for the transformation of bacteria for chemical methods: $CaCl_2$ method, Hanahan method, polyethylene glycol method [9].

In one study, gentle shaking has been found more efficient than heat shock transformation [5]. Thus, in our study we hypothesized that a combination of heat shock and the shaking would have a positive cumulative effect on the transformation efficiency. Thus we applied shaking at specific stages of the heat shock transformation. However heat shock transformation accompanied by vibration did not yield better or worse transformation efficiency. Shaking was positively affecting the transformation efficiency only when it is applied during incubation in the post heat shock stage. This latter effect is not new, as an incubation step before plating were found to be beneficial for the transformation of bacteria for chemical methods in general [9]. However in our experiment, incubation was done with or without vibration (control) thus emphasizing the important effect of vibrations induced during incubation.

In addition to the effect of bacterial growth phase and growth conditions, storage period and conditions and the type of the antibiotic, the size of the plasmid plays a significant role in transformation efficiency [11]. For example, according to Hanahan [12], transformation efficiency decreases linearly with increasing plasmid size. In our study we used a relatively bigger plasmid with a size of 5.4 kb which could reflect our lower transformation efficiencies (10^5–10^6) compared with the transformation of similar studies (typically 10^6–10^7) in which a plasmid vector with a size ranging from 2,53 kb is used in the transformation experiments [2, 8, 9], and [11].

Our results suggest some future investigations. First, the previous findings [5], regarding the effect of shaking have to be replicated if it is better than heat shock. Second our results should be studied by some other labs for replication. Third, in our study we used 10% glycerol in the final resuspension buffer of the $CaCl_2$ induction. But addition of glycerol in the growth media could be critical, as well. For example, in some bacteria (*Bacillus licheniformis),* the addition of glycerol in the growth media caused a decrease in C16:1 and C18:1 fatty acids thus leading to an increased permeability of cell membrane. It also caused an increase in C10:0 and C12:0 fatty acids, which may have an impact on the increased fluidity of cell membrane [13]. Thus, addition of glycerol into the growth media of *E. coli* during the competent cell preparation may have a positive effect.

Finally, further adjustments of our experimental conditions should be tested, especially different plasmid sizes and different amplitude and duration of vibrations during different stages of heat shock may increase the transformation efficiency. The heat shock is thought to cause two modifications: (1) the disruption of lipids and subsequent formation of pores on the bacterial surface, which helped DNA to cross outer membrane barrier, and (2) lowering of membrane potential, which facilitated DNA to cross inner membrane of *E. coli* [11]. Coupling heat shock with some other physical means such as vibration might have a facilitating effect on the transformation process eventually.

Compliance with Ethical Standards. Conflict of Interest: The authors declare that they have no conflict of interest.

References

1. Mandel, M., Higa, A.: Calcium-dependent bacteriophage DNA infection. J. Mol. Biol. **53**, 159–162 (1970)
2. Fregel, R., Rodríguez, V., Cabrera, V.M.: Microwave improved Escherichia coli transformation. Lett. Appl. Microbiol. **46**(4), 498–9 (2008)
3. Kurien, B.T., Scofield, R.H.: Polyethylene glycol-mediated bacterial colony transformation. Biotechniques **18**(6), 1023–6 (1995)
4. Li, X., Sui, X., Zhang, Y., Sun, Y., Zhao, Y., Zhai, Y., Wang, Q.: An improvised calcium chloride method preparation and transformation of competent cells. Afr. J. Biotechnol. **9**, 8549–8554 (2010)
5. Shanehbandi, D., Saei, A.A., Zarredar, H., Barzegari, A.: Vibration and glycerolmediated plasmid DNA transformation for Escherichia coli. FEMS Microbiol. Lett. **348**(1), 74–8 (2013)
6. Tu, Z., He, G., Li, K.X., Chen, M.J., Chang, J., Chen, L., Yao, Q., Liu, D.P., Ye, H., Shi, J., Wu, X.: An improvised system for competent cell preparation and high efficiency plasmid transformation using different Escherichia coli strains. Electron. J. Biotechnol. **8**, 113–120 (2005)
7. Sambrook, J., Russell, D.W.: Preparation and transformation of competent E. coli using calcium chloride. CSH Protoc. **1**, 116–118 (2006)
8. Chan, W.T., Verma, C.S., Lane, D.P., Gan, S.K.: A comparison and optimization of methods and factors affecting the transformation of Escherichia coli. Biosci. Rep. **33**(6)
9. Huff, J.P., Grant, B.J., Penning, C.A., Sullivan, K.F.: Optimization of routine transformation of Escherichia coli with plasmid DNA. Biotechniques **9**(5), 570-2, 574, 576-7 (1990)
10. Steele, C., Zhang, S., Shillitoe, E.J.: Effect of different antibiotics on efficiency of transformation of bacteria by electroporation. Biotechniques **17**(2), 360–5 (1994)
11. Chung, C.T., Niemela, S.L., Miller, R.H.: One-step preparation of competent Escherichia coli: transformation and storage of bacterial cells in the same solution. Proc. Nat. Acad. Sci. U.S.A. **86**(7), 2172–2175 (1989)
12. Hanahan, D.: Studies on transformation of Escherichia coli with plasmids. J. Mol. Biol. **166**(4), 557–580 (1983)
13. Du, G., Yang, G., Qu, Y., Chen, J., Lun, S.: Effects of glycerol on the production of poly (c-glutamic acid) by Bacillus licheniformis. Process Biochem. **40**, 2143–2147 (2005)

Detection of hrHPV DNA with Simulated HPV16 and HPV18 Typing Based on Real-Time PCR

Almedina Hajrović[1](✉), Arnisa Alibegović[2], Emir Šeherčehajić[1], Adisa Ramić[1], and Berina Hasanefendić[3]

[1] Faculty of Health Studies, University of Sarajevo, Sarajevo, Bosnia and Herzegovina
[2] Medical Biochemical Laboratory, Medlab, Gračanica, Bosnia and Herzegovina
[3] Institute for Clinical Biochemistry and Immunology, Clinical Center University of Sarajevo, Sarajevo, Bosnia and Herzegovina

Abstract. The complexity of molecular diagnosis of HPV that is directly related to cervical cancer is caused by a large number of genotypes, such as HPV genotypes 16 and 18, which show the greatest malignant potential and cause 70% of all cervical cancers. The objectives of this study are to determine the significance of Ct values detected by HPV DNA as well as intern control DNA in cervical smear samples and to evaluate the advantages and disadvantages of cervical cancer screening based on HPV DNA and to determine the incidence of HPV16, HPV18 and other mixed infections. In this report, the Abbott RealTime High-Risk HPV Diagnostic Test (Abbott Molecular Inc., IL, USA), for the presence of hrHPV infection, routinely tested a total of 1419 cervical samples of women with cytological changes on the cervix. The average age was 42.59 years. The prevalence of hrHPV infection was 29.81% (hrHPV16, 18 and others). Ct HPV18 showed a significant difference between age groups. Approximately constant values of Ct-IC samples of cervical swabs, Ct values of viral DNA could be considered as a parameter of disease progression/regression. A statistically significant correlation was found between Ct-IC and Ct-HPV16.

Keywords: HPV · rt-PCR · Cervical cancer

1 Introduction

The importance of molecular diagnosis of HPV lies in its direct association with cervical cancer, and the complexity is caused by a large number of genotypes, which need to be distinguished to explain the potential risks for the development of the disease and conduct screening programs [1, 2]. Preferably, the diagnostic method for HPV infection should allow simultaneous detection, genotyping, and quantification of HPV. There are various tests for detecting and genotyping HPV. These tests differ in goals, technologies, genotyping capabilities, as well as automation capacity. Genotyping plays a crucial role in determining the prevalence and relative degree of risk for each type of HPV [3, 4]. According to the oncogenic potential of HPV genotypes are classified into: low risk genotypes and high risk genotypes. HPV 16 and 18, as high-risk HPV subtypes, have

A. Badnjevic and L. Gurbeta Pokvić (Eds.): CMBEBIH 2021, IFMBE Proceedings 84, pp. 565–574, 2021.
https://doi.org/10.1007/978-3-030-73909-6_66

the highest propensity for persistence and the highest malignant potential, where they represent the most common cause of almost 70% of all cervical cancers [5, 11].

The discovery of the connection between HPV and cervical cancer conditioned the development of highly sensitive molecular tests for HPV detection (about 90%) and opened the possibility of its application in mass screening of the female population. HPV can only be reliably detected by DNA tests. HPV DNA testing is performed only in women older than 30 years, because in women younger than 30 years, despite the high prevalence of transient infections, there is a low prevalence of intraepithelial lesions [5, 10].

Although most commercially available HPV tests have excellent clinical sensitivity and specificity values, they are not suitable in a large screening program in low-resource countries, mainly because of their high cost [8, 9]. The use of low-cost cervical cell collection and transport devices and low-cost polymerase chain reaction (PCR) techniques to detect hrHPV infections are therefore suitable alternatives for developing countries. The genome of HPV, like most viruses, is not large enough or in sufficient quantity to be directly detected, so PCR is performed. Molecular methods based on the detection of virus nucleic acid can detect the virus and determine which genotype of HPV is involved. Virus typing determines whether a virus belongs to a group of high-risk HPV types [17]. DNA genotyping tests based on PCR using the target amplification technique can detect the presence of small amounts of virus and are considered the "gold standard" for virus detection [6, 7, 15].

2 Aims

1. To determine the significance of Ct values of detected HPV DNA as well as internal control DNA in cervical smear samples and to evaluate the advantages and disadvantages of HPV DNA based screening for cervical cancer.
2. To determine the frequencies: infections caused by HPV; infections caused by high-risk genotypes of HPV16, HPV18 and other hrHPVs and mixed infections caused by high-risk genotypes of HPV16, HPV18 and other hrHPVs.

3 Material and Methods

The research is a descriptive analytical, retrospective study. The study included 1419 women $\geq$ 30 years of age, in whom in the period from 02.10.2014 to 20.12.2017 year, based on cytological testing according to Papanicolaou (Papa), abnormal changes in the cervix were found. Data from the results of testing for the presence of HPV infection were taken from the available medical documentation. The clinical material analyzed was a cervical swab taken during regular gynecological examination of patients or before colposcopy, stored in a standardized viral transport medium Abbott Cervi-Collect Specimen Collection Kit (Abbott Molecular Inc., IL, USA) for testing Abbott RealTime high risk HPV test (Abbott Molecular Inc., IL, USA).

The study included women $\geq$30 years of age with cytologically diagnosed changes in the cervix, who were routinely analyzed by Abbott RealTime High Risk HPV for the presence of hrHPV infection. Excluded from the study were cervical swab specimens of women in whom the accompanying medical documentation before testing for the presence of hrHPV genotype infection was not properly completed, or in whom the result of cytological testing was not reported and women under 30 years of age.

Clinical material that met the inclusion criteria in the study was stored in appropriate aliquots required for hrHPV DNA extraction at +4 °C for up to 7 days, or longer at 20 °C/−70 °C until testing. Testing was performed using an automated m2000 system. The system consisted of m2000sp instruments for extraction and purification of DNA from cervical swab samples in the transport medium and preparation of PCR reactions and m2000rt for performing real-time PCR. Screening for the presence of HPV infection was performed by the RealTime high risk HPV method.

The specificity of the PCR reaction is given by modified GP5 + /6 + starting oligonucleotides (3 forward and 2 reverse starting oligonucleotides) which amplify the preserved part of the HPV L1 gene region and one pair of starting oligonucleotides for detection of human beta-globin gene as internal DNA controls. In addition to the initial oligonucleotides, the PCR mixture also contains specific probes for HPV-16, HPV-18, non-HPV16/18 genotypes and IC (internal control), which are labeled with 4 different fluorophores (fluorescent dyes, reporters).

The importance of internal control detection in each sample/control is in the validation of the whole process, ie evaluation of the presence of human cells potentially infected with the virus, quality and efficiency of DNA extraction process and efficiency of amplification and detection reaction. Detection is qualitative.

The use of multiple fluorescent dyes enabled multiplex analysis and obtaining three test results in one reaction: 1) detection of hrHPV, 2) genotyping of HPV-16 and 3) genotyping of HPV-18, as well as combinations of these three results. A positive result was set on the value of the PCR threshold cycle (Ct) determined based on the clinical performance of the test, as follows: for HPV genotype analysis it is up to Ct = 32, and for IC up to Ct = 35, and satisfied with the conditions for positive and negative control defined by the manufacturer. Statistical procedures were analyzed in IBM SPSS Statistics 23.0.

4 Results

The results of the study were processed in the total surveyed population of women over 30 years and according to individual age groups: I (30–39 years), II (40–49 years), III (50–59 years), IV (60–69 years) and V (70–79 years). Respondents were divided into 5 groups. The first age group (30–39 years) had the largest number of respondents of 639/1419 women (45.03%), followed by the second age group (40–49 years) which consisted of 461/1419 women (32.48%), while the smallest number of respondents had V group (70–79 years) of 11/1419 women, which was 0.77%. Of the total number of respondents, 423/1419 (29.81%) respondents had hrHPV infection. The share of individual types of infections in relation to the total examined population is shown in Fig. 1.

6.20% of subjects had HPV16 genotype infection, 1.55% HPV18 genotype infection and 17.89% other hrHPV genotypes. The incidence of mixed infections was as follows:

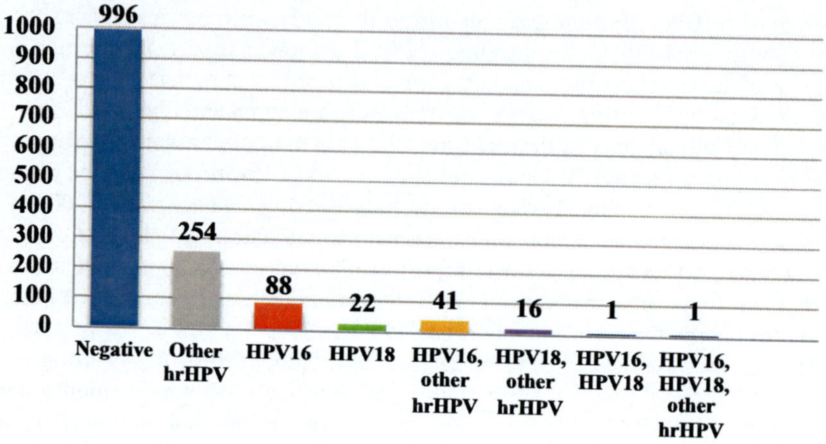

Fig. 1. Share of individual types of infections in the total examined population (n = 1419)

HPV16 + other hrHPV 2.88%, HPV18 + other hrHPV 1.12%, HPV16 + HPV18 and HPV16 + HPV18 + other hrHPV less than 1%.

The values of the age parameters of the subjects and Ct values for infections caused by HPV16, HPV18 and other hrHPV are shown in Table 1.

Table 1. Statistical values of parameters of infected respondents (n = 423/1419)

	Age	Ct-HPV16	Ct-HPV18	Ct-other hrHPV	Ct-intern control
MIN	30,00	11,23	12,62	10,15	17,76
MAX	79,00	31,53	32,75	35,08	31,23
MEAN	42,59	22,15	25,27	23,38	22,55
MEDIAN	41,00	21,78	26,65	23,45	22,26
SD	9,28	4,67	5,54	5,43	2,15

Table 1 shows the values of descriptive statistics for age and Ct parameter in infected subjects. The mean age of the study population was 42.59 years (SD ± 9.28). The youngest respondent was 30 years old and the oldest 79 years old.

In the I age group (30–39 years), 35.68% of respondents had hrHPV infection. In group I of subjects aged 30–39 years, 9.23% of subjects had infection with HPV16 genotype, 1.40% with HPV18 genotype, and 20.03% with other hrHPV genotypes. The incidence of mixed infections was as follows: HPV16 + other hrHPV 4.06%, HPV18 + other hrHPV 0.78% and HPV16 + HPV18 (<1%).

In the II age group (40–49 years), 111 (24.07%) subjects had hrHPV infection. In group II of subjects aged 40–49 years, 3.25% of subjects had infection with HPV16 genotype, 1.51% with HPV18 genotype, while the most common infection was with other hrHPV genotypes 15.40%. The incidence of mixed infections was as follows:

HPV16 + other hrHPV 2.38%, HPV18 + other hrHPV 1.30% and HPV16 + HPV18 + other hrHPV < 1%.

Out of the total number of respondents in the III age group (50–59 years), 25.00% had HPV infection. In group III of subjects aged 50–59 years, 4.66% of subjects had HPV16 genotype infection, 2.54% of HPV18 genotypes, and 13.98% of other hrHPV genotypes. The incidence of mixed infections was as follows: HPV16 + other hrHPV 1.69% and HPV18 + other hrHPV 2.11%.

Out of the total number of respondents in the IV age group (60–69 years), 29.16% of the respondents had hrHPV infection. In group IV of subjects aged 60–69 years, 1.38% of subjects had HPV16 genotype infection, and 27.77% of other hrHPV genotypes. HPV18 genotype infections as well as mixed infections were not reported.

Out of the total number of respondents in the V age group (70–79 years), 36.36% of respondents had hrHPV infection. In group V of subjects aged 70–79 years, 18.18% of subjects had infection with the HPV16 genotype and other hrHPV genotypes. HPV18 genotype infections as well as mixed infections were not reported (Table 2).

Table 2. Difference in the frequency of individual infections among groups with statistical significance

	Negative	Positive	HPV16	HPV18	Other hrHPV	All
I group (30–39 years)	411 (444.9) [2.44]	228 (197.87) [4.59]	59 (39.22) [9.98]	9 (10.70) [0.27]	128 (113.20) [1.94]	607
II group (40–49 years)	349 (323) [2.06]	111 (131.05) [3.07]	15 (28.56) [6.44]	7 (7,79) [0.08]	71 (82.43) [1.58]	442
III group (50–59 years)	178 (166.7) [0,76]	59 (68.01) [1.19]	11 (14.73) [0.95]	6 (4.02) [0.98]	33 (42.52) [2.13]	228
IV group (60–69 years)	51 (53.38) [0.11]	21 (22.28) [0.07]	1 (4,72) [2.93]	1 (1.29) [0.06]	20 (13.61) [3.00]	73
V group (70–79 years)	7 (8.78) [0.36]	4 (3.79) [0.01]	2 (0.78) [1.93]	1 (0.21) [2.94]	2 (2.24) [0.03]	12
All	**996**	**423**	**88**	**24**	**254**	**1758**

Table 3 Chi-square test shows that the frequency of positive findings is significantly different between the examined groups. The highest frequency of infections was observed in the group of the youngest respondents (aged 30–39). The frequency of infections decreases significantly in the groups of older women (IV and V). The statistical significance of these differences between age groups is $p = 0.00005$.

The frequency of mixed infections, as shown by the chi-square test, is not age-specific, although the highest number of mixed infections with HPV16 and other hrHPV is in the youngest subjects of groups I and II (Table 4). The incidence of other types of mixed infections is low in all examined groups and does not show a statistically significant difference between the groups. Determined by the chi-square test, mixed infections do not show significant differences between age groups and the p value is 0.1457.

Table 3. Difference in the frequency of mixed infections between groups with statistical significance

	Results of Chi-Square Test: 17.1023; p-value 0.14578				All
	HPV16, HPV18	HPV16, other hrHPV	HPV18, other hrHPV	HPV16, HPV18, other hrHPV	
I group (30–39 years)	1 (2.32) [0.75]	26 (19.99) [1.81]	5 (8.37) [1.35]	0 (2.32) [0.75]	32
II group (39–49 years)	0 (1.34) [0.09]	11 (11.51) [0.02]	6 (4.82) [0.29]	1 (1.34) [0.09]	18
III group (50–59 years)	0 (0.77) [0.07]	4 (6.66) [1.06]	5 (2.79) [1.75]	0 (0.77) [0.07]	9
IV group (60–69 years)	0 (0.28) [1.83]	0 (2.42) [0.84]	0 (1.01) [0.00]	0 (0.28) [1.83]	0
V group (70-79 years)	0 (0.28) [1.83]	0 (2.42) [0.84]	0 (1.01) [0.00]	0.(0.28) [1.83]	0
All	**1**	**41**	**16**	**1**	**59**

In assessing the differences between the mean values of individual HPV genotypes and Ct-internal control, the Kruskal-Wallis H test showed that there was no difference in the mean values of Ct-HPV16 between age groups, with $p = 0.866$. Likewise, Ctother HPV as well as Ct-internal controls did not show significant differences in mean values between age groups, with p values 0.5324 and 0.7868, respectively. Only mean Ct-HPV18 values showed a significant difference between age groups, with $p = 0.003$, determined by the Kruskal-Wallis H test. The mean values of Ct-HPV16 and Ct-HPV18 for the IV and V age groups were not included in the difference assessment due to the insufficient number of results that would allow statistical processing.

In the following work, the Pearson correlation coefficient was used to calculate the correlations between individual parameters in order to determine the possible existence of a relationship between the change in the value of Ct-internal control and HPV genotypes (Table 5).

A statistically significant correlation was found between Ct-IC and Ct-HPV16, as well as between Ct-IC and Ct-other hrHPV. All found correlations are positive, of different strength, with statistical significance.

The correlation between Ct-IC and Ct-other hrHPV is positive; $R = 0.331$, $R2 = 0.11$ with statistical significance $p < 0.01$.

Table 4. Difference between mean values of examined hrHPV genotypes and internal control, Kruskal - Wallis H test

		Ct-HPV16	Ct-HPV18	Ct-other hrHPV	Ct-intern control
Mean/average	I group (30–39 years)	22,0849	21,2427	23,4805	22,5178
	II group (39–49 years)	22,4933	27,8554	23,4231	22,6723
	III group (50–59 years)	22,1467	27,6891	23,8893	22,4620
	IV group (60–69 years)	/	/	21,7470	22,5862
	V group (70–79 years)	/	/	19,3300	22,4750
	H - statistic	0,2870	11,6000	2,1973	1,7212
	N	131	30	312	423
	p-value	0,8660	0,0030	0,5324	0,7868

Table 5. Pearson correlation between the examined parameters

		Ct-HPV16	Ct-HPV18	Ct-other hrHPV
Ct-intern control	R	0,421	0.339	0,331
	R^2	0,177	0.115	0,11
	P	$p < 0{,}000$	$p = 0{,}1227$	$p < 0.000$

5 Discussion

Our study included 1419 subjects with cytologically diagnosed changes in the cervix. The prevalence of hrHPV infections was 29.81 (423/1419). The total frequency of individual infections was 483 (34.03%), of which 131/483 (27.12%) subjects had HPV genotype infection, 40/483 (8.28%) subjects had HPV18 genotype infection, and others hrHPV genotypes 312/483 (64.59%) of the subject. Based on the findings, there is a significant difference ($p = 0.00005$) in the prevalence of infections caused by high-risk human papillomaviruses HPV16, HPV18 and other HPVs. Kegalj K. in the study indicated that all patients with a referral diagnosis of cervical cancer in situ, who were positive for HPV, had a high-risk type of HPV, while none was positive only for a low-risk type of HPV. In the group of patients with a referral diagnosis of anogenital warts, the largest percentage of samples was positive only for low-risk type of HPV compared to other referral diagnoses, and 29.6% had mixed high + low-risk HPV infection, while in the study [23] Vitković L et al. in their study of the distribution of high-risk types of HPV in relation to pathohistological findings of cervical biopsies examined in 56 women with suspicion of premalignant and malignant cervical lesions, due to suspicious colposcopic and cytological findings. 47 (84%) subjects had a positive HPV test and one or more types of HPV were proven to be present. The most common type of virus was HPV 16 [14].

In a previous study in BiH [20] on subjects with different cytological changes in the cervix (ASCUS to HSIL), the most common genotype of HPV was HPV16 detected in 32.6% of the subjects, followed by genotypes of HPV31 which had 14.3% subjects and HPV51 detected in 9.5% (14/147) subjects. The HPV18 genotype took the 4th place in terms of the frequency of infection in 7.5% (11/147) of the respondents.

The frequency of mixed infections, as shown by the chi-square test, is not age-specific (there is no significant difference between age groups, $p = 0.1457$), although the highest number of mixed infections with HPV16 and other hrHPV in the youngest subjects of groups I and II, 26/59 (44.06%) and 11/59 (18.64%), respectively. In study involving 20,017 women, including 785 women with abnormal cytological results, You W et al. found the presence of mixed infections in 17.8%, of which 98 had infection caused by two types, 37 cases had an infection caused by three HPV genotypes, 2 cases had an infection caused by four HPV genotypes, 2 cases with five HPV genotypes and 1 case caused by as many as six HPV genotypes. [19] Kleter B et al. demonstrated the presence of mixed infections in 11.8% of women with normal cytological results and ASCUS in 34.5% of women with mild and moderate dyskaryosis, and in 4.4% of women with cervical cancer [20]. They suggested that mixed infections were due to immune tolerance. An infected woman is likely to accumulate infections of different HPV genotypes, without an immune response and the disappearance of individual HPV genotypes [22, 23].

During real-time PCR, the Ct value is generally lower in samples of higher concentration of the target molecule (viral DNA or IC DNA).

Salimović-Bešić I et al. in their study found mixed infections with at least one highrisk or probably high-risk HPV genotype. It is assumed that in younger women, the prevalence of mixed infections is higher due to insufficient immune response and immature portion (vaginal part of the cervix), which makes it more susceptible to infections [16, 21].

van der Weele et al. in their study described the prevalence of the HPV18 genotype as the most prevalent genotype in cervical cancer. Of the 3282 subjects, 259 were positive for the HPV18 genotype (7.9%), indicating the clinical significance of HPV18 typing in screening tests. Accordingly, in our study, 40 subjects (2.82%) were positive for the HPV18 genotype, who had varying degrees of cervical cell abnormalities [22].

Vrdoljak-Mozetič D and co-workers included in their study subjects tested using a group-specific high-risk HPV DNA test (hybrid capture 2), and the HPV genotype was determined for positive findings using the INNO-LiPA HPV Genotyping v2 molecular method. The most common was the HPV16 genotype. Accordingly, disease progression was statistically significantly associated with peak HPV positivity ($p < 0.001$) and the presence of HPV16 ($p < 0.001$) [18].

How significant is the discovery of a cause-and-effect relationship between highrisk HPV genotypes and cervical cancer has led to the development of HPV vaccines, the two most common, HPV16 and HPV18 [12,13, 27]. Therefore, we can say that we have confirmed the working hypothesis, ie that the determination of HPV16 and HPV18 genotypes plays a significant role in predicting the potential coverage and effectiveness of primary prevention of cervical cancer in our area.

6 Conclusion

1. The used screening test for the detection of hrHPV DNA with simultaneous typing of HPV16 and HPV18, based on real-time PCR, could, based on the findings from our study, have a semi-quantitative character in addition to the qualitative one. Namely, with the discovery of an approximately constant Ct value of the internal control of cervical smear samples, the Ct values of the detected viral DNA could be considered as a parameter of disease progression/regression. With the use of adequate concentration standards, the method could fully take on a quantitative character, which, together with the cytological finding of changes in the cervix, would give a more complete insight into the prognosis of disease outcomes. By correlation, we determined the possible existence of a relationship between changes in the values of Ct-internal control and HPV genotypes. A statistically significant correlation was found between Ct-IC and Ct-HPV16, as well as between Ct-IC and Ct-other hrHPV. All found correlations are positive, of different strength, with statistical significance.
2. Analysis of the frequency of infections caused by HPV in 423/1419 women $\geq$ 30 years with abnormal cytological result of cervical smear showed a total share of 29.81% of positive results. By analyzing the frequency of individual infections caused by highrisk genotypes HPV16, HPV18 and other hrHPV in 423/1419 women $\geq$ 30 years, with abnormal cytological result of cervical swabs, we determined that the HPV16 genotype infection had 131/483 (27.12%) subjects, 40/483 (8.28%) subjects had the HPV18 genotype, and other hrHPV genotypes had 312/483 (64.59%) subjects compared to the total number of individual infections ($n = 483$). Mixed infections caused by high-risk genotypes HPV16, HPV18 and other hrHPV in 423/1419 women $\geq$ 30 years of age, with abnormal cytological result of cervical swabs were classified into groups. The incidence of mixed infections is low in all examined groups and does not show a statistically significant difference between the groups. It was found that mixed infections do not show significant differences between age groups, and $p = 0.145$.

References

1. Doorbar, J., Egawa, N., Griffin, H., Kranjec, C., Murakami, I.: Human papillomavirus molecular biology and disease association. Rev. Med. Virol. **25**(S1), 2–23 (2015)
2. Chow, L.T., Broker, T.R., Steinberg, B.M.: The natural history of human papillomavirus infections of the mucosal epithelia. AMPIS **118**, 422–49 (2010)
3. Paz-Zulueta, M., Alvarez-Paredes, L., Rodriguez Diaz, J.C., Paras-Bravo, P., Andrada Becerra, Ma.E., Rodriguez Ingelmos, J.M., Ruiz Garcia, M.M., Portilla, J., Santibanez, M.: Prevalence of high-risk HPV genotypes, categorized by their quadrivalent and ninevalent HPV vaccination coverage, and the genotype association with high-grade lesions. BMC Cancer **18**(112), 1–9 (2018)

4. Vince, A., Lepej, S.Z.: Diagnostic methods and techniques in cervicalcancer prevention Part II: Molecular diagnostics of HPV infection. Med. Glas. Ljek komore Zeničko-doboj kantona **7**, 18–25 (2010)
5. Snijders, P.J., Heideman, D.A., Meijer, C.J.: Methods for HPV detection in exfoliated cell and tissue specimens. Apmis **118**(6–7), 520–528 (2010)
6. Harari, A., Chen, Z., Burk, R.D.: Human papillomavirus genomics: past, present and future. Curr. Probl. Dermatol. **45**, 1–18 (2014)
7. Hadžisejdić, I., Grce, M., Grahovac, B.: Humani papiloma virus i karcinom cerviksa: mehanizmi karcinogeneze, epidemiologija, dijagnostika i profilaksa. Medicina fluminensis **46**(2), 112–123 (2010)
8. de Villiers, E.M., Fauquet, C., Broker, T., Bernard, H.U., zurHausena, H.: Classification of papillomaviruses. Virology **324**, 17–27 (2004)
9. Muñoz, N., Bosch, F.X., de Sanjosé, S., Herrero, R., Castellsagué, X., Shah, K.V., et al.: Epidemiologic classification of human papillomavirus types associated with cervical cancer. N. Engl. J. Med. **348**(6), 518–527 (2003)
10. Mirabello, L., Clarke, M.A., Nelson, C.W., Dean, M., Wentzensen, N., Yeager, M., Cullen, M., Boland, J.F., NCI HPV Workshop: Schiffman, M., Burk, R.D.: The intersection of HPV epidemiology, genomics and mechanistic studies of HPV-mediated carcinogenesis. Viruses **10**(80), 1–14 (2018)
11. Grahovac, B., Šimat, M., Kraševič, M.: Humani papiloma virus i karcinom cerviksa – imunopatogeneza i molekularna dijagnostika. Medix **58**, 67–71 (2005)
12. Kajitani, N., Satsuka, A., Kawate, A., Sakai, H.: Productive lifecycle of human papillomaviruses that depends upon squamous epithelial differentiation. Front. Microbiol. **24**(3), 152 (2012)
13. Vince, A., Lepej, S.Z.: Klinička značajnost molekularne analize humanih papiloma virusa. Infektološki glasnik **30**(3), 123–129 (2010)
14. Vitković, L., Perišić, Ž, Trajković, G., Mijović, M., Savić, S., Leštarević, S., Đerković, B.: Distribucija visokorizičnih tipova humanih papiloma virusa u odnosu na patohistološke nalaze cervikalnih biopsija kod žena. Praxis Medica **44**(4), 39–44 (2015)
15. World Health Organization. Human papillomavirus vaccines: WHO Position Paper, vol. 92, pp. 241–268 (2017)
16. Salimović-Bešić, I.: Određivanje onkogene aktivnosti visoko-rizičnih humanih papilomavirusa detekcijom HPVE6/E7 mRNA kod pacijentica sa prekanceroznim promjenama na grliću materice. Doktorska disertacija. Sarajevo (2013)
17. McGrae, S.L., Ferrante, J.M.: Update on prevention and screening of cervical cancer. World J. Clin. Ocol. **5**(4), 744–752 (2014)
18. Vrdoljak-Mozetič, D., Štemberger-Papić, S., Ostojić, D.V., Rubeša-Mihajlović, R., Dinter, M., Brnčić-Fischer, A., Kraševič, M.: Procjena citomorfologije, HPV statusa i HPV 16 genotipa u predikciji ishoda bolesti kod pacijentica s citološkim nalazom ASCUS i LSIL. Medicina fluminensis **52**(3), 377–389 (2016)
19. You, W., Li, S., Du, R., Zheng, J., Shen, A.: Epidemiological study of highrisk human papillomavirus infection in subjects with abnormal cytological findings in cervical cancer screening. Exp. Ther. Med. **15**, 412–418 (2018)
20. Kleter, B., van Doorn, L.J., Schrauwen, L., et al.: Development and clinical evaluation of highly sensitive PCR-reverse hybridization line probe assay for detection and identification of anogenital human papillomavirus. J. Clin. Microbiol. **37**, 2508–2517 (1999)
21. Salimović-Bešić, I., Poljak, M., Kocjan, B.: HPV genotypes in repeated Pap II smears of group of women in Slovenia. Med. Arh. **60**, 7–12 (2006)
22. van der Weele, P., Meijer, C.J.M.L., King, A.: High whole-genome sequence diversity of human papillomavirus type 18 isolates. Viruses **10**(68), 1–13 (2018)
23. Kegalj, K.: Kvalitativna detekcija humanih papilloma virusa (HPV) kombinacijom metoda lančane reakcije polimerazom (PCR) i reverzne transkriptaze. Završni rad. Osijek. (2017)

Comparison of Biofilm Category Determination Using TCP Method Depending on Signal Molecule Adherence

Faris Hrvat[2], Osman Hasanić[1](✉), Amina Aleta[2], Amel Spahić[3], Amra Džuho[1], and Mirsada Hukić[4,5]

[1] International Burch University, Sarajevo, Bosnia and Herzegovina
[2] University of Milan, Milano, Italy
[3] Mistral Technologies, Sarajevo, Bosnia and Herzegovina
[4] Institute for Biomedical Diagnostic and Research NALAZ, Sarajevo, Bosnia and Herzegovina
[5] Academy of Sciences and Arts of Bosnia and Herzegovina, Sarajevo, Bosnia and Herzegovina

Abstract. *Introduction.* Biofilms are defined as bacterial assemblies that are kept together by a biopolymer matrix. It can be composed of one or many bacterial species. Antibiotic treatments are unable to kill the bacteria in a biofilm, but may reverse the symptoms of a disease.

Methods. Microtiter plate assay is mainly used to analyze the biofilm formation by calculating the Optical Density Cutoff (ODc) and Optical Density (OD) values and describing the formation of biofilm belonging to one of the four categories. Biofilm structure can only be analyzed quantitatively, and it cannot be correlated with any parameters that can characterize its performance. The further analysis and understanding of the principles of biofilm forming and its characteristics is of great importance to reduce its negative influences on the health and environment. One of the main issues in biofilm formation analysis is the long time needed for the calculations to describe the formation of biofilm.

Results. This paper presents 2 biofilm calculators that were developed using Angular 5, and 192 samples were used to test the calculators.

Conclusion. The two developed biofilm calculators are of great importance, since they shorten the time of calculating tremendously, while not affecting the results.

Keywords: Biofilm · Biofilm classification · Biofilm calculator · Biofilm formation

1 Introduction

Biofilms have been defined as bacterial assemblies kept together by a self-produced biopolymer matrix. The matrix contains polysaccharides, microbial DNA and proteins that provide structural stability and protection to the biofilm. The mature biofilm may resemble primitive multicellular organism and contain water-filled channels. This assembly of the bacteria may consist of one or many species living in a social community. In

A. Badnjevic and L. Gurbeta Pokvić (Eds.): CMBEBIH 2021, IFMBE Proceedings 84, pp. 575–581, 2021.
https://doi.org/10.1007/978-3-030-73909-6_67

this communities bacteria communicate by synthesizing and reacting on quorum sensing molecules. Quorum sensing molecules are small peptides, usually called autoinducers, and the most famous one is N-acyl-l-homoserine lactones. Besides allowing communication between the bacteria, these molecules can influence the development of the biofilm and the interaction between quroum sensing and biofilm formation influences the pathogenesis of many bacterial infections [4, 6]. The properties of bacteria growing in biofilm are different when compared to the planktonic ones and this causes significant therapeutic and diagnostic issues. Biofilm producing bacteria causes chronic infections characterized by persistent inflammation and tissue damage [4]. The biofilm formation increases the ability of the bacteria to resist antibiotics and this antibiotic resistance causes a major threat to the disease management worldwide [8]. Antibiotic therapy can reverse the symptoms of disease cause by the bacteria growing in biofilm but is unable to kill the bacteria in the biofilm. Due to this antibiotic resistance biofilm infections show recurring symptoms even after cycles of antibiotic therapy. Several mechanisms have been proposed of biofilm resistance and two most common ones are antibiotic failure to penetrate full depth of the biofilm, since the polymers found in biofilms retard the diffusion of antibiotics and diffusion rate is slower within biofilms, and simply the bacterial cells exist in slow-growing state and are not susceptible to the antibiotics [1]. Besides being resistant to antibiotics bacteria in biofilm display resistance to disinfectants and host immune system.

Microtiter plate method (tissue culture plate assay) is mostly used for biofilm investigations in vitro. 96-well plate with U-shaped of V-shaped bottom is used for the biofilm analysis. The first step of biofilm formation is initial attachment of bacteria to the surface, depending on the ability of the bacteria to attach to the material. Even though, many different 96-well plates can be used, the one prepared for tissue culture allows the excellent cell attachment and proliferation. Another benefit of using tissue culture plates is that disruption and falling off the biofilm layer during the washing procedure is decreased [7].

Another method that is commonly used is tube method (TM). It is a qualitative assay that enables detection of biofilms. A biofilm has been formed, if a visible film can be seen on the walls of the tube [2].

To analyze the biofilm formation optical density (OD) of each 96-well plate's well is stained with crystal violet and measured using microtiter-plate reader. One of the problems suggested in this approach is that OD readers measure OD only at one point in the middle of the well, but if the thickness of the biofilm at that point differs from the rest of the well the measurement is questioned. However, resolubilization of the dye bound to the bacterial cells enables precise measurement of the biofilm production. The results are interpreted by using cut-off value. Cut-off value separates biofilm producing from biofilm nonproducing strains. Sometimes, this value is calculated by three standard deviations above the negative control that does not contain bacterial cells, while another method is to use result of positive control as a starting point for calculation.

However, the most usually the results obtained are averaged since all tested strains and negative controls are analyzed in triplicate or quadruplet. Also, the cut-off value (ODc) should be established by calculating three standard deviations (SD) above the mean OD of the negative control. Final OD values is expressed as average OD value reduced by Optical Density Cutoff (ODc) value. To provide easier understanding of the results the biofilm formation can be divided to categories: (1) no biofilm producer or non-adherent, (2) weak biofilm producer or slightly adherent, (3) moderate biofilm producer or moderately adherent, (4) strong biofilm producer or strongly adherent. Since many factors influence the biofilm production, the negative control as a reference for a calculation is a good start to avoid any inconveniences [7].

Biofilm structure can only be analyzed quantitatively, and it cannot be correlated with any parameters that can characterize its performance. The parameters that define biofilm structure are, however, straightforward. There are few computer software that quantify the parameters, but they do not interpret the obtained results [3]. The Biofilm classifier tool was developed to overcome certain difficulties in the analysis of the biofilm by the TCP method.

2 Material and Methods

2.1 Development of Biofilm Classifier

Biofilm classifier was developed in different stages, such as designing interface, so it is user-friendly and looking like 96-well plate, enabling inputs in boxes, and finally, applying formulas for biofilm forming category formations.

For development, Angular 5 was applied, using latest technologies for fastest and most responsive way that user can experience in that time. AngularJS used to be the prodigy of JavaScript frameworks, as it was first introduced in 2012 by Google. The structure, written in pure JavaScript, was designed to decouple the rationale of an application from the control of DOM (document object model), and went for dynamic page refreshes. Angular framework can have just a piece of the page constrained by AngularJS. This framework provided numerous incredible highlights that allowed the designer to efficiently build rich single-page application.

Biofilm classifier was tested on 192 samples provided by fellow researcher, sample results are used as an input model for biofilm classifier.

2.2 Biofilm Classifier Tool

Biofilm Classification Tool is available at biofilmclass.com and it is an online tool that uses two different approaches for classifying biofilm. First calculator calculates Optical Density for lower 4 quadruplets. Then we calculate Mean for upper four quadruplets in same row and compare to values found in table by using Optical Density Cutoff of lower four quadruplets.

Second calculator, by using TSB (A1-D1) or Negative Control we calculate ODc. After that, we find mean for every upper quadruplet and let it be our Mean 1, then find mean for lower quadruplets and name it Mean 2. Subtract our Mean 1 and Mean 2, and results that we get are values of our biofilm, which will be sorted by comparing it to table categories.

$$ODc = 3 * STDEV(neg\ control) + Mean\ OD(neg\ control)$$

The ODc value calculated using formulas above, was used for determining biofilm forming categories for the samples tested. The categories in which biofilm classifier could classify each samples are described in Table 1.

Table 1. Categorization of Biofilms

OD ≤ ODc	Non-adherent
ODc < OD ≤ 2 × ODc	Weakly adherent
2 × ODc < OD ≤ 4 × ODc	Moderately adherent
4 × ODc < OD	Strongly adherent

If optical density is lower or equal to optical density cut-off value, the biofilm classifier will classify biofilm as non-adherent, if optical density cut-off value is lower than optical density, and optical density is lower or equal to 2 times optical density cut-off value, biofilm classifier will categorize biofilm as weakly adherent, if 2 times optical density cut-off value is lower than optical density value, and optical density value is lower than 4 times optical density cut-off value, biofilm classifier will classify biofilm as moderately adherent and if 4 times optical density cut-off value is lower than optical density value biofilm classifier will classify biofilm as strongly adherent.

Biofilm Classifier Online Tool is represented as 96 well plate, A1 – H1, representing quadruplets just like 96-well plate used in laboratory. User can enter number between 2 and 24, to unlock quadruplet fields for inputs. Red fields represent Negative Control input fields. All other fields are inputs for bacterial values.

2.3 Examination of Biofilm Forming Capacity

The first approach uses the negative control of each dilution of the signaling molecule for the calculation of ODc. By using negative control in experiment, it is not expected to produce any results, since it is not treated by any bacteria. Control test is important because it helps the researchers to establish the baseline of the experiment by using comparation of results. Cutoff value is calculated by measuring three standard deviations plus mean of negative control. The second approach use TSB in quadruplet as the negative control and subtract the average of the ODc for each signaling molecule dilution from the ODc of the signaling molecule incubated with bacteria. This type of calculator works differently than first one. In first four rows, user enters negative control values. Next upper quadruplet calculates mean of entered values, so does the lower quadruplet. Next step is subtracting the mean result of upper quadruplet with mean result of lower quadruplet, and then comparing it with negative control through formula, calculating Optical Density Cutoff value.

2.4 Statistical Analysis of Biofilm Classifier

In order to determine the sensitivity, specificity and accuracy of the test, statistical analysis was performed using the MedCalc online software.

3 Results and Discussion

3.1 Statistical Analysis of Biofilm Classifier Tool

Previous studies of our fellow researcher provided samples used in the testing phase, the number of samples tested was 192. In order to determine sensitivity, specificity and accuracy of Biofilm Classifier Tool, results were tested using statistical calculator available on www.medcalc.org. The findings obtained were as follows: the mean sensitivity of the biofilm classifier is 100%, specificity and accuracy were as well 100%, as presented in Table 2. The precision of a test is its ability to correctly distinguish the cases. To measure the accuracy of a test, in all cases observed, the extent of the true positive and true negatives can be determined. The sensitivity of a test is its capacity to efficiently decide the cases. The specificity of a test is its capacity to correctly assess the cases found (Table 3).

Table 2. Specificity, sensitivity, and accuracy results of biofilm classifier

Type of testing	Non adherent	Weakly adherent	Moderately adherent	Strongly adherent
Specificity	100%	100%	100%	100%
Sensitivity	100%	100%	100%	100%
Accuracy	100%	100%	100%	100%
Total	100%	100%	100%	100%

Table 3. Result comparison of manual calculation and biofilm classifier

Type of calculator	Non adherent	Weekly adherent	Moderately adherent	Strongly adherent
Biofilm classifier 1	99	41	20	32
Biofilm classifier 2	99	41	20	32
Manual calculation	99	41	20	32

The table above presents a comparison of the results of manual calculation and calculation using the biofilm classifier. The results above show that the biofilm classifier gets just as accurate results as the manual calculation, without the possibility of bias. Of course, using a biofilm classifier makes computation time shorter and more reliable. Also, a different setting of the experiment may influence in the choice of one or another biofilm classifier, as stated in the previous sections of the scientific paper.

4 Discussion

The further analysis and understanding of the principles of biofilm forming and its characteristics is of great importance to reduce its negative influences on the health and environment. It is recognized that formation of biofilm is important in many bacteria causing diseases and the fact that these bacteria are resistant to most antibiotics makes the treatment of the various infections even more difficult. Furthermore, the biofilmproducing bacterial infections that may seem to be susceptible to the antibiotic treatment usually relapse after few week or months [5]. Another reason for further investigating the principles of biofilms is the development of the functional biomaterials. One of the most important aspects is determining the antibacterial methods for removal of biofilms from many different surfaces. To determine the microbial antibiotic susceptibility or assess the effectiveness of the antibiotic treatment the determining the relative reduction in bacteria due to the treatment and thus the analysis of biofilm forming capacity of different bacteria has to be done [8]. Several different methods for analyzing biofilms have emerged, but microtiter plate assay is most widely used one. Using this method biofilm structure can be analyzed quantitatively, since the parameters that define biofilm structure are straightforward. There are a few computer software that quantify the parameters, but they do not interpret the obtained results [3]. The two developed biofilm calculators are of great importance, since they shorten the time of calculating tremendously, while not affecting the results. As can be observed from the Table 2, specificity, accuracy, and sensitivity for both calculators were 100%. The main reason for obtaining the exact same results was the experimental setup: the samples used for testing these calculators had washed out negative control as well as adherence of tested substances was small, therefore the calculators gave the same outcome. The choice of calculator should be based on an individual's protocol of the experiment. Both calculators work on the same principle: the user inputs the variables obtained after the reading, and the calculators calculate the output of every well based on the four categories, as well as the mean ODc. The main difference between the two calculators is that one requires two negative controls in the experimental setup. The developed calculators will have a huge impact on prospective researches in this field, since it shortens the time of calculation and avoids possible bias.

5 Conclusion

This paper presents two biofilm calculators that were developed using Angular 5, and 192 samples were used to test the calculators. Both calculators had specificity, sensitivity, and accuracy of 100%. These types of calculators are very useful when doing biofilm experiments, since they shorten the time of calculation and reduce possible bias. The ease of use of the biofilm calculators and the short time needed for obtaining the results of the experiments improves the efficacy of the biofilm analyzing experiments. Therefore, these calculators are of huge importance for the biofilm formation analyzing experiments since they enable quick analysis of the results which directly influences the further improvement of the understanding of the biofilms that influence the health, environment, production of biomaterials and usage of the most optimal antibiotic treatment.

References

1. Costerton, J.W., Stewart, P.S., Greenberg, E.P.: Bacterial biofilms: a common cause of persistent infections. Science **284**(5418), 1318–1322 (1999)

2. Gordon, C., Simpson, A., Younger, J., Badour, L., Barret, F., Melton, D., Beachey, E.: Adherence of Coagulase-Negative Staphylococci to Plastic Tissue Culture Plates: a Quantitative Model for the Adherence of Staphylococci to Medical Devices (1985). Jcm.asm.org. https://jcm.asm.org/content/jcm/22/6/996.full.pdf. Accessed 1 Dec 2020
3. Beyenal, H., Lewandowski, Z., Harkin, G.: Quantifying biofilm structure: facts and fiction. Biofouling J. Bioadhesion Biofilm Res. **20**(1), 1–23 (2004)
4. Høiby, N., Bjarnsholt, T., Givskov, M., Molin, S., Ciofu, O.: Antibiotic resistance of bacterial biofilms. Int. J. Antimicrob. Agents **35**(4), 322–332 (2010)
5. Jefferson, K.K.: What drives bacteria to produce a biofilm? FEMS Microbiol. Lett. **236**(2), 163–173 (2004)
6. Nadell, C.D., Xavier, J.B., Levin, S.A., Foster, K.R.: The evolution of quorum sensing in bacterial biofilms. PLoS Biol **6**(1), e14 (2008)
7. Stepanović, S., Vuković, D., Hola, V., Bonaventura, G.D., Djukić, S., Ćirković, I., Ruzicka, F.: Quantification of biofilm in microtiter plates: overview of testing conditions and practical recommendations for assessment of biofilm production by staphylococci. Apmis **115**(8), 891–899 (2007)
8. Welch, K., Cai, Y., Strømme, M.: A method for quantitative determination of biofilm viability. J. Funct. Biomater. **3**(2), 418431 (2012)

Finite Element Analysis of Modified Hip Implant Surfaces

Aleksandra Vulović[1,2,3](✉) and Nenad Filipović[1,2,3]

[1] Faculty of Engineering, University of Kragujevac, Kragujevac, Serbia
{aleksandra.vulovic,fica}@kg.ac.rs
[2] Bioengineering Research and Development Center (BioIRC), Kragujevac, Serbia
[3] Steinbeis Advanced Risk Technologies Institute Doo Kragujevac, Kragujevac, Serbia

Abstract. Cementless hip implants are thought to be more durable than cemented hip implants. The primary problem with the cementless type is that unless the connection that forms between bone and implant is strong, the patient will require a revision surgery. A possible way to improve this connection is a surface modification of a hip implant, which would reduce micro movements between femoral bone and implant. The goal of the present study was to employ the numerical approach in order to analyze two hip implant surface modifications. Numerical approach consisted of the implementation of the finite element method (FEM) for the simulation of the bone-implant interaction and for the calculation of the shear stress values on the surface topographies of two hip implants.

Keywords: Finite element analysis · Hip implant · Shear stress · Surface modification · Titanium alloy

1 Introduction

Hip replacement procedure is one of the most commonly performed procedures in the world, with more than one million hip replacement surgeries performed annually [1]. When a hip joint is damaged and there is no available option for effective treatment, orthopedic surgeons need to replace that damaged part with an artificial one. This procedure involves the removal and replacement of the femoral head, in order to insert a metal stem into the hollow femoral bone.

Titanium and its alloys, especially Ti-6Al-4V, are considered to be a very good choice for hip implant material. That is due to the mechanical properties, mainly excellent biocompatibility and corrosion resistance. Ti-6Al-4V is for the cementless replacement components [2]. Although the hip replacement procedure maintains a minimum rate of early complications, they can still occur and are typically related to the aseptic loosening.

Aseptic loosening means that connection established between the femoral bone and the inserted implant is not strong. This can often happen with cementless hip implants, due to the way they are implanted. A possible way to reduce the aseptic loosening and need for revision surgery is to provide a better bone-implant fixation. This can be achieved by reducing the micro movements between the femoral bone and the hip

A. Badnjevic and L. Gurbeta Pokvić (Eds.): CMBEBIH 2021, IFMBE Proceedings 84, pp. 582–588, 2021.
https://doi.org/10.1007/978-3-030-73909-6_68

implant. According to the research [3, 4], this connection can be improved by using a hip implant with a rough surface. The effect of the implant surface modification on the bone-implant connection has been commonly analyzed using in vivo experiments. The obtained results have indicated the surface roughness of cylindrical implants can improve the bone – implant connection [5].

Numerical analysis (i.e. finite element analysis) provides a possibility to easily analyze numerous surfaces compared with in vivo studies. Not enough research has been performed in this area, but the published results are promising. Researchers have been able to employ numerical methods to optimize the groove angle on the femoral stem and the stem material based on the obtained stress values [6]. In our previous studies [7, 8], we have analyzed six implant surfaces that included cylindrical topographies. As a result of the application of the finite element method, we were able to obtain shear stress distribution on the surface of the hip implant topographies and perform assessment on which surface topography could potentially be used for hip implant.

The goal of the present study was to analyze two different modified hip implant surfaces in order to obtain shear stress distribution on the implant surface and compare the obtained results with our previous study. As in our previous study, the analysis was performed using the Finite Element Method, using the same boundary conditions.

The rest of the paper is organized as follows. Created models, their material properties and applied boundary conditions are described in Sect. 2. Section 3 covers the results of the finite element analysis and the discussion, while the conclusion is provided in Sect. 4.

2 Materials and Methods

2.1 Model Geometries

In order to carry out computational simulations of modified hip implant surface and its interaction with the femoral cortical bone, models had to be created. The analyzed models were developed utilizing a computer aided design (CAD) software. Both layers (hip implant and femoral cortical bone) for every model were exported from CAD software as.stp file which was then used for manual mesh creation.

Created hip implant models with are presented in Fig. 1. As it can be seen in this figure, the difference between the models was in the distance between half cylinder structures on the opposite sides of the model. Compared to our previous studies, in this one we did not position half cylinder topographies to be symmetric. Instead, we were interested to analyze what happens, if we position them as shown in Fig. 1.

The numbers of nodes and elements for each model (including femoral cortical bone and hip implant) are given in Table 1.

Table 1. Number of nodes and elements for each model

Model	Number of Nodes	Number of Elements
Model 1	58624	50460
Model 2	58562	50460

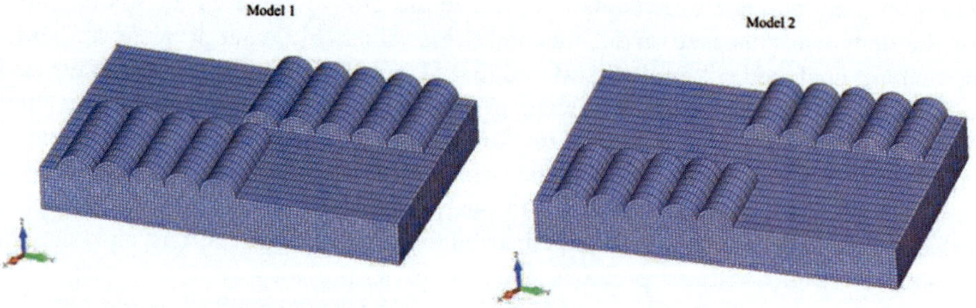

Fig. 1. Overview of the two hip implant models

2.2 Material Properties and Boundary Conditions

For the numerical analysis, two materials were used. Those materials were defined in order to describe behavior of femoral cortical bone and hip implant material (Ti6Al4V). Both materials were considered to be linear elastic, isotropic and homogeneous. Applied Young's modulus and Poisson ratio for both materials were adapted from the literature and the values are given in Table 2.

Table 2. Material properties applied for numerical analysis

Material	Young's modulus	Poisson ratio	Reference
Hip Implant (Ti6Al4V)	109 [GPa]	0.34	[9]
Femoral cortical bone	16.7 [GPa]	0.3	[10]

The constraints applied to the models were:

1) The upper surface of the cortical bone was fixed
2) The sideways of the implant were allowed to move in the y and z direction
3) The bottom surface of the implant was locked in the z direction
4) All the other elements were allowed to move in all three directions

The location where the force was applied can be seen in Fig. 2. In order to compare with previously obtained results, the value of the applied force was the same as in our previous studies [7, 8].

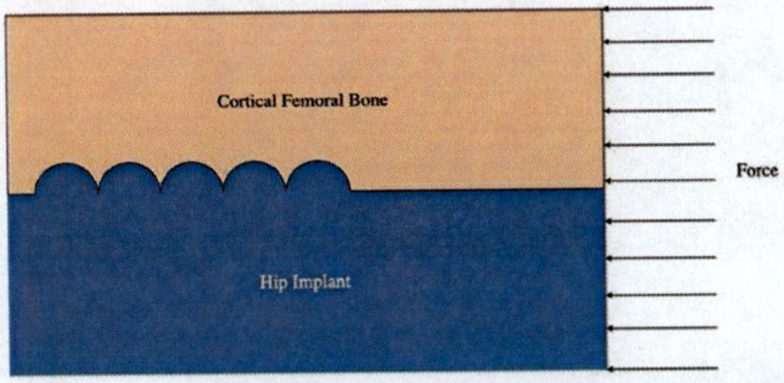

Fig. 2. Force application

Friction between cortical bone and hip implant was considered with a friction coefficient of 0.39 [9].

3 Results and Discussion

3.1 Model 1

The shear stress distribution for the first model can be seen in Fig. 3. The calculated shear stress values were in the range from 0.00463 to 7.299 MPa.

The highest values for the first implant model can be noted in the corners where the force was applied. The shear stress values in all other locations were lower than 7.299 MPa, and their maximum was around 6 MPa. Moving from location where the force was defined, it can be observed that the shear stress values on the smooth implant surface reduce, and around the middle of the model were almost 0 MPa. At the same time, it can be noticed that this was not the situation with rough implant surface where even the last cylindrical structure had the highest shear stress value of around 5 MPa.

3.2 Model 2

Figure 4 shows the shear stress distribution for the second model. In this case, shear stress values were in the range from 0.00697 to 7.283 MPa.

As in previous case, the highest values can be seen in corners where the force was applied. In other locations, the maximum value was around 6.5 MPa. The shear stress distribution for the second model was very similar to the first model as it can be observed of Fig. 3 and 4. The shear stress values on the smooth implant surfaces significantly reduced when moving towards the end of the model, and in this case the value started to get lower much closer to the area where the force was applied.

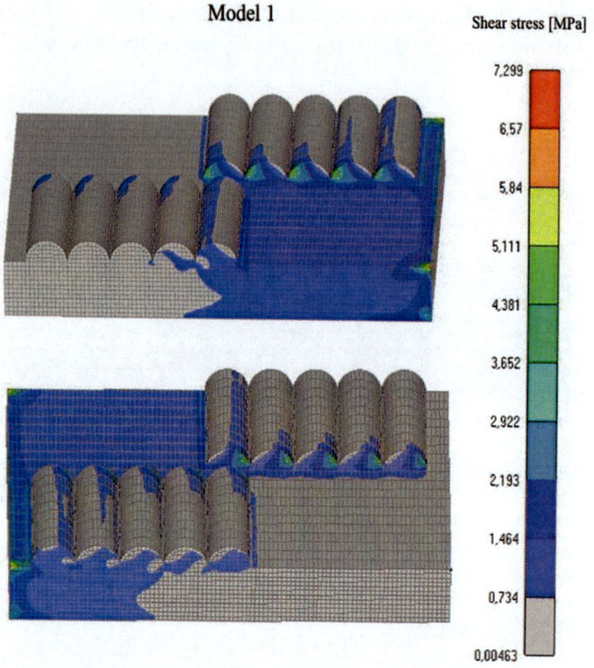

Fig. 3. Shear stress distribution – Model 1

3.3 Comparison of the Results

Comparing the previous results for models 1 and 2, it can be noticed that the shear stress values were in a similar range for both models. Our previous studies indicated that increased distance between the upper and bottom group of half cylinders leads to increased shear stress values, which in this case is not as clear considering very small difference. However, if we analyze the values without corner peaks, the same thing can be concluded - distance between half cylinders increased shear stress values on the rough implant surface.

According to [3] shear stress at the implant-bone interface should be minimized in order to improve bone ingrowth. Considering this, we can say that between these two models, model 2 would represent a more suitable choice.

Comparison with the best options from our previous studies is presented in Table 3.

Based on this comparison, we can say that the models presented in this paper would not represent a good option for hip implant surface topography considering only shear stress values. Even if we did not consider the values in the corners and compared the values from the rest of the models, conclusion would be the same.

It is important to note, that this study has a limitation in terms of applied force, which is dependent on the person's weight. The results presented here correspond to the force of a healthy person of an average weight.

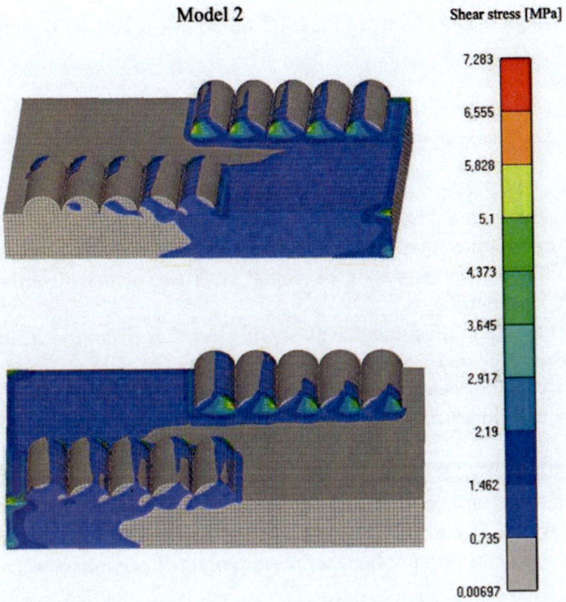

Fig. 4. Shear stress distribution – Model 2

Table 3. Shear stress results comparison with previously published results

Model 1	Model 2	Previous results [7]	Previous results [8]
7.299 MPa	7.283 MPa	4.145 MPa	3.209 MPa

4 Conclusion

Technological improvements during the previous decades have provided the opportunity to analyze a number of different surface topographies in a shorter amount of time, compared with in vivo studies. In order to improve bone-implant connection, shear stress should be minimized. Based on the shear stress parameter, the models presented here would not remain a suitable option for hip implant surface topography as we have obtained lower shear stress results in both of our previous studies. These preliminary results indicate that asymmetric surface topography should be avoided, at least for hip implants. However, we intend to analyze a few more asymmetric surface topographies before making the final decision.

Acknowledgement. This research is supported by the European Union's Horizon 2020 research and innovation programme under grant agreements No. 760921 and No. 952603. This article reflects only the author's view. The Commission is not responsible for any use that may be made of the information it contains. This research is also funded by Serbian Ministry of Education,

Science, and Technological Development [451–03-68/2020–14/200107 (Faculty of Engineering, University of Kragujevac)].

References

1. Shan, L., Shan, B., Graham, D., Saxena, A.: Total hip replacement: a systematic review and meta-analysis on mid-term quality of life. Osteoarthritis Cartilage **22**(3), 389–406 (2014)
2. Hu, C.Y., Yoon, T.-R.: Recent updates for biomaterials used in total hip arthroplasty. Biomater. Res. **22** (2018). article no. 33
3. Mattila, R.H., et al.: Bone attachment to glass-fibre- reinforced composite implant with porous surface. Acta Biomater. **5**, 1639–1646 (2009)
4. Halldin, A., Ander, M., Jacobsson, M., Hansson, S.: Simulation of the mechanical interlocking capacity of a rough bone implant surface during healing. BioMed Eng OnLine **14** (2015). article no. 45
5. Loberg, J.M.I., Hansson, S., Ahlberg, E.: Characterisation of titanium dental implants I: critical assessment of surface roughness parameters. Open Biomater. J. **2**, 1–8 (2010)
6. Noyama, Y., Nakano, T., Ishimoto, T., Sakai, T., Yoshikawa, H.: Design and optimization of the oriented groove on the hip implant surface to promote bone microstructure integrity. Bone **52**(2), 659–667 (2013)
7. Vulović, A., Filipović, N.: Computational analysis of hip implant surfaces. J. Serbian Soc. Comput. Mech. **13**(1), 109–119 (2019)
8. Vulović, A., Filipović, N.: Effect of hip implant surface modification on shear stress distribution. In: International Conference on Computational Bioengineering, pp. 151–159, Springer, Cham (2020)
9. Das, S., Sarangi, S.K.: Finite element analysis of femur fracture fixation plates. Int. J. Basic Appl. Biol. **1**(1), 1–5 (2014)
10. Aradhya, K.S.S., Doddamani, M.R.: Characterization of mechanical properties of SiC/Ti-6Al-4V Metal Matrix Composite (MMC) using finite element method. Am. J. Mater. Sci. **5**(3C), 7–11 (2015)

Genomic Alterations of KRAS and NRAS in B&H Colorectal and Non-small Cell Lung Cancer Patients

Lejla-Nur Smajović[1,2](✉), Dino Pećar[1], Lana Salihefendić[1], Altijana Hromić-Jahjefendić[2], and Rijad Konjhodžić[1]

[1] Alea Genetic Center, Olovska br. 67, 71000 Sarajevo, Bosnia and Herzegovina
[2] Genetics and Bioengineering Program, Faculty of Engineering and Natural Sciences, International University of Sarajevo, Hrasnička cesta br. 15, 71210 Sarajevo, Bosnia and Herzegovina
ahromic@ius.edu.ba

Abstract. Non-small cell lung cancer (NSCLC) and colorectal cancer (CRC) are two leading neoplasms in population of Bosnian cancer patients and a leading cause of death. RAS family of genes, in particular KRAS gene, is one of the most frequently mutated oncogene in these cancer types. This paper reports diagnostic samples from 36 CRC and 14 NSCLC Bosnian patients that were analyzed by next generation sequencing (NGS). Based on *Ion AmpliSeq* technology, NGS enabled molecular characterization of a panel of 22 hotspot regions of genes known to be involved in pathogenesis of NSCLC and CRC. We identified 33.3% of CRC patients harboring KRAS mutations and 11.1% patients harboring NRAS mutations. In NSCLC group, we identified 35.7% of KRAS-mutant patients. None of the NSCLC patients carried NRAS mutation. We found that CRC patients harbored 8 distinctive KRAS mutation subtypes, most of which were pathogenic. KRAS G13D was the most frequently occurring KRAS mutation subtype (33.3%) in CRC. 2 subtypes of NRAS mutations were detected, one of which was a well-documented pathogenic mutation. In NSCLC patients, we detected 3 KRAS mutant subgroups. KRAS G12C mutant was the most frequently detected KRAS mutation subtype in NSCLC patients (80%). Mean number of co-occurring aberrations per cancer type was 6.75 and 11 for CRC and NSCLC, respectively. Two most frequently co-occurring pathogenic aberrations were found in TP53 and PIK3CA genes. Thus, RAS-mutants in the population of B&H cancer patients are highly heterogeneous and distinctive molecular profiles have potentially different implications in cancer-associated pathways and therapeutic significance indicating the need for further investigation of KRAS subtype as a potent biomarker.

Keywords: KRAS mutations · NRAS mutations · Colorectal cancer · Non-small cell lung cancer · Next-generation sequencing

A. Badnjevic and L. Gurbeta Pokvić (Eds.): CMBEBIH 2021, IFMBE Proceedings 84, pp. 589–598, 2021.
https://doi.org/10.1007/978-3-030-73909-6_69

1 Introduction

At all levels of cells, gene products involved in proliferative, survival and differentiation pathways synthesized by proto-oncogenes are being expressed in tightly regulated manner. Proliferative signaling is a crucial component of healthy cellular metabolism and once proto-oncogenes are severely up-regulated as a result of mutations, they express abnormal proliferative activity and direct cell toward tumorigenesis [1, 2]. The most frequently mutated gene family in human cancer is the RAS family, present in more than 30% of human tumors with up to 3.4 million new cancer cases harboring some form of RAS mutation annually worldwide [3, 4]. Oncogenic RAS encompasses most of the original and some of the newly established hallmarks of cancer [5]. According to the Institute of Public health of Federation of Bosnia and Herzegovina, cancer is the second most common cause of morbidity in FB&H for an extended period of time (2014–2018), with the most lethal cancer types being malignant neoplasms of bronchus and lungs and malignant neoplasms of colon and rectum, both of which most commonly harbor RAS mutations [6]. Incidence rate of RAS oncogenic mutations in two of the most lethal and commonly occurring cancer types justify it as a crucially important molecular drug target.

However, there is a vast complexity among RAS mutations in mutational patterns not only because of varying RAS genes and cancer types, but because of the distinct patterns in RAS isoform mutated, position of mutation and the type of substitution all having implications in oncogenic potency, thus possible implications in treatment and prevention [7]. This oncogenic potency of RAS is manifested in a context-dependent manner and is dependable on subcellular, cellular and tissue environments in which RAS is present for determining its functional output [5].

In this study an overall aim is to analyze molecular profiles of the CRC and NSCLC patients of Bosnia and Herzegovina subjected to genomic profiling with special emphasis on KRAS and NRAS hotspot regions' mutational profiles to determine heterogeneity and genetic specificity of RAS mutations within B&H population. In order to analyze genetic specificity we focus on performing molecular-genetic profiling to: (a) determine KRAS and NRAS heterogeneity per cancer type, (b) compare relative frequencies of target mutations per cancer type, (c) determine pathogenicity status of specific mutations, and (d) determine most frequently co-occurring aberrations per mutation and cancer type in the population of Bosnian cancer patients. Molecular characterization has been generated by *Ion AmpliSeq* technology targeting hotspot regions of 22 genes known to be involved in pathogenesis of non-small cell lung cancer (NSCLC) and colorectal cancer (CRC). Our results show high RAS mutation heterogeneity within B&H population for both cancer types, with 8 and 3 distinctive mutation subtypes for CRC and NSCLC, respectively. Most detected mutations carry pathogenic status, and high incidence of other co-occurring pathogenic aberrations is observed. To our knowledge, this study provides the first overview of molecular characterization of RAS mutational patterns in cancer patients in the population of B&H.

2 Methods

2.1 DNA Extraction and Quantification

We extracted DNA from formalin-fixed and paraffin-embedded (FFPE) blocks of tumor biopsies and from buccal swabs, for determining genotype of cancerous tissue and *wild-type* noncancerous genotype, respectively. For the extraction of DNA from FFPE blocks we used *QIAamp DNA FFPE Tissue Kit*, and for extraction from buccal swab we used *QIAamp DNA Mini Kit*. Standard protocols suggested by kits' manufacturers were followed. The instrument used for quantification was the *Invitrogen Qubit 4* fluorometer and the assay used was *Qubit dsDNA HS Assay Kit* (in range of 10 pg/μl to 100 ng/μl). We followed standard protocol suggested for DNA quantification by *Thermo Fischer Scientific*.

2.2 NGS Library Preparation

NGS library was prepared with *Ion AmpliSeq Library Kit* 2.0 which contains necessary reagents for amplification, digestion and ligation of adapters to amplified regions of interest. *Ion AmpliSeq Colon and Lung Cancer Research Panel v2* was used which contains optimized primers for amplification of regions of interest targeted for sequencing. Amplification of DNA was used to perform a multiplex PCR reaction with 92 primer pairs. Followed by amplification is the ligation with adapters and barcodes of produced amplicons and partial digestion. After sequencing the instrument separated sequences from different patients according to unique barcode sequence. P1 adapter is necessary to ligate on amplified sequences of DNA so that the library can bind onto nanoparticle during clonal amplification. *Agencourt AMPure XP* reagent was used for purification of NGS library from primers' dimers, dNTPs and salts. Final concentration of library is important because optimal concentration used in final mix for chip loading is 100 pM and each library should be at optimal concentration before further usage.

2.3 Sample Loading and Initialization

Before sequencing the libraries were loaded on *Ion Torrent Chip* with *Ion Chef System*. Used kits for this procedure are: *Ion 510, Ion 520* and *Ion 530 Kit-Chef*. These kits contain all necessary reagents for chip loading for sequencing on *Ion S5 System*. With multi locus sequencing of genes, the *Ion AmpliSeq Colon and Lung Cancer Research Panel v2* targeted hotspot regions of 22 known genes associated with colon and lung neoplasm tissues, among which are KRAS and NRAS. *Torrent Browser* software was used to show results after the sequencing.

2.4 Analysis

The chi-square test was used to compare obtained overall frequencies. Statistical significance was defined as a p-value less than 0.05.

3 Results

In this study, total number of patients included in the study was 50. Out of 50 patients, 36 patients were diagnosed with CRC and 14 patients were diagnosed with NSCLC. Patients positive for RAS mutations detectable via panel (KRAS and NRAS mutations) were analyzed for genomic profiling and included in the study. We detected 15 CRC patients with at least one mutation in KRAS or NRAS gene, and 5 NSCLC patients with at least one KRAS or NRAS mutation. The mean age for CRC patients was 64.4 (range from 42 to 80 years). The predominant histology of CRC patients was *adenocarcinoma colonis* ($N = 10$). The mean age for NSCLC patients was 62 (range from 56 to 68 years). The predominant histology of NSCLC patients was *adenocarcinoma pulmonum* ($N = 4$).

3.1 KRAS Mutations in RAS⁺ CRC Patients

We determined that the frequency of RAS^+ CRC patients carrying at least one KRAS mutation is 80% ($N = 12$, out of 15 patients). The frequency of KRAS mutations in our sample is not statistically deviating from expected frequency of KRAS mutations present in CRC worldwide with value of $p = 0.06$ [4].

We determined that there were 8 different mutations occurring in KRAS gene among CRC patients. The most common KRAS subtype is mutation occurring at KRAS 25398281 C > T, detected in 4 patients (33.3%). Three subtypes of KRAS mutations (KRAS 25398284 C > T, KRAS 25378562 C > T, KRAS 25398285 C > T) were detected in 2 patients (16.7%), and other 4 KRAS mutation subtypes occurred in one patient each (8.3%) (see Table 1 and Fig. 1).

Table 1. KRAS mutations in CRC patients

	Gene	Position	Ref.	Var.	Patients w/ mut	% of patients w/mut.
1	KRAS	25398284	C	T	2	16.7
2	KRAS	25398284	C	A	1	8.3
3	KRAS	25378562	C	T	2	16.7
4	KRAS	25398285	C	T	2	16.7
5	KRAS	25398286	-	AGCTCT	1	8.3
6	KRAS	25380275	T	A	1	8.3
7	KRAS	25398279	C	T	1	8.3
8	KRAS	25398281	C	T	4	33.3

Abbreviations -, None; ref., Reference base; var., Variant detected; w/, With; mut., Mutation

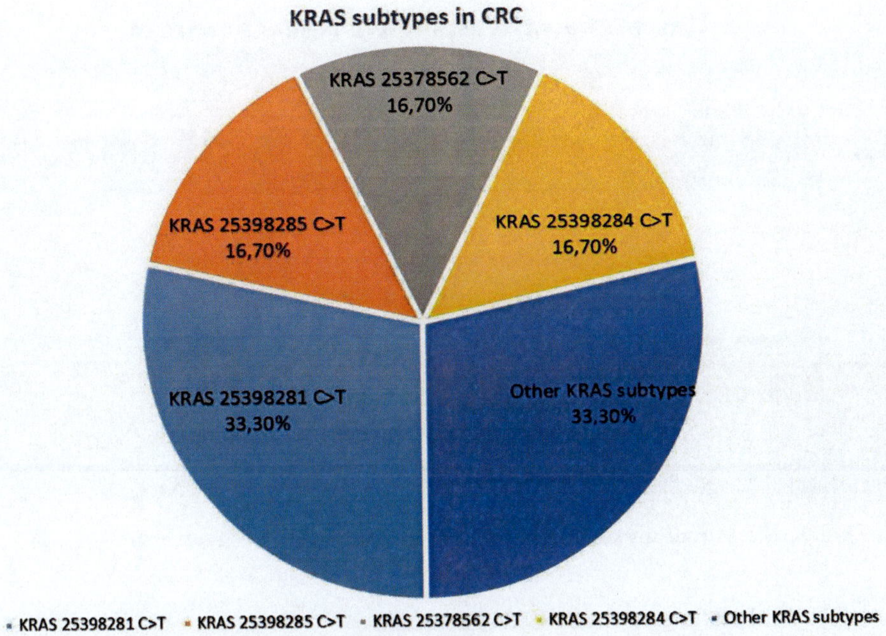

Fig. 1. Visual representation of KRAS subtypes in CRC patients

Out of 8 KRAS subtypes, 4 of them (50%) occurred as a single KRAS mutation, while 4 (50%) of the subtypes co-occurred with other KRAS mutation. Genomic profiling of 12 $KRAS^+$ patients diagnosed with colorectal cancer revealed 10 patients carrying only one KRAS mutation (83.3%), and 2 patients with two KRAS subtype mutations (16.7%). None of the patients carried more than two KRAS mutations.

We examined pathogenicity status of detected KRAS mutation in CRC patients according to ACMG classification of automated criteria and pre-existing submitted research about variant's pathogenicity [25]. All detected KRAS mutations were pathogenic except for one, which is not registered in online databases. Hence, 11 (91.7%) KRAS mutations were pathogenic according to ACMG classification. Further analysis of detected variants revealed that transversions accounted for 2 mutations (14.2%), while transitions accounted for 11 mutations (78.5%). Majority of KRAS mutations were single nucleotide polymorphisms accounting for 13 mutations (92.8%), with only one insertion (7.1%). All detected mutations of KRAS were heterozygous.

3.1.1 Co-occurring Aberrations of $KRAS^+$ CRC Patients

Out of the total $KRAS^+$ patients, 8 patients (66.7%) had at least one additional pathogenic mutation, with 7 of them (58.3%) having one and one of the patients (8.3%) having two co-occurring pathogenic aberrations.

Co-occurring pathogenic aberrations were detected in two genes: TP53 (50%) and PIK3CA (62.5%) (see Fig. 2).

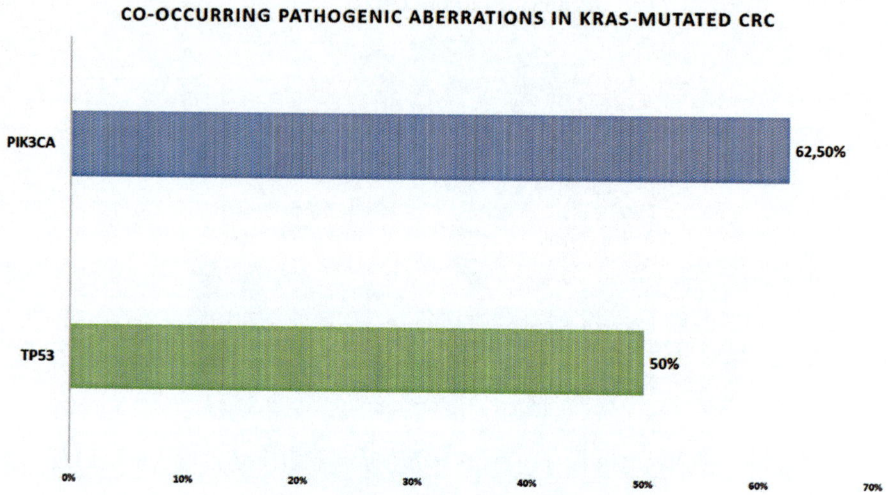

Fig. 2. Most frequently co-occurring pathogenic aberrations in KRAS+ CRC patients

3.2 NRAS Mutations in RAS+ CRC Patients

We detected two subtypes of NRAS mutations in CRC patients. The most common.

NRAS subtype is mutation occurring at NRAS 115256530 G > T, detected in 3 patients (75%). Other subgroup NRAS 115258718 G > T occurred in one patient (25%). According to ACMG classification [25], NRAS 115256530 G > T mutation subtype carries status of pathogenic mutation.

3.3 KRAS Mutations in RAS+ NSCLC Patients

We found that the frequency of KRAS mutations in our sample is not statistically deviating from expected frequency of KRAS mutations present in NSCLC, with observed value of $p = 0.39$ [4]. We determined that there were 3 different mutations occurring in KRAS gene among NSCLC patients. The most common KRAS subtype is mutation occurring at KRAS 25398285 C > A, detected in 4 patients (80%). One subtype of KRAS mutation (KRAS 25398284 C > A) was detected in 2 patients (40%) while one subtype of KRAS mutation (KRAS 25380358 G > A) was detected in 1 patient (20%) (see Fig. 3).

Genomic profiling of 5 KRAS+ patients diagnosed with NSCLC cancer revealed 3 patients carrying only one KRAS mutation (60%), and 2 patients with two KRAS subtype mutations (40%). None of the patients carried more than two KRAS mutations. Two detected KRAS mutations (66.7%) found in NSCLC patients were pathogenic according to ACMG classification [25]. Further analysis of detected variants revealed that transversions accounted for 6 mutations (85.7%), while transitions accounted for 1 mutation (14.3%). All detected mutations of KRAS were single nucleotide polymorphisms and their genotype was heterozygous.

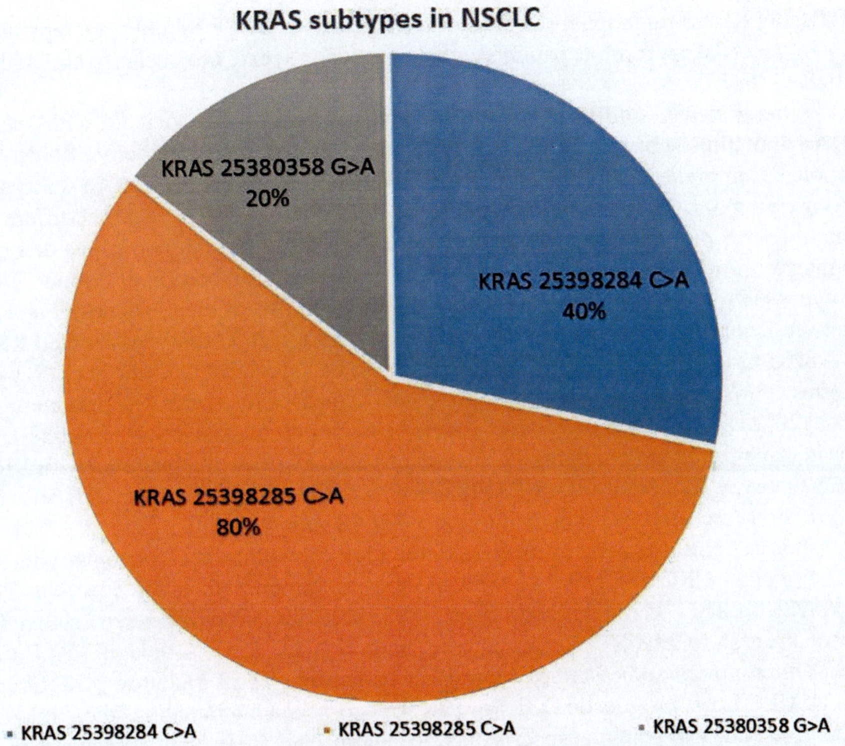

Fig. 3. Visual representation of KRAS subtypes in NSCLC patients

4 Discussion and Conclusion

We report the landscape of genomic alterations in a series of 15 RAS-mutant colorectal cancer patients and 5 KRAS-mutant non-small cell lung cancer patients with a distinctive genetic profile, and identifying most frequently occurring KRAS subtypes per cancer type along with most frequently co-occurring pathogenic aberrations in Bosnian cancer patients.

Downstream applications of such profiling may include identifying predictive and prognostic biomarkers within heterogeneous groups of KRAS mutation harboring patients. Accumulating literature suggests the prognostic role of KRAS mutations in both NSCLC and CRC is well supported and evident, regardless of the stage of cancer progression. $KRAS^+$ mutation status is correlated with significantly worse disease free survival and OS compared to *wildtype* patients [7–9]. Hence, it is an independent prognostic marker in CRC and NSCLC. Furthermore, $KRAS^+$ status was significantly associated with the absence of response to drug cetuximab and highly predictive selection criterion for treatment decision for patients with metastatic CRC and NSCLC, with benefits of cetuximab limited to patients with KRAS *wildtype* tumor status [10–12]. Another significant treatment implication of $KRAS^+$ status is its role as negative predictor of clinical benefit for tumor response in NSCLC patient treated with EGFR-TKIs [13,

14]. Using KRAS mutational status to restrict the use of EGFR inhibitor therapy would help to select the appropriate patients and ineffective and toxic treatments to non-suitable patients [15, 16].

As many newly studies report, differential outcomes are associated with specific KRAS mutation subtypes regarding amino acid and codon specificity [8]. In fact, National Comprehensive Cancer Network (NCCN) recommended KRAS subtype as the emerging new biomarker [17]. The reason for this is due to the fact that differential mutant KRAS proteins affect downstream signaling in a different way, hence therapeutic intervention may need to be re-directed towards taking into account tumour KRAS subtypes [18]. KRAS subtypes are heterogeneous in terms of their prognostic and predictive effect [19]. Most studied are the effects of KRAS G12C subtype (denoted KRAS 25398285 C > A in our study) which is proven to be associated with shorter OS, worse progression-free survival and adverse prognosis compared to other KRAS mutation subtypes [20, 21]. While not prognostic, it may be an important predictive biomarker as its covalent inhibitors are emerging [22]. Previous studied show that patients with KRAS G12V subtype (denoted KRAS 25398284 C > A in our study) also showed worse progression-free survival compared to other KRAS subtypes [18].

When it comes to concurrent mutations, they are common in patients with both KRAS-mutant CRC and NSCLC, with the most common pathogenic events in TP53 and PIK3CA. The commonly co-occurring TP53 aberrations have been proved to reduce responsiveness to EGFR TKIs in patients with NSCLC [18]. Although TP53 mutations are common, some studies report that concurrent TP53 mutation in KRASmutant NSCLC is not prognostic [23]. For PIK3CA co-occurring mutations there may be a clinical relevance as potential specific targeted approaches are in clinical evaluation [24].

In our study, in addition to reporting and selecting patients suitable for therapy based on $KRAS^+$ status, we also extend identification of patients based on the specificity of the genomic imprint of KRAS mutations which has implications in correlation to PFS and OS of both CRC and NSCLC. Thus, KRAS mutational subtype is of equal or even greater relevance than categorical identification of patients based on KRAS mutation presence.

Future therapeutic directions for managing neoplastic diseases should be redirected toward identifying tumor-specific and mutation-specific alterations for more successful treatment outcome and clinical decision-making process. Our study identifies importance of using the KRAS subtype as a potentially more beneficial biomarker for patients in B&H. However, this study implies a need for further research of KRAS subtypes as prognostic and predictive biomarkers and validation within population of Bosnian cancer patients over longer periods of time and with larger patients groups.

Compliance with Ethical Standards. Conflict of Interest: The authors declare that they have no conflict of interest.

Informed consent: The authors declare that no personal information was disclosed.

References

1. Macdonald, F., Ford, C.H.J., Casson, A.G.: Molecular Biology of Cancer, 2nd ed. Taylor & Francis e-Library (2005)

2. Divan, A., Royds, J.A.: Cancer biology and treatment. Oxford University Press, Oxford (2020)
3. Papke, B., Der, C.J.: Drugging RAS: know the enemy. Front. Can. Therapy **335**, 1158–1163 (2017). https://doi.org/10.1126/science.aam7622
4. Prior, I.A., Hood, F.E., Hartley, J.L.: The frequency of Ras mutations in cancer. Am. Assoc. Can. Res. (2020). https://doi.org/10.1158/0008-5472.CAN-19-3682
5. Pylayeva-Gupta, Y., Grabocka, E., Bar-Sagi, D.: RAS oncogenes: weaving a tumorigenic web. Nat. Rev. Cancer **11**, 761–773 (2011)
6. Statistički Godišnjaci (2019). ZZJZ FBiH. www.zzjzfbih.ba/statistickigodisnjaci/
7. Li, S., Balmain, A., Counter, C.M.: A model for RAS mutation patterns in cancers: finding the sweet spot. Nat. Rev. Cancer **18**, 767777 (2018)
8. Izar, B., Zhou, H., Heist, R., Azzoli, C., Muzikanski, A., et al.: The prognostic impact of KRAS, its codon and amino acid specific mutations, on survival in resected I lung adenocarcinoma. J. Thoracic Oncol. **9**(9), 1363–1369 (2014)
9. Nadal, E., Chen, G., Prensner, J.R., Shiratsuchi, H., Sam, C., Zhao, L., et al.: KRAS G12C mutation is associated with poor outcome in surgically resected lung adenocarcinoma. J. Thorcacic Oncol. **9**, 1512–1522 (2014)
10. Kim, J.H.K., Kim, H.S., Kim, B.J.: Prognostic value of KRAS mutation in advanced non-small-cell lung cancer treated with immune checkpoint inhibitors: a meta-analysis and review. Oncotarget **8**, 48248–58252 (2017). https://doi.org/10.18632/oncotarget.17594
11. Lievre, A., Bachet, J.B., Le Corre, D., Boige, V., Landi, B., et al.: KRAS mutation status is predictive of response to cetuximab therapy in colorectal cancer. Can. Res. **66**, 8 (2006)
12. Van Cutsem, E., Kohne, C.H., Hitre, E., Zaluski, J., et al.: Cetuximab and chemotherapy treatment for metastatic colorectal cancer. New Engl. J. Med. **360**(14), 1408–1417 (2009)
13. Mao, C., Qui, L., Liao, R., Du, F., et al.: KRAS mutations and resitance to EGFR-TKIs treatment in patients with non-small cell lung cancer: a meta- analysis of 22 studies. Lung Can. **69**(3), 272–278 (2010)
14. Roberts, P., Stinchcombe, T., Der, C.J., Socinski, M.A.: Personalized medicine in non-small-cell lung cancer: is KRAS a useful marker in selecting patients for epidermal growth factor receptor-targeted therapy? J. Clin. Oncol. **28**, 31 (2010)
15. Kris, M.G., Johnson, B.E., Berry, L.D., et al.: Using multiplexed Assays of oncogenic drivers in lung cancers to select targeted drugs. JAMA (2014). https://doi.org/10.1001/jama.2014.3741
16. Herbst, R.S., Schlessinger, J.: Small molecule combat cancercausing KRAS protein at last. Nature **575**, 294–295 (2019)
17. Eshraghi, A., Eshandbod, M., Eshraghi, A., et al.: The predictive value of KRAS and NRAS mutations in metastatic colorectal cancer. Med. J. Islamic Republic Iran **34**, 39 (2020)
18. Ihle, N.T., Byer, L.Y., Kim, E.S., et al.: Effect of KRAS oncogene substitutions on protein behaviour: implications for signalling and clinical outome. J. Nat. Can. Inst. **104**(3), 228–239 (2012)
19. Zer, A., Ding, K., Ming Lee, S., Glenwood, D.G., et al.: Pooled analysis of the prognostic and predictive value of KRAS mutation status and mutation subtype in patients with non-small cell lung cancer treated with epidermal growth receptor tyrosine kinase inhibitors. J. Thoracic Oncol. 312–33 (2016)
20. Schirripa, M., Nappo, F., Cremolini, C., Salvatore, L., et al.: KRAS G12C metastatic colorectal cancer: specific features of a new emerging target population. Clinical Colorectal Cancer **20** (2020)
21. Svaton, M., Fiala, O., Pesek, M., et al.: The prognostic role of KRAS mutation in patients with advances NSCLC treated with second- or third-line chemotherapy. Antican. Res. **36**, 1077–1082 (2016)
22. Cui, W., Franchini, F., Alexander, M., et al.: Real world outcomes in KRAS G12C mutation positive non-small cell lung cancer. Lung cancer **146**, 310–317 (2020)

23. Shepherd, FA., Lacas, B., Le Teuff, G., Hainaut, P., Jeanne, PA., Pignon, JP., et al.: Pooled analysis of the prognostic and predictive effects of TP53 computation status combined with KRAS or EFGR mutation in early stage resected nonsmall-cell lung cancer in four trials of adjuvant chemotherapy. J. Clin. Oncol. **35**, 2018-27
24. Horn, L., Cass, A.S.: Current landscape of personalized therapy. Thoracic Surgery Clinics 30 (2020)
25. Kopanos, C., Tsiolkas, V., Kouris, A., Chapple, C.E., Aguilera, M.A., Meyer, R, Massouras, A.: VarSome: The Human Genomic Variant Search Engine. Oxford Bioinfrmatics (2018). https://doi.org/10.1093/bioinformatics/bty897

Cardiovascular, Respiratory and Endocrine System Engineering

Incidence of Arterial Hypertension in Patients with Obstructive Sleep Apnea Treated at the Mostar Medical Sleep Center

Josip Lesko[1], Nikolina Obradović[2](✉), and Vana Turudić[2]

[1] Department of Otorinolaryngology, University Clinical Hospital Mostar, Mostar, Bosnia and Herzegovina

[2] Department of Family Medicine, DZ Mostar, Mostar, Bosnia and Herzegovina

Abstract. Background. Obstructive sleep apnea-syndrome and cardiovascular disease are a public health problem. There are a number of comorbidities associated with obstructive sleep apnea, yet it is most commonly associated with hypertension. The aim of this study was to determine the incidence of arterial hypertension in patients with apnea with respect to the degree of obstructive sleep apnea, and with regard to their age and sex, and to determine whether there is a difference between patients with obstructive sleep apnea suffering from arterial hypertension with respect to gender and age.

Methods. A total of 213 respondents who were admitted to the Center for Sleep Medicine Mostar participated in this cohort form of research. The study included those subjects, of both sexes and all age, who were diagnosed with obstructive sleep apnea. The polysomnographic device Respironics Alice LE (Philips-Respironics, Endhoven, the Netherlands) was used in the study to diagnose OSA. Data on arterial hypertension were obtained from discharge letters (anamnestic data). The parameter according to which we selected the study group is apnea / hypopnea index greater than 5.

The results: The study included 213 subjects, 163 subjects were men and 50 subjects were women. There was almost the same number of patients with severe apnea who had hypertension as those with severe apnea and no hypertension ($\chi^2 = 3.362$; $P = 0.118$). No statistically significant difference was found in OSA patients according to the degree of apnea and the sex of the patients ($\chi^2 = 5.1388$; $P = 0.076$). There was also no association between gender and the occurrence of hypertension in OSA patients ($P = 0.592$). The number of hypertensive OSA patients older than 50 years was higher compared to hypertensive OSA patients younger than 50 years ($P < 0.001$).

Conclusion. More than half of OSA patients who had arterial hypertension were older than 50 years.

Keywords: Obstructive sleep apnea · Arterial hypertension · Age · Sex · Comorbidities

A. Badnjevic and L. Gurbeta Pokvić (Eds.): CMBEBIH 2021, IFMBE Proceedings 84, pp. 601–608, 2021.
https://doi.org/10.1007/978-3-030-73909-6_70

1 Introduction

Obstructive sleep apnea-syndrome and cardiovascular disease are a public health problem [1]. Studies have shown that arterial hypertension and obstructive sleep apnea have a synergistic effect on the cardiovascular system [2].

Obstructive sleep apnea (OSA) is a multisystem breathing disorder associated with increased morbidity and mortality [3]. Clinical manifestations of apnea are post-awakening headache, daytime drowsiness, decreased cognitive performance, and snoring [4]. Obstructive sleep apnea (OSA) and obstructive sleep apnea / hypopnea syndrome (OSAHS) have long been considered predominantly male conditions [5]. Obstructive sleep apnea (OSA) is a sleep disorder caused by transient obstruction of the upper airway and results in intermittent hypoxia, sleep fragmentation, sympathetic nervous system activation, and arousal which can have an adverse effect on cardiovascular disease [6]. There are a number of comorbidities associated with obstructive sleep apnea, yet it is most commonly associated with hypertension [7, 8]. Respiratory events during sleep are associated with hypertensive peaks occurring in apnea and hypopnea [7]. The increased incidence of hypertension in OSA individuals has been well established, and there appears to be an association between OSA levels and blood pressure [9]. In addition to age and gender, a significant-elevated AHI (anea / hypopnea index) is one of the possible risks affecting the development of hypertension. It is considered to have a synergistic effect with age, obesity, nocturnal and diurnal hypoxia in the development of hypertension in these patients [10].

However, it is assumed that the increase in blood pressure in OSA patients occurs primarily due to excessive activity of the sympathetic nervous system. Intermittent nocturnal hypoxia and negative intrathoracic pressure lead to chemoreceptor activation and increased sympathetic activity that persists during wakefulness [11]. The basic therapy of obstructive sleep apnea is constant positive pressure therapy, CPAP. CPAP therapy, as well as other alternative therapies for apnea, has been shown to reduce sympathetic nervous system activity [12].

The aim of this study was to determine the incidence of arterial hypertension in patients with apnea considering the degree of obstructive sleep apnea, and their age and gender.

2 Subjects and Methods

The study included 213 subjects, 163 subjects were men and 50 subjects were women. All subjects were admitted to the Center for Sleep Medicine Mostar in the period from January 1, 2014 to February 22, 2019, and they underwent an all-night polysomnography. The study included those subjects who were diagnosed with obstructive sleep apnea. Respondents of both sexes and all age categories were included.

The polysomnographic device Respironics Alice LE (Philips-Respironics, Endhoven, the Netherlands) was used in the study to diagnose OSA. Data on arterial hypertension were obtained from discharge letters (anamnestic data). The parameter according to which we selected the study group is apnea / hypopnea index greater than 5. Within this group we observed the frequency of arterial hypertension considering to demographic data (age and sex), the degree of obstructive sleep apnea and accompanying comorbidities.

For statistical analysis of the obtained data was used the software systems SPSS for Windows (version 13.0, SPSS Incorporated Chicago, Illinois, USA) and Microsoft Excel (version Office 2010, Microsoft Corporation, Redmond, Washington, USA). The $\chi 2$ test was used to test the significance of the observed differences. The limit of statistical significance was set at $P < 0.05$. P values that could not be expressed to three decimal places are expressed as $P < 0.001$.

3 The Results

The study included 213 subjects, 163 subjects were men and 50 subjects were women.

There was almost the same number of patients with severe apnea who had hypertension as those with severe apnea and no hypertension. The largest number of hypertensive patients had a severe degree of OSA ($\chi^2 = 3.362$; $P = 0.118$) (Table 1).

Table 1. Prevalence of arterial hypertension in patients with apnea considering the degree of apnea.

	Mild OSA	Moderate OSA	Severe OSA	P	χ^2
	Number of patients %	Number of patients %	Number of patients %		
Hypertension	19 34,54	23 39,66	49 49,00	0,118	3,362
No hyper-tension	36 65,45	35 60,34	51 51,00		

In women, mild and moderate OSA were slightly more prevalent compared to severe OSA. In men, the most common was severe OSA, while mild and moderate OSA were less common ($\chi^2 = 5.1388$; $P = 0.076$) (Table 2).

Table 2. Representation of patients by sex and degree of apnea.

	Mild OSA	Moderate OSA	Severe OSA	p	χ^2
	Number of patients %	Number of patients %	Number of patients %		
Male	37 67,27	43 22,41	83 83	0,076	5,1388
women	18 32,72	15 25,82	17 17		

There was no statistically significant difference among the OSA patients examined in this study considering to gender and the presence of hypertension ($P = 0.592$) (Fig. 1).

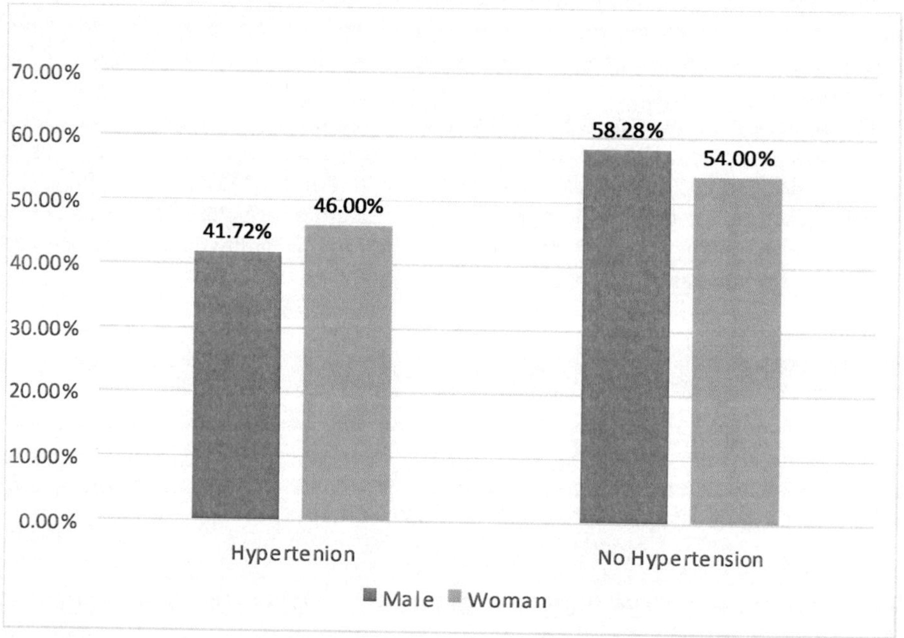

Fig. 1. Representation of patients with considering gender and the presence of hypertension in patients with obstructive sleep apnea

The number of hypertensive OSA patients older than 50 years is higher compared to hypertensive OSA patients younger than 50 years ($P < 0.001$) (Fig. 2).

Type 2 diabetes, depression, cardiac patients, hypothyroidism, and acromegaly were the most common other comorbidities (Table 3).

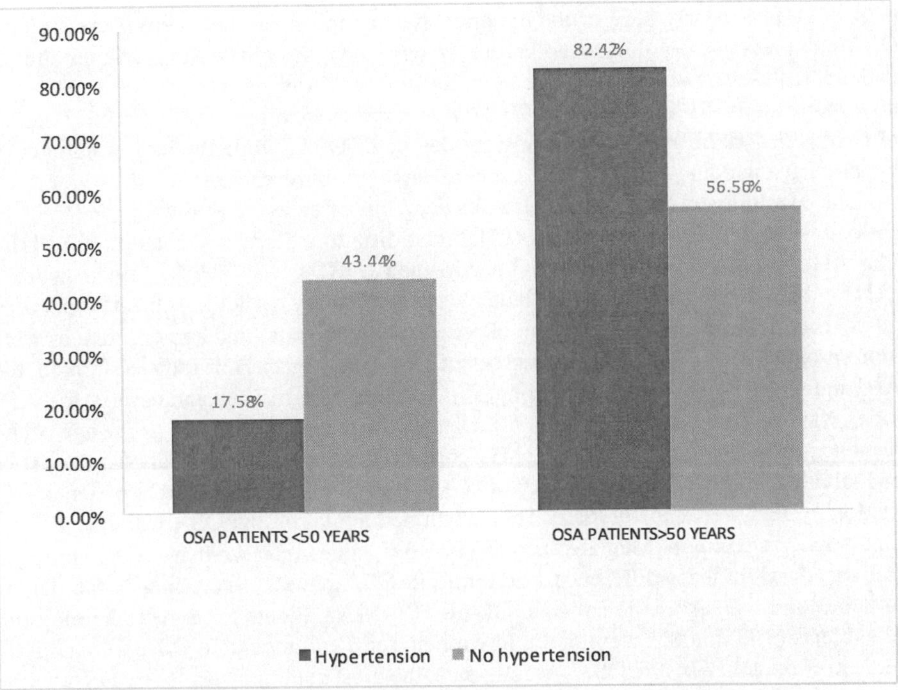

Fig. 2. Prevalence of arterial hypertension in patients with OSA considering age.

Table 3. Incidence of other comorbidities in OSA patients.

Other comorbidities	Number of patients	%
0- nothing special	159	74,60%
1- hypothyroidism	10	4,70%
2- type 2 diabetes	21	9,85%
3- depression	8	3,75%
4- cardiac patient	16	7,51%
5- acromegaly	6	2,81%

4 Discussion

According to the Wiscinsin Sleep cohort study, the prevalence ratio of hypertension during the four-year follow-up in patients with an AHI index of 0.1–4.9 compared with patients who did not have sleep-disordered breathing was 1.4. In those with an AHI index of 5.0–14.9, this ratio was higher and was 2.0, while in those with AHI $\geqslant$ 15.0, or severe apnea, this ratio was 2.9. This study showed a strong association between the degree of apnea and the development of hypertension [13]. In a study conducted by Hsu

H.C. et al. [14] nearly 80% of the hypertensive patients examined were found to have OSA that had not been diagnosed before. In our study was almost the same number of patients with severe apnea who have hypertension (49%) as those with severe apnea and no hypertension (51%). A total of 20.87% of hypertensive patients had a mild degree of OSA, 25.27% of them had a moderate degree of OSA, while the largest number of hypertensive patients, 53.84%, had a severe degree of hypertension.

The development of OSA as well as its degree is often associated with gender. Most patients with this disorder are men [15]. According to a study by Fietze I. et al. [16], The AHI index was significantly higher in men (AHI $\geq$ 15: 29.7%) than in women (AHI $\geq$ 15: 13.2%). Also, the AHI index was significantly higher in subjects older than 60 years compared to those less than 60 years of age. Gender and age differences were more pronounced in more severe degrees of OSA and higher AHI indices. In men, the AHI index grew continuously with age. In women, a significant increase in the AHI index was seen only after the age of 50. Gender differences decreased as the age of the respondents increased. Cai A. et al. [17] compared the epidemiological characteristics of individuals with varying degrees of OSA. According to their research, patients with moderate OSA were mostly male. They examined a total of 865 OSA patients, 79% of that number were men. Men also had higher AHI values than women. In our study, no statistically significant difference was found in OSA patients according to the degree of apnea and the sex of the patients. Of the 213 OSA patients examined, the majority of subjects (163 in total) were male. In women, mild and moderate OSA were slightly more prevalent than severe OSA. Among men, the most common was severe OSA, while mild and moderate OSA were less common. No statistically significant difference was found in this study with considering gender and the presence of hypertension in OSA patients, which may be due to the lack of data on the existence of arterial hypertension in the examined OSA patients. A 2017 study by Irene Cano Pumarega [18] examined the association between OSA and the risk of developing severe hypertension, taking into account the sex of the patient.

They used the Respirtory Disturbance Index (RDI), which is often used instead of the AHI index, to determine the degree of hypertension. According to their results, in men, RDI $\geq$14 (severe OSA) was significantly associated with severe hypertension. This association was not statistically significant in women. These results showed that moderate and severe OSA were associated with the development of more severe forms of hypertension in men but not in women. According to a study by Jenner and colleagues [19], of the 95 hypertensive subjects (56% were women), OSA was associated with the development of severe arterial hypertension in both sexes compared to those who were not OSA patients.

Age, however, according to our study, played the most significant role in the development of hypertension in OSA patients. There was a statistically significant difference which shows that the number of hypertensive OSA patients older than 50 years is higher compared to hypertensive OSA patients younger than 50 years. Vgontzas and colleagues [20] in the 2018 study also demonstrated a significant association between moderate to mild severe OSA and age with the onset of hypertension. The development of hypertension in patients aged 20 to 80 years was significantly higher compared with patients who were the same age but without OSA. However, according to the results of this study, the association between hypertension and OSA was significantly expressed before age 60, while the incidence of hypertension associated with moderate OSA in people older than 60 years was significantly lower.

Although hypertension is one of the most common comorbidities associated with OSA, there are other conditions whose possible cause is associated with the existence of OSA. According to Heinzer and colleagues [21] research, AHI values were significantly associated with diabetes in men and with depression in women. Mokhlesi B et al. [22] did a study in which they calculated the most common comorbidities of OSA. In their study, hypertension was the most common, while the second most common was type 2 diabetes, followed by cardiovascular disease, of which ischemic heart disease was the most common. In our study, type 2 diabetes, depression, cardiac patients, hypothyroidism, and acromegaly were the most common among OSA patients.

5 Conclusion

In our study, a larger number of respondents were men. Almost the same number patients were found with arterial hypertension as with those without hypertension. Although severe OSA was most prevalent in male patients, gender has shown to have no significant effect on the development of severe OSA. According to our results, the frequency of arterial hypertension does not depend on the sex of the patients. However, the frequency of arterial hypertension significantly depends on the age of the patient.

Also, other comorbidities such as type 2 diabetes, depression, cardiac patients, hypothyroidism, and acromegaly were common among OSA patients. Yet none of those comorbidities had a significantly high frequency.

Although our study did not confirm a significant incidence of arterial hypertension in OSA patients with considering patient gender and degree of apnea, the pathophysiological features of apnea, namely intermittent hypoxia leading to sympathetic activation and oxidative stress, which could lead over the years to an increase in arterial pressure [23], suggest that attention should certainly be paid to the possible association between these two clinical entities.

References

1. Halima, A.B., Aouadi, S., Bejjar, D., et al.: Hypertension and Atrial Fi-brillation: What is the prevalence of obstructive sleep apnea syndrome? Tunisia Med. **96**, 187–192 (2018)
2. Cai, A., Wang, L., Zhou, Y.: Hypertension and obstructive sleep apnea. Hypertens Res. **39**, 391–395 (2016)
3. Savini, S., Ciorba, A., Bianchini, C., et al.: Assessment of obstructive sleep apnoea (OSA) in children: an update. Acta Otorinolaryngol Ital. **39**, 289–297 (2019)
4. Maspero, C., Giannini, L., Galbiati, G., et al.: Obstructive sleep apnea syndrome: a literature review. Minerva Stomatol. **64**, 97–109 (2015)
5. Bonsignore, M.R., Saaresranta, T., Riha, R.L.: Sex differences in obstructive sleep apnea. Eur Respir Rev. **28**, 190030 (2019)
6. Cananzi, S.G., White, L.A., Barzeger, M., et al.: Obstructive sleep apnea intensifies stroke severity following middle cerebral aartery occlusion. Sleep Med. **67**, 278–285 (2020)

7. Bonsignore, M.R., Baiamonte, P., Mazzuca, E., et al.: Obstructive sleep apnea and comorbidities: a dangerous liaison. Multidiscip Respir Med. **14**, 8 (2019)
8. Ahmad, A.N., McLeod, G., Al Zahrani, N., et al.: Screening for High Risk of Sleep Apnea in an Ambulatory Care Setting in Saudi Arabia. Int. J. Environ. Crisp. Public Health. **16**, 459 (2019)
9. Salman, L.A., Shulman, R., Cohen, J.B.: Obstructive sleep apnea, hy-pertension, and cardiovascular risk: epidemiology, pathophysiology, and management. Curr. Cardiol. Rep. **22**, 6 (2020)
10. Natsois, G., Pastaka, C., Vavougios, G., et al.: Age, Body Mass Index, and Daytime and Nocturnal Hypoxia as Predictors of Hypertension in Patients With Obstructive Sleep Apnea. The Journal of Clinical Hy-pertension. **18**, 146–152 (2015)
11. Van Ryswyk, E.M., Mukherjee, S., Chai-Coetzer, C.L., et al.: Sleep disorders, including sleep apnea, and hypertension. Am. J. Hypertens. **31**, 857–864 (2018)
12. Lombardi, C., Pengo, M.F., Parati, G.: Systemic hypertension in ob-structive sleep apnea. J. Thorac. Dis. **10**, S4231–S4243 (2018)
13. Gonzaga, C., Bertolami, A., Bertolami, M., et al.: Obstructive sleep apnea, hypertension and cardiovascular diseases. J. Hum. Hypertens. **29**, 705–712 (2015)
14. Hsu, H.C., Chen, N.H., Ho, W.J., et al.: Factors associated with undi-agnosed obstructive sleep apnea among hypertensive patients: A multisite cross-sectional survey study in Taiwan. J. Clin. Nurs. **27**, 1901–1912 (2018)
15. Zhang, Z., Cheng, J., Yang, W., et al.: Gender differ-ences in clinical manifestations and polysomnographic findings in Chinese patients with obstructive sleep apnea. Sleep Breath. **24**, 1019–1026 (2020)
16. Fietze, I., Laharnar, N., Obst, A., et al.: Prevalence and association analysis of obstructive sleep apnea with gender and age differences - Results of SHIP-Trend. J. Sleep Res. **12770**, 5–6 (2018)
17. Cai, A., Zhou, Y., Zhang, J., et al.: Epidemiolog-ical characteristics and gender-specific differences of obstructive sleep apnea in a Chinese hypertensive population: a cross-sectional study. BMC Cardiovasc. Disord. **17**, 8 (2017)
18. Cano-Pumarega, I., Barbe, F., Esteban, A., et al.: Sleep apnea and hypertension: are there sex differences? The Victoria Sleep Cohort. Chest. **152**, 742–750 (2017)
19. Jenner, R., Fetureto-Borges, F., Costa-Hong, V., et al.: Association of obstructive sleep apnea with arterial riffness and nondipping blood pressure in patients with hypertension. J. Clin. Hypertens (Greenwich). **19**, 910–918 (2017)
20. Vgontzas, A.N., Li, Y., He, F., et al.: Mild-to-Moderate Sleep Apnea is associated with Incident Hy-pertension: Age Effect. Sleep **42**, 4 (2018)
21. Perger, E., Pengo, M.F., Lombardi, C.: Hypertension and atrial fibrilla-tion in obstructive sleep apnea: Is it a menopause issue? Graduation. **124**, 32–34 (2019)
22. Mokhlesi, B., Ham, S.A., Gozal, D.: The effect of sex and age on the comorbidity burden of OSA: an observational analysis from a large nationwide US health claims database. Eur. Respir. J. **47**, 1162–1169 (2016)
23. Maeder, M.T., Schoch, O.D., Rickley, H.: A clinical approach to obstructive sleep apnea as a risk factor for cardiovascular disease. Vasc Health Risk Manage. **12**, 85–103 (2016)

Frequency and Causes of Primary and Secondary Hyperparathyroidism in Patients Treated with Surgery at the University Clinical Hospital Mostar

Josip Lesko[1], Nikolina Obradović[2], and Vana Turudić[2(✉)]

[1] Department of Otorhinolaryngology, University Clinical Hospital Mostar, Mostar, Bosnia and Herzegovina

[2] Department of Family Medicine, DZ Mostar, Mostar, Bosnia and Herzegovina

Abstract. Background. The aim of this study was to investigate the causes of hyperparathyroidism and whether there is a difference between the sexes. An additional goal was to determine the incidence of primary and secondary hyperparathyroidism and the most common surgical method of treatment.

Subjects and methods: 56 subjects were included in the study, of which 49 were women and 7 were men with diagnosed hyperparathyroidism. Subjects were patients who were treated endocrinologically and had elevated serum PTH (>7 pmol/L) as well as elevated Ca values (>2.6 mmol/L). The parameters considered are: age, sex, primary or secondary hyperparathyroidism, the most common surgical method of treatment.

The results: The study included 56 respondents, of whom 49 women and 7 men. All respondents had a diagnosis of hyperparathyroidism. Statistically, a significantly higher proportion of female sex than men with primary hyperparathyroidism was diagnosed ($\chi^2 = 8.64$; s.s = 1; $P = 0.003$). The youngest patient was 40 years old and the oldest 92 years. The average age was 63 years.

Most of the subjects in our study who had hyperparathyroidism also had parathyroid adenoma ($\chi^2 = 6.74$; s.s. = 1; $P = 0{,}009$).

The most common surgical procedure was partial parathyroidectomy, with almost the same ratio between women and men. There were no differences in surgical procedures with respect to sex ($\chi^2 = 1.25$; s.s. = 1; $P = 0.262$).

Parathyroidectomy of the lower and upper left as well as the lower and upper right parathyroid glands was equally common in our study (Table 4).

Conclusion. The incidence of primary hyperparathyroidism was higher than secondary. Primary occured more frequently in females while secondary hyperparathyroidism occured equally in both sexes. The most common cause of primary hyperparathyroidism was parathyroid adenoma, and secondary chronic renal failure. The most common surgical procedure was partial parathyroidectomy. Parathyroidectomy of the lower and upper left as well as the lower and upper right parathyroid glands was equally common.

Keywords: Primary hyperparathyroidism · Secondary hyperparathyroidism · Adenoma · Chronic renal insufficiency · Parathyroidectomy

A. Badnjevic and L. Gurbeta Pokvić (Eds.): CMBEBIH 2021, IFMBE Proceedings 84, pp. 609–616, 2021.
https://doi.org/10.1007/978-3-030-73909-6_71

1 Introduction

During the fifth week of gestation, the lower parathyroid glands develop from the third and the upper from the fourth gill arch. Fetal parathyroid glands are functional, respond to blood calcium levels, and produce parathyroid hormone. Most people have 4 glands, while 2–5% have 5 parathyroid glands and more [1, 2].

The parathyroid glands respond to low serum calcium levels by releasing PTH which increases serum calcium levels by acting directly on the bones and kidneys, also stimulates osteoclasts to resorb bone and mobilizes calcium into the blood. In the kidneys, PTH acts to reduce calcium clearance and stimulates the synthesis of 1, 25-dihydroxy vitamin D, which stimulates calcium absorption in the gastrointestinal tract [3].

The most common disorder of the parathyroid gland is hyperparathyroidism, we distinguish primary, secondary and tertiary. Primary and tertiary are treated surgically, and secondary is usually treated with medication [2].

Primary hyperparathyroidism is the third most common endocrine disorder, occurring more frequently in women than in men in all age groups [4, 5]. Occurrence in pregnant women is rare, and since neonatal mortality is associated with maternal hyperparathyroidism, early diagnosis is essential. Treatment should be individualized and includes conservative treatment, parathyroidectomy in the second trimester, or in the early postpartum period [6].

Primary hyperparathyroidism occurs due to a primary defect in the parathyroid glands, so that elevated serum calcium does not inhibit PTH release. Primary hyperparathyroidism is caused by solitary parathyroid adenoma in 80% of cases, 4 gland hyperplasia in 10–15% of cases, multiple adenomas in 5% of cases, and parathyroid carcinoma in less than 1% of cases [7].

In addition to generalized fatigue and weakness, primary hyperparathyroidism may present with symptoms such as vomiting, abdominal pain, constipation, peptic ulcer, and pancreatitis. Hyperparathyroidism may be the most significant manifestation of endocrine diseases in patients with MEN-I syndrome. Symptoms by the renal system are manifested by polyuria, polydipsia, nephrolithiasis. Bone disease can manifest as bone pain and pathological fractures. Neuromuscular syndrome improves after parathyroidectomy in 80–90% of patients. Many of the psychiatric symptoms improved after parathyroidectomy. Cardiovascular symptoms may include hypertension and cardiac arrhythmias. Hypertension can occur in 50% of patients [2, 8].

Biochemical characteristics include elevated serum calcium levels with elevated or inappropriately normal PTH levels [9].

Surgery is the only definitive cure for primary hyperparathyroidism, avoiding long-term complications. Parathyroidectomy is indicated for symptomatic treatment or persistently elevated serum calcium. In some patients, existing comorbidities may prevent the need for surgery [3, 10].

Secondary hyperparathyroidism is most commonly caused by chronic renal failure. It is characterized by high serum PTH, glandular hyperplasia, and disbalance of mineral metabolism [11, 12].

These damaging effects may contribute to an increased risk of cardiovascular morbidity and mortality in patients with end-stage renal disease [13]. Calcium and phosphorus homeostasis is maintained through a complex connection between bones, intestines, kidneys, and parathyroid glands. PTH is probably the most important regulator of calcium metabolism [14].

In patients with damaged renal function, who are on dialysis, physiological control of mineral balance via parathyroid hormone through the kidney fails and hyperparathyroidism progresses. Ca-sensory receptor and vitamin D receptor abnormalities are associated with the pathogenesis of secondary hyperparathyroidism [15]. Abnormalities in renal tubular phosphate absorption lead to decreased phosphate excretion and hyperphosphatemia. Damage of renal conversion of 25-hydroxycolicalcaliferol to 1,25-dihydroxycholecalciferol also causes a decrease in intestinal calcium absorption. In combination, elevated serum phosphate levels and decreased vitamin D production result in decreased serum calcium levels and hypocalcemia. Stimulation of the parathyroid glands by chronic hypocalcemia results in hyperplasia of all glands [11, 12].

Early diagnosis of secondary hyperparathyroidism in chronic renal patients is crucial [16]. Treatment is medicamentous [2].

Parathyroidectomy may be useful for certain patients with severe secondary hyperparathyroidism and clinical symptoms [17].

Tertiary hyperparathyroidism occurs when the cause of stimulation is corrected, but the glands remain autonomically hyperfunctional [2]. It usually occurs in patients with chronic kidney disease after kidney transplantation, which is a therapeutic procedure in secondary hyperparathyroidism [18].

Unlike secondary, tertiary hyperparathyroidism often requires total parathyroidectomy [19].

2 Subjects and Methodes

Cohort form of research. The study included 56 subjects, 49 subjects were women and 7 subjects were men with diagnosed hyperparathyroidism hospitalized at the Clinic of Otorhinolaryngology and Maxillofacial Surgery and underwent surgery, partial or total parathyroidectomy at the University Clinical Hospital (SKB) Mostar in the period from January 1st 2009 to February 22nd 2019. Subjects were endocrinologically treated patients with elevated serum PTH (>7 pmol/L) and Ca (>2.6 mmol/L). Serum PTH and Ca values were obtained by taking serum samples from patients. The concentration of ionized calcium in serum was determined by ion selective electrode (ISE) potentiometry on a Rapidpoint400 analyzer manufactured by Siemens. PTH was determined on a Cobas e 411 automated analyzer (Roche Diagnoatics GmbH, Mannheim, Germany) by the ECLIA method (Electrochemiluminescence immunoassay electrostimulation method). Patient data were collected from the Hospital Information System (BIS) based on medical records in the form of a medical history. The parameters considered are: age, sex, primary or secondary hyperparathyroidism, the most common surgical method of treatment.

For statistical analysis of the obtained data was used the software systems SPSS for Windows (version 13.0, SPSS Incorporated Chicago, Illinois, USA) and Microsoft Excel (version Office 2010, Microsoft Corporation, Redmond, Washington, USA). The χ^2 test was used to test the significance of the observed differences. The limit of statistical significance was set at $P < 0.05$. P values that could not be expressed to three decimal places are expressed as $P < 0.001$.

3 Results

The study included 56 subjects, 49 subjects were women and 7 subjects were men with diagnosed hyperparathyroidism. The youngest patient was 40 years old and the oldest 92 years old. The average life expectancy was 63 years. There was a statistically significantly higher proportion of females compared to males with a diagnosis of primary hyperparathyroidism ($\chi^2 = 8.64$; s.s = 1; $P = 0.003$) (Table 1).

Table 1. Distribution of patients with primary and secondary hyperparathyroidism by gender.

	Primary hyperparathyroidism		Secondary hyperparathyroidism			
	Number of patiens	%	Number of patiens	%	P	χ^2
Male	4	8,0	3	50,0	0,003	8,64
Female	46	92,0	3	50,0		

Most of the subjects in our study who had hyperparathyroidism also had parathyroid adenoma ($\chi^2 = 6.74$; s.s. = 1; $P = 0{,}009$) (Table 2).

Table 2. Distribution of patients according to the etiology of hyperparathyroidism.

	Adenoma		Chronic renal failure			
	Number of patiens	%	Number of patiens	%	p	χ2
Male	4	8,16	3	42,86	0,009	6,74
Female	45	91,84	4	57,14		

The most common surgical procedure was partial parathyroidectomy, with almost the same ratio between women and men. There were no differences in surgical procedures with respect to sex ($\chi^2 = 1.25$; s.s. = 1; $P = 0.262$) (Table 3). Parathyroidectomy of the lower and upper left as well as the lower and upper right parathyroid glands was equally common in our study (Table 4).

Table 3. Frequency of partial and total parathyroidectomy

	Partial parathyroidectomy		Total parathyroidectomy			
	Number of patiens	%	Number of patiens	%	P	χ^2
Male	7	12,96	0	00,0	0,262	1,25
Female	47	87,04	2	100,0		

Table 4. Frequency of parathyroidectomy of individual glands.

Parathyroidectomy	Number of patiens	%	P	χ^2
Right lower parathyroid gland	9	16,07 (0.36)		
Right lower and upper parathyroid gland	19	33,92 (1.07)		
Left lower parathyroid gland	8	14,28 (0.82)	0,23406	5.565
Left lower and upper parathyroid gland	19	33,92 (1.07)		
Right and left lower parathyroid gland	1	1,78 (2.25)		

4 Discussion

In their study, Castellano et al. [20] found that the biochemical activity of primary hyperparathyroidism is gender independent, but the clinical picture is different, mainly due to the menopausal condition. Shah and coworkers [21] state that age and gender have a significant impact on the presentation of primary hyperparathyroidism. Bone pain and rickets were more common in children and adolescents, while kidney stones were more common in adults.

Bandeira et al. [22] claim that vitamin D deficiency or complete deficiency in primary hyperparathyroidism may contribute to the severity of the disease itself. Carsote et al. [23] also found that careful monitoring of 25-hydroxy vitamin D is crucial for a good outcome of parathyroidoidectomy.

In their study, Sharma et al. [24] state that one in every 20 patients with urolithiasis had primary hyperparathyroidism.

The data obtained in our study agree with the data provided by Jiang et al. [25] in their study. A larger number of patients had (72%) parathyroid adenoma as the cause of primary hyperparathyroidism, and in 9.3% of cases was parathyroid carcinoma. The incidence of parathyroid cancer was higher in men.

La et al. [26] conducted a study on the quality of sleep of patients who underwent parathyroidectomy for primary hyperparathyroidism and thyroidectomy for thyroid disease. Preoperatively, patients with primary hyperparathyroidism had poorer sleep quality, and after the procedure, their sleep quality improved, and there were no changes in patients with thyroid disease. In their study, Ma et al. [27] suggest that parathyroidectomy may beneficially modify calcium and phosphorus metabolism and bone mineral density in patients with secondary hyperparathyroidism undergoing hemodialysis. The question is when to perform parathyroidectomy in patients who are waiting for a kidney transplant and suffer from secondary hyperparathyroidism. Littbarski et al. [28] state that parathyroidectomy should be performed before transplantation, and if the procedure cannot be performed then as early as the first year after transplantation if possible. Also Parikh et al. [29] state that parathyroidectomy can lead to transient dysfunction of allogeneic kidney transplantation with possible recovery of graft function up to 12 months after parathyroidectomy.

Chudzinski and colleagues [30] in their study state that parathyroidectomy did not significantly impair renal transplant function, but there were transient decreased functions in the early postoperative period.

Konturek et al. [31] in their study state that partial parathyroidectomy is also a safe and effective treatment option for secondary hyperparathyroidism.

It remains unclear whether partial or total parathyroidectomy gives better results for secondary hyperparathyroidism. Isaksson et al. [32] investigated mortality, cardiovascular disorders, hip fractures, and the need for re-parathyroidectomy in patients on renal replacement therapy. They considered age, sex, cause of kidney disease, pre-existing cardiovascular disease, time on dialysis, kidney transplantation as well as medical treatment before parathyroidectomy. The study concludes that there is a higher cardiovascular risk after total parathyroidism than after partial. Filho et al. [33] state that parathyroidectomy increases the quality of life expectancy of hemodialysis patients with secondary hyperparathyroidism regardless of the type of procedure.

5 Conclusions

In this paper, we investigated the causes of hyperparathyroidism, and whether there is a difference between the sexes. We wanted to determine the frequency of primary and secondary hyperparathyroidism, and the most common surgical method of treatment. The study included 56 patients diagnosed with hyperparathyroidism. The average life expectancy was 63 years. There was a significantly higher proportion of females (92%) compared to males (8%) in patients with primary hyperparathyroidism, while in secondary hyperparathyroidism the number of patients was equal to both males and females. Research has shown that adenoma was the most common cause of hyperparathyroidism. The most common surgical procedure was partial parathyroidectomy. The study showed that parathyroidectomy of the lower and upper left as well as the lower and upper right parathyroid gland was equally common.

In our study, partial parathyroidectomy was performed in 87.04% of women and 12.96% of men, while total was performed only in women.

References

1. Van de Water, T.R., Staecker, H.: Otolaryngology Basic Science and Clinical Review 1th ed. Thieme (2005)
2. Lalwani, A.K.: Diagnosis & Treatment Otolaryngology Head and Neck surgery 3rd ed. The Mc Grow Hill Companies (2012)
3. Michels, T.C., Kelly, K.M.: Parathyroid Disordrs. Am Fam Physi. **88**, 249–257 (2013)
4. Fraser, W.D.: Hyperparathyroidism. Lancet. **11**, 145–158 (2009)
5. Miller, B.S., Dimick, J., Wainess, R., Burney, R.E.: Age- and sex-related incidence of surgically treated primary hyperparathyroidism. World J. Surg. **32**, 795–799 (2008)
6. Mokrysheva, N.G., Eremkina, A.K., Mirnaya, S.S., et al.: A case of pregnancy complicated by primary hyperparathyroidism due to a parathyroid adenoma. Am. J. Case Rep. **20**, 53–59 (2019)
7. Walker, M.D., Silverberg, S.J.: Primary hyperparathyroidism. Nat. Rev. Endocrinol. **14**, 115–125 (2018)
8. Flint, P.W., Haughey, B.H., Lund, V.J., et al.: Cummings Otolaryngology Head & Neck Surgery, 5th ed. Mosby Elsevier (2010)

9. Silverberg, S.J., Walker, M.D., Bilezikian, J.P.: Asymptomatic primary hyperparathyroidism. J. Clin. Densitom. **16**, 14–21 (2013)
10. Pasha, R.: Otolaryngology Head & Neck Surgery Clinical Reference Guide 1th ed. Singular/Thomson Learning (2000)
11. Pitt, S.C., Sippel, R.S., Chen, H.: Secondary and tertiary hyperparathyroidism. State Art Surg. Manage. Surg. Cli.n North Am. **89**, 1227–1239 (2009)
12. Cozzolino, M., Galassi, A., Conte, F., et al.: Treatment of secondary hyperparathyroidism: the clinical utility of etelcalcetide. Ther. Clin. Risk Manag. **13**, 679–689 (2017)
13. Nikodimopoulou, N., Liakos, S.: Secondary hyperparathyroidism and target organs in chronic kidney disease. Hippokratia. **15**, 33–38 (2011)
14. Perm, J.: Hyperparathyroidism of renal disease. Permanente J. **20**, 15–127 (2016)
15. Mizobuchi, M., Ogata, H., Koiwa, F.: Secondary hyperparathyroidism: pathogenesis and latest treatment. Ther. Apher. Dial. **22**, 229–235 (2018)
16. Saliba, W., El-Haddad, B.: Secondary hyperparathyroidism: pathophysiology and treatment. J. Am. Board. Fam. Med. **22**, 574–581 (2009)
17. Eidman, K.E., Wetmore, J.B.: The role of parathyroidectomy in the management of secondary hyperparathyroidism. Curr. Opin. Nephrol. Hypertens. **26**, 516–522 (2017)
18. Jamal, S.A., Miller, P.D.: Secondary and tertiary hyperparathyroidism. J. Clin. Densitom. **16**, 64–68 (2013)
19. Pitt, S.C., Panneerselvan, R., Chen, H., Sippel, R.S.: Secondary and tertiary hyperparathyroidism: the utility of ioPTH. Monitoring. World J. Surg. **34**, 1343–1349 (2010)
20. Castellano, E., Attanasio, R., Boriano, A., et al.: Sex Difference in the clinical presentation of primary hyperparathyroidism: influence of menopausal status. J. Clin. Endocrinol. Metab. **102**, 4148–4152 (2017)
21. Shah, V.N., Bhadada, S.K., Bhansali, A., et al.: Influence of age and gender on presentation of symptomatic primary hyperparathyroidism. J. Postgrad. Med. **58**, 107–111 (2012)
22. Bandeira, F., Caldas, G., Freese, E., et al.: Relationship between serum vitamin d status and clinical manifestations of primary hyperparathyroidism. Endocr Pract. **8**, 266–270 (2002)
23. Carsote, M., Paduraru, D.N., Nica, A.E., Valea, A.: Parathyroidectomy: is vitamin D a player for a good outcome? J. Med. Life. **9**, 348–352 (2016)
24. Sharma, S., Rastogi, A., Bhadada, S.K., et al.: Prevalence and predictors of primary hyperparathyroidism among patients with urolithiasis. Endocr. Pract. **23**, 1311–1315 (2017)
25. Jiang, T., Yao, X.A., Wei, B.B., Chang, H.: The influence of gender on clinical manifestations of primary hyperparathyroidism. Zhonghua Nei Ke Za Zhi. **57**, 753–755 (2018)
26. La, J., Wang, T.S., Hammad, A.Y., et al.: Parathyroidectomy for primary hyperparathyroidism improves sleep quality: a prospective study. Surgery. **161**, 25–34 (2017)
27. Ma, L., Zhao, S., Li, Z.: Effects of parathyroidectomy on bone metabolism in haemodialysis patients with secondary hyperparathyroidism. Scand. J. Clin. Lab. Invest. **77**, 527–534 (2017)
28. Littbarski, S.A., Kaltenborn, A., Gwiasda, J., et al.: Timing of parathyroidectomy in kidney transplant candidates with secondary hyperparathryroidism: effect of pretransplant versus early or late posttransplant parathyroidectomy. Surgery **16**, 373–380 (2018)
29. Parikh, S., Nagaraja, H., Agarwal, A., et al.: Impact of post-kidney transplant parathyroidectomy on allograft function. Clin. Transplant. **27**, 397–402 (2013)
30. Chudzinski, W., Wyrzykowska, M., Nazarewski, S., et al.: Does the Parathyroidectomy Endanger the Transplanted Kidney? Transplant Proc. **48**, 1633–1636 (2016)
31. Konturek, A., Barczyński, M., Stopa, M., Nowak, W.: Subtotal parathyroidectomy for secondary renal hyperparathyroidism: a 20-year surgical outcome study. Langenbeck's Archives Surg. **401**(7), 965–974 (2016). https://doi.org/10.1007/s00423-016-1447-7

32. Isaksson, E., Ivarsson, K., Akaberi, S., et al.: Total versus subtotal parathyroidectomy for secondary hyperparathyroidism. Surgery **165**, 142–150 (2019)
33. Filho, W.A., van der Plas, W.Y., Brescia, M.D.G., et al.: Quality of life after surgery in secondary hyperparathyroidism, comparing subtotalparathyroidectomy with total parathyroidectomy with immediate parathyroid autograft: Prospective randomized trial. Surgery **164**, 978–985 (2018)

Classification of Hypertension and Assessment of Cardiovascular Risk Using Digital Data and Biochemical Parameters

Čerkadžić Tarik, Talić Lejla, Zametica Rubina, Zećo Nadina, Zilkić Hasiba, and Zukić Džejla(✉)

Faculty of Pharmacy, University of Sarajevo, Sarajevo, Bosnia and Herzegovina
dzejlazukic@ffsa.unsa.ba

Abstract. Cardiovascular disease [CVD] is the leading cause of death worldwide. One of the major factors of CVD is hypertension. Blood pressure readings of ≥140/≥90 mmHg indicate hypertension and may lead to cardiovascular complications, if not treated. In this paper, we use patient information like blood pressure, age, gender, smoking, digital data like physical activity, BMI and sleep, as well as biochemical parameters [HDL-C, total cholesterol, creatinine, glucose] to classify the grades of hypertension and assess the total cardiovascular risk. We used machine learning algorithms that processed data from 300 individuals. Machine learning enables quick and accurate assessment of cardiovascular risk that can give promising results in medical practice Naïve Bayes [NB], Decision Tree [DT], Random Forest [RF] and k-Nearest Neighbor [k-NN] algorithms were used. The classifier based on Random Forest algorithms achieved the highest accuracy.

Keywords: Classification · Hypertension · Assessment · Cardiovascular risk · Biochemical parameters · Machine learning

1 Introduction

Hypertension is a persistent public health problem worldwide. Currently, about 1,13 billion people in the world are suffering from hypertension [1]. In high or raised blood pressure, blood flows through arteries at higher-than-normal pressures. Systolic pressure is generated by the pumping of blood by the ventricles, while diastolic pressure is generated between heartbeats [2]. In healthy individuals systolic blood pressure is less than or equal to 120 mm of mercury and the diastolic pressure is less than or equal to 80 mm of mercury. In hypertension, by definition, the systolic blood pressure values are ≥140 mmHg and/or diastolic blood pressure values are ≥90 mmHg in younger, middle-aged and older individuals, who are not taking antihypertensive drugs [3]. In the beginning phases of the disease, individuals may not feel any symptoms, which can result in late diagnosis of the disease. That can be a major threat to the overall health of patients, as hypertension is one of the leading risk factors for cardiovascular disease

A. Badnjevic and L. Gurbeta Pokvić (Eds.): CMBEBIH 2021, IFMBE Proceedings 84, pp. 617–629, 2021.
https://doi.org/10.1007/978-3-030-73909-6_72

(CVD). CVD is an umbrella term referring to heart and blood vessel diseases. Approximately 54% of strokes and 47% of coronary heart diseases worldwide, are attributable to high blood pressure [4]. The prevalence of hypertension increases with age and it is estimated that more than half of people over 65 years of age suffer from high blood pressure [4]. As the global population is aging, mortality and complication due to uncontrolled hypertension are to be expected. Other factors like lifestyle, stress and diet contribute to heart, brain, kidney and blood vessel changes, that can seriously affect the health of patients. Other risk factors include smoking, high Body Mass Index [BMI], dyslipidemia, glucose intolerance and family history [5]. The goal of a successful hypertension treatment is not only correcting the blood pressure value, but also decreasing all other risk factors that may affect a patient's health. Detection and treatment of hypertension may possibly make use of digital phenotyping [6]. Digital phenotyping is defined as "the moment-by-moment quantification of the individual level human phenotype *in situ* using data from personal digital devices". Digital data can give us accurate information about a patient's BMI, movement and sleep patterns, which will help health workers capture a bigger picture about the state of the patient [7]. It may also help with early detection of risk factors for cardiovascular disease. Traditionally, medical professionals use their knowledge and relevant guidelines to assess the total cardiovascular risk. That is not a reliable method in all cases, as patients may not be compliant or honest in all aspects. Some data sets are so massive that the human brain needs help finding patterns, and this is where machine learning is very helpful. Machine learning [referred to as ML later in text] stems from artificial intelligence. ML is a method of data analysis that automates analytical model building. In the past twenty years, expert systems based on machine learning methods, such as artifcial neural networks (ANNs) and fuzzy logic (FL) were used for the detection of different types of diseases [10, 11]. Expert systems based on neural networks and fuzzy logic have proven very successful in diagnosing various medical conditions and the integration of real-time transmission has allowed this system to be applied as a health monitor as well [11]. ANNs were even used for genome-enabled classifications with more than satisfactory results [12]. Recently, results demonstrated that the expert diagnostic system can correctly identify patients with asthma and chronic obstructive pulmonary disease [13]. That is very important for us to consider. Apart from that, the use of artificial neural networks has resulted in successful classification of different epileptic seizures, which with high accuracy, enables easier detection of certain epilepsy types [14]. This can also be very useful in terms of CVD. However, the most useful implementation of artificial intelligence in case of hypertension and CVD may be ML. Machine learning has been very successfully used in medicine. Prediction of medical device performance is one example [15]. Important factors such as performance of defibrillators and possible performance failures of the device which can affect performance have been resolved through ML [16]. That means that ML may be useful in not only prediction of outcome, but also correcting of the human error. That is why this method may be the most efficient and useful in case of CVD and hypertension.

Our goal in this paper is to identify and quantify risk factors and markers of CVD and hypertension using ML and in accordance to the European Society of Cardiology (ESC) and the European Society of Hypertension (ESH) Guidelines, SCORE Risk Charts (The European cardiovascular disease risk assessment model) and relevant guidelines [8, 9].

2 Methods and Materials

2.1 Data Collection

Parameters that were used for the classification of hypertension and assessment of total cardiovascular risk are the following: Age, Sex, Systolic blood pressure, Diastolic blood pressure, Creatinine, Glucose, Total Cholesterol, HDL Cholesterol, Physical Activity, Sleep, BMI and smoking. The research was done using a database that consisted of 300 samples. We analyzed data from 200 diseased [*Sample Group 1*] and 100 healthy individuals [*Sample group 2, control group*], looking at the age of the candidates, blood test parameters, blood pressured values and digital data that included sleep, physical activity and BMI, as well as smoking status. In Sample group 1, all individuals have high or elevated blood pressure values, suffer from kidney disease and/or diabetes, dyslipidemia and at have at least one other risk factor for development of cardiovascular complications [smoking, low physical activity, low sleep]. For healthy individuals, we selected individuals with normal or optimal blood pressure values, no relevant comorbidities [kidney disease, dyslipidemia and diabetes], and less than two other risk factors [smoking, low physical activity, low sleep, BMI]. We took into account the age and gender of patients, using that data to develop a fitting ML algorithm. The properties of the dataset used for the development of ML is presented in Table 1.

Table 1. Properties of the dataset used for development of ML

Class	Sample group 1	Control Group	Total
Male	100	52	152
Average age [male]	64,95	48,48	56,71
Female	100	48	148
Average age [female]	62.61	47,46	55,03
Total number of patients per class	200	100	300

2.2 Classification Methods

Age is an independent risk factor for hypertension and development of cardiovascular disease. In Sample group 1, the average age was 63,78. In Sample group 2 the average age was 47,99. Every patient above 55 years of age is at higher risk at developing cardiovascular complications [17]. The age distribution is presented in Figs. 1 and 2.

Another important parameter is gender. Men are at higher risk at developing CVD. Although men tend to develop coronary artery disease earlier in life, after age 65 the risk of heart disease in women is almost the same as in men [18]. In Sample Group 1, the number of male and female candidates is both 100. The number of females above 65 years of age is 51. These female patients of the same cardiovascular risk as males. Is the control group, 48 of the candidates are female, 52 of them are male. The number of women above 65 years of age in this group is 11.

Fig. 1. Age distribution across Sample Group 1

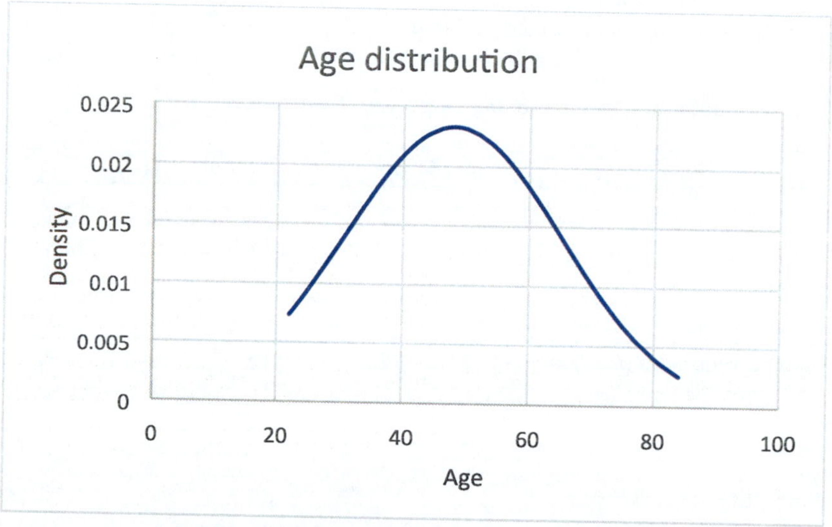

Fig. 2. Age Distribution across Sample Group 2

BMI, sleep and physical activity are classified as digital data derived from smart devices. We looked at the exercise the patients recorded manually and/or their smart device recorded for them. For adults, 18 years and older, the recommended amount of physical activity is 150–300 min of moderate-intensity aerobic physical activity or at least 75–150 min of vigorous-intensity aerobic physical activity in a sevenday period. Additional health benefits come from muscle-strengthening activities on 2 or more days in a week [19]. The categories for physical activity are the following: low- less than 150 min moderate-intensity activity/week; medium- 150–300 min moderate-intensity activity/week; high- ≤300 min moderate-intensity activity/week with additional muscle-strengthening activities [19]. Low physical activity poses a risk factor in the development of many diseases and is a general cardiovascular risk factor. Physical activity across both groups is shown in Fig. 3.

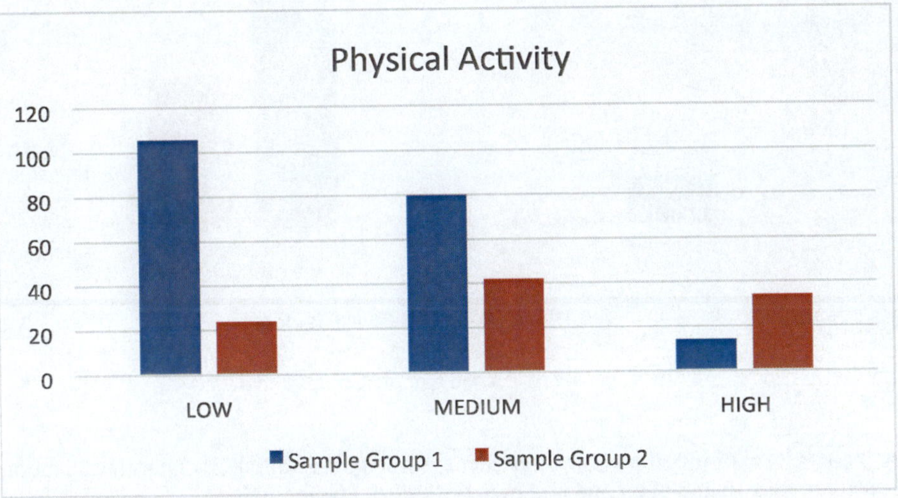

Fig. 3. Physical Activity in Sample Groups 1 and 2

As for BMI, the value is derived from the information about height and weight the patients entered into their smart devices. The formula for BMI is weight in kilograms divided by height in meters squared. It was calculated by their smart device. The BMI Categories are the following: Underweight = < 18.5, Normal weight = 18.5–24.9, Overweight = 25–29.9, Obesity = BMI of 30 or greater [20]. Underweight, overweight and obese patients are at higher risk of developing CVD [21]. BMI values of all candidates in the Control Group are normal. In Sample group 1, total of 12 candidates are underweight [6%] and 130 of them are overweight or obese [65%]. These patients are at a higher risk of developing CVD. Sleep is another risk factor we observed via smart devices. Smart phones and watches are able to record sleep patterns of users. Studies show that even small increases in blood pressure, particularly nighttime blood pressure levels, are associated with significant increases in CV complications [22]. The recommended amount of sleep for adults aged 18 to 65 is 7–9 h. For adults ≥65 years of age, the recommended amount is 7–8 h [23]. Habitual short sleep duration [less than 6 h] is associated with hypertension, especially during middle age and is a risk to developing CVD [22]. Hours of sleep in both groups is shown in Fig. 4.

For the classification of hypertension, we used the systolic and diastolic blood pressure values of patients. In accordance with most guidelines, hypertension be diagnosed when the systolic blood pressure [SBP] is ≥140 mm Hg and/or the diastolic blood pressure [DBP] is ≥90 mm Hg [24]. The classification of hypertension grades was made in accordance with the ESC/GHS Guidelines for the Management of Arterial Hypertension. The blood pressure category is defined according to seated clinic blood pressure and by

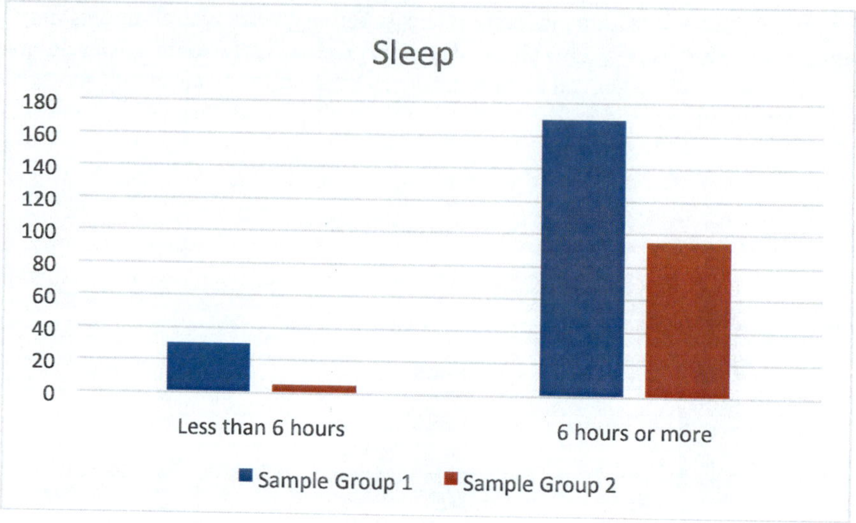

Fig. 4. Hours of sleep in Sample Groups 1 and 2

the highest level of blood pressure, whether systolic or diastolic [25]. This classification is used for ages 16 and older and is shown in Table 2. Hypertension has been noted in all candidates in Sample Group 1. The control group had optimal to normal blood pressure ranges.

Table 2. Classification of hypertension using seated clinic blood pressure values [25]

Category	Systolic [mmHg]		Diastolic [mmHg]
Optimal	<120	and	<80
Normal	120–129	and/or	80–84
High normal	130–139	and/or	85–89
Grade 1 hypertension	140–159	and/or	90–99
Grade 2 hypertension	160–179	and/or	100–109
Grade 3 hypertension	≥180	and/or	≥110
Isolated systolic hypertension	≥140	and	<90

Dyslipidemia, especially hypercholesterolemia increases fatty deposits in arteries and the risk of blockages. Patients with hyperlipidemia are at higher risk at developing CVD [26]. We measured the total Cholesterol and HDL-Cholesterol levels in the blood of all 300 patients. The normal range of total Cholesterol is 3,1–5,5 mmol/L, low levels are <3,1 mmol/L, while high levels are >5,5 mmol/L. The recommended range for HDLCholesterol is 1.03–1.55 mmol/L [27]. Figure 5 shows the incidence of dyslipidemia across all patients. No one in the Control Groups showed elevated lypid levels, while 79% of patients in Sample Group 1 had dyslipidemia, leaving them at a higher risk of developing CVD.

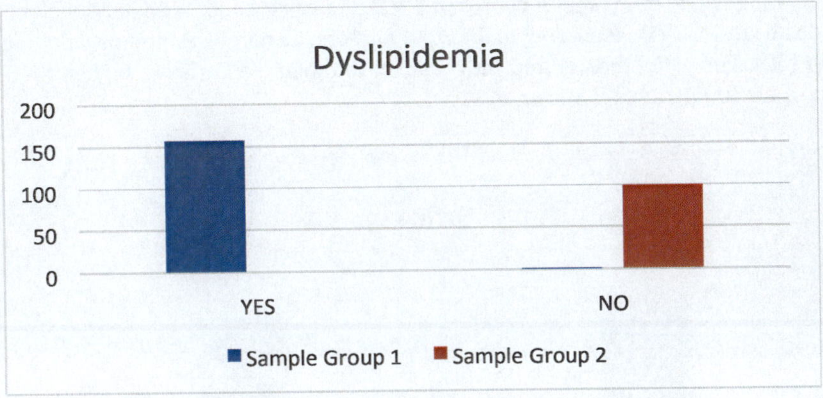

Fig. 5. Number of patients with and without dyslipidemia

Adults with diabetes are 2–4 times more likely to die from heart disease than non-diabetics [28]. The American Heart Association considers "diabetes to be one of the seven major controllable risk factors for cardiovascular disease" [28]. We used random plasma glucose tests for diabetes screening. In diabetics, the values of glucose in plasma are 11.1 mmol/L or more [29]. Diabetes incidence across both sample groups are shown in Fig. 6.

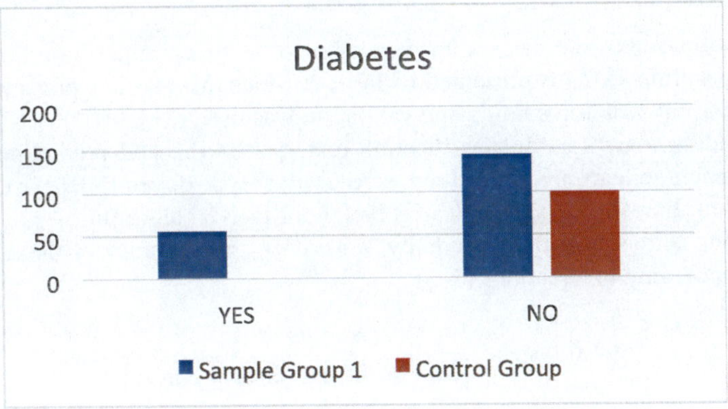

Fig. 6. Number of patients with and without Diabetes

For assessment of renal function, we used the serum creatinine test. Normal values of creatinine are: 62–115 μmol/L for men and 53–97 μmol/L for women [27]. Kidney failure could indicate and/or lead to cardiovascular complication. In those individuals, the creatinine levels are higher than normal. None of the patients in the Control group had elevated levels of creatinine. 1,5% of women and 6% of men in Sample Group 1 had normal levels of creatinine, the other 92,5% showed slight to very elevated levels of creatinine, indicating kidney disease.

Smoking is one of the major causes of CVD and causes approximately one of every four deaths from CVD, according to the 2014 Surgeon General's Report on smoking and health [30]. For patients who said they smoke, the total CVD risk is higher. Smoking status of all 300 patients is shown in Fig. 7.

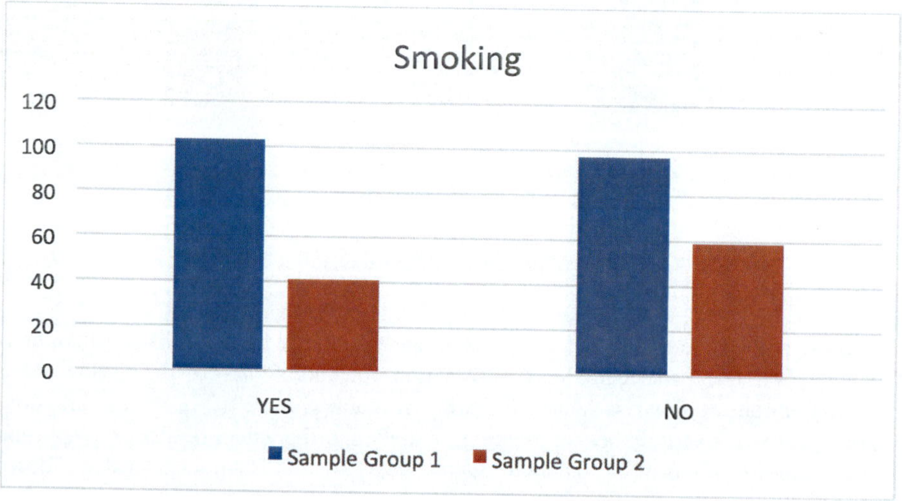

Fig. 7. Smoking Status in Sample Groups 1 and 2

The distribution of the dataset for development of the classifier based on machine learning algorithms [ML] is presented in Table 3. Since this is relatively small dataset, distribution of 70:30% for training and validation was used [31].

For building classifier for classification of hypertension and prediction of CVD, following machine learning algorithms were used: Naïve Bayes [NB], Decision Tree [DT], Random Forest [RF] and k-Nearest Neighbor [k-NN] algorithm.

According to these results, specificity, sensitivity and accuracy of the network are calculated according to equations [1–3]:

$$sensitivity = \frac{true\ positive}{[true\ positive + false\ negative]} \tag{1}$$

$$specificity = \frac{true\ negative}{[true\ negative + false\ positive]} \tag{2}$$

$$accuracy = \frac{[true\ negative + true\ positive]}{[true\ negative + true\ positive + false\ negative + false\ positive]} \tag{3}$$

Table 3. Dataset distribution

Total number of samples 300 samples [100%]	Healthy 100 samples [30%]	Disease 200 [70%]
Training dataset 210 samples [70%]	70 samples [70%]	140 samples [70%]
Training 147 samples [70%]	49 samples [70%]	98 samples [70%]
Testing 63 samples [30%]	21 samples [30%]	42 samples [30%]
Validation 90 samples [30%]	30 samples [30%]	60 samples [30%]

3 Results

Figure 10 presents the results of training of classifiers based on 4 different machine learning algorithms. As it can be seen from the figure, during training process, classifier based on Random Forest algorithms achieved the highest accuracy of 98.88%. The lowest accuracy was achieved by Naïve Bayes algorithm and it was 90.00% (Fig. 8).

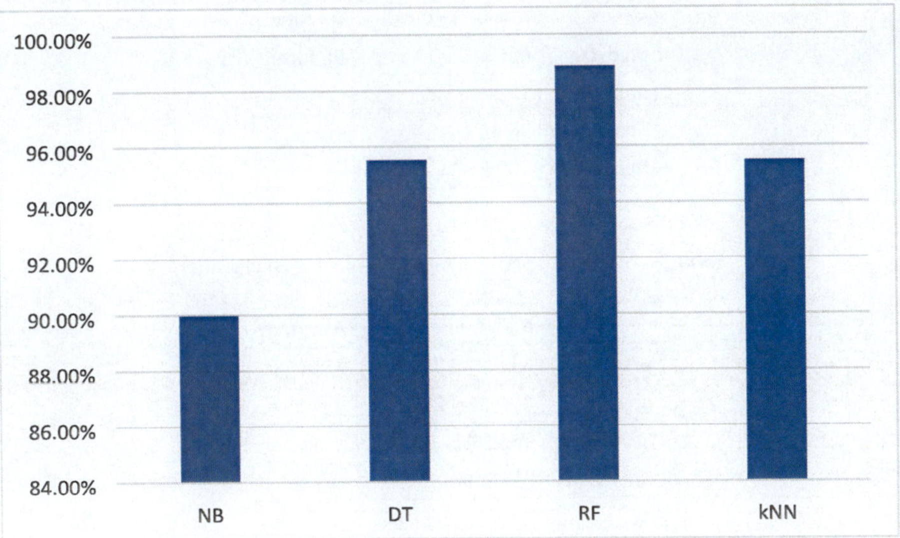

Fig. 8. Training performance of classifiers

Validation was performed on 90 samples. Each classifier was validated and the results of validation are presented in the Tables below (Tables 4, 5, 6 and 7).

Table 4. Validation performance of NB based classifier

	true 0	true 1	Accuracy
pred. 0	50	6	90.00%
pred. 1	3	31	
Sensitivity	94.33%		
Specificity	83.78%		

Table 5. Validation performance of DT based classifier

	true 0	true 1	Accuracy
pred. 0	58	1	95.55%
pred. 1	4	28	
Sensitivity	93.54%		
Specificity	96.55%		

Table 6. Validation performance of RF based classifier

	true 0	true 1	Accuracy
pred. 0	62	0	98.88%
pred. 1	1	27	
Sensitivity	98.41%		
Specificity	100%		

Table 7. Validation performance of k-NN based classifier

	true 0	true 1	Accuracy
pred. 0	63	3	95.50%
pred. 1	2	22	
Sensitivity	96.92%		
Specificity	88.00%		

4 Conclusion

Hypertension is one of the biggest risk factors for almost all different cardiovascular diseases in humans. Cardiovascular disease [ischaemic heart disease, stroke] is still the leading cause of death worldwide. Therefore, efficient and quick methods of assessment of cardiovascular risk linked to hypertension need to be created. We used machine learning in processing of our results. Machine learning enables that, instead of writing code, you feed data to the generic algorithm, and it builds logic based on the data given. That can be used to process data from every individual based on their dataset with high accuracy, as shown in the results. Classifier based on RF algorithms achieved the highest accuracy of 98.88%. Results and databases will be available to health professionals worldwide. If we can classify and predict risk factors of CVD and hypertension early on, we will be able to lower the chances of a lethal outcome. Especially in the modern age, this feature can be available on smart devices of patients and doctors. That means the best and most importantly, timely health care, that can improve and even save a life.

References

1. Brown, I.: Hypertension (2020). https://www.who.int/health-topics/hypertension/#tab=tab_2. Accessed 19 Nov 2020
2. The National Heart, Lung, and Blood Institute [NHLBI]: High Blood Pressure (2020). https://www.nhlbi.nih.gov/health-topics/high-blood-pressure. Accessed 19 Nov 2020
3. Ramzy, I.: Definition of hypertension and pressure goals during treatment [ESC-ESH Guidelines 2018] (2019). https://www.escardio.org/Journals/E-Journal-ofCardiology-Practice/Volume-17/definition-of-hypertension-and-pressure-goals-duringtreatment-esc-esh-guidelin. Accessed 19 Nov 2020
4. Wu, C.-Y., Hu, H.-Y., Chou, Y.-J., Huang, N., Chou, Y.-C., Li, C.-P.: High blood pressure and all-cause and cardiovascular disease mortalities in community-dwelling older adults. Medicine **94**(47) (2015). https://www.ncbi.nlm.nih.gov/pmc/articles/PMC5059018/
5. Tomić, M., Stepanović-Petrović, R., et al.: Farmakoterapija za farmaceute. Farmaceutski fakultet u Beogradu (2018)
6. Fagherazzi, G.: Deep digital phenotyping and digital twins for precision health: time to dig deeper. J. Med. Internet Res. **22**(3) (2020). https://pubmed.ncbi.nlm.nih.gov/32130138/. Accessed 19 Nov 2020
7. Lupton, D.: Digital health technologies and digital data: new ways of monitoring, measuring and commodifying human embodiment, health and illness (2016). https://www.researchgate.net/publication/305434534_Digital_Health_Technologies_and_Digital_Data_New_Ways_of_Monitoring_Measuring_and_Commodifying_Human_Embodiment_Health_and_Illness
8. Williams, B., Mancia, G., Spiering, W., et al.: ESC/ESH guidelines for the management of arterial hypertension: the Task Force for the management of arterial hypertension of the European Society of Cardiology [ESC] and the European Society of Hypertension [ESH]. Eur. Heart J. **39**(33), 3021–3104 (2018). https://academic.oup.com/eurheartj/article/39/33/3021/5079119
9. Piepoli, M.F., Hoes, A.W., Agewall, S., Albus, C., Brotons, C., Catapano, A. L., Cooney, M.T., Corrà, U., et al.: 2016 European Guidelines on cardiovascular disease prevention in clinical practice: The Sixth Joint Task Force of the European Society of Cardiology and Other Societies on Cardiovascular Disease Prevention in Clinical Practice [constituted by representatives of 10 societies and by invited experts]: Developed with the special contribution of the European Association for Cardiovascular Prevention & Rehabilitation [EACPR]. Eur. J. Prevent. Cardiol. **23**(11) (2016). https://pubmed.ncbi.nlm.nih.gov/27353126/
10. Hodzic, J., Gurbeta, L., Omanovic-Miklicanin, E., Badnjevic, A.: Overview of next-generation sequencing platforms used in published draft plant genomes in light of genotypization of immortelle plant [Helichrysium Arenarium]. Med Arch. **71**(4), 288–292 (2017). https://doi.org/10.5455/medarh.2017.71.288-292

11. Gurbeta, L., Maksimovic, M, Omanovic-Miklicanin, E., Badnjevic, A., Sejdic, E.: A telehealth system for automated diagnosis of asthma and chronical obstructive pulmonary disease. J. Am. Med. Inform. Assoc. (2018)
12. Catic, A., Gurbeta, L., Kurtovic-Kozaric, A., Mehmedbasic, S., Badnjevic, A.: Application of neural networks for classification of patau, edwards, down, turner and Klinefelter syndrome based on first trimester maternal serum screening data, ultrasonographic findings and patient demographics. BMC Med. Genomics **11**, 19 (2018). https://doi.org/10.1186/s12920-0180333-2
13. Badnjevic, A., Gurbeta, L., Custovic, E.: An expert diagnostic system to automatically identify asthma and chronic obstructive pulmonary disease in clinical settings. Sci. Rep. (2018)
14. Sarić, R., Jokić, D., Beganović, N., Pokvić, L.G., Badnjević, A.: FPGAbased real-time epileptic seizure classification using Artificial Neural Network. Biomed. Sig. Process. Control **62**, 102106 (2020). ISSN 1746-8094. https://doi.org/10.1016/j.bspc.2020.102106
15. Kovačević, Ž., Pokvić, L.G., Spahić, L., Badnjević, A.: Prediction of medical device performance using machine learning techniques: infant incubator case study Health and Technology. Health and Technology
16. Badnjevića, A., Pokvić, L.G., Hasičić, M., Bandić, L., Mašetić, Z., Kovačević, Ž., Kevrić, J., Hifziefendić, J., Pecchia, L.: Evidence-based clinical engineering: machine learning algorithms for prediction of defibrillator performance. Biomedical Signal Processing and Control, vol. 54, p. 101629, September 2019
17. World Heart Federation. [2019]. Risk factors - World Heart Federation. https://www.world-heart-federation.org/resources/risk-factors/. Accessed 24 Nov 2020
18. How Age and Gender Affect Your Heart. https://wa.kaiserpermanente.org/healthAndWellness/index.jhtml?item=%2Fcommon%2FhealthAndWellness%2Fconditions%2FheartDisease%2FageAndGender.html. Accessed 24 Nov 2020
19. World health organization: Physical activity and adults (2020). https://www.who.int/news-room/fact-sheets/detail/physical-activity. Accessed 24 Nov 2020
20. National Heart, Lung, and Blood Institute: Calculate Your BMI - Standard BMI Calculator. Nih.gov (2019). https://www.nhlbi.nih.gov/health/educational/lose_wt/BMI/bmicalc.htm. Accessed 25 Nov 2020
21. Park, D., Lee, J.-H., Han, S.: Underweight: another risk factor for cardiovascular disease? Medicine **96**(48) (2017). https://www.ncbi.nlm.nih.gov/pmc/articles/PMC5728753/
22. Calhoun, D.A., Harding, S.M.: Sleep and hypertension. Chest **138**(2), 434–443 (2010). https://www.ncbi.nlm.nih.gov/pmc/articles/PMC2913764/
23. National Sleep Foundation: National Sleep Foundation Recommends New Sleep Times|National Sleep Foundation. Sleepfoundation.org (2015). https://www.sleepfoundation.org/press-release/national-sleep-foundation-recommendsnew-sleep-times. Accessed 25 Nov 2020
24. Unger, T., Borghi, C., Charchar, F., Khan, N.A., Poulter, N.R., Prabhakaran, D., Ramirez, A., Schlaich, M., Stergiou, G.S., Tomaszewski, M., Wainford, R.D., Williams, B., Schutte, A.E.: 2020 international society of hypertension global hypertension practice guidelines. Hypertension **75**(6), 1334–1357 (2020). https://www.ahajournals.org/doi/10.1161/HYPERTENSIONAHA.120.15026
25. Williams, B., Mancia, G., Spiering, W., Agabiti Rosei, E., Azizi, M., Burnier et al.: 2018 ESC/ESH Guidelines for the management of arterial hypertension. European Heart Journal, 39(33), 3021–3104 (2018). https://academic.oup.com/eurheartj/article/39/33/3021/5079119. Accessed 25 Nov 2020
26. American Heart Association: Prevention and Treatment of High Cholesterol [Hyperlipidemia] (2017). https://www.heart.org/en/health-topics/cholesterol/prevention-and-treatment-of-high-cholesterol-hyperlipidemia

27. Čvorišćec, D., Čepelak, I.: Štrausova Medicinska biokemija. Medicinska Naklada, Zagreb (2009)
28. American Heart Association: Cardiovascular Disease and Diabetes (2015). https://www.heart.org/en/health-topics/diabetes/why-diabetes-matters/cardiovascular-disease--diabetes. Accessed 26 Nov 2020
29. Diabetes.co.uk.: Normal and Diabetic Blood Sugar Level Ranges - Blood Sugar Levels for Diabetes (2019). https://www.diabetes.co.uk/diabetes_care/bloodsugar-level-ranges.html. Accessed 26. Nov 2020
30. HHS/CDC: Smoking and Cardiovascular Disease (2014). https://www.cdc.gov/tobacco/data_statistics/sgr/50th-anniversary/pdfs/fs_smoking_CVD_508.pdf. Accessed 26 Nov 2020
31. Rau, H.-H., Hsu, C.-Y., Lin, Y.-A., Atique, S., Fuad, A., Wei, L.-M., Hsu, M.-H.: Development of a web-based liver cancer prediction model for type II diabetes patients by using an artificial neural network. Comput. Methods Programs Biomed. 125, 58–65 (2016). ISSN 0169–2607, https://doi.org/10.1016/j.cmpb.2015.11.009.

Finite Element Analysis of Patient-Specific Ascending Aortic Aneurysm

Smiljana Djorovic[1,2(✉)], Lazar Velicki[3,4], and Nenad Filipovic[1,2]

[1] Faculty of Engineering, University of Kragujevac, Sestre Janjic 6, 34 000 Kragujevac, Serbia
{smiljana,fica}@kg.ac.rs

[2] Bioengineering Research and Development Center (BioIRC), Prvoslava Stojanovica 6, 34 000 Kragujevac, Serbia

[3] Faculty of Medicine, University of Novi Sad, Hajduk Veljkova 3, 21 000 Novi Sad, Serbia
lazar@drvelicki.com

[4] Institute of Cardiovascular Diseases, Put doktora Goldmana 4, 21 204 Sremska Kamenica, Serbia

Abstract. This study investigated hemodynamics of the ascending aortic aneurysm caused by Bicuspid Aortic Valve (BAV) malformation, at the systole phase of cardiac cycle. The patient-specific numerical model had coupled computational fluid dynamics and finite element analysis simulation to analyze hemodynamics phenomena of ascending aortic aneurysm. Full patient-specific geometry was reconstructed based on Computed tomography (CT) scan images, in order to obtain the Three-Dimensional (3D) finite element mesh. The analysis which was performed using the possibilities of computational fluid dynamics, employed numeric methods and algorithms for the simulation of blood flow by solving the Navier-Stokes and continuity equations on computational meshes. The cardiac cycle was simulated for average blood properties and blood flow rate. The initial results for this single case study- the pressure, velocity field and shear stress distribution were quantified concerning anatomical patient's structures and showed alterations in dilated aortic area. The diseased

BAV produces an obstacle to blood flow and results in a decrease of blood flow in a whole flow regime which could cause huge energy loss during each heartbeat. This might induce high cyclic shear stress distribution and lead to other severe issues of the valve and surrounding aortic structures.

Keywords: Ascending aortic aneurysm · Finite element method · Computational modelling · Patient-specific geometry

1 Introduction

Bicuspid Aortic Valve (BAV) is the most common congenital heart disease (1–2% of the population) which reflects in fusion of two abnormal leaflets, in contrast to an initially healthy Tricuspid Aortic Valve (TAV). BAV is accompanied with an increased risk of aortic wall pathology, such as ascending aortic dilation (aneurysm), in up to 80% of cases [1]. While traditionally BAV aortopathy has been described as an intrinsic

A. Badnjevic and L. Gurbeta Pokvić (Eds.): CMBEBIH 2021, IFMBE Proceedings 84, pp. 630–637, 2021.
https://doi.org/10.1007/978-3-030-73909-6_73

aortopathy, recent advances in cardiovascular magnetic resonance have suggested that the flow pattern in the ascending aorta may be a major contributor to the aortopathy in BAV. In addition, abnormal aortic flow patterns in BAV may be partly responsible for the associated aortic dilation [2, 3]. Important factors in the development and progression of aortopathy in BAV patients include valve morphology and hemodynamic factors. Altered hemodynamic environment at the level of the ascending aorta can result in abnormal velocities and aortic wall shear stress (WSS), a wellknown stimulus leading to the expression of transcriptional factors implicated in vascular remodeling. Reported studies indicated altered WSS in the presence of BAVrelated flow abnormalities [2, 4]. These studies showed that BAV was associated with altered aortic flow patterns (nested flow, helical flow, or retrograde flow) and elevated or asymmetrically expressed WSS along the circumference of the aortic wall.

Recent findings in cardiovascular Magnetic Resonance Imaging (MRI) and Computational Fluid Dynamics (CFD) enabled the assessment of aortic valve morphology and Three-Dimensional (3D) blood flow velocity, in case of patients with normal and diseased aortic root. MRI and CFD can be used to calculate hemodynamic variables that explain the development and progression of aortic aneurysm as well as model the effects of surgery. In this study the CFD approach was performed with aim to examine the hemodynamic characteristics of patient-specific dilated ascending aorta, caused by BAV malformation. The main characteristics of this approach are presented in Table 1.

Table 1. Characteristics of CFD approach.

Definition	Application	Advantages	Disadvantages
Computational simulation of blood flow and calculation of hemodynamics in high spatial and temporal resolution. Examples of calculations include the Navier–Stokes equations for blood flow and other calculations of fluid dynamics	Examination of aneurysms and rupture risk. The design and evaluation of vascular devices, planning and predicting outcomes of vascular surgery [5]	Prediction of behavior of aneurysm, vascular device or outcome of surgery without subjecting the patient to that risk Calculation of pressures accounting wall motion via fluid–structure interaction equations [6] Hemodynamic measurements, such as velocities, pressures, shear stress, can be calculated from the anatomical and flow data	Computationally expensive with long simulations Compromised fidelity to reproduce *in vivo* hemodynamics due to assumptions concerning in-flow velocity profiles, blood rheology, choice of turbulence model and parameters, as well as the need for highquality data for geometry and flow boundary conditions [7] Limitations of modelling the vessel wall characteristic

In this study, the computational analysis of blood flow through the ascending aortic aneurysm was performed. The simulation included CFD analysis based on the Finite Element Method (FEM). The creation of a 3D patient-specific model of ascending aortic aneurysm, based on Computed Tomography (CT) scans is presented in the first part of the study. Creation of the model included segmentation od 2D imaging data. The analyzed patient suffered from the BAV (type 0), which had caused the ascending aortic aneurysm. The material properties and boundary conditions that were used for FE analyses are explained in the first part also. The second part covers discussion of results (velocity, pressure and shear stress distribution) for employed model. In the last part, main conclusions are given with plans for further improvements of presented computational analysis. Given the likely pathophysiological effects of the abnormal flow on aortic dilation in BAV disease, examining the flow patterns in the proximal dilated aorta after the bicuspid aortic valve may provide novel insights.

2 Materials and Methods

Aortic root and ascending aorta have a complex anatomy. Process of obtaining a full 3D patient-specific geometrical model includes several phases, that will we described. In the first phase of the model creation, segmentation and generation of the ascending aortic aneurysm geometry from the CT scans were performed. CT scanner was used to obtain a set of DICOM images which were imported in the segmentation software (*Materialise Mimics*). The software processes medical images and gives a highly accurate 3D models of the patients' anatomy. The resolution of CT scan images that were composed was 512 × 512 pixels with slice increment of 0.625 mm. Figure 1 shows one cross-section of modelled ascending aortic aneurysm in three planes (A-coronal, B-sagittal and C-transversal plane) where the BAV can be seen. Manipulation in three different planes enabled creation of more accurate 3D model of the patient's anatomy. The creation of the 3D model from 2D data was achieved by segmentation which was applied for each image.

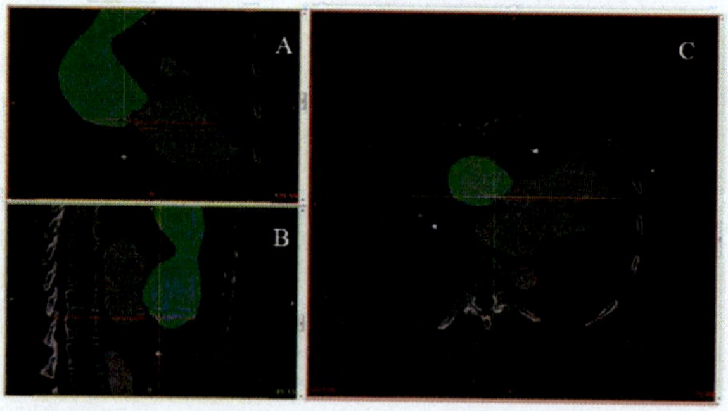

Fig. 1. Cross-sections of modelled ascending aorta in three planes (A - coronal, B - sagittal and C - transversal).

In the second phase, the initial surface mesh of obtained 3D model in the STL (stereolithography) format was optimized giving the smooth mesh and elements suitable for further computation. The final 3D surface finite element mesh of ascending aortic aneurysm is presented in Fig. 2 (anterior view). It can be seen that the patient has the ascending aortic aneurysm that is a common BAV consequence [8]. The created model started from the sinotubular junction and did not include aortic leaflets, sinuses and aortic annulus. The model contained the roots of arteries at the aortic arch, such as Brachiocephalic Trunk (BCT), left Common Carotid Artery (LCA) and Left Subclavian Artery (LSA), as well as the thoracic aorta. In addition, it can be seen that the aortic root dimensions are larger comparing to the normal aortic root which is caused by BAV type 0 malformation. In the area of aneurysmal dilatation its maximal diameter was approximately 80 mm.

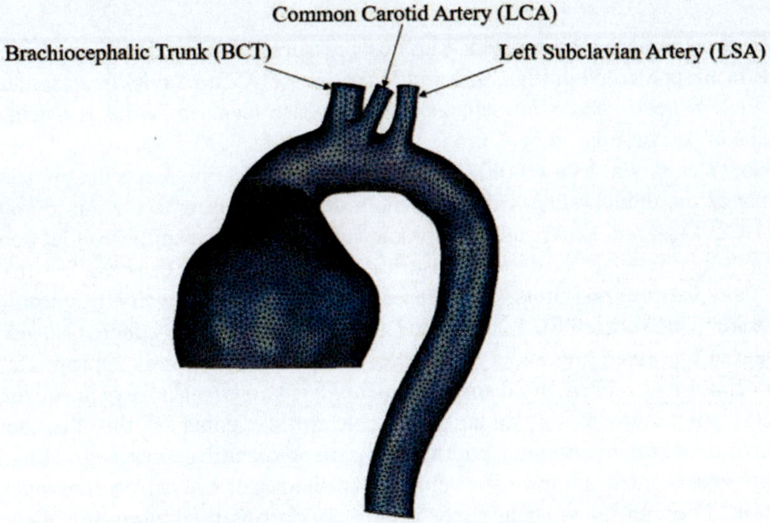

Fig. 2. Finite element mesh of the ascending aortic aneurysm.

3 Numerical Simulation

With aim to perform the computational simulation, the surface model was converted into the volumetric model, which consisted of the eight-node brick elements. The analyzed model, created from the eight-node brick elements, was imported into the inhouse software for further computational simulation of blood flow.

In order to perform analysis which reflects the realistic human blood flow we used average blood properties: a dynamic viscosity (μ) of 0,0035 Pa·s, and density (ρ) of 1,050 g/cm^3 [9]. Flow was considered to be laminar, homogeneous (Newtonian) and viscous incompressible. The code used for this complex time-dependent and full 3D fluid flow was validated using the analytical solution for shear stress and velocities through a 3D curved tube [10]. The 3D flow was governed by the Navier-Stokes equations and the continuity equation.

For this study the inlet boundary conditions in the ascending aorta were set as the mass flow boundary conditions with a pulsatile wave [11]. The outlet boundary conditions were given as the pressure boundary conditions, which reflected the external forces from outside the analysis domain. Also, zero-velocity at the rigid walls was prescribed.

4 Results and Discussion

The main fluid dynamic parameters, such as pressure, velocity field and shear stress distribution are presented in Figs. 3, 4 and 5, respectively. The model is presented from anterior view at peak systole moment, as the worst-case scenario, which is reached after about 0,3 s of the simulation.

Pressure (Units: Pa, 1 Pa = 0,0075 mm Hg) presented in Fig. 3, has the highest value at the inlet of the model, after which gradually decreases going to the aortic arch with roots of BCT, LCA and LSA, and the thoracic aorta. Due to the applied outlet boundary conditions, physiological resistance pressure at the arterial outlets can be noted. In case of velocity field distribution (Units: *m/s*) presented in Fig. 4, higher velocities dominate in the aortic arch with roots of BCT, LCA and LSA, while significantly decreased velocities are present at the ascending aneurysmal area. Such a distribution is appropriate to the boundary conditions and enlarged aortic diameter. It is known that size of aortic diameter is not sufficient parameter for evaluation of acute aortic events [12], thus the additional parameters, such as hemodynamic parameters, patient-specific geometry and blood flow need to be considered. In addition, the velocity distribution directly affects the shear stress distribution. Therefore, low shear stress (Units: *Pa*) is observed along the aneurysmal zone of ascending aorta (Fig. 5) where blood velocity has low values. It makes condition for bigger aneurysmal sac creation and its further growth, as it is known that the shear stress acts on the vessel wall and alters its properties. In contrary to the zones of low shear stress, the increased shear stress is located at the aortic arch with roots of BCT, LCA and LSA, as well as after the aneurysm.

Finally, it should be noted that this analysis was built on experiences of other groups having performed CFD simulations of blood flow in the human dilated ascending aorta [2–4, 13, 14]. CFD simulation was performed, with aim to investigate how the flow field is affected by ascending aortic aneurysm as a consequence of BAV, and by flow conditions.

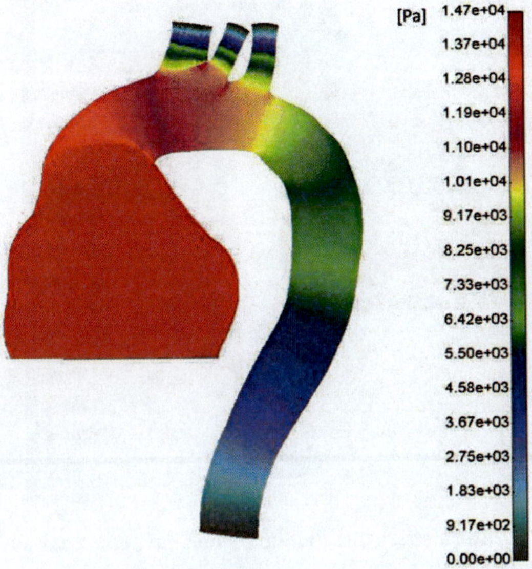

Fig. 3. Pressure field distribution (Units: Pa) at the peak systole.

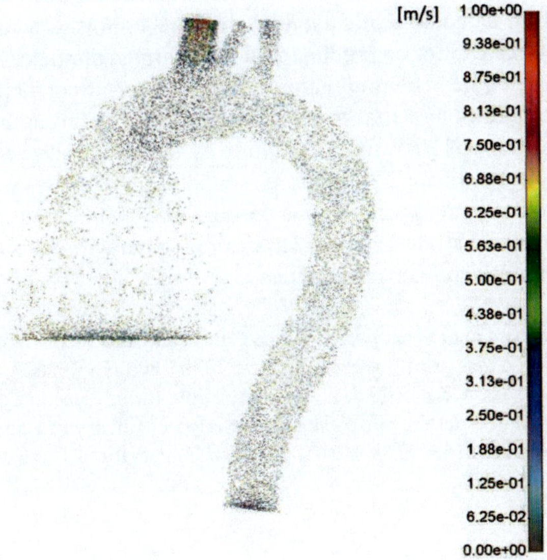

Fig. 4. Velocity field distribution (Units: m/s) at the peak systole.

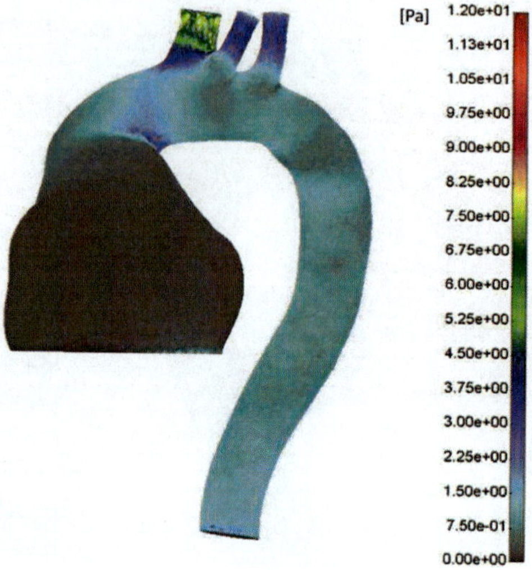

Fig. 5. Shear stress distribution (Units: Pa) at the peak systole.

5 Conclusions

In the past, much attention has been given to aortic diameter, but it has been shown to be a poor indicator of an acute aortic event in the BAV cohort. It is also known that the behavior of aneurysms cannot be predicted by their largest diameter alone. Their entire patient-specific geometry and hemodynamic conditions can affect their behavior and surrounding aortic structures. For that reason, the non-invasive computational modelling can improve assessment of BAV malformations by enabling risk prediction and monitoring of patients. As one of CFD limitations is modelling of vessel wall characteristics, the improved computational approach considering Fluid-Structure Interaction (FSI) will be objective of the future studies in order to examine characteristics of degraded aortic wall beside the hemodynamic characteristics.

Acknowledgment. This paper is supported by the European Union's Horizon 2020 research and innovation programme under grant agreements No 777204 and No 952603. This article reflects only the author's view. The Commission is not responsible for any use that may be made of the information it contains. The paper is also funded by Serbian Ministry of Education, Science, and Technological Development [451-03-68/2020-14/200107 (Faculty of Engineering, University of Kragujevac)].

Conflict of Interest Declaration. The authors declare no conflict of interest.

References

1. Ward, C.: Clinical significance of the bicuspid aortic valve. Heart **83**(1), 81–85 (2000)
2. Barker, A.J., Markl, M., Burk, J., Lorenz, R., Bock, J., Bauer, S., Schulz-Menger, J., von Knobelsdorff-Brenkenhoff, F.: Bicuspid aortic valve is associated with altered wall shear stress in the ascending aorta. Circ. Cardiovasc. Imaging **5**(4), 457–466 (2012)
3. Bissell, M.M., Hess, A.T., Biasiolli, L., Glaze, S.J., Loudon, M., Pitcher, A., Davis, A., Prendergast, B., Markl, M., Barker, A.J., Neubauer, S., Myerson, S.G.: Aortic dilation in bicuspid aortic valve disease: flow pattern is a major contributor and differs with valve fusion type. Circ. Cardiovasc. Imaging **6**(4), 499–507 (2013)
4. Barker, A.J., Markl, M.: The role of hemodynamics in bicuspid aortic valve disease. Eur. J. Cardiothorac. Surg. **39**, 805–806 (2012)
5. den Reijer, P.M., Sallee, D., III., van der Velden, P., Zaaijer, E.R., Parks, W.J., Ramamurthy, S., Robbie, T.Q., Donati, G., Lamphier, C., Beekman, R.P., Brummer, M.E.: Hemodynamic predictors of aortic dilatation in bicuspid aortic valve by velocity encoded cardiovascular magnetic resonance. J. Cardiovasc. Magn. Reson. **12**, 4 (2010)
6. Xiong, G., Figueroa, C.A., Xiao, N., Taylor, C.A.: Simulation of blood flow in deformable vessels using subject-specific geometry and spatially varying wall properties. Int. J. Numer. Method Biomed. Eng. **27**, 1000–1016 (2011)
7. Morris, P.D., Narracott, A., Tengg-Kobligk, H., Silva Soto, D.A., Hsiao, S., Lungu, A., et al.: Computational fluid dynamics modelling in cardiovascular medicine. Heart **102**, 18–28 (2016)
8. Tadros, T.M., Klein, M.D., Shapira, O.M.: Ascending aortic dilatation associated with bicuspid aortic valve: pathophysiology, molecular bioloy, and clinical implications. Circulation **119**, 880–890 (2009)
9. Gao, F., Watanabe, M., Matsuzawa, T.: Stress analysis in a layered aortic arch model under pulsatile blood flow. BioMed. Eng. Online **5**, 1–11 (2006)
10. Filipovic, N., Mijailovic, S., Tsuda, A., Kojic, M.: An implicit algorithm within the arbitrary Lagrangian-Eulerian formulation for solving incompressible fluid flow with large boundary motions. Comput. Methods Appl. Mech. Eng. **195**, 6347–6361 (2006)
11. Simao, M., Ferreira, J., Tomas, A., Fragata, J., Ramos, H.: Aorta ascending aneurysm analysis using CFD models towards possible anomalies. Fluids **2**(2), 31 (2017)
12. Michelena, H.I., Khanna, A.D., Mahoney, D., Margaryan, E., Topilsky, Y., Suri, R.M., et al.: Incidence of aortic complications in patients with bicuspid aortic valves. JAMA **306**, 1104–12 (2011)
13. Djorovic, S., Milosavljevic, A., Velicki, L., Filipovic, N.: Comparative finite element analysis of patient-specific tricuspid and bicuspid aortic valve. In: 17th International Conference on BioInformatics and BioEngineering (BIBE), Washington DC, Virginia, 23–25 October 2017, pp. 563–567 (2017)
14. Martin, C., Wei, S., Elefteriades, J.: Patient-specific finite element analysis of ascending aorta aneurysms. Integr. Cardiovasc. Physiol. Pathophysiol. **308**(10) (2015)

Particle Deposition in Respiratory Tract: Where are the Limits?

Amar Elezović[1](✉), Sandra Cvijić[2], Alisa Elezović[3], Saša Pilipović[1], and Jelena Parojčić[2]

[1] The Control Laboratory of the Agency for Medicinal Products and Medical Devices of Bosnia and Herzegovina, Sarajevo, Bosnia and Herzegovina
a.elezovic@almbih.gov.ba

[2] Department of Pharmaceutical Technology and Cosmetology, Faculty of Pharmacy, University of Belgrade, Belgrade, Serbia

[3] Department of Pharmaceutical Technology, Faculty of Pharmacy, University of Sarajevo, Sarajevo, Bosnia and Herzegovina

Abstract. In last two decades in silico biopharmaceutics modelling has become increasingly important in all stages of drug development. One of the widely used pharmacokinetic modelling software is Gastro Plus that, with its additional dosage routes modules, enables creation of pharmacokinetic models, amongst others, for pulmonary dosage route. Medicines that are applied via that route have specific methods for characterization and analysis. However, drug that is target of in silico method in Gastro Plus is not characterized in full extent with such parameters. In this work we created various models of size distributions and depositions of two pMDI preparations containing beclomethasone dipropionate in 4 respiratory compartments of Gastro Plus Pulmonary model. Those data were compared with results of in vivo studies and model with best correlation was obtained. Such model with defined cut-off limits between each compartment enables better prediction of pharmacokinetic properties in generated Gastro Plus in silico model.

Keywords: Lung deposition · In silico modeling · Aerodynamic assessment of particle size

1 Introduction

In silico biopharmaceutics modeling and simulation have become more popular and important in the last two decades. Before "in silico era", discovery and development of new drugs was extremely slow and expensive process. Use of in vivo and in vitro tests were the basis of data collection and proof for evaluation of new drug candidates. And, often, new drug candidates were proven to be unsuccessful, increasing general cost and time consumed in development of new drug. With the introduction of in silico methods, the process of new drug development was somewhat simplified and made to be more cost effective. As such, in silico methods are used in all stages of drug development, from lead identification and optimization, to preclinical and clinical studies [1]. Regulatory agencies have recognized that, and have begun to accept use of such methods in drug

A. Badnjevic and L. Gurbeta Pokvić (Eds.): CMBEBIH 2021, IFMBE Proceedings 84, pp. 638–644, 2021.
https://doi.org/10.1007/978-3-030-73909-6_74

development and regulatory submissions [2, 3]. One of the widely used pharmacokinetic modelling software is Gastro Plus (Simulations Plus Inc., USA) [4–6]. Besides modules for i.v. and oral routes of administration, it contains additional dosage routes modules which enable development of models for dermal, intraoral, pulmonary, ocular, and intramuscular routes of drug administration. Pulmonary dosage routes is described as The Pulmonary Compartmental Absorption & Transit (PCAT™) model that represents the lung/nose as a collection of the following compartments: an optional nose (containing the anterior nasal passages), extra-thoracic (naso- and oro-pharynx and the larynx), thoracic (trachea and bronchi), bronchiolar (bronchioles and terminal bronchioles) and alveolar-interstitial (respiratory bronchioles, alveolar ducts and sacs and interstitial connective tissue) [7]. Distribution of drug dose to those compartments is based on the Advanced ICRP 66 deposition model algorithms [8], Finlay deposition model [9] and option of custom model where user alone defines percentual deposition od drug in all compartments. Although mentioned models are long time foundation for measurement of particle deposition in the respiratory tract, both models originate from the field of radiological protection and have little or none "pharmaceutical background". As a consequence, input parameters in the PCAT model are not in line with Pharmacopoeial values and measurements that are used to characterize preparations for inhalation. There are two compendially accepted parameters that define quality and characteristics of inhaler products. First is Uniformity of delivered dose [10] which defines uniformity of dose emitted from inhaler and ensures that patient receives similar dose throughout the use of inhaler. Other is Aerodynamic assessment of fine particles [11]. In the latter test, aerosol particle size distribution is determined based on its aerodynamic diameter and test is performed on inertial impactors. Two most widely used are Andersen Cascade Impactors and Next Generation Impactor. They consist of series of stages with holes of decreasing diameter. As flow trough impactor is constant, speed of aerosol trough every stage is stepwise increased and, on every stage, smaller particles doe to increase of speed and inertia are deposited by impaction on the collection plates. After quantification of active substance on every plate, from obtained aerodynamic distribution parameters that can be used to characterize aerosol can be obtained. Fine particle mass or fraction is measure of mass or % from total dose that is composed of particles with aerodynamic diameter less than 5 μm. Particles with that size are considered to be sufficiently small to reach deeper part of lungs that are suitable for absorption or local effect of drug. Mass median aerodynamic diameter (MMAD) is diameter of particle below which 50% of the particle population lies on the basis of mass [12]. Geometric standard deviation (GSD) defines spread of distribution of particle size of aerosol. Although it is possible to enter data for mass median aerodynamic diameter and geometric standard deviations in particle size distribution section of the Gastro Plus, places for those data are not originally intended for those specific measurements and are designated for input of "Mean Particle Radius (um)" and "Standard Deviation". Also, MMAD and GSD do not completely characterize aerodynamic particle size distribution. Due to characteristic shape of particle size distributions od preparations for inhalations, fraction of particles with particles lager than 8 μm usually is not included in calculation of MMAD and GSD. That can be explained since such large particles do not enter lungs and stay in extrathoracic compartment, but since they usually consist for more than 30% of drug dose, they can have significant pharmacodynamic effect after being swallowed and absorbed through gastro-intestinal tract. Because of that, it would be useful to have possibility to incorporate data about aerodynamic particle size distribution in the PCAT model. The simplest solution is to define percentages of active substance that are deposited in each of the four compartments (Extra-thoracic, Thoracic, Bronchiolar and AlveolarInterstitial). Therefore, particle size cut-off values should be defined so that characterization of particle size distribution in the Gastro Plus software is more in line with standards and measurements in pharmaceutical industry.

2 Materials and Methods

Two pressurized metered dose inhalers (pMDI) with beclomethasone dipropionate as active substance were used in this study. Aerodynamic particle size distribution was determined on the Next Generation Impactor (Copley Scientific, Nottingham, UK) in accordance with the Ph.Eur. monograph Preparation for inhalation: Aerodynamic assessment of fine particles [11]. HPLC method for quantification was described elsewhere [13]. All calculations of particle size distributions were performed using the Copley Inhaler Testing Data Analysis Software (CITDAS) v. 3.10 (Copley Scientific, Nottingham, UK). In silico model was constructed using the Gastro Plus software v. 9.8 (Simulation Plus Inc., Lancaster, CA, USA) using "top down" approach based on published in vivo studies [14–16]. Model development will be subject of separate publication. Size limits for each compartment in the Gastro Plus pulmonary model were varied until the best correlation between in silico and in vivo data is obtained.

3 Results and Discussion

Results of aerodynamic assessment of fine particles of the two examined products are presented on Fig. 1.

It is evident that Product 1 have more desirable profile of particle sizes with smaller percentage coarse particles (in induction port) and higher percentage of small particles that can penetrate and be deposited in deeper parts of lungs. Deposition of those particles in the respiratory tract is governed by inertial impaction in the upper parts of respiratory tract and, as diameter of airways and speed of particles declines, other mechanisms gradually become more dominant (gravitational sedimentation and Brownian diffusion) [17]. Because of that, there is a general consensus that lung deposition of particles cannot be defined only by data obtained from inertial impactors (such as NGI) [18] and, consequently, lack of data that link location and extent of deposition in the respiratory tract to particle size distribution. Most frequently presented source that link aerodynamic particle size distribution to regional deposition in the respiratory tract, although highly criticized, is presented in the Table 1 [18].

Aerodynamic diameters provided on Table 1. Were used as starting point for comparison of effect of varying limits of particle sizes and quantity of deposition in each of the four respiratory compartments and was marked as base value (0). Values for limit sizes are then "moved" by 0.5 μm towards lower and higher values (points −0.5, +0.5

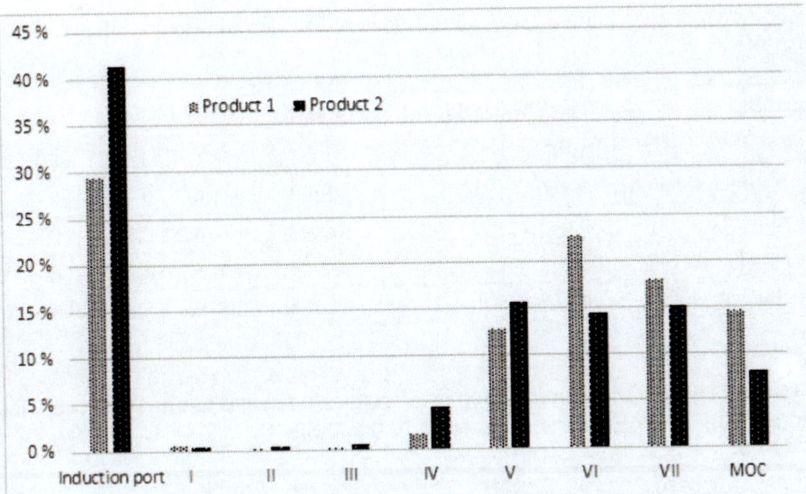

Fig. 1. Aerodynamic distribution of fine particles on stages of NGI

Table 1. Excerpt from the presentation of regional lung deposition relative to Andersen cascade Impactor (ACI) stage size range [18]

Equivallent to:	ACI Stage	Aerodynamic cut-off limit
Naso-pharynx	Preseparator	>10 μm
	Stage 0	9.0–10.0 μm
	Stage 1	5.8–9.0 μm
	Stage 2	4.7–5.8 μm
Trachea and primary bronchi	Stage 3	3.3–4.7 μm
Secondary bronchi	Stage 4	2.1–3.3 μm
Terminal bronchi	Stage 5	1.1–2.1 μm
Alveoli	Stage 6	0.65–1.1 μm
	Stage 7	0.43–0.65 μm

and +1). Also, limits marked as A1 and A2 were selected based on data from the reference [19]. In the pulmonary in silico model, created in the Gastro Plus, User Defined deposition model was selected and all selected groupings of active ingredient particles were entered (−0.5, 0, +0.5, +1, A1, A2). Correlation with in vivo data (Cp/t curve) was obtained for all groupings of particle size dimensions limits for Product 2. For Product 1 only one Cp/t point was usable and similarity was determined by calculating percentage prediction error value, PE% (PE% = (predicted value – measured value) × 100/ measured value). Values of particle size distribution in each compartment are presented in the Table 2.

Table 2. Aerodynamic particle size cut-off values for deposition in different respiratory compartments

Compartment	Particle size dimensions limit (μm)					
	−0,5	0	+0,5	+1	A1	A2
Alveolar-Interstitial (AI)	0–0.6	0–1.1	0–1.6	0–2.1	0–2	0–1
Bronchiolar (B)	0.6–3.2	1.1–3.7	1.6–4.2	2.1–4.7	2–4	1–3
Thoracic (T)	3.2–4.2	3.7–4.7	4.2–5.2	4.7–5.7	4–5	3–5
Extra-thoracic (ET)	4.2–∝	4.7–∝	5.2–∝	5.7–∝	5–∝	5–∝

Obtained values for the correlation in silico predicted and in vitro observed data and percentage prediction errors are presented in the Table 3.

Table 3. Influence of changing particle size limits on the correlation between in silico predicted and in vivo observed data

Particle size grouping	Product 1 percentage prediction error (%)	Product 2 R^2
−0,5	4,200	0,418
0	−2,580	0,877
+0,5	−0,940	0,984
+1	0,400	0,988
A1	0,440	0,990
A2	14,640	0,833

As can be seen in Table 3, variations of the Product 1 results range from 14.640% to 0.400%. Comparing size limits that have large errors, it can be concluded that limits for Alveolar-Interstitial compartments are set around 0–1 μm and 1–3 μm intervals. Compared with groupings with smaller error and better correlation, it can be concluded that this in silico model favors greater deposition in Alveolo-Interstitial and Bronchial compartment. Greatest error of 14,640% was also characterized with limits for Thoracic compartment of 3–5 μm. Lowest error was observed with particle size groupings +1 and A1 that are characterized with relatively higher values for particle size limits. Similar situation is with correlation factors obtained for Product 2. Lowest level of correlation was observed with groupings with relatively low particle size limits (−0.5, 0 and A2) and highest level of correlation was observed with groupings wilt relatively high particle size limits (A1). Two groupings with smallest errors are +1 and A1. They both have particle size limits for Alveolo-Interstitial compartment in similar ranges of around 0–2 μm. Grouping +1 have little bit wider limits for Bronchiolar compartment and limits for Thoracic compartment are moved by 0.7 μm. In relation to all groupings, grouping A1 have one of the greatest ranges for Alveolo-Interstitial compartment and narrower limits for Bronchiolar compartment. This indicates that employed in silico model favors greater deposition in the lower respiratory compartments where absorption is faster. That difference and increase in the rate of absorption is in this case "obtained" exclusively from Bronchiolar compartment that contains smaller particles that can be dissolved and absorbed faster. Thoracic and Extra-Thoracic compartments of grouping A1 are located in the "middle of the range" with several of groupings with higher and several groupings with lower limits.

4 Conclusions

There is a long discussion about usefulness of cascade impactor data in predicting deposition of inhaled particles in the respiratory tract. Many factors are opposed to affirmative conclusion and are more in line to view cascade impactors as primarily quality control tool in pharmaceutical industry. But, at the moment, there is no more suitable in vitro analytical technique for characterizing behavior of aerosol in conditions that approximately resemble respiratory tract. Also, increase of applications of in silico models in drug development and research will increase the need for quality and quantity of data that define drug under investigation. Present condition where usefulness of aerodynamic particle size data in in silico studies is generally limited to MMAD and GSD can present problem. Generally, in calculation of MMAD and GSD, sum of coarse particles that are captured in the induction port and preseparator are not included. Since those coarse particles are predominantly deposited in the oro-pharyngeal region and do not enter lower respiratory tract such logic is understandable, but those particles do compose part of the applied dose of the drug. They are usually swallowed and follow the path of orally applied drug. Even if active substance has very low oral bioavailability or study was conducted with oral application of charcoal, effect of that dose cannot be neglected since it "removes" part of the applied dose and can lead to overestimation of the deposited dose. Percentual distribution of active substance in each of the respiratory compartments could improve quality of results of in silico studies. Looking at the results, in comparison to +1 grouping, A1 grouping have much better correlation in Product 2 comparison and slightly higher level of error in comparison of Product 1. Together with high similarity with cut-off limits presented in reference [19] it can be concluded that grouping A1 is the most adequate to define deposition of beclomethasone dipropionate from pMDI products in the in silico model generated on the basis of before mentioned in vivo studies.

References

1. Baig, M.H., Ahmad, K., Adil, M., Khan, Z.A., Khan, M.I., Lohani, M., et al.: Drug discovery and in silico techniques: a mini-review. Enzyme Eng. **4**(1) (2014)
2. Valerio, L.G., Jr.: In silico toxicology models and databases as FDA Critical Path Initiative toolkits. Hum. Genomics **5**(3), 200–207 (2011)
3. Morrison, T.M., Pathmanathan, P., Adwan, M., Margerrison, E.: Advancing regulatory science with computational modeling for medical devices at the FDA's office of science and engineering laboratories. Front. Med. (Lausanne) **5**, 241 (2018)

4. Lin, W., Yan, J.-H., Heimbach, T., He, H.: Pediatric physiologically based pharmacokinetic model development: current status and challenges. Curr. Pharmacol. Rep. **4**(6), 491–501 (2018)
5. Al-Tabakha, M.M., Alomar, M.J.: In vitro dissolution and in silico modeling shortcuts in bioequivalence testing. Pharmaceutics **12**(1), 45 (2020)
6. Sjögren, E., Thörn, H., Tannergren, C.: In silico modeling of gastrointestinal drug absorption: predictive performance of three physiologically based absorption models. Mol. Pharm. **13**(6), 1763–1778 (2016)
7. SimulationPlus GastroPlus: Additional dosage routes 2020, 25 December 2020. https://www.simulations-plus.com/software/gastroplus/additionaldosage/
8. ICRP. Radiological Protection and Safety in Medicine Annals of the ICRP, vol. 26, no. 73. ICRP Publication (1996)
9. Finlay, W.H., Martin, A.R.: Recent advances in predictive understanding of respiratory tract deposition. J. Aerosol Med. Pulm. Drug Deliv. **21**(2), 189–206 (2008)
10. Europe1 Co. Preparation for inhalation. European Pharmacopoeia. 9.0 ed, pp. 875–880. Council of Europe, Strasbourg (2017)
11. Europe2 Co. Preparations for inhalation: Aerodynamic assessment of fine particles. European Pharmacopoeia. 9.0 ed, pp. 323–335. Council of Europe, Strasbourg (2017)
12. Lewis, D., Copley, M.: Inhaled product characterization. Pharm. Technol. **2011**(6) (2011)
13. Sahib, M.N., Darwis, Y., Khiang, P.K., Fung Tan, Y.T.: Aerodynamic characterization of beclomethasone dipropionate from Beclate-50 inhaler® by HPLC-UV. J. Liq. Chromatogr. Relat. Technol. **34**(8), 613–621 (2011)
14. Daley-Yates, P.T., Price, A.C., Sisson, J.R., Pereira, A., Dallow, N.: Beclomethasone dipropionate: absolute bioavailability, pharmacokinetics and metabolism following intravenous, oral, intranasal and inhaled administration in man. Br. J. Clin. Pharmacol. **51**(5), 400–409 (2001)
15. Harrison, L.I., Kurup, S., Wagner, C., Ekholm, B.P., Larson, J.S., Kaiser, H.B.: Pharmacokinetics of beclomethasone 17-monopropionate from a beclomethasone dipropionate extrafine aerosol in adults with asthma. Eur. J. Clin. Pharmacol. **58**(3), 197–201 (2002)
16. Singh, D., Piccinno, A., Borrill, Z., Poli, G., Acerbi, D., Meuleners, L., et al.: Tolerability of high cumulative doses of the HFA modulite beclomethasone dipropionate/formoterol combination inhaler in asthmatic patients. Pulm. Pharmacol. Ther. **21**(3), 551–557 (2008)
17. Darquenne, C.: Aerosol deposition in health and disease. J. Aerosol Med. Pulm. Drug Deliv. **25**(3), 140–147 (2012)
18. Dunbar, C., Mitchell, J.: Analysis of cascade impactor mass distributions. J. Aerosol Med.: Off. J. Int. Soc. Aerosols Med. **18**(4), 439–451 (2005)
19. Mitchell, J., Nagel, M.: Particle size analysis of aerosols from medicinal inhalers. KONA Powder Part. J. **22**, 32–65 (2004)

Epidemiologic Data of Adult Native Biopsy-Proven Renal Diseases in University Clinical Hospital Mostar

Davor Tomić[1](✉), Ivan Zeljko[2], Fila Raguž[3], Ivana Škoro[3], Slavica Ćorić[3], and Ivan Tomić[3]

[1] Department of Urology, University Hospital Mostar, Mostar, Bosnia and Herzegovina
[2] Community Health Center, Široki Brijeg, Bosnia and Herzegovina
[3] Department of Nephrology, University Hospital Mostar, Mostar, Bosnia and Herzegovina

Abstract. Renal biopsy is the most accurate method by which we analyze the pathological process in the kidney. Tissue analysis also provides insight into the prognosis and progression of the disease and evaluates the patient's therapeutic approach. The epidemiology of biopsy-proven renal diseases provides useful information about prevalence of renal diseases and its clinicopathological correlations. Data provided by renal biopsy registries could help better understanding the etiopathological aspects of these diseases. Today, many countries have epidemiological data on biopsy-proven renal disease, some of them Italy, Denmark, Brazil, Spain, Czech Republic and Saudi Arabia. The aim of this report is to show which renal diseases have been proven by biopsy at the Mostar University Clinical Hospital, determine their incidence and present the distribution of these diseases by age, sex and clinical presentation. This crosssectional study included 107 patients treated at the Internal Medicine Clinic with the dialysis center of the University Clinical Hospital Mostar who underwent renal biopsy due to renal pathology that could not have been differentiated differently. Data on total number of patients, their histological diagnosis, gender, age and clinical syndrome were collected and analyzed retrospectively. In all patients, light, immunofluorescence and electron microscopic analysis was performed. The median age of the patients was 48 years, and the majority of patients were men (51.4%). The most common clinical presentation was nephrotic syndrome (51.4%) followed by chronic nephritic syndrome (31.8%) and asymptomatic urinary abnormalities (16.8%). The most common disease presented by nephrotic syndrome was focal segmental glomerulosclerosis (34.5%), while the most common disease presented with chronic nephritic syndrome and asymptomatic urinary abnormalities was IgA glomerulonephritis (38.2% and 38.9%). The most common biopsy-proven renal disease in total was IgA glomerulonephritis (28.0%), followed by focal segmental glomerulosclerosis (23.4%). In men, the most common biopsy-proven renal diseases were found to be IgA glomerulonephritis (36.4%) and focal segmental glomerulosclerosis (16.4%), and in women focal segmental glomerulosclerosis (30.8%) and lupus nephritis (21.2%). This study showed that IgA glomerulonephritis is the most common biopsy-proven renal disease at Mostar University Hospital, and that the biopsy-proven renal disease is mostly affected by the male population. The most common clinical presentation of patients with renal biopsy indicated was found

A. Badnjevic and L. Gurbeta Pokvić (Eds.): CMBEBIH 2021, IFMBE Proceedings 84, pp. 645–654, 2021.
https://doi.org/10.1007/978-3-030-73909-6_75

to be nephrotic syndrome. Since all patients from Herzegovina gravitate to the University Clinical Hospital Mostar, this study represents epidemiological data of these diseases in Herzegovina.

Keywords: Biopsy-proven renal disease · Epidemiology · Glomerulonephritis · Renal biopsy · Nephrotic syndrome

1 Introduction

Renal biopsy is the most exact method by which we analyze the pathological process in the kidneys. Tissue analysis provides insight into the prognosis and progression of the disease and the therapeutic approach to the patient is assessed. Therefore, this method is irreplaceable in the assessment of renal disease [1]. The epidemiology of biopsy-proven renal diseases (BPRD) provides useful information about prevalence of renal diseases and its clinicopathological correlations. Data provided by renal biopsy registries could help better understanding the etiopathological aspects of these diseases. These data are the basis for future epidemiological studies aimed at identifying important risk factors in the development and progression of renal disease and can also help in the development of protocols for their prevention of these diseases [2, 3]. Today, many countries have epidemiological data on biopsyproven renal disease, and some of them are Italy, Denmark, Brazil, Spain, the Czech Republic, and Saudi Arabia [4–10].

This study will show which renal diseases were proven by biopsy at the University Clinical Hospital Mostar, determine their frequency and show the distribution of these diseases by age, sex and clinical presentation. Since all patients from Herzegovina gravitate to the University Clinical Hospital Mostar, this paper presents the epidemiological data of these diseases in Herzegovina.

2 Subjects and Methods

The research was conducted at the Clinic for Internal Medicine with the dialysis center of the University Clinical Hospital Mostar, and covers the period from January 1, 2012 to December 31, 2019. The study included 107 patients who were treated at the Clinic for Internal Medicine with a dialysis center at the University Clinical Hospital Mostar and who underwent a renal biopsy for the presence of renal pathology that could not be differentiated otherwise. The data collected for each patient were the date of renal biopsy, age, sex, serum creatinine and maximal 24-h proteinuria till the time of biopsy, as well as all other important laboratory findings and underlying conditions suggesting possible association with renal disease.

Pathohistological diagnoses were classified into following categories: IgA glomerulonephritis (IgAGN), focal segmental glomerulosclerosis (FSGS), nephroangiosclerosis (NAS), diabetic nephropathy (DMGN), amyloidosis and other renal diseases with organized deposits (AMORGDEP), membranous glomerulonephritis (MGN), minimal change disease (MCD), hereditary nephritis which includes Alport's syndrome and thin membrane disease (HERNEF), lupus nephritis (LN), pauci-immune glomerulonephritis (PCIMUNGN), membranoproliferative glomerulonephritis (MPGN), mezangioproliferative glomerulonephritis (MZPGN), nephronophthisis (NFNT) i C1Q nephropathy (C1QN).

According to the clinical presentation, which was also an indication for biopsy, the subjects were divided into the following categories: nephrotic syndrome (NS), asymptomatic urinary abnormalities (AUA) and chronic nephritic syndrome (CNS). Nephrotic syndrome was defined as proteinuria greater than 3.5 g/24 h. Asymptomatic urinary abnormalities were defined as haematuria or non-nephrotic proteinuria (proteinuria less than 3.5 g/24 h) or both with normal estimated glomerular filtration rate and no clinical symptoms. Chronic nephritic syndrome was defined as a permanent (longer than 6 months) decrease in the estimated glomerular filtration rate with non-nephrotic proteinuria (proteinuria less than 3.5 g/24 h) with or without hematuria.

According to age, patients were divided into 4 groups (16 to 29 years; 30 to 45 years; 46 to 60 years; >60 years). An encryption system has been introduced according to which it is not possible to determine the identity of persons whose anamnestic data were used in the research.

Biopsies were performed with a 16-gage (Tru-Cut) biopsy needle under the control of continuous ultrasound guidance. After the biopsy, one part of the tissue was fixed with formalin for 24 h, after which it was dehydrated with the help of ethanol. After these procedures, the tissue was embedded in paraffin and excised using a microtome. In order for the tissue to be seen as well as possible and for the cell structures to be able to differentiate as well as possible, it was necessary to stain the tissue, which was done with hematoxylin-eosin. The microscope slides thus prepared were examined under a light microscope (Celestron LABS CB2000CF). The second part of the tissue was prepared in a similar way, but was not stained with hematoxylin-eosin. Instead of that antibodies (IgG, IgM, IgA, C3, C1q, fibrinogen, albumin, kappa and lambda light chains) were used which were labeled with fluorescent dyes. This method is called immunofluorescence (the microscope used is Celestron LABS CB2000CF). The third part of the tissue was prepared so that after fixation it was excised by ultramicrotomy and examined on an electron microscope (JEOL 1400). Microscopic examination of the biopsy revealed all its pathohistological characteristics and made a final diagnosis of kidney disease.

The data of serum creatinine were collected from biochemical tests of venous blood. A blood sample was taken in a Vacutainer blood collection tube and sent for further analysis. 24-h proteinuria data were collected from biochemical analysis of urine that patients collected over 24 h in a specific vessel. The estimated glomerular filtration rate was calculated according to the MDRD GFR Equation. To calculate the estimated glomerular filtration rate data on the sex, age, race, and serum creatinine of the patients were needed and the equation goes like this: $eGFR = 186 \times [[\text{serum creatinine}]]^{\wedge} (-1{,}154) \times [[\text{year}]]^{\wedge} (-0{,}203) \times 1{,}212$ (for black race) $\times$ 0.742 (for women). According to the data on serum creatinine, 24-h proteinuria and the estimated glomerular filtration rate the clinical presentation was determined and based on that a renal biopsy was indicated.

Data were collected in an MS Excel database (version 11. Microsoft Corporation, Redmond, WA, USA), and the SPSS 20.0 statistical program (IBM Corp., Armonk, NY, USA) was used for statistical analysis. Data were processed by descriptive statistics methods, categorical variables were presented as frequency and percentage, while continuous variables were presented as arithmetic mean and standard deviation. The Hi square test was used to test the differences between the category variables, while the Student ttest for independent samples was used to analyze the differences between the continuous variables. The probability level of $p < 0.05$ was taken as statistically significant.

3 Results

During the study period (from 1 January 2012 to 31 December 2019), the study included 107 patients older than 15 years, including both sexes, who underwent a kidney biopsy. The majority of patients, 55 of them, were men while the rest were women (the ratio of male to female patients was 1.058: 1). The youngest patient was 16 and the oldest 95 years old. The median age of the patients was 48 (range 16–95 years) and was similar in men and women (Fig. 1).

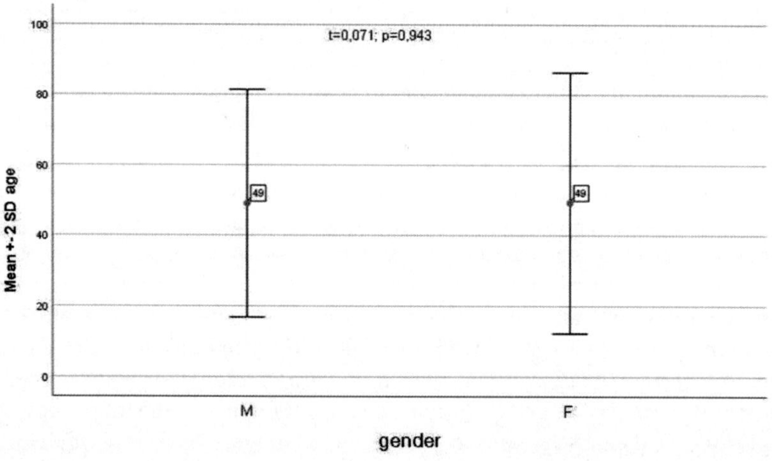

Fig. 1. The median age of male and female patients who underwent kidney biopsy

The most common clinical presentation, as well as an indication for kidney biopsy, was nephrotic syndrome (NS) (51.4%). It is followed by chronic nephritic syndrome (CNS) (31.8%), while the smallest number of patients, 16.8% of them, showed asymptomatic urine abnormalities (AUA). The clinical presentation was similar in both sexes, and there is no significant statistical difference between the clinical presentations with respect to gender (Fig. 2).

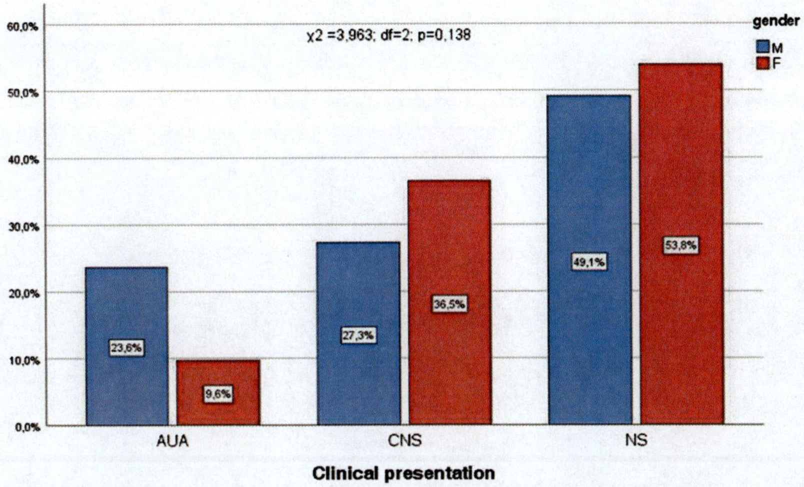

Fig. 2. Clinical presentation of patients who underwent renal biopsy according to gender

The most common biopsy-proven renal diseases were IgA glomerulonephritis (IgAGN) and focal segmental glomerulosclerosis (FSGS), which is statistically significant (Fig. 3).

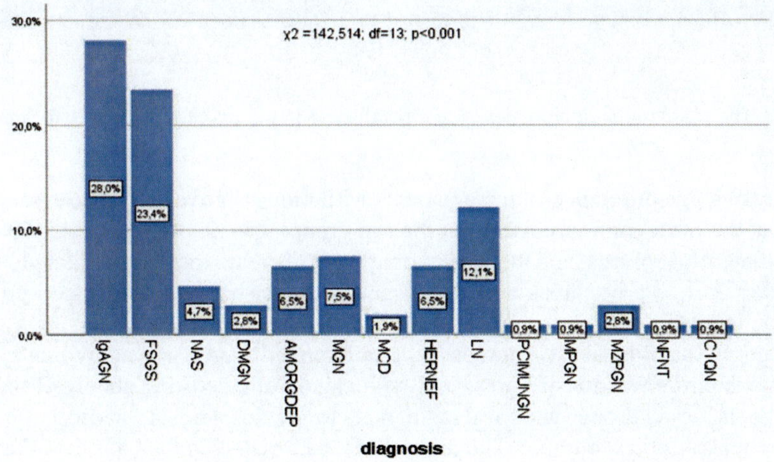

Fig. 3. Distribution and incidence of biopsy-proven renal disease

The distribution of biopsy-proven renal disease according to gender was also analyzed, and there were no significant statistical difference (Fig. 4 and Fig. 5).

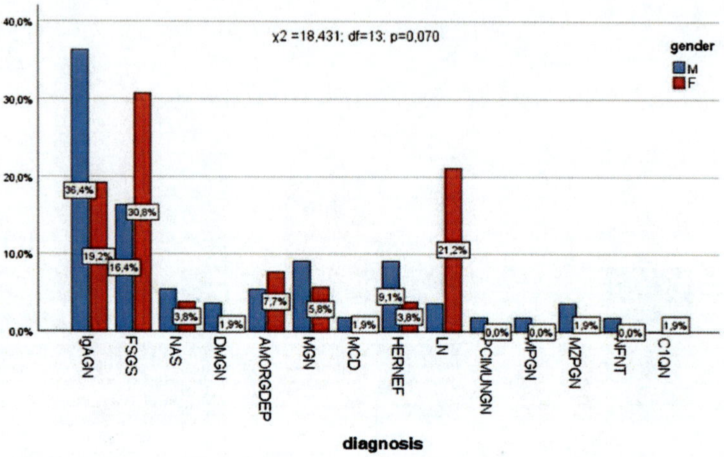

Fig. 4. Distribution and incidence of biopsy-proven renal disease according to gender

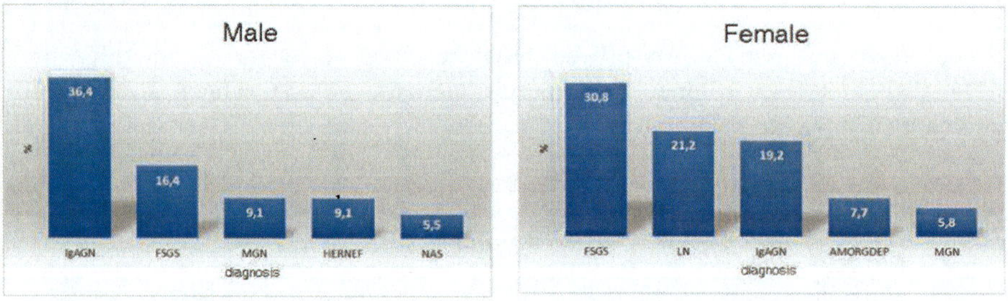

Fig. 5. The most common biopsy-proven renal disease in our patients according to gender

Analyzing age differences among patients with biopsy-proven renal disease, it was found that the most common disease in the age groups 16–29, 30–45 and 45–60 years is IgA glomerulonephritis, and in the age group over 60 years focal segmental glomerulosclerosis. There was no statistically significant difference between biopsy-proven renal disease and age (Table 1).

Regarding the clinical presentation of certain renal diseases proven by biopsy, IgA glomerulonephritis was mostly presented with asymptomatic urine abnormalities and chronic nephritic syndrome, and focal segmental glomerulosclerosis was most often presented with nephrotic syndrome. The differences were statistically significant (Table 2; Fig. 6).

Table 1. Distribution and incidence of biopsy-proven renal disease according age

	Age							
	16–29		30–45		46–60		>60	
	N		N		N		N	
IgAGN	5	29,4	10	41,7	11	28,9	4	14,3
FSGS	4	23,5	6	25,0	8	21,1	7	25,0
NAS	0	0,0	2	8,3	2	5,3	1	3,6
DMGN	0	0,0	1	4,2	1	2,6	1	3,6
AMORGDEP	0	0,0	0	0,0	1	2,6	6	21,4
MGN	1	5,9	1	4,2	4	10,5	2	7,1
MCD	0	0,0	0	0,0	0	0,0	2	7,1
HERNEF	1	5,9	3	12,5	2	5,3	1	3,6
LN	5	29,4	1	4,2	6	15,8	1	3,6
PCIMUNGN	0	0,0	0	0,0	1	2,6	0	0,0
MPGN	0	0,0	0	0,0	0	0,0	1	3,6
MZPGN	0	0,0	0	0,0	1	2,6	2	7,1
NFNT	1	5,9	0	0,0	0	0,0	0	0,0
C1QN	0	0,0	0	0,0	1	2,6	0	0,0

χ^2 =41,010; df=39; p=0,169 (Fisher's exact test)

Table 2. Distribution and incidence of biopsy-proven renal disease according to clinical presentation

	Clinical presentation					
	AUA		CNS		NS	
	N	%	N	%	N	%
IgAGN	7	38,9	13	38,2	10	18,2
FSGS	1	5,6	5	14,7	19	34,5
NAS	0	0,0	4	11,8	1	1,8
DMGN	0	0,0	0	0,0	3	5,5
AMORGDEP	0	0,0	1	2,9	6	10,9
MGN	0	0,0	2	5,9	6	10,9
MCD	0	0,0	0	0,0	2	3,6
HERNEF	6	33,3	0	0,0	1	1,8
LN	4	22,2	4	11,8	5	9,1
PCIMUNGN	0	0,0	1	2,9	0	0,0
MPGN	0	0,0	0	0,0	1	1,8
MZPGN	0	0,0	2	5,9	1	1,8
NFNT	0	0,0	1	2,9	0	0,0
C1QN	0	0,0	1	2,9	0	0,0

χ^2=47,480; df=26; p<0,001 (*Fisher's exact test)

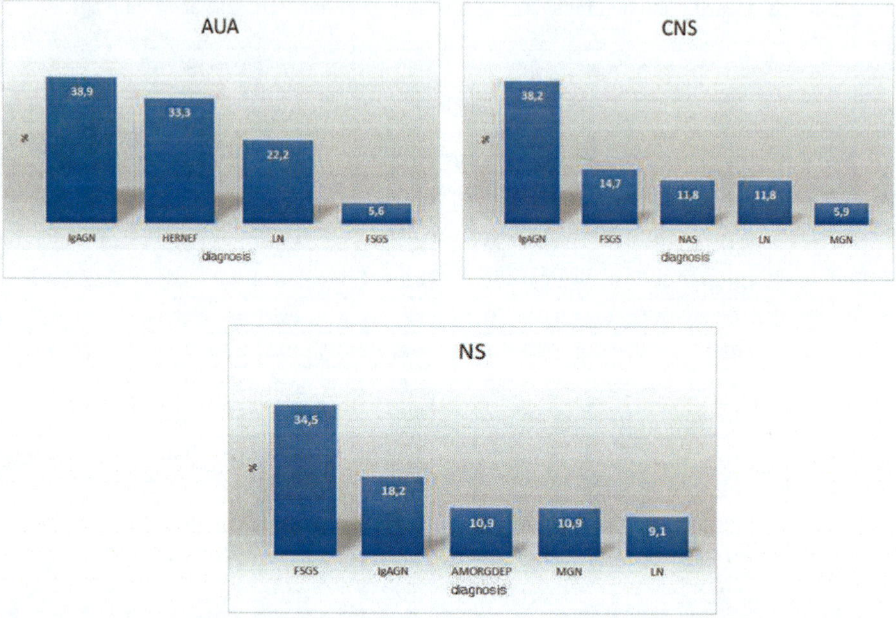

Fig. 6. The most common biopsy-proven renal disease in our patients according to clinical presentation

4 Discussion and Conclusion

Renal biopsy was performed in 107 patients in the period from January 1, 2012 to December 31, 2019 at the Clinic for Internal Medicine with the dialysis center of the University Clinical Hospital Mostar.

The results of this study show that men are more likely to suffer from kidney disease (51.4%). Such a trend can also be found in all available reports worldwide [2–14].

The most common clinical presentation, which is also an indication for renal biopsy, was nephrotic syndrome (51.4%) followed by chronic nephritic syndrome (31.8%) and asymptomatic urine abnormalities (16.8%). Similar results were shown in both the male and female populations, and there were no statistically significant differences between clinical presentations in relation to gender. These data are identical to most registry and database reports of other countries where most patients also presented with nephrotic syndrome, in contrast to Italy where the most common clinical presentation was asymptomatic urine abnormalities [2, 4, 7, 9, 11].

The study found that there were no statistically significant differences between biopsy-proven renal diseases with age. These data are similar to most papers which, like this one, include only the adult population [2, 4–6].

The main goal of this study was to show which renal diseases were proven by biopsy at the University Clinical Hospital Mostar, and it was found that the most common are IgA glomerulonephritis (28.0%) and focal segmental glomerulosclerosis (23.4%), which is statistically significant. Other countries, such as Croatia, Italy, Spain and the Czech

Republic, have similar data [2, 4, 5, 7, 9]. It is important to note that the distribution of biopsy-proven renal diseases depends on several factors. First of all, they depend on the age of the patients. Thus, registries and databases involving the paediatric population have more age-specific renal diseases such as minimal change disease [6, 12, 13]. Race is also an important factor to consider when comparing biopsy-proven renal diseases. The frequency and distribution of biopsy-proven renal disease also depends on the biopsy indications that differ from state to state. Although there are numerous published papers describing the frequency, pathohistological diagnosis, and clinical presentation of biopsy-proven renal disease from various kidney biopsy databases worldwide, it is not always easy to compare results mainly due to conflicting views regarding biopsy performance [2–14]. There are no generally accepted guidelines for kidney biopsy, so it is not surprising that the indications for biopsy differ significantly among different clinical centers, as well as among nephrologists themselves. Some centers perform biopsy only when they believe that pathohistological analysis could help modify therapy, while others have a relatively liberal biopsy policy. At the University Clinical Hospital Mostar, renal biopsy is recommended to all patients when there are abnormalities in the urine that suggest parenchymal kidney disease, and the patient has no contraindications for the procedure.

Regarding the distribution of kidney diseases proven by biopsy according to gender, it has been proven that in men the most common diseases are IgA glomerulonephritis (36.4%), focal segmental glomerulosclerosis (16.4%) and membrane glomerulonephritis (9.1%), and in women focal segmental glomerulosclerosis (30.8%), lupus nephritis (21.2%) and IgA glomerulonephritis (19.2%). There was no significant statistical difference between the distributions of biopsy-proven renal disease by gender. Comparing these data with the data from Horvatić's research from Croatia, it can be stated that they are quite similar in terms of males, while slight differences are seen in females where his research is dominated by hereditary kidney diseases including thin basement membrane disease and Alport's syndrome [2].

When we talk about the possible shortcomings of this study, it would primarily be the number of subjects who underwent a renal biopsy. Therefore, in future research on this topic, it would certainly be desirable to include a larger number of respondents or to extend the research to a longer period of time. Also, in future research it would be good to include some more parameters such as treatment regimen, monitoring treatment outcomes and response to therapy over a period of time, which could give an even clearer and more complete picture of clinical presentation and treatment arrangements.

This is the first study on the topic of biopsy-proven renal diseases conducted at the University Clinical Hospital Mostar. Since all patients from Herzegovina gravitate to the University Clinical Hospital Mostar, this paper presents the epidemiological data of these diseases in Herzegovina. It is important to emphasize that biopsy-proven kidney diseases are relatively rare and that there are many types of these diseases and this is the main reason why most nephrology centers around the world have the opportunity to see only a few cases of such diseases per year. It is for these reasons that such studies are very important because they provide an opportunity for physicians to learn more about these diseases and are the basis for future epidemiological studies aimed at identifying important risk factors in the development and progression of kidney disease and developing protocols for their prevention.

References

1. Vrhovac, B., Jakšić, B., Reiner, Ž., Vucelić, B.: INTERNA MEDICINA. 4. izd. Naklada Ljevak, Zagreb (2008)
2. Horvatic, I., Tisljar, M., Bulimbasic, S., Bozic, B., Galesic Ljubanovic, D., Galesic, K.: Epidemiologic data of adult native biopsy-proven renal diseases in Croatia. Int. Urol. Nephrol. **45**, 1577–1587 (2013)
3. Pesce, F., Pesce, F.P.: Worldwide distribution of glomerular diseases: the role of renal biopsy registers. Nephrol. Dial. Transplant. **25**, 334–336 (2010)
4. Schena, F.P.: Survey of the Italian Registry of Renal Biopsies. Frequency of the renal diseases for 7 consecutive years. The Italian Group of Renal Immunopathology. Nephrol. Dial. Transplant. **12**, 418–426 (1997)
5. Gesualdo, L., Palma, A.M., Morrone, L.F., Strippoli, G.F., Schena, F.P.: The Italian experience of the national registry of renal biopsies. Nephrol. Dial. Transpant. Kidney Int. **66**, 890–894 (2004)
6. Heaf, J., Løkkegaard, H., Larsen, S.: An overview on frequency of renal biopsy diagnosis in Brazil: clinical and pathological patterns based on 9,617 native kidney biopsies. Nephrol. Dial. Transplant. **25**, 490–496 (2010)
7. Rivera, F., López-Gómez, J.M., Pérez-García, R.: Frequency of renal pathology in Spain 1994–1999. Nephrol. Dial. Transplant. **17**, 1594–1602 (2002). Spanish Registry of Glomerulonephritis (2002)
8. Rivera, F., López-Gómez, J.M., Pérez-García, R.: Clinicopathologic correlations of renal pathology in Spain. Kidney Int. **66**, 898–904 (2004). Spanish Registry of Glomerulonephritis (2004)
9. Rychlík, I., Jancová, E., Tesar, V., Kolsky, A., Lácha, J., Stejskal, J., et al.: The Czech registry of renal biopsies. Occurrence of renal diseases in the years 1994–2000. Nephrol. Dial. Transplant. **19**, 3040–3049 (2004)
10. Huraib, S., Al Khader, A., Shaheen, F.A., Abu Aisha, H., Souqiyyeh, M.Z., Al Mohana, F., et al.: The spectrum of glomerulonephritis in Saudi Arabia: the results of the Saudi registry. Saudi J. Kidney Dis. Transplant. **11**, 434–441 (2000)
11. Fiorentino, M., Bolignano, D., Tesar, V., Pisano, A., Biesen, W., D'Arrigo, G., et al.: Renal biopsy in 2015 - from epidemiology to evidence-based indications. Am. J. Nephrol. **43**, 1–19 (2016)
12. Sarwal, D., D'Cruz, S., Singh Punia, R.P., Minz, R.W.: The spectrum of renal diseases observed in native renal biopsies in a single North Indian tertiary care center. Saudi J. Kidney Dis. Transplant. **30**, 492–500 (2019)
13. Onwubuya, I.M., Adelusola, K.A., Sabageh, D., Ezike, K.N., Olaofe, O.O.: Biopsy-proven renal disease in Ile-Ife, Nigeria: a histopathologic review. Indian J. Nephrol. **26**, 16–22 (2016)
14. Zhou, F.D., Zhao, M.H., Zou, W.Z., Liu, G., Wang, H.: The changing spectrum of primary glomerular diseases within 15 years: a survey of 3331 patients in a single Chinese centre. Nephrol. Dial. Transplant. **24**, 870–876 (2009)

Bio-micro/Nano Technologies

Fighting Cancer Using Nanoparticles – Diagnosis, Treatment and Monitoring

Selma Cifrić[1], Lemana Spahić Bećirović[1], Dina Osmanović[2], Emina Imamović[1], and Amar Deumić[1](✉)

[1] International Burch University, Sarajevo, Bosnia and Herzegovina
amar.deumic@stu.ibu.edu.ba
[2] Friedrich Schiller University, Jena, Germany

Abstract. Cancer is one of the most dangerous diseases in present times. Biological processes that lead to growth and spread of cancer occur at nanoscale level. Nanotechnology provides a novel approach in diagnosing and treatment of cancers. The main mission of nanotechnology in this field is tumor targeting, which enables differentiation between malignant and healthy cells, through passive and active targeting. Early detection of cancer is crucial for the adequate treatment of patients. Application of nanotechnology enables more precise and accurate diagnosis and continuous monitoring of the disease. The principle of nanoparticles as diagnostic tools is that they capture specific cancer biomarkers. In addition to standard imaging techniques (MRI) and computed tomography (CT), nanotechnology also enables development of novel imaging: PAT, Raman and multimodal imaging. Current therapeutic methods for fighting cancer are limited to chemotherapy, radiation and surgery. Application of nanotechnology offers advancements of current therapies with targeted drug delivery and therapy with less cytotoxicity for healthy cells. Few different types of anticancer therapies can be improved by including nanoparticles (NPs). Even though nanotechnology helps detect cancer at a very early stage and helps to monitor the disease with much greater precision than using standard methods, its potential toxicity should be fully investigated. This is done through both passive and active targeting. This review focuses on applications of nanoparticles in fighting cancer.

Keywords: Nanoparticles · Cancer treatment · Drug delivery · Diagnostics · Nanotechnology · EPR effect

1 Introduction

Nowadays, cancer has become one of the most serious diseases for humans. The disease develops and metastases rapidly, where patients have low survival rate and usually poor quality of life. Cancer is a name for many related diseases that are known for the abnormal growth of cells in body. Some cancers are malignant, which means that cancerous cells

L. Spahić Bećirović, D. Osmanović, E. Imamović and A. Deumić—These authors contributed equally to this work.

A. Badnjevic and L. Gurbeta Pokvić (Eds.): CMBEBIH 2021, IFMBE Proceedings 84, pp. 657–669, 2021.
https://doi.org/10.1007/978-3-030-73909-6_76

can travel and spread to the other parts of body. Many biological processes that are essential for human growth and survival occur at nano level, even the development of cancer. Cancer is caused by alterations in genes that are responsible for normal cell's function, their growth and reproduction [1, 2].

Nanoparticles are much smaller than cells in the human body and very similar in size to biological molecules (receptors, enzymes). These small nanoscale particles can easily travel through different parts of our body. This can enhance current methods for the detection of abnormal tissues and drug delivery [3, 4].

Application of nanodevices enables manipulation of cells and biomolecules, targeted drug delivery and detection of abnormal cells in real time [5]. Nanotechnology is highly sensitive and provides a novel approach in diagnosing and treating cancers. Nanotechnology has a potential to enhance specificity of the therapy, while it minimizes the toxicity for normal cells. The main mission of nanotechnology is tumor targeting, which enables differentiation between cancerous and normal cells [3, 6].

Tumor targeting is done through both passive and active targeting. Passive targeting is based on enhanced permeability and retention (EPR) effect. EPR effect tends to accumulate various biomolecules, nanoparticles, and drugs, in malignant tissues much faster compared to normal tissues. EPR effect is unique for tumor tissue. Since tumor mass grows quickly it develops a high density network of blood vessels, by a process known as angiogenesis. Angiogenesis creates many gaps that are responsible for the high accumulation of delivered drugs in tumor tissue [7, 8].

On the other hand, active tumor targeting involves recognition of receptors (proteins, antigens) on the surface of malignant cells by ligands attached to nanoparticles. Active targeting enhances drug accumulation and cell uptake, since surface molecules trigger the process of endocytosis [7, 9].

Various nanomaterials can be used for diagnostic purposes: nanoshells, nanotubes, quantum dots, superparamagnetic nanoparticles, nanowires, nanodiamonds, as well as for targeted therapy: gold and silica nanoparticles, carbon nanotubes, magnetic nanoparticles, organic nanomaterials (liposomes) and quantum dots [10].

In fighting cancer, the most important step is detection at its early stage. Nanotechnology is very sensitive; it can detect even a single cancer cell in vivo [10]. At the same time, nanoparticles can enhance drug delivery with reduced toxicity to normal tissues.

2 Objective and Methodology

The main purpose of this review paper is to summarize recent literature that gave us an insight in the beneficial use of nanoparticles in cancer diagnosis and treatment, especially focused on targeted drug delivery.

This review was based on papers found in relevant databases, for which selection criteria was:

- The paper must be in English,
- Availability of full text,
- The paper should include keywords: nanoparticles; nanotechnology; cancer treatment; diagnostics; drug delivery; EPR effect,

- Search through relevant databases: NCBI, Google Scholar, ResearchGate, Web of Science.

3 Current Application of NPs in Treatment and Diagnosis

The use of nanomaterials comes down to its potential in active and passive targeting (longer accumulation in the malignant tissue), to efficiently deliver drugs across biological barriers, and to act as a diagnostic tool.

This study investigates and shows the latest applications of nanotechnology for diagnosis, monitoring and treatment of cancer diseases.

3.1 Diagnosis and Monitoring

Early detection of cancer is crucial for the adequate treatment of patients. Application of nanotechnology enables more precise and accurate diagnosis and continual monitoring of the disease. For diagnosis of cancer, nanoparticles have usually been used for capturing circulating tumor cells (CTCs) and their DNA. They have the ability to recognize unique surface signatures of tumor cells [11–15].

Gold nanoparticles, superparamagnetic particles, nanotubes, quantum dots, polymer dots and other nanomaterials have been used for cancer detection [16–20].

The principle of nanoparticles as diagnostic tools is to capture specific cancer biomarkers (CTCs, surface proteins, CTCs DNA). The nanoparticle surface is usually covered with various ligands, antibodies, proteins, which can attach to and recognize malignant cells. Number of ligands on the surface can be enhanced, which improves the specificity and sensitivity of detection (Fig. 1) [3, 16].

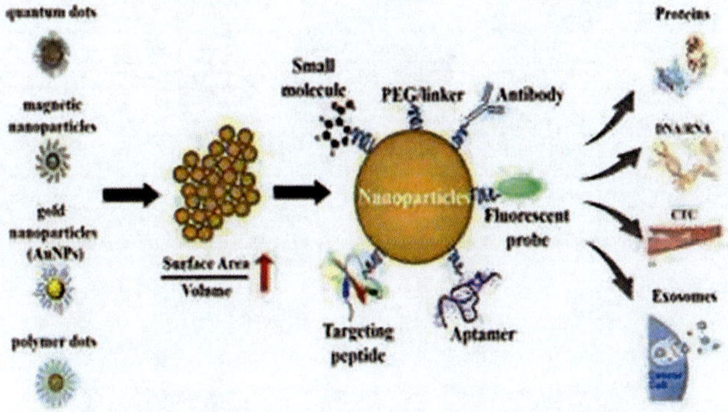

Fig. 1. Nanotechnology improves sensitivity and specificity of cancer detection [3]

A. Detection of CTCs from blood

Nano materials have large surface-to-volume ratio which enables efficient ligand binding. These moieties found on nanoparticles bind to surface proteins on CTCs, which enables their identification. Since cell surface protein markers vary among different types of cancer cells, the best option would be to look for EpCAM surface protein. This protein is highly expressed on the surface of CTCs of many cancer diseases. Therefore, anti-EpCAM molecules are attached to nano materials for accurate detection of CTCs [3].

Detection of CTCs in peripheral blood is a good prognostic factor for cancer behavior. Researchers have constructed gold nanoparticles to detect CTCs. These NPs are highly sensitive and allow noninvasive detection, and later destruction, of CTCs [14, 21, 22].

B. NP-mediated in vivo imaging

Current imaging techniques are not sensitive enough to detect small alterations in malignant tissues. Application of different nanotechnology-based contrast imaging agents (dendrimers, micelles, nanotubes) can improve sensitivity of current imaging techniques, magnetic resonance imaging (MRI) and computed tomography (CT) [23]. Nanotechnology also enables development of novel imaging platforms: photoacoustic tomography (PAT), Raman spectroscopic imaging, and multimodal imaging, that enhance detection of tumors [24–27]. Multimodal imaging is developed in order to be able to determine growth of tumor, before and intraoperatively. Multimodal imaging is improved by using Triple-modality nanoparticle (MRIphotoacoustic-Raman). It is administered intravenously. Nanoparticles are accumulated in a tumor tissue longer than standard contrast agents. MRI delineates the margins of the tumor, before the surgery. Photoacoustic imaging is used during the surgery to get a better 3D picture of the tumor and its location. While Raman acts as a guide for margin resection of tumor (Fig. 2) [28]. This type of imaging can be used for both monitoring and diagnostic purposes [29].

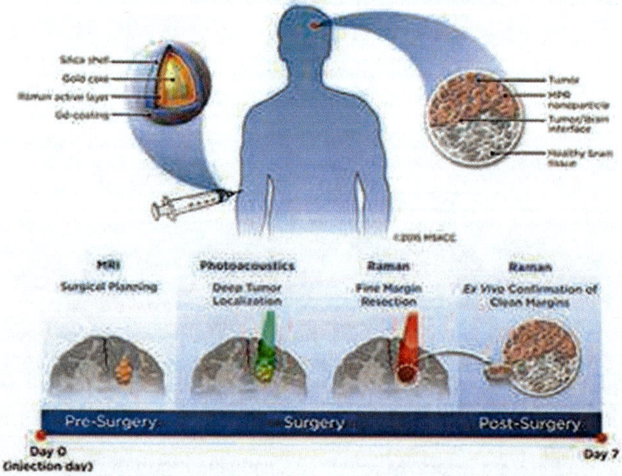

Fig. 2. Principle of triple-modality nanoparticle [29]

C. Quantum dot-based biosensors

Quantum dots are semiconductors whose size range from 2 to 10 nm in diameter. Quantum dots are made of 10–50 atoms. In response to applied energy, electrons, found in atoms of quantum dots, are moved to a higher state. During the stabilization state of electrons, the remaining energy is emitted as light of specific frequency. External source of energy is usually ultraviolet light [12].

QDs have important applications in the imaging process. It aids images to be visible for a longer period of time, since QDs are resistant to photobleaching [30].

Quantum Dot-based biosensors (QD) can be used for the detection of malignancies. They are resistant to photobleaching and degradation, and have high quantum yield. QD-based biosensors use a sandwich-type assay for the detection of cancer biomarkers, which can later be visualized by staining or more often by fluorescence [12, 30–33].

3.2 Treatment

Current therapeutic methods for fighting cancer are limited to chemotherapy, radiation and surgery. All of them can also be harmful to normal cells, or may not be completely efficient. Application of nanotechnology offers better efficiency of current therapies, targeted drug delivery and less toxicity to normal cells [28, 34–36]. Although nanoparticles are hundreds to thousands of times smaller than human cells, they can encapsulate various biomolecules and deliver them to target sites. Also, their large surface-to-volume ratio allows them to detect target cells by surface ligand binding [37–40]. Most of the therapeutic options are based on enhanced permeability and retention (EPR) effect, a unique property of cancer cells. It stated that biomolecules, nanoparticles, drugs, in tumor tissues accumulate in larger amounts than in normal tissues [7, 8, 41, 42].

A Gold nanoparticle for treatment

Gold nanoparticles are mostly used to improve specificity of drug delivery [43, 44]. Many ligands, together with therapeutic molecules, can be attached to its surface. These specific ligands allow them to circulate only in tumor tissue [45–48].

Destroying cancer cells with nanoparticles goes through the following steps: Each sphere of gold NPs is coated with gold. Lasers are used to stimulate gold nanoparticles to enter the cancer cells. The laser makes gold NPs to vibrate at high temperatures, which in the end kills cancer cells. In other words, gold NPs absorb energy from photons, and by converting it to heat, they destroy cancer cells [49].

Gold NPs have a high atomic number and good absorption of X-rays, which makes them suitable for the role of contrast agents. They can also be used for tumor-selective photothermal therapy and for improvement of pharmacokinetics (Fig. 3.) [48, 50, 51].

B NP-mediated Gene therapy

Gene therapy is an experimental therapy that uses genes to prevent or suppress various diseases. The most popular is gene silencing therapy, which uses smallinterfering RNAs (siRNAs). siRNAs are class of short double-stranded RNAs, whose function is to silence target genes through enzymatic cleavage [52].

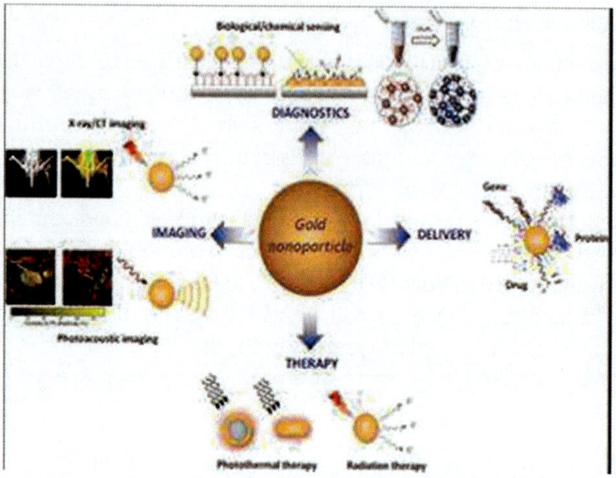

Fig. 3. Application of gold NPs in cancer treatment [51]

Although the gene therapy was discovered a few decades ago, their efficient application and development were slow due to challenges with DNAse degradation, toxicity and difficulty to enter into various cell types [23, 53, 54]. Recently, this type of gene therapy has shown to have extended half-lives when delivered via nanoparticles, either encapsulated or attached to the surface [55, 56]. This type of therapy is used to target "undruggable" cancer proteins. Nanotechnology addresses most of the issues of gene therapy and improves its efficiency [57–60].

C. CNP-mediated immunotherapy

Immunotherapy is transforming the field of oncology. Immunotherapy is a type of biological therapy which helps your immune system destroy cancer [61].

Nanomaterials can be engineered in that manner to improve the efficiency of immunotherapy by targeting specific immune processes [62–65]. The main goal of immunotherapy is to raise the number of T cells (to destroy cancer). This is done by delivering antigens that stimulate dendritic cell targets, and trigger prolonged immune stimulation (Fig. 4.). NPs can also be used to deliver immunostimulatory molecules combined with chemotherapy or radiotherapy [62, 66–68].

D NP-mediated chemotherapy

Chemotherapy is one of the oldest and best known treatments for cancer disease. It uses aggressive chemicals to kill cells that grow out of control. However, these aggressive chemicals often harm normal cells and tissues. Since chemotherapy is not very specific [29].

Application of nanotechnology in chemotherapy can enhance its delivery and selective targeting of malignant cells [69–71]. Enhanced specificity relies on passive targeting of tumor cells through EPR effect. Other factors that can influence NPmediated chemotherapy are pH, heat, composition of nanomaterials [29, 72, 73].

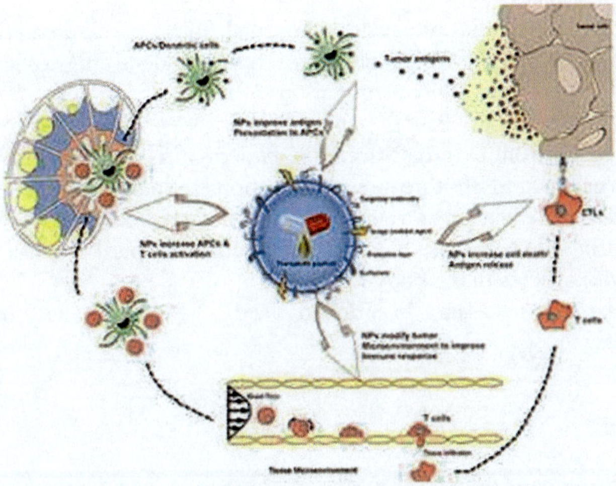

Fig. 4. Pathway of NP-mediated immunotherapy [29]

E NP-mediated radiotherapy

It is also possible to kill malignant cells by exposing them to ionizing radiation, which is known as radiotherapy. High- energy radiation (X-rays, gamma rays) kills cancer cells by damaging their DNA, which induces cell death – apoptosis. However, radiotherapy is very often toxic to all cells, not only cancer cells. It is always tricky to set up a dosage for radiotherapy, to be efficient enough and not too harmful to normal cells of the human body [29, 74–76].

Radiation therapy could become more specific and efficient by including nanoscale materials in the treatment process. Nanoparticles play a key role in enhancement of radiotherapy. Connection between nanoparticles and X-rays rely on inherent atomic properties used (high-Z atomic number) [74, 77–79].

There is another type of radiotherapy known as photodynamic therapy. It localizes superficial tumors using photosensitizer, which in turn enables nanoparticles to emit the visible light at the tumor site. This generates cytotoxic reactive oxygen that kills malignant cells [29].

3.3 Safety of Nanotechnology Used in Cancer Treatment

The main disadvantage or potential risk of using nanoparticles in fighting cancer is their toxicity. Nanomaterials differ in composition, charge, size etc. and therefore it is not possible to draw general conclusions about their toxicity [3, 23]. Other factors affecting the cytotoxicity include concentration of nanoparticles, whether they were in contact with cells for short or long time, as well as distribution of nanoparticles within the cells [30]. When assessing safety of nanoparticles, biodegradability should be taken into account [3].

Their small size allows them to enter the cells, while disrupting their normal function. Some of the toxic effects of NPs include inflammation, and disrupted oxidationreduction, both of which have a potential of causing abnormal function cell death or even cancer [23, 30, 74]. Research has shown that nanoparticles can induce epigenetic changes such as histone modifications and DNA methylation [30].

Side effects at the injection site of nanoparticles were also reported by several studies. These can range from skin discoloration to dermatitis and fibrosis. Anaphylactic reactions which can lead to life-threatening conditions are also possible [18]. Other side effects include fever, headache, nausea, pain in lower back and dry mouth [18, 48].

Some potential risks can be avoided by dilution of particles, as well as slow administration of particles in the body [18].

Further research is necessary in order to clarify the exact toxicity of NPs and its potential benefits.

4 Conclusion

As cancer is a lasting and rising concern in the modern world, constant advancements have to be made in the field of diagnosis and treatment. The wide application of nanoparticles for diagnostic purposes and treatment of cancer disease has been summarized in this review paper.

Application of nanomedicine in cancer treatment and diagnosis is very common in the last few years. Since nanoparticles are minimally invasive, they provide a great platform for the future development and production of diagnostic tools and drug delivery systems.

NP-mediated assays showed enhanced specificity and sensitivity compared to traditional techniques. Detection of cancer at a very early stage, as well as continual monitoring of disease, can improve the survival rate of cancer patients.

It is easy to conclude that nanotechnology helps in detection and monitoring of the disease with much greater precision than using standard methods. When toxicity is taken into account, there are a few aspects that need to be considered. First, they need to be biodegradable. There are no finite conclusions regarding their toxicity due to differences in charge and ionic constitution. Therefore, further research is necessary.

Conflict of Interest Declaration. Authors have nothing to declare.

References

1. Blackadar, C.B.: Historical review of the causes of cancer. World J. Clin. Oncol. **7**(1), 54 (2016)
2. Sarkar, S., Horn, G., Moulton, K., Oza, A., Byler, S., Kokolus, S., Longacre, M.: Cancer development, progression, and therapy: an epigenetic overview. Int. J. Mol. Sci. **14**(10), 21087–21113 (2013)
3. Zhang, Y., Li, M., Gao, X., Chen, Y., Liu, T.: Nanotechnology in cancer diagnosis: progress, challenges and opportunities. J. Hematol. Oncol. **12**(1), 137 (2019)
4. Maksimović, M., Omanović-Mikličanin, E., Badnjević, A.: Nanofood and Internet of Nano Things. Springer, Heidelberg (2019)

5. Badnjevic, A., Škrbić, R., Pokvić, L.G. (eds.): CMBEBIH 2019: Proceedings of the International Conference on Medical and Biological Engineering, Banja Luka, Bosnia and Herzegovina, 16–18 May 2019, vol. 73. Springer (2019)
6. Kumar, S., Kumari, P., Singh, R.: Emerging nanomaterials for cancer therapy. In: Nanoparticles in Medicine, pp. 25–54. Springer, Singapore (2020)
7. Torchilin, V.P.: Passive and active drug targeting: drug delivery to tumors as an example. In: Handbook of Experimental Pharmacology, vol. 197, pp. 3–53 (2010)
8. Fang, J., Nakamura, H., Maeda, H.: The EPR effect: unique features of tumor blood vessels for drug delivery, factors involved, and limitations and augmentation of the effect. Adv. Drug Deliv. Rev. **63**(3), 136–151 (2011)
9. Farokhzad, O.C., Langer, R.: Impact of nanotechnology on drug delivery. ACS Nano **27**(3), 16–20 (2009)
10. Jaishree, V., Gupta, P.D.: Nanotechnology: a revolution in cancer diagnosis. Indian J. Clin. Biochem. **27**(3), 214–220 (2012)
11. Jain, K.K.: Nanotechnology in clinical laboratory diagnostics. Clin. Chim. Acta **358**(1–2), 37–54 (2005)
12. Choi, Y.E., Kwak, J.W., Park, J.W.: Nanotechnology for early cancer detection. Sensors (Basel, Switzerland) **10**(1), 428–455 (2010)
13. Sun, N., Li, X., Wang, Z., Zhang, R., Wang, J., Wang, K., Pei, R.: A multiscale TiO2 nanorod array for ultrasensitive capture of circulating tumor cells. ACS Appl. Mater. Interfaces **8**(20), 12638–12643 (2016)
14. Li, Z., Wang, G., Shen, Y., Guo, N., Ma, N.: DNA-templated magnetic nanoparticle-quantum dot polymers for ultrasensitive capture and detection of circulating tumor cells. Adv. Funct. Mater. **28**(14), 1707152 (2018)
15. Wang, Z., Sun, N., Liu, H., Chen, C., Ding, P., Yue, X., Zou, H., Xing, C., Pei, R.: High-efficiency isolation and rapid identification of heterogeneous circulating tumor cells (CTCs) using dual-antibody-modified fluorescentmagnetic nanoparticles. ACS Appl. Mater. Interfaces **11**(43), 3958639593 (2019)
16. Jia, S., Zhang, R., Li, Z., Li, J.: Clinical and biological significance of circulating tumor cells, circulating tumor DNA, and exosomes as biomarkers in colorectal cancer. Oncotarget **8**(33), 55632–55645 (2017)
17. Saeed, A.A., Sánchez, J.L.A., O'Sullivan, C.K., Abbas, M.N.: DNA biosensors based on gold nanoparticles-modified graphene oxide for the detection of breast cancer biomarkers for early diagnosis. Bioelectrochemistry **118**, 91–99 (2017)
18. Czarniecki, M., Pesapane, F., Wood, B.J., Choyke, P.L., Turkbey, B.: Ultra-small superparamagnetic iron oxide contrast agents for lymph node staging of high-risk prostate cancer. Translational andrology and urology **7**(Suppl 4), S453 (2018)
19. Thapa, A., Soares, A.C., Soares, J.C., Awan, I.T., Volpati, D., Melendez, M.E., Fregnani, J.H.T.G., Carvalho, A.L., Oliveira, O.N., Jr.: Carbon nanotube matrix for highly sensitive biosensors to detect pancreatic cancer biomarker CA19-9. ACS Appl. Mater. Interfaces **9**(31), 25878–25886 (2017)
20. Gao, G., Jiang, Y.W., Jia, H.R., Yang, J., Wu, F.G.: On-off-on fluorescent nanosensor for Fe3+ detection and cancer/normal cell differentiation via silicon-doped carbon quantum dots. Carbon **134**, 232243 (2018)
21. Marta, T., Luca, S., Serena, M., Luisa, F., Fabio, C.: What is the role of nanotechnology in diagnosis and treatment of metastatic breast cancer? Promising scenarios for the near future. J. Nanomater. **2016** (2016)
22. Huang, Q., Wang, Y., Chen, X., Wang, Y., Li, Z., Du, S., Wang, L., Chen, S.: Nanotechnology-based strategies for early cancer diagnosis using circulating tumor cells as a liquid biops. Nanotheranostics **2**(1), 21 (2018)

23. Gmeiner, W.H., Ghosh, S.: Nanotechnology for cancer treatment. Nanotechnol. Rev. **3**(2), 111–122 (2014)
24. Shakiba, M., Ng, K.K., Huynh, E., Chan, H., Charron, D.M., Chen, J., Muhanna, N., Foster, F.S., Wilson, B.C., Zheng, G.: Stable J-aggregation enabled dual photoacoustic and fluorescence nanoparticles for intraoperative cancer imaging. Nanoscale **8**(25), 12618–12625 (2016)
25. Yang, R.M., Fu, C.P., Fang, J.Z., Xu, X.D., Wei, X.H., Tang, W.J., Jiang, X.Q., Zhang, L.M.: Hyaluronan-modified superparamagnetic iron oxide nanoparticles for bimodal breast cancer imaging and photothermal therapy. Int. J. Nanomed. **12**, 197 (2017)
26. lzoghby, A.O., Hemasa, A.L., Freag, M.S.: Hybrid proteininorganic nanoparticles: from tumor-targeted drug delivery to cancer imaging. J. Controlled Release **243**, 303–322 (2016)
27. Kang, T., Li, F., Baik, S., Shao, W., Ling, D., Hyeon, T.: Surface design of magnetic nanoparticles for stimuli-responsive cancer imaging and therapy. Biomaterials **136**, 98–114 (2017)
28. Harmsen, S., Wall, M.A., Huang, R., Kircher, M.F.: Cancer imaging using surface-enhanced resonance Raman scattering nanoparticles. Nature protocols **12**(7), 1400 (2017)
29. National Cancer Institute. https://www.cancer.gov/
30. Maysinger, D.: Nanoparticles and cells: good companions and doomed partnerships. Org. Biomol. Chem. **5**(15), 2335–2342 (2007)
31. Xi, J., Xie, C., Zhang, Y., Wang, L., Xiao, J., Duan, X., Ren, J., Xiao, F., Wang, S.: Pd nanoparticles decorated N-doped graphene quantum dots@ Ndoped carbon hollow nanospheres with high electrochemical sensing performance in cancer detection. ACS Appl. Mater. Interfaces **8**(34), 22563–22573 (2016)
32. Liang, L., Su, M., Li, L., Lan, F., Yang, G., Ge, S., Yu, J., Song, X.: Aptamer-based fluorescent and visual biosensor for multiplexed monitoring of cancer cells in microfluidic paper-based analytical devices. Sens. Actuators B: Chem. **229**, 347–354 (2016)
33. Yang, Y., Mao, G., Ji, X., He, Z.: DNA-templated quantum dots and their applications in biosensors, bioimaging, and therapy. J. Mater. Chem. B **8**(1), 9–17 (2020)
34. Sztandera, K., Gorzkiewicz, M., Klajnert-Maculewicz, B.: Gold nanoparticles in cancer treatment. Mol. Pharm. **16**(1), 1–23 (2018)
35. Rezvantalab, S., Drude, N.I., Moraveji, M.K., Güvener, N., Koons, E.K., Shi, Y., Lammers, T., Kiessling, F.: PLGA-based nanoparticles in cancer treatment. Front. Pharmacol. **9**, 1260 (2018)
36. Jose, J., Kumar, R., Harilal, S., Mathew, G.E., Prabhu, A., Uddin, M.S., Aleya, L., Kim, H., Mathew, B.: Magnetic nanoparticles for hyperthermia in cancer treatment: an emerging tool. Environ. Sci. Pollut. Res. **27**(16), 19214–19225 (2020)
37. Bae, K.H., Chung, H.J., Park, T.G.: Nanomaterials for cancer therapy and imaging. Mol. Cells **31**(4), 295–302 (2011)
38. Martínez-Carmona, M., Lozano, D., Colilla, M., Vallet-Regí, M.: Lectin-conjugated pH-responsive mesoporous silica nanoparticles for targeted bone cancer treatment. Acta Biomater. **65**, 393–404 (2018)
39. Khaydukov, E.V., Mironova, K.E., Semchishen, V.A., Generalova, A.N., Nechaev, A.V., Khochenkov, D.A., Stepanova, E.V., Lebedev, O.I., Zvyagin, A.V., Deyev, S.M., Panchenko, V.Y.: Riboflavin photoactivation by upconversion nanoparticles for cancer treatment. Sci. Rep. **6**, 35103 (2016)
40. Gu, T., Wang, Y., Lu, Y., Cheng, L., Feng, L., Zhang, H., Li, X., Han, G., Liu, Z.: Platinum nanoparticles to enable electrodynamic therapy for effective cancer treatment. Adv. Mater. **31**(14), 1806803 (2019)
41. Yuan, Y.G., Peng, Q.L., Gurunathan, S.: Silver nanoparticles enhance the apoptotic potential of gemcitabine in human ovarian cancer cells: combination therapy for effective cancer treatment. Int. J. Nanomed. **12**, 6487 (2017)

42. Khan, H., Ullah, H., Martorell, M., Valdes, S.E., Belwal, T., Tejada, S., Sureda, A., Kamal, M.A.: Flavonoids nanoparticles in cancer: treatment, prevention and clinical prospects. In: Seminars in Cancer Biology. Academic Press (2019)
43. Lin, W., Yao, N., Qian, L., Zhang, X., Chen, Q., Wang, J., Zhang, L.: pH-responsive unimolecular micelle-gold nanoparticles-drug nanohybrid system for cancer theranostics. Acta Biomater. **58**, 455–465 (2017)
44. Du, Y., Xia, L., Jo, A., Davis, R.M., Bissel, P., Ehrich, M.F., Kingston, D.G.: Synthesis and evaluation of doxorubicin-loaded gold nanoparticles for tumor-targeted drug delivery. Bioconjug. Chem. **29**(2), 420–430 (2018)
45. Manivasagan, P., Bharathiraja, S., Bui, N.Q., Jang, B., Oh, Y.O., Lim, I.G., Oh, J.: Doxorubicin-loaded fucoidan capped gold nanoparticles for drug delivery and photoacoustic imaging. Int. J. Biol. Macromol. **91**, 578–588 (2016)
46. Manivasagan, P., Bharathiraja, S., Bui, N.Q., Lim, I.G., Oh, J.: Paclitaxel-loaded chitosan oligosaccharide-stabilized gold nanoparticles as novel agents for drug delivery and photoacoustic imaging of cancer cells. Int. J. Pharm. **511**(1), 367–379 (2016)
47. Paciotti, G.F., Zhao, J., Cao, S., Brodie, P.J., Tamarkin, L., Huhta, M., Myer, L.D., Friedman, J., Kingston, D.G.: Synthesis and evaluation of paclitaxel-loaded gold nanoparticles for tumor-targeted drug delivery. Bioconjug. Chem. **27**(11), 2646–2657 (2016)
48. Jain, S., Hirst, D.G., O'Sullivan, J.M.: Gold nanoparticles as novel agents for cancer therapy. Br. J. Radiol. **85**(1010), 101–113 (2012)
49. Patra, H.K., Banerjee, S., Chaudhuri, U., Lahiri, P., Dasgupta, A.K.: Cell selective response to gold nanoparticles. Nanomed. Nanotechnol. Biol. Med. **3**(2), 111–119 (2007)
50. Debouttière, P.J., Roux, S., Vocanson, F., Billotey, C., Beuf, O., Favre-Réguillon, A., Lin, Y., Pellet-Rostaing, S., Lamartine, R., Perriat, P., Tillement, O.: Design of gold nanoparticles for magnetic resonance imaging. Adv. Funct. Mater. **16**(18), 2330–2339 (2006)
51. Her, S., Jaffray, D.A., Allen, C.: Gold nanoparticles for applications in cancer radiotherapy: mechanisms and recent advancements. Adv. Drug Deliv. Rev. **109**, 84–101 (2017)
52. Davis, M.E., Zuckerman, J.E., Choi, C.H.J., Seligson, D., Tolcher, A., Alabi, C.A., Yen, Y., Heidel, J.D., Ribas, A.: Evidence of RNAi in humans from systemically administered siRNA via targeted nanoparticles. Nature **464**(7291), 1067–1070 (2010)
53. Castro, J.E., Melo-Cardenas, J., Urquiza, M., Barajas-Gamboa, J.S., Pakbaz, R.S., Kipps, T.J.: Gene immunotherapy of chronic lymphocytic leukemia: a phase I study of intranodally injected adenovirus expressing a chimeric CD154 molecule. Cancer Res. **72**(12), 29372948 (2012)
54. Bartlett, D.W., Davis, M.E.: Impact of tumor-specific targeting and dosing schedule on tumor growth inhibition after intravenous administration of siRNA-containing nanoparticles. Biotechnol. Bioeng. **99**(4), 975–985 (2008)
55. Osman, G., Rodriguez, J., Chan, S.Y., Chisholm, J., Duncan, G., Kim, N., Tatler, A.L., Shakesheff, K.M., Hanes, J., Suk, J.S., Dixon, J.E.: PEGylated enhanced cell penetrating peptide nanoparticles for lung gene therapy. J. Controlled Release **285**, 3545 (2018)
56. Lin, G., Li, L., Panwar, N., Wang, J., Tjin, S.C., Wang, X., Yong, K.T.: Non-viral gene therapy using multifunctional nanoparticles: status, challenges, and opportunities. Coord. Chem. Rev. **374**, 133152 (2018)
57. Zylberberg, C., Gaskill, K., Pasley, S., Matosevic, S.: Engineering liposomal nanoparticles for targeted gene therapy. Gene Ther. **24**(8), 441452 (2017)
58. Chen, Z., Liu, F., Chen, Y., Liu, J., Wang, X., Chen, A.T., Deng, G., Zhang, H., Liu, J., Hong, Z., Zhou, J.: Targeted delivery of CRISPR/Cas9-mediated cancer gene therapy via liposome-templated hydrogel nanoparticles. Adv. Funct. Mater. **27**(46), 1703036 (2017)
59. Wang, Z., Chang, Z., Lu, M., Shao, D., Yue, J., Yang, D., Zheng, X., Li, M., He, K., Zhang, M., Chen, L.: Shape-controlled magnetic mesoporous silica nanoparticles for magnetically-mediated suicide gene therapy of hepatocellular carcinoma. Biomaterials **154**, 147–157 (2018)

60. Majidi, S., Zeinali Sehrig, F., Samiei, M., Milani, M., Abbasi, E., Dadashzadeh, K., Akbarzadeh, A.: Magnetic nanoparticles: applications in gene delivery and gene therapy. Artif. Cells Nanomed. Biotechnol. **44**(4), 1186–1193 (2016)
61. Goldberg, M.S.: Improving cancer immunotherapy through nanotechnology. Nat. Rev. Cancer **19**(10), 587–602 (2019)
62. Park, W., Heo, Y., Han, D.K.: New opportunities for nanoparticles in cancer immunotherapy. Biomater. Res. **22**, 24 (2018)
63. Min, Y., Roche, K.C., Tian, S., Eblan, M.J., McKinnon, K.P., Caster, J.M., Chai, S., Herring, L.E., Zhang, L., Zhang, T., DeSimone, J.M.: Antigen-capturing nanoparticles improve the abscopal effect and cancer immunotherapy. Nat. Nanotechnol. **12**(9), 877 (2017)
64. Chen, Q., Xu, L., Liang, C., Wang, C., Peng, R., Liu, Z.: Photothermal therapy with immune-adjuvant nanoparticles together with checkpoint blockade for effective cancer immunotherapy. Nat. Commun. **7**(1), 1–13 (2016)
65. Schmid, D., Park, C.G., Hartl, C.A., Subedi, N., Cartwright, A.N., Puerto, R.B., Zheng, Y., Maiarana, J., Freeman, G.J., Wucherpfennig, K.W., Irvine, D.J.: T cell-targeting nanoparticles focus delivery of immunotherapy to improve antitumor immunity. Nat. Commun. **8**(1), 1–12 (2017)
66. Rodell, C.B., Arlauckas, S.P., Cuccarese, M.F., Garris, C.S., Li, R., Ahmed, M.S., Kohler, R.H., Pittet, M.J., Weissleder, R.: TLR7/8-agonist-loaded nanoparticles promote the polarization of tumour-associated macrophages to enhance cancer immunotherapy. Nat. Biomed. Eng. **2**(8), 578588 (2018)
67. Wilson, D.R., Sen, R., Sunshine, J.C., Pardoll, D.M., Green, J.J., Kim, Y.J.: Biodegradable STING agonist nanoparticles for enhanced cancer immunotherapy. Nanomed. Nanotechnol. Biol. Med. **14**(2), 237–246 (2018)
68. Evans, E.R., Bugga, P., Asthana, V., Drezek, R.: Metallic nanoparticles for cancer immunotherapy. Mater. Today **21**(6), 673–685 (2018)
69. Zhao, C.Y., Cheng, R., Yang, Z., Tian, Z.M.: Nanotechnology for cancer therapy based on chemotherapy. Molecules (Basel, Switzerland) **23**(4), 826 (2018)
70. Zhang, Y., Cai, K., Li, C., Guo, Q., Chen, Q., He, X., Liu, L., Zhang, Y., Lu, Y., Chen, X., Sun, T.: Macrophage-membrane-coated nanoparticles for tumor-targeted chemotherapy. Nano Lett. **18**(3), 1908–1915 (2018)
71. Yong, T., Zhang, X., Bie, N., Zhang, H., Zhang, X., Li, F., Hakeem, A., Hu, J., Gan, L., Santos, H.A., Yang, X.: Tumor exosome-based nanoparticles are efficient drug carriers for chemotherapy. Nat. Commun. **10**(1), 1–16 (2019)
72. Shen, Z., Chen, T., Ma, X., Ren, W., Zhou, Z., Zhu, G., Zhang, A., Liu, Y., Song, J., Li, Z., Ruan, H.: Multifunctional theranostic nanoparticles based on exceedingly small magnetic iron oxide nanoparticles for T 1-weighted magnetic resonance imaging and chemotherapy. ACS Nano **11**(11), 10992–11004 (2017)
73. Pei, Q., Hu, X., Zheng, X., Liu, S., Li, Y., Jing, X., Xie, Z.: Lightactivatable red blood cell membrane-camouflaged dimeric prodrug nanoparticles for synergistic photodynamic/chemotherapy. ACS Nano **12**(2), 1630–1641 (2018)
74. Buzea, C., Pacheco, I.I., Robbie, K.: Nanomaterials and nanoparticles: sources and toxicity. Biointerphases **2**(4), MR17–MR71 (2007)
75. Song, G., Liang, C., Yi, X., Zhao, Q., Cheng, L., Yang, K., Liu, Z.: Perfluorocarbon-loaded hollow Bi2Se3 nanoparticles for timely supply of oxygen under near-infrared light to enhance the radiotherapy of cancer. Adv. Mater. **28**(14), 2716–2723 (2016)
76. Dou, Y., Guo, Y., Li, X., Li, X., Wang, S., Wang, L., Lv, G., Zhang, X., Wang, H., Gong, X., Chang, J.: Size-tuning ionization to optimize gold nanoparticles for simultaneous enhanced CT imaging and radiotherapy. ACS Nano **10**(2), 2536–2548 (2016)

77. Yi, X., Chen, L., Zhong, X., Gao, R., Qian, Y., Wu, F., Song, G., Chai, Z., Liu, Z., Yang, K.: Core–shell Au@ MnO 2 nanoparticles for enhanced radiotherapy via improving the tumor oxygenation. Nano Res. **9**(11), 3267–3278 (2016)
78. Du, J., Gu, Z., Yan, L., Yong, Y., Yi, X., Zhang, X., Liu, J., Wu, R., Ge, C., Chen, C., Zhao, Y.: Poly (Vinylpyrollidone)-and selenocysteine-modified Bi2Se3 nanoparticles enhance radiotherapy efficacy in tumors and promote radioprotection in normal tissues. Adv. Mater. **29**(34), 1701268 (2017)
79. Kwatra, D., Venugopal, A., Anant, S.: Nanoparticles in radiation therapy: a summary of various approaches to enhance radiosensitization in cancer. Transl. Cancer Res. **2**(4), 330–342 (2013)

Photopolymerization-Based Technologies for Microneedle Arrays Production

Merima Sirbubalo[1(✉)], Amina Tucak[1], Kenan Muhamedagić[2], Ognjenka Rahić[1], Ahmet Čekić[2], and Edina Vranić[1]

[1] Department of Pharmaceutical Technology, Faculty of Pharmacy, University of Sarajevo, Zmaja od Bosne 8, 71000 Sarajevo, Bosnia and Herzegovina
merima.sirbubalo@ffsa.unsa.ba

[2] Department of Machinery Production Engineering, Faculty of Mechanical Engineering, University of Sarajevo, Vilsonovo šetalište 9, 71000 Sarajevo, Bosnia and Herzegovina

Abstract. Recently, extensive research has been carried out on numerous manufacturing methods to produce the microneedle (MN) device suitable for painless and efficient transdermal drug delivery. However, the manufacturing process should be cost-effective and reproducible. Additive manufacturing technologies have, therefore, emerged as promising techniques for MN production, as they enable the use of a wide range of printable materials such as ceramics, liquids, metal, and plastic for MN production. This paper summarizes the potential and challenges of photopolymerization-based technologies, namely stereolithography (SLA), digital light processing (DLP), and two-photon polymerization (2PP) for the fabrication of MN arrays.

Keywords: Microneedles · Additive manufacturing · Transdermal drug delivery

1 Introduction

Microneedles (MNs) represent the concept of tiny, minimally invasive puncture devices of micron-sized dimensions designed to disrupt the outermost layer of the epidermis, the *stratum corneum* (SC), and to enhance the delivery of macromolecules, small drug molecules, or nanoparticles directly to the circulation of the skin [1]. MNs are manufactured in various shapes, dimensions (25–2000 μm in height, 50–250 μm of the base diameter, and 1–25 μm of the tip diameter), and densities (up to 20,000 MNs per cm^2) on adhesive membrane surface [2, 3]. In short, their height enables needles to physically pass through the epidermis and reach the capillary system of the skin without activating the nerve endings located in the dermis. At the same time, their base and tip diameters ensure that MNs remain unchanged during the application [4].

Generally, MNs are divided into five types (Fig. 1) [5]:

1. solid MNs – initially used in the skin pre-treatment, where MNs create aqueous microchannels in the SC, following by drug application in the form of a transdermal solution, cream, gel, or patch (Fig. 1a);

A. Badnjevic and L. Gurbeta Pokvić (Eds.): CMBEBIH 2021, IFMBE Proceedings 84, pp. 670–678, 2021.
https://doi.org/10.1007/978-3-030-73909-6_77

2. coated solid MNs – that deliver drug formulation, which is coated onto MN shafts (Fig. 1b);
3. hollow MNs – that enable the injection of liquid drug formulation into the skin (Fig. 1c);
4. dissolving MNs – that release drug incorporated into the MN matrix (Fig. 1d);
5. hydrogel MNs – that swell up in the skin and thus, release the drug formulation (Fig. 1d).

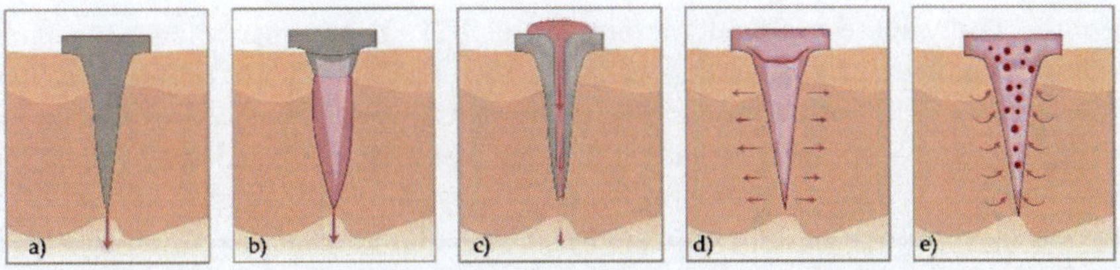

Fig. 1. MN types: a) solid MN, b) coated solid MN, c) hollow MN, d) dissolving MN, and e) hydrogel MN.

MNs can be made of various materials such as stainless steel [6], glass [7], sugar [8–10], polymers [11–13], and ceramics [14]. However, the success of the manufactured MN system highly depends to a large extent on its manufacturing technique, as the painless MN application and the efficient drug delivery into the skin circulation must be guaranteed. Furthermore, the production process itself should ensure the reproducible manufacturing of these microstructures [15, 16]. Up to now, the MN production has been reported using lithography methods [17], electrochemical and photochemical etching [17], laser cutting, laser ablation, and micromolding [18].

In addition, the literature describes the use of additive manufacturing (AM) for MN production [19]. This new, cutting-edge technology fabricates three-dimensional (3D) objects from a digital model generated using computer-aided design (CAD) software by fusing or depositing suitable materials such as ceramics, liquids, metal, plastic, powders or even living cells in layers [2, 19–21].

Although AM technologies include inkjet printing and fused deposition method (FDM), this paper will summarize the potential of photopolymerization-based techniques such as stereolithography (SLA), digital light processing (DLP), and two-photon polymerization (2PP) for the fabrication of MN arrays.

2 Photopolymerization-Based Techniques

3D printing technologies (3DP), which are based on the ability to selectively polymerize photo-sensitive polymers through laser emissions or projections of light are called photopolymerization-based technologies. The technologies are classified as SLA, DLP, and 2PP, depending on how the uncured polymer is exposed to the polymerization-inducing light. The curing process named photo-polymerization enables the manufacturing of MNs through the consecutive layer-wise polymerization of UVsensitive polymers – resins [2, 22].

To be appropriate for MN production, resins must be biocompatible and one has to know that material selection in 3DP of MN arrays is challenging. Up to now, approved medical grade resins are, respectively: dimethacrylate (DMA), polymethylmethacrylate (PMMA), and methyl methacrylate/acrylonitrile/butadiene styrene (MABS). Methacrylate-based proprietary photoresins and acrylic-based photoresins are commonly used resins in the SLA process [23].

2.1 Stereolithography

Stereolithography, more commonly referred to as SLA 3D printing, belongs to a family of additive manufacturing technologies known as vat photopolymerization [22]. Stereolithography is a liquid-based process that builds parts directly from CAD software. SLA uses a light source—UV laser to harden the photosensitive resin and achieve polymerization (Fig. 2). The manufacturing of 3D objects by stereolithography is based on the controlled solidification of a liquid resin by photo-polymerization. The process begins with a 3D CAD file. The file is digitally sliced into a series of parallel horizontal cross-sections which are then switched SLA devices and make one by one. UV laser draws the cross-section onto a bath of photopolymer resin which solidifies the crosssection. After the first layer, the platform is raised according to the layer thickness (typically about 0,05–0,15 mm) [23], the laser then solidifies the next cross-section, and the process is repeated until the whole part is complete. The resin that is not touched by the laser remains in the vat and can be reused.

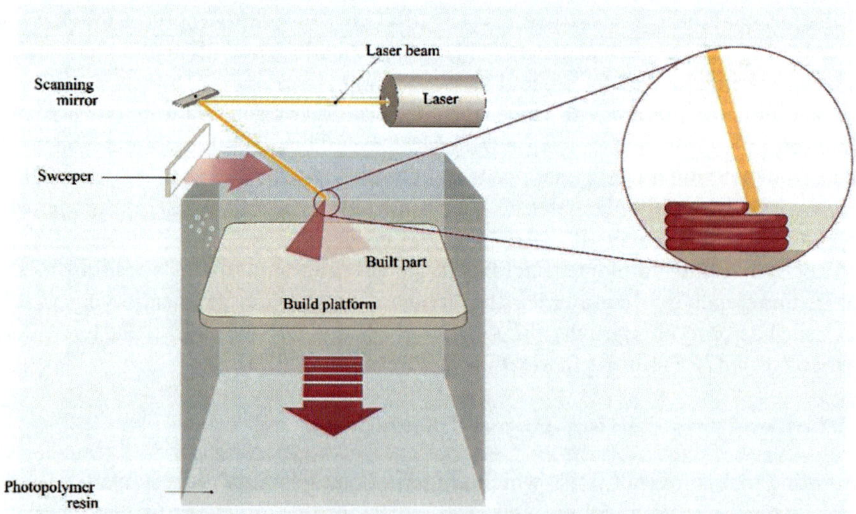

Fig. 2. The working principle of SLA technology

Based on different filling mechanisms, the SLA processes can be divided into two main categories: free surface and constrained surface (Fig. 3) [24].

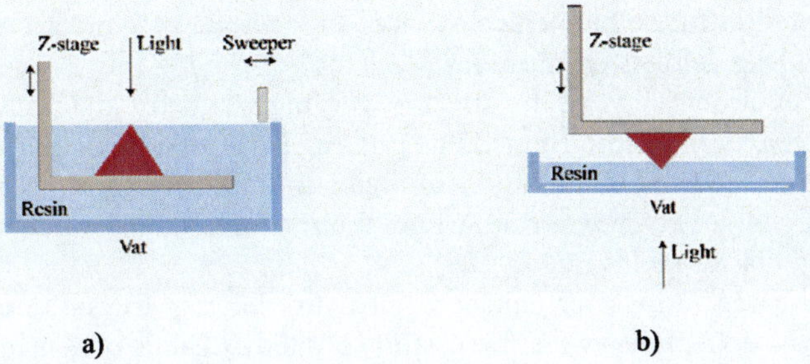

Fig. 3. Schemes of two types of stereolithography setups: (a) free-surface approach; (b) constrained-surface approach

In the free surface approach, a computer-controlled laser beam to draw a pattern, structures are built bottom-up from a support platform that rests just below the resin surface (Fig. 3a). Only a thin layer of resin is illuminated from above and cured on top of the structure as it is built in a layer-by-layer manner. The constrained surface approach (Fig. 3b) has a building platform, which can be suspended above the resin bath. Illumination from below, through the transparent floor, cures a layer of resin between the building platform and vat floor. This layer adheres to the platform, as the z-stage is raised by a defined distance. As each layer is cured, the building platform with the adhered part is elevated, and the part grows suspended from the platform downward [25].

After finishing the material polymerization, the platform rises out of the tank and the excess resin is drained. At the end of the process, the model is removed from the platform, washed of excess resin, and then placed in a UV oven for final curing. Post-print curing enables objects to reach the highest possible strength and become more stable.

To assess the possibility of photopolymer resin application in MNs manufacturing Pere et al. prepared MN arrays for transdermal delivery of insulin using a biocompatible Class 1 polymer as a resin. This study proved that a 3D printing stereolithographic technique can serve as an effective technology for the manufacturing of biocompatible MN patches with excellent mechanical strength and piercing capacity [26].

In research conducted by Economidou et al. MN arrays for intradermal insulin delivery were also prepared using a biocompatible Class 1 polymer. MNs patches of high quality and reproducibility featured two different needle shapes, a pyramid and a flat spear-shaped, were printed and then, subsequently coated with insulin–sugar films using inkjet printing. Authors reported superior skin penetration capacity of these MNs and in comparison with metallic MNs with the resembled geometry, SLA printed MNs required low forces to penetrate porcine skin varying (2–5 N). *In vivo* trials using diabetic mice were also performed and showed that SLA printed MNs facilitate rapid low glucose levels with longer duration compared to subcutaneous injections [15].

Farias et al. manufactured a custom 3D printed hollow MNe biodevice comprised of a reservoir chamber and an array of conical MN using SLA and the proprietary methacrylate-based Formlabs Clear photoresin® [27]. Yeung et al. went one step further and used SLA to fabricate a single-piece, multi-inlet, 3D microfluidic device with an embedded hollow MN array. Authors believe that the presented fabrication scheme can be adopted in future biomedical devices to facilitate new modes of operations for transdermal drug delivery applications [28].

Krieger et al. investigated the effect of varying geometry and print settings on the SLA printed MNs and demonstrated that printing MNs with good tip sharpness beyond the printer's defined resolution limits is possible if the correct print settings are chosen. They developed, a two-step "Print & Fill" fabrication method which allowed them to produce MN master molds rapidly [29].

Polymeric MN arrays for enhanced cisplatin delivery to A-431 epidermoid skin tumors for cancer treatment were successfully printed by Uddin et al. using SLA followed by inkjet coating. Rapid cisplatin release rates of 80–90% within 1 h were revealed using Franz cell diffusion studies and *in vivo* evaluation using Balb/c nude mice presented sufficient cisplatin permeabilization with high anticancer activity and tumor regression. This study demonstrates the potential for *in vivo* transdermal delivery of anticancer drugs using 3D printed MNs [30].

2.2 Digital Light Processing

Another photopolymerization-based technology which differs from SLA just in the used source of light is DLP. DLP is usually faster than SLA as high definition projector flashes the whole cross-section of the object at once, in the form of volumetric pixels (Fig. 4) [2].

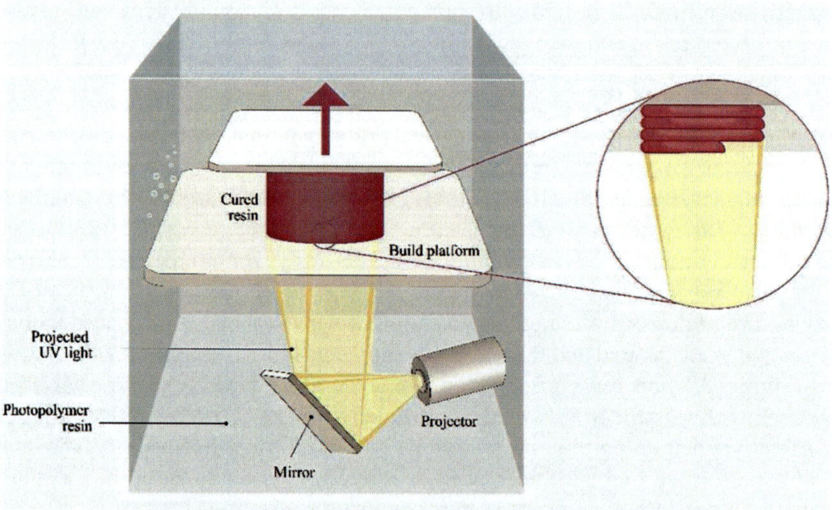

Fig. 4. The working principle of DLP technology

Gittard et al. successfully employed DLP to print solid MN array structures in various geometries out of an acrylate-based polymer for wound healing applications. In order to coat printed MNs with silver and zinc oxide to provide an antimicrobial effect, this research group used the pulsed laser deposition technique [31]. El-Sayed et al. also successfully used a desktop DLP 3D printer for MNs molds for nanoparticle delivery [32].

Lu et al. fabricated drug-loaded MN arrays for transdermal delivery of a chemotherapeutic drug using a microstereolithographic (DLP) apparatus. They incorporated dacarbazine into poly(propylene fumarate)/drug blends followed by photopolymerization through DLP that created the microneedle arrays. Results showed that dacarbazine can be released at a controlled rate for five weeks [33].

A very interesting approach was introduced by Lim et al. and his research group. They developed a personalized, dual-function MN splint via DLP 3D printer that can act as a splint to immobilize the affected trigger finger and deliver diclofenac through MN-assisted transdermal drug delivery for pain relief [34].

2.3 Two-Photon-Polymerization

MNs can be also manufactured using a 2PP printer, which is currently the highest resolution printer (structures with detail less than 100 nm). This specific technology enables a low-cost, layer-by-layer fabrication process of 3D structures from solid, liquid, or powder precursors in the microscale and nanoscale structures. Briefly, a drop of resin is placed on a glass substrate which is followed by focusing the laser beam directly inside a liquid resin droplet, which solidifies the focal point volume by two-photon absorption (Fig. 5) [35].

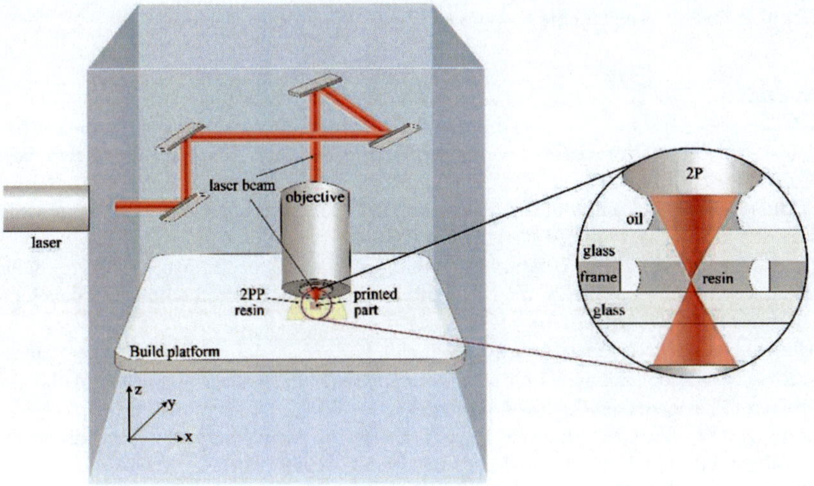

Fig. 5. The working principle of 2PP technology

Few research groups successfully demonstrated the fabrication of MNs using 2PP technology. Doraiswamy et al. were the first to report using 2PP to produce MNs from non-toxic and biologically inert Ormocer® (organically modified ceramic) materials. Obtained MNs exhibit appropriate mechanical properties, specifical penetration without fracture, against pork skin surfaces [36]. Gittard et al. suggested that 2PP can create MNs with a wide range of geometries while another research group printed ultrasharp polymer MNs via 2PP [37, 38]. Cordeiro et al. described an approach to fabricate high-quality MN array master templates using 2PP 3D [39].

Hollow MNs combined with internal laser-generated microchannels were developed by Trautmann et al. using 2PP technology. This novel hybrid approach combining 2PP printed microneedles with femtosecond laser generated microfluidic channels provides an important step towards versatile medical point-of-care systems [40].

Cylindrical, conical, and pyramidal MNs with excellent biocompatibility were successfully printed by 2PP technology and then endowed with magnetic properties through coating with iron in research conducted by Kavaldzhiev et al. [41].

3 Conclusion

Using photopolymerization-based 3DP techniques to fabricate MNs offers many advantages over traditional fabrication methods including versatility in terms of geometric complexity and resolution that these technologies provide, one-step process, the ability to rapidly optimize the geometry of MN, material, and other parameters of MNs. Comprehensive research, especially regarding stability and safety issues, is still necessary to make the photopolymerization-based 3DP techniques industrially feasible for microneedle production.

Conflict of Interest. The authors have no conflicts of interest to disclose.

References

1. Prausnitz, M.R.: Microneedles for transdermal drug delivery. Adv. Drug Deliv. Rev. **56**, 581–587 (2004)
2. Economidou, S.N., Lamprou, D.A., Douroumis, D.: 3D printing applications for transdermal drug delivery. Int. J. Pharm. **544**, 415–424 (2018)
3. Tucak, A., Sirbubalo, M., Hindija, L., Rahić, O., Hadžiabdić, J., Muhamedagić, K., et al.: Microneedles: characteristics, materials production methods and commercial development. Micromachines **11**, 961 (2020)
4. Alkilani, A.Z., McCrudden, M.T.C., Donnelly, R.F.: Transdermal drug delivery: Innovative pharmaceutical developments based on disruption of the barrier properties of the stratum corneum. Pharmaceutics **7**, 438–470 (2015)
5. Hong, X., Wu, Z., Chen, L., Wu, F., Wei, L., Yuan, W.: Hydrogel microneedle arrays for transdermal drug delivery. Nano-Micro Lett. **6**, 191–199 (2014)
6. Kaur, M., Ita, K.B., Popova, I.E., Parikh, S.J., Bair, D.A.: Microneedle-assisted delivery of verapamil hydrochloride and amlodipine besylate. Eur. J. Pharm. Biopharm. **86**, 284–291 (2014)
7. Gupta, J., Felner, E.I., Prausnitz, M.R.: Minimally invasive insulin delivery in subjects with type 1 diabetes using hollow microneedles. Diabetes Technol. Ther. **11**, 329–337 (2009)
8. Kolli, C.S., Banga, A.K.: Characterization of solid maltose microneedles and their use for transdermal delivery. Pharm. Res. **25**, 104–113 (2008)

9. Camović, M., Biščević, A., Brčić, I., Borčak, K., Bušatlić, S., Ćenanović, N., et al.: Acidresistant capsules with sugar microneedles for oral delivery of ascorbic acid. In: Badnjević, A., Škrbić, R., Gurbeta Pokvić, L. (eds.) CMBEBIH 2019. IFMBE Proceedings, pp. 749–753. Springer, Cham (2019)
10. Kathuria, H., Kang, K., Cai, J., Kang, L.: Rapid microneedle fabrication by heating and photolithography. Int. J. Pharm. **575**, 118992 (2020)
11. Park, J.H., Allen, M.G., Prausnitz, M.R.: Biodegradable polymer microneedles: fabrication, mechanics and transdermal drug delivery. J. Controlled Release **104**, 51–66 (2005)
12. Pérennès, F., Marmiroli, B., Matteucci, M., Tormen, M., Vaccari, L., Di Fabrizio, E.: Sharp beveled tip hollow microneedle arrays fabricated by LIGA and 3D soft lithography with polyvinyl alcohol. J. Micromech. Microeng. **16**, 473–479 (2006)
13. Park, J.H., Choi, S.O., Seo, S., Choy, Y.B., Prausnitz, M.R.: A microneedle roller for transdermal drug delivery. Eur. J. Pharm. Biopharm. **76**, 282–289 (2010)
14. Ovsianikov, A., Chichkov, B., Mente, P., Monteiro-Riviere, N.A., Doraiswamy, A., Narayan, R.J.: Two photon polymerization of polymer-ceramic hybrid materials for transdermal drug delivery. Int. J. Appl. Ceram. Technol. **4**, 22–29 (2007)
15. Economidou, S.N., Pere, C.P.P., Reid, A., Uddin, M.J., Windmill, J.F.C., Lamprou, D.A., et al.: 3D printed microneedle patches using stereolithography (SLA)for intradermal insulin delivery. Mater. Sci. Eng. C. **102**, 743–755 (2019)
16. Larrañeta, E., Lutton, R.E.M., Woolfson, A.D., Donnelly, R.F.: Microneedle arrays as transdermal and intradermal drug delivery systems: Materials science, manufacture and commercial development. Mater. Sci. Eng. R Rep. **104**, 1–32 (2016)
17. Ma, B., Liu, S., Gan, Z., Liu, G., Cai, X., Zhang, H., et al.: A PZT insulin pump integrated with a silicon microneedle array for transdermal drug delivery. Microfluid. Nanofluid. **2**, 417–423 (2006)
18. Zahn, J.D., Talbot, N.H., Liepmann, D., Pisano, A.P.: Microfabricated polysilicon microneedles for minimally invasive biomedical devices. Biomed. Microdevices **2**, 295–303 (2000)
19. Luzuriaga, M.A., Berry, D.R., Reagan, J.C., Smaldone, R.A., Gassensmitha, J.J.: Biodegradable 3D Printed Polymer Microneedles for Transdermal Drug delivery. Lab Chip **18**, 1223–1230 (2018)
20. Camović, M., Biščević, A., Brčić, I., Borčak, K., Bušatlić, S., Ćenanović, N., et al.: Coated 3D printed PLA microneedles as transdermal drug delivery systems. In: Badnjevic, A. (ed.) CMBEBIH 2019, IFMBE Proceedings, vol. 73, pp. 735–741. Springer (2019)
21. Alhnan, M.A., Okwuosa, T.C., Sadia, M., Wan, K.W., Ahmed, W., Arafat, B.: Emergence of 3D printed dosage forms: opportunities and challenges. Pharm. Res. **33**, 1817–1832 (2016)
22. ISO/ASTM52921 - 13. Standard Terminology for Additive Manufacturing—Coordinate Systems and Test Methodologies. https://www.astm.org/Standards/ISOASTM52921.htm. Accessed 22 Nov 2020
23. Rahmati, S., Ghadami, F.: Process parameters optimization to improve dimensional accuracy of stereolithography parts. Int. J. Adv. Des. Manuf. Technol. **7**, 59–65 (2014)
24. Schmidleithner, C., Kalaskar, D.M.: Stereolithography. 3D Print. https://doi.org/10.5772/intechopen.78147
25. Huang, J., Qin, Q., Wang, J.: A review of stereolithography: processes and systems. Processes **8**, 1138 (2020)
26. Pere, C.P.P., Economidou, S.N., Lall, G., Ziraud, C., Boateng, J.S., Alexander, B.D., et al.: 3D printed microneedles for insulin skin delivery. Int. J. Pharm. **544**, 425–432 (2018)
27. Farias, C., Lyman, R., Hemingway, C., Chau, H., Mahacek, A., Bouzos, E., et al.: Threedimensional (3D) printed microneedles for microencapsulated cell extrusion. Bioengineering **5**, 59 (2018)

28. Yeung, C., Chen, S., King, B., Lin, H., King, K., Akhtar, F., et al.: A 3D-printed microfluidic-enabled hollow microneedle architecture for transdermal drug delivery. Biomicrofluidics **13**, 064125 (2019)
29. Krieger, K.J., Bertollo, N., Dangol, M., Sheridan, J.T., Lowery, M.M., O'Cearbhaill, E.D.: Simple and customizable method for fabrication of high-aspect ratio microneedle molds using low-cost 3D printing. Microsyst. Nanoeng. **5**, 42 (2019)
30. Uddin, M.J., Scoutaris, N., Economidou, S.N., Giraud, C., Chowdhry, B.Z., Donnelly, R.F., et al.: 3D printed microneedles for anticancer therapy of skin tumours. Mater. Sci. Eng. C. **107**, 110248 (2020)
31. Gittard, S.D., Miller, P.R., Jin, C., Martin, T.N., Boehm, R.D., Chisholm, B.J., et al.: Deposition of antimicrobial coatings on microstereolithography-fabricated microneedles. Jom **63**, 59–68 (2011)
32. El-Sayed, N., Vaut, L., Schneider, M.: Customized fast-separable microneedles prepared with the aid of 3D printing for nanoparticle delivery. Eur. J. Pharm. Biopharm. **154**, 166–174 (2020)
33. Lu, Y., Mantha, S.N., Crowder, D.C., Chinchilla, S., Shah, K.N., Yun, Y.H., et al.: Microstereolithography and characterization of poly(propylene fumarate)-based drug-loaded microneedle arrays. Biofabrication **7**, 045001 (2015)
34. Lim, S.H., Ng, J.Y., Kang, L.: Three-dimensional printing of a microneedle array on personalized curved surfaces for dual-pronged treatment of trigger finger. Biofabrication **9**, 015010 (2017)
35. Sun, H.-B., Kawata, S.: Two-Photon photopolymerization and 3D lithographic microfabrication. APS **170**, 169–273 (2004)
36. Doraiswamy, A., Jin, C., Narayan, R.J., Mageswaran, P., Mente, P., Modi, R., et al.: Two photon induced polymerization of organic-inorganic hybrid biomaterials for microstructured medical devices. Acta Biomater. **2**, 267–275 (2006)
37. Aksit, A., Arteaga, D.N., Arriaga, M., Wang, X., Watanabe, H., Kasza, K.E., et al.: In-vitro perforation of the round window membrane via direct 3-D printed microneedles. Biomed. Microdevices **20**, 47 (2018)
38. Gittard, S.D., Ovsianikov, A., Chichkov, B.N., Doraiswamy, A., Narayan, R.J.: Two-photon polymerization of microneedles for transdermal drug delivery. Expert Opin. Drug Deliv. **7**, 513–533 (2010)
39. Cordeiro, A.S., Tekko, I.A., Jomaa, M.H., Vora, L., Mcalister, E., Volpe-zanutto, F., et al.: Two-photon polymerisation 3d printing of microneedle array templates with versatile designs: application in the development of polymeric drug delivery systems. Pharm. Res. **37**, 1–15 (2020)
40. Trautmann, A., Roth, G.L., Nujiqi, B., Walther, T., Hellmann, R.: Towards a versatile pointof-care system combining femtosecond laser generated microfluidic channels and direct laser written microneedle arrays. Microsyst. Nanoeng. **5**, 6 (2019)
41. Kavaldzhiev, M., Perez, J.E., Ivanov, Y., Bertoncini, A., Liberale, C., Kosel, J.: Biocompatible 3D printed magnetic micro needles. Biomed. Phys. Eng. Express **3**, 025005 (2017)

Gastric Parietal Cell Regeneration by Nano-Scaffolding in Hypochlorhydria and Achlorhydria Treatment

Merima Bukva, Evelina Pulo, Naida Omerović(✉), and Selma Škrbo

Faculty of Pharmacy, University of Sarajevo, Zmaja od Bosne 8, 71 000 Sarajevo, Bosnia and Herzegovina
naida.omerovic@ffsa.unsa.ba

Abstract. Hypochlorhydria and achlorhydria are conditions in which the production of hydrochloric acid (HCl) in the stomach is reduced or completely absent, mostly because of an underlying medical condition, such as gastric cancer. Subtotal or total gastrectomy is commonly performed in patients who suffer from gastric cancer. But, advances in the field of tissue engineering allowed the fabrication and regeneration of many tissues, therefore supporting the idea that hypochlorhydria and achlorhydria could be treated in that manner, especially by nano-scaffolding, the process of tissue regrowing using polymer fibres at a nanoscale. The main goal of this article is to present theoretical premises of tissue engineering and combine findings from multiple studies which, when formulated in a form composed of four steps (preparation of the polymeric biodegradable nano-scaffold, preparation and multiplication of gastric cells, integration of gastric cells into a nano-scaffold and transplantation of the nano-scaffolded tissue into the interior of the stomach), followed by the vascularisation process and biodegradation, gives a novel theoretical possibility when it comes to managing these patients. We gathered scientific articles, professional literature and electronic books by researching databases, such as PubMed and Google Scholar, which were used as main sources of information when creating this article. Additionally, we used Medical Subject Headings (MeSH) in order to make a more precise search. Further research is needed for this procedure to resuscitate and make treatment of hypochlorhydria and achlorhydria easier and more efficient.

Keywords: Parietal cells · Gastric · Achlorhydria · Nanomaterial · Tissue scaffolds · Tissue engineering

1 Introduction

The stomach is one of the earliest described organs by priests, physicians, and anatomists and was studied by alchemists, chemists, and physiologists [1–3]. It is a muscular, hollow organ located in the peritoneal cavity, in the left upper abdominal quadrant or the epigastric abdominal region. Formation of chyme, synthesis of proteins necessary for vitamin absorption, microbial defences, and propagation of the peristaltic reflex are the

A. Badnjevic and L. Gurbeta Pokvić (Eds.): CMBEBIH 2021, IFMBE Proceedings 84, pp. 679–689, 2021.
https://doi.org/10.1007/978-3-030-73909-6_78

main stomach functions [4]. It is comprised of four regions: cardia, fundus, body, and pylorus. The gastric wall consists of different tissue layers: mucosal layer, submucosa, muscularis externa, and adventitia/serosa. Gastric glands are lined with four specialized cells: surface mucous cells, acid-secreting parietal cells, pepsinogen-secreting chief or zymogenic cells, and enteroendocrine cells. They all contribute to independent functions [5, 6]. All these cells originate from stem cells residing in the isthmus region of gastric glands [7].

Three regulating molecules control the secretion of parietal cells: acetylcholine via M_3-muscarinic receptors, histamine via H_2-histamine receptors, and gastrin receptors. All receptors appear on the basal side of parietal cells and are in charge of a protein channel called H^+/K^+ ATPase (proton pump) on the luminal side of parietal cells [5]. In the resting state, proton pumps are contained within the membranes of intracellular tubulovesicles. However, during stimulation, tubulovesicles fuse with the apical membrane, relocating proton pumps to the apical surface of parietal cells [8, 9]. Proton pumps transport H^+ out of the cell into the gastric lumen and K^+ from the gastric lumen into the cell. Chloride ions follow the proton gradient via K^+/Cl^- channel into the gastric lumen, leading to the formation of hydrochloric acid (HCl). There is a constant basal level of acid production, even when parietal cells are not stimulated [10]. The basal level of acid secretion is caused by histamine and acetylcholine and is approximately 10% of the maximal acid output of stimulated parietal cells [11].

Disorders can occur at every step of HCl formation and lead to its hypersecretion or hyposecretion. Hypochlorhydria and achlorhydria are conditions in which HCl production in the stomach is reduced or completely absent, respectively. It is usually secondary to an underlying medical condition [12, 13]. Multiple disorders can cause achlorhydria, such as pernicious anaemia, *Helicobacter pylori* infection, hypothyroidism, or gastric cancer [14–16]. Depending on the primary cause, there are various symptoms of achlorhydria. Generally, patients develop epigastric pain, weight loss, heartburn, nausea, bloating, diarrhoea, abdominal pain, and vomiting. There is no specific treatment of achlorhydria. For patients who have *Helicobacter pylori* infection, the recommended treatment is its eradication. Other therapies improve the complications of achlorhydria that include replacement of calcium, vitamin D, iron, and vitamin B_{12} [17, 18]. Despite having a risk of developing into gastric adenocarcinoma and carcinoid tumour, achlorhydria has a good prognosis [19]. Gastric cancer is very common and one of the leading causes of death worldwide [20, 21]. The two main tumour sites of gastric adenocarcinoma are proximal (cardia) and distal (noncardia). Billroth and Schlatter performed the first subtotal and total gastrectomy more than a century ago [22, 23] and up to date, it has been the best procedure when it comes to patients who suffer from cancer or other problems that cannot be treated differently.

Different types of surgery can be performed to treat gastric cancer. The choice of procedure depends on tumour location and size, its stage, the volume of the remnant stomach, and surgeon's preference [24]. Up to date, radical surgery has been the most effective treatment for proximal and distal gastric cancer. Currently, two types of surgical procedures are available: total gastrectomy and proximal gastrectomy [25]. The main benefit of proximal gastrectomy is the preservation of the function of the remnant stomach, thus the operation is often recommended if cancer is only present in the lower part of the stomach [24]. On the other hand, long tumour-free resection margin can be achieved by performing total gastrectomy, therefore it is performed if cancer has spread throughout the stomach. Total gastrectomy is also advised if cancer is located near the oesophagus. Different studies have revealed that total gastrectomy showed similar outcomes compared with proximal gastrectomy. No significant difference was detected in the overall survival rate between the patients who underwent proximal or total gastrectomy [26–28]. However, the choice of surgical procedure for patients with proximal and distal gastric cancer remains controversial.

Gastric cancer spreads via lymphatics to the regional lymph nodes, so their dissection is beneficial and always proceeded in a subtotal or total gastrectomy. Different studies have shown that lymphadenectomy is associated with lower morbidity and mortality in patients after proximal or total gastrectomy [29, 30]. Endoscopic resection is being increasingly used for the treatment of early-stage gastric cancer, mainly because of its minimal invasiveness. The main advantages of endoscopic resection are the assessment of the depth of cancer invasion, as well as the degree of cancer differentiation and involvement of lymphatics, so the risk of lymph node metastasis can be predicted [31].

Another method that has recently gained popularity is laparoscopic, a procedure intended for gastric cancer patients with preoperative stage Ia diagnosis. This is a good substitute to gastrectomy, especially for patients with early-stage gastric cancer. Laparoscopic has significant advantages, such as less blood loss during the procedure, less postoperative pain, earlier recovery, and shorter hospital stay [32].

A few gastric replacement approaches have been proposed to improve the quality of patients' lives after gastrectomy. However, the optimal reconstruction procedure remains controversial [7]. Recent advances in the field of tissue engineering have allowed the fabrication of many tissues and organs, including urinary bladder, urethra, heart, and liver [33–37].

This paper aims to give the theoretical basis for the hypothetical procedure using tissue engineering guidelines and nano-scaffolding method when it comes to the treatment of hypochlorhydria and achlorhydria in patients suffering from gastric cancer or other diseases, which entail the loss of stomach tissue functioning. One of the examples of tissue engineering is nano-scaffolding, the process of tissue regrowing using polymer fibres at a nanoscale. This research gives an insight into the possible procedure that could be used as an aid for patients suffering from hypochlorhydria or achlorhydria, which are usually a result of destructible processes in the stomach, such as carcinomas.

2 Materials and Methods

When making a selection of studies and articles to use for this purpose, we gathered papers from verified sources of information, for example, databases, such as PubMed and Google Scholar, in which the given method showed a high success rate and the preconceived hypothesis of an author was confirmed. We combined several papers, whose selection was divided into multiple categories, considering the step of the theoretical method that we investigated at that moment, simultaneously following the criteria that all methods are suitable, so they can be interconnected and built on each other.

For example, when preparing the polymeric biodegradable nano-scaffold, polylactic acid (PLA) is a polymer chosen, based on the evidence in several studies that this material has good bioproperties and can be obtained from biomass. The phase separation method is a golden standard when making nanomaterials because a researcher can control the parameters, such as pore diameter, quite easily, which is extremely important for nanoscale materials. Preparation and multiplication of gastric cells were based on novel research from 2018, which was simple, yet it had a high success rate, as it combined discoveries published up to that year, along with the confirmation method that demonstrated that all types of gastric epithelial cells were present in the obtained sample. Finally, the integration of gastric cells into a nano-scaffold was based on another novel study from 2020, which constructed biocompatible urethral tissue substitute, which, with certain adjustments, might be a good basis for the gastric substitute usage.

3 Results

Tissue engineering is a rapidly growing field of biotechnology, which combines understandings and accomplishments in engineering, as well as material and life sciences with an ultimate goal of developing artificial tissue substitutes. The main idea behind the process is designing replacements, which can mimic cellular functions, such as proliferation, and ultimately restore, maintain, or improve tissue function of different human organs. It is a very complex process for which is necessary to strictly control all factors to obtain a product with high quality and usability. Mentioned techniques are recognised as an encouraging alternative for donor tissues, which are in high demand, as they can be very successfully replaced by synthetic scaffolds [38–40].

An ideal scaffold for the human tissue replacement should be mechanically stable, biocompatible, and with an appropriate architecture for cell repopulation. Therefore, required material is selected cautiously, taking into account all the desired properties, such as a wide range of pore distribution, as well as the porosity and surface-to-volume ratio, both of which should be high. When all these conditions are met, synthesised scaffolds are suitable for the cell attachment and proliferation, as they perfectly imitate natural extracellular matrix. Novel methods include principles of nanotechnology, as the nano-fibrous scaffolds have a very high surface-to-volume ratio, which enhances the cell adhesion better than the traditionally used materials. Nowadays, beyond everything, researchers tend to develop biodegradable materials, so scaffolds deteriorate some time after the transplantation [38, 40, 41]. Some of the current applications of nano-scaffolds include making substitutes for urethral tissue, bone tissue reparation, heart valve engineering, keratinocyte and myocyte support, skin reconstruction [41–44], and many more, all of which represent remarkable promotions in comparison with the traditionally used methods. The most commonly used methods when obtaining nano-fibrous materials are self-assembly, electrospinning, and phase separation. Each one has its advantages and disadvantages, so researchers select the one that suits the conditions of a study the best. For example, the self-assembly method assembles small-diameter nano-fibres, whereas electrospinning assembles large-diameter ones, when in comparison with the size range for the natural extracellular matrix. However, phase separation creates nano-fibres closest in pore size to the natural ones, so this method has the most potential in imitating a real tissue. Lately, methods such as three-dimensional (3D) printing have emerged, with a huge potential for future use [38, 39].

Polymers used in nano-scaffold formations are polyglycolic acid (PGA), PLA, its copolymer poly (D, L-lactide-co-glycolide) (PLGA), chitosan (CS), carboxymethylcellulose (CMC), polycaprolactone (PCL), collagen, or gelatin, but there are also some novel approaches, such as using carboxylated cellulose nanowhiskers and silver nanoparticle composites (CCNWs-AgNPs). The PLA has outstanding biocompatibility and biodegradability, as well as moderate mechanical and thermal properties, which can be upgraded by combining PLA with other polymers. Another polymer that is satisfactory for nano-scaffolds is CS, also with good biocompatibility and biodegradability, but with antimicrobial assets as well [40, 45]. Because nano-scaffolds are formations composed of nano-architecture polymeric foundation, together with the implanted cells, deoxyribonucleic acid (DNA), and protein components [46], it is necessary to explain in more detail all of the integral parts to create a complete picture of a nano-scaffold with properties customised for stomach tissue and parietal cell replacement – a topic our work is based upon.

Preparation of the gastric parietal cell nano-scaffolds has integral phases. Firstly, the biodegradable polymeric nano-scaffold foundation is designed and then developed. At the same time, parietal cells are isolated, multiplied, prepared, and then transplanted onto the scaffold. Scaffolds assembled this way are surgically built into the interior surface of the stomach. Next step is the vascularisation process of cells and biodegradation of the scaffold, while new active parietal cells remain on the gastric wall. Although presented simplified, the whole process of nano-scaffolding is complicated and requires detailed preparation and implementation of each step.

3.1 Preparation of the Polymeric Biodegradable Nano-Scaffold

Firstly, we select polymers that are highly biocompatible and biodegradable, such as PLA, but polymers such as cellulose derivates, chitin, PGA, or PCL can also be used. New polymer coatings with antimicrobial activity have been developed, as their activity depends on the presence of silver in the oxidized form [1]. Primarily, the design of a scaffold is made and adequate polymer (PLA) is chosen, because of its good bioproperties and renewable resource origin, since it is produced in the polymerization process of lactic acid gained from corn starch and sugarcanes. Four different PLA isomers exist: isotactic, syndiotactic, tactic, and isotactic stereoblock form, all of which differ in their mechanical properties and degradation time. This material is hydrophobic, and therefore soluble in organic solvents. The PLA polymers can be manufactured in two ways: biologically, which implies fermentation by lactic bacteria or fungi, and chemically, which implies using chemical reactions for polycondensation process, resulting in polymers of higher molecular weight and better control of final properties, such as crystallinity and chain orientation. The PLA has already been widely used in medicine as a material for stents, dialysis media, and drug delivery systems [38, 47].

The phase separation method is the best choice when making nano-scaffolds, which need to have precisely determined pore size and ultra-fine fibres. As a novelty, the method has been upgraded to be thermally induced, resulting in better control of the built material. The method itself uses the spinodal liquid-liquid phase separation, followed by the crystallization of the phase with a polymer. The process involves the following steps: polymer dissolution, phase separation and gelation, solvent extraction, freezing and freeze-drying. The critical stage is the gelation, on which the scaffold structure depends. The temperature should be low to create nano-scale fibres, along with keeping the polymer concentration optimal and choosing the right solvent. As a result of phase separation, a 3D continuous network of fibres with a size of 50–500 nm and 98% of porosity could be developed. To further increase the porosity, porogens, such as sugar and inorganic salt, could be added, creating macropores and allowing cells to be implanted in optimal conditions, which is necessary for the implantation of the previously multiplied parietal cells with an average size of 20 μm. By using the phase separation method, researchers are in complete control of parameters, such as pore diameter or pore interaction, all with simple equipment [38, 48].

3.2 Preparation and Multiplication of Gastric Cells

Gastric epithelial cells are obtained via the surgical gastrectomy method, based on the study by Qin and Pei [49]. Biopsies are taken from patients who are recovering from ulcers or gastritis, and for patients having a tumour, an obtained tissue should be at least 10 cm away from the invaded parts to be considered acceptable, as confirmed by pathological findings. Baratta et al. developed an isolation method for human gastric epithelial cells from surgically removed tissue, followed by their multiplication. Namely, gastric epithelial cells are released from the tissue and separated from the lamina propria by the enzymatic digestion of collagenase type IV. This enzyme also minimalises the damage of proteins and receptors present. Then, cell cultures were made by planting cells onto a plastic base, firstly without a biological matrix. These seedings represent primary cultures and it has been proven that multicellular aggregates obtained by collagenase activity (not single cells) have better chances and capacity to adhere to plastic. The procedure is done in the following way: surgical pieces are pinned onto the Petri dish and mucosa layers are removed using a sterile scalpel. Then, mucosa layers are moved to another Petri dish, cut into 1 mm^3 pieces, and washed in the cell culture media. Digestion solution is added and the mixture is incubated in a water bath at 37 °C with frequent manual agitation. Tubes are settled on ice for 2 min, followed by collecting the supernatant with cellular clumps and moving it to a new tube. For the undigested tissue, the process is repeated until all the obtained tissue decomposes. Cell suspensions are centrifuged at 200 g/5 min and washed three times with the phosphate buffered saline (PBS). Next, cell suspensions are seeded into plastic plates and left for 24 h to attach. After the time has expired, culture is rinsed with PBS a couple of times to remove cell debris and mucus, and serum-free culture medium is applied. The medium should be renewed every two days and, based on the study conducted by Schlaermann et al. [50], enriched with human epidermal growth factor, human noggin, human fibroblast growth factor, gastrin, nicotinamide, and N-acetylcysteine. Afterwards, the immunostaining method is utilized as a confirmation that the obtained cells are mucous, parietal, chief, and gastrin-secreting. After the making of colonies, a monolayer of gastric cells is obtained.

3.3 Integration of Gastric Cells into a Nano-Scaffold

Prepared nano-scaffolds need to have the ability to attract cells and allow their adhesion upon the implantation into the body organs, in this case, preserved stomach tissue. To allow the transmission of duplicated cells from their cultures onto the prepared polymer materials, the material needs to ideally mimic natural extracellular matrix, so cells can adhere and proliferate undisturbedly [46]. Modelled on the study by Wang et al. [41], the previously prepared cell cultures can be relocated to nano-scaffolds, which are prepared as follows: scaffolds are cut and sterilized with 75% ethanol and ultra-violet (UV) light for 30 min on each side. After seeding gastric cells into scaffolds, they are fixed with paraformaldehyde and incubated in ethanol aqueous solution, followed by the vacuum drying. Prepared nano-composites could be characterized via field emission scanning electron microscopy (FESEM), X-ray powder diffraction (XRD), transmission electron microscopy (TEM), Fourier-transform infra-red (FTIR) spectroscopy, or thermogravimetric analysis (TGA) [45].

3.4 Transplantation of the Nano-Scaffolded Tissue into the Interior of the Stomach

Nano-scaffold needs to have adequate properties so it can swimmingly integrate into the host tissue after the surgical transplantation of the prepared nano-composite. If the integration is appropriate, bioactive components and cells of the matrix move towards the stomach tissue that needs to regenerate and communicate with cells that survived [46]. The whole point of the gastric nano-scaffold transplantation is to restore the function of parietal cells so they can secrete HCl. Both polymeric scaffold and natural stomach tissue show a high degree of elasticity, so we can conclude that PLA scaffold is suitable to withstand motility and contractions of the stomach [51]. For transplanted cells to attach to the gastric interior wall, the vascularization process must occur. In tissue engineering, it is desirable to promote angiogenesis, e.g. the new blood vessel formation, so proangiogenic agents are incorporated, and methods, such as the implantation of endothelial progenitor cells that excrete angiogenic cytokines, are used [52]. The most commonly used angiogenic growth factors are fibroblast growth factor, interleukin-8, platelet-derived growth factor-BB, transforming growth factor-beta, tumour necrosis factor-alpha, or vascular endothelial growth factor. They can be applied individually or in a combination, and are usually released with a delay from a scaffold. Some studies have shown angiogenic effects of the gelatin hydrogel granules containing plasma growth factor mixture [53], and neovascularization ability of fibrin glue matrix [54–56]. The biggest problem that arises is the fact that vascularisation depends on the blood flow level of the host tissue. Therefore, approaches based on the creation of tubes and channels are being developed to allow microvascularization of all cells in a construct [52]. After some time, long enough to allow the transplanted cells to be vascularized and linked to the stomach tissue, the polymeric foundation component of a nano-scaffold must degrade and leave the body, so the additional surgical procedure is not needed. Biodegradable materials are defined as substances whose chemical and physical properties are decayed and which completely degrade when exposed to processes in the body, including microorganisms and aerobic or anaerobic activities [47]. The PLA is proven to be biodegradable, therefore it is a perfect material for the cultured parietal cell transplant utilization. As a result of all of the given phases (preparation of the polymeric biodegradable nano-scaffold, preparation and multiplication of gastric cells, integration of gastric cells into a nano-scaffold and transplantation of the nano-scaffolded tissue into the interior of the stomach), together with processes of vascularization and biodegradation, theoretically, we can obtain and transfer new parietal cells into the gastric wall to renew the stomach ability to produce HCl after the malign processes, which caused deterioration of stomach cells, including parietal ones.

4 Conclusion

Hypochlorhydria and achlorhydria are unavoidable accompanying elements to many destructible illnesses of the stomach, such as gastric cancer. The tumour removal, which is aggressive no matter if the procedure is done by radical surgery, laparoscopic, or any other method, is controversial, so there is a need for novel techniques. Destroyed tissue, which lost its function by removing active cells, needs to be revived to help a patient lead life as normal as it is possible. Approaches of tissue engineering have made it easier to treat destructed tissue of many organs, improving patients' lives, therefore the pressure to incorporate these methods into the treatment of gastric diseases is understandable. Using tissue engineering methods and seeding parietal cells into the polymer foundation, forming 3D nano-scaffolds, followed by the biodegradation of the material, with new parietal cells remaining on the gastric wall, allows easier and more efficient transplantation, which, together with necessary provascular factors, represents a sustainable substitute to the previously destroyed cells.

The proposed method is theoretical and hopefully can serve as a foundation for future studies and implementation of tissue engineering when dealing with gastrointestinal illnesses. So far, no method like this has been established. The main advantage in comparison to the existing procedures is that, by combining different approaches, a good starting point for this method is made, and afterwards, there is a possibility for the expansion and improvement. Today-existing methods are individual and refer only to the certain steps of the whole process. Following this study, the nano-scaffolding and integration of gastric cells could be automatised, which would facilitate and accelerate the process of providing assistance and medical care to patients.

Conflict of Interest. The authors have no conflicts of interest to disclose.

References

1. Baron, J.H.: The discovery of gastric acid. Gastroenterology **76**(5), 1056–1064 (1979)
2. Rosenfeld, L.: The last alchemist–the first biochemist: J.B. van Helmont (1577–1644). Clin. Chem. **31**(10), 1755–1760 (1985)
3. Modlin, I.M.: From prout to the proton pump: a history of the science of gastric acid secretion and the surgery of peptic ulcer. Surg. Gynecol. Obstet. **170**(1), 81–96 (1990)
4. Chaudhry, S.R., Liman, M.N.P., Peterson, D.C.: Anatomy, abdomen and pelvis, stomach. In: StatPearls [Internet]. Treasure Island (FL): StatPearls Publishing (2020)

5. Dockray, G.J.: Gastrin and gastric epithelial physiology. J. Physiol. **518**(Pt 2), 315–324 (1999)
6. Schubert, M.L.: Gastric acid secretion. Curr. Opin. Gastroenterol. **32**(6), 452–460 (2016)
7. Speer, A.L., Sala, F.G., Matthews, J.A., Grikscheit, T.C.: Murine tissue-engineered stomach demonstrates epithelial differentiation. J. Surg. Res. **171**(1), 6–14 (2011)
8. Mercer, D.W., Robinson, E.K.: Stomach. In: Townsend, C.M., (ed.) Sabiston textbook of surgery, 18th ed. Philadelphia: Saunders, pp. 1226–1238 (2008)
9. Urushidani, T., Forte, J.G.: Signal transduction and activation of acid secretion in the parietal cell. J. Membr. Biol. **159**(2), 99–111 (1997)
10. Silen, W.: Physiology of gastric function. In: Fischer, J.E., (ed.) Surgical basic science. St Louis (MO): Mosby, pp. 271–291 (1993)
11. Mulholland, M.W.: Gastric anatomy and physiology. In: Mulholland, M.W., (ed.) Greenfield's surgery: scientific principles and practice, 4th ed. Philadelphia: Lippincott Williams and Wilkins, pp. 712–720 (2006)
12. Li, Y., Xia, R., Zhang, B., Li, C.: Chronic atrophic gastritis: a review. J. Environ. Pathol. Toxicol. Oncol. **37**(3), 241–259 (2018)
13. Furuta, T., Baba, S., Yamade, M., Uotani, T., Kagami, T., Suzuki, T., et al.: High incidence of autoimmune gastritis in patients misdiagnosed with two or more failures of H. pylori eradication. Aliment. Pharmacol. Ther. **8**(3), 370–377 (2018)
14. Chen, C., Zheng, Z., Li, B., Zhou, L., Pang, J., Wu, W., et al.: Pancreatic VIPomas from China: case reports and literature review. Pancreatology **19**(1), 44–49 (2019)
15. Giusti, F., Marini, F., Brandi, M.L.: Multiple endocrine neoplasia type 1. In: GeneReviews® [Internet]. Seattle (WA): University of Washington (2005)
16. Belei, O.A., Heredea, E.R., Boeriu, E., Marcovici, T.M., Cerbu, S., Mărginean, O., et al.: Verner-Morrison syndrome. literature review. Rom. J. Morphol. Embryol. **58**(2), 371–376 (2017)
17. Tytgat, G.N.: Practical management issues for the Helicobacter pylori-infected patient at risk of gastric cancer. Aliment. Pharmacol. Ther. **12**(1), 123–128 (1998)
18. Glasbrenner, B., Eissele, R., Heeckt, P., Klatt, S., Beger, H.G., Adler, G.: Multicentric gastric carcinoid in autoimmune gastritis. Dtsch. Med. Wochenschr. **121**(8), 231–236 (1996)
19. Fatima, R., Aziz, M.: Achlorhydria. In: StatPearls [Internet]. Treasure Island (FL): StatPearls Publishing (2020)
20. Parkin, D.M., Bray, F.I., Devesa, S.S.: Cancer burden in the year 2000. The global picture. Eur. J. Cancer **37**(8), S4–66 (2001)
21. Parkin, D.M.: International variation. Oncogene **23**(38), 6329–6340 (2004)
22. Billroth, T.: Offenes Schreiben an Herrn Dr. L. Wittelshofer. Wien. Med. Wochenschr. **31**, 161–165 (1881)
23. Schlatter, K.: A unique case of complete removal of the stomach-successful esophago-enterostomy recovery. Med. Rec. **52**, 909–914 (1897)
24. Sugoor, P., Shah, S., Dusane, R., Desouza, A., Goel, M., Shrikhande, S.V.: Proximal gastrectomy versus total gastrectomy for proximal third gastric cancer: total gastrectomy is not always necessary. Langenbecks Arch. Surg. **401**(5), 687–697 (2016)
25. Yura, M., Yoshikawa, T., Otsuki, S., Yamagata, Y., Morita, S., Katai, H., et al.: Correction to: Oncological safety of proximal gastrectomy for T2/T3 proximal gastric cancer. Gastric Cancer **22**(4), 906 (2019)
26. Erturk, M., Çiçek, Y., Ersan, Y., Saribeyoğlu, K., Doğusoy, G., Erginoz, E.: Analysis of clinicopathological prognostic parameters in adenocarcinoma of the gastric cardia. Acta Chir. Belg. **103**(6), 611–615 (2003)
27. Smith, J., Brennan, M.: Surgical treatment of gastric cancer: proximal, mid, and distal stomach. Surg. Clin. North Am. **72**(2), 381–399 (1992)

28. Carey, S., Storey, D., Biankin, A.V., Martin, D., Young, J., Allman-Farinelli, M.: Long term nutritional status and quality of life following major upper gastrointestinal surgery – a cross-sectional study. Clin. Nutr. **30**(6), 774–779 (2011)
29. Seevaratnam, R., Bocicariu, A., Cardoso, R., Mahar, A., Kiss, A., Helyer, L.: A meta-analysis of D1 versus D2 lymph node dissection. Gastric Cancer **15**(1), S60–69 (2011)
30. Degiuli, M., Sasako, M., Calgaro, M., Garino, M., Rebecchi, F., Mineccia, M., et al.: Morbidity and mortality after D1 and D2 gastrectomy for cancer: interim analysis of the Italian gastric cancer study group (IGCSG) randomised surgical trial. Eur. J. Surg. Oncol. **30**(3), 303–308 (2004)
31. Gotoda, T., Jung, H.Y.: Endoscopic resection (endoscopic mucosal resection/ endoscopic submucosal dissection) for early gastric cancer. Dig. Endosc. **25**(1), 55–63 (2013)
32. Santoro, R.: Subtotal gastrectomy for gastric cancer. World J. Gastroenterol. **20**(38), 13667–13680 (2014)
33. Atala, A., Bauer, S.B., Soker, S., Yoo, J.J., Retik, A.B.: Tissue-engineered autologous bladders for patients needing cystoplasty. Lancet **367**(9518), 1241–1246 (2006)
34. Raya-Rivera, A., Esquiliano, D.R., Yoo, J.J., Lopez-Bayghen, E., Soker, S., Atala, A.: Tissue-engineered autologous urethras for patients who need reconstruction: an observational study. Lancet **377**(9772), 1175–1182 (2011)
35. Ott, H.C., Matthiesen, T.S., Goh, S.K., Black, L.D., Kren, S.M., Netoff, T.I., et al.: Perfusion-decellularized matrix: using nature's platform to engineer a bioartificial heart. Nat. Med. **14**(2), 213–221 (2008)
36. Baptista, P.M., Orlando, G., Mirmalek-Sani, S.H., Siddiqui, M., Atala, A., Soker, S.: Whole organ decellularization – a tool for bioscaffold fabrication and organ bioengineering. In: Conference Proceedings. Annual International Conference of the IEEE Engineering in Medicine and Biology Society. IEEE Engineering in Medicine and Biology Society. Annual Conference, pp. 6526–6529 (2009)
37. Uygun, B.E., Soto-Gutierrez, A., Yagi, H., Izamis, M.L., Guzzardi, M.A., Shulman, C., et al.: Organ reengineering through development of a transplantable recellularized liver graft using decellularized liver matrix. Nat. Med. **16**(7), 814–820 (2010)
38. Smith, L.A., Ma, P.X.: Nano-fibrous scaffolds for tissue engineering. Colloids Surf. B. Biointerfaces **39**(3), 125–131 (2004)
39. Yadid, M., Feiner, R., Dvir, T.: Gold nanoparticle-integrated scaffolds for tissue engineering and regenerative medicine. Nano. Lett. **19**(4), 2198–2206 (2019)
40. Li, W.J., Laurencin, C.T., Caterson, E.J., Tuan, R.S., Ko, F.K.: Electrospun nanofibrous structure: a novel scaffold for tissue engineering. J. Biomed. Mater. Res. **60**(4), 613–621 (2002)
41. Wang, Z., Hu, J., Yu, J., Chen, D.: Preparation and characterization of nano-laponite/PLGA composite scaffolds for urethra tissue engineering. Mol. Biotechnol. **62**(3), 192–199 (2020)
42. Holzapfel, B.M., Rudert, M., Hutmacher, D.W.: Gerüstträgerbasiertes knochen-tissue engineering. Orthopäde **46**, 701–710 (2017)
43. Wu, S., Duan, B., Qin, X., Butcher, J.T.: Living nano-micro fibrous woven fabric/hydrogel composite scaffolds for heart valve engineering. Acta Biomater. **51**, 89–100 (2017)
44. Yu, Q., Han, Y., Tian, T., Zhou, Q., Yi, Z., Chang, J., et al.: Chinese sesame stick-inspired nanofibrous scaffolds for tumor therapy and skin tissue reconstruction. Biomaterials **194**, 25–35 (2019)
45. Hasan, A., Waibhaw, G., Saxena, V., Pandey, L.M.: Nano-biocomposite scaffolds of chitosan, carboxymethyl cellulose and silver nanoparticle modified cellulose nanowhiskers for bone tissue engineering applications. Int. J. Biol. Macromol. **111**, 923–934 (2018)
46. Singh, D., Singh, D., Zo, S., Han, S.S.: Nano-biomimetics for nano/micro tissue regeneration. J. Biomed. Nanotechnol. **10**(10), 3141–3161 (2014)

47. Abhilash, M., Thomas, D.: Biopolymers for biocomposites and chemical sensor applications. In: Sadasivuni, K.K., Cabibihan, J.J., Ponnamma, D., Al-Maadeed, M.A.S.A., Kim, J., (eds.) Biopolymer composites in electronics. Amsterdam: Elsevier, pp. 405–435 (2017)
48. Vieira, W.A., Oberholzer, H.M., Pretorius, E.: The morphological effects of asthma, as well as conventional and alternative asthma therapies on parietal and chief cells in the stomach of BALB/c mice. Int. J. Morphol. **29**(4), 1341–1350 (2011)
49. Qin, J., Pei, X.: Isolation of human gastric epithelial cells from gastric surgical tissue and gastric biopsies for primary culture. Methods Mol. Biol. **1817**, 115–121 (2018)
50. Schlaermann, P., Toelle, B., Berger, H., Schmidt, S.C., Glanemann, M., Ordemann, J., et al.: A novel human gastric primary cell culture system for modelling Helicobacter pylori infection in vitro. Gut **65**(2), 202–213 (2016)
51. Greish, Y., Pulikkot, S., Mourad, A.H., Karam, S.: Biodegradable scaffolds for gastric tissue regeneration. In: Baino, F., (ed.) Scaffolds in tissue engineering – materials, technologies and clinical applications. London: IntechOpen, pp. 253–266 (2017)
52. Arkudas, A., Beier, J.P., Boccaccini, A.R., Horch, R.E.: Characterisation of vascularisation of scaffolds for tissue engineering. Mater. Sci. Tech. **31**(2), 180–187 (2015)
53. Matsui, M., Tabata, Y.: Enhanced angiogenesis by multiple release of platelet-rich plasma contents and basic fibroblast growth factor from gelatin hydrogels. Acta Biomater. **8**(5), 1792–1801 (2012)
54. Wilcke, I., Lohmeyer, J.A., Liu, S., Condurache, A., Krüger, S., Mailänder, P., et al.: VEGF(165) and bFGF protein-based therapy in a slow release system to improve angiogenesis in a bioartificial dermal substitute in vitro and in vivo. Langenbecks Arch. Surg. **392**(3), 305–314 (2007)
55. Mol, A., van Lieshout, M.I., Dam-de Veen, C.G., Neuenschwander, S., Hoerstrup, S.P., Baaijens, F.P., et al.: Fibrin as a cell carrier in cardiovascular tissue engineering applications. Biomaterials **26**(16), 3113–3121 (2005)
56. Mosesson, M.W.: Fibrinogen and fibrin structure and functions. J. Thromb. Haemost. **3**(8), 1894–1904 (2005)

Artificial Intelligence in Nanotechnology: Recent Trends, Challenges and Future Perspectives

Faris Hrvat[2], Amina Aleta[2], Amra Džuho[1](✉), Osman Hasanić[1], and Lemana Spahić Bećirović[1]

[1] Faculty of Engineering and Natural Sciences, Department of Genetics and Bioengineering, International Burch University, Sarajevo, Bosnia and Herzegovina
amra.dzuho@stu.ibu.edu.ba

[2] Faculty of Medicine, Medical Biotechnology and Molecular Medicine, University of Milan, Milan, Italy

Abstract. Nanoscience and nanotechnology are two overlapping areas of human activity, that take into consideration characteristics and utilization of nanoscale materials, and thus have applications in majority of science fields. Nanoscience and nanotechnology can be used in conjunction with Artificial Intelligence, since they are present in various fields ranging from medical diagnostics to robotics. Advancements in precision of medicine were highly influenced by nanomaterials. This paper is a review of artificial intelligence in nanotechnology, and was done through a literature review of 10 papers. According to the available literature on different platforms, it can be concluded that the concern for artificial intelligence in nanotechnology has been highly increased in the past decade. The paper presents how this combination of sophisticated techniques can be used for different medical purposes, and with different methodologies of both, artificial intelligence and machine learning.

Keywords: Artificial intelligence · Nanotechnology · Nanoscience · Nanomaterials · Machine learning · Medicine

1 Introduction

Even long before popularization of nanotechnology, people came across different nano-sized objects and the related processes of nano level and used them in practice. However, intuitive artifacts in nanotechnology developed spontaneously, without proper understanding of the nature of those objects and processes, people were subconsciously engaged in nanotechnology, without guessing that they had to deal with the nanoworld phenomena. Richard P. Feynman's 1959 discussion with the American Physical Society, "There's plenty of space at the edge," was the important event for nanotechnology. [1] This discussion led to other important events have put nanotechnology into the spotlight, introducing the development of scanning tunnel and atomic force microscopes and other groundbreaking discoveries [2, 3]. Like it is well known from history as can be Mendel called father of genetics, Feynman can be called father of nanotechnology. A number of

A. Badnjevic and L. Gurbeta Pokvić (Eds.): CMBEBIH 2021, IFMBE Proceedings 84, pp. 690–702, 2021.
https://doi.org/10.1007/978-3-030-73909-6_79

influential discoveries and inventions were made during the 1980s to the early 1990s, which created a fundamental impact on the farther development of nanotechnology. Since then, there has been a considerable intensification of nanotechnological research and design, the number of nanotechnology publications subjects are growing rapidly, practical application of nanotechnology is expanding, also project funding in nanotechnology is rising considerably, as is the number of organizations and countries are getting involved [4, 5].

Nanoscience and nanotechnology are two areas of human activity which are interwoven. The best description of nanoscience and nanotechnology is that they both deal with the characteristics and applications of extremely small materials and therefore, have utilization in almost all other scientific fields. Nano-dimensional material has different properties, sometimes quite the opposite, from the micro- or macro-dimensional properties of that same material. Nanoscience is therefore a science in which small-scale particles exhibit new physical properties, quantum effects, which depend on the particle size and are entirely different from matter in macro-dimension properties [6]. As it is known, nanoscience and nanotechnology can be very important in different fields, such as Artificial Intelligence (AI) field described in this paper.

Artificial intelligence has found its place in various areas of human action, such as medicine. [7–10] Application of nanotechnology in AI can be recognized in fields varying from medical diagnostic and patient care to robotics [11]. Integrating the link among current nano-sciences and AI can enhance research in these scientific fields which will have a major impact in our society. Whereas, numerous efforts have been making use of AI technologies in basic and applied nano science research, for example interpreting experimental techniques or assisting in the design of nano materials and instruments. Integrating the link among current nano-sciences and AI can enhance research in these scientific fields as well as provide a new generation of communication and information technologies which will have a major impact in our society. Whereas, numerous efforts have been making use of AI technologies in basic and applied nano science research, for example in interpretation of experiments or assisting design of nano materials and instruments. Current studies deal with the ways how computational power gain, resulting from nano-scientific advancements, will benefit AI in the near future. [12]. Machine learning methodologies are widely used for interpretation of demanding datasets and have proven highly efficient in treating many interacting variables in parallel. Machine learning methodologies are those of neural artificial networks (ANNs). Other biologically inspired AI methods use genetic algorithms and evolutionary algorithms for search and optimization related problems. [13]. AI and overall computer science will be able to evolve due to the increased computational power all aspects of nanotechnology offer in the field of implementation of learning methodologies. [14] The next sections discuss connection between AI and nanotechnology by using important features covering various uses and applications in medicine.

Nanomaterials have contributed, in all the medical stages, to the advancement of precision medicine [15]. New technologies for collecting omics data, like single molecule sequencing, allow for preservation of genetic context by increasing read lengths. Understanding of cancer biology is enabled by means of nanomedicine manufacturing techniques, that enable rationalization of targeted therapy design approaches using endogenous and exogenous stimuli for enhanced drug delivery. The advances contributed to theragnostic nanomedicine advancement, striving to combine a imaging agents and drugs along with patient treatment efficacy analysis [16].

Investing in nanotechnology has also been followed by criticism of its possible adverse effects on humans and the environment as all nanostructures behave differently from micro- and macro-structures in the environment, so it is difficult to determine their negative impact [17, 18]. The assets of applying nanotechnology and nanomaterials in different fields and industries mean however that many companies continue to invest in applications related to nanotechnology. As data gathered by Lux organization shows only in 2012 in whole world 18.5 billion was funded for nanotechnology researches and products [19]. All of these progression in nanotechnology conceptually based on Richard P. Feynman discussion which also led to discoveries and other events such as nanolithography, nanoscale chip computers, punctuate monitoring of atoms and further advancements [20].

2 Materials and Methods

This study was conducted through a literature review done in the period of April to June of 2020. The research was conducted using Google Scholar for screening and PubMed library with search terms "Artificial Intelligence and Nanotechnology in medicine"; "Artificial Intelligence and Nanomaterials in medicine". Selection was limited to English and Bosnian language literature, and the period of last 5 years. A further, in depth screening was done by independent reading of the paper abstract and title, where inclusion criteria's for choosing articles were: article was published in English or Bosnian language, article full text is publicly available for free, and ANN accuracy is provided in results. A total of 65 publications were excluded from the review because they, either lacked relevance to the investigated topic, or the approach of the paper was irrelevant for this review. The remaining 10 papers which were presented through Tables 1 and 2 in results section were included.

3 Results and Discussion

The articles that fulfilled inclusion criteria were grouped based on content in (1) Artificial Intelligence and nanotechnology in medicine, and (2) Artificial Intelligence and nanomaterials in medicine. The main findings of different papers are presented in a summarized form. Main characteristics of papers (topic of study, nanotechnology used and nanomaterial investigated, target molecule, AI methodology, and accuracy of prediction) are summarized in Tables 1 and 2.

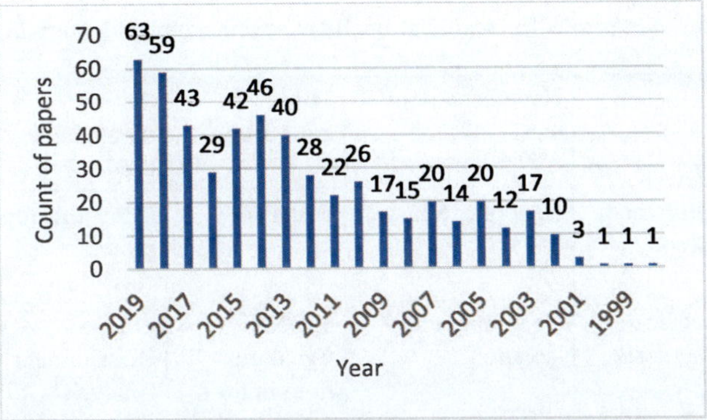

Fig. 1. Number of papers published on topic Artificial Intelligence in nanotechnology by PubMed database.

Figure 1 presents a graphical chart with number of papers published on topic "Artificial Intelligence in nanotechnology", from this figure we can conclude that interest in interconnection of these two fields has risen over the years. Data was obtained from PubMed search tool.

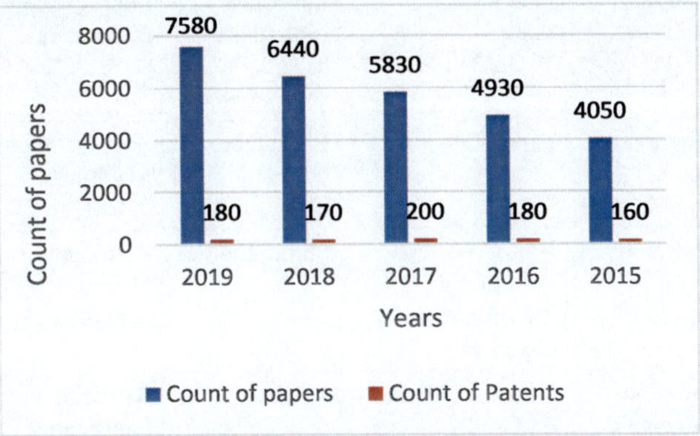

Fig. 2. Number of papers and patents published on topic Artificial Intelligence in nanotechnology by Google Scholar database. Figure 2 is a graphical chart with number of papers and patents published on topic "Artificial Intelligence in nanotechnology" in last five years, from figure above we can conclude that interest in interconnection of these two fields raised over 5 years. This data was obtained from Google Scholar database.

3.1 Artificial Intelligence and Nanotechnology in Medicine

The search using PubMed database yielded a total of 18 papers published on this topic since 2015. After reading the papers, 7 of them fulfilled the inclusion criteria and were considered relevant to this review.

Table 1. Relevant papers for Artificial Intelligence and nanotechnology in medicine.

Ref.	Topic	Nanotechnology used	Target molecule	AI methodology used	Accuracy
[21]	Whole-proteome identification	Nanopore Sensing	Proteins	Deep learning protein classifier	97%
[22]	Traumatic Brain Injury diagnosis	Nanomagnetic isolation	miRNA signatures from brain derived extracellular vesicles	Linear Discriminant Analysis (LDA)	AUC = 0.9
[23]	Computational sensing	Plasmonic nanosensors	Any molecule detectable by optical sensors	Feature selection using LASSO algorithm	Not discussed
[24]	Diagnosis and classification of 17 diseases	Nanoarray of gold nanoparticles	Exhaled molecules (volatile-organic compounds)	Artificial intelligence Nanoarray	86%
[25]	Determination of indication of estrogen induction	nanoLCms/ms	Plasma proteins	Support Vector Machine (SVM)	100%
[26]	Optimization of sizes of nanoparticles	Nanoprecipit tion in microfluidic channels	Stable Iodine 127I	Artificial Neural networks (ANN)	0.92
[27]	Differentiation between genetic mutations in breast cancer	Nanoarray based on Gold nanoparticles or Single Wall Carbon Nanotubes	Volatile organic compounds	Artificial intelligence Nanoarray	90.9%

Two intertwining fields that are important in achieving the goal of new treatments and customization of each patient's best treatment are artificial intelligence and nanotechnology. There are several paths of interconnection of AI and nanotechnology in medicine, and ones that were applied in the included papers will be covered in this section. The properties of nanoscale objects and materials, and those of the same chemical materials but produced with greater dimensions vary greatly. Also, another important concept of combination of nanotechnology and AI are nanosensors, based and developed with AI. The nanosensors intrinsic spectral, mechanical, and electrochemical properties allow an improved signal-to- noise performance ratio, making them desirable for sensing biomarkers of cancer [28]. Efficient nanosensors must be able to distinguish remarkably low concentrations of the molecular signature in an environment of many more interfering biological segments. Computational analysis plays a major role in making the change from rationally engineered sensors that track simple bio-markers to pattern-based nanosensors that require sophisticated methods for clustering and classifying large datasets. Main component analysis is a widely used technique for analyzing electronic nose data using linear combinations of input data features to generate a smaller number of new variables showing high variance. Instead, these are used to classify the data into different classes. Collision between fields of nanotechnology and AI enables better acquisition of patient data for targeted therapy. As it is known, every patient is different, not just by their apparent differences but also in their molecular profile and because of this each patient needs to be treated different. Differences within patients are especially evident in various forms of cancer, which are affected by aggregation of driver mutations, that lead to intra-tumor and inter-patient heterogeneities, making diagnosis and treatment more complex. AI algorithms' ability to process large datasets and identify complex patterns may be used for improving diagnosis and treatment outline [20, 24].

Nanopore sensing combined with a custom-made deep learning protein classifier has yielded accuracy of 97% when used for whole proteome identification, rendering combination of nano-based sequencing combination with AI a promising candidate for future development [21].

NanoLC ms/ms methodology utilizes Nanoscale liquid chromatography coupled to tandem mass spectrometry as a robust, highly sensitive system for protein analysis. A research performed by Schilling et al., combined this approach with a machine learning methodology of a support vector machine and yielded an accuracy of 100% in determination of sex-specific 17 β-estradiol-responsive expression patterns in white perch plasma proteins Eltd1 and Kng1 and made this E2 responsive proteins promising indicators of induction of estrogen [25].

Utilization of nanoprecipitation for nanosuspension of stable Iodine isotope ((127)I) is a common process done in microfluidic devices. As size optimization of these particles is highly important, ANN modelling was used for this specific purpose. This methodology provided insights in the relationship of input parameters, such as solvent and antisolvent flow rate, solvent temperature and surfactant concentration, to the output variable of nanoparticle size. The results indicated that antisolvent flow rate is inversely proportional to the size of nanoparticles and that low temperatures are inversely proportional to size of stable ((127)I) nanoparticles, whereas high temperatures are directly proportional. Also, surfactant concentration was impacted by solvent temperature, thereby the temperature had a huge effect on particle size [26].

The interaction of inherited and environmental risk factors for developing the continuous aggregation of both, genetic and epigenetic changes in breast cells, promotes breast cancer (BC). Gene signatures that are used as predictive biomarkers associated with targeted BC treatment involve hormone receptors, estrogen receptors (ER) or progesterone receptors (PgR), and epidermal growth factor receptors ErbB-1 (EGFR) and ErbB-2 (ERBB2, HER-2/ neu, HER2). By analyzing volatile organic compounds (VOCs), a promising approach to reliable, time-efficient and cost-effective molecular subtype classification of BC tumors is built on volatolomic studies. This offers a means of distinguishing among diverse molecular subtypes of BC which could be notably useful for patient risk evaluation, as well as identifying an effective targeted therapy for BC patients, especially in advanced disease where metastatic tumor biopsy is challenging. On the basis of Gold Nanoparticles (GNPs) and Single Wall Carbon Nanotubes (SWC-NTs) encapsulated with different organic layers, volatolomic signatures of breath VOCs based on genetic expression subtypes were determined. Analyzed the collective response from the array utilizing DFA to obtain specific volatolomic signatures of each subtype of BC. This research also reported accuracy of 90.9% by using AI for determination and prediction of subtypes mentioned. Also results of this article show that when several artificially intelligent nanoarray signals are combined into a single Discriminant function analysis (DFA) value, instead of a signal compound analysis or DFA analysis of the gas chromatography mass spectrometry (GC-MS) performance are more precise [27].

Another research was also based on volatolomic signatures of breath samples obtained from patients, with 17 different disease types in order to utensil machine learning system for detection and classification of patient diseases. In most cases, the binary classifiers of the breath samples obtained from the different disease states achieved very high accuracy, assisted by two separate validation methods, the cross-validation leaveone-out procedure, and the blind analysis of 14 samples for each classification. In through comparison, a minimum number of sensors were also used which reduced the risk of overfitting the tests. In contrast to the high accuracy attained among diseases, by research report 86%, the control sample classification excluded the possibility of chance and external bias. Of particular interest, findings from the artificially intelligent nanoarrays support the hypothesis represented in the research, that differences in pathophysiological processes are expressed in very similar patterns of breath. The fact that the artificially intelligent nanoarray does not rely on selective or a lock-and-key approach in detecting specific VOCs in the biomarkers could explain these observations.

The nanoarray is actually based on sensors that have different chemistries [29].

In field of medicine for decades, finding practical solutions for the detection of proteins, cancer biomarkers, viruses, and other small objects has been a major challenge for researchers worldwide. Mobile plasmonic readers can be a very breakthrough finding in this field, and intervene with AI can be even much more. Ballard, Z. S. et al. proposed computational sensing system that the plasmonic sensing community may typically use to design optimized Localized Surface Plasmon Resonance (LSPR) readout devices aimed at cost-effectiveness, mobility and robustness. This approach can be applied broadly to any LSPR sensor geometry, and can meet any set of effective design constraints. Beyond the scope of LSPR sensors, any type of biochemical sensor that operates via a spectrum change or shift can also use this computational sensing and design framework. The machine learning methodology in behind this research selects the optimum sub-set of LEDs from a predetermined library, and aims to generate the minimal model of refractive index prediction for a given plasmonic nano-structure design and nano-manufacturing process. This approach to computational sensing is especially useful in designing multi-spectral readers where the selection of optimum lowcost bands of illumination is crucial. This system is compatible with any arbitrary geometry of plasmonic sensors, and any lighting library based on user-defined design constraints. [30]

In the field of injuries medicine is limited by absence of accessible molecular biomarkers which represent this heterogeneous disease's pathophysiology, one of the injuries like that is brain injury. Approach of one of the included articles was based on multiple extracellular vesicle microRNA (EV miRNA) biomarkers which provide a more holistic image of the damaged and healing brain than one single biomarker is possible. Sometimes traditional approaches based on a single molecular biomarker are not precise enough and only offer a broad view of the healing brain. Since EVs package include plentiful biomarkers from the damaged and recovering brain, a panel of molecular biomarkers will be assessed to more accurately explain the complex states of the recovering brain. Several groups have evaluated several miRNAs or protein markers, and have united these different measures to diagnose traumatic brain injury (TBI) algorithmically. They have used quantitative polymerase chain reaction (qPCR) to measure the cycle threshold (Ct) values of EV miRNA biomarkers in each individual subject within a training cohort wT to develop Linear discriminant analysis (LDA)-based classification, for which the true state of each subject is known a priori. As an indication of the importance of this strategy, they assessed a training set consisting of a cohort of mice suffering a blast injury and a cohort of mice which is uninjured [22].

3.2 AI and Nanomaterials in Medicine

The search using PubMed database yielded a total of 17 papers published on this topic since 2015. After reading the papers, 3 of them were considered relevant to this review.

Table 2. Relevant papers for Artificial Intelligence and nanomaterials in medicine.

Ref.	Topic	Nanotechnology used	Target molecule
[31]	Detection of DNA base modifications	Oxford Nanopore Sequencing	DNA mutations
[23]	Base-calling	Oxford Nanopore Sequencing	DNA
[33]	Classification of Prostate cancer, Benign prostate hyperplasia and healthy subjects	Surface-enhanced Raman scattering using Silver nanoparticles as immunoprobes	PSA, PSMA, human kallikrein 2

Pattern recognition and classification algorithms can be used to differentiate between stable and diseased patients and to predict drug effectiveness in patients. Due to its high complexity, these analytical capabilities are especially important in the case of cancer. As well as this is one of the important facts in creation of nanomaterials for different types of sequencing or assays used in diseases detection. One of the breakthroughs in medicine and nanotechnology were nanoscale polymer capsules that can be designed to break down and release drugs at controlled rates, allowing differing releases in certain conditions, such as acid media, to promote tumor uptake versus normal tissue uptake. Some other computational methods assumed the optimal shape and size for binding nanoparticles and releasing drugs. For instance, the size and initial burst rates of polylactic glycolic acid (PLGA) nanoparticles were predicted using an ANN [24, 35].

Immunoassay has been the dominant tumor marker research tool in body fluids such as human serum, largely due to the highly precise molecular identification of an antigen's antibodies and epitopes. Radioimmunoassay, enzyme-linked immunosorbent assay (ELISA), fluoroimmune assay, and chemiluminescent immunoassay (CLIA) are among the most common forms of immunoassays. Such detection techniques have many benefits and thus are important in the clinical quantitative detection of tumor markers; nevertheless, they also suffer from some shortcomings, especially during the early cancer screening. Therefore, to satisfy the requirement of a high sensitivity, highthroughput, multi-analyte, and non-polluting tumor marker immunoassay, new immunoassay techniques are required. Surface enhanced Raman Scattering (SERS) spectroscopy displays excellent features with high sensitivity, a unique fingerprint and nondestructive data processing. One of the researches chosen for this review reports SERSbased immunoassay platform made of silver nanoparticles as immune samples, and SiC@Ag@Ag-NPs as a SERS immune substrate to detect various tumor markers. As well as they have implemented support vector machine (SVM) algorithms which is doing classification of healthy subjects, benign prostate hyperplasia (BPH) and patients with prostate cancer (PCa) based on immunoassay data. The designed immunoassay platform has outstanding detection performance of prostate-specific antigen (PSA), prostate-specific membrane antigen (PSMA), and human kallikrein 2 (hK2), such as low detection limitations (LODs) and large dynamic detection ranges. Additionally, in mentioned control experiments, the specificity of the immunoassay was validated with the usage of non-specific antigens and female serum samples. Results indicate that the use of various tumor markers (PSA, PSMA, and hK2) may enhance the diagnostic accuracy of PCa from 50% to 70% as reported, on the basis of the linear SVM classification model, as well as advanced classification for BPH and healthy subjects when compared to the results of using PSA alone [33].

As mentioned before, importance of nanotechnology in sequencing data is exceptional. One the articles included in review reports performance on Oxford Nanopore Technologies (ONT) in connection with neural network tools. ONT long-read sequencing is based on the following concept: transfer a single strand of DNA across a nanopore membrane and apply a difference in voltage across the membrane. The nucleotides present in the pore can influence the electrical resistance of the pore so current measurements will show the sequence of DNA bases passing through the pore over time. This signal of electric current is the raw data collected by an ONT sequencer. The process of transforming this raw signal into a DNA sequence is the basecalling for ONT devices. It's not a trivial problem because the electrical signals come from single molecules, which makes noisy and stochastic data. In addition to that, the electrical resistance of a pore is determined by the bases present in multiple nucleotides at the narrowest point of the pore, resulting in a wide number of possible states. This can be a very challenging fact for implementation machine learning inside these sequencers. Researchers reported that tests were done on ONT programs Albacore, Guppy, Scrappie and Flappie. Best results for the *K. Pneumoniae*, both in terms of read and consensus accuracy benchmarking package, was obtained using Guppy v2.2.3 with a custom model trained on data mainly from the same species. This exceptional performance is presumably from proper handling of Dcm methylation. Authors suggest that their results would be a more general trend than native DNA basecalling, since the patterns of modifying DNA can differ between taxa. Accuracy is the highest in cases when the model has been trained on native DNA from a certain species or is close enough to have similar DNA changes. Authors also reported that Sloika is ONT's neural network training toolkit that they were using doing development [23].

Another important approach of Artificial Intelligence in nanomaterial and medicine is DNA base modification. DNA base modifications are variants of standard nucleotides in a molecule of DNA by adding chemical groups to them. DNA-based modifications occur widely among different organisms and are involved in various biological processes such as genomic imprinting, inactivation of X-chromosome, genome stability, gene regulation, repression of transposable elements, aging, and carcinogenesis. DNA modifications, in particular DNA methylations, could be detected through both, short-read sequencing techniques and long-read sequencing. Concept studies of Oxford Nanopore sequencing techniques proposed the ability to detect DNA modifications, depending on the characteristics of the electrical signal when a transformed DNA molecule moves through nanopores. In one of the researches included in this review, authors have developed DeepMod to capture time-series characteristic of nanopore signals to detect DNA modification. It was firstly tested on *E. coli* organism and also on two other organisms, for evaluating performance and vaulability of developed deep neural network. DeepMod bridged the gap between both the rapid growth of Nanopore sequencing data and the growing need for genomic-scale detection of DNA modifications. DNA modifications are not compliant through the cycle of biological cells, and it may differ among types of cells. Along with Oxford Nanopore sequencing, DeepMod offers a valuable method to classify genomic-scale DNA modifications, with reported accuracy of 0.9–0.99%. However, if dissimilar aligners or improper variables are used with poor performance in alignment, the prediction of DeepMod will be affected as well [31].

4 Conclusion

As Faraday's colloidal "ruby" gold showed how gold nanoparticles can interact with light in different ways, an interest in nanotechnology spanned throughout the scientific community. But it wasn't until the discovery of nanomaterials, such as Buckminsterfullerene, 100 years after, that nanotechnology has started to make a huge impact in all fields of technology. If observed, from the medicinal point of view, the use of nanotechnology principles in microscopy has immensely impacted medicine. The 21st century marked a significant turning point for nanotechnology, as more investments were made here. First experimentations with AI in nanotechnology started in the 1960s but it was the MYCIN expert system from 1975 that proved the efficiency of expert systems in diagnosis. Henceforth, we can conclude that AI and nanotechnology gained on importance in medicine in a similar timeframe. Using these novel methodologies in conjunction is the goal of interdisciplinary research. This review covered the last five years of using AI in nanotechnology. A variety of nanotechnology and nanomaterials were used for different medicinal purposes, and combined with various AI/ML methodologies, ranging from simple algorithms to those of deep learning. What is evident, is that the topics that are trying to be solved by combining AI and nanotechnology are those of high importance in medicine, that have intrigued science for a long time. Delivery of personalized therapy using AI driven nanoparticles, identifications of proteome landscape, and AI driven, nanotechnology platform enabled genetics research are all of high importance. Breakthroughs in these fields could change the way in which medicine is practiced, and hinder old therapeutics, changing them with novel, sophisticated methods that will enhance the quality of life.

References

1. Drexler, K.E.: Nanotechnology: from Feynman to funding. Bull. Sci. Technol. Soc. **24**(1), 21–27 (2004). https://doi.org/10.1177/0270467604263113
2. Eigler, D.M., Schweizer, E.K.: Positioning single atoms with a scanning tunnelling microscope. Nature **344**(6266), 524–526 (1990). https://doi.org/10.1038/344524a0
3. Eigler, D.M., Weiss, P.S., Schweizer, E.K., Lang, N.D.: Imaging Xe with a lowtemperature scanning tunneling microscope. Phys. Rev. Lett. **66**(9), 1189 (1991). https://doi.org/10.1103/PhysRevLett.66.1189
4. Peterson, C.L.: Nanotechnology: from Feynman to the grand challenge of molecular manufacturing. IEEE Technol. Soc. Mag. **23**(4), 9–15 (2004). https://doi.org/10.1109/MTAS.2004.1371633
5. Tolochko, N.K.: History of nanotechnology. Encyclo. Life Support Syst. (EOLSS) (2009)
6. Miklicanin Omanovic, E., Badnjevic, A.: Uvod u nanotehnologiju (1st ed., pp. 7–34). Sarajevo: Univerzitet u Sarajevu, Poljoprivredno-prehrambeni fakultet (2017)
7. Catic, A., Gurbeta, L., Kurtovic-Kozaric, A., Mehmedbasic, S., Badnjevic, A.: Application of Neural Networks for classification of Patau, Edwards, down, turner and klinefelter syndrome based on first trimester maternal serum screening data, ultrasonographic findings and patient demographics. BMC Med. Genomics, **11**(19) (2018). https://doi.org/10.1186/s12920-018
8. Abdel-Ilah, L., Zukic, S., Veljovic, E., Badnjevic, A., Gurbeta, L.: Quantitative structure activity relationship and artificial neural networks in design of benzimidazoles as antiproliferative agents. Res. J. Pharm. Biol. Chem. Sci. (Res. J. Pharm., Biol. Chem. Sci.) (2019)
9. Kovačević, Z., Pokvić, L.G., Spahić, L., Badnjević, A.: Prediction of medical device performance using machine learning techniques: infant incubator case study. Health Technol. **10**, 151–155 (2020). https://doi.org/10.1007/s12553-019-00386-5
10. Spahić, L., Šehović, E., Šećerović, A., Đozić, Z., Smajlović-Skenderagić, L.: Lactose intolerance prediction using artificial neural networks. In: International Conference on Medical and Biological Engineering, pp. 505–510). Springer, Cham (2019). https://doi.org/10.1007/978-3-030-17971-7_75
11. Mulhall, D.: Our Molecular Future: How Nanotechnology, Robotics, Genetics and Artificial Intelligence Will Transform Our World. Prometheus Books, Buffalo (2010)
12. Ahmad, J., Akhter, S., Rizwanullah, M., Ahmed Khan, M., Pigeon, L.T., Addo, R., Amjad Kamal, M.: Nanotechnology based theranostic approaches in Alzheimer's disease management: current status and future perspective. Curr. Alzheimer Res. **14**(11), 1164–1181 (2017). https://doi.org/10.2174/1567205014666170508121031
13. McLarty, J., Wu, C., Wu, C.H., McLarty, J.W.: Neural networks and genome informatics (Methods in computational biology and biochemistry; v. 1) (1st ed.). Elsevier (2000)
14. Bishop, C.M.: Pattern Recognition and Machine Learning (Singapore: Springer) (2006)
15. Wang, Y., Sun, S., Zhang, Z., Shi, D.: Nanomaterials for cancer precision medicine. Adv. Mater. **30**(17), 1705660 (2018). https://doi.org/10.1002/adma.201705660

16. Lammers, T., Aime, S., Hennink, W.E., Storm, G., Kiessling, F.: Theranostic nanomedicine. Acc. Chem. Res. **44**(10), 1029–1038 (2011). https://doi.org/10.1021/ar200019c
17. Fischer, H.C., Chan, W.C.: Nanotoxicity: the growing need for in vivo study. Curr. Opin. Biotechnol. **18**(6), 565–571 (2007). https://doi.org/10.1016/j.copbio.2007.11.008
18. Fu, P.P., Xia, Q., Hwang, H.M., Ray, P.C., Yu, H.: Mechanisms of nanotoxicity: generation of reactive oxygen species. J. Food Drug Anal. **22**(1), 64–75 (2014)
19. Jia, L.: Global governmental investment in nanotechnologies. Curr. Nanosci. **1**(3), 263–266 (2005). https://doi.org/10.2174/157341305774642957
20. Thrall, J.H.: Nanotechnology and medicine. Radiology **230**(2), 315–318 (2004). https://doi.org/10.1148/radiol.2302031698
21. Ohayon, S., Girsault, A., Nasser, M., Shen-Orr, S., Meller, A.: Simulation of single-protein nanopore sensing shows feasibility for whole-proteome identification. PLoS Comput. Biol. **15**(5), (2019). https://doi.org/10.1371/journal.pcbi.1007067
22. Ko, J., Hemphill, M., Yang, Z., Sewell, E., Na, Y.J., Sandsmark, D.K., Haber, M., Fisher, S.A., Torre, E.A., Svane, K.C., Omelchenko, A., Firestein, B.L., Diaz-Arrastia, R., Kim, J., Meaney, D.F., Issadore, D.: Diagnosis of traumatic brain injury using miRNA signatures in nanomagnetically isolated brain-derived extracellular vesicles. Lab Chip, **18**(23), 3617–3630 (2018). https://doi.org/10.1039/c8lc00672e
23. Wick, R.R., Judd, L.M., Holt, K.E.: Performance of neural network basecalling tools for Oxford Nanopore sequencing. Genome Biol. **20**(1), 129 (2019). https://doi.org/10.1186/s13059-019-1727-y
24. Yamankurt, G., Berns, E.J., Xue, A., Lee, A., Bagheri, N., Mrksich, M., Mirkin, C.A.: Exploration of the nanomedicine-design space with high-throughput screening and machine learning. Nat. Biomed. Eng. **3**(4), 318–327 (2019). https://doi.org/10.1038/s41551-019-0351-1)
25. Schilling, J., Nepomuceno, A.I., Planchart, A., Yoder, J.A., Kelly, R.M., Muddiman, D.C., Reading, B.J.: Machine learning reveals sex-specific 17β-estradiol-responsive expression patterns in white perch (Morone Americana) plasma proteins. Proteomics, **15**(15), 2678–2690 (2015). https://doi.org/10.1002/pmic.201400606
26. Aghajani, M.H., Pashazadeh, A.M., Mostafavi, S.H., Abbasi, S., Hajibagheri-Fard, M.J., Assadi, M., Aghajani, M.: Size control in the nanoprecipitation process of stable iodine (127 I) using microchannel reactor—optimization by artificial neural networks. Aaps Pharmscitech, **16**(5), 1059–1068 (2015). https://doi.org/10.1208/s12249-015-0293-1
27. Barash, O., Zhang, W., Halpern, J.M., Hua, Q.L., Pan, Y.Y., Kayal, H., Khoury, K., Liu, H., Davies, M.P., Haick, H.: Differentiation between genetic mutations of breast cancer by breath volatolomics. Oncotarget, **6**(42), 44864–44876 (2015). https://doi.org/10.18632/oncotarget.6269
28. Kosaka, P.M., Pini, V., Ruz, J.J., Da Silva, R.A., González, M.U., Ramos, D., Tamayo, J.: Detection of cancer biomarkers in serum using a hybrid mechanical and optoplasmonic nanosensor. Nat. Nanotechnol. **9**(12), 1047 (2014). https://doi.org/10.1038/nnano.2014.250
29. Nakhleh, M.K., Amal, H., Jeries, R., Broza, Y.Y., Aboud, M., Gharra, A., Ivgi, H., Khatib, S., Badarneh, S., Har-Shai, L., Glass-Marmor, L., Lejbkowicz, I., Miller, A., Badarny, S., Winer, R., Finberg, J., Cohen-Kaminsky, S., Perros, F., Montani, D., Girerd, B., Haick, H.: Diagnosis and classification of 17 diseases from 1404 subjects via pattern analysis of exhaled molecules. ACS Nano, **11**(1), 112–125 (2017). https://doi.org/10.1021/acsnano.6b04930
30. Ballard, Z.S., Shir, D., Bhardwaj, A., Bazargan, S., Sathianathan, S., Ozcan, A.: Computational sensing using low-cost and mobile plasmonic readers designed by machine learning. ACS Nano **11**(2), 2266–2274 (2017). https://doi.org/10.1021/acsnano.7b00105
31. Liu, Q., Fang, L., Yu, G., Wang, D., Xiao, C.L., Wang, K.: Detection of DNA base modifications by deep recurrent neural network on oxford nanopore sequencing data. Nat. Commun. **10**(1), 2449 (2019). https://doi.org/10.1038/s41467-019-10168-2

32. Zhou, L., Liu, Y., Wang, F., Jia, Z., Zhou, J., Jiang, T., et al.: Classification analyses for prostate cancer, benign prostate hyperplasia and healthy subjects by SERS-based immunoassay of multiple tumour markers. Talanta **188**, 238–244 (2018). https://doi.org/10.1016/j.talanta.2018.05.070
33. Baghaei, B., Saeb, M.R., Jafari, S.H., Khonakdar, H.A., Rezaee, B., Goodarzi, V., Mohammadi, Y.: Modeling and closed-loop control of particle size and initial burst of PLGA biodegradable nanoparticles for targeted drug delivery. J. Appl. Polym. Sci. **134**, 45145 (2017). https://doi.org/10.1002/app.45145

Tolerance Assays Performed in Animal Models During the Evaluation of Nanoparticles for Ocular Drug Delivery

Naida Omerović[1(✉)], Selma Škrbo[1], and Edina Vranić[2]

[1] Department of Clinical Pharmacy, Faculty of Pharmacy, University of Sarajevo, Zmaja od Bosne 8, 71 000 Sarajevo, Bosnia and Herzegovina
naida.omerovic@ffsa.unsa.ba

[2] Department of Pharmaceutical Technology, Faculty of Pharmacy, University of Sarajevo, Zmaja od Bosne 8, 71 000 Sarajevo, Bosnia and Herzegovina

Abstract. Nanotechnology is one of the most advanced technologies regarding the development of formulations for ocular drug delivery. Nanoparticles have shown great efficacy in the treatment of different ocular diseases, but the question arises as to whether nanoparticles are tolerable to the ocular surface of the animal and human eye, because that request needs to be fulfilled during the evaluation of their properties. The aim of this research is to provide a short overview of the tolerance assays performed in animal models during the evaluation of nanoparticles for ocular drug delivery. The following databases were searched: PubMed, Web of Science, Scopus, Science Direct, EBSCO. Both abstracts and full texts were included in the search and there were no exclusion criteria regarding the year when a certain article was published. Results showed that rabbits, rats and guinea pigs are typically used animal models for the evaluation of ocular tolerance of nanosystems, because of the precision of the animal-to-human extrapolation. The Draize test is the most commonly used test when assessing ocular tolerance in animal models. Other methods to evaluate ocular tolerance are the grading system of the macroscopic signs for the colloidal systems tested, the Nelson's classification for squamous metaplasia or the clinical recovery regarding ocular irritation. *In vivo* studies conducted in animals have shown that nanoparticles usually cause no irritation or inflammation, because of their small size. Still, further research needs to be carried out to get firmer confirmation about the safety of these nanoformulations in humans.

Keywords: Nanoparticles · Ocular drug delivery · Tolerance assays · Animal models

1 Introduction

Ocular drug application can be achieved using different dosage forms, such as eye drops or ointments. These conventional dosage forms are safe and noninvasive but are inconsistent concerning bioavailability, solubility and pericorneal drug retention. In the last

A. Badnjevic and L. Gurbeta Pokvić (Eds.): CMBEBIH 2021, IFMBE Proceedings 84, pp. 703–717, 2021.
https://doi.org/10.1007/978-3-030-73909-6_80

few decades, nanoformulations have been intensively analysed regarding drug delivery to anterior and posterior parts of the eye. One of the most important steps in the evaluation of nanoparticles for ocular drug delivery is checking whether they cause irritation or inflammation because the tolerability of certain formulation impacts patients' adherence [1].

An animal model usually used in tolerance assays is rabbit. There are many benefits of using rabbits for this kind of analysis because they are very cheap and sensitive to irritating substances, have big eyes and appropriate anatomy and physiology for the animal-to-human extrapolation. Guinea pigs and rats are also commonly used [2].

Tolerability factors limit the use of many nanomaterials in ocular drug delivery [3]. For example, some polymers have both mucoadhesive and viscosity-modifying properties. If a drug solution becomes too viscous after the addition of a polymer, it will irritate the eye, consequently causing the tear production and increasing the drug clearance. The aim should be to enhance mucoadhesiveness without significantly changing the viscosity [4]. Chitosan (CS) is often used as a polymer in ophthalmic formulations. This is a cationic polysaccharide, so it reacts with the negatively charged mucus and conjunctiva of the eye. It is only soluble in acids, so it needs to be neutralized to be formulated into an ophthalmic formulation. Galactosylated chitosan (GC) and thiolated chitosan (TC), derivatives of CS, have better solubility properties for ocular drug delivery, so they are preferred in practice [5, 6]. Many polymers, such as poly(caprolactone) (PCL) and poly(lactic acid) (PLA), have also been thoroughly studied regarding the incorporation into nanotechnology-based ocular dosage forms [7]. Some of the surfactants used as stabilising agents caused ocular irritation when used above an optimal concentration. But, Kolliphor P188 showed no irritation signs even at the highest concentration, Tween 80 did not cause irritation signs up to the concentration of 0.05% and sodium dodecyl sulphate (SDS) caused severe inflammation. Therefore, once a nanosystem is prepared, the excess surfactant should be removed [8]. Timolol maleate-loaded polymeric nanoparticles containing flaxseed gum (FX) and CS were also evaluated. The pH value of the FX and CS solutions was 6.78 ± 0.02 and 6.42 ± 0.04, respectively. The near-neutral pH value implied that there would not be any irritation signs. Results of morphological studies showed that nanoparticles of size less than 100 nm and spherical or ovoid shape with no sharp edges did not cause any irritation signs in the cul-de-sac [9].

In this paper, we aim to provide a short overview of the methodology of conducting tolerance assays during the evaluation of nanoparticles for ocular drug delivery, with an emphasis on *in vivo* animal models.

2 Materials and Methods

The following databases were searched: PubMed, Web of Science, Scopus, Science Direct, EBSCO. Search terms included: nanoparticles, ocular drug delivery, tolerance assays, animal models. Various combinations of Boolean terms (AND/OR/NOT) were used to capture all relevant articles but also to make sure our search is limited only to the papers that help answer our research question. Both abstracts and full texts were included in the search. There were no exclusion criteria regarding the year when a certain article was published.

3 Results

In Table 1. we present results of notable studies conducted in animal models, which investigated ocular tolerance of certain nanosystems.

Table 1. Results of notable studies conducted in animal models, which investigated ocular tolerance of certain nanosystems

Study	Animal model	Nanosystem	Notes
[10]	New Zealand white rabbits	Pranoprofen-loaded nanoparticles with poly(lactic-co-glycolic acid) (PLGA) as a polymer and Carbomer 934 as a hydrogel	Compared with the pranoprofen solution and eye drops, nanoparticles were preferred because they showed no signs of ocular irritation
[11]	New Zealand white rabbits	Silver nanoparticles	Conjunctival redness, discharge, and oedema were monitored and no signs of irritation to the cornea, iris, or conjunctiva were found 24, 48, and 72 h post-application
[12]	Rabbits	Fullerene nanoparticles	No signs of corneal opacity, iris anomaly, or chemosis were found after 24 h, but conjunctival redness and blood vessel hyperemia were observed one hour post-application. Thus, long-term tolerance studies need to be carried out
[13]	Rabbits	Multi-walled carbon nanotubes (MWCNTs) of two different sizes	Reversible conjunctival redness and discharge were found, indicating low ocular toxicity of MWCNTs

(*continued*)

Table 1. (*continued*)

Study	Animal model	Nanosystem	Notes
[14]	Rabbits	MWCNTs and single-walled carbon nanotubes (SWCNTs)	Only one of the MWCNTs showed a very weak and reversible acute ocular irritation potential, causing conjunctival redness and blood vessel hyperemia one hour post-application
[15]	Rabbits Sprague Dawley rats	Graphene oxide nanoparticles	Graphene oxide did not cause acute ocular irritation in rabbits, but short-term repeated exposure of Sprague Dawley rats to this substance caused reversible ocular damage because of the oxidative stress occurrence
[16]	Rabbits	Indomethacin-PCL nanoparticles	Optimal ocular tolerance was revealed, especially in the cornea, conjunctiva, and iris, with a possibility of the conjunctival hyperemia occurring later. Cationic polymer-coated nanocapsules as ocular drug carriers should generally be considered safe [17]
[18]	Rabbits	Poly(methylmethacrylate-co-sulphopropylmethacrylate) (PMS) nanoparticles	When formulated with arecaidine propargyl ester and aceclidine, no signs of ocular irritation were shown
[19]	Rabbits	Carteolol-PCL nanocapsules	No signs of ocular irritation were shown

(*continued*)

Table 1. (*continued*)

Study	Animal model	Nanosystem	Notes
[20]	Rabbits	Acyclovir-PLA and poly(ethylcyanoacrylate) (PECA) nanospheres	No signs of discomfort or inflammation were noted. Insignificant conjunctival hyperemia was noted 10 min post-application, but no signs of inflammation were observed six hours post-application. Both PLA and PECA nanospheres were thought to be safe in ocular drug delivery
[21–24]	Rabbits	Eudragit-based nanoparticles	Nanoparticles did not cause any ocular irritation, indicating their safe use in ocular drug delivery
[25]	Wistar rats	Ganciclovir-bovine serum albumin (BSA) nanoparticles	Results showed that prolonged ocular presence of these nanoparticles was well-tolerated

The evaluation criteria of the ocular irritation according to the Draize test are divided into four categories presented in Table 2.

Table 2. The evaluation criteria of the ocular irritation according to the Draize test [26–28]

Score	Reaction
0–3.9	Nonirritant
4–8.9	Slightly irritant
9–12.9	Moderately irritant
13–16	Seriously irritant

In Table 3. we present results of notable studies in which the Draize test was used to evaluate ocular irritation of various nanosystems.

Table 3. Results of notable studies in which the Draize test was used to evaluate ocular irritation of various nanosystems

Study	Animal model	Nanosystem	Notes
[29]	Guinea pigs	Silver nanoparticles	An irritation score of 1 was noted for the conjunctiva after 24 h. It is also well-known that long-term exposure of eyes to silver could cause argyrosis [30, 31]
[32]	Rabbits	Ocular supersaturated self-nanoemulsifying drug delivery systems (SSNEDDS) of econazole nitrate with and without hydroxypropylmethylcellulose (HPMC) as a precipitation inhibitor	Rabbits were monitored for any ocular anomaly directly after the exposure and one hour post-application for three days. Both SSNEDDS-A and SSNEDDS-A2 produced a nonirritant effect to the eyes. They scored 0.111 ± 0.055 and 0.091 ± 0.087, respectively. Both of these formulations were well-tolerated and did not show any evidence of toxic or inflammatory effects on the ocular surface
[33]	New Zealand white rabbits	Naringenin-loaded sulphobutylether (SBE)-cyclodextrin (CD)-CS nanoparticles (Nag-SBE-CD-CS-NPs)	Both single-dose and long-term ocular irritation of Nag-SBE-CD-CS-NPs was studied. Before the test was conducted, both eyes were verified to be without inflammation. For the assessment of long-term ocular irritation, the right eye was treated with the formulation twice per day for seven days. This formulation was proven to be nonirritant to the eyes, because scores were 0 and 1.67 for single-dose and long-term ocular irritation, respectively. It can be concluded that Nag-SBE-CD-CS-NPs could be considered safe in ocular drug delivery
[34]	New Zealand white rabbits	Thermoresponsive gel with curcumin-loaded BSA nanoparticles (Cur-BSA-NPs)	Changes in the cornea, iris, and conjunctiva, secretion, and chemosis caused by this formulation were compared with the control. No lesion formation was observed during the study. The Cur-BSA-NPs gel did not irritate the eyes, because the irritation score was 0. This formulation could be considered safe for ocular use

(*continued*)

Table 3. (*continued*)

Study	Animal model	Nanosystem	Notes
[35]	Rabbits	Timolol maleate-loaded gelatin nanoparticles	Eight rabbits were grouped into three groups (three for each of two gelatin-nanoparticle formulations (F1: amino groups: glutaraldehyde (NH_2:GA) 1:2, F2: NH_2:GA 1:1) and two for eye drops). Time points were one hour, 24, 48, and 72 h, 7, 14, and 21 days post-application. Cornea, iris, conjunctiva, pupil, and anterior chamber were checked for any signs of ocular irritation. None of the gelatin-nanoparticle formulations irritated the ocular structures, whereas eye drops caused minor irritation in one rabbit's eye in the first hour. This minor irritation caused by timolol maleate eye drops was probably because of the preservatives used in this dosage form [36]

The evaluation criteria of the ocular irritation according to the grading system of the macroscopic signs for the colloidal systems tested are divided into three categories presented in Table 4.

Table 4. The evaluation criteria of the ocular irritation according to the grading system of the macroscopic signs for the colloidal systems tested [37]

Grade	Discomfort	Cornea	Conjunctiva	Discharge	Lids
0	No reaction	No alterations	No alterations	No discharge	No swelling
1	Blinking	Mild opacity	Mild hyperemia, mild oedema	Mild discharge without moistened hair	Mild swelling
2	Enhanced blinking, intense tearing, vocalisations	Intense opacity	Intense hyperemia, intense oedema, haemorrhage	Intense discharge with moistened hair	Obvious swelling

In Table 5. we present results of notable studies in which the grading system of the macroscopic signs for the colloidal systems tested was used to evaluate ocular irritation of various nanosystems.

Table 5. Results of notable studies in which the grading system of the macroscopic signs for the colloidal systems tested was used to evaluate ocular irritation of various nanosystems

Study	Animal model	Nanosystem	Notes
[38]	New Zealand white rabbits	Acyclovir-PLA nanospheres	These nanospheres caused no tissue alteration or ocular inflammation in the rabbits' eyes. Irritation scores of conjunctival congestion, swelling, and discharge, as well as iris hyperemia and corneal opacity, were 0
[39]	Rabbits	Tetramethyl rhodamine (TAMRA)-gelatin nanoparticles	A slight conjunctival redness was observed after four hours, but after 16 h, no difference was noted. Previous studies showed that the rabbits' eyes were more sensitive and needed more time for the epithelial repair compared with the human eyes [40], and in this study, the application of gelatin nanoparticles was safe and caused no irritation to the rabbits' eyes. It is reasonable to expect that positively charged gelatin nanoparticles would also be well-tolerated in the human eyes

Blinking test can also be used to evaluate ocular tolerance of nanosystems [41]. In Table 6. we present results of notable studies in which the blinking test was used to evaluate ocular irritation of various nanosystems.

Nelson's classification system, which evaluates the morphology of conjunctival ocular surface and the degree of squamous metaplasia, is presented in Table 7.

Table 6. Results of notable studies in which the blinking test was used to evaluate ocular irritation of various nanosystems

Study	Animal model	Nanosystem	Notes
[42]	Rabbits	Amikacin-loaded nanosuspension prepared with two bioadhesive, positively charged polymers, Eudragit RS 100 and Eudragit RL 100	Rabbits were forced to blink once to spread formulations (simulated tear fluid, amikacin eye drops, amikacin-loaded nanosuspension) uniformly on the cornea. Frequency of rabbits' blinking five minutes post-application was monitored. The normal blinking rate was 1–2 per minute. Results showed that the ocular irritation caused by these nanoparticles was almost the same compared with the control
[43]	New Zealand white rabbits	Cyclophosphamidepoly(butylcyanoacry late) (PBCA) nanospheres	No irritation effects on the iris, cornea, and conjunctiva were shown, suggesting their good ocular tolerance. Blinking rate post-application revealed significant differences because cyclophosphamide solution was shown to be irritant to the cornea, whereas nanospheres did not cause a significant increase in blinking rate. Blinking rate caused by drug-loaded PBCA nanospheres was higher compared with the PBCA nanospheres, which could be associated with the amount of cyclophosphamide incorporated into nanospheres

Table 7. The evaluation criteria of the ocular irritation according to the Nelson's classification for squamous metaplasia

Grade	Feature
0	>500 goblet cells/mm^2; small, round epithelial cells with large nuclei
1	100–500 goblet cells/mm^2
2	
3	<100 goblet cells/mm^2; large, polygonal epithelial cells with small nuclei

In Table 8. we present results of notable studies in which the Nelson's classification for squamous metaplasia was used to evaluate ocular irritation of various nanosystems.

Table 8. Results of notable studies in which the Nelson's classification for squamous metaplasia was used to evaluate ocular irritation of various nanosystems

Study	Animal model	Nanosystem	Notes
[45]	Rabbits	Fluoresceinamine-labelled hyaluronic acid-CS nanoparticles	Acute ocular tolerance was evaluated 24 h post-application and compared with the prior-application values. Five rabbits received fluorescent nanoparticles or fluorescent conjugate in the right eye, 12 applications total. Twenty-four hours post-application, evaluation of the ocular tolerance of these formulations was conducted and it included clinical signs, such as ocular discomfort, irritation, discharge, eyelid swelling, and presence of corneal and/or conjunctival alterations. Clinical signs in all rabbits at all time points scored grade 0 because there was no ocular discomfort, irritation, discharge, or eyelid swelling. After six hours, all of the nanoparticle-treated eyes scored grade 1 mucus-like discharge. No oedema, redness, or corneal vascularization were noted. Six days prior-application, conjunctival impression cytology (CIC) samples showed normal epithelial cells and abundant goblet cells, which were evenly distributed (grade 0). Scattered polymorphonuclear cells were present in all CIC samples. Twenty-four hours post-application, CIC samples from control eyes showed no difference when compared with the samples taken six days prior-application. The CIC samples for nanoparticle-exposed and conjugate-exposed eyes showed a normal distribution and morphology of epithelial cells, but there was an increased presence of polymorphonuclear cells. Rabbits showed no signs of ocular discomfort or irritation after exposure to these nanoparticles

(continued)

Table 8. (*continued*)

Study	Animal model	Nanosystem	Notes
[46]	Albino New Zealand rabbits	CS-BSA nanoparticles labelled with fluorescein isothiocyanate	The CS-BSA nanoparticles were applied to the right eye every 30 min for six hours, whereas the left eye was not treated and served as a control. The CIC samples were collected from both control and experimental eyes of all rabbits, six days before the beginning of the study and 24 h post-application. The CIC samples were analysed in a masked fashion according to the Nelson's classification. Grades 0 and 1 resemble normal morphology and grades 2 and 3 abnormal morphology. Grades were assigned for nuclear morphology, nuclear or cytoplasmic ratio, metachromasia, and goblet cell density. In CIC samples collected six days before the beginning of the study, animals showed no clinical signs and the ocular surface structures were normal, thus the score of mean clinical macroscopic signs was 0. For both control and experimental eyes, Nelson's grade was 0 to 1, because of the normal morphology of the conjunctival epithelia regarding goblet and nongoblet epithelial cells. Twenty-four hours post-application, CIC samples were obtained from both control and experimental eyes and between them, no difference was observed, thus Nelson's grade was 0 to 1, too. Ocular irritation was analysed using the grading system of the macroscopic signs for the colloidal systems tested, three, six and 24 h post-application. Rabbits showed no signs of discomfort 24 h post-application. Nanoparticle-treated eyes had a clinical macroscopic sign score of 1, because of the mucus-like discharge six hours post-application, but it was minimal and probably caused by the ability of CS-BSA nanoparticles to aggregate. It can be concluded that the ocular surface of rabbits showed no signs of inflammation and that these nanoparticles were well-tolerated

The evaluation of clinical recovery regarding ocular irritation is presented in Table 9.

Table 9. The evaluation of clinical recovery regarding ocular irritation [47]

Grade	Description
0	Absent
1	Mild
2	Moderate
3	Severe

Ocular irritation potential of clindamycin-PLGA nanoimplants was evaluated in rabbits by scoring the clinical parameters from 0 to 3. *In vivo*, PLGA degrades into lactic acid and glycolic acid that are utilized in the Krebs cycle and converted into water and carbon dioxide. These nanoimplants, with or without clindamycin, caused no inflammation and were biocompatible and well-tolerated [47].

4 Conclusion

Ocular tolerance of nanosystems must be evaluated, primarily in animal models and afterwards in humans. By conducting tolerance assays, one can conclude whether a nanosystem used for ocular drug delivery is safe or not for animals and humans. These nanoformulations usually show great ocular tolerance in animals, with no signs of irritation or inflammation, when ocular tolerance is assessed via the Draize test, or according to the grading system of the macroscopic signs for the colloidal systems tested, the Nelson's classification for squamous metaplasia or the clinical recovery regarding ocular irritation. Still, it is crucial to precisely extrapolate results from animals to humans, so these nanosystems could be used for the treatment of ocular diseases in humans. Further research needs to be carried out to get firmer confirmation about the safety and efficacy of these nanoformulations in humans.

Conflict of Interest. The authors have no conflicts of interest to disclose.

References

1. Omerović, N., Vranić, E.: Application of nanoparticles in ocular drug delivery systems. Health Technol. **10**(1), 61–78 (2020)
2. Roggeband, R., York, M., Pericoi, M., Braun, W.: Eye irritation responses in rabbit and man after single applications of equal volumes of undiluted model liquid detergent products. Food Chem. Toxicol. **38**(8), 727–734 (2000)
3. Del Amo, E.M., Rimpelä, A.K., Heikkinen, E., Kari, O.K., Ramsay, E., et al.: Pharmacokinetic aspects of retinal drug delivery. Prog. Retin. Eye Res. **57**(1), 134–185 (2017)
4. Peterson, G.I.: Biodegradable shape memory polymers in medicine. Adv. Healthc. Mater. **6**(21), 1700694 (2017)

5. Zhao, R., Li, J., Wang, J., Yin, Z., Zhu, Y., Liu, W.: Development of timolol-loaded galactosylated chitosan nanoparticles and evaluation of their potential for ocular drug delivery. AAPS PharmSciTech **18**(4), 997–1008 (2017)
6. Quinones, J.P., Peniche, H., Peniche, C.: Chitosan based self-assembled nanoparticles in drug delivery. Polymers (Basel) **10**(3), 235 (2018)
7. Da Silva, G.R., Lima, T.H., Orefice, R.L., Silva-Cunha, A., Zhao, M., Behar-Cohen, F.: In vitro and in vivo ocular biocompatibility of electrospun poly(caprolactone) nanofibers. Eur. J. Pharm. Sci. **73**(1), 9–19 (2015)
8. Leonardi, A., Bucolo, C., Romano, G.L., Platania, C.B.M., Drago, F., Puglisi, G., et al.: Influence of different surfactants on the technological properties and in vivo ocular tolerability of lipid nanoparticles. Int. J. Pharm. **470**(1–2), 133–140 (2014)
9. Mittal, N., Kaur, G.: Investigations on polymeric nanoparticles for ocular delivery. Adv. Polym. Technol. **2019**(1), 1–14 (2019)
10. Abrego, G., Alvarado, H., Souto, E.B., Guevara, B., Bellowa, L.H., Parra, A., et al.: Biopharmaceutical profile of pranoprofen-loaded PLGA nanoparticles containing hydrogels for ocular administration. Eur. J. Pharm. Biopharm. **95**(Pt B), 261–270 (2015)
11. Kim, J.S., Song, K.S., Sung, J.H., Ryu, H.R., Choi, B.G., Cho, H.S., et al.: Genotoxicity, acute oral and dermal toxicity, eye and dermal irritation and corrosion and skin sensitisation evaluation of silver nanoparticles. Nanotoxicology **7**(5), 953–960 (2013)
12. Ema, M., Matsuda, A., Kobayashi, N., Naya, M., Nakanishi, J.: Dermal and ocular irritation and skin sensitization studies of fullerene C60 nanoparticles. Cutan. Ocul. Toxicol. **32**(2), 128–134 (2013)
13. Kishore, A.S., Surekha, P., Murthy, P.B.: Assessment of the dermal and ocular irritation potential of multi-walled carbon nanotubes by using in vitro and in vivo methods. Toxicol. Lett. **191**(2–3), 268–274 (2009)
14. Ema, M., Matsuda, A., Kobayashi, N., Naya, M., Nakanishi, J.: Evaluation of dermal and eye irritation and skin sensitization due to carbon nanotubes. Regul. Toxicol. Pharmacol. **61**(3), 276–281 (2011)
15. Wu, W., Yan, L., Wu, Q., Li, Y., Li, Q., Chen, S., et al.: Evaluation of the toxicity of graphene oxide exposure to the eye. Nanotoxicology **10**(9), 1329–1340 (2016)
16. Calvo, P., Alonso, M.J., Vila-Jato, J.L., Robinson, J.R.: Improved ocular bioavailability of indomethacin by novel ocular drug carriers. J. Pharm. Pharmacol. **48**(11), 1147–1152 (1996)
17. Calvo, P., Vila-Jato, J.L., Alonso, M.J.: Evaluation of cationic polymer-coated nanocapsules as ocular drug carriers. Int. J. Pharm. **153**(1), 41–50 (1997)
18. Langer, K., Mutschler, E., Lambrecht, G., Mayer, D., Troschau, G., Stieneker, F., et al.: Methylmethacrylate sulfopropylmethacrylate copolymer nanoparticles for drug delivery, part III: evaluation as drug delivery system for ophthalmic applications. Int. J. Pharm. **158**(1), 219–231 (1997)
19. Marchal-Heussler, L., Sirbat, D., Hoffman, M., Maincent, P.: Poly(ε-caprolactone) nanocapsules in carteolol ophthalmic delivery. Pharm. Res. **10**(1), 386–390 (1993)
20. Fresta, M., Fontana, G., Bucolo, C., Cavallaro, G., Giammona, G., Puglisi, G.: Ocular tolerability and in vivo bioavailability of poly(ethylene glycol) (PEG)-coated polyethyl-2-cyanoacrylate nanosphere-encapsulated acyclovir. J. Pharm. Sci. **90**(3), 288–297 (2001)
21. Bucolo, C., Maltese, A., Puglisi, G., Pignatello, R.: Enhanced ocular anti-inflammatory activity of ibuprofen carried by an Eudragit RS100® nanoparticle suspension. Ophthalmic Res. **34**(1), 319–323 (2002)
22. Bucolo, C., Maltese, A., Maugeri, F., Busa, B., Puglisi, G., Pignatello, R.: Eudragit RL100 nanoparticle system for the ophthalmic delivery of cloricromene. J. Pharm. Pharmacol. **56**(7), 841–846 (2004)

23. Pignatello, R., Bucolo, C., Ferrara, P., Maltese, A., Puleo, A., Puglisi, G.: Eudragit RS100 nanosuspensions for the ophthalmic controlled delivery of ibuprofen. Eur. J. Pharm. Sci. **16**(1–2), 53–61 (2002)
24. Pignatello, R., Bucolo, C., Spedalieri, G., Maltese, A., Puglisi, G.: Flurbiprofen-loaded acrylate polymer nanosuspensions for ophthalmic application. Biomaterials **23**(15), 3247–3255 (2002)
25. Merodio, M., Irache, J.M., Valamanesh, F., Mirshahi, M.: Ocular disposition and tolerance of ganciclovir-loaded albumin nanoparticles after intravitreal injection in rats. Biomaterials **23**(7), 1587–1594 (2002)
26. Liu, Z., Li, J., Nie, S., Guo, H., Pan, W.: Effects of Transcutol P on the corneal permeability of drugs and evaluation of its ocular irritation of rabbit eyes. J. Pharm. Pharmacol. **58**(1), 45–50 (2006)
27. Li, J., Liu, H., Liu, L.L., Cai, C.N., Xin, H.X., Liu, W.: Design and evaluation of a brinzolamide drug-resin in situ thermosensitive gelling system for sustained ophthalmic drug delivery. Chem. Pharm. Bull. **62**(10), 1000–1008 (2014)
28. Bozdag, S., Gumus, K., Gumus, O., Unlu, N.: Formulation and in vitro evaluation of cysteamine hydrochloride viscous solutions for the treatment of corneal cystinosis. Eur. J. Pharm. Biopharm. **70**(1), 260–269 (2008)
29. Maneewattanapinyo, P., Banlunara, W., Thammacharoen, C., Ekgasit, S., Kaewamatawong, T.: An evaluation of acute toxicity of colloidal silver nanoparticles. J. Vet. Med. Sci. **73**(11), 1417–1423 (2011)
30. Drake, P.L., Hazelwood, K.J.: Exposure-related health effects of silver and silver compounds: a review. Ann. Occup. Hyg. **49**(7), 575–585 (2005)
31. Lansdown, A.B.: A pharmacological and toxicological profile of silver as an antimicrobial agent in medical devices. Adv. Pharmacol. Sci. **2010**(1), 910686 (2010)
32. El Kasabgy, N.A.: Ocular supersaturated self-nanoemulsifying drug delivery systems (SSNEDDS) to enhance econazole nitrate bioavailability. Int. J. Pharm. **460**(1–2), 33–44 (2014)
33. Zhang, P., Liu, X., Hu, W., Bai, Y., Zhang, L.: Preparation and evaluation of naringenin-loaded sulfobutylether-cyclodextrin/chitosan nanoparticles for ocular drug delivery. Carbohydr. Polym. **149**(1), 224–230 (2016)
34. Lou, J., Hu, W., Tian, R., Zhang, H., Jia, Y., Zhang, J., et al.: Optimization and evaluation of a thermoresponsive ophthalmic in situ gel containing curcumin-loaded albumin nanoparticles. Int. J. Nanomedicine **9**(1), 2517–2525 (2014)
35. Shokry, M., Hathout, R.M., Mansour, S.: Exploring gelatin nanoparticles as novel nanocarriers for timolol maleate: augmented in-vivo efficacy and safe histological profile. Int. J. Pharm. **545**(1–2), 229–239 (2018)
36. Singh, J., Chhabra, G., Pathak, K.: Development of acetazolamide-loaded, pH-triggered polymeric nanoparticulate in situ gel for sustained ocular delivery: in vitro, ex vivo evaluation and pharmacodynamic study. Drug Dev. Ind. Pharm. **40**(9), 1223–1232 (2014)
37. Organization for Economic Co-Operation and Development. Test Guideline 405: Acute Eye Irritation/Corrosion (2002)
38. Giannavola, C., Bucolo, C., Maltese, A., Paolino, D., Vandelli, M.A., Puglisi, G., et al.: Influence of preparation conditions on acyclovir-loaded poly-d, l-lactic acid nanospheres and effect of PEG coating on ocular drug bioavailability. Pharm. Res. **20**(4), 584–590 (2003)
39. Tseng, C.L., Chen, K.H., Su, W.Y., Lee, Y.H., Wu, C.C., Lin, F.H.: Cationic gelatin nanoparticles for drug delivery to the ocular surface: in vitro and in vivo evaluation. J. Nanomater. **2013**(6), 1–11 (2013)
40. Gokce, E.H., Sandri, G., Egrilmez, S., Bonferoni, M.C., Guneri, T., Caramella, C.: Cyclosporine A-loaded solid lipid nanoparticles: ocular tolerance and in vivo drug release in rabbit eyes. Curr. Eye Res. **34**(11), 996–1003 (2009)

41. Wei, G., Ding, P., Zhao, H., Zheng, J.: Effects of penetration enhancers on the permeability of timolol maleate through isolated rabbit cornea. J. Chin. Pharm. Sci. **37**(6), 430–433 (2002)
42. Sharma, U.K., Verma, A., Prajapati, S.K., Pandey, H., Pandey, A.C.: In vitro, in vivo and pharmacokinetic assessment of amikacin sulphate laden polymeric nanoparticles meant for controlled ocular drug delivery. Appl. Nanosci. **5**(1), 143–155 (2015)
43. Salgueiro, A., Egea, M.A., Espina, M., Valls, O., Garcia, M.L.: Stability and ocular tolerance of cyclophosphamide-loaded nanospheres. J. Microencapsul. **21**(2), 213–223 (2004)
44. Enriquez de Salamanca, A., Diebold, Y., Calonge, M., García-Vazquez, C., Callejo, S., Vila, A., et al.: Chitosan nanoparticles as a potential drug delivery system for the ocular surface: toxicity, uptake mechanism and in vivo tolerance. Investig. Ophthalmol. Vis. Sci. **47**(4), 1416–1425 (2006)
45. Nelson, J.D.: Impression cytology. Cornea **7**(1), 71–81 (1988)
46. Contreras-Ruiz, L., de la Fuente, M., García-Vázquez, C., Sáez, V., Seijo, B., Alonso, M., et al.: Ocular tolerance to a topical formulation of hyaluronic acid and chitosan-based nanoparticles. Cornea **29**(5), 550–558 (2010)
47. Fernandes-Cunha, G.M., Gouvea, D.R., Fulgêncio Gde, O., Rezende, C.M., da Silva, G.R., Bretas, J.M., et al.: Development of a method to quantify clindamycin in vitreous humor of rabbits' eyes by UPLC-MS/MS: application to a comparative pharmacokinetic study and in vivo ocular biocompatibility evaluation. J. Pharm. Biomed. Anal. **102**(1), 346–352 (2015)

Development of Microfluidic Lab-on-Chip System for Cultivation of Cells and Tissues

Nevena Milivojević[1(✉)], Dalibor Nikolić[1,2], Dragana Šeklić[1], Živana Jovanović[3], Marko Živanović[1,2], and Nenad Filipović[2,3]

[1] Institute of Information Technologies, University of Kragujevac, Kragujevac, Serbia
[2] BioIRC - Bioengineering Research and Development Center, Univesity of Kragujevac, Kragujevac, Serbia
[3] Faculty of Engineering, University of Kragujevac, Kragujevac, Serbia

Abstract. Lab-on-chip systems are microfluidic devices that can be used as powerful tools for research in fields of molecular biology and bioengineering. Novel chip systems such as miniaturized microfluidic human tissue and organ models are powerful platforms to *ex vivo* study functional units of any important biological and physiological parameters of their *in vivo* counterparts. Also, development of microfluidic system with mathematical modeling described fluid dynamics in a chip can be used in real-time monitoring of the transport, efficacy, and cytotoxicity of potential drugs on the same platform.

We have developed a Chip system that is suitable for cultivation of cells or primitive tissues. The sandwich chip design is composed of acrylic plates - material that is transparent for easy optical following of cells growing. It is manufactured by milling processes performed on in-house developed 3D CNC router – 3D CNC milling machine and by the FDM 3D printing process on Creality 3D CR-10max printer. The main components are: i) bottom plate with cylindrical main chamber for cells growing and two supplying channels; ii) top plate with other half of the channel as well as two holes for tubing fittings.

Connection of Chip to a peristaltic pump gives a microfluidic system physiologically relevant microenvironment for cell growth. Fluidic control mimics conditions in real *in vivo* systems and leads to successful cell proliferation.

In near future, organ-on-chip systems promise to be powerful platforms for *ex vivo* studies of functional units of some organs, for specific biological processes, for examination of tumor with its microenvironment. Also, these small devices can be used for personalized/precision medicine and drug screening for direct treatment decision-making. The chip model that we developed can be used for successful growth of cells or smaller tissues in conditions as suitable as in a living system.

Keywords: Lab-on-chip · Microfluidics · Tissue engineering

1 Introduction

Lab-on-chip systems, which have recently emerged as powerful tools in research, are microfluidic devices that aim to recreate relevant physiological features [1]. 'Engineered

A. Badnjevic and L. Gurbeta Pokvić (Eds.): CMBEBIH 2021, IFMBE Proceedings 84, pp. 718–725, 2021.
https://doi.org/10.1007/978-3-030-73909-6_81

tissue' culture on a chip is a promising technology with relative simplicity of sample acquisition and availability [2]. Device mathematical modeling describes fluid dynamics chip that can be used in real-time monitoring of the transport, efficacy and cytotoxicity of potential drugs on the same platform [1, 2]. The most common microfluidic devices are fabricated in the elastomer poly (dimethylsiloxane) (PDMS) and operated using variable-control syringe pumps [3]. Integration of fine fluidic control in platform with cell-laden chamber compartments inside developed a new approach for tissue mimicry [4]. Organ-on-a-chip studies have shown the importance of producing physiologically relevant microenvironments with respect to static flow well-plate systems [5].

Recent observations constitute a proof of concept that miniaturized microfluidic human tissue and organ models are powerful platforms to study *ex vivo* functional units of some organs (e.g., liver), specific biological processes (e.g., angiogenesis), tumor with its microenvironment and for drug preclinical testing [6, 7]. Future work on treatment application on these devices will offer such opportunities, minimizing sample volumes and maximizing control over the manipulation of particles suspended in microfluidic system [2, 5]. Organ-on-chip systems promise to be powerful enablers of personalized/precision medicine and drug screening in order to direct treatment decision-making [1, 2]. Animal models can satisfy requirement and provide an *in vivo* setting to study treatment, but they have their own complications, they are expensive, ethically questionable and fail to provide a precise recapitulation of results to human physiology [1, 7]. *Ex vivo* assays, such as Organ-on-a-chip, that predict therapy response may fill these gaps [8].

Novel chip systems can be tailored to reconstruct important biological and physiological parameters of their *in vivo* counterparts. Furthermore, multiple organon-a-chip models can be connected in a similar manner in which they are arranged *in vivo*. So, they can provide analyzes of multi organ interactions making a body-on-achip system comparable to an entire human body [9, 10].

In this paper we presented a chip produced for growing cells in conditions similar to physiological ones. With connection to a peristaltic pump, a controlled flow was obtained, which enables successful growth and proliferation of cells in the system.

2 Methodology

The chip is sandwich designed, composed of two acrylic plates 5 mm thick. Selected material is transparent for microscopic following of cell growth. The channels are designed and milled on the inner side of both panels. The SolidWorks 2020 [11] CAD software package was used to design and build the 3D model of the chip. The main components are: bottom plate (A – Fig. 1), dimensions 30 mm by 70 mm and 5 mm thick. The cylindrical main chamber for cells cultivation with a diameter of 16 mm and a depth of 0.8 mm as well as two supplying channels (3 – Fig. 1) with half of the cylindrical cross-section with diameter 0.25 mm are milled in this plate. The other half of the channel is milled in the top plate (B – Fig. 1) as well as two holes for tubing fittings. At the end of each supplying channels tubing fittings (C – Fig. 1) are attached and the purposes are to provide an easy connection between the chip and the rest of the experimental equipment (e.g. peristaltic pump). Those fittings are manufactured by the FDM 3D printing process from PLA material. Creality 3D CR-10max printer was used, with a high-quality printing setup (layer thickness 0.1 mm). Also, the plates are drilled holes by diameter 3 mm (4 – Fig. 1) for easy bolting them.

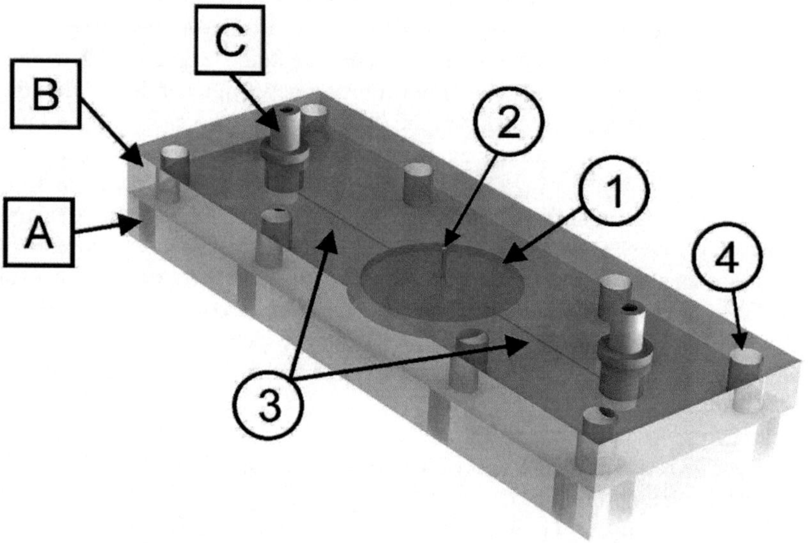

Fig. 1. Chip: A - bottom part; B - top part; C - tube fitting; 1 - main chamber; 2 - air pillar, for removing bubbles; 3 - supply channels

Milling processes are performed on in-house developed 3D CNC router – 3D CNC milling machine. This machine is designed for very precise work with an installed 2.2 KW water cooling milling head. This head provides max. 24000 RPM. For the axial precise (with error of 0.01 mm per 100 mm) and robust operation of the machine, a hybrid system was used. A highly precise encoder was added to each stepper motor, and hybrid step drivers with feedback coupling were used to control them. In this way, stable operation is ensured and the problem of skipping steps that occurs in machines with classic stepper motors is overcome. For milling larger chambers and openings on plates, a milling machine with 2 blades and a diameter of 2 mm was used, while for small supply channels, a flute with a ball end and diameter of 0.25 mm was used. With a cutting depth of 0.25 mm is ensured a semicylindrical cross-section profile of supplying channels (Fig. 2).

Cell line HeLa (Human cervix epitheloid carcinoma) used for seeding in Chip was obtained from ATCC–American Type Culture Collection, Manassas, VA, USA. The cells were propagated and maintained under optimum conditions and standard protocols in DMEM medium (Dulbecco's Modified Eagle Medium, Gibco, USA) supplemented with 10% fetal bovine serum (Gibco Invitrogen, USA) and antibiotics 100 IU/ml penicillin and 100 μg/ml streptomycin (Sigma Aldrich). Experiments were performed in incubator in a humidified atmosphere of 5% CO_2 at 37 °C. Cells were seeded at a concentration of 200,000 cells per ml and were in static conditions for 24 h while adhering to the base (both in the chip and in the 6-well). The seeded cells in the 6-well served as a control

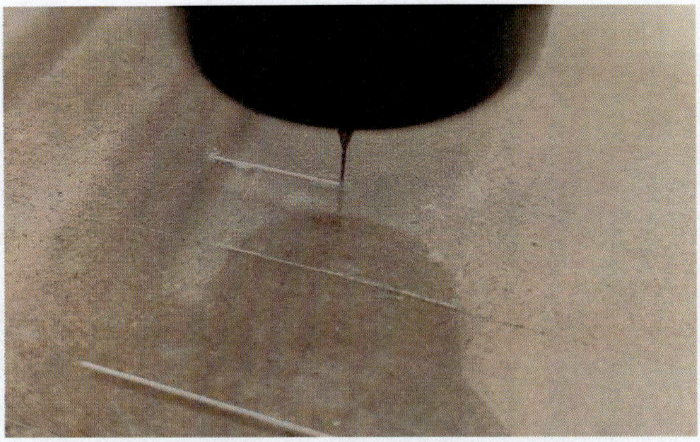

Fig. 2. Milling process, ball end flute 0.25 mm

group for comparison with the group under dynamic conditions, which represents a microfluidic chip system. After 24 h incubation, the cells in the chip were switched on medium flow, and their proliferation was observed for 24 and 48 h, respectively.

3 Results

We have developed a microfluidic Lab-on-chip system that is suitable for cultivation of cells or primitive tissues. Figure 3 shows computer designed model of our Chip, and on Fig. 4 there is a real model after milling process.

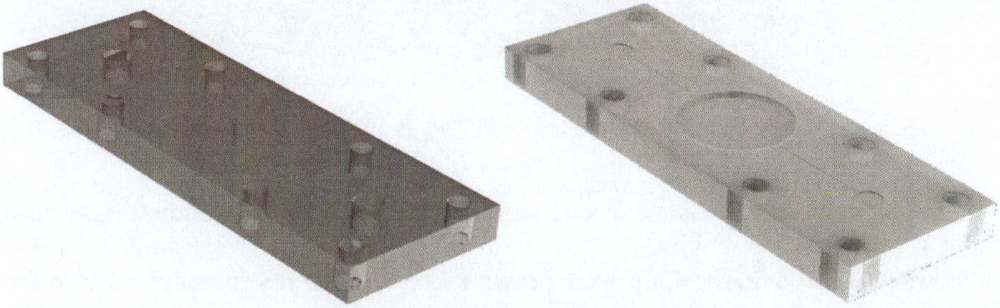

Fig. 3. Computer designed model

When the Chip is connected to a peristaltic pump, we get a microfluidic system with a physiologically relevant microenvironment for cells or tissues (Fig. 5). Peristaltic pump provides fine fluidic control that mimics conditions in real *in vivo* systems, and thus leads to successful cell proliferation.

After seeding the cells in the Chip and in the presence of controlled flow of 0.03 ml/min medium, the cells are successfully cultivated. After 24 and 48 h of flow-controlled Chip experiment in incubator, the cells in the chip survived and divided undisturbed.

Fig. 4. Real model

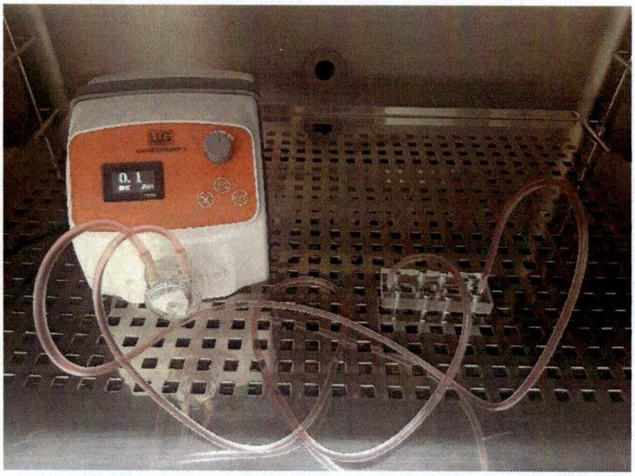

Fig. 5. Chip connected to a peristaltic pump with flow

Compared to the static model such as seeding cells in 6 wells (oftenly used *in vitro* studies), cells from the chip are cultivated in conditions similar to physiological. We concluded that cells on the Chip were grown more successfully in comparison to the static conditions (6-well). Figure 7a shows the cells from the static 6-well, while Fig. 7b shows the cells after growing under dynamic conditions in the Chip system.

4 Discussion

Lab-on-Chip is a system that is primarily intended to replace *in vivo* experiments due to its simplification, shortening of testing time, and ethical issues regarding the use of a large number of animals. Our chip model can be used for successful growth of cells or smaller tissues in conditions as suitable as in a living system.

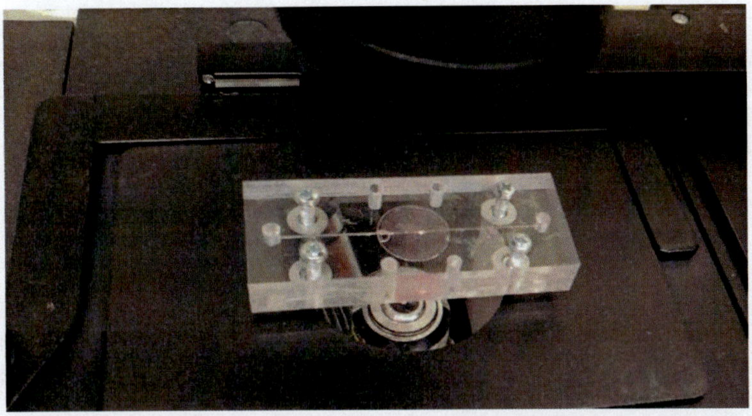

Fig. 6. Imaging of cells in Chip on microscope

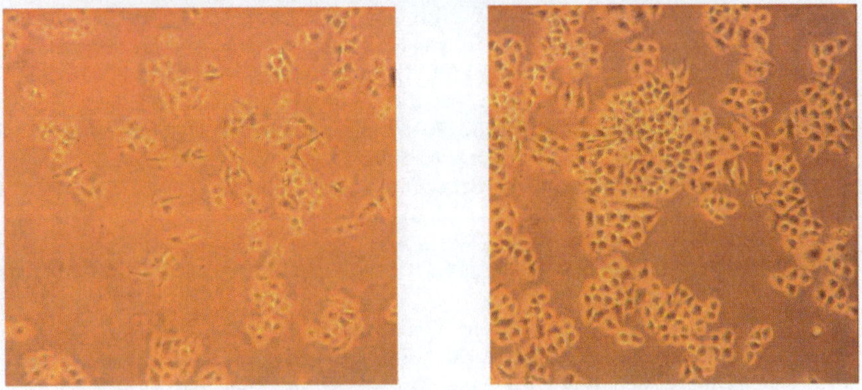

Fig. 7 a. Cells seeded in the 6-well b. Cells in the Chip

By further system developing we aim to create a Lab-on-chip microfluidic device that can be used for various purposes. More improved Chip systems will be manufactured by inserting multiple channels for different fluids, by inserting more chambers where several types of cells will be seeded, with deeper central chamber that will create the conditions for the cultivation of 3D cells seeded in hydrogels or various tissues.

5 Conclusion

We have developed functional lab-on-chip system and were able to cultivate cells in such conditions that successfully mimics growing cells as *in vivo* systems. With further work we will improve microfluidic devices in order to use them in several directions for research in the field of molecular biology and bioengineering.

Acknowledgment. This research is supported by the European Union's Horizon 2020 research and innovation programme under grant agreement No 952603 (SGABU). This article reflects only the author's view. The Commission is not responsible for any use that may be made of the information it contains. Also, the research is funded by Serbian Minis- try of Education, Science, and Technological Development [451-03-68/202014/200378 (Institute for Information Technologies, University of Kragujevac)].

Conflict of Interest Declaration. The authors declare that they have no conflict of interest.

References

1. Trujillo-de Santiago, G., Flores-Garza, B.G., Tavares-Negrete, J.A., Lara- Mayorga, I.M., González-Gamboa, I., Zhang, Y.S., Rojas-Martínez, A., Ortiz-López, R., Álvarez, M.M.: The tumor-on-chip: recent advances in the development of microfluidic systems to recapitulate the physiology of solid tumors. Mater. (Basel) **12**(18), 2945 (2019). https://doi.org/10.3390/ma12182945. Review
2. Bower, R., Green, V.L., Kuvshinova, E., Kuvshinov, D., Karsai, L., Crank, S.T., Stafford, N.D., Greenman, J.: Maintenance of head and neck tumor on-chip: gateway to personalized treatment? Future Sci OA. **3**(2), FSO174 (2017). https://doi.org/10.4155/fsoa-2016-0089
3. Ahmed, N., Sukovich, D., Abate, A.R.: Operation of droplet-microfluidic devices with a lab centrifuge. Micromach. (Basel) **7**(9), 161 (2016). https://doi.org/10.3390/mi7090161
4. Gumuscu, B., Albers, H.J., van den Berg, A., Eijkel, J.C.T., van der Meer, A.D.: Compartmentalized 3D tissue culture arrays under controlled mi- crofluidic delivery. Sci. Rep. **7**(1), 3381 (2017). https://doi.org/10.1038/s41598-017-01944-5
5. Cunha-Matos, C.A., Millington, O.R., Wark, A.W., Zagnoni, M.: Real-time assessment of nanoparticle-mediated antigen delivery and cell re- sponse. Lab Chip **16**(17), 3374–81 (2016). https://doi.org/10.1039/c6lc00599c
6. Carvalho, M.R., Barata, D., Teixeira, L.M., Giselbrecht, S., Reis, R.L., Oliveira, J.M., Truckenmüller, R., Habibovic, P.: Colorectal tumor-on-a-chip system: a 3D tool for precision onco-nanomedicine. Sci. Adv. **5**(5), eaaw1317 (2019). https://doi.org/10.1126/sciadv.aaw1317
7. Nguyen, M., De Ninno, A., Mencattini, A., Mermet-Meillon, F., Forna-baio, G., Evans, S.S., Cossutta, M., Khira, Y., Han, W., Sirven, P., Pelon, F., Di Giuseppe, D., Bertani, F.R., Gerardino, A., Yamada, A., Descroix, S., Soumelis, V., Mechta-Grigoriou, F., Zalcman, G., Camonis, J., Martinelli, E., Businaro, L., Parrini, M.C.: Dissecting effects of anticancer drugs and cancer-associated fibroblasts by on-chip reconstitution of immunocompetent tumor microenvironments. Cell Rep. **25**(13), 3884–3893 (2018). https://doi.org/10.1016/j.celrep.2018.12.015
8. Meijer, T.G., Naipal, K.A., Jager, A., van Gent, D.C.: Ex vivo tumor culture systems for functional drug testing and therapy response prediction. Future Sci. OA. **3**(2), FSO190 (2017). https://doi.org/10.4155/fsoa-2017-0003
9. Baghban, R., Roshangar, L., Jahanban-Esfahlan, R., Seidi, K., Ebrahimi- Kalan, A., Jaymand, M., Kolahian, S., Javaheri, T., Zare, P.: Tumor micro- environment complexity and therapeutic implications at a glance. Cell Commun. Sign. **18**(1), 59 (2020). https://doi.org/10.1186/s12964-020-0530-4. Review

10. Zhang, Y.S., Aleman, J., Shin, S.R., Kilic, T., Kim, D., Mousavi Shaegh, S.A., Massa, S., Riahi, R., Chae, S., Hu, N., Avci, H., Zhang, W., Silvestri, A., Sana- tiNezhad, A., Manbohi, A., De Ferrari, F., Polini, A., Calzone, G., Shaikh, N., Alerasool, P., Budina, E., Kang, J., Bhise, N., Ribas, J., Pourmand, A., Skardal, A., Shupe, T., Bishop, C.E., Dokmeci, M.R., Atala, A., Khademhosseini, A.: Multisensor-integrated organs-onchips platform for automated and continual in situ monitoring of organoid behaviors. Proc. Natl. Acad. Sci. **114**(12), E2293–E2302 (2017). https://doi.org/10.1073/pnas.1612906114
11. Solidworks: Computer Aided Engineering CAD – Solid - Works Documentation, Dassault Systèmes, Providence, RI, USA (2020)

Electrospinning and Electrospun Nanofibrous Materials – Promising Scaffolds in Tissue Engineering

Katarina Virijević[1(✉)], Jelena Grujić[1], Mihajlo Kokanović[2], Dalibor Nikolić[1,2], Marko Živanović[1,2], and Nenad Filipović[2,3]

[1] Institute for Information Technologies, University of Kragujevac, Kragujevac, Serbia
katarina.virijevic@uni.kg.ac.rs
[2] BioIRC - Bioengineering Research and Development Centre, Kragujevac, Serbia
[3] Faculty of Engineering, University of Kragujevac, Kragujevac, Serbia

Abstract. Electrospinning is a highly used technique in the tissue engineering field, particularly for the development of tissue supportive scaffolds. Use of electrospun derived nano and microfibers is of crucial importance for extracellular matrix mimicking and cell adhesion and proliferation. This paper is focused on the electrospinning process and its ideal conditions for producing scaffolds for tissue engineering purpose. With use of a series of biocompatible polymers and solvents, solutions were tested in various electrospinning settings in order to produce microscale fibers. The production of these fibers is directly related to the electrospinning parameters in use, hence the importance of optimizing this process. A series of scaffold samples were built from various solutions in different parameter conditions. The scaffolds were analyzed with microscopy for fiber diameter measurement. Within this paper, we analyzed influence of input parameters on fiber output diameter which is of particular importance for cell adhesion. Input to the system are material properties of the tested solutions, needle electrode dimensions, the flow rate of the solution, the distance between electrodes and the applied voltage.

Keywords: Electrospinning · Nanofibers · Polymers

1 Introduction

Tissue degeneration, damage, and deterioration occur practically in every organism, so the focus of progressive methods of healing and restoring are the main problems for today's ediie.

Tissue engineering is a wide field based on medical, biological, and bioengineering approach. In the last ten years, all these segments evolved and get into the focus of numerous scientific communities. One of the main goals of tissue engineering is to join the damaged parts of tissue that can be performed by healing, replacing and regeneration. This refers not only to parts of damaged organs, but to the whole organs as well by

A. Badnjevic and L. Gurbeta Pokvić (Eds.): CMBEBIH 2021, IFMBE Proceedings 84, pp. 726–733, 2021.
https://doi.org/10.1007/978-3-030-73909-6_82

utilization of biocompatible materials applied in accordance with appropriate methods and efficient techniques.

In this study, we used electrospinning procedure for obtaining the most optimal nanofibers that are used in scaffold production and tissue engineering. Electrospinning has gained popularity, particularly in recent years for production of fibrous scaffolds from different polymers with high porosities, large surface area to volume ratios, and nano to micro-sized fiber diameters [1].

Features characteristics determine the success of the practical scaffold application in medicine and tissue engineering. In order to gain the most optimal results, obtained scaffolds must possess similar properties to the natural extracellular matrix structure that supports cell attachment and proliferation [2]. Moreover, from a biological aspect, almost all the tissues and organs, such as bone, blood vessel, nerve, skin, and other composite constructions are consisted and arranged into a fibrous structure with fiber dimensions down to nanometer scale [3].

The concept of electrospinning is based on the application of various polymer solutions. Electrostatic forces used in this process produce ultrafine nanofibers with a different range of fiber diameters. The method considers using various synthetic, natural polymers, biopolymers, bioactive molecules, nanoparticles, and other compatible materials. Their application is known in the fields of biosensors, chemical filtrations, cosmetics, wound healing, and many other [4]. On the one hand, the main goal of electrospinning process is to form desirable nanofiber structures. Referring to these particular requirements, researchers are conducting experiments in different conditions trying to achieve the most favorable structures. Thus, it can be said that scaffold structures are predefined by the used methods. On the other hand, scaffolds features, i.e. nanofiber structures are predefined by characteristics of used polymers and solvents.

In this study we used chemically synthesized polymers: polycaprolacton (PCL) and poly (ethyleneglycol) (PEG).

PCL is known for its biomedical application in scaffold production for fibroblasts and osteoblasts growth support, and has been widely used in long-term implants, in tissue engineering and controlled drug delivery applications [5]. According to the literature data, PCL and PEG are known as co-friendly copolymers, easy produceable, biodegradable, biocompatible and with sturdy function in tissue engineering [6]. Previous study reported importance of PEG as type of hydrophilic polymer used for biomedical applications because of its hydrophilicity, lack of toxicity, and absence of antigenicity and immunogenicity [5].

As polymer-based fibers are the product of electrospinning, the main aim of this study is to optimize the parameters of electrospinning process in order to obtain the appropriate scaffolds for further using in tissue engineering.

2 Materials and Methods

2.1 Materials

Poly (ε-caprolacton) (PCL) (average Mn ~ 80,000, 440744-500G) and poly (ethylene glycol) (PEG) (average Mn ~ 4,000, 81240-1 kg) were purchased from Sigma-Aldrich.

Chloroform ($CHCl_3$) was purchased from Macron Fine Chemicals (4444-25), while *N,N*-Dimethylformamide (DMF, 99.5%) from Fisher Scientific (D/3841/17). In this study all the chemicals were used without further purification.

2.2 Methods

2.2.1 Preparation of Electrospinning Solution

The examined solutions were prepared by dissolving the PCL: PEG in volume ratios (70:30) in solvents $CHCl_3$: DMF (mixed in the volume ratio of 70:30). The range of concentrations polymer solutions was 1727%. After the preparation, solutions were stirred at 1000 rpm for 24 h at room temperature. After 24 h a blend solutions were completely dissolved and ready for electrospinning. The process was carried out by using vertical position setup. When the solutions were prepared, they were placed in the 10 mL syringes which with 21G needle with applied voltage of 15 kV. Applied flow rate was 1.5 mL/h^{-1}. Duration of the process was 30 s for each run. Aluminium plate was used as a collector. This procedure was completed according to protocol already described [4].

2.2.2 Electrospinning

Electrospinning equipment used in this study is our in-house produced device. This apparatus consists of: syringe pump, with attached needle (NIPRO, Japan, size 21G), collector plate and high voltage supplier. The whole process is controlled *via* our in-house produced computer software.

Firstly, room humidity and temperature were checked after which the polymer solution was placed in syringe with attached needle. Electric field was initiated by application of appropriate voltage. As the intensity of the electric field increases, the solution drops on the top of the needle extended and resulted in creation of shape known as the Taylor cone. As already known, electrostatic forces induced formation of nanofibers, while solvents from polymer solution evaporated under the influence of these forces. The obtained fibrous scaffolds were collected on the aluminum plate collector (Fig. 1) and then transported to the Petri dishes. Minimum of 10 micrograms of nanofibers per sample were analyzed on the inverted microscope (Delta Optical Genetic PRO microscope) at 100x magnification.

2.2.3 Fibers Morphology Characterization and Statistical Analysis

The obtained fibers morphology was analyzed with software package Image J that was used for analysis of desired image, allowing measuring, calibrating, and editing the image. The morphology of fibers, consisted of various polymer solutions, was analyzed measuring fiber diameters on the obtained image. The results were obtained in excel software and evaluated with statistical software package SPSS (Chicago, IL), using the one-way ANOVA test to highlight the statistical significance. The data were expressed as the mean of diameter fibers ± standard error.

Fig. 1. Electrospinning setup

3 Results and Discussion

In Table 1 it is presented the input parameters for electrospinning procedure. It is very important to analyze all important parameters in regard for optimization of fiber formation. Thus, we prepared various concentrations of PCL: PEG (mixed in 70:30% ratio) in $CHCl_3$: DMF solvent mixture (70:30% ratio). After many attempts, we found optimal parameters for production of appropriate scaffolds that have been intensively analyzed by microscopy and consequent image processing. For this kind of chemical polymers, it was shown that room temperature conditions are enough fair for good production of fibers. As the concentration of mixed chemicals increased, their viscosity also increased, thus it was needed to apply much greater force in pumping process. For example, for 17% solution the flow rate was 0.5 mL/h, but for more viscous 27% the flow rate needed to be increased to 1 mL/h. In the same manner, as the concentration/viscosity increases the applied voltage needed for fiber production also increased (from 10 to 15 kV). Under these conditions the spinning time was fixed to 30 s and the electrode distance was 12 cm.

The intensive image processing enabled us to understand the fiber nature in regards of their diameter (Fig. 2). Our results shown that only fibers under 1 μm length produced scaffolds that could be considered for further *in vitro* analyzing (Fig. 3). The thick fibers produce scaffolds that are not appear similar to extracellular matrices.

When we analyze the influence of chemical composition on fiber formation, our conclusion is that solutions with concentration up to 21% produce thin extracellular matrix similar fibers. On the other hand, by using of these concentrations we use lesser force in pumping process and lesser voltages. Concentrations lower than 17% produced

Table 1. Input parameters for electrospinning process for polymer solutions

PCL:PEG	CHCl3:DMF	t [°C]	Q [ml/h]	U [kV]	t [sec]	L [cm]
70:30	70:30					
17%		24	0.5	10	30	12
18%		24	0.5	12	30	12
19,20%		24	0.6	11	30	12
21%		24	0.6	15	30	12
22,23%		24	0.7	15	30	12
24%		24	01	12	30	12
25%–27		24	01	15	30	12
		24	1	15	30	12

t – Temperature
Q – Flow rate
U – Voltage
L – The distance between needle and collector plate

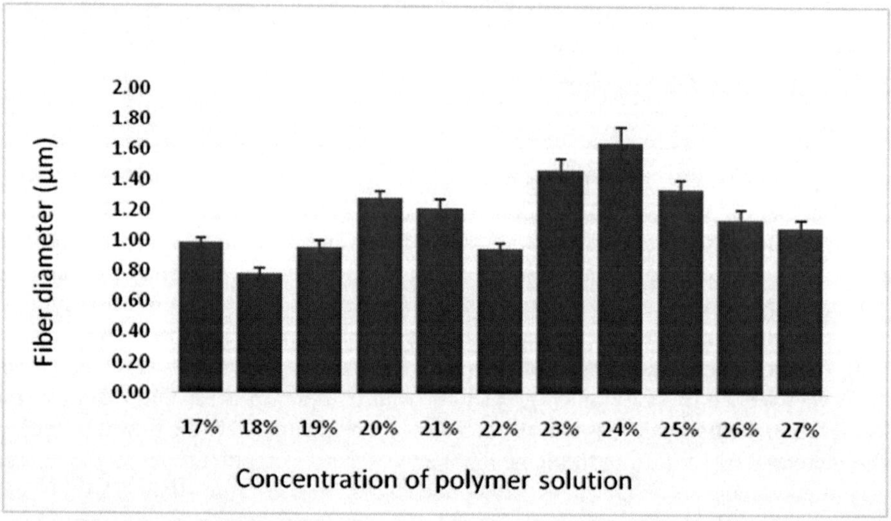

Fig. 2. The mean values of fibers diameter shown as mean ± standard error of 50 measurements per micrograph

poorly spinnable fibers not easy for manipulation. As the concentration increases, the diameter distribution is moved towards 2 and even 2.5 μm fibers (Fig. 4).

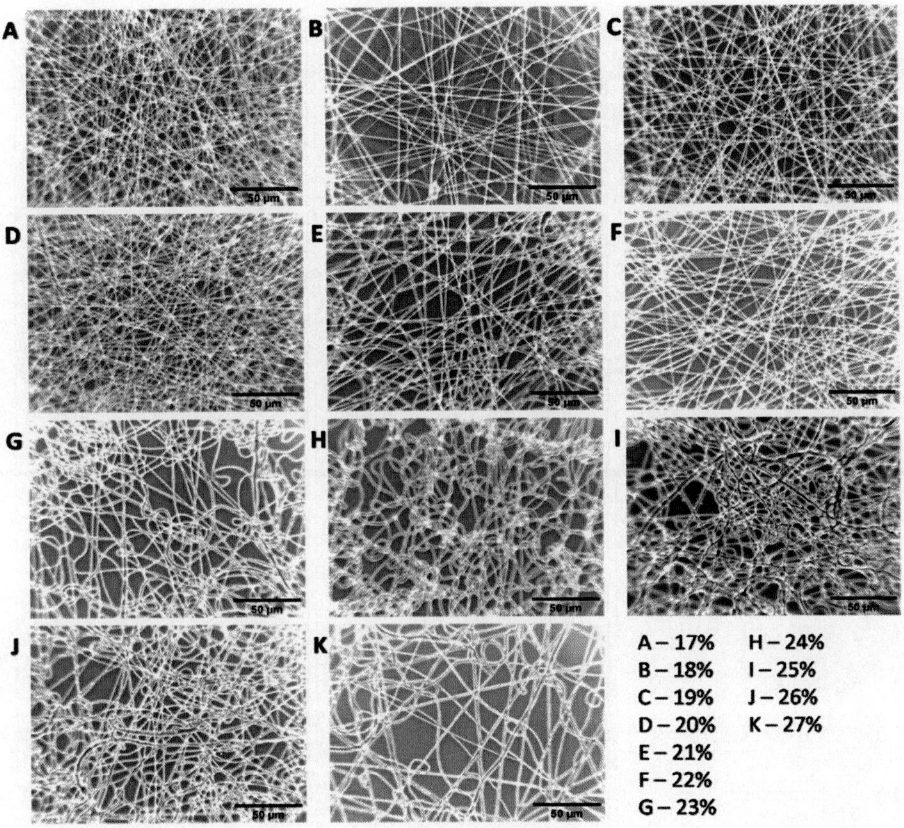

Fig. 3. Representative micrographs of synthesized fibers of various polymer concentrations

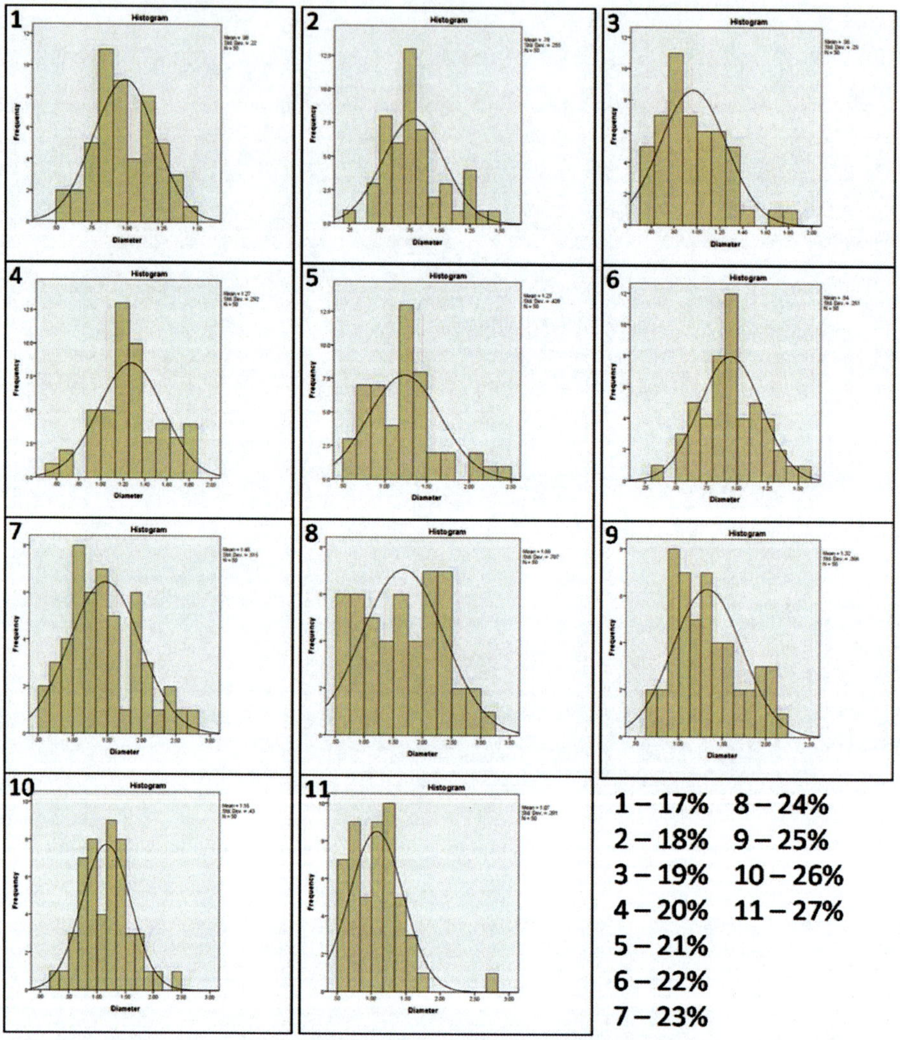

Fig. 4. Distribution of electrospun PCL/PEG fibers

4 Conclusions

Optimization of many experimental parameters for creating a scaffold that can be used in tissue engineering is a demanding process. Scaffolds made with desired properties will be used for cell seeding prior to create primitive tissues *in vitro*. We analyzed fiber diameter parameters for many combinations of PCL/PEG polymers. With a suitable scaffold, cell seeding can take place with aims towards developing an artificial tissue. Our conclusion is that use of 17 to 21% PCL: PEG solutions are optimal for producing the most appropriate and extracellular matrix like fibers.

Acknowledgment. This research is supported by the European Uio's Horizo 2020 research and innovation programme under grant agreement No 952603 (SGABU). This article reflects only the author's view. The Commission is not responsible for any use that may be made of the information it contains. Also, the research is funded by Serbian Ministry of Education, Science, and Technological Development [451-03-68/202014/200378 (Institute for Information Technologies, University of Kragujevac)].

Conflict of Interest Declaration. The authors have no conflict of declaration.

References

1. Vasita, R., Katti, D.S.: Nanofibers and their applications in tissue engineering. Int. J. Nanomed. **1**(1), 15–30 (2006)
2. Bui, H.T., Chung, O.H., Dela Cruz, J., Park, J.S.: Fabrication and characterization of electrospun curcumin-loaded polycaprolactone-polyethylene glycol nanofibers for enhanced wound healing. Macromol. Res. **22**, 1288–1296 (2014)
3. Liu, H., Ding, X., Zhou, G., Li, P., Wei, X., Fan, Y.: Electrospinning of nanofibers for tissue engineering applications. Biocompatibility and Toxicity of Nanobiomaterials, p. 495708 (2013)
4. Živanović, M.N.: Use of electrospinning to ehae the versatility of drug delivery. In: Lai, W.F. (eds) Systemic Delivery Technologies in Anti-Aging Medicine: Methods and Applications. Healthy Ageing and Longevity, vol. 13. Springer, Cham (2020). https://doi.org/10.1007/978-3-030-54490-4_14
5. Koupaei, N., Karkhaneh, A., Daliri, J.M.: Preparation and characterization of (PCL-crosslinked-PEG)/hydroxyapatite as bone tissue engineering scaffolds. J. Biomed. Mater. Res. Part A **103**(12), 3919–3926 (2015)
6. Danafar, H.: Preparation and characterization of PCL-PEG-PCL copolymeric nanoparticles as polymersomes for delivery hydrophilic drugs. Iran. J. Pharm. Sci. **14**(2), 21–32 (2018)

The Application of Nanotechnology in Constructing Scaffolds for Bone Tissue Engineering

Naida Mirvic(✉), Edina Vranic(✉), Jasmina Hadziabdic, Alisa Elezovic, and Lamija Hindija

Department of Pharmaceutical Technology, Faculty of Pharmacy, University of Sarajevo, Zmaja od Bosne 8, 71000 Sarajevo, Bosnia and Herzegovina
{naidamirvic,edina.vranic}@ffsa.unsa.ba

Abstract. The bone tissue regeneration represents one of the biggest challenges in nowadays medicine, firstly due to a complex structure of human bones and wide range of diseases that can occur on that level. Bone tissue engineering and nanotechnology approach reveals a brand new, revolutionary era for treatment of bone diseases and regeneration of bone structures. Targeted drug delivery with the use of optimal dosage and reduced incidence of side effects and infections, and improved therapeutic outcome can be achieved by applying nanocarriers. The formulations based on the nanotechnology principles will enable minimally invasive therapy and better uptake of therapeutic agents from the bone cells. Among various types of nanoparticles, the most used in bone tissue engineering and bone diseases therapy are liposomes, polymeric nanoparticles, dendrimers, silica nanoparticles, carbon nanotubes, quantum dots, gold and bioceramic nanoparticles. Limitations of regenerative strategies include mismatch of cell proliferation and differentiation, insufficient mechanical strength of the scaffold and inadequate synthesis of extrinsic factors necessary for osteogenesis. Given these limitations, nanotechnology offers new solutions, attributed to advantages of nanoparticles as drug carriers which are the result of their size that is similar to the size of natural bone structures.

Scaffolds are extremely important in providing architectural support for cell growth. They are biomaterials used as a replacement and guide for tissue repair and regeneration in the form of bulk materials or with the incorporation of nanomaterials. Our aim is to provide an overview of potential application of nanotechnology in bone tissue engineering and manufacturing bio-mimetic scaffolds that can be able to replace native tissue while minimizing complications. In this article we will discuss about the latest developments in the context of scaffold construction combined with nanotechnology, as well as the potential to encourage osteogenic differentiation of such formulations.

Keywords: Bone tissue · Scaffolds · Nanotechnology · Regenerative medicine · Tissue engineering · Nanoparticles

A. Badnjevic and L. Gurbeta Pokvić (Eds.): CMBEBIH 2021, IFMBE Proceedings 84, pp. 734–743, 2021.
https://doi.org/10.1007/978-3-030-73909-6_83

1 Introduction

According to recent research, bone has an impact on the activity of many other organs of the human body, but bones are also affected by other organs, which shows the complexity and dynamic nature of bone tissue [1]. The inorganic matrix is mostly occupied by hydroxyapatite, while collagen type 1 is found in the organic part, as well as other non-collagenous proteins, such as osteopontin, osteocalcin, and proteoglycans [2]. It is important to mention signaling molecules, such as TGF-β and IGF1, as noncollagenous proteins within the bone matrix. The mineral phase of bone, in the form of modified hydroxyapatite crystals, makes up 2/3 of the total bone mass. Based on the above, it can be concluded that the skeleton is a metabolically active organ, in which the organization of minerals and organic components determines the performance of mechanical function, which is enabled by a combination of dense, compact and trabecular bone, reinforced in places exposed to stress [3].

Osteochondral and bone defects are one of the main causes of disability worldwide. Bone lesions result from trauma or tissue degeneration and cause loss of joint mobility, severe pain, and deformity. Limitations of regenerative strategies include mismatch of cell proliferation and differentiation, insufficient mechanical strength of the scaffold and inadequate synthesis of extrinsic factors necessary for osteogenesis. Given these limitations, nanotechnology offers new solutions, producing materials that mimic the structures naturally present in bone. Due to their size similar to bone and cartilage structures, nanostructures play a very important role in bone tissue engineering [4, 5]. Nanoparticles (NPs) are used to model scaffold properties, which will result in superior mechanical properties and implant osteointegration, osteoconduction, and osteoinduction. If applied topically, they enhance tissue regeneration and reduce the risk of infections [5, 6].

The main goal of tissue engineering and regenerative medicine (TERM) is to create a functional replacement for damaged or diseased tissues with the connection of biological properties with engineering. The strategy used by TERM combines three main elements - scaffolds, stem cells and growth factors, in order to build the structure of the designed tissue. The role of the scaffold is to support cell growth and tissue formation, which is why they are seeded with stem cells that have the potential for differentiation and the formation of new tissue. Growth factors are also very important, as their function is to regulate the process of differentiation and proliferation [6].

From a clinical point of view, the reconstruction of bone damage depends on a number of factors, such as the size and area of damage, and mechanical load. Using bone tissue engineering, biomaterials are designed and constructed to increase cell growth affinity, improve integration, and encourage the growth and deposition of new bone tissue. Harmful tissue is removed, the cells needed for regeneration are then introduced into the configuration, which optimizes their survival in the environment. Tissue engineering offers advantages over cell transplantation, as it designs and develops three-dimensionally organized functional tissue [7, 8].

2 Scaffolds and Composites

The scaffolds are 3D matrices with a porous architecture capable of stimulating bone recovery and the growth of new bone tissue. The main goal of scaffold-based strategies is to obtain an implant capable of replacing native tissue, which will also support the tissue regeneration process, while minimizing complications. To achieve this goal, scaffolds must meet the following criteria [6, 9, 10]:

- Three-dimensionality and high porosity,
- Biocompatibility,
- Controlled rate of degradation and resorption,
- Surface chemistry should allow cell binding, proliferation and differentiation, and be similar to tissues at the site of implantation,
- Possibility of delivery of physico-chemical and mechanochemical signs, providing mechanical strength and support.

In general, scaffolds can be divided into 3 categories: metallic (alkali metal or alloy scaffolds), inorganic (bioactive glass and hydroxyapatite), and organic (polymerbased scaffolds) [9].

Nanoscaffolds are used as materials for growth in tissue engineering, as a system of local delivery of drugs, genes, but also for the regulation of cellular processes. They are typically composed of electrospun nanofibers, and are designed to be bio-mimicking, specifically in terms of representing the structure and function of the extracellular matrix. This technology could be incorporated into implants to increase bone integration and prevent infection as the leading cause of implantation failure [11].

Composites represent a combination of two or more materials with different properties, where each of them shows certain advantages and specific disadvantages. The combinations may be in the form of co-polymers, polymer-polymer mixtures or polymer-ceramics. Co-polymers are composed of two or more monomers, and an example is PLGA composed of poly lactide and polyglycolide. Polymer-polymer mixtures are composed of two different polymers, such as a PLGA-polyphosphazene mixture, wherein the polyphosphazines release only neutral or basic products. In this way, problems related to the formation of acid degradation products of PLGA, which cause necrosis and implant failure, are prevented. Polymer-ceramic composites are completely biomimetic, and the incorporation of bioceramic particles, as well as bioactive glass particles, carbon nanotubes, improves the mechanical characteristics of the scaffold [12].

2.1 Incorporation of Liposomes in Scaffolds

Due to their structure, liposomes have the ability to encapsulate both lipophilic and hydrophilic agents, with lipophilic being incorporated into the lipid bilayer and hydrophilic into the aqueous space [13]. Liposomes have enabled efficient drug delivery, while reducing side effects, all due to their benefits that include biocompatibility, biodegradability, non-toxicity and non-immunogenicity; possibility of protection of encapsulated drugs from the external environment; reduced toxicity due to reduced exposure of surrounding tissues to the drug; flexibility in pairing with site-specific ligands, to achieve active targeting [14].

In general, unmodified scaffolds show reduced ability to promote tissue regeneration and, consequently, to treat disease. Therefore, liposome-scaffold composite systems are designed to take advantage of both components, namely liposome biocompatibility and scaffold mechanical strength. Liposomes allow the scaffold to carry drugs, and when liposomes are modified on the surface of the implant or mixed into the scaffold matrix, there is the possibility of their concentration at a specific site [9].

With the aim of designing a scaffold that mimics the extracellular matrix, Mohammadi and associates developed a BMP-2 peptide encapsulated in liposomes covalently bound to scaffold nanofibers. The scaffold is made of PLLA nanofibers, obtained by electrospinning, coated with hydroxyapatite (HA). In order to improve the osteogenic differentiation of MSCs, the aim was to improve the efficiency of liposome uptake by modifying their surface area with BMP-2 peptide. It has been hypothesized that the electrostatically charged BMP-2 peptide may behave as a target ligand, in order to improve the insertion of nanoparticles into the MSC. Accordingly, two liposome formulations were prepared: BMP-2 peptide encapsulation and BMP-2 peptide surface modification. The results of the study showed that the electrostatically immobilized BMP-2 peptide acted as a ligand that enhances cellular internalization of liposomes, and osteogenic differentiation of MSCs, with an increase in ALP activity, compared to the encapsulated peptide. Liposomes filled with BMP-2 peptide and labeled with FAM (fluorescein amidites) were covalently bound to the scaffold and a sustained release of up to 20% encapsulated BMP-2 was observed over 6 days [15].

In a study by Sarkar and Bose liposomes filled with curcumin as an antioxidant, anti-inflammatory, anticancer and osteogenic agent were incorporated in a 3D printed calcium phosphate scaffold. The results showed that after 60 days, 17% of curcumin was released from the liposome formulation, while curcumin itself showed 30.9% release. Based on the above, it can be concluded that curcumin encapsulated in liposomes shows more controlled and sustainable release compared to curcumin itself. Liposomal curcumin released from 3D printed scaffolds showed significant cytotoxicity to in vitro osteosarcoma cells, and increased cell viability and osteoblast proliferation [16].

2.2 Silica Nanomaterials as Scaffolds

Silica nanomaterial (SNM) scaffolds can stimulate the growth of bone cells and enable the simultaneous delivery of biologically active molecules encapsulated in mesopores. They are good scaffolds due to their high mechanical strength, low cytotoxicity, and ability to stimulate proliferation and osteogenic differentiation of progenitor cells. The ability to deliver small biomolecules such as dexamethasone, growth factors, vitamins and mineral ions depends on the morphology, porosity and crystallinity of SNM and their composites with other polymeric materials. Controlled release of rhBMP-2 can be achieved by loading on mesoporous SNMs used as scaffolds [17].

The use of dexamethasone in combination with FGF-18 in the scaffold could be an improved method of stimulating the formation of new bone. The second most important application of SNM-based scaffolds is the delivery of antitumor drugs in the treatment of osteosarcoma, where controlled drug release is achieved with SNM uniform pore sizes [17].

Yao and associates developed a 3D nanofibrous gelatin scaffold (GF) with incorporated mesoporous silica nanoparticles, for dual delivery of BMP-2 and deferoxamine (DFO). Deferoxamine is a hypoxia-mimetic drug, which activates hypoxia-inducible factor-1 α (HIF-1α), which in turn leads to angiogenesis. The sustained release system of BMP-2 was achieved by encapsulation in highly porous MSN, and the relatively short-term release of DFO was achieved by covalent conjugation with chitosan, with reduced cytotoxicity and extended half-life. Both components were incorporated into a porous 3D nanofibrous scaffold that served as a biomimetic osteogenic microenvironment. It has been shown that these components are released at different rates while achieving their angiogenic and osteogenic effects, as well as that DFO significantly improved the osteogenic differentiation induced by BMP-2. Significantly higher amounts of BMP-2 are released from the GF/MSN/BMP2/CTS scaffold compared to GF/MSN-BMP-2, so it can be concluded that coating the surface of MSN and GF scaffold with chitosan allows to protect proteins from rapid degradation. It is similar in the case of deferoxamine where covalent conjugation of deferoxamine with chitosan achieved sustainable DFO release over a period of 10 days [18].

2.3 Carbon Nanotubes (CNT) Combined with the Scaffolds

The high strength, elasticity and resistance of carbon nanotubes make them a very important material in the fabrication of composite scaffolds in bone tissue engineering. They form strong bonds with the composite, which helps to efficiently transfer the charge, and improves the strength of the matrix material. On the other hand, the 3D porous structure of CNT has a highly specific surface area, which leads to increased protein adsorption and increased cell adhesion [19].

Tanaka and coworkers created a 3D block structure composed of MWCNT with maximum mechanical strength and monitored efficiency as a scaffold material for bone regeneration. MWCNT blocks containing 5 μg rhBMP-2 were implanted in the mouse back muscle and evaluated with microcomputed tomography (μCT) 3 weeks after implantation. Ectopic new bone formation was confirmed on a positive control of PET (polyethylene terphthalate)-reinforced collagen scaffolds with BMP-2, and similar ectopic formation was observed on MWCNT blocks with rhBMP-2. In an in vitro study, MWCNT blocks did not show toxicity to three representative cell types [20].

2.4 Bioceramics as Scaffolds

Bone-bioactive nanoceramics, including calcium phosphate (CPC), nHA, and bioactive glass are used to regenerate bone tissue, where nanoceramics with smaller pore sizes could better promote osteochondral formation, and consequently both osteogenesis and vascularization. Nanostructured CPC is a synthetic material for bone graft and was first approved by the FDA in 1996 for the repair of craniofacial defects [7].

In addition to mechanical support, scaffolds based on hydroxyapatite continuously release bioactive elements such as calcium, phosphorus, important for tissue regeneration [9]. Bioactive glass nanomaterials can bind to the bone surface and convert to an apatite-like bone mineral deposited on the surface, which increases bone graft bioactivity, osteogenic differentiation, and cell mineralization. However, the disadvantages of these nanomaterials are poor flexibility, fragility and poor mechanical strength, which limits their application as scaffold materials [7].

Chen and associates prepared a biomimetic scaffold composite of collagen and biphasic calcium phosphate nanoparticles (BCP-NP) with controlled release of dexamethasone and controlled pore structures for bone tissue engineering. Dexamethasone (DEX), as a low molecular weight osteoinductive factor, was incorporated into BCPNPs during their synthesis and hybridized to collagen scaffolds, whose pore structure was controlled. Scaffold composites showed high mechanical strength, sustained release of dexamethasone, interconnected pore structures, and good biocompatibility. Scaffolds also promoted osteogenic differentiation of hMSCs when used for 3D culture of said cells. Subcutaneous implantation of the composite into athymic nude mice as models resulted in facilitated, ectopic bone tissue regeneration [21].

2.5 Nanopolymeric Scaffolds

Polymers can be both bioactive and biodegradable, making them easily workable and adaptable, which is why they are often used to make scaffolds. As the most commonly used organic nanomaterial, nanostructured polymers typically exhibit a onedimensional size of less than 100 nm, but the dimensions can be adjusted according to different applications. Polymer nanomaterials used for nanoscaffolds include poly (1lactide) (PLLA), poly (lactide-co-glycolide) (PLGA), poly (ε-caprolactone) (PCL), silk, and chitosan. They are also well known for their biocompatibility and resorption by natural metabolic pathways, without causing side effects, and have also shown improved osteoconductivity and osseointegration [7]. Besides the above mentioned advantages, synthetic polymers as materials show disadvantages in terms of weaker interaction with cells compared to natural polymers, which show better bioactivity properties due to their intrinsic nature. Although polymers are the most promising materials for application in BTE, they show low rigidity and cannot completely recapitulate the mechanical properties of normal bone tissue [7, 12].

Chitosan (CS) enhances osteoblast adhesion, proliferation, differentiation, and mineralization. It is also used in BTE, as it can be constructed to form 3D scaffolds with different pore structures, as well as composites with a number of materials, including ceramics and polymers. Chitosan acts as a template for biomineralization due to the presence of - NH2 and C = O functional groups. Chitosan-based biocomposite scaffolds show improved biomineralization potential, compared to chitosan itself [10].

Gharib and associates developed chitosan-nanohydroxyapatite scaffolds as bone grafts in rabbits with induced osteoporosis. Two groups of rabbits were treated: group A (osteoporosis without graft material) and group B (osteoporosis with CS (nHA)). In group A, the cups recovered spontaneously without the use of material, while in group B the cups were filled with CS/nHA. The effects were evaluated with SEM and EDXA, where SEM revealed greater trabecular bone thickness compared to the regular bone surface, and uniform lacunae of osteocytes in group B, compared with group A. EDXA analysis showed a higher ratio of Ca: P in group B. In conclusion of this study, it is stated that scaffolds based on CS/nHA are effective materials for hard tissue engineering, with interconnected pores and mechanical strength. They enable migration, growth, and proliferation resulting in good integration with surrounding tissue, and rapid bone recovery [22].

Injectable Hydrogels

An important class of polymers used in BTE are hydrogels, and they represent hydrophilic polymer networks that can absorb water. This property allows cells to adhere, multiply and differentiate. Natural (agarose, alginate and gelatin) and synthetic (polyvinyl alcohol) hydrogels can mimic ECM topography and deliver bioactive molecules [12].

Hydrogels are used as scaffolds in bone and cartilage engineering primarily because of their structural similarities to the extracellular matrix, as well as because of the porous framework that allows cell transplantation and proliferation. These are 3D networks that swell in an aqueous medium and provide an adequate environment similar to ECM, which further leads to easier migration, adhesion, proliferation and differentiation of chondrocytes and osteoprogenitor cells into osteoblasts. Injectable hydrogels can replace surgery, with a minimally invasive method of injection, and can take the desired shape as they can coincide with irregular defects [23].

Filling hydrogels with bioactive agents is a method by which it is possible to expand their application. However, for this application, the challenge is to achieve local application and slow release. Liu and associates developed adhesive liposomes (A-LIP) filled with BMP-2, followed by incorporation into PEG hydrogels based on the coordinated cross-linking principle of thiolated polyethylene glycol (SH-PEG) and Ag +, and produced an injectable, antibacterial, multifunctional drug delivery system. This formulation can be injected locally into the osteoporotic fracture and the bone marrow cavity, after which adhesive liposomes are released, adhered to the site of damage and they promote bone reconstruction. In vitro studies have shown that the formulation significantly promoted osteogenic differentiation, without much impact on cell proliferation. Compared with the classical lipo-hydrogel, A-LIP-PEG showed better in vivo adhesion, better osteogenic differentiation and consequently faster local bone remodeling in osteoporotic fractures in rats [24].

A synergistic approach to incorporating osteoconductive nanomaterials built on hydrophilic, biocompatible scaffolds, specifically incorporated with biochemical signs for direct bone regeneration, is promising for future generations of therapies to reverse bone loss that occurs in osteoporosis [25]. One of the unresolved barriers is the creation of vascularized and innervated bone tissue. Imitation of bone tissue, including cells and ECM requires a combination of nanotechnology with advanced fabrication techniques, such as multimaterial bioprinting. The challenges of such an approach are the different mechanical properties of relatively hard materials, which are suitable for modeling ECMs and hydrogels used in bioprinting. With the advancement of high-resolution printing techniques in the future, there will be great advances in the field of hierarchical bone tissue bioprinting [26]. Table 1 shows the scaffolds and composites with proven efficacy in various studies.

Table 1. The scaffolds and composites with proven efficacy in various studies.

Scaffolds and composites	Applications and results of the studies	References
BMP-2 peptide encapsulated in liposomes covalently bound to scaffold nanofibers	Scaffold that mimics the extracellular matrix; a sustained release of up to 20% encapsulated BMP-2 was observed over 6 days	[15]
Liposomes filled with curcumin, incorporated in a 3D printed calcium phosphate scaffold	Liposomal curcumin released from 3D printed scaffolds showed significant cytotoxicity to in vitro osteosarcoma cells	[16]
3D nanofibrous gelatin scaffold (GF) with incorporated mesoporous silica nanoparticles, for dual delivery of BMP-2 and deferoxamine (DFO)	The sustained release system of BMP-2 was achieved by encapsulation in highly porous MSN, and the relatively short-term release of DFO was achieved by covalent conjugation with chitosan, achieving angiogenic and osteogenic effects	[18]
3D block structure composed of MWCNT blocks containing 5 μg rhBMP-2	Ectopic new bone formation	[20]
Biomimetic scaffold composite of collagen and biphasic calcium phosphate nanoparticles (BCP-NP)	High mechanical strength, interconnected pore structures, and good biocompatibility	[21]
Chitosan-nanohydroxyap-atite scaffolds	Enable migration, growth, and proliferation resulting in good integration with surrounding tissue, and rapid bone recovery	[22]

3 Conclusion

Bone regeneration is a great challenge for tissue engineering, which is a consequence of the complexity of the structure of the skeletal system and the great mechanical stress to which the bones are exposed. By applying nanotechnology, it is possible to overcome the limitations of current strategies, since the materials in the nanoscale are structures similar to natural bone structures. Nanoparticles are used in the construction of bio-mimicking scaffolds and nanocomposites, and it has been proven that such systems more effectively encourage the formation of new bone compared to materials in the microscale. Nanotechnology offers improved efficiency and longevity of implants, efficient regeneration of bone structures, reducing the risk of developing infections. However, in order to understand the safety profile and long-term toxic effects in vivo, further research is needed.

Conflict of Interests. The authors have no conflicts of interest to disclose.

References

1. Silva, R.F., Sasso, G.R., Sasso-Ceri, E., Simoes, M.J., Cerri, P.S.: Biology of bone tissue: structure, function and factors that influence bone cells. Biomed. Res. Int. **2015**, 1–17 (2015)
2. Brunetti, G., Colaianni, G., Colucci, S., Grano, M.: Anatomy and physiology of skeletal tissue: the bone cells. In: Lenzi, A., Migliaccio, S. (eds) Multidisciplinary Approach to Osteoporosis, pp. 1–23. Springer International Publishing, Cham (2018)
3. Sims, N.A., Martin, T.J.: The osteoblast lineage: its actions and communication mechanisms. In: Bilezikian, J.P., Clemens, T.L., Martin, T.J., Rosen, C.J. (eds) Principles of Bone Biology, Fourth edition. Elsevier Science, London (2020)
4. Filippi, M., Born, G., Felder-Flesch, D., Scherberich, A.: Use of nanoparticles in skeletal tissue regeneration and engineering. Histol. Histopathol. **35**(4), 331–350 (2020)
5. Walmsley, G.G., McArdle, A., Tevlin, R., Momeni, A., Atashroo, D., Hu, M.S., et al.: Nanotechnology in bone tissue engineering. Nanomed. Nanotechnol. Biol. Med. **11**(5), 1253–1263 (2015)
6. Vieira, S., Vial, S., Reis, R.L., Oliveira, J.M.: Nanoparticles for bone tissue engineering. Biotechnol. Prog. **33**, 590–611 (2017)
7. Lowe, B., Guastaldi, F., Muller, M.L., Gootkind, F., Troulis, M.J., Ye, Q.: Nanobiomaterials for bone tissue engineering. In: Choi, A., Ben-Nissan, B. (eds) Marine-Derived Biomaterials for Tissue Engineering Applications, pp. 81–97. Springer, Singapore (2019)
8. Vacanti, J.P., Vacanti, C.A.: The history and scope of tissue engineering. In: Lanza, R., Langer, R., Vacanti, J. (eds) Principles of Tissue Engineering: Fourth edition, pp. 3–8. Elsevier Inc., Cambridge (2014)
9. Cheng, R., Liu, L., Xiang, Y., Lu, Y., Deng, L., Zhang, H., et al.: Advanced liposome-loaded scaffolds for therapeutic and tissue engineering applications. Biomaterials **232**, 119706 (2020)
10. Saravanan, S., Leena, R.S., Selvamurugan, N.: Chitosan based biocomposite scaffolds for bone tissue engineering. Int. J. Biol. Macromol. **93**, 1354–1365 (2016)
11. Kwaees, T.A., Pearce, A., Ring, J., Sutton, P., Charalambous, P.: Nanotechnology and its applications in knee surgery. In: Jackson, M.J., Ahmed, W. (eds.) Micro and Nanomanufacturing Volume II, pp. 35–53. Springer, Cham (2018)
12. Roseti, L., Parisi, V., Petretta, M., Cavallo, C., Desando, G., Bartolotti, I., et al.: Scaffolds for bone tissue engineering: state of the art and new perspectives. Mater. Sci. Eng. **78**, 1246–1262 (2017)
13. Bhatia, S.: Nanoparticles types, classification, characterization, fabrication methods and drug delivery applications. In: Natural Polymer Drug Delivery Systems, pp. 33–93. Springer International Publishing, Cham (2016)
14. Kapoor, B., Singh, S.K., Gulati, M., Gupta, R., Vaidya, Y.: Application of liposomes in treatment of rheumatoid arthritis: quo vadis. Sci. World J. **2014**, 17 (2014)
15. Mohammadi, M., Alibolandi, M., Abnous, K., Salmasi, Z., Jaafari, M.R., Ramezani, M.: Fabrication of hybrid scaffold based on hydroxy apatite biodegradable nanofibers incorporated with liposomal formulation of BMP-2 peptide for bone tissue engineering. Nanomed. Nanotechnol. Biol. Med. **14**(7), 1987–1997 (2018)
16. Sarkar, N., Bose, S.: Liposome-encapsulated curcumin-loaded 3D printed scaffold for bone tissue engineering. ACS Appl. Mater. Interfaces. **11**(19), 17184–17192 (2019)
17. Eivazzadeh-Keihan, R., Chenab, K.K., Taheri-Ledari, R., Mosafer, J., Hashemi, S.M., Mokhtarzadeh, A., et al.: Recent advances in the application of mesoporous silica-based nanomaterials for bone tissue engineering. Mater. Sci. Eng. C **107**, 110267 (2020)
18. Yao, Q., Liu, X., Selvaratnam, B., Koodali, R.T., Sun, H.: Mesoporous silicate nanoparticles/3D nanofibrous scaffold-mediated dual-drug delivery for bone tissue engineering. J. Control. Release **279**, 69–78 (2018)

19. Pei, B., Wang, W., Dunne, N., Li, X.: Applications of carbon nanotubes in bone tissue regeneration and engineering: superiority, concerns, current advancements, and prospects. Nanomaterials **9**, 1501 (2019)
20. Tanaka, M., Aoki, K., Haniu, H., Kamanaka, T., Takizawa, T., Sobajima, A., et al.: Applications of carbon nanotubes in bone regenerative medicine. Nanomaterials **10**, 659 (2020)
21. Chen, Y., Kawazoe, N., Chen, G.: Preparation of dexamethasone-loaded biphasic calcium phosphate nanoparticles/collagen porous composite scaffolds for bone tissue engineering. Acta Biomater. **67**, 341–353 (2018)
22. Gharib, H., El-Sawa, A.A., Karam, S.S., Hamza, S.A.: Scanining electron microscopic evaluation of chitosan nanohydroxyapatite scaffold for socket healing in rabbits with induced osteoporosis. Alexandria Dent. J. **45**(1), 27–33 (2020)
23. Liu, M., Zeng, X., Ma, C., Yi, H., Ali, Z., Mou, X., et al.: Injectable hydrogels for cartilage and bone tissue engineering. Bone Res. **5**, 17014 (2017)
24. Liu, L., Xiang, Y., Wang, Z., Yang, X., Yu, X., Lu, Y., et al.: Adhesive liposomes loaded onto an injectable, self-healing and antibacterial hydrogel for promoting bone reconstruction. NPG Asia Mater. **11**, 81 (2019)
25. Barry, M., Pearce, H., Cross, L., Tatullo, M., Gaharwar, A.K.: Advances in nanotechnology for the treatment of osteoporosis, current osteoporosis reports. Curr. Osteoporos. Rep. **14**(3), 87–94 (2016)
26. Hill, M.J., Qi, B., Bayaniahangar, R., Araban, V., Bakhtiary, Z., Doschak, M.R., et al.: Nanomaterials for bone tissue regeneration: updates and future perspectives. Nanomedicine **14**(22), 2987–3006 (2019)

Silver Nanoparticles Biosynthesis from Cinnamon (*Cinnamomum Verum*) and Cloves (*Syzygium Aromaticum*)

Ermina Krndžija, Amina Stambolić(✉), and Enisa Omanović-Mikličanin

Faculty of Agriculture and Food Sciences, University of Sarajevo, Zmaja od Bosne 8, 71000 Sarajevo, Bosnia and Herzegovina
{a.stambolic,e.omanovic-miklicanin}@ppf.unsa.ba

Abstract. Nanotechnology represents the molecular production or creation of objects at the level of atoms, by grouping individual atoms and molecules. Dried plant material is cheap and easily available, and contains the necessary active substances for nanoparticle synthesis. Synthesis pathways such as this one do not involve the use of toxic substances and do not result in their production, so they belong to the green synthesis of nanoparticles. This paper describes the green method of obtaining silver nanoparticles (AgNPs) using plant extracts, namely cinnamon (Cinnamomum verum) and cloves (Syzygium aromaticum). After their synthesis, the obtained nanoparticles were characterized by UV/Vis spectroscopy. Morphological analysis of nanoparticles using a transmission electron microscope (TEM) was also performed.

Keywords: Nanotechnology · Nanoparticles · Silver nanoparticles · Green synthesis · UV/Vis spectroscopy · TEM analysis

1 Introduction

The meaning of "nanotechnology" varies from field to field and country to country and is widely used as a "catch all" description for anything very small. The most common commonly it is defined as understanding, control, and restructuring of matter on the order of nanometers (i.e., less than 100 nm) to create materials with fundamentally new properties and functions [1]. Nanotechnology has almost unlimited potential in diverse applications. The era of nanotechnology has been named the second Industrial Revolution [2].

Involving green chemistry and green engineering principles in nanotechnology, green nanotechnology has been created. Green nanotechnology enables full potential of nanotechnology applications without concern about nanoparticle safety, impact on the environment and consequences on human health [2]. The appliance of green chemistry principles and green engineering in the development of nanomaterials and in the design of production techniques, enable green nanotechnology to create sustainability alongside the protected environment and human health [3–5].

A. Badnjevic and L. Gurbeta Pokvić (Eds.): CMBEBIH 2021, IFMBE Proceedings 84, pp. 744–747, 2021.
https://doi.org/10.1007/978-3-030-73909-6_84

Silver has been used in numerous applications because it retains antimicrobial activity against pathogens. It is used in the form of pure free metal or as a compound. Due to the production of products with potential impact on the environment while synthesizing AgNPs, the development of a more ecologically safe green methods is progressing as an important branch of nanotechnology [6, 7]. Nanoparticles synthesized using green technology have diverse natures, greater stability and appropriate dimensions since they are synthesized using a one-step procedure. Utilization of plants as a green source for nanoparticles synthesis is most commonly practiced since plants are the hub of a wide range of phytochemicals which can act as reducing and stabilizing agent for NPs synthesis. There is no need to use high pressure, energy, temperature, and toxic chemicals in green synthesis of nanoparticles [8–10].

2 Materials and Method

Ground cinnamon and cloves required for obtaining the necessary extracts were obtained from commercial sources. The extracts were prepared by boiling 2,5 g of ground plant material in 100 mL of distilled water. After cooling to room temperature, the extracts were filtered through Whatman No.1 filter paper. Silver nanoparticles were then prepared by adding 1 mL of extract to 50 mL of $AgNO_3$ c = 1 mM solution, with constant stirring at 650 rpm. The resulting solutions were then left to rest for 12 h to ensure silver nanoparticle synthesis.

Silver nanoparticle synthesis was confirmed by UV/Vis spectroscopy by identifying a stable peak of absorbance at 430 nm which is in accordance with silver nanoparticles having a peak in the 400–500 nm range. Redistilled water was used as the blank probe. The shape and size of the synthesized silver nanoparticles were characterized by a transmission electron microscope. The samples for TEM analysis were prepared by dripping the AgNPs solution onto a carbon coated copper mesh grid. Excess liquid was removed with filter paper.

3 Results and Discussion

In this paper, a AgNPs synthesis using the extracts of cinnamon and cloves is observed. A silver nitrate solution was used as the source of silver ions. The silver ions were reduced to elementary silver by the bioactive components of previously mentioned extracts, mainly by eugenol. Nanoparticles prepared this way were stable enough, so the introduction of additional stabilizers wasn't needed. Cinnamon and cloves extracts, after mixing with the silver nitrate solution were incubated at room temperature for 12 h during which the color of the solution changed from yellow to brown with a noticable increase of color intensity. This color change indicates the reduction of silver ions and the aglomeration of the formed silver atoms.

The unique property of silver nanoparticles is that they can exhibit surface plasma resonance (SPR) in interaction with visible light photons, resulting in a darker color of the nanoparticles. SPR peaks and wavelengths have been shown to be sensitive to the size and shape of nanoparticles, and their height provides concentration information.

For the AgNPs sythesized using the cloves extract the peak of absorbance was observed at 430 nm. Nanoparticles obtained from the cinnamon extract showed the same peak of absorbance. The solution spectra were recorded again after 7 days. The peaks retained their position which proved the stability of the nanoparticles without any additional stabilizers. Based on these findings, it can be concluded that substances contained in plant extracts also serve as stabilizers, and not only reducing agents. According to previous research, it has been proven that terpenoids are often associated with nanoparticles. Terpenoids are a class of diverse organic polymers synthesized in plants from five-carbon isoprene units that exhibit strong antioxidant activity. Eugenol, the major terpenoid of *Cinnamomum verum* and *Syzygium aromaticum* extracts, was found to play a major role in the bioreduction of $AgNO_3$ to nanoparticles. When it comes to the stability of the obtained nanoparticles even after a certain period, it is considered that the phytochemicals present in the essential oil *Cinnamomum verum* and *Syzygium aromaticum* act on the surface of the nanoparticles and lead to their stabilization and preservation.

3.1 TEM Analysis

Based on the obtained images, we find that the silver nanoparticles obtained from *Cinnamomum verum* extract are quite small, their dimensions are 8.20 nm, 12.08 nm, 14.07 nm, 17.7 nm. The particles did not agglomerate, they are quite scattered throughout the sample, they are mostly spherical in shape, although hexagonal shapes are also noticeable.

On the other hand, nanoparticles obtained from the *Syzygium aromaticum* extract, are much more stable, i.e. they are extremely noticeable and are double the size of the ones obtained from the *Cinnamomum verum* extract: 24.50 nm, 30.40 nm, 32.99 nm, 39.72 nm. In this case, hexagonal, spherical and cubic shapes predominate although elongated shapes are also visible. So, what we can conclude is that nanoparticles obtained from the cloves extract have better visual characterstics. In fact, the *Syzygium aromaticum* extract had a much darker color from the beginning compared to *Cinnamomum verum extract*, although an equal amount of sample was taken to prepare them. We attribute this phenomenon to the fact that *Syzygium aromaticum* contains greater amounts of bioactive substances that participate in this process of reduction of Ag^{+}ions, so that they later contributed to the formation of more concentrated nanoparticles.

4 Conclusion

Plant extracts of *Cinnamomum verum* and *Syzygium aromaticum* can be used as reducing agents for the production of stable silver nanoparticles as a method of green nanoparticle synthesis. The presence of synthesized silver nanoparticles was confirmed within 30 min, primarily visually, by observing the solution color change from yellow to brown. Prolonging the incubation time, there was an increase in the concentration of nanoparticles, which was observed by the increase of color intensity. It was further proven by UV/Vis spectroscopy where the absorption maximum was in the 400–500 nm range. The obtained silver nanoparticles proved to be extremely stable without the use of external artificial stabilizers, demonstrated by the fact that the maximum absorbtion wavelength stayed the same during seven days. Images obtained by TEM showed that the dimensions of *Cinnamomum verum* nanoparticles were almost half the size (8.20 nm, 12.08 nm, 14.07 nm, 17.7 nm), compared to the dimensions of *Syzygium aromaticum* nanoparticles (24, 50 nm, 30.40 nm, 32.99 nm, 39.72 nm). TEM obtained images show that the shapes of Cinnanomum verum nanoparticles are exclusively spherical, whereas in Syzygium *aromaticum* nanoparticles, hexagonal, spherical shapes, cubic and elongated nanoparticles can be distinguished.

References

1. Sanchez, F., Sobolev, K.: Nanotechnology in concrete – a review. Constr. Build. Mater. **24**, 2060–2071 (2010)
2. Maksimovic, M., Omanovic-Miklicanin, E., Badnjevic, A.: Nanofood and Internet of Nano Things-For the Next Generation of Agriculture and Food Sciences. Springer Nature, Switzerland AG (2019). ISBN 978-3-030-15053-2
3. Maksimovic, M., Omanovic-Miklicanin, E.: Towards green nanotechnology: maximizing benefits and minimizing harm. In: CMBEBiH 2017, pp. 164–170 (2017). https://doi.org/10.1007/978-981-10-4166-2_26
4. Nath, D., Banerjee, O.: Green nanotechnology – a new hope for medical biology. Environ. Toxicol. Pharmacol. **36**, 997–1014 (2013)
5. Bhattacharya, S., Kumar Agarwal, A., Chanda, N., Pandey, A., Kumar Sen, A.: Environmental Chemical and Medical Sensors. Springer Nature, Singapore Pte. Ltd. (2018)
6. Kazlagić, A., Omanović-Mikličanin, E., Hamidović, S.: Synthesis, characterization and antimicrobial activity of silver nanoparticles. In: Karabegović, I. (ed.) New Technologies, Development and Application II. NT 2019. Lecture Notes in Networks and Systems, vol. 76. Springer, Cham (2020)
7. Kazlagić, A., Lagumdzija, A., Borovac, B., Abud, O.A., Hamidović, S., Omanović-Mikličanin, E.: Green Synthesis and characterization of silver nanoparticles using fresh leaf extract of Aloe vera barbadensis Miller, Aloe vera and Sempervivum tectorum and Its Antimicrobial Activity Studies. In: Brka, M., et al. (Eds.) AgriConf 2019. IFMBE Proceedings, vol. 78, pp. 335–341 (2020). https://doi.org/10.1007/978-3-030-400491_42
8. Zhang, Y., Yang, D., Kong, Y., Wang, X., Pandoli, O., Guo, G.: Synergetic antibacterial effects of silver nanoparticles aloe vera prepared via a green method. Nano Biomed. Eng. **2**(4), 252–257 (2010)
9. Happy, A., Soumya, M., Kumar, S.V., Rajeshkumar, S., Sheba, R.D., Lakshmi, T., Nallaswamy, V.D.: Phyto-assisted synthesis of zinc oxide nanoparticles using Cassia alata and its antibacterial activity against Escherichia coli. Biochem. Biophys. Rep. **17**, 208–211 (2019)
10. Joghee, S., Ganeshan, P., Vincent, A., et al.: Ecofriendly biosynthesis of zinc oxide and magnesium oxide particles from medicinal plant Pisonia grandis R. Br. leaf extract and their antimicrobial activity. BioNanoScience **9**, 141–154 (2019)

Medical Physics, Biomedical Imaging and Radiation Protection

Breast Lesions Detection Using FADHECAL and Multilevel Otsu Thresholding Segmentation in Digital Mammograms

Saifullah Harith Suradi[1](✉), Kamarul Amin Abdullah[1,2], and Nor Ashidi Mat Isa[3]

[1] School of Medical Imaging, Faculty of Health Sciences, Universiti Sultan Zainal Abidin (UniSZA), 21300 Kuala Nerus, Terengganu, Malaysia

[2] Medical Imaging Research Group (MIRG), Faculty of Health Sciences, Universiti Sultan Zainal Abidin (UniSZA), 21300 Kuala Nerus, Terengganu, Malaysia

[3] School of Electrical and Electronic Engineering, Universiti Sains Malaysia, 14300 Nibong Tebal, Penang, Malaysia

Abstract. Breast cancer is the most common cause of mortality among women. Early detection plays an important role to improve survival rates. Digital mammograms can be used to detect breast lesions within the breast tissue. However, digital mammograms have a limitation of low contrast images due to the low exposure factors used. As a result, the extraction of breast lesions using the region of interest (ROI) tool will be difficult and, thus, lead to misclassification. This paper presents a novel technique to detect breast lesions in digital mammograms, known as Fuzzy Anisotropic Diffusion Histogram Equalization Contrast Adaptive Limited (FADHECAL) incorporated with Multilevel Otsu Thresholding Segmentation. FADHECAL will enhance the breast lesions by reducing the image noise while preserving the details. Multilevel Otsu Thresholding Segmentation detects the breast lesions using the ROI tool at different intensity levels. The performance of FADHECAL incorporated with Multilevel Otsu Thresholding Segmentation has been tested on 115 digital mammograms from the Mammographic Image Analysis Society (MIAS) database with the abnormal conditions. The efficiency of the proposed technique is 94.8%, and the error rate is 5.2%. In conclusion, FADHECAL incorporated with the Multilevel Otsu Thresholding Segmentation has provided sufficient detection of breast lesions with the appropriate quality of the digital mammograms.

Keywords: Breast cancer · Digital image processing · Histogram equalization · Mammography · Morphological operations

1 Introduction

In most developed countries, breast cancer is one of the main causes of increased mortality among women. According to the American Cancer Society (ACR), in 2020, a total of 276,480 new cases of breast cancer are expected to be diagnosed among women in

A. Badnjevic and L. Gurbeta Pokvić (Eds.): CMBEBIH 2021, IFMBE Proceedings 84, pp. 751–759, 2021.
https://doi.org/10.1007/978-3-030-73909-6_85

the United States [1]. This expectation can cause the mortality of about 42,170 women. However, from 2013 to 2017, a total death rate of breast cancer has decreased by 1.3% per year [1]. One of the main factors is the early screening and detection of breast cancer.

Digital mammograms can be used to screen and diagnose breast cancer at an early stage. Digital mammograms, a low dose of X-ray imaging, are able to demonstrate the soft tissue of the breast. However, the low exposure factors used in digital mammograms have led to low contrast appearances and thus, reduces breast lesions' detectability and image details. The low contrast appearances are mainly attributed to image noise [2]. Image enhancement techniques are used to emphasize the image details and enhance any suspected irregular tissues in digital mammograms. Therefore, using an appropriate image enhancement technique is necessary to improve the image quality in digital mammograms and reduce the image noise while preserving the image quality.

When using computer-aided diagnosis (CAD), breast lesions detection is a significant indication to improve the image quality of digital mammograms for clinical diagnosis. Image segmentation is used to perform and detect affected areas, especially at the breast lesions regions in the digital mammograms [3]. The main limitation of the CAD is the difficulty to extract the breast lesions using the ROI tool from low contrast appearances and thus, lead to misclassification [4, 5]. Therefore, an accurate segmentation will greatly affect the performance to detect breast lesions in digital mammograms. In previous studies, various image segmentation techniques have been introduced, such as Double Thresholding [6], Graph Cuts [7], and Marker Controlled Watershed Transform [3]. However, Otsu thresholding segmentation is one of the most commonly used due to fast and straightforward implementation [8].

This paper proposes a novel technique to detect breast lesions known as Fuzzy Anisotropic Diffusion Histogram Equalization Contrast Adaptive Limited (FADHECAL) incorporated with Multilevel Otsu Thresholding Segmentation method in digital mammograms.

2 Material and Methods

2.1 Image Dataset

Digital mammograms used in this study were retrieved from the Mammographic Image Analysis Society (MIAS) database [9]. The database consists of 322 digital mammograms. Of these, 115 abnormal digital mammograms were selected. Each digital mammogram has a size of 1024 × 1024 pixels. The radiologists marked the location of abnormalities with centroid and radius in the digital mammograms. Table 1 shows selected digital mammograms with different background tissues and abnormalities.

Table 1. Selected digital mammogram with three types of background tissues and two types of abnormalities (benign and malignant).

Condition of breast lesions	Background tissue	Number of digital mammograms
Benign	Fatty	21
	Glandular	19
	Dense	23
Malignant	Fatty	19
	Glandular	17
	Dense	16

2.2 Fuzzy Anisotropic Diffusion Histogram Equalization Contrast Adaptive Limited (FADHECAL) and Multilevel Otsu Thresholding Segmentation Technique

The combination of Fuzzy Anisotropic Diffusion Histogram Equalization Contrast Adaptive Limited (FADHECAL) enhancement and Multilevel Otsu Thresholding Segmentation technique can be summarized in the flowchart shown in Fig. 1. The first step, the original digital mammogram images were pre-processed and enhanced using FADHECAL. Then, the enhanced digital mammograms were segmented by Multilevel Otsu Thresholding Segmentation to detect the breast lesions within the ROI. These breast lesions were then analysed and measured based on area, perimeter, circularity, and solidity. In the final step, the contours were created on the enhanced digital mammogram images by referring to the boundaries of the ROI and thus, produce the output detection of breast lesions in digital mammograms.

2.3 Pre-processing and FADHECAL

The pre-processing of digital mammograms is an initial process to remove the unwanted region, such as the unwanted labels, to improve the contrast appearances. The removal of unwanted labels is to clearly demonstrate the breast tissue regions, which later will facilitate the segmentation process.

FADHECAL is an improvement of the Contrast-Limited Adaptive Histogram Equalization (CLAHE) technique with a combination of fuzzy clipped inference system (FCIS) and Anisotropic Diffusion Filter (ADF). This FADHECAL technique consists of a fuzzy clipped inference system that automates the clip-limit selection for the enhancement process. This automatic selection can replace the previous method of selecting the clip-limit manually.

The anisotropic diffusion filter (ADF) is used to eliminate the noise and preserve the edges of image details in digital mammograms. Equation (1) describes the ADF model.

$$\frac{\partial I(x, y)}{\partial t} = div(g(|\nabla I(x, y)|)\nabla I(x, y)) \tag{1}$$

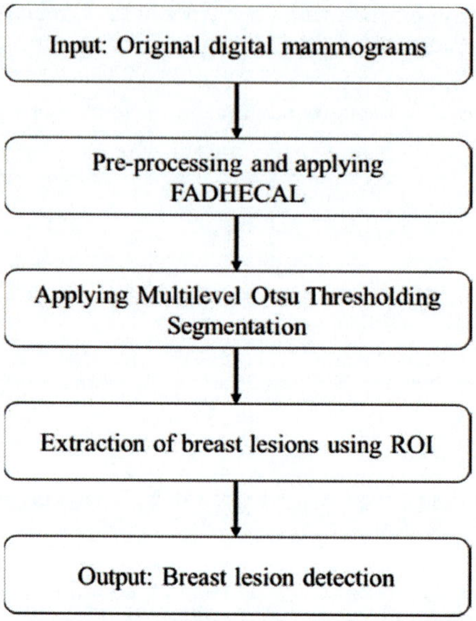

Fig. 1. The flowchart of combination FADHECAL and Multilevel Otsu Thresholding Segmentation technique.

Where $div(\cdot)$ and ∇ represent the divergence and the gradient operators respectively while $g(|\nabla I(x, y)|)$ is a decreasing function known as the edge stopping function that has an important role in controlling the diffusion process.

Fuzzy Clipped Interface System (FCIS) has been set up with two input measures, which include the contrast (C) and the entropy (E). The output measure of FCIS is the Fuzzy Histogram Clipped Limit (FHCL). The triangular membership function is used to construct the rules of FCIS and classify the FHCL into low clip limit (CL1) and high clip limit (CL2). The output of each rule is aggregated into a single fuzzy set for overall FCIS output. The result of the defuzzification of FCIS is 0.0185. As a result, the FHCL is 0.0185 for the given values (C = 0.325) and (E = 0.444). The $f\beta$ varies between 0 and 0.1. The FADHECAL method uses Eq. (2) to calculate the new clip limit that depends on the C and E of the input images.

$$FADHECAL = \left[\frac{\varphi}{256}\right] + \left[FHCL.\left(\varphi - \left[\frac{\varphi}{256}\right]\right)\right] \tag{2}$$

Where FHCL is the fuzzy histogram clip limit (ranges from 0 to 0.1) and [.] denotes the truncating value to the nearest integer. φ is the product of block size. The value of 256 denotes the pixel intensity values range from 0 to 255.

2.4 Multiple Otsu Thresholding Segmentation

Image segmentation is used to extract the breast lesions within the ROI for further analysis [10]. Otsu thresholding segmentation is one of the most commonly used threshold-based segmentation technique [8]. The thresholding segmentation technique is a fast and straightforward implementation based on a threshold value obtained from the image histogram [11]. The threshold value in a histogram separates intensities of pixels into two sections: foreground and background. Global thresholding segmentation converts the image by replacing all pixels in the input image based on a threshold value. After the threshold values have been chosen, the procedure is followed with the binarization process. Binarization is a process of converting pixel images into binary images. Binary image is an image that perceives two values for every pixel, 0 and 1. These pixels represent only two colours that are contained in a binary image, which is black and white. Equation (3) describes the thresholding operation.

$$g(x, y) = \begin{cases} 0, \; if \; (x, y) < T \\ 1, f \; (x, y) \geq T \end{cases} \quad (3)$$

Where, $g(x, y)$ is the pixel intensity, T as the threshold value, and (x, y) is the position of the threshold value point. If the input image's pixels have values larger than the threshold, then the pixels are converted to '1' or white. Otherwise, the pixels are converted to '0', which is black.

Otsu thresholding segmentation is an effective technique that automatically selects the threshold and works effectively on a region-based segmentation technique [12]. This Otsu thresholding segmentation chooses the optimal threshold value by searching across the whole range of the image's pixel values until the intra-class variances reach their maximum using Eq. (4).

$$\sigma_B^2(T) = max_{0<K<L-1}\sigma_B^2(T) \quad (4)$$

Where, σ_B between class variances, L intensity levels [0, .., L−1].

The segmentation process must satisfy certain conditions to obtain the region of breast masses. The selection region of interest for breast masses in digital mammograms can identify the types of breast lesions. Digital mammograms often abrupt changes in the intensity peak values [7]. Thus, it may lead to misclassifying the breast lesions region due to different types of breast tissue densities. To overcome this problem, using Multilevel Otsu Thresholding Segmentation can segment the region of breast lesions at different intensities levels. The best Multilevel Otsu Thresholding Segmentation values lie between $0.75 \leq T \geq 0.85$.

2.5 Region of Interest (ROI) Extraction

Region of Interest (ROI) extraction is a crucial detail for digital mammogram analysis. The method used to extract the breast lesions within the ROI of blob properties is known as morphological operations. In this study, the use of morphological operations was to help structure elements and measure the shape of the image. It could refine the characteristics of the ROI and help to remove the unwanted regions [13]. The blob properties segmented from digital mammograms were calculated based on areas, perimeter, circularity, and solidity. These features were used to measure the ROI of blob properties as described below;

a) Areas
The area is used to measure based on the number of pixels in shape occupied by breast lesions. The white pixels represent the ROI of breast lesions.
b) Perimeter
The perimeter is used to measure the length of a region boundary.
c) Circularity
The circularity is used to measure the roundness, which excludes local irregularities can be obtained as the ratio area of an object to the area of a circle with the same convex perimeter. Equation (5) describes the circularity.

$$Circularity = \frac{4\pi.Area}{(Convex.perimeter)^2} \tag{5}$$

d) Solidity
The solidity is used to measure the density of ROI for breast lesions. The solidity measurement obtains from the ratio of the ROI area to the area of ROI convex. Equation (6) describes solidarity.

$$Solidity = \frac{Area}{Convex\ area} \tag{6}$$

3 Results and Discussion

The performance of FADHECAL incorporated with Multilevel Otsu Thresholding Segmentation technique has been tested on 115 digital mammograms extracted from the MIAS database. The input or original digital mammograms in Fig. 2(a) have shown low contrast appearances of dense tissue in the breast. The enhancement of digital mammograms using FADHECAL in Fig. 2(b) has improved the images with appropriate noise reduction while preserving the details. The interpretation of breast lesions has improved as the details of dense tissue in the breast are highlighted and preserved.

Figure 2(c) has shown the ROI of blob properties segmented using Multilevel Otsu Thresholding Segmentation to analyse and identify the pattern of breast lesions. This segmentation method is very significant to identify the ROI of breast lesions. Although, the breast lesions are superimposed on the tissue with the same intensity level. Figure 2(d) has shown the detection of ROI for breast lesions contoured onto the digital mammograms to locate the breast lesions.

The method has successfully detected the breast lesions with verification of ground truth from the MIAS database. The proposed method's efficiency and error were measured by the following Eq. (7) and (8), respectively.

$$Efficiency\ percentage = \frac{Number\ of\ true\ identification\ of\ breast\ lesions}{Total\ of\ abnormal\ digital\ mammograms} \times 100\% \tag{7}$$

$$Error\ percentage = \frac{Number\ of\ false\ identification\ of\ breast\ lesions}{Total\ of\ abnormal\ digital\ mammograms} \times 100\% \tag{8}$$

The proposed method successfully identified 109 breast lesions among 115 abnormal digital mammograms. Figure 3 shows the performance of FADHECAL incorporated

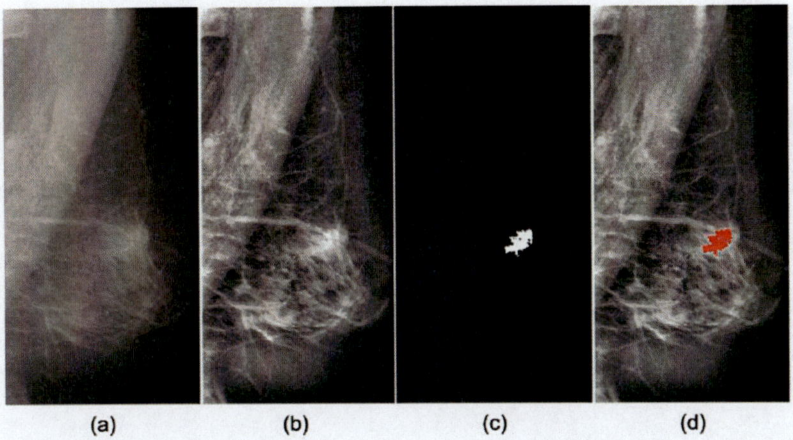

Fig. 2. (a) Original digital mammogram images, (b) FADHECAL enhancement process, (c) Multilevel Otsu Thresholding Segmentation process, and (d) ROI of breast lesions.

with Multilevel Otsu Thresholding Segmentation. Another six digital mammograms were incorrectly identified due to the low intensity and breast density. Breasts with high density are harder to interpret as the dense tissue overshadows the tumor regions [14]. Thus, the segmentation of ROI for breast lesions become challenging due to white intensities of dense tissue, and breast lesions appear approximately the same. The efficiency percentage of the proposed method is 94.8%, and the error rate percentage is 5.2%.

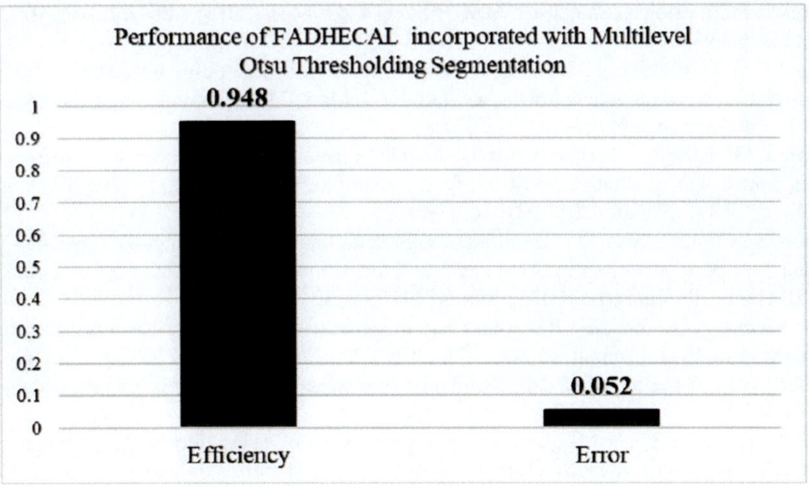

Fig. 3. The performance of proposed method (FADHECAL) incorporated with Multilevel Otsu Thresholding Segmentation.

4 Conclusion

A novel technique of FADHECAL incorporated with the Multilevel Otsu Thresholding Segmentation has shown appropriate results to detect breast lesions for digital mammograms with an efficiency percentage of 94.8%. Therefore, this proposed method can improve the detection of breast lesions with a better quality of digital mammograms.

Ethics Approval. Not required.

Conflict of Interest. The authors declare no conflict of interest.

Acknowledgments. Declared none.

References

1. American Cancer Society: Cancer Facts & Figures 2020, pp. 1–52 (2020)
2. Lee, S., Jin Park, S., Jeon, J.M., et al.: Noise removal in medical mammography images using fast non-local means denoising algorithm for early breast cancer detection: a phantom study. Optik (Stuttg) **180**, 569–575 (2019). https://doi.org/10.1016/j.ijleo.2018.11.167
3. Chattaraj, A., Das, A., Bhattacharya, M.: Mammographic image segmentation by marker controlled watershed algorithm. In: Proceedings of 2017 IEEE International Conference on Bioinformatics and Biomedicine (BIBM 2017), January 2017, pp. 1000–1003 (2017). https://doi.org/10.1109/BIBM.2017.8217793
4. Sridhar, B., Reddy, K.V.V.S., Prasad, A.M.: Detection of masses in mammogram based on non-linear filtering techniques. J. Med. Bioeng. **4**, 430–435 (2015). https://doi.org/10.12720/jomb.4.6.430-435
5. Sreeja, G.B., Rathika, P., Devaraj, D.: Detection of tumours in digital mammograms using wavelet based adaptive windowing method. Int. J. Mod. Educ. Comput. Sci. **4**, 57–65 (2012). https://doi.org/10.5815/ijmecs.2012.03.08
6. Samir, M., Alaa, A., Hassan, E., Mohaammed, T.: Breast cancer detection with mammogram segmentation: a qualitative study. Int. J. Adv. Comput. Sci. Appl. **8**, 117–120 (2017). https://doi.org/10.14569/ijacsa.2017.081016
7. Don, S., Choi, E., Min, D.: Breast mass segmentation in digital mammography using graph cuts. In: Communications in Computer and Information Science (CCIS), vol. 206, pp. 88–96 (2011). https://doi.org/10.1007/978-3-64224106-2_12
8. Makandar, A., Halalli, B.: Threshold based segmentation technique for mass detection in mammography. J. Comput. **11**, 472–478 (2016). https://doi.org/10.17706/jcp.11.6.463-4712
9. Clark, A.F.: The mini-MIAS database of mammograms (2012). https://peipa.essex.ac.uk/info/mias.html
10. Senthilkumaran, N., Vaithegi, S.: Image segmentation by using thresholding techniques for medical images. Comput. Sci. Eng. Int. J. **6**, 1–13 (2016). https://doi.org/10.5121/cseij.2016.6101
11. Bhargavi, K., Jyothi, S.: A survey on threshold based segmentation technique in image processing. Int. J. Innov. Res. Dev. **3**, 234–39 (2014)
12. Khan, A.K.: Wavelet based automatic lesion detection using improved OTSU method, vol. 2, pp. 119–127 (2014)

13. Guzmán-Cabrera, R., Guzmán-Sepúlveda, J.R., Torres-Cisneros, M., et al.: Digital image processing technique for breast cancer detection. Int. J. Thermophys. **34**, 1519–1531 (2013). https://doi.org/10.1007/s10765-012-1328-4
14. Duffy, S.W., Morrish, O.W.E., Allgood, P.C., et al.: Mammographic density and breast cancer risk in breast screening assessment cases and women with a family history of breast cancer. Eur. J. Cancer **88**, 48–56 (2018). https://doi.org/10.1016/j.ejca.2017.10.022

Estimation of Effective Doses to Patients in Whole Body Computed Tomography with Automatic Tube Current Modulation Systems

Adnan Beganović[1](✉), Samra Stabančić-Dragunić[2], Senad Odžak[3], Amra Skopljak-Beganović[1], Rahima Jašić[1], and Irmina Sefić-Pašić[4]

[1] Radiation Protection and Medical Physics Department, Clinical Centre of Sarajevo University, Sarajevo, Bosnia and Herzegovina

[2] Institute for Occupational Health and Sports Medicine, Zenica, Bosnia and Herzegovina

[3] Faculty of Science, University of Sarajevo, Sarajevo, Bosnia and Herzegovina

[4] Clinic of Radiology, Clinical Centre of Sarajevo University, Sarajevo, Bosnia and Herzegovina

Abstract. Computed tomography (CT) is a digital radiological modality which has been classified as the largest source of medical radiation exposure. In this study, we estimated the effective dose to patients undergoing CT examinations using information obtained from DICOM images, including tube current, patient size and positioning, collimation etc. The custom code was used to calculate the conversion factors between an effective dose and air kerma length product. Their values are 0.0232 mSv/(mGy cm) and 0.0263 mSv/(mGy cm), for male and female patients, respectively. The effective dose for the whole-body CT is estimated to be 8.7 mSv.

Keywords: Computed tomography · Effective dose · Tube current modulation

1 Introduction

Computed tomography (CT) is a digital radiological modality that has certainly made the biggest change in diagnostic radiology since the discovery of x-rays. Previous planar radiological methods hid the third dimension of the human body, so a large number of morphological changes in the body could not be noticed. Despite its benefits, CT examination is always accompanied by radiation exposure of the patient.

Concerns about radiation risk due to CT examinations are rising in both the medical physics community and the general public. Cohort studies investigating stochastic effects of radiation exposure from CT scans are being still being published [1, 2]. CT imaging and interventional procedures have been classified as the largest source of medical radiation exposure. The effective dose in CT-guided interventions can easily exceed 100 mSv [3]. National Council on Radiation Protection and Measurements (NCRP) Report 184 estimates that effective dose per capita in the United States of America is 1.45 mSv [4, 5].

A. Badnjevic and L. Gurbeta Pokvić (Eds.): CMBEBIH 2021, IFMBE Proceedings 84, pp. 760–767, 2021.
https://doi.org/10.1007/978-3-030-73909-6_86

The value has not changed much since 2010, indicating that the rapid increase in collective effective dose from CT examinations that happened since the introduction of multi-slice CT scanners has stopped [6]. We could argue that the same trend is being witnessed in other nations on a similar healthcare level.

Many institutions are actively developing dose control programs to make more responsible use of computed tomography and to avoid unnecessary exposure to ionizing radiation [7–9]. Such programs automatically extract values relevant for direct assessment of patient exposure from patient dosimetry reports, such as volume computed tomography air kerma index C_{VOL} or CT air kerma-length product $P_{\mathrm{KL,CT}}$, and monitor changes in their value over time.

However, these values are related to measurements on the dosimetry phantom and do not provide information on the actual patient dose, and therefore cannot be compared to doses in other diagnostic modalities or different patient sizes.

One way to solve this problem is to convert C_{VOL} to organ doses using conversion factors labeled h. These factors have been shown to be exponentially dependent on body diameter. For organs that are fully covered by the primary radiation beam, the h factors do not depend much on the scan parameters and the CT device model [10]. These properties suggest the need to incorporate conversion factors h into direct patient exposure assessment programs to allow control of organ doses for individual patients. Several research groups have examined how changes in tube current affect dose and conversion factors [11–18].

There are significant differences in the implementation of tube current modulation (TCM) technology among different manufacturers of CT devices. Some manufacturers aim to provide a unique image noise for all body regions and patient sizes, while others aim to reduce the dose for the patient by allowing slightly higher noise for larger patients [19]. New technologies allow organ-based TCM which is capable of anterior and total dose reduction with minimal loss of image quality [20].

The dependence of the h factor on the modulation schemes was investigated in the study of Li et al. which focused on examinations of the thorax, abdomen, and pelvis, which are examinations in which TCM is most commonly used. In the conclusion of this paper, it was pointed out that ignoring the modulation of the anode current when estimating the dose to a patient can lead to errors reaching values in the order of 200% [10].

However, the dose estimate also depends on the selected modulation strength, the choice of scanning protocol, such as minimum and maximum anode current, and the current modulation method itself, which is based on the patient's scans and is a trade secret from the CT manufacturer. Taking all the above into account, it is impossible to give a universal way of estimating the effective dose and the dose to individual organs using the mean value C_{VOL}.

2 Materials and Methods

In this study, the effective dose and the dose to individual organs in wholebody CT were estimated using tube current values extracted from the header of DICOM images. In the process of data collection, a software has been designed that can extract tube current values from tomographic images, apply conversion factors, and use readily available Monte

Carlo data from CTVoxDos software to estimate the effective dose to the patient. CTVoxDos software dose estimations are based on the scheme of Schlattl et al. [15, 21, 22]. In this study we used voxel computational phantoms RCP-AM and RCP-AF, which correspond to an average male and female body, respectively. Body mass and height of RCP-AM is 73 kg and 177.6 cm, respectively, while for RCP-AF these values are 60 kg and 168.4 cm. To examine the influence of these values on E we followed this basic idea:

- obtain conversion factors $k_i(z_i)$ between the effective dose per slice $E_i(z_i)$ and C_{VOL} from CTVoxDos;
- obtain $I_i(z_i)$ from DICOM images for a group of patients that performed whole-body CT;
- interpolate values of $I_i(z_i)$ to match positions z_i on the phantom, and adjust for different body parts proportions;
- calculate $C_{\mathrm{VOL},i}$ from tube current using ${}_{\mathrm{n}}C_{\mathrm{VOL}}$;
- multiply the obtained value with slice width Δz and divide with pitch factor p;
- calculate the effective dose as:

$$E = \frac{{}_{\mathrm{n}}C_{\mathrm{VOL}} \Delta z}{p} \sum_i k_i(z_i) I_i(z_i) \quad (1)$$

Although there might be other convenient solutions to get the values of tube current from DICOM images, we used the command-line software dicom2 (Sebastien Barre, 2007), as an accessory for our own code. Visual Basic in Microsoft Access was used to run dicom2, store collected data, perform necessary calculations, as well as to organize results.

3 Results and Discussion

Table 1 shows the height (m), body mass (kg) and age of the patients included in this study. Twenty-eight out of 53 patients (52.8%) were female and 25 male (47.2%). Average height of patients was 1.72 m ($\sigma = 0.10$ m) with body mass of 74 kg ($\sigma = 16$ kg). Most of the patients were normally sized according to the WHO classification of body mass index. Height of patients does not differ significantly from computational phantom size. According to the two-tailed one sample t-test, p-values for males and females are 0.5379 and 0.1422, respectively. Body mass, however, is slightly higher, with p-values equal to 0.0003 and 0.0214 for male and female patients, respectively.

The median age of the female patient group was 53 years (IQR = 28y) which did not significantly differ (Mann-Whitney U test, $p = 0.358$) from the median age of male patients, which was 57 years (IQR = 14y).

Within the statistical data processing, the analysis of the correlation between certain quantities was performed. As the body mass index increases, so does the anode current, and thus the C_{VOL} and $P_{\mathrm{KL,CT}}$ (Table 2). The obtained results were expected and confirm that the software that modulates the anode current adjusts the scan parameters to match patient body size.

The average value of C_{VOL} is 3.0 mGy (IQR = 2.0 mGy), and the tube current read for images is 65.8 mA (IQR = 45.6 mA). The statistically significant correlation

Table 1. Height, body mass and age of the patients included in this study

Sex	N	Height (m)		Body mass (kg)		Age (years)	
		$\bar{x}$	σ	$\bar{x}$	σ	$\tilde{x}$	IQR
Male	25	1.78	0.08	84	13	57	14
Female	28	1.66	0.07	66	13	53	28
Total	53	1.72	0.10	74	16	54	25

Table 2. Value of tube current read from images, C_{VOL} and $P_{KL,CT}$ as indicated by the PET-CT for different BMI groups

BMI (kg/m^2)	I (mA)		C_{VOL} (mGy)		$P_{KL,CT}$ (mGycm)	
	$\tilde{x}$	IQR	$\tilde{x}$	IQR	$\tilde{x}$	IQR
<20.0	30.0	14.6	1.4	0.5	141.6	61.3
20.0–24.9	53.6	19.1	2.5	0.8	261.7	93.9
25.0–29.9	84.0	35.9	3.6	1.5	406.4	180.7
≥30.0	141.1	26.8	5.8	2.8	616.9	307.3
Total	65.8	45.6	3.0	2.0	314.6	225.3

was found between tube current I and C_{VOL}, which is shown in the Fig. 1. The figure confirms the theoretical linear relationship between the tube loading (product of anode current and exposure time, P_{It}) and air kerma, which is in this case represented by C_{VOL}.

Using software developed for this study, data on scanning parameters were obtained that cannot be read directly from the device console or the process would take a very long time. Values of table position z and anode current I were loaded from the header of a series of DICOM files (tomographic images). A series of transformations have been performed for z to match specific positions on the computational phantom and adjust for different body parts proportions.

The final results (z vs. median I) are presented in the figure, where shaded areas represent the interval between 1st and 3rd quartile (Fig. 2).

The anode current is greatest in the pelvic region when the pelvic and hip bones are in the beam. The current is the lowest in the head region. However, we notice that the value of I increases unexpectedly near the scalp. The reason for this is the inability of the software to estimate the adequate value of the anode current when single projection radiograph (SPR or scout) does not cover the entire patient head. Due to the lack of input data at the beginning of the scan, software resorts to a solution that will certainly give a picture of satisfactory quality. Hence, the tube current is increased to its maximum permissible value. Figure 3 shows the effective dose E to patients after whole-body CT vs. $P_{KL,CT}$ for males and females. The conversion factor k can be read from the equations describing the linear regression curves. For male patients $k = 0.0232$ mSvmGy^{-1}cm^{-1}, and for female patients $k = 0.0263$ mSvmGy^{-1}cm^{-1}.

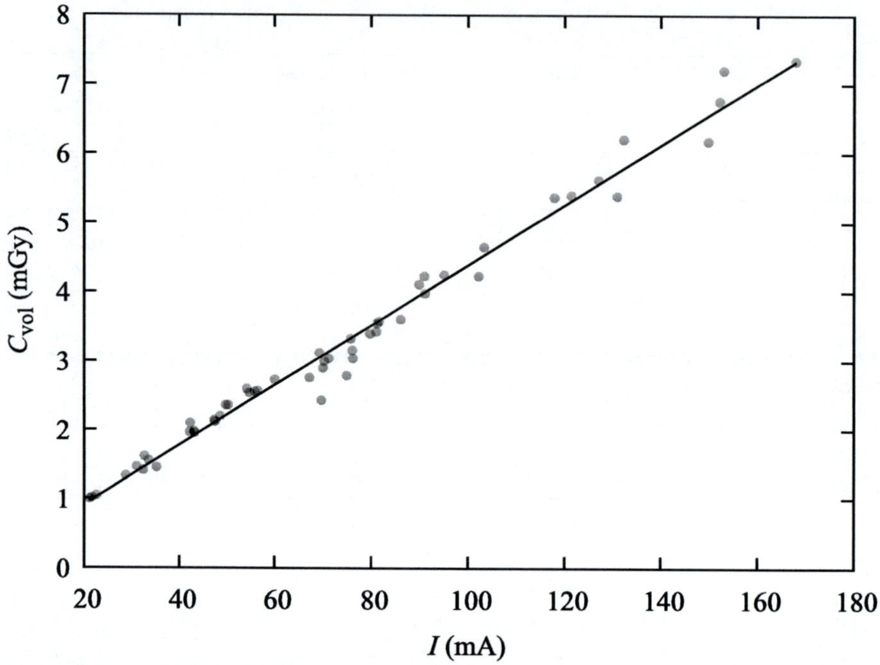

Fig. 1. Value of tube current I vs. C_{VOL}. Linear correlation is confirmed ($a = 0.0434 \pm 0.0008$, $b = 0.04 \pm 0.06$, $R^2 = 0.984$)

The slope or linear regression coefficients (a) can be used to convert $P_{KL,CT}$ to effective dose E after whole-body CT. For male patients, it has the value of 0.0232 $mSvmGy^{-1}cm^{-1}$, and for females 0.0263 $mSvmGy^{-1}cm^{-1}$. We should have in mind that this method is valid only for one protocol of one CT manufacturer. There is no universal conversion factor for all protocols and all CT device manufacturers [23]. It should be noted that the tube current values drop to minimum in the leg region. Hence, the average C_{VOL} is lower than the value observed in the regions of radiosensitive organs. Depending on the protocol, these differences can have significant impact on the overall conversion factor.

Other methods of obtaining the conversion factor can be found in the literature for devices that use anode current modulation. In the study of Kawaguchi et al. for CT devices Toshiba Aquilion 64 and Aquilion RXL, the doses obtained did not differ significantly from the doses obtained via $P_{KL,CT}$ conversion factor [24].

To estimate the final value of effective dose to patients undergoing whole body CT, we used the obtained conversion factors, multiplied them with average $P_{KL,CT}$ for male and female patients, and calculated the average between two. Finally, the effective dose for whole-body CT is estimated to be 8.7 mSv, which is comparable to values found in the literature [25–27].

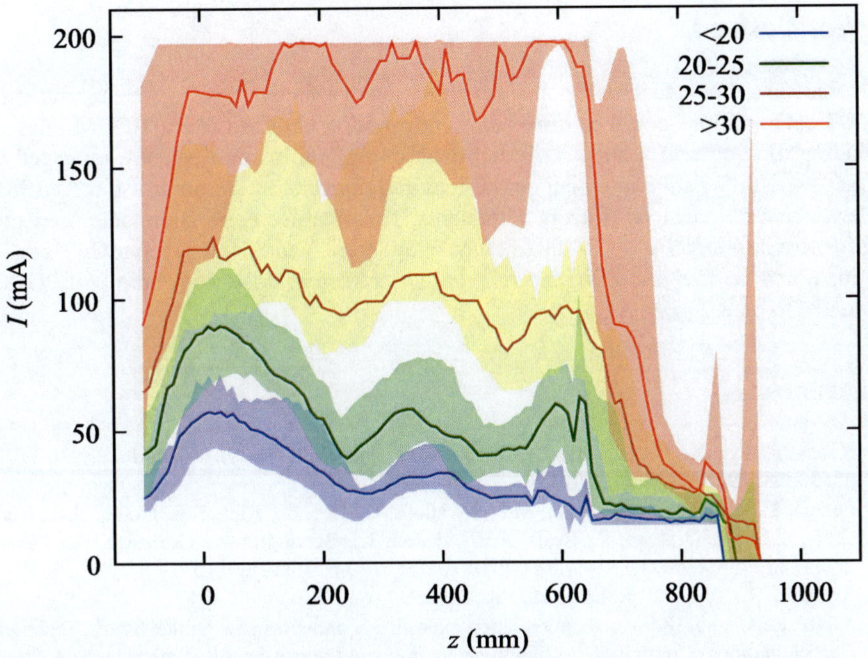

Fig. 2. Change of median tube current I with table position z for different patient BMI (kgm^{-2}). Shaded areas represent the interval between 1st and 3rd quartile.

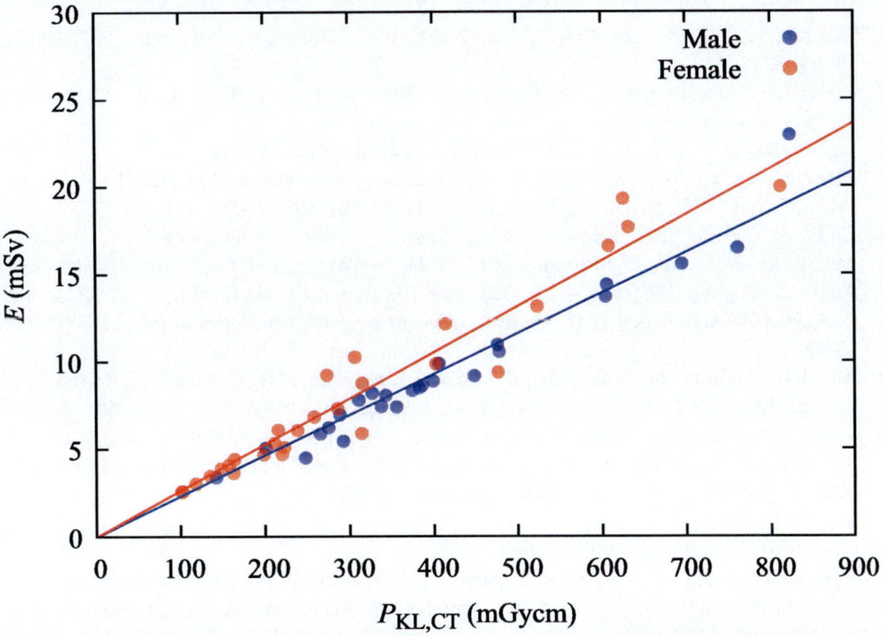

Fig. 3. Effective dose E to patients after whole-body CT vs. $P_{\mathrm{KL,CT}}$ for male and female patients (Male: $a = 0.0232 \pm 0.0005$, $R^2 = 0.991$; Female: $a = 0.0263 \pm 0.0007$, $R^2 = 0.984$)

4 Conclusions

This study has shown that more accurate estimation of effective dose to patients undergoing CT examinations could be done using information obtained from DICOM images, including tube current, patient size and positioning, collimation etc. We managed to extract necessary data and calculate the required conversion factors using the custom code which ran images taken from 53 patients. The obtained conversion factor between an effective dose and $P_{KL,CT}$ is 0.0232 $mSvmGy^{-1}cm^{-1}$ and 0.0263 $mSvmGy^{-1}cm^{-1}$, for male and female patients, respectively. The effective dose for whole body CT is estimated to be 8.7 mSv.

References

1. Huang, R., Liu, X., He, L., Zhou, P.K.: Dose-Response **18**(2) (2020). https://doi.org/10.1177/1559325820923828. PMID: 32425727
2. Pearce, M.S., Salotti, J.A., Little, M.P., McHugh, K., Lee, C., Kim, K.P., Howe, N.L., Ronckers, C.M., Rajaraman, P., Craft, A.W., Parker, L., Berrington de González, A.: Lancet **380**(9840), 499 (2012). https://doi.org/10.1016/S0140-6736(12)60815-0
3. Arellano, R., Yang, K., Rehani, M.: Eur. Radiol. 1–6 (2020)
4. NCRP, Report no. 184 – medical radiation exposure of patients in the United States. Technical Report, National Council on Radiation Protection and Measurements, Bethesda, MD (2019)
5. Mettler, F.A., Mahesh, M., Bhargavan-Chatfield, M., Chambers, C.E., Elee, J.G., Frush, D.P., Miller, D.L., Royal, H.D., Milano, M.T., Spelic, D.C., Ansari, A.J., Bolch, W.E., Guebert, G.M., Sherrier, R.H., Smith, J.M., Vetter, R.J.: Radiology **295**(2), 418 (2020). https://doi.org/10.1148/radiol.2020192256. PMID: 32181730
6. Einstein, A.J.: Radiology 295(2), 428 (2020). https://doi.org/10.1148/radiol.2020200200. PMID: 32186459
7. Cook, T.S., Zimmerman, S., Maidment, A.D., Kim, W., Boonn, W.W.: J. Am. Coll. Radiol. **7**(11), 871 (2010)
8. Li, X., Zhang, D., Liu, B.: Am. J. Roentgenol. **196**(6), W781 (2011)
9. Christianson, O., Li, X., Frush, D., Samei, E.: Med. Phys. **39**(11), 7131 (2012)
10. Li, X., Segars, W.P., Samei, E.: Phys. Med. Biol. **59**(16), 4525 (2014)
11. Huda, W., Ogden, K.M., Khorasani, M.R.: Radiology **248**(3), 995 (2008)
12. van Straten, M., Deak, P., Shrimpton, P.C., Kalender, W.A.: Med. Phys. **36**(11), 4881 (2009)
13. Angel, E., Yaghmai, N., Jude, C.M., DeMarco, J.J., Cagnon, C.H., Goldin, J.G., McCollough, C.H., Primak, A.N., Cody, D.D., Stevens, D.M., et al.: AJR. Am. J. Roentgenol. **193**(5), 1340 (2009)
14. Angel, E., Yaghmai, N., Jude, C.M., DeMarco, J.J., Cagnon, C.H., Goldin, J.G., Primak, A.N., Stevens, D.M., Cody, D.D., McCollough, C.H., et al.: Phys. Med. Biol. **54**(3), 497 (2009)
15. Schlattl, H., Zankl, M., Becker, J., Hoeschen, C.: Phys. Med. Biol. **55**(20), 6243 (2010)
16. Khatonabadi, M., Zhang, D., Mathieu, K., Kim, H.J., Lu, P., Cody, D., DeMarco, J.J., Cagnon, C.H., McNitt-Gray, M.F.: Med. Phys. **39**(8), 5212 (2012)
17. Khatonabadi, M., Kim, H.J., Lu, P., McMillan, K.L., Cagnon, C.H., DeMarco, J.J., McNitt-Gray, M.F.: Med. Phys. **40**(5), 051903 (2013)
18. Rupcich, F., Badal, A., Kyprianou, I., Schmidt, T.G.: Med. Phys. **39**(9), 5336 (2012)
19. Kalra, M.K., Maher, M.M., Toth, T.L., Schmidt, B., Westerman, B.L., Morgan, H.T., Saini, S.: Radiology **233**(3), 649 (2004). https://doi.org/10.1148/radiol.2333031150. URL https://bit.ly/Q2x7Mr

20. Mussmann, B., Mørup, S., Skov, P., Foley, S., Brenøe, A., Eldahl, F., Jørgensen, G., Precht, H.: Radiography (2020). https://doi.org/10.1016/j.radi.2020.04.011
21. Schlattl, H., Zankl, M., Becker, J., Hoeschen, C.: Phys. Med. Biol. **57**(20), 6309 (2012)
22. Fehrmann, M.L., Schegerer, A., Werncke, T., Schlattl, H.: Radiat. Prot. Dosimetry **190**(1), 71 (2020). https://doi.org/10.1093/raddos/ncaa070
23. Christner, J.A., Kofler, J.M., McCollough, C.H.: Am. J. Roentgenol. **194**(4), 881 (2010)
24. Kawaguchi, A., Matsunaga, Y., Kobayashi, M., Suzuki, S., Matsubara, K., Chida, K.: Radiat. Prot. Dosimetry **167**(4), 562 (2014). https://doi.org/10.1093/rpd/ncu340
25. Alkhybari, E.M., McEntee, M.F., Brennan, P.C., Willowson, K.P., Kench, P.L.: J. Med. Imag. Radiat. Oncol. **63**(3), 291 (2019). https://doi.org/10.1111/1754-9485.12857
26. Sabri, A.S.A., Wong, J.H.D.: J. Phys.: Conf. Ser. **1248**, 012006 (2019). https://doi.org/10.1088/1742-6596/1248/1/012006
27. Sato, T., Saito, Y., Saito, N., Ooba, M.: Nihon Hoshasen Gijutsu Gakkai zasshi **73**(4), 258—266 (2017). https://doi.org/10.6009/jjrt.2017_jsrt_73.4.258

Local Diagnostic Reference Levels in Emergency Computed Tomography of the Head

Taa Awad-Dedić[1], Lejla M. Čiva[2](✉), Adnan Beganović[3], Mustafa Busuladžić[1], Edis Đedović[4], and Sandra Vegar-Zubović[5]

[1] Faculty of Medicine, University of Sarajevo, Sarajevo, Bosnia and Herzegovina
[2] Medical School, University Sarajevo School of Science and Technology, Sarajevo, Bosnia and Herzegovina
lejla.civa@ssst.edu.ba
[3] Radiation Protection and Medical Physics Department, Clinical Centre of Sarajevo University, Sarajevo, Bosnia and Herzegovina
[4] Medical Physics and Radiation Protection Department, University Clinical Centre of Tuzla, Tuzla, Bosnia and Herzegovina
[5] Clinic of Radiology, Clinical Centre of Sarajevo University, Sarajevo, Bosnia and Herzegovina

Abstract. Computed tomography (CT) is a widely used modality in diagnostic radiology. It has found its way into emergency medicine where it represents an indispensable diagnostic tool. The main disadvantage of CT is the high radiation dose delivered to patients, which is why a lot of emphases is given to the development of optimised scanning protocols. Aim of this study was to assess the patient doses in the form of air kerma length product ($P_{KL,CT}$ or DLP) and volume computed tomography air kerma dose index (C_{VOL} or CTDIvol), for the most common head CT procedures, and to compare them to different diagnostic reference levels (DRLs). The results have shown that patient doses for non-contrast CT of the head are in line with the DRLs in many countries. The results from this paper will provide a basis for the optimisation of surveyed radiological procedures.

Keywords: Head CT · Radiation dose · DLP · CTDI · Emergency medicine

1 Introduction

Despite magnetic resonance imaging (MRI) has outperformed computed tomography (CT) in representing neuroanatomy and pathology, CT studies in neuroimaging are a mainstay for several reasons. New diagnostic methods such as CT angiography (CTA) and CT perfusion, as well as the application of helical and multidetector CTs (MDCT) lead to an increase in the number of CT examinations. In the setting of acute trauma, CT allows rapid assessment of brain scans and provides patients to receive adequate care, for which it remains the modality of choice for preliminary evaluation. However, its widespread use leads to increasing patient dose and represents the main disadvantage of this diagnostic modality [1–4]. It is becoming crucial to develop new techniques in order to optimise current CT imaging protocols. The aim of this study was to assess the

A. Badnjevic and L. Gurbeta Pokvić (Eds.): CMBEBIH 2021, IFMBE Proceedings 84, pp. 768–776, 2021.
https://doi.org/10.1007/978-3-030-73909-6_87

patient doses expressed in form of dose indices – air kerma length product ($P_{KL,CT}$ or KLP, commonly abbreviated as DLP) and pitch-corrected volume CT air kerma index (C_{VOL}, commonly abbreviated as CTDIvol) for several head CT procedures and to compare them to diagnostic reference levels (DRLs) of surveyed countries.

2 Materials and Methods

In this retrospective study, all the data was provided by OpenREM (The Royal Marsden NHS Foundation Trust, London, United Kingdom), a dose management program, which collected the information about the CT scans at the Emergency Department of the Clinical Centre of Sarajevo University in the period from April to August of 2020.

OpenREM is a free, open-source application designed for patient dose monitoring. It collects radiation dose structured reports (RDSR) from different diagnostic modalities, such as CT, fluoroscopy, planar X-ray, and mammography. It displays exposure data, charts for visualisation, dose maps, allows export of data etc.

The study included adult patients who underwent non-contrast CT of the head (NC-CTH) or contrast-enhanced CT of the head (CE-CTH), as well as CT angiography of the head (CTAH). A total of 911 cases were selected. The collected data contained the following information: type of examination, date and time of the procedure, patient age and gender, exposure time, scan length, slice thickness, collimation width, pitch, C_{VOL}, $P_{KL,CT}$, tube potential (kV), maximum and mean tube current (mA), and rotation time. All patients were scanned on CT model Toshiba Astelion 16 (Toshiba, Tokyo, Japan), which takes 16 cross-sectional images per rotation. The minimum slice thickness is 0.5 mm.

3 Results

Data were collected and statistically analysed for a total of 911 adult patients. The number of patients in individual diagnostic procedures is summarised in Table 3. The most common diagnostic procedure was NC-CTH (834 patients), followed by CE-CTH

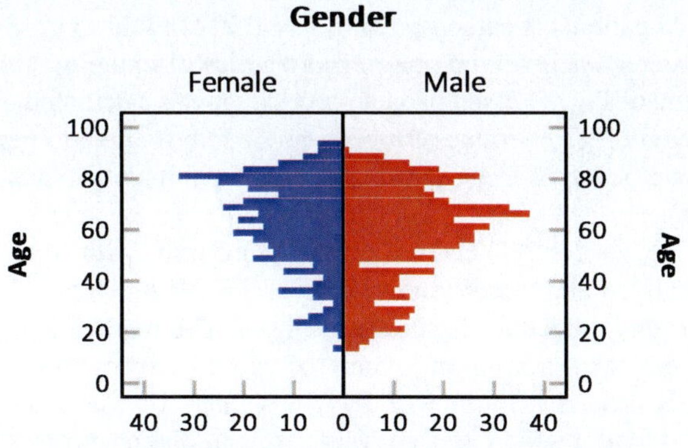

Fig. 1. Age distribution of patients relative to gender

Table 1. Number of patients, values of scanning parameters and median volume computed tomography (CT) air kerma index (C_{VOL}) for CT angiography of the head (CTAH), non-contrast CT of the head (NC-CTH) and contrast enhanced CT of the head (CE-CTH)

	Examination type		
	CTAH	NC-CTH	CE-CTH
Number of patients	21	834	56
Tube potential (kV)	120	120	120
Average tube current (mA)	200	200	200
Rotation time (s)	0.75	0.75	0.75
Collimation (mm)	16	16	16
Pitch	0.69	0.69	0.69
Slice thickness[a] (mm)	3	3	3
Reconsturcted slice[b] (mm)	1	3 & 1	3 & 1
Median scan length (cm)	202	193	193
Median C_{VOL} (mGy)	49.8	49.8	49.8

[a]Slice thickness used in noise calculations, not thickness of reconstructed images
[b]Slice thicknesses for two reconstructions – soft for brain and sharp for bone

(56 patients) and CTAH (21 patients). Table 3 shows values of scan parameters for different examination protocols and C_{VOL}. It has to be noted that most parameters, including average tube current (mA), had constant values in all examination protocols. Hence, the median C_{VOL} is the same for all examinations.

The histogram of gender and age distribution is presented in Fig. 1. Of the total number of patients, 508 or 55.8% were men while 403 or 44.2% were women. The number of male patients is significantly higher (one-sample binomial test, $p = 0.001$). Majority of patients were 65 years or older (45.8%). However, this is not uniform across different examination types. CE-CTH and CTAH are more common in the group of 35.0–64.9-year-old patients (Pearson χ^2 test, $p = 0.014$) (Table 1).

Diagnostic procedures involved one or more series of imaging, and for each of the cases the total value of $P_{KL,CT}$, used in data processing, was calculated. The figure shows differences in median $P_{KL,CT}$ values for each procedure relative to gender. In the case of NC-CTH, a lower value of $P_{KL,CT}$ was observed for female patients (Mann-Whitney U test, $p < 0.001$).

The median $P_{KL,CT}$ for NC-CTH was 968.1 mGycm, with a minimum recorded value of 848.5 mGycm and a maximum value of 1390.7 mGycm. The distribution of $P_{KL,CT}$ values for this procedure is shown in Fig. 3. The median $P_{KL,CT}$ for CE-CTH was 1891.5 mGycm, with a minimum value of 923.4 mGycm and a maximum value of 2147.2 mGycm. The distribution of $P_{KL,CT}$ values for this diagnostic procedure is shown in Fig. 4. The median $P_{KL,CT}$ value for the diagnostic procedure of CTAH was 2533.5 mGycm, with a minimum value of 2008.0 mGycm and maximum value of

Table 2. Minimum, 25th percentile, median, mean, 75th percentile and maximum value of total air kerma length product ($P_{KL,CT}$) for computed tomography angiography of the head (CTAH), non-contrast computed tomography of the head (NC-CTH) and contrast enhanced computed tomography of the head (CE-CTH)

	$P_{KL,CT}$ (Gycm)			
	CTAH	NC-CTH	CE-CTH	
			1 series	2 series
Minimum	2.01	0.85	0.92	1.58
25th percentile	2.26	0.94	1.01	1.85
Median	2.53	0.97	1.03	1.91
Mean	2.96	0.97	1.07	1.91
75th percentile	3.41	1.00	1.10	2.00
Maximum	4.96	1.39	2.40	2.15

4956.4 mGycm. The distribution of $P_{KL,CT}$ values for this diagnostic procedure is shown in Fig. 5.

Minimum, 25th percentile, median, mean, 75th percentile and maximum value of total $P_{KL,CT}$ for CTAH, NC-CTH and CE-CTH are shown in Table 2. The values of the 75th percentile are of special importance because they are taken to determine the local diagnostic reference level (LDRL) [5–7]. Table 3 shows the same statistical parameters for C_{VOL}.

4 Discussion

In the study, C_{VOL} and $P_{KL,CT}$ are used as dose descriptors of patient doses. Although other descriptors could be used to express patient doses, use of these quantities allows comparison of obtained results to results in other studies and national diagnostic reference levels [5, 8].

In a brief overlook of the scanning parameters used for imaging procedures, it is indicative that all imaging protocols are done with the same tube potential and current, rotation time, collimation, pitch and slice thickness used for noise calculations. All parameters have constant values, including tube current which is set to the fixed value of 200 mAs. This will inevitably lead to constant values of C_{VOL}. Indeed, the vast majority of patients are exposed to 49.8 mGy, exactly. However, some of them (less than 5%) received doses above this value. The most likely reason for these unusual discrepancies is the off-centre positioning of patients which leads to the use of larger field-of-view (FOV). This may cause the change of beam conditions (increased scatter) or even the use of inappropriate bow-tie filter. This needs to be further investigated with controlled tests using a phantom (Fig. 2).

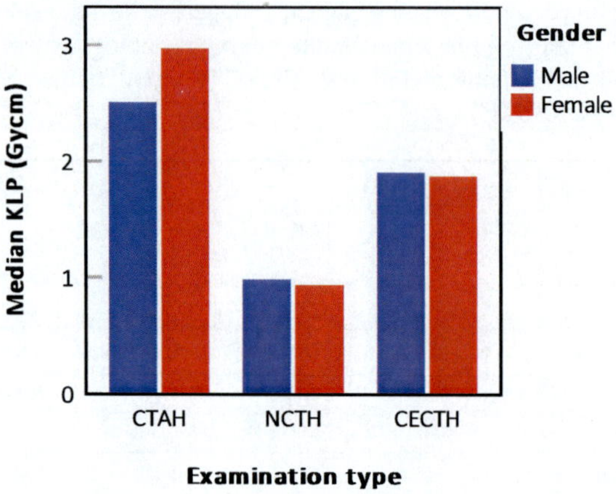

Fig. 2. Median air kerma–length product (KLP) values for computed tomography angiography of the head (CTAH), non-contrast computed tomography of the head (NCCTH) and contrast enhanced computed tomography of the head (CE-CTH), classified by gender

NC-CTH is technologically one of the less demanding procedures in CT. It is the most useful study for the evaluation of patients after trauma, but it is also routinely acquired for patients with neurological symptoms in the emergency department [9]. In this study, the median $P_{KL,CT}$ for the NC-CTH was slightly higher for the male than for female patients (Mann Whitney U test, $p < 0.001$). CT scanning parameters were not gender-dependent, and the only difference in the scan procedure (and therefore in $P_{KL,CT}$ values) may be the scan length which depends on the patient's anatomy. Scan length was slightly larger for male patients included in the study. However, although significant, the actual difference in $P_{KL,CT}$ between two genders is very small – 4.7%. The median $P_{KL,CT}$ NC-CTH was 970 mGycm, which is slightly higher than those reported in similar studies such as Trop et al. (740 mGycm), Hidajat et al. (587 mGycm) and Shrimpton et al. (787 mGycm) [10–13]. It should be noted that the reported results are lower than the national DRL value in Bosnia and Herzegovina, as well as in the European Union. In the regulations of the State Regulatory Agency for Radiation and Nuclear Safety of Bosnia and Herzegovina, which dates back to 2011, the DRL for CT head examinations is set to 1050 mGycm [14, 15]. We could argue that optimisation with the aim to reduce the dose in this procedure could be done. In this particular case, we the reconstructed slice thickness for soft tissue images could be increased to 5 mm. This action would lead to noise reduction and allow change of the preselected value of current-exposure time product P_{It}. This change could go along with the reduction of tube potential to 110 kV or similar. However, the technical parameters are within those recommended by the American Association of Physicists in Medicine (AAPM) for similar CT scanners [16]. It is plausible that slightly longer scan lengths lead to increased values of $P_{KL,CT}$.

The 75th percentile of $P_{KL,CT}$ for NC-CTH in our study was 1,000 mGycm, and is lower compared to the national DRLs in Japan and France, but also higher than those in Germany and Australia. Another study from Nigeria in 2015 conducted to set up a national DRL, analyzed data from three clinical centres, and reported 985 mGycm

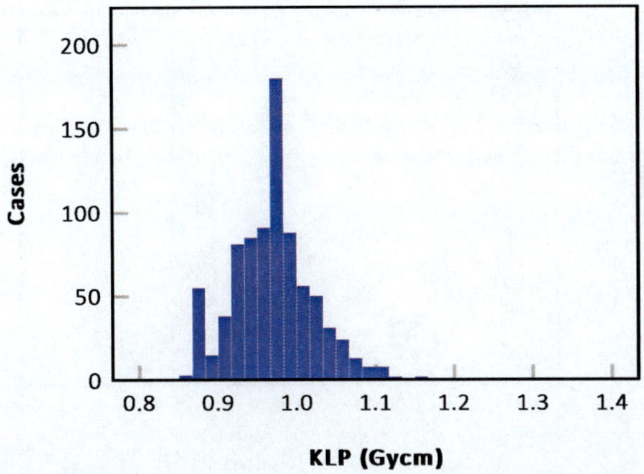

Fig. 3. Histogram of air kerma–length product (KLP or $P_{KL,CT}$) distribution for noncontrast computed tomography of the head

Table 3. Minimum, 25th percentile, median, mean, 75th percentile and maximum value of volume computed tomography air kerma index (C_{VOL}) for computed tomography angiography of the head (CTAH), non-contrast computed tomography of the head (NC-CTH) and contrast enhanced computed tomography of the head (CE-CTH)

	C_{VOL} (mGy)		
	CTAH	NC-CTH	CE-CTH
Minimum	49.8	49.8	49.8
25th percentile	49.8	49.8	49.8
Median	49.8	49.8	49.8
Mean	50.9	49.9	49.9
75th percentile	49.8	49.8	49.8
Maximum	68.5	56.4	51.3

as a recommended value for NC-CTH. $P_{KL,CT}$ values for the same procedure varied significantly from centre to centre, due to different scanning protocols [17].

CE-CTH adds a diagnostic value in examinations of patients with acute nontraumatic symptoms referable to the brain. Common indications for CE-CTH include headache, dizziness, altered mental status and seizure [18]. Two peaks in the histogram of CE-CTH $P_{KL,CT}$ indicate that this procedure could be done in two different ways. The most common one is in combination with NC-CTH (2 series), and the other with CE-CTH only (1 series). All statistical parameters for examinations in 1 series only are very close to the values of those in NC-CTH. The 75th percentile of $P_{KL,CT}$ for 2-series CE-CTH is 2,000 mGycm.

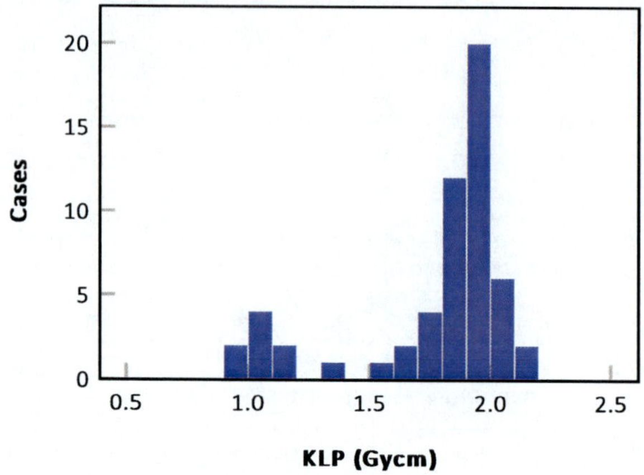

Fig. 4. Histogram of air kerma–length product (KLP or $P_{KL,CT}$) for contrast enhanced computed tomography of the head

CTAH is an examination done to investigate the vascular structure of the head and neck. The images provide information on arteriovenous malformation, bleeding in the brain, brain tumour, stroke, narrowed or blocked carotid arteries or vertebral artery in the neck. For the group of patients who underwent CTAH, the median $P_{KL,CT}$ was approximately 2,500 mGycm. However, it should be emphasised that only 21 patient underwent this diagnostic procedure. Although the number of patients is small, the human anatomy in the head and neck region is consistent. Hence, LDRLs could be defined as 75th percentile or 3rd quartile of $P_{KL,CT}$. If rounded to two significant figures, the LDRL for CATH is set to 3,400 mGycm. This value is comparable to diagnostic reference levels in the USA, UK, Australia, France, Germany and Japan [19–24].

When compared to NC-CTH and CE-CTH, CATH has the highest reported doses, although the scanning parameters are practically the same, including the value of C_{VOL}. Again, the discrepancies are due to differences in scanning length in CTAH the scanned volume includes neck arteries. This leads to higher patient doses, as observed in the study.

The ICRP gives a recommendation for institutions which report $P_{KL,CT}$ values that are significantly lower than the national DRL. The recommendation consists of revising the scanning protocols due to the possibility of reduced image quality at the expense of radiation protection. The optimisation process is time-consuming, and it is very risky to introduce large dose changes without a detailed examination of the impact of image quality. Optimising a protocol can take several years [25]. Setting the DRLs for CE-CTH and CTAH is not standard practice at the national level. Several reports state that DRLs are de-termined for the most common procedures in CT scanning which explains the unavailability of data. However, one study from India processed data from several clinical centres and determined $P_{KL,CT}$ values for multiphase CT sequences, and then compared them with several other studies. Multiphase CT sequences involve pre-contrast and post-contrast admission series of scans. The values of the 75th percentile of $P_{KL,CT}$ were 2,040 mGycm, which is consistent with our values for CE-CTH, where the value of the 75th percentile was 2,000 mGycm [26]. Approximate values (2130 mGycm) were also recorded in a 2013 University of California study [27].

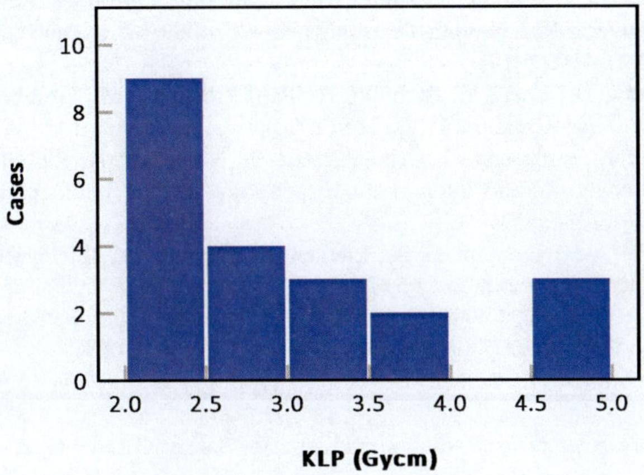

Fig. 5. Histogram of air kerma–length product (KLP or $P_{KL,CT}$) for computed tomography angiography of the head

5 Conclusion

The rapid development of CT scanners leads to their widespread use. Computed tomography is an indispensable diagnostic tool, especially in emergency medicine, where the information they provide could mean a difference between life and death. This convenience comes with a cost – the increase of patient doses. Our study aimed to check radiation doses and compare them to DRLs. The study included non-contrast and contrast-enhanced CT of the head, as well as CT angiography of the head. DRL data were consistent with those found in the literature. The obtained results will provide a basis for the optimisation of these procedures. The limitation of this study, in terms of radiation dose analysis, is the fact that the data was collected from one health facility only, and may not provide true insight on a precise range of C_{VOL} or $P_{KL,CT}$ values across the country.

References

1. Coles, J.: Br. J. Anaesth. **99**(1), 49 (2007)
2. Ibrahim, M., Parmar, H., Christodoulou, E., Mukherji, S.: Am. J. Neuroradiol. **35**(4), 619 (2014)
3. Currie, S., Saleem, N., Straiton, J.A., Macmullen-Price, J., Warren, D.J., Craven, I.J.: Postgrad. Med. J. **92**(1083), 41 (2016)
4. Kalender, W.A., Buchenau, S., Deak, P., Kellermeier, M., Langner, O., van Straten, M., Vollmar, S., Wilharm, S.: Phys. Med. **24**(2), 71 (2008)

5. Vañó, E., Miller, D., Martin, C., Rehani, M., Kang, K., Rosenstein, M., Ortiz-López, P., Mattsson, S., Padovani, R., Rogers, A., et al.: Ann. ICRP **46**(1), 1 (2017)
6. Martin, C.J., Vano, E.: J. Radiol. Prot. **38**(1) (2017)
7. Ghetti, C., Ortenzia, O., Palleri, F., Sireus, M.: Radiat. Prot. Dosimetry **175**(1), 38 (2017)
8. Kanal, K.M., Butler, P.F., Sengupta, D., Bhargavan-Chatfield, M., Coombs, L.P., Morin, R.L.: Radiology **284**(1), 120 (2017)
9. Callen, A., Chow, D., Chen, Y., Richelle, H., Pao, J., Bardis, M., Weinberg, B., Hess, C., Sugrue, L.: Am. J. Neuroradiol. **41**(2), 213 (2020)
10. Torp, C., Olerud, H., Einarsson, G., Grøn, P., Leitz, W., Servomaa, A.: Radiological protection of patients in diagnostic and interventional radiology, nuclear medicine and radiotherapy (2001)
11. Hidajat, N., Wolf, M., Nunnemann, A., Liersch, P., Gebauer, B., Teichgräber, U., Schröder, R.J., Felix, R.: Radiology **218**(2), 395 (2001)
12. Shrimpton, P., Hillier, M., Lewis, M., Dunn, M.: Doses from computed tomography (CT) examinations in the UK-2003 review, vol. 67. NRPB Chilton (2005)
13. Menzel, H., Schibilla, H., Teunen, D., et al.: European Commission 16262, Luxembourg (2000)
14. State Regulatory Agency for Radiation and Nuclear Safety, Official Gazette of BiH12 (13) (2011)
15. Suljić-Beganović, F., Tanković, A., Talajić, M., Lutvica, S., Muhamedagić, L.: BiltenLjekarske komore **19**, 165 (2015)
16. A.A. of Physicists in Medicine: Routine adult head CT (2016). https://www.aapm.org/pubs/CTprotocols. Accessed 25 Dec 2020
17. Garba, I., Engel-Hills, P., Davidson, F., Tabari, A.: Radiat. Prot. Dosimetry **165**(1–4), 98 (2015)
18. Shuaib, W., Tiwana, M., Chokshi, F., Johnson, J., Bedi, H., Khosa, F.: Ir. J. Med. Sci. (1971-) **184**(3), 631 (2015)
19. ACR-AAPM-SPR: ACR-AAPM-SPR practice parameter for diagnostic reference levels and achievable doses in medical x-ray imaging. Technical Report. American College of Radiology (2014)
20. Lee, K.L., Beveridge, T., Sanagou, M., Thomas, P.: J. Med. Radiat. Sci. **67**(1), 5 (2020)
21. Shrimpton, P., Hillier, M., Lewis, M., Dunn, M.: Review Archived, pp. 9–22 (2011)
22. Schegerer, A., Loose, R., Heuser, L.J., Brix, G.: RöFo-Fortschritte auf dem Gebietder Röntgenstrahlen und der bildgebenden Verfahren **191**(08), 739 (2019)
23. Kanda, R., Akahane, M., Koba, Y., Chang, W., Akahane, K., Okuda, Y., Hosono, M.: Jpn. J. Radiol. 1–8 (2020)
24. Roch, P., Aubert, B.: Radiat. Prot. Dosimetry **154**(1), 52 (2013)
25. Greenwood, T.J., Lopez-Costa, R.I., Rhoades, P.D., Ramírez-Giraldo, J.C., Starr, M., Street, M., Duncan, J., McKinstry, R.C.: Radiographics **35**(5), 1539 (2015)
26. Varghese, B., Kandanga, I., Puthussery, P., Vijayan, D., Babu, S.H., Aneesh, M., Noufal, M., Binu, E., Babu, A.C., James, S.M., et al.: Indian J. Radiol. Imaging **28**(2), 250 (2018)
27. Smith-Bindman, R., Moghadassi, M., Wilson, N., Nelson, T.R., Boone, J.M., Cagnon, C.H., Gould, R., Hall, D.J., Krishnam, M., Lamba, R., et al.: Radiology **277**(1), 134 (2015)

Lattice Boltzmann Simulation of Fluid Flow Between Two Rotating Cylinders and Application in Biomedicine

Tijana Djukic[1,2](✉) and Nenad Filipovic[3]

[1] Institute for Information Technologies, University of Kragujevac, Jovana Cvijica bb, 34000 Kragujevac, Serbia
tijana@kg.ac.rs

[2] Bioengineering Research and Development Center, BioIRC, Prvoslava Stojanovica 6, 34000 Kragujevac, Serbia

[3] Faculty of Engineering, University of Kragujevac, Sestre Janjic 6, 34000 Kragujevac, Serbia

Abstract. In many processes in industrial, technical, biological and biomedical systems it is necessary to analyze the flow between coaxial cylinders. This phenomenon was analyzed in literature – analytically, experimentally and numerically. In this paper, numerical simulations of fluid flow between two concentric rotating cylinders are performed using a special formulation of the LB (lattice Boltzmann) method for moving walls, called the mass-conserved volumetric lattice Boltzmann (abbreviated MCVLB) method. Overall three different kinematic cases were considered - the inner and outer cylinder rotated with equal angular velocities, the inner cylinder was at rest and the cylinders rotated with equal and opposite angular velocities. The results of simulations are compared with other numerical results and analytical solutions to verify the accuracy of the proposed model. Finally, a simulation of fluid flow through the semicircular canals of the inner ear is performed to demonstrate the possibilities of application of the proposed method in biomedical engineering.

Keywords: Lattice Boltzmann method · Moving boundaries · Validation of results · Vestibular system

1 Introduction

The analysis of flow through straight or curved pipes or between coaxial cylinders can be useful in many fields, including industrial, technical, biological and biomedical systems. Simulation of time-periodic fluid flow is important because it provides insight into the parameters of oscillatory flow occurring in diverse fields, e.g. during blood flow in arteries or the motion of endolymph within the inner ear. The rotational fluid flow was examined analytically in literature [1]. The authors developed an approximate analytical method to analyze the steady state and transient motion of an incompressible, Newtonian fluid between two concentric, rotating, rigid spheres. This problem was also analyzed experimentally in article [2]. The authors developed an experimental setup

A. Badnjevic and L. Gurbeta Pokvić (Eds.): CMBEBIH 2021, IFMBE Proceedings 84, pp. 777–783, 2021.
https://doi.org/10.1007/978-3-030-73909-6_88

with two concentric cylinders, with a gap between them that contained the fluid. The cylinders were closed by a top and bottom plate that rotate with the cylinders. They used ultrasound Doppler velocimetry to measure the parameters of the flow. Numerical simulations of fluid flow between two rotating cylinders were also performed in literature [3, 4]. Two numerical methods – DPD (dissipative particle dynamics) and SPH (smoothed particle hemodynamics) were used and the results were compared with the known, time-dependent analytical solution. In this paper, a special formulation of the LB (lattice Boltzmann) method for moving walls will be used to perform simulations of fluid flow and the results will be compared with results presented in literature [3, 4].

2 Numerical model

LB method observes the fluid as a set of particles that are fixed to a Cartesian mesh. These particles can move in prescribed directions defined with vectors $\overline{\xi}_i$ that represent the abscissae of the chosen lattice structure. For a two-dimensional problem, there are overall 9 vectors defining the behavior of particles in the considered node and 8 adjacent nodes. A specific distribution function f defined identically for each particle depends on the state of the particular particle and its closest neighboring particles. The dynamics of motion of all particles is modeled through their mutual collisions and further propagation in the domain.

In literature it was demonstrated that this numerical model can be used to model the fluid flow at the macro scale level [5, 6]. LB method was successfully applied to the modeling of motion of LDL (low-density lipoprotein) particles [7] through the bloodstream, CTCs (circulating tumor cells) through microfluidic chip [8] and RBCs (red blood cells) through the vein plexus of zebrafish [9].

The entire LB method can be ultimately represented using two equations, representing the collision and propagation steps.

$$\overline{f}_i^{out}(\mathbf{x},t) = \overline{f}_i^{in}(\mathbf{x},t) - \frac{1}{\overline{\tau}}\left(\overline{f}_i^{in}(\mathbf{x},t) - f_i^{(0)}(\rho,\mathbf{u})\right) \tag{1}$$

$$\overline{f}_i^{in}\left(\mathbf{x}+\overline{\xi}_i, t+1\right) = \overline{f}_i^{out}(\mathbf{x},t) \tag{2}$$

In these equations, two distribution functions are defined $-\overline{f}_i^{in}$ and $\overline{f}_i^{out}$, that represent values of the distribution function before and after collision, and $\overline{\tau}$ represents the relaxation time.

In the standard LB method, the boundary condition (BC) used for fixed walls is called bounce-back and using this BC, the nonslip boundary condition is imposed. This approach is preferred when the solid-fluid boundary is made of straight lines and the walls are fixed. But this approach cannot be used if the solid-fluid boundary is complex with curved boundaries and/or if the boundary is not fixed. In this case, a different approach must be applied for simulating moving boundaries. In this paper, an approach developed by Yu et al. [10] that is called the mass-conserved volumetric lattice Boltzmann (abbreviated MCVLB) method is used for numerical simulations.

This is an adaptation of the standard LB method that enables the introduction of predefined velocity of the boundary nodes.

An additional quantity called fluid volume portion is defined in the MCVLB method. This quantity is used to define whether a particular node (or more precisely cell) in the lattice mesh is occupied completely by solid, by fluid or partially by fluid and solid. The occupation of solid volume is defined as:

$$P(\mathbf{x},t) = \frac{\Delta V_s(\mathbf{x},t)}{\Delta V} \tag{3}$$

where the volume that solid occupies within the observed cell is denoted by $\Delta V_s(\mathbf{x},t)$ and ΔV is the overall cell volume, which is taken to be unity. For cells that are completely occupied by fluid, this parameter is equal to 0, for cells that are completely occupied by solid, this parameter is equal to 1 and $0 < \mathrm{P} < 1$ is valid for cells that are partially occupied by solid and partially by fluid.

Now it is possible to define the fluid volume portion:

$$\Delta V_f(\mathbf{x},t) = [1 - P(\mathbf{x},t)]\Delta V \tag{4}$$

In MCVLB method, the adapted distribution function n_i is related to the distribution function in standard LB method by the following equation:

$$f_i(\mathbf{x},t) = n_i(\mathbf{x},t)/\Delta V_f(\mathbf{x},t) \tag{5}$$

MCVLB method consists of three steps. The first step considers the collision of the particles. This step is different from the standard collision step (defined in Eq. (1)) in that the collisions of particles with moving solid boundary are dealt with corrections included in the equation:

$$n_i^{out}(\mathbf{x},t) = n_i(\mathbf{x},t) - \frac{1}{\tau}\left(n_i(\mathbf{x},t) - n_i^{(0)}(\mathbf{x},t)\right) \tag{6}$$

where the superscript out again denotes the values of distribution function after collision and $n_i^{(0)}$ represents the equilibrium particle distribution function, that is given by:

$$n_i^{(0)}(\mathbf{x},t) = N\omega_i\left[1 + \frac{\boldsymbol{\xi}_i \cdot \mathbf{U}}{c_s^2} + \frac{(\boldsymbol{\xi}_i \cdot \mathbf{U})^2}{2c_s^4} + \frac{\mathbf{U} \cdot \mathbf{U}}{2c_s^2}\right] \tag{7}$$

where $N(\mathbf{x},t) = \sum_i n_i(\mathbf{x},t)$ represents the constant related to the LB method, defined using the relation $c_s^2 = 1/3$.

The value of velocity $\mathbf{U}$ in Eq. (7) is calculated such that the effect of moving boundary is included in the calculation. The macroscopic velocity $\mathbf{u}$ that is used in standard LB method is increased for a velocity correction ($\mathbf{U} = \mathbf{u} + \delta\mathbf{u}$), where the correction $\delta\mathbf{u}$ is given by:

$$\delta\mathbf{u} = \tau P(\mathbf{x},t)\mathbf{u}_B(\mathbf{x},t) + \frac{\tau\sum_j P(\mathbf{x} + \boldsymbol{\xi}_j, t)n_j(\mathbf{x},t)\mathbf{u}_b(\mathbf{x} + \boldsymbol{\xi}_j, t)}{N(\mathbf{x},t)} \tag{8}$$

where $\mathrm{u}_B(\mathbf{x},t)$ is the prescribed velocity of the boundary in considered lattice cell.

The second step of the MCVLB method considers the propagation of the particles. Again, this step is different from the standard propagation step (defined in Eq. (2)), because the interaction of the fluid particles with solid boundaries is modeled using a special type of bounce-back boundary condition:

$$n_i^{in}(\mathbf{x},t) = [1 - P(\mathbf{x},t)]n_i^{out}\left(\mathbf{x} - \boldsymbol{\xi}_i\Delta t,\, t\right) + P\left(\mathbf{x} + \boldsymbol{\xi}_{i*}\Delta t,\, t\right)n_{i*}^{out}(\mathbf{x},t) \tag{9}$$

where the index $i*$ corresponds to the direction opposite to the i-th direction, i.e. $\overline{\xi}_{i*} = -\overline{\xi}_i$. It should be noted that for cells with P = 0, i.e. for cells that are entirely occupied by fluid, Eq. (8) is identical to the one used for propagation in standard LB method (Eq. (2)).

The third step of the MCVLB method considers mass conservation of the fluid, when the solid boundary is moving and meanwhile changing the fluid volume portion of the cells in the lattice mesh. In this paper, the solid boundary is considered to be rigid and moving along with the fluid. Therefore the mentioned third step does not have to be included in the numerical simulations within this paper.

Macroscopic quantities are calculated using the components of the particle distribution function. The equations used to define pressure and macroscopic velocity are given by:

$$p = c_s^2 \frac{\sum_i n_i(\mathbf{x},t)}{1 - P(\mathbf{x},t)} \tag{10}$$

$$\mathbf{u} = \frac{\sum_i \boldsymbol{\xi}_i n_i(\mathbf{x},t)}{\sum_i n_i(\mathbf{x},t)} \tag{11}$$

Equations (6) and (9) are solved in iterations, for all nodes in the lattice mesh, until the steady state is reached.

3 Results and discussion

3.1 Fluid Flow Between Two Rotating Cylinders

The numerical model presented in Sect. 2 is used to simulate flow in a domain consisting of fluid that is located between two concentric cylinders that rotate with equal angular velocity, in the same direction. This example was used in literature [3] to validate the results of the DPD method against analytical solutions. A similar example was also used in literature [11] to test the accuracy of the boundary conditions of LB method. The same simulation setup as in article [3] was used in this paper. The radius of the outer cylinder was 2.14 times greater than the radius of the inner cylinder. The fluid is initially at rest and after overall 50,000 iterations the LB simulation reaches the steady state. The results of LB simulations are compared with DPD results and analytical solutions for several moments in time during simulation and these results for the distribution of nondimensional velocity along the radial direction are shown in Fig. 1A. Full lines in Fig. 1A represent analytical solutions, while the discrete data symbols show numerical

A B

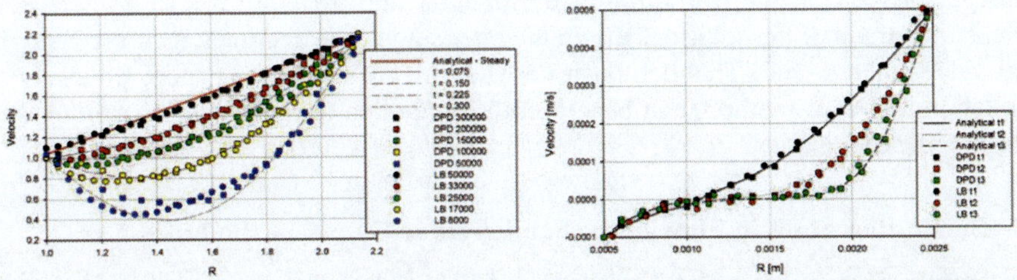

Fig. 1. Distribution of velocity along the radial direction and comparison with analytical solution and DPD results presented in literature [3, 4]. A – The cylinders rotate with equal angular velocities in the same direction; B - The cylinders rotate with equal angular velocities in the opposite direction (moments in time are defined as follows t1 = 4.3216×10^{-2} s, t2 = 9.6111×10^{-2} s, t3 = 24.1381×10^{-2} s).

results using DPD and LB. In this case, following the representation used in literature [3], the nondimensional quantities are represented on the graph.

Three different cases were analyzed using a similar simulation setup in literature [4]. In these simulations the radius of the outer cylinder was 5 times greater than the radius of the inner cylinder. The difference was in the kinematic conditions. Namely, the inner cylinder rotated with equal angular velocity as the outer cylinder in case 1, in case 2 the inner cylinder was at rest and in case 3 the inner cylinder rotated with equal and opposite angular velocity of the outer cylinder. LB simulation for case 3 was compared in this paper with DPD results and analytical solution and the distribution of velocity along the radial direction for this case is shown in Fig. 1B. Again, full lines in Fig. 1B represent analytical solutions, while the discrete data symbols show numerical results using DPD and LB.

The velocity distribution field, together with velocity streamlines and vectors for the 3 mentioned cases is shown in Fig. 2.

A B C

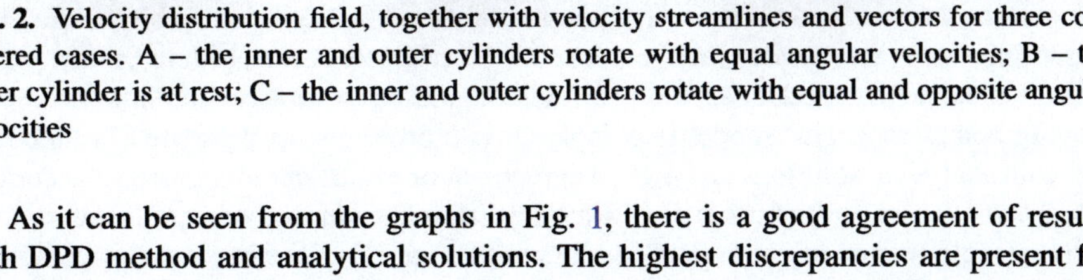

Fig. 2. Velocity distribution field, together with velocity streamlines and vectors for three considered cases. A – the inner and outer cylinders rotate with equal angular velocities; B – the inner cylinder is at rest; C – the inner and outer cylinders rotate with equal and opposite angular velocities

As it can be seen from the graphs in Fig. 1, there is a good agreement of results with DPD method and analytical solutions. The highest discrepancies are present for the early phase of the simulations in Fig. 1A, but this is due to the different nature of both methods and to the resolution of the domain that probably has to be greater to increase accuracy of the solution. But in the later stages, the accuracy increases and the final values match well. This demonstrates that the LB method, or more precisely its adaptation MCVLB method, can be successfully applied to model the motion of fluid with moving boundaries.

3.2 Simulation of Fluid Flow in the Semicircular Canals of the Inner Ear

The presented numerical model can be used to simulate the process of fluid flow in many biomedical systems. In this paper, the semicircular canals (SCC) of the inner ear are taken as an example. The SCCs are part of the vestibular system responsible for the balance of the human organism. They are filled with a fluid called endolymph, whose characteristics are similar to water. During the motion of the head, there is a flow of this fluid caused by the motion of the walls that further causes the human to sense the movements. The rotation of the SCCs is simulated using the numerical model presented in Sect. 2 and the change of velocity in the SCCs during rotation is shown in Fig. 3.

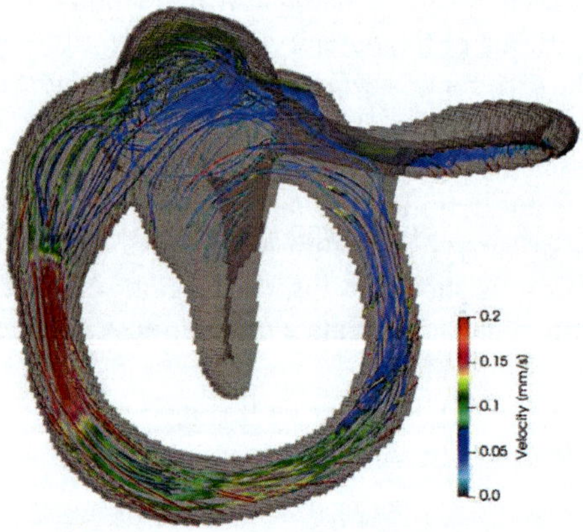

Fig. 3. Velocity streamlines in the semicircular canals of the inner ear during rotation of the whole domain

4 Conclusion

LB method was applied in literature to model diverse phenomena, e.g. to model the behavior of harbour seal vibrissae [12], to model flows through permeable media [13], to investigate the pore-scale reactive transport in carbonate rocks [14] etc. In this paper, a discrete numerical model based on LB method is used to simulate the fluid flow with moving boundaries. This model is compared with other results presented in literature and with analytical solutions and a good agreement of results demonstrated its accuracy. Finally, this model can be further used to model different phenomena in biomedicine. One such example (the endolymph flow in semicircular canals of the inner ear) is presented in this paper.

Acknowledgements. This research is funded by Serbian Ministry of Education, Science, and Technological Development [451-03-68/2020-14/200378 (Institute for Information Technologies, University of Kragujevac)]. This research is also supported by the European Union's Horizon 2020 research and innovation programme under grant agreement No. 952603. This article reflects only the author's view. The Commission is not responsible for any use that may be made of the information it contains.

References

1. Gagliardi, J.C., Nigro, N.J., Elkouh, A.F., Yang, J.-K., Rodriguez, L.S.J.: Study of the axially symmetric motion of an incompressible viscous fluid between two concentric rotating spheres. J. Eng. Math. **24**, 1–23 (1990)
2. Ern, P., Wesfreid, J.E.: Flow between time-periodically co-rotating cylinders. J. Fluid Mech. **397**, 73–98 (1999)
3. Haber, S., Filipovic, N., Kojic, M., Tsuda, A.: Dissipative particle dynamics simulation of flow generated by two rotating concentric cylinders: Boundary conditions. Phys. Rev. E **74**, 046701 (2006)
4. Filipovic, N., Ivanovic, M., Kojic, M.: A comparative numerical study between dissipative particle dynamics and smoothed particle hydrodynamics when applied to simple unsteady flows in microfluidics. Microfluid. Nanofluid. **7**, 227–235 (2009)
5. Đukić, T.: Modeling motion of deformable body inside fluid and its application in biomedical engineering. Ph.D. dissertation, Faculty of Engineering, University of Kragujevac, Kragujevac (2015)
6. Malaspinas, O.P.: Lattice Boltzmann method for the simulation of viscoelastic fluid flows. Ph.D. dissertation, Switzerland (2009)
7. Filipovic, N., Zivic, M., Obradovic, M., Djukic, T., Markovic, Z., Rosic, M.: Numerical and experimental LDL transport through arterial wall. Microfluid. Nanofluid. **16**(3), 455–464 (2014)
8. Djukic, T., Topalovic, M., Filipovic, N.: Numerical simulation of isolation of cancer cells in a microfluidic chip. J. Micromech. Microeng. **25**(8), 084012 (2015)
9. Djukic, T., Karthik, S., Saveljic, I., Djonov, V., Filipovic, N.: Modeling the behavior of red blood cells within the caudal vein plexus of zebrafish. Front. Physiol. **7**, 455 (2016). https://doi.org/10.3389/fphys.2016.00455
10. Yu, H., Chen, X., Wang, Z., Deep, D., Lima, E., Zhao, Y., Teague, S.D.: Mass-conserved volumetric lattice Boltzmann method for complex flows with willfully moving boundaries. Phys. Rev. E **89**, 063304 (2014)
11. Chang, C., Liu, C.-H., Lin, C.-A.: Boundary conditions for lattice Boltzmann simulations with complex geometry flows. Comput. Math. Appl. **58**(5), 940–949 (2009)
12. Morrison, H., Brede, M., Dehnhardt, G., Leder, A.: Simulating the flow and trail following capabilities of harbour seal vibrissae with the lattice Boltzmann method. J. Comput. Sci. **17**, 394–402 (2016)
13. Boroni, G., Silin, N., Clausse, A.: A Python implementation in graphic processing unit of a lattice Boltzmann model for unstable three-dimensional flows in immersed permeable media. Phys. Fluids **32**, 127107 (2020)
14. An, S., Erfani, H., Hellevang, H., Niasar, V.: Lattice-Boltzmann simulation of dissolution of carbonate rock during CO2-saturated brine injection. Chem. Eng. J. **408**, 127235 (2021)

Influence of Random Vibration on Semicircular Canals During Exposure to Whole Body Vibrations

Igor Saveljic[1,3](✉), Slavica Macuzic Saveljic[2], and Nenad Filipovic[2,3]

[1] Institute for Information Technology, University of Kragujevac, 34000 Kragujevac, Serbia
isaveljic@kg.ac.rs

[2] Faculty of Engineering, University of Kragujevac, 34000 Kragujevac, Serbia

[3] Bioengineering Research and Development Center, 34000 Kragujevac, Serbia

Abstract. Vibrations and vehicles reduce the comfort of drivers and passengers, causing stress, fatigue and discomfort. Chronic exposure to whole body vibration can affect the lumbar spine, the gastrointestinal system, the peripheral veins and the vestibular system. Until recently, it was possible to study the semicircular canals geometry only in cadaveric specimens, but now, with the development of high-resolution computed tomography and magnetic resonance imaging, it has become possible to mathematically model their geometry. The semicircular canals (SCC), as a part of vestibular system, are responsible for sensing angular head motion in three-dimensional space and for providing neural inputs to the central nervous system. In this study, one male subject was exposed to root-mean-square WBV acceleration level of 0.8 m/s^2 while seated on an electrohydraulic vibration simulator with multi-axial excitation. The movements recorded on the head of the examinee were transferred to a 3D model. Numerical analysis provides fluid velocity generated in semicircular canals.

Keywords: Numerical analysis · Random vibration · Semicircular canals · Whole body vibration

1 Introduction

The human vestibular system is a sensory and equilibrium system that manages and controls the human sense of balance and movement. The semicircular canals are three semicircular, interconnected tubes located in the innermost part of each ear, the inner ear. The three canals are the horizontal, superior and posterior semicircular canals. They are part of the vestibular system (Fig. 1) located in the inner ear. The SCC are filled with endolymph, a fluid with properties similar to water. When the head is rotated, inertia induces a flow of the endolymph. This flow deflects a gelatinous structure called cupula which obstructs the whole lumen at one end of each SCC. The cupular deflection stimulates hair cells which lead to the perception of angular motion. Any perturbation of the endolymph flow may cause vertigo and balance disorders.

Balance disorders can occur at any age.

A. Badnjevic and L. Gurbeta Pokvić (Eds.): CMBEBIH 2021, IFMBE Proceedings 84, pp. 784–791, 2021.
https://doi.org/10.1007/978-3-030-73909-6_89

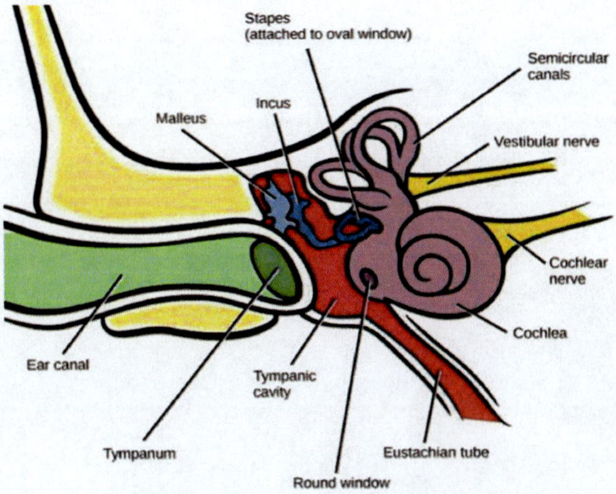

Fig. 1. The vestibular system.

Exposure to vibrations affects a person in different ways, starting from ordinary disturbances, to reduced work performance and damage to health. A guideline in defining human tolerance to whole-body vibration as an international standard ISO 2631-1 (1997) [1] was adopted today. ISO 2631-5 (2004) [2] is used to assess exposure to high levels of vibration and shock. ISO 2631-4 (2001) [3] is used to define different methods for measuring periodic, random and transient vibrations. The vibrations that are absorbed lead to muscle contractions that can cause muscle fatigue, especially at resonant frequencies. Vertical vibrations in the range of 5–10 Hz cause resonance in the thoracic abdominal system, from 4–8 Hz in the spinal part, from 20–30 Hz in the area of the head and neck and from 60–90 Hz in the area of the eyeballs.

The aim of this paper is to show how exposure to vibrations of different amplitudes affects the fluid located in the semicircular channels.

2 Materials and Methods

In order to provoke excitation of different amplitudes and frequencies, an electrohydraulic pulsator HP-2007 was used, which contains a car seat on which the subjects were exposed to vibrations. This device has the ability to excite vibrations in two directions. On the head of the subjects via plastic helmet was mounted a system of three-axis accelerators where the accelerator AC102-1A, weight of 90 g, sensitivity of 100 mV/g, and frequency range of 0.5–15.000 Hz was used. In order to determine head displacements, accelerations was measured on the head using three-axis accelerometers. The 01dB-Metravib NetdB PRO-132 acquisition systems was. The measurement scheme is shown in Fig. 2. The parameters of the experiment were: the sampling rate 51.200 Hz, the block duration 80 ms, the number of samples 4.096, the number of averaging 2, and sampling step $\Delta t = 0.0195$ ms. Signal overlap was 75%. The frequency step was $\Delta f = 0.390625$ Hz, and the bandwidth was 39 Hz.

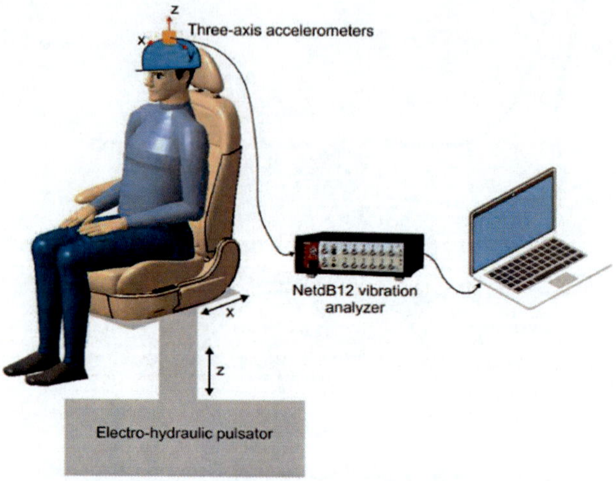

Fig. 2. Position of the subjects on the electro-hydraulic pulsator.

One male healthy subject (aged 37 years, 188 cm tall and 95 kg mass) was exposed to a random stimulus to both single-axis vibration in vertical and fore and aft. The vibration amplitude was 0.8 m/s^2. The excitation frequency range was 0.5 Hz–30 Hz. There was one seating angle of 90° of inclination. The subject rested his hands on thighs. Position of hands does not have effect on vibration transmission through human body [4, 5].

Figure 3 shows input boundary condition applied to model - r.m.s. acceleration measured on subject head.

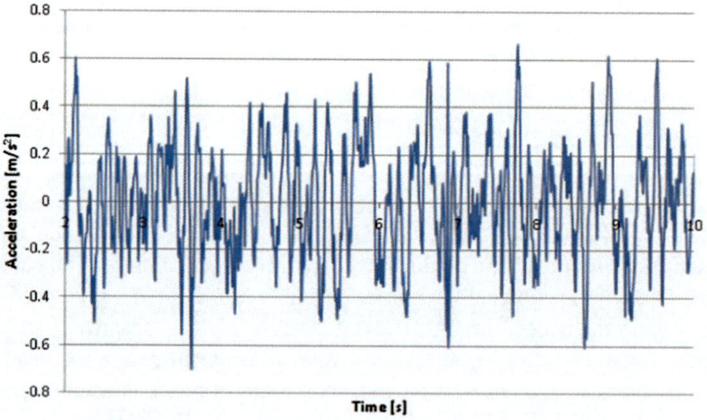

Fig. 3. R.m.s. acceleration measured on subject head.

Healthy human right membranous labyrinths based on the morphologically descriptive model were reconstructed using the CT scans. The high resolution CT data were read into Mimics 17.0 (Materialise Inc., Leuven, Belgium) visualization software, where the images were segmented by thresholding to obtain 3D model (Fig. 4).

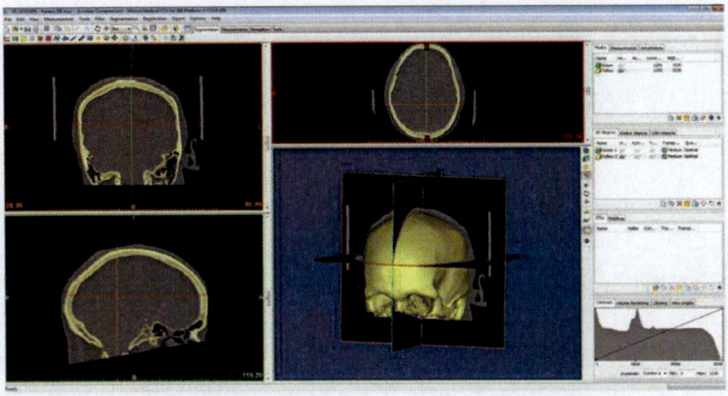

Fig. 4. Automatic segmentation of CT scans.

The computational model consists of the utricular cavity, the horizontal canal (HC), the anterior canal (AC), the posterior canal (PC), and their ampullae (Fig. 5) [6].

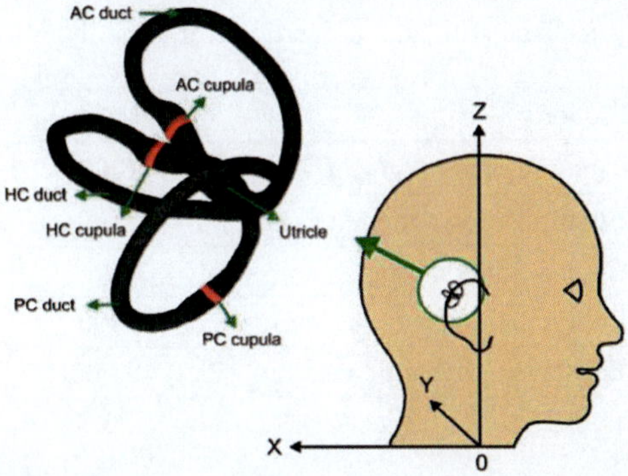

Fig. 5. The reconstructed model of a healthy human membranous (the semicircular canals) labyrinth and position in head

The endolymph flow were described by the conservation equations of mass, momentum, and energy. An Arbitrary Lagrangian Eulerian (ALE) approach is adopted, and the endolymph is assumed to be a slightly compressible Newtonian fluid with constant properties. All the governing equations are listed below. The continuity equation is approximated as [7]:

$$v_{i,i} + \frac{p}{\lambda} = 0 \tag{1}$$

where λ is a selected large number, the penalty parameter. Substituting the pressure p from Eq. 1 into the Navier-Stokes equations it can be obtained

$$p\left(\frac{\partial v_i}{\partial t} + \partial v_{i,k} v_k\right) - \lambda v_{j,ij} - \mu v_{i,kk} - f_i^V = 0 \tag{2}$$

Then the FE equation of balance becomes

$$M\dot{V} + (K_{vv} + K_{vv}^{\lambda})V = F_v + F_\lambda \tag{3}$$

where:

$$\left[K_{KJ}^{\lambda}\right]_{i,k} = \lambda \int_V N_{K,i} N_{J,k} dV, \quad (F_\lambda)_{Ki} = \lambda \int_S N_K v_{j,j} n_i dS \tag{4}$$

Fluid-structure interaction for cupula deformation and endolymph flow is implemented. Cupula was modeled as elastic 3D membrane with brick finite element and endolymph domain as 3D 8–node finite elements. The values for the endolymph conductivity, reference density, and viscosity are taken from measured data provided by [8]. The physical and structural properties of the endolymph and cupula are shown in Table 1. The meshing was conducted primarily with Femap v.10. Linear tetrahedron was used as the final element.

Table 1. The physical and structural properties of the endolymph and cupula [8].

Property	Value
Cupula density (kg/m^3)	1000
Cupula Young's modulus (Pa)	5
Cupula Poisson ratio	0.45
Endolymph density (kg/m^3)	1000
Endolymph dynamic viscosity (Pa·s)	0.000852

3 Results

Numerical simulation was performed using PAKFS solver. Three value of acceleration was investigated. The results of velocities changes in the channels of the vestibular system are shown in the following figures. Figure 6 shows velocity changes through the SCC geometry due to a head acceleration of 0.40 mm/s^2. It can be noticed that the maximum velocity of 4.1e−03 mm/s is recorded in the PC duct, while the maximum velocities in the HC duct and the AC duct are 3.35e−03 mm/s and 2.10e−03 mm/s, respectively. In AC cupula and HC cupula are relatively low fluid velocities, about 2.5e−04 mm/s, while a value of 1.2e−03 mm/s is recorded in the PC cupula. Ultricle is also characterized by low fluid velocity.

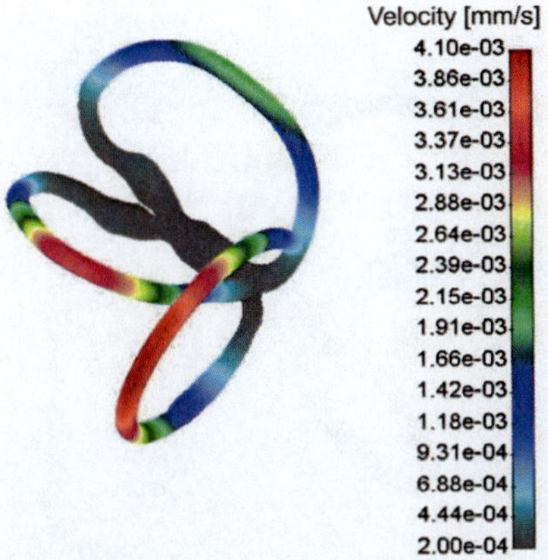

Fig. 6. Velocity changes through the SCC geometry due to a head acceleration of 0.40 mm/s^2

Figure 7 shows velocity changes through the SCC geometry due to a head acceleration of 0.55 mm/s^2. Unlike the previous picture, the highest recorded velocity of 4.1e03 mm/s is in the AC cupula, the HC chupula and Ultricle. Observing the PC duct, the highest recorded velocity is 6.7e−04 mm/s, while in the HC duct it was measured 9.4e04 mm/s, and in the AC duct the velocity was 2.2e−03 mm/s. The velocity in the PC cupula is 1.95e−03 mm/s.

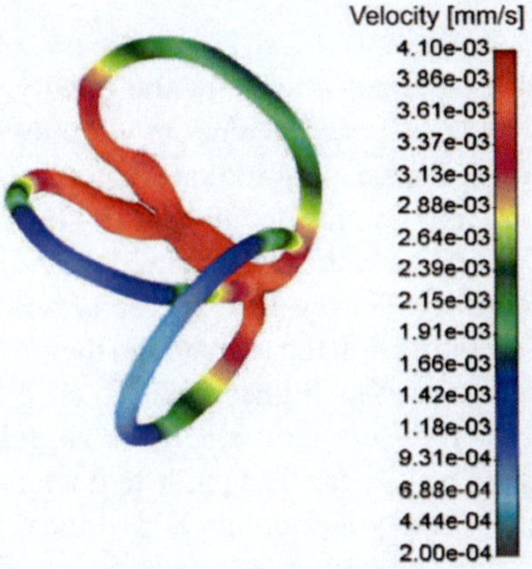

Fig. 7. Velocity changes through the SCC geometry due to a head acceleration of 0.55 mm/s^2

Figure 8 shows velocity changes through the SCC geometry due to a head acceleration of 0.60 mm/s^2. Also, this case differs from the previous two. The highest velocity of 4.1e−03 mm/s was recorded in the PC duct, while in the HC duct the velocity is slightly lower and amounts to 3.75e−03 mm/s. In the AC duct, the fluid velocity is 2.75e−03 mm/s. Observing the AC and HC cupula, the maximum velocity is around 6.9e−04 mm/s, while the Ultricle has a slightly lower maximum speed of 5.5e−04 mm/s. The PC cupula records a fluid velocity value of 1.95e−03 mm/s.

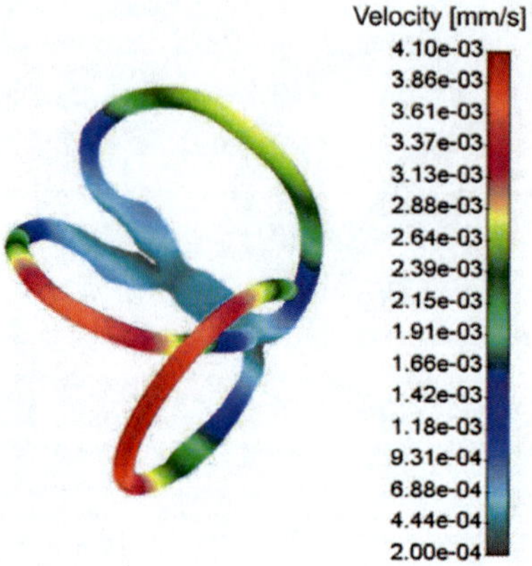

Fig. 8. Velocity changes through the SCC geometry due to a head acceleration of 0.60 mm/s^2

The previous figures clearly shows how the change in acceleration to which the subject's head is exposed affects the change in fluid velocity in certain parts of the semicircular canal. The presented numerical simulation can provide doctors very valuable information on how different amplitudes of vibrations to which a person is exposed at his workplace, have an impact on his balance.

4 Conclusion

In this paper we presented numerical model for the endolymph flow in semicircular canals when the subject's head is exposed to random vibrations in two directions. Three value of acceleration was investigated – 0.4 mm/s^2, 0.55 mm/s^2 and 0.6 mm/s^2. For the first case of head acceleration maximum velocity of 4.1e−03 mm/s is recorded in the PC duct, while the maximum velocities in the HC duct and the AC duct are 3.35e−03 mm/s and 2.10e−03 mm/s, respectively. For the second case of head acceleration the highest recorded velocity of 4.1e−03 mm/s is in the AC cupula, the HC chupula and Ultricle. For the third case of head acceleration The highest velocity of 4.1e03 mm/s was recorded in the PC duct, while in the HC duct the velocity is slightly lower and amounts to 3.75e−03 mm/s. The obtained results can be a guide to doctors when there are problems with the patient's balance. By applying numerical methods, all physical quantities of interest can be determined: fluid velocity, cupula pressure, etc. The finite element method demonstrates once again that this numerical method is a robust technique to obtain fast and reliable results even in the biomechanical field.

Acknowledgments. This research is supported by the European Union's Horizon 2020 research and innovation programme under grant agreement No 952603. This article reflects only the author's view. The Commission is not responsible for any use that may be made of the information it contains. Also, this research is funded by Serbian Ministry of Education, Science, and Technological Development [451-03-68/2020-14/200378 (Institute for Information Technologies, University of Kragujevac)].

References

1. ISO 2631-1, 1997: Mechanical vibration and shock - evaluation of human exposure to whole-body vibration - part 1: general requirements (1997)
2. ISO 2631-5, 2004: Mechanical vibration and shock - evaluation of human exposure to whole-body vibration - part 5: method for evaluation of vibration containing multiple shocks (2004)
3. ISO 2631-4, 2001: Mechanical vibration and shock- evaluation of human exposure to whole-body vibration - part 4: guidelines for the evaluation of the effects of vibration and rotational motion on passenger and crew comfort in fixed-guideway transport systems. Texas Tech University (2001)
4. Demic, M., Lukic, J.: Investigation of the transmission of fore and aft vibration through the human body. Appl. Ergon. **40**, 622–629 (2009)
5. Mandapurama, S., Rakheja, S., Boileau, P.E., Maeda, P.E.: Apparent mass and head vibration transmission responses of seated body to three translational axis vibration. Int. J. Ind. Ergon. **42**(3), 268–277 (2012)
6. Tascioglu, A.B.: Brief review of vestibular anatomy and its higher order projections. Neuroanatomy **4**, 24–27 (2005)
7. Kojic, M., Filipovic, N., Stojanovic, B., Kojic, N.: Computer Modeling in Bioengineering - Theoretical Background Examples and Software. Wiley, Chichester (2008)
8. Hain, T.C., Helminski, J.O.: Anatomy and physiology of the normal vestibular system. In: Herdman, S.J. (ed.) Vestibular Rehabilitation. F.A. Davis Company, Philadelphia (2007)

A Novel Implementation of Road Mapping from Digital Subtraction Angiography Images

Eleonora Tiribilli[1], Ernesto Iadanza[1], Cosimo Lorenzetto[2], Leonardo Manetti[2], and Leonardo Bocchi[1](✉)

[1] Department of Information Engineering, University of Florence, Via di Santa Marta 3, 50139 Florence, Italy
leonardo.bocchi@unifi.it

[2] Epica Imaginalis, Via Rodolfo Morandi 13/15, 50019 Sesto Fiorentino, Italy

Abstract. Road Mapping (RM) is a variant of the commonly used Digital Subtraction Angiography (DSA). In this work we propose a Road Mapping algorithm in which the mask is extracted by means of vessels segmentation from DSA images. The algorithm extracts a two-dimensional map consisting of two values (black and white), to be superimposed on the live fluoroscopic image. A dedicated phantom has been created to develop the RM algorithm, simulating a circulatory system. A prototype algorithm is developed in Matlab R2018b in order to reach some preliminary results and compare different techniques. A Qt graphical interface is also developed for a fast and effective use of this new RDA approach to real veterinary fluoroscopies. Two workflows are presented, both succeeding in performing segmentation of the vessel. Their performances are compared and discussed.

Keywords: DSA · Road mapping · Fluoroscopy · Segmentation · Radiology

1 Introduction

Road Mapping (RM) is a variant of the commonly used Digital Subtraction Angiography (DSA). It used in interventional radiology as a navigation support procedure, during insertion of surgical material inside the patient. The first patents for this technique date back to the 1980s [1]. It is now used in the vascular field in many settings, such as angioplasty, embolization, insertion of stents or cava filters. It is also employed in carrying out guided biopsies, as well as for interventional operations in the orthopedic and in neurological field to evaluate and treat aneurysms [2, 3]. RM is performed starting from fluoroscopic images, acquired after the injection of the contrast agent. A mask of contrasted anatomical structures extracted from fluoroscopic images is saved. This mask is then used to perform a DSA with the current frame in live fluoroscopy. A fluoroscopic series in which the vessel is white and the remaining portion is gray is generated, due to the cancellation of all the structures present in both images [4].

In this work we propose a Road Mapping algorithm in which the mask is extracted by means of vessels segmentation from DSA images. The algorithm extracts a two-dimensional map consisting of two values (black and white), to be superimposed on

A. Badnjevic and L. Gurbeta Pokvić (Eds.): CMBEBIH 2021, IFMBE Proceedings 84, pp. 792–799, 2021.
https://doi.org/10.1007/978-3-030-73909-6_90

the live fluoroscopic image. The interest in this approach stems from the fact that the mask could be treated as a second layer superimposed on fluoroscopies. This allows the operator to set transparency, color and display mode as desired. In addition, the physician has the possibility, during the clinical practice, to see the inside of the vascular structure of interest and to move the surgical tools more accurately, by having a superimposed view of the vessels boundaries. Furthermore, extracting a mask by segmentation opens the door to the possibility of automatically studying some of the characteristics of the district under examination.

2 Materials and Methods

2.1 Phantom

A dedicated phantom has been created to develop the RM algorithm. To simulate a circulatory system, a circuit of plastic pipes connected to a small re-circulation pump has been designed and built (Fig. 1). The circuit is fixed with a silicone rubber casting, which has a double task: preserving the circuit from unwanted movements, creating an absorption layer to simulate soft tissues interaction with X-rays. Two bones are inserted in the phantom to simulate the absorption of hard tissues. The acquisitions are made during a water injection into the plastic pipes, using an automatic injection pump (SyringePumpTWO NE-1000 - Kf Technology).

2.2 Matlab Prototype

A prototype algorithm is developed in Matlab R2018b in order to reach some preliminary results and compare different techniques. First, fluoroscopic images require a pre-processing to reduce spatial and temporal noise. Three different spatial filters are evaluated: median, medium value, Gaussian smoothing. The temporal filter applied on the fluoroscopic series is an Exponential Weighted Moving Average (EWMA) with $\alpha = 0.4$. The fluoroscopic series must be integrated to obtain a single mask. Two approaches to this problem are identified for integrating frames: before thresholding or after binarization. The first solution consists in a pixel to pixel comparison that identifies the darker pixels in each of the processed fluoroscopic frames, herein referred to as Gray Level Integration. The second one consists in an integration of binary images previously segmented, through pixel-to-pixel logical operations, herein referred to as Binary Integration. It makes a logical OR between the current frame and an image, obtained as the result of a logical AND executed on a specific interval preceding the current frame. The amplitude of this interval defines how long the pixel must remain stable during fluoroscopic series. This allows the reduction of some noise contributions.

Segmentation is carried out by thresholding with two different techniques:

- Global thresholding: threshold value identified by the Otsu method [5].
- Adaptive thresholding: threshold calculated for each pixel of the image, based on pixel neighborhood information.

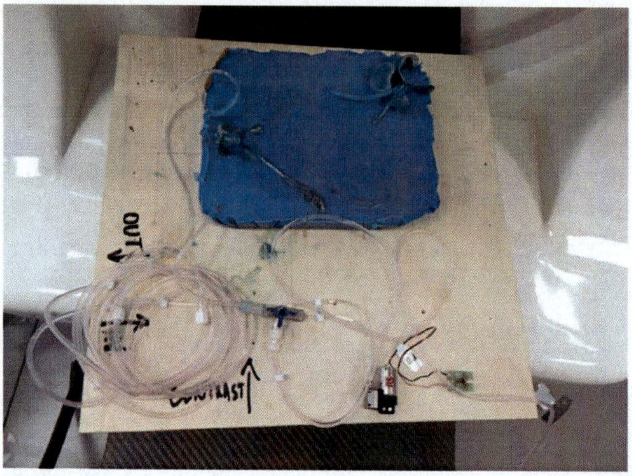

Fig. 1. Dedicated phantom created to develop the RM algorithm

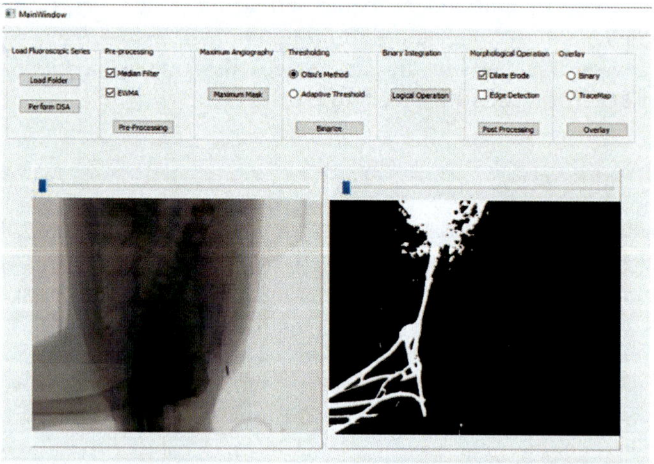

Fig. 2. Graphic User Interface developed to tests and compare different processing methods on real fluoroscopic studies

The binary image resulting from segmentation is subjected to morphological operations. A morphological opening operation is applied to remove noise granules: an erosion followed by an expansion with a structuring element 5 × 5. The binary mask is finally superimposed on the fluoroscopic frame. Overlay takes place with two different types of visualization: the first involves coloring binarized vessels in blended red, the second computes and visualizes only the boundaries of the mask.

The edge is obtained by eroding the image through a 3 × 3 structuring element. The result is subtracted from the original binary image.

2.3 Graphical User Interface (GUI)

A Qt [6] graphical interface is developed for a fast and effective use of this new RM approach to real veterinary fluoroscopies. The interface is also created with the scope of assessing the different filtering and screening techniques in an immediate and more intuitive way, observing the image resulting from each of the steps that compose the algorithm, allowing for optimizing the parameters, such as kernel sizes.

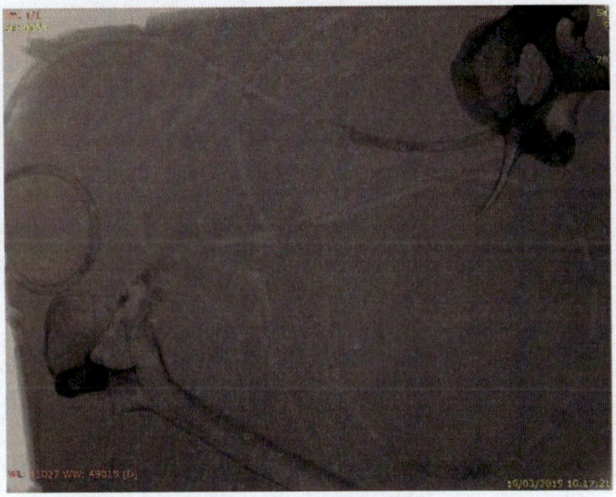

Fig. 3. Phantom fluoroscopy

The main functions to be tested are implemented in C++ by means of Qt framework and OpenCV [7] libraries. The GUI is designed to let the user select the functions to be applied, while examining the result in real time. The application, shown in Fig. 2 uses two display windows, one for the original image and the other to view processing results.

The interface is composed by a block dedicated to pre-processing functions, one to address thresholding operations, another to the apply morphological algorithms on binary images and a last one that includes the functions necessary to superimpose the mask on the fluoroscopic image.

3 Results

The algorithm's performance is first assessed on the phantom, and then on real veterinary studies by means of the developed GUI. In order to evaluate the performance of the Matlab prototype, two acquisitions of the phantom previously described are considered:

- fluoroscopy during automatic water injection (Fig. 3).
- fluoroscopy during water injection, with a bag of physiological solution placed between the source and the phantom, providing an uneven background.

The analysis of these fluoroscopic series allows to study and evaluate the results given by the different processing steps.

a) *Pre-processing:* Among the three spatial filters analyzed, the most satisfactory results for these acquisitions are obtained with the median. The medium value filter does not completely suppress Gaussian-type noise. A larger kernel improves noise suppression, but at the expenses of greater image degradation. Also, it is unable to remove the so-called "salt and pepper" noise. In this case, the large deviation of the noise value from the neighborhood pixel value disturbs the average value. Indeed, the "salt and pepper" noise operates as an outlier in the distribution of pixels; for this reason it is necessary to treat it using a more robust statistical measurement, compared to the outliers. In this sense, the median filter is more effective, and manages to preserve high-frequency details, lowering the image degradation. Finally, the spatial smoothing filter provides an image in which the abrupt transitions of intensity are reduced and the irrelevant details are eliminated (i.e. those with dimensions comparable with the mask's ones). This lowers down the high frequency noise levels of the image, although blurring the outlines. The temporal filter also decreases the noise on the image but partially compromises the temporal resolution.
b) *Mask:* The global Otsu method manages to find an adequate threshold to extract the flow inside the tubes when the intensity of the pixels of interest is homogeneous (first acquisition). Instead, if the gray levels of interest show uneven values (second acquisition), the method fails to segment all the water bubbles. The adaptive thresholding method extracts a complete path of water flowing into the pipes even in the presence of the bag of physiological solution. This happens at the cost of increasing the noise granules present on the image, which must be removed by post-processing step.
c) *Integration:* Gray Level Integration, integrating the gray-level fluoroscopic frame by looking for the darkest value of each pixel in the series, is the fastest method, allowing to threshold a single frame. However, Binary Integration removes small grains of noise included in the thresholding of some scattered frames.
d) *Post-Processing:* The morphological opening operation provides a solution for making the contours of the segmented vessel more homogeneous, eliminating small interruptions and subtle protuberances. It also removes isolated points on the mask, which can be considered noise granules.
e) *Overlay:* The two masks are superimposed on real fluoroscopy according to the two different modalities. Figure 4a and b show an example of the results obtained by the Matlab prototype, compared with the standard procedure (Fig. 5).

The optimal workflow in terms of image quality is based on Binar Integration and it's the following:

- performing a DSA without registration
- median filtering with a filter of kernel 5×5
- temporal filtering with the EWMA and $\alpha = 0.5$
- Gaussian adaptive thresholding with 55 pixel kernel

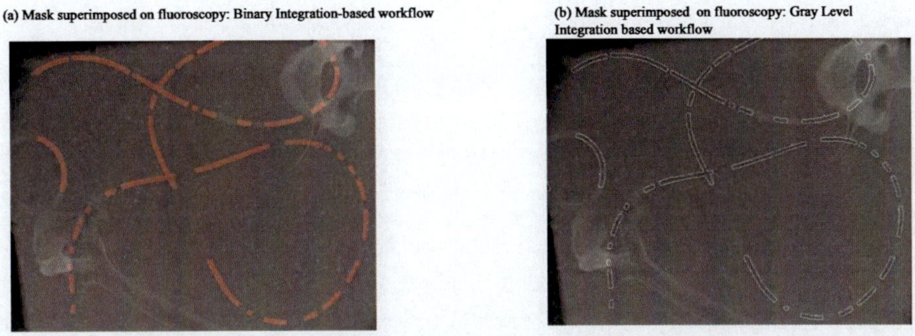

Fig. 4. Results obtained on phantom fluoroscopy, during water injection.

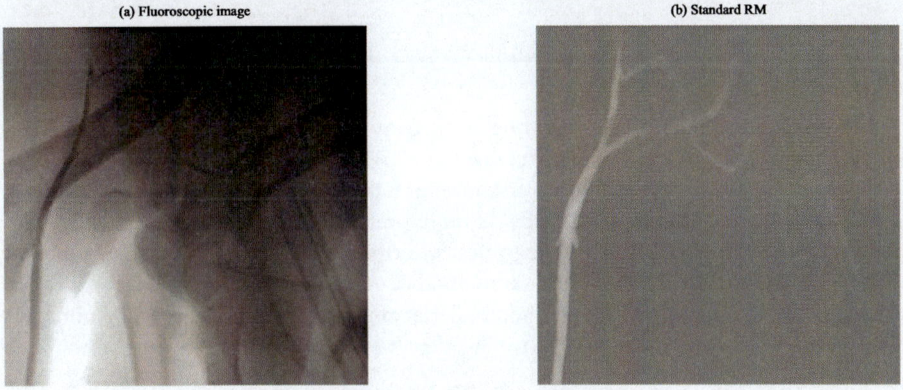

Fig. 5. Standard RM obtained as subtraction

- integration of binary frames through a Boolean logic operation applied on an interval of five frames
- post-processing on the binary image with morphological opening.

This procedure requires 817 ms for processing 300 frames, obtainig the binary mask shown in Fig. 6a. A strong reduction of computational time (350 ms) can be reached by the following steps, using Gray Level Integration:

- run a DSA without registration
- integration of DSA frames into a single image, identifying for each pixel the darkest value in the series
- median filtering with a 5×5 kernel filter
- EWMA temporal filter with $\alpha = 0.5$
- thresholding with the Otsu method
- post-processing on the binary image through morphological opening.

(a) Binary Integration-based workflow of our RM algorithm. It offers better precision at the expense of slightly higher computational time

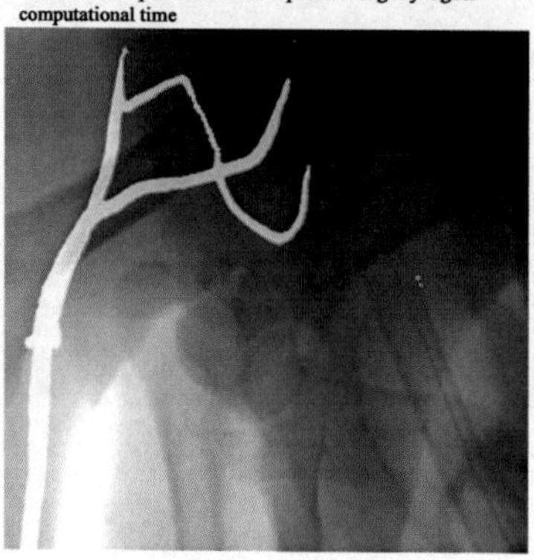

(b) Gray Level Integration-based workflow of our RM algorithm: superimposition with only boundaries. It offers shorter computational time but some points outside the vessel are included in the mask

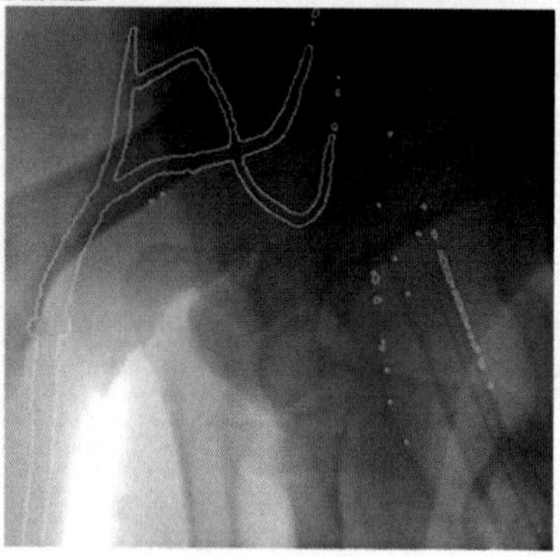

Fig. 6. Results obtained with workflows presented in this article

Figure 6b shows the image resulting from Gray Level Integration-based workflow, only the mask boundaries are superimposed.

Ultimately, these two methods both succeed in performing segmentation of the vessel: Binary Integration-based workflow is more precise but with a slightly higher computational cost, while Gray Level Integration workflow is faster but with less satisfactory results. The higher precision of the first method depends on the integration of the frame after thresholding that allows to remove small movement artifacts present in DSA images.

4 Conclusions

In general, compared to classical Road Mapping performed by subtraction, Road Mapping performed by segmentation, presented in this work, provides the possibility to view the map during the intervention directly on the fluoroscopic image and not as a DSA image. For an implementation in a real settings an appropriate risk management process will be needed [8]. The flow that involves thresholding all the frames allows to obtain, in practice, a real-time segmentation of the vessel. This opens the door to future developments that allow automatic extraction of information on the binary mask. For example, evaluation of the diffusion speed of the contrast medium in the various districts. The map of the spread of the contrast agent, also, gives the possibility to extract data concerning the dimensions of the contrasting vessels and districts, or their orientation. The analysis of images of segmented vessels is overall of primary importance in the field of biomedical images processing and allows various future applications.

References

1. Wagner, U., Limburgerhof, H.J., Henrici, K., Kuessner, H., Volkamer, K., Fuerst Ernst, I.: Dynamic coronary road mapping (1982)
2. Rossitti, S., Pfister, M.: 3D road-mapping in the endovascular treatment of cerebral aneurysms and arteriovenous malformations. Interv. Neuroradiol. **15**, 283–290 (2017)

3. Canbay, A., Malfertheiner, P., Pech, M., Kandulski, A., Obst, W., Weigt, J.: Road map fluoroscopy successfully guides endoscopic interventions in the esophagus. Endosc. Int. Open **05**(07), E608–E612 (2017)
4. Strother, C., Stieghorst, M., Turski, P., Mistretta, C., Crummy, A., Lieber-man, R.: Digital subtraction angiography road map. Am. J. Roentgenol. **139**(6), 1233–1234 (2013)
5. Smith, P., Reid, D.B., Environment, C., Palo, L., Alto, P., Smith, P.L.: A threshold selection method from gray-level histograms. IEEE Trans. Syst. **9**, 62–66 (1979)
6. Qt Documentation. https://doc.qt.io. Accessed 3 Feb 2020
7. OpenCV documentation index. https://docs.opencv.org. Accessed 3 Feb 2020
8. Bambi, F., et al.: Analysis and management of the risks related to the collection, processing and distribution of peripheral blood haematopoietic stem cells. Blood Transfus. **7**(1), 3–17 (2009)

How Differently Generated Clinical Tasks Affect the Observer Performances in CT Images Analysis

Edis Đedović[1](✉), Azra Gazibegović-Busuladžić[2], and Mustafa Busuladžić[3]

[1] Medical Physics and Radiation Protection Department, University Clinical Centre of Tuzla, Tuzla, Bosnia and Herzegovina

[2] Faculty of Science, University of Sarajevo, Sarajevo, Bosnia and Herzegovina

[3] Faculty of Medicine, University of Sarajevo, Sarajevo, Bosnia and Herzegovina

Abstract. The objective assessment of diagnostic image quality parameters cannot be considered as a means of assessing the performance of the entire imaging chain. It is performed without taking into account any clinical task under consideration, observer performances or intra and inter-observer variations and viewing conditions.

The aim of this study is, for the given scanning and reconstruction parameters and given viewing conditions, to consider the observer performances, estimated using the non-prewhitening (NPW) model observer method.

For obtained results for the task transfer function (TTF) and 2D noise power spectrum (NPS) the detectability indices d', and area under curve (*AUC*) values of NPW model observer have been determined for the objects of designer, Gaussian and flat contrast-profile of the diameters of 3 mm, 5 mm, 7 mm and 10 mm and with the peak contrast against background of 15 Hounsfield units (HU).

This study showed that the shape of contrast profiles significantly determines the extent to which NPS will affect AUC values. Because of this influence, the corresponding AUC values of the larger objects of flat or Gaussian (SD = 1) contrast profile can be smaller compared to AUC values of smaller objects of Gaussian (SD = 2 or 3) contrast profiles.

Keywords: Computed tomography (CT) · Model observer · Detectability index

1 Introduction

The receiver operating characteristic (ROC) analysis is a fundamental way of assessing performances of an observer (radiologist), and thus the entire imaging chain [1–3]. Because of its robustness, the ROC analysis serves as the gold standard in assessing the performances - from x-ray generator and detector across the data processing and post-processing system to image display monitor and the observer. The complexity of the ROC method and the great variety of clinical questions to be answered are the reasons why the assessment of an imaging system is limited to only the small number of simple and specific clinical questions. The ROC method poses a binary problem for an

A. Badnjevic and L. Gurbeta Pokvić (Eds.): CMBEBIH 2021, IFMBE Proceedings 84, pp. 800–808, 2021.
https://doi.org/10.1007/978-3-030-73909-6_91

observer whose task is to determine the presence or the absence of a signal in a region of interest (ROI) within an image. It is common for the signal to be presented as a low-contrast object of a certain dimension embedded on a uniform background. In order to increase the statistical power of the ROC method several additional ROC paradigms are introduced [4]. To avoid the learning effect - the observer always expects the signal to appear at a specific location within the image, the so-called localization ROC (LROC) analysis is introduced in which the observer has the task not only to determine whether the signal is present in the image or not but also to determine the location within the image where the signal appears [5]. Free ROC (FROC) analysis sets to an observer the task of marking all areas in the image that he/she considers to contain a signal [6]. An even more effective method of the ROC analysis is the so-called rating method, in which the observer assigns a rating for each individual case depending on how likely it is to contain a signal [4]. In each of these approaches, the sensitivity (*Se*) - a contribution of positively identified cases in relation to the total number of positive cases (true positive fraction) and specificity (*Sp*) - a contribution of negatively identified cases in relation to the total number of negative cases (true negative fraction) are chosen as a measure of the observer performance. Based on the obtained values for *Se* and *Sp*, the ROC curve is obtained on the graph whose x - axis is reserved for the values of 1 - *Sp*, and the y - axis for the values of *Se*. The area under curve (*AUC*) has more advantages over pair (*Se, Sp*) in describing observer performance and it is used, in addition to the detection index, d′, and the percentage of correct cases, PC, as the basic figure of merit (FOM) [7]. The objective assessment of diagnostic image quality parameters, expressed in the terms of physical metrics through modulation transfer function (MTF), noise power spectrum (NPS), signal to noise ratio (SNR) and noise equivalent quanta (NEQ) cannot be considered as a means of assessing the performance of the entire imaging chain. The article by Leh-Nien D Loo et al., from 1984, is the first work in which the authors are trying to make a connection between objective image quality parameters and observer performances [8]. Lehn-Nien D Loo et al. work, as well as many others since then, has contributed to the development of various model-observer methods for assessing the observer performances based on physical metrics [9–13].

A model-observer can be used to assess the image quality of diagnostic images. The mathematical model observer takes into account the physical parameters (MTF and NPS) as well as the specific clinical task for modeling the response and predicting the performances of a human observer.

The aim of this study is, for the given scanning and reconstruction parameters and given viewing conditions, to consider the observer performances, estimated using the non-prewhitening (NPW) model observer method, in the case of the different clinical tasks (the objects of different sizes generated by using different contrast profile functions) [14].

2 Methods and Materials

2.1 Theoretical Considerations

The NPW model observer method can serve in modeling the observer response to a stimuli in the form of diagnostic image of different values of image quality parameters (MTF, NPS) taking into account the specific clinical task under consideration. Detection index, d', given in the form

$$d'_{NPW} = \frac{\iint |W(u,v)|^2 TTF^2(u,v)dudv}{\sqrt{\iint |W(u,v)|^2 TTF^2(u,v)NPS(u,v)dudv}} \tag{1}$$

represents the basic FOM in the case of the NPW model observer [14]. *W (u, v)* represents the task function that takes into account different aspects of the clinical task to be solved: size, shape, contrast and contrast profile of an object; *TTF (u, v)* is a so called task transfer function - *MTF* function of the systems that use iterative reconstruction algorithms that take into account the influence of an object's contrast on the resolution; *NPS* represents the noise power spectrum obtained from the image [14–17].

The task function, *W (u, v),* is obtained as the Fourier transform of an image of the object, i.e.

$$W(u,v) = FT\{c(r)\}. \tag{2}$$

In this work the images of the objects are generated using three different functions to obtain the objects' contrast – profiles: designer, Gaussian and flat profile. The designer contrast – profile of an object is obtained as [14]

$$c(r) = C\left[1 - \left(\frac{2r}{D}\right)^2\right]^n, \tag{3}$$

the Gaussian contrast – profile as

$$c(r) = \frac{c}{2}\left(1 - erf\left(\frac{r-D}{n}\right)\right), \tag{4}$$

and the flat contrast – profile as [14]

$$c(r) = C\prod\left(\frac{rD}{2}\right), \tag{5}$$

where *c(r)* is a value of the signal at a radial distance *r*; *C* is the peak contrast of the signal against the background expressed in the Hounsfield units (HU); *D* is the signal (object) diameter; *n* is a blur factor which is, in the case of designer contrast-profile, equal to 1 and in the case of Gaussian contrast-profile it corresponds to the profile standard deviation with the value equal to 1, 2 and 3 (Gaussian1, Gaussian2 and Gaussian3) chosen in this work; *erf* is the Gaussian error function; Π is rect function. To make the connection with the ROC analysis, on the basis of d'_{NPW} values, the AUC values can be calculated as

$$AUC = \frac{1}{\sqrt{2}}\Phi(d'), \tag{6}$$

where $\Phi(x) = \int_{-\infty}^{x} \phi(y)dy$ is a Gaussian cumulative probability distribution function and $\phi(x) = \frac{1}{\sqrt{2\pi}}e^{\frac{-x^2}{2}}$ is a Gaussian probability density function.

2.2 Materials

For the analysis the images of the Philips medical systems brilliance 16 series performance phantom obtained on the Siemens Somatom Sensation 64 CT scanner were used. The imaging and reconstruction parameters are presented in Table 1. The NPSs were obtained from the five consecutive images of the homogenous water-filled part of the phantom. The images of the phantom's module intended to determine the accuracy of CT numbers for water, acryl, teflon and polyethylene material, are used to obtain TTFs on the teflon disk for three consecutive slices. Both NPS and TTF were obtained by using imQuest v7.1 software [18]. The volume computed tomography dose index (C_{VOL}) values were obtained on the basis of data from the CT scanner system.

For obtained results for TTF and 2D NPS the detectability indices d′, and AUC values of NPW model observer have been determined for the objects with designer, Gaussian and flat contrast-profile of the object's diameters of 3 mm, 5 mm, 7 mm and 10 mm and with the peak contrast against background of 15 HU.

Table 1. Imaging and reconstruction protocols

	Protocol's number												
	1	2	3	4	5	6	7	8	9	10	11	12	13
Tube voltage (kV)	80	80	80	100	100	100	120	120	120	120	140	140	140
Quality reference (mAs)	135	150	165	135	150	165	135	150	165	180	135	150	165
C_{VOL} (mGy)	2.57	2.68	2.83	5.22	5.80	5.96	8.91	9.62	10.36	10.91	12.56	13.62	14.65
Rotation time (s)	0.5												
Beam width (mm)	32 x 0.6												
FOV (mm)	330												
Pitch	1.4												
Recon. slice thickness (mm)	5												
Recon.kernel	B20												

3 Results

An example of the contrast – profiles of the 3 mm diameter object is presented on the Figs. 1 and 2 as the images and graph, respectively.

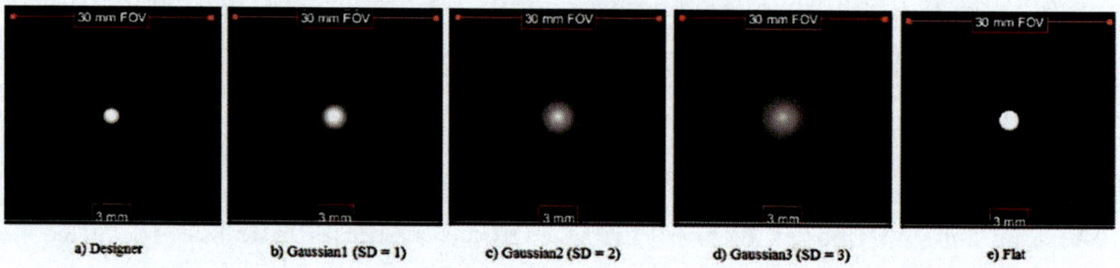

Fig. 1. Different contrast – profile images for the object of 3 mm diameter

The AUC vs C_{VOL} values in the case of the objects of 3 mm, 5 mm and 7 mm diameters are presented on the Fig. 3 a), b) and c), respectively, for four different contrast – profiles.

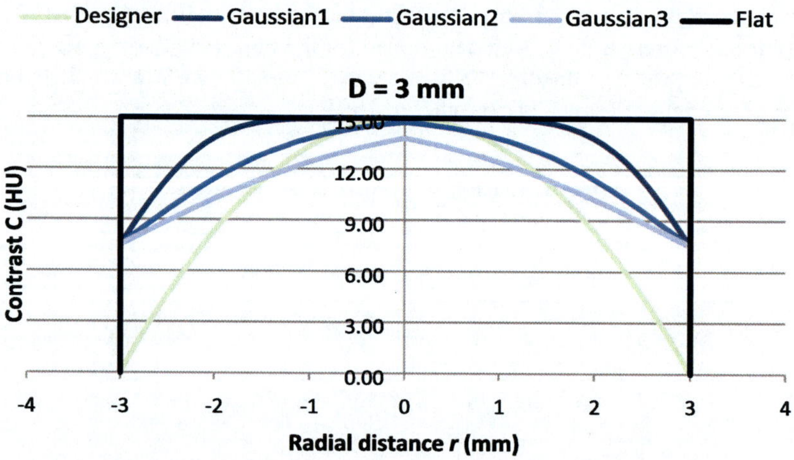

Fig. 2. Graphical representation of the contrast – profiles

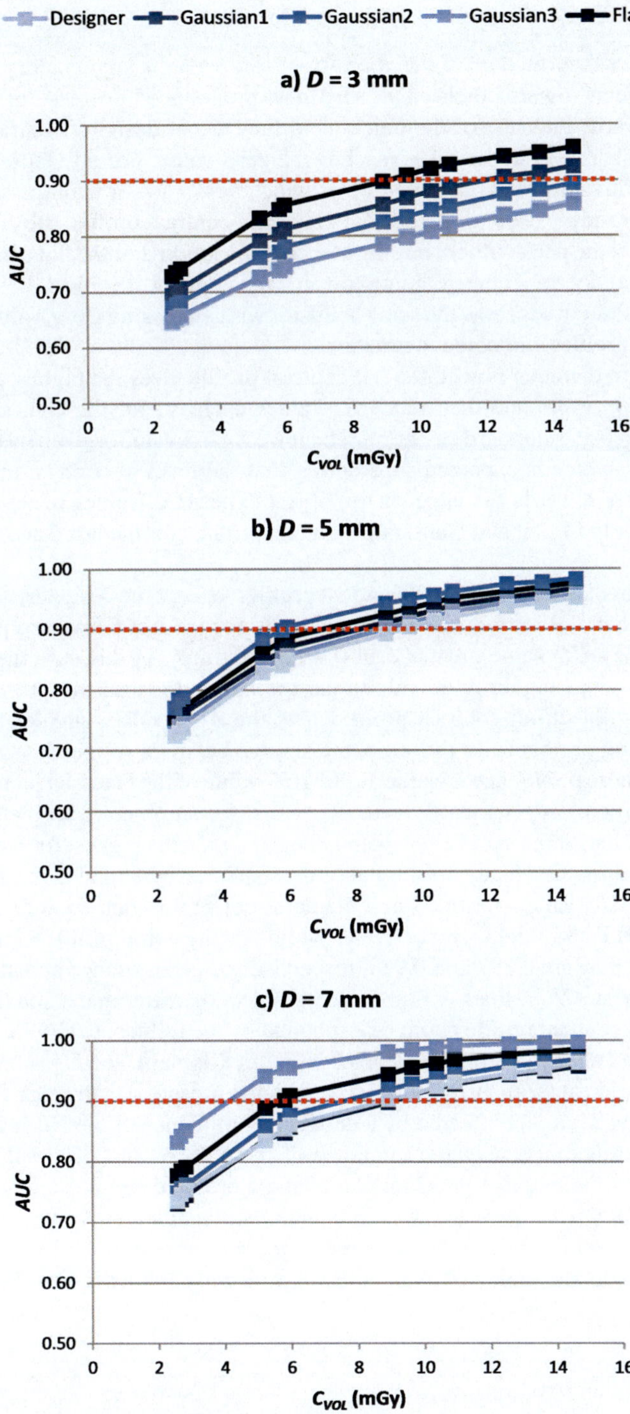

Fig. 3. *AUC* vs C_{VOL} values for different object's diameters and contrast - profiles

4 Discussion

The Fig. 2 shows that the type of the contrast – profile affects significantly the contrast level of the objects towards their edge. The most pronounced decrease in the contrast level going towards the edge of the object is in the case of designer contrast profile in which the contrast level at the edge reaches the zero value. For all Gaussian contrast profiles, the contrast level at the edge of the object is 50% lower than the set value (15 HU). It is interesting to note that with the Gaussian contrast profile, the contrast level decreases even at the peak with increasing standard deviation. For the flat contrast profile the contrast level does not change going towards the edge of the object.

The *AUC* value curves show the same behavior with increasing C_{VOL} values observed for all contrast profiles and object diameters.

For the 3 mm diameter object, the flat contrast profile gives the highest *AUC* values for a given C_{VOL} value and then the *AUC* values decrease for the Gaussian contrast profile with increasing standard deviation. The lowest *AUC* value is given by the designer contrast profile, which is expected considering that the level of contrast in this profile decreases rapidly towards the edge of the object. The *AUC* values of 0.9 and higher were obtained only for flat and Gaussian1 contrast profiles for the last 6 and 3 protocols, respectively.

For the 5 mm diameter object, all contrast profiles, except the designer, give the *AUC* values greater than 0.9 for the last 7 protocols, and for the last 6 imaging protocols all profiles have the *AUC* value greater than 0.9. Interestingly, in this case the Gaussian2 contrast profile gives higher *AUC* value than the flat and Gaussian1 profiles. For the 7 mm diameter object, all contrast profiles give the *AUC* values greater than 0.9 for the last 7 imaging protocols. In this case, the strange behavior of the Gaussian contrast profile is even more pronounced. Namely, the *AUC* value of the Gaussian contrast profile increases with increasing standard deviation. Thus, for an object with the diameter of 7 mm, we have that, at the same C_{VOL} value, the *AUC* value is highest for the Gaussian3 profile, followed by the values for the flat, Gaussian2, designer and Gaussian1 profile. Moreover, the *AUC* values for the 5 mm diameter object for Gaussian2 and Gaussian3 profiles are 0.911 and 0.943, respectively, while for the same profiles for the 7 mm diameter object they are 0.897 and 0.919, respectively. Such a strange and at first glance illogical change in *AUC* values with increasing object diameter and standard deviation of the Gaussian contrast profile could be explained by the influence of NPS whose peak values are shifted towards lower frequencies in Fourier domain (see Fig. 4.). This means that the information at the edge of the objects, which correspond to the high frequency in the Fourier domain, are not affected by a destructive influence of the NPS. The positive effect of the absence of a negative influence of the NPS on the *AUC* values was not canceled even by the fact that the Gaussian contrast profile decreases the contrast level in the peak itself. On the other hand, objects with the flat contrast profile, since they do not change the level of contrast towards the edge, correspond to the low frequency in the Fourier domain and their *AUC* values are significantly influenced by the NPS.

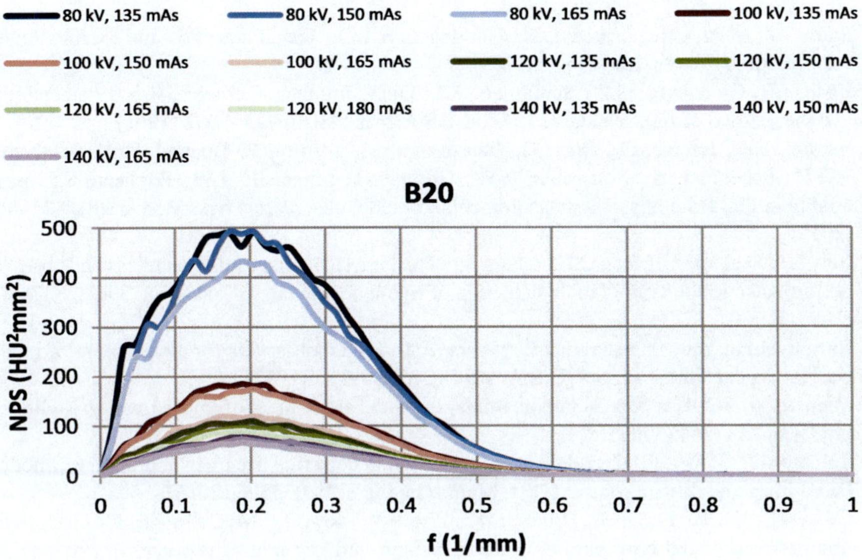

Fig. 4. The shapes of NPS for all imaging protocols

5 Conclusion

To ensure image quality that will provide reliable diagnostic information for a wide range of clinical tasks, it is not enough to simply consider exposure parameters, applied reconstruction kernels, and doses. The presented considerations show that it is necessary to take into account all the specifics of a clinical task in order to select the adequate imaging protocols. This study showed that the shape of the contrast profile significantly determines the extent to which NPS of the smooth B20 kernel will affect AUC values. Because of this, a larger object diameter is not a guarantee for obtaining the higher AUC value. Smaller objects with Gaussian contrast profiles of larger standard deviation (SD = 2, 3) can give higher AUC values than objects of larger diameters and of flat or Gaussian (SD = 1) contrast profile. The reason for this is that smaller objects as well as information on the edges of Gaussian (SD = 2, 3) contrast profiles are in the high frequency domain. Because of this, their AUC values are less exposed to the negative influence of NPS whose peak values are shifted towards the low frequency domain. On the other side, larger objects of flat and Gaussian (SD = 1) contrast profiles are in the low frequency domain where the influence of NPS on AUC values is more significant.

References

1. Boyle, J.A.: Introduction to medical decision making. Arthritis Rheum. **12**(5), 535–536 (1969)
2. Metz, C.E.: ROC methodology in radiologic imaging. Invest Radiol. **21**(9), 720–733 (1986)
3. Metz, C.E.: Basic principles of ROC analysis. Semin. Nucl. Med. **8**(4), 283–298 (1978)
4. Chakraborty, D.P.: Observer Performance Methods for Diagnostic Imaging: Foundations, Modeling, and Applications with R-Based Examples. CRC Press, Taylor & Francis Group, Boca Raton (2017)

5. Starr, S.J., Metz, C.E., Lusted, L.B., Goodenough, D.J.: Visual detection and localization of radio-graphic images. Radiology **116**(3), 533–538 (1975)
6. Egan, J.P., Greenburg, G.Z., Schulman, A.I.: Operating characteristics, signal detectability and the method of free-response. J. Acoust. Soc. Am. **33**(8), 993–1007 (1961)
7. Verdun, F.R., Racine, D., Ott, J.G., Tapiovaara, M.J., Toroi, P., Bochud, F.O., Veldkamp, W.J.H., Schegerer, A., Bouwman, R.W., Giron, I.H., Marshall, N.W., Edyvean, S.: Image quality in CT: from physical measurements to model observers. Phys. Med. **31**(8), 823–843 (2015)
8. Loo, L.N.D., Doi, K., Metz, C.E.: A comparison of physical image quality indices and observer performance in the radiographic detection of nylon beads. Phys. Med. Biol. **29**(7), 837–856 (1984)
9. Barrett, H.H., Yao, J., Rolland, J.P., Myers, K.J.: Model observers for assess-ment of image quality. Proc. Natl. Acad. Sci. U.S.A. **90**(21), 9758–9765 (1993)
10. Vennart, W.: ICRU report 54: medical imaging—the assessment of image quality. Radiography **3**(3), 243–244 (1997)
11. Solomon, J., Samei, E.: Correlation between human detection accuracy and observer model-based image quality metrics in CT. J. Med. Imaging **3**(3), 035506 (2016)
12. Gong, H., Yu, L., Leng, S., Dilger, S.K., Zhou, W., Ren, L., McCollough, C.H.: Correlation between model observers in uniform background and human observer in patient liver background for a low-contrast detection task in CT. In: Medical Imaging: Image Perception, Observer Performance, and Technology Assessment. Proceedings of SPIE, p. 105770M. International Society for Optics and Photonics (2018)
13. Solomon, J., Samei, E.: What observer models best reflect low-contrast detectability in CT? In: Medical Imaging: Image Perception, Observer Performance, and Technology Assessment. Proceedings of SPIE, p. 94160I (2015)
14. Samei, E., Bakalyar, D., Boedeker, K.L., Brady, S., Fan, J., Leng, S., Myers, K.J., Popescu, L.M., Ramirez-Giraldo, J.C., Ranallo, F., Solomon, J., Vaishnav, J., Wang, J.: Performance evaluation of computed tomography systems: summary of AAPM task group 233. Med. Phys. **46**(11), e735–e756 (2019)
15. Richard, S., Li, X., Yadava, G., Samei, E.: Predictive models for observer performance in CT: applications in protocol optimization. In: Medical Imaging: Physics of Medical Imaging. Proceedings of SPIE, p. 79601H (2011)
16. Richard, S., Husarik, D.B., Yadava, G., Murphy, S.N., Samei, E.: Towards task-based assessment of CT performance: system and object MTF across different reconstruction algorithms: towards task-based assessment of CT performance. Med. Phys. **39**(7), 4115–4122 (2012)
17. Chen, B., Christianson, O., Wilson, J.M., Samei, E.: Assessment of volumetric noise and resolution performance for linear and nonlinear CT reconstruction methods: assessing CT noise and resolution for nonlinear reconstruction. Med. Phys. **41**(7), 071909 (2014)
18. https://deckard.mc.duke.edu/~samei/tg233.html

Radiation Exposure of Patients in Mammography: An Overview of the 15-Year Practice

Hanka Bečirović[1(✉)], Suad Kunosić[2], Minela Halilović[3], Ivan Lasić[4], Stipe Galić[4], and Jasmina Bajrović[5]

[1] Faculty of Science, University of Sarajevo, Sarajevo, Bosnia and Herzegovina
[2] Faculty of Natural Sciences and Mathematics, University of Tuzla, Tuzla, Bosnia and Herzegovina
[3] Faculty of Health Studies, University of Sarajevo, Sarajevo, Bosnia and Herzegovina
[4] Medical Physics and Radiation Protection Centre, University Clinical Hospital Mostar, Mostar, Bosnia and Herzegovina
[5] Clinic of Radiology, Clinical Centre of Sarajevo University, Sarajevo, Bosnia and Herzegovina

Abstract. Mammography is a non-invasive X-ray examination of the breast. The primary goal of mammography is the early detection of breast cancer, which is associated with better survival. About a third of all malignant tumours in women are breast cancers. Over the past couple of decades, the imaging techniques used in mammography changed from film-screen to computed radiography and full-field digital mammography, and finally to digital breast tomosynthesis. Generally, the image quality has improved, but this might have come with the cost of increased radiation dose. In this paper, we analysed the change of patient doses for different imaging techniques over the last 15 years.

Keywords: Mammography · Imaging · Mean glandular dose · Patient dose · Digital breast tomosynthesis

1 Introduction

Mammography is a radiological modality for examining the breast. Images are produced using an X-ray device equipped with a specially designed X-ray tube that emits low-energy X-rays. The first X-rays were used in the diagnosis of breast changes in 1913 when the German surgeon Soloman made radiographic images of surgically removed breasts and compared them with operative and histological findings. The first significant radiological findings of healthy and diseased breasts were published by Warren Warren in 1930. Gross constructs a mammogram with a molybdenum anode, which leads to significantly better image quality. The most significant advancement in mammography in recent years has been the digitization of detectors via computerized radiography or mammography (CR) and plate detectors [1, 2]. Mammography is used as one of the best ways to detect breast cancer early in women over the age of 40. Early mammograms performed to detect breast cancer can reduce the risk of death in women. Small

A. Badnjevic and L. Gurbeta Pokvić (Eds.): CMBEBIH 2021, IFMBE Proceedings 84, pp. 809–816, 2021.
https://doi.org/10.1007/978-3-030-73909-6_92

differences in soft tissue density must be seen on the mammogram, so the low-contrast resolution is very important. Low energy radiation is used to achieve satisfactory contrast. The sharpness of the image is important due to the need to see small calcifications. The image noise should be reduced to increase the visibility of microcalcifications [3]. Mammography is performed from two projections, craniocaudal (breast top view) (CC) and mediolateral (side view breast) (MLO) projections are accepted as standard (Fig. 1). Breast compression plays an important role in image quality [4].

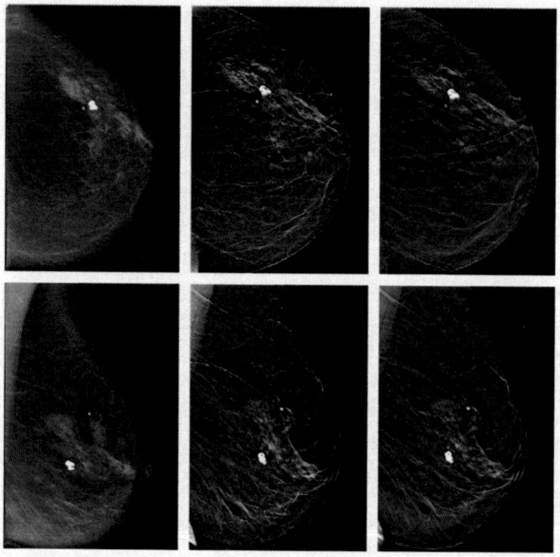

Fig. 1. Mammography projections, cranio-caudal (top) and mediolateral oblique (bottom) in computed radiography, full-field digital mammography and digital breast tomosynthesis (left to right)

During the mammography examination, the patient is exposed to ionizing radiation. As a measure of exposure, the mean absorbed dose to the mammary gland, i.e., the mean or average glandular dose (AGD), is used. The mean glandular dose for a standard breast is calculated by the formula:

$$D_G = gcsK_i \tag{1}$$

Where K_i is the incident air kerma, g is the granularity factor, c is the granularity correction factor and s is the target and filter combination factor. The values of the conversion factor g were obtained by numerical Monte Carlo simulations for breasts of different thicknesses, T, and filter half-values of the radiation beam, HVL, and molybdenum target and filter. The calculation was performed under the assumption that the ratio of glandular (glandular) and adipose (adipose) tissue in the breast is uniform. The Monte Carlo simulation was repeated for other target and filter combinations. Dance et al. published the final values of factors g, c, and s for conventional mammography in 2009 and then extended them to other breast imaging modalities [5–8]. Dosimetry protocol for digital breast tomosynthesis (DBT) has been developed as an extension of the method used for 2D imaging. The average glandular dose $\bar{D}_G$ for breast dosimetry in DBT is estimated using:

$$D_G = gcsK_iT \quad (2)$$

where factor T is a correction factor for DBT. The factor T has to be determined by integrating angular dependent tomography conversion factors, t, over the angles used for the DBT exposure, and factors are given in as a function of breast thickness for the two DBT systems. This method is now official in Europe [7, 9]. For contrast-enhanced digital mammography (CEDM), Dance et al. published conversion factors for the high-energy spectra from a tungsten target at tube voltage range 40–50 kV, filtered by 0.28, 0.30 and 0.32 mm of copper [8].

2 Materials and Methods

The study included patient doses collected at different time intervals. The first set of data included 55 patients that underwent film-screen mammography (FSM) examination in 2006. The second set (51 patients) originates from 2018 when CR system was used on the same mammography unit. Finally, the set from 2020 consists of dosimetry data from full-field digital mammography (FFDM) and DBT as collected by patient dosimetry management software OpenREM (The Royal Marsden NHS Foundation Trust, London, UK). Patients included in the study underwent a common mammography examination with CC and MLO projections.

The mammography unit used for FSM and CR images was Siemens Mammomat 3000 Nova (Siemens, Erlangen, Germany), while FFDM and DBT examination was conducted on Hologic 3Dimensions (Marlborough, Massachusetts, USA). Due to the expected log-normal distribution of the data, the reported central tendency value for dosimetry results was median ($\tilde{x}$), while the interquartile range was used as a measure of variability. The arithmetic mean ($\bar{x}$) and standard deviation (σ) were used only in cases when the observed variable is normally distributed. The significance level used in the study was $\alpha = 0.05$.

3 Results

Table 1 shows type (right or left, CC and MLO projections) and a number of analyzed mammography projections (N), and corresponding median values of tube potential (U), tube current-exposure time product (P_{It}), breast thickness (T), compression force (F), incident air kerma (K_i), and mean glandular dose per projection (D_G). Distributions of U P_{It}, and F differ significantly between imaging modalities (Kruskal-Wallis test, $p <$ 0.001), while distributions of breast thicknesses remain the same (Kruskal-Wallis test, $p = 0.397$).

Table 1. Type and number of analyzed mammography projections (N), and corresponding median values of tube potential (U), tube current-exposure time product (P_{It}), breast thickness (T), compression force (F), incident air kerma (K_i), and average glandular dose per projection (D_G).

Imaging		Mammography projection			
		R-CC	L-CC	R-MLO	L-MLO
FSM	N	55	55	55	55
	U (kV)	30	30	30	30
	P_{It} (mAs)	24.9	20.9	32.9	30.0
	T (mm)	57	54	61	60
	F (N)	.	.	.	.
	K_i (mGy)	4.84	3.82	6.26	5.82
	D_G (mGy)	0.791	0.670	1.00	0.913
CR	N	51	51	51	51
	U (kV)	30	30	30	30
	P_{It} (mAs)	28.8	31.1	45.0	47.4
	T (mm)	50	54	64	64
	F (N)	110	110	100	100
	K_i (mGy)	3.17	3.78	5.38	5.82
	D_G (mGy)	0.681	0.766	0.995	1.07
FFDM	N	293	294	293	294
	U (kV)	29	29	30	30
	P_{It} (mAs)	132	135	160	159
	T (mm)	54	54	60	62
	F (N)	83.5	91.8	123	118
	K_i (mGy)	5.37	5.77	7.65	8.23
	D_G (mGy)	1.45	1.54	1.90	1.98
DBT	N	116	114	115	115
	U (kV)	31	32	33	33
	P_{It} (mAs)	65.8	66.1	71.6	72.5
	T (mm)	54	56	62	63
	F (N)	91.6	90.0	112	107
	K_i (mGy)	6.64	6.65	8.23	8.83
	D_G (mGy)	1.91	1.96	2.29	2.38

Figure 2 shows how average glandular dose per projection (D_G) changes with breast thickness. There is a clear positive correlation between two variables ($p < 0.001$). Figure includes exponential regression lines ($y = \alpha e^{bx}$) which represent experimental data well. For FSM, CT, FFDM and DBT the coefficients of determination (R^2) are 0.746, 0.828, 0.625 and 0.902, respectively.

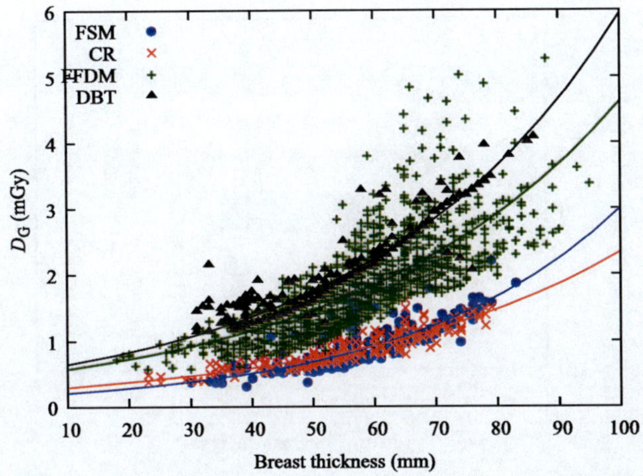

Fig. 2. Dependence of average glandular dose per projection (D_G) on breast thickness for different imaging modalities.

The total D_G, averaged over two exposed breasts, is presented in Fig. 3. The boxplot diagram shows the differences in distributions and median values for different imaging modalities. The median D_G for FSM and CR is 1.7 mGy and 1.8 mGy, respectively. The distribution of D_G is the same across these two modalities (Mann-Whitney U test, $p = 0.711$). The median average glandular dose in FFDM is 3.4 mGy, which is significantly higher than FSM or CR ($p < 0.001$). The reported median of D_G in DBT is 4.4 mGy, which is the imaging modality with the highest patient dose ($p < 0.001$).

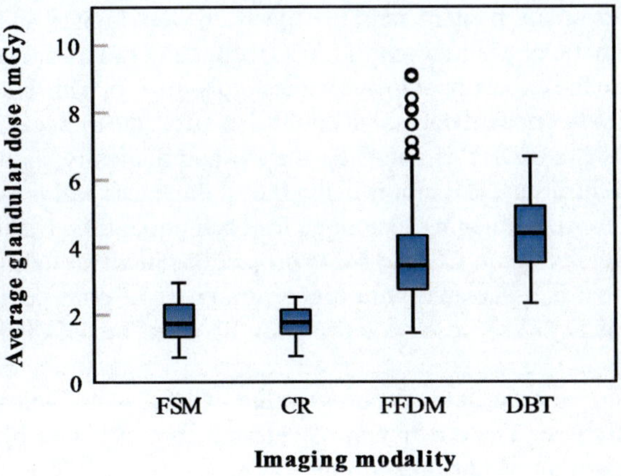

Fig. 3. Total average glandular dose for different imaging modalities

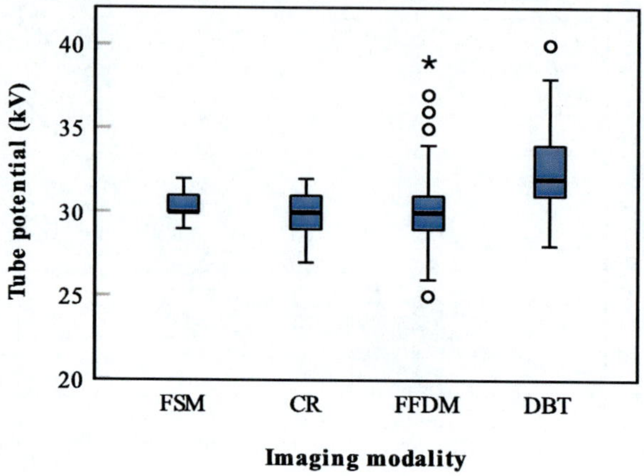

Fig. 4. Tube potential (kV) used in different imaging modalities

4 Discussion

Results indicate that the most common tube potential used in FSM and CR is 30 kV. The practice in the evaluated mammography room was to select tube potential manually, which was a common practice for the x-ray unit model used in the study. Pre-programmed protocols, available at the unit, were seldomly used.

The example of a good radiographic technique in the European guidelines on quality criteria for diagnostic radiographic images recommends 28 kV to be used in both CC and MLO projections for standard-sized breasts [10]. This is something we find in scientific literature too [11, 12]. Thus, tube potential used in the selected patients was higher than optimal. Hence, photon energies in the spectrum were higher, and therefore, more penetrating. Automatic exposure control (AEC) will detect more radiation in a shorter time, and terminate the exposure at lower values of P_{It} (mAs). This chain of events will eventually lead to lower patient doses, but could also affect the image quality. The digital unit used for FFDM and DBT is choosing the P_{It} automatically, as well, but also the tube potential and filtration. The unit is using breast thickness and pre-exposure data to determine the appropriate imaging technique for each projection. Mean glandular dose per projection differs between CC and MLO projections in all modalities. As expected, MLO projection that is associated with higher thickness of compressed tissue results in higher values of D_G. The positive correlation between breast thickness and D_G is expected [13–16].

It is interesting, however, how reported value of D_G in DBT closely matches the chosen regression curve. The coefficient of determination R^2 is as higher than 0.9. It seems that the algorithm used to determine P_{It}, and hence the dose, heavily relies on measured breast thickness.

Doses received by patients in observed cases match the doses reported in other studies. James et al. reported values of 2.1 mGy in FFDM, and 2.5 mGy in DBT for 63 mm compressed breast size [17]. Olgar et al. report (1.84, 1.81, 1.97. and 1.91) mGy for RCC, LCC, RMLO and LMLO, respectively. Doses in DBT were higher, (2.62,

2.45, 2.62 and 2.64) mGy for RCC, LCC, RMLO and LMLO, respectively [18]. Older studies report doses for FSM. Kruger and Schueler calculated mean glandular dose on the sample of 24,471 mammograms to be 2.6 mGy [19].

5 Conclusion

Mammography plays a central role in the early detection of breast cancer. During the mammography, patients are exposed to harmful X-rays. Understanding the doses and associated risk from radiation exposure is a cornerstone of radiation protection of patients. In this study, we analyzed mean glandular doses for patients that underwent digital breast tomosynthesis and full-field digital mammography and compared them to the older dataset of doses received by patients in film-screen and computed radiography mammography.

References

1. Gold, R.H., Bassett, L.W., Widoff, B.E.: Highlights from the history of mammography. Radiographics **10**(6), 1111 (1990)
2. Monnin, P., Gutierrez, D., Bulling, S., Guntern, D., Verdun, F.: A comparison of the performance of digital mammography systems. Med. Phys. **34**(3), 906–914 (2007)
3. Dance, D., Christofides, S., Maidment, A., McLean, I., Ng, K.: International Atomic Energy Agency (IAEA): IAEA (2014
4. Ciraj-Bjelac, O., Avramova-Cholakova, S., Beganovic, A., Economides, S., Faj, D., Gershan, V., Grupetta, E., Kharita, M., Milakovic, M., Milu, C., et al.: Image quality and dose in mammography in 17 countries in Africa, Asia and Eastern Europe: results from IAEA projects. Euro. J. Radiol. **81**(9), 2161 (2012)
5. Dance, D., Skinner, C., Young, K., Beckett, J., Kotre, C.: Additional factors for the estimation of mean glandular breast dose using the UK mammography dosimetry protocol. Phys. Med. Biol. **45**(11), 3225 (2000)
6. Dance, D., Young, K., Van Engen, R.: Further factors for the estimation of mean glandular dose using the United Kingdom, European and IAEA breast dosimetry protocols. Phys. Med. Biol. **54**(14), 4361 (2009)
7. Dance, D., Young, K., Van Engen, R.: Estimation of mean glandular dose for breast tomosynthesis: factors for use with the UK, European and IAEA breast dosimetry protocols. Phys. Med. Biol. **56**(2), 453 (2010)
8. Dance, D., Young, K.: Estimation of mean glandular dose for contrast enhanced digital mammography: factors for use with the UK, European and IAEA breast dosimetry protocols. Phys. Med. Biol. **59**(9), 2127 (2014)
9. van Engen, R.E., Bosmans, H., Bouwman, R.W., Dance, D.R., Heid, P., Lazzari, B., Marshall, N.W., Phelan, N., Schopphoven, S., Strudley, C., Thijssen, M., Young, K.C.: Protocol for the Quality Control of the Physical and Technical Aspects of Digital Breast Tomosynthesis Systems (EUREF 2018)
10. Carmichael, J.H.E., Maccia, C., Moores, B.M., Oestmann, J.W., Schibilla, H., Teunen, D., Van Tiggelen, R., Wall, B. (eds.), European guidelines on quality criteria for diagnostic radiographic images (EUR 16260 EN) (European Commision, 1996)
11. Hendrick, R.E., Berns, E.A.: Optimizing techniques in screen-film mammography. Radiol. Clin. North Am. **38**(4), 701 (2000)

12. Satoh, K., Kuwabara, T., Yasuda, H., Arakawa, S.: Medical imaging: physics of medical imaging. (Int. Soc. Optics Photon.), **6510**, 65104 (2007)
13. Hendrick, R.E.: Radiation doses and cancer risks from breast imaging studies. Radiology **257**(1), 246 (2010)
14. Chevalier, M., Morán, P., Ten, J.I., Fernández Soto, J.M., Cepeda, T., Vañó, E.: Patient dose in digital mammography. Med. Phys. **31**(9), 2471 (2004)
15. Hendrick, R.E., Pisano, E.D., Averbukh, A., Moran, C., Berns, E.A., Yaffe, M.J., Herman, B., Acharyya, S., Gatsonis, C.: Comparison of acquisition parameters and breast dose in digital mammography and screen-film mammography in the American college of radiology imaging network digital mammographic imaging screening trial. Am. J. Roentgenol. **194**(2), 362 (2010)
16. Fausto, A.M., Lopes, M., de Sousa, M., Furquim, T.A., Mol, A.W., Velasco, F.G.: Optimization of image quality and dose in digital mammography. J. Digit. Imaging **30**(2), 185 (2017)
17. James, J.R., Pavlicek, W., Hanson, J.A., Boltz, T.F., Patel, B.K.: Breast radiation dose with CESM compared with 2D FFDM and 3D tomosynthesis mammography. Am. J. Roentgenol. **208**(2), 362 (2017)
18. Olgar, T., Kahn, T., Gosch, D.: Fortschritte auf dem Gebiet der Röntgenstrahlenund bildgebenden Verfahren, **184**(10), 911 (2012)
19. Kruger, R.L., Schueler, B.A.: A survey of clinical factors and patient dose in mammography. Med. Phys. **28**(7), 1449 (2001)

Evaluation of the Effectiveness of Protective Aprons in the Primary and Scattered Radiation X-Ray Beam

Amra Skopljak-Beganović[1], Lejla M. Čiva[2], Edis Đedović[3], Selma Zulić Hrelja[4], Azra Gazibegović-Busuladžić[4], and Adnan Beganović[1](✉)

[1] Radiation Protection and Medical Physics Department, Clinical Centre of Sarajevo University, Sarajevo, Bosnia and Herzegovina

[2] Medical School, University Sarajevo School of Science and Technology, Sarajevo, Bosnia and Herzegovina

[3] Medical Physics and Radiation Protection Department, University Clinical Centre of Tuzla, Tuzla, Bosnia and Herzegovina

[4] Faculty of Science, University of Sarajevo, Sarajevo, Bosnia and Herzegovina

Abstract. Occupational radiation protection is important for the safe use of radiation in medicine. In some situations, medical professionals are required to use personal protective equipment, such as lead aprons, thyroid shield collars and glasses. Weight of lead aprons is an issue interventionalists face when performing long procedures. A lot of vendors market "lightweight" aprons that are made of alternative materials which supposedly provide the same attenuation power as the lead. In this paper, we propose a method to test protective garments, in both direct and scattered radiation beam, simultaneously, in a consistent and repeatable manner. The obtained results were compared to measurements in other studies, including those done in the clinical environment.

Keywords: Radiation protection · Apron · Lead equivalence · Quality control

1 Introduction

Occupational exposure to ionising radiation comes as a precondition for the use of radiation in medical purposes. For this reason, ensuring occupational radiation protection is important for the safe use of radiation in medicine [1]. Workers are commonly protected by structural shielding (walls, barriers, protective glass etc.). However, whenever they are exposed to unshielded scattered radiation, medical professionals are required to use personal protective equipment, such as lead aprons, thyroid shield collars and glasses. Dosimeters measuring personal dose equivalent, $H_{\mathrm{P}}(d)$ are used to estimate the effective dose received by occupationally exposed workers [2–7]. They are commonly worn under the lead apron. Sometimes the effective dose might be underestimated when measured with a single dosimeter. In such cases, the second dosimeter may be included to be worn near the neck, over the protective garment [8].

A. Badnjevic and L. Gurbeta Pokvić (Eds.): CMBEBIH 2021, IFMBE Proceedings 84, pp. 817–825, 2021.
https://doi.org/10.1007/978-3-030-73909-6_93

Weight of lead apron is an issue interventionalists face when performing long procedures in radiology, cardiology, cardiac surgery etc. [9]. A lot of vendors market "lightweight" aprons that are made of alternative materials which supposedly provide the same attenuation power as the lead [10]. Hence, many authors suggest that the manufacturer stated lead equivalence of protective aprons should be checked and recorded along with the details of associated quality assurance (QA) [11, 12].

In this paper, we provide a solution for type testing of lead aprons in the direct and scattered beam of X-ray radiation. Measurements were done on cutouts with varying thicknesses in order to examine the effectiveness of different lead aprons. Our results are compared with those in available relevant literature with a reflection on the protocol and procedure differences.

2 Materials and Methods

The effectiveness of lead-equivalent material used in the production of protective aprons was estimated by measuring incident air kerma, K_i, in both direct and scattered radiation beam, with and without layers of protective material. Corresponding measurements have been performed using the setup shown in Figs. 1 and 2. The protective layers (filters) are placed close to the detector, which matches the method explained in IEC 61331:2014 standard for determination of attenuation properties of materials used for protective devices against diagnostic medical X-radiation [13]. It should be noted that "lead-free" or "lead-reduced" materials, if used, could introduce forward-scatter radiation in the beam, which would lead to false results (increased K_i).

The source of X-rays was a standard radiography unit, that was able to provide exposure with different tube parameters. During the experiment, tube voltage was set to different values, namely (60, 81, and 102) kV. A 20 cm-thick PMMA phantom, which is a material similar to the human tissue, was placed 80 cm from the source. The exposure time-current product was held fixed at 25 mAs. Generally, these values are typical for those used in diagnostic and interventional radiology. The size of the X-ray beam was 20×20 cm^2 at 80 cm distance form the X-ray tube focus.

Piranha (RTI, Mölndal, Sweden), a multi-functional meter for regular quality control of radiological equipment, was used to simultaneously measure K_i in two different points. The main detector body was placed in the primary beam, on top of the PMMA phantom. It provided measurements of K_i, but it was also used as a triggering device for the second detector – Piranha Dose Probe, a solidstate detector specially designed for low air kerma rate measurements. Both detectors require no corrections for pressure and temperature. Piranha Dose Probe measured scattered radiation $\left(K_s^1\right)$ at 1 m distance from the phantom centre, laterally.

Protective material cutouts (5×5 cm^2) were placed on both detectors. The thickness of protection material (T) was changed subsequently, simulating different lead apron thickness. Four values of T were chosen: 0 mmPb (no lead filter), 0.25 mmPb, 0.5 mmPb and 0.75 mmPb. When cutouts are not available, one could cover the detector with different parts of untouched apron. The back is usually 0.25 mmPb, front 0.5 mmPb, and 0.75 mmPb could be obtained with overlapping these two layers.

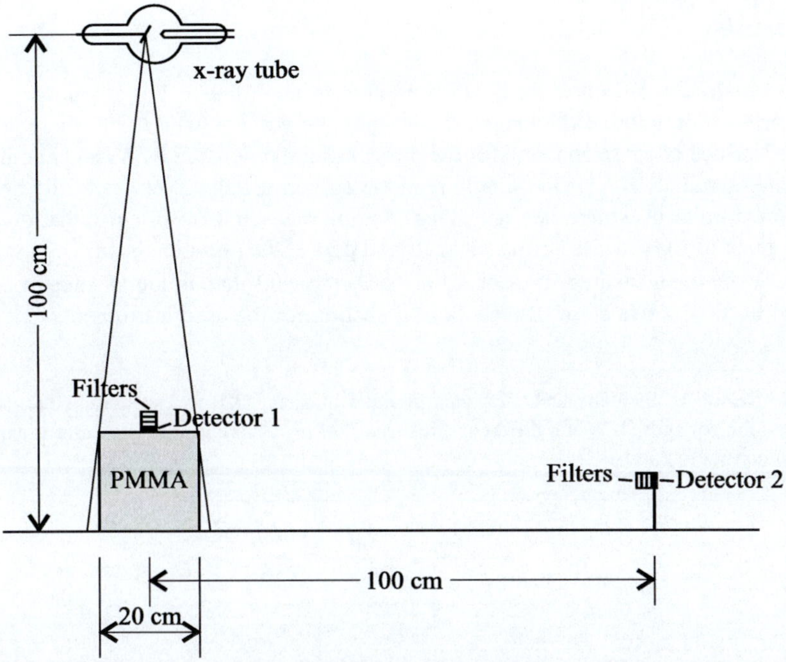

Fig. 1. Measurement setup. Detector 1 and 2 are solid-state detectors with lead backing, measuring incident air kerma only. Filters are layers of protective material.

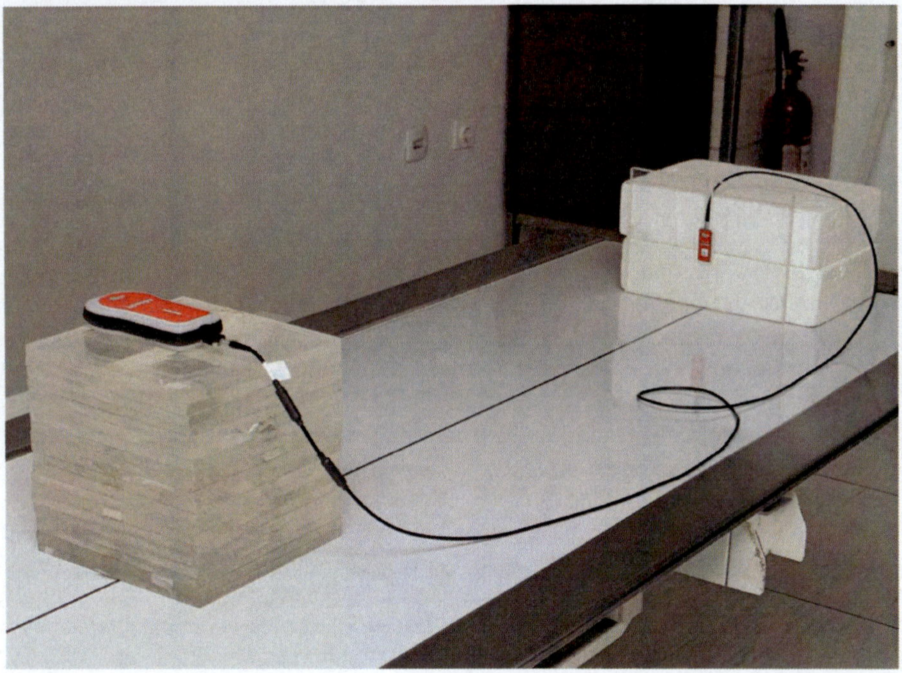

Fig. 2. Measurement photos. Body of Piranha detector is placed on top of the PMMA phantom, while Piranha Dose Probe 1 m away from the centre of phantom, laterally.

3 Results

Results obtained within this study are presented in the Table 1. For any chosen value of the protection material thickness, *T*, and the tube voltage values, U_P, there are two obtained values of air kerma, one for the direct radiation denoted by K_i and another for the scattered radiation K_s^1. One should read the air kerma values very carefully because of different units in which they are given. Kerma measured for direct radiation beam (K_i) is given in μGy, while kerma for scattered part of the radiation beam $\left(K_s^1\right)$ is given in nGy. When the measured value of K_s^1 is 0 nGy it means that its actual value is smaller than 0.1 nGy, which is below threshold of detection for the used instrument.

Table 1. Results of incident air kerma measurement in direct (*K*i) and scattered beam (K_s^1) at different tube voltages (*UP*) and different thickness *T* of protective layers with current-exposure time product set to 25 mAs

U_P	*T* (mmPb)	*Ki* (μGy)	$\left(K_s^1\right)$ (nGy)
60	0.00	626.7	452.1
kV	0.25	20.08	0
	0.50	2.750	0
	0.75	0.5142	0
81	0.00	1198	1095
kV	0.25	105.0	64.45
	0.50	27.09	0
	0.75	9.260	0
102	0.00	2031	2191
kV	0.25	296.4	222.9
	0.50	104.0	64.18
	0.75	44.87	0

Figures of corresponding normalised measured air kerma values for the direct (K_i) and the scattered radiation beam $\left(K_s^1\right)$ as function of the tube voltage (U_P) are given in Fig. 3 and Fig. 4, respectively. Each obtained value of K_i and K_s^1 at different thickness of the protection material *T* is divided by the corresponding kerma value in the case the protection material was not used at all – $K_{i,}0$ and $K_{s,}^1 0$ for direct and scattered air kerma, respectively. Ratios $K_i/K_{i,}0$ and $K_s^1 / K_{s,}^1 0$ account for normalised values for direct and scattered radiation, denoted as ${}_nK_i$ and ${}_nK_s^1$, respectively.

The exponential regression lines ($y = ae^{-bx}$) seen in Fig. 3 and Fig. 4 do not provide seamless fit to experimental data. However, they represent approximation of theoretical exponential attenuation of monochromatic narrow beam through material. The fitting parameter $a = 1.00$ in all cases, while *b* depends on U_P and beam type. In the primary beam, *b* is equal to (13.7 ± 0.3, 9.6 ± 0.5 and 7.4 ± 0.6) mm^{-1}, while in scattered beam *b* has values of (26.0 ± 0.3, 11.4 ± 0.2 and 9.0 ± 0.5) mm^{-1}, for (60, 81 and 102) kV, respectively.

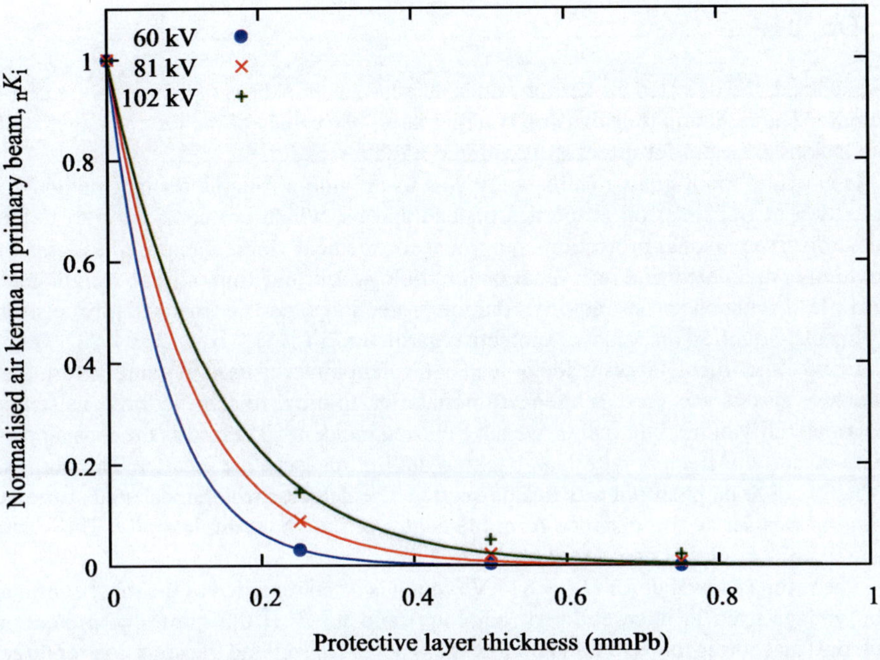

Fig. 3. Transmission of lead apron sheets with different nominal thickness in primary beam with different tube voltages. Curves represent exponential regression lines ($y = ae - bx$)

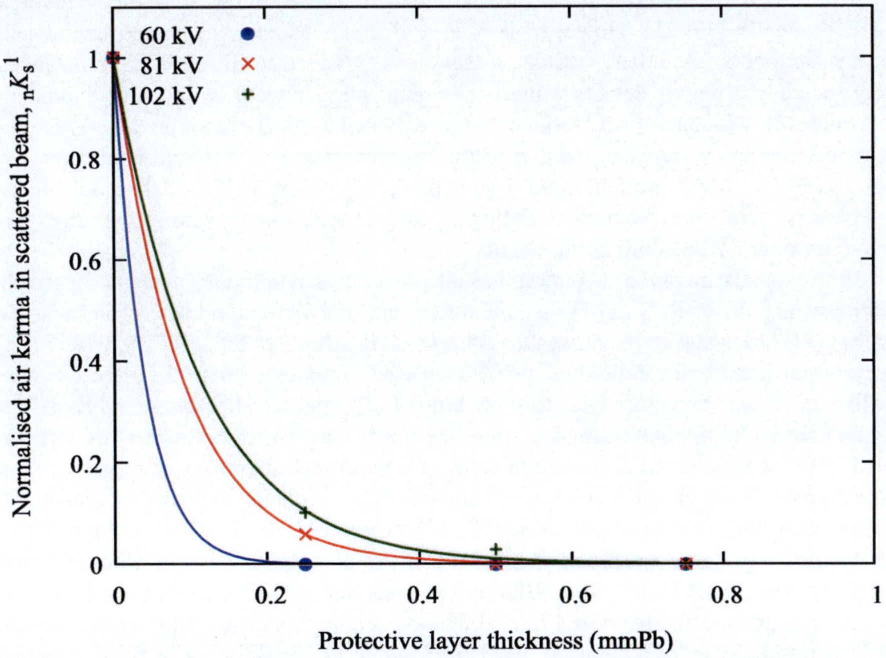

Fig. 4. Transmission of lead apron sheets with different nominal thickness in scattered beam with different tube voltages. Curves represent exponential regression lines ($y = ae - bx$)

4 Discussion

As expected, the detected air kerma values decrease as thickness of protection material increase. The radiation transmission is larger at higher values of the tube voltage. Both conclusions are valid for direct and scattered radiation beam.

One of the main goals of this study was to provide a reliable testing method for evaluation of transmission properties of lead aprons, which are considered to be the most effective personal protection equipment for medical workers exposed to ionising radiation in diagnostic and intervenational radiology. Introduction of light-weight lead-free or lead-reduced aprons require radiation protection experts or medical physicists to pay special attention on personal protective garments [14, 15].

In our experiment, layers made of lead equivalent material used in manufacturing of protective aprons was used as attenuation material. In order to achieve the consistency and reputability of measurements, we have chosen incident air kerma as the most appropriate quantity. Detectors with lead-backing did not measure backscatter. The purpose of the 20 cm-thick phantom was to side-scatter. The detector which measured scattered radiation was set at 1 m distance from the centre of the phantom, laterally. This setup ensured uniform field of scattered radiation.

The results show that for $U_P = 81$ kV (which is often regarded as the most common tube voltage used in interventional radiology) and for $T = 0.5$ mmPb of protection material (that corresponds to standard thickness of lead apron) the transmission for direct radiation beam was 2.3%, while for the scattered beam it was close to zero. Overall, the results correspond to those found in other studies [12, 16].

There is a lot of scientific papers devoted to the influence of radiation on the medical worker's health [17–21]. However, it is very difficult to compare experimental data to measurements in clinical settings, as they are collected in different circumstances. This includes different detectors used in monitoring, as well as different diagnostic procedures in which medical workers can be exposed to ionisation radiation. There are even differences in the considered procedure performances in different countries [18–20]. Also, one of the important issues is how medical professionals used their dosimeters and personal protection equipment. Following above mentioned reasoning, we can expect some deviations when comparing results.

Professionals involved in interventional procedures commonly wear two thermo-luminescence dosimeters (TLD) — one under and one above the lead apron (near the neck) [17]. The quantity they measure is personal dose equivalent, $H_P(10)$, which is an operational quantity for individual monitoring used for assessment of the effective dose. Unlike solid-state detectors with lead backing, TLDs measure backscattered radiation. As we know, effective and equivalent doses are not directly measurable. For this purpose, the ICRP has introduced and defined a set of operational quantities, such as personal dose equivalent which can be measured, and which are intended to provide a reasonable estimate for the protection quantities [22, 23]. According to the ICRP and the ICRU, the whole-body doses are reported in terms of the personal dose equivalent H_P (10). By using conversion factors and different methods, the personal dose equivalent gives an estimate of the effective dose [22, 23]. Niklason's methodology, which relies on two TLD measurements, is commonly used to estimate the effective dose from radiation sources in interventional radiology.

If that method is not possible to apply for effective dose calculation, measured personal dose equivalent $H_P(10)$ is regarded as the effective dose. More information on the method of effective dose calculations can be found in [17, 22]. Keeping in mind all differences in two considered approaches and uncertainty in TLD dosimetry it was reasonable to expect that results obtained in the study and this investigation differs quantitatively [17].

In the study conducted by Bašić et al., most of the exposed medical workers in interventional radiology (physicians, radiographers and nurses) received the yearly mean effective dose below 2 mSv [17]. Because of the limitation of the measuring instrument (TLD dosimeter worn by professionals) at low dose values, we considered only cases when the above-the-apron dose was above 10 mSv. Results are compared to our investigation. For corresponding medical professionals, the mean value ($\pm$ standard deviation) of the received dose under the lead apron was (1.3 ± 0.3) mSv, while the corresponding mean dose detected above the lead apron was (18.7 ± 1.8) mSv. Ratio of these values is 0.071.

In our measurement for the tube voltage of 102 kV and thickness of the protection material of 0.5 mm (which corresponds to thickness of the most often used lead apron in interventional radiology), the ratio is 0.0293, while for thinner protection material (0.25 mm) the ratio is 0.1017. Comparing these values with those presented in paper by Bašić et al. one can see a significant deviation [17]. There are a few reasons for that. First of them was previously mentioned uncertainties regarding the the limitations of the TLD dosimeter. Average annual dose reported in paper was 1.3 mSv, which is approximately about 0.1 mSv per one month [17]. This dose, however, is near the detection threshold for TLD dosimeter. At these values, the displayed value may differ from the real situation by a factor of 2. Also, natural radiation can influence the measured dose value. Contribution of this kind of radiation varies from to place, and can also vary depending on the season, especially in Sarajevo region [17]. All these factors influence dose values read by TLD system. Beside above-mentioned reasons, there are a few more things that can influence measurement results such as: different angles at which radiation falls on dosimeter, air temperature inside the room and its correlation with dose, manner in which medical professionals wear measuring devices, etc. Because of these factors, TLD dosimeters are constructed to show higher value of doses to avoid any underestimation. This is a conservative approach which is characteristic for any radiation protection measurements, including thermoluminescence dosimetry.

The detection threshold problem can be confirmed by additional analysis of the results. If the average above-the-apron dose (18.7 mSv) is used with corresponding transmission data from this study, the dose under the lead apron is estimated to be 0.548 mSv for $T = 0.25$ mmPb and 0.0381 mSv for $T = 0.5$ mmPb. Both calculated values are significantly below the threshold of TLD dosimeter. Calculated values are well below 1.7 mSv reported in paper by Bašić et al. [17].

5 Conclusions

Labelling of personal protective garments in terms of lead equivalence, while composite or lead-free materials are used, means that special attention is required by radiation protection experts or medical physicists. This study demonstrated a reliable testing method for evaluation of transmission properties of protective aprons.

Incident air kerma values have been measured in direct and scattered beam of X-ray radiation, simultaneously. Measurements were repeated for different tube voltages and with varying thickness of protection materials. Due to changes in radiation energy, the decreasing rate is more significant for scattered radiation. Doses that could not be detected are below detecting threshold of the used measurement devices.

We compare our results to those available in scientific literature to some extent, as well as to real data from measurements in clinical environment in Bosnia and Herzegovina, where doses were measured using a pair of TLDs worn by medical professionals. It is important to stress out that workplace conditions can greatly influence the final result and apparent efficiency of the protection material. More precisely, for the tube voltages commonly used in interventional radiology (not more than 100 kV) and the 0.5 mmPb-thick apron, only 3% of the incident radiation penetrate throughout protection. It seems that standard requirements for protection material are enough to prevent medical workers from unwanted effects of the ionising radiation on their health. We can conclude that lead aprons should be worn and stored properly in order to protect professionals from both direct and scattered radiation.

Acknowledgements. This paper is dedicated to the memory of Dr. Davorin Samek who sadly passed away on Thursday, 21st May, 2020. He was a full professor at the University of Sarajevo teaching general physics and biophysics. He was our teacher and collaborator.

References

1. IAEA, Radiation protection and safety in medical uses of ionizing radiation: Specif. Safe. Guide No. SSG-46 (International Atomic Energy Agency, 2018)
2. ICRP, ICRP Publication 103: Recommendations of the International Commision on Radiological Protection (International Commision on Radiological Protection, 2008)
3. IAEA, Radiological protection of patients in diagnostic and interventional radiology, nuclear medicine and radiotherapy: proceedings of an international conference held in Málaga, Spain, 26–30 March 2001 (International Atomic Energy Agency, Vienna, 2001)
4. IAEA, radiation protection and safety of radiation sources: international basic safety standards. (Int. Atom. Energy Agency, Vienna, 2014)
5. ICRP, ICRP Publication 60: Recommendations of the International Commision on Radiological Protection (International Commision on Radiological Protection, 1991)
6. ICRP, ICRP Publication 27: Recommendations of the International Commision on Radiological Protection (International Commision on Radiological Protection, 1977)
7. ICRU, Fundamental Quantities and Units for Ionizing Radiation (Report 60) (Oxford University Press, Oxford, 2011)
8. von Boetticher, H., Lachmund, J., Hoffmann, W.: Effective dose estimation in diagnostic radiology with two dosimeters: impact of the 2007 recommendations of the ICRP. Health Phys. **95**(3), 337 (2008)

9. Klein, L.W., Tra, Y., Garratt, K.N., Powell, W., Lopez-Cruz, G., Chambers, C., Goldstein, J.A.S.: For cardiovascular angiography, interventions, catheterization and cardiovascular interventions **86**(5), 913 (2015)
10. Bartal, G., Sailer, A.M., Vano, E.: Should we keep the lead in the aprons? Tech. Vasc. Interv. Radiol. **21**(1), 2–6 (2018)
11. Finnerty, M., Brennan, P.: Protective aprons in imaging departments: manufacturer stated lead equivalence values require validation. Euro. Radiol. **15**(7), 1477 (2005)
12. Mori, H., Koshida, K., Ishigamori, O., Matsubara, K.: Evaluation of the effectiveness of X-ray protective aprons in experimental and practical fields. Radiol. Phys. Technol. **7**(1), 158 (2014)
13. IEC, IEC 61331-1:2014: Protective devices against diagnostic medical X-radiation - Part 1: Determination of attenuation properties of materials (International Electrotechnical Commission, Geneva, Switzerland, 2014)
14. Muir, S., McLeod, R., Dove, R.: Light-weight lead aprons—light on weight, protection or labelling accuracy? Australas. Phys. Eng. Sci. Med. **28**(2), 128 (2005)
15. Lichliter, A., Weir, V., Heithaus, R.E., Gipson, S., Syed, A., West, J., Rees, C.: Clinical evaluation of protective garments with respect to garment characteristics and manufacturer label information. J. Vasc. Intervent. Radiol. **28**(1), 148–155 (2017)
16. Christodoulou, E.G., Goodsitt, M.M., Larson, S.C., Darner, K.L., Satti, J., Chan, H.P.: Evaluation of the transmitted exposure through lead equivalent aprons used in a radiology department, including the contribution from backscatter. Med. Phys. **30**(6), 1033 (2003)
17. Bašić, B., Beganović, A., Skopljak-Beganović, A., Samek, D.: Occupational exposure doses in interventional procedures in Bosnia and Herzegovina. Radiat. Protectiondosimetry **144**(1–4), 501 (2011)
18. Yun, J.W., Cha, E.S., Ko, S., Lee, W.J.: Work practices and radiation exposure among male radiologic technologists assisting fluoroscopically guided interventional procedures. Radiat. Protect. Dosim. **176**(4), 418–424 (2017)
19. Chida, K., Kaga, Y., Haga, Y., Kataoka, N., Kumasaka, E., Meguro, T., Zuguchi, M.: Occupational dose in interventional radiology procedures. Am. J. Roentgenol. **200**(1), 138–141 (2013)
20. Sánchez, R.M., Vano, E., Fernández, J.M., Rosales, F., Sotil, J., Carrera, F., García, M.A., Soler, M.M., Hernández-Armas, J., Martínez, L.C., et al.: Staff doses in interventional radiology: a national survey. J. Vascu. Intervent. Radiol. **23**(11), 1496–1501 (2012)
21. Elshami, W., Abuzaid, M., Piersson, A.D., Mira, O., AbdelHamid, M., Zheng, X., Kawooya, M.: Occupational dose and radiation protection practice in uae: a retrospective cross-sectional cohort study (2002--2016). Radiat. Protect. Dosim. **187**(4), 426 (2019)
22. Niklason, L.T., Marx, M.V., Chan, H.P.: The estimation of occupational effective dose in diagnostic radiology with two dosimeters. Health Phys. **67**(6), 611–615 (1994)
23. Harrison, C., Streffer.: Radiation protection dosimetry **127**(1–4), 12 (2007)

Use of a Smaller Size Phantom When Measuring Scatter Radiation in Diagnostic and Interventional Radiology

Amra Skopljak-Beganović[1], Lejla M. Čiva[2], Rahima Jašić[1](✉), Branka Metlić[1], Alma Pašić-Alić[3], and Davorin Samek[4]

[1] Radiation Protection and Medical Physics Department, Clinical Centre of Sarajevo University, Sarajevo, Bosnia and Herzegovina

[2] Medical School, University Sarajevo School of Science and Technology, Sarajevo, Bosnia and Herzegovina

[3] Faculty of Science, University of Sarajevo, Sarajevo, Bosnia and Herzegovina

[4] Veterinary Faculty, University of Sarajevo, Sarajevo, Bosnia and Herzegovina

Abstract. Regular control of X-ray installations includes radiation protection tests. Scattered radiation is measured in a standardised environment, using phantoms that simulate patients. In this paper, we investigated the possibility to use smaller size phantoms when measuring scatter radiation in diagnostic and interventional radiology. We examined how ambient dose equivalent is changing with different characteristics of the primary beam and phantom sizes. The source of radiation was a radiographic X-ray unit equipped with the possibility to change the exposure time and current individually. Measurements were done using three ionisation chambers. During the measurement procedure distance from the source of scattered radiation to the detector was changed. Two phantoms characterised by different size were used. Inconsistency was noted between different instruments during the measurements at distances close to the centre of the primary field. This can be attributed either to different dimensions of ionisation chambers or to non-uniformly scattered radiation. Other results are in accordance with the expectations. Use of smaller-sized phantoms is possible with the use of appropriate correction factors.

Keywords: Diagnostic and interventional radiology · Scattered radiation · Phantoms · Ionisation chamber · Ambient dose equivalent

1 Introduction

As a part of regular control of units and facilities in diagnostic and interventional radiology, it is necessary to carry out verification of radiation protection measures. The use of ionising radiation for medical purposes has increased as a result of the development of medical diagnostics and treatment. Consequently, exposed workers should employ protective measures and use personal dosimeters. X-ray images are acquired by the transmission of ionising radiation through the human body which relies on the attenuation coefficients.

A. Badnjevic and L. Gurbeta Pokvić (Eds.): CMBEBIH 2021, IFMBE Proceedings 84, pp. 826–832, 2021.
https://doi.org/10.1007/978-3-030-73909-6_94

One of the most prominent modes of interaction of radiation with matter is the Compton effect that produces scattered radiation. Furthermore, it reduces the image quality and spatial resolution and contributes to staff and patient exposure [1–3].

Measuring of scattered radiation is still an interesting topic for many researchers [4–7]. Today, the primary focus is in on optimisation of radiation protection in interventional radiology [8]. Even for a common workload, dose to the eye lens of interventionalist could exceed the dose limit of 20 mSv per year [9]. Hence, reliable measurement of ambient dose equivalent is an important task for medical physicists or radiation protection experts. At the same time, they are facing ever-changing challenges in other domains of their work, such as extensive quality control (QC) of imaging equipment, optimisation of diagnostic procedures, etc. All these tests require the use of advanced electronics (dosimeters, laptops, etc.) and supporting phantoms for quality control. Carrying this equipment becomes a challenge, especially when QC testing is done outside of the base institution.

Aim of this paper is to investigate possibilities of use of smaller-sized phantom for measurement of radiation scatter in diagnostic and interventional radiology.

2 Materials and Methods

In this study, we explored the dependence of the ambient dose equivalent on the characteristics of the primary beam and the phantom size, for different xray exposure parameters and detector to phantom distances. Source of ionising radiation was a common X-ray tube that is used within standard radiography. Measurements were done using three ionisation chambers – Victoreen 450B (Victoreen LLC, OH, US), Ludlum 9DP (Ludlum Measurements INC, TX, US) and TBM-IC-MARK V (Direct Scientific, MA,US). Corresponding measurements have been performed using the setup shown in Fig. 1.

During the measurement procedure distance from the source of scattered radiation to the detector was changing from 0.5 m to 1.5 m. Two different phantoms were used; one 20×20 cm^2 and the other 30×30 cm^2, both 20 cm thick. Measurements at 1 m distance were repeated for different values of tube voltage (U_P) that ranged from 40 to 120 kV, which are in standard use in diagnostic and interventional radiology. X-ray field size was constant – 20×20 cm^2.

3 Results

Dosimetric values are represented in terms of ambient dose equivalent at 10 mm depth per unit tube current-exposure time product or shortly $H*(10)/P_{It}$, expressed in terms of μSv/(mAs).

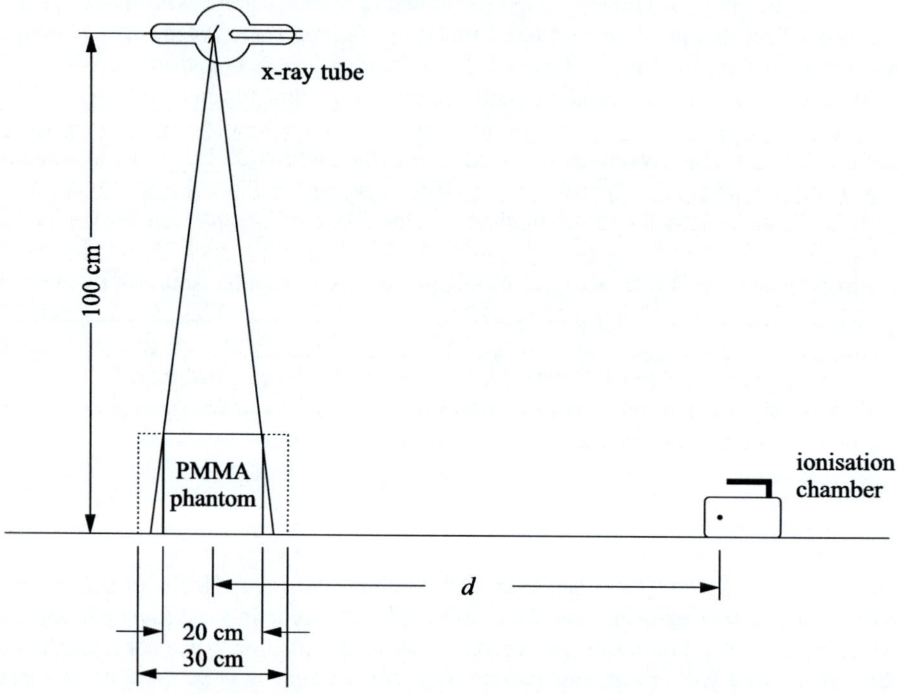

Fig. 1. Measurement setup. Ionisation chamber is placed at various distances 9(*d*) from the centre of 20 cm-thick PMMA phantom which is placed at 80 cm from X-ray tube focus.

Figure 2 shows how $H*(10)/P_{\text{It}}$ depends on tube voltage. The form of regression line is αU_P^b, which is common for a relationship between incident air kerma K_i and anode potential difference U_P [10]. While $\alpha \approx 0$, values of b change depending on detector model and phantom size. For phantom size 20×20 cm^2, *b* has values of 2.9 ± 0.2, 3.0 ± 0.2 and 3.4 ± 0.2 for ionisation chambers TBM-IC-MARK V, Ludlum 3DP and Victoreen 450B, respectively. Similarly, for 30×30 cm^2 phantom size, *b* fits to values 3.58 ± 0.04, 3.91 ± 0.6 and $4.27.0 \pm 0.02$ for same ionisation chambers, respectively.

Figures 3 and 4 show how $H*(10)/P_{\text{It}}$ change with distance from two phantoms, 20×20 cm^2 and 30×30 cm^2, respectively.

Table 1 shows average values, standard deviations and coefficient of variations for measurements with TBM-IC-MARK V, Ludlum 3DP and Victoreen 450B at 1 m distance from the phantom centre for different values of tube voltages and phantom sizes. Table 1 also illustrates the differences in scatter between two phantom sizes. The ratio of measured values in 20×20 cm^2 i 30×30 cm^2 is shown in last 4 rows of table.

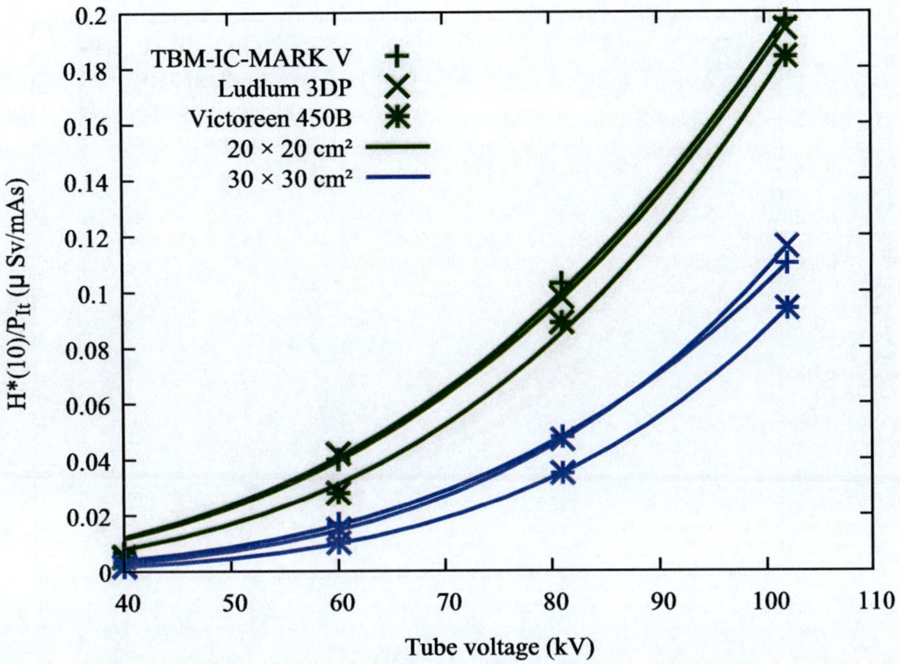

Fig. 2. Dependence of the value of the ambient dose equivalent per unit of tube current time product on the value of the selected tube voltage for two phantom dimensions. The regression lines are in form of. αU_P^b

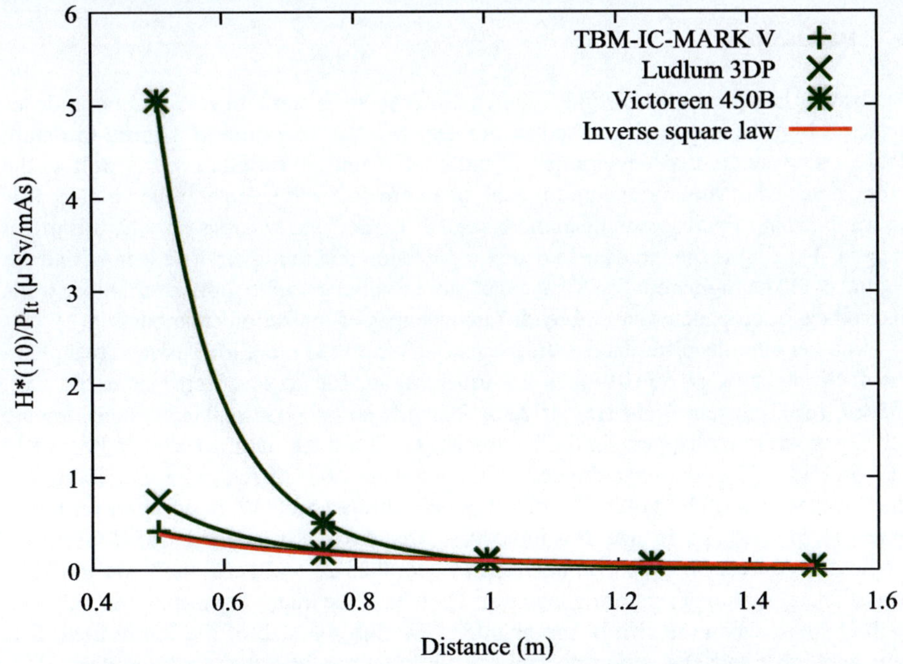

Fig. 3. Dependence of measured ambient dose equivalent at 10 mm depth per unit tube current-exposure time product $H*(10)/P_{\text{It}}$ on distance from 20 × 20 cm^2 PMMA phantom.

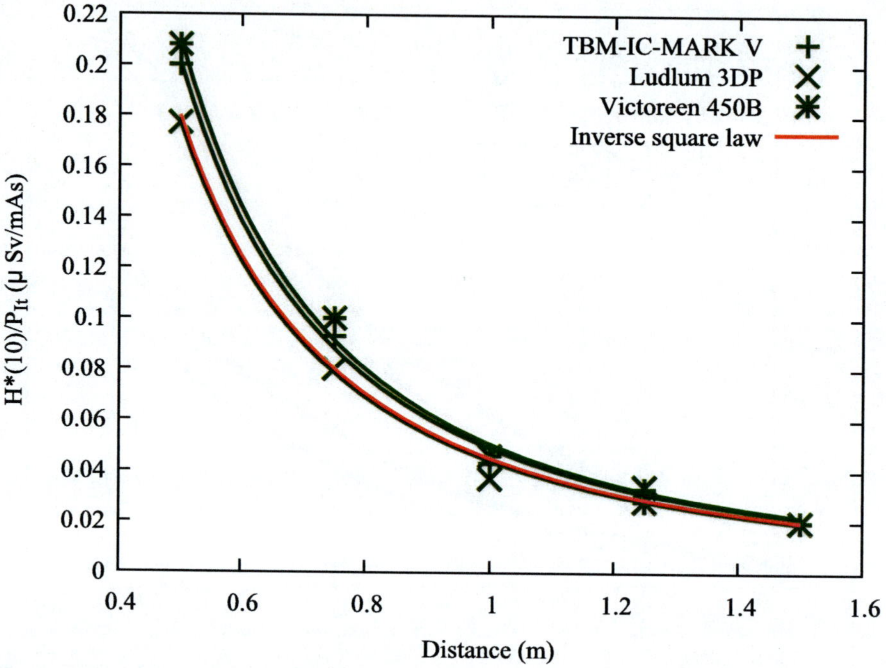

Fig. 4. Dependence of measured ambient dose equivalent at 10 mm depth per unit tube current-exposure time product $H*(10)/P_{\mathrm{It}}$ on distance from $30 \times 30\ \mathrm{cm}^2$ PMMA phantom.

4 Discussion

Analysing data presented in Fig. 2 indicate that there is some deviation among doses detected by different devices used in our experiment. To examine it more in detail, Table 1 shows standard deviations (σ) and coefficients of variation (c) as well as the average values of measurements for each phantom size and selected tube voltage. It is noticeable that c decreases with an increase of U_{P}. There are two reasons why this might happen. The first is the increase in detector precision due to higher dose values, and the second could be higher energy X-rays that more easily penetrate through chamber walls and reduce discrepancies caused by different shapes of ionisation chambers.

Another obvious difference is the intensity of scattered radiation between two phantoms. On average, $H*(10)/P_{\mathrm{It}}$ is 2.3 times higher for $20 \times 20\ \mathrm{cm}^2$ phantom size. Indeed, for a constant field size of $20 \times 20\ \mathrm{cm}^2$, larger phantoms have considerable self-absorption that happens in the 5 cm margin around the field, which can be clearly seen in Fig. 1. This causes reduction of the intensity of scattered radiation, and hence the lower values of $H*(10)/P_{\mathrm{It}}$ for same exposure parameters in comparison to the phantom $20 \times 20\ \mathrm{cm}^2$ in size. It is most likely that the measured values of $H*(10)/P_{\mathrm{It}}$ would also decrease if we used phantoms smaller than $20 \times 20\ \mathrm{cm}^2$, and this would be caused by reduction of scattering material. Therefore, the intensity of scattered radiation is the highest when the size of the phantom matches the size of the X-ray field. It is noticeable that self-absorption decreases with higher values of the tube voltage. This

Table 1 Mean value ($\bar{x}$), standard deviation (σ) and coefficient of variation (c) of measurements by 3 types of dosimeters (TBM-IC-MARK V, Ludlum 3DP and Victoreen 450B) at 1 m distance from the phantom centre for different values of tube voltages (UP) and phantom sizes, and the ratio of measurements.

Phantom size	U_P (kV)	$H^*(10)/P_{It}$ $\bar{x}$ (μSv/mAs)	σ (μSv/mAs)	c
20 × 20 cm²	40	0.0057	0.0009	0.16
	60	0.036	0.008	0.21
	81	0.097	0.007	0.074
	102	0.192	0.007	0.036
30 × 30 cm²	40	0.0023	0.0009	0.39
	60	0.014	0.004	0.26
	81	0.044	0.008	0.18
	102	0.11	0.01	0.11
Ratio	40	2.4		
	60	2.7		
	81	2.2		
	102	1.8		

happens due to greater penetration of the X-ray radiation at higher voltages. $H*(10)/P_{It}$ as a function of distance from phantom needs a special consideration. Results are presented in Figs. 3 and 4. As it is well known, inverse square law is stating that a dose is inversely proportional to the square of the distance from the source of radiation. For parameters chosen in our investigation that kind of dependence is presented by the solid red curve in both diagrams. The other curves present measured dependence for dose readout on different measurement instruments (ionisation chambers) used in our study (TBM-IC-MARK, Ludlum 9DP and Victoreen 450B). Analysis of the presented curves leads to the conclusion that, in the case of the larger sized phantom, no significant deviation from the inverse square law is noticeable for any detector i.e. mentioned inverse square law is approximately satisfied (Figure 4). On the other hand, for smaller sized phantom dose distribution as a function of distance do not follow inverse square law, and the corresponding deviation is greatest for readouts from Ludlum 3DP and Victoreen 450B ionization chambers (Fig. 3).

We could argue that this behaviour is mostly due to inhomogeneity of the beam scattered on a small phantom. Sharp edges of the phantom scatter the radiation "unpredictably," and this becomes clearly noticeable at distances $d < 1$ m. The other option is that the size and shape of ionisation chambers play a significant role at short distances.

Nevertheless, it is interesting to see that the inverse square law provides a reasonable approximation of the dose reduction with the distance from the source, although it is clear that the examined scattered radiation does not originate from a single point-source.

5 Conclusion

In this paper, we presented results of the investigation on characteristics of scattered X-ray beam depending on dimensions of the phantom, distance from the phantom and X-ray tube voltage used in the experiment. Inverse-square law defined for point radiation sources fails at smaller distances when the phantom is at distances less than 1 m from ionisation chambers. This is more prominent when the scattering phantom has the same dimensions as field size. The inhomogeneities of scattered X-ray beam cause unpredictable readouts that depend on the precise position of the ionisation chamber and its size. Other results are in accordance with our expectations and theoretical predictions. At the same values of the current–exposure time product, ambient dose equivalent increases with the voltage as a power of 2 or more. When phantom larger than the field of size is used, the amount of scattered radiation is smaller due to self-absorption which becomes a significant factor. The self-absorption decreases at higher tube voltages, as higher energies of X-rays lead to higher penetration. In conclusion, the use of smaller size phantoms for the purpose of measuring scatter radiation in diagnostic and interventional radiology is possible only with the use of appropriate correction factors.

Acknowledgements. This paper is dedicated to the memory of Dr. Davorin Samek who sadly passed away on Thursday, 21st May 2020. He was a full professor at the University of Sarajevo teaching general physics and biophysics. He was our teacher and collaborator.

References

1. Abrantes, A., Rebelo, C., Sousa, P., Rodrigues, S., Almeida, R., Pinheiro, K., Aevedo, K., Ribeiro, L.: Health Manage. **17**(1) (2017)
2. Nguyen, M.K., Truong, T.T., Morvidone, M., Zaidi, H.: Scattered radiation emission imaging: principles and applications. Int. J. Biomed. Imaging (2011)
3. IAEA, radiation protection and safety of radiation sources: international basic safety standards. Int. Atom. Energy Agency, Vienna (2014)
4. McCaffrey, J., Tessier, F., Shen, H.: Medical physics **39**(7Part1), 4537 (2012)
5. Vlachos, I., Tsantilas, X., Kalyvas, N., Delis, H., Kandarakis, I., Panayiotakis, G.: Measuring scatter radiation in diagnostic X rays for radiation protection purposes. Radiat. Protect. Dosim. **165**(1–4), 382 (2015)
6. Fetterly, K.A., Schueler, B.A., Grams, M.P., Sturchio, G.M.: Estimating head and neck tissue dose from x-ray scatter to physicians performing x-ray guided cardiovascular procedures: a phantom study. J. Radiol. Protect. **37**(1), 43 (2016)
7. Lalrinmawia, J., Pau, K.S., Tiwari, R.C.: Radiol. Phys. Technol. **12**(3), 312 (2019)
8. Leyton, F., Nogueira, M.S., Saad, J., dos Santos, J.A., Vano, E., Oliveira, M.A., Ubeda, C.: Scatter radiation dose at the height of the operator's eye in interventional cardiology. Radiat. Meas. **71**, 349 (2014). https://doi.org/10.1016/j.radmeas.2014.07.015.0, http://www.sciencedirect.com/science/article/ pii/S1350448714002182. Proceedings of the 17th Solid State Dosimetry Conference (SSD17)
9. ICRP, ICRP Publication 103: Recommendations of the International Commision on Radiological Protection (International Commision on Radiological Protection (2008)
10. IAEA, Diagnostic radiology physics: a handbook for teachers and students. Int. Atom. Energy Agency, Vienna, (2014)

COVID-19 Related Research

The Most Significant Biomarkers and Specific Antibodies for the Early Diagnosis and Monitoring in COVID-19 Patients

Merima Bukva(✉), Minela Islamović, Selma Jusufović, Enisa Njemčević, Neven Meseldžić, and Tamer Bego

Faculty of Pharmacy, University of Sarajevo, Zmaja od Bosne 8, 71000 Sarajevo, Bosnia and Herzegovina
merimabukva@ffsa.unsa.ba

Abstract. As the COVID-19 ravages the world, the efforts to identify this infection more quickly and efficiently grow in significance every day. In given study, we have gathered the most specific biomarkers discovered by now for the COVID-19 infection diagnosis and disease progress monitoring, thus giving basic guidelines for the admitted patients' care. Divided into sections: inflammatory, hematologic, liver biomarkers and specific antibodies, we have consolidated the findings of different indicators of disease such as cytokine storm markers, C-reactive protein, ferritin, blood components (lymphocytes, neutrophils, thrombocytes), D-dimer, glycosylated hemoglobin, alanine and aspartate aminotransferase, lactate dehydrogenase, creatinine, together with the specific COVID-19 antibodies, all with an aim of making a foundation for quicker reaction of healthcare professionals when it comes to these patients.

Keywords: COVID-19 · Biomarkers · Cytokine storm · Inflammation · Antibodies

1 Introduction

Coronavirus Disease 2019 (COVID-19) was first discovered in Wuhan, China, in December 2019 and it has become a major health problem. The microorganism that causes this infection is coronavirus named SARS-CoV-2. The most common clinical symptoms that occur are fever, cough, shortness of breath, other breathing difficulties, headache, fatigue, muscle pain, diarrhea, vomiting and loss of sense of smell and taste [1]. Tissue injury in infected patients who develop acute respiratory syndrome is associated with prolonged hospitalization and more severe clinical features, which is explained by the fact that SARS-CoV-2 binds to angiotensin-converting enzyme 2 (ACE2) receptor. Stated receptor is expressed in different human cells including liver and bile duct cells and attaching to it causes abnormal activity of liver enzymes [2]. Simultaneously, deregulation of immune response happens, which causes the state of hyperinflammation followed by the high levels of C-reactive-protein (CRP), interleukin-6 (IL-6), and the

A. Badnjevic and L. Gurbeta Pokvić (Eds.): CMBEBIH 2021, IFMBE Proceedings 84, pp. 835–846, 2021.
https://doi.org/10.1007/978-3-030-73909-6_95

possibility of the cytokine storm development [3]. Furthermore, there is deregulation of coagulopathy process and patients with this condition are at higher risk of developing more severe infection [4]. Serology tests are a good option for detection of the antibodies that are developed due to the COVID-19 infection. These assays can detect IgG and IgM antibodies and be used for an estimation of the duration of immunity (Table 1) [3].

Table 1. Normal and COVID-19 patients' biomarkers values [21, 65–72]

	Normal values	COVID-19 infection values
Inflammatory biomarkers		
IL-1β (pg/mL)	0.5–12	5.0
IL-6 (pg/mL)	0–16.4	74.2
IL-8 (pg/mL)	24.4–35.9	51.7
IL-10 (pg/mL)	8.5–16.7	5.0–20.1
TNFα (pg/mL)	0–8.1	7.5–28.7
IFN γ IP-10 (pg/mL)	<2.0	203.7–939.4
GM-CSF (pg/mL)	26.3–63.8	420
MCP-1 (pg/mL)	69.5–175.2	293.2–1207.5
CRP (mg/dL)	<5.0	65–169
Ferritin (μg/L)	30–400	745–2733
Hematologic biomarkers		
Lymphocytes (cells/μL)	900–4000	700–1250
Neutrophils (cells/μL)	1500–8000	2855–6725
Trombocytes (cells/μL)	$130–400 * 10^3$	155–242
PT (s)	10.0–13.0	prolongation > 3 s
D-dimer (ng/mL)	<232	330–985
HbA1c (%)	<6	≥9 increases the risk of hospitalization
Liver biomarkers		
ALT (U/L)	10–50	24–58
AST (U/L)	18–39	34–68
LDH (U/L)	135–225	251–481
Creatinin (mg/dL)	0.60–1	0.9–1.3
Specific COVID-19 antibodies		
IgM (mg/dL)	40–230	↑
IgG (mg/dL)	700–1600	↑

All of the stated markers can be used for public health, risk management and academic perspective. Thus, the aim of this article is to give a condensed insight into the most used biomarkers for the COVID-19 detection, diagnosis, disease and therapy monitoring and the prediction of severity and outcome of the infection.

2 Inflammatory Biomarkers

2.1 Cytokine Storm

Cytokine storm is a high inflammatory state and, by the recent studies, it is connected to the development and a more severe clinical picture of the COVID-19. We can observe it as a complex network of overlapping pathways and cross-talks between fragments of the immune system [5]. In the normal conditions, inflammation is a useful and necessary tool for fighting foreign organisms in the body. However, disturbance in these regulatory mechanisms leads to organs damage, intracorporeal chaos and ultimately, death. [6, 7]. Virus is first attacked by the mast cells which are widespread in the respiratory system, followed by the releasing of many proinflammatory cytokines. Infected cell activates B cells, natural killer cells, dendritic cells, macrophages. Rapid activation of the Th cells and inflammatory monocytes occurs, huge number of cytokines are produced and released, leading up to the cytokine storm state managed by the positive feedback loop. All listed elements circulate in the blood, easily reaching distant places of the organism and inducing signs of inflammation - fever, pain, redness, vascular hyperpermeability and swelling. When not under control, they generate a major damage in the body [6, 8–11].

The greatest issue that takes place within the cytokine storm state is that it provokes higher severity of the COVID-19 disease - people are more likely to die or to develop irreversible impairment of the body organs, especially lungs, where most commonly acute respiratory distress syndrome (ARDS) and pulmonary edema occur. [6, 9]. The most important markers whose levels point to the severity of clinical picture and mortality of patients infected with COVID-19 will be further discussed.

Interleukins

Interleukins are regulating factors of inflammatory process and immune cell differentiation and activation. When bilateral lobular pneumonia is present in patients with COVID-19, levels of IL-1β, IL-7, IL-8 and IL-9 are always increased, as they are released from the damaged cells. IL-1 has a role in T-cell immunity and IL-2 secretion. Subspecies IL-1α and IL-1β maintain acute-phase signaling and transfer of immune cells to the place of inflammation. They are also responsible for the cytokine production [12]. IL-4 is a suppressor of inflammation which is mostly detectable in the late phase of the disease [8].

Probably the most important role belongs to the IL-6. IL-6 is a proinflammatory cytokine that is secreted by different kinds of cells - macrophages, neutrophils, dendritic cells and lymphocytes. The intensity of the inflammation is increased as IL-6 arrives to liver and encourages the releasing of C-reactive protein, fibrinogen and serum amyloid A. Furthermore, IL-6 helps the differentiation of naïve CD4 T cells to the effector and helper cells, promotes Th7, CD8 T cells, T-follicular helper cells, plasma cells, IL-21 and B cells differentiation, consequently controlling the levels of immunoglobulins. Together with the D-dimer, it can be used as a predicting indicator for the disease severity and the need for ventilation support. It can also be used for the predicament of the requirement of ICU admission [5, 8, 10].

IL-8, together with MCP-1, is an intermediate in the transport of the neutrophils and macrophages to the location of inflammation. It is a strong chemoattractant which prevents cell death, therefore extending neutrophils impact [12, 13]. IL-10 is responsible for the differentiation and activation of the CD8 T cells and Th cells. The level of IL-10 is increased on the second week following typical symptoms as one of the last efforts of the body to control the cytokine storm [8]. IL-17 induces G-CSF, IL-1β, IL6, TNFα, IL-8, MIP2A, MIP3A, IP10. Together with the IL-22, it induces mucosal peptides that have antimicrobial properties. IL-22 encourages mucin, fibrinogen and serum amyloid A excretion, finally causing lung edema [14]. Macrophages produce IL-18 as an inactive precursor. IL-18 binds to IL-18R and by acting on helper, cytotoxic T cells and NK cells induces the production of IFN-γ [15].

Tumor Necrosis Factor

Tumor necrosis factor (TNF) is a proinflammatory and pyrogen cytokine [16]. In most cases, elevation of the TNF levels in blood is the first sign of the infection present in the body [17]. It is mostly produced by the macrophages, monocytes, neutrophils, but also muscle, adipocyte and astrocyte cells. TNF-α levels increase early on, and stay increased throughout the infection process. In the lungs, it induces hyaluronan-synthase-2 which induces fluid permeability and decrease in ventilation. TNF-α also causes the apoptosis of lung cells [8, 13]. It may contribute to the development of the lymphocytopenia, often connected to the COVID-19 state [18]. Higher levels of TNF are associated with the severity of cytokine storm and overall, clinical picture of the patient. Together with the IL-6, it represents the most common blood-marker determined in the COVID-19 patients [19].

Interferon Gamma Inducible Protein 10

Chemokines, including Interferon gamma inducible protein 10 (IFN γ IP-10) or CXCL10, have chemoattractant effect to the immune system cells and take part in an innate immunity. IFN γ IP-10 is produced by the macrophages and NK cells as a response to the infection. Also, it can be produced by the helper and cytotoxic T cells. After the contact with the virus, IFN-γ IP-10 can slow down its multiplication and simulate cells to express antiviral elements, along with regulating apoptosis. It is likely that IP-10 helps the transfer of the macrophages into the lungs [13, 16]. Together with IL-6, IFN-γ IP-10 is one of the indicators of probability of ICU admission of the patient, as blood levels of IFN-γ IP-10 are a good predictor of the COVID-19 outcome. [6, 8].

Granulocyte-Macrophage Colony-Stimulating Factor

Granulocyte-macrophage colony-stimulating factor (GM-CSF) is produced by the Th1 cells, induced by the IL-6. Helper cells differentiate to the toxic Th1 cells which then secrete GM-CSF. Said factor stimulates the inflammation state and the cascade of events which leads up to the cytokine releasement and the symptoms of the COVID-19 [13, 17]. Furthermore, GM-CSF is discovered to be important factor in lung alveoli where it has a role in the development of alveolar macrophages. For the most part, it is expressed by the myeloid cells, and during immune reaction it has a role in polarizing the same myeloid cells to proinflammatory forms and helping them reach the inflammation sites [20].

Monocyte Chemoattractant Protein-1

Monocyte chemoattractant protein-1 (MCP-1) is a chemokine compound that helps macrophage migration in the body, therefore takes part in causing uncontrolled macrophage activation syndrome (MAS). It exclusively attracts monocytes and basophils. Elevation in the levels of MCP-1 are connected to the probability of venous thrombosis ie. coagulation, which on the other hand is closely related to the COVID19 infection. In the serum of the critically ill patients, MCP-1 is much higher when in comparison to the severe ill ones [19, 21].

2.2 C-reactive Protein

C-reactive protein (CRP) is well known acute inflammatory protein. It increases up to 1,000-fold at sites of infection or inflammation. CRP is produced in liver hepatocytes. Also, it can be synthesized in smooth muscle cells, macrophages, endothelial cells, lymphocytes and adipocytes [22]. Mudatsir et al. observed that inflammatory markers, including elevated levels of CRP are found significantly higher in severe and mild COVID-19 patients [23]. Increased CRP level during the first 48 h of hospitalization is a good predictor of respiratory decline, [24]. Mickael et al. state that CRP correlates with CT scans that showed pneumonia, therefore CRP could be a good predictor for pneumonia due to COVID-19 infection where a CT scan is not available [25]. It was confirmed that CRP levels are indicative of different aspect of COVID-19 severity, oxygen requirement, presence of acute respiratory distress and other severe clinical features [26, 27].

2.3 Ferritin

Ferritin is a form of the total body iron stores. Small amounts of ferritin are secreted into the plasma within normal range, but increased free serum ferritin correlates with presence of ongoing inflammation in COVID-19 patients [28]. In the classification of patients with the COVID-19, study results showed that severe patients had significantly higher levels of ferritin, between 1.5–5 times higher than in ones with moderate symptoms. As for non-survivors, ferritin levels were 3–4 times higher compared to survivors. Ferritin synthesis is tightly regulated by cytokines, as one of their pathways regulates it indirectly through increasing NO and stimulating the secretion of ferritin from hepatocytes [29, 30].

3 Hematologic Biomarkers

3.1 Lymphocytes

One significant sign of COVID-19 infection is a lymphopenia occurrence, particularly important in severe clinical picture, as it is followed by the decaying of lung function and the need for oxygen supply and assisted ventilation. The assumption is that lymphopenia indicates the interstitial lymphocytic pneumonia, ending with severe alveolar damage, as well as hyaline membrane and pneumocytes damage [31]. Cytotoxic lymphocites draining in the early stages results in a stronger progression of the infection [32]. In a meta-analysis by Huang et al. [33], the total number of lymphocites and subsets was counted in the hospitalized COVID-19 patients. In comparison to the mild manifestation of the disease, numbers were notably reduced in patients with severe disease. Helper, regulatory and suppressor T cells are below normal numbers, where extremely low helper T cells indicate poor prognosis. In contrary, naïve helper cells are reported as increased in several COVID-19 cases, and regulatory T cells are visibly damaged in more difficult cases [34, 35]. Also, the life cycle of NK-cells has been investigated with a conclusion that NK cells decrease in number, but are highly activated and stay that way until the patients' condition improves [36].

3.2 Neutrophils

Neutrophil numbers are increased in the early stages of the disease, followed by the abnormal cell shape and granulation presence. Also, immature granulocyte forms are frequently found [37, 38]. Neutrophilia most frequently develops in patients in ICUs. Increased neutrophil-to-lymphocyte ratio in early stages indicates more serious clinical picture and patients show neutrophil infiltration in the pulmonary vessels, as well as neutrophil-caused mucositis. Considering that the neutrophils take part in killing microorganisms that enter the body, their prolonged activity causes tissue damage and might result in cytokine storm [39].

3.3 Trombocytes

Thrombocytopenia neatly follows the pathological picture of the COVID-19 patients. In one of the first ever COVID studies by Yang et al. [40], it was found that 20.7% patients had low thrombocytes - a marker that was associated to the much higher risk of mortality, as non-survivors had lower platelet counts. It is considered that the cause of given condition lies in COVID and cytokine storm - caused damage to the bone marrow and therefore hematopoietic cells, destruction of the thrombocytes by novel circulating antibodies, or the initial platelet aggregation forming thrombi, which then causes tissue damage and increased thrombocytes utilization [41]. Another factor which is proven to correlate with the disease progression is the prothrombin time (PT). It gradually increases and can be detected as prolonged, so it is recommended for PT to be monitored as early as possible so coagulopathic complications such as thromboembolism can be detected on time, as it has been proven in several studies already [42–44].

3.4 D-dimer

D-dimer is a product formed due to clot decomposition and it can be measured in a blood sample. It has high sensitivity and low specificity for the detection of pulmonary embolism or deep vein thrombosis [45]. A link between D-dimer and survival rate due COVID-19 infection has been established. Namely, D-dimer values greater than

11.0 μg/mL are associated with the increased mortality [46]. D-dimer is an excellent prognostic marker and a marker which indicates whether a COVID-positive patient needs to take anticoagulant therapy [47].

3.5 Glycosylated Hemoglobin

Glycosylated hemoglobin (HbA1c) is the form of hemoglobin binded with glucose that is used as a test for detecting Diabetes Mellitus, as it shows levels of glucose in the blood for the past 3 months [48]. Result of different studies showed that patients with higher HbA1c ($\geq$8%) had poorer outcomes in relation to COVID-19, which suggest that hypergycaemia is a good biomarker for the severity, but there are no proofs until now that higher levels of HbA1c can have causative effects [49, 50].

4 Liver Biomarkers

4.1 Alanine and Aspartate Aminotransferase

Alanine-aminotransferase (ALT) and aspartate-aminotransferase (AST) are enzymes found in the cytosol of hepatocytes. Liver damage during COVID-19 infection can occur due to the use of drugs, liver hypoxia and reaction of the immune system. Gu et al. performed a study in which ALT activity was elevated 21.2% patients who had.

COVID-19 infection [51]. Further, it is proven that higher ALT activity is associated with a poorer prognosis of COVID-19 infection and significantly higher ALT activity was reported in the elderly and men population, which is explained by the fact that ACE2 receptor expression is higher in the male population [52, 53].

AST can be used as a prognostic marker of myocardial injury or liver injury [54]. Lee and co-workers report that increased AST levels have been linked to increased inflammatory parameters and hypoxia in COVID-19 infection [55]. Ye et al. state that lower levels of AST activity are associated with a higher survival rate [56]. An increase in both AST and ALT is associated with a higher risk of mortality and they can be used as prognostic markers [57].

4.2 Lactate Dehydrogenase

Lactate dehydrogenase (LDH) is an important enzyme in the anaerobic metabolic pathway. It belongs to the class of oxidoreductases with a main role in the conversion of lactate to pyruvate, reducing NAD+ to NADH [58]. Zhou et al. reported that serum LDH levels are elevated in patients in whom COVID-19 infection has caused a severe clinical state. These patients have a higher risk of developing respiratory distress syndrome [59]. Scientists state that the ratio of LDH and leukocyte levels can be a prognostic and detection marker when it comes to COVID infection. Still, due to the number of false negative results when it comes to PCR testing, this ratio may be an alternative [60]. Salinas et al. report that LDH activity is an excellent predictor of death in COVID-19 patients [61].

4.3 Creatinine

Creatinine is a product of the breakdown of creatine and phosphocreatine in the muscles. It is produced in equal proportions in the body every day and excreted from by the kidneys [60]. COVID-19 patients have significantly higher creatinine values that might be associated with a fatal outcome [62]. Alvarez-Belon et al. state that it is necessary to measure the value of serum creatinine in COVID-19 infection for the disease predicament in COVID-19 patients every 24 h or more often if it is necessary [61].

5 Specific COVID-19 Antibodies

The immune system helps body fight the infections. IgG and IgM antibodies are used as biomarkers for detecting COVID-19 antigenes - coronavirus spike-S and nucelocapsid-N proteins. ELISA and chemiluminescence methods are shown to be consistent with nucleic acid detection assays, eliminating a possible cross reaction with other coronaviruses and avoiding false-negative results. S-specific IgM antibodies are detectable from day 4 onward, with IgM antibody titres increasing over time and peaking at approximately day 20. As for S-specific IgG antibodies - they are detectable from day 7 onwards, peaking at approximately day 25 and maintaining high level after 4 weeks of infection [63, 64].

6 Conclusion

As the COVID-19 ravages the world, the efforts to identify this infection more quickly and efficiently grow in significance every day. In given study, we have gathered the most specific biomarkers discovered by now for the COVID-19 infection diagnosis and disease progress monitoring, thus giving basic guidelines for the admitted patients' care. With this paper, our goal is to facilitate and accelerate the treatment process of every patient battling a COVID-19 infection by recognizing the presence of the disease and its severity, because only if acted upon in a timely manner does the patient have a great chance of survival.

Conflict of Interest. The authors have no conflicts of interest to disclose.

References

1. Ouassou, H., Kharchoufa, L., Bouhrim, M., et al.: The pathogenesis of coronavirus disease 2019 (COVID-19): evaluation and prevention. J. Immunol. Res. **2020**, 1357983 (2020)
2. Ghoda, A., Ghoda, M.: Liver injury in COVID-19 infection: a systematic review. Cureus **12**(7), 9487 (2020)
3. Feng, X., Li, S., Sun, Q., Zhu, J., et al.: Immune-inflammatory parameters in COVID-19 cases: a systematic review and meta-analysis. Front. Med. (Lausanne). **9**(7), 301 (2020)
4. Luan, Y.Y., Liu, Y., Liu, X.Y., et al.: Coronavirus disease 2019 (COVID-19) associated coagulopathy and its impact on outcomes in Shenzhen, China: A retrospective cohort study. Thromb. Res. **195**, 62–68 (2020)

5. Mahmudpour, M., Roozbeh, J., Keshavarz, M., Farrokhi, S., Nabipour, I.: COVID-19 cytokine storm: the anger of inflammation. Cytokine **133**, 155151 (2020)
6. de la Rica, R., Borges, M., Gonzalez-Freire, M.: COVID-19: in the eye of the cytokine storm. Front. Immunol. **11**, 558898 (2020)
7. Nile, S.H., Nile, A., Qiu, J., Li, L., Jia, X., Kai, G.: COVID-19: Pathogenesis, cytokine storm and therapeutic potential of interferons. Cytokine Growth Factor Rev. **53**, 66–70 (2020)
8. Fara, A., Mitrev, Z., Rosalia, R.A., Assas, B.M.: Cytokine storm and COVID-19: a chronicle of pro-inflammatory cytokines. Open Biol. **10**(9), 200160 (2020)
9. Jose, R.J., Manuel, A.: COVID-19 cytokine storm: the interplay between inflammation and coagulation. Lancet. Respir. Med. **8**(6), e46–e47 (2020)
10. Kempuraj, D., et al.: COVID-19, mast cells, cytokine storm, psychological stress, and neuroinflammation. Neuroscientist. **26**(5–6), 402–414 (2020)
11. Tang, Y., Liu, J., Zhang, D., Xu, Z., Ji, J., Wen, C.: Cytokine storm in COVID-19: the current evidence and treatment strategies. Front. Immunol. **11**, 1708 (2020)
12. Song, P., Li, W., Xie, J., Hou, Y., You, C.: Cytokine storm induced by SARS-CoV-2. Clin. Chim. Acta. **509**, 280–287 (2020)
13. Pelaia, C., Tinello, C., Vatrella, A., De Sarro, G., Pelaia, G.: Lung under attack by COVID-19induced cytokine storm: pathogenic mechanisms and therapeutic implications. Ther. Adv. Respir. Dis. **14**, 1753466620933508 (2020)
14. Wu, D., Yang, X.O.: TH17 responses in cytokine storm of COVID-19: an emerging target of JAK2 inhibitor Fedratinib. J. Microbiol. Immunol. Infect. **53**(3), 368–370 (2020)
15. Jamilloux, Y., et al.: Should we stimulate or suppress immune responses in COVID-19? Cytokine and anti-cytokine interventions. Autoimmun Rev. **19**(7), 102567 (2020)
16. Coperchini, F., Chiovato, L., Croce, L., Magri, F., Rotondi, M.: The cytokine storm in COVID19: An overview of the involvement of the chemokine/chemokine-receptor system. Cytokine Growth. Factor Rev. **53**, 25–32 (2020)
17. Wang, J., Jiang, M., Chen, X., Montaner, L.J.: Cytokine storm and leukocyte changes in mild versus severe SARS-CoV-2 infection: Review of 3939 COVID-19 patients in China and emerging pathogenesis and therapy concepts. J. Leukoc Biol. **108**(1), 17–41 (2020)
18. Iannaccone, G., et al.: Weathering the cytokine storm in COVID-19: therapeutic implications. Cardiorenal Med. **10**(5), 277–287 (2020)
19. Pearce, L., Davidson, S.M., Yellon, D.M.: The cytokine storm of COVID-19: a spotlight on prevention and protection. Expert Opin. Ther. Targets. **24**(8), 723–730 (2020)
20. Lang, F.M., Lee, K.M., Teijaro, J.R., Becher, B., Hamilton, J.A.: GM-CSF-based treatments in COVID-19: reconciling opposing therapeutic approaches. Nat. Rev. Immunol. **20**(8), 507–514 (2020)
21. Chen, Y., Wang, J., Liu, C., Su, L., Zhang, D., Fan, J., Yang, Y., Xiao, M., Xie, J., Xu, Y., Li, Y., Zhang, S.: IP-10 and MCP-1 as biomarkers associated with disease severity of COVID-19. Mol Med. **26**(1), 97 (2020)
22. Sproston, N.R., Ashworth, J.J.: Role of C-reactive protein at sites of inflammation and infection. Front. Immunol. **9**, 754 (2018)
23. Mudatsir, M., Fajar, J.K., Wulandari, L., et al.: Predictors of COVID-19 severity: a systematic review and meta-analysis. F1000Res **9**, 110 (2020)
24. Mueller, A.A., Tamura, T., Crowley, C.P., et al.: Inflammatory biomarker trends predict respiratory decline in COVID-19 patients. Cell Rep Med **1**(8), 100144 (2020)
25. Mickael, T., Ahmed, M., Rahul, M., et al.: Determining extent of COVID-19 pneumonia on CT based on biological variables. Respir. Med. **5**(175), 106206 (2020)
26. Keddie, S., Ziff, O., Chou, M.K.L., et al.: Laboratory biomarkers associated with COVID-19 severity and management. Clin. Immunol. **221**, 108614 (2020)

27. Chen, Z., Zhang, F., Hu, W., Chen, Q., Li, C., Wu, L., Zhang, Z., Li, B., Ye, Q., Mei, J., Yue, J.: Laboratory markers associated with COVID-19 progression in patients with or without comorbidity: a retrospective study. J. Clin. Lab. Anal. e23644 (2020)
28. Domellöf, M., Dewey, K.G., Lönnerdal, B., Cohen, R.J., Hernell, O.: The diagnostic criteria for iron deficiency in infants should be reevaluated. J. Nutr. **132**(12), 3680–3686 (2002)
29. Dahan, S., et al.: Ferritin as a marker of severity in COVID-19 patients: a fatal correlation. Isr. Med. Assoc. J. **8**(22), 429–434 (2020)
30. Gómez-Pastora, J., Weigand, M., Kim, J., Wu, X., Strayer, J., Palmer, A.F., Zborowski, M., Yazer, M., Chalmers, J.J.: Hyperferritinemia in critically ill COVID-19 patients - Is ferritin the product of inflammation or a pathogenic mediator? Clin. Chim. Acta. 509, 249–251 (2020)
31. De Rosa, F.G., Lupia, T., Corcione, S.: COVID-19: where have the lymphocytes gone? Intern. Med. J. **50**(11), 1436–1437 (2020)
32. Zheng, M., et al.: Functional exhaustion of antiviral lymphocytes in COVID-19 patients. Cell Mol. Immunol. **17**(5), 533–535 (2020)
33. Huang, W., et al.: Lymphocyte subset counts in COVID-19 patients: a meta-analysis. Cytometry A. **97**(8), 772–776 (2020)
34. Qin, C., Zhou, L., Hu, Z., et al.: Dysregulation of immune response in patients with COVID-19 in Wuhan, China. Clin Infect Dis. (2020)
35. Cossarizza, A., De Biasi, S., Guaraldi, G., et al.: SARS-CoV2, the virus that causes COVID-19. Cytometry **97**(4), 340–343 (2020)
36. Jiang, Y., et al.: COVID-19 pneumonia: CD8+ T and NK cells are decreased in number but compensatory increased in cytotoxic potential. Clin. Immunol. **218**, 108516 (2020)
37. Salib, C., Teruya-Feldstein, J.: Hypersegmented granulocytes and COVID-19 infection. Blood. **135**(24), 2196 (2020)
38. Zini, G., Bellesi, S., Ramundo, F., d'Onofrio, G.: Morphological anomalies of circulating blood cells in COVID-19. Am. J. Hematol. **95**(7), 870–872 (2020)
39. Tomar, B., Anders, H.J., Desai, J., Mulay, S.R.: Neutrophils and neutrophil extracellular traps drive necroinflammation in COVID-19. Cells **9**(6), 1383 (2020)
40. Yang, X., et al.: Thrombocytopenia and its association with mortality in patients with COVID-19. J. Thromb. Haemost. **18**(6), 14691472 (2020)
41. Finelli, C., Parisi, S.: The clinical impact of COVID-19 epidemic in the hematologic setting. Adv. Biol. Regul. **77**, 100742 (2020)
42. Long, H., et al.: D-dimer and prothrombin time are the significant indicators of severe COVID-19 and poor prognosis. Biomed. Res. Int. **2020**, 6159720 (2020)
43. Wang, L., He, W.B., Yu, X.M., Hu, D.L., Jiang, H.: Prolonged prothrombin time at admission predicts poor clinical outcome in COVID-19 patients. World J. Clin. Cases. **8**(19), 43704379 (2020)
44. Iba, T., Levy, J.H., Connors, J.M., Warkentin, T.E., Thachil, J., Levi, M.: The unique characteristics of COVID-19 coagulopathy. Crit. Care. **24**(1), 360 (2020)
45. Bounds EJ, Kok SJ. D Dimer. [16.8.2020]. In: StatPearls [Internet]. Treasure Island (FL): StatPearls Publishing; 2020.
46. Kermali, M., Khalsa, R.K., Pillai, K., et al.: The role of biomarkers in diagnosis of COVID-19 - a systematic review. Life Sci. **1**(254), 117788 (2020)
47. Vidali, S., Morosetti, D., Cossu, E., et al.: D-dimer as an indicator of prognosis in SARS-CoV-2 infection: a systematic review. ERJ Open Res. **6**(2), 00260–02020 (2020)
48. Sherwani, S.I., Khan, H.A., Ekhzaimy, A., Masood, A., Sakharkar, M.K.: Significance of HbA1c test in diagnosis and prognosis of diabetic patients. Biomark Insights. **11**, 95–104 (2016)
49. Wang, F., Yang, Y., Dong, K., et al.: Clinical characteristics of 28 patients with diabetes and Covid-19 in Wuhan China. Endocr. Pract. **26**(6), 668–674 (2020)

50. Merzon, E., Green, I., Shpigelman, M., Vinker, S., Raz, I., Golan-Cohen, A., Eldor, R.: Haemoglobin A1c is a predictor of COVID-19 severity in patients with diabetes. Diabetes Metab. Res. Rev. e3398 (2020)
51. Gu, X., Li, X., An, X., et al.: Elevated serum aspartate aminotransferase level identifies patients with coronavirus disease 2019 and predicts the length of hospital stay. J. Clin. Lab. Anal. **34**, e2339 (2020)
52. Sharma, A., Jaiswal, P., Kerakhan, Y., et al.: Liver disease and outcomes among COVID-19 hospitalized patients - a systematic review and meta-analysis. Ann Hepatol. **16**, S1665–2681(20)30188–5 (2020)
53. Saini, R.K., Saini, N., Ram, S., Soni, S.L., et al.: COVID-19 associated variations in liver function parameters: a retrospective study. Postgrad. Med. J. 2020–138930 (2020)
54. Shen, J., Zhang, J., Wen, J., Ming, Q., et al.: Correlation of serum alanine aminotransferase and aspartate aminotransferase with coronary heart disease. Int. J. Clin. Exp. Med. **8**(3), 4399–404 (2015)
55. Li, X., Zhang, Z.C., Zhang, P.L.: Severe COVID-19 patients with liver injury: a seven-case series. Eur. Rev. Med. Pharmacol. Sci. **24**(14), 7855–7860 (2020)
56. Ye, L., Chen, B., Wang, Y., et al.: Prognostic value of liver biochemical parameters for COVID-19 mortality. Ann. Hepatol. **3**, S1665–2681(20)30204–0 (2020)
57. Chen, L.Y., Chu, H.K., Bai, T., et al.: Liver damage at admission is an independent prognostic factor for COVID-19. J. Dig. Dis. **21**(9), 512–518 (2020)
58. Klein, R., Nagy, O., Tóthová, C., Chovanová, F.: Clinical and Diagnostic Significance of Lactate Dehydrogenase and Its Isoenzymes in Animals. Vet. Med. Int. **2020**(15), 5346483 (2020)
59. Zhou, Y., Ding, N., Yang, G., et al.: Serum lactate dehydrogenase level may predict acute respiratory distress syndrome of patients with fever infected by SARS-CoV-2. Ann. Transl. Med. **8**(17), 1118 (2020)
60. Serin, I., Sari, N.D., Dogu, M.H., et al.: A new parameter in COVID-19 pandemic: initial lactate dehydrogenase (LDH)/Lymphocyte ratio for diagnosis and mortality. J. Infect. Public Health. **13**(11), 1664–1670 (2020)
61. Salinas, M., Blasco, Á., Santo-Quiles, et al.: Laboratory parameters in patients with COVID19 on first emergency admission is different in non-survivors: albumin and lactate dehydrogenase as risk factors. J. Clin. Pathol. **6**, jclinpath-2020–206865 (2020)
62. Ouyang, L., Gong, Y., Zhu, Y., et al.: Association of acute kidney injury with the severity and mortality of SARS-CoV-2 infection: a meta-analysis. Am. J. Emerg. Med. **2**, S07356757(20)30783-X (2020)
63. Ma, H., Zeng, W., He, H., et al.: Serum IgA, IgM, and IgG responses in COVID-19. Cell. Mol. Immunol. **17**, 773–775 (2020)
64. Zhong, L., et al.: Detection of serum IgM and IgG for COVID-19 diagnosis. Sci. China Life Sci. **63**(5), 777–780 (2020). https://doi.org/10.1007/s11427-020-1688-9
65. Marklund, E., et al.: Serum-IgG responses to SARS-CoV-2 after mild and severe COVID-19 infection and analysis of IgG non-responders. PLoS ONE **15**(10), e0241104 (2020)
66. Inciardi, R.M., et al.: Characteristics and outcomes of patients hospitalized for COVID-19 and cardiac disease in Northern Italy. Eur. Heart J. **41**(19), 1821–1829 (2020)
67. Klok, F.A., Kruip, M.J.H.A., van der Meer, N.J.M., Arbous, M.S., Gommers, D.A.M.P.J., Kant, K.M., Kaptein, F.H.J., van Paassen, J., Stals, M.A.M., Huisman, M.V., Endeman, H.: Incidence of thrombotic complications in critically ill ICU patients with COVID-19. Thromb Res. **191**, 145–147 (2020)
68. Gonzalez-Quintela, A., Alende, R., Gude, F., Campos, J., Rey, J., Meijide, L.M., FernandezMerino, C., Vidal, C.: Serum levels of immunoglobulins (IgG, IgA, IgM) in a general adult population and their relationship with alcohol consumption, smoking and common metabolic abnormalities. Clin. Exp. Immunol. **151**(1), 42–50 (2008)

69. Kinra, P., Dutta, V.: Serum TNF alpha levels: a prognostic marker for assessment of severity of malaria. Trop. Biomed. **30**(4), 645–653 (2013)
70. Blot, M., Bour, J.B., Quenot, J.P., Bourredjem, A., Nguyen, M., Guy, J., Monier, S., Georges, M., Large, A., Dargent, A., Guilhem, A., Mouries-Martin, S., Barben, J., Bouhemad, B., Charles, P.E., Chavanet, P., Binquet, C., Piroth, L.: LYMPHONIE study group. The dysregulated innate immune response in severe COVID-19 pneumonia that could drive poorer outcome. J. Transl. Med. **18**(1), 457 (2020)
71. O'Neill, C.M., et al.: Circulating levels of IL-1B+IL-6 cause ER stress and dysfunction in islets from prediabetic male mice. Endocrinology **154**(9), 3077–3088 (2013)
72. Kleiner, G., Marcuzzi, A., Zanin, V., Monasta, L., Zauli, G.: Cytokine levels in the serum of healthy subjects. Mediators Inflamm. **2013**, 434010 (2013)

Investigations of Degradation of Virus Spread by Physical Techniques

Baki Karaböce[1](✉), Ahmet Baş[2], Ahsen Aydın Böyük[3], Mihli Nur Bülün[3], and Kadir Ak[3]

[1] TÜBİTAK UME, Istanbul, Turkey
baki.karaboce@tubitak.gov.tr
[2] Internal Medical Sciences, İstanbul University Cerrahpaşa, Istanbul, Turkey
[3] Biomedical Faculty, Kocaeli University, İzmit, Turkey

Abstract. The COVID-19 pandemic, which emerged in 2019, spread rapidly, exposing healthcare professionals and the healthcare system in particular and the entire society in general, to face a deadly virus. Learning how the virus spread will enable the development of physical methods that can prevent both COVID19 and future pandemics. Pandemics emerges when viruses are able to survive outside the host for contamination to occur. Viruses are known to spread extensively through aerosols and droplets because such environments provide conditions in which the virus can survive for a long time. Temperature and humidity parameters of aerosols and droplets also affect the survival of viruses. This article aims to analyze aspects related to destroying environments where viruses can live using infrared sources.

Keywords: Virus spread · Isotonic solution · Degradation of virus · Infrared · Temperature measurements

1 Introduction

Viruses cause about 60% of human infections and 1.5 million deaths occur worldwide each year [1]. Outbreaks such as the Severe Acute Respiratory Syndrome Coronavirus (SARS-CoV), which emerged in 2002–2003, followed by SARS CoV-2, and COVID-19, which appeared in December 2019, occurred during the winter months. This is an indication of the increase in various respiratory tract infections in the winter season [2–5]. The most important mechanism explaining this effect is cold and humid air. This mechanism is the effect of temperature and humidity on respiratory virus stability and transmission rates, and the damage caused by cold and dry air to respiratory system mucosa in winter [6–10]. Therefore, respiratory viral infections are predominantly transmitted indoors (crowded indoor spaces such as homes, schools, social facilities, work offices and hospitals) and facilitate disease morbidity and mortality [11]. Influenza, Corona and Respiratory Syncytial Virus (RSV) are known as three main types of viruses that cause epidemics. Such respiratory viral infections can theoretically be transmitted by direct/indirect contact, by droplet route (by coughing, sneezing or speaking) at a short

A. Badnjevic and L. Gurbeta Pokvić (Eds.): CMBEBIH 2021, IFMBE Proceedings 84, pp. 847–857, 2021.
https://doi.org/10.1007/978-3-030-73909-6_96

distance and by aerosol and secretion (or fomites) at a long distance [12–17]. The residence time of these viruses on the contacted surface also varies depending on the values of the environment such as temperature and humidity, and whether the surface is porous or non-porous or surface properties (plastic, metallic, paper, wooden, etc.) [18, 19]. The longer the virus remains on the surface, the longer the contamination time and the greater the number of contaminated people, resulting in an accelerated spread [20, 21].

Respiratory transmitted viral diseases (SARS, MERS, and COVID-19) undergo constantly mutation [22, 23]. Even if preventive vaccines or drugs are developed for the virus, when the virus is mutated, drug resistance may develop or the vaccine may lose its effectiveness, and drug studies may be required. Even worse, more-virulent viruses may arise in the future that is difficult to vaccinate and treat [24]. For this reason, due to the slow development of vaccines and drugs in the prevention of viral diseases, physical methods targeting the transmission routes of all viruses are needed in order to prevent or at least reduce the virus transmission causing major epidemics worldwide [25, 26]. The most known of these methods is Ultraviolet (UV) radiation, which is used to inactivate viruses and bacteria. UV have been used in indoor air and surface sterilization for many years with their mechanism of action on virus DNA. For example, UVC radiation (100 to 280 nm) is highly effective in disinfecting air and surfaces for viral sterilization [27, 28]. But ultraviolet radiation (UV) cannot be used directly in public areas for sterilization due to reasons such as the detrimental mechanism of action on DNA is caused by skin cancer, squamous cell carcinoma, basal cell carcinoma on human, causing vision loss by damaging the retina or causing cataracts, causing hair loss by damaging the keratin fiber structure in the hair, damaging the human immune system, etc. [29–34]. For all these reasons, there has been a need to develop physical systems that are harmless to human health in order to reduce virus spread. This paper presents the studies about disinfection with infrared radiation (IR) targeting droplets (>10 μm diameter) and aerosols (<5 μm diameter), which are the carriers of viruses transmitted by air in public areas.

1.1 Theoretical Background

Among the environmentally friendly technologies, IR technology applications have attracted considerable attention [35]. Such as chemical-free technology, rapid heating, uniform surface heating, and energy efficiency, IR heating has numerous advantages [35–44]. If the absolute temperature of an object is above zero (0 °K or −273 °C), it emits thermal radiation in the form of electromagnetic waves [36]. Visible light (400 nm–700 nm) and IR radiation (700 nm–1 mm) wavelengths fall within this range of thermal radiation. Infrared radiation are trapped in the electromagnetic spectrum between the visible region and microwaves and they have a warming effect on molecules [45, 46]. Depending on the spectrum band, IR radiation is divided into three different part:

- 0.75 μm to 1.4 μm is near IR radiation (NIR),
- 3 μm to 8 μm is medium IR radiation (MIR) and
- 15 μm to 1000 μm is far IR radiation (FIR)

The black body is an ideal object, absorbing all incoming radiation and radiating maximum. The radiant energy emitted by the heat source is proportional to the emission and surface temperature of the material. According to Stefan-Boltzmann law, the emitted energy flux q″ (W/m^2) is a multiple of the temperature T (K) of the emitter [47].

$$q'' = \sigma T \tag{1}$$

The Stefan-Boltzmann constant are;

$$\sigma = 5.669\,x^{-8}\mathrm{W/m\,K} \tag{2}$$

The amount of energy transmitted, reflected and absorbed varies depending on the wavelength of the IR energy, surface conditions and material properties of the materials. If the temperature of the object increases, the spectrum shifts to short wavelength and the total energy increases. This results in the Wien displacement curve. The peak of a wave is formulated by this Wien's law: [48].

$$\lambda_{peak} = C/\mathrm{T} \tag{3}$$

$$C = 2897.8\ \mu\mathrm{mK} \tag{4}$$

Generally, air and gases absorb very little IR energy, so IR does not heat the air and highly efficient heat transfer is achieved without contact between the heated materials.

1.2 Materials and Methods

As it was stated that viruses cannot survive in high temperatures and without host environment, conducting the virus inactivation experiments in isotonic solution that best simulates human fluid secretion is first step before further investigations. In the experiments, it was aimed to examine the evaporation rate of isotonic water sprayed on different surfaces under laboratory conditions without IR source and to have information about the evaporation rate. The isotonic water used is OSEL 0.9%. Plexiglass, paper, wooden and metal surfaces with a surface area of (5 × 5) *cm* were selected. The purpose of using these surfaces is that they are porous and non-porous surfaces that are constantly in contact in daily life and the survival time of viruses on different surfaces is different. Isotonic water was sprayed through TLC (Thin Layer Chromatography) spray kit to ensure homogeneous distribution on the surfaces. In order to observe the effect of different amounts of liquid, 10 mg, 30 mg, 50 mg and 100 mg liquids were sprayed onto the surfaces.

In order to observe the evaporation rate depending on the time, isotonic solution sprayed materials were placed on the sensitive balance (Sartorius-ME235S Semi-Micro Balance) and the weight change was recorded. During the experiment, the surface temperatures of the materials were measured with a thermal imager (Optris PI). The obtained temperature and weight data were observed at 30-s intervals and recorded in the table created in Excel. Time dependent temperature and weight change graphs were created by using this data.

The temperature distribution of the surfaces of IR sources with different wavelengths was determined by using a thermal camera. Kuas ISP Basic450 has a power of 450 W/h,

FIR and dimensions of 600 mm × 600 mm × 25 mm. Kuas ISP Basic200 has a power of 200 W/h, FIR and dimensions of 300 mm × 300 mm × 25 mm. Kuas Hybridboard Basic 600 has a power of 600 W/h, FIR and dimensions of 600 mm × 600 mm × 40 mm. Kuas I-PIPE has a power of 1000 W/h, NIR and dimensions of 113.5 cm × 20 cm. In the experiment, the surfaces of the sources were divided into areas of 10 cm × 10 cm and the temperature values of these areas were measured.

An experiment was carried out in order to observe the temperature distribution of IR sources on black absorbent plates according to the distance. Three identical black plates with a surface area of 60 cm × 60 cm were heated at different distances at 50 cm, 100 cm, 150 cm, 200 cm and 250 cm with different IR as seen Fig. 1.

Fig. 1. Surface temperature distribution measurements of IR surfaces

Droplet evaporation experiments were carried out under the effect of IR sources. The distance between the IR sources and the surfaces with droplets are 0.5 m, 1 m and 1.3 m respectively. Different amounts of isotonic water solution (10 mg, 30 mg, 50 mg and 100 mg) were used to spray different materials. The temperature and weight change data were taken at 30 s intervals were recorded.

A cabin was designed and produced for determination of humidity dependence of droplet evaporation by IR. Test Chamber from Weiss Technik was also used in the measurements. Humidity and temperature values were adjusted between 30% RH and 90% RD and 20 °C and 50 °C. Results show the effect of different humidity values on the virus degradation.

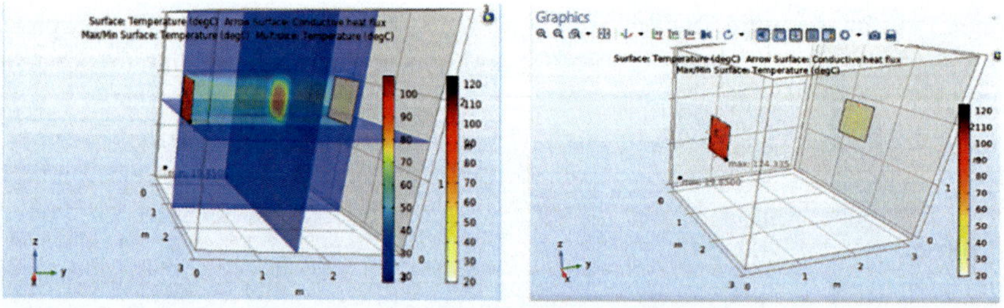

Fig. 2. Comsol multiphysics simulation

A simulation was performed with a heat transfer module in the Comsol Multiphysics program in order to simulate the surface temperature of the plates heated by IR radiation as seen in Fig. 2. The properties of the materials used in the experiment, the IR source and the environment (heat capacity, density, thermal conductivity, relative permeability, heat emission on the surface, etc.) were defined on the materials. The simulation and the experiments were compared and a consistency was observed.

2 Results

As can be seen from the graphs in Fig. 3, the isotonic solution was completely removed from the surface in shorter times by applying IR radiation. NIR reach the 58 °C on the surface and remove the isotonic droplet within minutes. While the isotonic droplet in an amount of 100 mg, stays 80 min. on the surfaces of metal, paper, Plexiglas and wood, application of NIR, MIR and FIR decrease the time of stay of the solution on the

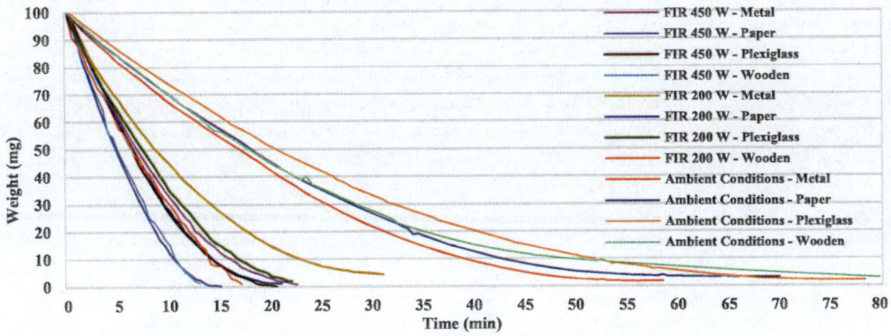

(a)

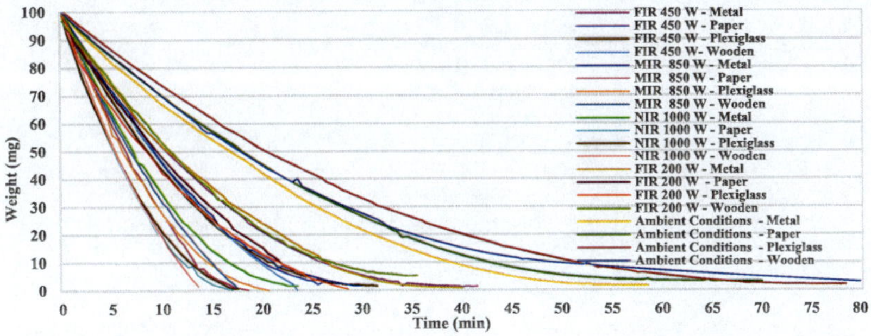

(b)

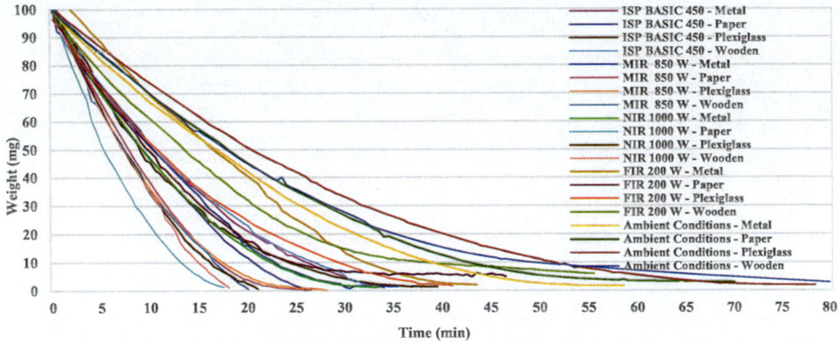

(c)

Fig. 3. Time-dependent decrease of 100 mg isotonic droplet under the effect of IR sources and ambient conditions on surfaces at (a) 0.5 m, (b) 1.0 m and (c) 1.3 m.

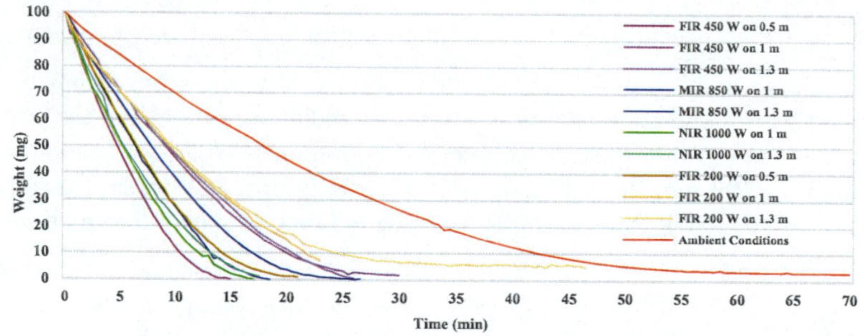

(a)

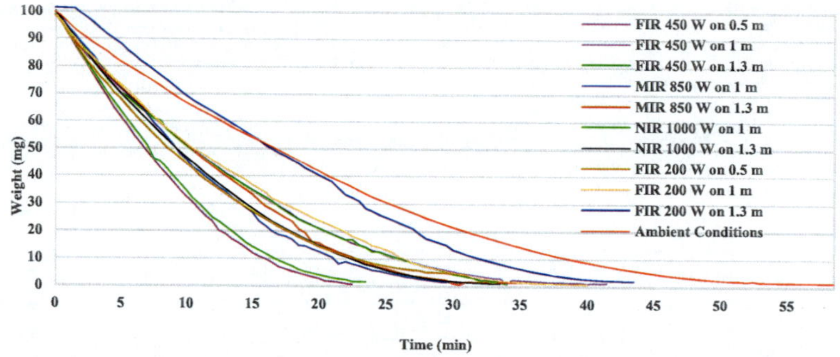

(b)

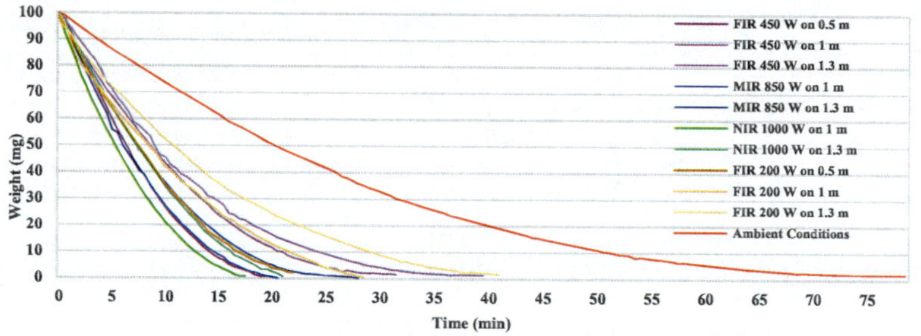

(c)

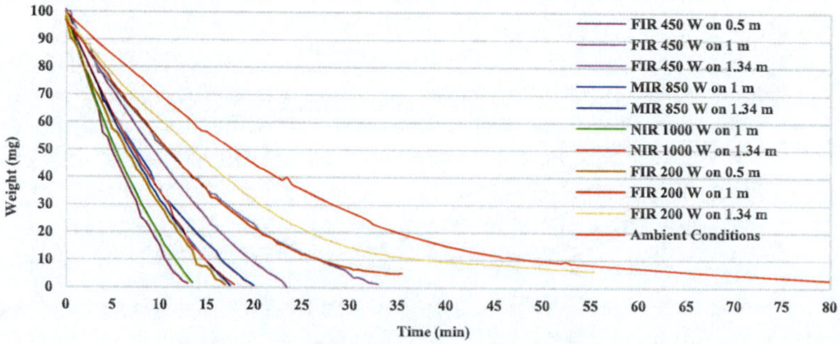

(d)

Fig. 4. Time-dependent decrease of 100 mg isotonic droplet under the effect of IR sources and ambient conditions on surfaces (a) paper (b) metal (c) pleksiglas and (d) wooden.

surfaces as can be seen in in Fig. 3. NIR, MIR and FIR (very healthy for body) reaches the temperature 58 °C, 47 °C and 38 °C on the surfaces correspondingly at 0,5 m.

In addition, the disappearance time of isotonic droplet on paper and wood surfaces was shorter than metal and plexiglass surfaces, and IR sources also affected the disappearance of liquids on these surfaces more as can be seen in Fig. 4. The reason for this is that the surfaces are porous or non-porous, as we mentioned in the previous sections. During the experiments, ambient temperature and humidity were measured and recorded continuously around 23 °C and 50% RH and as a result, no significant temperature increase was observed in ambient conditions. The IR effect can be expected to have a similar effect on virus-added isotonic water droplets as the virus can live without host environment like solution, droplet etc. As a result, decrease in the number and the the quality of droplets that carry viruses can lead to decrease of the virus spread and transmission. It can be noted that I-PIPE and MIR source was not applied at 0.5 m, which would cause high temperature increase and damage the measuring instruments.

Different amount of isotonic droplets change the evaporation time as can be seen in Fig. 5. Reduction of different amounts of isotonic droplets in Environmental chamber at 80% RH can be seen in Fig. 6.

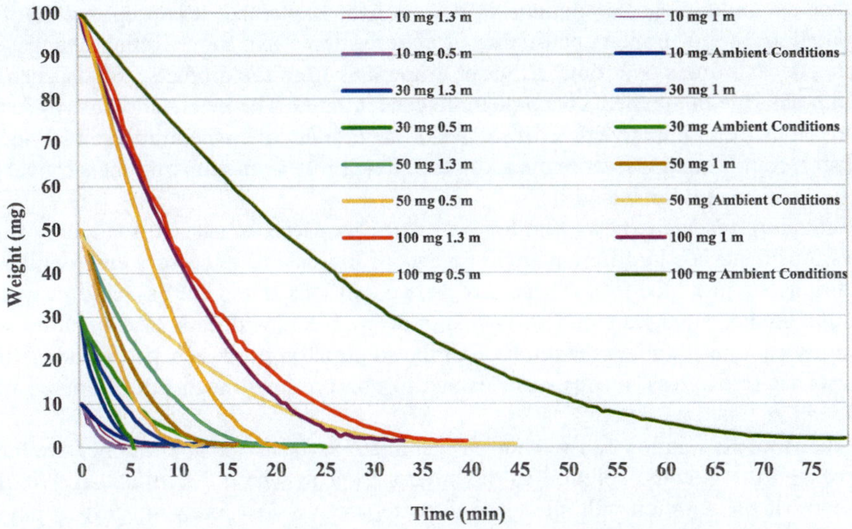

Fig. 5. Decrease of different amount of isotonic droplets under the effect of IR source.

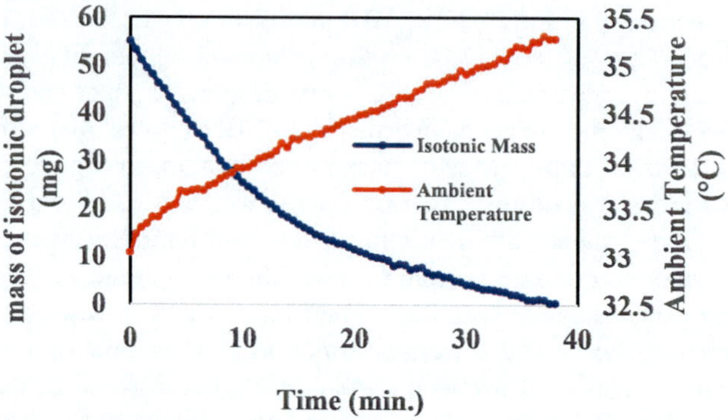

Fig. 6. Reduction of different amount of isotonic droplets in environmental chamber

3 Discussion

The transmission of infectious diseases requires that pathogens remain infectious in the environment, outside the host. Time-dependent decay of viruses is known; however, the mechanism underlying the relationship has not been explained. It is proposed that the evaporation rate of droplets consisting of culture media and the resulting changes in solute concentrations over time; as water evaporates from the droplets, solutes such as sodium chloride in the media become more concentrated. The inactivation of viruses is governed by the cumulative dose of solutes or the product of concentration and time, as in disinfection kinetics. Evaporation kinetics play a role in modulating the survival of microorganisms in droplets [49].

Literature research shows that human enteric viruses can survive for a long time on various materials in different environments as long as the necessary environmental conditions are provided [50]. These favorable conditions stated are basically temperature and humidity parameters. Our goal is to determine how different temperature and humidity parameters affect the duration of the solution on porous (paper, wooden) and non-porous (plexiglass, metal) surfaces and to observe how much the IR source will differentiate this time.

The isotonic solution can remain on inanimate surfaces for hours. Since we have determined the isotonic solution as the environment in which the virus can live, the presence of the solution will allow the virus to survive. Our study suggests a physical disinfection system to reduce the risk of contamination to patients from inanimate surfaces. The IR heating system placed in the environment, aiming to directly heat the objects without affecting the ambient temperature, will ensure that this solution is quickly destroyed and the risk of contamination will be minimized.

4 Conclusion

Respiratory viral infections are predominantly transmitted indoors (crowded indoor spaces such as homes, schools, social facilities, work offices and hospitals) and facilitate disease morbidity and mortality. Influenza, Corona and Respiratory Syncytial Virus

(RSV) are known as three main types of viruses that cause epidemics. Learning how these types of viruses spread is one of the most important issues because of the world-wide epidemic. Such respiratory viral infections can theoretically be transmitted by direct/indirect contact, by droplet route (by coughing, sneezing or speaking) at a short distance and by aerosol and secretion (or fomites) at a long distance. Air flow in the measurement environment is another parameter that must be taken into account for deeper research.

Non-enveloped and enveloped phages like MS2 and Phi6 were also explored as viral aerosol models [51]. MS2 is a single-stranded RNA phage of the Leviviridae family and Phi6 is a segmented double-stranded RNA phage of the Cystoviridae family. Studies are going on the virus models. MS2 and Phi6 virus models are used to characterize the effect of IR on virus spread. IR sources with different wavelengths (and temperatures) show different effects on virus spread depending on the application times and temperatures.

Applications of IR panels in the places where virus availability and activation are high look very attractive tool. Moreover the healthy uses of FIR is well known. But it must be characterized by means of metrological approach. Effect of IR sources must be further investigated since they may not only causing the temperature increase and extermination but may also weaken the viruses.

References

1. Boone, S.A., Gerba, C.P.: Significance of fomites in the spread of respiratory and enteric viral disease. Appl. Environ. Microbiol. **73**(6), 1687–1696 (2007)
2. Razzini, K., et al.: SARS-CoV-2 RNA detection in the air and on surfaces in the COVID-19 ward of a hospital in Milan, Italy. Sci. Total Environ. **742**, 140540 (2020)
3. Dowell, S.F., et al.: Severe acute respiratory syndrome coronavirus on hospital surfaces. Clin. Infect. Dis. **39**(5), 652–657 (2004)
4. Fisman, D.N.: Seasonality of infectious diseases. In: Annual Review of Public Health (2007)
5. Altizer, S., Dobson, A., Hosseini, P., Hudson, P., Pascual, M., Rohani, P.: Seasonality and the dynamics of infectious diseases. Ecol. Lett. (2006)
6. Lowen, A.C., Mubareka, S., Steel, J., Palese, P.: Influenza virus transmission is dependent on relative humidity and temperature. PLoS Pathog. (2007)
7. Chan, K.H., Peiris, J.S.M., Lam, S.Y., Poon, L.L.M., Yuen, K.Y., Seto, W.H.: The effects of temperature and relative humidity on the viability of the SARS coronavirus. Adv. Virol. (2011)
8. Casanova, L.M., Jeon, S., Rutala, W.A., Weber, D.J., Sobsey, M.D.: Effects of air temperature and relative humidity on coronavirus survival on surfaces. Appl. Environ. Microbiol. (2010)
9. Ma, Y., et al.: Effects of temperature variation and humidity on the death of COVID-19 in Wuhan, China. Sci. Total Environ. (2020)
10. Dowell, S.F., Shang Ho, M.: Seasonality of infectious diseases and severe acute respiratory syndrome - What we don't know can hurt us. Lancet Infectious Diseases (2004)
11. Abad, F.X., Pinto, R.M., Bosch, A.: Survival of enteric viruses on environmental fomites. Appl. Environ. Microbiol. **60**(10), 3704–3710 (1994)
12. Otter, J.A., Donskey, C., Yezli, S., Douthwaite, S., Goldenberg, S.D., Weber, D.J.: Transmission of SARS and MERS coronaviruses and influenza virus in healthcare settings: the possible role of dry surface contamination. J. Hospital Infect. (2016)
13. Ge, Z., Yang, L., Xia, J., Fu, X., Zhang, Y.: Possible aerosol transmission of COVID-19 and special precautions in dentistry. J. Zhejiang Univ. Sci. B (2020)

14. World Health Organization: Modes of transmission of virus causing COVID-19: implications for IPC precaution recommendations. Geneva World Heal. Organ. (2020)
15. Cai, J., Sun, W., Huang, J., Gamber, M., Wu, J., He, G.: Indirect virus transmission in cluster of COVID-19 cases, Wenzhou, China, 2020. Emerg. Infect. Dis. (2020)
16. Shereen, M.A., Khan, S., Kazmi, A., Bashir, N., Siddique, R.: COVID-19 infection: Origin, transmission, and characteristics of human coronaviruses. J. Adv. Res. (2020)
17. Morawska, L., Cao, J.: Airborne transmission of SARS-CoV-2: the world should face the reality. In: Environment International (2020)
18. Kramer, A., Schwebke, I., Kampf, G.: How long do nosocomial pathogens persist on inanimate surfaces? A systematic review. BMC Infect. Dis. **6**, 1–8 (2006)
19. Davis, R.E., Dougherty, E., McArthur, C., Huang, Q.S., Baker, M.G.: Cold, dry air is associated with influenza and pneumonia mortality in Auckland, New Zealand. Influenza Other Respi. Viruses (2016)
20. van Doremalen, N., et al.: Aerosol and surface stability of SARS-CoV-2 as compared with SARS-CoV-1. N. Engl. J. Med. (2020)
21. Tellier, R.: Review of aerosol transmission of influenza A virus. Emerg. Infect. Dis. **12**(11), 1657–1662 (2006)
22. Benvenuto, D., Giovanetti, M., Ciccozzi, A., Spoto, S., Angeletti, S., Ciccozzi, M.: The 2019-new coronavirus epidemic: Evidence for virus evolution. J. Med. Virol. (2020)
23. Sanjuán, R., Domingo-Calap, P.: Mechanisms of viral mutation. Cell. Mol. Life Sci. **73**(23), 4433–4448 (2016). https://doi.org/10.1007/s00018-016-2299-6
24. Clavel, F., Hance, A.J.: HIV drug resistance. N. Engl. J. Med. (2004)
25. Darnell, M.E.R., Subbarao, K., Feinstone, S.M., Taylor, D.R.: Inactivation of the coronavirus that induces severe acute respiratory syndrome, SARS-CoV. J. Virol. Methods (2004)
26. Abhimanyu, Coussens, A.K.: The role of UV radiation and Vitamin D in the seasonality and outcomes of infectious disease. Photochem. Photobiol. Sci. (2017)
27. Inagaki, H., Saito, A., Sugiyama, H., Okabayashi, T., Fujimoto, S.: Rapid inactivation of SARS-CoV-2 with Deep-UV LED irradiation. Emerg. Microbes Infect. (2020)
28. Houser, K.W.: Ten Facts about UV radiation and COVID-19. LEUKOS J. Illuminat. Eng. Soc. North Am. (2020)
29. Behar-Cohen, F., et al.: Ultraviolet damage to the eye revisited: Eye-sun protection factor (E-SPF®), a new ultraviolet protection label for eyewear. Clin. Ophthalmol. (2014)
30. Kim, Y., He, Y.Y.: Ultraviolet radiation-induced non-melanoma skin cancer: regulation of DNA damage repair and inflammation. Genes Diseases (2014)
31. Šebetic, K., Masnec, I.S., Èavka, V., Biljan, D., Krolo, I.: UV damage of the Hair. Coll. Antropol. (2008)
32. Levy, S.B.: UV filters. In: Handbook of Cosmetic Science and Technology, 3rd edn. (2009)
33. Ravanat, J.L., Douki, T.: UV and ionizing radiations induced DNA damage, differences and similarities. Radiat. Phys. Chem. (2016)
34. Duthie, M.S., Kimber, I., Norval, M.: The effects of ultraviolet radiation on the human immune system. Br. J. Dermatol. (1999)
35. Kastberger, G., Stachl, R.: Infrared imaging technology and biological applications. In: Behavior Research Methods, Instruments, and Computers (2003)
36. Z. Pan, G. Atungulu, and X. Li, "Infrared heating," *Resour. Eng. Technol. Sustain. World*, 2013.
37. Vaidyanathan, J.S., Krishnamurthy, K.: Infrared Heating for Decontamination. In: *I*nnovative Food Processing Technologies (2020)
38. Tanaka, F., Verboven, P., Scheerlinck, N., Morita, K., Iwasaki, K., Nicolaï, B.: Investigation of far infrared radiation heating as an alternative technique for surface decontamination of strawberry. J. Food Eng. (2007)

39. Ramaswamy, R., Krishnamurthy, K., Jun, S.: Microbial decontamination of food by infrared (IR) heating. In: Microbial Decontamination in the Food Industry: Novel Methods and Applications (2012)
40. Staack, N., Ahrné, L., Borch, E., Knorr, D.: Effect of infrared heating on quality and microbial decontamination in paprika powder. J. Food Eng. (2008)
41. Kaczmarek, M., Nowakowski, A.: Active IR-thermal imaging in medicine. J. Nondestr. Eval. **35**(1), 1–16 (2016). https://doi.org/10.1007/s10921-016-0335-y
42. Demirci, A., Ngadi, M.O.: Microbial decontamination in the food industry: Novel methods and applications (2012)
43. Wilson, S.A., Okeyo, A.A., Olatunde, G.A., Atungulu, G.G.: Radiant heat treatments for corn drying and decontamination. J. Food Process. Preserv. 2017
44. Bunaciu, A.A., Fleschin, S., Aboul-Enein, H.Y.: Cancer diagnosis by ft-Ir Spectrophotometry. Rev. Roum. Chim. (2015)
45. Buijs, H.: Infrared spectroscopy. In: Springer Handbooks (2006)
46. Sankaran, S., Ehsani, R.: Introduction to the electromagnetic spectrum. In: Imaging with Electromagnetic Spectrum: Applications in Food and Agriculture (2014)
47. Howell, J.R., Mengüç, M.P., Siegel, R.: Thermal radiation heat transfer, 6th edn. (2015)
48. Lahiri, B.B., Bagavathiappan, S., Jayakumar, T., Philip, J.: Medical applications of infrared thermography: a review. Infrared Phys. Technol. (2012)
49. Lin, L., Marr, L.C.: Humidity-dependent decay of viruses, but not bacteria, in aerosols and droplets follows disinfection kinetics. Environ. Sci. Technol. **54**(2), 1024–1032 (2020)
50. Sabino, C.P., et al.: Light-based technologies for management of COVID-19 pandemic crisis. J. Photochem. Photobiol. B Biol. **212** (2020)
51. Verreault, D., Moineau, S., Duchaine, C.: Methods for sampling of airborne viruses. Microbiol. Mol. Biol. Rev. **72**, 413–444 (2008). https://doi.org/10.1128/MMBR.00002-08

A Review of Novel Methods for Diagnosing COVID-19

Tarik Abdukić, Tamer Bego, Neven Meseldžić, Matej Planinić, Evelina Pulo(✉), and Faruk Šehić

Faculty of Pharmacy Sarajevo, University of Sarajevo, Sarajevo, Bosnia and Herzegovina
evelinapulo@ffsa.unsa.ba

Abstract. SARS-CoV-2 is part of a diverse family of viruses, which mainly affect the respiratory system, central nervous, hepatic, and gastrointestinal systems. The symptoms are fever, cough, shortness of breath, muscle pain, confusion, headache, sore throat, rhinorrhea, chest pain, diarrhea, and nausea and vomiting. The COVID-19 upheaval has certainly become the most significant issue around the world, and its results have left nobody immaculate.

Rapid discoveries, especially in the terms of diagnosing COVID-19 can assist with controlling the infection spread just as additional flare-ups of this infectious diseases. Real-Time Polymerase Chain Reaction (RT-PCR) is a golden standard when it comes to diagnosing COVID-19 infection. Novel methods have been developed to overcome time-consuming problems including RT-LAMP and AI-LAMP.

On the other hand, various antigen tests are being currently developed. One of the novel ones is the ultrasensitive antigen-based test for SARS-CoV-2 spike protein detection. Furthermore, the Panbio test detects nucleocapsid protein of SARS-CoV-2 in nasopharyngeal samples. Next, Field-Effect Transistor (FET) based biosensors are very attractive due to their high sensitivity and instant measurements with small amounts of needed analytes.

Moreover, the peak of rapid testing is a device that can detect COVID-19 in exhaled breath, making it the fastest and easiest test to run. Up to this day, "Breathonix" and "Scentech Medical" companies alongside professor Niann Sun developed breath analyzers that are still in the testing phase.

In this paper, we review novel methods for diagnosing COVID-19, including the RT-PCR. The advantages, as well as disadvantages of the mentioned methods, are included in this paper. Data on COVID-19 are evolving rapidly. Considering science and technology's swift development, some of the specifics in this review could be changed as more studies become available.

Keywords: COVID-19 · Diagnosing · Novel methods · RT-PCR- · RT-LAMP · AI-LAMP · Spike protein detection · Panbio · FET · Breathonix · Scentech Medica

1 Introduction

The swift spread of COVID-19 caused by severe acute respiratory syndrome Coronavirus (SARS-CoV-2), first lead to pronouncing it as a general well being crisis of

A. Badnjevic and L. Gurbeta Pokvić (Eds.): CMBEBIH 2021, IFMBE Proceedings 84, pp. 858–866, 2021.
https://doi.org/10.1007/978-3-030-73909-6_97

worldwide concern (PHEIC) on 30 January 2020., and after that on 11 March 2020. the World Health Organization (WHO) declared COVID-19 a pandemic [1]. Transmission of Covid-19 happens from individual-to-individual, either through direct transmission or contact transmission [2, 3].

This virus is part of a diverse family of viruses, consisting of four viral genera (alpha-, beta-, gamma-, and delta-coronaviruses) [4]. They infect different body systems mainly affecting the respiratory system, central nervous, hepatic, and gastrointestinal systems [5]. The clinical features of COVID-19, as determined from 99 patients in Wuhan, China are fever, cough, shortness of breath, muscle pain, confusion, headache, sore throat, rhinorrhea, chest pain, diarrhea, and nausea and vomiting [6]. Currently, the number of patients with COVID-19 is rapidly rising to nearly 57 million leaving more than 1,4 million dead [7]. It is certain that the COVID-19 upheaval has become the most significant issue around the world, and its results have left nobody immaculate. Rapid discoveries, especially in the terms of diagnosing COVID-19 can assist with controlling the infection spread just as additional flare-ups of this infectious diseases.

2 Real-Time Polymerase Chain Reaction (RT-PCR)

The method used to test patients for SARS-CoV-2 is based on reverse transcription quantitative Real-Time Polymerase Chain Reaction (RT-PCR), which is performed after RNA extraction [8, 9] Real-time RT-PCR detects amplified SARS-CoV-2 genome in upper and lower respiratory specimens. This method is considered as the 'gold standard' due to its benefits as a specific, simple qualitative assay that has adequate sensitivity for diagnosing early infection [10–12]. In principal the viral RNA is converted to cDNA (complementary DNA) using reverse transcriptase (RNA dependent DNA polymerase enzyme). Then, cDNA is amplified in a polymerase chain reaction (PCR) in three steps: 1. denaturation of cDNA at 95 °C, 2. annealing of the primers and probe to the respective denatured cDNA strands at 60 °C, and last one is extension or synthesis of RNA copies by DNA polymerase at 72 °C. The TaqMan probe is used for the quantification of RNA copies by producing fluorescence signal during amplification cycles [13]. Commercially available RT-PCR diagnostics for COVID-19 are either one-step which utilizes only one tube and buffers for RT and PCR steps, and in two-step kit assay RT and PCR are performed separately in different tubes [14].

In general, RT-PCR offers extremely sensitive, specific, and quantitative detection of the SARS-CoV-2 RNA. However, this method is complex, expensive, slow to deliver and its application is intended for qualified clinical laboratory personnel and medium- or high-complexity laboratories The main problem is that, some studies suggest a high false-negative rate of RT-PCR diagnostics for COVID-19 [6, 15]. Speed, simplicity are some of the problems that have been resolved by using reverse transcription loop-mediated isothermal amplification.

2.1 Rapid Point-of-Care Detection (RT-LAMP) and (AI-LAMP)

RT-LAMP recognizes SARS-CoV-2 in clinical swab tests without the RNA extraction step. The process is exactly like RT-PCR but the LAMP strategy requires just one steady temperature, therefore the presentation gadget can be less expensive and more modest than a warm cycler making it ideal to use in poorer countries without high complexity laboratories. In the event that the format is RNA, the amplification response can be cultivated in one stage by adding a reverse transcriptase. The LAMP-based technique can identify SARS-CoV-2 qualities ORF8 and N straightforwardly from pharyngeal swab tests.

The test shows no cross-reactivity when tried on 20 other respiratory microbes, and it is multiple times quicker and less expensive than a routine PCR test. RT-LAMP is a simple to carry out test, which requires no special equipment and can be performed on locations such as railroad stations, air terminals, or medical clinics. The analysis was conducted through fluorescence. To screen the fluorescence signal advancement, the examples could be investigated at 5 min, 15 min, and 25 min time focus during RT-LAMP with minimum additional time required.

To sum up, while PCR is a golden standard, it has its cons. To run this test, one needs a molecular biological laboratory facility with access to highly specialized instruments along with highly trained workers. It is also time-consuming since it needs RNA extraction and purification before the test is run. RT-LAMP, on the other hand, does not require this step, thus is faster to run. Moreover, this test is conducted at a constant temperature of 65 °C, bypasses DNA denaturation step by heat, and does not require highly specialized equipment to be run, making it perfect for mass testing [16]. Among various molecular diagnostic sciences, LAMP innovation gives a promising way to deal with the fast and dependable identification in asset restricted settings using Artificial Intelligence-Assisted Loop-Mediated Isothermal Amplification (AILAMP). The use of this novel LAMP measure might be especially valuable for recognizing COVID-19 cases with low popular burdens and when testing upper respiratory plot examples (nasal or oral swabs) from patients [17].

3 Novel Antigen-Based Methods of COVID-19 Detection

The ongoing global Coronavirus pandemic has increased the need for less time consuming, more cost-effective, rapid, easy-to-use, and more accurate antigen-based tests for clinical application. These tests should be able to help clinicians to track sick and recovered patients, providing a clearer picture of this pandemic [18]. Although earlier this year, the WHO stated that it did not recommend the use of antigen-based tests for patient care, because of their usual low detection sensitivity (more false-negative results) and not high enough detection specificity (more false-positive results) for accurate diagnosis, their performance, and potential diagnostic utility are highly encouraging [3].

In this section, several novel antigen-based tests for SARS-CoV-2 detection will be presented and reviewed.

3.1 Ultrasensitive Antigen-Based Test for SARS-CoV-2 Spike Protein Detection

This diagnostic test is based on a method that uses a combination of sandwich enzyme-linked immunosorbent assay (ELISA) and thio-nicotinamide adenine dinucleotide (thio-NAD) cycling, for detection of the specific spike surface glycoprotein. The role of the spike protein, as a major transmembrane protein, is to facilitate the binding of the virus to the host cell receptor via angiotensin-converting enzyme 2 receptor (ACE2) as the main receptor. To target the specific spike protein, the method uses antispike protein antibodies which grant high specificity in detection of SARS-CoV-2 virus [20–22].

The principle of this method is based on measuring SARS CoV-2 spike proteins (detected at the attomole level) by a sandwich ELISA in combination with thio-NAD cycling using alkaline phosphatase, androsterone derivates (specifically 17β-methoxy-5β-androstan-3α-ol 3-phosphate), 3α-hydroxysteroid dehydrogenase, and its coenzymes (NADH and thio-NAD). Accumulated thio-NADH, during the cycling reaction, is measured directly at an absorbance of 405 without interference from other cofactors [23, 24].

Experimental results showed that the limit of detection (LOD) of the test is 2,3 $\times 10^{-18}$ moles/assay, meaning that if the virus has approximately 25 spike proteins on its surface, this method should detect on the order of 10^{-20} moles of virus/assay, which correlates to ~10 copies of the virus RNA/assay. When comparing the detection sensitivity, ultrasensitive ELISA method can approximate to that of PCR-based assays, with their average virus RNA loads being ~ 10 copies per swab. It is important to note that the test is able to distinguish SARS-CoV-2 from SARS-CoV, while using both the control antigen (spike protein of SARS-CoV), and our antigen of interest. The study also found a correlation between limit of detection (LOD) and the period of thio-NAD cycling, as decreasing the detection time decreased the detection sensitivity. The best detection sensitivity data was obtained from combining ELISA with 60 min of thioNAD cycling, while sufficient sensitivity can be achieved within the 10 min window [25, 26].

This test presents an easy-to-use, rapid, ultrasensitive, and specific test which can be performed via microplate reader and it allows simultaneous multiple measurements. Scientists are currently working on an automatic version for this test, so it can be implemented in various places as a point-of-care technique [25].

3.2 Panbio Rapid Antigen-Based Test for SARS-CoV-2 Detection

This test represents a lateral flow test that uses qualitative membrane-based immunoassay which detects nucleocapsid protein of SARS-CoV-2 in nasopharyngeal samples. Lateral flow assays are disposable devices, usually in the form of plastic cassettes similar to pregnancy tests. The antigen component is captured and bound to dedicated antibodies that could be either fluorescent- or colloidal gold- labeled. Detection is indicated via visible lines appearing on the test strip or through the appearance of fluorescence, which is detected by an immunoflorescence analyzer [27].

Results from a study that tested 255 nasopharyngeal samples (184 of which were symptomatic), using the rapid antigen-based test in comparison to RT-PCR, showed that the detection sensitivity (86,5%) and specificity were highest in samples of patients with high viral loads, within the first 7 days after the onset of symptoms. On the other hand, there was a statistically significant difference in sensitivity and specificity in samples from patients after 7 days of the onset of symptoms and samples from asymptomatic patients as well [28].

This technique definitely shows usability as a point-of-care method because it's cheap, rapid, easily mass-produced, user-friendly, and cost-effective. Panbio test enables rapid identification of moderately to highly infected individuals, and could also be used for screening in high-risk areas to quickly identify people and prevent transmission [28]. It's important to note that more studies have to be carried out for a better assessment of the potential usefulness of this test.

3.3 Field-Effect Transistor-Based Biosensor for SARS-CoV-2 Detection

Among other available diagnostic methods, Field-Effect Transistor (FET) based biosensors are very attractive due to their characteristics, including high sensitivity and instant measurements with small amounts of needed analytes. This sensor is fabricated from coating graphene sheets of the field-effect transistor along with a dedicated antibody, which is used as the sensing area, against SARS CoV-2 surface spike glycoprotein [29, 30].

Graphene has proven to be a practical material for sensing platforms, because of it's remarkable assets, such as high electronic conductivity, large specific area, and high carrier mobility [31].

For a functionalized detection platform, the SARS-CoV-2 spike antibody was immobilized into onto the biosensor through 1-pyrenebutyric acid N-hydroxysuccinimide ester (PBASE), which represents an interface coupling agent. Furthermore, this device is able to detect SARS-CoV-2 spike protein in various samples, such as transport medium used for nasopharyngeal swabs, cultured SARS-CoV-2 virus, and clinical samples as well, which proves the potential clinical and point-of-care applicability of the method. The limit of detection for the FET sensor is 1 fg/mL, and it also has the ability to differentiate the antigen protein of SARS-CoV-2 from those of MERS-CoV. This technology proves to be a simple, rapid, highly sensitive platform in the detection of SARS-CoV-2 virus in various samples, and with proper adaptation, it can be useful in the detection of other emerging viruses as well [32].

4 Diagnosis of SARS COVID-19 by Breath Analysis

A quick, non-invasive test to detect infection would be a big step forward, there is a high chance that this disease can be detected in exhaled breath. Currently, the world's scientific research centers are working on developing a device that works on this principle [33].

4.1 60-second Breath Test to Detect COVID-19 ("Breathonix")

"Breathonix" is a company that has developed a COVID-19 one-minute detection test, which does not require medically trained personnel to use it. "Breathonix" is useful for mass testing especially in places with mass gatherings. This company believes that this test could be found at every airport, bus, tram, etc. [33].

It works on the principle of detecting volatile organic compounds that are present in a person's breath. Volatile organic compounds can be used as markers for diseases such as COVID-19. The person must blow into a disposable mouthpiece connected to a high-precision breath analyzer. Exhaled breath is collected and stored in a mass spectrometer.

The software analyzes volatile organic compounds and generates results in less than a minute. The disposable mouthpiece does not allow saliva to enter the machine, this prevents cross-contamination from occurring. Clinical research is still ongoing, and more testing is needed to advance technology [33].

4.2 New Sensors that Detect Corona Virus Particles in the Breath

Professor Niann Sun created a gas sensor for various molecules in the air. It is a device that can probe the chemistry of airborne pathogens, including the SARS-CoV-2 coronavirus. He is currently working on developing a breath analysis device that will give results in a matter of second [34].

The device uses electrochemical sensors consisting of a special material that has cavities that are the same shape and size as the crown structures of the proteins that cover the SARS-CoV-2 virus, only these proteins can fill cavities in that material. The sensors captures these particles from the air, then the material reacts with the proteins and sends an electrical signal indicating that the virus is present. These sensors have shown same efficacy as the RT-PCR [34].

4.3 Scentech Medical Breath Analyzer

Scentech Medical has begun the early stages of testing for their breath analyser along with Meir Medical Center in Kfar Saba. Although the companies do not want to reveal how the technology works, they explained that it is a combination of hardware and software that enables the detection of volatile organic compounds in a few seconds. The process uses gas chromatography, then mass spectrometry, a technique and the ReCIVA breath collector. These techniques can analyze up to 8000 volatile organic compounds that are presented with each breath. The idea is to use an analytical elimination process to separate known from unknown compounds and then narrow the process so that coronavirus biomarkers can be found [34].

5 Conclusion

More than ever before coordinated global effort is required for rapid discoveries, especially in the terms of diagnosing COVID-19 which can assist with controlling the infection spread just as additional flare-ups of this infectious diseases. Realtime RT-PCR is still considered the 'gold standard' when it comes to diagnosing.

COVID-19 despite its flaws. In order to improve its performance, Loop-mediated Isothermal Amplification (RT-LAMP) and Artificial Intelligence-Assisted Loop-Mediated Isothermal Amplification (AI-LAMP) are novel methods that have been developed based on the RT-PCR method. On the other hand, more accurate antigen-based tests for the clinical application have been developed including the Ultrasensitive Antigen-based Test for SARS-CoV-2 spike protein detection and Field-effect Transistor (FET) based biosensors which showed outstanding performance. However, the most recent trend in the scientific community is developing a device that would be able to detect COVID19 in exhaled breath. Up to this day, "Breathonix" and "Scentech Medical" companies alongside professor Niann Sun developed breath analysers that are still in the testing phase. Data on COVID-19 are evolving rapidly. Some of the specifics in this review could be changed as more studies become available. Some referenced manuscripts are preprints and have not been peer-reviewed yet.

Conflicts of Interest. The authors declare no conflict of interest. The funders had no role in the design of the study; in the collection, analyses, or interpretation of data; in the writing of the manuscript, or in the decision to publish the results.

References

1. WHO Director-General's opening remarks at the media briefing on COVID-19 - 11 March 2020. In: Who.int. (2020). https://www.who.int/director-general/speeches/detail/who-director-general-s-opening-remarks-at-the-media-briefing-on-covid-19---11-march-2020. Accessed 11 Dec 2020
2. Guo, H., Zhou, Y., Liu, X., Tan, J.: The impact of the COVID-19 epidemic on the utilization of emergency dental services. J. Dental Sci. **15**, 564–567 (2020). https://doi.org/10.1016/j.jds.2020.02.002
3. Meng, L., Hua, F., Bian, Z.: Coronavirus disease 2019 (COVID-19): emerging and future challenges for dental and oral medicine. J. Dent. Res. **99**, 481–487 (2020). https://doi.org/10.1177/0022034520914246
4. Chen, Y., Liu, Q., Guo, D.: Emerging coronaviruses: genome structure, replication, and pathogenesis. J. Med. Virol. **92**, 2249 (2020). https://doi.org/10.1002/jmv.26234
5. Jaimes, J., Millet, J., Stout, A., et al.: A tale of two viruses: the distinct spike glycoproteins of feline coronaviruses. Viruses **12**, 83 (2020). https://doi.org/10.3390/v12010083
6. Chen, N., Zhou, M., Dong, X., et al.: Epidemiological and clinical characteristics of 99 cases of 2019 novel coronavirus pneumonia in Wuhan, China: a descriptive study. Lancet **395**, 507–513 (2020). https://doi.org/10.1016/s0140-6736(20)30211-7
7. WHO Coronavirus Disease (COVID-19) Dashboard. In: Covid19.who.int. (2020). https://covid19.who.int/. Accessed 11 Dec 2020
8. Chu, D., Pan, Y., Cheng, S., et al.: Molecular diagnosis of a novel coronavirus (2019nCoV) causing an outbreak of pneumonia. Clin. Chem. **66**, 549–555 (2020). https://doi.org/10.1093/clinchem/hvaa029
9. Corman, V., Landt, O., Kaiser, M., et al.: Detection of 2019 novel coronavirus (2019nCoV) by real-time RT-PCR. Eurosurveillance (2020). https://doi.org/10.2807/1560-7917.es.2020.25.3.2000045
10. Shen, M., Zhou, Y., Ye, J., et al.: Recent advances and perspectives of nucleic acid detection for coronavirus. J. Pharm. Anal. **10**, 97–101 (2020). https://doi.org/10.1016/j.jpha.2020.02.010
11. Wan, Z., Zhang, Y., He, Z., et al.: A melting curve-based multiplex RT-qPCR assay for simultaneous detection of four human coronaviruses. Int. J. Mol. Sci. **17**, 1880 (2016). https://doi.org/10.3390/ijms17111880
12. Noh, J., Yoon, S., Kim, D., et al.: Correction to: Simultaneous detection of severe acute respiratory syndrome, Middle East respiratory syndrome, and related bat coronaviruses by real-time reverse transcription PCR. Adv. Virol. **163**, 819 (2017). https://doi.org/10.1007/s00705-017-3677-6
13. Afzal, A.: Molecular diagnostic technologies for COVID-19: limitations and challenges. J. Adv. Res. **26**, 149–159 (2020). https://doi.org/10.1016/j.jare.2020.08.002
14. Carter, L., Garner, L., Smoot, J., et al.: Assay techniques and test development for COVID-19 diagnosis. ACS Central Sci. **6**, 591–605 (2020). https://doi.org/10.1021/acscentsci.0c00501

15. Liu, R., Han, H., Liu, F., et al.: Positive rate of RT-PCR detection of SARS-CoV-2 infection in 4880 cases from one hospital in Wuhan, China, from Jan to Feb 2020. Clin. Chim. Acta **505**, 172–175 (2020). https://doi.org/10.1016/j.cca.2020.03.009
16. Mautner, L., Baillie, C., Herold, H., et al.: Rapid point-of-care detection of SARS-CoV-2 using reverse transcription loop-mediated isothermal amplification (RT-LAMP). Virol. J. (2020). https://doi.org/10.1186/s12985-020-01435-6
17. Rohaim, M., Clayton, E., Sahin, I., et al.: Artificial intelligence-assisted loop mediated isothermal amplification (AI-LAMP) for rapid detection of SARS-CoV-2. Viruses **12**, 972 (2020). https://doi.org/10.3390/v12090972
18. Udugama, B., Kadhiresan, P., Kozlowski, H., et al.: Diagnosing COVID-19: the disease and tools for detection. ACS Nano **14**, 3822–3835 (2020). https://doi.org/10.1021/acsnano.0c02624
19. Advice on the use of point-of-care immunodiagnostic tests for COVID-19. In: Who.int. (2020). https://www.who.int/news-room/commentaries/detail/advice-on-the-use-of-point-of-careimmunodiagnostic-tests-for-covid-19. Accessed 12 Dec 2020
20. Wu, A., Peng, Y., Huang, B., et al.: Genome composition and divergence of the novel coronavirus (2019-nCoV) originating in China. Cell Host Microbe **27**, 325–328 (2020). https://doi.org/10.1016/j.chom.2020.02.001
21. Watabe, S., Kodama, H., Kaneda, M., et al.: Ultrasensitive enzyme-linked immunosorbent assay (ELISA) of proteins by combination with the thio-NAD cycling method. Biophysics **10**, 49–54 (2014). https://doi.org/10.2142/biophysics.10.49
22. Wan, J., Xing, S., Ding, L., et al.: Human IgG neutralizing monoclonal antibodies block SARS-CoV-2 infection. SSRN Electr. J. (2020). https://doi.org/10.2139/ssrn.3613271
23. Ito, E., Kaneda, M., Kodama, H., et al.: Immunoreactive insulin in diabetes mellitus patient sera detected by ultrasensitive ELISA with thio-NAD cycling. Biotechniques (2015). https://doi.org/10.2144/000114355
24. Nakatsuma, A., Kaneda, M., Kodama, H., et al.: Detection of HIV-1 p24 at attomole level by ultrasensitive ELISA with thio-NAD cycling. PLoS ONE **10**, e0131319 (2015). https://doi.org/10.1371/journal.pone.013131
25. Kyosei, Y., Namba, M., Yamura, S., et al.: Proposal of de novo antigen test for COVID19: ultrasensitive detection of spike proteins of SARS-CoV-2. Diagnostics **10**, 594 (2020). https://doi.org/10.3390/diagnostics10080594
26. Ke, Z., Oton, J., Qu, K., et al.: Structures and distributions of SARS-CoV-2 spike proteins on intact virions. Nature **588**, 498–502 (2020). https://doi.org/10.1038/s41586-020-2665-2
27. Dinnes, J., Deeks, J., Adriano, A., et al.: Rapid, point-of-care antigen and molecular-based tests for diagnosis of SARS-CoV-2 infection. Cochrane Database of Systematic Reviews (2020). https://doi.org/10.1002/14651858.cd013705
28. Linares, M., Pérez-Tanoira, R., Carrero, A., et al.: Panbio antigen rapid test is reliable to diagnose SARS-CoV-2 infection in the first 7 days after the onset of symptoms. J. Clin. Virol. **133**, 104659 (2020). https://doi.org/10.1016/j.jcv.2020.104659
29. Janissen, R., Sahoo, P., Santos, C., et al.: InP nanowire biosensor with tailored biofunctionalization: ultrasensitive and highly selective disease biomarker detection. Nano Lett. **17**, 5938–5949 (2017). https://doi.org/10.1021/acs.nanolett.7b01803
30. Liu, J., Chen, X., Wang, Q., et al.: Ultrasensitive monolayer MoS2 field-effect transistor based DNA sensors for screening of down syndrome. Nano Lett. **19**, 1437–1444 (2019). https://doi.org/10.1021/acs.nanolett.8b03818
31. Geim, A., Novoselov, K.: The rise of graphene. Nat. Mater. **6**, 183–191 (2007). https://doi.org/10.1038/nmat1849
32. Seo, G., Lee, G., Kim, M., et al.: Correction to rapid detection of COVID-19 causative virus (SARS-CoV-2) in human nasopharyngeal swab specimens using field-effect transistor-based biosensor. ACS Nano **14**, 12257–12258 (2020). https://doi.org/10.1021/acsnano.0c06726

33. 60-second breath test to detect COVID-19. In: 60-second breath test to detect COVID-19 (2020). https://news.nus.edu.sg/60-second-breath-test-to-detect-covid-19_20201026094346785/. Accessed 13 Dec 2020
34. These new sensors can detect coronavirus particles on your breath, instantly. In: News.northeastern.edu. (2020). https://news.northeastern.edu/2020/07/17/could-this-new-gas-sensor-help-researcherstest-regularly-for-the-coronavirus-in-the-air/. Accessed 13 Dec 2020
35. Aiming for 10-second result, breathalyzer detector for COVID begins early tests. In: Timesofisrael.com. (2020) https://www.timesofisrael.com/israeli-startups-breathalyzer-detector-for-covid-19-beginsearly-tests/. Accessed 13 Dec 2020

Machine Learning Techniques for Predicting Outcomes of COVID-19 for Patients with preexisting Chronic Diseases

Belmina Pramenković, Džejna Prasko(✉), Evelina Pulo, Ines Rončević, Rasema Ramić, and Adna Rakovac

Faculty of Pharmacy, University of Sarajevo, Sarajevo, Bosnia and Herzegovina
dzejnaprasko@ffsa.unsa.ba

Abstract. COVID-19 was officially confirmed during December 2019 in the city of Wuhan, China, while the first case of the disease was recorded on 17 November 2019. The World Health Organization has declared a pandemic due to the rapid spread of this disease. It is believed, due to worldwide population aging, that middle-aged and geriatric patients who suffer from chronic diseases are more prone to respiratory failure and having a poorer outcome caused by COVID-19. This paper presents the association of certain chronic diseases such as diabetes mellitus, COPD, hypertension, asthma, and others with COVID-19. Testing was done on 400 samples who were positive for the virus, 250 samples were sick of some of the listed diseases, while 150 were healthy. Machine learning algorithms use computational methods to "learn" information directly from data without relying on a predetermined equation as a model. The algorithms adaptively improve their performance as the number of samples available for learning increases. Specific parameters for different diseases were used in the paper. Based on the results presented in the paper, we concluded that chronic diseases greatly affect the number of people infected with COVID-19.

Keywords: COVID-19 · Machine learning · Prediction · Chronic diseases

1 Introduction

The rapid spread of COVID-19 caused by severe acute respiratory syndrome Coronavirus 2 (SARS-CoV-2), first lead to pronouncing it as a general well being a crisis of worldwide concern (PHEIC) on January 30, 2020, and after that, on March 11, 2020, the World Health Organization (WHO) declared COVID-19 a pandemic [1]. COVID19 has resulted in extensive morbidity and mortality in 217 countries and regions worldwide. At present, the number of patients with COVID-19 is rapidly rising to nearly 57 million leaving more than 1,4 million dead [2]. The COVID-19 upheaval has certainly become the most significant issue around the world, and its results have left nobody immaculate. The weight of COVID-19 stretches out far beyond that of contagious disease because of its

A. Badnjevic and L. Gurbeta Pokvić (Eds.): CMBEBIH 2021, IFMBE Proceedings 84, pp. 867–882, 2021.
https://doi.org/10.1007/978-3-030-73909-6_98

affection on every sphere of society, the entire health system as well as its ability to increase the overall mortality through the impact on non-communicable diseases.

Chronic diseases, also known as non-communicable illnesses (NCDs), strikingly diabetes, cardiovascular diseases, malignant growths, and persistent respiratory diseases are the main sources of death worldwide, influencing a greater number of individuals every year than all different causes joined. NCDs are liable for over 70% of all deaths all over the world [3]. It is believed, due to worldwide population aging, that middleaged and geriatric patients who suffer from chronic diseases are more prone to respiratory failure and having a poorer outcome caused by COVID-19 [4, 5]. Confirming this, 96.2% of patients who died in Italian hospitals had comorbidities. The comorbidities were mostly NCDs, such as hypertension, diabetes, and chronic lung disease, especially chronic obstructive pulmonary disease [6, 7]. A combination of aging alongside chronic diseases and COVID-19 represents a scenario that can rapidly lead to hospitalization, intensive care, or even hospital readmissions. This might explain the increased death rates in some countries, especially those with a high number of old-aged citizens [8].

Developing tools for prediction of outcome of COVID-19 can be extremely extensive and challenging, especially when the association between input and target values is non-linear and depending on multiple factors [9]. Machine learning methods such as Artificial Neural Networks (ANNs)are promising tools for overcoming these difficulties.

Recent studies identified that Machine Learning (ML) is a promising technology as it results in better scale-up, speed-up processing power, it is reliable and even able to outperform humans in specific healthcare tasks [10]. Machine learning evolved from different techniques in Artificial Intelligence (AI) areas and has been used in data mining. It allows the system to learn useful structural patterns and models from training data. A machine learning approach usually consists of two main phases: the training phase and the decision making phase. During the training phase, ML methods are applied to learn the system model using the training dataset. During the decisionmaking phase, the system can obtain the estimated output for each new input based on the trained model [11]. Computer based systems in healthcareare are able to process large set of information very quickly so faster and more accurate diagnosis can be established and it could be used to deduce the possible disease for patient [12].

Ever since the COVID-19 outbreak, machine learning techniques have been used in the taxonomic classification of COVID-19 genomes, CRISPR-based COVID-19 detection assay, survival prediction of severe COVID-19 patients, and discovering potential drug candidates against COVID-19 [13–16].

In this paper, we will give the effects of preexisting chronic conditions: diabetes, asthma, hypertension, and chronic obstructive pulmonary disease in COVID-19 patients and its outcomes by using artificial neural networks.

2 Materials and Methods

The Maslach Burnout Inventory Human Services Survey (MBI-HSS) was distributed to a target population of health-system (pharmacists) to assess study participants for burnout, which is characterized by feelings of emotional exhaustion, depersonalization, and reduced personal accomplishment. Electronic questionnaire regarding burnout was used, and distributed via e-mail. Descriptive statistics were used to assess MBI-HSS scores and risk factors associated with burnout in this cross-sectional study. The dataset for this study consisted of 400 samples. Out of overall number, 250 samples are with Covid-19, and 150 is for healthy class of patients.

2.1 Chronic Obstructive Pulmonary Disease - COPD

Chronic obstructive pulmonary disease (COPD) is a disease in which there is a slow progression of irreversible bronchial obstruction caused by chronic inflammation of the airways [17]. By 2020, chronic respiratory diseases will become the world's biggest cause of mortality [18]. A major challenge in chronic disease management is the number of patients with chronic respiratory diseases who are inaccurately diagnosed or misdiagnosed for having other respiratory diseases such as the common cold [19].

In our research, we have taken samples of 400 patients who were separated into two groups. The first group consisted of 320 healthy patients and the second group of 80 patients with COPD and COVID-19. The research showed that 20,3% of respondents had some symptoms of COPD and 79,7% of respondents didn't show any symptoms of COPD.

The clinical diagnosis of COPD is based on spirometry results. This test is performed on people who have symptoms of COPD including chronic cough, breathlessness, wheezing or sputum production, and/or a history of risk factors such as current or previous tobacco smoking and/or occupational exposure [20]. Spirometry is one of the most commonly used methods for testing lung function and involves measuring the volume of air and/or the speed at which air can be exhaled/inhaled. The usual parameters that accompany and are important for the diagnosis of COPD are FEV1/FVC relation [20]. The FEV1/FVC relation showed that all respondents with COPD had this relation $< 70\%$ [20], while all the healthy respondents had a relation higher than 70%.

Cigarette smoking is a big risk factor, although only about 15% of smokers develop COPD [22]. In our research from 80 patients who have COPD 53% of them are smokers.

The other most common symptoms of COPD are chronic cough, wheezing, and sputum [19, 22]. In our research, 53% of respondents with COPD had a chronic cough. Whistling in the lungs may be noticeable in patients and it was in 54% of our respondents. The third symptom, sputum, was present in 56,1% of patients with COPD [22]. These three symptoms come alone or in a combination of two or all three.

2.2 Diabetes Mellitus

Diabetes is one of the world's leading diseases today, in 2015 there were 415 milion patients diagnosed with diabetes, the expentance rate is 642 milions by 2040 [23]. This condition is the cause of morbidity, mortality as well as macrovascular and microvascular complications. There is a close relationship between diabetes and infections, however, influenza and pneumonia are particularly common and dangerous in the elderly with type 2 diabetes mellitus [24]. The outbreak of the Covid-19 epidemic resulted in many more severe cases. Scientists have spent a lot of time identifying risk factors for infection and severe outcomes, as well as understanding their mechanisms. The most common comorbidities that occurred with more severe outcomes were diabetes and cerebrovascular disease. Studies have shown that the risk of serious complications and death from coronavirus in people with diabetes is 50% higher than in people without diabetes [25].

In our study, we estimated 250 patients. We had two groups of patients. One group consisted of 150 healthy and the other of 100 patients with diabetes and Covid-19. The research showed that 60% of them had diabetes, and 40% of people did not have diabetes.

The study showed that the parameters that differed the most in people with diabetes compared to those without diabetes were: the absolute number of lymphocytes in the peripheral blood, absolute number of neutrophils, IL-6, serum ferritin, CRP, D-dimer, HbA1C. The absolute number of lymphocytes in the peripheral blood was significantly lower in the range of 0.41 × 109/L to 0.89 × 109/L, while in healthy people the range was from 1.1 × 109/L to 3,2 × 109/L. The absolute neutrophil counts were elevated (2.3 × 109/L - 6.52 × 109/L), while in healthy individuals the ratio was (1.8 × 109/L - 6.3 × 109/L). Levels of inflammation-related biomarkers such as CRP, IL-6, and serum ferritin are much higher in patients with DM compared to individuals without DM. CRP was higher than 11.3 mg/L up to 93 mg/L, in contrast to healthy individuals with CRP < 8mg/L. Serum ferritin was high (164 ng / ml - 1146.2 ng / ml), compared to healthy respondents (21.8 ng / ml - 275 ng / ml). IL-6 was >7.3 pg/ml while in healthy individuals the upper limit was 2.9 pg/ml. The D-dimer was higher than 0.74 μg/L up to 1,89 μg/L, while in healthy people it is less than 0.5 μg/L. For diseases that can cause a cytokine storm, IL-6 is a very good predictor of disease severity and prognosis, and the expression time is longer than other cytokines [26].

If HbA1C (glycated hemoglobin) levels are > 9% the risk is increased by 60% of hospitalizations and complications associated with pneumonia during bacterial infection [27]. Also, in diabetic samples, the possibility of hypercoagulation is increased compared to healthy samples [28].

Patients with diabetes required an increased dose of insulin after the patient was infected with SARS-CoV-2, indicating that the virus affects the patient's glucose metabolism. Inadequate glucose regulation will exacerbate diabetes, and this will affect the severity of pneumonia. Patients with diabetic complications showed a higher mortality rate. All this indicates that diabetes is one of the risk factors for a poor outcome of COVID-19.

2.3 Asthma

Asthma is a chronic disease characterized by widespread narrowing of the tracheobronchial tree [29].

In our research, we estimated 400 patients. The research showed that 62,5% of respondents have some kind of asthma and 37,5% of respondents didn't show any symptoms. For the confirmation of existing asthma, we used 6 different pieces of information that are used to diagnose. The first diagnostic test, FEV1/FVC relation [22, 31] showed that all respondents with asthma had this relation <70% (21), while all the healthy respondents had the relation higher than 70%. FEV1/FVC relation is used for the most precise diagnostic test [29].

The three most common symptoms of asthma are cough, wheeze, and dyspnea. [22, 29, 31] In our study, 54% of respondents with asthma had a cough. The wheeze in the lungs can be perceptible in patients and it was in 56% of our respondents. In other patients with asthma, there is shown a phenomenon of "quite lungs" [22]. The third symptom, dyspnoea, was present in 56,4% of patients with asthma. These three symptoms come alone or in a combination of two or all three [30]. These symptoms are not confident results as the FEV1/FVC relation or the paradoxical pulse.

All respondents with asthma had present tachycardia in the range of 160–200 beats per minute while healthy respondents had normocapnia in the range of 60–100 beats per minute [22, 31].

Besides the FEV1/FVC relation, the paradoxical pulse is another diagnostic test that confirmed the presence of asthma. Paradoxical pulse is a state where the systolic pressure is decreased in inspired by >10mmHg [22]. Our study showed that 73,2% of patients with asthma had a paradoxical pulse. In recent decades, the prevalence of asthma has been as well as the costs for asthma care [32].

2.4 Hypertension

Hypertension is the primary cause of cardiovascular disease, with systolic blood pressure ≥130 mmHg and diastolic ≥89 mmHg [33].

In our research, we estimated 400 patients. We had two groups of patients, the first group included 250 patients which had hypertension with COVID-19, and the second control group, 150 patients were healthy, without hypertension and COVID-19. As parameters in determining hypertension were included gender, age, medications, comorbidities, pressure value [34]. Comorbidities in the second patient group were: chronic kidney disease, coronary heart disease, diabetes, or no additional disease. Most patients had diabetes as an additional disease [35]. Their blood pressure was also measured by nurses using a mercury BP monitor, twice per day (in the morning and the afternoon) after admission [33]. The pressure value in the healthy group was in the optimal range ≤ 130/ ≤ 89 mmHg and in the second group, pressures were manifested ≥ 140/ ≥ 90 mmHg [3].

Research showed that 75% of the subjects had hypertension and COVID-19, and 35% were completely healthy. Of 75% of patients, 25% (average age 64) were taking hypotension drugs: angiotensin-converting enzyme inhibitors (ACEI) and angiotensin II receptor blockers (ARB), while the rest 50% of patients (average age 64), we're taking other antihypertensive drugs. The results showed a lower risk of death from COVID19 in patients receiving ACEI / ARB compared to patients not receiving ACEI/ARB [36].

To conclude, we included two groups of patients in our research. The first group was sick individuals, while the participants of the second group were healthy individuals. The first group was made of 250 persons and the second group was made of 150 persons. Based on these groups and regarding asthma we had 170 individuals positive on COVID-19, while 80 of them who didn't have asthma were negative on COVID-19 (Fig. 1).

The second research was regarding diabetes. 101 individuals, had TYPE 1 DIABETES and TYPE 2 DIABETES which were positive on COVID-19, while 149 individuals were negative with the fact that they did not have diabetes (Fig. 2).

The third research was regarding hypertension. Regarding hypertension we had 151 individuals positive on COVID-19, while 99 individuals without hypertension were negative on COVID-19 (Fig. 3).

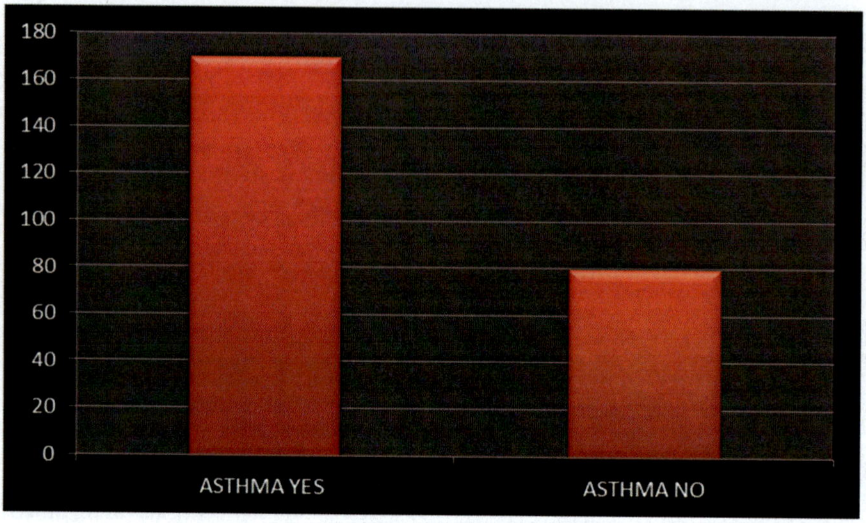

Fig. 1. Number of patients with and without asthma

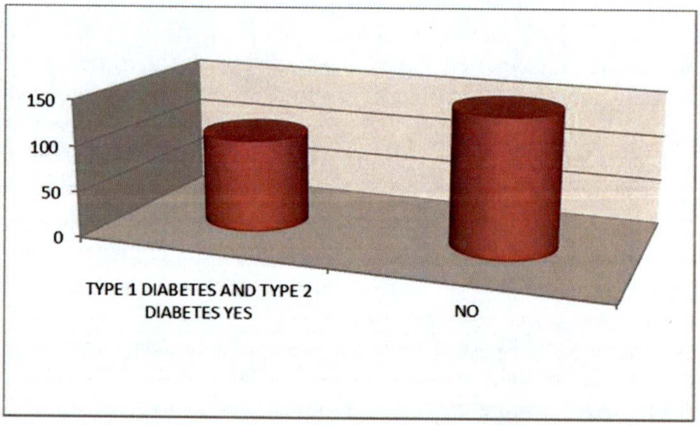

Fig. 2. Number of patients with Type 1 and Type 2 Diabetes

The last research was regarding COPD. Regarding COPD we had 84 individuals positive for the virus, while 166 do not have COPD and were negative on COVID-19 (Fig. 4).

This data is used for creating an automated model for determining correlation between patients with chronical diseases and COVID-19 and its' outcomes using appropriate neural network architectures.

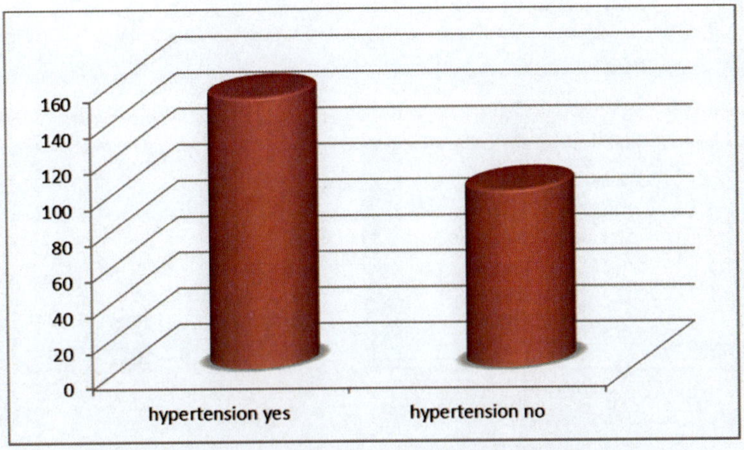

Fig. 3. Number of patients with and without hypertension

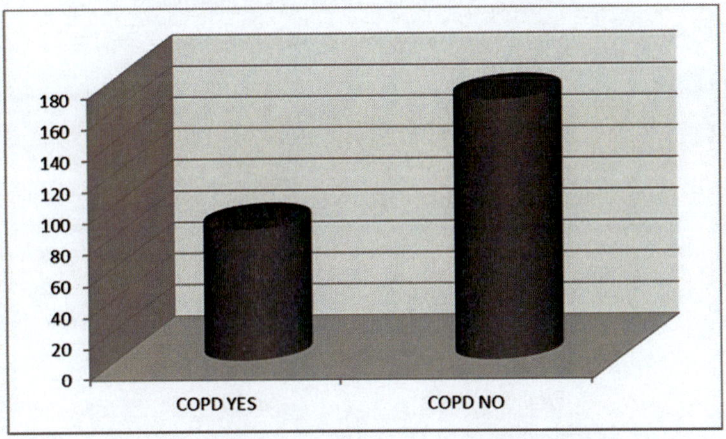

Fig. 4. Number or patients with and without COPD

3 Results

Figure 5 presents the results of training of classifiers based on 4 different machine learning algorithms (Tables 1, 2, 3, 4, 5, 6, 7, 8, 9, 10, 11, 12, 13, 14, 15, 16, Figs. 5, 67, 8 and 8).

a)

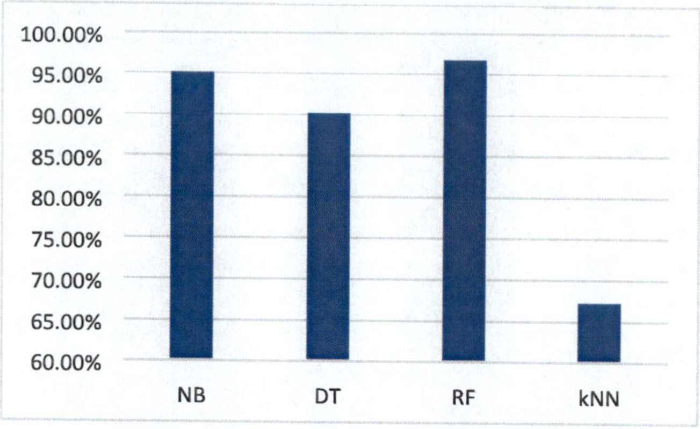

Fig. 5. Training performance of classifiers

Table 1. Performance of NB based classifier

	True 0	True 1	Accuracy
Pred. 0	23	1	95.08%
Pred. 1	2	35	
Total samples			

Table 2. Performance of DT based classifier

	True 0	True 1	Accuracy
Pred. 0	24	2	90.16%
Pred. 1	4	31	
Total samples			

Table 3. Performance of RF based classifier

	True 0	True 1	Accuracy
Pred. 0	29	0	96.72%
Pred. 1	2	30	
Total samples			

Table 4. Performance of k-NN based classifier

	True 0	True 1	Accuracy
Pred. 0	16	11	67.21%
Pred. 1	9	25	
Total samples			

b)

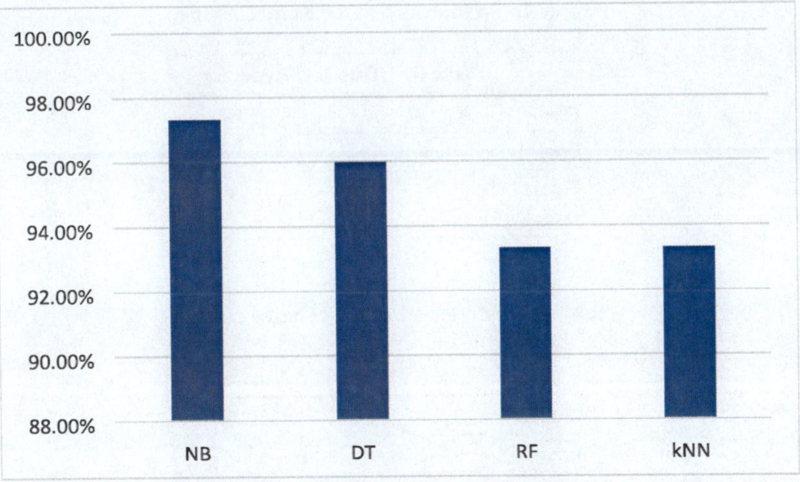

Fig. 6. Training performance of classifiers

Table 5. Performance of NB based classifier

	True 0	True 1	Accuracy
Pred. 0	27	1	97.33%
Pred. 1	1	46	
Total samples			

Table 6. Performance of DT based classifier

	True 0	True 1	Accuracy
Pred. 0	29	1	96.00%
Pred. 1	2	43	
Total samples			

Table 7. Performance of RF based classifier

	True 0	True 1	Accuracy
Pred. 0	26	1	93.33%
Pred. 1	4	44	
Total samples			

Table 8. Performance of k-NN based classifier

	True 0	True 1	Accuracy
Pred. 0	22	0	93.33%
Pred. 1	5	48	
Total samples			

c)

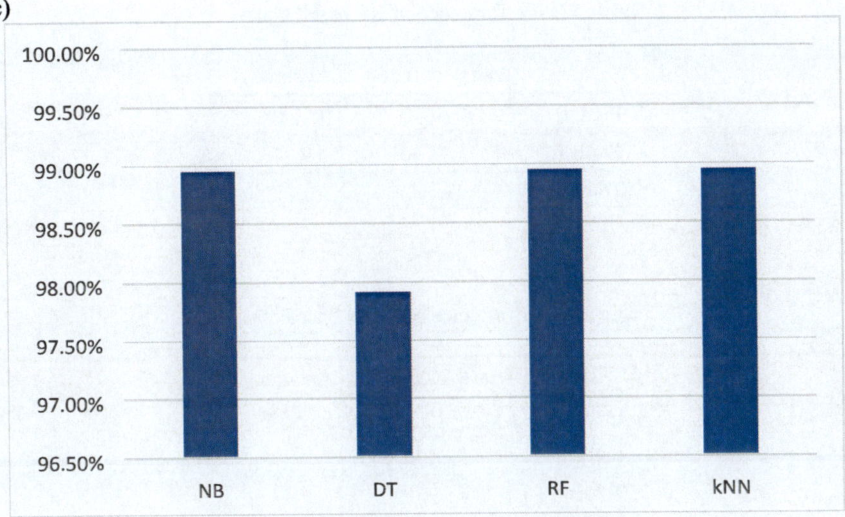

Fig. 7. Training performance of classifiers

Table 9. Performance of NB based classifier

	True 0	True 1	Accuracy
Pred. 0	52	1	98.95%
Pred. 1	0	43	
Total samples			

Table 10. Performance of DT based classifier

	True 0	True 1	Accuracy
Pred. 0	49	2	97.91%
Pred. 1	0	45	
Total samples			

Table 11. Performance of RF based classifier

	True 0	True 1	Accuracy
Pred. 0	59	0	98.95%
Pred. 1	1	36	
Total samples			

Table 12. Performance of k-NN based classifier

	True 0	True 1	Accuracy
Pred. 0	53	0	98.95%
Pred. 1	1	42	
Total samples			

d)

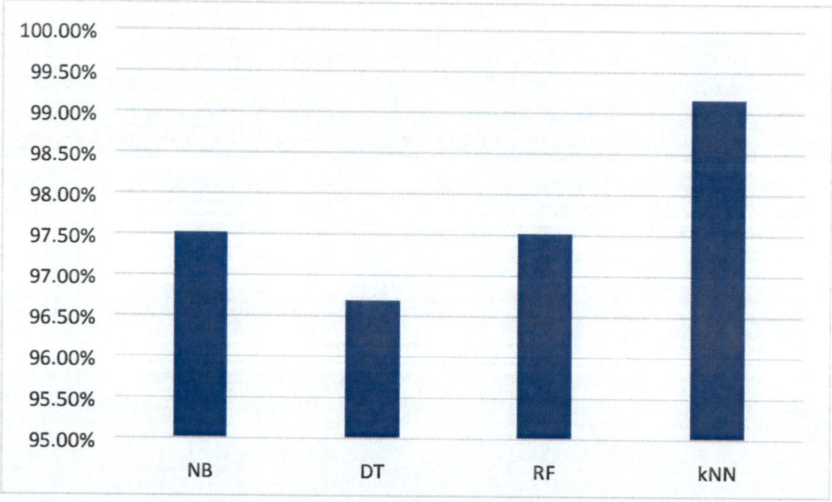

Fig. 8. Training performance of classifiers

Table 13. Performance of NB based classifier

	True 0	True 1	Accuracy
Pred. 0	26	1	97.52%
Pred. 1	2	92	
Total samples			

Table 14. Performance of DT based classifier

	True 0	True 1	Accuracy
Pred. 0	20	2	96.69%
Pred. 1	2	97	
Total samples			

Table 15. Performance of RF based classifier

	True 0	True 1	Accuracy
Pred. 0	29	0	97.52%
Pred. 1	3	89	
Total samples			

Table 16. Performance of k-NN based classifier

	True 0	True 1	Accuracy
Pred. 0	26	0	99.17%
Pred. 1	1	94	
Total samples			

4 Conclusion

The aim of this paper was to give the effects of preexisting chronic conditions: diabetes, asthma, hypertension, and chronic obstructive pulmonary disease in COVID-19 patients and its outcomes by using artificial neural networks.

References

1. WHO Director-General's opening remarks at the media briefing on COVID-19, 11 March 2020. In: Who.int (2020). https://www.who.int/director-general/speeches/detail/whodirector-general-s-opening-remarks-at-the-media-briefing-on-covid-19---11-march-2020. Accessed 12 Dec 2020
2. WHO Coronavirus Disease (COVID-19) Dashboard. In: Covid19.who.int (2020). https://covid19.who.int/. Accessed 12 Dec 2020
3. Disease burden and mortality estimates. In: World Health Organization (2020). https://www.who.int/healthinfo/global_burden_disease/estimates/en/. Accessed 12 Dec 2020
4. Yang, X., et al.: Clinical course and outcomes of critically ill patients with SARS-CoV-2 pneumonia in Wuhan, China: a single-centered, retrospective, observational study. Lancet Resp. Med. **8**, 475–481 (2020). https://doi.org/10.1016/s22132600(20)30079-5
5. Liu, K., et al.: Clinical characteristics of novel coronavirus cases in tertiary hospitals in Hubei Province. Chin. Med. J. **133**, 1025–1031 (2020). https://doi.org/10.1097/cm9.0000000000000744
6. Kluge, H., et al.: Prevention and control of non-communicable diseases in the COVID-19 response. Lancet **395**, 1678–1680 (2020). https://doi.org/10.1016/s0140-6736(20)31067-9
7. Thakur, J.: Novel coronavirus pandemic may worsen existing global noncommunicable disease crisis. Int. J. Noncommun. Diseases **5**, 1 (2020). https://doi.org/10.4103/jncd.jncd_2_20
8. Dyer, O.: Covid-19: Pandemic is having "severe" impact on non-communicable disease care, WHO survey finds. BM J. m2210 (2020). https://doi.org/10.1136/bmj.m2210
9. Catic, A., Gurbeta, L., Kurtovic-Kozaric, A., Mehmedbasic, S., Badnjevic, A.: Application of Neural Networks for classification of Patau, Edwards, Down, Turner and Klinefelter Syndrome based on first trimester maternal serum screening data, ultrasonographic findings and patient demographics. BMC Med. Genomics (2018). https://doi.org/10.1186/s12920-018-0333-2
10. Davenport, T., Kalakota, R.: The potential for artificial intelligence in healthcare. Future Healthc. J. **6**, 94–98 (2019). https://doi.org/10.7861/futurehosp.6-2-94
11. Xie, J., Yu, F., Huang, T., Xie, R., Liu, J., Wang, C., Liu, Y.: A survey of machine learning techniques applied to software defined networking (SDN): research issues and challenges. IEEE Commun. Surv. Tutor. **21**, 393–430 (2019). https://doi.org/10.1109/comst.2018.2866942
12. Badnjevic, A., Gurbeta, L., Cifrek, M., Pecchia, L.: Pre-classification process symptom questionnaire based on fuzzy logic for pulmonary function test cost reduction. In: IFMBE Proceedings, pp. 608–616 (2017). https://doi.org/10.1007/978-981-10-4166-2_92
13. Randhawa, G., Soltysiak, M., El Roz, H., de Souza, C., Hill, K., Kari, L.: Machine learning using intrinsic genomic signatures for rapid classification of novel pathogens: COVID-19 case study. PLoS ONE **15**, e0232391 (2020). https://doi.org/10.1371/journal.pone.0232391
14. Metsky, H., Freije, C., Kosoko-Thoroddsen, T., Sabeti, P., Myhrvold, C.: CRISPR-based surveillance for COVID-19 using genomically-comprehensive machine learning design (2020). https://doi.org/10.1101/2020.02.26.967026

15. Yan, L., Zhang, H., Goncalves, J., Xiao, Y., Wang, M., Guo, Y., Sun, C., Tang, X., Jin, L., Zhang, M., Huang, X., Xiao, Y., Cao, H., Chen, Y., Ren, T., Wang, F., Xiao, Y., Huang, S., Tan, X., Huang, N., Jiao, B., Zhang, Y., Luo, A., Mombaerts, L., Jin, J., Cao, Z., Li, S., Xu, H., Yuan, Y.: A machine learning-based model for survival prediction in patients with severe COVID-19 infection (2020). https://doi.org/10.1101/2020.02.27.20028027
16. Ge, Y., Tian, T., Huang, S., Wan, F., Li, J., Li, S., Yang, H., Hong, L., Wu, N., Yuan, E., Cheng, L., Lei, Y., Shu, H., Feng, X., Jiang, Z., Chi, Y., Guo, X., Cui, L., Xiao, L., Li, Z., Yang, C., Miao, Z., Tang, H., Chen, L., Zeng, H., Zhao, D., Zhu, F., Shen, X., Zeng, J.: A data-driven drug repositioning framework discovered a potential therapeutic agent targeting COVID-19 (2020). https://doi.org/10.1101/2020.03.11.986836
17. Mannino, D., Gagnon, R., Petty, T., Lydick, E.: Obstructive lung disease and low lung function in adults in the United States. Arch. Internal Med. **160**, 1683 (2020). https://doi.org/10.1001/archinte.160.11.1683
18. Gurbeta, L., Badnjevic, A., Maksimovic, M., Omanovic-Miklicanin, E., Sejdic, E.: A telehealth system for automated diagnosis of asthma and chronical obstructive pulmonary disease. J. Am. Med. Inform. Assoc. **25**, 1213–1217 (2018). 10. 1093/jamia/ocy055
19. Badnjevic, A., Gurbeta, L., Custovic, E.: An expert diagnostic system to automatically identify asthma and chronic obstructive pulmonary disease in clinical settings. Sci. Rep. (2018). https://doi.org/10.1038/s41598-018-30116-2
20. Overview | Chronic obstructive pulmonary disease in over 16s: diagnosis and management | Guidance | NICE. In: Nice.org.uk (2020). https://www.nice.org.uk/guidance/NG115. Accessed 14 Dec 2020
21. Ugrešić, N., Stepanović-Petrović, R., Savić, M.: Farmakoterapija za farmaceute, pp. 173–198 (2017)
22. MSD priručnik dijagnostike i terapije: Kronična opstruktivna plućna bolest. In: Msd-prirucnici.placebo.hr. (2020). https://www.msd-prirucnici.placebo.hr/msdprirucnik/pulmologija/kronicna-opstruktivna-plucna-bolest. Accessed 14 Dec 2020
23. Sejdinović, D., Gurbeta, L., Badnjević, A., Malenica, M., Dujić, T., Čaušević, A., Bego, T., Mehmedović, L.: Classification of prediabetes and type 2 diabetes using artificial neural network. In: IFMBE Proceedings, pp. 685–689 (2017). https://doi.org/10.1007/978-981-10-4166-2_103
24. Hussain, A., Bhowmik, B., do Vale Moreira, N.: COVID-19 and diabetes: knowledge in progress. Diabetes Res. Clin. Pract. **162**, 108142 (2020). https://doi.org/10.1016/j.diabres.2020.108142
25. Muniangi-Muhitu, H., Akalestou, E., Salem, V., et al.: Covid-19 and diabetes: a complex bidirectional relationship. Front. Endocrinol. (2020). https://doi.org/10.3389/fendo.2020.582936
26. Guo, W., Li, M., Dong, Y., et al.: Diabetes is a risk factor for the progression and prognosis of COVID -19. Diabetes/Metab. Res. Rev. (2020). https://doi.org/10.1002/dmrr.3319
27. Singh, A., Gupta, R., Ghosh, A., Misra, A.: Diabetes in COVID-19: Prevalence, pathophysiology, prognosis and practical considerations. Diabet. Metab. Syndr. Clin. Res. Rev. **14**, 303–310 (2020). https://doi.org/10.1016/j.dsx.2020.04.004
28. Sardu, C., Gargiulo, G., Esposito, G., et al.: Impact of diabetes mellitus on clinical outcomes in patients affected by Covid-19. Cardiovasc. Diabetol. (2020). https://doi.org/10.1186/s12933020-01047-y
29. Yoon, H.: Evidence-based pharmacologic therapy of bronchial asthma. J. Korean Med. Assoc. **47**, 568 (2004). https://doi.org/10.5124/jkma.2004.47.6.568
30. Global Initiative for Asthma - Global Initiative for Asthma - GINA. In: Global Initiative for Asthma – GINA (2020). https://ginasthma.org/. Accessed 15 Dec 2020

31. Zhang, D., Shen, X., Qi, X.: Resting heart rate and all-cause and cardiovascular mortality in the general population: a meta-analysis. Can. Med. Assoc. J. **188**, E53–E63 (2015). https://doi.org/10.1503/cmaj.150535
32. Badnjevic, A., Gurbeta, L., Cifrek, M., Marjanovic, D.: Classification of asthma using artificial neural network. In: 2016 39th International Convention on Information and Communication Technology, Electronics and Microelectronics (MIPRO) (2016). https://doi.org/10.1109/mipro.2016.7522173
33. COVID-19 and Hypertension: What We Know and Don't Know - American College of Cardiology. In: American College of Cardiology (2020). https://www.acc.org/latest-incardiology/articles/2020/07/06/08/15/covid-19-and-hypertension. Accessed 16 Dec 2020
34. Rouette, J., Suissa, K., Azoulay, L.: Re. Association of inpatient use of angiotensin converting enzyme inhibitors and angiotensin II receptor blockers with mortality among patients with hypertension hospitalized with COVID-19. Epidemiology (2020). https://doi.org/10.1097/ede.0000000000001250
35. Flack, J., Adekola, B.: Blood pressure and the new ACC/AHA hypertension guidelines. Trends Cardiovasc. Med. **30**, 160–164 (2020). https://doi.org/10.1016/j.tcm.2019.05.003
36. Bakris, G., Ali, W., Parati, G.: ACC/AHA Versus ESC/ESH on Hypertension Guidelines. J. Am. Coll. Cardiol. **73**, 3018–3026 (2019). https://doi.org/10.1016/j.jacc.2019.03.507

Genetic Predisposition – Impact on the COVID-19 Infection Severity

Ines Rončević(✉), Valida Abdagić, Amar Kolašinac, Denisa Tahirović, Indira Zuko, and Tamer Bego

Faculty of Pharmacy, University of Sarajevo, Zmaja od Bosne 8, 71 000 Sarajevo, Bosnia and Herzegovina
inesronceviс@ffsa.unsa.ba

Abstract. Since the end of 2019, world deals with the new infection called COVID-19 caused by SARS-CoV-2. SARS-CoV-2 binds to angiotensin-converting enzyme 2 (ACE) 2 receptors and enter through cell membranes of the lungs, kidneys, intestines, heart and other organs. It causes different clinical features among the population. It has been observed that there are young people with no comorbidities that will develop severe COVID-19 infection. It has been proposed that these severe COVID-19 infections can be determined by genetic predisposition. Main genes that have been analyzed are ACE polymorphisms, TMPSRSS2, Human Leukocyte Antigen (HLA), ADAM17 gene and Heme-oxygenase-1-enzymes polymorphism. These genes can be associated with development of COVID-19 infection and its severity. The frequency of these genes varies when it comes to different populations all around the globe. Also it is important to consider the interaction between genes and environment.

Keywords: SARS-CoV-2 · COVID-19 · Polymorphism · Gene

1 Introduction

In the end of 2019, a new cause of viral infection appeared in the city of Wuhan in China. It was named SARS-CoV-2. It caused the unknown viral pneumonia and it was transmissed very easily [1]. There are more than 77.00 milion cases of COVID-19 infection and more than 1.7 milion deaths [2]. The disease which is caused by SARS-CoV-2 is named COVID-19 which means coronavirus disease 2019. Since then, the virus, SARS-CoV-2, spread around the world and caused pandemic [1]. Spike proteins of the coronavirus, specifically SARS-CoV-2, are composed of homotrimers of S proteins that direct the virus to host receptors. Interestingly, the spike-receptor-binding domain (RBD) has a similarity of only 40% when it comes to the amino acid sequence, with other SARS-CoV viruses [3]. It binds to the angiotensin-converting enzyme 2 (ACE2) receptor in order to enter the cell. This receptor is a glycoprotein that can be found on the cell membranes of the lungs, kidneys, intestines and heart [4].

The most common types of genetic variations in the human genome are single nucleotide polymorphisms (SNPs). SNPs in genes regulate DNA mismatch repair, cell

A. Badnjevic and L. Gurbeta Pokvić (Eds.): CMBEBIH 2021, IFMBE Proceedings 84, pp. 883–891, 2021.
https://doi.org/10.1007/978-3-030-73909-6_99

cycle regulation, immunity and metabolism. They can be associated with genetic susceptibility with diseases such as cancer [5]. Guo et al. state that misens mutation of ACE2 receptor may affect the clinical course of COVID-19 infection [6]. For example ACE1 I/D polymorphisms associated with ACE2 enzyme function have also been isolated. Both ACE1 and ACE2 polymorphisms have been associated with the development of hypertension and it is one of the risks for the development of a severe clinical manifestation in COVID-19 patients [7]. At the same time, studies showed the association of ACE2 and TMPRSS2 polymorphisms with susceptibility to COVID-19 infection and the severity of disease and the outcome of the disease itself [8]. Also genome wide association studies showed that there is significant association between genetic variants and severe COVID-19. For example variations at two loci in the human genome were associated with increased risk of respiratory failure. One is within the ABO blood type cluster on chromosome 9 and the other on position 3p21.3 [9]. Also some polymorphisms of human leukocyte antigen (HLA) genotypes have been associated with a higher risk of developing COVID-19 infection [10]. The importance of isolating gene variants that lead to a severe infection is reflected in the fact that young individuals can develop a severe disease even if they did not have any comorbidity. At the same time, it is important to consider the interaction between genes and environment [11] (Fig. 1).

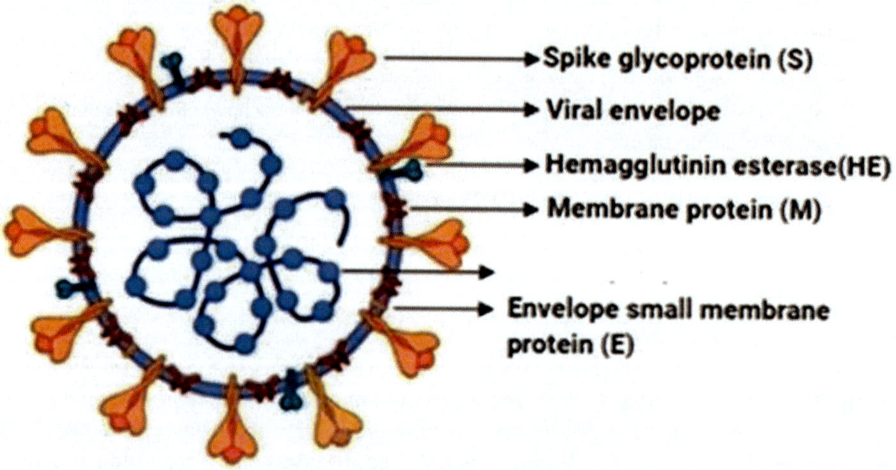

Fig. 1. Structure of SARS-CoV-1

2 Materials and Methods

This is a literary work that includes the results of previous scientific research. For the purpose of writing this paper, various medical journals, books, manuals, articles, as well as medical databases (Medline) and Internet search engines were used as data sources.

3 Results

See Table 1.

Table 1. Presentation of genes that can impact COVID-19 infection severity

Gene	Polymorphism	Impact on COVID-19 infection	Source
ACE	DD polymorphisms	Associated with ARDS	Guo et al.
HLA	HLA-A*11:01 and B*44	Associated with more severe clinical features	Warren et al. Correale et al. Tomita et al.
ADAM17	Should be more investigated	Associated with higher ACE expression which is a risk factor for severe clinical features	Brest et al.
HO-1	Longer (GT)n repeats in the HMOX1 gene	Associated with comorbidities that contribute to severe clinical features	Bao et al.

3.1 ACE Polymorphisms

It has been observed that ACE gene that contains an insertion/deletion (I/D) polymorphism (rs4646994) is associated with higher serum ACE levels, obesity, hypertension and increased cardiovascular risk. This all clinical conditions are correlated with more aggressive COVID-19. Also, the ACE DD polymorphism is associated with mortality when it comes to acute respiratory distress syndrome (ARDS) [12]. The reason why that happens is explained by the fact that DD carriers and I/D carriers have higher mean ACE activity levels than II genotype individuals. The racial difference of this ACE gene polymorphism has been also well established. The highest frequencies of D allele have African Americans. In Europe, populations in Italy, France and Spain have a high frequency of D allele. The frequency of D allele comes up to 8287% there [13]. Gomez et al. state that D allele was significantly higher in hypoxemic that in non-hypoxemic patients. They did not find the difference for DD-frequencies between COVID-19 and controls. However they find significant higher risk for sever form of COVID in males. It was explained it by the fact that D-allele was a risk for hypertension in their male patients [14]. Adamzik et. al reported that mortality due to acute respiratory distress syndrome (ARDS) was increased in their study in patients with ACE DD genotype compared with I allele and the ACE I/D polymorphism. It was an independent prognostic factor for 30-days survival. DD genotype can be a used as a predictor for mortality outcome [15]. COVID-19 infection can cause ARDS and emergency rooms all around the world are dealing with it [16]. Wuster et al. reported that five polymorphisms (rs4240157, rs6632680, rs4830965, rs1476524, and rs2048683), with minor alleles that associated with higher tissue expression of ACE2. They were associated with increased need for hospitalization due to COVID-19 [17]. Conclusion was that poorer outcome of

COVID-19 infection can be connected with hypertension that is determined with ACE polymorphisms [18].

3.2 TMPRSS2 Gene

TMPRSS2 gene is gene that produces fusion between TMPRSS2 and ERG (ETS related genes). This fusion results in one of the most frequent genetic lesions in prostate cancer patients. It is androgen responsive gene so the fusion results in androgen dependent transcription of ERG in prostate tumor cells. So from said above it can be hypothesized that males might have higher TMPRSS2 expression in the lungs. This higher expression can improve the ability of SARS-CoV-2 to enter the cells [19]. Viral entry requires not only binding to ACE receptor but it also primes the virus spike protein by the transmembrase serine 2 (TMPRSS2). It has been observed that TMPRSS2 expression in lung cells may vary across different populations. This could be implied in susceptibility to coronavirus and influenza virus infections. It has been confirmed that single nucleotide polymorphisms (SNPs) have been linked to higher expression of TMPRSS2. It is associated with an increased susceptibility to influenza virus infection [20]. Because of the location of TMPRSS2 which is on 21q22.3, it can be speculated that individuals with Down syndrome can be at higher risk for COVID-19 infection. Oncogenic roles of this gene can be linked to poor outcomes with COVID-19 infection [21]. In the other hand, infants express TMPRSS2 at very low levels in epithelial lineages. Children exhibited similarly low levels of this gene in secretory and alveolar epithelial cells. Adult subjects had higher TMPRSS2 expression in secretory, ciliated and AT1 cells. TMPRSS2 expression is associated with aging and this can connected with COVID-19 infection severity [22].

3.3 Human Leukocyte Antigen

Human Leukocyte Antigen (HLA) includes a large set of genes that have important actions in immune responses against viral infections and they are encoded by Major Histocompatibility Complex (MHC) [23, 24]. HLA genes have extreme diversity and several thousand already reported polymorphisms. These MHC molecules are receptors for viral peptides and they also present peptides to the virus-specific cytotoxic T lymphocytes. When it comes to HLA genes, it is well known that they are crucial for MHC peptide interactions [24]. Warren et al. using in silica method confirmed that HLA alleles A*11:01, C*04:01, DQA1*01:02 may be prognostic indicators for poor outcome [25]. Nguyen et al. showed that there is a link between HLA-B*46.01 allele and that this allele could be associated with more severe infection. They also state that there is high degree of similarity between the two across HLA types when it comes to comparing SARS-CoV-1 and SARS-CoV-2 peptide [20]. Correale et al. found that HLA-A*25, B*08, B*44, B*15:01, B*51, and C*01 showed positive correlation with COVID-19 incidence rate. They also concluded that potential genetic determinants are HLA-C*1 and B*44 alleles for the identification of individual at risk but it is also said that more case-controls studies are needed [26]. Tomita et al. showed in theirs silico study possible association between HLA-A*02:01 and increased risk for COVID-19. This allel,

HLA-A*02:01 has relatively lower capacity to present SARS-CoV-2 antigens to T cells compared with HLA-A*11:01 and HLA-A*24:02 [27].

3.4 ADAM17 Gene

The ADAM17 locus is on chromosome 2p25.1. It presents two clusters and three unique SNPs that result in strong differences in terms of allelic profiles between Asian and European populations [28]. It has been observed that down regulation of ADAM17 inhibition can impair immune response and increase susceptibility to infections. This down regulation causes imbalance in the release of CD14, CD16 via monocytes, natural killer cells and T cells. Also the ADAM17 is connected with hypertension [29]. There are many genetic polymorphisms that influence ADAM17 expression and they can contribute to the modulation of ACE2 shedding intensity. ACE2 also regulates blood pressure [28]. Many reports suggest that hypertension is associated with severe COVID-19 infection and mortality rate [30]. In the same time, ADAM17 has important role in chronicle obstructive pulmonary disease (COPD). It is mediating inflammatory responses and it induces pro-inflammatory molecules secretion such as CXL8 and interleukin-6 cytokine receptor. Reduction of expression of these gene can contribute to anti-inflammation process and tissue remodeling when it comes to the lungs [31]. COPD is associated with higher fatality rate due to SARS-CoV-2 infection [32].

3.5 Heme-Oxygenase-1-Enzymes Polymorphism

SARS-CoV-2 has ability to bind porphyrins with even stronger affinity than ACE2 receptors. It results in up regulation of free heme and severe reactive oxygen species (ROS) formation and decreased levels of heme-oxygenase-1enzymes (HO-1). Free hem and iron release contribute to production of ROS. This production of ROS inhibits HO-1 enyzmes that were previously active. Deficiency of human HO-1 normally causes early death as result of iron toxicity, renal failure and nephrotic proteinuria. The same clinical issues have been observed with COVID-19. Catalase is a vital hemoprotein that can be potentially affected by SARS-CoV-2. When catalase is not present, there is no conversion of hydrogen peroxide to water and oxygen which then contributes to ROS production [33]. Many polymorphisms of HO-1 have been reported and they can be connected with developing several diseases [34, 35].

Short alleles of GT repeat are associated with greater inducibility of HO-1. This greater inducibility of HO-1 leads to increased cytoprotection and reduced inflammation [33].

Bao et al. estimate that persons carrying longer (GT)n repeats in the HMOX1 gene promoter can have higher risk of type 2 diabetes [36]. It has been established that diabetes is one of the comorbidities that are associated with morbidity and mortality of COVID-19. Hypertension, obesity and cardiovascular diseases that usually come with diabetes alter the ACE2 expression, immune deregulation, alveolar and endothelial dysfunction and increase the systemic coagulation. This all can be a reason why people with diabetes are at higher risk for developing severe COVID-19 infection [37]. It can be assumed that COVID-19 patients with longer GT sequences are at higher risk for developing complications of COVID-19 [33].

4 Discussion

All genes that are mentioned, ACE polymorphisms, TMPRSS2 gene, HLA gene and even ADAM17 gene can probably be associated to predisposition and severity of COVID-19 infection. For instance, it was mentioned that Italian population has high frequency of D allele that contributes to ACE polymorphism [13]. Benetti et al. reported that variants namely c.2158A > G p.(Asn720Asp), c.1166C > A p.(Pro389His), and c.1051C > G p.(Leu351Val), one of which polymorphic c.2158A > G p.(Asn720Asp), moderately expressed in the Italian and European non-Finnish populations. At the same time these variants can be found with a very low allele frequency in the Eastern Asia population. This results in differences of the clinical state due to bonding SARS-CoV-2 to ACE2 receptor [37]. It is well known that Italy was facing an explosion of COVID-19 cases and since February 2020, it had much higher case fatality rate (CFR) than the other countries. Giangreco et al. tried to explain it by the fact that Italians have different social behavior but this hypothesis could not explain the CFR. There can be an impact of genetic predisposition [38].

TMPRSS2 gene has lower expression in infants and children population. Low mortality and morbidity among children infected with severe respiratory syndrome due to SARS-CoV-2 has already been observed. Reports from China said that children usually have mild symptoms. Mortality rates are much higher in adult population, even up to 3,7% in the USA [39]. TMPRSS2 can be a target for pharmacological treatment. It has been observed that inhibition of TMPRSS2 activity using camostat mesylate in human lung cells in vitro, has showed protective effects against SARS-CoV-2. There are several differences between TMPRSS2 polymorphisms. Several polymorphisms have been isolated such as rs75603675; c.23G > T p.Gly8Val and rs12329760; c.589G > A p.Val197Met. They can also contribute to COVID-19 infection severity [40].

For HLA genes, further investigation could be needed in order to determinate which polymorphisms can be used as genetic markers [26]. It has been observed that HLA-B*46:01 allele carriers can be at risk of being attacked by SARS-CoV-2. On the other hand, HLA-B*15:03 allele can promote cross-protective T-cell based immunity. In the same time, HLA polymorphisms can cause production of anti-drug antibodies. For example, HLA-DQA1 may induce production of anti-drug antibody against anti-TNF like infliximab and adalimumab. Infliximab and adalimumab are used as an options in treatment of COVID-19 [41].

ADAM17 gene should be more investigated because it has impact on ACE2 receptor [27]. Late results show that ADAM17 inhibition by decreasing ACE2 shedding may have protective effects on COVID-19 infection. It could possibly reduce viral load [42].

Patients who have genetic polymorphism of the CYP12D6 and the HO-1 GT allele make prognosis challenging and these polymorphisms should be more observed. In general, in steady state expression of HMOX1 GENE that is coding HO-1, is quite low. This HO-1 activity has protective actions in lungs and this gene can activated by different stimuli. SARS-CoV-2 can decrease HO-1 activity due to ROS production and this can contribute to the COVID-19 infection severity [43].

Severity of COVID-19 infection due these polymorphisms can be higher but in the same time there might be treatment failure [33].

5 Conclusion

It is important to discover genetic predisposition that can contribute to COVID-19 infection severity. Differences between children, adults and elderly populations can be explained by genetic predisposition. Also treatment failure can be connected with polymorphisms. D allele polymorphisms of ACE2 contribute to more severe clinical features of COVID-19 infection. Expression of TMPRSS2 gene in cells can also impact COVID-19 infection severity. Some of HLA polymorphisms are associated with COVID-19 infection severity such as HLA-A*11:01 and B*44. Carriers wih longer GT sequences of HO-1 genes can develop more severe infection. This all shows that impact of genetic predisposition is huge. In the same time, interaction between genes and environment should not be forgotten.

References

1. Hu, B., Guo, H., Zhou, P., et al.: Characteristics of SARS-CoV-2 and COVID-19. Nat. Rev. Microbiol. **6**, 1–14 (2020)
2. Worldmeter. Coronavirus cases (2020). https://www.worldometers.info/coronavirus/country/iran/. Accessed 21 Dec 2020
3. Cascella, M., Rajnik, M., Cuomo, A., et al.: Features, evaluation, and treatment of coronavirus. In: StatPearls [Internet]. Treasure Island (FL). StatPearls Publishing (2020). https://www.ncbi.nlm.nih.gov/books/NBK554776
4. Neerukonda, S.N., Katneni, U.: A review on SARS-CoV-2 virology, pathophysiology, animal models, and anti-viral interventions. Pathogens **29**;9(6), 426 (2020)
5. Deng, N., Zhou, H., Fan, H., Yuan, Y.: Single nucleotide polymorphisms and cancer susceptibility. Oncotarget. 7;8(66), pp. 110635–110649 (2017)
6. Guo, X., Chen, Z., Xia, Y., et al.: Investigation of the genetic variation in ACE2 on the structural recognition by the novel coronavirus (SARS-CoV-2). J. Transl. Med. **18**, 321 (2020)
7. Delanghe, J.R., Speeckaert, M.M., De Buyzere, M.L.: ACE polymorphism and COVID-19 outcome. Endocrine **70**(1), 13–14 (2020). https://doi.org/10.1007/s12020-020-02454-7
8. Hou, Y., Zhao, J., Martin, W., et al.: New insights into genetic susceptibility of COVID-19: an ACE2 and TMPRSS2 polymorphism analysis. BMC Med. **18**, 216 (2020)
9. Jacob, C.O.: On the genetics and immunopathogenesis of COVID-19. Clin. Immunol. **220**, 108591 (2020)
10. LoPresti, M., Beck, D.B., Duggal, P., et al.: Role of host genetic factors in coronavirus susceptibility: review of animal and systematic review of human literature. Am. J. Hum. Genet. **107**(3), 381–402 (2020)
11. Darbeheshti, F., Rezaei, N.: Genetic predisposition models to COVID-19 infection. Med. Hypotheses **142**, 109818 (2020)
12. Bellone, M., Calvisi, S.L.: ACE polymorphisms and COVID-19-related mortality in Europe. J. Mol. Med. (Berl). **98**(11), 1505–1509 (2020)
13. Zheng, H., Cao, J.J.: Angiotensin-converting enzyme gene polymorphism and severe lung injury in patients with coronavirus disease 2019. Am. J. Pathol. **190**(10), 2013–2017 (2020)
14. Gómez, J., Albaiceta, G.M., García-Clemente, M., et al.: Angiotensin-converting enzymes (ACE, ACE2) gene variants and COVID-19 outcome. Gene **15**(762), 145102 (2020)
15. Adamzik, M., Frey, U., Six, S., et al.: ACE I/D but not AGT (-6)A/G polymorphism is a risk factor for mortality in ARDS. Eur. Resp. J. **29**(3), 482–488 (2007)

16. Fan, E., Beitler, J.R., Brochard, L., et al.: COVID-19-associated acute respiratory distress syndrome: is a different approach to management warranted? Lancet Respir Med. **8**(8), 816–821 (2020)
17. Wooster, L., Nicholson, C.J., Sigurslid, H.H., et al.: Polymorphisms in the ACE2 Locus Associate with Severity of COVID-19 Infection. medRxiv 2020.06.18.20135152 (2020)
18. Asselta, R., Paraboschi, E.M., Mantovani, A., et al.: ACE2 and TMPRSS2 variants and expression as candidates to sex and country differences in COVID-19 severity in Italy. Aging (Albany NY). **12**(11), 10087–10098 (2020)
19. Mollia, V., Rizzo, A., Massari, F.: The pivotal role of TMPRSS2 in coronavirus disease 2019 and prostate cancer. Future Oncol. **16**(27) (2020)
20. Hou, Y., Zhao, J., Martin, W., et al.: New insights into genetic susceptibility of COVID-19: an ACE2 and TMPRSS2 polymorphism analysis. BMC Med. **18**(1), 216 (2020)
21. Schuler, B.A., Christian Habermann, A., Plosa, E.J., et al.: Vanderbilt COVID-19 Consortium Cohort. In: Webber, S.A., Banovich, N.E., Kropski, J.A., Sucre, J.M.S. (eds.) HCA Lung Biological Network. Agerelated expression of SARS-CoV-2 priming protease TMPRSS2 in the developing lung. bioRxiv [Preprint]. 23:2020.05.22.111187 (2020)
22. Iturrieta-Zuazo, I., Rita, C.G., García-Soidán, A., et al.: Possible role of HLA class-I genotype in SARS-CoV-2 infection and progression: a pilot study in a cohort of Covid-19 Spanish patients. Clin Immunol. **219**, 108572 (2020)
23. Debnath, M., Banerjee, M., Berk, M.: Genetic gateways to COVID-19 infection: implications for risk, severity, and outcomes. FASEB J. **34**(7), 8787–8795 (2020)
24. Warren, R.L., Birol, I.: Retrospective in silico HLA predictions from COVID-19 patients reveal alleles associated with disease prognosis. medRxiv. 2:2020.10.27.20220863 (2020)
25. Nguyen, A., David, J.K., Maden, S.K., et al.: Human leukocyte antigen susceptibility map for severe acute respiratory syndrome coronavirus 2. J. Virol. **94**(13), e00510-20 (2020)
26. Correale, P., Mutti, L., Pentimalli, F., et al.: HLA-B*44 and C*01 Prevalence Correlates with Covid19 Spreading across Italy. Int. J. Mol. Sci. **21**(15), 5205 (2020)
27. Tomita, Y., Ikeda, T., Sato, R., et al.: Association between HLA gene polymorphisms and mortality of COVID-19: an in silico analysis. Immun. Inflamm. Dis. **8**, 684–694 (2020)
28. Brest, P., Refae, S., Mograbi, B., et al.: Host polymorphisms may impact SARS-CoV-2 infectivity. Trends Genet. **36**(11), 813–815 (2020)
29. de Queiroz, T.M., Lakkappa, N., Lazartigues, E.: ADAM17-mediated shedding of inflammatory cytokines in hypertension. Front Pharmacol. **29**(11), 1154 (2020)
30. Cook, T.M.: The importance of hypertension as a risk factor for severe illness and mortality in COVID-19. Anaesthesia **75**(7), 976–977 (2020)
31. Zipeto, D., Palmeira, J.D.F., Argañaraz, G.A., Argañaraz, E.R.: ACE2/ADAM17/TMPRSS2 interplay may be the main risk factor for COVID-19. Front Immunol. **7**(11), 576745 (2020)
32. Alqahtani, J.S., Oyelade, T., Aldhahir, A.M., et al.: Prevalence, severity and mortality associated with COPD and smoking in patients with COVID-19: a rapid systematic review and metaanalysis. PLoS One **15**(5), e0233147 (2020)
33. Fakhouri, E.W., Peterson, S.J., Kothari, J., et al.: Genetic polymorphisms complicate covid-19 therapy: pivotal role of HO-1 in cytokine storm. Antioxidants (Basel) **9**(7), 636 (2020)
34. Exner, M., Schillinger, M., Minar, E., et al.: Heme oxygenase-1 gene promoter microsatellite polymorphism is associated with restenosis after percutaneous transluminal angioplasty. J. Endovasc. Ther. **8**(5), 433–440 (2001)
35. Bao, W., Song, F., Li, X., et al.: Association between heme oxygenase-1 gene promoter polymorphisms and type 2 diabetes mellitus: a HuGE review and meta-analysis. Am. J. Epidemiol. **172**(6), 631–636 (2010)
36. Erener, S.: Diabetes, infection risk and COVID-19. Mol. Metab. **39**, 101044 (2020)

37. Benetti, E., Tita, R., Spiga, O., et al.: ACE2 gene variants may underlie interindividual variability and susceptibility to COVID-19 in the Italian population. Eur. J. Hum. Genet. **28**, 1602–1614 (2020)
38. Giangreco, G.: Case fatality rate analysis of Italian COVID-19 outbreak. J. Med. Virol. **92**(7), 919–923 (2020)
39. Hon, K.L.E., Leung, K.K.Y.: Pediatric COVID-19: what disease is this? World J. Pediatr. **16**(4), 323–325 (2020). https://doi.org/10.1007/s12519-020-00375-z
40. Baughn, L.B., Sharma, N., Elhaik, E., et al.: TMPRSS2 in SARS-CoV-2 Infection. Mayo Clin. Proc. **95**(9), 1989–1999 (2020)
41. Raslan, M.A., Alshahawey, M., Shehata, E.M., et al.: Does human leukocyte antigen gene polymorphism affect management of COVID-19 Patients? A review article. Sci. J. Genet. Gene Ther. **6**(1), 001–003 (2020)
42. Palau, V., Riera, M., Soler, M.J.: ADAM17 inhibition may exert a protective effect on COVID-19. Nephrol. Dial Transplant. **35**(6), 1071–1072 (2020)
43. Dattilo, M.: The role of host defences in Covid 19 and treatments thereof. Mol. Med. **26**(1), 90 (2020)

COVID-19 Diagnostic Approaches: An Overview

Zejneba Jassin(✉), Amir Heric, Amar Mujkic, and Ena Baralic

Department of Genetics and Bioengineering, International Burch University, Sarajevo, Bosnia and Herzegovina
zejneba.jassin@stu.ibu.edu.ba

Abstract. In the midst of the global COVID-19 pandemic, diagnostic approaches have played a critical role in tracking and fighting the virus. In the light of this, the state-of-the-art biosensing devices and procedures being repurposed for COVID-19 diagnosis are critically evaluated. Various COVID-19 diagnostic approaches are constantly being tested and optimized in order to ensure accurate diagnosis and reduce false results. This review spotlights diagnostic strategies such as: (1) Reverse transcription polymerase chain reaction (RT-PCR), (2) Enzyme-linked immunosorbent test (ELISA), (3) Chemiluminescent Immunoassay (CLIA), (4) Lateral flow immunoassay (LFIA), (5) Clustered Regularly Interspersed Palindromic Repeats (CRISPR-Cas13), (6) Clustered Regularly Interspersed Palindromic Repeats (CRISPR-Cas12a), (7) Loop-mediated Isothermal Amplification (LAMP-based colorimetric method), (8) Point-of-care tests (POC- RT-PCR), (9) Point-of-care tests (POC- Isothermal DNA amplification), (10) reactive oxygen species (ROS), (11) Chest imaging using computerized tomography (CT), (12) X-ray radiography (CXR), (13) Field-effect transistor (FET) in order to address different biomarkers correlated with COVID-19. The overview of these diagnostic approaches is given in terms of their strengths, limitations, time per analysis, costs, and availability. Finally, conclusions on diagnostic tests are highlighted.

Keywords: COVID-19 · SARS-CoV-2 · Viral diagnosis · Diagnostic approaches · Overview

1 Introduction

The novel virus known as a coronavirus (Sars-CoV-2) caused "unidentified viral pneumonia" that emerged in China on December 31, 2019 [1, 2]. The province of Hubei, Wuhan City dealt first with this unfamiliar disease and it began to spread to the rest of the world.

SARS-Cov-2 presents similar sequence homology with the SARS-CoV virus which emerged in 2003 [1, 3, 4], however, many significant differences and similarities are seen in contrast to the current virus. SARS-CoV-2 replicates predominantly in the upper and lower respiratory tracts and is primarily transmitted from asymptomatic and symptomatic infected subjects through droplets and aerosols [5, 6]. Also, case fatality rates for this virus are lower than those of MERS-CoV and SARS-CoV viruses.

While SARS-CoV-2 currently shares 79% of its genome with SARS-CoV, it seems to be even more communicable [5]. The baseline reproduction number (R0) is calculated

A. Badnjevic and L. Gurbeta Pokvić (Eds.): CMBEBIH 2021, IFMBE Proceedings 84, pp. 892–906, 2021.
https://doi.org/10.1007/978-3-030-73909-6_100

to be between 2 and 3 and the median incubation period is 6 days (range 2–14 days) [6]. Around 80% of infections with COVID-19 are mild or asymptomatic; 15% are serious, requiring additional oxygen; and 5% are vital, requiring mechanical ventilation. Breath shortages occur in 18 percent of COVID-19 patients [7].

Over 1.8 million new COVID-19 cases and 39 000 new deaths have been reported to the WHO on the 16th of August, bringing the total number to 21.2 million confirmed COVID-19 cases, including 761 000 deaths [8]. The current prevalence of infection fatality in all age groups is between 0.5% and 1%, and the rate increases after the sixth decade of life [6]. The fatality of COVID-19 is the product of several mechanisms, including comorbidity complications (e.g., cardiovascular disorders), massive alveolar damage, fulminant myocarditis, progressive respiratory failure, and in critical cases, cytokine storms [9–11]. The vaccines that are currently being produced have been quickly monitored and it is therefore important to store them at even colder temperatures, some as low as −80°C. It is still unknown exactly how stable some of them are under the best possible conditions. Even at 90% efficiency, it can be reduced to about 70% of its potency and below if the conditions are less favourable [12].

Early diagnostic procedures are of great importance for fighting the pandemic so infected cases can be effectively treated. Accurate and fast diagnostic approaches contribute towards the reduction in health care expenses caused by hospitalizations and therapies. As the novel virus appeared, so did the diagnostic and therapeutic approaches. Currently, the gold standard for diagnosing COVID-19 is based on a molecular test of the reverse transcription-polymerase chain reaction (RT-PCR), aimed at detecting the RNA of the virus in respiratory samples such as nasopharyngeal swabs or bronchial aspirate [13, 14]. Even though this method is used as a gold standard, it has its limitations and can produce false-negative results.

Extraordinary instances of rapid, early-stage testing have taken place in a few countries; South Korea can check thousands of people per day using vigorous marketing strategies allowed by the drive-through and 'phone booth' tests [15]. This country has performed over 1.5 million COVID-19 tests, which may contribute to the fact that they have one of the lower COVID-19 mortality rates globally. On the other hand, in other countries, there have been major supply shortages of specialized swabs, face masks, and reagents, which are all materials needed by laboratory personnel to perform tests [16]. Due to this, researchers are striving to develop rapid, sensitive testing approaches that do not require numerous reagents or expensive laboratory equipment [17].

Since professionals are adapting to circumstances in real life while fighting this pandemic, the usage of other diagnostic approaches is not uncommon. Therefore, this paper presents an overview of diagnostic approaches for COVID-19.

2 Materials and Methods

The review presented in this paper is based on X papers that were selected as a result of the search strategy. A manual search was conducted in Google Scholar and NCBI databases. Gray literature from the following databases was also taken into account: World Health Organization (WHO), Centers for Disease Control and Prevention (CDC), NIH National Library of Medicine, PubMed, and AAAS Science Advances, Food and

Drug Administration (FDA). The search was done without limits of timeframe and using descriptors: diagnostic, test, SARS-CoV-2, COVID-19, biosensors. Two authors performed all the literature selection steps individually and then discussed the differences and edited with the other two authors.

All found papers and sources were checked for eligibility based on defined criteria which were:

1. Paper is written in the English language;
2. Paper presents an evaluation of any diagnostic method for COVID-19;
3. Paper presents any diagnostic method for COVID-19;
4. Paper is reporting data on the accuracy of the test (sensitivity and/or specificity);
5. Paper presents biosensing approach in COVID-19 diagnosis;

After the selection of the papers, data extraction was performed. From all selected papers, following information was extracted: general study details, diagnostic approach details, biomarkers, procedure, time per test ratio, sensitivity, costs, availability, biosensing principle, strengths and limitations.

3 Results

Table 1 presents the list of COVID-19 recent diagnostic approaches and gives an overview of their strengths and limitations (Tables 2 and 3).

Table 1. Overview of recent diagnostic approaches for COVID-19

No.	Name of the diagnostic approach	SARS-CoV-2 biomarkers	Procedure
1	RT-PCR	S gene, E genes, N genes, RdRp, (ORF1a/b) genes [18]	A specific gene of interest is amplified: through heating and cooling steps, small amounts of template DNA and primers specific to the gene of interest are cycled, to which the primers bind to and amplify the gene into millions of copies. An additional step is done in RT-PCR, where RNA molecules are reverse transcribed into DNA through the effects of reverse transcriptase [16, 19]

(*continued*)

Table 1. (*continued*)

No.	Name of the diagnostic approach	SARS-CoV-2 biomarkers	Procedure
2	Enzyme-linked immunosorbent test (ELISA)	Antibodies against S, N proteins [20–23]	ELISA immobilizes antibodies to the plate. Patient antibodies are specific to the captured antigen in the sample that is bound to the immobilized captured antigen on the plate. An enzyme-labeled detection antibody that is unique to each of the isotypes of the antibody is added to and directly binds to the captured patient antibodies [19, 24]
3	Chemiluminescent Immunoassay (CLIA)	Antibodies against S, N proteins [20, 25]	Analytes are detected through the chemical reaction of chemical probes to antibodies which generates light [26]
4	Lateral flow immunoassay (LFIA)	Antibodies against S and N proteins [20]	LFIA is based on immune complex reactions. The sample analyte is placed on a pad. The reaction of bioconjugate and antigen forms an immune complex, which moves through the LFIA strip and binds with the complementary antibody [27, 28]
5	CRISPR-Cas13	S and Orflab genes [17, 29]	Cas13 effector protein combines with crRNA to produce an rRNA-guided RNA-targeting CRISPR effector complex that cleaves ssRNA instead of DNA to produce a signal [30, 31]
6	CRISPR-Cas12a	S, N genes [29]	This assay performs simultaneous reverse transcription and isothermal amplification using loop-mediated amplification (RT-LAMP)14 for RNA isolated from universal transport medium (UTM), nasopharyngeal or oropharyngeal swabs, accompanied by Cas12 detection of predefined coronavirus sequences, after which the detection of the virus is verified by reporter molecule cleavage [29]

(*continued*)

Table 1. (*continued*)

No.	Name of the diagnostic approach	SARS-CoV-2 biomarkers	Procedure
7	LAMP-based colorimetric method	S, E, and N genes [32, 33]	The loop-mediated isothermal amplification (LAMP) approach operates on the theory of nucleic acid amplification. Specifically designed primers are used to amplify nucleic acid sequence under isothermal conditions (usually 60–65 °C) [32, 34–36]
8	POC – test (Point-ofcare) RT-PCR	N2, E, RNA dependent RNA polymerase (RdRp) [19, 37]	Processing of the specimen, nucleic acid extraction, RT-PCR, and the detection of amplified genes are all done in a single cartridge [38]
9	POC – test (Point-of-care) Isothermal DNA amplification	N [19]	The test relies on isothermal amplification on nasal swabs and detects the SARS-CoV-2N gene. The test also has a POC test cartridge that can form a connected diagnostic platform with a mobile phone, giving the patients access to information about their health [19, 39]
10	ROS detector system (Reactive Oxygen Species)	H_2O_2/ROS [40]	Sputum samples are applied to a device, where an electrochemical sensor detects the intensity of reactive oxygen species. The software analyzes the data and reports the results [40, 41]
11	CT scan	Ground glass opacities, crazypaving pattern, and consolidation [18]	CT is combined with an electronic detector array that uses ionizing radiation or x-rays to record a density pattern and produce an image of a "slice" or "cut" of a particular tissue. Inside the detector, the x-ray beam rotates around the target, so that several x-ray projections pass through the object [42]
12	X-ray radiography	Irregular masses [43–47]	Using a relatively small dose of ionizing radiation, chest x-rays create images of the interior of the chest [48]
13	FET	Sars-Cov-2 Spike Protein [49]	Nasopharyngeal swab specimens are applied to the FET. Graphene sheets are coated with antibody against SARS-CoV-2 spike protein to produce the sensor [49]

Table 2. Characteristics of recent diagnostic approaches for COVID-19

No.	Name of diagnostic approach	Time per test ratio	Sensitivity	Costs	Availability
1	RT-PCR	Their results may take from 4–6 h up to 3 days [16, 24]	40%–90% [19, 50, 51]	$50–$200 [19]	Hospitals, drive-through clinics, academic institutions [16, 19]
2	Enzyme-linked immunosorbent test (ELISA)	100–160 min [19, 20, 52]	50%–98% [20, 53–57]	Antibody lab test is around $50–$150 [19]	Hospitals, drive-through clinics, academic institutions [16, 19]
3	Chemiluminescent immunoassay (CLIA)	30 min [57, 58]	73, 3%–98,6% [25, 54, 57, 59, 60]	Antibody lab test is around $50–$150 [19, 58]	Clinical laboratories [30]
4	Lateral flow immunoassay (LFIA)	5–20 min [57, 58, 61, 62]	65%–96% [27, 54, 57, 61, 63]	Antibody lab test is around $50–$150 [19]	Care homes, universities, NHS staff, public health teams, and schools [64]
5	CRISPR-Cas13	~1 h [65]	96%–97% [66]	Under $40USD [67]	Laboratories [65]
6	CRISPR-Cas12a	45 min [18]	90% [18]	Estimated to be ~$6 [29, 68]	Home, Small clinics Pending clinical verification [29, 68]
7	LAMP-based colorimetric method	20–50 min [32, 33, 69]	95% [41]	~$25 [70]	This technique holds promise for use outside of a central laboratory (in-field detection) by staff without special training and without need of advanced equipment [33]
8	POC – test (Point-of-care) RT-PCR	~ 40 min [19]	100% [19]	$20 [71]	Physician offices and clinics [19]
9	POC – test (Point-of-care) Isothermal DNA amplification	~20 min [19]	1–2X LoD 95% 5X LoD 100% 10X LoD 100% 50X LoD 100% [19]	/	Certified laboratories [72]

(continued)

Table 2. (*continued*)

No.	Name of diagnostic approach	Time per test ratio	Sensitivity	Costs	Availability
10	ROS detector system (Reactive Oxygen Species)	Less than 30 s [41]	97% [40]	/	Laboratories, clinics [40]
11	CT scan	~30 min [72, 73]	92% [32]	$437–$2,239 [74]	Centralized diagnostic services, hospitals, clinics [74, 75]
12	X-ray radiography	5 s [76]	94.38% [77]	$41–$285 [74]	Healthcare infrastructure, hospitals [44, 78]
13	FET	10 min [79]	/	/	In the developmental stage [49]

Table 3. Overview of biosensing principles in recent diagnostic approaches for COVID-19

No.	Name of diagnostic approach	Biosensing principle	Strengths/Limitations
1	RT-PCR	The viral RNA molecule is converted into cDNA by reverse transcriptase. Afterward, the DNA molecule is amplified, and the results of the reaction are detected [80]	– Poor accuracy – High sensitivity, the possibility of a large-scale operation – Professional technicians needed, difficult data analysis, expensive, false-negative (asymptomatic) or false-positive results, global shortages of swabs and personal protective equipment – During insufficient shipping, genetic content can be denatured, bringing in false negatives as well – The quality and quantity of reagents used by different manufacturers of qRT-PCR kits may also affect consistency between the results – Restricted due to lack of flexibility, and lacks of functional reliability [13, 16, 32, 81–83]

(*continued*)

Table 3. (*continued*)

No.	Name of diagnostic approach	Biosensing principle	Strengths/Limitations
2	Enzyme-linked immunosorbent test (ELISA)	The enzyme is used to detect binding antigen or antibody [84]	– Likely to be false-positive rather than true-positive, even when the high-specificity test is used – The basic advantages of serological tests are their faster turn-around time, high output, less workload – The precision of the test, reliability, simplicity, speed, flexibility, and cost will determine the scope of impact [85–88]
3	Chemiluminescent immunoassay (CLIA)	Immunoreaction is combined with a chemiluminescent system to detect analytes [26]	– Larger throughput than ELISA – Doesn't require reagents for stopping the enzymatic reaction – Shorter incubation steps [19]
4	Lateral flow immunoassay (LFIA)	Analyte moves through different zones of the polymeric strip on which molecules that can interact with the analyte are placed [27]	– Quick, point of care approach – Simple to use – Results are often false negative – Verification of positive or doubtful results needed – More scientific research is needed on this test [62, 89]
5	CRISPR-Cas13	crRNA-Cas13a complex binds specifically to target nucleic acid and cleaves it, alongside nontarget RNA associated with a fluorescent reporter [90]	– Rapid identification of a single shift in a single molecule in a DNA or RNA sequence – Low false-negative and false-positive results – CRISPR has an increased advantage in testing as it can detect multiple variants of SARS-CoV-2 [17, 29, 65]
6	CRISPR-Cas12a	Complex crRNA-Cas12a binds to target DNA and generates ssDNA cleavage that is associated with a fluorescent reporter [90]	– Simple, rapid, ultrasensitive, precise – gives consistent results, onepot, and visual detection – Detection findings can be directly visualized by the naked eye, greatly simplifying the detection process and removing the need for separate detection based on lateral flow – Low false-negative and false-positive results [17, 18, 29, 68]

(*continued*)

Table 3. (*continued*)

No.	Name of diagnostic approach	Biosensing principle	Strengths/Limitations
7	LAMP-based colorimetric method	Target nucleic acid is amplified at a constant temperature, with the help of 4–6 specially designed primers [38]	– Simple, rapid amplification, simple operation, easy detection and will be useful in developing nations for rapid and accurate clinical diagnosis of SARS-CoV [35]
8	POC – test (Point-of-care) RT-PCR	Real-time PCR molecular test is combined with the software that allows the user to see amplification curve and results for N2 and E biomarkers [19, 37]	– Designed not to require complex machinery or devices and can preferably be used by individuals in the home environment. Therefore, anyone can be checked anywhere or anytime [19]
9	POC – test (Point-of-care) Isothermal DNA amplification	Target genes are amplified at a constant temperature. After completion of the test, results are instantly sent to the user's smartphone via a software platform [19, 39]	– Designed not to require complex machinery or devices and can preferably be used by individuals in the home environment. Therefore, anyone can be checked anywhere or anytime – With isothermal amplification, genes in a multiplex are more difficult to detect [19]
10	ROS detector system (Reactive Oxygen Species)	Detects overproduction of ROS by mitochondria, caused by lung cell dysfunction [40]	– Portable, simple, rapid [41]
11	CT scan	CT uses ionizing radiation or x-rays, together with an electronic detector array to record a density pattern and produce an image of a tissue [42]	– Anomalies in the CT scans may be linked with severity and prognosis of the disorder – With CT scans, not all cases can be perfectly detected as some cases may be asymptomatic or not yet be visible as affecting the lungs
			– The combined method with visualization, using CT and molecular therapy will also help to better track and managing of COVID-19 [3, 79, 91–93]
12	X-ray radiography	X-ray beam that passes through the body is either absorbed or scattered by internal tissues; the remaining X-rays is passed to a detector [94]	– Available and accessible, reduced logistical difficulties and cost - Inconclusive results due to inability to distinguish between COVID 19 and other pneumonia inducing pathogens [43–45, 78, 95, 96]
13	FET	Conjugation of SARS-CoV-2 spike protein to a graphene sheet [49]	– Highly sensitive, no sample pretreatment is required, no cross reactivity with MERS-CoV [49]

4 Conclusion

Accurate diagnosis is a crucial step in the fight against the COVID-19 pandemic, as it provides us with methods for timely patients screening and management of viral outbreaks. RT-PCR molecular tests are regarded as a gold standard, as they provide relatively accurate results. However, they are not easily scalable for mass population testing due to several reasons such as: the need for trained personnel, high turnaround time or kits that may affect consistency between the results. Lack of flexibility as well as specimen collection and processing are found to infect medical personnel and the cost is relatively high. Nevertheless, point of care (POC) tests are being developed as they make the process of data acquisition and sharing more efficient but also prove to be more affordable. On the other hand, serological tests have been used because they complement RT-PCRs disadvantages, but they should not be used as a standalone test to diagnose or screen or as a surveillance tool in presymptomatic and asymptomatic populations. Their sensitivity highly varies in regards to the manufacturer, and the day from which the symptoms onset. Due to the need for effective mass population testing, LAMP tests based on isothermal amplification of the target genes are being developed as they are relatively cheap and can be performed by non-trained personnel making them the perfect candidates for testing in developing countries. In some cases, LAMP is combined with Cas12a proteins as it simplifies the detection process and gives consistent results. Due to the emergence of new variants of Sars-Cov-2, more attention should be put toward CRISPR-Cas13a, as it can distinguish between different variants of the virus. On another note, newly developed biosensors as the ROS detector system and FET show promising results. Particularly the nanotechnology-based biosensor FET which can accurately detect the presence of the virus even in individuals with a small viral load. Unfortunately, they are still in the developmental phase and more research is needed. When it comes to assessing the progression of the COVID-19 in symptomatic cases, CT and X-ray methods are useful but they should never be done as standalone tests, as they can't reveal the cause of infection if it is even present. Furthermore, combination with molecular test is desirable but it will require several days to get all the results and the costs won't be affordable to all, which is a major drawback. In our best knowledge, tracking the patient's history of symptoms will be the best way in order to decide which type of the test should be obtained. All in all, proper population diagnostic is a must in the current pandemic, but also in every possible future viral outbreak that may come.

Conflict of Interests. The authors have nothing to declare.

References

1. Ai, T., Yang, Z., Hou, H., Zhan, C., Chen, C., Lv, W., Tao, Q., Sun, Z., Xia, L.: Correlation of chest CT and RT-PCR testing in coronavirus disease 2019 (covid-19) in china: a report of 1014 cases. Radiology **296**(2), E32–E40 (2020)
2. Rada, G., Verdugo-Paiva, F., Avila, C., Morel-Marambio, M., Bravo-Jeria, R., Pesce, F., Madrid, E., Izcovich, A.: Evidence synthesis relevant to COVID-19: a protocol for multiple systematic reviews and overviews of systematic reviews. Medwave **220**(3), e7868 (2020)

3. Vellingiri, B., Jayaramayya, K., Iyer, M., Narayanasamy, A., Govindasamy, V., Giridharan, B., Ganesan, S., Venugopal, A., Venkatesan, D., Ganesan, H., et al.: COVID-19: a promising cure for the global panic. Sci. Total Environ. **825**, 138277 (2020)
4. Saif, L.: Animal coronaviruses: what can they teach us about the severe acute respiratory syndrome? Revue scientifique et technique-Office international des epizooties **23**(2), 643–660 (2004)
5. Jiang, F., Deng, L., Zhang, L., Cai, Y., Cheung, C.W., Xia, Z.: Review of the clinical characteristics of coronavirus disease 2019 (COVID-19). J. Gen. Intern. Med. **4**, 1–5 (2020)
6. Salzberger, B., Buder, F., Lampl, B., Ehrenstein, B., Hitzenbichler, F., Holzmann, T., Schmidt, B., Hanses, F.: Epidemiology of SARS-CoV-2. Infection **8**, 1–7 (2020)
7. Dashraath, P., Jeslyn, W.J., Karen, L.M., Min, L.L., Sarah, L., Biswas, A., Choolani, M.A., Mattar, C., Lin, S.L.: Coronavirus disease 2019 (COVID-19) pandemic and pregnancy. Am. J. Obstet. Gynecol. **222**(6), 521–531 (2020)
8. World Health Organization: Coronavirus disease (COVID-19): weekly epidemiological, update 1 (2020)
9. Mahmoudi, M.: Emerging biomolecular testing to assess risk of mortality from COVID-19 infection. Mol. Pharm. **18**(2), 476–482 (2020)
10. Mehta, P., McAuley, D.F., Brown, M., Sanchez, E., Tattersall, R.S., Manson, J.J., Collaboration, H.A.S., et al.: COVID-19: consider cytokine storm syndromes and immunosuppression. Lancet (London, England) **395**(10229), 1033 (2020)
11. Ruan, Q., Yang, K., Wang, W., Jiang, L., Song, J.: Clinical predictors of mortality due to COVID-19 based on an analysis of data of 150 patients from Wuhan, China. Intensive Care Med. **46**(5), 846–848 (2020)
12. Gruenwald, H.: Covid-19 and vaccination
13. Tang, Y.-W., Schmitz, J.E., Persing, D.H., Stratton, C.W.: Laboratory diagnosis of COVID19: current issues and challenges. J. Clin. Microbiol. **58**(6), e00512–e00520 (2020)
14. Qiu, G., Gai, Z., Tao, Y., Schmitt, J., Kullak-Ublick, G.A., Wang, J.: Dual-functional plasmonic photothermal biosensors for highly accurate severe acute respiratory syndrome coronavirus 2 detection. ACS Nano **14**(5), 5268–5277 (2020)
15. Santiago, I.: Trends and innovations in biosensors for COVID-19 mass testing. ChemBioChem **21**, 1–11 (2020)
16. Morales-Narvaez, E., Dincer, C.: The impact of biosensing in a pandemic outbreak: COVID-19. Biosens. Bioelectron. **163**, 112274 (2020)
17. Zhang, F., Abudayyeh, O.O., Gootenberg, J.S.: A protocol for detection of covid-19 using crispr diagnostics (2020)
18. Cui, F., Zhou, H.S.: Diagnostic methods and potential portable biosensors for coronavirus disease 2019. Biosens. Bioelectron. **165**, 112349 (2020)
19. Ravi, N., Cortade, D.L., Ng, E., Wang, S.X.: Diagnostics for SARS-CoV-2 detection: a comprehensive review of the FDA-EUA COVID-19 testing landscape. Biosens. Bioelectron. **1**(165), 112454 (2020)
20. Tantuoyir, M.M., Rezaei, N.: Serological tests for COVID-19: potential opportunities. Cell Biol. Int. **45**(4), 740–748 (2020)
21. Touma, M.: COVID-19: molecular diagnostics overview. J. Mol. Med. **98**(7), 947–54 (2020)
22. Sapkal, G., Shete-Aich, A., Jain, R., Yadav, P.D., Sarkale, P., Lakra, R., Baradkar, S., Deshpande, G.R., Mali, D., Tilekar, B.N., Majumdar, T.: Development of indigenous IgG ELISA for the detection of anti-SARS-CoV-2 IgG. Indian J. Med. Res. **151**(5), 444 (2020)
23. Zhang, L., Guo, H.: Biomarkers of COVID-19 and technologies to combat SARS-CoV-2. Adv. Biomark. Sci. Technol. **2**, 1–23 (2020)
24. Weissleder, R., Lee, H., Ko, J., Pittet, M.J.: COVID-19 diagnostics in context. Sci. Transl. Med. **12**(546), eabc1931 (2020)

25. Infantino, M., Grossi, V., Lari, B., Bambi, R., Perri, A., Manneschi, M., Terenzi, G., Liotti, I., Ciotta, G., Taddei, C., Benucci, M.: Diagnostic accuracy of an automated chemiluminescent immunoassay for anti-SARS-CoV-2 IgM and IgG antibodies: an Italian experience. J. Med. Virol. **92**(9), 1671–5 (2020)
26. Xiao, Q., Xu, C.: Research progress on chemiluminescence immunoassay combined with novel technologies. TrAC Trends Anal. Chem. **124**, 115780 (2020)
27. Borse, V.B., Konwar, A.N., Jayant, R.D., Patil, P.O.: Perspectives of characterization and bioconjugation of gold nanoparticles and their application in lateral flow immunosensing. Drug Deliv. Transl. Res. **10**(4), 878–902 (2020)
28. Koczula, K.M., Gallotta, A.: Lateral flow assays. Essays Biochem. **60**(1), 111–20 (2016)
29. Broughton, J.P., Deng, X., Yu, G., Fasching, C.L., Servellita, V., Singh, J., Miao, X., Streithorst, J.A., Granados, A., Sotomayor-Gonzalez, A., et al.: CRISPR–Cas12-based detection of SARS-CoV-2. Nat. Biotechnol. **38**(7), 870–874 (2020)
30. Nicol, T., Lefeuvre, C., Serri, O., Pivert, A., Joubaud, F., Dubée, V., Kouatchet, A., Ducancelle, A., Lunel-Fabiani, F., Le Guillou-Guillemette, H.: Assessment of SARS-CoV-2 serological tests for the diagnosis of COVID-19 through the evaluation of three immunoassays: two automated immunoassays (Euroimmun and Abbott) and one rapid lateral flow immunoassay (NG Biotech). J. Clin. Virol. **129**, 104511 (2020)
31. East-Seletsky, A., O'Connell, M.R., Knight, S.C., Burstein, D., Cate, J.H., Tjian, R., Doudna, J.A.: Two distinct RNase activities of CRISPR-C2c2 enable guide-RNA processing and RNA detection. Nature **538**(7624), 270–273 (2016)
32. Benzigar, M.R., Bhattacharjee, R., Baharfar, M., Liu, G.: Current methods for diagnosis of human coronaviruses: pros and cons. Anal. Bioanal. Chem. **20**, 1–20 (2020)
33. Lamb, L.E., Bartolone, S.N., Ward, E., Chancellor, M.B.: Rapid detection of novel coronavirus (COVID19) by reverse transcription-loop-mediated isothermal amplification. Available at SSRN 3539654 (2020)
34. Pyrc, K., Milewska, A., Potempa, J.: Development of loop-mediated isothermal amplification assay for detection of human coronavirus-NL63. J. Virol. Methods **175**(1), 133–136 (2011)
35. Thai, H.T., Le, M.Q., Vuong, C.D., Parida, M., Minekawa, H., Notomi, T., Hasebe, F., Morita, K.: Development and evaluation of a novel loop-mediated isothermal amplification method for rapid detection of severe acute respiratory syndrome coronavirus. J. Clin. Microbiol. **42**(5), 1956–61 (2004)
36. Augustine, R., Hasan, A., Das, S., Ahmed, R., Mori, Y., Notomi, T., Kevadiya, B.D., Thakor, A.S.: Loop-mediated isothermal amplification (Lamp): a rapid, sensitive, specific, and cost-effective point-of-care test for coronaviruses in the context of COVID-19 pandemic. Biology **9**(8), 182 (2020)
37. Loeffelholz, M.J., Alland, D., Butler-Wu, S.M., Pandey, U., Perno, C.F., Nava, A., Carroll, K.C., Mostafa, H., Davies, E., McEwan, A., Rakeman, J.L.: Multicenter evaluation of the cepheid Xpert Xpress SARS-CoV-2 test. J. Clin. Microbiol. **58**(8), e00926-20 (2020)
38. Huang, W.E., Lim, B., Hsu, C.C., Xiong, D., Wu, W., Yu, Y., Jia, H., Wang, Y., Zeng, Y., Ji, M., Chang, H.: RT-LAMP for rapid diagnosis of coronavirus SARS-CoV-2. Microbial Biotechnol. **13**(4), 950–961 (2020)
39. Diagnostic testing at the speed of life | Cue (2020). https://www.cuehealth.com/#product. Accessed 26 Dec 2020
40. Miripour, Z.S., Sarrami-Forooshani, R., Sanati, H., Makarem, J., Taheri, M.S., Shojaeian, F., Eskafi, A.H., Abbasvandi, F., Namdar, N., Ghafari, H., Aghaee, P.: Real-time diagnosis of reactive oxygen species (ROS) in fresh sputum by electrochemical tracing; correlation between COVID-19 and viral-induced ROS in lung/respiratory epithelium during this pandemic. Biosens. Bioelectron. **165**, 112435 (2020)
41. Xu, L., Li, D., Ramadan, S., Li, Y., Klein, N.: Facile biosensors for rapid detection of COVID-19. Biosens. Bioelectron. **170**, 112673 (2020)

42. Caldemeyer, K.S., Buckwalter, K.A.: The basic principles of computed tomography and magnetic resonance imaging. J. Am. Acad. Dermatol. **41**(5), 768–71 (1999)
43. Yin, Z., Wu, M., Wu, Z.: Evaluation of deep learning-based approaches for COVID-19 classification based on chest X-ray images. SIViP **7**, 1–8 (2021)
44. Li, A.C., Lee, D.T., Misquitta, K.K., Uno, K., Wald, S.: COVID-19 detection from chest radiographs using machine learning and convolutional neural networks. medRxiv (2020)
45. Han, Y., Chen, C., Tewfik, A.H., Ding, Y., Peng Y.: Pneumonia detection on chest X-ray using radiomic features and contrastive learning. arXiv preprint arXiv:2101.04269 (2021)
46. Afifi, A., Hafsa, N.E., Ali, M.A., Alhumam, A., Alsalman, S.: An ensemble of global and local attention based convolutional neural networks for COVID-19 diagnosis on chest X-ray images. Symmetry **13**(1), 113 (2021)
47. Siddiqui, S.Y., Abbas, S., Khan, M.A., Naseer, I., Masood, T., Khan, K.M., Al Ghamdi, M.A., Almotiri, S.H.: Intelligent decision support system for COVID-19 empowered with deep learning. CMC-Comput. Mater. Continua **66**(2), 1719–1732 (2021)
48. Ypsilantis, P.P., Montana, G.: Learning what to look in chest X-rays with a recurrent visual attention model. arXiv preprint arXiv:1701.06452 (2017)
49. Seo, G., Lee, G., Kim, M.J., Baek, S.-H., Choi, M., Ku, K.B., Lee, C.-S., Jun, S., Park, D., Kim, H.G., et al.: Rapid detection of COVID-19 causative virus (SARS-CoV-2) in human nasopharyngeal swab specimens using field-effect transistor-based biosensor. ACS Nano **14**(4), 5135–5142 (2020)
50. Fda.gov (2020). https://www.fda.gov/media/134922/download. Accessed 26 Dec 2020
51. Miller, T.E., Garcia Beltran, W.F., Bard, A.Z., Gogakos, T., Anahtar, M.N., Astudillo, M.G., Yang, D., Thierauf, J., Fisch, A.S., Mahowald, G.K., Fitzpatrick, M.J.: Clinical sensitivity and interpretation of PCR and serological COVID-19 diagnostics for patients presenting to the hospital. FASEB J. **34**(10), 13877–13884 (2020)
52. Lionex.de (2021). https://lionex.de/wpcontent/uploads/2020/05/COVID-19-ELISA-Human-IgG_EN_Instructions-for-use-rev.-0.pdf. Accessed 31 Jan 2021]
53. Open.fda.gov (2021). https://open.fda.gov/apis/device/covid19serology/. Accessed 31 Jan 2021
54. Bastos, M.L., Tavaziva, G., Abidi, S.K., Campbell, J.R., Haraoui, L.P., Johnston, J.C., Lan, Z., Law, S., MacLean, E., Trajman, A., Menzies, D.: Diagnostic accuracy of serological tests for COVID-19: systematic review and meta-analysis. BMJ **370**, m2516 (2020)
55. Gutiérrez-Cobos, A., de Frutos, S.G., García, D.D., Lara, E.N., Carrión, A.Y., García-Rodrigo, L.F., Torres, A.M., Domingo, L.C.: Evaluation of diagnostic accuracy of 10 serological assays for detection of SARS-CoV-2 antibodies. Eur. J. Clin. Microbiol. Infect. Dis. **24**, 1–7 (2020)
56. Krüttgen, A., Cornelissen, C.G., Dreher, M., Hornef, M., Imöhl, M., Kleines, M.: Comparison of four new commercial serologic assays for determination of SARS-CoV-2 IgG. J. Clin. Virol. **128**, 104394 (2020)
57. Kontou, P.I., Braliou, G.G., Dimou, N.L., Nikolopoulos, G., Bagos, P.G.: Antibody tests in detecting SARS-CoV-2 infection: a meta-analysis. Diagnostics **10**(5), 319 (2020)
58. Vashist, S.K.: In vitro diagnostic assays for COVID-19: recent advances and emerging trends. Diagnostics **10**(4), 202 (2020)
59. Ma, H., Zeng, W., He, H., Zhao, D., Yang, Y., Jiang, D., Zhou, P., Qi, Y., He, W., Zhao, C., Yi, R.: COVID-19 diagnosis and study of serum SARS-CoV-2 specific IgA, IgM and IgG by a quantitative and sensitive immunoassay. MedRxiv (2020)
60. Soleimani, R., Khourssaji, M., Gruson, D., Rodriguez-Villalobos, H., Berghmans, M., Belkhir, L., Yombi, J.C., Kabamba-Mukadi, B.: Clinical usefulness of fully automated chemiluminescent immunoassay for quantitative antibody measurements in COVID-19 patients. J. Med. Virol. **93**(3), 1465–1477 (2020)

61. Li, Z., Yi, Y., Luo, X., Xiong, N., Liu, Y., Li, S., Sun, R., Wang, Y., Hu, B., Chen, W., Zhang, Y.: Development and clinical application of a rapid IgM-IgG combined antibody test for SARS-CoV-infection diagnosis. J. Med. Virol. **92**(9), 1518–1524 (2020)
62. Kubina, R., Dziedzic, A.: Molecular and serological tests for COVID-19 a comparative review of SARS-CoV-2 coronavirus laboratory and point-of-care diagnostics. Diagnostics **10**(6), 434 (2020)
63. Adams, E.R., Ainsworth, M., Anand, R., Andersson, M.I., Auckland, K., Baillie, J.K., Barnes, E., Beer, S., Bell, J.I., Berry, T., Bibi, S.: Antibody testing for COVID-19: a report from the National COVID Scientific Advisory Panel. Wellcome Open Res. **5**(139), 139 (2020)
64. Deeks, J.J., Raffle, A.E.: Lateral flow tests cannot rule out SARS-CoV-2 infection. BMJ **371**, m4787 (2020)
65. Sherlock Biosciences receives FDA emergency use authorization for CRISPR SARS-CoV-2 rapid diagnostic. Sherlock Biosciences Sherlock.bio (2021). https://sherlock.bio/sherlock-biosciences-receives-fdaemergency-use-authorization-for-crispr-sars-cov-2-rapid-diagnostic/. Accessed 31 Jan 2021
66. Patchsung, M., Jantarug, K., Pattama, A., Aphicho, K., Suraritdechachai, S., Meesawat, P., Sappakhaw, K., Leelahakorn, N., Ruenkam, T., Wongsatit, T., Athipanyasilp, N.: Clinical validation of a Cas13-based assay for the detection of SARS-CoV-2 RNA. Nat. Biomed. Eng. **4**(12), 1140–9 (2020)
67. Joung, J., Ladha, A., Saito, M., Segel, M., Bruneau, R., Huang, M.L., Kim, N.G., Yu, X., Li, J., Walker, B.D., Greninger, A.L.: Point-of-care testing for COVID-19 using SHERLOCK diagnostics. MedRxiv (2020)
68. Ding, X., Yin, K., Li, Z., Lalla, R.V., Ballesteros, E., Sfeir, M.M., Liu, C.: Ultrasensitive and visual detection of SARS-CoV-2 using all-in-one dual CRISPR-Cas12a assay. Nat. Commun. **11**(1), 1 (2020)
69. Shirato, K., Semba, S., El-Kafrawy, S.A., Hassan, A.M., Tolah, A.M., Takayama, I., Kageyama, T., Notomi, T., Kamitani, W., Matsuyama, S., Azhar, E.I.: Development of fluorescent reverse transcription loop-mediated isothermal amplification (RT-LAMP) using quenching probes for the detection of the Middle East respiratory syndrome coronavirus. J. Virol. Methods **1**(258), 41–8 (2018)
70. Oxford researchers develop portable COVID-19 test costing less than $25. Fierce-Biotech (2021). https://www.fiercebiotech.com/medtech/oxford-researchers-develop-portable-covid-19-test-costing-less-than25. Accessed 31 Jan 2021
71. Cepheid charging four times more than it should for coronavirus COVID-19 tests | MSF. Médecins Sans Frontières (MSF) International (2021). https://www.msf.org/diagnostic-company-cepheid-charging-more-itshould-covid-19-tests. Accessed 31 Jan 2021
72. Fda.gov (2021). https://www.fda.gov/media/138826/download. Accessed 31 Jan 2021
73. Scan, P.: Preparing for a CT Scan, WakeMed Health & Hospitals, Raleigh & Wake County, NC. Wakemed.org (2021). https://www.wakemed.org/care-and-services/imaging-services/ct-scan/preparing-for-a-ctscan. Accessed 31 Jan 2021
74. Gooch, K.: The out-of-pocket costs of X-rays, CT scans across 3 states: 4 things to know. Beckershospitalreview.com (2021). https://www.beckershospitalreview.com/finance/the-out-of-pocket-costs-of-x-rays-ctscans-across-3-states-4-things-to-know.html#:~:text=Researchers%20found%20that%20across%20hospitals,3. Accessed 31 Jan 2021
75. Ai, T., Yang, Z., Hou, H., Zhan, C., Chen, C., Lv, W., Tao, Q., Sun, Z., Xia, L.: Correlation of chest CT and RT-PCR testing for coronavirus disease 2019 (COVID-19) in China: a report of 1014 cases. Radiology **296**(2), E32–E40 (2020)
76. Panwar, H., Gupta, P.K., Siddiqui, M.K., Morales-Menendez, R., Singh, V.: Application of deep learning for fast detection of COVID-19 in X-Rays using nCOVnet. Chaos Solitons Fractals **138**, 109944 (2020)

77. Ijitee.org (2021). https://www.ijitee.org/wp-content/uploads/Souvenir_Volume-9_Issue-6_April_2020.pdf. Accessed 31 Jan 2021
78. Ahmad, F., Farooq, A., Ghani, M.U.: Deep Ensemble model for classification of novel coronavirus in chest X-ray images. Comput. Intell. Neurosci. **12**, 2021 (2021)
79. Ji, T., Liu, Z., Wang, G., Guo, X., Lai, C., Chen, H., Huang, S., Xia, S., Chen, B., Jia, H., Chen, Y.: Detection of COVID-19: A review of the current literature and future perspectives. Biosens. Bioelectron. **166**, 112455 (2020)
80. Wong, M.L., Medrano, J.F.: Real-time PCR for mRNA quantitation. Biotechniques **39**(1), 75–85 (2005)
81. Long, C., Xu, H., Shen, Q., Zhang, X., Fan, B., Wang, C., Zeng, B., Li, Z., Li, X., Li, H.: Diagnosis of the coronavirus disease (COVID-19): rRT-PCR or CT? Eur. J. Radiol. **126**, 108961 (2020)
82. Li, Y., Yao, L., Li, J., Chen, L., Song, Y., Cai, Z., Yang, C.: Stability issues of RT-PCR testing of SARS-CoV-2 for hospitalized patients clinically diagnosed with COVID-19. J. Med. Virol. **92**(7), 903–908 (2020)
83. Suo, T., Liu, X., Feng, J., Guo, M., Hu, W., Guo, D., Ullah, H., Yang, Y., Zhang, Q., Wang, X., Sajid, M.: ddPCR: a more accurate tool for SARS-CoV-2 detection in low viral load specimens. Emerg. Microbes Infect. **9**(1), 1259–1268 (2020)
84. Singh, R.S., Singh, T., Pandey, A.: Microbial enzymes—an overview. Adv. Enzyme Technol. **1**, 1–40 (2019)
85. Gurbuz, M.: Molecular and serological tests for COVID-19. Eurasian J. Pulmonol. **22**(4), 29 (2020)
86. Peeling, R.W., Wedderburn, C.J., Garcia, P.J., Boeras, D., Fongwen, N., Nkengasong, J., Sall, A., Tanuri, A., Heymann, D.L.: Serology testing in the COVID-19 pandemic response. Lancet Infect. Dis. 20 (2020)
87. COVID WT, Montrose TI: A pandemic response update for the Board of Health
88. Chertow, D.S.: Next-generation diagnostics with CRISPR. Science **360**(6387), 381–382 (2018)
89. Wise, J.: COVID-19: safety of lateral flow tests questioned after they are found to miss half of cases. BMJ **371**, m4744 (2020)
90. Xiang, X., Qian, K., Zhang, Z., Lin, F., Xie, Y., Liu, Y., Yang, Z.: CRISPR-Cas systems based molecular diagnostic tool for infectious diseases and emerging 2019 novel coronavirus (COVID-19) pneumonia. J. Drug Target. **28**(7–8), 727–731 (2020)
91. Wang, D., Hu, B., Hu, C., Zhu, F., Liu, X., Zhang, J., Wang, B., Xiang, H., Cheng, Z., Xiong, Y., et al.: Clinical characteristics of 138 hospitalized patients with 2019 novel coronavirus–infected pneumonia in Wuhan, China. JAMA **323**(11), 1061–1069 (2020)
92. Jin, Y.-H., Cai, L., Cheng, Z.-S., Cheng, H., Deng, T., Fan, Y.-P., Fang, C., Huang, D., Huang, L.-Q., Huang, Q., et al.: A rapid advice guideline for the diagnosis and treatment of 2019 novel coronavirus (2019-nCoV) infected pneumonia (standard version). Mil. Med. Res. **7**(1), 4 (2020)
93. Fang, Y., Zhang, H., Xie, J., Lin, M., Ying, L., Pang, P., Ji, W.: Sensitivity of chest CT for covid-19: comparison to RT-PCR. Radiology **296**(2), E115–E117 (2020)
94. 2021. https://www.fda.gov/radiationemitting-products/medical-imaging/medical-x-ray-imag-ing#:~:text=CT%2C%20radiography%2C%20and%20fluoroscopy%20all,computer%20screen)%20for%20recording%20or. Accessed 31 Jan 2021
95. Sakib, S., Siddique, M.A., Khan, M.M., Yasmin, N., Aziz, A., Chowdhury, M., Tasawar, I.K.: Detection of COVID-19 disease from chest X-ray images: a deep transfer learning framework. medRxiv (2020)
96. Jain, R., Gupta, M., Taneja, S., Hemanth, D.J.: Deep learning based detection and analysis of COVID-19 on chest X-ray images. Appl. Intell. **51**, 1690–1700 (2020)

First Report on Public Opinion Regarding COVID-19 Vaccination in Bosnia and Herzegovina

Faris Hrvat[2], Amina Aleta[2], Amra Džuho[1], Osman Hasanić[1], and Lemana Spahić Bećirović[1](✉)

[1] Faculty of Engineering and Natural Sciences, Department of Genetics and Bioengineering, International Burch University, Sarajevo, Bosnia and Herzegovina
lemana.spahic@ibu.edu.ba

[2] Faculty of Medicine, Medical Biotechnology and Molecular Medicine, University of Milan, Milano, Italy

Abstract. Background: COVID-19 has become an emerging health concern since February 2020, and is still affecting both health and economy in the entire world. World scientists are providing joint efforts for better understanding of the virus and the pharmaceutical companies are racing to produce the COVID-19 vaccine which would relieve the stress posed on the world. Social implications of COVID-19 are immense and comparable to health implications. Given the divided public opinion on vaccination for other infectious diseases in general it is interesting to investigate what do people think about COVID-19 vaccine.

Methods: Data was acquired in Bosnia and Herzegovina during the period from June to July of 2020 by means of a public questionnaire which consisted of 13 questions. A total of 1237 was provided by 420 male and 817 female participants in the age range of 16–79.

Result: Out of all respondents 679 would approve administration of the vaccine, while 558 would not. The results presented in this paper show that most of the respondents who are against vaccination belong to less educated part of the community, while higher educated respondents are more optimistic regarding the vaccine effects.

Conclusion: Most vaccination opponents have little to no connection with medical sciences indicating that their attitude towards vaccination is based upon news repositories and social network campaigns which are often filled with false information. As the questionnaire was presented to the public in the period when the effects of COVID-19 were less pronounced than today, it is possible that the public opinion in the future will be shifted to vaccination approval by the general public.

Keywords: COVID-19 · Vaccination · Pandemic · Survey

1 Introduction

Two centuries have passed since Edward Jenner performed first vaccination in 1796 [12] and the consequences of this attempt have been argued extensively. Although this event

A. Badnjevic and L. Gurbeta Pokvić (Eds.): CMBEBIH 2021, IFMBE Proceedings 84, pp. 907–920, 2021.
https://doi.org/10.1007/978-3-030-73909-6_101

has been marked as the start of the modern vaccination, Chinese [21] were performing vaccination 7 centuries before Jenner and the Ancient Greeks [28] were aware that the survivors of plague epidemic were resistant to the new infection. These cases are the evidence that immunity to certain infectious disease has been of human interest for a long time, but it is important to mention that immunology as a separate entity from microbiology began in 20th century [20].

Vaccines are nowadays used as an effective way of improving health of the communities by most usually routinely administering vaccines to the children against major diseases such as tuberculosis, hepatitis B, pertussis, diphtheria, tetanus, polio, crumple, rubella and mumps. [1] Up to now 27 diseases are preventable by using immunization [31]. Vaccines are classified into two groups: (1) live attenuated vaccines [4] which represent the weakened versions of the pathogen and (2) subunit vaccines [18] toxoid vaccines consisting of inactivated toxins [6], conjugate vaccines [8] and carbohydrate vaccines [3]. The latter contains adjuvant substances that enhance magnitude and modify the quality of the immune response [25].

The vast majority of the vaccines are developed empirically [26] which means that little or zero understanding of the immunological mechanisms of their protective immunity induction. In contrast, the need for novel vaccine development against global pandemics such as human immunodeficiency virus (HIV) pandemic has placed the emphasis on the understanding of immunologic mechanisms for inducing protective immunity [25]. Even though vaccination has improved public health the administration of vaccines is far from universal worldwide. The developing world is still susceptible to diseases that can be prevented by vaccines. In addition, the rest of the world has been faced with many obstacles connected with the vaccination despite of HIV, SARS and bird flu pandemics which renewed the importance of the vaccination [14, 20].

According to the data published by World Health Organization (WHO), COVID19 was characterized as a pandemic on 11th March, 2020. At that point, there were 4970 confirmed cases and 313 deaths, globally. Nowadays, there are more than 56,623,643 confirmed cases and more than 1,355,963 deaths globally (until 20th November). In order to ensure the efficient immunization, vaccination rate should be 95% and upward on local and global level respectively [29, 32, 33].

The vaccine development against COVID-19 was triggered and accelerated after the genetic sequence of SARS-CoV-2 was published on 11 January 2020. The high range of humanitarian and economic impact of pandemic has hastened the vaccine development and the first vaccine candidates entered human clinical testing on 16 March 2020. 115 vaccine candidates were presented on April 8 2020. 78 candidates were confirmed as active, while 37 were not confirmed as active. 73 out of 78 confirmed active candidates are at exploratory or preclinical stages and the most advanced candidates have moved into clinical development. Range of technology platforms being evaluated for COVID-19 vaccine include nucleic acids (DNA and RNA), peptide, viral vector, virus-like particle, recombinant protein, live attenuated virus and inactivated virus approach. Some of these technologies are not the basis for licensed vaccine already being used in the practice, but are used in oncology. These techniques are being investigated in order to speed up the vaccine development and manufacture. The technique based on nucleic acids is more flexible in antigen manipulation and the development is much faster, therefore mRNA-based vaccine entered clinical testing 2 months after genetic sequence was revealed [17, 27]. DNA and RNA vaccines do not require culture or fermentation, but are developed using synthetic processes which makes them optimum choice in high-speed development [16, 17] In addition, nucleic acid vaccine elicit CD8 + cytotoxic T cell responses in addition to CD4 + T cell response and antibody and therefore has a key role in virus elimination. DNA vaccines are more stable than mRNA, but mRNA is non-integrating and carries no risk of insertional mutagenesis [27]. Viral vector vaccines are characterized by high level of protein expression and long-time stability as well as inducing strong immune response. In spite of high number of vaccine candidates, public information on the specific SARS-CoV2 antigen used in vaccine is limited. Peptide-based vaccines are the simplest form that can be readily validated and rapidly manufactured. These vaccines may be produced as peptides plus adjuvant mixtures or only peptides can be delivered by a nanocarrier or encoded by nucleic acid vaccine formulation [27]. Mostly, the aim is to induce neutralizing antibodies against the viral spike (S) protein, and thus preventing uptake by the ACE2 receptor [16, 27]. S protein is also glycosylated and that property makes it more organized as opposed to non-glycosylated counterparts as indicated by computational analysis. So, the glycosylation of vaccine candidates might also be considered in following cases: nucleic acid vaccine is produced in situ inside the patient and may carry native glycosylation and in recombinant subunits vaccines glycoengineering protocols may be applied [27]. When it comes to ethical acceptability of accelerating vaccine development, two requirements should be primarily considered – social value and scientific validity. Social value of an effective and safe COVID-19 vaccine would be seen in the reduction of disease, the range of mortality and morbidity and the economic and social costs of the global pandemic. Conversely, the social value would depend on the public confidence in the effectiveness and safeness of the vaccine candidates. Rushing forward could cause lack of public confidence about benefits of particular vaccine candidates and weaken any effort made during the development, especially if rushing gives inadequate or unreliable data. Public could get a mistaken view of how protected they are and become distracted from important health measures [11]. Finally, vaccine developers, funders, regulators, public health bodies and governments should all be open for international collaboration and technology transfer in order to ensure that COVID-19 vaccine quickly moves forward and is provided in sufficient quantities to all affected areas [24].

Several surveys have been conducted worldwide to assess the willingness of the population to get future COVID-19 vaccine. Lazarus, *et al.* (2020), reported survey results of the probability of vaccine acceptance from a population of 19 countries. Chinese population gave the highest number of positive responses (88.6%), while respondents from Poland gave the highest number of negative responses (27.3%). Out of all questioned respondents, 71.5% showed a willingness to get a vaccine if it is proven to be safe and effective and 64% showed a willingness to get a vaccine if their employer recommends it. The overall conclusion was that the current level of willingness to accept the vaccine, in most of the 19 countries surveyed, is insufficient for the acquirement of community immunity [15]. Another study, reported by Feleszko, *et al.* (2020), showed that 28%

of the adult population in Poland was against the COVID-19 vaccine, while 37% supported vaccination. Significantly fewer respondents supported the COVID-19 vaccine when compared to the 78% of the population that supported vaccination in 2018. Also, the vaccine skepticism was very low in China (26%) and very high in Czech Republic (43%) and Turkey (44%) [9].

Most activity concerning COVID-19 vaccine development is seen in North America (36 developers of the active vaccine candidates), followed by China (14), Europe (14). Effort being made on a global scale in response to the COVID-19 pandemic is remarkable, with an indication that vaccines could be available by early 2021. Such rapid development of the vaccine would represent change in the traditional developmental pathway [16]. It will be of great importance to ensure good coordination of manufacturing and supplying the vaccine, since many of the lead developers are small and inexperienced in large-scale vaccine development [16].

Despite the vaccine development for COVID-19 takes place in a world that has increased scientific knowledge including more understanding of genomics, structural biology and other branches which may lead to new way of vaccine development the trust in the vaccination differs among the population [17]. In this crisis of confidence in vaccines exposure to criticism of vaccination, misinformation and "antivax" activists all of which is highly accessible through social networks. What is more, a study in England stated that among almost half of the population there is certain dose of mistrust against a future COVID-19 vaccine [30]. In Bosnia and Herzegovina since the time of the former Yugoslavia mandatory vaccination program against 10 infectious diseases is present. The health insurance system fully covers the basic need of the vaccination for paediatric population despite various political and socioeconomic crises. If we take in consideration that vaccine coverage in 2017 in Federation of Bosnia and Herzegovina was low for MMR1 and MMR2 vaccine, being 62.7% and 77.4%, respectively, we may start to question how ready is the population to adopt new vaccine for COVID-19. It is thought that the one of the reasons for low trust in vaccination is paper published by Andrew Wakefield that associated vaccines with autism. The paper was rejected but due to the availability of the information using the social networks and media it has done an enormous harm to the public health [2].

According to Hrvat et al., the vaccination awareness of residents of Canton Sarajevo is as follows: not all the respondents have the same vaccines administered, and the populations opinion for vaccination is divided [13]. This paper presents results of investigation of public opinion regarding COVID-19 vaccination that was conducted in Bosnia and Herzegovina.

2 Methods

The study provides information on adult vaccination assessment experience, health insurance coverage, awareness of vaccine-preventable diseases and potential SARSCoV-2 vaccines, and more. Complete set of questions in survey is summarized in table below. The data was collected by means of an online questionnaire. A questionnaire was conducted using forms from Google. The number of individuals responding was 1237. The questionnaire was conducted anonymously, and in the results section, the results obtained are summarized. Questionnaire and graphical representation of data are available in supplementary material.

3 Results and Discussion

As vaccine development in the world is picking up its pace, it is important to keep records of public opinion of residents of a country in order to know what is their attitude towards vaccination and will the health care sector have difficulties in administering the vaccines to the population. The results of our 13-question questionnaire, which aimed to assess the public opinion, are presented in the table below. Since the beginning of COVID-19 pandemic, the rapid development of a vaccine for its prevention is of a global importance [11]. By applying knowledge from SARS and Middle East respiratory syndrome (MERS) vaccine development, first potential COVID-19 vaccine appeared within only a few weeks after the outbreak and entered human clinical testing on 16 March 2020. Since then, a fair amount of other COVID-19 vaccine candidates has been proposed. In order to keep track of every vaccine and its phase of development, the Coalition for Epidemic Preparedness Innovations (CEPI) together with global health authorities provided a database that includes all programs of vaccine development reported through the World Health Organization's (WHO's) authoritative [16, 24].

Currently, the database includes 115 vaccine candidates, out of which 78 are confirmed as active. For the remaining 37, the developmental status cannot be determined (Fig. 1). Most of the active projects are at the time in exploratory or preclinical stage, but the most advanced candidates are already in clinical development, such as mRNA-1273 from Moderna, Ad5-nCoV from CanSino Biologicals and INO4800 from Inovio (Table 1) [16].

SARS and MERS vaccine studies are probably the base used to select an appropriate target antigen and vaccine platform. All selected antigens and platforms that are tested for SARS and MERS in clinical and preclinical stages are shown in Table 2. Most current attempts for COVID-19 vaccine development are based on inducing neutralizing antibodies against the viral spike (S) protein, which could prevent the uptake through the human ACE2 receptor. However, it is unknown how different forms of the S protein used in different vaccine candidates relate to each other or to the epidemiology of the disease [16, 24].

Even with advanced technology and novel platforms, vaccine development is a long and expensive process. In order to produce a licensed vaccine, it usually takes multiple candidates and many years. But with a great need for the COVID-19 vaccine, many steps in the development need to be executed in parallel without confirming successfulness of another step [17]. Despite the urgent need of the vaccine, there are some challenges that need to be considered. First, it's the optimization of the antigen design which is important for optimal immune response. Second, from the experience with SARS and MERS vaccine studies, there are concerns about the vaccine having potential to worsen lung disease, either by the antibody-dependent enhancement or directly. Consequently, it will be of great importance to rigorously monitor the testing of the vaccine and for the test to be performed in a suitable animal model. Third challenge to be considered is the fact that the duration of immunity is still unknown. It is uncertain whether a single dose of the vaccine will be sufficient. Considering all challenges, gathering of important information and careful evaluation should be well defined [17].

Table 1. Candidate vaccines in clinical stage of evaluation (Taken and adapted from WHO's authoritative and continually updated list)

Vaccine candidate	Platform	Current phase of evaluation	Developer
Adenovirus Type 5 Vector	Non-replicating Viral Vector	*Phase 2 (ChiCTR2000031781) *Phase 1 (ChiCTR2000030906)	CanSino Biological Inc./Beijing Institute of Biotechnology
DNA plasmid vaccine Electroporation Device	DNA	*Phase 1 (NCT04336410)	Inovio Pharmaceuticals
Inactivated	Inactivated	*Phase 1 (ChiCTR2000031809)	Beijing Institute of Biological Products/Wuhan Institute of Biological Products
Inactivated + alum	Inactivated	*Phase 1	Sinovac
LNP encapsulated mRNA	RNA	*Phase 1 (NCT04283461)	Moderna/NIAID

All of the results can also be used for comparative approaches as well as statistical comparison for upcoming studies.

In Table 3 education levels of the respondents is presented, indicating that most of them have either High School Diploma or are holders of a BSc diploma, as well as results of health insurance respondents. 97% of respondents are covered by health insurance, while only 3% does not have health insurance. The type of vaccines and the number of participants who obtained them during their lifetime are presented in table above. As can be seen from the data, all respondents did not receive each and every vaccine. The vaccine obtained by most respondents was BCG, while Hib was the least received.

Through results we presented and collected data about the number of respondents who were tested for presence of SARS-CoV-2, as it can be seen that the majority of respondents did not have contact with COVID-19 up to that time. Vaccination awareness presented through data shows that, 68% of respondents were not against vaccination while 32% were against vaccination, or as can been seen from Table 3, 320 out of 1237 are against vaccination while 847 out of 1237 are for vaccination, while collecting data researchers wanted to collect information about vaccination awareness of possible SARS-CoV-2 vaccine. Results collected presents, 45% of respondents were not against possible SARS-CoV 2 vaccination while 55% were against possible vaccination, or as can been seen from Table 3, 679 out of 1237 are against vaccination while 558 out of 1237 are for possible vaccination.

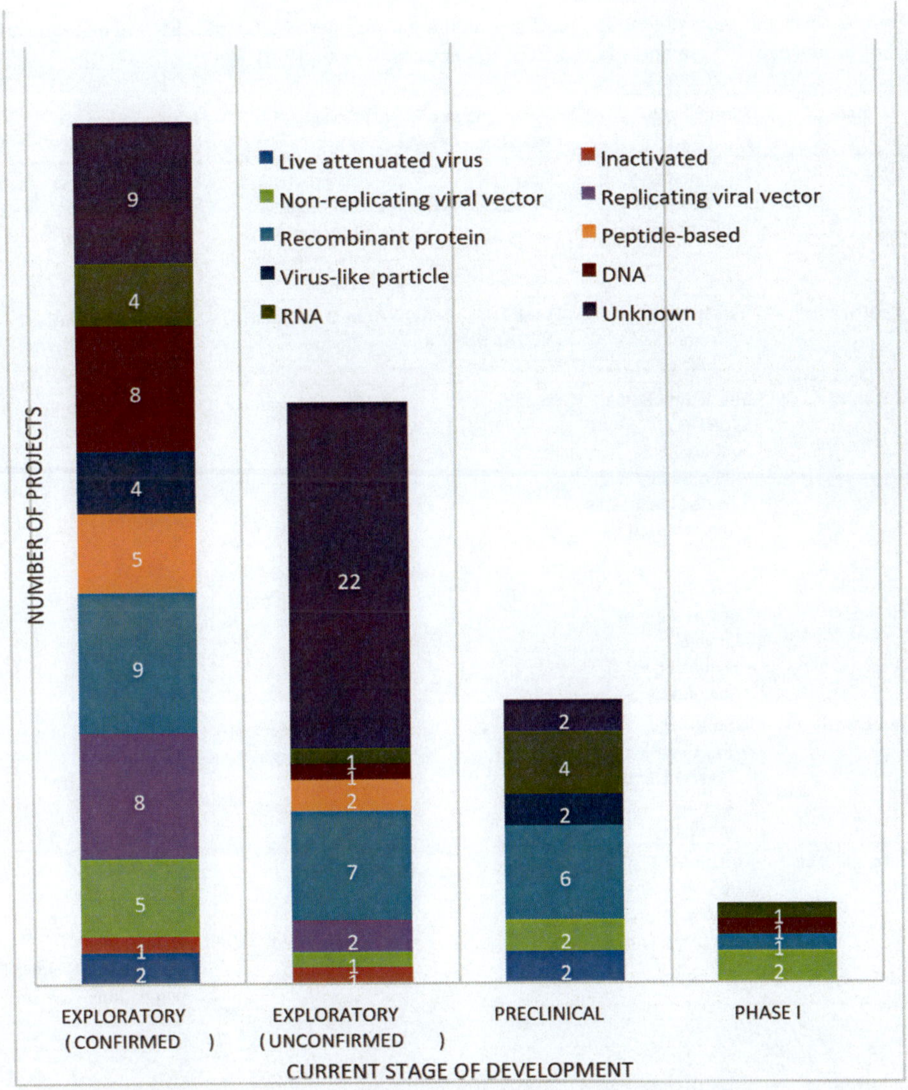

Fig. 1. Analysis of vaccine projects based on technology platform

As it can be seen in Table 3, 679 individuals that were questioned were against the COVID-19 vaccine. Several reasons have been proposed for their vaccine rejection and some of them are the short time frame in which the vaccine would be developed, the belief that the only aim of vaccination is to increase the income of the pharmaceutical companies, the COVID-19 is similar to seasonal flu so the vaccination is not needed, low trust in medical institutions in Bosnia and Herzegovina, the vaccine would not be mandatory, possible side-effects of the vaccination, and possible high cost of the vaccine. If we take in consideration a higher number of individuals that are against the

Table 2. All selected antigens and platforms that are tested for SARS and MERS in clinical and preclinical stages (Taken and adapted from Prompetchara et al., [24])

Vaccine platform	Immunogen	Current phase of evaluation	Advantage	Disadvantage
DNA	Full-length Spike – IM follow by electroporation	Phase I, II (NCT03721718)	– Fast production – Manipulative – Both T and B cells response induced	– Lower immune response when compared to live vaccines
Viral vector	Full-length Spike – Vector used: ChAd or MVA	Phase I (NCT0339957 8, NCT0361591 1)	– Great immune induction	– Possible TH2 bias – Chance of different immune responses
Subunit	Full-length Spike, S1, RDB, nucleocapsid – Formulated with various adjuvants and/or fused with Fc	Preclinical	– Humoral immune response induction – Consistent production – High safety profile	Cost effectiveness
Virus like particles	RDB, S or Coexpressing of S1, M, and E – Produced in baculovirus	Preclinical	– Preserve virus structure – Displays multimeric antigen	– Require optimal conditions
Inactivated	Whole virus – Inactivated by Formaldehyde or gamma irradiation	Preclinical	– Fast development – Neutralizing Ab induction – Preserve virus structure	– Possible TH2 bias – Can cause hypersensitivity
Live attenuated virus	Mutant MERSCoV and SARS-CoV or recombination with other live attenuated virus	Preclinical	– T and B cells response induction – Site directed mutagenesis	– Can reverse to virulent state – Not for sensitive and immunocompromised population

vaccination, we may propose that some of the arguments are due to the campaigns against the vaccination. Anti-vaccine campaigns propose several arguments and some of them are that side effects of the vaccines are more likely to happen than to be infected with a vaccine-preventable disease whose prevalence is minimal and that unvaccinated persons are unlikely to be infected due to the herd immunity. Another possible argument that is responsible for vaccination to be rejected by certain groups of people is that vaccines may not be as safe as it was taught since there is a belief that vaccine for measles, mumps and rubella (MMR) may cause autism. Besides these possible reasons a study that proved that Rotashield vaccine for rotavirus, that was withdrawn from the use, caused acute intussusception may decrease the willingness for the vaccination. In addition to negative effects on trust in immunization by community these factors increase the pressure on

Table 3. Results obtained from the questionnaire.

Gender?	
Male	420
Female	817
Age?	
>20	182
21–27	555
28–34	212
35–41	134
42–50	88
<50	66
Education level	
Dr.	30
Mr.	167
B.Sc.	514
High school diploma	477
Elementary school diploma	8
Other	41
Employment status	
Employed	563
Unemployed	198
Student	445
Retiree	23
Not active	8
Do you have health insurance?	
Yes	1195
No	42
Did you have or currently have disease from which you could vaccinate?	
Tuberculosis	15
Hepatitis B	16
Diphtheria or Tetanus	32
Child paralysis	7
Hemophilus Influenzae type B	50
Measles or mumps	86
No	1094
Which vaccines did you receive?	
BCG	1066

(*continued*)

Table 3. (*continued*)

HBV	928
DTaP	965
IPV	934
HiB	641
MRP	962
dT	917
Do you have chronic medical diseases?	
Asthma or chronic lung diseases	58
Liver diseases	9
Heart diseases	21
Diabetes	21
Immunological diseases	72
No	1087
Have you been issued a decision for mandatory self-isolation due to COVID-19 by any of the competent institutions?	
Yes	50
No	1187
Have you been tested for presence of SARS-CoV 2?	
Yes	75
No	1162
If you have been tested, what was the outcome?	
Positive after testing by qPCR method	5
Positive after testing by rapid antibody method	0
Negative after testing by qPCR method	58
Negative after testing by rapid antibody method	30
Are you against vaccination?	
Yes	390
No	847
If vaccine against SARS-CoV 2 become available on market would you undergo vaccination?	
Yes	558
No	679

the governments to increase the regulation of vaccine production therefore increasing the cost of the vaccine production [5]. Refusal or acceptance with certain doubts about the vaccination and its usefulness and safety has been recognized as one of the most important health problems in the nowadays world. Besides being reluctant for the already existent and applicable vaccines significant portions of populations around the world are also reluctant about the COVID-19 vaccines despite the high risk connected with the pandemic. As an example, 25% of people in 5 surveys in France were against the future vaccine for COVID-19. Some of the reasons for their refusal were safety of a vaccine developed in such a short time and in the middle of pandemic. Assurance in vaccination depends on different factors such as trust in health care workers and system, science, social context as well as political context. Confidence also depends on the patient and his/her relationship with the health care workers [30].

According to the study by Larson, et al., in which 67 countries were involved in questionnaire about the vaccination [14], European region showed the highest negative responses for vaccine importance, safety and effectiveness. One of the countries that were reported to have high importance and safety-related scepticism and least confidence in vaccines is Bosnia and Herzegovina with 38.3%, while in the results of this questionnaire 68% were against the vaccination. According to these results, it can be concluded that the trend of vaccine scepticism increased over the years. If the results of the public opinion about the novel COVID-19 vaccination from this study are compared with the results of the study that reported 73.9% of people from Denmark, France, Germany, Italy, Portugal, the Netherland, and the UK are willing to get vaccine, it can be stated that population in Bosnia and Herzegovina show less confidence in vaccination. According to these results, the public should get reassured that the vaccine will be safe and effective, even though it is produced in short time. Otherwise, the risk of losing the public trust in the vaccine will increase [22].

Vaccine development is a stepwise and slow process that starts with experiments that try to mimic natural protection, followed by animal studies, studies for finding small doses, safety and immunogenicity evaluation and finally large-scale randomized controlled clinical trials. Through all mentioned steps, efficacy is tested after which vaccine license is attained. Average time for proper vaccine development and investigation is 10 years or more [11]. In the past, rapidly developed vaccines resulted in unexpected events and deaths. Hence, fast development of COVID-19 vaccine is highly likely to increase the risks related to delivering a licensed vaccine. Finding the best way to accelerate vaccine development without compromising any ethical concerns and scientific norms is of great need. Primary goal for vaccines that are supposed to be tested on otherwise healthy people is of course safety [16, 27]. In order to minimize the safety concerns, COVID-19-specific animal models are being developed, such as ACE2-transgenic mice, hamsters, ferrets and non-human primates [16].

Recently, there has been an urgent approval for the use of two types of vaccines within the United States and the European Union. The manufacturers of these two vaccines are Pfizer-BioNTech and Moderna. Approval for the emergency use of these vaccines has been issued by the Food and Drug Administration (FDA) and the European Medicines Agency (EMA). The U.S. on Dec. 11th, 2020, released the first Emergency Use Authorization (EUA) by FDA, for the 2019 coronavirus disease prevention vaccine

(COVID-19) for individuals 16 years of age and older due to extreme acute respiratory syndrome coronavirus 2 (SARS-CoV-2). The U.S. on 18th December 2020, has given an Emergency Use Authorization (EUA) by FDA, for the second 2019 coronavirus disease prevention vaccine (COVID-19) caused by extreme coronavirus 2 acute respiratory syndrome (SARS-CoV-2). The emergency use authorization makes it possible to administer the Moderna COVID-19 vaccine in the U.S. for use. The EU on 21st December approved administration of Pfizer-BioNTech and Moderna vaccines for disease prevention of SARS-CoV-2 [7, 19, 23].

4 Conclusion

Since the first vaccine of Edward Jenner came to the market, world has recognized their medical potential and up to this year, there is a total of 27 diseases that can be prevented by immunization. The onset of COVID-19 pandemic was pronounced by WHO on March 11th 2020, and since then, large pharma companies worldwide have begun to develop various vaccine types. However, raising vaccination controversies and conspiracy theories have impacted the public opinion regarding vaccination against COVID-19. This research and accompanying questionnaire were designed with an aim of assessing public opinion of the population of Bosnia and Herzegovina it comes to vaccination against COVID-19. The questionnaire was filled in by a total of 1237 people. Most of the participants were in the young, 21–27 years, age group. This was mainly due to online method of collecting answers which is more suitable for younger population. Out of the total number, only 5 participants tested positive for COVID-19, while 1162 were never tested at all. Out of 1237 participants, 679 would approve administering of the vaccine, while 558 would not take the vaccine. Out of all respondents, most of them who were against vaccination do not have a high education degree and their explanations for vaccination disapproval are mostly similar. Most opponents have low to none relation with medical sciences, and believe in inaccurate educational materials, that spread disinformation, making uneducated people believe in conspiracy theories, causing them to become opponents of vaccination. Concerns are mainly in the field of safety and effectiveness and general lack of confidence in our healthcare system. Hence, general population should be supplied with appropriate scientific research materials for education, to reduce disinformation possibility. As this questionnaire was concluded with July 2020, after the first wave of infection has relieved its grasp, it is likely that the ratios would be tilted towards vaccination optimism due to the new wave of infection taking its toll.

Conflict of Interests. Authors declare no conflict of interests.

Supplementary Material

Supplementary material available upon request.

References

1. Anderson, E.J., Daugherty, M.A., Pickering, L.K., Orenstein, W.A., Yogev, R.: Protecting the community through child vaccination. Clin. Infect. Dis. **67**(3), 464–471 (2018)
2. Arapović, J., Sulaver, Ž, Rajič, B., Pilav, A.: The 2019 measles epidemic in Bosnia and Herzegovina: what is wrong with the mandatory vaccination program? Bosn. J. Basic Med. Sci. **19**(3), 210 (2019)
3. Astronomo, R.D., Burton, D.R.: Carbohydrate vaccines: developing sweet solutions to sticky situations? Nat. Rev. Drug Discov. **9**(4), 308–324 (2010)
4. Belshe, R.B., Edwards, K.M., Vesikari, T., Black, S.V., Walker, R.E., Hultquist, M., Kemble, G., Connor, E.M.: Live attenuated versus inactivated influenza vaccine in infants and young children. N. Engl. J. Med. **356**(7), 685–696 (2007)
5. Bloom, D.E., Canning, D., Weston, M.: The value of vaccination. World Econ. **6**(3), 15 (2005)
6. Dhillon, S., Pace, D.: Meningococcal quadrivalent tetanus toxoid conjugate vaccine (MenACWY-TT; Nimenrix®): a review. Drugs **77**(17), 1881–1896 (2017)
7. EMA recommends first COVID-19 vaccine for authorisation in the EU. European Medicines Agency (2020). https://www.ema.europa.eu/en/news/emarecommends-first-covid-19-vaccineauthorisation-eu
8. Ewald, H., Briel, M., Vuichard, D., Kreutle, V., Zhydkov, A., Gloy, V.: The clinical effectiveness of pneumococcal conjugate vaccines: a systematic review and meta-analysis of randomized controlled trials. Deutsches Ärzteblatt International **113**(9), 139 (2016)
9. Feleszko, W., Lewulis, P., Czarnecki, A., Waszkiewicz, P.: Flattening the curve of COVID-19 vaccine rejection—a global overview. Available at SSRN (2020)
10. Grady, C., Shah, S., Miller, F., Danis, M., Nicolini, M., Ochoa, J., Taylor, H., Wendler, D., Rid, A.: So much at stake: Ethical trade-offs in accelerating SARS-CoV-2 vaccine development. Vaccine **38**(41), 6381–6387 (2020)
11. Graham, B.S.: Rapid COVID-19 vaccine development. Science **368**(6494), 945–946 (2020)
12. Henderson, D.A.: Edward Jenner's vaccine. Public Health Rep. **112**(2), 116 (1997)
13. Hrvat, F., Cifric, S., Hasanic, O., Aleta, A.: Vaccination awareness among population in Canton Sarajevo, Bosnia and Herzegovina. Int. J. Sci. Basic Appl. Res. (IJSBAR) **51**(2), 179–190 (2020)
14. Larson, H.J., De Figueiredo, A., Xiahong, Z., Schulz, W.S., Verger, P., Johnston, I.G., Cook, A.R., Jones, N.S.: The state of vaccine confidence 2016: global insights through a 67-country survey. EBioMedicine **12**, 295–301 (2016)
15. Lazarus, J.V., Ratzan, S.C., Palayew, A., Gostin, L.O., Larson, H.J., Rabin, K., Kimbell, S., El-Mohandes, A.: A global survey of potential acceptance of a COVID-19 vaccine. Nat. Med. **27**(2), 225–228 (2020)
16. Le, T.T., Andreadakis, Z., Kumar, A., Roman, R.G., Tollefsen, S., Saville, M., Mayhew, S.: The COVID-19 vaccine development landscape. Nat. Rev. Drug Discov. **19**(5), 305–306 (2020)
17. Lurie, N., Saville, M., Hatchett, R., Halton, J.: Developing Covid-19 vaccines at pandemic speed. N. Engl. J. Med. **382**(21), 1969–1973 (2020)
18. Mascola, J.R., Snyder, S.W., Weislow, O.S., Belay, S.M., Belshe, R.B., Schwartz, D.H., Clements, M.L., Dolin, R., Graham, B.S., Gorse, G.J., Keefer, M.C., McElrath, M.J., Walker, M.C., Wagner, K.F., McNeil, J.G., McCutchan, F.E., Burke, D.S.: Immunization with envelope subunit vaccine products elicits neutralizing antibodies against laboratory-adapted but not primary isolates of human immunodeficiency virus type 1. J. Infect. Dis. **173**(2), 340348 (1996)
19. Moderna COVID-19 vaccine (2020). https://www.fda.gov/emergency-preparedness-and-response/coronavirus-disease-2019-covid-19/moderna-covid-19-vaccine

20. Morgan, A.J., Parker, S.: Translational mini-review series on vaccines: the Edward Jenner Museum and the history of vaccination. Clin. Exp. Immunol. **147**(3), 389–394 (2007)
21. Needham, J.: China and the Origins of Immunology. University of Hong Kong, Centre of Asian Studies (1980)
22. Neumann-Böhme, S., Varghese, N.E., Sabat, I., Barros, P.P., Brouwer, W., van Exel, J., Schreyögg, J., Stargardt, T.: Once we have it, will we use it? A European survey on willingness to be vaccinated against COVID-19. Eur. J. Health Econ. **21**(7), 977–982 (2020)
23. Pfizer-BioNTech COVID-19 Vaccine (2020). https://www.fda.gov/emergency-preparedness-and-response/coronavirus-disease-2019-covid-19/pfizer-biontech-covid-19-vaccine
24. Prompetchara, E., Ketloy, C., Palaga, T.: Immune responses in COVID-19 and potential vaccines: lessons learned from SARS and MERS epidemic. Asian Pac. J. Allergy Immunol. **38**(1), 1–9 (2020)
25. Pulendran, B., Ahmed, R.: Immunological mechanisms of vaccination. Nat. Immunol. **12**(6), 509 (2011)
26. Rueckert, C., Guzmán, C.A.: Vaccines: from empirical development to rational design. PLoS Pathog. **8**(11), e1003001 (2012)
27. Shin, M.D., Shukla, S., Chung, Y.H., Beiss, V., Chan, S.K., Ortega-Rivera, O.A., Wirth, D.M., Chen, A., Sack, M., Pokorsk, J.K., Steinmetz, N.F.: COVID-19 vaccine development and a potential nanomaterial path forward. Nat. Nanotechnol. **15**(8), 646–655 (2020)
28. Sistrunk, K.C.: Hit me with your best shot: a critical analysis of the resistance to vaccine utilization (2019)
29. UNICEF: Znanje, stavovi i prakse u vezi sa imunizacijom dece u Srbiji. UNICEF, Beograd (2018)
30. Verger, P., Dubé, E.: Restoring confidence in vaccines in the COVID-19 era. Expert Rev. Vaccines **19**(11), 991–993 (2020)
31. WHO (2020). https://www.who.int/ith/ITHChapter6.pdf. Accessed 22 Nov 2020
32. WHO (2020). https://covid19.who.int/. Accessed 20 Nov 2020
33. WHO (2020). https://www.who.int/news/item/29-062020covidtimeline. Accessed 20 Nov 2020

Implementation of Service Robots for Space Disinfection in Medical Institutions: A Review of Control of Corona Virus Infection

Isak Karabegović[1](✉), Ermin Husak[2], Safet Isić[3], Lejla Banjanović-Mehmedović[4], and Almir Badnjević[5,6]

[1] Academy of Sciences and Arts, Sarajevo, Bosnia and Herzegovina
ikarabegovic@anubih.ba
[2] University of Bihać, Bihać, Bosnia and Herzegovina
[3] University "Džemal Bijedić" of Mostar, Mostar, Bosnia and Herzegovina
[4] University of Tuzla, Tuzla, Bosnia and Herzegovina
[5] Verlab Ltd Sarajevo, Medical Device Verification Laboratory, Sarajevo, Bosnia and Herzegovina
[6] Faculty of Pharmacy, University of Sarajevo, Sarajevo, Bosnia and Herzegovina

Abstract. Last year and part of this year, we have witnessed a large number of people being infected, and subsequently died, as a result of the corona virus COVID-19. The physical environment plays an important role in transmitting the infection. We must urgently find a solution and answer to all situations in society that have a detrimental effect on human health. When it comes to virus transmission, it is necessary to disinfect both the body and the room. The development of new technologies leads to the development of service robots that are implemented for disinfection in the medical institutions. The paper presents a review of the implementation of service robots that use ultraviolet light for disinfection in medical institutions. UV disinfection of surface with service robots is one of the most promising new technologies for several reasons, of which some are: faster treatment in the room, higher productivity, highly effective treatments, whole room treatment, one room treatment, etc. In addition to service robots for indoor disinfection, service robots for disinfection of outdoor surfaces have also been developed. New materials and technologies are now being introduced into medical facilities in an attempt to reduce the ecological pathway of disease transmission.

Keywords: Infection · COVID 19 · Service robots · Medical institutions · Disinfection

1 Introduction

The application of mobile robots is increasing every day. It can be argued that there is no segment of human life in which mobile robots are not included. Service robots find application in: medicine, defence, agriculture, construction, logistics, rescue and security, professional cleaning, inspection and maintenance, space exploration, education,

A. Badnjevic and L. Gurbeta Pokvić (Eds.): CMBEBIH 2021, IFMBE Proceedings 84, pp. 921–928, 2021.
https://doi.org/10.1007/978-3-030-73909-6_102

household, etc. Service robots and artificial intelligence coexist and therefore it is difficult to imagine that today's robot it is not some kind of artificial intelligence. It is believed that robots and artificial intelligence will indeed prolong human life and improve the quality of life in general. Mobile robots in medicine have found application in many areas of medicine, including: performing sensitive surgeries, replacing missing limbs, rehabilitation therapy for patients who have had a stroke, patient visits, serving patients with food and drugs, remote patient treatment, remote monitoring of surgical procedures in medical facilities, disinfection of surgical rooms, etc. [1–6]. Mobile robotics can currently provide a number of benefits. The development of new technologies, primarily sensory and information technologies that are widely represented in robotic technologies, has resulted in the development of different robotic systems designed for different medical applications. Since the beginning of the COVID-19 epidemic, medical experts have warned people to take all necessary precautions and follow medical advice to stay safe. The World Health Organization (WHO) has officially declared a COVID-19 pandemic due to the speed with which it has spread across China as well as across borders, causing multiple incidents in all countries of the world, including Bosnia and Herzegovina, as shown in Fig. 1.

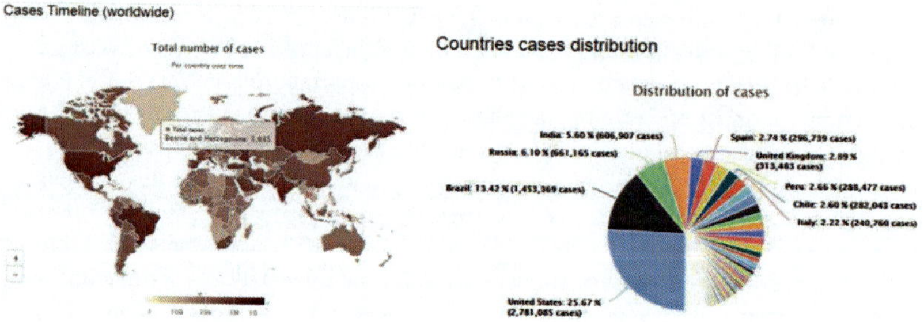

#	Country, Other	Total Cases	New Cases	Total Deaths	New Deaths	Total Recovered	Active Cases	Serious, Critical	Tot Cases/ 1M pop	Deaths/ 1M pop	Total Tests	Tests/ 1M pop	Population
84	Gabon	5,513		42		2,508	2,963	10	2,477	19	38,692	17,387	2,225,384
85	Guinea	5,404		33		4,346	1,025	24	412	3	14,407	1,097	13,129,739
86	Bulgaria	5,154		232		2,722	2,200	32	742	33	144,369	20,779	6,947,970
87	Bosnia and Herzegovina	4,788	+182	189	+1	2,516	2,084	4	1,459	58	95,885	29,227	3,280,646

Fig. 1. The World Health Organization (WHO) review of the expansion of the pandemic on 27 June 2020 with reference to Bosnia and Herzegovina [4]

Even today, the complications in healthcare sector due to the widespread of infection are present [7] and the number of people infected with coronavirus in the world, including Bosnia and Herzegovina, is constantly increasing. The application of mobile robots minimizes contact between people and provides disinfection, sterilization and support in hospitals and similar facilities, such as quarantine. This will result in minimizing the life-threatening situations of medical staff and physicians who are actively involved in managing the treatment of patients with COVID-19.

2 Distribution of Service Robots in Medicine

We can acquire a real insight into the implementation of service robots in medical institutions if we conduct an analysis of the distribution of service robots in the last ten years. The International Federation of Robotics (IFR), the United Nations Economic Commission for Europe (UNECE) and the Organization for Economic Cooperation and Development (OECD) have adopted an introductory system for classification of service robots in medicine by category and type of interaction, so service robots in medicine have the following classification [1–4] (Table 1):

Table 1. Classification of professional service robots in medicine

II. Service robots for professional use	
40–43	Medical robotics
40	Diagnostic systems
41	Robot assisted surgery or therapy
42	Rehabilitation systems
43	Other medical robots

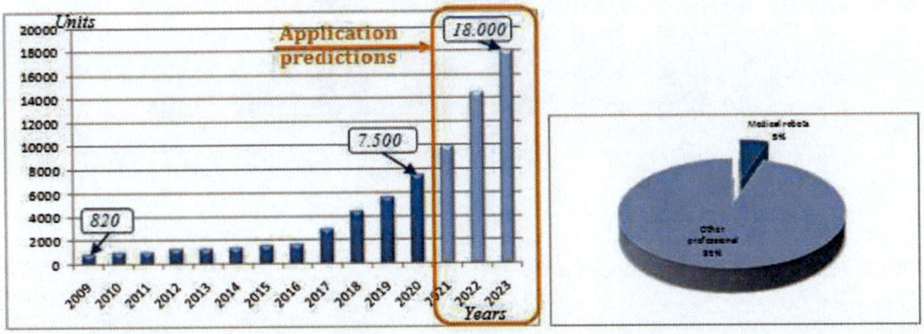

a) b)

Fig. 2. a) Implementation of service robots in medical institutions on an annual basis for the period 2009–2019 and estimated implementation by 2023; b) implementation of service robots for medical services in 2019

The statistical data for the analysis of the trend of distribution and implementation of service robots in medical institutions were obtained from the International Federation of Robotics (IFR), the UN Economic Commission for Europe (UNECE) and the Organization for Economic Co-operation and Development (OECD) [6, 8–17]. Figure 2 shows the trends of the total application of service robots in medical institutions by years for the period 2009–2019, as well as estimates of implementation to 2023.

The trend of implementation of service robots for medical services in medical institutions is shown in Fig. 2, which shows that implementation of service robots for medical services in 2019 amounted to 5% of the total amount of service robots for professional services [4]. This is a low percentage and we believe that it will have a growing trend in the coming years due to the fact that our population is getting older and will need more services performed by robots to free people that serve the elderly. In addition, there has been a high investment in the development of robotic technology and continuous development of new service robots for medical services to perform various tasks. Another conclusion can be made based on the trend of sales and implementation of service robots for medical services shown in the Fig. 2.

As can be seen, the implementation of service robots for medical services in 2009 amounted to about 820 units, and in just ten years the trend increased to about 7.500 service robot units in 2020, which is an increase of about seven times. It is estimated that the trend will continuously grow in the coming years, so that in 2023 the implementation of about 18.000 robot units is expected. It can be concluded that the implementation of service robots for medical services has been growing exponentially since 2016. Of the total 5% of service robots for medical services implemented in 2019, in relation to their implementation in other areas, they were implemented in different medical services, as shown in Fig. 3 [1, 5].

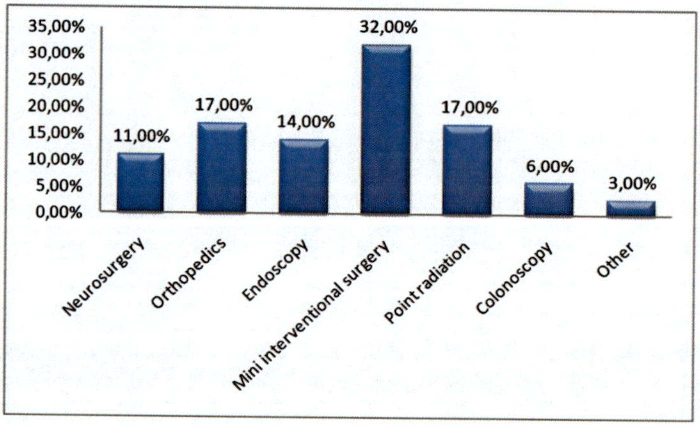

Fig. 3. Implementation of service robots for medical services by areas

Based on Fig. 2, we can conclude that the highest number of service robot implementations in medicine is as follows: small interventional surgeries 32%, orthopaedics 17%, neurosurgery 11%, endoscopy 14%, spot radiation 17%, colonoscopy 6%, etc. The most relevant useful factors are: high quality work and productivity, reduction of manual work, increased safety or risk avoidance, increased operational usability, temporary flexibility, new, previously available facilities, conditions, etc. The following Table 2 shows an assessment of the relevance of factors for the types of service robots used in medicine.

The most relevant useful factors are: high quality of work and productivity, reduction of manual work, increased safety or risk avoidance, increased operational usability, temporary flexibility, new previously available facilities, conditions, etc. Table 2 shows an assessment of factor relevance for types of service robots used in medicine [1–3].

Table 2. Assessment of factor relevance for types of service robots in medicine

40–43	Medical robotics	High quality labour productivity	Reduction of manual labour	Increasing security, avoiding risk
40	Diagnostic systems	• •		•
41	Robot assisted surgery or therapy	• •		•
42	Rehabilitation systems	•		•
43	Other medical robots			

The table shows an estimate of the factor relevance for each type of service robot, with a degree of relevance marked from 0 (*not relevant*) to two points (*high relevance*).

3 Implementation of Service Robots for Disinfection in Medical Institutions

It is well known that disinfection is necessary and required in medical institutions – hospitals. We are currently witnessing a pandemic caused by the COVID-19 virus. For years, hospitals have accepted ultraviolet light as a cleaning agent for decontamination of rooms in medical institutions. The three main types of UV ultraviolet rays are UVA, UVB and UVC. Because UVC rays have the shortest wavelength, and therefore the highest energy, they are capable of killing bacteria and viruses, also known as pathogens. UVC light has a wavelength between 100 and 280 nm (nm). It is extremely effective in decontamination, because it destroys the molecular bonds that hold the DNA of viruses and bacteria, including “superbugs”, which have developed stronger resistance to antibiotics [18–25]. UV photons penetrate cell membranes and cause thymine molecules to bind to DNA and RNA, preventing replication. Bacteria and viruses have known but limited repair mechanisms that are easily overcome. Unlike antibiotics and chemical disinfectants, bacteria and viruses have not been shown to systematically develop further resistance. However, this broad-spectrum light also poses a health hazard. It is linked to diseases such as skin cancer and cataract, and people cannot be present in the room when they are used. A group of researchers is working on research with narrow spectrum UVC rays with a wavelength of 207–222 nm (nm), because this type of UVC light kills bacteria and viruses without penetrating the farthest cell layer of human skin. Many studies are currently in progress in regard to UVC light, so a 2017 study showed that 222 nm UVC light killed methicillin-resistant Staphylococcus aureus (MRSA) bacteria just as effectively as 254 nm UVC light, which would be toxic for people. This study was repeated in 2018 on the H1N1 virus, and narrow-spectrum UVC light was again found to be effective in removing the virus. This has particularly important implications for public health, as the possibility of nontoxic UV overhead lighting in public spaces could drastically reduce disease transmission. Numerous design solutions for disinfection robots have been developed [25]. Figure 4 shows the different constructions of patient room disinfection in a medical facility.

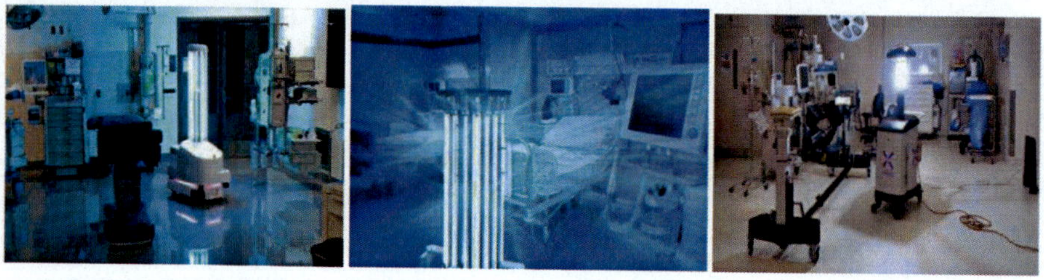

Fig. 4. Different constructions of service robots for disinfection in medical institutions [26, 27]

Service robots for disinfection have a number of advantages, to name just a few: faster in-room treatment, higher productivity, easier integration into the EVS cleaning protocol, highly effective treatments, highly effective pathogen elimination, whole room treatment, one room treatment, disinfection of all surfaces in the room, etc. Service robots have shown great efficiency in disinfecting the entire room during a corona virus pandemic. We can secure maximum disinfection in less than 10 min, which allows us to disinfect more rooms in a short period of time. Service robots for disinfection of rooms in medical institutions are very practical because they perform disinfection quickly, and they are safe and easy to use. One service robot can perform disinfection of 64 rooms in medical institutions per day.

4 Conclusion

We have seen that virus infections are the cause of death of a large number of people (today it is the COVID 19 virus, and tomorrow it will be another virus). We see that the physical environment can play a significant role in disease transmission than previously thought. We need to respond to such phenomena and find solutions to protect ourselves from the spread of infection now and in the future. The development of new technologies, including robotic technologies, has encouraged the development of service robots for disinfection in medical institutions. The paper presents certain constructions of service robots for disinfection, which proved well at the time of corona virus infection. It has been proven that UV disinfection of surfaces with service robots is one of the most promising new technologies. We have witnessed that service robots are used to disinfect open surfaces with certain disinfectants. In the coming period, the development of autonomous systems of service robots for disinfection of both indoor and outdoor areas is expected.

References

1. Karabegović, I., Doleček, V.: Service Robotic. Technical fakulty, Bihać

2. Karabegović, I., Doleček, V.: The role of service robots and robots systems in the treatment of patients in Medical Institutions. In: Hadžikadić, M. (ed.) Advanced Technologies, Systems and Applications, pp. 9–27. Springer, Deutschland (2016)
3. Karabegović, I., Doleček, V., Karabegović, E.: World service robot distribution and application, In: Proceedings of 12th International Conference, Research and Development in Mechanical Industry, Serbia, Vrnjačka Banja, pp. 22–30, 14–17 September (2012)
4. The World Health Organization (WHO) (2020). https://www.worldometers.info/coronavirus/worldwide-graphs/; https://www.worldometers.info/coronavirus/#countries. Accessed 27 June 2020
5. World Robotics 2020-Service Robots, pp. 11–12. The International Federation of Robotics, Statistical Department, Frankfurt am Main (2020)
6. Karabegović, I., Felić, M., Đukanović, M.: Design and application of service robots in assisting patients and rehabilitations of patients. Int. J. Eng. Technol. **13**(02), 11–17 (2013)
7. Badnjević, A., Pokvić, L.G., Džemić, Z., et al.: Risks of emergency use authorizations for medical products during outbreak situations: a COVID-19 case study. BioMed. Eng. OnLine **19**, 75 (2020). https://doi.org/10.1186/s12938-020-00820-0
8. World Robotics 2019-Service Robots, pp. 11–12. The International Federation of Robotics, Statistical Department, Frankfurt am Main (2019)
9. World Robotics 2018-Service Robots, pp. 11–16. The International Federation of Robotics, Statistical Department, Frankfurt am Main (2018)
10. World Robotics 2017-Service Robots, pp. 12–18. The International Federation of Robotics, Statistical Department, Frankfurt am Main (2017)
11. World Robotics 2016-Service Robots, pp. 1–5. The International Federation of Robotics, Statistical Department, Frankfurt am Main (2016)
12. World Robotics 2015-Service Robots, pp. 22–26. The International Federation of Robotics, Statistical Department, Frankfurt am Main (2015)
13. World Robotics 2014-Service Robots, pp. 20–24. The International Federation of Robotics, Statistical Department, Frankfurt am Main, German (2014)
14. World Robotics 2013-Service Robots, pp. 18–21. The International Federation of Robotics, Statistical Department, Frankfurt am Main (2013)
15. World Robotics 2012-Service Robots (15–18). The International Federation of Robotics, Statistical Department, Frankfurt am Main (2012)
16. World Robotics 2011-Service Robots, pp. 14–16. The International Federation of Robotics, Statistical Department, Frankfurt am Main (2011)
17. World Robotics 2010-Service Robots, pp. 13–14. The International Federation of Robotics, Statistical Department, Frankfurt am Main (2010)
18. Reed, N.G.: The history of ultraviolet germicidal irradiation for air disinfection. Public Health Rep. **125**(1), 15–27 (2010)
19. Bolton, J.R., Cotton, C.A.: The Ultraviolet Disinfection Handbook. American Water Works Association, Denver (2008)
20. Otter, J.A., Yezli, S., Salkeld, J.A., French, G.L.: Evidence that contaminated surfaces contribute to the transmission of hospital pathogens and an overview of strategies to address contaminated surfaces in hospital settings. Am. J. Infect. Control **41**(5), S6–S11 (2013)
21. Huslage, K., Rutala, W.A., Gergen, M.F., Sickbert-Bennett, E.E., Weber, D.J.: Microbial assessment of high-, medium-, and low-touch hospital room surfaces. Infect. Control Hosp. Epidemiol. **34**(2), 211–212 (2013)
22. Hayden, M.K., Blom, D.W., Lyle, E.A., Moore, C.G., Weinstein, R.A.: Risk of hand or glove contamination after contact with patients colonized with vancomycin-resistant enterococcus or the colonized patients' environment. Infect. Control Hosp. Epidemiol. **29**(2), 149–154 (2008)

23. Anderson, J., Gergen, M.F., Smathers, E., Sexton, D.J., Chen, L.F., Weber, D.J., Rutala, W.A., CDC Prevention Epicenters Program: Decontamination of targeted pathogens from patient rooms using an automated ultraviolet-C-emitting device. Infect. Control Hosp. Epidemiol. **34**(5), 466–471 (2013)
24. Simmons, S., Morgan, M., Hopkins, T., Helsabeck, K., Stachowiak, J., Stibich, M.: Impact of a multi-hospital intervention utilising screening, hand hygiene education and pulsed xenon ultraviolet (PX-UV) on the rate of hospital associated methicillin resistant Staphylococcus aureus infection. J. Infect. Prev. **14**(5), 172–174 (2013)
25. Karabegović, I., Đukanović, M.: The tendency of development and application of service robots for defense rescue and security. Int. J. Adv. Eng. Res. Sci. **4**(9), 063–068 (2017)
26. Does UV light kill germs? MUVGI-8 (UV-C) against COVID-19 (2021). https://thalheimerkuhlung.com/en/does-uv-light-kill-germs-muvgi-8-uv-c-against-covid-19/
27. The Xenex LightStrike Germ-Zapping Robot (2021). https://lsdc.net/lsdc-news/xenexlightstrike-germ-zapping-robot/

Author Index

A. Badnjevic and L. Gurbeta Pokvić (Eds.): CMBEBIH 2021, IFMBE Proceedings 84, pp. 929–933, 2021.
https://doi.org/10.1007/978-3-030-73909-6